# Methods in Enzymology

Volume 82

STRUCTURAL AND CONTRACTILE PROTEINS

Part A

Extracellular Matrix

# METHODS IN ENZYMOLOGY

EDITORS-IN-CHIEF

Sidney P. Colowick     Nathan O. Kaplan

*Methods in Enzymology*

Volume 82

# Structural and Contractile Proteins

Part A

*Extracellular Matrix*

EDITED BY

## Leon W. Cunningham

DEPARTMENT OF BIOCHEMISTRY
VANDERBILT UNIVERSITY
NASHVILLE, TENNESSEE

## Dixie W. Frederiksen

DEPARTMENT OF BIOCHEMISTRY
VANDERBILT UNIVERSITY
NASHVILLE, TENNESSEE

1982

**ACADEMIC PRESS**
*A Subsidiary of Harcourt Brace Jovanovich, Publishers*

New York   London
Paris   San Diego   San Francisco   São Paulo   Sydney   Tokyo   Toronto

COPYRIGHT © 1982, BY ACADEMIC PRESS, INC.
ALL RIGHTS RESERVED.
NO PART OF THIS PUBLICATION MAY BE REPRODUCED OR
TRANSMITTED IN ANY FORM OR BY ANY MEANS, ELECTRONIC
OR MECHANICAL, INCLUDING PHOTOCOPY, RECORDING, OR ANY
INFORMATION STORAGE AND RETRIEVAL SYSTEM, WITHOUT
PERMISSION IN WRITING FROM THE PUBLISHER.


ACADEMIC PRESS, INC.
111 Fifth Avenue, New York, New York 10003


*United Kingdom Edition published by*
ACADEMIC PRESS, INC. (LONDON) LTD.
24/28 Oval Road, London NW1 7DX


Library of Congress Cataloging in Publication Data
Main entry under title:

Structural and contractile proteins.

   (Methods in enzymology ; v. 82)
   Includes bibliographical references and index.
   Contents: pt. A. Extracellular matrix.
   1. Proteins. I. Colowick, Sidney P. II. Kaplan,
Nathan O. III. Series.
QP601.M49 vol. 82 [QP551] 574.19'25s 81-20622
ISBN 0-12-181982-5 (v. 82) [574.19'245] AACR2


PRINTED IN THE UNITED STATES OF AMERICA

82 83 84 85   9 8 7 6 5 4 3 2 1

# Table of Contents

CONTRIBUTORS TO VOLUME 82. . . . . . . . . . . . . . . . . . . . . . ix

PREFACE . . . . . . . . . . . . . . . . . . . . . . . . . . . . . . . . xiii

VOLUMES IN SERIES . . . . . . . . . . . . . . . . . . . . . . . . . . xv

## Section I. Collagen

### A. The Multiple Types and Forms of Collagen

| | | |
|---|---|---|
| 1. Collagen: An Overview | EDWARD J. MILLER AND STEFFEN GAY | 3 |
| 2. Preparation and Characterization of the Different Types of Collagen | EDWARD J. MILLER AND R. KENT RHODES | 33 |
| 3. Preparation and Characterization of Invertebrate Collagens | LOUANN W. MURRAY, J. HERBERT WAITE, MARVIN L. TANZER, AND PETER V. HAUSCHKA | 65 |
| 4. Preparation and Characterization of Procollagens and Procollagen–Collagen Intermediates | HELENE SAGE AND PAUL BORNSTEIN | 96 |
| 5. Characterization of Fibrous Forms of Collagen | BARBARA BRODSKY AND ERIC F. EIKENBERRY | 127 |
| 6. Solid State NMR Studies of Molecular Motion in Collagen Fibrils | D. A. TORCHIA | 174 |
| 7. Characterization of Soluble Collagens by Physical Techniques | ARTHUR VEIS | 186 |

### B. Biosynthesis of Procollagen

| | | |
|---|---|---|
| 8. Preparation and Translation of Interstitial (Calvaria) Procollagen mRNA | JANET M. MONSON | 218 |
| 9. Cell-Free Synthesis of Noninterstitial (CHL Cell) Procollagen Chains | MICHAEL A. HARALSON | 225 |
| 10. Posttranslational Enzymes in the Biosynthesis of Collagen: Intracellular Enzymes | KARI I. KIVIRIKKO AND RAILI MYLLYLA | 245 |
| 11. Posttranslational Enzymes in the Biosynthesis of Collagen: Extracellular Enzymes | DARWIN J. PROCKOP AND LEENA TUDERMAN | 305 |

## C. Preparation of Other Proteins Containing Collagen-like Sequences

| | | |
|---|---|---|
| 12. C1q | K. B. M. Reid | 319 |
| 13. Acetylcholinesterase | Terrone L. Rosenberry, Philip Barnett, and Carol Mays | 325 |

## D. Characterization of Special Features in the Chemistry of Collagens

| | | |
|---|---|---|
| 14. Hydroxylysine Glycosides | William T. Butler | 339 |
| 15. Asparagine-Linked Glycosides | Charles C. Clark | 346 |
| 16. Covalent Cross-Links in Collagen | Nicholas D. Light and Allen J. Bailey | 360 |
| 17. Determination of 3- and 4-Hydroxyproline | Richard A. Berg | 372 |
| 18. Urinary Metabolites of Collagen | Jere P. Segrest | 398 |
| 19. Estimation of the Size of Collagenous Proteins by Electrophoresis and Gel Chromatography | Ralph J. Butkowski, Milton E. Noelken, and Billy G. Hudson | 410 |

## E. Proteolysis of Collagen

| | | |
|---|---|---|
| 20. Vertebrate Collagenases | Edward D. Harris, Jr. and Carol A. Vater | 423 |
| 21. Bacterial Collagenase | Beverly Peterkofsky | 453 |

## F. Immunology of Collagen

| | | |
|---|---|---|
| 22. Antibodies to Collagens and Procollagens | Rupert Timpl | 472 |

## G. Interaction of Cells with Collagen

| | | |
|---|---|---|
| 23. Cell–Collagen Interactions: Overview | Frederick Grinnell | 499 |
| 24. Fibroblast Adhesion to Collagen Substrates | Hynda K. Kleinman | 503 |
| 25. Platelet–Collagen Adhesion | Samuel A. Santoro and Leon W. Cunningham | 509 |
| 26. Hepatocyte–Collagen Adhesion | Björn Öbrink | 513 |
| 27. Biosynthetic Matrices from Cells in Culture | Carl G. Hellerqvist | 530 |
| 28. Ultrastructural Studies of Cell–Collagen Interactions | Frederick Grinnell and Marylyn Hoy Bennett | 535 |

# TABLE OF CONTENTS

| | | |
|---|---|---|
| 29. Collagen as a Substrate for Cell Growth and Differentiation | STEPHEN C. STROM AND GEORGE MICHALOPOULOS | 544 |

## Section II. Elastin

### A. Elastin Structure and Biosynthesis

| | | |
|---|---|---|
| 30. Elastin Structure and Biosynthesis: An Overview | JUDITH ANN FOSTER | 559 |
| 31. Elastin Isolation and Cross-Linking | MERCEDES A. PAZ, DAVID A. KEITH, AND PAUL M. GALLOP | 571 |
| 32. Proteolysis of Insoluble Elastin | P. J. STONE, C. FRANZBLAU, AND H. M. KAGAN | 588 |
| 33. Primary Structure of Insoluble Elastin | RASHID A. ANWAR | 606 |
| 34. Biosynthesis of Insoluble Elastin in Cell and Organ Cultures | CARL FRANZBLAU AND BARBARA FARIS | 615 |
| 35. Lysyl Oxidase: Preparation and Role in Elastin Biosynthesis | HERBERT M. KAGAN AND KATHLEEN A. SULLIVAN | 637 |
| 36. Isolation of Soluble Elastin from Copper-Deficient Chick Aorta | ROBERT B. RUCKER | 650 |
| 37. Production and Isolation of Soluble Elastin from Copper-Deficient Swine | L. B. SANDBERG AND T. B. WOLT | 657 |
| 38. Isolation of Soluble Elastin-Lathyrism | CELESTE B. RICH AND JUDITH ANN FOSTER | 665 |
| 39. Characterization of Soluble Peptides of Elastin by Physical Techniques | DAN W. URRY | 673 |
| 40. Biosynthesis of Soluble Elastin in Organ and Cell Culture | JOEL ROSENBLOOM | 716 |
| 41. Cell-Free Translation of Elastin mRNAs | JUDITH ANN FOSTER, CELESTE B. RICH, STEVEN B. KARR, AND ALAN PRZYBYLA | 731 |

### B. Immunology of Elastin

| | | |
|---|---|---|
| 42. Antibodies to Insoluble and Solubilized Elastin | ROBERT P. MECHAM AND GARY LANGE | 744 |
| 43. Antibodies to Desmosine | BARRY C. STARCHER | 759 |

| | | |
|---|---|---|
| 44. Preparation of Antiserum to Tropoelastin | JUDITH ANN FOSTER, CELESTE B. RICH, SU CHEN WU, AND MARIE-THERESE MARZULLO | 762 |

## Section III. Proteoglycans

| | | |
|---|---|---|
| 45. Proteoglycans: Isolation and Characterization | VINCENT C. HASCALL AND JAMES H. KIMURA | 769 |

## Section IV. Other Connective Tissue Proteins

| | | |
|---|---|---|
| 46. Fibronectin: Purification, Immunochemical Properties, and Biological Activities | ERKKI RUOSLAHTI, EDWARD G. HAYMAN, MICHAEL PIERSCHBACHER, AND EVA ENGVALL | 803 |
| 47. Laminin | RUPERT TIMPL, HEILWIG ROHDE, LEILA RISTELI, URSULA OTT, PAMELA GEHRON ROBEY, AND GEORGE R. MARTIN | 831 |
| 48. Structural Glycoproteins | LADISLAS ROBERT AND MADELEINE MOCZAR | 839 |

| | |
|---|---|
| AUTHOR INDEX . . . . . . . . . . . . . . . . . . | 853 |
| SUBJECT INDEX . . . . . . . . . . . . . . . . . | 885 |

# Contributors to Volume 82

Article numbers are in parentheses following the names of contributors.
Affiliations listed are current.

RASHID A. ANWAR (33), *Department of Biochemistry, University of Toronto, Toronto M5S 1A8, Canada*

ALLEN J. BAILEY (16), *Agricultural Research Council, Meat Research Institute, Langford, Bristol BS18 7DY, England*

PHILIP BARNETT (13), *Department of Neurology, College of Physicians and Surgeons, Columbia University, New York, New York 10032*

MARYLYN HOY BENNETT (28), *Department of Cell Biology, The University of Texas Health Science Center at Dallas, Dallas, Texas 75235*

RICHARD A. BERG (17), *Department of Biochemistry, College of Medicine and Dentistry of New Jersey–Rutgers Medical School, Piscataway, New Jersey 08854*

PAUL BORNSTEIN (4), *Departments of Biochemistry and Medicine, University of Washington, Seattle, Washington 98195*

BARBARA BRODSKY (5), *Department of Biochemistry, College of Medicine and Dentistry of New Jersey–Rutgers Medical School, Piscataway, New Jersey 08854*

RALPH J. BUTKOWSKI (19), *Department of Biochemistry, University of Kansas Medical Center, Kansas City, Kansas 66103*

WILLIAM T. BUTLER (14), *Institute of Dental Research, University of Alabama in Birmingham, Birmingham, Alabama 35294*

CHARLES C. CLARK (15), *Connective Tissue Research Institute, University City Science Center, Philadelphia, Pennsylvania 19104*

LEON W. CUNNINGHAM (25), *Department of Biochemistry, Vanderbilt University School of Medicine, Nashville, Tennessee 37232*

ERIC F. EIKENBERRY (5), *Department of Biochemistry, College of Medicine and Dentistry of New Jersey–Rutgers Medical School, Piscataway, New Jersey 08854*

EVA ENGVALL (46), *La Jolla Cancer Research Foundation, 2945 Science Park Road, La Jolla, California 92037*

BARBARA FARIS (34), *Department of Biochemistry, Boston University School of Medicine, Boston, Massachusetts 02118*

JUDITH ANN FOSTER (30, 38, 41, 44), *Department of Biochemistry, University of Georgia, Athens, Georgia 30602*

C. FRANZBLAU (32, 34), *Department of Biochemistry, Boston University School of Medicine, Boston, Massachusetts 02118*

PAUL M. GALLOP (31), *Departments of Biological Chemistry and Oral Biology, Harvard Schools of Medicine and Dental Medicine, Children's Hospital Medical Center, Boston, Massachusetts 02115*

STEFFEN GAY (1), *Department of Medicine and Institute of Dental Research, University of Alabama in Birmingham, Birmingham, Alabama 35294*

FREDERICK GRINNELL (23, 28), *Department of Cell Biology, University of Texas Health Science Center at Dallas, Dallas, Texas 75235*

MICHAEL A. HARALSON (9), *Department of Pathology, Vanderbilt University School of Medicine, Nashville, Tennessee 37232*

EDWARD D. HARRIS, JR. (20), *Department of Medicine, Dartmouth-Hitchcock Medical Center, Hanover, New Hampshire 03755*

VINCENT C. HASCALL (45), *Laboratory of Biochemistry, National Institute of Dental Research, National Institutes of Health, Bethesda, Maryland 20205*

PETER V. HAUSCHKA (3), *Children's Hospital Medical Center and Harvard School of Dental Medicine, Boston, Massachusetts 02115*

EDWARD G. HAYMAN (46), *La Jolla Cancer Research Foundation, 2945 Science Park Road, La Jolla, California 92037*

CARL G. HELLERQVIST (27), *Department of Biochemistry, Vanderbilt University School of Medicine, Nashville, Tennessee 37232*

BILLY G. HUDSON (19), *Department of Biochemistry, University of Kansas Medical Center, Kansas City, Kansas 66103*

H. M. KAGAN (32, 35), *Department of Biochemistry, Boston University School of Medicine, Boston, Massachusetts 02118*

STEVEN B. KARR (41), *Department of Biochemistry, University of Georgia, Athens, Georgia 30602*

DAVID A. KEITH (31), *Department of Oral and Maxillofacial Surgery, Harvard School of Dental Medicine, Laboratory of Human Biochemistry, Children's Hospital Medical Center, Boston, Massachusetts 02115*

JAMES H. KIMURA (45), *Laboratory of Biochemistry, National Institute of Dental Research, National Institutes of Health, Bethesda, Maryland 20205*

KARI I. KIVIRIKKO (10), *Department of Medical Biochemistry, University of Oulu, SF-90220 Oulu 22, Finland*

HYNDA K. KLEINMAN (24), *Laboratory of Developmental Biology and Anomalies, National Institute of Dental Research, National Institutes of Health, Bethesda, Maryland 20205*

GARY LANGE (42), *Pulmonary Division, Department of Medicine, Washington University School of Medicine at the Jewish Hospital of St. Louis, St. Louis, Missouri 63110*

NICHOLAS D. LIGHT (16), *Department of Animal Husbandry, University of Bristol, Langford, Bristol BS18 7DU, England*

GEORGE R. MARTIN (47), *Laboratory of Developmental Biology and Anomalies, National Institute of Dental Research, National Institutes of Health, Bethesda, Maryland 20205*

MARIE-THERESE MARZULLO (44), *Department of Biochemistry, University of Georgia, Athens, Georgia 30602*

CAROL MAYS (13), *The Division of Arteriosclerosis and Metabolism, Mount Sinai School of Medicine, New York, New York 10029*

ROBERT P. MECHAM (42), *Pulmonary Division, Department of Medicine, Washington University School of Medicine at the Jewish Hospital of St. Louis, St. Louis, Missouri 63110*

GEORGE MICHALOPOULOS (29), *Department of Pathology, Jones Center for Cancer Research, Duke University Medical Center, Durham, North Carolina 27710*

EDWARD J. MILLER (1, 2), *Department of Biochemistry and Institute of Dental Research, University of Alabama in Birmingham, Birmingham, Alabama 35294*

MADELEINE MOCZAR (48), *Laboratoire de Biochimie du Tissu Conjonctif, Groupe de Recherche No. 40 du CNRS, Faculté de Médecine, Université Paris Val de Marne, 8 rue du Général Sarrail, 94010 Créteil Cedex, France*

JANET M. MONSON (8), *Department of Biochemistry and Biophysics, University of California, San Francisco, California 94143*

LOUANN W. MURRAY (3), *Department of Pediatrics, Division of Medical Genetics, UCLA-Harbor Hospital, Torrance, California 90509*

RAILI MYLLYLÄ (10), *Department of Medical Biochemistry, University of Oulu, SF-90220 Oulu 22, Finland*

MILTON E. NOELKEN (19), *Department of Biochemistry, University of Kansas Medical Center, Kansas City, Kansas 66103*

BJÖRN ÖBRINK (26), *Department of Medicine and Physiological Chemistry, University of Uppsala, The Biomedical Center, S-751 23 Uppsala, Sweden*

URSULA OTT (47), *Medizinische Universitäts-Klinik, D-3550 Marburg/Lahn, Federal Republic of Germany*

MERCEDES A. PAZ (31), *Department of Oral Biology, Harvard School of Dental Medicine, Laboratory of Human Biochemistry, Children's Hospital Medical Center, Boston, Massachusetts 02115*

BEVERLY PETERKOFSKY (21), *Laboratory of Biochemistry, Division of Cancer Biology* and Diagnosis, National Cancer Institute, National Institutes of Health, Bethesda, Maryland 20205

MICHAEL PIERSCHBACHER (46), La Jolla Cancer Research Foundation, 2945 Science Park Road, La Jolla, California 92037

DARWIN J. PROCKOP (11), Department of Biochemistry, College of Medicine and Dentistry of New Jersey-Rutgers Medical School, Piscataway, New Jersey 08854

ALAN PRZYBYLA (41), Department of Biochemistry, University of Georgia, Athens, Georgia 30602

K. B. M. REID (12), M.R.C. Immunochemistry Unit, Department of Biochemistry, University of Oxford, Oxford OX1 3QU, England

R. KENT RHODES (2), Department of Biochemistry, University of Alabama in Birmingham, Birmingham, Alabama 35294

CELESTE B. RICH (38, 41, 44), Department of Biochemistry, University of Georgia, Athens, Georgia 30602

LEILA RISTELI (47), Department of Medical Biochemistry, University of Oulu, SF-90220 Oulu 22, Finland

LADISLAS ROBERT (48), Laboratoire de Biochimie du Tissu Conjonctif, Groupe de Recherche No. 40 du CNRS, Faculté de Médecine, Université Paris Val de Marne, 8 rue du Général Sarrail, 94010 Créteil Cedex, France

PAMELA GEHRON ROBEY (47), National Eye Institute, National Institutes of Health, Bethesda, Maryland 20205

HEILWIG ROHDE (47), Max-Planck-Institut für Biochemie, Martinsried, D-8033 Munich, Federal Republic of Germany

TERRONE L. ROSENBERRY (13), Department of Pharmacology, Case Western Reserve University, Cleveland, Ohio 44106

JOEL ROSENBLOOM (40), Department of Histology, School of Dental Medicine, The University of Pennsylvania, Philadelphia, Pennsylvania 19104

ROBERT B. RUCKER (36), Department of Nutrition, University of California, Davis, Davis, California 95616

ERKKI RUOSLAHTI (46), La Jolla Cancer Research Foundation, 2945 Science Park Road, La Jolla, California 92037

HELENE SAGE (4), Department of Biochemistry, University of Washington, Seattle, Washington 98195

L. B. SANDBERG (37), Department of Pathology of the University of Utah Medical Center and Veterans Administration Hospital, Salt Lake City, Utah 84132

SAMUEL A. SANTORO (25), Division of Laboratory Medicine, Barnes Hospital and Washington University, St. Louis, Missouri 63110

JERE P. SEGREST (18), Departments of Pathology, Biochemistry and Microbiology, The Comprehensive Cancer Center, The Institute of Dental Research and Diabetes Research and Training Center, University of Alabama in Birmingham Medical Center, Birmingham, Alabama 35294

BARRY C. STARCHER (43), Division of Nutrition and Foods, Department of Home Economics, University of Texas at Austin, Austin, Texas 78712

P. J. STONE (32), Department of Biochemistry, Boston University School of Medicine, Boston, Massachusetts 02118

STEPHEN C. STROM (29), Department of Pathology, Jones Center for Cancer Research, Duke University Medical Center, Durham, North Carolina 27710

KATHLEEN A. SULLIVAN (35), Department of Biochemistry, University of Massachusetts Medical Center, Worcester, Massachusetts 01605

MARVIN L. TANZER (3), Department of Biochemistry, University of Connecticut Health Center, Farmington, Connecticut 06032

RUPERT TIMPL (22, 47), Max-Planck-Institut für Biochemie, Martinsried, D-8033 Munich, Federal Republic of Germany

D. A. TORCHIA (6), Laboratory of Biochemistry, National Institute of Dental Research, National Institutes of Health, Bethesda, Maryland 20205

LEENA TUDERMAN (11), *Department of Biochemistry, College of Medicine and Dentistry of New Jersey-Rutgers Medical School, Piscataway, New Jersey 08854*

DAN W. URRY (39), *Laboratory of Molecular Biophysics, University of Alabama in Birmingham, Birmingham, Alabama 35294*

CAROL A. VATER (20), *Department of Medicine, Dartmouth-Hitchcock Medical Center, Hanover, New Hampshire 03755*

ARTHUR VEIS (7), *Department of Oral Biology, Northwestern University Dental School, Chicago, Illinois 60611*

J. HERBERT WAITE (3), *Division of Orthopedic Surgery, University of Connecticut Health Center, Farmington, Connecticut 06032*

T. B. WOLT (37), *Department of Pathology of the University of Utah Medical Center and Veterans Administration Hospital, Salt Lake City, Utah 84132*

SU CHEN WU (44), *Department of Biochemistry, University of Georgia, Athens, Georgia 30602*

# Preface

Although interest in the predominant components of connective tissue, collagen and mucopolysaccharides, developed early in the history of biochemistry, progress in understanding the biosynthesis and physiological roles of these and other components was delayed by their structural complexity, their high molecular weight and in some cases insolubility, and by the then prevailing lack of understanding of the more general processes of protein synthesis, addition of heteropolysaccharides, secretion of proteins, etc. During the past fifteen years great progress has been made in all of these areas, not least in our understanding of the underlying special features of the chemistry of collagens and proteoglycans. These developments have combined to make possible spectacular improvements in our understanding of the biochemistry of the extracellular matrix during the past decade.

In view of the relatively limited treatment of extracellular matrix components in earlier volumes of *Methods in Enzymology*, it seemed desirable to attempt to collect in one volume the central methodology current in this field.

An emphasis on specific tissues represents a novel approach in the *Methods in Enzymology* series. We have attempted to use it in this volume, "Structural and Contractile Proteins" both in Part A, "Extracellular Matrix" and in Part B, "Contractile Apparatus and the Cytoskeleton."

It is our intent in Part A to focus on the isolation, characterization, and in a few instances, the function, of the predominant and well characterized components of the extracellular matrix. Often the procedures involved are those more generally applied in protein chemistry and have been covered in depth in other volumes of the *Methods in Enzymology* series. In such instances, we have attempted to emphasize the special chatacteristics of matrix components, anticipating ancillary use of the more general and detailed discussions of earlier volumes, as needed. On the other hand, we have attempted to present information sufficiently close to the leading edge of new understanding that in a few cases some apparent overlap occurs between related sections. Here the possible importance of minor variations in procedure has led us to accept this. We have also included a few brief reviews of major topics as an aid to understanding nomenclature and the other complexities that typically arise in a rapidly moving investigative area.

Our principal goal has been to create, through the participation of leading investigators in the field, a single volume which includes directly usable descriptions of the preparation of well defined constituents of the extracellular matrix and of the specialized techniques which have been developed to characterize them. If we have been successful, this volume should serve as an essential handbook to all of us who share a common interest in and appreciation for the central physiological importance of the matrix in multicellular animals.

We wish to emphasize our indebtedness to each contributor to this volume, and especially to those who gave us invaluable advice in planning and organization, particularly Drs. E. J. Miller and Judith Foster. We also wish to express our appreciation for the organizational and secretarial skills of Mrs. Marlene Jayne. The most helpful and pleasant cooperation of the staff of Academic Press is gratefully acknowledged.

LEON W. CUNNINGHAM<br>
DIXIE W. FREDERIKSEN

# METHODS IN ENZYMOLOGY

EDITED BY

### Sidney P. Colowick and Nathan O. Kaplan
VANDERBILT UNIVERSITY
SCHOOL OF MEDICINE
NASHVILLE, TENNESSEE

DEPARTMENT OF CHEMISTRY
UNIVERSITY OF CALIFORNIA
AT SAN DIEGO
LA JOLLA, CALIFORNIA

I. Preparation and Assay of Enzymes
II. Preparation and Assay of Enzymes
III. Preparation and Assay of Substrates
IV. Special Techniques for the Enzymologist
V. Preparation and Assay of Enzymes
VI. Preparation and Assay of Enzymes (*Continued*)
    Preparation and Assay of Substrates
    Special Techniques
VII. Cumulative Subject Index

# METHODS IN ENZYMOLOGY

EDITORS-IN-CHIEF

Sidney P. Colowick     Nathan O. Kaplan

VOLUME VIII. Complex Carbohydrates
*Edited by* ELIZABETH F. NEUFELD AND VICTOR GINSBURG

VOLUME IX. Carbohydrate Metabolism
*Edited by* WILLIS A. WOOD

VOLUME X. Oxidation and Phosphorylation
*Edited by* RONALD W. ESTABROOK AND MAYNARD E. PULLMAN

VOLUME XI. Enzyme Structure
*Edited by* C. H. W. HIRS

VOLUME XII. Nucleic Acids (Parts A and B)
*Edited by* LAWRENCE GROSSMAN AND KIVIE MOLDAVE

VOLUME XIII. Citric Acid Cycle
*Edited by* J. M. LOWENSTEIN

VOLUME XIV. Lipids
*Edited by* J. M. LOWENSTEIN

VOLUME XV. Steroids and Terpenoids
*Edited by* RAYMOND B. CLAYTON

VOLUME XVI. Fast Reactions
*Edited by* KENNETH KUSTIN

VOLUME XVII. Metabolism of Amino Acids and Amines (Parts A and B)
*Edited by* HERBERT TABOR AND CELIA WHITE TABOR

VOLUME XVIII. Vitamins and Coenzymes (Parts A, B, and C)
*Edited by* DONALD B. MCCORMICK AND LEMUEL D. WRIGHT

VOLUME XIX. Proteolytic Enzymes
*Edited by* GERTRUDE E. PERLMANN AND LASZLO LORAND

VOLUME XX. Nucleic Acids and Protein Synthesis (Part C)
*Edited by* KIVIE MOLDAVE AND LAWRENCE GROSSMAN

VOLUME XXI. Nucleic Acids (Part D)
*Edited by* LAWRENCE GROSSMAN AND KIVIE MOLDAVE

VOLUME XXII. Enzyme Purification and Related Techniques
*Edited by* WILLIAM B. JAKOBY

VOLUME XXIII. Photosynthesis (Part A)
*Edited by* ANTHONY SAN PIETRO

VOLUME XXIV. Photosynthesis and Nitrogen Fixation (Part B)
*Edited by* ANTHONY SAN PIETRO

VOLUME XXV. Enzyme Structure (Part B)
*Edited by* C. H. W. HIRS AND SERGE N. TIMASHEFF

VOLUME XXVI. Enzyme Structure (Part C)
*Edited by* C. H. W. HIRS AND SERGE N. TIMASHEFF

VOLUME XXVII. Enzyme Structure (Part D)
*Edited by* C. H. W. HIRS AND SERGE N. TIMASHEFF

VOLUME XXVIII. Complex Carbohydrates (Part B)
*Edited by* VICTOR GINSBURG

VOLUME XXIX. Nucleic Acids and Protein Synthesis (Part E)
*Edited by* LAWRENCE GROSSMAN AND KIVIE MOLDAVE

VOLUME XXX. Nucleic Acids and Protein Synthesis (Part F)
*Edited by* KIVIE MOLDAVE AND LAWRENCE GROSSMAN

VOLUME XXXI. Biomembranes (Part A)
*Edited by* SIDNEY FLEISCHER AND LESTER PACKER

VOLUME XXXII. Biomembranes (Part B)
*Edited by* SIDNEY FLEISCHER AND LESTER PACKER

VOLUME XXXIII. Cumulative Subject Index Volumes I-XXX
*Edited by* MARTHA G. DENNIS AND EDWARD A. DENNIS

VOLUME XXXIV. Affinity Techniques (Enzyme Purification: Part B)
*Edited by* WILLIAM B. JAKOBY AND MEIR WILCHEK

VOLUME XXXV. Lipids (Part B)
*Edited by* JOHN M. LOWENSTEIN

VOLUME XXXVI. Hormone Action (Part A: Steroid Hormones)
*Edited by* BERT W. O'MALLEY AND JOEL G. HARDMAN

VOLUME XXXVII. Hormone Action (Part B: Peptide Hormones)
*Edited by* BERT W. O'MALLEY AND JOEL G. HARDMAN

VOLUME XXXVIII. Hormone Action (Part C: Cyclic Nucleotides)
*Edited by* JOEL G. HARDMAN AND BERT W. O'MALLEY

VOLUME XXXIX. Hormone Action (Part D: Isolated Cells, Tissues, and Organ Systems)
*Edited by* JOEL G. HARDMAN AND BERT W. O'MALLEY

VOLUME XL. Hormone Action (Part E: Nuclear Structure and Function)
*Edited by* BERT W. O'MALLEY AND JOEL G. HARDMAN

VOLUME XLI. Carbohydrate Metabolism (Part B)
*Edited by* W. A. WOOD

VOLUME XLII. Carbohydrate Metabolism (Part C)
*Edited by* W. A. WOOD

VOLUME XLIII. Antibiotics
*Edited by* JOHN H. HASH

VOLUME XLIV. Immobilized Enzymes
*Edited by* KLAUS MOSBACH

VOLUME XLV. Proteolytic Enzymes (Part B)
*Edited by* LASZLO LORAND

VOLUME XLVI. Affinity Labeling
*Edited by* WILLIAM B. JAKOBY AND MEIR WILCHEK

VOLUME XLVII. Enzyme Structure (Part E)
*Edited by* C. H. W. HIRS AND SERGE N. TIMASHEFF

VOLUME XLVIII. Enzyme Structure (Part F)
*Edited by* C. H. W. HIRS AND SERGE N. TIMASHEFF

VOLUME XLIX. Enzyme Structure (Part G)
*Edited by* C. H. W. HIRS AND SERGE N. TIMASHEFF

VOLUME L. Complex Carbohydrates (Part C)
*Edited by* VICTOR GINSBURG

VOLUME LI. Purine and Pyrimidine Nucleotide Metabolism
*Edited by* PATRICIA A. HOFFEE AND MARY ELLEN JONES

VOLUME LII. Biomembranes (Part C: Biological Oxidations)
*Edited by* SIDNEY FLEISCHER AND LESTER PACKER

VOLUME LIII. Biomembranes (Part D: Biological Oxidations)
*Edited by* SIDNEY FLEISCHER AND LESTER PACKER

VOLUME LIV. Biomembranes (Part E: Biological Oxidations)
*Edited by* SIDNEY FLEISCHER AND LESTER PACKER

VOLUME LV. Biomembranes (Part F: Bioenergetics)
*Edited by* SIDNEY FLEISCHER AND LESTER PACKER

VOLUME LVI. Biomembranes (Part G: Bioenergetics)
*Edited by* SIDNEY FLEISCHER AND LESTER PACKER

VOLUME LVII. Bioluminescence and Chemiluminescence
*Edited by* MARLENE A. DELUCA

VOLUME LVIII. Cell Culture
*Edited by* WILLIAM B. JAKOBY AND IRA H. PASTAN

VOLUME LIX. Nucleic Acids and Protein Synthesis (Part G)
*Edited by* KIVIE MOLDAVE AND LAWRENCE GROSSMAN

VOLUME LX. Nucleic Acids and Protein Synthesis (Part H)
*Edited by* KIVIE MOLDAVE AND LAWRENCE GROSSMAN

VOLUME 61. Enzyme Structure (Part H)
*Edited by* C. H. W. HIRS AND SERGE N. TIMASHEFF

VOLUME 62. Vitamins and Coenzymes (Part D)
*Edited by* DONALD B. MCCORMICK AND LEMUEL D. WRIGHT

VOLUME 63. Enzyme Kinetics and Mechanism (Part A: Initial Rate and Inhibitor Methods)
*Edited by* DANIEL L. PURICH

VOLUME 64. Enzyme Kinetics and Mechanism (Part B: Isotopic Probes and Complex Enzyme Systems)
*Edited by* DANIEL L. PURICH

VOLUME 65. Nucleic Acids (Part I)
*Edited by* LAWRENCE GROSSMAN AND KIVIE MOLDAVE

VOLUME 66. Vitamins and Coenzymes (Part E)
*Edited by* DONALD B. MCCORMICK AND LEMUEL D. WRIGHT

VOLUME 67. Vitamins and Coenzymes (Part F)
*Edited by* DONALD B. MCCORMICK AND LEMUEL D. WRIGHT

VOLUME 68. Recombinant DNA
*Edited by* RAY WU

VOLUME 69. Photosynthesis and Nitrogen Fixation (Part C)
*Edited by* ANTHONY SAN PIETRO

VOLUME 70. Immunochemical Techniques (Part A)
*Edited by* HELEN VAN VUNAKIS AND JOHN J. LANGONE

VOLUME 71. Lipids (Part C)
*Edited by* JOHN M. LOWENSTEIN

VOLUME 72. Lipids (Part D)
*Edited by* JOHN M. LOWENSTEIN

VOLUME 73. Immunochemical Techniques (Part B)
*Edited by* JOHN J. LANGONE AND HELEN VAN VUNAKIS

VOLUME 74. Immunochemical Techniques (Part C)
*Edited by* JOHN J. LANGONE AND HELEN VAN VUNAKIS

VOLUME 75. Cumulative Subject Index Volumes XXXI, XXXII, and XXXIV-LV (in preparation)
*Edited by* EDWARD A. DENNIS AND MARTHA G. DENNIS

VOLUME 76. Hemoglobins
*Edited by* ERALDO ANTONINI, LUIGI ROSSI-BERNARDI, AND EMILIA CHIANCONE

VOLUME 77. Detoxication and Drug Metabolism
*Edited by* WILLIAM B. JAKOBY

VOLUME 78. Interferons (Part A)
*Edited by* SIDNEY PESTKA

VOLUME 79. Interferons (Part B)
*Edited by* SIDNEY PESTKA

VOLUME 80. Proteolytic Enzymes (Part C)
*Edited by* LASZLO LORAND

VOLUME 81. Biomembranes (Part H: Visual Pigments and Purple Membranes, I)
*Edited by* LESTER PACKER

VOLUME 82. Structural and Contractile Proteins (Part A: Extracellular Matrix)
*Edited by* LEON W. CUNNINGHAM AND DIXIE W. FREDERIKSEN

VOLUME 83. Complex Carbohydrates (Part D) (in preparation)
*Edited by* VICTOR GINSBURG

VOLUME 84. Immunochemical Techniques (Part D) (in preparation)
*Edited by* JOHN J. LANGONE AND HELEN VAN VUNAKIS

VOLUME 85. Structural and Contractile Proteins (Part B: The Contractile Apparatus and the Cytoskeleton) (in preparation)
*Edited by* DIXIE W. FREDERIKSEN AND LEON W. CUNNINGHAM

VOLUME 86. Prostaglandins and Arachidonate Metabolites (in preparation)
*Edited by* WILLIAM E. M. LANDS AND WILLIAM L. SMITH

# Section I

# Collagen

A. The Multiple Types and Forms of Collagen
*Articles 1 through 7*

B. Biosynthesis of Procollagen
*Articles 8 through 11*

C. Preparation of Other Proteins Containing Collagen-like Sequences
*Articles 12 and 13*

D. Characterization of Special Features in the Chemistry of Collagens
*Articles 14 through 19*

E. Proteolysis of Collagen
*Articles 20 and 21*

F. Immunology of Collagen
*Article 22*

G. Interaction of Cells with Collagen
*Articles 23 through 29*

## [1] Collagen: An Overview

By EDWARD J. MILLER and STEFFEN GAY

The protein recognized as collagen occurs in all but the simplest of organisms. In organisms representing the more advanced phyla, collagen is prevalent in a number of disparate tissues and organs. It could therefore be inferred that collagen molecules originate biologically as products of a number of cell types. Experimental verification for this inference has been obtained in numerous studies over the past few years. Newly synthesized collagen molecules are, for the most part, utilized in the construction of extracellular supporting elements. The ultimate and classic functional role of collagen is achieved by these elements, which serve to maintain the physical integrity of the different tissues and organs. Nevertheless, the supporting elements exhibit a great deal of morphological and architectural variability. They are as diverse as the relatively massive striated fibers in tendons and ligaments, the thin and often nonstriated fibrils observed in portions of hyaline cartilages, the delicate reticular networks prevalent in several connective tissues, and the apparently amorphous aggregates that occur in a variety of thin membranous sheets commonly recognized as basement membranes. Furthermore, in certain locations the collagenous supporting elements may actually incorporate or be intimately associated with a variety of extracellular matrix materials. These include amorphous or crystalline deposits of calcium phosphate salts, proteoglycan aggregates, and specific macromolecular glycoproteins, such as fibronectin.

Information concerning the biological and architectural diversity of the collagenous supporting elements has been complemented by studies detailing the molecular heterogeneity of molecules comprising the elements. In this regard, two developments have been particularly significant, i.e., the acquisition of definitive data indicating the presence of genetically and structurally distinct collagens within the same organism[1] and the detection of procollagen.[2,3] Research on the biology and chemistry of collagen in the intervening years has been largely dominated by studies leading to the amplification and refinement of the concepts generated by these initial observations. With respect to genetically determined heterogeneity, it now seems certain that collagen molecules in a given organism

[1] E. J. Miller and V. J. Matukas, *Proc. Natl. Acad. Sci. U.S.A.* **64**, 1264 (1969).
[2] D. L. Layman, E. B. McGoodwin, and G. R. Martin, *Proc. Natl. Acad. Sci. U.S.A.* **68**, 454 (1971).
[3] G. Bellamy and P. Bornstein, *Proc. Natl. Acad. Sci. U.S.A.* **68**, 1138 (1971).

collectively represent the products of one or perhaps several multigene families. As of this writing, nine different collagen chains that are utilized in the formation of various types of molecules have been identified and characterized at least with respect to general compositional features. Additional chains have been identified, and their characterization is proceeding in a number of laboratories. Therefore, the actual size of the collagen gene pool can only be estimated in conservative terms at this time. It should be noted that the term "collagen chain" in the present context refers to one member of a set of homologous polypeptide chains, each of which exhibits a molecular weight ($M_r$) of about 95,000 and each of which is characterized by the presence of the typical Gly-X-Y repeating sequence throughout 90% or more of its entire length. The term, then, does not include polypeptides known to contain relatively short collagen-like regions, such as the chains of C1q and the chains of the tail region of the acetylcholinesterase molecule.

Current evidence further indicates that the collagen chains are synthesized in the form of precursor (preprocollagen) chains. Although many details concerning the reactions involved in the processing of preprocollagen to procollagen and the processing of procollagen to collagen remain to be elucidated, the extant data indicate that not all types of procollagen molecules are processed in an equivalent fashion or to the same extent. Thus, supporting elements derived from one type of collagen may be comprised largely, if not exclusively, of collagen molecules, whereas the supporting elements derived from an alternate collagen type may be comprised of partially processed procollagen molecules or procollagen molecules. Since the chemical properties of the precursor-specific regions of procollagen molecules differ substantially from those in the main body of the molecule, it seems obvious that the general chemical properties of the supporting elements will be dependent on the relative proportions of procollagen, partially processed procollagen, and collagen molecules used in constructing the aggregates. In this regard, then, considerations relevant to the molecular heterogeneity of collagen molecules cannot be confined solely to genetically determined heterogeneity, but should also recognize heterogeneity stemming from the presence of various forms of the different types of molecules.

The scope of this overview does not permit a comprehensive survey and evaluation of the copious body of literature currently available on the biochemistry, biosynthesis, and physiological significance of the various types of collagens. We have therefore confined the discussion to those topics that are likely to be of significant value in orienting readers using a volume devoted largely to methodological aspects. References to additional recent reviews of the topics discussed are supplied in the appropriate locations in the following text.

Nomenclature

The development of the nomenclature system for the collagens and their constituent chains has been the subject of brief comments in recent years.[4,5] This discussion is intended to serve as a means of summarizing current usage of the nomenclature system. In general, the system is founded on the traditional use of the lower case Greek letter $\alpha$ to designate single-chain components observed on denaturation of a collagen preparation. Additional letters such as $\beta$, $\gamma$, and $\delta$, have been reserved, respectively, to denote covalently linked dimers, trimers, and tetramers of the $\alpha$ chains. As a result of this established pattern, the various chains identified as the primary constituents of collagen molecules have been, for the most part, designated by the letter $\alpha$ plus an additional notation that serves to distinguish each chain. Thus, when it became apparent in the early 1960s that the monomer collagen molecules extracted from certain tissues consisted of two distinct $\alpha$ chains, these were designated $\alpha 1$ and $\alpha 2$.[6] The subsequent finding that collagen molecules extracted from hyaline cartilages were comprised of another type of chain[1,7,8] prompted use of the notation $\alpha 1(II)$ for the latter chain, and it was proposed that $\alpha 1(I)$ be used for the closely related chain in the previously described heteropolymer.[1] The rationale for these assignments was the apparent high degree of structural homology between $\alpha 1(II)$ and $\alpha 1(I)$ when these chains were compared with $\alpha 2$. This has subsequently been verified in numerous sequencing studies, the results of which have been compiled in reviews.[9,10] These developments eventually led to a general consensus stipulating that the native heteropolymer be referred to as type I collagen and that the native molecule comprised of $\alpha 1(II)$ chains be referred to as type II collagen. Similar considerations were applied on detection of the $\alpha 1(III)$ chain,[11] the isolation of native molecules comprised of this chain,[12,13] and designation of these molecules as type III collagen. On discovery of molecules apparently comprised of $\alpha 1(I)$ chains it was proposed that this molecular form

---

[4] E. J. Miller, *Mol. Cell. Biochem.* **13**, 165 (1976).
[5] P. Bornstein and H. Sage, *Annu. Rev. Biochem.* **49**, 957 (1980).
[6] K. A. Piez, M. S. Lewis, G. R. Martin, and J. Gross, *Biochim. Biophys. Acta* **53**, 596 (1961).
[7] R. L. Trelstad, A. H. Kang, S. Igarashi, and J. Gross, *Biochemistry* **9**, 4993 (1970).
[8] E. J. Miller, *Biochemistry* **10**, 1652 (1971).
[9] P. P. Fietzek and K. Kühn, *Int. Rev. Connect. Tissue Res.* **7**, 1 (1976).
[10] P. Bornstein and W. Traub, in "The Proteins" (H. Neurath and R. L. Hill, eds.), 3rd ed., Vol. 4, p. 411. Academic Press, New York, 1979.
[11] E. J. Miller, E. H. Epstein, Jr., and K. A. Piez, *Biochem. Biophys. Res. Commun.* **42**, 1024 (1971).
[12] E. Chung and E. J. Miller, *Science* **183**, 1200 (1974).
[13] E. H. Epstein, Jr., *J. Biol. Chem.* **249**, 3225 (1974).

TABLE I
CURRENT NOMENCLATURE FOR THE COLLAGENS AND THEIR
CONSTITUENT CHAINS

| Collagens | Constituent chain(s)[a] | Molecular species |
|---|---|---|
| Type I | $\alpha 1(I)$,[1] $\alpha 2(I)$[b] | 1 |
| Type II | $\alpha 1(II)$[1] | |
| | $\alpha 1(II)$Major, $\alpha 1(II)$Minor[15] | Possibly 2 |
| Type III | $\alpha 1(III)$[11] | 1 |
| Type I-trimer | $\alpha 1(I)$ | 1 |
| Type IV | C, D[20–22] | |
| | 100K, 70K[23] | |
| | 100K, 70K-II[24] | |
| | $\alpha''(IV)$, $\alpha'(IV)$[25] | Possibly 2 |
| | (P3, P1), P2[26] | |
| | $\alpha 1(IV)$, $\alpha 2(IV)$[27,28] | |
| Type V | A, B[32] | |
| | $\alpha A$, $\alpha B$, $\alpha C$[31,33,34] | At least 2 |
| | $\alpha 2(V)$, $\alpha 1(V)$, $\alpha 3(V)$[5] | |

[a] Superscript numbers refer to text footnotes.
[b] The $\alpha 2$ chain is now frequently designated as $\alpha 2(I)$ to signify its presence in the type I collagen molecule.

of collagen be designated type I-trimer collagen,[14] a term that has since gained wide acceptance.

The four types of molecules mentioned above are often referred to collectively as the interstitial collagens. The nomenclature currently used for the interstitial collagens is summarized in Table I. It is important to note that, as far as current evidence is concerned, the terms type I, type III, and type I-trimer each denote a single molecular species of collagen.[4] However, recent sequence data strongly suggest that there may be at least two distinct chains within the population of chains originally designated as $\alpha 1(II)$.[15] Thus, there is the distinct possibility that more than one molecular species is actually extant in the population of molecules currently referred to as type II collagen.

Concomitant with the development of information concerning the interstitial collagens, it became clear that collagen resident in a number of basement membrane structures differed substantially from any of the interstitial collagens. Initial studies designed to characterize the basement membrane collagen molecules suggested that they were comprised of a single unique $\alpha$ chain,[16] and it was proposed that this chain be designated

[14] R. Mayne, M. S. Vail, and E. J. Miller, *Proc. Natl. Acad. Sci. U.S.A.* **72**, 4511 (1975).
[15] W. T. Butler, J. E. Finch, Jr., and E. J. Miller, *J. Biol. Chem.* **252**, 639 (1977).
[16] N. A. Kefalides, *Biochem. Biophys. Res. Commun.* **45**, 226 (1971).

as $\alpha 1(IV)$.[17] Accordingly, basement membrane collagen molecules have frequently been referred to as type IV collagen. The initial conclusions with respect to the nature and number of the chains comprising the collagen molecules of basement membranes were disputed in subsequent studies,[18,19] the results of which suggested the presence of multiple components as the primary constituents of basement membrane collagen molecules.

Some of the apparent conflicts in these studies have been resolved in studies showing that basement membrane collagen molecules contain at least two distinct chains[20-28] that in native conformation exhibit a relatively high degree of susceptibility to proteolysis[20,21,26] compared to other known collagen chains. A plethora of terms have been used to designate the two basement membrane chains. These are listed for reference purposes in Table I along with appropriate citations. At the moment, definitive information concerning the distribution of these chains in native molecules is lacking. Nevertheless, the different rates at which the two chains are degraded on exposure of basement membrane collagen to proteolytic enzymes,[29] as well as the varying ratios of the chains observed in precipitates formed from basement membrane collagen preparations,[28,30] strongly suggest that the chains reside in different molecules.

An additional development to be reckoned with in this context is the discovery of what is frequently referred to as AB collagen or type V collagen. Initial studies on type V collagen indicated the presence of two distinct chains, designated $\alpha A$ and $\alpha B$[31] or A and B.[32] Further studies revealed that certain preparations of type V collagen contain yet another chain, designated $\alpha C$.[33,34] More recently, it has been proposed that the

[17] N. A. Kefalides, *Int. Rev. Connect. Tissue Res.* **6**, 63 (1973).
[18] J. R. Daniels and G. H. Chu, *J. Biol. Chem.* **250**, 3531 (1975).
[19] T. Sato and R. G. Spiro, *J. Biol. Chem.* **251**, 4062 (1976).
[20] T. F. Kresina and E. J. Miller, *Biochemistry* **18**, 3089 (1979).
[21] S. Gay and E. J. Miller, *Arch. Biochem. Biophys.* **198**, 370 (1979).
[22] S. N. Dixit, *FEBS Lett.* **106**, 379 (1979).
[23] A. J. Bailey, T. J. Sims, V. C. Duance, and N. D. Light, *FEBS Lett.* **99**, 361 (1979).
[24] H. Sage, R. G. Woodbury, and P. Bornstein, *J. Biol. Chem.* **254**, 9893 (1979).
[25] R. W. Glanville, A. Rauter, and P. P. Fietzek, *Eur. J. Biochem.* **95**, 383 (1979).
[26] R. Timpl, P. Bruckner, and P. Fietzek, *Eur. J. Biochem.* **95**, 255 (1979).
[27] E. Crouch, H. Sage, and P. Bornstein, *Proc. Natl. Acad. Sci. U.S.A.* **77**, 745 (1980).
[28] K. Tryggvason, P. G. Robey, and G. R. Martin, *Biochemistry* **19**, 1284 (1980).
[29] V.-J. Uitto, D. Schwartz, and A. Veis, *Eur. J. Biochem.* **105**, 409 (1980).
[30] P. G. Robey and G. R. Martin, *Collagen Rel. Res.* **1**, 27 (1981).
[31] R. E. Burgeson, F. A. El Adli, I. I. Kaitila, and D. W. Hollister, *Proc. Natl. Acad. Sci. U.S.A.* **73**, 2579 (1976).
[32] E. Chung, R. K. Rhodes, and E. J. Miller, *Biochem. Biophys. Res. Commun.* **71**, 1167 (1976).
[33] R. A. Brown, C. A. Shuttleworth, and J. B. Weiss, *Biochem. Biophys. Res. Commun.* **80**, 866 (1978).

αA, αB, and αC chains of type V collagen be designated as α2(V), α1(V), and α3(V), respectively.[5] The various terms used for the chains of type V collagen are also summarized in Table I. There is evidence for the presence of molecules with the chain composition $[\alpha 1(V)]_2 \alpha 2(V)$[31,35,36] and $[\alpha 1(V)]_3$[37,38] in various preparations of type V collagen. There undoubtedly exist other molecular species in which at least one of the constituent chains is α3(V). Therefore, the term type V collagen likewise denotes more than one molecular species of collagen.

In addition to the collagens and chains discussed so far, recent evidence has been offered for the existence of a series of cysteine-rich collagenous fragments derived from two unique high molecular weight aggregates,[39,40] and at least three additional collagen chains have been tentatively identified in extracts of young cartilages.[41] The molecular organization of these components is currently not known, and their relationship with the other collagens and their constituent chains has not been determined. The collagens represented by these components are therefore not presently classified within the framework of the nomenclature system described above.

It is clear that the current nomenclature system developed for the collagens and their constituent chains is not without its ambiguities and inconsistencies. Most notable among these is use of the term type I collagen to specify a single molecular species, whereas similar terms, such as type II, type IV, and type V, most likely denote a mixture containing more than one molecular species. Moreover, identification of the individual chains as α1(I), α1(II), α1(III), α1(IV), and α1(V) as well as α2(I), α2(IV), and α2(V) more than likely does not specify the degree of structural homology between the various chains as envisioned in the original use of the α1(I) and α1(II) designations. These developments, however, are to be expected in a rapidly expanding field, and it is likely that a truly satisfactory and comprehensive nomenclature system cannot be adopted in the absence of more detailed structural analyses of all extant collagen chains

---

[34] H. Sage and P. Bornstein, *Biochemistry* **18**, 3815 (1979).
[35] H. Bentz, H. P. Bächinger, R. Glanville, and K. Kühn, *Eur. J. Biochem.* **92**, 563 (1978).
[36] C. Welsh, S. Gay, R. K. Rhodes, R. Pfister, and E. J. Miller, *Biochim. Biophys. Acta* **625**, 78 (1980).
[37] R. K. Rhodes and E. J. Miller, *Biochemistry* **17**, 3442 (1978).
[38] M. A. Haralson, W. M. Mitchell, R. K. Rhodes, T. F. Kresina, R. Gay, and E. J. Miller, *Proc. Natl. Acad. Sci. U.S.A.* **77**, 5206 (1980).
[39] D. K. Furuto and E. J. Miller, *J. Biol. Chem.* **255**, 290 (1980).
[40] J. Risteli, H. P. Bächinger, J. Engel, H. Furthmayr, and R. Timpl, *Eur. J. Biochem.* **108**, 239 (1980).
[41] R. E. Burgeson and D. W. Hollister, *Biochem. Biophys. Res. Commun.* **87**, 1124 (1979).

as well as more complete information concerning the molecular species formed by these chains.

The nomenclature generally used for procollagen and intermediate forms of these molecules generated in the processing of procollagen to collagen is based on the system established for the various collagens with additional notations specifying the characteristic features of the procollagen and their intermediates. In early studies on procollagen, the term was used to denote a precursor form of the collagen molecule, and the constituent chains of the procollagen molecule were designated with the prefix "pro," e.g., pro-$\alpha$1 and pro-$\alpha$2. At the same time, p-collagen was used to denote a modified (truncated or partially processed) form of the precursor molecule, the constituent chains of which were referred to as p-$\alpha$1 and p-$\alpha$2. With advancing information concerning the nature of the procollagen molecule and the existence of various types of procollagen molecules, a more precise set of terms has come into general use. Thus, procollagen type I designates the intact precursor of the type I molecule, and use of this term indicates the presence of a complete precursor-specific domain at both the $NH_2$- and COOH-terminal extremities of the molecule. The constituent chains of procollagen type I are designated pro-$\alpha$1(I) and pro-$\alpha$2(I). The term pN-collagen type I designates a form of the precursor from which the COOH-terminal precursor-specific domain has been removed. The constituent chains of this form of the precursor are designated pN-$\alpha$1(I) and pN-$\alpha$2(I). By analogy with pN-collagen type I, pC-collagen type I denotes a form of the precursor that is devoid of the $NH_2$-terminal precursor-specific domain. The constituent chains of the latter form of the precursor are designated pC-$\alpha$1(I) and pC-$\alpha$2(I). These terms, which are now in general use to denote procollagen type I, its intermediate forms, and the chains of the respective molecules, are summarized in Table II. An entirely analogous system is used for other procollagens.

Preparation of Collagens and Procollagens

In the preparation of collagens and procollagens, the investigator is presented with several different options. The ultimate strategy chosen for a given preparation will, in large part, be determined by the type and form of collagen to be obtained as well as the anticipated yield of the desired protein. Our purpose here is not to present a multitude of procedural details, since these will be covered at length in the ensuing contributions. Rather, we intend to introduce these topics by providing a concise compendium of the concepts governing the design and use of techniques for the isolation of the various collagens and their respective precursor forms.

TABLE II
CURRENT NOMENCLATURE FOR THE VARIOUS FORMS OF
PROCOLLAGEN I AND THEIR CONSTITUENT CHAINS[a]

| Procollagen form | Constituent chains |
|---|---|
| Procollagen type I (intact precursor) | pro-$\alpha$1(I), pro-$\alpha$2(I) |
| pN-collagen type I (precursor devoid of COOH-terminal sequences) | pN-$\alpha$1(I), pN-$\alpha$2(I) |
| pC-collagen type I (precursor devoid of NH$_2$-terminal sequences) | pC-$\alpha$1(I), pC-$\alpha$2(I) |

[a] The nomenclature used for other types of procollagen, intermediate forms, and their constituent chains follows the outline given here for procollagen type I.

For the sake of clarity, preparation of the collagens and procollagens will be treated separately even though similar approaches are employed in the isolation of these molecules.

*Collagens*

In the preparation of the various collagens, initial concerns involve the choice of a convenient source as well as the use of appropriate preliminary procedures designed to optimize the yield of collagen during extraction. Although type I collagen can be prepared from virtually any tissue or organ, the most convenient sources for the collagen are dermis, tendon, and bone of relatively young organisms. Each of these tissues contains an abundance of collagen. The fibers of tendon and bone are derived almost exclusively from type I collagen molecules,[4] which considerably simplifies the procedures required to purify type I collagen extracted from these sources. Moreover, prior to extraction of dermis or tendon, the tissues generally need only be dispersed in order to achieve relatively good yields of collagen. Effective yields of native bone collagen are obtained, however, only if cross-linking has been inhibited in experimental animals by administration of a lathyrogen and the bones are subsequently demineralized.[42]

Type II collagen is most conveniently prepared from specimens of hyaline cartilage, such as mammalian articular cartilage, epiphysial growth plate, the chick sternum, and bovine nasal septum.[4] These tissues are somewhat less "collagen-rich" than the tissues commonly used in the preparation of type I collagen. Nevertheless, their fibrous elements are formed largely from type II collagen molecules. Induction of a state of

[42] E. J. Miller, G. R. Martin, K. A. Piez, and M. J. Powers, *J. Biol. Chem.* **242**, 5481 (1967).

lathyrism in experimental animals has a dramatic effect on the extractability of type II collagen,[1,7,8,43] which is similar to that noted above for bone collagen. In addition, extraction of type II collagen is generally facilitated by preliminary extraction procedures designed to eliminate proteoglycan molecules present in the tissues.[44,45]

There apparently exist no tissues or organs in which the prevalence of type III collagen approaches that of type I collagen in tendon and bone or that of type II collagen in hyaline cartilages. Nevertheless, the dermis of relatively young organisms[12,13,46,47] constitutes a good source of type III collagen. It has also been prepared in good yields from major vessels,[12,48] a smooth muscle cell tumor,[12] lung parenchymal tissue,[49] and cirrhotic liver.[50] Although type III collagen molecules are apparently cross-linked through mechanisms common to the other known collagens,[51-53] there have been no studies demonstrating enhanced extractability of this collagen in lathyrism. However, it has been observed that prior treatment of major vessels with elastase facilitates the subsequent solubilization of collagen,[54] suggesting that the yields of type III collagen from vascular tissue might be increased by use of this protocol. Type I-trimer collagen apparently is present in minimal amounts in several connective tissues and has been prepared from dermis[55] and dentin[56] as well as a polyoma virus-induced murine tumor.[57]

The relatively large and accessible basement membrane structures, such as the lens capsule, the glomerular basement membrane, and Descemet's membrane, have been the traditional sources of type IV collagen

[43] B. D. Smith, G. R. Martin, E. J. Miller, A. Dorfman, and R. Swarm, *Arch. Biochem. Biophys.* **166**, 181 (1975).
[44] E. J. Miller, *Biochemistry* **11**, 4903 (1972).
[45] E. J. Miller and L. G. Lunde, *Biochemistry* **12**, 3153 (1973).
[46] P. H. Byers, K. H. McKenney, J. R. Lichtenstein, and G. R. Martin, *Biochemistry* **13**, 5243 (1974).
[47] R. Timpl, R. W. Glanville, H. Nowack, H. Wiedemann, P. P. Fietzek, and K. Kühn, *Hoppe-Seyler's Z. Physiol. Chem.* **356**, 1783 (1975).
[48] E. H. Epstein, Jr. and N. H. Munderloh, *J. Biol. Chem.* **250**, 9304 (1975).
[49] B. D. McLees, G. Schleiter, and S. R. Pinnell, *Biochemistry* **16**, 185 (1977).
[50] J. M. Seyer, E. T. Hutcheson, and A. H. Kang, *J. Clin. Invest.* **59**, 241 (1977).
[51] A. J. Bailey and T. J. Sims, *Biochem. J.* **153**, 211 (1976).
[52] K. Fujii, M. L. Tanzer, B. V. Nusgens, and C. M. Lapiere, *Biochem. Biophys. Res. Commun.* **69**, 128 (1976).
[53] W. Henkel, J. Rauterberg, and R. W. Glanville, *Eur. J. Biochem.* **96**, 249 (1979).
[54] B. Faris, R. Moscaritolo, A. Levine, R. Snider, R. Goldstein, and C. Franzblau, *Biochim. Biophys. Acta* **534**, 64 (1978).
[55] J. Uitto, *Arch. Biochem. Biophys.* **192**, 371 (1979).
[56] M. Wohllebe and D. J. Carmichael, *Eur. J. Biochem.* **92**, 183 (1978).
[57] L. Moro and B. D. Smith, *Arch. Biochem. Biophys.* **182**, 33 (1977).

molecules.[17] The latter may also be prepared from extracts of complex organs, such as the human placenta,[20,23-25] as well as murine basement membrane-containing tumor.[58,59] Since the tumor may be grown in animals during the induction of lathyrism, the subsequent extraction of intact basement membrane collagen molecules is considerably facilitated. Whole placenta,[34,37] the amnion and chorion,[31,34,35] as well as cornea,[36] have been useful in preparing relatively large amounts of type V collagen.

Although a variety of solvents have been used to extract or disperse collagen from tissues,[10] the solvents commonly used in the preparation of the various collagens are neutral salt solutions ($0.15-2.0\,M$ NaCl) or $0.5\,M$ acetic acid. When employed at low temperatures, these solvents dissociate collagenous fibrous elements into native, largely monomeric molecules. In contrast, use of denaturing solvents generally results in extraction of large amounts of aggregates that are difficult to resolve with present techniques. Moreover, many of the approaches recently developed to separate the various collagens once they have been brought into solution are applicable only for the native proteins. In general, neutral salt solvents preferentially extract newly synthesized molecules that are as yet retained within the fibers only through noncovalent bonds. The efficacy of dilute acid solvents stems largely from their low pH, which induces swelling of the fibers and promotes the dissociation of intermolecular cross-links containing aldimine bonds. Intermolecular cross-links containing the ketoamine configuration are much less labile.[60] Therefore, collagen fibers in which the latter form of covalent cross-linking is prevalent, such as in bone, cartilage, and a number of other tissues, exhibit little tendency to dissolve on exposure to dilute acid solvents. As noted above, effective extraction of collagens from these tissues as native intact molecules requires measures to inhibit the formation of covalent cross-links.

Under the best conditions, extracts of collagen, particularly those obtained with dilute acid solvents, commonly contain aggregates of covalently cross-linked molecules. The aggregates may be effectively eliminated, however, by selective salt precipitation from $0.5\,M$ acetic acid at relatively low NaCl concentrations.[61] Although the helical region of most collagen molecules is resistant to nonspecific proteolysis, the short nonhelical sequences at both the $NH_2$- and COOH-terminal ends of the molecules are subject to proteolysis by tissue proteases during extraction or

---

[58] R. W. Orkin, P. Gehron, E. B. McGoodwin, G. R. Martin, T. Valentine, and R. Swarm, *J. Exp. Med.* **145**, 204 (1977).

[59] R. Timpl, G. R. Martin, P. Bruckner, G. Wick, and H. Wiedemann, *Eur. J. Biochem.* **84**, 43 (1978).

[60] A. J. Bailey, S. P. Robins, and G. Balian, *Nature (London)* **251**, 105 (1974).

[61] G. Chandrakasan, D. A. Torchia, and K. A. Piez, *J. Biol. Chem.* **251**, 6062 (1976).

purification, or both. It has been recommended that protease inhibitors be included in the extracting solvents and that the extracted collagen be purified rapidly in order to avoid proteolytic alteration of the nonhelical ends.[61]

A somewhat different approach to the extraction of collagen has frequently been used in recent years and involves the deliberate use of proteolytic enzymes to facilitate dissolution of collagen fibers. Use of this approach has been stimulated by the desire to evaluate the nature of the collagens in a wide variety of tissues where use of more conventional extraction procedures is not productive. This situation arises when the molecules are intermolecularly cross-linked by acid-stable bonds and inhibition of cross-linking is not feasible owing to ethical or practical considerations. The rationale for this approach rests on the observations that exposure of collagen molecules to proteolytic enzymes under nondenaturing conditions has relatively little effect on the overall structural features of the molecules, yet releases small noncollagenous sequences and serves to depolymerize the molecules.[62] Additional studies revealed that the susceptible peptide bonds are located in the $NH_2$-[63] and COOH-[64,65] terminal nonhelical ends of the molecules and that these regions contain sites for both intramolecular[66] and intermolecular[67-69] cross-link formation. Limited proteolysis, then, facilitates the extraction of otherwise insoluble collagen pools by selectively cleaving peptide bonds in cross-linking sequences, thereby rendering the molecules soluble largely as monomers. Pepsin is the most commonly used enzyme for this purpose chiefly because it is available in pure form from commercial sources and can be employed in acidic solvents in which the monomer molecules are readily dissolved. Although limited proteolysis with pepsin has been extremely useful in preparing relatively large amounts of the various collagens in essentially monomeric form from a number of animal and human tissues, the procedure has its inherent limitations. Its most obvious limitation is that the molecules are obtained with altered nonhelical extremities; this effectively precludes subsequent studies designed to evaluate the structure and function of these regions. In addition, the portion of type IV collagen molecules homologous to the major helical domain in other colla-

[62] M. P. Drake, P. F. Davison, S. Bump, and F. O. Schmitt, *Biochemistry* **5**, 301 (1966).
[63] P. Bornstein, A. H. Kang, and K. A. Piez, *Biochemistry* **5**, 3803 (1966).
[64] M. Stark, J. Rauterberg, and K. Kühn, *FEBS Lett.* **13**, 101 (1971).
[65] J. Rauterberg, P. P. Fietzek, F. Rexrodt, U. Becker, M. Stark, and K. Kühn, *FEBS Lett.* **21**, 75 (1972).
[66] P. Bornstein and K. A. Piez, *Biochemistry* **5**, 3460 (1966).
[67] E. J. Miller, *Biochem. Biophys. Res. Commun.* **45**, 444 (1971).
[68] A. H. Kang, *Biochemistry* **11**, 1828 (1972).
[69] W. Henkel, J. Rauterberg, and T. Stirtz, *Eur. J. Biochem.* **69**, 223 (1976).

gens or procollagens contains pepsin-susceptible bonds;[70] this results in considerable fragmentation of the molecules isolated in this fashion.[20,21]

Regardless of the method chosen to extract collagen molecules, the extract is likely to contain a variety of impurities. The nature and proportion of the latter substances are dependent on the source of the extract. In general, native collagen molecules present in solution at or near neutral pH can be quantitatively precipitated at high (3.5–4.0 $M$ NaCl) or low (below 0.15 $M$ NaCl) salt concentrations. When present in dilute acid solvents, the molecules can be precipitated by the addition of NaCl to a concentration of 1.0–2.0 $M$. Historically, repeated precipitation from neutral-salt and dilute-acid solvents under the conditions outlined above has constituted the preferred method for recovery and purification of extracted collagen molecules. Use of the indicated precipitation procedures, however, does not discriminate between the various types of collagen molecules that might be present in a given extract. Consequently, these procedures have essentially been replaced in recent years by more selective precipitation techniques designed to separate the collagens on the basis of their different solubility properties. The latter approach was introduced in studies on the collagens extracted from chick sternal cartilages.[71] These studies showed that the small amount of type I collagen present in neutral salt extracts could be effectively precipitated from solution at a NaCl concentration of about 2.2 $M$, whereas the type II collagen present in the extracts remained in solution under these conditions and could subsequently be precipitated on increasing the NaCl concentration to 4.4 $M$. Selective salt precipitation from both neutral salt solvents and dilute acid solvents has subsequently been employed to achieve at least initial resolution and purification of the collagens encountered in extracts of various tissues.

In addition to selective salt precipitation, which is performed at relatively low temperatures, preliminary results indicate that type IV[72] as well as type V[73] collagen molecules fail to precipitate under conditions that promote the formation of native heat gels on the part of the types I, II, and III collagens. This approach provides an alternative method of achieving initial resolution of the collagens present in complex mixtures but has the disadvantage that the nonprecipitating collagens must be exposed in solution to temperatures approaching 37°, where some degree of denaturation is possible.

Selective salt precipitation represents a relatively rapid and convenient

[70] D. Schwartz and A. Veis, *FEBS Lett.* **85**, 326 (1978).
[71] R. L. Trelstad, A. H. Kang, B. P. Toole, and J. Gross, *J. Biol. Chem.* **247**, 6469 (1972).
[72] R. L. Trelstad and K. R. Lawley, *Biochem. Biophys. Res. Commun.* **76**, 376 (1977).
[73] R. L. Trelstad and A. C. A. Carvalho, *J. Lab. Clin. Med.* **93**, 499 (1979).

method for the separation and purification of the collagens present in a given mixture. Nevertheless, as might be expected, the degree of purity attained by selective salt precipitation is often less than ideal. This may be due to the presence of relatively large amounts of noncollagenous contaminants as well as the failure to achieve quantitative precipitation of the various collagens at the appropriate salt concentrations. The latter factor, then, results in contamination of collagen precipitating at relatively high salt concentrations with some collagen, which theoretically should be removed from solution at lower salt concentrations. This problem is most frequently encountered when one of the molecular species of collagen present in the mixture represents a high proportion of the total collagen.[38]

In an attempt to overcome these problems and achieve greater purification of the isolated collagens, methods have been developed for chromatographing the collagens under nondenaturing conditions. Thus, application of type II[8] and type IV[59] collagen preparations to DEAE-cellulose under conditions in which the native collagens are not retained by the column has proved to be an effective means of removing noncollagenous contaminants and of resolving type IV collagen from other apparently collagenous contaminants. It has also been shown that preparations of type V collagen can be successfully chromatographed and resolved from other collagens on columns of DEAE-[35] or CM-[36] cellulose. Moreover, chromatography of type IV collagen on CM-cellulose[20,21,25] is useful in resolving this collagen from other potential contaminating collagens. And finally, zone precipitation chromatography employing stepwise decreases in ionic strength at neutral pH to resolubilize collagens precipitated on the column appears to hold some promise in facilitating critical separations of the collagens, particularly when collagens present in relatively small samples are to be resolved.[74] This procedure relies on the well-known solubility properties of the various collagens at different salt concentrations, and its utility will be dependent on the ability to achieve quantitative resolubilization of a given type of collagen in the appropriate eluent. Although the range of separations attainable in any single chromatographic procedure appear, at this time, to be somewhat limited, these procedures represent effective means of achieving final purification of specific collagen fractions obtained by selective salt precipitation.

*Procollagens*

Since the methods currently employed to isolate procollagens are equally applicable to the isolation of procollagen intermediates, the term "procollagen" is used here to denote the various molecular species of

[74] H. P. Ehrlich, *Prep. Biochem.* **9**, 407 (1979).

apparently intact procollagen molecules as well as their respective intermediates. In general, the strategies devised to isolate procollagens are patterned after those found advantageous in isolating the collagens. For reasons already discussed above, neutral salt solvents containing protease inhibitors are the preferred solvents for extraction of native procollagens and have been employed to extract these proteins from a variety of sources including fresh tissue preparations,[46,47] tissues maintained in culture,[75] as well as the cell layers of cultured cells.[76] In an exception to this general rule, procollagen type IV molecules are most readily extracted from fresh tissue preparations,[59] cultured tissues,[28] and cultured cells[77] in dilute acid solvents. Procollagens present in the media of cultured cells are conveniently recovered by precipitation with ammonium sulfate.[76] Procollagens obtained from these sources may generally be resolved by selective salt precipitation under conditions identical to those used for precipitation of the various collagens.[47] This technique, however, does not differentiate between the individual collagens and their respective procollagens. Therefore, depending on the source of protein, precipitates obtained by this procedure may contain a mixture of a given collagen plus one or more of its precursor forms, necessitating an additional purification step for recovery of the procollagen molecules. In general, this is most conveniently achieved by chromatography of the mixtures on DEAE-cellulose under conditions in which native collagen molecules are not retained on the column while various forms of the precursor molecules are retained owing to their greater negative charge at slightly alkaline pH. Using this approach, satisfactory resolution of the molecular species present in relatively simple mixtures has been attained. Thus, DEAE-cellulose chromatography has proved to be useful in resolving procollagen types I and III from each other and their respective collagens,[46] in separating type III collagen from pN-collagen type III,[47] and in resolving mixtures of pC-collagen type I and procollagen type I.[78] The limitations of this method, however, lie in the incomplete resolution attained when chromatography of more complex mixtures are attempted. In this regard, it has been shown that type I collagen and pC-collagen type I as well as pN-collagen type I and procollagen type I coelute when chromatographed on DEAE-cellulose.[79] This technique, then, allows only partial resolution of a given procollagen, its potential intermediate forms, and collagen. Isolation and purification of these molecules is thus dependent on the use

[75] J. M. Monson, E. M. Click, and P. Bornstein, *Biochemistry* **14,** 4088 (1975).
[76] R. L. Church, J. A. Yaeger, and M. L. Tanzer, *J. Mol. Biol.* **86,** 785 (1974).
[77] E. Crouch, H. Sage, and P. Bornstein, *Proc. Natl. Acad. Sci. U.S.A.* **77,** 745 (1980).
[78] J. Uitto, J. R. Lichtenstein, and E. A. Bauer, *Biochemistry* **15,** 4935 (1976).
[79] E. Crouch and P. Bornstein, *Biochemistry* **17,** 5499 (1978).

of various systems in which discrete and potentially resolvable mixtures of the molecular species are present.

Procollagens may also be extracted from tissues in denaturing solvents.[80-82] Extraction under these conditions may result in the presence of a large proportion of high molecular weight aggregates in the extract.[80] Moreover, the procollagens are present in denatured form. Nevertheless, it has been shown that mixtures of procollagen types I and III can be resolved in denatured form by velocity sedimentation in sucrose gradients.[81] The resolving power of this technique is likely due to the presence of interchain disulfide linkages at both ends of the procollagen type III molecule, which considerably restricts the hydrodynamic volume of these molecules in the denatured state relative to that of procollagen type I molecules containing interchain disulfide linkages only in the COOH-terminal region. Moreover, owing to the presence of interchain disulfide linkages, reacquisition of native conformation on the part of denatured procollagen molecules is a relatively rapid and efficient process.[81] This factor, then, would allow the use of fractionation procedures dependent on the presence of native conformation, even though the molecules are obtained originally in denatured form.

It should be further noted that procollagens may be resolved from collagens by covalent chromatography on columns containing activated thiol groups[83] or by adsorption chromatography on porous glass beads.[84] Selective retention of procollagen on thiol columns, however, requires that the molecules be reduced under denaturing conditions prior to application to the column. Moreover, neither of these methods is capable of resolving the various types of procollagens or the various intermediate forms of the precursors that might be present. Such procedures may prove to be valuable, however, as initial steps in the resolution of complex mixtures, particularly when the molecular species of procollagen are present in relatively small proportions.

Characterization of Collagens and Procollagens

After preparation of a collagen or procollagen, the protein may be characterized by a variety of physical and chemical techniques. Studies of this nature have led to considerable insight into the structural features of collagen and procollagen molecules, their chain composition, as well as

[80] U. Becker, R. Timpl, O. Helle, and D. J. Prockop, *Biochemistry* **15**, 2853 (1976).
[81] L. I. Fessler and J. H. Fessler, *J. Biol. Chem.* **254**, 233 (1979).
[82] J. Uitto, R. E. Allan, and K. L. Polak, *Eur. J. Biochem.* **99**, 97 (1979).
[83] K. Angermann and H.-J. Barrach, *Anal. Biochem.* **94**, 253 (1979).
[84] S. Gerard and W. M. Mitchell, *Anal. Biochem.* **96**, 433 (1979).

the nature of their constituent chains. In addition, fragmentation of the chains by chemical and enzymic means, resolution of the cleavage products by standard chromatographic procedures, and application of automated sequence techniques to the isolated peptides have led to significant advances in the determination of the primary structure of the chains. The information derived from these studies will be summarized here with particular emphasis on the compositional features of the various $\alpha$ chains and the general structural features of procollagen molecules as determined by studies on collagen as well as procollagen molecules.

*Collagens*

The constituent chains of most of the known collagens can be resolved by chromatography on CM-cellulose under denaturing conditions. Depending on the collagen, its source, and its mode of preparation, molecular sieve chromatography is often used in conjunction with CM-cellulose chromatography to remove cross-linked components that are incompletely resolved from the primary chain constituents during ion-exchange chromatography. This approach has been widely used to resolve and purify the constituent chains derived from preparations of type I, type II, and type III collagens.[4,85] It has also been found useful in resolving the pepsin-resistant regions derived from each of the constituent chains of type IV collagen.[20-26] The chains present in molecules comprising type V collagen preparations, however, are not readily resolved by CM-cellulose chromatography but may be resolved by chromatography on phosphocellulose,[37] DEAE-cellulose,[35] or hydroxyapatite,[86] or by use of sequential separations on phosphocellulose and CM-cellulose.[34] In general, the ion-exchange properties of the individual collagen chains are indicative of their respective identities. Thus, the chromatographic procedures can be used not only for preparative purposes but for analytical purposes. It should be noted, however, that the resolving power of any single chromatographic procedure is limited. Therefore, the results obtained in both preparative and analytical approaches are more reliable provided the specific collagen under investigation has been carefully purified prior to denaturation.

A somewhat more versatile analytical procedure is sodium dodecyl sulfate (SDS)–polyacrylamide gel electrophoresis employing either cylindrical[87,88] or slab[89,90] gels. Although the various collagen $\alpha$ chains are

[85] E. J. Miller, in "The Methodology of Connective Tissue Research" (D. A. Hall, ed.), p. 197. Joynson-Bruvvers, Oxford, 1976.
[86] B.-S. Hong, P. F. Davison, and D. J. Cannon, *Biochemistry* 18, 4278 (1979).
[87] H. Furthmayr and R. Timpl, *Anal. Biochem.* 41, 510 (1971).
[88] T. Hayashi and Y. Nagai, *J. Biochem.* 86, 453 (1979).
[89] F. W. Studier, *J. Mol. Biol.* 79, 237 (1973).

virtually identical in molecular weight, a number of the chains exhibit characteristic mobilities when electrophoresed in these systems.[31,33,37,88,91] This factor allows the resolution and identification of the chains present in relatively simple mixtures. As pointed out with respect to chromatographic techniques, however, the resolving power of polyacrylamide gel electrophoresis is likewise limited, and the most accurate and reliable results are likely to be obtained after electrophoresis of the denaturation products derived from highly purified collagen preparations. Estimates of chain ratios in collagen preparations based on ion-exchange or electrophoresis profiles have been valuable in conjunction with other data in assessing the chain composition of the various collagens.

The compositional features of the individual collagen α chains are presented in Table III. Although interspecies differences for a given type of chain are generally minimal, the data presented here are derived exclusively from studies on human collagens in order to allow chain–chain comparisons in the absence of species differences. Moreover, since the majority of the human collagens can be obtained in good yields only through limited proteolysis to facilitate their solubilization, all the values listed in Table III are for chains derived from collagens brought into solution by means of limited pepsin digestion. This information represents the most complete set of data currently available for each of the chains.

Inspection of these data reveals, as expected, that the number of glycyl residues in each chain is highly conserved at approximately one-third of the total amino acid residues. Other features, however, clearly serve to differentiate the individual chains as well as groups of chains. With respect to the latter point, the chains derived from types IV and V collagens contain much less alanine than do the chains derived from types I, II, and III collagens. The decreased alanine content in the type IV chains is accompanied by a sharp increase in the total number of the larger hydrophobic amino acids. In the type V chains, however, the decreased alanine content is associated with a less dramatic increase in the number of the larger hydrophobic residues plus a moderate increase in the total number of the basic amino acid residues. These data thus strongly suggest that these collagen chains may be classified, for the moment at least, into three general groups of chains, i.e., the interstitial collagen chains containing relatively large amounts of alanine, the type IV chains characterized by a large number of the larger hydrophobic amino acids, and the type V chains containing intermediate quantities of the larger hydrophobic residues as well as somewhat elevated levels of the basic amino acids.

---

[90] B. U. Steinmann and A. H. Reddi, *Biochem. J.* **186**, 919 (1980).
[91] B. Sykes, B. Puddle, M. Francis, and R. Smith, *Biochem. Biophys. Res. Commun.* **72**, 1472 (1976).

TABLE III
AMINO ACID COMPOSITION OF THE HUMAN COLLAGEN CHAINS (RESIDUES/1000 TOTAL RESIDUES)

| Amino acid | α1(I)[a] | α2(I)[a] | α1(II)[b] | α1(III)[c] | C or α1(IV)[d] | D or α2(IV)[d] | αA or α2(V)[e] | αB or α1(V)[e] | αC or α3(V)[f] |
|---|---|---|---|---|---|---|---|---|---|
| 3-Hydroxyproline | 1 | 1 | 2 | 0 | 1 | 1 | 3 | 5 | 1 |
| 4-Hydroxyproline | 108 | 93 | 97 | 125 | 122 | 110 | 106 | 110 | 91 |
| Aspartic acid | 42 | 44 | 43 | 42 | 45 | 49 | 50 | 49 | 42 |
| Threonine | 16 | 19 | 23 | 13 | 19 | 30 | 29 | 21 | 19 |
| Serine | 34 | 30 | 25 | 39 | 38 | 30 | 34 | 23 | 34 |
| Glutamic acid | 73 | 68 | 89 | 71 | 78 | 65 | 89 | 100 | 97 |
| Proline | 124 | 113 | 120 | 107 | 85 | 73 | 107 | 130 | 98 |
| Glycine | 333 | 338 | 333 | 350 | 334 | 324 | 331 | 332 | 330 |
| Alanine | 115 | 102 | 103 | 96 | 30 | 47 | 54 | 39 | 49 |
| Half-cystine | 0 | 0 | 0 | 2 | 0 | 2 | 0 | 0 | 1 |
| Valine | 21 | 35 | 18 | 14 | 33 | 27 | 27 | 17 | 29 |
| Methionine | 7 | 5 | 10 | 8 | 15 | 14 | 11 | 9 | 8 |
| Isoleucine | 6 | 14 | 9 | 13 | 32 | 38 | 15 | 17 | 20 |
| Leucine | 19 | 30 | 26 | 22 | 52 | 56 | 37 | 36 | 56 |
| Tyrosine | 1 | 4 | 2 | 3 | 5 | 7 | 2 | 4 | 2 |
| Phenylalanine | 12 | 12 | 13 | 8 | 27 | 36 | 11 | 12 | 9 |
| Hydroxylysine | 9 | 12 | 20 | 5 | 50 | 36 | 23 | 36 | 43 |
| Lysine | 26 | 18 | 15 | 30 | 6 | 7 | 13 | 14 | 15 |
| Histidine | 3 | 12 | 2 | 6 | 6 | 6 | 10 | 6 | 14 |
| Arginine | 50 | 50 | 50 | 46 | 22 | 42 | 48 | 40 | 42 |
| Total | 1000 | 1000 | 1000 | 1000 | 1000 | 1000 | 1000 | 1000 | 1000 |
| Gal-hydroxylysine | 1 | 1 | 4 | — | 2 | 2 | 3 | 5 | 7 |
| Glc-Gal-hydroxylysine | 1 | 2 | 12 | — | 44 | 29 | 5 | 29 | 17 |

[a] Chains derived from placental type I collagen.
[b] Chain derived from infant epiphysial type II collagen.
[c] Data from Chung and Miller.[12]
[d] Data from Kresina and Miller.[20]
[e] Data from Chung et al.[32] and Rhodes and Miller.[37]
[f] Data from Sage and Bornstein.[34]

Additional correlations between the groups of chains are possible. For instance, the percentage of hydroxylated prolyl and lysyl residues in the type IV chains is significantly higher than in the other chains. The percentage of hydroxylated prolyl residues in type IV chains derived from bovine lens capsules is even higher,[21] indicating a wide variability with respect to prolyl hydroxylation as a function of source for the $\alpha1(IV)$ and $\alpha2(IV)$ chains. In addition, the percentage of glycosylated hydroxylysine residues tends to be quite high in the type IV chains, although this feature is apparent in the $\alpha1(II)$ and $\alpha1(V)$ chains as well. In general, the posttranslational modifications of a given collagen chain exhibit a certain degree of variability and are dependent on the source of the collagen. This is particularly true with respect to lysyl hydroxylation in the $\alpha1(I)$ and $\alpha2(I)$ chains[42] and, as pointed out above, for prolyl hydroxylation in the $\alpha(IV)$ and $\alpha2(IV)$ chains. The extent to which these modifications occur, then, is a relatively poor criterion for chain identification.

As noted above, the availability of techniques specifically and reproducibly to fragment the various collagen chains has considerably facilitated their further characterization. The CNBr cleavage products have been of great value in this regard. The CNBr peptides derived from the various chains can, in general, be isolated and purified by employing molecular sieve and ion-exchange chromatography including chromatography on CM-cellulose, phosphocellulose, and sulfonated polystyrene beads.[45,92] Utilizing these techniques, the CNBr cleavage products derived from the interstitial collagen $\alpha$ chains have been extensively characterized.[4,9,10] More recent reports concerning the isolation and characterization of the CNBr cleavage products derived from the $\alpha1(V)$,[93] $\alpha1(IV)$,[94] and $\alpha2(IV)$[95] chains have appeared. Although the various collagen chains exhibit a great deal of homology, the CNBr cleavage products derived from each chain represent a unique set of peptides. Therefore, chromatography of the major peptides derived from a given chain on CM-cellulose not only provides a means of preparing the peptides, but also serves as a means of identifying the chain based on the characteristic CM-cellulose elution profile.[4] Sodium dodecyl sulfate–polyacrylamide gel electrophoresis in cylindrical[43] and slab[24] gels represents an alternative and highly useful analytical procedure for resolving and identifying the CNBr cleavage products of the various chains. In addition, recent studies have shown that high-performance liquid chromatography may be a useful

---

[92] E. J. Miller, *Biochemistry* **10**, 3030 (1971).
[93] R. K. Rhodes and E. J. Miller, *J. Biol. Chem.* **254**, 12084 (1979).
[94] S. N. Dixit and A. H. Kang, *Biochemistry* **18**, 5686 (1979).
[95] S. N. Dixit and A. H. Kang, *Biochemistry* **19**, 2692 (1980).

analytical technique in distinguishing between the CNBr cleavage products derived from the different collagens.[96]

It has also been shown that treatment of $\alpha 1$(III) chains[97] as well as $\alpha 1$(IV) and $\alpha 2$(IV) chains[24] with a mast cell protease generates from each chain a unique set of peptide fragments, which when resolved by SDS–polyacrylamide gel electrophoresis on slab gels are indicative of the chain of origin. Similarly, it has been shown that *Staphylococcus aureus* V8 protease may be used to generate a unique set of cleavage products derived from partially processed pro-$\alpha 1$(IV) and pro-$\alpha 2$(IV) chains.[30] Use of these enzymes, then, clearly provides a useful alternative approach to attain specific cleavage products of the various chains.

*Procollagens*

Considerable information concerning the nature of procollagen molecules has been gained by visualization of segment-long-spacing crystallites formed by the molecules[98] as well as by studies on the products derived from procollagen molecules on exposure to an animal collagenase.[99] The constituent chains of various procollagen preparations have been resolved and purified by DEAE-cellulose chromatography followed by preparative gel electrophoresis,[75] by DEAE-cellulose chromatography,[47] by CM-cellulose chromatography in conjunction with molecular sieve chromatography,[80] and by preparative gel electrophoresis.[30] In general, fragments representing portions of the precursor-specific regions of pro-$\alpha$ chains may be recovered by chromatography of their CNBr cleavage products or their bacterial collagenase digestion products.[80] In a somewhat different approach, precursor-specific fragments of procollagen type I generated in the biological processing of the procollagen have been successfully isolated from the media of cultured chick embryo tendons.[100,101]

For the purposes of the present discussion, the structural features of the procollagen molecule are illustrated diagrammatically in Fig. 1. This model has been derived largely from studies on the pro-$\alpha 1$(I) chain and therefore more specifically depicts the structural features of procollagen

---

[96] C. Black, D. M. Douglas, and M. L. Tanzer, *J. Chromatogr.* **190**, 393 (1980).
[97] H. Sage, E. Crouch, and P. Bornstein, *Biochemistry* **18**, 5433 (1979).
[98] M. L. Tanzer, R. L. Church, J. A. Yaeger, D. E. Wampler, and E.-D. Park, *Proc. Natl. Acad. Sci. U.S.A.* **71**, 3009 (1974).
[99] P. H. Byers, E. M. Click, E. Harper, and P. Bornstein, *Proc. Natl. Acad. Sci. U.S.A.* **72**, 3009 (1975).
[100] B. R. Olsen, N. A. Guzman, J. Engel, C. Condit, and S. Aase, *Biochemistry* **16**, 3030 (1977).
[101] D. M. Pesciotta, M. H. Silkowitz, P. P. Fietzek, P. N. Graves, R. A. Berg, and B. R. Olsen, *Biochemistry* **19**, 2447 (1980).

FIG. 1. An approximate scale drawing of the procollagen I molecule illustrating the presence of five molecular domains, labeled A–E. Regions of helicity are depicted as rectangular bars. In globular or nonhelical domains, the individual chains are drawn with solid lines representing pro-α1(I) chains and a dashed line depicting the pro-α2(I) chain. Arrows indicate the cleavage sites of procollagen peptidases.

type I molecules. However, current data indicate that the model is valid, in general terms, for all types of procollagen molecules.

As illustrated in Fig. 1, the procollagen type I molecule may be divided into five major molecular domains, labeled A, B, C, D, and E. At the $NH_2$-terminal end of the molecule, region A represents a globular domain to which each pro-α1(I) chain contributes 86 amino acid residues.[102,103] The pro-α1(I) chain at this point is relatively rich in acidic and hydrophobic residues and contains ten cysteinyl residues,[102,103] none of which are involved in interchain disulfide bonding.[80,99] Region B comprises a triple-helical domain extending for 48 residues and characterized by the presence in the pro-α1(I) chain of 16 contiguous Gly-X-Y triplets, which are relatively rich in proline and hydroxyproline.[102] Region C is a short nonhelical domain that joins the small helical domain in region B with the large helical domain in region D. Insofar as pro-α1(I) is concerned, region C contains a sequence of 21 amino acids.[102,104,105] This sequence, similar to that in region A, is characterized by an irregular placement of glycyl residues but does not contain cysteine. Within region C, however, is the Pro-Gln bond (at residues 5 and 6), which is cleaved by an amino procollagen peptidase during the conversion of procollagen type I to type I collagen.[104] The precursor-specific domain at the $NH_2$-terminal portion of the pro-α1(I) chain thus contains a total of 139 amino acid residues.

This information concerning the $NH_2$-terminal region of pro-α1(I) has been obtained, for the most part, in studies on pN-collagen type I extracted from the dermis of dermatosparactic animals in which a genetic

---

[102] D. Hörlein, P. P. Fietzek, E. Wachter, C. M. Lapiere, and K. Kühn, *Eur. J. Biochem.* **99**, 31 (1979).

[103] H. Rohde, E. Wachter, W. J. Richter, P. Brückner, O. Helle, and R. Timpl, *Biochem. J.* **179**, 631 (1979).

[104] B. Hörlein, P. P. Fietzek, and K. Kühn, *FEBS Lett.* **89**, 279 (1978).

[105] J. Rauterberg, R. Timpl, and H. Furthmayr, *Eur. J. Biochem.* **27**, 231 (1972).

defect in the conversion of procollagen to collagen results in the accumulation of this form of the precursor within the extracellular spaces of the tissue. Far less extensive data are currently available for the homologous portion of the pro-α2(I) chain. This is due largely to the fact that pN-α2(I) chains derived from dermatosparactic skin provide relatively poor yields of the precursor-specific peptide.[80] Nevertheless, amino acid analyses on a truncated form of the peptide have established that it contains collagen-like sequences in approximately the same location as determined by the precursor-specific region of pN-α1(I).[106] Moreover, studies on metabolically labeled procollagen type I have shown that the precursor-specific region of pN-α1(I) and pN-α2(I) are approximately the same size.[107]

Region D (Fig. 1) constitutes the major helical domain of the procollagen molecule as well as the main body of the collagen molecule produced on removal of the precursor-specific regions. Extensive sequence studies on the α1(I) chain from a number of species have shown that, in this region, the chain is characterized by a series of Gly-X-Y triplets extending throughout 1014 amino acid residues.[9,10,108] Comparable data exist for the α2(I) chain indicating that it likewise exhibits a repetitive Gly-X-Y triplet structure extending throughout 1014 residues.[108]

Region E at the COOH-terminal portion of the procollagen type I molecule constitutes the second major globular domain of the molecule. Current data indicate that the general structural and chemical features of this domain closely resemble those of region A. There are differences, however, in that each COOH-terminal precursor-specific peptide exhibits an apparent $M_r$ of about 33,000,[100] suggesting that the COOH-terminal globular region is considerably larger than its counterpart at the opposite end of the molecule. Moreover, the COOH-terminal precursor-specific peptides are linked through disulfide bonds,[78,99,100] and contain at least one mannose-rich, asparagine-linked oligosaccharide moiety per chain.[100,109,110] Initial studies on the primary structure of the COOH-terminal precursor-specific portion of the pro-α1(I) chain have been reported.[110] As judged from sequence studies on the COOH-terminal region of the α1(I) chain,[111,112] a carboxy procollagen peptidase cleaves the

[106] U. Becker, O. Helle, and R. Timpl, *FEBS Lett.* **73**, 197 (1977).
[107] B. D. Smith, K. H. McKenney, and T. J. Lustberg, *Biochemistry* **16**, 2980 (1977).
[108] H. Hoffmann, P. P. Fietzek, and K. Kühn, *J. Mol. Biol.* **125**, 137 (1978).
[109] C. C. Clark, *J. Biol. Chem.* **254**, 10798 (1979).
[110] A. M. Showalter, D. M. Pesciotta, E. F. Eikenberry, T. Yamamoto, I. Pastan, B. DeCrombrugghe, P. P. Fietzek, and B. R. Olsen, *FEBS Lett.* **111**, 61 (1980).
[111] P. P. Fietzek, F. W. Rexrodt, P. Wendt, M. Stark, and K. Kühn, *Eur. J. Biochem.* **30**, 163 (1972).
[112] J. Rauterberg, P. Fietzek, F. Rexrodt, U. Becker, M. Stark, and K. Kühn, *FEBS Lett.* **21**, 75 (1972).

procollagen type I molecule at a point located 25 amino acid residues from the end of the major helical domain. This is likewise illustrated diagrammatically in Fig. 1. The conversion of procollagen type I to type I collagen, then, results in removal of approximately one-third of the molecular mass of the procollagen molecule. The major product of these reactions, the collagen molecule, retains short nonhelical sequences located at both ends of the molecule.

From the description given above, it can be discerned that the procollagen molecule is a relatively immense and complex structure. Allowing for some uncertainty concerning the actual size of the COOH-terminal precursor-specific portion of the pro-$\alpha$1(I) and pro-$\alpha$2(I) chains, it can be estimated that each of the three constituent chains of procollagen type I exhibits an $M_r$ of about 140,000. The structural and chemical features of procollagen type II remain relatively undefined owing largely to the lack of a convenient source for preparing chemical quantities of the precursor. However, studies on procollagen type II synthesized by chondrocytes in culture have revealed that the molecule contains both $NH_2$- and COOH-terminal precursor-specific regions and that the constituent pro-$\alpha$1(II) chains are of approximately the same size as pro-$\alpha$1(I) or pro-$\alpha$2(I) chains.[113-115] These studies also showed that the precursor-specific portions of procollagen type II clearly resemble the homologous regions of procollagen type I with respect to cysteine content and the presence of interchain disulfide bonds in the COOH-terminal region. In addition, both the $NH_2$- and COOH-terminal precursor-specific segments of procollagen type II appear to contain mannose-rich oligosaccharide chains.[115]

Somewhat more detailed information is available concerning the procollagen III molecule. Studies on a form of this precursor, pN-collagen type III, extracted from bovine fetal skin[116,117] have revealed the presence of an $NH_2$-terminal precursor-specific segment that closely resembles the one that has been established for the homologous portion of procollagen type I. The pN-collagen type III segment thus contains a cysteine-rich globular domain (region A, Fig. 1), a relatively short helical domain (region B, Fig. 1), and a small nonhelical domain intervening between the end of the helical domain and the beginning of the sequence represented by $\alpha$1(III) chains (initial portion of region C, Fig. 1). In contrast to the pN-

---

[113] A. H. Merry, R. Harwood, D. E. Wooley, M. E. Grant, and D. S. Jackson, *Biochem. Biophys. Res. Commun.* **71**, 83 (1976).

[114] J. Uitto, H.-P. Hoffmann, and D. J. Prockop, *Arch. Biochem. Biophys.* **179**, 654 (1977).

[115] N. A. Guzman, P. N. Graves, and D. J. Prockop, *Biochem. Biophys. Res. Commun.* **84**, 691 (1978).

[116] H. Nowack, B. R. Olsen, and R. Timpl, *Eur. J. Biochem.* **70**, 205 (1976).

[117] P. Bruckner, H. P. Bächinger, R. Timpl, and J. Engel, *Eur. J. Biochem.* **90**, 595 (1978).

collagen I segment, however, the pN-collagen III segment contains interchain disulfide bonds that are apparently localized at the junction of regions B and C. Very little is currently known about the COOH-terminal precursor-specific domain of procollagen type III. Data on procollagen type III extracted from cultured chick embryo blood vessels indicate, however, that this region is characterized by the presence of interchain disulfide bonds and closely resembles the homologous portions of procollagens type I and II with respect to all other general properties.[81] Procollagen type I-trimer has been detected in the medium of cultured epithelioid amniotic fluid cells.[79] Its properties were consistent with the presence of structural features similar to procollagen type I.

Relatively little is known, likewise, concerning the structural features of procollagen type IV molecules. They have been detected in several *in vitro* culture systems as relatively high molecular weight molecules the constituent chains of which are linked through disulfide bonding. On reduction, estimates for the $M_r$ of the pro-$\alpha$ chains have ranged from 140,000 in earlier studies[118,119] to somewhat higher values approaching 180,000 in more recent investigations.[28,30,120,121] These differences in molecular weight estimates for the pro-$\alpha$ chains of procollagen type IV could very well be due to the unique characteristics of the individual systems employed in the various studies. On the other hand, the differences more than likely reflect increasing sophistication in the use of techniques designed to preserve the structural features of the procollagen molecules during their isolation and purification. Thus, even if one allows for some imprecision in the molecular weight estimates for these relatively large components, these data strongly suggest that the constituent pro-$\alpha$ chains of procollagen type IV are considerably larger than the pro-$\alpha$ chains of procollagens types I, II, III, and I-trimer. In addition, treatment of procollagen type IV molecules with pepsin under conditions that result in complete degradation of the precursor-specific regions of other procollagens has a relatively small effect on procollagen type IV molecules. In this regard, the constituent chains of pepsin-treated procollagen IV molecules are approximately the same size as pro-$\alpha$1(I) chains,[120,122] remain cross-linked through disulfide bonds,[122] and exhibit compositional features indicative of the presence of rather lengthy noncollagenous domains.[122] These findings are consistent with observations indicating that collagen

[118] M. E. Grant, N. A. Kefalides, and D. J. Prockop, *J. Biol. Chem.* **247**, 3545 (1972).
[119] M. E. Grant, R. Harwood, and I. F. Williams, *Eur. J. Biochem.* **54**, 531 (1975).
[120] R. R. Minor, C. C. Clark, E. L. Strause, T. R. Koszalka, R. L. Brent, and N. A. Kefalides, *J. Biol. Chem.* **251**, 1789 (1976).
[121] J. G. Heathcote, C. H. J. Sear, and M. E. Grant, *Biochem. J.* **176**, 283 (1978).
[122] E. Crouch and P. Bornstein, *J. Biol. Chem.* **254**, 4197 (1979).

molecules extracted from lens capsules, under conditions where minimal amounts of pepsin are used to facilitate extraction, still retain globular domains at both ends of the molecules as ascertained by the appearance of segment-long-spacing crystallites in the electron microscope.[123] The results of the latter studies also indicated that the length of the major helical domain in these preparations of type IV collagen was quite similar to that known for other collagens. Cumulatively, then, the results outlined above for procollagen type IV molecules certainly suggest a model similar to that illustrated in Fig. 1 for procollagen type I. Procollagen type IV molecules, however, may have more extended globular domains with unique structural features that bestow a certain degree of resistance to proteolysis to at least portions of these globular domains.

As of this writing, there exist no published reports describing the properties of procollagen type V molecules. One might anticipate, however, that procollagen type V molecules exhibit the same general structural features as described above for other known procollagens.

Processing of Procollagens to Collagens

Various aspects of procollagen biosynthesis have recently been evaluated in a number of reviews.[124-126] We present here a brief discussion of current information on procollagen processing in order to develop further the theme concerning the structural and chemical heterogeneity of fibrous elements formed by the different collagens. It should be emphasized that what are now referred to as procollagen molecules probably do not contain the initial translation products of collagen mRNA. In this regard, studies on the cell-free translation products of chick embryo calvaria mRNA have revealed the presence of components somewhat larger than pro-$\alpha$ chains. In one study it was estimated that the $M_r$ of the putative prepro-$\alpha$1(I) chain exceeded that of pro-$\alpha$1(I) by about 5,500,[127] whereas this difference was estimated to approach 20,000 for both the prepro-$\alpha$1(I) and prepro-$\alpha$2(I) chains in another study.[128] The reason for this apparent discrepancy is not obvious, but may be related to problems encountered in estimating the molecular weight of relatively large components contain-

---

[123] D. Schwartz and A. Veis, *Eur. J. Biochem.* **103**, 29 (1980).
[124] D. J. Prockop, R. A. Berg, K. I. Kivirikko, and J. Uitto, in "Biochemistry of Collagen" (G. N. Ramachandran and A. H. Reddi, eds.), p. 163. Plenum, New York, 1976.
[125] J. H. Fessler and L. I. Fessler, *Annu. Rev. Biochem.* **47**, 129 (1978).
[126] R. Harwood, *Int. Rev. Connect. Tissue Res.* **8**, 159 (1979).
[127] R. D. Palmiter, J. M. Davidson, J. Gagnon, D. W. Rowe, and P. Bornstein, *J. Biol. Chem.* **254**, 1433 (1979).
[128] L. Sandell and A. Veis, *Biochem. Biophys. Res. Commun.* **92**, 554 (1980).

ing both collagenous and noncollagenous domains as a function of electrophoretic mobility. Since the prepro-α1(I) chains detected in these studies contained an initiator methionyl residue and were converted to components resembling pro-α1(I) chains on incubation with microsomal membrane preparations, the results certainly suggest the presence of an unusually lengthy signal sequence at the $NH_2$ terminus of the preprocollagen chain. Alternatively, these results may indicate the presence of an additional transient precursor domain,[128] the function of which remains to be determined. Clearly, further work will be required to resolve the issues raised by these observations.

The processing of procollagen to collagen has been studied most extensively by means of pulse-chase experiments in culture systems in which the predominant procollagen synthesized is procollagen type I. These studies have revealed that conversion of procollagen type I to type I collagen is a multistep process involving initial removal of the $NH_2$-terminal precursor-specific region yielding pC-collagen type I, which is subsequently converted to type I collagen on removal of the COOH-terminal precursor-specific region.[129,130] Removal of the $NH_2$-terminal propeptides apparently involves rapid and perhaps simultaneous cleavage of the susceptible peptide bond in each chain, whereas removal of the COOH-terminal propeptides appears to be a slower process allowing the detection of pC-collagen type I intermediates in which cleavage has occurred in only one or two of the constituent chains.[131] Current data further indicate that removal of the $NH_2$- and COOH-terminal propeptides requires the activity of separate procollagen peptidases,[132] and that the cleavage reactions yield essentially intact $NH_2$- and COOH-terminal propeptides.[133] The ultimate metabolic fate of the propeptides is presently not known. However, it is of interest that the $NH_2$-terminal globular domain derived from pN-α1(I) or pN-α1(III) effectively inhibits procollagen synthesis when added to cultured fibroblasts[134] as well as the synthesis of pro-α chains when added to preparations of procollagen mRNA,[135] suggesting a regulatory role for at least portions of the propeptides in collagen synthesis.

[129] L. I. Fessler, N. P. Morris, and J. H. Fessler, *Proc. Natl. Acad. Sci. U.S.A.* **72**, 4905 (1975).
[130] J. M. Davidson, L. S. G. McEneany, and P. Bornstein, *Biochemistry* **14**, 5188 (1975).
[131] J. M. Davidson, L. S. G. McEneany, and P. Bornstein, *Eur. J. Biochem.* **81**, 349 (1977).
[132] M. K. K. Leung, L. I. Fessler, D. B. Greenberg, and J. H. Fessler, *J. Biol. Chem.* **254**, 224 (1979).
[133] N. P. Morris, L. I. Fessler, and J. H. Fessler, *J. Biol. Chem.* **254**, 11024 (1979).
[134] M. Wiestner, T. Krieg, D. Hörlein, R. W. Glanville, P. Fietzek, and P. K. Müller, *J. Biol. Chem.* **254**, 7016 (1979).
[135] L. Paglia, J. Wilczek, L. D. deLeon, G. R. Martin, D. Hörlein, and P. Müller, *Biochemistry* **18**, 5030 (1979).

Indirect evidence indicates that procollagen type I is, for the most part, rapidly processed *in vivo* in a manner analogous to that observed *in vitro*. In this regard, procollagen type I or its intermediates are not generally present in significant quantities in extracts of normal tissues.[107] Moreover, failure to remove the $NH_2$-terminal precursor-specific domain from a substantial proportion of newly synthesized procollagen type I molecules in dermatosparactic animals leads to pronounced fragility of the dermis. This apparently is due to the inability of pN-collagen type I molecules to form the relatively massive, well structured fibers characteristically formed by type I molecules.[136] A similar situation prevails in the Ehlers–Danlos type VII syndrome, in which failure to complete the conversion of procollagen to collagen is due to an amino acid substitution at the $NH_2$-terminal cleavage site in the pro-$\alpha 2$(I) chain.[137] This is consistent with an autosomal dominant mode of inheritance for the syndrome. On the other hand, indirect immunofluorescence tests using antibodies specific for determinants in the $NH_2$-terminal precursor-specific region of pN-collagen type I generally reveal a moderate amount of staining in sections derived from normal tissues known to contain large quantities of type I collagen.[138,139] Although alternative explanations are possible, these results suggest that not all procollagen type I molecules are completely converted to type I collagen.

The conversion of procollagen type II to type II collagen has been studied in culture systems employing embryonic chick sterna as well as chondrocytes.[82,140] These studies showed that the processing of procollagen type II proceeds in a fashion quite similar to that previously observed for procollagen type I. Nevertheless, evaluation of the nature of the procollagen type II intermediates present at various intervals during pulse-chase experiments suggested that the $NH_2$- and COOH-terminal precursor-specific regions of procollagen type II molecules were removed more or less at random with no clear preference for initial removal of the $NH_2$-terminal domain.[82] Additional analogy between the processing of procollagen types I and II molecules is suggested by the observations that a partially purified amino procollagen peptidase is active in removing the $NH_2$-terminal precursor-specific domain from both procollagens.[141]

In contrast to the *in vitro* studies on the processing of procollagen types I and II, similar studies on the processing of procollagen type III in vari-

---

[136] C. M. Lapiere and B. Nusgens, *Biochim. Biophys. Acta* **342**, 237 (1974).
[137] B. Steinmann, L. Tuderman, L. Peltonen, G. R. Martin, V. A. McKusick, and D. J. Prockop, *J. Biol. Chem.* **255**, 8887 (1980).
[138] G. Wick, H. Furthmayr, and R. Timpl, *Int. Arch. Allergy Appl. Immunol.* **48**, 664 (1975).
[139] G. Wick, H. Nowack, E. Hahn, R. Timpl, and E. J. Miller, *J. Immunol.* **117**, 298 (1976).
[140] J. Uitto, *Biochemistry* **16**, 3421 (1977).
[141] L. Tuderman, K. I. Kivirikko, and D. J. Prockop, *Biochemistry* **17**, 2948 (1978).

ous systems have shown either no conversion to procollagen intermediates or collagen[97] or conversion only to intermediates that remained somewhat larger than type III molecules.[81,142,143] In one of the latter studies,[81] characterization of the apparent final product in procollagen type III processing suggested that it represented pN-collagen type III or a molecule closely resembling this form of the precursor. These results could be interpreted as indicating that the enzymes required for procollagen type III processing are generally unavailable in *in vitro* culture systems. This is a likely possibility, since it has been shown that a preparation of the amino procollagen peptidase capable of converting pN-collagen type I to type I collagen exhibits no activity when pN-collagen type III is used as substrate, indicating that processing of the respective procollagens requires different enzymes at least for removal of the $NH_2$-terminal precursor-specific domain.[144] The *in vitro* results are, however, entirely consistent with the observations that relatively large quantities of procollagen type III intermediates can be recovered in neutral salt extracts of normal tissues[46,116,117,144] and that antibodies specific for the $NH_2$-terminal globular domain of pN-collagen type III stain tissue sections with approximately the same pattern and intensity as that observed for antibodies to type III collagen.[145] The results obtained *in vitro*, then, more than likely accurately reflect limited processing of procollagen III molecules *in vivo*. It is clear, however, that some procollagen III molecules are processed to collagen, since extracts containing relatively high proportions of procollagen type III intermediates likewise contain type III molecules. Moreover, the $\alpha 1(III)$ chains derived from the latter molecules exhibit an $NH_2$-terminal sequence highly homologous to that of $\alpha 1(I)$ chains,[146] suggesting that the amino procollagen peptidase cleavage site is highly conserved in procollagens type I and III.

*In vitro* studies on procollagen type IV processing have yielded results similar to those obtained in studies on procollagen type III processing. The results have thus indicated either no conversion of procollagen IV molecules to smaller molecular weight products in rat parietal yolk sac[120] and lens capsule,[121] or conversion only to products in which the constituent chains remain substantially larger than collagen $\alpha$ chains in chick lens capsule[118] and rat glomerulus.[119] These results could likewise be interpreted as indicating a relative lack of enzymes for procollagen type IV

---

[142] B. Goldberg, *Proc. Natl. Acad. Sci. U.S.A.* **74**, 3322 (1977).
[143] H. F. Limeback and J. Sodek, *Eur. J. Biochem.* **100**, 541 (1979).
[144] A. Lenaers and C. M. Lapiere, *Biochim. Biophys. Acta* **400**, 121 (1975).
[145] H. Nowack, S. Gay, G. Wick, U. Becker, and R. Timpl, *J. Immunol. Methods* **12**, 117 (1976).
[146] R. W. Glanville and P. P. Fietzek, *FEBS Lett.* **71**, 99 (1976).

processing in the various culture systems. Nevertheless, more recent *in vitro* studies on procollagen type IV biosynthesis in cultures of a murine basement membrane tumor as well as yolk sac have shown that the constituent pro-$\alpha$1(IV) and pro-$\alpha$2(IV) chains in the labeled precursor molecules exhibit an apparent $M_r$ of 185,000 and 170,000, but that the chains are present in extracts of the tumor collagen as components with an apparent $M_r$ of 160,000 and 140,000, respectively.[28] It has subsequently been demonstrated by amino acid analyses that each of the latter chains is likely to contain relatively extensive noncollagenous domains.[30] Current data, then, strongly support the notion that procollagen type IV processing results in limited alteration of the precursor molecules and that procollagen-like molecules containing extensive noncollagenous domains are incorporated into the matrices containing this collagen.

Since it may be assumed that the products of procollagen processing are ultimately utilized in the formation of the supporting elements or functional aggregates in various connective tissues, the foregoing discussion concerning the chemical properties of the procollagens and their processing indicates that the nature of the functional aggregates formed in any given location will be profoundly influenced not only by the type of collagen involved, but by the extent to which the precursor molecules have been processed. In this regard, it seems fairly obvious that fibers derived from type I molecules, for instance, will differ markedly with respect to chemical properties from fibers comprised largely of partially processed procollagen type III molecules. Moreover, based on the results cited above, it would appear that the extent of procollagen processing controls, to some degree, the morphological features of the resulting aggregates. Thus, type I molecules are generally involved in the formation of relatively large fibrous elements, and these elements cannot be formed efficiently by pN-collagen type I molecules.[136] In accordance with these observations, partially processed procollagen type III molecules are deposited, for the most part, in fine reticular networks[145] whereas the partially processed procollagen type IV molecules of basement membranes likewise appear to form only fine filamentous elements.[147] At this time, one cannot rule out the possibility that incomplete processing of procollagen type II molecules is responsible for the formation of relatively small fibrils often observed in hyaline cartilage matrices.

Summary and Perspectives

In the preceding sections we have outlined the development of information concerning the molecular heterogeneity of collagen. This hetero-

---

[147] N. A. Kefalides, R. Alper, and C. C. Clark, *Int. Rev. Cytol.* **61**, 167 (1979).

geneity is ultimately determined by the genome, but is amplified by posttranslational events that allow for wide variations in the extent to which different types of procollagen molecules are processed prior to incorporation into their respective fibrous elements. The extent to which procollagen molecules are processed more than likely specifies the general properties of the fibrous elements formed at any given site. The general properties of the fibrous elements may, in turn, be of great importance with respect to specific interactions with other matrix macromolecules as well as with cells. These possibilities will undoubtedly be scrutinized closely in future investigations.

The above considerations as well as information concerning the distribution and localization of the collagens in various connective tissues[4,148] strongly support the notion that each of the collagens has evolved in order to accommodate specific physiological requirements of organisms of increasing complexity. In this regard, it is likely that the multiple genes for collagen synthesis represent an informational multigene family, or perhaps a number of families, the existence of which allows the selective synthesis of discrete, albeit functionally related, proteins.[149] At present, however, data on the location of the various collagen genes within the genome are lacking, the control mechanisms for collagen synthesis have not been determined, and the physiological significance of the various collagens have not been defined in molecular terms. It is anticipated that future endeavors will be productive in providing considerable insight into these and other factors related to the molecular biology of the most prevalent protein in vertebrate organisms.

Acknowledgments

Investigations in the author's laboratory are supported by N.I.H. grants DE-02670 and HL-11310 and a grant from the Osteogenesis Imperfecta Foundation.

[148] S. Gay and E. J. Miller, "Collagen in the Physiology and Pathology of Connective Tissue." Fischer, Stuttgart, 1978.
[149] L. Hood, J. H. Campbell, and S. C. R. Elgin, *Annu. Rev. Genet.* **9**, 305 (1975).

## [2] Preparation and Characterization of the Different Types of Collagen

*By* EDWARD J. MILLER AND R. KENT RHODES

Preparation of Collagens

The preparation of a collagen sample generally involves several different steps. These include acquisition and preliminary processing of an appropriate tissue or organ, extraction of the collagen, and its purification. The latter process requires not only elimination of the noncollagenous components that are present in the extract, but may also require selective removal of alternative collagen types. This can usually be accomplished by a judicious use of selective precipitation techniques accompanied ultimately by one or two chromatographic steps. Given the extreme diversity of the tissues or organs in which collagen occurs as well as the multiplicity of collagen types that may be present in a given source, there understandably exists no single standard or preferred method for the preparation of collagen. Accordingly, the following discussion will present details of methods generally applicable to the preparation of native collagen from a variety of sources. Unless otherwise indicated, all procedures are conducted at relatively low temperatures, in the range of 4–8°. This minimizes bacterial growth, enhances the solubility of native collagens, and ensures the retention of native conformation on the part of the solubilized collagens.

*Acquisition and Preliminary Processing of Tissues*

In preparing collagen from the tissues of small laboratory animals, the increased extractability of native collagen achieved through inhibition of cross-linking often justifies the added expense and time involved in rendering the animals lathyritic. Young rats (100–150 g)[1] and newly hatched chicks[2] can be conveniently made lathyritic by being fed ad libitum a standard diet containing the lathyrogen $\beta$-aminopropionitrile (fumarate salt) at a level of 1 g per kilogram of diet for 3–4 weeks. At this interval the symptoms of lathyrism are clearly evident as skeletal deformities and continued feeding of the lathyrogen results in high mortality. Similar results have been obtained with mice fed a diet containing the lathyrogen

---

[1] P. Bornstein and K. A. Piez, *Biochemistry* **5**, 3460 (1966).
[2] E. J. Miller, G. R. Martin, K. A. Piez, and M. J. Powers, *J. Biol. Chem.* **242**, 5481 (1967).

at a level of 4 g per kilogram of diet for 2 weeks.[3] Although β-aminopropionitrile has been the most commonly used lathyrogen, more recent results suggest that a related compound, α-aminoacetonitrile, is somewhat more effective in inhibiting cross-linking of connective tissue components and considerably less toxic when administered to young chicks.[4] The activity of lathyrogens with respect to cross-linking may be ascribed to an irreversible inhibition of lysyl oxidase.[5] The latter enzyme is a copper-containing[5] amine oxidase[6] that catalyzes the oxidative deamination of certain lysyl and hydroxylysyl residues in collagen molecules forming reactive aldehyde precursors for intra- and intermolecular cross-links. The administration of certain aminothiol compounds, such as D-penicillamine, to young rats at a level of 4 g per kilogram of diet for 2 weeks likewise enhances the extractability of collagen.[7] The effect of D-penicillamine is most pronounced on collagen present in soft tissues, particularly dermis, and has been attributed to the ability of the compound to form moderately stable complexes with the aldehyde precursors of the cross-links[8] as well as its ability to inactivate lysyl oxidase by chelating copper.[9]

In general, tissues commonly used in the preparation of various collagens (Table I) can be excised and cleaned of adherent structures by standard surgical procedures. An exception to this rule are the basement membranes, other than the lens capsule. Owing to their small size, basement membrane structures generally must be isolated by a sequence of procedures basically designed to dissociate the membranes from cells and other connective tissue elements. Glomerular basement membrane is, for instance, conventionally isolated by first gently forcing slices of kidney cortex through a 115-mesh sieve[10] or passing a homogenate of whole kidney through a double layer of cheesecloth.[11] The resultant filtrates are passed through tandemly arranged 80- and 150-mesh sieves, which are subsequently washed with a large volume of 0.15 $M$ NaCl.[10,11] The glomeruli collected on the 150-mesh sieve are suspended in 1.0 $M$ NaCl and disrupted by sonication[10] or homogenization,[11] and the glomerular

[3] R. W. Orkin, P. Gehron, E. B. McGoodwin, G. R. Martin, T. Valentine, and R. Swarm, *J. Exp. Med.* **145,** 204 (1977).
[4] J. A. Foster, C. B. Rich, N. Berglund, S. Huber, R. P. Mecham, and G. Lange, *Biochim. Biophys. Acta* **587,** 477 (1979).
[5] R. C. Siegel, S. R. Pinnell, and G. R. Martin, *Biochemistry* **9,** 4486 (1970).
[6] P. C. Trackman and H. M. Kagan, *J. Biol. Chem.* **254,** 7831 (1979).
[7] M. E. Nimni, *J. Biol. Chem.* **243,** 1457 (1968).
[8] K. Deshmukh and M. E. Nimni, *J. Biol. Chem.* **244,** 1787 (1969).
[9] M. E. Nimni, K. Deshmukh, and N. Gerth, *Nature (London)* **240,** 220 (1972).
[10] R. G. Spiro, *J. Biol. Chem.* **242,** 1915 (1967).
[11] J. R. Daniels and G. H. Chu, *J. Biol. Chem.* **250,** 3531 (1975).

TABLE I
MAJOR VERTEBRATE TISSUES USED IN THE
PREPARATION OF THE VARIOUS COLLAGENS

| Tissue | Prevalent collagen(s)[a] |
|---|---|
| Bone | Type I |
| Tendon | Type I |
| Hyaline cartilage | Type II |
| Basement membranes | Type IV |
| Dermis | Types I and III |
| Cirrhotic liver | Types I and III |
| Vessel wall | Types I, III, and V |
| Uterine wall | Types I, III, and V |
| Amnion and chorion | Types I, III, and V |
| Placental villi | Types I, III, V, and IV |

[a] The molecular species of collagen listed here represents the most abundant collagen(s) present in the tissue and those that can conveniently be obtained in relatively large quantities on extraction of the indicated tissue. Where more than one molecular species is prevalent, the order of listing indicates relative abundance.

basement membrane is collected by centrifugation and washed in distilled water. The isolated glomeruli and basement membranes are inspected by phase contrast microscopy in order to monitor the purity of these fractions. In a recent modification of this procedure, the protease inhibitors (ethylenedinitrilo)tetraacetic acid (EDTA), diisopropyl fluorophosphate (DFP), $N$-ethylmaleimide (NEM), and $\epsilon$-aminocaproic acid ($\epsilon$-ACA) were added to all solvents in order to minimize proteolysis of the basement membrane components.[12] Similar homogenization and sieving techniques have been used to isolate intact brain[13] and retinal[14] microvessels as well as renal tubules and glomeruli.[15] In further developments, it has been shown that ultrastructurally pure basement membrane can be isolated from each of these preparations by sequential extraction at room temperature with water containing 0.1% sodium azide, 1.0 $M$ NaCl containing DNase, and 4% sodium deoxycholate containing 0.1% sodium azide.[16]

[12] J. W. Freytag, M. Ohno, and B. G. Hudson, *Biochem. Biophys. Res. Commun.* **72**, 796 (1976).
[13] K. Brendel, E. Meezan, and E. C. Carlson, *Science* **185**, 953 (1974).
[14] E. Meezan, K. Brendel, and E. C. Carlson, *Nature (London)* **251**, 65 (1974).
[15] E. Meezan, K. Brendel, J. Ulreich, and E. C. Carlson, *J. Pharmacol. Exp. Ther.* **187**, 332 (1973).
[16] E. Meezan, J. T. Hjelle, K. Brendel, and E. C. Carlson, *Life Sci.* **17**, 1721 (1975).

The method chosen to disperse various tissues prior to extraction will in large part be determined by the nature of the tissue and the quantity of the tissue that is available. Mineralized tissues, such as bone and dentin, are readily fragmented by use of a stainless steel mortar and pestle. Cartilaginous tissues are far less brittle and do not readily disperse on homogenization. They can, nevertheless, be cut into relatively thin slices with a scalpel blade or a butcher's meat slicer. All other tissues from which collagen is commonly extracted can be conveniently dispersed in a mechanical meat grinder followed by homogenization with a Waring blender or Polytron homogenizer if necessary. In some cases, tissues have been frozen in liquid nitrogen and subsequently pulverized by milling.

After dispersal of the tissue, it is frequently advisable to perform a series of preliminary extractions designed to remove selectively substances that may either inhibit efficient extraction of the collagen or render purification of the extracted collagen more difficult. The nature and source of the tissue generally dictates the type of solvent used in preliminary extractions. Mineralized tissues are rapidly decalcified on extraction with 0.5 $M$ EDTA, pH 7.4,[17] and somewhat more slowly decalcified on extraction with 0.5 $M$ acetic acid.[2] If the tissues are derived from lathyritic animals, some collagen may also be solubilized during decalcification with 0.5 $M$ acetic acid. The collagen may, however, be conveniently recovered by enclosing the specimens in dialysis bags during decalcification.[2] Many cartilaginous structures contain appreciable quantities of proteoglycan components that combine with collagen molecules in solution at acid pH to form insoluble aggregates. It is therefore advantageous to extract the proteoglycans prior to extraction of the collagen. This can be accomplished by exposing slices of cartilage to solutions of 4.0 $M$ guanidine-HCl, 0.05 $M$ Tris, pH 7.5, at room temperature.[18] This procedure was originally designed to facilitate extraction of proteoglycan components from cartilage[19] and results in the solubilization of little, if any, of the collagen from normal cartilages. The cartilage slices are then extensively washed with water, at which time they are comprised largely of collagen fibrils.

Other tissues used in the preparation of collagen are, as a rule, extracted initially with water or neutral salt solvents (0.15–1.0 $M$ NaCl) to remove soluble noncollagenous substances. This process, however, may also result in the simultaneous solubilization of substantial amounts of collagen, particularly when the tissue is derived from a relatively young and/or lathyritic organism. Moreover, tissues with large amounts of vas-

[17] A. Veis and A. Perry, *Biochemistry* **6**, 2409 (1967).
[18] E. J. Miller and L. G. Lunde, *Biochemistry* **12**, 3153 (1973).
[19] S. W. Sajdera and V. C. Hascall, *J. Biol. Chem.* **244**, 77 (1969).

culature often require repeated preliminary extractions to remove relatively high levels of blood-borne contaminants, thereby increasing the likelihood that certain portions of the collagen molecules will be degraded through endogenous protease activity. These problems can generally be avoided by performing the initial extractions with 4.5 $M$ NaCl, 0.05 $M$ Tris, pH 7.5, under which conditions all the known collagens are insoluble (see below), and by adding protease inhibitors. It is recommended that the latter include at least one inhibitor for each of the four recognized classes of proteinases,[20] i.e., an inhibitor of metalloproteinases, such as EDTA (20.0 m$M$); an inhibitor of serine proteinases, such as DFP (1.0 m$M$) or phenylmethanesulfonyl fluoride (PMSF) (1.0 m$M$); an inhibitor of sulfhydryl proteinases, such as NEM (2.0 m$M$); and an inhibitor of carboxylproteinases, such as pepstatin (1 $\mu$g/ml).

*Extraction of Collagen*

For the extraction of collagen, the three following solvent systems have been found to be generally useful and convenient.

1. A neutral salt solvent, 1.0 $M$ NaCl, 0.05 $M$ Tris, pH 7.5. The concentration of NaCl in neutral salt solvents may be reduced to as low as 0.15 $M$. However, the efficiency of extraction is increased at higher salt concentrations. Owing to the variety of tissue proteinases that may be present and active in the neutral pH range, it is advisable to include a number of inhibitors (EDTA, DFP, NEM, and pepstatin) in the solvent in order to minimize proteolytic degradation of nonhelical or other protease-sensitive regions in the collagen molecules. Of the solvent systems discussed here, neutral salt solvents exhibit the least capacity to solubilize collagen. In this regard, these solvents essentially extract molecules that are not bound within the fibers through covalent intermolecular cross-links. In the tissues of normal organisms, the proportion of these molecules is usually quite small but may be substantially increased when cross-linking is inhibited.

2. A dilute organic acid solvent, 0.5 $M$ acetic acid. The pH of this solvent is approximately 3.0. However, the solvent exhibits a somewhat increased capacity to induce swelling of most tissues if the pH is adjusted to 2.5 with 1.0 $N$ HCl prior to use. In this pH range, the addition of pepstatin, leupeptin, and NEM should allow effective inhibition of endogenous proteinases, although PMSF[3,21] as well as EDTA and $\alpha,\alpha'$-dipyridyl[21] have been used for this purpose. Dilute acid solvents are capable of solubilizing non-cross-linked molecules but also solubilize fibers in

[20] A. J. Barrett, *Fed. Proc., Fed. Am. Soc. Exp. Biol.* **39**, 9 (1980).
[21] P. Dehm and N. A. Kefalides, *J. Biol. Chem.* **253**, 6680 (1978).

which aldimine-containing cross-links are prevalent owing to the lability of this type of cross-link at acid pH. In practical terms, however, this effect of dilute acid solvents is limited to a portion of the collagen in the dermis and some tendons of relatively young organisms, since the more stable keto-amine form of cross-link is prevalent in the fibers of virtually all other tissues.

3. Acetic acid, 0.5 $M$, containing pepsin (EC 3.4.23.1, pepsin A, Worthington Biochemical Corp.). This solvent system can likewise be used effectively in the pH range of 2.5–3.0. The amount of pepsin required to attain effective solubilization of the collagen in a given tissue may be variable. In general, however, the enzyme is added to the solvent in sufficient quantities to achieve a 1 : 10 ratio between the weight of enzyme and dry weight of the tissue to be extracted.[22] Since the efficacy of this solvent system lies in the ability selectively to degrade those portions of the native collagen molecule from which cross-links originate, it is the most versatile and widely applicable solvent system for the extraction of native collagen. However, the molecules are recovered with modified nonhelical extremities, and in certain instances even more extensive degradation may occur, particularly on exposure of type IV molecules to pepsin.[23–26]

Based on the considerations outlined above, a reasonably good choice of solvent or solvents to use in the extraction of collagen can be made in most cases. Prior to the actual extraction procedure, it is advisable to wash the dispersed tissue briefly with water containing protease inhibitors in order to remove relatively high levels of NaCl that may have accumulated during preliminary extractions. The tissue may then be suspended in approximately 20 volumes of the solvent of choice. Extraction is generally carried out for intervals of 12–24 hr with continuous vigorous stirring. The extraction mixture may then be filtered through several layers of cheesecloth to achieve initial separation of the collagen solution and insoluble tissue residue. The latter may be resuspended in fresh solvent and reextracted several times although the yield of soluble collagen diminishes markedly in successive extracts. Solutions of collagen attained in this manner can be centrifuged in a fixed-angle rotor at forces between 40,000 and 50,000 $g$ for 1–2 hr. This is often sufficient to clarify the solution and remove suspended material especially, if the extract is derived from a tissue that has been thoroughly cleansed during the preliminary extraction

[22] E. J. Miller, *Biochemistry* **11**, 4903 (1972).
[23] D. Schwartz and A. Veis, *FEBS Lett.* **85**, 326 (1978).
[24] T. F. Kresina and E. J. Miller, *Biochemistry* **18**, 3089 (1979).
[25] S. Gay and E. J. Miller, *Arch. Biochem. Biophys.* **198**, 370 (1979).
[26] R. Timpl, P. Bruckner, and P. Fietzek, *Eur. J. Biochem.* **95**, 255 (1979).

intervals. Should the solution remain turbid following centrifugation, however, as described above, it is necessary to centrifuge the extract a second time at forces approaching 100,000 $g$ for 2–3 hr in an ultracentrifuge. Failure to attain clear solutions of collagen during ultracentrifugation is often due to the high viscosity of relatively concentrated collagen solutions, and the samples should be appropriately diluted with solvent and centrifuged again.

*Purification of Collagen (Initial Steps)*

Initial purification of the extracted collagen is achieved in a series of precipitation steps in which the collagen is alternatively precipitated from neutral and acidic solvents. The principles governing these steps are as follows.

1. Native collagen molecules present in solution at or slightly above neutral pH are quantitatively precipitated at a NaCl concentration of 4.0 $M$ or at low ionic strength.
2. Native collagen molecules present in solution at acid pH are quantitatively precipitated at a NaCl concentration of 2.0 $M$.

Accordingly, collagen extracted in a neutral salt solvent, such as 1.0 $M$ NaCl, 0.05 $M$ Tris, pH 7.5, is most conveniently recovered by increasing the NaCl concentration of the solution to 4.0 $M$. This can be accomplished by the slow addition of appropriate quantities of crystalline NaCl with constant stirring. The rapidity with which the precipitate forms as well as its consistency is dependent on the concentration of collagen in the extract. At concentrations of collagen approaching 1 mg/ml or greater, the precipitate is visible immediately and generally forms heavy aggregates that readily fall to the bottom of the vessel on cessation of stirring. When precipitates are formed from more dilute solutions, they may not be visible for an hour or more and often remain dispersed as particles suspended in the solution. In any event, it is generally advisable to allow the precipitate to form and settle without stirring over a 12–24-hr interval to ensure optimum recovery of the collagen. The precipitates may then be recovered by centrifugation at 35,000 $g$ for 1 hr, at which time the precipitate will form a compact pellet. The pellets can then be resuspended in 0.5 $M$ acetic acid and redissolved in this solvent by stirring. In general, collagen extracted in neutral salt solvents readily redissolves in 0.5 $M$ acetic acid at a concentration of 1 mg/ml or less within a few hours. In some cases, it may be necessary to dialyze the suspension versus 0.5 $M$ acetic acid in order to remove NaCl that may have been entrapped within the pellets derived from heavy precipitates. Material that does not redissolve in 0.5 $M$ acetic acid should be removed by centrifugation. The collagen present in solution in 0.5 $M$ acetic acid is then reprecipitated by the addition of crystalline NaCl to a concentration of 2.0 $M$. The precipitates obtained by this procedure are recovered after an appropriate interval by centrifugation. At this point, the protein may be recycled through neutral salt and dilute acid solvents as often as deemed necessary. In practice, however, two complete cycles are ordinarily sufficient to obtain a preparation in which contamination by noncollagenous proteins is less than 5% as judged by amino acid analyses. Ultimately the protein is dissolved in 0.5 $M$ acetic acid, dialyzed extensively versus this solvent, and lyophilized or stored as a frozen solution.

In a modification of the procedure, collagen extracted in 1.0 $M$ NaCl, 0.05 $M$ Tris, pH 7.5, can be initially precipitated by dialysis versus 0.02 $M$ dibasic sodium phosphate. This procedure requires more time than when the protein is precipitated by the addition of NaCl to 4.0 $M$, but affords the opportunity of recovering the collagen from a slightly alkaline (pH 8.2) solution of low ionic strength.

Collagen extracted in 0.5 $M$ acetic acid with or without pepsin is recovered and purified by alternate precipitations from acidic and neutral salt solvents in virtually the same manner as described above for collagen initially solubilized in neutral salt solvents. The first precipitate is obtained by the addition of sufficient NaCl to attain a concentration of 2.0 $M$. Alternatively, this precipitate can be obtained by dialysis of the extract versus 0.02 $M$ dibasic sodium phosphate. On redissolving collagen precipitated from acid solvents in neutral salt solvents, care must be taken to maintain the pH at 7.5 owing to the tendency of acid adsorbed to the pellets to overwhelm the buffering capacity of the neutral salt solvent when the pellets are dispersed. The pH can readily be controlled by dropwise addition of a solution of 1.0 $N$ NaOH. Even under the best of conditions, collagen initially brought into solution in dilute acid solvents is often difficult to redissolve in a neutral salt solvent such as 1.0 $M$ NaCl, 0.05 $M$ Tris, pH 7.5. For collagen extracted in 0.5 $M$ acetic acid, relatively poor solubility in neutral salt solvents generally reflects the presence of native cross-linked aggregates of type I collagen, which will not redissolve in neutral salt solvents. These components may therefore be eliminated at this step in the purification procedure by discarding the material that remains insoluble on exposure to the neutral salt solvent. Alternatively, the high molecular weight aggregates may be selectively precipitated from the 0.5 $M$ acetic acid extract by adding NaCl to a concentration of 0.6 $M$.[27]

When dealing with collagen extracted in 0.5 $M$ acetic acid containing pepsin, relative insolubility in neutral salt solvents is likely to be due to the presence of type III collagen molecules, which do not readily dissolve

[27] G. Chandrakasan, D. A. Torchia, and K. A. Piez, *J. Biol. Chem.* **251**, 6062 (1976).

in these solvents, particularly when derived from chick[28] or rat[29] tissues. The reason for this refractory behavior on the part of chick or rat type III molecules is not readily apparent, since no difficulty in solubilizing the human[30,31] or calf[32] type III molecules in neutral salt solvents has been encountered. In any event, the possibility that some type III collagen remains undissolved at this step should be entertained. This fraction of type III collagen molecules can, however, be purified by repeated precipitation from acid solvents. Even though some type III collagen may fail to redissolve in neutral salt solvents, collagen extracted with 0.5 $M$ acetic acid containing pepsin should be exposed to neutral salt solvents at some interval during purification in order to inactivate any pepsin that may have coprecipitated with the collagen in initial precipitations.

An alternative approach to achieving initial purification of the collagen present in extracts of various tissues is chromatography of the extracts on columns of DEAE-cellulose.[33] For this purpose, the extracts are dialyzed extensively versus 0.2 $M$ NaCl, 0.05 $M$ Tris, pH 7.5, and any material precipitating at this step is removed by centrifugation. Aliquots (50 ml) of the dialyzed extracts are then applied at a flow rate of 100 ml/hr to a 2.5 × 15 cm column of DEAE-cellulose (DE-32, Whatman) equilibrated with the same buffer. After application of the sample, elution with starting buffer is continued until no further ultraviolet-absorbing material is eluted from the column. The column is then washed with 1.0 $M$ NaCl, 0.05 $M$ Tris, pH 7.5, to elute retained proteins. Since the relatively basic collagen molecules are not retained on DEAE-cellulose under these conditions, they can be effectively resolved from acidic contaminants by this procedure.

*Purification of Collagen (Selective Salt Precipitation of Collagen Types)*

Selective salt precipitation currently constitutes the most widely used method for recovery of the individual collagen types that may be present in extracts of various tissues. The collagens may be selectively precipitated from neutral salt solvents or dilute acid solvents. Indeed, selective precipitation from both types of solvent may actually be necessary to resolve the molecular species of collagen present in complex mixtures. In general, the results achieved by selective salt precipitation are reproducible and the collagens are purified and fairly well resolved when the tech-

---

[28] H. Herrmann and K. von der Mark, *Hoppe-Seyler's Z. Physiol. Chem.* **356**, 1605 (1975).
[29] J. M. Seyer, *Biochim. Biophys. Acta* **629**, 490 (1980).
[30] E. Chung and E. J. Miller, *Science* **183**, 1200 (1974).
[31] E. H. Epstein, Jr., *J. Biol. Chem.* **249**, 3225 (1974).
[32] T. Fujii and K. Kühn, *Hoppe-Seyler's Z. Physiol. Chem.* **356**, 1793 (1975).
[33] E. J. Miller, *Biochemistry* **10**, 1652 (1971).

nique is performed properly and under appropriate conditions. Although these factors have not been systematically evaluated, some empirical observations are of value in this context.

In this regard, the precipitates may be formed and collected in the manner described for the initial purification of collagen with the exception that NaCl is added to the solutions to discrete concentrations at which the various collagens are known to precipitate. Owing to the critical nature of the separations to be achieved, however, it is advisable to deliver NaCl to the solutions in a much more controlled fashion than is possible by the addition of the crystalline salt in order to avoid localized concentration gradients that may result in premature precipitation of a portion of one or more of the molecular species of collagen present in the solution. This can generally be accomplished by extensively dialyzing the solution versus solvent containing NaCl at the desired concentration. A more rapid and equally suitable approach is the dropwise addition of a concentrated NaCl solution during constant stirring of the collagen solution. In the latter approach, the concentrated NaCl solution can be delivered from a burette, which facilitates control of the rate at which drops are delivered and affords the means of readily determining the amount of solution added.

The best results will be obtained when the precipitates are derived from clear solutions of collagen. In addition, the formation of relatively large heavy precipitates should be avoided. For reasons that are not readily apparent, such precipitates tend to contain a portion of collagen molecules that normally would not precipitate until much higher salt concentrations are attained. Therefore, as a general rule, the precipitates should be derived from solutions containing collagen at concentrations of 0.5 mg/ml or less. Sufficient time should also be allowed for the precipitate to form at a given salt concentration, and an interval of 24 hr can be considered appropriate for this purpose. And finally, the precipitate obtained at a given salt concentration should be suspended and washed in solvent at a slightly higher salt concentration in order to remove other molecular species that may have coprecipitated.

Table II summarizes the results that may be expected in experiments designed to precipitate selectively the collagens from neutral salt solvents as well as dilute acid solvents.[3,24,30,31,34-42] Inspection of these data reveals

[34] R. K. Rhodes and E. J. Miller, *Biochemistry* **17**, 3442 (1978).
[35] K. Tryggvason, P. G. Robey, and G. R. Martin, *Biochemistry* **19**, 1284 (1980).
[36] J. Risteli, H. P. Bächinger, J. Engel, H. Furthmayr, and R. Timpl, *Eur. J. Biochem.* **108**, 239 (1980).
[37] R. W. Glanville, A. Rauter, and P. P. Fietzek, *Eur. J. Biochem.* **95**, 383 (1979).
[38] D. K. Furuto and E. J. Miller, *J. Biol. Chem.* **255**, 290 (1980).
[39] R. L. Trelstad, A. H. Kang, B. P. Toole, and J. Gross, *J. Biol. Chem.* **247**, 6469 (1972).

TABLE II
SELECTIVE SALT PRECIPITATION OF THE COLLAGENS

| Concentration of NaCl (M) | Collagen(s) precipitating from | | Primary references[c] |
|---|---|---|---|
| | Neutral salt solvent[a] | Dilute acid solvent[b] | |
| 0.7 | | Types I, II, III | 34 |
| 1.2 | | Types IV, V | 24, 34, 35 |
| | | 7 S | 36 |
| 1.5–1.8 | Type III | | 30, 31 |
| | Type IV | | 3, 37 |
| 1.8 | | Cys-rich aggregate | 38 |
| 2.0 | 7 S | | 36 |
| 2.2–2.5 | Type I | | 30, 31, 39 |
| 3.5 | Type I | | 40 |
| | Type I-trimer | | 41 |
| 4.0 (>4.0) | Type II | | 39 |
| | Type I-trimer | | 41 |
| | Type V | | 42 |

[a] Most commonly employed solvent: 1.0 M NaCl, 0.05 M Tris, pH 7.5.
[b] Most commonly employed solvent: 0.5 M acetic acid, pH adjusted to 2.5–3.0.
[c] Numbers refer to text footnotes.

several interesting points of practical importance. In this regard, all the collagens isolated thus far precipitate from dilute acid solvents in a more narrow range of NaCl concentrations than when precipitated from neutral salt solvents. In addition, certain collagens are not necessarily precipitated in a quantitative fashion at a given salt concentration. Thus, it has been observed that the bulk of type I collagen present in neutral salt solvents can be precipitated at 2.5 M NaCl, but that a portion of this collagen remains in solution until a NaCl concentration of 3.5 M is attained.[40] Similar observations have been made with respect to the precipitation behavior of type I-trimer collagen, a portion of which can be precipitated at 3.5 M NaCl with the remainder precipitating at 4.0 M NaCl.[41] Moreover, selective precipitation from either neutral salt solvents or dilute acid solvents alone is not an adequate procedure for resolving

---

[40] C. Welsh, S. Gay, R. K. Rhodes, R. Pfister, and E. J. Miller, *Biochim. Biophys. Acta* **625**, 78 (1980).
[41] L. Moro and B. D. Smith, *Arch. Biochem. Biophys.* **182**, 33 (1977).
[42] R. E. Burgeson, F. A. El Adli, I. I. Kaitila, and D. W. Hollister, *Proc. Natl. Acad. Sci. U.S.A.* **73**, 2579 (1976).

complex mixtures of the collagens such as those encountered in extracts of placental tissue that contain types I, III, IV, and V collagens as well as the high molecular weight, cysteine-rich collagenous aggregates.[36,38] The data summarized in Table II do, however, indicate that a combination of procedures, e.g., initial fractionation from dilute acid solvents followed by selective precipitation from neutral salt solvents, provides a reasonable approach to resolving even the most complex mixtures. It is of further interest to note that the type IV procollagen-like molecules extracted from basement membrane-containing tumors grown in lathyritic mice[3,43] exhibit the same precipitation behavior as type IV collagen molecules brought into solution from human placental tissue by means of limited proteolysis with pepsin.[37] Similarly, pN-collagen type III has been shown to precipitate selectively with type III collagen.[44] These observations indicate that the structural features responsible for the selective precipitation behavior of procollagen molecules reside largely in their lengthy pepsin-resistant collagenous domain and suggest that selective salt precipitation may be a useful general approach to resolving mixtures of the procollagens as well.

In addition to selective salt precipitation, other approaches to the isolation of specific types of collagen have proved to be feasible in a limited number of cases. In this regard, mixtures of types I and III collagens have been resolved by allowing the denatured proteins to renature at room temperature in water.[45] Under these conditions, helical structure is rapidly reestablished on the part of type III molecules, since the constituent $\alpha 1$(III) chains remain attached and aligned in the denatured state through interchain disulfide linkages. Rapid re-formation of native type III molecules then leads to their selective precipitation from the aqueous medium before the chains of type I molecules reassociate to form native molecules. In yet another approach, the resistance of 7 S collagen to degradation with bacterial collagenase affords the means of isolating the latter collagen in essentially pure form by treating mixtures of collagen containing the 7 S protein with the enzyme followed by dialysis and molecular sieve chromatography of the incubation mixture to remove the degradation products of alternative collagens.[36]

*Purification of Collagen (Chromatographic Techniques)*

Chromatography of the collagens under nondenaturing conditions provides the means of achieving final purification and resolution of the collagens that may not be readily or completely separated using selective

[43] R. Timpl, G. R. Martin, P. Bruckner, G. Wick, and H. Wiedemann, *Eur. J. Biochem.* **84**, 43 (1978).
[44] R. Timpl, R. W. Glanville, H. Nowack, H. Wiedemann, P. P. Fietzek, and K. Kühn, *Hoppe-Seyler's Z. Physiol. Chem.* **356**, 1783 (1975).
[45] J. Chandrarajan, *Biochem. Biophys. Res. Commun.* **83**, 180 (1978).

salt precipitation techniques. Although chromatography of the native collagens has been used primarily in the preparation of types IV and V collagens, the methods are generally applicable, within limits, to the preparation of other collagens as well.

CM-CELLULOSE CHROMATOGRAPHY (ACID pH)

CM-cellulose chromatography at acid pH has been useful in the resolution of type IV collagen from relatively acidic noncollagenous contaminants as well as other collagens. The following procedure is adapted from previously published methods.[24,37] In addition, we supply here several details of the methods generally applicable in ion-exchange chromatography of native as well as denatured collagen preparations. These details, which need not be repeated in subsequent sections, concern the preparation of the stationary phase, packing of the column, a description of the instruments used in chromatography, and dissolution of the sample in starting buffer.

*Stock Solutions*

Sodium acetate, 0.4 $M$ ($Na^+$), pH 4.8: Dissolve 54.7 g of $NaC_2H_3O_2 \cdot 3 H_2O$ in approximately 600 ml of distilled water, add 15.4 ml of glacial acetic acid, and dilute to 1 liter.

Urea, 8.0 $M$, deionized (480.5 g of reagent grade urea per liter in distilled water). The solution is deionized by passage through a mixed bed resin [AG 501-X8(D), 20–50 mesh, Bio-Rad Laboratories].

NaCl, 1.0 $M$, (58.4 g of NaCl per liter in distilled water)

*Gradient Solutions.* These are prepared from stock solutions.

Starting buffer: 0.04 $M$ ($Na^+$) sodium acetate containing 2.0 $M$ urea, pH 4.8

Limit buffer: 0.04 $M$ ($Na^+$) sodium acetate containing 2.0 $M$ urea and 0.4 $M$ NaCl, pH 4.8

*Procedure.* The stationary phase, CM-cellulose (CM-32 or CM-52, Whatman), should be precycled through acidic and basic solvents as specified by the supplier. Ultimately the particles are equilibrated with starting buffer. This is most readily accomplished by washing the particles several times with the stock solution, 0.4 $M$ ($Na^+$) sodium acetate, pH 4.8, followed by several additional washes with the starting buffer. Throughout the precycling and equilibration procedures, it is advisable to pour off and discard suspensions of fine particles that do not settle rapidly with the bulk of the particles. Removal of the finer particles in this fashion generally allows the use of higher flow rates during subsequent chromatographic procedures.

With respect to column parameters, a suitable design is one with an inside diameter of 2.5 cm equipped with adjustable flow adapters that permit upward elution in a bed height of about 10 cm. If the column is not used in a cold room, it should be jacketed and attached to a refrigerated circulating water bath to allow pouring and subsequent use at temperatures between 4° and 8°. The column is most conveniently packed by the slow addition of a 50% slurry of CM-cellulose in starting buffer, at the same time withdrawing excess buffer from the bottom of the column at a flow rate approximating that to be used in subsequent chromatographic procedures (100 ml/hr). The addition of slurry is stopped when a bed height of 10–11 cm has been attained, and the top flow adapter is replaced to a position where it gently compresses the top of the CM-cellulose bed. The column packing should then be evaluated by pumping starting buffer upward through the bed at a flow rate of 100 ml/hr. If this flow rate can be sustained for several minutes with little or no contracture on the part of the CM-cellulose bed, the column may be considered ready for use.

During chromatography, the column is eluted with a linear gradient delivered to the bottom of the column by a peristaltic pump. The gradient is, in turn, delivered to the pump from a dual-chamber constant-level device containing 500 ml of starting buffer in the mixing chamber and 500 ml of limit buffer in the reservoir chamber. This apparatus is easily constructed by connecting two aspirator bottles of equal size with a piece of flexible plastic tubing. The mixing chamber is placed on a magnetic stirrer to allow continuous stirring during delivery of the gradient to the pump. The column effluent should be monitored continuously by means of a UV-detector equipped with a flow cell and capable of monitoring in the range of 220–240 nm. A continuous record of absorbance versus time of elution may be generated by a recorder, and fractions of suitable volume are collected in an automatic fraction collector.

The sample to be chromatographed (20–50 mg) is dissolved in starting buffer at a concentration of 1 mg/ml. This may require several hours of stirring if the sample has been lyophilized, an an appreciable quantity of sample may fail to redissolve under these conditions. Consequently, it is best to retain collagen to be chromatographed in this fashion in solution in 0.5 M acetic acid and dialyze the solution versus starting buffer. Any material precipitating at this step should be removed by centrifugation.

The sample is then pumped into the bottom of the column. After application of the sample, the column is eluted initially with starting buffer until all unretained material has been washed through the column, at which time elution with the gradient is initiated. Figure 1 illustrates the typical results obtained in this procedure when chromatographing a preparation of type IV collagen derived from human placenta by means of limited digestion with pepsin. The material applied to the column is eluted in essentially three peaks containing (in the order of elution): unretained

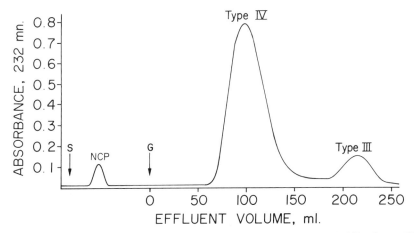

FIG. 1. Carboxymethyl (CM) cellulose chromatography of pepsin-solubilized type IV collagen from human placenta. The sample represents 50 mg of type IV collagen purified by selective salt precipitation prior to chromatography on a 2.5 × 10 cm column of CM-cellulose. Elution was achieved at a flow rate of 100 ml/hr in 0.04 $M$ (Na$^+$) sodium acetate containing 2.0 $M$ urea, pH 4.8, with a linear gradient of 0.0 to 0.4 $M$ NaCl over a total volume of 1 liter. The column was maintained at 8° with a refrigerated circulating bath. S, the point of sample application; G, the initiation of the gradient; NCP, noncollagenous proteins.

noncollagenous proteins, type IV collagen, and other molecular species of collagen that are more strongly retained on the column. The relative proportions of material eluting in the three indicated peaks as well as the types of collagen present in the third peak are dependent on how well the sample has been purified prior to chromatography.

CM-CELLULOSE CHROMATOGRAPHY (ALKALINE pH)

CM-cellulose chromatography at slightly alkaline pH has been used in the purification of one of the molecular species of type V collagen, i.e., molecules with the chain composition $[\alpha1(V)]_2\alpha2(V)$.[40]

*Stock Solutions*

Tris, 0.2 $M$, pH 8.0: Dissolve 24.2 g of Tris in 900 ml of distilled water, adjust the pH to 8.0 by the addition of distilled 6 $N$ HCl, and dilute to 1 liter with water.
Urea, 8.0 $M$, deionized
NaCl, 2.5 $M$
*Gradient Solutions.* These are prepared from stock solutions
Starting buffer: 0.02 $M$ Tris containing 6.0 $M$ urea, pH 8.0
Limit buffer: 0.02 $M$ Tris containing 6.0 $M$ urea and 0.25 $M$ NaCl, pH 8.0

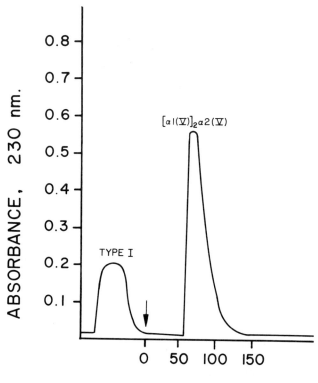

FIG. 2. CM-cellulose chromatography of pepsin-solubilized type V collagen from human placenta. The sample represents 25 mg of type V collagen chromatographed at 4° on a 2.5 × 10 cm column. Elution was achieved at 150 ml/hr in 0.02 $M$ Tris containing 6.0 $M$ urea, pH 8.0, using a linear gradient of 0.0 to 0.25 $M$ NaCl over a total volume of 400 ml. The arrow designates the initiation of the gradient.

*Procedures.* For chromatography in this system, the sample (up to 50 mg) is dissolved in or dialyzed into starting buffer at a concentration of 1 mg/ml. The sample is then applied to a 2.5 × 10 cm column of CM-cellulose previously equilibrated with starting buffer. After sample application, the column is washed with starting buffer until all unbound material has been eluted. At this time, gradient elution is initiated at a flow rate of 150 ml/hr using 200 ml each of starting and limit buffers in the respective chambers of a constant-level device.

Chromatography under these conditions results in an elution profile similar to that depicted in Fig. 2. Depending on the sample and its mode of preparation and purification, the initial peak of unretained substances may contain a mixture of types I, III, and IV collagens, and the retained protein eluted with the gradient in the second peak is comprised of molecules with the chain composition $[\alpha 1(V)]_2\alpha 2(V)$.[40]

DEAE-CELLULOSE CHROMATOGRAPHY

Preparations of type V collagen may also be successfully chromatographed and purified on DEAE-cellulose.[46]

*Stock Solutions*

> Tris, 0.5 M, and 0.2 M NaCl, pH 8.6: Dissolve 60.6 g of Tris in 900 ml of distilled water, add 11.7 g of NaCl, adjust the pH to 8.6 by the addition of distilled 6 N HCl, and dilute to 1 liter with water.
> Urea, 8.0 M, deionized
> NaCl, 1.0 M

*Gradient Solutions.* These are prepared from stock solutions.
> Starting buffer: 0.05 M Tris containing 0.02 M NaCl and 2.0 M urea, pH 8.6
> Limit buffer: 0.05 M Tris containing 0.32 M NaCl and 2.0 M urea, pH 8.6

*Procedures.* For chromatography in this system, the sample is dissolved in or dialyzed into starting buffer at a concentration approximating 1 mg/ml. The sample is then applied to a 2.5 × 12 cm column of DEAE-cellulose (DE-52, Whatman) equilibrated with starting buffer and maintained at 15°. After sample application, elution is initiated with starting buffer and continued until unretained proteins have washed through the column. Gradient elution is then initiated utilizing a flow rate of 100 ml/hr with a total gradient volume of 1 liter.

Under these or similar chromatographic conditions, type IV collagen is unretained by the DEAE-cellulose column,[43] whereas molecules with the chain composition $[\alpha 1(V)]_2\alpha 2(V)$ are eluted shortly after initiation of the gradient and more acidic noncollagenous materials are eluted near the end of the gradient.[46]

Characterization of Collagens

Preparations of collagen are commonly characterized by a variety of physical and chemical techniques. The physical techniques utilized in characterizing preparations of soluble collagen are described in this volume [7]. Approaches utilized in elucidating some of the special features of the collagens are likewise described in subsequent contributions.

[46] H. Bentz, H. P. Bächinger, R. Glanville, and K. Kühn, *Eur. J. Biochem.* **92**, 563 (1978).

Our treatment here is thus confined to the chromatographic and electrophoretic techniques useful in isolating and characterizing the primary chain constituents of the various collagens. In general, these procedures are performed at elevated temperatures at which the soluble collagens are denatured or they are performed at room temperatures in solvents in which renaturation of a previously denatured collagen preparation cannot occur.

*Chromatography of Collagens (Denaturing Conditions)*

MOLECULAR SIEVE CHROMATOGRAPHY

Chromatography of denatured collagen preparations on molecular sieve columns provides a reasonably efficient method of evaluating the molecular weight distribution of the denaturation products and isolating the primary chain constituents. The eluents used in this type of chromatography are 1.0 $M$ calcium chloride, 0.05 $M$ Tris, pH 7.5[47]; or 2.0 $M$ guanidine-HCl, 0.05 $M$ Tris, pH 7.5.[48] The latter solvent exhibits a somewhat increased capacity to solubilize collagen preparations with a relatively high proportion of aggregates among the denaturation products. In general, a stationary phase of agarose beads with a molecular weight operating range of about $10^6$–$10^4$ for random coil polypeptides (BioGel A-5m, 200–400 mesh, Bio-Rad Laboratories) is suitable for the separations to be achieved. The gel is equilibrated with and swollen for a suitable interval (1–2 days) in the solvent of choice and then poured as a 50% slurry into a simple glass column with an inside diameter of 1.5 cm and sufficient length to allow a bed height of 150–160 cm. During packing of the stationary phase, it is advisable to control the flow rate of excess solvent exiting from the bottom of the column at 10–15 ml/hr. A column of these dimensions can easily accommodate a sample of up to 50 mg of denatured collagen dissolved in and applied to the column in 2.5–5.0 ml of eluent. For larger samples, the column diameter can be increased to 2.0 or 2.5 cm and the sample be applied in 10–20 ml of eluent. Prior to application to the column, the sample should be suspended in the appropriate amount of eluent and dissolved by warming the suspension to 45° for 0.5 hr. Any undissolved material can be removed by centrifugation at room temperature. The sample is then applied to the column and eluted at a flow rate of 10–15 ml/hr at room temperature. Figure 3 illustrates the typical results to be expected on chromatography of a denatured collagen preparation on BioGel A-5m. In this particular example, the elution pattern is that obtained for a preparation of type I collagen extracted originally in 0.5

[47] K. A. Piez, *Anal. Biochem.* **26**, 305 (1968).
[48] E. Chung, E. M. Keele, and E. J. Miller, *Biochemistry* **13**, 3459 (1974).

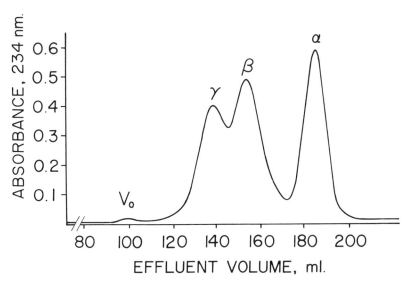

FIG. 3. Molecular sieve chromatography of a denatured type I collagen preparation on a 1.5 × 155 cm column of BioGel A-5m. The sample represents 30 mg of acid-soluble guinea pig dermal collagen. The column was eluted with 2.0 $M$ guanidine-HCl, 0.05 $M$ Tris, pH 7.5, at a flow rate of 13 ml/hr.

$M$ acetic acid from guinea pig dermis. The denaturation products are eluted in the order of decreasing molecular weight and exhibit a logarithmic relationship between elution volume and molecular weight. Although mixtures of $\beta$-, $\gamma$-, and higher molecular weight components are generally not well resolved by molecular sieve chromatography, the $\alpha$ chains present in a denatured collagen preparation are well resolved from the other components, as shown in Fig. 3.

ION-EXCHANGE CHROMATOGRAPHY

Ion-exchange chromatography of denatured collagen preparations is the most suitable method of resolving and purifying the constituent primary chains in a given collagen preparation. In these applications, columns essentially identical to those described above for ion-exchange chromatography of native collagen preparations are used. During chromatography of denatured collagens, however, the columns are maintained at 40–45°, and all buffers used in equilibrating the stationary phases, dissolving the samples, and developing the columns are deaerated prior to use and likewise maintained at 40–45°.

Of the approaches currently used in ion-exchange chromatography of denatured collagens, chromatography on CM-cellulose represents the most versatile and widely applicable technique. The following description of the technique is adapted from previous accounts.[49,50]

*Stock Solutions*

Sodium acetate, 0.6 $M$ ($Na^+$), pH 4.8: Dissolve 81.6 g of $NaC_2H_3O_2 \cdot 3 H_2O$ in approximately 600 ml of distilled water, add 23.0 ml of glacial acetic acid, and dilute to 1 liter.
Urea, 8.0 $M$, deionized
NaCl, 1.0 $M$

*Gradient Solutions.* These are prepared from stock solutions.

Starting buffer: (a) 0.06 $M$ ($Na^+$) sodium acetate, pH 4.8[49]; (b) 0.02 $M$ ($Na^+$) sodium acetate containing 1.0 $M$ urea[33]
Limit buffer: (a) 0.06 $M$ ($Na^+$) sodium acetate containing 0.1 $M$ NaCl[49]; (b) 0.02 $M$ ($Na^+$) sodium acetate containing 1.0 $M$ urea and 0.1 $M$ NaCl[33]

*Procedure.* As noted, the gradient solutions used for CM-cellulose chromatography of denatured collagens may be prepared without urea. The addition of urea, however, often enhances the solubility of collagen preparations in the starting buffer and it has been used in the development of CM-cellulose columns at concentrations as high as 6.0 $M$.[51] In the presence of urea, the affinity of the denatured collagen components for the stationary phase tends to be reduced. Accordingly, the concentration of buffer salts may be reduced as in the examples given above in order to allow the same degree of retention for all components in the presence of urea. In any given instance, the precise composition of the starting buffer can be adapted to achieve optimum solubilization of the sample. The sample to be chromatographed can be suspended in starting buffer and immediately warmed to 40–45° for several minutes. Material that fails to dissolve under these conditions within an hour should be removed by centrifugation at room temperature. Alternatively, the sample may be dissolved in a solvent such as 0.5 $M$ acetic acid at 4° and subsequently dialyzed versus starting buffer in the cold. The latter method of achieving dissolution of the sample in starting buffer requires more time but generally results in solubilization of a larger proportion of the sample. For the sake of convenience, the volume of the sample solution should be minimized even though almost any reasonable volume of sample can be applied to the column. Moreover, the concentration of collagen in starting buffer need not be carefully controlled because solutions of denatured

---

[49] K. A. Piez, E. A. Eigner, and M. S. Lewis, *Biochemistry* **2**, 58 (1963).
[50] E. J. Miller, in "The Methodology of Connective Tissue Research" (D. A. Hall, ed.), p. 197. Joynson-Bruvvers, Oxford, 1976.
[51] H. Sage and P. Bornstein, *Biochemistry* **18**, 3815 (1979).

TABLE III
SUMMARY OF CONDITIONS EMPLOYED IN THE CHROMATOGRAPHY OF
DENATURED COLLAGEN PREPARATIONS ON CM-CELLULOSE

| Sample (mg) | Column[a] | | | Gradient volume (ml)[b] | Flow rate (ml/hr) |
|---|---|---|---|---|---|
| | I.D. (cm) | L (cm) | A (cm$^2$) | | |
| 200–300 | 2.5 | 10 | 4.9 | 2000 | 200 |
| 100–200 | 1.8 | 10 | 2.5 | 1000 | 200 |
| 50–100 | 1.5 | 8 | 1.8 | 700 | 150 |
| 25–50 | 1.3 | 8 | 1.3 | 500 | 150 |
| 5–25 | 0.9 | 7 | 0.6 | 250 | 100 |

[a] I.D., inside diameter; L, length; A, cross-sectional area of the column.
[b] Total volume of gradient, distributed equally as starting and limit buffers.

collagen in the range of 2–5 mg/ml are relatively nonviscous and exhibit little resistance to flow through the confines of the column. After application of the sample to the column, the collagen components are eluted with a linear salt gradient over the range of 0.0 to 0.1 $M$ NaCl. The gradient is fabricated by placing appropriate amounts of starting and limit buffer in the respective chambers of a constant-level device. Table III summarizes the precise conditions for CM-cellulose chromatography of denatured collagens. As indicated in the table, the amount of sample that can be successfully chromatographed in a single run as well as the size of the gradient to be employed and flow rate can be correlated with the cross-sectional area of the column.

A representative CM-cellulose elution pattern obtained during chromatography of a denatured type I collagen preparation is presented in Fig. 4A. It is clear from this illustration that the primary chain constituents of the type I molecule, $\alpha 1(I)$ and $\alpha 2(I)$, are not generally well resolved from higher molecular weight components by CM-cellulose chromatography. Prior chromatography of the sample on a molecular sieve column to isolate the $\alpha$ chains and subsequent rechromatography of the latter on CM-cellulose, however, allows the recovery of the chains as illustrated in Fig. 4B. A similar sequence of the chromatographic steps may be used in the isolation and purification of the primary chain constituents of the other types of collagen. With reference to Fig. 4B, the chromatographic properties of chains derived from the other collagens are as follows: $\alpha 1(II)$ chains elute in virtually the same position as $\alpha 1(I)$ chains; $\alpha 1(III)$ chains elute just prior to $\alpha 2(I)$ chains; $\alpha 1(IV)$ and $\alpha 2(IV)$ chains derived from

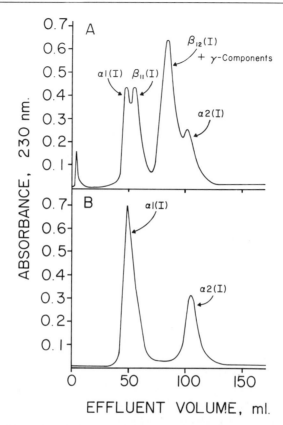

FIG. 4. (A) CM-cellulose chromatography of a denatured type I collagen preparation employing a 0.9 × 7 cm column. The sample represents 25 mg of acid-soluble guinea pig dermal collagen. Elution was achieved at a flow rate of 100 ml/hr in 0.02 $M$ (Na$^+$) sodium acetate containing 1.0 $M$ urea, pH 4.8, by employing a linear gradient from 0.0 to 0.1 $M$ NaCl in a total gradient volume of 250 ml. (B) CM-cellulose chromatography of the $\alpha$ chains isolated from a denatured preparation of type I collagen by molecular sieve chromatography. The sample represents 20 mg of $\alpha$ chains, and chromatography was performed as described in (A).

pepsin-solubilized type IV collagen elute in positions virtually coincident with $\alpha$1(I) and $\alpha$2(I), respectively; and the chains of type V collagen chromatograph as a poorly resolved doublet peak in a position approximating that observed for $\alpha$2(I).

Since the $\alpha$1(V) and $\alpha$2(V) chains (B and A chains or $\alpha$B and $\alpha$A chains, respectively) in type V collagen preparations are not suitably resolved by chromatography on CM-cellulose, two alternative procedures have been devised specifically for the resolution of these chains. In one procedure, the chains are applied to and eluted from a 1.5 × 9 cm column of phosphocellulose (P-11, Whatman, dry particles collected in the 140–230-mesh range).[34] Elution is achieved at a flow rate of 100 ml/hr in 0.06 $M$ (Na$^+$) diabasic sodium phosphate, pH 6.3, employing a linear gradient from 0.0 to 0.2 $M$ NaCl in a total gradient volume of 400 ml. Under these conditions, the $\alpha 1(V)$ and $\alpha 2(V)$ chains are eluted as two completely resolved peaks after the delivery of slightly more than one-half of the gradient. In a further approach, these chains may be resolved by chromatography on a 2.5 × 11 cm column of DEAE-cellulose (DE-52, Whatman).[46] In this case, elution is achieved in 0.025 $M$ Tris containing 4.0 $M$ urea, pH 8.6, using a linear gradient from 0.0 to 0.1 $M$ NaCl in a total gradient volume of 1 liter. Under these conditions, $\alpha 2(V)$ chains are not retained on the column and $\alpha 1(V)$ chains are eluted after delivery of about two-thirds of the total gradient.

### Chromatography of Peptides

Fragmentation of the various collagen $\alpha$ chains by cleavage at methionyl residues with CNBr constitutes the most widely employed method of recovering the chains as a series of discrete peptides. The latter, which represent suitable starting materials for the determination of primary structure, can be isolated and purified by a combination of ion-exchange and molecular sieve chromatography. In addition, the unique set of CNBr peptides derived from each type of chain serves as a definitive means of identifying the respective chains.

#### CM-Cellulose Chromatography (pH 3.6, Linear Gradients)

Although it may often be advisable initially to resolve the CNBr peptides derived from a given chain by molecular sieve chromatography,[52] initial resolution of the peptides can also be conveniently achieved by CM-cellulose chromatography. The method for CM-cellulose chromatography of the CNBr peptides described here is adapted from the technique originally used in chromatographing the peptides derived from $\alpha 1(I)$ chains.[53] This approach has subsequently been used in the initial resolution of the CNBr cleavage products of several other chains as well.

### Stock Solutions

Citrate, 0.1 $M$, pH 3.6: Dissolve 11.05 g of sodium citrate, Na$_3$C$_6$H$_5$O$_7$ · 2 H$_2$O, in approximately 600 ml of distilled water; add 13.1 g of citric acid, H$_3$C$_6$H$_5$O$_7$ · H$_2$O; and dilute to 1 liter with water.

NaCl, 1.0 $M$

---

[52] E. J. Miller, *Biochemistry* **10**, 3030 (1971).
[53] W. T. Butler, K. A. Piez, and P. Bornstein, *Biochemistry* **6**, 3771 (1967).

*Gradient Solutions.* These are prepared from stock solutions.

Starting buffer: (a) 0.02 $M$ citrate containing 0.04 $M$ NaCl, pH 3.6[53]; (b) 0.02 $M$ citrate containing 0.01 $M$ NaCl, pH 3.6[54]; (c) 0.02 $M$ citrate, pH 3.6[52]; (d) 0.005 $M$ citrate containing 0.02 $M$ NaCl, pH 3.6[55]

Limit buffer: (a) 0.02 $M$ citrate containing 0.14 $M$ NaCl, pH 3.6[53]; (b) 0.02 $M$ citrate containing 0.16 $M$ NaCl, pH 3.6[54]; (c) 0.02 $M$ citrate containing 0.14 $M$ NaCl, pH 3.6[52]; (d) 0.005 $M$ citrate containing 0.18 $M$ NaCl, pH 3.6[55]

*Procedure.* As indicated above, the precise composition of the gradient solutions used for CM-cellulose chromatography of the CNBr peptides may be varied to accommodate the nature and number of the peptides expected among the cleavage products of a given chain. The most appropriate gradient conditions allowing adequate resolution of the majority of the peptides can often be discerned following one or two preliminary runs. Since chromatography is performed at 40–45°, all buffers used in equilibrating and eluting the column should be deaerated. The peptides derived from the various collagen chains are readily dissolved in the starting buffer of choice. As a rule, concentrations of peptides in starting buffer of about 10 mg/ml are desirable in order to minimize the volume of sample to be applied to the column. It is further advisable to warm the samples to 40–45° for a few minutes prior to application to the column to ensure that all polypeptide constituents in the samples are present in random coil form. After application of the sample to the column, the peptides are eluted with a linear salt gradient over the range established by the gradient solutions employed. In general, the conditions chosen for CM-cellulose chromatography of the CNBr peptides (i.e., column dimensions, total gradient volume, and flow rate) are dependent on the amount of sample to be chromatographed as detailed in Table III for CM-cellulose chromatography of the denatured collagen preparations.

Figure 5 illustrates an application of this technique and shows the elution profile of the CNBr peptides derived from the $\alpha1(V)$ chain ($\alpha B$ chain or B chain) when chromatographed on CM-cellulose. The individual peptides are numbered in the order of elution from the column and are, for the most part, fairly well resolved under the conditions specified in the legend to the figure. The peptide or peptides recovered in various portions of the CM-cellulose effluent can be further purified and resolved by rechromatography in one or more of the systems described below.

[54] R. K. Rhodes and E. J. Miller, *J. Biol. Chem.* **254**, 12084 (1979).

[55] M. A. Haralson, W. M. Mitchell, R. K. Rhodes, T. F. Kresina, R. Gay, and E. J. Miller, *Proc. Natl. Acad. Sci. U.S.A.* **77**, 5206 (1980).

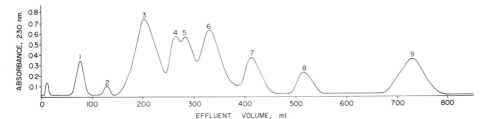

FIG. 5. CM-cellulose elution profile of the CNBr peptides derived from the human α1(V) chain. The sample represents 50 mg of peptides chromatographed on a 1.5 × 9 cm column. Elution was achieved at a flow rate of 100 ml/hr in 0.005 $M$ citrate containing 0.02 $M$ NaCl, pH 3.6, using a linear gradient of 0.02 to 0.18 $M$ NaCl over a total volume of 1 liter.

CM-CELLULOSE CHROMATOGRAPHY (pH 4.8, CONCAVE GRADIENTS)

In certain instances, it has been found advantageous to rechromatograph peptides, which are initially recovered as described above, on CM-cellulose utilizing modified gradient conditions.[48,53]

*Stock Solutions*

Sodium acetate, 0.6 $M$ (Na$^+$), pH 4.8, prepared as described for chromatography of denatured collagens
NaCl, 1.0 $M$

*Gradient Solutions.* These are prepared from stock solutions.
Starting buffer: (a) 0.02 $M$ (Na$^+$) sodium acetate, pH 4.8; (b) 0.03 $M$ (Na$^+$) sodium acetate, pH 4.8; (c) 0.05 $M$ (Na$^+$) sodium acetate, pH 4.8[48]
Limit buffer: starting buffer (a), (b), or (c), containing 0.12 $M$ NaCl[48]

*Procedure.* Peptide samples to be rechromatographed in this system are dissolved in the appropriate starting buffer and applied to a 2.5 × 10 cm column of CM-cellulose previously equilibrated with starting buffer. Elution is then achieved at a flow rate of 200 ml/hr and at 40–45° by delivery of a concave salt gradient formed by placing 825 ml of starting buffer in the mixing chamber and 500 ml of limit buffer in the reservoir chamber of a constant-level device.[48] Rechromatography of the relatively acidic peptides is most conveniently performed using starting and limit buffers with 0.02 $M$ (Na$^+$) sodium acetate, and the more basic peptides may be rechromatographed using buffers at the indicated higher concentrations. When rechromatographed under these conditions, certain peptides, particularly those derived from α1(I) chains, tend to be resolved into two or as many as three different peaks each of which exhibit the same amino acid composition.[53] The precise reason for the chromatographic heterogeneity observed with respect to these peptides is not currently known, although it was proposed to result from the establishment of an equilibrium between the free acid and lactone forms of COOH-terminal homoserine residues.[53]

PHOSPHOCELLULOSE CHROMATOGRAPHY

Phosphocellulose chromatography is useful as an additional ion-exchange procedure for the purification of the relatively acidic CNBr peptides.[1,56]

*Stock Solutions*

> Sodium acetate, 0.1 $M$ ($Na^+$), pH 3.8: Dissolve 13.6 g of $NaC_2H_3O_2 \cdot 3 H_2O$ in approximately 600 ml of distilled water, add 57.2 ml of glacial acetic acid, and dilute to 1 liter with water.
> NaCl, 1.0 $M$

*Gradient Solutions.* These are prepared from stock solutions.

> Starting buffer: 0.001 $M$ ($Na^+$) sodium acetate, pH 3.8
> Limit buffer: 0.001 $M$ ($Na^+$) sodium acetate containing 0.3 $M$ NaCl, pH 3.8

*Procedure.* Peptides are made ready for chromatography in this system by dissolving in starting buffer, and the solution is applied to a 1.8 × 8 cm column of phosphocellulose (P-11, Whatman). The particles of phosphocellulose are prepared for chromatography by sieving and equilibration with starting buffer as noted above for the use of phosphocellulose columns in the separation of collagen chains. Elution is achieved at a flow rate of 100 ml/hr and at 40–45° by means of a linear salt gradient over a total gradient volume of 500 ml. When required, the resolving power of the phosphocellulose procedure may be increased by lowering the slope of the gradient. This may be accomplished either by decreasing the concentration of NaCl in the limit buffer or by increasing the total gradient volume.[52]

MOLECULAR SIEVE CHROMATOGRAPHY

Used in conjunction with ion-exchange techniques, molecular sieve chromatography provides an excellent means of resolving and purifying the CNBr peptides, at the same time affording the means of making reasonably accurate estimates of molecular weight.[47] For these purposes, the majority of the peptides derived from the various collagen chains can be chromatographed on columns of BioGel A-1.5m (200–400 mesh) or BioGel P-150 (100–200 mesh), both obtainable from Bio-Rad Laboratories. For peptides in the molecular weight range of 2000 or less, a more suitable stationary phase is BioGel P-6. Columns measuring 1.5 × 120–

---

[56] E. J. Miller, J. M. Lane, and K. A. Piez, *Biochemistry* **8**, 30 (1969).

160 cm are packed with these or comparable media and eluted under the same conditions as noted for the molecular sieve chromatography of collagen chains. Molecular weight estimates for the peptides are made by reference to standard curves constructed by plotting elution volume versus the logarithm of the molecular weight for a series of well-characterized collagen peptides.

*Polyacrylamide Gel Electrophoresis*

Polyacrylamide gel electrophoresis is commonly used in the identification and characterization of the different collagen types. It has proved to be a useful technique in studies on the chains as well as the peptides derived from the chains and has the advantages of being a relatively rapid, highly sensitive technique that can be used to analyze several samples simultaneously. Two systems are widely used. Both are adaptations of previously published techniques, one a continuous buffer system[57] and the second a discontinuous system.[58,59] Each is equally adaptable to either disc or slab gel methodology.

CONTINUOUS BUFFER SYSTEM

*Solutions*

    A. 30% acrylamide, 0.08% $N,N'$-methylene bisacrylamide. Dissolve 30.0 g of acrylamide and 0.8 g of bisacrylamide in approximately 50 ml of distilled water. Dilute to a final volume of 100 ml.

    B. 0.6 $M$ ($PO_4^{-3}$) sodium phosphate, pH 7.2. Dissolve 115.8 g of $Na_2HPO_4 \cdot 7\ H_2O$ and 23.1 g of $NaH_2PO_4 \cdot H_2O$ in approximately 750 ml of distilled water. Dilute to a final volume of 1 liter.

    C. 10% (w/v) sodium dodecyl sulfate (SDS)

    D. 8.0 $M$ urea, deionized

    E. 0.45% (w/v) ammonium persulfate, 0.3% (v/v) $N,N,N',N'$-tetramethylethylenediamine (TEMED), 0.6% (w/v) SDS. Dissolve 45 mg of $(NH_4)_2S_2O_8$ in 3 ml of distilled water. Add 30 $\mu$l of TEMED and 0.6 ml of solution C, and dilute to 10 ml. This solution should be made fresh just prior to use and is enough to prepare up to 60 ml of gel solution.

    F. Sample buffer: 0.01 $M$ ($PO_4^{3-}$) sodium phosphate, pH 7.2, 0.2% SDS, 5.0 $M$ urea, 0.01% bromophenol blue. Combine 1.7 ml of solution B, 2 ml of solution C, 62.5 ml of solution D, and then

[57] H. Furthmayr and R. Timpl, *Anal. Biochem.* **41**, 510 (1971).
[58] U. K. Laemmli, *Nature (London)* **227**, 680 (1970).
[59] F. W. Studier, *J. Mol. Biol.* **79**, 237 (1973).

dissolve 10 mg of bromophenol blue in the mixture. Dilute to 100 ml with distilled water.
G. Water-saturated isobutanol. Mix 50 ml of distilled water and 50 ml of isobutanol with vigorous stirring in a tightly capped bottle. The less dense isobutanol will rise to the surface when mixing stops and the phases are allowed to separate.
H. Electrode buffer: $0.1\ M$ ($PO_4^{3-}$) sodium phosphate, pH 7.2, 0.1% SDS. Dissolve 1.0 g of SDS in 167 ml of solution B and dilute to 1 liter with distilled water.
I. Staining and fixing solution. Dissolve 2.5 g of Coomassie Brilliant Blue-R250 in a solution composed of 200 ml of distilled water, 500 ml of absolute methanol, and 100 ml of glacial acetic acid. Stir a minimum of 2 hr, at which time the volume should be brought up to 1 liter with water and the solution is filtered through qualitative grade filter paper.
J. Destaining solution. Mix 100 ml of absolute methanol, 100 ml of glacial acetic acid, and 800 ml of distilled water.

*Gel Preparation.* Prepare gels of the desired acrylamide concentration by mixing thoroughly solutions A, B, E, and distilled water in the proportions given in Table IV. As a general rule collagen $\alpha$ chains and the $\beta$ and $\gamma$ components are electrophoresed using 5% gels. Depending on their size, peptides derived from the chains may be electrophoresed on the higher percentage gels. Disc gels should be polymerized in $0.5 \times 7.5$ cm glass tubes at a height of 6 cm. Slab gels should be approximately 10 cm in length. Immediately after delivery of the gel solution into the polymerization vessels, water-saturated isobutanol (solution G) should be gently layered on the upper gel surface to a height of 3–4 mm. This will exclude air that may interfere with polymerization and produce a flat surface on the gel. The gel should be allowed to polymerize for approximately 1 hr.

TABLE IV
PROPORTIONS OF STOCK SOLUTIONS USED IN THE PREPARATION OF POLYACRYLAMIDE GELS OF DIFFERENT PERCENTAGES

|  | Percentage of polyacrylamide (final concentration) | | | |
| --- | --- | --- | --- | --- |
|  | 5 | 10 | 15 | 20 |
| Solution A | 1 | 2 | 3 | 4 |
| Solution B | 1 | 1 | 1 | 1 |
| Solution E | 1 | 1 | 1 | 1 |
| Distilled water | 3 | 2 | 1 | 0 |

When polymerization is complete, remove the isobutanol and rinse the gel surface with a small amount of electrode buffer (solution H). Place the gels in the electrophoresis apparatus and add a sufficient quantity of solution H to cover the bottom 3–4 cm of the gels. Care should be taken to ensure that no air bubbles are trapped at the bottom of the gels. A sufficient amount of solution H is then added to the top buffer reservoir to provide a layer of solvent the height of which exceeds the upper gel surfaces by 1–2 cm.

*Preparation and Application of Samples.* Dissolve the samples in solution F at a concentration of 4 mg/ml and denature by heating for 0.5 hr at 50°. Reduction of disulfide bonds may be accomplished by including 5% 2-mercaptoethanol in the sample buffer at the time of denaturation. Samples of 10–50 $\mu$g (2.5–12.5 $\mu$l) are applied by gently layering the sample solution under the electrophoresis buffer onto the tops of the gels. This is readily accomplished by use of a pipette or microliter syringe, since the high urea concentration in the sample buffer provides a sufficiently dense solution, which remains in position at the top of the gels.

*Electrophoresis.* For electrophoresis, connect the power supply so that the bottom chamber serves as the anode. Activate the power supply and adjust to 3 mA/gel or 15 V for disk and slab gels, respectively. These power settings are used for approximately 10 min, until the bromophenol blue has penetrated the gel. At this time, the power settings are adjusted to 6 mA/gel or 35 V and electrophoresis is continued until the tracking dye reaches the end of the gel. The power supply is then disconnected, and the gels are removed from the tubes or plates and stained in an excess of solution I for 1–4 hr. Disc gels can be freed from their tubes by expressing water between the gel and the wall of the tube using a syringe and a small-diameter hypodermic needle. After staining, the gels should be destained by diffusion in solution J and stored in 7% acetic acid.

DISCONTINUOUS BUFFER SYSTEM

*Solutions*

A. 30% acrylamide, 0.8% $N,N'$-methylene bisacrylamide. For preparation, see solution A of continuous buffer system.
B. Separating gel buffer: 2.25 $M$ Tris, 0.6% SDS, pH 8.8. Dissolve 272.6 g of Tris and 6.0 g of SDS in 700 ml of distilled water. Adjust to pH 8.8 with concentrated HCl and dilute to 1 liter with water.
C. Stacking gel buffer: 0.125 $M$ Tris, 0.2% SDS, pH 6.8. Dissolve 15.1 g of Tris and 2.0 g of SDS in 700 ml of distilled water. Adjust to pH 6.8 with concentrated HCl and dilute to 1 liter with water.
D. Sample buffer: 0.063 $M$ Tris, 2% SDS, 10% glycerol, 0.01% bromophenol blue, pH 6.8. Dissolve 0.76 g of Tris, 2.0 g of SDS, 10

ml of glycerol, and 0.1 g of bromophenol blue in 70 ml of distilled water. Adjust to pH 6.8 with concentrated HCl and dilute to 100 ml.

E. 0.45% ammonium persulfate and 0.15% (v/v) TEMED. Dissolve 45 mg of $(NH_4)_2S_2O_8$ in approximately 8 ml of distilled $H_2O$. Add 15 μl of TEMED and dilute to 10 ml. This solution should be prepared fresh daily and is enough to prepare up to 60 ml of gel solution.

F. Electrode buffer (5 ×): 0.25 $M$ Tris, 0.5% SDS, 1.92 $M$ glycine, pH 8.3. Dissolve 30.3 g of Tris, 5.0 g of SDS, and 144.1 g of glycine in 700 ml of distilled water. Adjust to pH 8.3 with concentrated HCl and dilute to 1 liter. The electrode buffer is prepared by making a 1 : 5 dilution of the stock 5 × buffer with distilled water.)

G, I, and K. These solutions are the same as for the continuous buffer system.

*Gel Preparation.* Electrophoresis in the discontinuous buffer system involves use of a stacking gel on top of the separating gel. To prepare the separating gel, select the desired acrylamide concentration and mix solutions A, B, E, and distilled water in the proportions indicated in Table IV. Deliver the gel into the polymerization vessels and overlay with solution H. Allow the polymerization to go to completion for approximately 1 hr. Remove solution H and rinse the gel surface with $H_2O$, finally removing as much water as possible with a Pasteur pipette.

Prepare the stacking gel by mixing 1 part solution A, 2.7 parts solution C, 5 parts solution E, and 1.3 parts distilled water. This produces a stacking gel of 3% acrylamide. Layer the gel solution on top of the separating gel to a height of 1.0 cm. Again overlay the gel with solution H and allow 1 hr for polymerization.

After polymerization remove solution H and rinse the gel with the diluted electrode buffer prepared from solution F. Place the gels in the electrophoresis apparatus containing the 1 × buffer, then add additional buffer to the upper chamber in sufficient quantity to cover the gels.

*Preparation and Application of Samples and Electrophoresis.* Samples should be dissolved at 4 mg/ml in solution D, heat-denatured at 50° for 0.5 hr, and, if desired, reduced by the addition of 2-mercaptoethanol to give a concentration of 5%. Apply the samples by layering the more dense sample buffer solution on the tops of the gels under the electrophoresis buffer. Connect the apparatus to the power supply and electrophorese at 6 mA/gel for disc gels or 35 V for slab gels until the tracking dye reaches the end of the separating gel. The gels are then stained and destained following the procedure described for the continuous buffer system.

In some respects collagen chains and peptides behave anomalously when electrophoresed in SDS–polyacrylamide gels. They migrate more

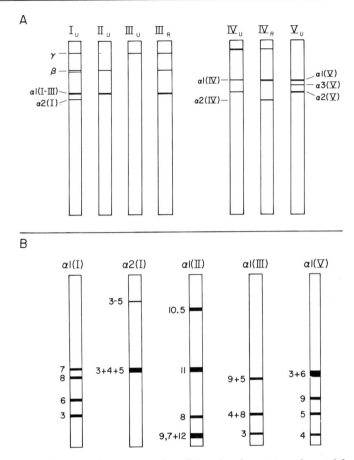

FIG. 6. (A) Diagrammatic representation of the migration pattern observed for components of denatured collagen preparations when electrophoresed in 5% SDS–polyacrylamide gels. The collagen type in each lane is designated by a roman numeral at the top of the gel, with U and R indicating unreduced or reduced samples, respectively. (B) Diagrammatic representation of the migration patterns for the major CNBr peptides derived from various collagen $\alpha$ chains. The patterns are for chains of human origin, and the peptides are numbered according to previous designations.[54,61] The exact relationships between the peptides will vary slightly depending on the acrylamide concentration and the buffer system employed.

slowly than globular proteins of comparable size, and often collagen chains of the same size do not migrate at the same rate. This is illustrated in Fig. 6A, which summarizes diagrammatically the results observed when different collagen preparations are denatured and electrophoresed in 5% polyacrylamide cylindrical or slab gels using the methodology outlined above. As indicated in Fig. 6A, the constituent $\alpha1(I)$ and $\alpha2(I)$ chains of type I collagen exhibit different mobilities, and these chains are well resolved from the $\beta$ and $\gamma$ components present in a preparation of acid-extracted type I collagen. The electrophoresis profiles of denatured types II and III collagens are much simpler owing to the presence of only one constituent chain in these collagens. The $\alpha1(II)$ chain migrates, as shown, with a mobility essentially identical to that of $\alpha1(I)$, and the disulfide-linked $\alpha1(III)$ chains migrate in the position of $\gamma$ components. In reduced samples of type III collagen, however (sample $III_R$), the $\alpha1(III)$ chains migrate with a mobility similar to $\alpha1(I)$ and $\alpha1(II)$ chains. Migration of the $\alpha1(III)$ chains is considerably delayed if the samples are reduced after 1 hr of electrophoresis by allowing buffer containing a reducing agent to diffuse into the top of the gels.[60] This approach, then allows resolution of the $\alpha1(III)$, $\alpha1(I)$, and $\alpha2(I)$ chains in samples containing a mixture of types I and III collagens. Preparations of pepsin-solubilized type IV collagen are comprised largely of disulfide-bonded aggregates that barely penetrate the gels and migrate less rapidly than $\gamma$ components. On reduction, the constituent $\alpha1(IV)$ and $\alpha2(IV)$ chains are well resolved, the former migrating slightly slower than $\alpha1(I)$ chains and the latter migrating slightly faster than $\alpha2(I)$ chains. As also shown in Fig. 6A, the constituent chains of type V collagen preparations are fairly well resolved on polyacrylamide gels.

Figure 6B summarizes the results to be expected when the CNBr peptides derived from various chains are electrophoresed in SDS–polyacrylamide gel systems.[54,61] In general, the electrophoretic mobility of the individual peptides can be roughly correlated with the logarithm of the molecular weight. Of greater practical value, however, is the observation that the CNBr peptide pattern for each chain is unique, a factor that allows the various chains to be identified as a function of their respective CNBr cleavage products.

Acknowledgments

Investigations in the authors' laboratory are supported by National Institutes of Health grants DE-02670, HL-11310, GM-27993, and a grant from the Osteogenesis Imperfecta Foundation.

[60] B. Sykes, B. Puddle, M. Francis, and R. Smith, *Biochem. Biophys. Res. Commun.* **72**, 1472 (1976).
[61] E. J. Miller, *Mol. Cell. Biochem.* **13**, 165 (1976).

## [3] Preparation and Characterization of Invertebrate Collagens

*By* LOUANN W. MURRAY, J. HERBERT WAITE, MARVIN L. TANZER, and PETER V. HAUSCHKA

Most invertebrate collagens resemble either the interstitial or basement membrane collagens of vertebrates. Standard techniques of acid extraction, sometimes with the aid of pepsin digestion, and differential salt precipitation can be used to obtain both interstitial and basement membrane collagens from many invertebrate animals.[1-3] Other chapters in this volume describe suitable methods for the extraction and purification of these collagen types. The properties of invertebrate collagens have been reviewed.[4-7]

Using a method of collagen classification based on plotting ratios of total hydroxylic, hydrophobic, and polar amino acids onto a triangular coordinate grid, Matsumura[8] found that the 97 vertebrate and invertebrate collagens that were examined fell into one of four catgories: (*a*) interstitial; (*b*) basement membrane; (*c*) *Ascaris* cuticle; and (*d*) *Lumbricus* cuticle. Using published amino acid analyses and plotting them according to Matsumura's technique, we found that cuticle collagen from the nematode *Caenorhabditis elegans* is located very close to that of the nematode *Ascaris* (Fig. 1). However, polychaete cuticle collagens are located near to, but distinctly separate from, oligochaete cuticle collagens and constitute a fifth class of collagen. The nematode, polychaete annelid, and oligochaete annelid cuticle collagens have distinctive chemical and physical properties that allow them to be considered separately from each other and from interstitial and basement membrane collagens. Techniques for the extraction, purification, and analysis of these three types of cuticle collagen are presented in this chapter.

[1] D. Ashhurst and A. Bailey, *Eur. J. Biochem.* **103**, 75 (1980).
[2] A. Nordwig and U. Hayduk, *J. Mol. Biol.* **44**, 161 (1969).
[3] C. Hung, R. J. Butkowski, and B. G. Hudson, *J. Biol. Chem.* **255**, 4964 (1980).
[4] E. Adams, *Science* **202**, 591 (1978).
[5] M. L. Tanzer, *Trends Int. Biochem. Soc.* **3**, 15 (1978).
[6] R. Garrone, *Frontiers Matrix Biol.* **5**, (1978).
[7] S. Kimura, *Hikaku Kagaku* **21**, 62 (1975).
[8] T. Matsumura, *Int. J. Biochem.* **3**, 265 (1972).

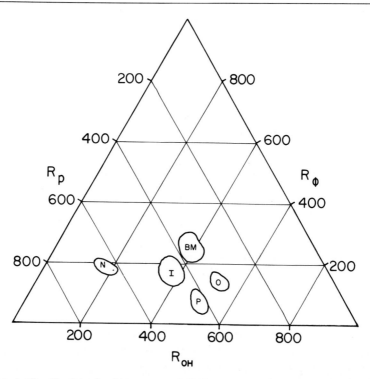

FIG. 1. Classification of collagen types. Collagens can be categorized by calculating the ratios of polar (p), hydroxylic (OH) and hydrophobic ($\phi$) amino acids to total amino acids and graphing the resulting ratios ($R$ values) onto a triangular, coordinate grid.[3] Cuticle collagens from nematodes cluster in the area labeled "N." Cuticle collagens from oligochaete annelid worms cluster in the area labeled "O." Cuticle collagens from polychaete annelid worms cluster in the area labeled "P." All other invertebrate collagens for which $R$ values have been calculated[3] (see the Appendix) cluster in either the interstitial area (I) or the basement membrane area (BM). Interestingly, the more recently discovered vertebrate type V collagen [R. E. Burgeson, F. A. El Adli, I. I. Kaitils, and D. W. Hollister, *Proc. Natl. Acad. Sci. U.S.A.* **73**, 2579 (1976)], a component of basement membranes, has $R$ values that locate it in the interstitial region of the graph. For formulas and the calculated $R$ values that were used to make this graph, see the Appendix.

## Extraction and Purification of Polychaete Cuticle Collagen

*Extraction of Acid-Soluble Cuticle Collagen*

### Reagents

Neutral pH salt buffer containing protease inhibitors: 0.05 $M$ Tris-HCl, pH 7.5; 1.0 $M$ NaCl; 0.001 $M$ phenylmethanesulfonyl fluoride

(PMSF), dissolved in a small volume of 2-propanol or DMSO; 0.01 $M$ $N$-ethylmaleimide (NEM); 0.025 $M$ ethylenediaminetetraacetic acid (EDTA)

NaCl, 20% (w/v)

Acetic acid, 0.1 $M$

*Procedures.* Polychaete cuticle collagens have been most frequently extracted from two species of marine worms, *Nereis virens* and *Nereis japonica*, obtainable seasonally from bait dealers located near the ocean. They are commonly known as sandworms, clamworms, and ragworms. After two rapid cycles of freezing and thawing in tap water, cuticles are easily freed of adhering tissue by scraping them under ice water. Worms that have been stored for extended periods at $-20°$ need only one cycle of thawing. Unless otherwise stated, all work is performed at 4°.

Collagen is extracted from the cuticles with either neutral pH salt buffer or dilute acid. The former solubilization is evidently produced by as yet undefined proteolytic reactions during extraction, as virtually no collagen is solubilized by neutral salt buffer containing protease inhibitors. It should be noted that, in contrast, virtually all collagen in earthworm cuticles is solubilized by neutral pH salt buffer containing protease inhibitors; different procedures are necessary for its extraction and purification (see below).

The preparation of acid-soluble cuticle collagen (ASCC) is outlined in Fig. 2. Cleaned cuticles are stirred for 24–48 hr in neutral pH salt buffer containing protease inhibitors. The cuticles are then washed several times in cold distilled water and extracted for 24–48 hr with 0.1 $M$ acetic acid, which almost completely solubilizes the cuticles. The acid extract is centrifuged at 40,000 $g$ for 1 hr to pellet chaetae and insoluble residue. The collagen in the supernatant is precipitated, while continuously stirring, by the slow addition of 20% NaCl to a final concentration of 2.5% NaCl. After 4 hr, the precipitated collagen is pelleted by centrifugation at 12,000 $g$ for 30 min. The pellet is redissolved in 0.1 $M$ acetic acid, and the cycle of centrifugation, precipitation, centrifugation, and solubilization is twice repeated. As a final step, the ASCC is centrifuged for 2 hr at 40,000 $g$, poured into sterile bottles, and stored at 4°; it remains stable for at least a year.

Kimura and Tanzer[9] have included a step involving pepsin digestion in the purification of cuticle collagen. Collagen is dissolved in 0.1 $M$ acetic acid and incubated with pepsin (enzyme : collagen ratio of 3 : 100) at 4° with continuous stirring for 48 hr. The mixture is dialyzed against two to three changes of 0.02 $M$ Na$_2$HPO$_4$; this precipitates the collagen, which is collected by low speed centrifugation and redissolved in 0.1 $M$ acetic acid.

[9] S. Kimura and M. L. Tanzer, *Biochemistry* **16**, 2554 (1977).

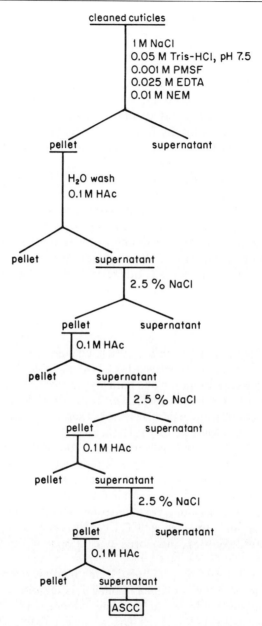

FIG. 2. Extraction scheme for acid-soluble cuticle collagen (ASCC).

More recent experiments suggest that in most cases the pepsin treatment is superfluous.

*Extraction of Neutral pH Salt-Soluble Cuticle Collagen*

  *Reagents*

  Neutral pH salt buffer: 0.05 $M$ Tris-HCl, pH 7.5; 1.0 $M$ NaCl
  NaCl, 20% (w/v)
  Acetic acid, 0.1 $M$
  $Na_2HPO_4$, 0.02 $M$

  *Procedure.* Neutral pH salt-soluble cuticle collagen (NSCC) is prepared in the absence of protease inhibitors (Fig. 3). Cleaned cuticles are stirred in neutral pH salt buffer for 24–48 hr. After 48 hr, as much as 40% of the cuticle collagen is solubilized, presumably by proteolysis. The responsible proteases have not been identified. The NSCC extract is centrifuged at 40,000 $g$ for 1 hr to pellet chaetae and undissolved material. Collagen is precipitated from the supernatant solution by dialysis against 0.02 $M$ $Na_2HPO_4$ and harvested by centrifugation at 12,000 $g$ for 1 hr. The pellet is redissolved in 0.1 $M$ acetic acid and purified by successive salt precipitation as for ASCC.

*Purification of Subunits*

Although preparations of native ASCC appear homogeneous as judged by sedimentation equilibrium and velocity centrifugation studies,[10,11] denatured ASCC consists of two distinct subunits, designated A and B chains.[9] Cuticle collagen is easily denatured owing to a low melting temperature ($T_d \leq 31°$); thus incubation for 5 min at 40° assures virtually complete denaturation.

GEL FILTRATION

  *Reagents*

  Denaturing buffer: 1.2 $M$ CaCl, 0.05 $M$ Tris pH 7.5

  *Procedure.* Denatured ASCC is dissolved in denaturing buffer at 40° and applied to a 1.5 × 180 cm column of BioGel A-5m (200–400 mesh) eluted with the same buffer at a flow rate of 6.5 ml/hr at room temperature. A single peak, containing both the A and B subunits, is evident by monitoring fractions at 220 or 230 nm.

[10] S. Kimura, *Nippon Suisan Gakkaishi* **37**, 419 (1971).
[11] H. Mizuno, N. Onda, T. Saito, N. Iso, and H. Ogawa, *Nippon Suisan Gakkaishi* **45**, 193 (1979).

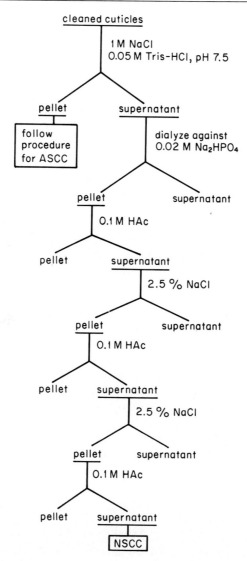

FIG. 3. Extraction scheme for neutral pH salt-soluble cuticle collagen (NSCC).

## Ion Exchange Chromatography

*Reagents*

Acetate buffer No. 1: 0.04 $M$ sodium acetate, pH 4.8
Acetate buffer No. 2: 0.04 $M$ sodium acetate, pH 4.8; 0.1 $M$ NaCl

*Procedure.* The subunits can be separated on CM-cellulose as originally described by Piez et al.[12] Lyophilized cuticle collagen is denatured in acetate buffer No. 1 by warming to 40° for 5 min. Undissolved material is removed by centrifugation at 40,000 $g$ for 1 hr using a rotor that has been warmed to 40° by immersion in hot water. The supernatant material is applied to a column (2.5 × 20 cm) of Whatman CM-52 equilibrated at 40° with acetate buffer No. 1. Elution is accomplished by a linear gradient of acetate buffers Nos. 1 and 2 using a total volume of 1300 ml at a flow rate of 120 ml/hr. As exhibited in Fig. 4, a complete separation of subunits A and B is afforded in *N. virens* cuticle collagen, but the separation in *N. japonica* is much less marked. Peak pooled fractions can be desalted on Sephadex G-25 eluted with 0.1 $M$ acetic acid and subsequently lyophilized.

## Sodium Dodecyl Sulfate-Polyacrylamide Gel Electrophoresis (SDS PAGE)

*Reagents*

Acrylamide stock: 16 g of acrylamide; 0.43 g of methylene bis-acrylamide; $H_2O$ to 100 ml
Sodium dodecyl sulfate (SDS): 10% (w/v) SDS in $H_2O$ to 100 ml
Tris-acetate stock: 0.4 $M$ Tris with glacial acetic acid to pH 6.6
Tetraethylmethylenediamine (TEMED)
Ammonium persulfate (APS): 10% (w/v) in $H_2O$, made fresh
Running buffer: 250 ml of Tris-acetate stock; 740 ml of $H_2O$; 10 ml of 10% SDS
Sample buffer: 0.05 $M$ sodium phosphate buffer, pH 7.4; 20% glycerol; 2% mercaptoethanol; 2% SDS; 0.033% phenol red as tracker dye
Coomassie stain: 0.1% (w/v) Coomassie Blue R-250; 10% glacial acetic acid; 25% isopropanol
Destaining solution: 10% isopropanol, 10% glacial acetic acid

*Procedure.* Electrophoresis of annelid cuticle collagens presents special problems because of the enormous size of the subunit chains. Polyacrylamide gels of 4% (w/v) can be either purchased (Biophore) precast from Bio-Rad (Richmond, California) or made by mixing 10 ml of acrylamide

---

[12] K. A. Piez, A. Eiger, and M. S. Lewis, *Biochemistry* **2**, 58 (1963).

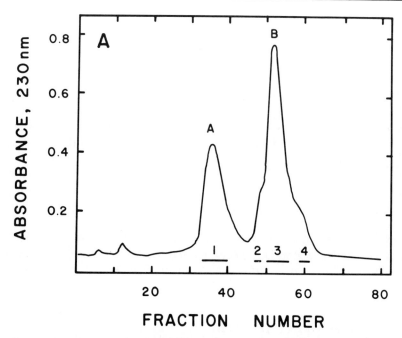

Fig. 4. (A) Elution pattern of approximately 50 mg of acid-soluble cuticle collagen from *Nereis virens* on a column of CM-cellulose at 40°. Bar 1 contains the A chain, and bar 3 contains the B chain. The shoulder preceding the B chain peak (bar 2) contains B'. The tail on the B chain peak (bar 4) consists of degradation products. (B) Elution pattern of approximately 35 mg of acid-soluble cuticle collagen from *N. japonica* on a column of CM-cellulose at 40°. Bar 1 contains mostly the A chain with two minor components, A' and A". Bar 2 contains only B chain. Figures are reprinted, with permission, from Kimura and Tanzer.[9]

stock, 0.4 ml of 10% SDS, 20 ml of Tris-acetate stock, 0.020 TEMED, 0.3 ml of 10% APS, and distilled water to 40 ml. The mixture is deaerated under vacuum, and poured into clean glass tubes (O.D. 7 mm; I.D. 5.5 mm; length 125 mm) that have Parafilm-sealed or tightly stoppered ends. Poured gels are immediately overlaid with distilled water and left to polymerize for at least 45 min. Precast gels are pre-electrophoresed with running buffer at 3 mA per gel overnight. Lyophilized cuticle collagen samples (about 40 μg) are dissolved in 40 μl of sample buffer, heated at 100° for 2 min, cooled to room temperature and applied to gels that are then run at 8 mA/gel. Precast Biophore gels are run for 8–12 hr whereas self-made gels are run for 6–8 hr.

After electrophoresis, the gels are removed from the glass tubes and stained in Coomassie stain for 1–2 hr, and destained for 24 hr with 1–2 changes of destaining solution. Inclusion of an old polyurethane sponge

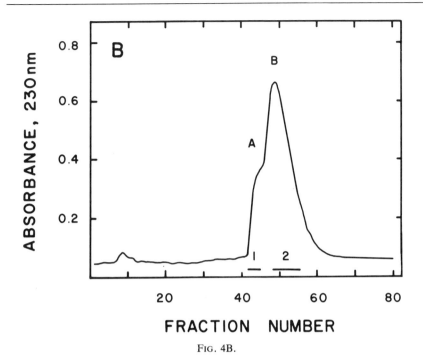

FIG. 4B.

(foam plastic bottle stopper) in the destaining solution will soak up the eluted Coomassie Blue dye and minimize the need for changing the destaining solution. The gels are stored at 4° to eliminate further destaining.[13]

As indicated, precast Biophore gels require longer equilibration and running times than the self-made gels. This is due to the necessity of introducing running buffer into the gels and to the greater degree of cross-linking that exists in the Biophore gels. In general, although both types of gels offer adequate separation of the subunits, precast gels give crisper resolution but are costlier than the laboratory versions. Protein mobility tends to be linear as a function of the log of the molecular weight on self-made gels, from 185,000 to 600,000, using vertebrate type I collagen standards, whereas on the precast gels polypeptide mobility is not linear with the log of the molecular weight.

SDS-PAGE of cuticle collagens using a Tris-glycine (pH 8.4) running buffer and 4% acrylamide gels was found to give a rapid (3 hr) and adequate separation of A and B subunits.[14] The mobility of A and B was reversed, and the resolution was no better than with the Tris-acetate system reported here.

[13] P. Davison, *J. Biol. Chem.* **253**, 5635 (1978).
[14] J. H. Waite, M. L. Tanzer, and J. R. Merkel, *J. Biol. Chem.* **255**, 3596 (1980).

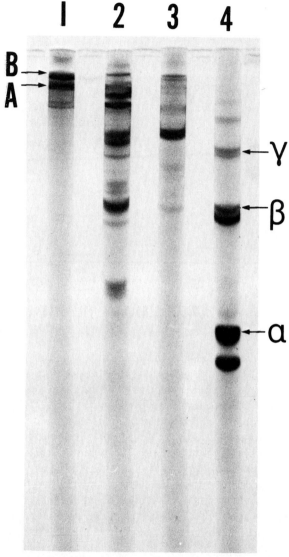

Fig. 5. Sodium dodecyl sulfate-polyacrylamide gel electrophoresis of *Nereis virens* cuticle collagen on 4% acrylamide self-made gels run in Tris-acetate buffer at 6 mA/gel for 6 hr as described in the text. Lane 1: Acid-soluble cuticle collagen (ASCC); A and B chains are indicated. Minor bands are A" and A' +B'. Lane 2: Neutral pH salt-soluble cuticle collagen (NSCC). In addition to A and B chains, a number of fragments with apparent molecular weights of 125,000 to 400,000 are evident. Lane 3: ASCC treated with clostridial collagenase for 22 hr.[15] The major band is the $M_r$ 300,000 collagenase-resistant fragment. Lane 4: Type I collagen from calf skin. The $\alpha$, $\beta$, and $\gamma$ bands are indicated.

SDS-PAGE most clearly demonstrates that polychaete cuticle collagen is composed of two subunits, which occur in a 1:2 ratio (Fig. 5). Preparations of ASCC contain minor components as well, designated A', B', and A", comprising about 10% of the total applied sample. These faster migrating proteins are thought to be degradation products of A and B chains because of their similarity to A and B chains in regard to mobility on CM-cellulose and amino acid composition.[15] NSCC consists of a variety of lower molecular weight collagenous proteins, ranging from about 125,000 to 400,000 $M_r$, in addition to A and B chains (Fig. 5). Although the pattern of bands from NSCC on SDS-PAGE is fairly consistent from one preparation to the next, the total yield and amount of protein in each band varies a great deal. This variation seems to be related to the length of time of storage of the worms at $-20°$ prior to extraction.

Properties of Polychaete Cuticle Collagen

*Physical Properties*

Many of the physical properties of native polychaete cuticle collagen have been defined (Table I). There is, however, still some ambiguity about the sedimentation coefficient and the intrinsic viscosity; $s^°_{20,w}$ is highly dependent upon both the collagen concentration and the buffer used. The 6.7 S value was obtained by both Kimura[10] and Mizuno *et al.*[11] using 0.15 M sodium citrate, whereas a value of 4.4 S was observed using 0.1 M acetic acid. The calculation of intrinsic viscosities is known to be dependent upon shear rate as well as upon collagen concentration. Kimura's low [$\eta$] values[10] reflect extrapolations based on only three shear rates, all of which were too high (140–1400 sec$^{-1}$). Using a rotational viscometer operated at very low rates of shear (5–15 sec$^{-1}$), Mizuno *et al.*[11] extrapolated the [$\eta$] value of 126 dl/g from plots of $\eta$ sp/$c$ vs $c$ and [$\eta$] $\dot\gamma$ vs $\dot\gamma$, where $\eta$ sp is specific viscosity, $c$ is concentration (g/dl), and [$\eta$] $\dot\gamma$ is intrinsic viscosity at each $\dot\gamma$ (rate of shear, sec$^{-2}$). All workers agree that the molecular weight of native cuticle collagen lies between $1.65 \times 10^6$ and $1.70 \times 10^6$.

As noted above, results obtained from SDS-PAGE and ion exchange chromatography of denatured collagens point to the presence of two dissimilar subunits, A and B, which have similar molecular weights of about 500,000 and occur in a 1:2 ratio, respectively.[9] Table II lists a few properties of the subunits. The two sedimentation coefficients represent two components detected in 2 M KCNS. These have calculated molecular weights of 300,000 and 600,000, respectively. Unfortunately, this is in

---

[15] S. Kimura and M. L. Tanzer, *Biochemistry* **16**, 2554 (1977).

## TABLE I
### Physical Properties of Polychaete Cuticle Collagens

| Property | Method | Value | Reference[a] |
|---|---|---|---|
| Sedimentation coefficient, $s^°_{20,w}$ | Sedimentation velocity | 6.7 S[b] | 10, 11 |
| | | 4.4 S[b] | 11 |
| Partial specific volume, $\overline{V}$ | Amino acid composition | 0.69 ml/g[b] | 10, 11 |
| Intrinsic viscosity, $[\eta]$ | Capillary viscometry | 70 dl/g[b] | 10 |
| | | 68 dl/g[c] | 10 |
| | | 62 dl/g[d] | 10 |
| | Zimm viscometry | 126 dl/g[b] | 11 |
| Frictional ratio, $f/f_0$ | Viscometry | 8.9 dl/g | 10 |
| Optical rotation, $[\alpha]_D$ | Polarimetry | $-390°$ [b,c] | 10 |
| Average particle length | Flow birefringence | 9000 Å[b,e] | 10 |
| Average particle diameter | Flow birefringence | 16.2 Å[b] | 10 |
| Molecular weight, $M$ | Sedimentation equilibrium | $1.65 \times 10^6$ [b] | 11 |
| | Sedimentation velocity | $1.70 \times 10^6$ [b] | 10 |
| | Light scattering | $1.66 \times 10^6$ [b] | 11 |
| | Sedimentation velocity and viscosity | $1.70 \times 10^6$ [b] | 10 |

[a] Numbers refer to text footnotes.
[b] *Nereis japonica* (previously *Neanthes diversicolor*).
[c] *Marphysa sanguinea*.
[d] *Lumbriconereis heteropoda*.
[e] The value of 9000 Å was obtained from a preparation that probably contained a mixture of whole and degraded molecules, as protease inhibitors were not used, and should be considered a minimal value. Based on values of molecular weight for *Nereis* sp., cuticle collagen subunits[9] and on measurements of segment-long-spacing (SLS),[33] a value of 15,000 Å for the particle length of *Nereis* sp. cuticle collagen would be more accurate.

## TABLE II
### Physical Properties of Polychaete Cuticle Collagen Subunits

| Property | Method | Value | Reference[a] |
|---|---|---|---|
| Sedimentation coefficient, $s^°_{20,w}$ | Sedimentation velocity | 6 S | 10 |
| | | 8.3 S | |
| Intrinsic viscosity, $[\eta]$ | Capillary viscometry | 1.3 dl/g[b] | 9, 10 |
| | | 1.4 dl/g[c] | |
| Molecular weight, $M$ | Gel filtration | 470,000[b,c] | 9 |
| | Viscometry | 480,000[b,c] | 10 |
| | | 462,000[b] | 10 |
| | | 513,000[c] | 9 |
| | Osmotic pressure | 430,000 | 10 |

[a] Numbers refer to text footnotes.
[b] *Nereis japonica*.
[c] *Nereis virens*.

TABLE III
THERMAL STABILITY OF CUTICLE COLLAGENS[a]

| Property | Temperature | Species |
|---|---|---|
| Denaturation temperature, $T_d$ | 29 | Nereis japonica |
| | 31 | Marphysa sanguinea |
| | 29 | Lumbriconereis heteropoda |
| Shrinkage temperature, $T_s$ | 43 | Nereis japonica |
| | 46 | Marphysa sanguinea |
| | 46 | Lumbriconereis heteropoda |

[a] Data from Kimura.[10]

direct contrast to results obtained using other techniques, e.g., SDS-PAGE and gel filtration, that suggest both subunits to be of similar molecular weight. Sedimentation equilibrium centrifugation could resolve this dilemma, but has yet to be performed.

*Stability*

Table III lists the denaturation and shrinkage temperatures of three cuticle collagens. *Nereis virens* cuticle collagen exhibits remarkable stability to digestion by clostridial collagenases,[14,15] the cause of which is not known (see Fig. 5, gel 3).

*Chemical Composition*

*Amino Acids.* The chemical composition of cuticle collagens of four species of polychaete worm is given in Table IV. Analyses of additional cuticle collagens have appeared, but the purity of these collagens is not defined.[16] The cuticle collagens all have glycine as approximately one-third of all the residues. The other abundant residues are glutamic acid (115–133 residues/1000) and 4-hydroxyproline (147–170 residues/1000). Totally absent are cystine and hydroxylysine as well as the aldimine and aldol condensation product cross-links commonly associated with collagens.[17]

The composition of subunits A and B of *Nereis* sp. cuticle collagen is listed in Table V. The A chain has greater amounts of aspartic and glutamic acids, valine, methionine, isoleucine, leucine, and phenylalanine and lesser amounts of 4-hydroxyproline, threonine, and serine. The A

[16] M. F. Voss-Foucart, S. Barzin, and C. Toussaint, *Arch. Zool. Exp. Gen.* **118**, 457 (1978).
[17] T. Housley and M. L. Tanzer, unpublished observation.

## TABLE IV
### CHEMICAL COMPOSITION OF POLYCHAETE CUTICLE COLLAGENS[a]

| Amino acid | Nereis japonica[b] | Nereis virens[b] | Marphysa sanguinea[c] | Lumbriconereis heteropoda[c] |
|---|---|---|---|---|
| 3-Hydroxyproline | 5.9 | 3.9 | 14 | 6.1 |
| 4-Hydroxyproline | 170 | 147 | 169 | 160 |
| Aspartic acid | 12 | 10 | 17.2 | 25.1 |
| Threonine | 43 | 43 | 37.8 | 42.8 |
| Serine | 61 | 69 | 60 | 56 |
| Glutamic acid | 124 | 133 | 117 | 115 |
| Proline | 30 | 51 | 38.1 | 28 |
| Glycine | 344 | 347 | 348 | 318 |
| Alanine | 93 | 88 | 82 | 100 |
| Cystine | 0 | 0 | — | — |
| Valine | 8.4 | 7 | 9.1 | 12.7 |
| Methionine | 1.1 | 0.8 | 1.2 | 0.4 |
| Isoleucine | 7.3 | 5.0 | 7.1 | 12.7 |
| Leucine | 19 | 18 | 11.4 | 28.5 |
| Tyrosine | 0.7 | 0.4 | 1.4 | 3.4 |
| Phenylalanine | 2.9 | 2.2 | 3.1 | 5.8 |
| Hydroxylysine | 0 | 0 | 0 | 0 |
| Lysine | 0.7 | 0.6 | 2.1 | 10.3 |
| Histidine | 0.2 | 0.3 | 0.8 | 2.1 |
| Arginine | 75 | 76 | 80.7 | 73.1 |
| Hexosamine | 0.3 | ND[d] | 0.6 | 0.7 |
| Hexose[e] | 2.3 | ND | 1.9 | 2.9 |

[a] Residues/1000.
[b] Kimura and Tanzer.[9]
[c] Kimura.[10]
[d] ND, not determined.
[e] Galactose, mannose, and glucose, expressed as percentage of weight.

chain is also more highly glycosylated. The presence of only one methionine residue in the B chain has been exploited by Kimura and Tanzer[9] in the preparation of two cyanogen bromide peptides from B chain.

*Carbohydrates.* All polychaete cuticle collagens exhibit some degree of glycosylation by neutral and amino sugars (Tables IV and V). The nature of the known linkages between the carbohydrate and protein moieties is summarized in Table VI. The acidic disaccharide 6-$O$-$\alpha$-D-glucuronosylmannose is novel for collagens. Up to 80% of the carbohydrate in *Nereis* sp. cuticle collagens is presumed to occur as saccharide units that are O-glycosidically linked to threonine and serine residues.[18]

[18] R. G. Spiro and V. D. Bhoyroo, *J. Biol. Chem.* **255**, 5347 (1980).

TABLE V
CHEMICAL COMPOSITION OF POLYCHAETE CUTICLE COLLAGEN SUBUNITS[a,b]

|  | Nereis japonica | | Nereis virens | |
| --- | --- | --- | --- | --- |
| Amino acid | A chain | B chain | A chain | B chain |
| 3-Hydroxyproline | 5.9 | 5.5 | 4.9 | 2.9 |
| 4-Hydroxyproline | 162 | 179 | 128 | 151 |
| Aspartic acid | 10 | 8.9 | 12 | 7.2 |
| Threonine | 33 | 44 | 35 | 45 |
| Serine | 57 | 61 | 53 | 74 |
| Glutamic acid | 131 | 125 | 146 | 129 |
| Proline | 19 | 24 | 51 | 50 |
| Glycine | 360 | 351 | 361 | 351 |
| Alanine | 93 | 97 | 85 | 94 |
| Cystine | 0 | 0 | 0 | 0 |
| Valine | 9.7 | 5.1 | 9.3 | 4.1 |
| Methionine | 1.3 | 0.3 | 0.9 | 0.2 |
| Isoleucine | 12 | 6.5 | 9.0 | 4.1 |
| Leucine | 27 | 15 | 27 | 13 |
| Tyrosine | 0.1 | 0.1 | 0.2 | 0.1 |
| Phenylalanine | 4.2 | 0.7 | 4.7 | 0.5 |
| Hydroxylysine | 0 | 0 | 0 | 0 |
| Lysine | 0.1 | 0.1 | 0.1 | 0.1 |
| Histidine | 0.1 | 0.1 | 0.1 | 0.1 |
| Arginine | 75 | 77 | 73 | 74 |
| Galactose[c] | 1.5 | 10 | 11 | 6.2 |
| Glucose[c] | 1.6 | 1.4 | 1.1 | 0.6 |
| Mannose[c] | 3.9 | 1.0 | 1.8 | 0.4 |

[a] As residues/1000.
[b] Data from Kimura and Tanzer.[9]
[c] Monosaccharides were calculated on the basis of a mean amino acid residue weight of 92.

TABLE VI
DISTRIBUTION OF O-GLYCOSIDICALLY LINKED CARBOHYDRATE UNITS IN THE CUTICLE COLLAGEN OF Nereis Virens[a]

| Carbohydrate | Units/1000 total amino acid residues |
| --- | --- |
| Galactose | 0.7 |
| (Galactose)$_2$ | 1.6 |
| (Galactose)$_3$ | 1.5 |
| Glucuronosylmannose | 1.9 |
|  | 5.7 |

[a] Data from Spiro and Bhoyroo.[18]

Electron Microscopy of Annelid Cuticle Collagen

The cuticles of annelids have been well studied by electron microscopy using standard thin-sectioning methods. One of the most puzzling aspects of the ultrastructural organization of the cuticle is the lack of banding pattern in the cuticle fibers.[19-25] All interstitial collagens, including annelid gut[26] and brain sheath collagen,[27] exhibit a characteristic periodicity when viewed by electron microscopy. It has been theorized that the polysaccharide content of annelid cuticles is responsible for the lack of visible periodicity.[28] If annelid cuticles are stained for polysaccharide by fixing with tris-1-aziridinylphosphine oxide (TAPO) followed by the Thiery silver staining method, periodicities of 62.5 and 185 Å can be seen, indicating that there is an ordered array of at least the carbohydrate moieties along the cuticle fibers.[29]

*Native Fibers*

*Reagents*

Fixative: 1% glutaraldehyde in 0.1 $M$ sodium phosphate buffer, pH 7.2

Amylase: 0.1% bacterial amylase in 0.2 $M$ sodium phosphate buffer, pH 5.4

Trypsin: 0.1% trypsin in 0.2 $M$ sodium phosphate buffer, pH 7.0

NaCl, 0.9%

Phosphate-buffered saline: 0.1 $M$ NaCl in 0.1 $M$ sodium phosphate buffer, pH 7.4

PTA, pH 7.0: 2% phosphotungstic acid in $H_2O$, to pH 7.0 with NaOH

*Procedure.* Cuticular fibers can be isolated and negatively stained for examination in the electron microscope. One method, used for earthworm cuticle fibers, involves fixation and enzymic cleaning of the cuticle before

[19] C. Ruska and H. Ruska, *Z. Zellforsch. Mikrosk. Anat.* **53**, 759 (1961).
[20] J. Gross, in "Comparative Biochemistry" (M. Florkin and H. S. Mason, eds.), Vol. 5, p. 307. Academic Press, New York, 1963.
[21] R. Coggeshall, *J. Cell Biol.* **28**, 95 (1966).
[22] K. M. Rudall, in "Treatise on Collagen" (B. S. Gould, ed.), Vol. 2, p. 83. Academic Press, New York, 1968.
[23] V. Storch and U. Welsch, *Z. Morphol. Tiere* **66**, 310 (1970).
[24] D. Goodman and W. B. Parrish, *J. Morphol.* **135**, 71 (1971).
[25] M. B. Mathews, "Connective Tissue; Macromolecular Structure and Evolution." Springer-Verlag, Berlin and New York, 1975.
[26] B. Bacetti, *J. Cell Biol.* **34**, 885 (1967).
[27] O. Hermans, *J. Ultrastruct. Res.* **30**, 255 (1970).
[28] M. D. Maser and R. V. Rice, *J. Biol. Chem.* **18**, 569 (1963).
[29] W. Djaczenko and C. C. Cimmino, *J. Biol. Chem.* **57**, 859 (1973).

isolation of the cuticular fibers.[30] Whole earthworm cuticles, prepared by the freeze-thaw method outlined above, are placed in fixative for 15 min. Cuticles are incubated for 18 hr in amylase, followed by incubation for 1 hr in trypsin, all at room temperature. The cuticles are washed in 0.9% NaCl for 2 hr, then mechanically dispersed. In our laboratory, cleanly dissected sandworm cuticles are thoroughly washed in ice water, then mechanically dispersed at 4° in either water or phosphate-buffered saline by homogenizing small pieces of cuticle in a Dounce homogenizer with a B pestle. No prior fixation or enzymic digestion is performed, resulting in better preservation of the interfibrillar connections and faster preparation time but also contributing to more contamination by nonfibrillar components of the cuticle. The fiber suspension is diluted with water, dropped onto either a Formvar- or a carbon-coated grid, washed with water, and negatively stained with PTA, pH 7.0. The fibers are characteristically nonstriated, average 200 nm in diameter, and are composed of fibrils that are about 20-40 nm in diameter (Fig. 6).

*Reconstituted Fibers*

*Reagents*

NaCl, 0.1 $M$
NaCl, 0.35 $M$
NaCl, 0.5 $M$
$Na_2HPO_4$, 0.0001 $M$ (pH 7.55)
$Na_2HPO_4$, 0.01 $M$ (pH 9.05)

*Procedure.* Fibers that are morphologically similar to native fibers can be reconstituted from acid solutions of either ASCC or NSCC from *N. virens.* One milliter of ASCC or NSCC at 0.5 mg/ml in 0.1 $M$ acetic acid is dialyzed for 3 days at 4° against 200 ml of either water or one of the NaCl or $Na_2HPO_4$ solutions listed above. The resulting thick gel is dispersed by mild homogenization, and the suspension is negatively stained as for native fibers. Although no gross morphological differences can be detected between fibers prepared by any of these methods and native fibers, it is not known whether the organization of the molecules within the reconstituted fibers is the same as in the native fibers.

*Segment-Long-Spacing Crystallites (SLS)*

*Reagents*

ATP solution: 0.8% adenosine triphosphate, 0.1 $M$ NaCl, 0.01 $M$ acetic acid, 0.02% sodium azide, as an antibacterial agent

[30] L. V. Zuccarello, *Cell Tissue Res.* **201**, 459 (1979).

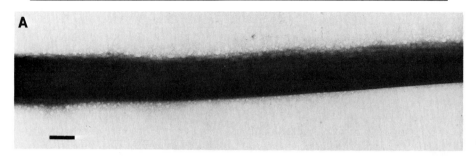
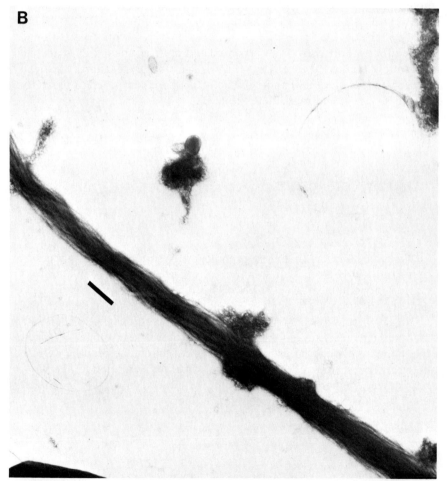

FIG. 6. *Nereis virens* native cuticle fibers, prepared without the use of enzymes, as described in the text, and negatively stained. (A) Native fiber, 270 nm in diameter. (B) Native fiber, 140 nm in diameter, in a more open configuration, showing interweaving of the 30–40 nm fibrils. ×73,700; bar represents 0.1 μm.

PTA: Phosphotungstic acid, 2% in $H_2O$, pH 1.5
UrAc: Uranyl acetate 2% in $H_2O$, pH 4.5

*Procedure.* Interstitial and basement membrane collagens will form segment-long-spacing crystallites (SLS), a side-by-side aggregate of collagen molecules in exact parallel register, upon dialysis of an acid solution of collagen against an acidic ATP solution.[31,32] Examination of SLS that have been positively stained with phosphotungstic acid or uranyl acetate can provide useful information about the distribution of the charged amino acids along the collagen molecules.

Preparations of earthworm cuticle collagen that had been mechanically sheared to the size of interstitial collagen were reported to form very low yields of SLS,[28] but other workers attributed those SLS to contaminating gut collagen and claimed no success in attempts to produce SLS.[22] Using the following method, it is possible to obtain good yields of authentic SLS from NSCC from *N. virens.*[33] Several enzymic digests of ASCC will also form SLS. However, undegraded ASCC molecules do not line up in exact parallel register, but will instead form banded ordered aggregates (BOA) that average 2100 nm in length.[33] The BOA probably represent parallel molecules that, for unknown reasons, are slightly staggered rather than exactly aligned.

To make SLS from *Nereis* cuticle collagen, 0.5 ml of a 1 mg/ml solution of NSCC in 0.1 $M$ acetic acid is mixed in a previously boiled dialysis bag with 0.5 ml of ATP solution. The dialysis bag is placed in a sterile test tube containing 3 ml of ATP solution and allowed to sit at 4° for 1 week before electron microscopic examination. Satisfactory SLS can also be prepared by mixing 0.5 ml of NSCC directly with 3.5 ml of ATP solution and allowing it to sit at 4° for 1 week. These long time periods result in better quality images of the crystallites. The resulting precipitate is gently resuspended, diluted 1 : 10 with ATP solution, dropped onto a carbon-coated grid, rinsed with water, stained for 2–10 min with PTA or UrAc, water-rinsed, and air-dried. Grid preparation is done at 4°, and the grid is not allowed to dry until the last step. The resulting SLS from NSCC will vary in length from 300 nm to 1500 nm, reflecting the molecular lengths of the starting materials. An example of the type of SLS that can be expected from NSCC of *N. virens* is seen in Fig. 7.

Extraction and Purification of Oligochaete Cuticle Collagen

The absence of typical banded collagen fibrils in electron micrographs of cuticles of the common earthworm or "nightcrawler," *Lumbricus ter-*

---

[31] R. R. Bruns and J. Gross, *Biochemistry* **12**, 808 (1973).
[32] R. Timpl, G. B. Martin, P. Bruckner, G. Wick, and H. Wiedemann, *Eur. J. Biochem.* **84**, 43 (1978).
[33] L. W. Murray and M. L. Tanzer, *Proc. Int. Congr. Biochem. 11th*, p. 213 (1979).

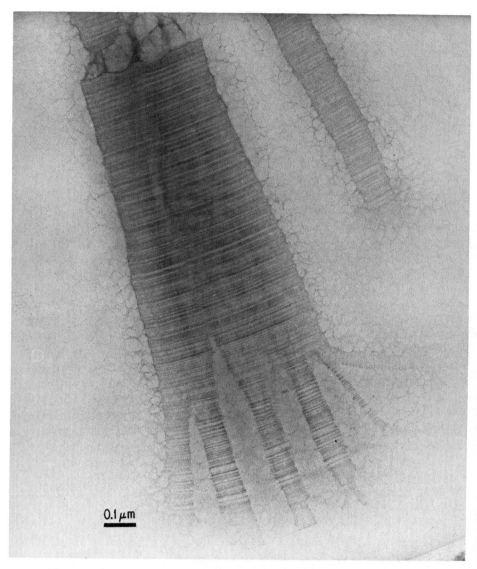

FIG. 7. Segment-long-spacing crystallites (SLS) from neutral pH salt-soluble cuticle collagen (NSCC) of *Nereis virens*. This 1580-nm SLS is probably composed of shorter molecular fragments as well as whole molecules. ×110,000.

*restris*, prompted early attempts to isolate and characterize this unusual collagen. Two procedures were developed. One, by Maser and Rice,[34] was based on acetic acid extraction. A second, somewhat gentler method of Josse and Harrington,[35] utilizing neutral salt solubilization and ammonium sulfate precipitation, is described below.

*Reagents and Materials*

NaCl, 0.5 $M$

NaCl, 0.2 $M$

Ammonium sulfate, saturated aqueous solution at 5°, pH 8

Earthworms, 100 large *L. terrestris*, generally obtainable as "nightcrawlers," which can be shipped airmail by fishing bait suppliers

*Procedure.* Live earthworms are washed in distilled water, placed by the handful in stainless steel beakers, and rapidly frozen by cooling the beaker in a solid $CO_2$–isopropanol bath. The blocks of frozen worms are then thawed in large vessels of distilled water at 5°, after which the cuticles are easily removed with forceps and minced with a scalpel. All subsequent steps are performed at 5°. Inclusion of protease inhibitors in the washing and extracting steps would be a modern modification worthy of consideration. In contrast to results from polychaete worms, the cuticle collagen of earthworms is almost totally soluble in neutral pH salt solutions in the presence of protease inhibitors (L. W. Murray, unpublished observation). The cuticles are washed by stirring in 5 liters of distilled water for 1 hr and should appear opalescent or translucent white with no trace of red or brown color. Disintegration of the cuticles into a thick slurry is achieved by grinding in ice water with a ceramic mortar and pestle. Excessive shearing forces, such as those produced by high speed homogenizers, may have adverse effects on the highly asymmetric earthworm cuticle collagen molecule. The cuticle slurry is suspended in 100 ml of 0.5 $M$ NaCl and stirred for 48 hr. The extract is centrifuged, and the resultant viscous supernatant, containing about 0.7 mg of hydroxyproline-rich protein per worm, is subjected to ammonium sulfate fractionation.

To 100 ml of extract supernatant, 25 ml of saturated ammonium sulfate are added over a 10-min period. After an additional 10 min of stirring, the suspension is centrifuged and the pellet is dissolved in 50 ml of 0.5 $M$ NaCl. A second precipitation is carried out by adding 12.5 ml of saturated ammonium sulfate, and after stirring and centrifugation a white pellet of earthworm cuticle collagen is obtained. The final yield is 0.1–0.3 mg of collagen per cuticle. This pellet is dissolved in 30 ml of 0.2 $M$ NaCl and centrifuged at 90,000 $g$ for 1 hr; the supernatant liquid is stored at 5°. The

[34] M. D. Maser and R. V. Rice, *Biochim. Biophys. Acta* **63**, 255 (1962).
[35] J. Josse and W. F. Harrington, *J. Mol. Biol.* **9**, 269 (1964).

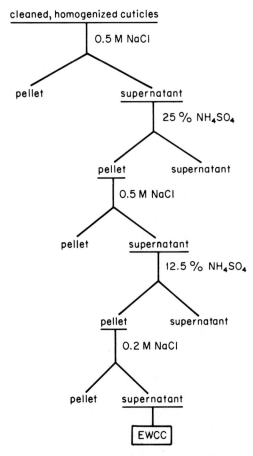

FIG. 8. Extraction scheme for earthworm cuticle collagen (EWCC).

final collagen concentration should be about 1 mg/ml. Such stock solutions are stable for up to 4 months, showing no change in sedimentation, viscosity, or optical rotatory properties.[35] The evidence for collagen purity is that additional precipitation steps with either ammonium sulfate or organic solvents and column chromatography on ion exchange resins or Sephadex fail to alter the hydroxyproline:protein ratio of earthworm cuticle collagen. See Fig. 8 for a summary of the extraction procedure.

Properties of Earthworm Cuticle Collagen

*Physical Characteristics*

The distinguishing features of this protein are its immense molecular weight[34,35] and high hexose content.[35] Thorough investigation of the phys-

TABLE VII
PHYSICAL PROPERTIES OF OLIGOCHAETE AND NEMATODE CUTICLE COLLAGENS

| Property | Lumbricus terrestris collagen[a] | Pheretima communissima native[b] | Ascaris lumbricoides Native[a,c] | Ascaris lumbricoides Subunit[c-e] | Panagrellus silusiae subunit[f] |
|---|---|---|---|---|---|
| Molecular weight | | | | | |
| Native | 1,900,000 | 1,690,000 | 900,000 | — | — |
| 5 $M$ GuHCl | 600,000 | — | 1,100,000 | — | — |
| Subunit | — | — | — | 52,000 | 89,000 to 110,000 |
| $s^\circ_{20,w}$ | | | | | |
| Native | 7.0 S | 5.19 S | 7.0 S | 2.8 S | — |
| 5 $M$ GuHCl | 5.3 S | — | 9.6 S | — | — |
| $[\eta]$ | | | | | |
| Native | 62 dl/g | 109 dl/g | 12.3 dl/g | 1.0 dl/g | — |
| 5 $M$ GuHCl | 4.3 dl/g | — | 2.9 dl/g | — | — |
| $T_m(^\circ C)$ | 22° | — | 52.8° | 34.7° | — |

[a] Josse and Harrington.[35]
[b] Utiyama et al.[42]
[c] Hauschka and Harrington.[48]
[d] McBride and W. F. Harrington.[49]
[e] Evans et al.[43]
[f] Leuschner and Pasternak.[44]

ical properties of the earthworm cuticle collagen were carried out by Josse and Harrington[35] (see Table VII). The sedimentation coefficient is highly dependent upon concentration, with an extrapolated value of $s^\circ_{20,w}$ = 7.0 S. In contrast, the smaller vertebrate tropocollagen molecule has a sedimentation coefficient of only 2.8 S to 3.3 S.

Intrinsic viscosity $[\eta]$ is strongly influenced by molecular asymmetry (axial ratio), and thus the 62 dl/g value for earthworm cuticle collagen reflects the tremendous length of this rodlike protein[35] (see Table VII). The molecular diameter and length calculated from $[\eta]$ are 17 Å and 9400 Å, respectively. Josse and Harrington[35] pointed out that this is a *minimum* estimate of length because of the type of viscometer employed and the known shear gradient dependence of $[\eta]$ for highly asymmetric particles. The actual molecular length probably approaches 18,000 Å. Using the Scheraga–Mandelkern equation with the measured values of $[\eta]$ and $s^\circ_{20,w}$, a molecular weight of 1,900,000 is calculated for earthworm cuticle collagen.[35] Thus this molecule is some six times larger than vertebrate tropocollagen.

A sharp collagen → gelatin transition at 22° is observed for the earthworm protein in 0.2 $M$ NaCl, pH 6, using viscometry and optical rotatory dispersion.[35] The melting transition is accompanied by a drop in mo-

TABLE VIII
AMINO ACID COMPOSITION OF OLIGOCHAETE AND NEMATODE CUTICLE COLLAGENS[a]

| Amino acid | Lumbricus terrestris[b] | Ascaris lumbricoides Native[c] | Ascaris lumbricoides B subunit[d] | Caenorhabditis elegans[e] |
|---|---|---|---|---|
| 3-Hyp | 2 | — | 0 | — |
| 4-Hyp | 160 | 19 | 22 | 60 |
| Asp | 58 | 64 | 58 | 97 |
| Thr | 49 | 16 | 15 | 32 |
| Ser | 88 | 17 | 17 | |
| Glu | 81 | 72 | 52 | 100 |
| Pro | 9 | 291 | 376 | 127 |
| Gly | 348 | 261 | 268 | 277 |
| Ala | 100 | 78 | 67 | 161 |
| ½-Cys | 0 | 16 | 31 | 18 |
| Val | 16 | 27 | 13 | 28 |
| Met | 0 | 4 | 3 | 0 |
| Ile | 15 | 10 | 7 | 12 |
| Leu | 27 | 20 | 17 | 16 |
| Tyr | 1 | 4 | 2 | 5 |
| Phe | 6 | 10 | 8 | 10 |
| His | 1 | 10 | 7 | 10 |
| Hyl | 0 | — | 0 | 0 |
| Lys | 16 | 49 | 38 | 27 |
| Arg | 23 | 32 | 23 | 19 |
| Trp | — | — | 2 | — |

[a] Data are expressed in residues/1000.
[b] Adams.[4]
[c] Josse and Harrington.[35]
[d] Evans et al.[43]
[e] Ouazana and Gibert.[45]

lecular weight to roughly one-third that of the original collagen (600,000). The physical properties of the denatured protein ($[\eta] = 4.5$ dl/g, $s^\circ_{20,w} = 5.1$ S)[35] are consistent with values expected for single, random-coil, gelatin molecules containing some 6000 amino acid residues. Adams[4] has pointed out that this may be the largest single polypeptide chain found in nature.

## Chemical Characteristics

Earthworm cuticle collage is rich in hydroxylated amino acid residues, with the highest reported values of hydroxyproline and serine, and a very high threonine content[35] (see Table VIII). In contrast to vertebrate collagens, the hydroxyproline of earthworm cuticle collagen is restricted to the X position in the Gly-X-Y triplet sequence.[36] Remarkably, this sequence anomaly was predicted from thermal stability and renaturation characteristics[37] before chemical confirmation was obtained.

Earthworm cuticle collagen contains 12-14% D-galactose by weight, most of which is in di- and trisaccharide units attached glycosidically to serine and threonine residues of the protein.[38] This stands in sharp contrast to the vertebrate collagens, for which hydroxylysine is the site of glycosidic linkage to galactose. An interesting susceptibility of both earthworm cuticle collagen and gelatin to periodate[39] raises the possibility that some galactose residues are involved in structural stabilization. In the presence of 2 m$M$ NaIO$_4$ (2 mol of IO$_4^-$ per mole of D-galactose) the reduced viscosity of earthworm cuticle collagen drops precipitously from 74 dl/g to 40 dl/g within 20 min at 19°. Similar rapid changes occur in the gelatin chains, with periodate consumption paralleling the 50% decrease in viscosity. It is not clear whether oxidation of galactose residues shifts the $T_m$ of the collagen, or whether peptide bond scission occurs under these mild conditions. Early suggestions by Grassman and Kühn[40] that hexoses might be involved in collagen cross-linking were never substantiated by periodate oxidation studies with vertebrate collagens,[41] but the hypothesis may merit more thorough investigation with some of the unusual invertebrate collagens, such as that from earthworm cuticle.

Few other annelids in the class Oligochaeta have been studied. Maser and Rice[34] performed some experiments on collagen from the South American giant earthworm *Rhinodrilus fafneri* in conjunction with work on *Lumbricus terrestris*. Utiyama *et al.*[42] studied the hydrodynamic properties of cuticle collagen from the earthworm *Pheretima communissima* (see Table VII). The extremely high intrinsic viscosity measured at zero shear, 109 dl/g, suggests that a collagen triple-helical rod about 18,000 Å in length is needed to accommodate the measured molecular weight of 1,690,000.[42]

Extraction and Purification of Native *Ascaris* Cuticle Collagen

*Ascaris lumbricoides* (var. *suis*) is the common parasitic roundworm of the hog intestinal tract. *Ascaris* cuticle collagen ($M_r$ 900,000)[35] and its

[36] A. Goldstein and E. Adams, *J. Biol. Chem.* **243**, 3550 (1968).
[37] N. V. Rao and W. F. Harrington, *J. Mol. Biol.*, **21**, 5771 (1966).
[38] L. Muir and Y. C. Lee, *J. Biol. Chem.* **245**, 502 (1970).
[39] P. V. Hauschka, unpublished observations.
[40] W. Grassmann and K. Kühn, *Hoppe-Seyler's Z. Physiol. Chem.* **301**, 1 (1955).
[41] R. J. Schlueter and A. Veis, *Biochemistry* **3**, 1657 (1964).
[42] H. Utiyama, K. Sakato, K. Ikehara, T. Setsuiye, and M. Kurata, *Biopolymers* **12**, 53 (1973).

small subunit chains ($M_r$ 52,000)[43] have been widely studied. The isolation of these proteins is described below. Other nematodes whose collagens are of interest are *Panagrellus silusiae*[44] and *Caenorhabditis elegans*.[45] The latter species has been intensively investigated by Brenner and his colleagues[46] as a model for cell differentiation and development in multicellular organisms, and a variety of interesting cuticle mutants ("rollers") were recently described.[47]

*Reagents and Materials*

NaCl 0.5 $M$
NaCl, 0.2 $M$
Ammonium sulfate, saturated aqueous solution at 5°, pH 8
*Ascaris lumbricoides* (var. *suis*), 100 live adult worms, about 20 cm, generally available from a local slaughterhouse

*Procedure.* Following established procedures[35,48] identical to those described above for earthworm cuticle, *Ascaris* specimens are washed, rapidly frozen, thawed in distilled water, and then slit lengthwise with a scalpel. Slitting is greatly facilitated by a "tunnel" (6 mm wide × 6 mm high × 24 mm long), easily constructed by gluing plastic strips on a 15 × 30 cm sheet of Lucite. A slot in the roof of the tunnel accommodates a No. 10 scalpel blade, which slices the body wall as the worm is drawn through. The cuticles are scraped with the back of a scalpel blade and then washed in water. It is advisable to autoclave all discarded waste tissue, which may contain millions of viable eggs; some individuals develop severe allergic reactions after handling *Ascaris,* so gloves should be worn and cleanliness observed. All procedures are carried out at 5°.

Washed cuticles are ground with a mortar and pestle to form a thick slurry and then stirred several hours in 100 ml of 0.5 $M$ NaCl to remove most noncollagenous proteins. The cuticles are collected by centrifugation and then reduced to a fine suspension in 0.5 $M$ NaCl with a motor-driven TenBroeck glass homogenizer. The volume is brought to 100 ml of 0.5 $M$ NaCl, and extraction proceeds by stirring for 48 hr. After centrifugation, the supernatant is subjected to two 20% ammonium sulfate precipitations as described for earthworm cuticle collagen above. The pellet can be saved for subunit preparation. The yield of native *Ascaris* cuticle collagen is 0.2–0.7 mg per worm. See Fig. 9 for a summary of the extraction procedure.

[43] H. J. Evans, C. E. Sullivan, and K. A. Piez, *Biochemistry* **15**, 1435 (1976).
[44] J. Leushner and J. Pasternak, *J. Exp. Zool.* **204**, 155 (1978).
[45] R. Ouazana and M.-A. Gibert, *C. R. Hebd. Seances Acad. Sci. Ser. D* **288**, 911 (1979).
[46] S. Brenner, *Genetics* **77**, 71 (1974).
[47] G. N. Cox, J. S. Laufer, M. Kusch, and R. S. Edgar, *Genetics* **95**, 317 (1980).
[48] P. V. Hauschka and W. F. Harrington, *Biochemistry* **9**, 3734 (1970).

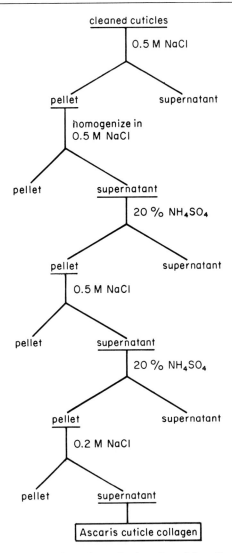

Fig. 9. Extraction scheme for *Ascaris* cuticle collagen.

## Preparation of Subunit Gelatin Chains from *Ascaris* Cuticle Collagen

Two variations based on the original method of McBride and Harrington[49] yield large quantities of reduced, carboxymethylated collagen subunits (RCM-*Ascaris*)[48] or reduced, methylated subunits (RMe-*Ascaris*).[43] In the method of Hauschka and Harrington,[48] cuticles previ-

[49] O. W. McBride and W. F. Harrington, *Biochemistry* **6**, 1484 (1967).

ously extracted with 0.5 $M$ NaCl are dissolved in 1% 2-mercaptoethanol in 0.5 $M$ NaCl, pH 8, and the reduced protein is twice precipitated with ammonium sulfate at 30% saturation. Carboxymethylation is performed in 5 $M$ guanidine-HCl–Tris-EDTA, as described, yielding 10–20 mg of RCM-*Ascaris* per worm.[48] Chromatography at 40° on CM-cellulose and DEAE-cellulose showed approximately equal quantities of *two* RCM-*Ascaris* gelatin chain species. Later work by Evans *et al.*[43] showed that whole *Ascaris* cuticle is composed of *three* species of gelatin subunit. Presumably one of these three was lost in the pre-extraction of native collagen or in the ammonium sulfate steps of the former procedure.[48] The method of Evans *et al.*[43] is described below.

*Reagents and Materials*

*Ascaris* cuticles, 2 g wet weight, water washed
Tris–urea–EDTA: 0.58 $M$ Tris-HCl, pH 8.6; 8 $M$ urea; 0.2% Na$_2$EDTA
2-Mercaptoethanol
Methyl-$p$-nitrobenzenesulfonate, 1 $M$ in acetonitrile
BioGel P-2 (100–200 mesh) in a 3 × 45 cm column equilibrated at 25° with 0.1% acetic acid for desalting after reduction and methylation
Phosphocellulose (Whatman P-1) in a jacketed 18 × 25 mm column equilibrated at 40° with 0.01 $M$ sodium formate, pH 3.5. An 800 ml of gradient from 0.05 $M$ NaCl to 0.45 $M$ NaCl in the column buffer is used to elute the gelatin chains
BioGel A-0.5m (200–400 mesh) in a 2.1 × 120 cm column equilibrated with 1.0 $M$ CaCl$_2$, 0.05 $M$ Tris-HCl, pH 7.5

*Procedure.* Cuticles (2 g) are suspended in 8 ml of Tris-urea-EDTA, and dissolution is achieved by adding 0.08 ml of 2-mercaptoethanol. After 3 hr at 25°, the small amount of insoluble material is removed by passage through a 5-$\mu$m Millipore filter, and methylation is carried out at 40°.[50] Desalting on BioGel P-2 is followed by lyophilization. Chromatography on phosphocellulose resolves three principal gelatin species (A, B, and C) and some proteolytic breakdown products of the C chain. After removal of these smaller peptides by gel filtration (BioGel A-0.5m), rechromatography on phosphocellulose produces homogeneous fractions of A, B, and C. The average yield of these three chains is in the ratio 1 : 2.7 : 1.9.[43]

Properties of Native *Ascaris* Cuticle Collagen and Its Subunits

*Physical Characteristics*

Native *Ascaris* cuticle collagen is considerably larger than the tropocollagen of vertebrate species. Hydrodynamic studies[35] revealed

[50] R. L. Heinrickson, *Biochem. Biophys. Res. Commun.* **41**, 967 (1970).

$s^\circ_{20,w}$ = 7.0 S and [$\eta$] = 12.3 dl/g, with calculated dimensions of 18 Å by 4100 Å. In keeping with the abundance of cysteine-derived disulfide cross-links in *Ascaris* collagen, the molecular weight of 900,000 is unchanged by denaturation in 5 $M$ guanidine-HCl, even though this solvent abolishes all triple-helical collagen conformation as judged by optical rotary dispersion[35] (see Table VII). The extremely high thermal denaturation temperature of 52.8°[35,48] and the rapid regain of collagen conformation after melting[35,48,49,51] are other interesting properties bestowed by the disulfide cross-links.

The subunit structure of *Ascaris* cuticle collagen is unusual. Each native molecule comprises some 15–20 polypeptide chains of molecular weight 52,000.[35,43,49] The individual chains are capable of *intra*molecular collagen triple-helix formation by back-folding,[49,52] and thus may be cross-linked into larger aggregates.[51] Whether or not the A, B, and C chains of Evans *et al.*[43] are present in constant proportions in each 900,000-dalton native *Ascaris* collagen molecule has not been established.

*Chemical Characteristics*

In addition to its very high proline and low hydroxyproline and glycine proportions, *Ascaris* cuticle collagen contains the highest half-cysteine and lysine, and among the lowest serine and threonine contents of any collagen reported[35] (see Table VIII). Interestingly, this anomalous collagen was isolated in an era when one principle for collagen purity was the *absence* of cystine, a criterion that has since evaporated with the advent of vertebrate procollagen studies. *Ascaris* cuticle collagen contains up to 36% proline.[43] A substantial fraction of proline apparently occupies the Y position in the Gly-X-Y triplet sequence because of stability properties[38] and because RCM-*Ascaris* collagen is a good substrate for vertebrate prolyl hydroxylase.[4,53] There is almost no carbohydrate in *Ascaris* cuticle collagen.

## Cuticle Collagens from Other Nematode Species

Cuticles from adult *C. elegans* (1 mm body length) have been isolated in quantity by sonication,[44] followed by washing in 1% SDS (4 hr, 25°) to remove internal tissue and cellular debris.[45] Extraction for 1 hr in 5% trichloroacetic acid at 90° yielded a collagenous material with the composition shown in Table VIII. In view of the harsh conditions employed and of the problem of selective extraction of certain collagen subclasses in hot acid,[4,54] it would be useful to apply *Ascaris* methods to *C. elegans* cuticles.

[51] P. V. Hauschka and W. F. Harrington, *Biochemistry* **9**, 3754 (1970).
[52] N. V. Rao and W. F. Harrington, *Biochemistry* **9**, 3714 (1970).
[53] D. Fujimoto and D. J. Prockop, *J. Biol. Chem.* **243**, 4138 (1968).
[54] D. Fujimoto and E. Adams, *Biochem. Biophys. Res. Commun.* **17**, 437 (1964).

From the high half-cysteine content, one would predict that direct reduction and methylation[43] of *C. elegans* would yield the subunit gelatin chains. Subunits have been obtained from *P. silusiae* by extraction under reducing conditions.[44]

The distinctions for annelids and nematodes between the unusual ectodermally derived cuticle collagens and the mesodermal collagens of body wall, muscle, and intestinal basement membrane have been reviewed by Adams.[4]

## Appendix

### Calculation of $R$ Values, Using Amino Acid Analyses Expressed as Residues/1000

$$aa_p = Asp + Glu + Hyl + Lys + His + Arg$$
$$aa_{OH} = Hyp + Thr + Ser + Tyr + Hyl$$
$$aa_\phi = Val + Met + Leu + Ile + Tyr + Phe$$

$$R_p = \frac{aa_p \times 1000}{aa_p + aa_{OH} + aa_\phi}$$

$$R_{OH} = \frac{aa_{OH} \times 1000}{aa_p + aa_{OH} + aa_\phi}$$

$$R_\phi = \frac{aa_\phi \times 1000}{aa_p + aa_{OH} + aa_\phi}$$

The resulting coordinates ($R_p$, $R_{OH}$, $R_\phi$) from each collagen analysis allow it to be plotted in a triangular coordinate space (see Fig. 1). Matsumura[8] plotted the results from 30 invertebrate and 67 vertebrate collagen analyses to conclude that there were four basic classes of collagen. However, the data available to him in the nematode, annelid, and basement membrane collagen groups were limited. The $R$ values in the following tabulation are calculated from more recent analyses and were used in making Fig. 1. It can be seen that polychaete annelid cuticle collagen forms a fifth class of collagens. Matsumura's values are included for comparison.

| Species | Collagen type or tissue source | $R_\phi$ | $R_{OH}$ | $R_p$ | Reference[a] |
|---|---|---|---|---|---|
| *Ascaris lumbricoides* | Nematode cuticle | 177–209 | 156–181 | 636–642 | 8 |
| *Caenorhabditis elegans* | Nematode cuticle | 168 | 231 | 600 | 45 |
| *Ascaris lumbricoides* | Nematode cuticle, B chain | 176 | 197 | 626 | 43 |
| *Lumbricus* sp. | Oligochaete cuticle | 117–157 | 507–574 | 309–337 | 8 |
| *Lumbricus terrestris* | Oligochaete cuticle | 119 | 551 | 330 | b |
| | Whole cuticle | 155 | 529 | 315 | 38 |
| | Salt extract of cuticle | 128 | 531 | 341 | 38 |
| | Acid extract of cuticle | 114 | 559 | 326 | 38 |
| *Marphysa sanguinea* | Polychaete cuticle | 62 | 530 | 409 | 10 |
| *Lumbriconereis heteropoda* | Polychaete cuticle | 110 | 480 | 410 | 10 |
| *Nereis japonica* | Polychaete cuticle | 120 | 440 | 450 | 11 |
| | Polychaete cuticle, ASCC | 74 | 523 | 402 | 9 |
| | Polychaete cuticle, NSCC | 73 | 530 | 400 | 9 |
| *N. virens* | Polychaete cuticle | 59 | 517 | 424 | 18 |
| | Polychaete cuticle, ASCC | 65 | 512 | 423 | 9 |
| | Polychaete cuticle, SSCC | 69 | 500 | 430 | c |
| *Locusta migratoria* | Insect ejaculatory duct | | | | |
| | 2.4 *N* NaCl ppt | 208 | 340 | 453 | 1 |
| | 4.0 *M* NaCl ppt | 166 | 375 | 459 | 1 |
| Invertebrate | Interstitial | 140–226 | 296–403 | 418–521 | 8 |
| Vertebrate | Interstitial | 147–217 | 298–397 | 441–538 | 8 |
| | Type I, $\alpha$1 | 147 | 391 | 462 | d |
| | Type I, $\alpha$2 | 214 | 353 | 433 | d |
| | Type II | 166 | 355 | 479 | d |
| | Type III | 150 | 408 | 442 | d |
| | Type V, $\alpha$A | 191 | 339 | 470 | d |
| | Type V, $\alpha$B | 183 | 344 | 473 | d |
| | Type V, $\alpha$A, skin | 179 | 391 | 429 | e |
| | Type V, $\alpha$B, skin | 180 | 374 | 446 | e |
| | Type V, $\alpha$A, placenta | 193 | 370 | 437 | f |
| | Type V, $\alpha$A, calvaria | 193 | 367 | 439 | f |
| | Type V, $\alpha$B, placenta | 176 | 369 | 455 | f |
| | Type V, $\alpha$B, calvaria | 178 | 361 | 461 | f |
| | Type V, $\alpha$B, cartilage | 176 | 363 | 461 | f |
| | Type V, $\alpha$A, placenta | 185 | 367 | 449 | g |
| | Type V, $\alpha$B, placenta | 173 | 355 | 472 | g |
| Dog | Type IV, basement membrane | 232–249 | 394–415 | 353–362 | 8 |

(*continued*)

| Species | Collagen type or tissue source | $R_\phi$ | $R_{OH}$ | $R_p$ | Reference[a] |
|---|---|---|---|---|---|
| *Ascaris suum* | Nematode basement membrane | 253 | 358 | 389 | 3 |
| *Homo sapiens* | Type IV, placenta | 260 | 372 | 360 | h |
| | Type IV, procollagen | 286 | 341 | 373 | i |
| | Type IV, pepsin-treated procollagen | 237 | 390 | 374 | i |
| Mouse | Type IV, EHS sarcoma | 241 | 379 | 379 | 34 |
| Cow | Type IV, lens capsule | 222 | 408 | 370 | j |
| Cow | Type IV, lens capsule | 213 | 404 | 384 | k |
| Cow | Type IV, lens capsule | 247 | 399 | 354 | l |

[a] Numbers refer to text footnotes.
[b] A. Goldstein and E. Adams, *J. Biol. Chem.* **245**, 5478 (1970).
[c] L. Murray and T. Housley, unpublished observations.
[d] R. Burgeson, F. A. ElAdli, I. I. Kaitils, and D. W. Hollister, *Proc. Natl. Acad. Sci. U.S.A.* **73**, 2579 (1976).
[e] E. Chung, R. K. Rhodes, and E. J. Miller, *Biochem. Biophys. Res. Commun.* **71**, 1167 (1976).
[f] R. K. Rhodes and E. J. Miller, *Biochemistry* **17**, 3442 (1978).
[g] H. Bentz *et al.*, *Eur. J. Biochem.* **92**, 563 (1978).
[h] H. Sage and P. Bornstein, *J. Biol. Chem.* **255**, 9893 (1980).
[i] E. Crouch and P. Bornstein, *J. Biol. Chem.* **254**, 4197 (1979).
[j] D. Schwartz and A. Veis, *Eur. J. Biochem.* **103**, 29 (1980).
[k] S. Dixit, *FEBS Lett.* **85**, 153 (1978).
[l] P. Dehm and N. Kefalides, *J. Biol. Chem.* **253**, 6680 (1978).

## [4] Preparation and Characterization of Procollagens and Procollagen–Collagen Intermediates

*By* HELENE SAGE and PAUL BORNSTEIN

Collagen is now known to exist as a family of structurally related proteins comprising at least five genetically distinct types.[1] Procollagen is defined as the soluble biosynthetic precursor of collagen. It can be isolated from tissues and from systems capable of synthesizing the protein *in vitro* as a molecule containing both N- and C-terminal extensions, or in various intermediate forms that occur as a result of specific proteolytic processing, leading to the final product, collagen (for extensive reviews of the

[1] P. Bornstein and H. Sage, *Annu. Rev. Biochem.* **49**, 957 (1980).

chemistry and biology of collagen and procollagen, see Fessler and Fessler[2] and Bornstein and Traub[3]).

Although procollagen intermediates (termed p-collagens) had been described earlier, recovery of intact procollagen was accomplished only after the inclusion of protease inhibitors during purification. Type I procollagen is composed of two identical pro$\alpha$1 chains and one pro$\alpha$2 chain and is accordingly described by the formula [pro$\alpha$1(I)]$_2$pro$\alpha$2 (for collagen nomenclature, see Bornstein and Sage[1] and this volume [1]). It contains globular domains at both the N and C termini, termed extensions or propeptides, and has a molecular weight[3a] of approximately 450,000; after disulfide bond reduction, each chain migrates on SDS-PAGE with an $M_r$ of about 150,000. Procollagen containing the C-terminal propeptides (p$_C$-collagen) is generated following proteolytic scission of the N-propeptides and has an apparent $M_r$ of 130,000 after reduction of disulfide bonds. Procollagen from which the C-propeptides have been removed (p$_N$-collagen) is not disulfide bonded and has an apparent $M_r$ of 120,000. It should be noted that procollagen chains migrate anomalously on SDS-PAGE compared with globular proteins and that their hybrid structure imposes further difficulties in estimation of molecular weights. It is now known that the helical domain of the pro$\alpha$1(I) chain contains 1014 amino acids; the N-terminal domain, 155 amino acids[4]; and the C-terminal domain, 272 amino acids.[5]

The structure of type II procollagen is analogous to that of type I with the exception that only one chain type is present, pro$\alpha$1(II). Type III procollagen similarly is composed of only one chain type [pro$\alpha$1(III)], but interchain disulfide bonds are present in both the C- and N-propeptide domains. The structure of type IV procollagen appears to be quite different from that of types I–III; a much more limited type of processing may

[2] J. H. Fessler and L. I. Fessler, *Annu. Rev. Biochem.* **47**, 129 (1978).
[3] P. Bornstein and W. Traub, in "The Proteins" (H. Neurath and R. L. Hill, eds.), 3rd ed., Vol. 4, pp. 411–632. Academic Press, New York, 1979.
[3a] Abbreviations: $M_r$, molecular weight; PAGE, polyacrylamide gel electrophoresis; SDS, sodium dodecyl sulfate; PMSF, phenylmethanesulfonyl fluoride; EDTA, ethylenediaminetetraacetic acid; NEM, $N$-ethylmaleimide; PCMB, $p$-chloromercuribenzoate; DTT, dithiothreitol; PBS, phosphate-buffered saline (0.15 $M$ NaCl, 4 m$M$ KH$_2$PO$_4$, 10 m$M$ Na$_2$HPO$_4$, pH 7.5); DMEM, Dulbecco's modified Eagle's medium; $\beta$-APN, $\beta$-aminopropionitrile fumarate; DEAE, diethylaminoethyl; CM, carboxymethyl; PMB, $p$-mercuribenzoate; w/v, weight to volume ratio; v/v, volume to volume ratio; $p$-HMB, $p$-hydroxymercuribenzoate; DFP, diisopropylfluorophosphate; BSA, bovine serum albumin; Tris-saline, 0.15 $M$ NaCl, 50 m$M$ Tris-HCl, pH 7.5; HEPES, $N$-2-hydroxyethylpiperazine-$N'$-2-ethanesulfonic acid.
[4] D. Hörlein, P. P. Fietzek, E. Wachter, C. M. Lapiere, and K. Kühn, *Eur. J. Biochem.* **99**, 31 (1979).
[5] F. Fuller and H. Boedtker, *Biochemistry* **20**, 996 (1981).

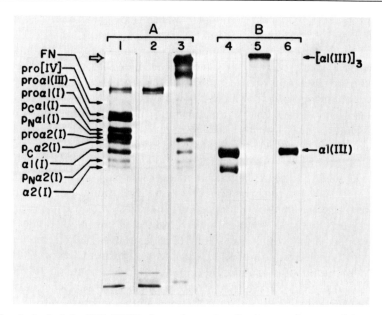

FIG. 1. Analysis by SDS–PAGE of smooth muscle cell culture medium containing procollagen and collagen chains. (A) lane 1: [$^3$H]proline-labeled protein from bovine aortic smooth muscle cell culture medium was isolated by trichloroacetic acid precipitation and resolved on a 6%/10% SDS–polyacrylamide slab gel in the presence of 50 m$M$ dithiothreitol. Visualization was by fluorescence autoradiography. Lane 2: protein in lane 1 after digestion with bacterial collagenase. Only fibronectin (FN) remains uncleaved. Lane 3: lane 1, unreduced. Non-disulfide-bonded chains ($p_N\alpha 1$, $\alpha 1$, $p_N\alpha 2$ and $\alpha 2$) are observed. (B) Protein in lane 1 was digested with pepsin. Types I and III collagens were separated by fractional salt precipitation and purified by CM-cellulose chromatography, prior to SDS-PAGE; lane 4: type I collagen, reduced or unreduced; lane 5: type III collagen, unreduced; lane 6: lane 5, reduced. Fibronectin, procollagen, p-collagen, and collagen chains are identified; white arrow indicates position of migration of the intact procollagen molecule.

occur, which is compatible with the presumed absence of discrete, globular N- and C-propeptides in this disulfide-linked molecule[6,7] and with the existence of discontinuities in the helical body of the molecule.[8]

Figure 1, lane 1, represents an autoradiographic analysis of the high molecular weight, proline-labeled proteins in smooth muscle cell culture medium and illustrates the relative mobilities on SDS-PAGE of some of the procollagen, p-collagen, and collagen chains. Prior to reduction, these proteins migrate as shown in lane 3. The position of migration of $p_N$-

---

[6] K. Tryggvason, P. G. Robey, and G. R. Martin, *Biochemistry* **19**, 1284 (1980).
[7] E. Crouch and P. Bornstein, *J. Biol. Chem.* **254**, 4197 (1979).
[8] D. Schuppan, R. Timpl, and R. W. Glanville, *FEBS Lett.* **115**, 297 (1980).

collagen and collagen chains, which lack interchain disulfide bonds, is unchanged. After bacterial collagenase digestion, only fibronectin remains uncleaved (lane 2).

This chapter describes specific methodology for the purification of the five different procollagen types both from tissues and from sources that synthesize these proteins *in vitro*. In addition, we discuss several procedures for assessment of purity and for characterization of procollagen chains. Aspects of the preparation of preprocollagen[9] and the purification of procollagen extension peptides[4,10] will not be considered here.

## Isolation of Procollagens Synthesized *in Vitro*

*1. Choice of Organ or Cell Culture System*

In Table I are listed several systems that have been used successfully for preparation of radiolabeled procollagens *in vitro*.[11-17] In general, mesenchymal cells, such as skin and tendon fibroblasts, and aortic smooth muscle cells secrete a major portion of radiolabeled protein as types I and III procollagens.[18] Epithelial cells or tissues, such as amniotic fluid cells,[7] lens capsule,[19] kidney glomerulus,[20] and rat parietal yolk sac endoderm,[15] have proved to be good sources of type IV procollagen. In addition, certain tumors such as the EHS murine sarcoma[21] and a human fibrosarcoma cell line HT 1080 (obtainable from the American Type Culture collection),[16] are advantageous as sources of type IV procollagen owing to their high growth rate and longevity in culture. Type II procollagen has been isolated most successfully from sternae in organ culture and from

---

[9] R. D. Palmiter, J. M. Davidson, J. Gagnon, D. W. Rowe, and P. Bornstein, *J. Biol. Chem.* **254**, 1433 (1979).
[10] B. R. Olsen, N. A. Guzman, J. Engel, C. Condit, and S. Aase, *Biochemistry* **16**, 3030 (1977).
[11] L. Peltonen, A. Palotie, T. Hayashi, and D. J. Prockop, *Proc. Natl. Acad. Sci. U.S.A.* **77**, 162 (1980).
[12] K. Alitalo, M. Kurkinen, A. Vaheri, T. Krieg, and R. Timpl, *Cell* **19**, 1053 (1980).
[13] H. Sage, E. Crouch, and P. Bornstein, *Biochemistry* **18**, 5433 (1979).
[14] B. A. Booth, K. L. Polak, and J. Uitto, *Biochim. Biophys. Acta* **607**, 145 (1980).
[15] C. C. Clark, E. A. Tomichek, T. R. Koszalka, R. R. Minor, and N. A. Kefalides, *J. Biol. Chem.* **250**, 5259 (1975).
[16] K. Alitalo, A. Vaheri, T. Krieg, and R. Timpl, *Eur. J. Biochem.* **109**, 247 (1980).
[17] C. Kumamoto and J. H. Fessler, *Proc. Natl. Acad. Sci. U.S.A.* **77**, (1980).
[18] P. Dehm and D. J. Prockop, *Biochim. Biophys. Acta* **264**, 375 (1972).
[19] G. J. Heathcote, C. H. J. Sear, and M. E. Grant, *in* "Biology and Chemistry of Basement Membranes" (N. A. Kefalides, ed.), p. 335. Academic Press, New York, 1978.
[20] M. E. Grant, R. Harwood, and I. F. Williams, *Eur. J. Biochem.* **54**, 531 (1975).
[21] R. W. Orkin, P. Gehron, E. B. McGoodwin, G. R. Martin, T. Valentine, and R. Swarm, *J. Exp. Med.* **145**, 204 (1977).

TABLE I
SYNTHESIS OF PROCOLLAGENS in Vitro

| Procollagen type | Source (quantity) | Metabolic label (hr) | Yield[a,b] | Reference |
|---|---|---|---|---|
| I | Human fibroblasts (10–150 cm$^2$ flasks containing 10$^8$ cells) | 800 $\mu$Ci of [$^{14}$C]amino acid mixture (24) | 2 mg at 2 × 10$^6$ cpm/mg | 11 (see also this volume [11]) |
| III | Human amniotic epithelial cells | 10 $\mu$Ci/ml [5-$^3$H]proline | 3.3 pg/cell (13.7% of total protein in medium and cell layer) | 12 |
| III | Human embryonic fibroblasts | 10 $\mu$Ci/ml [5-$^3$H]proline | 0.69 pg/cell (3.1% of total protein in medium and cell layer) | 12 |
| III | Bovine aortic endothelial cells (10–100 mm dishes containing 75 × 10$^6$ cells) | 2.5 mCi [5-$^3$H]proline (24) | 6.8 × 10$^6$ dpm in procollagen (3.1% of dpm in medium protein) | 13 |
| I and III | Human skin fibroblasts (1 confluent 75 cm$^2$ flask) | 30 $\mu$Ci of [$^3$H]proline (20) | Medium: 2.2 × 10$^5$ dpm [$^3$H]Hyp/mg cell protein | 14 |
|  |  |  | Cell layer: 3.1 × 10$^4$ dpm [$^3$H]Hyp/mg cell protein | 14 |
| IV | Embryonic rat parietal yolk sacs (10) | 10 $\mu$Ci of [$^{14}$C]proline (4) | 6% of total newly synthesized protein | 15 |
| IV | HT 1080-human fibrosarcoma | 5 $\mu$Ci/ml L-[U-$^{14}$C]proline | Medium: 2.3 × 10$^4$ cpm in procollagen/10$^6$ cells (1.7% of total medium protein) | 16 |
|  |  |  | Cell layer: 2.3 × 10$^3$ cpm in procollagen/10$^6$ cells (0.1% of total medium protein) | 16 |
| V | Chick embryo crops (50) | 750 $\mu$Ci of [$^3$H]proline (5) | 6% of total radioactive soluble collagens | 17 |

[a] In many cases it is not possible to calculate specific activities of procollagens because of their recovery in microgram or submicrogram quantities.
[b] Yields refer to procollagen in the culture medium unless specified otherwise.

matrix-free chondrocytes.[22] These cells, however, can undergo phenotypic modulation in culture and will cease production of type II procollagen in favor of types I and III[23] (see Bornstein and Sage[1] for review).

We have chosen to describe smooth muscle cell cultures in detail

[22] J. Uitto, *Biochemistry* **16**, 3421 (1977).
[23] P. Benya, S. Padilla, and M. Nimni, *Biochemistry* **16**, 865 (1977).

because the cells are easily propagated in tissue culture, they produce relatively large amounts of types I and III procollagen,[24] and they also synthesize small but detectable amounts of types IV and V procollagen.[25]

## Types I and III Procollagen

*Isolation of Smooth Muscle Cells.*[26] Fresh bovine aortas were handled under sterile conditions and processed within 2–3 hr after slaughter. A piece of tissue (approximately 1 cm$^2$) comprising both intimal and medial layers, but excluding the adventitia, which is rich in fibroblasts, was excised from the lumen of the vessel and finely minced. The tissue was then transferred to a 100-mm plastic tissue culture dish, and a drop of Waymouth's medium containing mycostatin (100 units/ml), penicillin G (100 units/ml), streptomycin sulfate (100 µg/ml), and either 10% (v/v) fetal calf or adult bovine serum was placed on top of each small piece. The dish was incubated for 1 hr at 37° in a mixture of 5% $CO_2$ and 95% air. Within this time, cells attach to the plastic surface. Ten milliliters of complete Waymouth's medium were added, and the dish was incubated for 3–4 days at 37°. At this point the healthy explants were refed and the contaminated ones were discarded (we find with bovine tissues that from 10–50% of the preparations can become infected). One week after the initial plating, the cells were trypsinized (1 ml of 0.25% trypsin plus 1 ml of 0.02% EDTA, w/v) and seeded into flasks (approximately $0.5 \times 10^6$ cells per 25-cm$^2$ flask). They were maintained in Waymouth's or in DMEM containing antibiotics and 10% serum and were subcultured weekly at a split ratio of 1:3.

*Metabolic Labeling.* Cultures of smooth muscle cells which had just reached confluence were used between the second and eighth passage. The cells were preincubated at 37° for 1 hr in serum-free DMEM containing 50 µg of sodium ascorbate (freshly made solution added immediately before use) and 64 µg of β-APN fumarate per milliliter. The medium was removed and replaced with fresh preincubation medium containing 25 µCi of [2,3-$^3$H]proline (35 Ci/mmol; New England Nuclear) per milliliter. Five milliliters were used per 100-mm dish or 75-cm$^2$ flask and 10 ml per 135-mm plate. After 24 hr at 37°, the culture medium was removed from the cells with a siliconized Pasteur pipette and transferred to plastic centrifuge tubes containing a freshly prepared mixture of protease inhibitors, to produce a final concentration in the medium of 0.2 m$M$ PMSF (stock solution: 0.2 $M$ in absolute ethanol), 10 m$M$ NEM (added in solid form

---

[24] J. M. Burke, G. Balian, R. Ross, and P. Bornstein, *Biochemistry* **16**, 3243 (1977).
[25] R. Mayne, M. S. Vail, and E. J. Miller, *Biochemistry* **17**, 446 (1978).
[26] R. Ross, *J. Cell Biol.* **50**, 172 (1971).

with vortexing), and 2.5 m$M$ EDTA (stock solution: 0.25 $M$ EDTA, disodium salt, 50 m$M$ Tris-HCl, pH 7.5). Cell layers were rinsed with 1–2 ml of PBS containing 0.2 m$M$ PMSF, and this rinse was added to the radiolabeled culture medium. The medium was clarified in a clinical centrifuge at 400 $g$ to remove cellular debris (higher centrifugal force will result in cell lysis and subsequent release of proteases) and was transferred to a siliconized glass flask. The solution was cooled to 4°, and solid ammonium sulfate (Ultrapure) was added over a period of 2–3 hr to a final concentration of 20% (w/v). The suspension was stirred gently for 16 hr at 4°, and the resulting precipitate, which contained >90% of the collagenous proteins in the culture medium, was isolated by centrifugation in a refrigerated centrifuge at 48,000 $g$ at 4° for 30 min.

*Ion-Exchange Chromatography.* The proteins precipitated in 20% ammonium sulfate were dissolved in buffer at 4° (6 $M$ deionized urea, 50 m$M$ Tris-HCl, pH 8.0, containing 0.2 m$M$ PMSF and 2.5 m$M$ EDTA) by agitation with a rubber policeman (approximately 2 ml per pellet). This solution was dialyzed against the same buffer at 4° and then centrifuged at 48,000 $g$ for 20 min at 4°. The supernatant (approximately 10 ml) was loaded onto a 2.0 × 22 cm DEAE-cellulose column (DE-52, Whatman) at 30 ml/hr, and at least one column volume of buffer was pumped through prior to inception of a linear gradient of 0 to 200 m$M$ NaCl in a total of 400 ml of column buffer. Three-milliliter fractions were collected into siliconized tubes, and the column was monitored by counting 10–25 $\mu$l of each fraction in 5–10 ml of Aquasol (New England Nuclear). Appropriate fractions were pooled in the presence of 10 $\mu$g of pepstatin A (Peninsula Laboratories; stock solution: 100 mg/ml in dimethyl sulfoxide) per milliliter and dialyzed extensively against 0.1 $M$ acetic acid at 4°. The solutions were lyophilized in plastic tubes and stored in desiccated form at −20°.

Procollagens were identified by mobility before and after disulfide bond reduction on SDS-polyacrylamide slab gels.[27] All electrophoreses described in this chapter were run at room temperature at a constant current of 30–35 mA. The gels were processed for fluorescence autoradiography using standard techniques.[28,29] Figure 2 is a representative chromatograph of a 20% ammonium sulfate precipitate of bovine smooth muscle cell culture medium after fractionation on DEAE-cellulose. Proteins in the peak fractions have been identified by their position of migration on SDS-PAGE (inset). Collagen containing $\alpha$1 and $\alpha$2 chains does not bind to DEAE-cellulose at pH 8.0; similarly, bovine type I p$_C$-collagen is eluted prior to the start of the gradient (peak I). However, in other

[27] U. K. Laemmli, *Nature (London)* **227**, 680 (1970).
[28] W. M. Bonner and R. A. Laskey, *Eur. J. Biochem.* **46**, 83 (1974).
[29] R. A. Laskey and A. D. Mills, *Eur. J. Biochem.* **56**, 335 (1975).

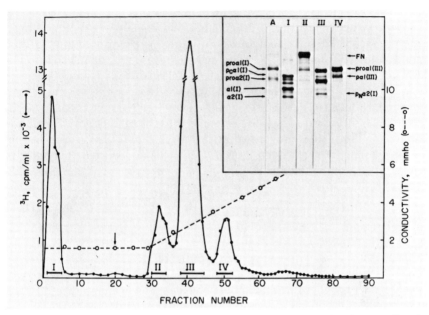

FIG. 2. Fractionation of procollagens from smooth muscle cell culture medium by DEAE-cellulose chromatography. [$^3$H]Proline-labeled protein was precipitated from culture medium by 20% (w/v) ammonium sulfate and chromatographed on DEAE-cellulose in 6 $M$ urea, 50 m$M$ Tris-HCl, pH 8.0, buffer containing 0.2 m$M$ PMSF and 2.5 m$M$ EDTA at 4°. Gradient elution was performed from 0 to 200 m$M$ NaCl over a total of 400 ml. Peaks containing labeled protein were pooled as indicated by Roman numerals. *Inset:* SDS-PAGE of DEAE-cellulose chromatographic fractions. Proteins were resolved on a 6% polyacrylamide slab gel in the presence of 50 m$M$ DTT and visualized by fluorescence autoradiography. Roman numerals refer to pooled peaks, and A represents the material prior to chromatography. Fibronectin (FN) and types I and III procollagen, p-collagen, and collagen chains are identified.

species, such as rat, chick, and human, type I $p_C$-collagen is eluted in the position of peak II, and fibronectin and type I procollagen are often eluted together. Figure 2 also illustrates that type I $p_N$-collagen coelutes with intact type I procollagen (peak III), while type III pro- and p-collagen (peak IV) are clearly separated from other proteins. The starting material from which the chromatogram in Fig. 2 was generated contained $3.7 \times 10^7$ cpm, obtained from six 135-mm plates ($1.2 \times 10^7$ cells), and, by scanning densitometry, 94% of the recovered counts per minute were in procollagen.

COMMENT. Since fractionation on DEAE-cellulose is performed under conditions in which procollagen retains its native conformation, it is critical, after removal of the labeled culture medium from the cells, to perform all procedures at 0–4° to prevent denaturation of the protein. We have achieved higher recoveries of procollagen by using either plastic or, preferably, siliconized glass containers and pipettes, by pretreating dialysis tubing with 1% BSA in PBS, and by using DEAE-cellulose which has not been regenerated at acidic pH. Preferential loss of type I procollagen in type I/III mixtures due to aggregation (upon prolonged dialysis) has been reported.[30] Optimal recoveries of both types I and III procollagen were observed with repeated use of the same DEAE-cellulose or when new batches of the resin were first exposed to 20% fetal calf serum in the starting buffer.[30] In addition, Booth et al.[14] have reported that human fibroblasts produce procollagen optimally when cultured in the presence of serum.

*Type II Procollagen*

Type II procollagen has been prepared from chick sternal cartilages[31] or from matrix-free cartilage cells.[32] Sternae were removed from 17-day-old embryonic chicks, washed with modified Krebs medium II, and digested with 4 mg of purified bacterial collagenase (Worthington) and 10 mg of trypsin in 3 ml of Eagle's minimum essential medium for 2 hr at 37° by shaking in an atmosphere of 95% air/5% $CO_2$. The cells were filtered through lens paper and centrifuged at $600\,g$ at room temperature for 6 min; they were subsequently resuspended in, and washed three times with, modified Krebs medium containing 10% fetal calf serum. The reported yield was 4 to 6 × $10^6$ cells/sternum.[32]

Matrix-free cartilage cells were then preincubated at 37° for 15 min in modified Krebs medium ($10^7$ cells/ml) supplemented with 20% fetal calf serum and 50 µg each of β-APN-HCl and ascorbic acid per milliliter; 12 µCi of [$^{14}$C]proline in preincubation medium was then added to 3.8 × $10^8$ cells, and labeling proceeded for 2 hr.[22] The culture medium was processed and chromatographed on DEAE-cellulose as described above except that (a) proteins were precipitated using 176 mg/ml ammonium sulfate (30% of saturation); (b) the buffer used for DEAE-cellulose chromatography was 2 $M$ urea, 0.025 $M$ Tris-HCl, pH 7.5, containing 1 m$M$ $Na_2$ EDTA; and (c) gradient elution was performed from 0 to 300 m$M$ NaCl over 600 ml. Type II procollagen, chromatographed under these conditions, eluted as a sharp peak within the first 200 ml of the gradient, and it migrated on SDS-PAGE similarly to type I procollagen before reduction and to the proα1(I) chain after reduction.[22]

[30] J. Uitto, B. A. Booth, and K. L. Polak, *Biochim. Biophys. Acta* **624**, 545 (1980).
[31] P. K. Müller and O. Jamhawi, *Biochim. Biophys. Acta* **365**, 158 (1974).
[32] P. Dehm and D. J. Prockop, *Eur. J. Biochem.* **35**, 159 (1973).

## Type IV Procollagen

Type IV procollagen has been prepared and characterized predominantly from three biosynthetic systems: the EHS sarcoma (a transplantable murine tumor),[21] human amniotic fluid cells,[7] and rat parietal yolk sac.[15] However, the first two are not readily available and the third does not produce sufficiently large quantities of the protein to permit detailed characterization.[6,15] In confirmation of a recent report,[16] we have found that cells from a human fibrosarcoma, HT 1080 (American Type Culture Collection, CCL 121) produce relatively large quantities of type IV procollagen to the exclusion of other procollagen types. These cells were maintained in DMEM containing antibiotics and 10% fetal calf or calf serum and were observed to have a minimum doubling time of 12–18 hr.

Procedures for isolation of type IV procollagen are the same as those just described for types I and III. However, a distinctive feature of type IV procollagen was its failure to bind to DEAE-cellulose at pH 8.0 and low ionic strength.[7] The protein can therefore be recovered in the initial elution peak of a chromatogram of HT 1080 culture medium (see Fig. 2 for comparison). This procollagen is disulfide-bonded and after reduction migrates on SDS-PAGE as a closely-spaced doublet between fibronectin and pro$\alpha$1(I) (see Fig. 1, lane 1).

## Type V(AB) Procollagen

The methodology for isolation and partial characterization of type V procollagen has been described by Kumamoto and Fessler.[17] Crops were excised from fifty 19-day-old chick embryos and incubated in DMEM[33] supplemented with 64 $\mu$g/ml $\beta$-APN, 100 $\mu$g/ml ascorbate, and 10% heat-inactivated fetal calf serum. Labeling was performed using 50 $\mu$Ci of [$^3$H]proline per milliliter for 5 hr, with a change of medium every 1.5–2 hr. Crops were then homogenized at 4° in 1 $M$ NaCl, 50 m$M$ Tris-HCl, 10 m$M$ EDTA, pH 7.5, containing 1.25 mg of NEM and 150 $\mu$g of PMSF per milliliter, and the homogenate was clarified by centrifugation at 40,000 rpm for 15 min. To 1 volume of the supernatant was added 0.5 volume of 6 $M$ urea, 50 m$M$ Tris-HCl, and 0.3% Triton X-100 buffer, pH 7.5; this solution was dialyzed at 4° against the column buffer (4 $M$ urea, 50 m$M$ Tris-HCl, 75 m$M$ NaCl, 0.1% Triton X-100, at pH 7.8) for 3 hr. The dialysis buffer was supplemented with EDTA (4 m$M$), PMSF (10 $\mu$g/ml), and NEM (250 $\mu$g/ml) for the first hour. The sample was then centrifuged briefly (15,500 rpm for 20 min) to remove any precipitated material and was chromatographed on a DEAE-cellulose column (16 × 140 mm;

---

[33] L. I. Fessler and J. Fessler, *J. Biol. Chem.* **249**, 7637 (1974).

DE-52, Whatman) at 4° using a linear gradient of 75–275 m$M$ NaCl in 100 ml, followed by a 1 $M$ NaCl wash.

The major peak, which was eluted at a conductivity of approximately 6.4 mmho/cm, contained, by SDS-PAGE analysis after reduction, two bands identified as proα1(V) (or proαB) and proα2(V) (or proαA). When this protein was subjected to SDS-PAGE in the absence of reduction, two disulfide-linked oligomers were present ([proα1(V)]$_2$proα2 and [proα1(V)]$_3$), as well as smaller amounts of proα1(V) and proα2(V). The second peak, eluted from DEAE-cellulose in the 1 $M$ NaCl step, was shown to consist of partially processed type V procollagen intermediates. After reduction of disulfide bonds, their order in increasing mobility on SDS-PAGE was pα1(V), fα1(V), and fα2(V). The designation fα was used by Kumamoto and Fessler[17] to describe biosynthetically processed type V chains that are not disulfide-bonded and are larger than the corresponding α1(V) and α2(V) chains obtained by pepsin treatment (see discussion under Identification of Procollagens and Criteria of Purity).

Isolation of Procollagens from Tissues

Table II includes a partial list of tissues that have been utilized as sources of procollagen.[21,34–36] For the isolation of some procollagen types it is advantageous to use radiolabeled proteins synthesized *in vitro*, because (*a*) the purification scheme is often less complex owing to the presence of fewer labeled contaminating proteins; (*b*) the extraction procedures tend to be gentler and more efficient, reducing the likelihood of partial proteolysis; and (*c*) there is often an increased yield of intact procollagen relative to p-collagen. However, disadvantages of biosynthetic systems include lower yields of procollagens and p-collagens, and the additional expense and time involved in tissue or organ culture. For these reasons we describe methods for extraction of procollagens from several tissues capable of yielding relatively large amounts of protein.

*Types I and III Procollagen*

Modification of an earlier procedure for the isolation of types I and III pro- and p-collagen from rat skin[37] has resulted in an efficient method for extraction of these precursors with enhanced recoveries.[34] Fifty

[34] B. D. Smith, K. H. McKenney, and T. J. Lustberg, *Biochemistry* **16**, 2980 (1977).
[35] K. Angermann and H.-J. Barrach, *Anal. Biochem.* **94**, 253 (1979).
[36] R. Timpl, R. W. Glanville, H. Nowack, H. Wiedemann, P. P. Fietzek, and K. Kühn, *Hoppe Seyler's Z. Physiol. Chem.* **356**, 1783 (1975).
[37] P. H. Byers, K. H. McKenney, J. R. Lichtenstein, and G. R. Martin, *Biochemistry* **13**, 5243 (1974).

TABLE II
ISOLATION OF PROCOLLAGENS FROM TISSUES

| Procollagen type | Source (quantity) | Yield[a] (mg) | Reference[b] |
|---|---|---|---|
| I (procollagen and p-collagen) | Rat skins (from 50 rats), after colchicine injection | 25 | 34 |
| II (procollagen and $p_C$-collagen) | Swarm chondrosarcoma (500 g from 30 rats)[c] | 2.5 | 35 |
| III ($p_N$collagen) | Fetal bovine skin (1 kg)[c] | 50–100 | 36 |
| III (procollagen and p-collagen) | Rat skins (from 50 rats), after colchicine injection | 36 | 34 |
| IV (procollagen) | Murine EHS sarcoma (1 kg)[c] | 50–100[d] | 21 |

[a] After chromatographic purification.
[b] Numbers refer to text footnotes.
[c] Wet weight.
[d] Approximation after salt fractionation followed by several chromatographic procedures.

Sprague-Dawley rats (75 g each) were injected intraperitoneally with colchicine (5 mg/animal) 30 min prior to sacrifice. All subsequent procedures were performed at 4°. The skins were ground twice in an electric grinder, homogenized in a Polytron tissue homogenizer until the tissue was finely dispersed, and extracted for 30 min in 2 liters of 150 m$M$ NaCl, 50 m$M$ Tris-HCl, 20 m$M$ EDTA, 10 $\mu M$ PMSF, 10 m$M$ PMB, pH 7.4. Particulates were initially removed by filtration through a laboratory towel, and further clarification was accomplished by passage of the solution through coarse paper and Celite (Fischer Hi-Flow Super Cel). Solid ammonium sulfate (Schwarz-Mann, Ultra Pure Grade) was then added slowly to 20% of saturation (11.2 g/100 ml), and the precipitated procollagens were isolated by centrifugation (15,000 $g$ for 20 min). The pellets were subsequently washed twice and resuspended in 600 ml of the original extraction buffer. After stirring for several hours at 4°, the solution was clarified by centrifugation at 70,000 $g$ for 15 min.

At this point, type III collagen and precursors were selectively precipitated by addition of 10 g NaCl per 100 ml of solution followed by gentle stirring for 1–3 hr. The precipitate was collected by centrifugation (70,000 $g$ for 15 min). Type I collagen and precursor molecules were then precipitated from the supernatant by addition of NaCl to a final concentration of 20 g/100 ml; this material was also collected by centrifugation. The precipitate corresponding to each procollagen type was separately dissolved in approximately 500 ml of column buffer 1 (200 m$M$ NaCl, 50 m$M$ Tris-HCl, pH 7.6) and dialyzed at 4° against the same buffer (the inclusion of protease inhibitors as previously described is recommended during this step). The solutions were subsequently chromatographed on a DEAE-cellulose column (DE-52, Whatman; 4 × 25 cm) to remove acidic proteins and proteoglycans.[38] Protein that did not bind to the DEAE-cellulose was then dialyzed against column buffer 2 (20 m$M$ NaCl, 50 m$M$ Tris-HCl, 2 $M$ urea, pH 8, containing 2.5 m$M$ EDTA and 0.2 m$M$ PMSF). Each sample was centrifuged briefly after dialysis and was then applied to another DEAE-cellulose column (2.5 × 10 cm) equilibrated in buffer 2 at 4–8°. Collagens did not bind to the DEAE-cellulose, and types I and III procollagen were eluted at a flow rate of 30–50 ml/hr with a linear gradient of 20–200 m$M$ NaCl in a total volume of 600 ml. The column effluent was monitored at 230 nm, and 12-ml fractions were collected. Peak-containing fractions were pooled in the presence of pepstatin (1 μg/ml), dialyzed at 4° against 0.1 $N$ acetic acid, and lyophilized. Storage in a desiccated chamber at −20° is recommended.

With rat skin as a source of procollagen, the 10% NaCl precipitate contained approximately 90% type III pro- and p-collagens, while the 20% NaCl precipitate was composed predominantly of type I collagenous protein (approximately 70%). Type I precursors (which eluted first) were clearly resolved from type III precursors by DEAE-cellulose chromatography; type III collagen precursors were always in excess over those of type I. However, it has been noted that the position of elution of pro- and p-collagens from DEAE-cellulose will vary with the species from which the procollagen was derived.[34] In this regard, the precursors of neither rat type I nor rat type III collagens (i.e., pro-, $p_C$- and $p_N$collagens) were resolved from one another on DEAE-cellulose, whereas at least partial resolution has been achieved in the bovine (see Fig. 2), chick,[39] and human systems.[40]

Fetal skin also contains relatively large amounts of type III procollagen. A procedure for the extraction of this protein, which in calf occurs mostly in the form of $p_N$collagen, has been described.[36] The procedure has the advantage of allowing selective recovery of type III $p_N$-collagen without contamination by type I pro- and p-collagens. Approximately 1 kg (wet weight) of fetal bovine skin was homogenized and extracted for 2 days in 2.5 liters of 1 $M$ NaCl, 50 m$M$ Tris-HCl, pH 7.4, containing 10 m$M$ EDTA, PMSF (3 mg/l), and PCMB (3 mg/liter). All procedures were performed at 4°. Collagenous protein was precipitated from the clarified extract by addition of solid ammonium sulfate to 45% saturation, and after centrifugation the insoluble material was suspended in 1 liter and dialyzed

[38] E. J. Miller, *Biochemistry* **10**, 1652 (1971).
[39] J. M. Davidson, L. S. G. McEneany, and P. Bornstein, *Biochemistry* **14**, 5189 (1975).
[40] E. Crouch and P. Bornstein, *Biochemistry* **17**, 5499 (1978).

against buffer containing 0.5 M NaCl, 50 mM Tris-HCl, pH 7.4. The solution was again clarified by centrifugation, and solid NaCl was added to the supernatant to a final concentration of 1.7 M. After 6–16 hr the precipitate was collected by centrifugation (18,000 g for 30 min) and redissolved in buffer. Precipitation at 1.7 M NaCl was repeated twice.

The final precipitate, resuspended in buffer, was subsequently dialyzed against 200 mM NaCl, 30 mM Tris-HCl, 2 M urea, pH 7.4 (we recommend inclusion of the protease inhibitors PMSF and EDTA) and chromatographed over a DEAE-cellulose column (2.5 × 20 cm), as described above, to remove contaminating acidic components. The final separation of type III collagen and $p_N$collagen was performed on DEAE-cellulose at low ionic strength as described above for the rat types I and III procollagens.

*Type II Procollagen*

The Swarm chondrosarcoma, a transplantable tumor maintained by serial passage in rats,[41] has been shown to be a useful source of type II procollagen.[35] Tumors were injected subcutaneously into Sprague–Dawley rats, and the animals were then maintained on a commercial diet supplemented with 0.3% β-APN for approximately 4 weeks prior to sacrifice. Excised tumors (500 g) were frozen in liquid $N_2$ and pulverized using a meat grinder. All procedures were carried out at 4°. The tissue was homogenized in a Waring blender in 1.5 liters of PBS containing 20 mM EDTA, 1 mM NEM, and 10 μM PMSF and centrifuged for 30 min at 30,000 g. The supernatant was saved, and the insoluble material was suspended in 1 liter of $H_2O$, rehomogenized, and stirred for 1 hr in the presence of 2% Triton X-100 and the protease inhibitors mentioned above. The buffer was then adjusted to 150 mM NaCl, 50 mM Tris-HCl, pH 7.5 and the suspension was stirred overnight. After centrifugation, this and the previous supernatant were combined, and collagenous protein was precipitated using a 30% saturation of ammonium sulfate (170 g/liter). Insoluble material was collected, washed twice with buffer containing 20% NaCl, and extracted three times with buffer containing 1 M NaCl. At this point, the extraction solutions were clarified by centrifugation and the solubilized proteins were reprecipitated by addition of NaCl to 20% (w/v). These precipitates were collected by centrifugation (30,000 g for 1 hr) and were subsequently dissolved in and dialyzed against 200 mM NaCl, 50 mM Tris-HCl, pH 7.5, containing the protease inhibitors, prior to chromatography on a DEAE-cellulose column (5 × 15 cm) as described

---

[41] B. D. Smith, G. R. Martin, E. J. Miller, A. Dorfman, and R. Swarm, *Arch. Biochem. Biophys.* **166**, 181 (1975).

above for removal of acidic components. The fraction that did not bind to the DEAE-cellulose was subsequently dialyzed against 2 $M$ urea, 50 m$M$ Tris-HCl, pH 8.0, containing the same protease inhibitors and was chromatographed on a second DEAE-cellulose column (5 × 10 cm) equilibrated with the same buffer. Linear gradient elution was performed from 0 to 100 m$M$ NaCl in a total of 1800 ml. SDS-PAGE of material that eluted as a single peak within the gradient indicated the presence of type II procollagen, $p_C$collagen, and collagen $\alpha$ chains.

COMMENT. Preparation of type II procollagen has also been accomplished using isolated chick embryo sternae.[31] The procedure described above could be applied to large amounts of this tissue (100–250 sternae).

*Type IV Procollagen*

The EHS sarcoma, a transplantable murine tumor that produces a copious extracellular matrix *in vivo* resembling authentic basement membrane in several respects, has been used to good advantage as a source of type IV procollagen.[21,42] Tumors were grown and routinely passaged in mice (Swiss–Webster or C57/BL) by subcutaneous implantation of minced tumor tissue. These tumors when grown in intramuscular sites produce significantly less matrix but contain a proportionately greater number of cells, making the intramuscular tumors more desirable for *in vitro* synthetic studies using labeled amino acids.[6] Mice were made lathyritic approximately 2 weeks prior to sacrifice by inclusion of $\beta$-APN in their diet (4 g/kg ground rat chow and 0.1% in the drinking water). Tumors were excised, dissected free of capsular material, and disrupted with an Ultraturrax homogenizer in a minimum of 20 volumes of 25% NaCl, 50 m$M$ Tris-HCl, pH 7.4, containing 50 $\mu$g of $p$-HMB and 50 $\mu$g of PMSF per milliliter. All procedures were conducted at 4°. The insoluble material was centrifuged, resuspended in the same buffer, and recentrifuged. The pelleted residue was extracted in 0.5 $M$ acetic acid overnight, and the suspension was subsequently clarified at 10,000 $g$. A second acetic acid extraction can be performed. The combined supernatant protein, which accounted for as much as 40% of the total collagenous protein in the tissue, was precipitated by addition of NaCl to a final concentration of 10% and was subsequently collected by centrifugation. The precipitated material was redissolved in 0.5 $M$ acetic acid and dialyzed against 10% NaCl, 50 m$M$ Tris-HCl, pH 7.5, containing protease inhibitors. At this point any type I collagen that might have been present remained in solution, while the type IV procollagen was insoluble. The resulting precipitate was col-

---

[42] R. Timpl, G. R. Martin, P. Bruckner, G. Wick, and H. Wiedemann, *Eur. J. Biochem.* **84**, 43 (1978).

lected by centrifugation, dissolved in and dialyzed against 0.5 $M$ acetic acid, and lyophilized.

The lyophilized material was dissolved in 0.01 $M$ acetic acid (0.5–2 g in 200–400 ml) and dialyzed against 2 $M$ urea, 50 m$M$ Tris-HCl, pH 8.6, containing 0.5 $M$ NaCl. After centrifugation, the supernatant was diluted with an equal volume of the above buffer, which lacked NaCl. The resulting solution was applied to a DEAE-cellulose column (2.5 × 25 cm) equilibrated in the same buffer containing 0.25 $M$ NaCl, and that fraction which did not bind to the DEAE-cellulose was dialyzed against the 2 $M$ urea, 50 m$M$ Tris-HCl buffer, pH 8.6, without NaCl. Protein in this solution was then chromatographed over a second DEAE-cellulose column of the same dimensions, equilibrated in the buffer without NaCl. Elution was performed using a linear gradient from 0 to 300 m$M$ NaCl in 1600 ml of column buffer. Of the type IV procollagen, 60–80% was eluted from the second DEAE-cellulose column before initiation of the gradient. This material was pooled, dialyzed against 0.05 $M$ acetic acid, and lyophilized. Amino acid analysis indicated a highly purified type IV procollagen.[42]

COMMENT. Type IV procollagen has also been prepared from embryonic mouse parietal yolk sac using procedures similar to those described for the EHS sarcoma tissue.[6]

Purification of Procollagens and Procollagen Chains

In some instances it may be possible to isolate essentially radiochemically pure native procollagen by extraction and salt precipitation, especially from biosynthetic systems that produce only one procollagen type. However, for use in structural or immunological studies, further purification of most preparations of native procollagen is usually necessary. In addition, chromatography under denaturing conditions must be employed to separate individual procollagen chains.

*Ion-Exchange Chromatography*

Monson et al.[43] described a procedure for the separation of pro$\alpha$1(I) and pro$\alpha$2(I) which required prior reduction and alkylation of the procollagen. The sample (5 mg or less) was dissolved at 23° in 20 ml of 8 $M$ urea, 0.5 $M$ Tris-HCl, pH 8.5, containing 10 m$M$ EDTA, 1 m$M$ DFP, 1 m$M$ benzamidine, and 0.001% pepstatin. Alternatively, samples in column buffer, extraction buffer, or dilute acetic acid could be dialyzed into the above buffer. The solution was clarified by centrifugation and reduced under $N_2$ for 4 hr by the addition of 0.05 $M$ DTT at 23°. The pH was then

[43] J. M. Monson, E. M. Click, and P. Bornstein, *Biochemistry* **14**, 4088 (1975).

adjusted to 8.0 with NaOH, and iodoacetic acid was added to a final concentration of 0.11 $M$. The sample was placed in the dark for 40 min at 23°, after which the reaction was terminated by adding an excess of 2-mercaptoethanol.

The reduced and alkylated samples were subsequently dialyzed against column buffer (2 $M$ urea, 25 m$M$ Tris-HCl, pH 8.5, 0.1% 2-mercaptoethanol, 1 m$M$ benzamidine, 1 m$M$ PMSF, and 2 m$M$ EDTA) (using three changes of buffer in less than 24 hr). Procollagen was denatured by heating at 40° for 20 min and centrifuged (39,000 $g$ for 20 min) at 40° to remove insoluble material. The sample was chromatographed on a DEAE-cellulose column (2.5 × 15 cm) at 40° using a flow rate of 220 ml/hr, and procollagen chains were eluted with a linear gradient from 0 to 250 m$M$ NaCl in a total of 1200 ml. Ten-milliliter fractions were collected, and the column was monitored by counting small aliquots of each fraction in 10-ml of Aquasol or by absorbance at 230 nm. Two peaks were eluted within the gradient, the first of which corresponded to pro$\alpha$2(I) and the second, to pro$\alpha$1(I).

CM-cellulose chromatography has also been utilized for the separation of type I pro- and $p_C$collagen chains.[34] Preparations of salt-extracted type I procollagen which had been isolated initially using DEAE-cellulose chromatography (described under Isolation of Procollagen from Tissues) were dissolved at approximately 20 mg/5 ml in 8 $M$ urea (deionized), 40 m$M$ Tris-HCl, pH 8, and were reduced for 1 hr at room temperature by addition of 2-mercaptoethanol to a final concentration of 1 m$M$. The sample was then dialyzed against column buffer consisting of 8 $M$ urea, 50 m$M$ sodium acetate, pH 4.8 and, after complete equilibration in this buffer, was chromatographed on a CM-cellulose column (2 × 7 cm; CM-52, Whatman) at 25°. Procollagen chains were eluted at a flow rate of 70 ml/hr using a linear gradient from 0 to 60 m$M$ NaCl in a total volume of 400 ml. The following chains were resolved, in order of increasing ionic strength: $p_N\alpha1$(I), pro$\alpha$1(I), $\alpha$1(I), $p_C\alpha1$(I), and $p_N\alpha2$(I). $p_C\alpha2$(I) coeluted with pro$\alpha$2(I) at the end of the gradient.

COMMENT. These chains can be dialyzed at 4° against either 0.1 $N$ acetic acid in the presence of 0.5 $\mu$g/ml pepstatin or 1 m$M$ ammonium bicarbonate containing 0.2 m$M$ PMSF, 2.5 m$M$ EDTA, and 10 m$M$ NEM, prior to lyophilization. Recoveries from ion-exchange columns usually range from 50 to 85% of the applied counts per minute. We have observed enhanced recoveries when the following precautions were taken: (*a*) pretreatment of the ion-exchange resin with an expendable solution of collagen; (*b*) washing of the dialysis tubing (which had been boiled and stored in the presence of an antibacterial agent such as octanoic acid) with a 1%

solution of type I collagen or BSA; (c) use of an ultrapure grade of urea immediately after deionization and of freshly prepared solutions of protease inhibitors; and (d) collection of column fractions in siliconized glass tubes.

CM-cellulose chromatography, conducted under conditions usually employed for the separation of collagen chains, has also been applied to procollagen chains with varying degrees of success. Resolution of pro- and p-collagen chains and collagen $\alpha$ chains of a single type is often incomplete, and the position of elution has been found to be highly dependent on the species of origin of the protein. However, additional purification from noncollagenous components and other procollagens was achieved for human type IV procollagen.[7] The lyophilized sample was dissolved (approximately 100 $\mu$g/1-2 ml) in 0.1 $M$ acetic acid containing 1 $\mu$g of pepstatin per milliliter and dialyzed against the column buffer (6 $M$ urea, 40 m$M$ sodium acetate, pH 4.8) at room temperature. Alternatively, samples in other buffers can be dialyzed directly into the column buffer. After clarification by centrifugation, the sample was heated at 42° for 30 min and applied to a CM-cellulose column (0.9 × 10 cm) at 42°. Type IV procollagen chains were eluted as a single peak using a linear gradient from 0 to 100 m$M$ NaCl in a total volume of 200 ml of the column buffer. Under these conditions separation of pro$\alpha$1(IV) and pro$\alpha$2(IV) was not achieved.

*Molecular Sieve Chromatography*

Molecular exclusion chromatography has been shown to be very effective for the purification of procollagen from non-disulfide-bonded aggregates, collagen $\alpha$ chains, and noncollagenous contaminants. However, owing to the tendency of collagenous proteins to aggregate in nondissociating buffers and to adsorb to solid supports such as Sephadex, procollagen must be chromatographed after denaturation. Depending on the molecular size of the contaminants to be removed, the unfolded procollagen can be chromatographed either unreduced (as a triple-stranded $\gamma$ component) or after disulfide bond reduction (as pro- or p-collagen chains).

Lyophilized procollagen was dissolved at a concentration of 1-3 mg/ml in a maximum total volume of 3 ml of 0.1 $M$ acetic acid containing 1 $\mu$g of pepstatin per milliliter at 4°. Alternatively, 0.2 to 1 × $10^6$ cpm/ml are required for detection and subsequent analysis. The sample was dialyzed at room temperature against 1 $M$ CaCl$_2$ (prepared by adsorption on activated charcoal), 50 m$M$ Tris-HCl, pH 7.5, containing 0.2 m$M$ PMSF and 2.5 m$M$ EDTA and was subsequently clarified by centrifugation. The solution was heated at 45° for 15 min and applied to a 6% agarose column (BioGel A-5m, 200–400 mesh, Bio-Rad Laboratories; 1.5 × 165 cm) at a flow rate of 10 ml (3 fractions) per hour. If procollagen is to be chromatographed in the reduced state, samples are first reduced and alkylated as described in the preceding subsection on ion-exchange chromatography, followed by dialysis against the column buffer. As collagenous proteins display higher hydrodynamic volumes relative to globular proteins, it is recommended that the column be calibrated with type I collagen molecular weight standards: $\alpha$ (95,000), $\beta$ (190,000), and $\gamma$ (285,000).

## Affinity Chromatography

Additional purification of native procollagens after DEAE-cellulose chromatography has been achieved using a concanavalin A-Sepharose column.[30] Type I procollagen was precipitated from DEAE-cellulose column fractions using 50% ethanol and dialyzed against 0.4 $M$ NaCl, 50 m$M$ Tris-HCl, pH 7.5, containing 10 m$M$ NEM and 0.3 m$M$ PMSF. The sample was made 5 m$M$ in $CaCl_2$ and was applied to a concanavalin A column (Pharmacia; 1.5 × 3.0 cm) equilibrated with the above buffer. After the baseline absorbance returned to the initial level, 0.5 $M$ $\alpha$-methyl-D-mannoside in the same buffer was passed through the column. After five fractions were collected (1 ml each), procollagen was eluted using a buffer containing 1 $M$ $CaCl_2$, 50 m$M$ Tris-HCl (pH 7.5). This procedure was found to be particularly useful in removing contaminating nonglycoprotein material from type III procollagen preparations.[30]

In certain systems such as rat, pro- and p-collagens and/or collagens are not sufficiently resolved on DEAE-cellulose. Separation has been achieved, however, using covalent chromatography on activated thiol-Sepharose 4B, a technique that allows disulfide-bonded components (pro- and $p_C$collagen) to be resolved from those lacking cysteine ($p_N$-collagen and collagen).[35,44] Twenty milligrams of type II procollagen were dissolved in 5 ml of 0.3 $M$ NaCl, 1 $M$ urea, 1 m$M$ EDTA, and 100 m$M$ Tris-HCl, pH 8.0. Dithiothreitol was added to a final concentration of 5 m$M$, and the solution was heated at 60° for 30 min. The sample was centrifuged briefly, and DTT was removed by passage over a Sephadex G-25 column (2.5 × 35 cm) equilibrated in the above buffer. The reduced procollagen was then mixed by gentle shaking with 5 g of activated thiol-Sepharose 4B (Pharmacia; preswollen and washed with the same buffer) for 2 hr at 42°. The suspension was poured into a prewarmed 1.6 × 20 column, and unbound type II $p_N$collagen and collagen were eluted with the starting buffer at a maximum flow rate of 100 ml/hr. The column was washed

[44] B. C. Sykes, *FEBS Lett.* **61**, 180 (1976).

briefly with starting buffer containing 1 $M$ NaCl and was subsequently equilibrated in starting buffer containing 5 m$M$ DTT. The flow was stopped for 40 min to allow complete sulfhydryl exchange, and upon resumption of flow, type II pro- and p$_C$collagens were eluted from the column. This procedure could also be applied to type I procollagen preparations, but not to those containing type III, as disulfide bonds link both the pro$\alpha$ and $\alpha$ chains of this collagen.

*Sedimentation Velocity Centrifugation*

This technique has been successfully adapted to the purification of several procollagen types.[33,45] The method can be used under both native and denaturing conditions, permits high recoveries of radiolabeled protein, and, when used in conjunction with limited protease treatment and SDS-PAGE, provides detailed information about the structure of procollagen types.

Type V procollagen isolated by DEAE-cellulose chromatography was further purified by sedimentation velocity centrifugation under nondenaturing conditions.[17] Column fractions that were eluted at low ionic strength (see preceding discussion of type V procollagen isolation) were dialyzed at 4° against 7% NaCl, 50 m$M$ Tris-HCl, 0.1% Triton X-100, pH 7.5, and procollagens were precipitated from this solution by the addition of carrier collagen (types I and III at 0.1 mg/ml) and acetic acid to a final concentration of 0.5 $M$. The precipitates were dissolved in 4 $M$ urea, 0.5 $M$ Tris-HCl, 0.1% Triton X-100, pH 8.5, and dialyzed first against 1 $M$ NaCl, 50 m$M$ Tris-HCl, 0.1% Triton X-100, pH 7.5, followed by dialysis against the same buffer containing 0.5 $M$ NaCl and 2 $M$ urea (buffer a). A linear sucrose gradient (3.5 ml; 5 to 20% w/v) was prepared in buffer a and placed over a 60% sucrose pad; 200 $\mu$l of the sample were applied onto the gradient and centrifuged at 4° (56,000 rpm for 19 hr) in a Beckman SW60 rotor. Type V procollagen sedimented as a single peak approximately 1.2 times faster than native type I procollagen.

Sedimentation velocity centrifugation under denaturing conditions has been used to separate chick types I and III procollagen, which in the native state do not resolve by this technique or by DEAE-cellulose chromatography.[45] A concentrated protein fraction containing radiolabeled types I and III procollagen was dialyzed against 6 $M$ urea, 50 m$M$ Tris-HCl, 100 m$M$ NaCl, 0.1% Triton X-100, pH 7.5, at 20°. After denaturation at 40° for 30 min, the sample was loaded onto a 5 to 20% (w/v) linear sucrose gradient in the same buffer and centrifuged in an SW60 Beckman rotor (56,000 rpm for 19 hr) at 22°. Proteins in peak-

[45] L. I. Fessler and J. H. Fessler, *J. Biol. Chem.* **254**, 233 (1979).

containing fractions were identified by SDS-PAGE, and by comparison of sedimentation coefficients with those of native and denatured standard procollagens.

### Identification of Procollagens and Criteria of Purity

The species from which the procollagen was purified, as well as the extent of posttranslational modifications, which include prolyl and lysyl hydroxylation and glycosylation, are at least partially responsible for the behavior of procollagen (both in native and denatured states) on ion-exchange chromatography. Because of the variability in elution position for any given procollagen type, it is not sufficient to identify a procollagen using this criterion alone. For this reason, other techniques that depend on different characteristics of the protein are commonly used. These include mobility on SDS-PAGE before and after reduction and after exposure to bacterial collagenase, amino acid composition, peptide mapping (usually one-dimensional) using a variety of chemical and enzymic degradations, immunochemical identification using specific antibodies to procollagen types, and analysis of SLS-crystallite patterns, which are a direct consequence of the primary sequence and are therefore specific for each procollagen type. The latter two procedures are discussed in detail in chapters [5] and [22], respectively, of this volume. We have chosen to describe several straightforward techniques that are particularly useful for the characterization of procollagen chains.

### Analysis by SDS-PAGE

Although collagenous proteins migrate anomalously on SDS gels (this volume [19]), their migration relative to one another and to other proteins is reasonably constant and can provide a great deal of information, especially in conjunction with the use of reducing agents and bacterial collagenase.[46] Figure 1 (lanes 1 and 3) shows the trichloroacetic acid (TCA)-precipitable proteins that are present in smooth muscle cell culture medium, and lane 2 shows the same proteins after digestion with bacterial collagenase. Two separate 1-ml aliquots of radiolabeled culture medium, which was removed from the cells in the presence of protease inhibitors and clarified (as described earlier), were chilled on ice in plastic microfuge tubes. To each tube was added 50 $\mu$g of BSA (stock solution: 1% by weight in Tris-saline), 10 $\mu$l of pepstatin and cold 100% TCA to a final concentration of 10% (v/v). After 1–3 hr on ice, the samples were pelleted in a microfuge and the supernatants were removed by careful aspiration. To one tube was added 50 $\mu$l of 1 $N$ NaOH plus 100 $\mu$l of collagenase

---

[46] B. Peterkofsky, *Arch. Biochem. Biophys.* **152**, 318 (1972).

buffer (1 $M$ HEPES, pH 7.4, 50 m$M$ CaCl$_2$, 2 m$M$ PMSF), and the contents were vortexed. Water containing 10 m$M$ NEM was added to produce a final volume of 1 ml, and the solution was vortexed and clarified in a microfuge. All buffers for this procedure must be sterilized by filtration (0.45 μm, Nalgene) before use. After the pH was checked (readjusted to pH 7.4 with 0.2 $N$ NaOH if necessary), 5–10 μl of the solution were counted in 5 ml of Aquasol. Approximately 100,000 cpm were pipetted into each of two new microfuge tubes, and the volume was adjusted to 1 ml (control) or to 950 μl (sample to be digested with enzyme) with a 1:10 dilution (using H$_2$O containing 10 m$M$ NEM) of collagenase buffer. Bacterial collagenase (Advance Biofactures, Form III, 100 μg/vial) was dissolved at 0° by injecting 1 ml of 0.33 $M$ calcium acetate, 0.025 $M$ Tris-HCl buffer, pH 7.4, through the rubber stopper, and the mixture was gently tipped back and forth until the protein appeared to be in solution; 50 μl of collagenase solution were added to the sample, and both tubes were incubated at 37° for 2 hr with intermittent vortexing. The samples were then chilled to 0° and precipitated using 10% TCA as described above, omitting the addition of BSA. All three samples [the original TCA-precipitated material (1), incubated control (2), and collagenase-treated sample (3)] were subsequently dissolved in 5 μl of SDS-PAGE sample buffer,[27] heated at 100° for 2 min, and spun briefly (5 sec) in a microfuge. Aliquots (5 μl) of each sample were then counted in a scintillation counter to assess the percentage of collagenase-sensitive counts. Half of sample 1 was run on SDS-PAGE unreduced (Fig. 1, lane 3); the remainder of sample 1 and sample 3 were reduced by the addition of 5 μl of 500 m$M$ DTT and resolved by SDS-PAGE to produce the patterns shown in Fig. 1. From these patterns it can be concluded that smooth muscle cells synthesize a high proportion of types I and III (and possibly type IV) procollagen.

*Amino Acid Composition*

In Table III are listed representative amino acid compositions[7,36,37,43,47] for several pro- and p-collagen chain types, as well as for the type IV procollagen molecule, which, in this preparation, contained both the proα1(IV) and proα2(IV) chains. In general, procollagens can be distinguished from collagens by ($a$) a lower content of the four amino acids that are most characteristic of collagen (Gly, Pro + Hyp, and Hyl); ($b$) an increase in Cys; ($c$) an increase in acidic residues; and ($d$) generally elevated levels of Ile + Leu, His, and Tyr. Although not measured by amino acid analysis, tryptophan has been identified in several procollagens by incorporation of the radiolabeled amino acid. Types II and IV procolla-

---

[47] J. Uitto, H.-P. Hoffmann, and D. J. Prockop, *Arch. Biochem. Biophys.* **179**, 654 (1977).

TABLE III
AMINO ACID COMPOSITION OF SEVERAL PROCOLLAGEN TYPES AND DERIVATIVE CHAINS[a]

| Amino acid | proα1(I)[b] | [proα1(II)]$_3$[c] | p$_C$α1(III)[d] | p$_N$α1(III)[e] | [proα1(IV)]$_2$proα2[f] |
|---|---|---|---|---|---|
| 3-Hyp | NR[g] | NR | NR | NR | 6.7 |
| 4-Hyp | 68 | 77 | 113 | 105 | 93 |
| Asp | 61 | 71 | 61 | 57 | 54 |
| Thr | 31 | 40 | 18 | 21 | 30 |
| Ser | 44 | 42 | 46 | 51 | 49 |
| Glu | 98 | 96 | 80 | 81 | 91 |
| Pro | 100 | 95 | 100 | 108 | 69 |
| Gly | 274 | 280 | 323 | 303 | 269 |
| Ala | 107 | 88 | 65 | 82 | 49 |
| Cys/2 | 8.9 | NR | 8 | 8.7 | 13 |
| Val | 25 | 28 | 23 | 23 | 36 |
| Met | 11 | 9.6 | 6 | 5.4 | 11 |
| Ile | 22 | 20 | 17 | 21 | 32 |
| Leu | 34 | 34 | 23 | 26 | 63 |
| Tyr | 10 | 6.7 | 9 | 8.7 | 15 |
| Phe | 17 | 18 | 13 | 12 | 33 |
| Hyl | 5.6 | 19 | 5 | 4.3 | 33 |
| Lys | 32 | 24 | 34 | 33 | 20 |
| His | 7.2 | 5.4 | 11 | 8.4 | 13 |
| Arg | 45 | 48 | 45 | 42 | 37 |

[a] Values are expressed as residues per 1000.
[b] Chick bone.[43]
[c] Chick embryo cartilage.[47]
[d] Rat skin.[37]
[e] Fetal bovine skin.[36]
[f] Human amniotic fluid cells.[7]
[g] NR, not reported.

gens exhibit the highest levels of Hyl and are also more highly glycosylated than the other types. A distinguishing feature of type III procollagen is its Gly content, which is the highest of all the procollagens, whereas type IV procollagen has the lowest levels of Gly, Ala, and Arg. Other distinguishing characteristics of collagen types have been described in this volume [2].

## Peptide Mapping

One very reliable approach to the identification and characterization of procollagens utilizes specific and limited proteolytic or nonenzymic cleavage of the protein, in conjunction with a method for resolving the cleavage products, usually SDS-PAGE or ion-exchange chromatography. These procedures are relatively quick, require very small amounts of protein, and provide a sound basis for identification of a procollagen type.

*Pepsin Treatment.* As summarized earlier, pepsin removes the globular N- and C-propeptides and the short, nontriple helical extensions (telopeptides) from native procollagen molecules, leaving the triple-helical region, which is composed of three collagen $\alpha$ chains, intact. Therefore, native type I procollagen [pro$\alpha$1(I)]$_2$pro$\alpha$2(I) is cleaved to native collagen [$\alpha$1(I)]$_2\alpha$2(I). If the molecule is denatured, then the chains are rapidly degraded to small fragments. Figure 1B illustrates the results of pepsin treatment of type I procollagen (lane 4) and of type III procollagen (lanes 5 and 6). Both preparations were isolated after DEAE-cellulose chromatography of smooth muscle cell culture medium protein (see lane 1). The characteristics of each collagen type after purification from pepsin-digested tissues have been discussed in this volume [2]. In this section we describe the methods for cleavage of procollagen with pepsin and the identification of the cleavage products.

A lyophilized sample containing $^3$H-labeled procollagen was dissolved at 100–500 $\mu$g/ml in 0.5 $M$ acetic acid containing 100 $\mu$g of type I carrier collagen per milliliter. This and all subsequent procedures were performed at 4°. Alternatively, samples dissolved in other buffers can be dialyzed directly into 0.5 $M$ acetic acid at 4°. The aliquot should contain 2.5 to 5 × $10^5$ cpm in procollagen. Pepsin (Worthington, 2 × recrystallized) was dissolved in $H_2O$ at 10 mg/ml and added to the procollagen to produce a final concentration of 10–50 $\mu$g/ml or a 1 : 10 enzyme to substrate ratio by weight. The reaction mixture was stirred at 4° for 21 hr and was subsequently centrifuged (48,000 $g$) for 30 min. Solid NaCl was added to the clarified pepsin supernatant to a final concentration of 1.8 $M$, and the collagen that precipitated after 3–8 hr was collected by centrifugation and redissolved in 1–3 ml of 1 $M$ NaCl, 50 m$M$ Tris-HCl, pH 7.5, containing a twofold molar excess of pepstatin over pepsin. The solution was dialyzed against the same buffer without pepstatin and was subsequently clarified by centrifugation. Solid NaCl was added to a final concentration of 4.5 $M$, and the precipitated collagens were collected by centrifugation, dissolved in and dialyzed against 0.1 $M$ acetic acid, and lyophilized.

Approximately 25 to 100 × $10^3$ cpm were loaded on an SDS-gel both in the presence and in the absence of reducing agents, as described above for analysis by SDS-PAGE. Pepsin treatment of type I procollagen produced two nonreducible bands (Fig. 1, lane 4) corresponding to $\alpha$1 and $\alpha$2 chains of type I collagen ($M_r$ 95,000). Similar treatment of type II procollagen produced a single, nondisulfide-bonded $\alpha$ chain, and a triple helical disulfide-bonded molecule was seen after digestion of type III procollagen; the latter, after reduction, migrated like an $\alpha$ chain (Fig. 1, lanes 5 and 6). Pepsin treatment of type IV procollagen can produce several components, the exact nature of which differs with the species, tissue of origin, and extent of digestion[7,48]; this procedure is therefore not recommended as a means of identifying this procollagen type. Pepsin digestion of type V procollagen ([pro$\alpha$1(V)]$_2$pro$\alpha$2(V)) yielded two nondisulfide-bonded components of $M_r$ 135,000 ($\alpha$1) and 105,000 ($\alpha$2) on SDS-PAGE.[17]

COMMENT. Uitto et al.[30] have indicated that optimal conditions for complete digestion of types I and III procollagen from cell culture material were 3 hr at either 4° or 15° using 100 $\mu$g of pepsin per milliliter. As found previously by Burke et al.,[24] pepsin can selectively cleave type III collagen, particularly when incubation times are increased, leading to an underestimation of the amount of type III procollagen in type III/I mixtures.

CNBr Cleavage. The specific cleavage of methionyl peptide bonds by CNBr has been used to generate limited, one-dimensional peptide maps that are unique for each procollagen chain type[1] (for a review of this technique, see also Piez[49] and Chapter [2] of this volume). The peptides have been resolved either on CM-cellulose[50] or by SDS-PAGE.[24]

In Figure 3 are shown the cleavage products of types I and III procollagen. Lyophilized procollagen, fractionated on DEAE-cellulose, was dissolved at 1 mg/ml (for nonradioactive samples) or at 1 to 2.5 × 10$^5$ cpm/ml in a siliconized, heavy-walled centrifuge tube in 70% formic acid (98–100%, Pierce Chemical Co.), and the solution was aerated with N$_2$ for several minutes. Clear crystals of fresh CNBr (Pierce Chemical Co.) were then added to a final concentration of 20 mg/ml. The solution was again flushed with N$_2$ (1–2 min), tightly stoppered, vortexed until the CNBr was dissolved, and incubated at 30° for 4–5 hr with intermittent vortexing. To terminate the reaction, 4 ml of H$_2$O were added, and the solution was frozen in a Dry Ice–ethanol bath and lyophilized in a flask that was immersed in the bath and covered with aluminum foil. When the sample was dry, it was resuspended in 2 ml of H$_2$O, frozen, and lyophilized again. It was then dissolved in SDS-PAGE sample buffer containing 50 m$M$ DTT and resolved on a 12.5% separating gel in conjunction with a 5% stacking gel.[27] Approximately 5–10 $\mu$g of protein per band (or 50–200 $\mu$g per sample) or 1 to 2.5 × 10$^5$ cpm of radioactive protein were loaded in each well prior to electrophoresis.

Alternatively, CNBr peptides were chromatographed on a CM-cellulose column (9 × 120 mm) at 45° in a 20 m$M$ sodium formate, 30 m$M$

[48] E. C. Crouch, H. Sage, and P. Bornstein, *Proc. Natl. Acad. Sci. U.S.A.* **77**, 745 (1980).
[49] K. A. Piez, *in* "Biochemistry of Collagen" (G. N. Ramachandran and A. H. Reddi, eds.), pp. 1–44. Plenum, New York, 1976.
[50] J. R. Lichtenstein, P. H. Byers, B. D. Smith, and G. R. Martin, *Biochemistry* **14**, 1589 (1975).

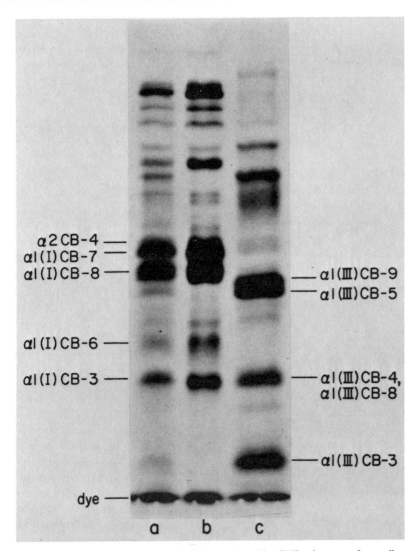

FIG. 3. Analysis by SDS-PAGE of peptides produced by CNBr cleavage of procollagen. [$^3$H]Proline-labeled procollagens from monkey skin fibroblast cell culture medium were fractionated on DEAE-cellulose prior to cleavage with CNBr. Cleavage products were resolved on a 12.5% polyacrylamide gel in the presence of 50 m$M$ DTT and were visualized by fluorescence autoradiography. Lane a: protein prior to DEAE-cellulose chromatography; lane b: type I procollagen; lane c: type III procollagen. Major cleavage products are identified. Reprinted from J. M. Burke, G. Balian, R. Ross, and P. Bornstein, *Biochemistry*, **16**, 3243 (1977), with permission.

NaCl, pH 3.8 buffer, and were resolved using a linear NaCl gradient from 30 to 140 m$M$ in a total volume of 100 ml, at a flow rate of 15 ml/hr. The peptide patterns that result from this cleavage are sufficiently specific for each procollagen type to permit direct comparison of the proteins in question. Comparisons, however, should be made among procollagen types from the same species.

COMMENT. In some instances, to effect more complete cleavage of certain procollagens (especially type IV procollagen), reduction of disulfide bonds with or without alkylation, as already described in this chapter, is recommended.

*Mast Cell Protease Cleavage.* Mast cell protease, an enzyme with chymotrypsin-like specificity, has been purified from rat skeletal muscle and extensively characterized.[51] It was described as producing limited but highly specific cleavages in denatured procollagen and collagen chains, which could then be compared by SDS-PAGE.[13,52] In addition, the enzyme has been used to characterize native procollagen types.[48] Examples of digestions of both native and denatured substrates are shown for types I, III, and IV procollagens in Fig. 4.

Lyophilized, purified procollagen that has been reduced and alkylated (as described earlier) was dissolved at 1 mg/ml (for nonradioactive samples) or at $5 \times 10^5$ cpm per 100 $\mu$l of Tris-saline buffer at 4°. Alternatively, better solubilization was achieved by initial dissolution at the same concentration in 0.1 $M$ acetic acid followed by dialysis against Tris-saline. Owing to the absence of protease inhibitors, it is recommended that the dialysis step be accomplished as quickly as possible (four changes in a maximum of 24 hr). The sample was transferred to a microfuge tube and clarified by centrifugation; 25-$\mu$l aliquots were removed to separate microfuge tubes, and the samples were denatured at 45° for 10 min. To several tubes were then added equal aliquots of mast cell protease (1 mg/ml in 0.5 $M$ Tris-HCl, pH 7.5) to produce a 1:100 enzyme to substrate ratio by weight, or 0.2–1.0 $\mu$g of protease per radioactive sample. It is recommended that the enzyme be assayed immediately before use to determine the amount of active protease present.[51] To one tube was added an aliquot of Tris-saline equal to that used for the enzyme. Because these digests are not carried to completion, the samples were incubated at 37° for several time intervals. We have found a span of 2 hr in 15- or 30-min increments to be useful. To terminate each incubation, the tubes were chilled on ice. When all digestions were completed, the samples were equilibrated at room temperature, and 25 $\mu$l of 2 × SDS-PAGE sample buffer containing

[51] R. G. Woodbury, M. Everitt, Y. Sanada, N. Katunuma, D. Lagunoff, and H. Neurath, *Proc. Natl. Acad. Sci. U.S.A.* **75**, 5311 (1978).
[52] H. Sage, R. G. Woodbury, and P. Bornstein, *J. Biol. Chem.* **254**, 9893 (1979).

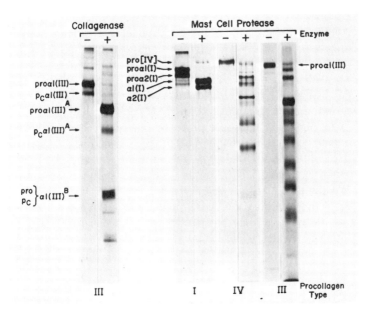

FIG. 4. Characterization of procollagens by cleavage with vertebrate collagenase and mast cell protease. Purified [³H]proline-labeled procollagens were incubated with enzyme as described in the text, and the cleavage products were resolved on 7.5%/12.5% or 10% (mast cell protease) SDS-polyacrylamide gels in the presence of 50 mM DTT and visualized by fluorescence autoradiography. All procollagens were cleaved as native procollagen molecules with the exception of type III procollagen, which was incubated with mast cell protease after denaturation. Intact procollagen chains and some cleavage products are identified.

50 mM DTT[27] was added to each microfuge tube. The samples were heated for 2–5 min at 100°, spun briefly in a microfuge (5 sec), and loaded onto a 5% stacking–10% separating SDS-polyacrylamide slab gel. The cleavage patterns generated for each procollagen type are unique[1]; however, it is recommended that comparisons be made among procollagens from the same species.

Mast cell protease cleaves native types I and III procollagens to collagens (Fig. 4, I). The only native procollagen (or collagen) that is cleaved within the triple-helical region is type IV procollagen[48,52] (Fig. 4, IV). Presumably such cleavages occur at discontinuities in the helix that exist in this procollagen.[8] These differences are useful for characterizing procollagen types. In order to cleave native substrates, the denaturation step described above was omitted and incubation in the presence of enzyme was performed at room temperature for 24 hr.

*Digestion with Vertebrate Collagenase.* Vertebrate collagenase produces one specific cleavage at the same location in each chain in the native procollagen or collagen molecule, resulting in the production of two fragments that comprise approximately 75% ("A" fragment) and 25% ("B" fragment) of the original triple-helical sequence (for a review, see Bornstein and Traub[3]). It has been found, however, that the highly purified enzyme from human skin does not cleave types IV and V (pro)collagen.[7,53] Collagenase digestion has been used to characterize procollagen types because the fragments are readily identifiable on the basis of their size and distribution of disulfide bonds.[30] Figure 4 illustrates the cleavage of type III pro- and $p_C$collagen with human skin collagenase (source: Dr. A. Eisen, Washington University, St. Louis, Missouri). Because $p_C$collagen lacks the N-propeptides, the A fragment ($p_C\alpha 1(III)^A$) is correspondingly smaller than that generated from intact procollagen (pro$\alpha 1(III)^A$) and migrates similarly to an $\alpha 1(III)$ chain on SDS-PAGE. However, since both pro- and $p_C$collagen contain the C-propeptides, the B fragment is the same size for both forms. In the case of type I procollagen, both the A and B fragments are seen as doublets corresponding to the pro$\alpha 1$ and pro$\alpha 2$ chain, respectively.

Lyophilized native procollagen was dissolved at 1 mg/ml or $2 \times 10^5$ cpm/100 $\mu$l in 0.1 $N$ acetic acid at 4° and dialyzed into 100 m$M$ HEPES buffer, pH 7.4, containing 5 m$M$ CaCl$_2$, and 0.2 m$M$ PMSF. The sample was clarified by centrifugation, and equal aliquots of 25 $\mu$l each were placed into two microfuge tubes; 2–5 $\mu$l of collagenase solution (representing approximately a final 1 : 100 active enzyme to substrate ratio by weight) was added to one tube, and an equal aliquot of buffer was added to the control. Samples were incubated for 24 hr at room temperature, diluted with 25 $\mu$l of 2 $\times$ SDS-PAGE sample buffer containing 50 m$M$ DTT, and heated for 2–5 min at 100°. The samples were then spun briefly in a microfuge and resolved on an SDS-slab gel composed of a 4% stacking gel and a composite 7.5%/12.5% separating gel.

*Two-Dimensional Peptide Mapping Using $^{125}I$-Labeled Procollagen.* A method has been described that adapts a previously developed two-dimensional peptide mapping technique using radioiodinated protein[54] to procollagen and collagen.[55] This method offers two major advantages: it is very sensitive, requiring less than 5 $\mu$g of protein, and it does not require prior purification of the protein. Several representative maps are shown in Fig. 5; they are specific for each procollagen type and can distinguish sequence differences between procollagen intermediates (panels A and B).

[53] H. Sage and P. Bornstein, *Biochemistry* **18**, 3815 (1979).
[54] J. H. Elder, R. A. Pickett, J. Hampton, and R. A. Lerner, *J. Biol. Chem.* **252**, 6510 (1977).
[55] H. Sage, P. Pritzl, and P. Bornstein, *Collagen Rel. Res.* **1**, 3 (1981).

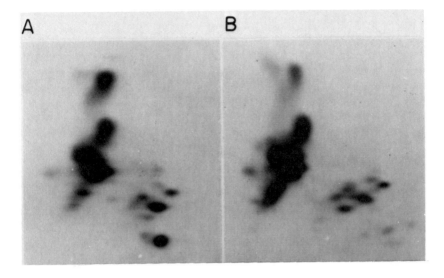
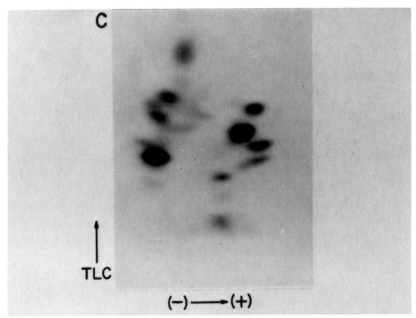

FIG. 5. Two-dimensional peptide mapping of $^{125}$I-labeled procollagens. Procollagens were initially resolved by SDS-PAGE under reducing conditions, and the protein-containing bands, as visualized by Coomassie Blue staining, were cut from the gel and radioiodinated. Washed gel slices were then incubated with proteinase K, and the resulting cleavage products were subjected to thin-layer electrophoresis followed by chromatography on precoated cellulose plates. Directions of electrophoresis, $(-) \rightarrow (+)$, and chromatography (TLC) are indicated. Origin is in lower left corner in each map. (A) proα1(I), human fibroblasts; (B) p$_C$α1(I), human fibroblasts; (C) type IV procollagen [proα1(IV) + proα2(IV)], human amniotic fluid cells.

Procollagen samples were resolved by SDS-PAGE, usually under reducing conditions, and the gels were stained with Coomassie blue. The band of interest was excised from the gel, washed with vigorous shaking in four 1-ml aliquots of 25% isopropanol followed by three 1-ml aliquots of 10% methanol, and dried by lyophilization. Bolton–Hunter reagent ($[^{125}I]p$-hydroxyphenylpropionic acid, $N$-hydroxysuccinimide ester, approximately 4000 Ci/mmol, New England Nuclear), supplied by the manufacturer in aliquots of 0.25 mCi per combi vial and dissolved in benzene, was evaporated to dryness under a stream of $N_2$. The dried gel slice, containing 5–10 $\mu$g of protein, was rehydrated in 250 $\mu$l of 0.1 $M$ calcium acetate buffer, pH 8.5, and transferred to the combi vial with an additional 250 $\mu$l of buffer. The vial was capped and shaken vigorously at 0° for 2 hr; 100 $\mu$l of 0.2 $M$ glycine in the calcium acetate buffer were then added, and the solution was agitated at 0° for another 30 min. The gel slice was rinsed three times with 10% methanol and subsequently placed in a microtiter plate with small holes drilled through the wells and the cover. Dialysis was performed against 10% methanol at room temperature, prior to lyophilization.

A solution of proteinase K (EM Biochemicals) was prepared at a concentration of 40 $\mu$g/ml in 0.05 $M$ ammonium bicarbonate, pH 8.0. The $^{125}$I-labeled gel slice was transferred to a siliconized glass tube and incubated with 0.5 ml of the enzyme solution at 37° for 18 hr. The reaction was terminated by transfer of the solution to a siliconized 3-ml Pyrex tube and lyophilization.

The lyophilized sample was dissolved in 10 $\mu$l of buffer (acetic acid–formic acid–$H_2O$, 15:5:80), centrifuged briefly, and counted in a $\gamma$ counter; $2.5 \times 10^5$ cpm were spotted in the lower left corner of a 10 × 10 cm precoated cellulose plate (without fluorescent indicator, EM Biochemicals). A tracer dye (2% Orange G and 1% acid fuschin in the same buffer) was applied in the upper right corner, and high voltage electrophoresis was conducted at 15° at 400 V (15–20 mA) for approximately 1 hr, until the dye reached a preset mark. After the plate was thoroughly dried at room temperature, thin-layer chromatography was conducted in the second dimension in a tank equilibrated with butanol–pyridine–acetic acid–$H_2O$ (32.5:25:5:20) containing 7% 2,5-diphenyloxazole (w/v). $\epsilon$-DNP lysine was spotted at the solvent level, and the run was terminated when the dye was 5.5 cm from the bottom of the plate and the solvent was near the top. The plate was dried at room temperature and subsequently exposed to Kodak BB-5 X-ray film in a cassette with a DuPont Chronex Hiplus intensifying screen at $-70°$ for 4–8 hr.

These peptide maps should be compared with those of other procollagen chain types produced using the same technique. The sensitivity of the method would predict that minor differences in maps of the same procollagen would occur. However, we have found a high degree of reproducibility among maps of the same protein produced at different times.

COMMENT. V-8 protease from *Staphylococcus aureus* (Miles Laboratories) will also cleave protein in gel slices and has been used to generate unique one-dimensional peptide maps for type IV[12] and type V procollagen.[17]

### Acknowledgments

We thank colleagues who supplied us with procedures and experimental data prior to publication. Special appreciation is due Drs. C. Kumamoto and J. Fessler for making their manuscript on type V procollagen available to us. We also thank Dr. S. Schwartz and S. Wang for their procedure for isolation of bovine aortic smooth muscle cells, Dr. G. Raugi for the data shown in Fig. 2, P. Pritzl for the $^{125}$I-peptide maps, and D. Stewart for typing the manuscript.

Contributions of experimental data from this laboratory were supported in part by NIH grant AM 11248, DE 02600, and HL 18645. H. S. was a recipient of a fellowship from R. J. Reynolds Industries, Inc.

## [5] Characterization of Fibrous Forms of Collagen[1]

### By BARBARA BRODSKY and ERIC F. EIKENBERRY

In higher animals collagen forms a major structural component of the connective tissues, including tendon, bone, cartilage, skin, vascular tissues, and basement membranes. Within a given organism, several genetically distinct types of collagen are found. These genetic types have been characterized biochemically, and each kind of tissue is found to have a characteristic composition of collagen types.[2] The most prevalent type, designated type I, is found as the nearly exclusive collagenous component of tendon and bone, whereas in many other tissues, such as skin, vascular tissues, and reticular tissues, type I is found together with a lesser amount of type III collagen. Type II collagen is the major component of cartilage. These three interstitial collagens, types I, II, and III, form characteristic periodic fibrils and are considered to occur largely in fibrous form in the extracellular matrix. Other genetic types, such as type IV collagen in basement membranes and type V collagen, are apparently not found as typical collagen fibrils.

[1] This research was supported by U.S. Public Health Service Research Grant AM 19626 from the National Institute of Arthritis, Metabolism, and Digestive Diseases.
[2] E. J. Miller and S. Gay, this volume [1].

Collagen molecules are composed of three polypeptide chains, designated α-chains, with repeating sequences of the form Gly-X-Y along their length, where X and Y are any amino acid residues. In type I collagen there are two distinct kinds of chains, α1(I) and α2, which occur in a ratio of two to one, respectively. Each chain has 1014 residues present as repeating tripeptide sequences. Type II and type III collagen molecules each contain only one kind of chain, α1(II) and α1(III), respectively. The different types of α1 chains are homologous, especially in the locations of charged amino acid residues, but substitutions are present in as many as 40% of the nonglycine residues.[3,4] In addition to these variations between chains of different genetic type, collagens from different animals are found to contain substitutions in their α1(I) and α2 chains.

The length of a collagen molecule is about 3000 Å, as demonstrated by physical-chemical studies of collagen in solution and by direct measurements of segment-long-spacing forms of collagen (see below) in electron micrographs.[5,6] The conformation of the collagen molecule was determined to be triple-helical from amino acid sequence characteristics and from X-ray data. The rise per residue in each left-handed helical chain is about 2.9 Å, and three of these helices are supercoiled in a right-handed fashion about a central axis to give the final structure. The topic of the collagen triple-helical conformation has been covered in several review articles.[7,8] High-angle X-ray patterns of very high resolution and quality were reported for stretched rat tail tendon, and these data have resulted in the most refined triple-helical model yet presented.[9]

Other physicochemical properties that have been found to be characteristic of the triple-helical conformation in solution include distinctive circular dichroism and optical rotatory dispersion spectra, and a sharp melting transition just above the body temperature of the organism.[10,11] Types II and III collagens have amino acid sequence characteristics similar to those in type I collagen, with glycine as every third residue and a large content of imino acids.[8] The native molecules of these types also have spectra and melting curves similar to those found for type I colla-

[3] W. T. Butler, J. E. Finch, Jr., and E. J. Miller, *Biochemistry* **16**, 4981 (1977).
[4] P. P. Fietzek *in* "Collagen in Health and Disease" (M. Jayson and J. Weiss, eds.). Churchill-Livingstone, Edinburgh, in press.
[5] H. Boedtker and P. Doty, *J. Am. Chem. Soc.* **78**, 4267 (1956).
[6] F. O. Schmitt, J. Gross, and J. H. Highberger, *Exp. Cell Res. Suppl.* **3**, 306 (1955).
[7] G. N. Ramachandran *in* "Biochemistry of Collagen" (G. N. Ramachandran and A. H. Reddi, eds.), p. 103. Academic Press, New York, 1976.
[8] P. Bornstein and W. Traub *in* "The Proteins" (H. Neurath and R. L. Hill, eds.), 3rd ed., Vol. 4, p. 411. Academic Press, New York, 1979.
[9] R. D. B. Fraser, T. P. MacRae, and E. Suzuki, *J. Mol. Biol.* **129**, 463 (1979).
[10] T. Hayashi, S. Curran-Patel, and D. J. Prockop, *Biochemistry* **18**, 4182 (1979).
[11] J. P. Carver and E. R. Blout, *in* "Treatise on Collagen" (G. N. Ramachandran, ed.), Vol. 1, p. 441. Academic Press, New York, 1967.

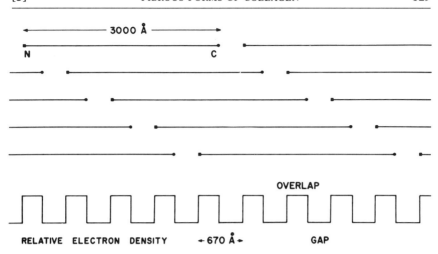

FIG. 1. Schematic illustration of the staggered arrangement of collagen molecules that gives rise to the D-periodic gap:overlap structure. For clarity, the molecules in this structure are shown in two dimensions. The N- and C-telopeptides are represented, respectively, as squares and circles.

gen.[12-14] Such observations suggest that types II and III collagen molecules have a triple-helical conformation similar to that determined for type I collagen. High-angle X-ray patterns have not been obtained for genetic types II and III, so it is possible that their triple-helical structures differ in some features from that of type I collagen.

Although the large central portions of these collagen chains contain glycine as every third residue, the regions at each end of the chain, called telopeptides, do not show this sequence regularity. For example, the bovine $\alpha1(I)$ chain has 1014 residues with glycine at every third position, but has 16 residues at the N-terminal end (N-telopeptide) and 25 residues at its C-terminal end (C-telopeptide) without this regularity.[4,8] The large central portion can adopt a triple-helical conformation, but the conformation of the small telopeptides is unknown. Calculations on the N-telopeptide amino acid sequence have led to a proposed model for this region's secondary structure.[15]

X-Ray diffraction and electron microscopy on collagen fibers in tissues such as tendons, demonstrate a repeating unit of length 670 Å (called D) in the axial fibril direction. This repeating unit is accounted for by the widely accepted one-dimensional model, shown in Fig. 1, in which adjacent molecules, 3000 Å in length, are staggered by 670 Å (or a multiple of 670 Å)

[12] E. J. Miller, *Biochemistry* **10**, 1652 (1971).
[13] T. Fujii and K. Kühn, *Hoppe-Seyler's Z. Physiol. Chem.* **356**, 1793 (1975).
[14] P. P. Fietzek and B. Brodsky, unpublished results.
[15] D. L. Helseth, J. H. Lechner, and A. Veis, *Biopolymers* **18**, 3005 (1979).

with respect to one another. This D stagger leaves a gap between the ends of linearly adjacent molecules because the molecular length is not an exact multiple of the D period, which results in a gap region and an overlap region in each D repeat. The gap region occupies more than half of each D period, and in such a region only 80% as many collagen chains are present as in the overlap region. The existence of these gap and overlap regions is the major low-resolution feature of collagen fibrils and dominates both the image of negatively stained electron micrographs and the low-resolution X-ray pattern. If there were no telopeptides, the gap : overlap ratio would be 0.68 : 0.32, but the presence of telopeptides reduces the size of the gap region. The length of the gap region cannot be calculated a priori because the conformations, and thus the axial extensions, of the N- and C-telopeptides are not known.

In addition to their collagen composition, different connective tissues are characterized by specific noncollagenous components.[16] Collagen comprises as much as 90% of the dry weight of most tendons, but other tissues, such as skin, contain larger amounts of noncollagenous components, such as proteoglycans, glycoproteins, lipids, elastin, and cellular material. Cartilage may contain very large amounts of proteoglycan, ranging from 3% to over 40% of the dry tissue weight, depending on the cartilage type. These components are undoubtedly important in determining the properties of tissues, but it is not clear whether they either influence the formation of collagen fibrils or interact specifically with collagen fibrils after they are formed.

With the knowledge that different connective tissues contain both different collagenous and noncollagenous components, it is necessary to look beyond determining the structure of one collagen in one tissue and to consider whether there are differences in collagen fibril structure in different tissues that are related to the composition and function of the tissue. Specifically the following questions will be addressed:

1. What are the methods of preparation of fibrous forms of collagen, and what are the techniques that are useful for investigating the structure of these forms, both in native tissues and as reprecipitated fibers?
2. Is the structure of type II collagen fibrils different from that of type I?
3. What role does type III collagen play in collagen fibril structure?
4. How does the fibril structure in developing tissues differ from that in mature tissues?

[16] M. B. Mathews, "Connective Tissue: Macromolecular Structure and Evolution." Springer-Verlag, Berlin and New York, 1975.

5. What alterations in fibril structure are found in connective tissue diseases?

Preparation and Investigation of Fibrous Forms of Collagen

Interstitial collagens, types I, II, and III, are found in fibrous form in a variety of tissues. The structure of these fibers and their associations with other tissue components may be studied directly in tissues by light microscopy, electron microscopy, X-ray diffraction, or nuclear magnetic resonance. Alternatively, collagen may be solubilized and purified from tissues, and the purified collagen reprecipitated into a variety of fibrous forms for investigation. Studies of these synthetic fibrous forms, which include native-type fibrils, SLS (segment long spacing), and FLS (fibrous long spacing) aggregates, have given information on the nature of the collagen molecules, the forces involved in forming aggregates, and the ability of purified collagen molecules to determine the fibril structure found in native tissues. As yet, collagens type IV and type V have not been observed in naturally occurring periodic fibers and are difficult to prepare in reprecipitated fibrils by the techniques used for other collagen types.[17]

In this section, we will consider the preparation of fibrous specimens from tissues and from purified collagen and the examination of these specimens by microscopy and by X-ray diffraction. Nuclear magnetic resonance studies are considered separately.[18]

Electron Microscopy

Electron microscopy has been a widely used technique in the study of the structure of collagen fibrils. This technique is applicable to a variety of specimen preparations, including intact tissues, aqueous suspensions of tissue components, and suspensions of reprecipitated fibers. Examples of specific protocols for preparations for electron microscopy will be given here, and references are given to other procedures.

*Suspensions of Collagen Fibrils*

TISSUES

Small pieces of tissue can be mechanically disrupted in 10 or more volumes of buffered physiological saline using a tissue grinder (e.g., Polytron). A drop of the resulting suspension is placed on a carbon-coated

[17] P. Bornstein and H. Sage, *Annu. Rev. Biochem.* **49,** 957 (1980).
[18] D. A. Torchia, this volume [6].

Parlodion film, supported on a standard 200- or 400-mesh grid, to adhere material to the film. At the end of 1 min, the excess liquid is run off the grid with bibulous paper. The adhered material is then stained, either positively or negatively.

A suitable positive stain for collagenous material is exposure first to 0.4% phosphotungstic acid (unbuffered) for 5 min, followed by rinsing the grid in water, followed by exposure to 1% uranyl acetate (unbuffered) for 5 min and again rinsing in water. This procedure stains both negatively and positively charged groups and gives a characteristic axial banding pattern consisting, in type I collagen, of about 12 bands per D repeating period.[19] The positions of these bands have been demonstrated to correlate very well with the positions of the charged residues in the amino acid sequence.[20-23]

Negative staining, or negative contrasting, of collagen may be achieved by treating the adhered specimen for 1 min with 1% sodium phosphotungstate, pH 7.0, after which the excess reagent is run off the grid without rinsing. The contrasting agent infiltrates the open volume of the specimen and highlights the stain-excluding regions. In collagen fibrils, a densely stained band of length approximately 0.6 D is seen, which is interpreted to be the gap region, together with a stain-excluding band of length 0.4 D, which is interpreted to be the overlap region. Exact measurement of these regions is difficult owing to variations in staining and to the presence in some fibrils of strongly stain excluding regions at the ends of the gap, which may be caused by telopeptides. Measurement of fibril diameters can also be made on these specimens, although some flattening of the cylindrical fibrils would be expected.

### Reprecipitated Forms of Collagen

Collagen may be solubilized in neutral salt or acid solutions, or by pepsin digestion, and then purified in the native triple-helical form by a variety of methods.[24] Collagen isolated and purified by any of these methods will have the entire triple-helical regions of the molecules intact, but portions of the N- or C-telopeptide may be removed by proteolytic degradation, even in the neutral salt or acid extraction procedures.[25] It appears

---

[19] R. R. Bruns and J. Gross, *Biopolymers* **13**, 931 (1974).
[20] B. B. Doyle, D. J. S. Hulmes, A. Miller, D. A. D. Parry, K. A. Piez, and J. Woodhead-Galloway, *Proc. R. Soc. London Ser. B.* **187**, 37 (1974).
[21] K. M. Meek, J. A. Chapman, and R. A. Hardcastle, *J. Biol. Chem.* **254**, 10710 (1979).
[22] A. G. Walton, *Biomed. Mater. Symp.* **5**, 409 (1974).
[23] P. P. Fietzek and K. Kühn, *Mol. Cell. Biochem.* **8**, 141 (1975).
[24] E. J. Miller, *Mol. Cell. Biochem.* **13**, 165 (1975).
[25] P. P. Fietzek and K. Kühn, *Int. Rev. Connect. Tissue Res.* **7**, 1 (1976).

to be especially difficult to retain the entire C-telopeptide,[26] though some studies have indicated intact ends in a rapid acid extraction of rat tail tendons.[27]

Purified collagen can be precipitated into different forms, and each of three main forms, SLS, FLS, and native-type fibrils, will be discussed.

*Segment Long Spacing (SLS).* Although not actually a fibrous structure, the SLS form of collagen has been exceptionally useful in helping to characterize the structure of the collagen molecule. The SLS form is obtained as a precipitate upon dialyzing a solution of collagen, at a minimum concentration of 50 μg/ml in 0.01 N acetic acid, against a solution of 0.2% adenosine triphosphate (ATP) in 0.01 N acetic acid. The best aggregates are produced when the ATP is introduced very slowly, an effect that may be enhanced by placing the ATP as a concentrated solution in a second dialysis bag, and not stirring, thereby making the transport of ATP dependent upon diffusion. The precipitate is placed on the grid and stained positively with phosphotungstic acid and uranyl acetate by the procedure outlined above.

Examination of this specimen reveals aggregates 3000 Å long and of varying widths, with the best specimens found as ribbons of width much greater than the molecular length. Fibrous aggregates of such SLS fragments may also be observed. These segments represent molecules lined up in parallel and aggregated laterally in exact register,[6] presumably by the binding of negatively charged phosphate groups to two or more collagen molecules at identical positively charged residues. The positive stain produces a banding pattern which, for type I collagen, has been shown to correspond precisely to the amino acid sequence.[20,28,29] Comparison of SLS aggregates from a variety of animals, including invertebrates, has shown that the distribution of these charged residues has been conserved with evolution.[29]

The SLS crystallites formed from type II collagen have a staining pattern resembling that of type I collagen, but significant differences are found in several regions.[30,31] Several differences in the banding patterns, and thus in the distribution of charged residues were also seen between SLS forms of type III collagen and those of type I collagen.[32,33] SLS

[26] U. Becker, R. Timpl, and K. Kühn, *Eur. J. Biochem.* **28**, 221 (1972).
[27] G. Chandrakasan, D. A. Torchia, and K. A. Piez, *J. Biol. Chem.* **251**, 6062 (1976).
[28] J. A. Chapman, *Connect. Tissue Res.* **2**, 137 (1974).
[29] R. R. Bruns and J. Gross, *Biochemistry* **12**, 808 (1973).
[30] R. L. Trelstad, A. H. Kang, and J. Gross, *Biochemistry* **9**, 4933 (1970).
[31] M. Stark, E. J. Miller, and K. Kühn, *Eur. J. Biochem.* **27**, 192 (1972).
[32] H. Wiedemann, E. Chung, T. Fujii, E. J. Miller, and K. Kühn, *Eur. J. Biochem.* **51**, 363 (1975).
[33] J. Rauterberg and D. B. Von Bassewitz, *Hoppe-Seyler's Z. Physiol. Chem.* **356**, 95 (1975).

aggregates formed from procollagen have led to the visualization of the N- and C-propeptide domains.[34]

*Fibrous Long Spacing (FLS).* The FLS fibers will form when equal parts of a 0.1% solution of collagen in 0.1% acetic acid and a 0.1% solution of chondroitin sulfate in 0.05% acetic acid are mixed, brought to pH 6.5 by the addition of sodium acetate, and dialyzed overnight against tap water.[35] Long flexible negatively charged polymers other than chondroitin sulfate will also interact with collagen in this way to give FLS fibers. These aggregates may be adhered to a support film and stained by the procedures outlined above.

Positively stained FLS fibers, first reported by Highberger *et al.*,[36] appear in electron micrographs with repeating periods of about 2600 Å, 1700 Å, or 900 Å, with a center of symmetry in the banding pattern of each repeating unit. At least four distinct forms of FLS fibers have been observed: FLS I, FLS II, FLS III, FLS IV. All forms have symmetrical banding patterns and thus involve antiparallel relationships between molecules as well as axial stagger between them. The molecular arrangement in each of the FLS forms has been deduced by simulation of the observed positively stained banding pattern using staggered antiparallel arrangements of the $\alpha 1(I)$ sequence, and by using the negatively stained pattern.[37] The occurrence of a pseudo-center of symmetry in the distribution of unpaired positive charges in a collagen molecule suggests that an antiparallel arrangement may be favorable for bridging of these charges by long negatively charged molecules and for minimizing the electrostatic repulsions between unpaired negatively charged residues, which do not show this symmetric arrangement.[37]

*Reprecipitated Native-type Fibrils.* Collagen in solution can re-form fibrils that closely resemble native fibrils under a variety of conditions. Some common procedures to form such native-type fibrils follow.

Dialysis against tap water. A collagen solution at a concentration greater than 0.5 mg/ml is dialyzed first against 0.225 $M$ sodium citrate, pH 3.7, and then against tap water, with several changes, in the cold.[38] If the protein concentration is high enough, a gel forms, which is a suitable specimen both for X-ray diffraction studies (see below) and for electron microscopic observations.

[34] M. L. Tanzer, R. L. Church, J. A. Yaeger, D. E. Wampler, and E.-D. Park, *Proc. Natl. Acad. Sci. U.S.A.* **71**, 3009 (1974).
[35] B. R. Olsen, *Z. Zellforsch. Mikrosk. Anat.* **59**, 199 (1963).
[36] J. H. Highberger, J. Gross, and F. O. Schmitt, *J. Am. Chem. Soc.* **72**, 3321 (1950).
[37] B. B. Doyle, D. W. L. Hukins, D. J. S. Hulmes, A. Miller, and J. Woodhead-Galloway, *J. Mol. Biol.* **91**, 79 (1975).
[38] C. Tkocz and K. Kühn, *Eur. J. Biochem.* **7**, 454 (1969).

Dialysis against neutral phosphate. A solution of collagen in 0.5 N acetic acid upon dialysis at 4° against 0.02 M sodium phosphate, pH 7.4, will, after several changes of buffer, yield a precipitate of fibrils. This procedure is frequently used in purification of collagen, since it separates denatured material that does not precipitate from native molecules that form fibrils.

Neutral salt precipitation. A solution of collagen in 0.5 N acetic acid is dialyzed at 4° against 0.15 M NaCl for 24 hr, at which time the outer dialyzate is neutralized with 1 N NaOH. Dialysis is then continued for 2 days in the cold to produce a fibrous precipitate.[39]

Heat gelation. An acidic solution of collagen is neutralized with sodium hydroxide and brought to 37°, which causes a gel to form.[40] These gels also are suitable for X-ray studies. The original collagen solution may be in neutral solution before heating,[41] but it is difficult to keep the protein in solution before initiating fibril formation unless it is very dilute and kept cold.

*Thin Sections*

Collagenous tissues or reprecipitated fibers of collagen may be prepared for electron microscopy by fixing, staining, dehydrating, embedding, and sectioning the specimen according to a variety of procedures. A particular procedure that we have used successfully is described. More-detailed discussions of such procedures are to be found in Hayat.[42]

Very small bits of tissue are fixed in 3% glutaraldehyde, 0.2 M sodium phosphate, pH 7.2, for 2 hr or more. The tissue is washed in several changes of the same buffer, treated with 1% osmium tetroxide in the same buffer for 1 hr in the cold, and again washed in buffer. The tissue pieces are dehydrated successively in 70%, 80%, and 95% acetone for 10 min in each step and then treated three times with pure acetone for 30 min each time. The specimens are soaked in the complete embedding medium for 24 hr and embedded in a fresh change of medium at 60° for 3 days. The composition of the embedding medium by weight is: Epon-812, 10 parts; nadic methyl anhydride (NMA), 10.2 parts; dodecenyl succinic anhydride (DDSA), 4.2 parts; and 2,4,6-tri(dimethylaminomethyl)phenol (DMP-30), 0.5 part. The complete mixture can be stored in the freezer for several months. The sections are mounted on support grids and stained with lead

[39] R. R. Bruns, R. L. Trelstad, and J. Gross, *Science* **181**, 269 (1973).
[40] T. R. Oegema, J. Laidlaw, V. C. Hascall, and D. Dziewiatkowski, *Arch. Biochem. Biophys.* **170**, 698 (1975).
[41] G. C. Wood and M. K. Keech, *Biochem. J.* **75**, 588 (1960).
[42] M. A. Hayat, "Principles and Techniques of Electron Microscopy," Vol. 1. Van Nostrand-Reinhold, New York, 1970.

citrate for 1–5 min. This stain is made by dissolving 0.02 g of lead citrate in 10 ml of freshly boiled distilled water in a screw-top tube and adding 0.1 ml of 10 $N$ NaOH.[43]

In longitudinal sections, collagen fibrils have a clear $D$ periodicity. Thin section preparations always result in positive staining, and a series of bands, related to groups of charged residues, can be seen in each repeating unit. A directionality of each fibril can be assigned because of the asymmetry of the banding pattern. In transverse sections, the diameters of fibrils and their organization in the tissue can be seen. These features vary markedly in different kinds of tissues, in immature tissues, and in some diseased tissues.

*Freeze Fracture*

Freeze-fracture preparation of connective tissues for electron microscopy has recently provided some interesting new views of fibril structure and morphology. In these micrographs, one can see the collagen fibrils clearly, and striations that appear to represent molecules or groups of molecules are seen along the fibrils.[44,45] Significant differences in fibril structure can be seen in different tissues, and in tissues treated with denaturing agents, by this method.

For freeze fracture, specimens are fixed for 2–4 hr in 2% glutaraldehyde, 0.1 $M$ sodium cacodylate, pH 7.3, glycerinated for 60 min in 30% glycerol in saline, and frozen in liquid Freon 22 at $-150°$. Fracturing is performed at $-110°$ in, e.g., the Balzers 360 $M$ unit; after this, the specimen is replicated with platinum and carbon and is transferred to a grid support for observation in the electron microscope. A full discussion of freeze-fracture techniques is given by Koehler.[46]

*Labeling with Conjugated Antibodies*

Specific antibodies directed against collagen or against selected portions of the collagen molecule or its precursors have proved to be very useful in the localization of particular collagen determinants within tissue structures.[47] The antibody molecules are rendered visible by conjugation with ferritin, which is easily seen in the electron microscope, or with a

[43] J. H. Venable and R. Coggeshall, *J. Cell Biol.* **25**, 407 (1965).
[44] A. Ruggeri, F. Benazzo, and E. Reale, *J. Ultrastruct. Res.* **68**, 101 (1979).
[45] D. G. Rayns, *J. Ultrastruct. Res.* **48**, 59 (1974).
[46] J. K. Koehler, in "Principles and Techniques of Electron Microscopy" (M. A. Hayat, ed.), Vol. 2, p. 53. Van Nostrand-Reinhold, New York, 1972.
[47] G. Wick, B. R. Olsen, and R. Timpl, *Lab. Invest.* **39**, 151 (1978).

peroxidase, the enzymic activity of which may be used to produce an electron dense image.[48,49] The conjugation reactions are typically carried out using glutaraldehyde or a water-soluble carbodiimide as a crosslinking reagent. The resolution of procedures using peroxidase-labeled antibodies is less than those using ferritin because of the tendency of the identifiable products of the enzymic reaction to diffuse away from the site of antigen–antibody binding. Techniques using labeled antibodies may be able to answer critical questions concerning the distribution of the different genetic types or of precursor molecules within fibrillar structures.

*Degree of Structural Preservation in Electron Microscopy*

The procedures required to prepare a specimen for electron microscopy, and the high vacuum and electron beam damage encountered during observation, raise the possibility that the structure seen may not exactly correspond to the native structure. Values of the $D$ period obtained by electron microscopy vary from 500 to 690 Å, 640 Å being the typical value reported. These variations in the $D$ value may be a result of distortion during specimen preparation or observation, but in some instances may reflect true differences in the $D$ period in different tissues.[50]

Since the diameters and distribution of fibrils in tissues have not been reported by any technique other than electron microscopy, it is hard to evaluate the preservation of these features. Some observations show an internal consistency that suggests good preservation, such as the observation that fibril diameters are multiples of 80 Å.[51] X-Ray diffraction studies indicate that electron microscopic preparative procedures may not preserve lateral order within fibrils and may lead to shrinkage of fibril diameters (vide infra).

Light Microscopy Using Labeled Antibodies

The specificity of the antigen–antibody reaction has also been of value in localizing collagenous components within fibril structures at the level of resolution accessible to the light microscope. The antibody molecules are rendered visible by conjugation to peroxidase enzymes, as mentioned in the preceding section, or to fluorescent molecules. Antibodies to different genetic types of collagen and to procollagens have been visualized by

[48] Y. Kishida, B. R. Olsen, R. A. Berg, and D. J. Prockop, *J. Cell Biol.* **64**, 331 (1975).
[49] S. Gay, R. Gay, and B. R. Olsen, in "Electron Microscopy in Human Medicine" (J. V. Johannessen, ed.), p. 243. McGraw-Hill, New York, 1978.
[50] M. H. Flint and M. J. Merrilees, *Histochem. J.* **9**, 1 (1977).
[51] D. A. D. Parry and A. S. Craig, *Nature (London)* **282**, 213 (1979).

immunofluorescence in different tissues, and have been used to determine the location of different components.[52] The fluorescent antibody technique is very versatile, but caution is required in interpreting the results obtained. The reaction of an antibody with a tissue component indicates its presence, but the lack of reactivity may not necessarily mean the absence of that component, since the component may be inaccessible or unreactive for other reasons. However, the use of this approach together with biochemical analysis and other structural methods can provide important information concerning the presence and accessibility of components to these antibodies and the distribution of components in tissues.

X-Ray Diffraction

X-Ray diffraction can be used to probe structure of collagenous tissues at a level overlapping that achieved by biochemical analysis and that of electron microscopy, since it is capable of giving information on both molecular and fibril parameters.

*Specimen Preparation*

Both native tissues and fibers reconstituted from purified collagen can be examined by X-ray diffraction. A specimen of dimensions approximately $1 \times 1 \times 4$ mm is convenient to work with in that it can be easily held and stretched, but considerably smaller pieces can be used successfully. A suitable tissue specimen can be dissected from various sources, clamped in an enclosed specimen cell with Mylar windows, and stretched, if possible, to improve orientation. Alternatively, the specimen may be sealed in a thin-walled capillary tube.

A reconstituted native-type fiber suitable for X-ray diffraction studies can be formed either by dialysis of a solution of collagen in citrate buffer against tap water or by heat gelation, as described in a preceding section.[53] Fiber formation is carried out in small-diameter dialysis tubing, after which the tubing is cut away from the gel and partially dried. The gel is oriented by stretching, hanging with attached weights or rolling on a hydrophobic surface. This oriented, partially dried fiber is then clamped in an enclosed specimen cell as described above.

Diffraction has a distinct advantage in comparison with electron microscopy in that it may be carried out on intact, hydrated specimens. The experimental arrangement for X-ray diffraction, using either an enclosed

[52] H. Nowack, S. Gay, G. Wick, V. Becker, and R. Timpl, *J. Immunol. Methods* **12**, 117 (1976).
[53] E. Eikenberry and B. Brodsky, *J. Mol. Biol.* **144**, 397 (1980).

cell or a sealed capillary tube, permits the specimen to be kept continuously hydrated by contact with water vapor or by immersion in buffer.

*Applications of X-Ray Diffraction to Collagenous Structures*

The general configuration of an X-ray diffraction experiment suitable for use with connective tissues is shown in Fig. 2. A source of X-rays, either a tube stand or a rotating anode generator, illuminates the entrance aperture of an X-ray optical system, which may be either of a focusing type (as shown in Fig. 2) or of a nonfocusing type. Diffraction at high angles is most conveniently recorded on a camera with a specimen to film distance of 3-5 cm, and for this type of camera an optical system consisting of simple pinhole collimation provides adequate definition of the beam. At lower angles, longer specimen to film distances and more precise X-ray optical systems are required. The focusing optical system shown schematically in Fig. 2 consists either of two bent glass mirrors positioned orthogonally or of a bent glass mirror followed by a bent quartz crystal monochromator, again positioned orthogonally. With either optical system, the radiation is brought to a focus at the detector plane. The radiation passes through the specimen, which is enclosed in a watertight cell as described above and impinges on the beamstop in front of the detector plane. The detector may be either X-ray film or a one- or two-dimensional position-sensitive detector connected to a computer.[54] The discussion below assumes that film is used as the detector. The path between the specimen and the detector is evacuated or filled with helium to eliminate air scatter.

The specimen causes X-rays to be scattered away from the incident beam and to fall on the film in a characteristic pattern. The angle through which the radiation is scattered to reach a particular point on the film is the scattering angle, designated $2\theta$. X-ray diffraction is sensitive, almost exclusively, to that part of a structure that is repeated in an ordered array. Such an ordered array defines sets of parallel planes within the specimen and the distance, $D$, between the sets of planes may be calculated from the positions of an intensity maximum on the film using Bragg's law [Eq. (1)].

$$2 D \sin \theta = n \lambda \tag{1}$$

where $\lambda$ is the wavelength of the radiation (1.54 Å for copper radiation) and $n$ is the integer order of the reflection. As may be seen from this equation, the level of resolution of features in the specimen is inversely related to the scattering angle, with diffraction at very low angles relating to broad features (large $D$) and diffraction at high angles relating to the

[54] A. Gabriel and Y. Dupont, *Rev. Sci. Instrum.* **43**, 1600 (1972).

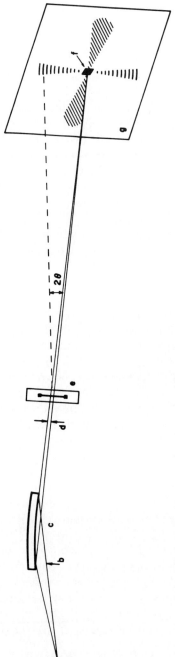

FIG. 2. Diagram of an X-ray diffraction apparatus suitable for examining collagenous structures. A source of X-radiation, a, illuminates the collimating slit, b, of a bent plate mirror optical system, c, where only one mirror is shown for clarity. After passing through guard slits, d, the convergent beam passes through the specimen, which is mounted in an enclosed specimen cell, e; the direct beam is then absorbed by a beam stop, f, which is located in front of the film plane, g. A diffracted ray, scattered by the specimen through an angle $2\theta$, with respect to the center of the incident beam, is shown by the dashed line.

finest features. Diffraction from connective tissues at high angles yields a distribution of intensity near the meridian together with other features of the pattern that can be interpreted with high precision in terms of the parameters of the triple helix.[9] Diffraction at low angles carries information on the distribution of groups of residues along the fibril axis, the lateral packing of molecules within the fibrils, and the packing of fibrils within the tissues.

With a specimen-to-film distance of 16 cm, well oriented connective tissues give a characteristic pattern (such as that shown in Fig. 4A for rat tail tendon). A set of meridional reflections are seen parallel to the fiber axis. These reflections are successive orders ($n = 1, 2, \ldots$) of the fundamental repeating unit in the axial direction in collagen, and application of Bragg's law shows the repeating unit, $D$, to be 670 Å in length. To record the lowest meridional orders of collagen (as shown in Fig. 4B), a camera with mirror-monochromator optics and a specimen to film distance of almost 1 meter was used. In the equatorial direction, perpendicular to the fiber axis, is a group of intensity maxima arising from the lateral packing of the collagen molecules in the fiber.

*Data Processing*

CALCULATION OF INTEGRATED INTENSITIES

Integrated intensities, proportional to the corrected total net blackening of the film for each reflection, are derived from one-dimensional or two-dimensional microdensitometric data. The two methods arrive at similar estimates in most cases.

Two-dimensional microdensitometric data consists of a stream of multibyte records on magnetic tape, one byte for each pixel in the original X-ray pattern. Tape format errors are corrected, and the content of each recorded image is printed as a contour map to permit location of the X-ray pattern within the data stream. Pattern location and recognition are further aided by production of overprint density maps and of cross-sections along arbitrary lines specified by the experimenter (Fig. 3). From these, the centers of the reflections are located, the reflections are indexed, and the $D$-periodicity is determined using a linear regression procedure to make optimal use of the data. Integration is performed within a template chosen by the experimenter. The template is an arbitrary polygon, specified by its vertices, chosen on the basis of the density maps and cross-sections so that the template just encloses each reflection. The integrated intensity is the sum of the optical densities of each pixel within the template corrected for film background estimated on the template perime-

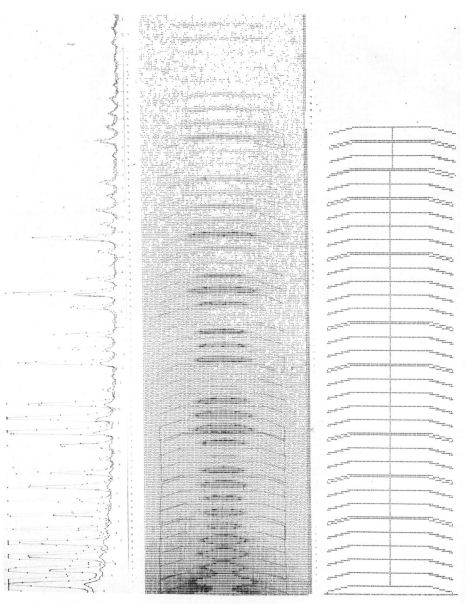

FIG. 3. *Left:* Line printer display of a cross-section of a two-dimensional microdensitometric scan of an X-ray diffraction pattern of rat tail tendon on X-ray film. Three cross-sections are represented, each parallel to the meridian, but slightly displaced from each other to permit exploration of the widths of the meridional reflections. The bottom order is the 9th and the top is the 49th. *Center:* Line printer over-strike density map of the same rat tail tendon data. The stronger reflections are clearly visible. The templates used for integration have been drawn in. *Right:* Templates used for integration of the net intensity of the meridional reflections shown above.

ter. Different templates are chosen for different parts of the image to compensate for differences in the shapes of reflections. This method of background subtraction is considerably superior to methods using only a single value, because the background is not at all uniform across the film.

A density saturation correction is applied to each pixel in the image to account for the decreased response of the optical density of the film at higher, but nonsaturating, X-ray exposures. This correction is significant at optical densities greater than about 1.2, whereas the saturating optical density of the Optronics microdensitometer is 3.0. We assume that the films within one pack (usually two or three films are used for any given exposure) are identical except that the exposure of the second film is attenuated by a factor, $k$, with respect to the exposure of the first film at every point in the image. The film saturation kinetics on exposure to X-rays is given by

$$D = D_{sat}(1 - e^{bE}) \qquad (2)$$

where $D$ is optical density of the film, $D_{sat}$ is the effective maximum optical density of the film, $E$ is the exposure, and $b$ is a constant to accommodate the choice of units for $E$. By making a graph of $D_1$ (observed density on film 1) as a function of $D_2$ (observed density on film 2) for a large number of pixels encompassing a full range of density values, one may determine $D_{sat}$. The value of $D_{sat}$ may be used to correct any observed optical density, $D_{obs}$, to the value that would have been observed had the film been perfectly linear, $D_{lin}$, according to Eq. (3).

$$D_{lin} = D_{sat} \ln(1 - D_{obs}/D_{sat}) \qquad (3)$$

The value of $k$ used to scale the data of films 2 and 3 to the scale of film 1 is also determined during this procedure.

Since the above procedures are complex, relative integrated intensities are also estimated from one-dimensional microdensitometer scans as being proportional to the areas under the peaks in a highly magnified scan of the meridian. Relative peak areas are approximated by height times width at half-height. In general, these two methods agree fairly well, except in regions where the background drops steeply as a function of distance from the center of the pattern.

CORRECTION OF INTEGRATED INTENSITIES

Integrated intensities are corrected for absorption, polarization, fibrillar disorientation, beam convengence, and path length in the film according to Eq. (4).

$$\text{Corrected intensity} = \text{observed intensity} \frac{C_3}{C_1 \cdot C_2 \cdot C_4} \qquad (4)$$

$C_1$ of Eq. (4) is the absorption correction, given by

$$C_1 = T^{\left[\frac{1}{2}\left(\frac{1}{\cos 2\theta} - 1\right)\right]}$$

where $T$ is the specimen transmittance at zero scattering angle, $2\theta$. $C_2$ is the polarization correction, given for mirror-quartz monochromator optics by

$$C_2 = \frac{1 + \cos^2 2\theta \cdot \cos^2 2\theta_Q}{1 + \cos^2 2\theta_Q}$$

or for double mirror optics by

$$C_2 = 1 + \cos^2 2\theta$$

where $2\theta_Q = 26° 42'$, the Bragg angle for the $10\bar{1}1$ planes of quartz. $C_3$ is the correction for specimen disorientation and beam convergence given by Wang and Worthington.[55,56]

$$C_3 = \frac{1}{R}\left(R^2 + \frac{h^2\omega^2}{D^2} + \frac{h^2\epsilon^2}{D^2}\right)^{1/2} \exp\left(\frac{\Delta^2}{R^2 + \frac{h^2\omega^2}{D^2} + \frac{h^2\epsilon^2}{D^2}}\right)$$

where $r$ = average radius of the diffracting fibrils, $R$ = radius of disk in reciprocal space such that $rR \simeq 1$, $h$ = order number, $\omega$ = mean angle of disorientation (estimated from the film), $D$ = axial period, $\epsilon$ = mean angle of beam convergence in the camera, $\Delta$ = deviation of sphere of reflection (Ewald sphere) from the center of the disk. $C_4$ is the correction for oblique passage of the X-rays through the film at large scattering angles, given by

$$C_4 = 1/\cos 2\theta$$

The net effect of these corrections is to increase substantially the estimated intensities of the higher orders while leaving the low orders relatively unaffected. After correction, which is done separately for the two camera systems that we use, data sets are merged by finding a common factor among overlapping sets of reflections, and the data are scaled arbitrarily to set the fifth order equal to 500 or the first order equal to 1000.

In a different approach to making the required corrections, Fraser et al.[57] have developed a procedure for directly mapping fiber diffraction data onto a central section of reciprocal space, thus preserving all the data in a two-dimensional form for further analysis, such as computation of integrated intensities. This contrasts with the procedure outlined above,

[55] S. K. Wang and C. R. Worthington, *Biophys. J.* **15**, 322a (1975).
[56] B. Crist and C. R. Worthington, *J. Appl. Crystallogr.* **13**, 585 (1980).
[57] R. D. B. Fraser, T. P. MacRae, A. Miller, and R. J. Rowlands, *J. Appl. Crystallogr.* **9**, 81 (1976).

which yields lists of intensities together with their associated positions in reciprocal space.

*Interpretation of X-Ray Data*

The positions of the intensity maxima on the film are directly interpretable in terms of repeating distances within the specimen. Thus, values of the meridional $D$ period and measures of the lateral intermolecular spacing derived from the positions at the equatorial intensity maxima can be compared directly for different specimens. The meridional intensities are related to the axial electron density distribution within the $D$ period, but part of the information necessary to reconstruct the electron density—the phases—is not present in the pattern. If one has both the intensities and the phases, one can reconstruct the electron density distribution within the $D$ period in the specimen. Approaches to obtaining phases have included calculations from models based on amino acid sequence data and triple-helical structure parameters[58] and experimental determination from heavy atom derivatives.[59] Direct methods of phase determination may also be applicable to this system, but have not yet been successful.

If one has the phases for the reflections, then an axial electron density distribution in one dimension can be synthesized using the Fourier transformation [Eq. (5)].

$$\rho(x) = \frac{1}{D} \sum_h F_h e^{-2\pi i h x} e^{-Bh^2} \qquad (5)$$

where $D$ = the length of the unit cell, $x$ = fractional coordinate within the unit cell, $h$ = order number, $F_h = |F_h|e^{i\phi_h}$, $|F_h|^2 = I_h$, the corrected observed intensity, $\phi_h$ = phase angle of the $h$th order, and $B$ = an artificial temperature coefficient used to dampen the Fourier series.

The difference between two structures can be visualized using the difference Fourier synthesis

$$\Delta\rho(x) = \rho_1(x) - \rho_2(x) \simeq \frac{1}{D} \sum_h (|F_{h_1}| - |F_{h_2}|) e^{i\phi_h} e^{-2\pi i h x} e^{-Bh^2} \qquad (6)$$

Since we do not have absolute intensity data, the coefficient $A$ is arbitrarily chosen to minimize the squared difference averaged over the unit cell:

$$\int_0^1 [\Delta\rho(x)]^2 dx = \text{minimum} \qquad (7)$$

[58] D. J. S. Hulmes, A. Miller, S. W. White, and B. Brodsky-Doyle, *J. Mol. Biol.* **110**, 543 (1977).
[59] R. H. Stinson, M. W. Bartlett, T. Kurg, P. R. Sweeny, and R. W. Hendricks, *Biophys. J.* **26**, 209 (1979).

The phases in this case are chosen to correspond to one intensity set or the other, the assumption being made that the differences between the two sets are small, and hence this choice is of insignificant consequence.

### Collagen Fibril Structure in Type I-Containing Tissues

Tendons and bones are tissues in which type I collagen is the only major collagenous component.[60] Tendons are the best-studied connective tissues in terms of fibril structure, and among them, rat tail tendon has been the prototype. We will first consider the structure of rat tail tendon, and then compare the structure of other tendons with this. Bone has noncollagenous components very different from those of tendon, and its structure will be discussed. Then the structure of fibers reconstituted from type I collagen will be reviewed.

### Collagen Fibril Structure in Rat Tail Tendons

Rat tail tendons are composed of very well-oriented collagen fibrils, and contain largely type I collagen, with less than 10% of the dry weight present as proteoglycans, lipids, elastic fibers, and other components. These properties make this tissue an excellent specimen for the study of collagen structure, especially by X-ray diffraction.

*Electron Microscopy*

Rat tail tendon has been characterized by electron microscopy both on mechanically dispersed suspensions and by thin sectioning of embedded material.[61] In suspensions, rat tail tendon gives a well-banded positive or negative image, with a $D$ period in the range of 600–700 Å. The positively stained bands within each $D$ period have been correlated with the amino acid sequence, and a very good correspondence was found within the resolution of the micrograph.[20,21] From negatively stained specimens, the gap : overlap ratio was found to be close to 0.58 : 0.42.[61]

On embedded specimens, a very detailed analysis was carried out on the diameters of collagen fibrils in tail tendons of rats of different ages.[62,63] Parry and Craig found a relatively narrow unimodal distribution of fibril diameters in fetal and newborn specimens and observed a spreading of the distribution to a very broad bimodal one in mature rats. Their studies on fetal and newborn rat tails and on other immature tissues with a narrow

---

[60] E. H. Epstein and N. H. Munderloh, *J. Biol. Chem.* **250**, 9304 (1975).
[61] B. R. Olsen, *Z. Zellforsch. Mikrosk. Anat.* **59**, 184 (1963).
[62] D. A. D. Parry and A. S. Craig, *Biopolymers* **16**, 1015 (1977).
[63] D. A. D. Parry and A. S. Craig, *Biopolymers* **17**, 843 (1978).

distribution of fibril diameters indicated that the diameters were discrete multiples of 80 Å, an observation that was interpreted in terms of discrete microfibril units.[51] Electron micrographs of transverse sections have not produced evidence of a highly ordered lateral arrangement of molecules in fibrils, although such a crystalline array is indicated by X-ray studies on rat tail tendon.[64] A 38 Å repeating unit within sector-shaped domains of fibrils has been visualized, and this may reflect some feature of the original lateral order, as discussed in the next section.[65]

In longitudinal sections of embedded rat tail tendon specimens, adjacent fibrils often run in opposite directions, and it has been estimated that approximately half of the fibrils are oriented in each direction.[62,66] Electron microscopy on freeze-fracture preparations of rat tail tendon shows fibrils with striations running almost parallel to the fibril axis (see Fig. 7A).[44] Similar observations of longitudinal striations have been reported in negatively stained fibrils.[61] It is thought that these striations indicate the orientation of the collagen molecules within the specimens (cf. below).

*X-Ray Diffraction*

Medium- and low-angle X-ray patterns of rat tail tendons show a series of Bragg reflections parallel to the fiber axis, called meridional reflections, which give information on the axial fibril structure, and some maxima perpendicular to the fiber axis, equatorial reflections, which are related to the lateral packing of molecules in fibrils (Fig. 4). A detailed account of all aspects of X-ray diffraction from rat tail tendon was given by Miller.[64]

MERIDIONAL DIFFRACTION: AXIAL FIBRIL STRUCTURE

The spacing of the meridional reflections corresponds to 670 Å, the $D$ axial stagger of collagen molecules in fibers (Table I). The integrated intensities of 52 meridional orders for rat tail tendon are given in Table II. For the low orders, one observes a pattern of strong odd orders (orders 1, 3, 5, 9) and weak even orders (orders 2, 4, 6), a feature suggesting that, at low resolution, the structure within a $D$ period corresponds to a rectangular function of electron density with roughly half at high density and half at a lower density. Such a pattern is consistent with the gap : overlap feature of collagen fibrils, and a ratio of 0.53 : 0.47 for the gap : overlap ratio best fits the X-ray data.[67] The intensities of the higher orders are more difficult

---

[64] A. Miller, in "Biochemistry of Collagen" (G. N. Ramachandran and H. Reddi, eds.), p. 85. Academic Press, New York, 1976.
[65] D. J. S. Hulmes, J.-C. Jesoir, A. Miller, C. Berthet-Colominas, and C. Wolff, *Proc. Natl. Acad. Sci. U.S.A.* **78**, 3567 (1981).
[66] R. Gieseking, *Z. Zellforsch. Mikrosk. Anat.* **58**, 160 (1962).
[67] S. G. Tomlin and C. R. Worthington, *Proc. R. Soc. London, Ser. A* **235**, 189 (1956).

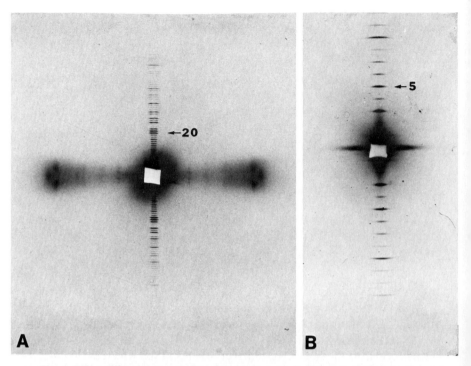

FIG. 4. X-Ray diffraction by rat tail tendon, immersed in buffered physiological saline and slightly stretched: (A) recorded on a camera with a specimen-to-film distance of 15.94 cm (the 20th meridional order is indicated); (B) recorded on a camera with a specimen to film distance of 86.8 cm (the 5th meridional order is indicated).

to interpret since a phase set is not known, but one can make comparisons of the observed intensities with those predicted for model structures.

Hulmes et al.[58] have constructed models of the collagen structure using the amino acid sequence, triple-helical parameters, and the $D$ stagger. A major problem in forming such model structures is the lack of knowledge about the telopeptide conformation. By adjusting the conformation of the N- and C-telopeptides to get the best fit between the calculated and observed X-ray data, a model was found that predicted the observed meridional X-ray intensity pattern well.[58,68] A similar analysis has been done using neutron diffraction.[68] The intensities of twelve meridional orders from the neutron diffraction pattern of rat tail tendon were collected for a range of concentration ratios of $H_2O : D_2O$, and the intensities predicted for the model were calculated. The residue translation for both the N- and

[68] D. J. S. Hulmes, A. Miller, S. W. White, P. A. Timmins, and C. Berthet-Colominas, *Int. J. Biol. Macromol.* **2**, 338 (1980).

TABLE I
THE $D$ PERIODICITY OF THE MERIDIONAL REFLECTIONS
IN VARIOUS CONNECTIVE TISSUES

| Tissue | $D$ spacing (Å) |
|---|---|
| Tendons | |
|   Rat tail | |
|     Stretched | 670–675;[a] 671[b] |
|     Unstretched | 675 |
|   Rat Achilles | 670 |
|   Calf Achilles | 667; 670[b] |
|   Rabbit leg | 675 |
|   Lamb leg | 669[a] |
|   Kangaroo tail | 673[b] |
|   Chicken leg | 674 |
|   17-Day embryonic chick leg | 669 |
|   1-Week rat tail | 671 |
| Skins | |
|   Lamb | 652[a] |
|   Rat | 657[a] |
|   Human | 655 |
|   Dog | 658 |
|   Calf | 656 |
|   Tadpole, frog, chicken, mice | 645–655[c] |
| Reticulum | |
|   Calf spleen | 659 |
| Cartilage and notochord | |
|   Human costal cartilage | 670[d] |
|   Pig ear treated with chrondroitinase | 667 |
|   Lamprey notochord | 668 |
| Reprecipitated fibers | |
|   Fiber of rat tail tendon collagen | 667 |
|   Fiber of calf skin type I collagen | 670 |
|   Fiber of type III collagen, salt extracted from calf skin | 669 |
|   Fiber of 85% I, 15% III collagen | 674 |

[a] Brodsky et al.[101]
[b] Tomlin and Worthington.[67]
[c] Stinson and Sweeny.[102]
[d] Bordas et al.[93]

C-telopeptides were then varied to find the best agreement with the neutron data, and the results agreed well with those found for the X-ray data. Since both the X-ray and neutron meridional intensities predicted from this model structure were in good agreement with the observed ones, the model appears to be a reasonable one, and the phases from this model are a feasible preliminary set. The phases for this model also received some experimental support from isomorphous replacement studies.[59]

## TABLE II
### Corrected Relative Intensities of the Meridional Reflections for Different Tendons, Scaled So That the Fifth Orders Are Equal to 500

| | Mammalian | | | | | Avian | |
|---|---|---|---|---|---|---|---|
| Order | Rat tail tendon[a] | Rat Achilles tendon[a] | Calf Achilles tendon[a] | Lamb leg tendon[a] | Rabbit leg tendon[a] | Chicken leg tendon[a] | Turkey leg tendon[b] |
| 1  | 14,265 | 12,594 | 7253 | 10,236 | —     | 11,589 | 10,420 |
| 2  | 245    | 225    | 84   | 151    | 283   | 89     | 94     |
| 3  | 1,162  | 1,070  | 779  | 1,395  | 1,153 | 1,180  | 1,115  |
| 4  | 132    | 85     | 48   | 97     | 119   | 58     | 73     |
| 5  | 500    | 500    | 500  | 500    | 500   | 500    | 500    |
| 6  | 180    | 148    | 116  | 149    | 212   | 125    | 83     |
| 7  | 142    | 152    | 226  | 285    | 235   | 321    | 313    |
| 8  | 54     | 85     | 106  | 143    | 102   | 33     | 83     |
| 9  | 445    | 502    | 376  | 582    | 361   | 352    | 386    |
| 10 | 106    | 97     | 98   | 152    | 156   | 133    | 208    |
| 11 | 69     | 74     | 63   | 118    | 109   | 76     | 135    |
| 12 | 160    | 129    | 176  | 262    | 224   | 107    | 229    |
| 13 | 7      | 4      | 6    | 7      | 45    | 8      | 21     |
| 14 | 29     | 29     | 29   | 56     | 54    | 58     | 125    |
| 15 | 23     | 17     | 40   | 113    | 119   | 31     | 73     |
| 16 | 19     | 25     | 40   | 96     | 156   | 53     | 73     |
| 17 | 36     | 29     | 33   | 71     | 82    | 58     | 115    |
| 18 | 13     | 15     | 35   | 81     | 75    | 30     | 42     |
| 19 | 13     | 14     | 8    | 15     | 101   | 24     | 31     |
| 20 | 155    | 153    | 175  | 439    | 619   | 204    | 323    |
| 21 | 113    | 102    | 125  | 319    | 574   | 152    | 219    |
| 22 | 52     | 44     | 62   | 135    | 215   | 54     | 52     |
| 23 | 3      | 2      | 13   | 5      | 36    | 0      | 0      |
| 24 | 2      | 0      | 3    | 0      | 23    | 8      | 10     |
| 25 | 62     | 32     | 79   | 171    | 336   | 58     | 94     |
| 26 | 16     | 2      | 16   | 20     | 127   | 34     | 31     |
| 27 | 44     | 25     | 26   | 47     | 167   | 74     | 104    |
| 28 | 1      | 0      | 4    | 0      | 34    | 0      | 0      |
| 29 | 18     | 6      | 16   | 21     | 131   | 28     | 21     |
| 30 | 50     | 27     | 28   | 43     | 185   | 32     | 21     |
| 31 | 14     | 0      | 6    | 5      | 61    | 36     | 21     |
| 32 | 10     | 7      | 11   | 9      | 61    | 41     | 42     |
| 33 | 10     | 0      | 7    | 2      | 49    | 14     | 10     |
| 34 | 60     | 31     | 31   | 50     | 158   | 32     | 31     |
| 35 | 3      | 11     | 4    | 0      | 27    | 0      | 0      |
| 36 | 12     | 5      | 12   | 10     | 51    | 5      | 0      |
| 37 | 17     | 5      | 17   | 31     | 107   | 49     | 42     |
| 38 | 28     | 3      | 8    | 13     | 70    | 19     | 21     |
| 39 | 3      | 6      | 7    | 5      | 60    | 43     | 31     |
| 40 | 14     | 4      | 7    | 5      | 70    | 7      | 0      |

TABLE II (continued)

|  | Mammalian | | | | | Avian | |
| --- | --- | --- | --- | --- | --- | --- | --- |
| Order | Rat tail tendon[a] | Rat Achilles tendon[a] | Calf Achilles tendon[a] | Lamb leg tendon[a] | Rabbit leg tendon[a] | Chicken leg tendon[a] | Turkey leg tendon[b] |
| 41 | 46 | 4 | 20 | 32 | 165 | 52 | 42 |
| 42 | 15 | — | — | — | 62 | 19 | — |
| 43 | 16 | — | — | — | 37 | 7 | — |
| 44 | 13 | — | — | — | 54 | 6 | — |
| 45 | 9 | — | — | — | 51 | 0 | — |
| 46 | 5 | — | — | — | 42 | 8 | — |
| 47 | 5 | — | — | — | 47 | 0 | — |
| 48 | 16 | — | — | — | 1 | 45 | — |
| 49 | 16 | — | — | — | 58 | 1 | — |
| 50 | 0 | — | — | — | 59 | 0 | — |
| 51 | 6 | — | — | — | 35 | 11 | — |
| 52 | 62 | — | — | — | 199 | 40 | — |

[a] B. Brodsky and E. Eikenberry, *Biopolymers*, in press.
[b] Berthet-Colominas et al.[77]

Thus, the meridional diffraction is derived from the axial fibril structure. The data give the axial stagger of molecules in fibrils and some information on low-resolution features such as the gap : overlap ratio, but at this time information on the higher-resolution structure is limited to finding good agreement with model structures. Even though the intensity data are not now directly interpretable, it is possible to make comparisons of different patterns, to see whether there are variations in structure and, if so, at what level of resolution.

EQUATORIAL DIFFRACTION: LATERAL FIBRIL STRUCTURE

In the equatorial direction, perpendicular to the tendon axis, there is a set of sharp maxima, overlaying some diffuse scatter (Fig. 4A).[69] The strongest maxima are located at positions corresponding to 39, 25, 19, 13.7, 13.3, and 12.6 Å (Table III). The presence of these sharp reflections indicates the existence of an ordered lateral packing of molecules in some fibrils. A number of models have been proposed for this lateral packing, and the two that give the best agreement with observed data are the microfibril model and the quasi-hexagonal model.

The microfibril model postulates the existence of cylindrical subunits composed of five molecules, each staggered by $D$ with respect to its

[69] A. Miller and J. S. Wray, *Nature (London)* **230**, 437 (1971).

## TABLE III
### Positions of Equatorial Diffraction Maxima in Hydrated Connective Tissues and Reprecipitated Fibers

| Tissues | Positions (Å) |
|---|---|
| **Sets of sharp reflections** | |
| Rat tail tendon | 12.64(equ.), 13, 28, 13.69, 17.48(equ.), 18.90, 26.51, 38.00[a] |
| Rat achilles tendon | 12.4(equ.), 13.5, 17.3(equ.), 18.8, 25.0, 26.2, 37.8 |
| Rabbit leg tendon | 13.8, 20, 25, 38 |
| **Diffuse maxima** | |
| *Tendons* | |
| Rat Achilles | 14.8 |
| Calf Achilles | 14.9; 14.7[b] |
| Lamb leg | 14.8 |
| Kangaroo tail | 14.7[b] |
| Turkey leg (uncalcified) | 15.1[c] |
| Chicken leg (uncalcified) | 15.3; 15.3[d] |
| 17-Day embryonic chick leg | 15.4 |
| *Skins* | |
| Rat | 15.4 |
| Lamb | 15.3 |
| Human | 15.5 |
| Cat | 15.0 |
| Calf | 15.2 |
| Chicken | 14.8 |
| *Cartilage* | |
| Annulus fibrosus | 16[e] |
| Nucleus pulposus (type II) | 16[f] |
| Bovine nasal | 16 |
| Rabbit knee articular | 16.5 |
| Lamprey notochord | 15.8 |
| *Bone* | |
| Rat tibia (demineralized) | 15.2; 15.32[g] |
| Chicken leg (demineralized) | 15.3[h] |
| Bovine dentine | 15.4[g] |
| Rat dentine | 15.2[i] |
| *Reticulin* | |
| Calf spleen reticulum | 15.5 |
| *Reconstituted fibers* | |
| Rat tail tendon collagen (acid extract) | 15.5 |
| Calf skin type I collagen (salt extract) | 16.10 |
| Calf skin type I collagen (pepsin extract) | 17.0 |
| Calf skin type III (salt extract) | 15 |
| Embryonic chick leg tendon (acid extract) | 15.9 |

[a] A. Miller and D. A. D. Parry, *J. Mol. Biol.* **75**, 441 (1973).
[b] R. Bear, *Adv. Protein Chem.* **7**, 69 (1952).
[c] Berthet-Colominas et al.[77]
[d] E. D. Eanes, D. R. Lundy, and G. N. Martin, *Calcif. Tissue Res.* **6**, 239 (1970).
[e] Berthet et al.[95]
[f] Grynpas et al.[96]
[g] E. P. Katz and S. Li, *Biochem. Biophys. Res. Commun.* **46**, 1368 (1972).
[h] E. D. Eanes and E. Miller, *Arch. Biochem. Biophys.* **129**, 769 (1969).
[i] Katz and Li.[86]

neighbor. Such subunits would be consistent with the observation of units running axially in the fibrils in negative-stained and freeze-fractured electron micrographs and has the appealing feature of having all molecules in equivalent positions. Other arguments, involving small units seen during fibril formation and the location of cross-links, have been put forth in favor of such a microfibril model.[70] To fit the observed X-ray reflections, the microfibrils of diameter 38 Å, would be arranged on a square lattice, with a unit cell of side 2 × 38 Å, consisting of four microfibril units.[64]

Hulmes and Miller[71] proposed a quasi-hexagonal model, constructed to give an excellent fit to the X-ray data. In a true hexagonal packing scheme, the distances between planes of adjacent molecules would be identical in the three major directions. In the quasi-hexagonal model, the three interplanar distances are slightly different, at positions 13.7, 13.3, and 12.6 Å, corresponding to the three strongest equatorial and near-equatorial reflections. The molecules are tilted with respect to the fiber axis when viewed along the 13.7 and 13.3 Å intermolecular planes, but not along the 12.6 Å plane, resulting in the observed off-equatorial and equatorial positions for these reflections. The density predicted by this quasi-hexagonal model is in excellent agreement with experimental data whereas the microfibril model predicts a much lower density. The X-ray pattern predicted by the quasi-hexagonal model was calculated recently and was shown to be similar to that observed.[72] The two models discussed need not be exclusive, since it is possible that the five stranded microfibril units could be compressed so as to fit on the proposed quasi-hexagonal lattice.[70,72a]

This ordered lateral array of molecules in the fibrils, as indicated by X-ray diffraction, has not been successfully visualized by electron microscopy of transverse sections. Disruption of the lateral structure could occur during specimen preparations for electron microscopy. The preservation of the lattice X-ray reflections after each stage of specimen preparation for electron microscopy has been systematically examined.[65,73] Fixation and staining of specimens results in the intensification of row lines near 40 Å and 20 Å in the X-ray pattern, although the original pattern, with additional sharp reflections, was not preserved. Such procedures, leading to the presence of these two strong row lines in the final embedded specimens, resulted in electron micrographs which, in trans-

[70] B. L. Trus and K. A. Piez, *Nature* (*London*) **286**, 300 (1980).
[71] D. J. S. Hulmes and A. Miller, *Nature* (*London*) **282**, 878 (1979).
[72] A. Miller and D. Tocchetti, *Int. J. Biol. Macromol.* **3**, 9 (1981).
[72a] Recently, data from a rat tail tendon pattern intensified by staining was shown to be consistent with a quasi-hexagonal, but not a tetragonal model. R. D. B. Fraser and T. P. MacRae, *Int. J. Biol. Macromol.* **3**, 193 (1981).
[73] J.-C. Jesoir, A. Miller, and C. Berthet-Colominas, *FEBS Lett.* **113**, 238 (1980).

verse sections, showed a 38 Å repeat distance between radial layers within sector-shaped domains of the fibril.[65] This repeating unit probably corresponds to part of the original lattice structure, although partial disruption may have occurred. A systematic approach such as this is a good way to evaluate discrepancies between X-ray and electron microscopic observations, and to arrive at a better understanding of the structure.

Collagen Fibril Structure Variations in Tendons

All tendons are similar in having type I collagen as their major component and in having a small amount of noncollagenous material. However, different tendons from a given animal can vary in their mechanical role, in the diameters of collagen fibrils, and in the amounts and kinds of proteoglycan present.[74,75] And, in different animals, the type I collagen chains show amino acid substitutions. For the amino acid sequences that have been determined, approximately 7% of the nonglycine positions differ between $\alpha 1(I)$ chains of calf and rat, and about 10% between calf and chick. For the $\alpha 2$ chain, where less of the sequences for different animals is known, there is even greater variation, with about 15% of the positions differing between calf and rat (among 96 nonglycine residues) and about 23% between calf and chick (among 223 nonglycine residues).[4,8] Since there are anatomical differences among tendons in one animal and biochemical differences in the type I collagen between different animals, it is worth considering whether these result in detectable variations in collagen fibril structure.

Electron micrographs of various tendons have been observed. No differences in the axial banding pattern for either positively stained or negatively stained fibrils have been reported. Differences have been noted in the diameters of fibrils in different tendons, although all tendons appear to have a narrow unimodal distribution of fibril diameters at early steps of development, changing to a broader distribution at maturity.[74]

Differences were found in a detailed comparison of the X-ray patterns of a variety of tendons.[76] All tendons give well oriented patterns, and the meridional D spacings vary between 667 and 675 Å (Table I). Meridional intensities for up to 52 orders are given in Table II for a variety of tendons.

The patterns of two different tendons from rat were compared. The

[74] D.A.D. Parry, G. R. C. Barnes, and A. S. Craig, *Proc. R. Soc. London Ser. B* **203**, 305 (1978).
[75] G. C. Gillard, M. J. Merrilees, P. G. Bell-Booth, H. C. Reilly, and M. H. Flint, *Biochem. J.* **163**, 145 (1977).
[76] B. Brodsky and E. Eikenberry, *Biopolymers,* in press.

meridional intensities of the rat Achilles and rat tail tendon were very similar for 52 orders, suggesting a high degree of similarity in their axial fibril structure.

A comparison of the pattern of tendons of different mammals (rat, calf, rabbit, lamb) indicated similar meridional patterns, but variations were seen in the relative intensities of the 14th–18th orders. These variations were small but reproducible. Such changes in observed relative intensities could be related to changes in the collagen itself, such as the amino acid sequence variations in the triple-helical or telopeptide regions, or could be due to small differences in periodic binding of noncollagenous components. In calculations of meridional intensities from models, the intensities of the 11th–19th orders were reported to be markedly affected by changes in telopeptide conformation, suggesting that this could be the explanation of the observed differences.[58]

Significant differences were seen between avian and mammalian meridional patterns, both in the low (2nd–8th) and the higher (29th–52nd) orders. A stronger 7th order and weaker 2nd, 4th, and 6th orders were seen in all avian patterns (Table II), an observation that is consistent with a gap:overlap ratio closer to 0.5:0.5 than the ratio found in mammalian tendons. Calculations show a 0.525:0.475 ratio best describes the pattern from chicken tendon, whereas 0.530:0.470 best fits the rat pattern. Significant differences were also seen for the highest orders recorded. Rat, calf, lamb, and rabbit tendons have a characteristic pattern of relative intensities for the 29th–52nd orders, whereas chicken, turkey, and quail tendons have a different and distinct pattern for these orders (Fig. 5; Table II). The larger number of substitutions in amino acid sequence that are found between chicken and calf type I collagen chains, as compared to those found between calf and rat, may lead to significant electron density differences between the two classes of animals and thus cause observed higher meridional intensity differences. This correlation of both low- and high-order intensities in a variety of mammalian and avian tendons suggests that the collagen molecules themselves, rather than noncollagenous components, which could vary among tissue types, are responsible for the distinctive features seen.[76]

The equatorial diffraction of various tendons indicates a diversity in lateral packing of collagen molecules in fibrils. Some specimens of rat Achilles tendon gave lattice reflections very similar to those seen for rat tail tendon, whereas rabbit tendon and some specimens of quail leg and wing tendons gave similar discrete reflections but lacked the strong equatorial 12.6 Å peak (Table III). Many tendons gave only a broad equatorial maximum, with no discrete diffraction peaks. Jesoir *et al.*,[73]

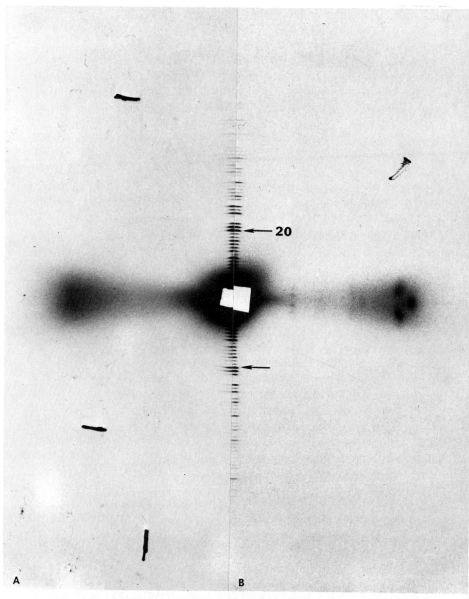

FIG. 5. Split comparison of X-ray diffraction patterns of (A) chicken leg tendon and (B) rat tail tendon. The specimens were immersed in physiological buffer during the X-ray exposure. The patterns were recorded on a camera with a 15.94 cm specimen-to-film distance, and the 20th orders are indicated. The differences in the relative intensities of the 29th–52nd orders of these two patterns can be seen.

have found that fixing and staining a variety of tendons, which showed no lattice reflections in their original pattern, resulted in row lines reminiscent of those seen in stained rat tail tendon, which are intensifications of the original lattice reflections. This suggests that some of the lateral order found in native rat tail tendon may be present in all tendons, but in less crystalline form.

Tendons with only a broad equatorial maximum show considerable variation in its position, from 14.4 to 15.4 Å. These maxima in quail leg and chicken wing tendons were near 14.4 Å, those of lamb and calf tendons were near 14.7 to 14.9 Å, and those of chicken leg and turkey leg tendons were near 15.1 to 15.4 Å.[76,77] Such differences may reflect variations in the degree of order or in the spacing between collagen molecules. The larger spacing in chicken leg and turkey leg tendon may be related to the observation that these tendons calcify in older animals.[77]

Collagen Fibril Structure in Bone

Bone, like tendon, contains type I as its only significant collagen type. The amino acid sequences of the $\alpha 1(I)$ and $\alpha 2$ chains of bone appear to be the same as in tendon, but there is an increase in the posttranslational modification of lysyl to hydroxylysyl residues in bone.[78] The lysyl residue in the N-telopeptide region is significantly more hydroxylated in bone than in tendon or skin, and this may result in differences in cross-linking and glycosylation as well. Bone differs from tendon in being calcified, and as much as 75% of its dry weight may be hydroxyapatite.[79] In addition, proteoglycan, glycoprotein, and other proteins may comprise 5–50% of the total organic matter, depending on the type of bone.[79] Leg tendons of turkeys and chickens become calcified with age, and calcified turkey tendon has been used as a model system for calcification.[80]

Electron microscopy of thin sections of bone show well banded collagen fibrils, with a clear $D$ repeat and often with adjacent fibrils in register.[81] The distribution of diameters may be unimodal about a mean of 800 Å, although there is variation with age and with types of bone. At early stages of mineralization, the hydroxyapatite is seen clearly in association

[77] C. Berthet-Colominas, A. Miller, and S. W. White, *J. Mol. Biol.* **134**, 431 (1979).

[78] M. J. Barnes, B. J. Constable, L. F. Morton, and P. M. Royce, *Biochem. J.* **139**, 461 (1974).

[79] G. M. Herring, *in* "The Biochemistry and Physiology of Bone" (G. H. Bourne, ed.), 2nd ed., Vol. 1, p. 127. Academic Press, New York, 1972.

[80] M. U. Nylen, D. B. Scott, and V. C. Moseley, *in* "Calcification in Biological Systems" (R. F. Sognnaes, ed.), p. 129. Am. Assoc. Advan. Sci., Washington, D.C., 1960.

[81] D. Cameron, *in* "The Biochemistry and Physiology of Bone" (G. H. Bourne, ed.), 2nd ed., Vol. 1, p. 191. Academic Press, New York, 1972.

with the collagen fibrils and is present at regular intervals of 600–700 Å. Glimcher and Krane[82] proposed that the hydroxyapatite is present in the gap regions of the fibril. Similar electron micrographs, where the dense mineral appears to fill the gap region, were seen for calcifying turkey leg tendon.[80]

The hydroxyapatite gives a strong crystalline high-angle X-ray pattern, and at low angles the diffuse scatter from the mineral is very strong. These features dominate the X-ray pattern of bone, although some 670 Å reflections can be seen.[83] These 670 Å reflections may be due to the presence at regular intervals of hydroxyapatite crystals.[83] Unlike calcified bone, where mineral scattering dominates, demineralized bone proves to be a good specimen for X-ray diffraction studies of collagen fibril structure. Well oriented patterns of hydrated and dry demineralized rat, fish, chicken, human, and turtle bones have been obtained.[82,84,85] The meridional $D$ period of such specimens is close to 670 Å, and the meridional intensities resemble those of tendons, although detailed comparisons have not yet been reported.

The equatorial diffraction of demineralized bone consists of a broad maximum near 15.1–15.4 Å (Table III).[85,86] This position is larger than seen for the diffuse maxima of mammalian tendon patterns, suggesting a larger intermolecular spacing. For example, the spacing for rat leg bone (15.2 Å) is greater than that seen for the rat Achilles tendon diffuse maximum (14.7 Å). For chick bone, the intermolecular spacing appears to be at the same position, 15.3 Å, as observed for chicken leg tendon, a tissue that calcifies in older animals.[85]

The collagen fibril structure in decalcified bone appears similar to that in noncalcifying tendon, except for a larger intermolecular spacing. Since the genetic type of collagen is the same in both tendon and bone, it is possible that cross-links or another component keeps the molecules farther apart in bone collagen fibrils than in noncalcifying tendons.

Detailed X-ray and neutron diffraction studies have been carried out on calcifying turkey leg tendon.[77,87] These studies suggest that the mineral occupies the gap region in collagen fibrils and that this mineral deposit grows in an orderly way during calcification.

---

[82] M. J. Glimcher and S. M. Krane, *in* "Treatise on Collagen" (B. S. Gould, ed.), Vol. 2B, p. 68. Academic Press, New York, 1968.

[83] A. Engstrom, *in* "The Biochemistry and Physiology of Bone" (G. H. Bourne, ed.), 2nd ed., Vol. 1, p. 237. Academic Press, New York, 1972.

[84] M. Grynpas, *Nature (London)* **265**, 381 (1977).

[85] D. L. Broek, E. F. Eikenberry, P. P. Fietzek, and B. Brodsk, *in* "First International Conference on the Chemistry and Biology of Mineralized Connective Tissues" (A. Veis and W. Butler, eds.), 1981, in press.

[86] E. P. Katz and S. Li, *J. Mol. Biol.* **80**, 1 (1973).

[87] S. W. White, D. J. S. Hulmes, A. Miller, and P. A. Timmins, *Nature (London)* **266**, 421 (1977).

## Reconstituted Fibers of Type I Collagen

As discussed above, solutions of collagen can be precipitated by a variety of procedures into fibers with a $D$ periodicity and an appearance like that of native fibrils. Comparison of fibers reconstituted from purified collagen with fibers in native tissues can indicate whether the purified collagen itself has the information necessary to determine the axial and lateral fibril structure and fibril diameter. If the reconstituted fiber does not reproduce the features of native fibrils, it suggests that some *in vivo* condition or component may play a role in determining these structural features.

Comparisons of electron micrographs of reconstituted fibers formed by any of the procedures discussed above in the section on preparation of fibrous forms of collagen with fibers in native tissues, shows these specimens to be indistinguishable in both the negative and positive staining patterns and in the axial repeat. The diameters of reconstituted fibrils do not have the same distribution as that seen in the native tissue from which the collagen was extracted. This is not surprising, since the rate of fibril formation *in vitro* and the diameters of the fibrils formed are affected by such factors as pH, temperature, ionic strength, and the presence of noncollagenous components, e.g., proteoglycans.

X-Ray patterns of oriented hydrated and dry fibers reconstituted from rat tail tendon collagen are very similar to those obtained from native rat tail tendon.[53,88] The meridional D spacing is close to that in native tendon (Table I), and the meridional intensities of the 1st to 25th orders are similar, except for some small variations in the relative intensities of the 14th and 18th orders.[53] A difference electron density distribution was calculated, as described in the section on X-ray diffraction, using the observed intensities from the rat tail tendon pattern and the reprecipitated fiber pattern with the phases calculated by Hulmes *et al.*[58] (Fig. 6). No significant differences in electron density were seen.[53]

For the hydrated reconstituted fiber, a broad equatorial maximum was seen at 15.5 Å, a position larger than that seen for tendon, suggesting that there may be a looser packing of molecules in reconstituted fibrils. No sharp lattice reflections were observed, but upon fixation and staining, weak row lines reminiscent of those seen in fixed and stained rat tail tendon patterns were seen.[53]

These X-ray diffraction and electron microscopic studies confirm that the information for the axial structure is present in purified collagen. Some of the purified collagen used for such studies may have undergone partial degradation of its telopeptide regions, but remaining intact molecules may form the best fibrils and be both the most strongly diffracting objects and

---

[88] R. W. G. Wyckoff and R. B. Corey, *Proc. Soc. Exp. Biol. Med.* **34**, 285 (1936).

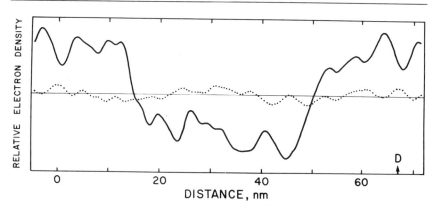

FIG. 6. Fourier synthesis of the relative electron density in one $D$ period of collagen using the intensity data for rat tail tendon shown in Table II and the phases calculated by Hulmes *et al.*[58] Note the prominent rectangular wave function of the density distribution, indicative of the gap:overlap structure illustrated in Fig. 1. The dotted line shows the difference in electron density between rat tail tendon and a synthetic fiber of purified rat tail tendon collagen, using the phases as above. Reproduced from Eikenberry and Brodsky.[53]

the good fibrils seen in the electron micrographs. The determinants of lateral order within the fibril and the fibril diameters are not clear. It is possible that information present in the original collagen molecule was lost during extraction—for example, by partial loss of the telopeptides. Perhaps more probable is the need for some condition or component present in the native tissue that determines fibril diameters and regulates the lateral aggregation of molecules into fibrils.

### Collagen Fibril Structure in Type II-Containing Tissues

Type II is the major collagenous component of cartilage and notochord.[2,17] A small amount of several minor collagen types may also be found in hyaline cartilage.[89] The amount of proteoglycan present depends on the type of cartilage, varying from over 50% of the dry weight in bovine nasal cartilage, to less than 5% of the dry weight in fibrocartilage, such as intervertebral disk.[90] This amount of proteoglycan is always much greater than that found in tendon.

Electron Microscopy

Collagen fibrils in cartilage have a distinctive appearance in electron micrographs of thin sections, being of small diameter and randomly dispersed throughout the matrix. In the very thin fibrils (100–200 Å) that are

[89] R. E. Burgeson and D. W. Hollister, *Biochem. Biophys. Res. Commun.* **84**, 627 (1979).
[90] A. Serafini-Francassini and J. W. Smith, "The Structure and Biochemistry of Cartilage." Churchill-Livingstone, Edinburgh, 1974.

usually present, no banding pattern can be seen, but in the somewhat larger fibrils (about 500 Å in diameter) found in some cartilage, the banding pattern resembles that of tendon fibrils, but is less well defined.[90] It has been suggested that the proteoglycan obscures the collagen fibrils so that they cannot be visualized clearly; it is also possible that the proteoglycan prevents the thin units from aggregating into larger fibrils. Extraction of proteoglycan from cartilage by either enzymic treatment or guanidinium hydrochloride results in well-banded fibrils, with a clear periodicity.[91] However, in the sterna of nanomelic chicks, where 90% of the proteoglycan is missing, the collagen fibrils are still randomly oriented, thin, and poorly banded.[92]

X-Ray Diffraction

Cartilage in general has proved to be a poor specimen for X-ray diffraction, giving only weak patterns, with poorly oriented equatorial maxima and a few weak meridional reflections.[93,94] Good X-ray patterns have been reported for annulus fibrosus,[95] which contains about half type I and half type II collagen, and for dry nucleus pulposus fibers,[96] which contain either all type II collagen, all type I collagen, or a defined mixture. Recently, a partial set of meridional intensity data was obtained on human articular cartilage using synchrotron radiation.[96a] We have obtained good patterns from lamprey notochord.[94] Those tissues that diffract well have a low proteoglycan content and well oriented collagen fibrils, suggesting that the high proteoglycan content and the poor collagen orientation may prevent diffraction by most cartilages. Preliminary experiments on removing proteoglycan from cartilage by treatment with guanidinum hydrochloride or chondroitinase has resulted in improved X-ray patterns.[94]

The X-ray patterns obtained have led to some information on the collagen fibril structure in type II-containing tissues. The $D$ period is near 670 Å, and the limited data available suggest that there are significant differences in the meridional intensities from those observed for tendons. The equatorial diffuse maximum is found near 16–16.7 Å for these various type II-containing tissues (Table III), a position considerably larger than

[91] K. Kuhn and K. von der Mark, in "Collagen-Platelet Interaction" (H. Gastpar, ed.), p. 123. Verlag, New York, 1978.
[92] J. P. Pennypacker and P. F. Goetinck, *Dev. Biol.* **50**, 35 (1976).
[93] J. Bordas, J. Woodhead-Galloway, and D. W. L. Hukins, *Biochem. Biophys. Res. Commun.* **84**, 627 (1978).
[94] B. Childs, T. Doering, E. Eikenberry, and B. Brodsky, unpublished results.
[95] C. Berthet, D. J. S. Hulmes, and P. A. Timmins, *Science* **199**, 547 (1978).
[96] M. Grynpas, D. Kirschner, and D. R. Eyre, *Biochim. Biophys. Acta* **626**, 346 (1981).
[96a] C. Berthet-Colominas, C. Mason, A. Miller, and D. Tochetti, in "Biology of Collagen" (A. Viidik and J. Vuust, eds.), p. 53. Academic Press, New York, 1980.

found for any type I-containing tissue.[95,96] This larger intermolecular spacing could be related to the amino acid sequence of type II collagen, to the increased glycosylation,[97] or to some interaction with proteoglycans in tissues.

### Reconstituted Fibers of Type II Collagen

Purified type II collagen can be reprecipitated into fibrils by the techniques discussed in the section on preparation of fibrous collagen. In electron micrographs these fibrils show a greater polymorphism of structure than is seen for any other type of collagen. Small fibrils with poorly defined or no banding are seen, as well as very wide clearly banded fibrils, and composite fibrils formed from units of various sizes staggered by 90 Å with respect to one another.[39,98] Symmetric forms have also been reported.[20] The failure always to form small fibrils *in vitro* could be due either to alterations in the type II molecules, such as partial loss of the telopeptides, or to the need for proteoglycan to restrict the diameters.

The banding pattern of the reconstituted type II collagen after positive or negative staining cannot be distinguished from that of type I collagen. There are several differences visible between the SLS patterns of type I and type II collagens, but corresponding differences are not apparent in the native-type patterns. X-Ray patterns of reconstituted type II collagen fibers have not yet been obtained.

Thus, the electron microscopic and X-ray evidence suggest that the axial and lateral structure of collagen fibrils and their diameters may be different in tissues with type II collagen than found in tendon. However, at this time, it is difficult to determine whether these variations in fibril structure are due to the presence of type II rather than type I collagen, or if they are due to the proteoglycans found in these type II-containing tissues.

### Collagen Fibril Structure in Tissues with Type III Collagen

Unlike type I and type II collagens, type III collagen has not been found as the major collagenous component of a tissue. Immunofluorescent observations indicated that the reticular networks of spleen and liver reacted with antibodies directed against type III collagen and type III procollagen, but did not react with antibodies against type I collagen.[52,99]

[97] R. Spiro, *J. Biol. Chem.* **244**, 602 (1969).
[98] B. B. Doyle, D. J. S. Hulmes, A. Miller, D. A. D. Parry, K. A. Piez, and J. Woodhead-Galloway, *Proc. R. Soc. London, Ser. B* **186**, 67 (1974).
[99] S. Gay, P. P. Fietzek, K. Remberger, M. Eder, and K. Kühn, *Klin. Wochenschr.* **53**, 205 (1975).

However, analysis of these tissues by comparison of cyanogen bromide peptides and by sodium dodecyl sulfate gel electrophoresis indicated that type I is the major collagen present and that type III is a minor, but significant, component.[60,100] The immunofluorescent observations were originally interpreted as suggesting that only type III collagen was present in such tissues, whereas, because of the biochemical data, it is now thought that the observations indicate inaccessibility of type I collagen to antibodies directed against it.

Thus type III is always associated with type I collagen as the major collagen type, and it has been suggested that all connective tissues except for tendon, bone, and cartilage contain type III collagen in significant amounts.[60] It is not known whether type III collagen is present in separate fibrils or in mixed fibrils with type I collagen. If type III and type I collagens are present in the same fibril, the two types could be randomly mixed or each type could be in some specific location. An additional distinctive feature of type III-containing tissues is the presence of type III pN-collagen. Unlike the other genetic types, type III procollagen is incompletely cleaved, resulting in the presence of significant amounts of type III pN-collagen in tissues, perhaps in the fibers themselves.

Because type III collagen is never found alone in a tissue, one can study its properties in native tissues only by examining those that contain both type I and type III collagens, such as skin. One can examine the properties of fibrils formed from purified type III collagen, but at this time we cannot be sure that such pure type III fibrils exist *in vivo*.

Electron Microscopy

Extensive electron microscopic observations have been made on skin, which contains both type I and type III collagen. The axial banding pattern of fibrils in skin, using either positive or negative staining, is indistinguishable from that found in tendon. If there were fibrils of type III collagen alone, one might expect to see fibrils with a significantly shorter gap : overlap ratio, since both the C-telopeptide and the N-telopeptide of type III collagen are longer than those found in type I collagen.[4] However, these have not been seen.

If there is a set of fibrils containing only type I collagen and a separate set containing only type III collagen, one might expect to see two distinct populations of fibril sizes, with, perhaps, type III fibrils having smaller diameters than type I fibrils. The diameters of the fibrils in skin are near 1000 Å, with a unimodal distribution of diameters.[74] Rat spleen reticulum, with about 80% type I and 20% type III collagen, contains a very narrow distribution of fibril diameters, clustered around 325 Å.[100] In fact, skin, in

[100] B. Brodsky, H.-T. Chen, and E. Eikenberry, manuscript in preparation.

general, is found to have a unimodal distribution of fibril diameters, whereas mature tendon has a bimodal distribution. Thus there is no evidence for two distinct populations of fibrils of different sizes in tissues containing type III collagen. This suggests that both type III and type I collagens are included in this unimodal population of fibrils, and supports the idea that they are found mixed in fibrils. It is interesting, although perhaps coincidental, that the proportion of type III collagen present in the rat spleen reticulum, is just sufficient to coat the surface of a 325 Å diameter cylinder. Such a model, with the type III collagen coating a fibril, would explain the immunofluorescent observation of antibody reactivity with type III, but not type I, collagen in such tissues. Another alternative is that one of the collagen types is not present in fibrous form. Clearly, electron microscopy using labeled antibodies to each of the involved components could do much to clarify the distribution of type I collagen, type III collagen, and type III pN-collagen in tissues.

Freeze-fracture electron microscopic studies on various tissues show subunits, which may be molecules or groups of molecules, running along the fibril. In cartilage and tendon, these subunits appear to be almost parallel to the fiber axis, whereas in skin, aorta, and other tissues, these units appear to run at a significant angle with respect to the fibril axis (Fig. 7).[44,45] Ruggeri[44] has reported an angle near 18° for this tilting and has suggested that it occurs generally in tissues that contain type III collagen.

### X-Ray Diffraction

Low- and medium-angle X-ray patterns have been recorded on skin,[101,102] spleen reticulum,[100] and aorta,[103] tissues containing both type III and type I collagens. Complete data sets were recorded for skins (see Fig. 9A), and the features were compared with those of tendon patterns. The $D$ period for skins was found to be significantly shorter than that in tendons (Table I). This was found for a large variety of skins in the hydrated state, and the difference was shown not to be due to the increased stretching of tendons, since unstretched tendons gave a repeat of 670–675 Å. The meridional intensities for 27 orders of lamb skin showed small differences from those of lamb tendon, but, since only one data set has been collected, this result is preliminary. The equatorial maxima of skins appeared consistently at larger distances than found in tendons (Table III). For example, calf tendon has a broad equatorial maximum at 14.8 Å, whereas calf skin has its maximum at 15.2 Å.

[101] B. Brodsky, E. Eikenberry, and K. Cassidy, *Biochim. Biophys. Acta* **621**, 162 (1980).
[102] R. H. Stinson and P. R. Sweeny, *Biochim. Biophys. Acta* **621**, 158 (1980).
[103] N. Roveri, A. Ripamonti, S. Garbisa, and D. Volpin, *Connect. Tissue Res.* **5**, 249 (1978).

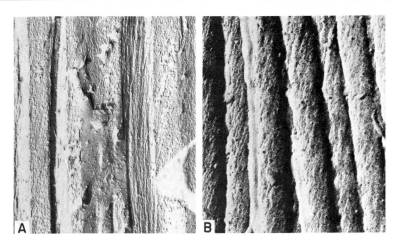

FIG. 7. Electron micrographs of freeze fracture specimens of (A) rat tail tendon and (B) skin. Reproduced from Ruggeri et al.[44]

The meridional and equatorial reflections are always spread over an arc-shaped area. The meridional arcing and part of the equatorial spread is due to disorientation of fibrils within the specimen. But in the rat tail tendon, as was pointed out by Miller,[64] the fanning of the 9.5 Å layer line and splitting of the sharp equatorial reflections on the equator can arise only from the tilting of molecules. The angles of meridional and equatorial spread are difficult to measure, but our estimates on tendons indicate that the amount of equatorial arcing is about the same as the meridional arcing, whereas in skins it is about 5–12° greater. For example, in lamb leg tendon, the equatorial arcing is 2° greater than the meridional arcing, a value not different from zero within experimental error, while the equatorial arcing is 11° greater than the meridional for lamb skin. This excess equatorial arc is a measure of tilting of molecules in the fibril.

A decreased $D$ period and increased tilting in skin would be consistent with the image of skin seen by Ruggeri in freeze-fracture pictures, where the subunits are tilted with respect to the fibril axis. Tilting of molecules by 12° with respect to the fiber axis would decrease the $D$ period by cos 12°, so that a 670 Å value would decrease to 655 Å.

The X-ray pattern of spleen reticulum also had a decreased $D$ period (Table I) of about 659 Å, an increased equatorial spread, and an increased intermolecular spacing compared with tendons. These features were all similar to those found in skin.[100]

The studies on dried aorta suggested some altered meridional intensities,[103] but more data are needed on this tissue before one can make detailed comparisons.

Reconstituted Fibers Containing Type III Collagen

Purified type III collagen can be reprecipitated to form native-type fibrils, with a clear $D$ periodicity. Electron microscope observations on fibers formed with type III collagen suggested that its presence leads to fibrils of smaller diameter than formed from type I collagen.[104] Although the SLS banding pattern of type III collagen can be distinguished from that of type I collagen, the positively stained reprecipitated type III native-type fibril cannot be distinguished from type I fibrils.[32] The negatively stained fibrils also resemble those of type I, with a similar gap : overlap ratio. A smaller gap : overlap ratio might have been expected because of the longer telopeptides of type III collagen, but it is possible that partial cleavage of the telopeptides had occurred during purification of the material.

Reconstituted fibers composed of type III collagen, isolated by salt extraction of calf skin, were oriented and examined by X-ray diffraction.[101] These patterns had a normal meridional $D$ spacing, near 668 Å, not the decreased $D$ period found in skin. The meridional intensities for these reconstituted type III fibrils appeared to be very different from those found for type I fibrils, but the pattern was too weak for calculation of intensity values. The $D$ period and meridional intensities of fibers reconstituted from 85% type I collagen and 15% type III collagen were similar to those found for a type I reconstituted fiber, but there was no direct evidence that type III was present in fibrils in this case. The intermolecular spacings of these reconstituted fibril containing type III collagen were similar to reconstituted fibers formed from only type I collagen (Table III). Thus, the decreased $D$ period and increased intermolecular spacing found in the X-ray patterns of skin and spleen reticulum were not observed in these reconstituted fibers containing type III collagen. It is possible that the type III collagen was altered during purification; or that the conditions for reprecipitation were not suitable to lead the *in vivo* arrangement of components; or that other tissue components are involved in the determination of these features.

In summary, a number of distinctive features were found for fibrils in tissues with type III collagen. Most striking are the decreased $D$ period, the increased equatorial spread, the tilting of molecules with respect to the fibrils axis, and a decreased fibril size of reprecipitated type III fibers. It is not clear at this point whether these features are caused by the presence of type III collagen or are related to some other factor found in *in vivo* situations.

[104] C. M. Lapiere, B. Nusgens, and G. E. Pierard, *Connect. Tissue Res.* 5, 21 (1977).

## Collagen Fibril Structure in Developing Connective Tissues

Embryonic or developing connective tissues show both biochemical and morphological differences from mature tissues. There is more hydroxylysine and glycosylation in embryonic tendon, bone, and skin, and this seems to decrease with age.[78] Embryonic skin contains more type III collagen than mature skin, and embryonic tendon may contain an increased amount of $\alpha 1(I)$ trimer.[105] In addition to these differences in collagen composition, there are changes in other components, a larger amount of cellular material, fibronectin, and proteoglycan being found in embryonic tissues.

Collagen fibrils in immature tissues all appear to be of small and uniform diameter,[74,106] and they also are argyrophilic, with a distinctive silver staining reaction. Fibrils appear to be on a disordered lattice, with adjacent fibers separated by a distance comparable to the fibril diameter.

X-Ray diffraction patterns have been obtained on embryonic chick leg tendons of varying ages.[107] The meridional $D$ spacing and the intermolecular spacing were similar to those found in mature chicken leg tendon patterns. The meridional intensities were similar except for a significant decrease in the intensity of the first order in the embryonic tendon, which is consistent with a partial filling of the gap region. In the embryonic patterns several very low-angle equatorial maxima were present that are never seen in patterns of mature tendons (Fig. 8). The positions and intensities of these maxima were consistent with diffraction from cylinders of uniform diameter. Calculation of fibril diameters indicated an increase with age, going from 460 Å in 13-day embryo tendons to 580 Å in 19-day embryo tendons. These data suggest that the fibrils of larger diameter and more loosely packed than has been visualized by electron microscopy.[106]

Thus, collagen fibrils in immature tissues have characteristic biochemical and morphological features, and it may now be possible to examine how the components in the tissue determine fibril structure and lead to their distinctive appearance and distribution.

## Collagen Fibril Structure in Diseases of Connective Tissues

Investigations are being carried out on diseases of connective tissue to determine the basic molecular defect and the way in which this defect

---

[105] S. A. Jimenez, R. I. Bashey, M. Benditt, and R. Yankowski, *Biochem. Biophys. Res. Commun.* **78**, 1354 (1977).
[106] S. Fitton Jackson, *Proc. R. Soc. London, Ser. B* **144**, 556 (1956).
[107] E. Eikenberry, B. Brodsky, and D. A. D. Parry, *Int. J. Biol. Macromol.*, submitted.

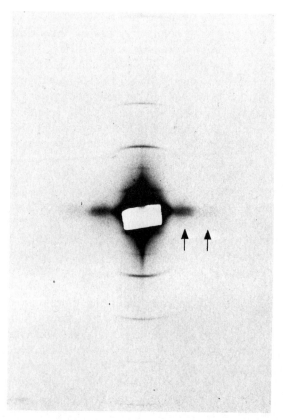

FIG. 8. Low-angle X-ray diffraction of 17-day embryonic chick leg tendon. The low-angle equatorial maxima, which are present only in specimens of immature tendons, are indicated by arrows.

causes tissue abnormalities and symptoms.[108,109] The best characterized case is dermatosparaxis, in which a defect in procollagen N-proteinase activity results in the accumulation of pN-collagen in the skin. Other biochemical alterations in composition have also been found, such as abnormal type I : type III ratios in the skins of patients with Ehlers–Danlos type IV and in some osteogenesis imperfecta patients. The relationship between such compositional alterations and the defective tissue properties must be considered. The molecular defect could alter the collagen fibril structure and thus alter tissue properties, or another more indi-

[108] R. R. Minor, *Am. J. Pathol.* **98**, 227 (1980).
[109] P. Bornstein and P. H. Byers, "Collagen Metabolism." Upjohn, 1980.

rect mechanism could be involved. Electron microscopy and X-ray diffraction are both methods of investigating possible abnormalities in fibril structure, and thus may be useful in bridging the gap between the molecular etiology and the defective tissue structure.

### Electron Microscopy

Dermatosparaxis has been the best characterized collagen disease from a structural viewpoint. Collagen fibrils in the skins of dermatosparactic calves and lambs show very abnormal fibril cross sections, with thin irregular branches.[110] In longitudinal sections, or on fibrils from suspensions, the $D$ period is clearly visible, but some twisting of fibrils is seen. Collagen fibrils of tendons from dermatosparactic animals, which contain less pN-collagen than do skins, show less distortion of fibril cross sections and less twisting. Labeling experiments were done on dermatosparactic skin with ferritin-conjugated antibodies directed against the N-propeptide, and electron micrographs clearly indicated that pN-collagen is incorporated into fibrils.[47] Thus, in dermatosparaxis, abnormal fibril morphology appears to be the result of a molecular defect and could account for the observed skin fragility.

Fibers in other connective tissue diseases, such as in several of the Ehlers–Danlos syndromes, look normal in transmission electron micrographs, but show some distortions in scanning electron microscopy.[111] Alternations in fibril diameters and distributions and other irregularities have been reported from electron micrographs of a variety of animals with connective tissue abnormalities.[109] However, interpretation of such abnormal fibril features in terms of molecular defects has proved to be difficult. As discussed earlier, it is not yet known what determines the diameter of collagen fibrils, so it is difficult to hypothesize what defect results in abnormal diameters.

### X-Ray Diffraction

X-Ray diffraction can be done on hydrated specimens, and gives specific information on fibril structure; so, it is a useful technique for investigating alterations in collagen fibril structure. However, in general, only tendons have been examined by low- and medium-angle X-ray diffraction and, in connective tissue diseases, it is often other tissues that are

---

[110] M. Fjolstad and O. Helle, *J. Pathol.* **112**, 183 (1975).
[111] C. M. Black, L. J. Gathercole, A. J. Bailey, and P. Beighton, *Br. J. Dermatol.* **102**, 85 (1980).

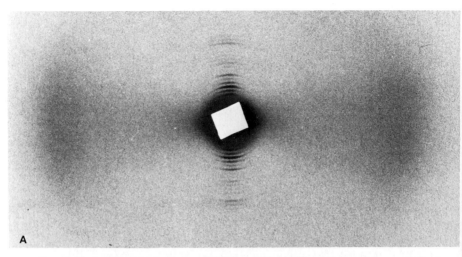

FIG. 9. X-ray diffraction patterns of (A) normal lamb skin; (B) dematosparactic lamb skin; (C) dog skin with abnormal fibril morphology; (D) skin from a biopsy sample from a patient with osteogenesis imperfecta. All specimens were immersed in physiological buffer while the patterns were being recorded, and a 15.94 cm specimen-to-film distance was used in all cases. The meridional $D$ period in (A) in 652 Å; all pictures are to the same scale.

most affected. However, with the ability to get good medium- and low-angle patterns from skin, demineralized bone, aorta, and other tissues, it may be possible to extend X-ray studies to a variety of diseased tissues. A limitation to this approach arises from the loss of the meridional X-ray diffraction pattern upon disruption of long range axial order in fibrils, which occurs in some disease states.

A detailed X-ray investigation comparing the collagen fibril structure in tendon and skin of dermatosparactic lambs with that in normal tissues was carried out.[112] The dermatosparactic tendons, with about 13% pN-collagen, gave a pattern indistinguishable from that of normal lamb tendons, whereas the affected skin gave extremely weak meridional diffraction, indicating a lack of long-range axial order (Fig. 9B). Electron micrographs of suspensions of this skin do show a well defined $D$ period in the fibrils, giving the appearance of long-range axial order. It may be that the original disordered structure becomes ordered during specimen preparation, or perhaps that the electron microscopic image appears more ordered because it is two-dimensional. Both drying the dermatosparactic skin and treating it with disulfide bond-reducing agents resulted in improved meridional diffraction. It is likely that reduction results in a loss of

[112] K. Cassidy, E. Eikenberry, and B. Brodsky, *Lab. Invest.* **43**, 542 (1980).

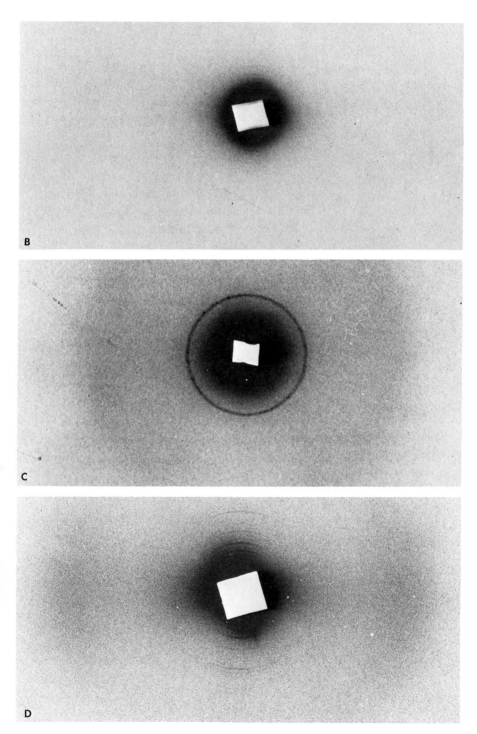

FIG. 9B, C, and D.

the native conformation of the N-propeptide, which then allows rearrangement of molecules to a more ordered state.

X-Ray diffraction was done on skins where altered fibril diameter distributions were seen by electron microscopy.[113,114] In these skins, strong lipid diffraction was often observed, sometimes in the form of a powder pattern of crystalline material (Fig. 9C). X-Ray patterns of several skin biopsies from patients with osteogenesis imperfecta gave patterns indistinguishable from those of normal skin (Fig. 9D), suggesting that largely normal collagen fibril morphology is present in these skins.

The ability to examine skin by medium- and low-angle X-ray diffraction has allowed the examination of diseased skins, and the acquisition of information to complement the electron microscopic observations of fibril structure abnormalities. It is hoped that the ability to look at other tissues and the increased knowledge of the biochemical abnormalities may make X-ray diffraction a useful tool in relating such defects to alterations in fibril structure.

Conclusion

Collagen fibers can be studied in tissues or as fibers reprecipitated from solutions of purified collagen. The biochemical data establishing differences in the collagenous and noncollagenous components of various connective tissues has clarified the need to compare structures of fibrils in different tissues and of fibers reprecipitated from different genetic types of collagen. Such investigations have been carried out using electron microscopy and X-ray diffraction on both native tissues and reprecipitated fibers.

From electron microscopy, one gets an image of the fibril showing the gap:overlap regions (by negative staining) or the distribution of charged residues (by positive staining). Examination of various tissues and reprecipitated forms has led to the following conclusions.

1. The SLS banding patterns of different genetic types of collagen can be distinguished, but no differences can be seen in the native-type fibrils reprecipitated from purified type I, type II, or type III collagens.
2. Freeze-fracture preparations of cartilage or tendon show fibers with striations running parallel to the fiber axis, whereas skin, aorta, and other type III-containing tissues have striations tilted at an angle to the fibril axis.

[113] K. Cassidy, P. Byers, K. Holbrook, E. Eikenberry, and B. Brodsky, unpublished results.
[114] R. Minor and K. Cassidy, unpublished results.

3. The diameters of fibrils are characteristic of each kind of tissue. In skin, there is a unimodal distribution of fibril diameters, while in mature tendon, a bimodal distribution is present. The diameters of reprecipitated native-type fibrils depend on the conditions of fibril formation and the presence of other components.
4. Fibrils in developing tissues are of small diameter with a narrow distribution of diameters, unlike those in mature tissues.
5. In diseased tissues, fibrils may appear disrupted, not cylindrical, and the distribution of fibril diameters may be different from normal.

X-Ray diffraction studies give information on the axial fibril structure and the lateral packing of molecules in the fibril. Examination of various tissues and reprecipitated fibers has indicated that the X-ray patterns vary from the well-studied one of rat tail tendon in the following features.

1. The decreased $D$ period and increased equatorial spread observed in skin specimens is consistent with the tilting of molecules by about 12° with respect to the fibril axis.
2. There are differences in the degree of order of the lateral packing of molecules, and the spacing between molecules varies in different tissues, being smallest for noncalcifying tendons (14.4–14.9 Å), followed by skin, bone, and calcifying avian tendons (about 15.3 Å), and being largest for cartilage (about 16 Å).
3. There are differences in the relative meridional intensities for tendons from different animals, especially between avian and mammalian ones. Apparently, type II-containing tissues have very different meridional intensities from type I tissues, but more data are needed to confirm this.
4. Developing tendons show a decreased first-order meridional intensity and the presence of several low-angle equatorial maxima, related to the presence of small uniform cylinders. These maxima can be used to determine the fibril diameter.
5. Diseased tissues may show a dramatic loss of diffraction, due to loss of long-range order, or other unusual features, such as the appearance of new crystalline lipid rings. Some tissues from patients or animals with connective tissue diseases give normal diffraction patterns, suggesting that the fibril structure is not involved in the abnormality.

These electron microscopic and X-ray observations clearly indicate differences in fibril structure and diameters in different tissues, but do not clarify whether such differences are due directly to the presence of different genetic types of collagen, or to the different noncollagenous components. The factors determining the diameters of fibrils are still unclear. The challenge will be to determine more precisely the fibril structures present in different tissues and to relate these structures on the one hand to the biochemical composition of each tissue and on the other hand to the mechanical and functional needs of the tissue.

## [6] Solid State NMR Studies of Molecular Motion in Collagen Fibrils

*By* D. A. TORCHIA

Nuclear magnetic resonance (NMR) studies of molecular motion within collagen fibrils are briefly described in this chapter. Solid state NMR methods, when used in conjunction with samples that contain $^{13}$C- or $^{2}$H-enriched amino acid residues, provide information about the rate and angular range of molecular motion at specific sites in collagen. In Vols. 27, 49, and 61 of this series NMR studies of soluble proteins were reviewed. Because solid state and solution NMR methodology differ, I will first present background information about solid state NMR.

### $^{13}$C and $^{2}$H Nuclear Magnetic Resonance in Solids

#### $^{13}$C–$^{1}$H Magnetic Double Resonance in Solids

High-resolution $^{13}$C spectra of solutions normally exhibit a single resonance for each magnetically distinct carbon when the $^{13}$C–$^{1}$H scalar coupling ($J$ coupling) is suppressed by resonant irradiation of the protons; this is called scalar decoupling.[1,2] Since $^{13}$C–$^{1}$H scalar couplings are usually less than 200 Hz, a rotating frame field of 0.7 G ($\gamma H_2/2\pi$ = 3 kHz) is sufficient for scalar decoupling. In addition to the scalar interaction, there is also an orientation-dependent dipole–dipole interaction between the $^{13}$C and hydrogen nuclei. This static dipolar interaction is much stronger than the scalar interaction.[3,4] For example, the rms dipolar coupling between the

[1] T. C. Farrar and E. D. Becker, "Introduction to Pulse and Fourier Transform NMR Methods." Academic Press, New York, 1971.
[2] G. C. Levy and G. L. Nelson, "Carbon-13 Nuclear Magnetic Resonance for Organic Chemists." Wiley (Interscience), New York, 1972.
[3] A. Abragam, "The Principles of Nuclear Magnetism." Oxford Univ. Press, London and New York, 1961.
[4] C. P. Slichter, "Principles of Magnetic Resonance." Springer-Verlag, Berlin and New York, 1978.

peptide carbonyl carbon and the two nearest protons has a value of 2.2 kHz, and the coupling between the peptide $\alpha$-carbon and the directly bonded $\alpha$-proton has a value of about 11 kHz. The static dipole–dipole coupling between $^{13}C$ nuclei is small compared with the $^{13}C$–$^1H$ dipolar interaction because the natural abundance of $^{13}C$ is low (1.1%) and the carbon magnetogyric ratio is one-fourth that of the proton. Unlike the scalar interaction, the $^{13}C$–$^1H$ dipolar interaction depends on the orientation of the C—H bond axis with respect to the external field. Thus, isotropic rotational diffusion, characterized by a correlation time $\tau \leq 10^{-8}$ sec, averages the dipolar interaction and results in small $^{13}C$ linewidths ($\leq 10$ Hz).[5,6] If motion is anisotropic[7] and any correlation time exceeds $10^{-5}$ sec, or if rotational diffusion is absent, as in a rigid lattice, the linewidth is so large that observation of the $^{13}C$ signal is normally not possible.

A method for eliminating the line broadening that arises from heteronuclear dipolar interactions was first suggested by Bloch.[8] This dipolar decoupling technique is similar to that used for scalar decoupling, except that a much larger field ($\gtrsim 12$ G, $\gamma H_2/2\pi > 50$ kHz) is applied to the protons,[8,9] because the decoupling field must exceed the strong $^{13}C$–$^1H$ and $^1H$–$^1H$ dipolar couplings. As is the case with scalar decoupled spectra in solution, spin-lattice relaxation times ($T_1$) and nuclear Overhauser enhancements (NOE) can be derived from dipolar decoupled spectra and used to obtain information about molecular motions on a time scale of microseconds to nanoseconds.[10-14] In addition, the observed lineshape often provides information about the residual chemical shift anisotropy. Analysis of the lineshape in these cases provides information about molecular motions in the millisecond time scale.

While dipolar decoupling greatly extends the range of application of the $^{13}C$ NMR, sensitivity is sometimes a problem. This is particularly true in rigid solids, where $^{13}C$ $T_1$ values are very large and the time required to acquire a signal is prohibitively long. Pines et al.[9] have shown that cross-polarization produced by a matched Hartmann–Hahn[15] contact provides the sensitivity enhancement that is needed for studying samples with large

---

[5] I. Solomon, *Phys. Rev.* **99**, 559 (1955).
[6] D. Doddrell, V. Glushko, and A. Allerhand, *J. Chem. Phys.* **56**, 3683 (1972).
[7] D. E. Woessner, *J. Chem. Phys.* **37**, 647 (1962).
[8] F. Bloch, *Phys. Rev.* **111**, 841 (1958).
[9] A. Pines, M. G. Gibby, and J. S. Waugh, *J. Chem. Phys.* **59**, 569 (1973).
[10] M. G. Gibby, A. Pines, and J. S. Waugh, *Chem. Phys. Lett.* **16**, 296 (1972).
[11] M. Mehring, *NMR: Basic Princ. Prog.* **11**, 1 (1976).
[12] H. W. Spiess, *NMR: Basic Princ. Prog.* **15**, 55 (1978).
[13] J. Schaefer and E. O. Stejskal, *Top. Carbon-13 NMR Spectrosc.* **3**, 284 (1979).
[14] D. A. Torchia and D. L. VanderHart, *Top. Carbon-13 NMR Spectrosc.* **3**, 325 (1979).
[15] S. R. Hartmann and E. L. Hahn, *Phys. Rev.* **128**, 2042 (1962).

$^{13}$C $T_1$ values. After the $^{13}$C polarization is established, the Hartmann–Hahn contact is broken, and the $^{13}$C free-induction decay is detected with dipolar decoupling.[9] The proton-enhanced $^{13}$C spectrum is the Fourier transform of the free-induction decay. Since the proton magnetogyric ratio is four times that of $^{13}$C, a single contact provides up to a fourfold enhancement of $^{13}$C signal intensity over the $90°-t$ pulse sequence. A signal can also be acquired much faster when the proton $T_1$ values are much smaller than those of $^{13}$C.

One caveat regarding sensitivity enhancement is that $^{13}$C $T_1$ values can be significantly shorter than $^1$H $T_1$ values. This has been shown to be true in hydrated systems (including collagen) where there is spin diffusion between water protons and solute protons.[16]

## $^2$H Nuclear Magnetic Resonance in Solids

The deuteron has spin 1, and therefore has a quadrupole moment.[3,4,17] The quadrupole interaction is dependent upon the orientation of the deuterium chemical bond axis with respect to the external magnetic field. Therefore, in a solid each orientation of a deuterium bond axis gives rise to a resonance at a different frequency resulting in a powder lineshape that covers a frequency range of up to 300 kHz. Because of this large spread of resonance frequencies, spectrometer recovery time exceeds the inhomogeneous spin–spin relaxation time $T_2^*$, and only a small fraction of the free induction decay following a 90° pulse is detected. This problem can be circumvented using the pulse sequence $90°_{\pm x} - t_1 - 90°_y - t_2$ which produces an echo at the end of $t_2$.[18,19] The echo is identical to the free induction decay following a 90° pulse if $t_1 + t_2$ is small compared with the homogeneous spin–spin relaxation time, $T_2$.[20] The Fourier transform of the echo is the $^2$H NMR signal in the frequency domain. Analysis of the lineshape provides information about molecular motion of the deuterium–carbon bond axis.

It is desirable to obtain $^2$H NMR spectra at high magnetic fields in order to maximize sensitivity and to minimize spectrometer recovery problems. A high power probe and amplifier and a fast transient recorder (~500 kHz window) are also required.

[16] L. W. Jelinski, C. E. Sullivan, and D. A. Torchia, *J. Magn. Reson.* **41**, 133 (1980).
[17] H. H. Mantsch, H. Saito, and I. C. P. Smith, *Prog. Nucl. Magn. Reson. Spectrosc.* **11**, 24 (1977).
[18] I. Solomon, *Phys. Rev.* **110**, 61 (1958).
[19] J. H. Davis, K. R. Jeffrey, M. Bloom, M. I. Valic, and T. P. Higgs, *Chem. Phys. Lett.* **42**, 390 (1976).
[20] H. W. Spiess and H. Silescue, *J. Magn. Res.* **42**, 381 (1981).

Preparation of $^{13}$C- and $^2$H-Labeled Collagen Fibrils

The low natural abundance of $^{13}$C (1.1%) and $^2$H (0.016%) has often been a deterrent to obtaining NMR spectra of these nuclei. Their low natural abundance is an advantage if $^{13}$C- or $^2$H-enriched amino acid residues are incorporated into a protein, since resonances of the enriched nuclei are readily assigned. Collagen, containing amino acid residues enriched 15- to 80-fold with $^{13}$C or $^2$H, has been obtained as follows[21-25]: Calvaria parietal bones from 25 dozen 17-day-old chick embryos are preincubated at 38–39° in a humidified 5% $CO_2$/air atmosphere for two 2-hr periods in order to deplete the free amino acid pool. Both incubations use 250 ml of Eagle's minimal essential medium, deficient in the amino acid of interest supplemented, per 100 ml, with the following: 10 mg of glycine, 10 mg of L-proline, 5–20 mg of the $^{13}$C- or $^2$H-labeled amino acid of interest, 5 mg of L-ascorbic acid (sodium salt), 10–50 $\mu$Ci of the $^{14}$C- or $^3$H-labeled amino acid, and 5000 units of penicillin/streptomycin. After the two 2-hr preincubation periods, the calvaria are incubated for two 48-hr periods in 250 ml of the above media, to which 5 mg of $\beta$-aminopropionitrile fumarate are added per 100 ml. ($\beta$-Aminopropionitrile fumarate inhibits collagen cross-linking, and thus produces lathyrism.)

After incubation, the calvaria are pooled, washed five times with a total of 1 liter of cold water, homogenized in 100 ml of 1 $M$ NaCl, 0.05 $M$ Tris (pH 7.4), and stirred overnight. The soluble collagen is separated from the ground bones by centrifugation and is then precipitated by the addition of 20% (w/w) solid NaCl. The precipitate is collected by centrifugation, dissolved in 0.5 $M$ acetic acid, and dialyzed versus 0.5 $M$ acetic acid. The dialysis bag contents are clarified by centrifugation for 90 min at 45,000 rpm (Beckman Ti-50 rotor) and then precipitated by dialysis against successive changes of 0.02 $M$ $Na_2HPO_4$. The fibrils are collected by centrifugation, and their morphology is checked by electron microscopy. Collagen concentrations and melting temperatures are determined after dissolution of these fibrils into 0.5 $M$ acetic acid and dialysis versus 0.5 $M$ acetic acid. Typically, 10–20 mg of collagen are obtained from 25 dozen chick embryos.

The percentage incorporation of $^{13}$C or $^2$H is determined by gas chromatography–mass spectroscopy analysis of the $N$-acetylmethyl ester

---

[21] D. A. Torchia and D. L. VanderHart, *J. Mol. Biol.* **104**, 315 (1976).
[22] L. W. Jelinski and D. A. Torchia, *J. Mol. Biol.* **133**, 45 (1979).
[23] L. W. Jelinski and D. A. Torchia, *J. Mol. Biol.* **138**, 255 (1980).
[24] L. W. Jelinski, C. E. Sullivan, and D. A. Torchia, *Nature (London)* **284**, 531 (1980).
[25] L. W. Jelinski, C. E. Sullivan, L. S. Batchelder, and D. A. Torchia, *Biophys. J.* **15**, 515 (1980).

TABLE I
LEVELS OF $^{13}$C- AND $^{2}$H-LABELED AMINO ACIDS
INCORPORATED INTO CHICK CALVARIA
COLLAGEN[a]

| Amino acid | Incorporation (%) |
|---|---|
| [1-$^{13}$C]Glycine | 48 |
| [2-$^{13}$C]Glycine | 54 |
| [3,3,3-$^{2}$H$_3$]Alanine | 14 |
| [methyl-$^{13}$C]Methionine | 85 |
| [6-$^{13}$C]Lysine | 66 |
| [5-$^{13}$C]Glutamic acid | 15 |
| [$^{2}$H$_{10}$]Leucine | 35 |

[a] Natural abundance levels are 1.1% for $^{13}$C and 0.016% for $^{2}$H.

derivatives of the protein hydrolyzates.[22] Table I lists the levels of enrichment obtained for various amino acid residues in chick calvaria collagen.

## Motion of the Polypeptide Backbone in the Collagen Fibril

Collagen fibrils enriched with [2-$^{13}$C]glycine (α-carbon label), [1-$^{13}$C]glycine (peptide carbonyl label), and [3,3,3-$^{2}$H$_3$]alanine (methyl label) were studied to obtain information about reorientation of the collagen backbone.

$^{13}$C spectra of collagen labeled with [2-$^{13}$C]glycine are shown in Fig. 1 for fibrils (a) and for the triple helix in solution (b). High power decoupling is required to obtain these spectra, since the diffusion coefficient for end-over-end tumbling of the molecule, $R_2$, is small ($R_2 \leq 10^3$ sec$^{-1}$) in solution and zero in the fibrils. $^{13}$C spin lattice relaxation times, $T_1$, and $^{13}$C-$\{^1$H$\}$ nuclear Overhauser enhancements were measured for the glycine α-carbon using standard pulse sequences, and the results are tabulated in Table II. These data and dipolar relaxation theory[5,7] provide information about reorientation of the glycine C$^\alpha$—H$^\alpha$ bond axes in collagen once a model for molecular motion is assumed. In solution, we take a rigid ellipsoid of revolution as a model for the collagen molecule. The dimensions of the model ellipsoid are specified by the condition that the axial ratio and volume of the ellipsoid equal those of a 1.5 nm × 300 nm cylinder.[26,27]

[26] P. H. von Hippel, in "Treatise on Collagen" (G. N. Ramachandran, ed.), Vol. 1, p. 253. Academic Press, New York, 1967.
[27] G. C. Fletcher, Biopolymers 15, 2201 (1976).

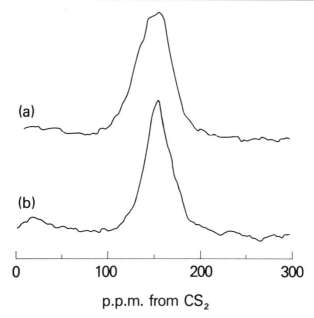

FIG. 1. $^{13}$C NMR spectra of [2-$^{13}$C]glycine-labeled chick calvaria collagen as fibrils and in solution. Curve a: Dipolar decoupled spectrum of collagen fibrils in equilibrium with 0.02 $M$ Na$_2$HPO$_4$, 4096 acquisitions, $t = 1$ sec, 16°. Curve b: Dipolar decoupled spectrum of collagen in solution, 13 mg/ml in 0.1 $M$ acetic acid, 5361 acquisitions, $t = 1$ sec, 16°. The above spectra were obtained at 15.09 MHz using a (90° − $t$) pulse sequence, with a nuclear Overhauser enhancement.

Analysis of the NMR data[22] yields values of $R_1$ (the diffusion coefficient for reorientation about the long axis of the molecule) of $1 \times 10^7$ sec$^{-1}$ from the $T_1$ and $6.3 \times 10^7$ sec$^{-1}$ from the NOE. A value of $R_1$ of $1.7 \times 10^6$ sec$^{-1}$ is calculated from theory for rigid body rotational diffusion of the ellipsoid.[7,22,28] The calculated value of $R_1$ is about an order of magnitude smaller than the values of $R_1$ obtained from the NMR data. This result suggests that, in addition to rigid body reorientation, fast ($R > 10^7$ sec$^{-1}$) small amplitude reorientation, such as would result from torsion about the long axis, also takes place in the collagen molecule.

The $T_1$ and NOE values obtained for collagen fibrils are similar to the values of these parameters obtained in solution, a result that immediately suggests that the collagen backbone reorients in the fibril. In view of the fact that the molecules are well ordered in the direction parallel to the long axis of the molecule, we assume that reorientation in the fibril is around the long axis. As will be seen, $^{13}$C and $^2$H NMR lineshapes of collagen

[28] F. Perrin, *J. Phys. Radium* **7**, 1 (1936).

TABLE II
$T_1$ AND NUCLEAR OVERHAUSER ENHANCEMENT (NOE) VALUES
MEASURED FOR COLLAGEN LABELED WITH [2-$^{13}$C]GLYCINE AND
[1-$^{13}$C]GLYCINE

|  | $T_1$(sec) | | NOE | |
| --- | --- | --- | --- | --- |
|  | Solution | Fibrils | Solution | Fibrils |
| [2-$^{13}$C]Glycine | 0.03 | 0.13 | 1.44 | 1.74 |
| [1-$^{13}$C]Glycine | 0.54 | 2.0 | 1.52 | 1.60 |

fibers exclude free rotation about the long axis. The simplest model for reorientation that is compatible with the NMR data is a two-site model in which the molecule is assumed to jump between two azimuthal orientations separated by an angle $\Delta\phi$. For simplicity, we assume that the residence time in each orientation is the same, and that the time required to jump from one orientation to the other is negligible compared to the residence time. Analysis of the NMR data[21,22] shows that $\Delta\phi \sim 30°$ if $R_j \sim 10^8$ sec$^{-1}$ ($R_j^{-1}$ is the residence time). The value of $\Delta\phi$ is greater than 30° if $R_j$ differs significantly from $10^8$ sec$^{-1}$.

The spectral parameters obtained for collagen fibrils labeled with [1 − $^{13}$C]glycine support the conclusions about backbone reorientation just discussed. The small $T_1$ value and the NOE value (Table II) (which is between the extreme values obtained for very fast or very slow motion) provide additional evidence for backbone reorientation having $R \sim 10^8$ sec$^{-1}$. The carbonyl signal of the fibrils has the lineshape of an axially asymmetric chemical shift tensor and a width of 100 ppm. The linewidth is about two-thirds of the 150 ppm powder linewidth observed for the peptide carbonyl of polycrystalline glycylglycine (Fig. 2). The observation of a smaller linewidth and an axially asymmetric lineshape in the spectrum of the collagen fibrils confirms that (*a*) rapid reorientation, $R \geqslant 10^3$ sec$^{-1}$, of the collagen backbone takes place in the fibrils; and (*b*) the motion is not axially symmetric.

At present, it is not possible to obtain an estimate of $\Delta\phi$ from an analysis of the glycine carbonyl lineshape, because the orientation of the peptide carbonyl chemical shift tensor has not been determined. It is possible to obtain $\Delta\phi$ from an analysis of a deuterium lineshape because it is known that the deuterium field gradient tensor is axially symmetric and that the symmetry axis is along the C–D bond axis.[17] The deuterium spectrum of polycrystalline [3,3,3-$^2$H$_3$]alanine (Fig. 3a) shows an axially symmetric powder pattern with a quadrupolar splitting of 39 kHz. This pattern is a motionally averaged lineshape resulting from threefold jumps

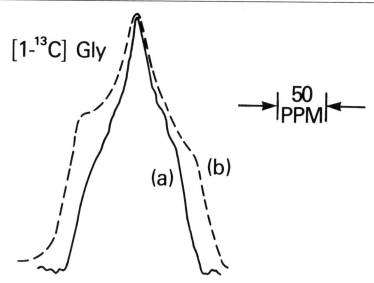

FIG. 2. Comparison of the $^{13}$C NMR spectrum of [1-$^{13}$C]glycine-labeled chick calvaria collagen fibrils with the spectrum of polycrystalline glycylglycine 10% labeled at the peptide carbonyl carbon. Curve a: Dipolar decoupled (90° − $t$) spectrum of reconstituted collagen fibrils in equilibrium with 0.02 $M$ Na$_2$HPO$_4$, 8163 acquisitions, $t$ = 6 sec, 16°, 15.09 MHz. Curve b: Proton-enhanced spectrum of glycylglycine, dry powder at 20°, 74 MHz.

of the methyl group around the C$^\alpha$—C$^\beta$ bond axis of alanine. As a consequence of this motion, the symmetry axis of the averaged field gradient tensor is the C$^\alpha$—C$^\beta$ bond axis. The lineshape of frozen [3,3,3-$^2$H$_3$]alanine-labeled collagen fibrils (Fig. 3b) is similar to that of the polycrystalline amino acid. In contrast, the spectrum of the alanine-labeled collagen at 18° (Fig. 3c) shows a markedly different lineshape and contains only 40% of the signal intensity observed for the frozen fibrils. Both the loss in signal intensity and the observed lineshape are a consequence of reorientation of the alanine C$^\alpha$—C$^\beta$ bond axis. In order to calculate the fibril lineshape, we again assume that azimuthal reorientation of the collagen backbone takes place by means of a two-site jump. Since the $^{13}$C relaxation data show that the reorientation is fast ($R \geq 10^7 \gg \Delta\nu_q$) the principal axes, principal frequencies, and lineshape can be readily calculated as a function of $\Delta\phi$ for the two-site model.[24,25,29] The calculated lineshape that best matches the experimental spectrum is obtained using $\Delta\phi$ = 30°. This value agrees with the value of $\Delta\phi$ obtained from the analysis of the $^{13}$C relaxation data.

[29] G. Soda and T. Chiba, *J. Chem. Phys.* **50**, 439 (1969).

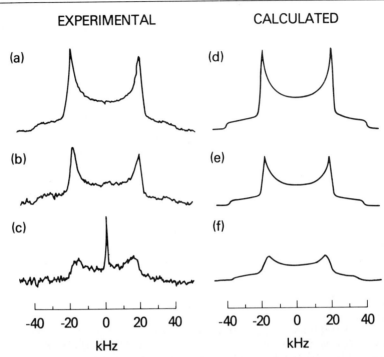

FIG. 3. Comparison of experimental and calculated $^2$H NMR spectra of [3,3,3-$^2$H$_3$]alanine and [3,3,3-$^2$H$_3$]alanine-labeled collagen. The experimental spectra (a–c) were obtained in quadrature using the solid echo pulse sequence ($90°_{\pm x} - t_1 - 90°_y - t_2 - T$) with $t_1$ and $t_2$ equal to 50 and 59 μsec, respectively. (a) [3,3,3-$^2$H$_3$]Alanine as a polycrystalline amino acid, 256 accumulations, 0.25 sec repetition rate, 18°. (b) [3,3,3-$^2$H$_3$]Alanine-labeled collagen fibrils, in equilibrium with excess 0.02 $M$ Na$_2$HPO$_4$, 0.25 sec repetition rate, $-18°$. (c) [3,3,3-$^2$H$_3$]Alanine-labeled collagen fibrils in equilibrium with excess 0.02 $M$ Na$_2$HPO$_4$, 1.67 × 10$^5$ accumulations, 0.25 sec repetition rate, 18°. Calculated spectra (d–f) of [3,3,3-$^2$H$_3$]alanine and [3,3,3-$^2$H$_3$]alanine-labeled collagen. Lorentzian line broadening was employed. (d) Polycrystalline [3,3,3-$^2$H$_3$]alanine, $\Delta\nu_q$ = 38.8 kHz, line broadening 1.3 kHz. (e) Frozen [3,3,3-$^2$H$_3$]alanine-labeled collagen $\Delta\nu_q$ = 37.3 kHz, line broadening = 1.5 kHz. (f) Spectrum of [3,3,3-$^2$H$_3$]alanine-labeled collagen fibrils at 18°, calculated assuming that the alanine C$^\alpha$—C$^\beta$ bond axes undergo jump diffusion between two equally populated azimuthal orientations. The angles $\theta$ between the C$^\alpha$—C$^\beta$ bond axes and the helix axis are 69° and 85°. The angle $\Delta\phi$ through which each alanyl C$^\alpha$—C$^\beta$ bond axis is carried by reorientation of the peptide backbone is 30°. The spectra calculated for $\theta$ = 69° and $\theta$ = 85° were summed with equal weight, $\Delta\nu_q$ = 37.3 kHz, line broadening = 2.5 kHz.

The value of $\Delta\phi$ obtained from this analysis is model dependent, and the value of 30° is almost certainly an underestimate of the true range of reorientation. Larger values of $\Delta\phi$ must be used to account for the data if the two sites have unequal populations or if there is a continuous range of azimuthal orientations rather than only two.

Another point worth noting is that $^{13}$C NMR intensity measurements show that at least 90% of the glycine residues in collagen fibrils undergo reorientation. The deuterium spectra support this conclusion, since if more than 5% of the $C^\alpha$—$C^\beta$ alanine bond axes were rigid, sharp components would be observed at $\pm$ 18–19 kHz, as in the spectra (Fig. 3a,b) of the alanine powder and the frozen fibrils.

## Motions of the Polypeptide Side Chains in the Collagen Fibril

These results show that the azimuthal orientation of a collagen molecule in the fibril is not fixed, suggesting that the collagen side chains (whose interactions are responsible for fibril assembly) also reorient. This possibility has been investigated[23,25] by analyzing NMR spectra of collagen containing side chains specifically labeled with $^{13}$C or $^2$H. Spectra of collagen samples labeled at terminal carbon atoms in the side chains are shown in Fig. 4, where each is compared with the spectrum of the corresponding labeled amino acid as a crystalline powder. In every case the spectrum obtained for the labeled collagen has a significantly smaller linewidth than is observed in the spectrum of the crystalline powder. This result shows that reorientation of the labeled carbon groups has averaged the static chemical shift powder patterns. In the case of lysine and methionine, the linewidths in the fibrils are narrow, showing that the approximately 30 ppm static shift anisotropies of the labeled carbons are averaged by nearly isotropic reorientation. In the case of glutamic acid, the 150 ppm static anisotropy is averaged to about 20 ppm in the fibrils. If the orientation of the chemical shift tensor of the glutamic acid carboxyl carbon is similar to tensor orientations measured for model carboxylic acids, rotation around all three glutamic acid side-chain bond axes is required to account for the observed linewidth.[23]

The $^2$H spectrum of the leucine-labeled collagen fibrils (Fig. 5b) differs significantly from the spectrum of the polycrystalline amino acid (Fig. 5a). The $^2$H spectrum observed for the methyls of polycrystalline [$^2$H$_{10}$]leucine is similar to that observed for [3,3,3-$^2$H$_3$]alanine since threefold methyl reorientation occurs in both cases. Because of the methyl reorientation, the symmetry axis of the averaged field gradient tensor is along the leucine $C^\gamma$—$C^\delta$ bond axis of each methyl group. Reorientation of the leucine $C^\gamma$—$C^\delta$ bond axes causes the spectrum observed for leucine in collagen fibrils at 18° to differ from the amino acid spectrum. X-Ray structures of various crystalline peptides containing leucine show that only two side-chain conformations, $\chi_1 = 180°$, $\chi_2 = 60°$ and $\chi_1 = 300°$, $\chi_2 = 180°$, are found.[30,31] The orientations of the $C^\gamma$—$C^\delta$ bond axes change by 109.5°

[30] E. Benedetti, *Pept. Proc. Am. Pept. Symp. 5th, 1977*, p. 257 (1977).
[31] J. Janin, S. Wodak, M. Levitt, and B. Maigret, *J. Mol. Biol.* **125**, 357 (1978).

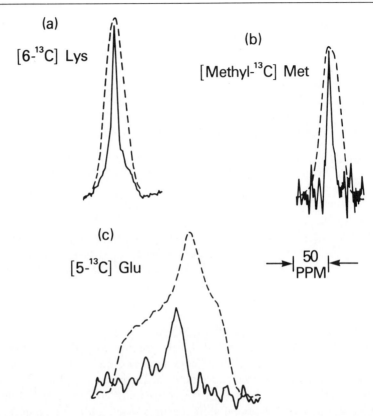

FIG. 4. Comparison of [13]C NMR spectra of collagen labeled at the terminal carbons of lysine, methionine, and glutamic acid with the [13]C NMR spectra of the corresponding powdered amino acids. (a) ——, Dipolar decoupled ($90° - t$) spectrum of [6-[13]C]lysine-labeled collagen fibrils in equilibrium with 0.02 $M$ Na$_2$HPO$_4$, 8192 acquisitions, 1 sec repetition rate, with NOE, 16°, 15.09 MHz. ----, Proton-enhanced spectrum of DL-[6-[13]C]lysine sodium phosphate salt, 4 msec contact time, 1 sec repetition rate, ambient temperature 74 MHz. (b) ——, Dipolar decoupled ($90° - t$) difference spectrum of [methyl-[13]C]methionine-labeled collagen fibrils, in equilibrium with 0.02 $M$ Na$_2$HPO$_4$, 6 sec repetition rate with NOE, 16°, 15.09 MHz. This spectrum was obtained by subtracting a spectrum of unlabeled collagen fibrils from the spectrum of labeled collagen in natural abundance. ----, Proton-enhanced spectrum of L-[methyl-[13]C]methionine (9 atom % [13]C) powder, 3031 acquisitions, 1 msec contact time, 1 sec repetition rate, −99°, 15.09 MHz. (c) ——, Dipolar decoupled ($90° - t$) spectrum of [5-[13]C]glutamic acid-labeled collagen fibrils, in equilibrium with 0.02 $M$ Na$_2$HPO$_4$, from which the natural abundance background was subtracted, 11,491 acquisitions, 6 sec repetition rate, with NOE, 16°, 15.09 MHz. ----, Proton-enhanced spectrum of DL-[5-[13]C]glutamic acid, (90 atom % [13]C) powder, 1024 acquisitions, 1 msec contact time, 3 sec repetition rate, −41°, 15.09 MHz.

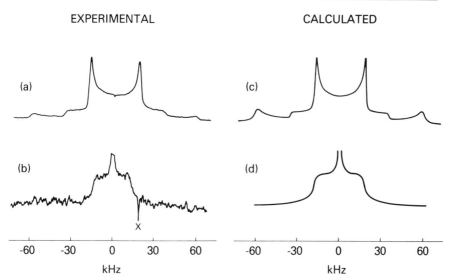

FIG. 5. Comparison of experimental and calculated $^2$H NMR spectra for [$^2$H$_{10}$]leucine and [$^2$H]leucine-labeled collagen. These spectra were obtained at 33.8 MHz using the pulse sequence $(90°_{\pm x} - t_1 - 90°_y - t_2 - T)$, with $t_1$ and $t_2$ set at 50 and 59 μsec, respectively. (a) [$^2$H$_{10}$]leucine as a polycrystalline amino acid, 1024 accumulations, 0.1 sec repetition rate, 18°. (b) [$^2$H]leucine-labeled collagen fibrils, in equilibrium with excess 0.02 $M$ Na$_2$HPO$_4$, 2.12 × 10$^5$ accumulations, 0.1 sec repetition rate, + 18°. Calculated spectra (c, d) of [$^2$H$_{10}$]leucine and [$^2$H]leucine-labeled collagen. Lorentzian line broadening was employed. (c) Calculated spectrum of polycrystalline [$^2$H$_{10}$]leucine, $\Delta\nu_q = 35.2$ kHz for the methyl groups $\Delta\nu_q = 117$ kHz for the nonmethyl deuterons, line broadening equals 1.1 kHz for methyl deuterons and 4.4 kHz for the nonmethyl deuterons. (d) Calculated spectrum of [$^2$H]leucine-labeled collagen fibrils at + 18°. The calculated spectrum is obtained assuming that the leucine side chain jumps between the two conformations discussed in the text with a rate equal to 6 × 10$^5$ sec$^{-1}$. A Lorentzian signal with 670 Hz linewidth was superimposed in the center of the spectrum to simulate the signal from deuterated water.

when the side-chain conformation changes from one of these states to the other. If the two side-chain conformations have equal populations in collagen and the rate of jumping between the two conformations is 6 × 10$^5$ sec$^{-1}$, one calculates the theoretical lineshape in Fig. 5d. The calculated spectrum is an accurate representation of the shape of the experimental spectrum. Although the calculated spectrum is broader than the experimental spectrum, this discrepancy is expected since backbone reorientation of collagen was not included in the theoretical analysis and would, if included, reduce the width of the calculated spectrum.

It is evident from these experiments that the amplitudes of side-chain reorientations within collagen fibrils are large and that the rates of these motions are greater than 10$^3$ sec$^{-1}$. The $T_1$ and NOE values measured for the $^{13}$C labeled side chains show that some of the side-chain motions have rates that exceed $10^9$ sec$^{-1}$.

Taken together, the NMR data for the labeled collagens show that the intermolecular interactions that occur in collagen fibrils take place in fluid-like domains at the molecular surfaces. We anticipate that our quantitative understanding of these interactions will be further refined by studies of the temperature dependence of $^2$H lineshapes.

## [7] Characterization of Soluble Collagens by Physical Techniques

*By* ARTHUR VEIS

Paper after paper describes the collagen molecule as a long rigid rod with a well defined axial ratio. Sequence analyses have brought us to the stage of knowing, at least for the interstitial collagens, the polypeptide chain sequences, which, with the known helix parameters, provide us with a detailed knowledge of the distribution of charged, polar and hydrophobic side chains along the helix. Why then have an extensive chapter on the characterization of soluble collagen molecules? The facts are that unperturbed collagen solutions rarely contain monomers; that the rigidity of the triple-helix region has not been experimentally determined; and that the conformations of the amino- and carboxyl-terminal regions of the molecules, either completed or in procollagen form, have not been worked out. Furthermore, the systems under the most intensive study at this time, the mechanism of fibrillogenesis and the role of collagen in platelet aggregation and activation, are highly dependent on the state of aggregation of the collagen used in *in vitro* experiments. Finally, there is the intrinsic interest of understanding the nature and properties of systems of highly asymmetric macromolecules.

The ensuing discussion begins with a theoretical description of the solution properties of asymmetric particles, using both the thermodynamic and hydrodynamic approaches. A number of excellent, very detailed discussions of the pertinent methods for measurement of the relevant properties are available, particularly in "Methods in Enzymology" Vols. 25–27, 47–49, and 61 on enzyme structure edited by C. H. W. Hirs and S. N. Timasheff. Rather than repeat these discussions on instrumentation and methodologies, the major point of the present chapter is to examine the nature of the data generated and the specific implications for the study of collagen and to suggest which methods of investigation might be selected to answer particular types of questions. The two features that make collagen distinctive are the rodlike character of the molecule and its highly self-interactive nature. We shall thus emphasize an analysis of those methodologies of special use in the characterization of the essential parameters of rodlike molecules and then conclude with a few specific examples.

## The Solution Behavior of Asymmetric Particles

### Basic Thermodynamic Theory

The fundamental approach to the thermodynamics of any binary system is through the use of Raoult's law and the Gibbs–Duhem equation, which link the activities ($a_i$) and chemical potentials ($\mu_i$) of the components of the mixture to their mole fraction ($N_i$) and in a binary solution enable one to express the activity of component 1, the solvent, in terms of the mole fraction of solute, component 2. In its usual form even this most basic expression is an approximation [Eq. (1)] involving a series expansion of $a_1$ in terms of $N_2$.

$$\ln a_1 = -[N_2 + \tfrac{1}{2}N_2^2 + \tfrac{1}{3}N_2^3 + \cdots] \tag{1}$$

This formulation is more useful in terms of the difference in chemical potential of component 1, $\mu_1$, between the solution and pure component 1 in its standard state, $\mu_1^0$, and the weight concentration, $g$, of component 2 [Eq. (2)].

$$\mu_1 - \mu_1^0 = RT\left[\frac{\overline{V}_1^0}{M_2}g + \frac{\overline{V}_1^0}{2\,M_2^2}g^2 + \frac{\overline{V}_1^0}{3\,M_2^3}g^3 + \cdots\right] \tag{2}$$

This difference is directly accessible to measurement by any one of the colligative properties, such as the osmotic pressure $\pi$ [Eq. (3)].

$$\pi/g = RT\,[A + Bg + Cg^2 + \cdots] \tag{3}$$

In an ideal solution, $\pi$ is given by the first term, in which $A = 1/M_2$, where $M_2$ is the particle molecular weight. Departures from ideality are thus dealt with by the remaining terms of the power series, where $B$, $C$, $\cdots$ are called the second, third, etc., virial coefficients, respectively.

It is obvious that a plot of $\pi/g$ vs $g$ should, at sufficiently low concentrations where the approximations apply, yield a straight line, the $g = 0$ intercept of which provides the value of $1/M_2$, and hence the number average molecular weight of component 2. In a system such as type I collagen, where the monomer value of $M_2$ is well known, $M_2$ is of interest as one measure of solution heterogeneity or aggregation.

Of equal interest in producing insight into the behavior of the solution is the value of the second virial coefficient, $B$. In operational terms the value of $B$ represents the sum of all interactions leading to deviation from ideal behavior. A zero value for $B$, pseudo ideality or the $\theta$ condition,[1] does not in fact signify truely ideal behavior, merely a balancing of favorable attractive and unfavorable repulsive interactions. An important contribution to $B$ that cannot be ignored in collagen solutions is the excluded volume. In essence, Eqs. (1) and (2) assume that all volume elements in a solution are equally accessible to solvent. If a solute particle is larger than a single solvent volume element, then a certain number of volume elements are excluded from occupancy by a second solute molecule. This excluded volume, $u$, is a function of the size, shape, and rigidity of the solute molecules and, in general, does not correspond to the physical volume of the molecule except in very concentrated solutions.

It can be shown that $B$ is related to $u$ by Eq. (4).

$$B = N_{av} u/2 M_2^2 \tag{4}$$

For rigid rodlike molecules, which can sweep out a large number of orientations around their center mass, $u$ is given by Eq. (5).

$$u = \frac{1}{2} dL^2 = \frac{2 L M_2 v_2}{dN_{av}} \tag{5}$$

where $L$ is the rod length, $d$ the diameter, $M_2$ the rod molecular weight, $v_2$, the partial specific volume in cm$^3$/g, and $N_{av}$, Avogadros' number. $B$ thus can readily be computed from Eq. (6).

$$B = (Lv_2)/dM_2 \tag{6}$$

For monomeric collagen, where $L$ is 300 nm, $d = 1.5$ nm, $v_2 = 0.72$ cm$^3$/g, and $M_2 = 290,000$, $B$ has the value of $4.96 \times 10^{-4}$ mol-cm$^3$/g$^2$. According to Eq. (3), $\pi/g$ at a concentration of 0.01 g/ml (10 g/liter) would have the value

$$\pi/g = RT [3.45 \times 10^{-6} + 5 \times 10^{-6}]$$

Thus, at a 1% concentration $\pi/g$ would be 144% too large. Inspection of Eq. (6) shows that an end-linked dimer, $L = 600$ nm, $d = 1.5$ nm, $M_2 = 6 \times 10^5$, would also have a $B$ value of $5 \times 10^{-4}$ mol-cm$^3$/g$^2$, but substitution into Eq. (3) yields a correction term 300% larger than the intercept value. A side-by-side aggregate, on the other hand, considered as a cylinder of uniform density, would have an axial ratio of 199 and $B = 1.42 \times 10^{-4}$ cm$^3$-mol/g$^2$, yielding a relative increase of only 82% at 0.01 g/ml. Table I summarizes these values and shows that in order to

---

[1] P. J. Flory and W. R. Krigbaum, *J. Chem. Phys.* **18**, 1086 (1950).

TABLE I
EXCLUDED VOLUME SECOND VIRIAL COEFFICIENTS FOR COLLAGEN MONOMER
AND DIMERS OF DIFFERENT STRUCTURE

| Model | $L/D$ | $B$ (cm$^3$-mol/g$^2$) | Percentage error in $\pi/g$ at 1% concentration | Concentration at which measurement must be made to be within 10% of correct value (g/liter) |
|---|---|---|---|---|
| Monomer | 200 | $4.96 \times 10^{-4}$ | 145 | 0.0695 |
| End-linked dimer | 363 | $4.50 \times 10^{-4}$ | 257 | 0.038 |
| Side-by-side dimer | 119 | $1.42 \times 10^{-4}$ | 82 | 0.121 |

achieve a result within 10% of the correct value via a single low-concentration measurement, very low concentrations of collagen must be used. In the ideal case, the osmotic pressure developed by a monomeric collagen solution at such a concentration, at 25°, would be only $6 \times 10^{-2}$ mm H$_2$O. In practical terms, pressures of ~1 cm H$_2$O are the minimum useful for reliable measurement. Boedtker and Doty[2] used a range from 3 to 7 g/liter for measurements on ichthyocol. The error in extrapolation was large with $M_n = 310,000 \pm 50,000$ and $B = 2.3 \, (\pm 1) \times 10^{-4}$ cm$^3$-mol/g$^2$.

All methods of calculation based ultimately on mole fractions provide number average molecular weights, $M_n$. A solution containing 80% by weight monomer and 20% dimer, would have $M_n = 3.19 \times 10^5$, a result essentially not significantly different from the monomer value of $2.85 \times 10^5$. Thus, measurements based on the colligative properties are not sufficiently sensitive to be used for determination of molecular weight heterogeneity.

Molecular weight determinations are usually made under conditions whereby intermolecular interactions are swamped out to the maximum extent by appropriate choice of solvent composition and temperature. If knowledge of intermolecular interactions is the goal, then conditions may be selected so as to accentuate the value of $B$, such as by adjusting pH away from the isoelectric point at low ionic strength. Repulsive electrostatic interactions near pH 3 can increase the value of $B$ for collagen by more than one order of magnitude, so that the concentration dependence of any of the colligative properties can be measured with relative ease in readily accessible concentration ranges.

An important question, even under conditions of minimum intermolecular interaction, is: At what concentration do molecular domains over-

---

[2] H. Boedtker and P. Doty, *J. Am. Chem. Soc.* **28**, 4267 (1956).

lap? that is: At which weight concentration are molecules constrained to exist in a region smaller than $u$? For collagen monomer rods, this concentration is 2.35 g/liter; for end-linked dimer, this value is 1.38 g/liter. These concentrations are just at the lower limit of the range accessible to most thermodynamically based methods of solution analysis.

## Hydrodynamic Properties

### Translational Motion

The movement of particles through a viscous medium is regulated by both the viscosity of the medium and the frictional resistance of the particle itself. The same frictional factors are involved regardless of the driving force; gravitational, electrostatic, or chemical potential. The frictional resistance of a particle is very much dependent upon particle shape.

Consider a particle moving at constant velocity $v$ under an applied force $F$. At this steady state of flow, where there is no particle acceleration, the frictional resistance force must be equal to $F$. Under the conditions of low velocity that are usual during measurement, the frictional resistance force is directly proportional to the velocity, or

$$F = f_t v \tag{7}$$

where $f_t$ is called the translational frictional coefficient, the parameter of direct interest. Any solid object can be modeled as an assembly of small unit structures packed together. Only those units, $u$, in the surface of the object contribute to $f_t$, with $f_t = \sum_u f_u n_u$. The number of surface units, and their individual contributions to resistance, $f_u$, is obviously a direct function of particle shape. Stokes[3] examined this problem long ago for the case of spheres of radius $R$, and determined the familiar result

$$f_0 = 6\pi\eta R \tag{8}$$

where $\eta$ is the viscosity of the viscous medium (ml$^{-1}$ $t^{-1}$). A number of workers have determined the translational frictional coefficients for other structures. Most pertinent to our interest is the Perrin[4] equation for prolate ellipsoids of revolution, Eq. (9).

$$f_t = 6\pi\eta R \frac{(1 - (d/L)^2)^{1/2}}{(d/L)^{2/3} \ln\left\{\frac{1 + [1 - (d/L)^2]^{1/2}}{d/L}\right\}} \tag{9}$$

In Eq. (9) $R$ is the radius of a sphere equal in volume to that of the ellipsiod of resolution with semi-axes $L/2$, $d/2$, $d/2$. For collagen

[3] G. Stokes, *Trans. Cambridge Phil. Soc.* **8**, 287 (1849); **9**, 8 (1856).
[4] F. Perrin, *J. Phys. Radium* [7] **7**, 1 (1936).

## TABLE II
CALCULATED TRANSLATIONAL FRICTIONAL COEFFICIENTS, SEDIMENTATION COEFFICIENTS, AND DIFFUSION COEFFICIENTS FOR TYPE I COLLAGEN

| Model | $L/D$ | $f_t/f_0$ | $f_t$ (g/sec) | $S_{calc}$ (sec$^{-1}$) | $D$ (cm$^2$/sec) | $S/D$ (cm$^{-2}$) |
|---|---|---|---|---|---|---|
| Monomer | 200 | 5.72 | $4.72 \times 10^{-7}$ | $2.95 \times 10^{-13}$ | $8.68 \times 10^{-8}$ | $3.39 \times 10^{-6}$ |
| End-linked dimer | 363 | 7.74 | $8.06 \times 10^{-7}$ | $3.46 \times 10^{-13}$ | $5.14 \times 10^{-8}$ | $6.73 \times 10^{-6}$ |
| Side-by-side dimer | 119 | 4.43 | $5.17 \times 10^{-7}$ | $5.39 \times 10^{-13}$ | $8.00 \times 10^{-8}$ | $6.73 \times 10^{-6}$ |

monomer, with $L = 300$ nm and $d = 1.5$ nm, $f_t = 5.7 f_0$, or $f_t = 4.72 \times 10^{-2}$ g/sec. Values for $f_t$ and $f_t/f_0$ for the dimer models of Table I are shown in Table II.

Two of the principal procedures based on translational movement are sedimentation velocity and translational diffusion measurements. Both sedimentation and diffusion coefficients are related to $f_t$ by the well-known phenomenological equations

$$S = \frac{M(1 - \bar{v}_2\rho)}{f_t}; \qquad D = \frac{kT}{f_t} \qquad (10)$$

Ideal values for monomer, end-to-end dimer, and side-by-side dimer models, computed from $f_t$ are given in Table II. It is evident that sedimentation velocity measurements can discriminate quite well between the various aggregate states. On the other hand, the translational diffusion coefficient is less useful, although it is dependent on shape and aggregate weight.

The computations implicit in Table II refer to the situation of infinite dilution. No adequate theory for the dependence of $S$ on concentration has yet been developed. Three factors appear to be important: the density of the solution, the effective viscosity of the solution, and the possibility of a backflow of solvent which, in a closed ultracentrifuge cell, is displaced by downward sedimentation of the macromolecules. Empirically, the relationship of Eq. (11)

$$S = S_o/(1 + kg) \qquad (11)$$

has been found to be applicable to most proteins, and plots of $1/S$ vs $g$ for collagen solutions are linear in the range 0.5 to 2.0 g/liter.[2,5,6] The value of the slope of such a plot is very much dependent on the solution conditions. Since it is likely that the concentration dependence of any form of lengthened aggregate will be greater than that of monomer, the sedimentation velocity of dimer in a mixture will be slowed and the Johnston–

---

[5] M. P. Drake and A. Veis, *J. Biol. Chem.* **238**, 2003 (1963).
[6] B. Öbrink, *Eur. J. Biochem.* **25**, 563 (1972).

Ogston effect accentuated.[7] Thus, while the presence of end-linked dimer in solution could probably be identified qualitatively, quantitation of the amount would require very careful analysis and the dimer sedimentation coefficient would be difficult to extrapolate accurately to zero concentration. Experimental reports of the dimer $S$ value should be treated with caution.

The driving force for translational diffusion is the difference in chemical potential of solute along the concentration gradient. $D$ thus has the same form of concentration dependence as the thermodynamically based colligative properties, that is, a typical virial expansion form, Eq. (12).

$$D = \frac{RT}{N_{av}f_t}\left[1 + g\left(\frac{d \ln y_2}{dg}\right)\right] = \frac{RT}{N_{av}f_t}[1 + g(BM + 2CM^2g + \cdots)] \quad (12)$$

If $f_t$ were concentration independent, by Eq. (12) $D$ would increase at increasing concentration. However, $f_t$ is concentration dependent and for rodlike particles becomes larger as $g$ increases. Hence, in practice for collagen solutions, $D$ actually decreases with increasing concentration. Fortunately, plots of $D$ vs $g$, in the range of 0.5 to 3 g/liter are linear.[6] Unfortunately accurate diffusion coefficients are difficult to obtain in transport and boundary spreading experiments, and extrapolations are subject to considerable error. Frequently, a better molecular weight extrapolation can be achieved by plotting $(S_{obs}/D_{obs})_g$ vs $g$, the zero concentration value yielding a weight average value, $M_w$, according to Eq. (13).

$$M_w = (S/D)^0 [RT/(1 - \bar{v}\rho_0)] \quad (13)$$

In a heterogeneous system containing collagen monomers, dimers, etc., a sedimentation velocity run provides a reasonably good value of $S$ for the monomer component. The diffusion coefficient determined from any sort of boundary spreading or mass transport experiment is a weight average value, hence $D_{w,apparent}$ will be smaller than its correct value. In the case of 80% monomer–20% end-linked-dimer the $M_w$ determined from Eq. (13) using the monomer component $S$ value would have been 9% high, based on the calculated values for each component as in Table II.

*Viscosity*

A pure liquid moving through a tube or along a container will change from near zero velocity at the wall to the bulk phase velocity as one moves into the liquid. The viscosity of the liquid describes the resistance of the liquid to the resultant shear stress. The viscosity coefficient, $\eta$, in

---

[7] V. P. Johnston and A. G. Ogston, *Trans. Faraday Soc.* **42**, 789 (1946).

steady-state nonturbulent flow, is the tangential force, $F$, required to maintain a unit velocity gradient or rate of shear. The shear stress is given by $F/A$, dynes/cm$^2$, and the shear gradient $G$ by $dv/dr$, sec$^{-1}$ where $v$ is the velocity in cm/sec and $r$ the distance of separation of areas $A$. Thus

$$\eta = \frac{F}{AG}, \frac{\text{dynes-sec}}{\text{cm}^2} \quad (14)$$

and the work, $w$, expended per unit time is

$$dw/dt = (G^2\eta)V \quad (15)$$

where $V$ is the volume of liquid. The work required to establish a shear gradient in a liquid containing large particles is invariably greater than that required to maintain the same shear stress in the pure liquid as a result of changes in flow patterns in the liquid and additional interactions. Theoretically, the problem has been handled by considering the inverse problem, the flow of particles through a stationary liquid, which obviously involves the shape- and surface area-dependent translational frictional coefficient already alluded to in Eqs. (7)–(9).

The usual approach has been to consider the increase in viscosity coefficient, $\eta$, for a solution relative to that of the pure solvent, $\eta_0$. Basic definitions for the relative, specific, and reduced viscosities are, respectively,

$$\eta_{\text{rel}} = \eta/\eta_0; \quad \eta_{\text{sp}} = \eta_{\text{rel}} - 1; \quad \eta_{\text{red}} = \eta_{\text{sp}}/g \quad (16)$$

where $\eta$ is the macroscopic viscosity of the solution and $\eta_0$ is the pure solvent viscosity. The relative viscosity is always greater than 1. Einstein showed that, if $\Phi$ is the solute volume fraction

$$\eta_{\text{sp}} = \nu\Phi \quad (17)$$

where $\nu$ is a shape-dependent coefficient called the viscosity increment. For spheres, $\nu = 2.5$ at small $\Phi$ ($<0.03$). Upon extrapolation of $\eta_{\text{red}}$ vs $g$ to infinite dilution the intercept, $\lim_{g \to 0} \eta_{\text{red}}$ is defined as the intrinsic viscosity, related only to the solute particle shape and size. The intrinsic viscosity, $[\eta]$, determined with $g$ in units of g/100 ml has the units of $dL/g$. Since $\nu = 1/\Phi\,(\eta_{\text{sp}})$,

$$\nu = \frac{100[\eta]}{\bar{v}_2} \quad (18)$$

At very low shear gradients, where particle orientation is not a factor, the viscosity increment can be approximated by the Simha[8] equation

[8] R. Simha, J. Phys. Chem. **44**, 25 (1940).

TABLE III
CALCULATED HYDRODYNAMIC PARAMETERS FOR COLLAGEN AND MODEL DIMERS

| Model | $L$ (Å) | $L/D$ | $\nu$ | $[\eta]$ (ml/g) | $f_{r,b}$ (cm$^2$-g sec$^{-1}$) | $\theta_b$ (sec$^{-1}$) | $\tau_a$ (sec) |
|---|---|---|---|---|---|---|---|
| Monomer | 3000 | 200 | 2050 | 1455 | $5.15 \times 10^{-17}$ | 785 | $6.37 \times 10^{-4}$ |
| End-linked dimer | 5520 | 363 | 6054 | 4298 | $2.89 \times 10^{-16}$ | 140 | $3.57 \times 10^{-3}$ |
| Side-by-side dimer | 3000 | 119 | 807 | 572 | $5.68 \times 10^{-17}$ | 711 | $7.03 \times 10^{-4}$ |

for elongated, prolate ellipsoids of revolution, similar to the Perrin derivation, with the result that, for rods with $L/d > 100$

$$\nu = \frac{(L/d)^2}{15[\ln (2L/d) - 1.5]} + \frac{(L/d)^2}{5[\ln (2L/d) - 0.5]} + 14/15 \quad (19)$$

Using the same $L/d$ values as in Tables I and II, the values of $\nu$ and $[\eta]$ in (ml/g) are shown in Table III for the monomer and dimer models. The viscosity increment is even more sensitive to $L/d$ than the frictional coefficient, and this is reflected in the calculation that a solution containing 80% monomer and 20% end-linked dimer would have a weight average intrinsic viscosity of 2024 ml/g, 39% greater than that of a monomeric solution.

The Simha equations are valid only for the case where the particle distribution in space is independent of the shearing forces in the moving liquid, that is, where orienting forces are less than those of random Brownian movement and no barriers to internal rotation exist. These conditions apply only at low shear gradients and at high dilution.

The proper selection of measurement conditions can be made through consideration of the rotational diffusion coefficient, $\theta$. At some high shear gradient a system of asymmetric rodlike particles will have an average preferred axial orientation, $\phi$, different from that at zero shear. If the shear force is brought to zero, the particles will return, under the random Brownian motion of the solvent, to a uniform distribution in space. The time required for this orientation "relaxation" is dependent on the internal frictional resistance of the particle, rotating about its own axes of revolution. For an ellipsoid with semiaxes $a$, $b$ and $c$, each with different lengths, three different frictional coefficients, $f_{r,a}$, $f_{r,b}$, and $f_{r,c}$ would be required to describe the rate of relaxation. The corresponding rotary diffusion coefficents $\theta_a$, $\theta_b$, $\theta_c$ are defined by

$$\frac{\delta \rho(\phi)}{\delta t} = \theta \frac{\delta^2 \rho(\theta)}{\delta \phi^2} \quad (20)$$

where $\rho(\phi)$ describes the particle distribution function at any time $t$ after release from the orienting stress. The rotational diffusion coefficients (sec$^{-1}$) are thus the counterparts of the Fick's law translational $D$ and

$$\theta_a = \frac{kT}{f_{r,a}}; \quad \theta_b = \frac{kT}{f_{r,b}}; \quad \theta_c = \frac{kT}{f_{r,c}} \tag{21}$$

For our model highly elongated ellipsoid of revolution, the semiminor axes are equivalent, $b = c$, and the only motion of special import is rotation about $b$ or $c$. The relaxation time for orientation of the long axis $a$, $\tau_a$, is therefore given by

$$\tau_a = \frac{1}{\theta_b + \theta_c} = \frac{1}{2\theta_b} = \frac{f_{r,b}}{2kT} \tag{22}$$

Perrin[9] determined that for ellipsoids of revolution with $a > 5b$

$$f_{r,b} = \frac{16 \pi \eta_0 a^3}{3[(2 \ln 2a/b) - 1]} \tag{23}$$

Frequently, the formulations of Burgers [Eq. (24)][10]

$$f_{r,b} = \frac{16 \pi \eta_0 a^3}{3[(2 \ln 2a/b) - 1.6]} \tag{24}$$

or of Broersma [Eq. (25)][11]

$$f_{r,b} = \frac{16 \pi \eta_0 a^3}{3 \left[ 2 \ln 2a/b - 3.14 + 14 \left( \frac{1}{\ln 2a/b} - 0.28 \right)^2 \right]} \tag{25}$$

are referred to rather than Eq. (23). For $a/b = 200$, the bracketed terms in the denominators of Eqs. (23), (24), and (25) are, respectively, 10.98, 10.38, and 9.02. The Burgers and Broersma equations are modified to model the molecule as a cylinder, rather than an ellipsoid of revolution, in which $a$ and $b$ are half rod length and diameter, whereas in the Perrin equation the equivalent ellipsoid of equal volume with $a/b = (2/3)^{1/2} L/d$ should be used. Accordingly, $[2 \ln a/b - 1]$ would have a value of 9.18. These numerical calculations based on the model systems are shown to emphasize that all such calculations of the frictional coefficient can be treated only as approximations to about 10%. Most important is internal consistency in model dimensions and choice of very similar equations.

The rotational frictional coefficient, and $\theta_b$, are obviously particularly sensitive to particle length because of the $a^3$ term. Thus direct measurement of $\theta_b$ is the most sensitive way to determine particle length. For the

[9] F. Perrin, *J. Phys. Radium* [7] **5**, 497 (1934).
[10] J. M. Burgers, *Verh. K. Ned. Akad. Wet.* **16**, 113 (1938).
[11] S. Broersma, *J. Chem. Phys.* **32**, 1626 (1960).

present, however, consider only the viscosity effects of the shear gradient, $\beta$, and the rotatory diffusion relaxation times. The shear gradient must be such that particles can achieve random orientation during measurement, and this is true only when $\beta/\theta_b \ll 1$. Table III shows the values of $f_{r,b}$, $\theta_b$, and $\tau_a$ calculated from Eqs. (21)–(23). Clearly, viscometers with shear gradients of 100 sec$^{-1}$ cannot provide data devoid of orientation effects. Even though many of the data in the literature based on capillary viscometry appear to be in the correct range, these data are useless for interpretation of molecular parameters because they typically use shear gradients of 200 sec$^{-1}$ or more. Moreover, since the end-linked dimer has such a long relaxation time, its presence will be consistently underestimated. As shown by Ananthanarayanan and Veis[12] shear gradients >25 sec$^{-1}$ will produce erroneous results even for essentially monomeric collagen preparations. Except for qualitative studies of the denaturation or degradation of collagen to low-viscosity forms, capillary viscometers should not be used. Couette-type viscometers, in which shear gradients of ~1 sec$^{-1}$ can be attained, should be utilized.

Procollagens produced in organ or cell culture systems, and type IV collagens extracted from basement membranes or basement membrane tumors without enzymic degradation[13,14] retain their amino- and carboxyl-terminal globular regions. These molecules must be modeled as dumbbells, two globular regions with a rigid joining rod. The most convenient method for modeling this complex system is the procedure of Bird *et al.*,[15] which treats the structure as a set of beads with various kinds of edges and surfaces. Each surface element in this system has a characteristic shielding coefficient, modifying the frictional coefficient. One determines the block geometries, the distribution of sections about the center of gravity of the particle, and the individual contribution to the shielding coefficient and viscosity. For rigid dumbbells, the viscosity is always higher than that of a rod of equivalent uniformly distributed mass. For example, a "procollagen" dumbbell from anterior lens capsule basement membrane collagen with a single rodlike triple helical section and two globular end regions has been estimated to have an intrinsic viscosity of 1520 ml/g, to be compared with a value of 1380 ml/g determined for monomer rods by the same method.[13]

Relaxation Processes

Flow birefringence, electric birefringence, and dielectric dispersion measurements all provide information on the rotary diffusion coefficient by the change in particle orientation upon imposition, termination, or

[12] S. Ananthanarayanan and A. Veis, *Biopolymers* **11**, 1365 (1972).
[13] A. Veis and D. Schwartz, *Collagen Rel. Res.* **1**, 269 (1981).

alteration of the orienting force. Quasi-elastic laser light scattering, in which Brownian motion of the solute particles causes a spectral broadening of the scattered light, provides data on the translational diffusion coefficient, with secondary contributions from $\theta$.

Within the past decade no studies of the flow birefringence of collagen solutions have been reported. The method, described in detail by Kasai and Oosawa,[16] will not be dealt with here. On the other hand, a number of studies have dealt with electric birefringence and the related dielectric constant and dielectric relaxation phenomena. Fundamentally, both depend on the existence of a dipole moment on the solute molecule. Hence these topics are considered together.

The magnitude of the dipole moment of a collagen molecule at time $t$, $\mu(t)$, will depend in part on the presence and magnitude of an external electric field $E$. The major contribution to $\mu$ derives from the distribution of polar and ionic groups along the rod, producing a permanent dipole moment not dependent on the applied field. In aqueous solution, there is a microcounterion ionic atmosphere, which, since it may be distorted in distribution under an applied field, produces an induced moment for $E \neq 0$. Finally, perturbation in the distribution of electrons relative to their atomic nuclei at $E \neq 0$ may produce a small induced moment. When a field $E$ is applied, all permanent and induced dipoles tend to become oriented in the direction of the field and a polarization or displacement current, $C$, is produced. This current (Fig. 1) consists of a small, but instantaneous, change when $E$ is turned on owing to the rapid electronic distortions by the applied field, followed by an exponential rise as the larger dipoles orient in the direction of the field. The reverse effects appear when $E$ is turned off. The value of $C$ is related to the mean molecular moment, $m(t)$ of the ensemble of molecules at $t$, and is proportional to $E$, the proportionality constant being the molecular polarizability $\alpha$, comprising the sum of polarizabilities of the previously mentioned permanent and induced moment terms. If the moment of the ensemble of molecules at $t = 0$ ($E = 0$) is $m(0)$ and the dipole moments of the individual molecules $\mu(0)$, and if the ensemble has a moment $m(t)$ at time $t$, then we can define the correlation function,[17] $\Phi$ [Eq. (26)], such that

$$\Phi = \frac{\langle m(t) \cdot \mu(0) \rangle}{\langle m(0) \cdot \mu(0) \rangle} \quad (26)$$

$\Phi$ is the probability that, if the moment is $m(0)$ at $t = 0$, for molecules with

---

[14] R. Timpl, J. Ristelli, and M. P. Bachinger, *FEBS Lett.* **101**, 265 (1979).
[15] R. Bird, O. Hassager, R. Armstrong, and C. Curtiss, "Dynamics of Polymeric Liquids," pp. 648–656. Wiley, New York, 1977.
[16] M. Kasai and F. Oosawa, this series, Vol. 26, p. 289.
[17] R. H. Cole, *J. Chem. Phys.* **42**, 637 (1965).

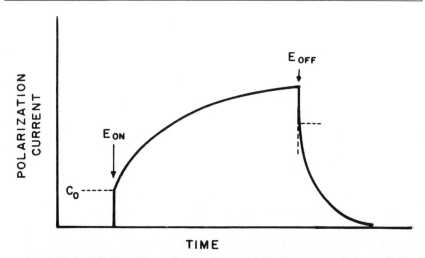

FIG. 1. Dielectric relaxation. When the external field $E$ is turned on, a fast induced moment is formed, producing the essentially instantaneous displacement or polarization current $C_0$. Orientation of the permanent dipoles then occurs over time. At $E_{OFF}$, an instantaneous drop equivalent to $C_0$ is seen, followed by an exponential decay in orientation of the permanent dipoles. It is this decay that yields $\tau$ in Eq. (27).

dipole moments $\mu(0)$, the moment at $t$ will be $m(t)$. If the decay is exponential, then

$$\Phi = e^{-t/\tau} \tag{27}$$

where $\tau$ is the relaxation time.

If $\epsilon_0$ is the dielectric constant at low applied frequencies of a periodic field $E(t) = E_0 \exp(i\omega t)$ and $\epsilon_\infty$ the dielectric constant at high frequency, then it can be shown that

$$\epsilon^* = \epsilon_\infty + \frac{(\epsilon_0 - \epsilon_\infty)}{1 + i\omega\tau} \tag{28}$$

where $\epsilon^*$ is a complex quantity with both real and imaginary components and $\omega$ is the angular frequency, $2\pi\nu$. Debye, in the classic work on dielectric constant,[18] showed that Eq. (28) can be separated into real ($\epsilon'$) and imaginary ($\epsilon''$) parts.

$$\epsilon' = \epsilon_\infty + \frac{(\epsilon_0 - \epsilon_\infty)}{1 + (\omega\tau)^2} \tag{29}$$

$$\epsilon'' = (\epsilon_0 - \epsilon_\infty) \frac{\omega\tau}{1 + (\omega\tau)^2} \tag{30}$$

[18] P. Debye, "Polar Molecules." Dover, New York, 1929.

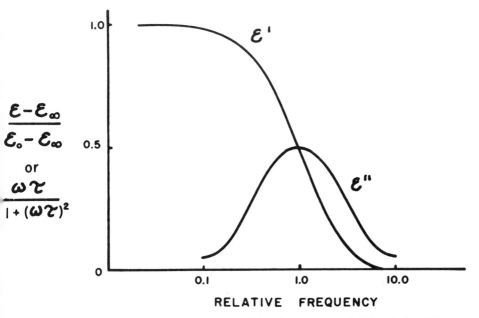

Fig. 2. The frequency dependence of the dielectric constant and dielectric loss. The curve labeled $\epsilon'$ represents the relative value of the dielectric constant between high and low frequency limits. The curve labeled $\epsilon''$ is the relative dielectric loss term. The point of intersection is given by Eq. (31).

Equation (30) represents an energy absorption term and is also called the dielectric loss term, while Eq. (29), the real part, is the observed dielectric constant. Figure 2 shows plots of Eqs. (29) and (30) as a function of $\omega$. It is evident that the maximum in the dielectric loss corresponds to the midpoint of the dielectric dispersion curve, at the point where

$$\epsilon'' = (\epsilon_0 - \epsilon_\infty)/2 \tag{31}$$

At this point, the critical relaxation frequency $\nu_c$ is related to the relaxation time by

$$\tau = 1/(2\pi\nu_c) \tag{32}$$

and $\tau$, in turn, is related to $\theta$ by[19]

$$\tau = 1/6\theta \tag{33}$$

For collagen monomer, where $\theta = 785$ sec$^{-1}$ according to the computations in Table III, $\tau$ has the value $2.12 \times 10^{-4}$ sec or 212 $\mu$sec and $\nu_c$ is thus 750 Hz, a value much lower than for most proteins.

[19] H. Benoit, Ann. Phys. (Leipzig) 6, 561 (1951).

Measurements of the dielectric constant are usually made in bridge systems, where the capacitance and conductance of a cell filled with solution or solvent is balanced with standard capacitors, etc. Such bridges work most efficiently at higher frequencies (>100 kHz) than the critical relaxation frequency for collagen. Thus special cells designed to minimize electrode polarization must be used.[20,21] Obviously, collagen end-linked dimers with lower $\theta$ would require still lower frequencies to demonstrate their dielectric relaxation.

Theoretically, the dipole moment of collagen could be determined from the low-frequency dielectric increment,

$$\Delta\epsilon_0 = \epsilon_0 - \epsilon_{\text{solvent}} \qquad (34)$$

However, as pointed out earlier, the dipole moment of a molecule such as collagen is a complex function of ionic side chain distributions and the ionic atmosphere. Thus, the magnitude of $\mu$ will be very dependent on pH, ionic strength, and solvent composition, and experimental determinations of this type appear to be essentially irrelevant unless one were to undertake the task of computing the expected permanent dipole moment from side-chain distributions determined from sequence analysis, and then specifically inquire about the counterion cloud distribution effects on the induced moment.

If $n$ is the index of refraction of the solution at zero electric field, the change in $n$, $\Delta n$, occasioned by imposition of the field $E$ is related to $E$ by the Kerr constants (Eqs. (35) and (36)].

$$B = \Delta n/\lambda E^2 = \Phi/2\pi l E^2 \qquad (35)$$
$$K = \Delta n/nE^2 \qquad (36)$$

The term $\Phi$ corresponds to that defined in Eq. (27) and describes the average molecular orientation at time $t$, $\lambda$ is the wavelength of light used, and $l$ is the length of light path between the electrodes. As in the case of the polarization current depicted in Fig. 1, the electrically induced birefringence can be treated in terms of rise and decay portions.

During the birefringence buildup portion, one must consider both the permanent and induced dipole moments. Benoit[19] determined that

$$\frac{\Delta n}{\Delta n_\infty} = 1 - \frac{3r}{2(r+1)} e^{-2\theta t} + \frac{r-2}{2(r+1)} e^{-6\theta t} \qquad (37)$$

where $r = (\mu/kt)^2/[(\alpha_\parallel - \alpha_\perp)/kT]$. The numerator of $r$ relates to the permanent dipole moment, the denominator to the excess polarizabilities

---

[20] H. P. Schwan and C. D. Ferris, *Annu. Conf. Enzymes Med. Biol. 16th*, p. 84. Institute of Electrical and Electronic Engineers, Baltimore (1963).

[21] M. Hanss and J.-C. Bernengo, *Biopolymers* **12**, 2151 (1973).

over solvent polarizability of the macromolecule in directions parallel to and perpendicular to the principal particle molecular axis. In Eq. (37) $\Delta n_\infty$ is the birefringence at saturation, high $E$, where maximal orientation is achieved. The interpretation of the buildup portion is complicated by the possibility that the induced moment may have a sign opposite to that of the permanent dipole. On the other hand, the decay portion of the curve, beginning with maximal birefringence $\Delta n_0$ at the time, $t_0$, at which the field is turned off, is given by Equation (38)[19]

$$\Delta n = \Delta n_0 e^{-6\theta t} \tag{38}$$

A plot of ln $\Delta n/\Delta n_0$ vs $t$ for a monodisperse solution of particles yields a straight line. In a polydisperse solution of sufficient dilution, the decay of birefringence is the sum of a set of the independent decays of each component. If the $\theta_i$'s are too close, a curved plot will be found. In a monomer–dimer mixture a set of two straight lines will appear only if $\theta_1 > 5\theta_2$. In the case of an 80% monomer–20% end-linked dimer mixture with the $\theta$ values 785 and 140 as computed in Table III, the data plotted according to Eq. (38) would appear as the upper line of Fig. 3, with an apparent monomer $\tau$ of 240 $\mu$sec, and a readily discernible second component. A monomer–end-linked dimer mixture in the amounts of 95%–5% also yields a plot that, over three periods of $\tau$, exhibits curvature clearly showing the presence of the dimer component. The value of $\tau$ abstracted from the initial slope is 235 $\mu$sec. This method rivals low-shear gradient viscometry in its ability to detect the presence of end-linked aggregates. On the other hand, with the extreme sensitivity of $\theta$ to length, side-by-side dimers with $\theta$ values comparable to the monomer length cannot easily be distinguished from the monomer birefringence decay, yet the presence of such an aggregate could easily be distinguished by viscosity.

O'Konski and Haltner[22] suggested that, if one rapidly reversed the polarity of the orienting pulse in a birefringence apparatus, one could distinguish between the rapidly induced component of the dipole and the permanent molecular dipole moment. If field reversal causes no change in birefringence, the moment is entirely attributable to a fast induced dipole moment. If both permanent and induced dipole moments are present, the birefringence drops and then returns, according to Eq. (39) deduced by Bernengo et al.[23] from the work of Tinoco and Yamaoka,[24]

$$\frac{\Delta n}{\Delta n_0} = 1 + \frac{3r}{r+1}[e^{-2\theta t} - e^{-6\theta t}] \tag{39}$$

---

[22] C. T. O'Konski and A. J. Haltner, *J. Am. Chem. Soc.* **79**, 5634 (1957).
[23] J.-C. Bernengo, D. Herbage, and B. Rous, *Colloq. Int. CNRS* **287**, 41 (1978).
[24] I. Tinoco and K. Yamaoka, *J. Phys. Chem.* **63**, 423 (1959).

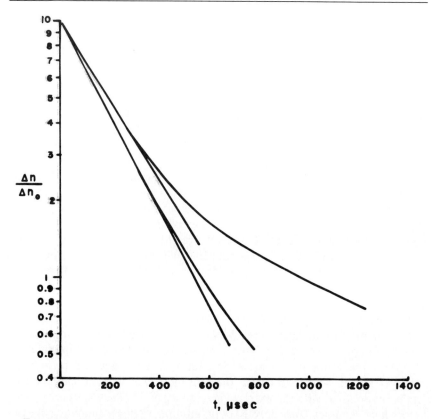

FIG. 3. The electric birefringence decay computed for hypothetical 95% monomer–5% end-linked dimer mixture (lower set of lines) and 80% monomer–20% end-linked dimer (upper set of lines) mixture. The limiting slopes are given by the straight lines in each case; for the 95% monomer case the slope is indistinguishable from the monomer value.

where $r = \mu^2/(\Delta\alpha_E kT)$ and $\Delta\alpha_E$ is the difference in polarizability along the major and minor axis of the particle, Equation (39) produces the change in birefringence shown in Fig. 4. The minimum in birefringence is characterized by a time $t_m$ such that

$$t_m = \frac{\ln 3}{4\theta} \qquad (40)$$

Equation (39) and the reversed-field method provide a very "pure" value for $\theta$. Still another approach is to use the sine wave field of increasing frequency.[25] In this case (Fig. 5), just as in dielectric dispersion measure-

[25] G. B. Thurston and D. I. Bowling, *J. Colloid Interface Sci.* **30**, 34 (1969).

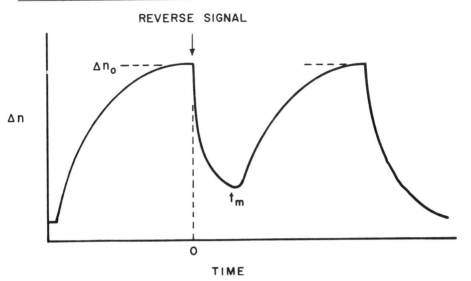

FIG. 4. Reversed pulse birefringence signal. After buildup of the initial orientation, the orienting voltage is reversed in polarity. The decrease in birefringence after this time indicates the presence of permanent dipoles. The position of the minimum $t_m$ provides a value for $\tau$ according to Eq. (40).

ments a frequency, $\nu_c$, is reached where reorientation of the particles cannot follow the field and the decay of $\Delta n$ can be used in Eq. (32) to yield $\tau_s$ and $\theta$.

Before the development of lasers, the classical approach of Rayleigh light scattering, via measurements of intensity of scattered light as a function of angle, was able to provide weight-average molecular weights and z-average estimates of particle radius of gyration. The best works of this type related to collagen were probably those of Boedtker and Doty[2] and of Öbrink.[6] Such measurements showed the presence of aggregates in solution and, through the second virial coefficient, the complex nature of intermolecular interactions. Although the fact that Rayleigh scattering depends on concentration or refractive index fluctuations was understood from the start, and that such movement through solution was in turn related to the translational and rotational frictional coefficients, no use could be made of these understandings until the high-intensity monochromatic, coherent light of lasers and means for the analysis of the spectral purity of light scattered from such beams was available. Pecora[26] was the first to provide a detailed theory of the spectrum of scattered

[26] R. Pecora, *J. Chem. Phys.* **40**, 1604 (1964).

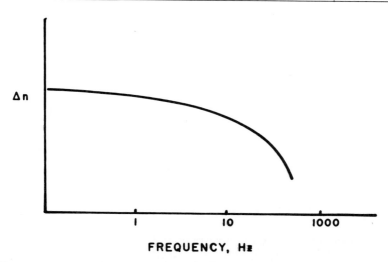

FIG. 5. The frequency dependence of electric birefringence using sine wave pulses of increasing frequency. At very low frequency, with the field frequency much longer than $\tau$, $\Delta n$ remains near its maximum value. Dipoles cannot keep up with the field as the sine wave frequency, at constant amplitude, is increased, hence $\Delta n$ decreases. When pulses have a frequency faster than $\tau$, $\Delta n$ drops to zero.

light from macromolecules. The applications since that time have multiplied at an enormous rate. French et al.[27] were the first to apply the approach of quasi-elastic laser light scattering to collagen. More recently, the best examples of this approach are in the works of Fletcher,[28] Thomas and Fletcher,[29] and Silver et al.[30] There have been further studies on the topic since these papers and one can expect major emphasis on this method in the future. There has been continued improvement in instrumentation, and laser light scattering apparatus and autocorrelators are commerically available.

In spite of its importance, discussion of quasi-elastic light scattering has been reserved to last in this section so that by this point the reader should have a firm grasp on the nature of the translational and rotational diffusion coefficients, the key parameters in data interpretation, and the basic information to be derived from relaxation approaches. To interpret the scattering data one needs in addition a reasonable grasp of the scattering theory and of the nature of the autocorrelation functions. A brief exposition of the theory follows, based primarily on the approaches

[27] M. J. French, J. C. Angus, and A. G. Walton, Biochim. Biophys. Acta 251, 320 (1971).
[28] G. C. Fletcher, Biopolymers 15, 2201 (1976).
[29] J. C. Thomas and G. C. Fletcher, Biopolymers 18, 1333 (1979).
[30] F. H. Silver, K. H. Langley, and R. L. Trelstad, Biopolymers 18, 2523 (1979).

taken by Cummins and Swinney[31] and Berne and Pecora.[32] These major works are highly recommended starting points for one seriously entering this field.

In classical light scattering the basic premise is that the scattered light has the same frequency as the incident light, $\nu_s = \nu_0$. Since the scattering particles in a fluid medium are in constant motion, this is not true and the movement causes a Doppler shift in frequency so that $\Delta\nu = \nu_s - \nu_0 \neq 0$ and

$$\Delta\nu/\nu_0 = vn/c \tag{41}$$

In Eq. (41) $n$ is the refractive index of the medium, $c$ the speed of light, and $v$ the velocity of the scattering particle. Obviously $vn/c$ is exceedingly small. The root mean square displacement of a particle undergoing random Brownian motion in a time $\tau$ (the Einstein one-dimensional random walk problem), $\langle x^2 \rangle^{1/2}$, is related to $\tau$ by

$$\langle x^2 \rangle = 2D\tau \tag{42}$$

The velocity of the scattering particle is thus

$$v \approx \langle x^2 \rangle^{1/2}/\tau \tag{43}$$

and, hence,

$$\Delta\nu \approx 2(n/\lambda_0)^2 D \tag{44}$$

with $\lambda_0$ the wavelength of the incident light *in vacuo*. For collagen monomer with $D$, the translational diffusion coefficient, taken from Table II, helium–neon laser light at 633 nm, and $n = 1.33$ for water, $\Delta\nu \approx 75$ Hz, a very small shift indeed, requiring a very high-resolution low-frequency detection system. Spectrum analyzers presently in use allow measurements from about 50 Hz to 500 kHz, so that collagen monomers come just at the lower limit for study.

The photomultiplier tube utilized for analysis receives the light scattered on its photosensitive cathode, producing a photocurrent. This photocurrent can be analyzed as a function of time or of its frequency. The electric field $E(t)$ of the light falling on the photocathode at time $t$ is related to the intensity of the light $I(t)$ by

$$I(t) = |E(t)|^2 \tag{45}$$

If $\sigma$ is the quantum efficiency of the photodetector, then the probability of photoelectron emission $W^{(1)}(t)$ is

[31] H. Z. Cummins and H. L. Swinney, *Prog. Opt.* **8**, 133 (1970).
[32] B. J. Berne and R. Pecora, "Dynamic Light Scattering: With Application to Chemistry, Biology and Physics." Wiley (Interscience), New York, 1976.

$$W^{(1)}(t) = \sigma I(t) \tag{46}$$

and the photocurrent at time $t$, $i(t)$ can be expressed as

$$i(t) = e\gamma W^{(1)}(t) = e\gamma\sigma I(t) \tag{47}$$

where $e$ is the electron charge and $\gamma$ is the photomultiplier gain. In the time domain, the probability that one photoelectron will be emitted at $t$ and another at $(t + \tau)$ will be

$$W^{(2)}(t, t + \tau) = \sigma^2 I(t) I(t + \tau) \tag{48}$$

If the system is stationary, that is, if its scattering properties are invariant over time, then $\langle i(t) \rangle = \langle i \rangle = e\gamma\sigma \langle I \rangle$ so that

$$W^{(2)}(t, t + \tau) = \sigma^2 \langle I \rangle^2 g^{(2)}(\tau) \tag{49}$$

where $g^{(2)}(\tau)$ is the normalized second-order correlation function

$$g^{(2)}(\tau) = \langle I(t) I(t + \tau) \rangle / \langle I \rangle^2 \tag{50}$$

The actual parameter measured is not the scattered intensity $I$ but the photocurrent $i$, and the autocorrelation function for $i$, $C_i(\tau)$ is

$$C_i(\tau) = \langle i(t) i(t + \tau) \rangle \tag{51}$$

The photocurrent consists of photoelectrons arising from shot noise, randomly emitted electrons, and those from scattered light so that

$$C_i(\tau) = e \langle i \rangle \delta(\tau) + \langle i \rangle^2 g^{(2)}(\tau) \tag{52}$$

The first term contains a delta function, $\delta(\tau)$, such that $\delta(\tau) = 1$ at $\tau = 0$ and zero at all other $\tau$. The electric field autocorrelation function

$$C_E(\tau) = \langle I \rangle g^{(1)}(\tau) \tag{53}$$

In homodyne or self-beating detection systems where the photocathode senses only the scattered light, and for the case where the scatterers act independently and the electric field is Gaussian and random, then Mandel[33] has shown that

$$g^{(2)}(\tau) = 1 + |g^{(1)}(\tau)|^2 \tag{54}$$

The normalized first-order correlation function is the factor of interest, since, as will be shown below, it is this function that is directly related to $D$ and $\theta$. For the homodyne system, combining Eqs. (52) and (54) relates the photocurrent correlation to the first-order correlation function, Eq. (55).

$$C_i^{\text{homo}}(\tau) = e \langle i \rangle \delta(\tau) + \langle i \rangle^2 + \langle i \rangle^2 |g^{(1)}(\tau)|^2 \tag{55}$$

[33] L. Mandel, *Prog. Opt.* **2**, 181 (1963).

The first term is the shot noise; the second is the dc component of the current, which is constant.

In the heterodyne system, where the scattered light is mixed with incident light, it can be shown that

$$C_i^{het}(\tau) = ei_{in}\delta(\tau) + i_{in}^2 + i_{in}\langle i_s \rangle [e^{i\omega_{in}\tau}g^{(1)}(\tau) + e^{-i\omega_{in}\tau}g^{(1)*}(\tau)] \quad (56)$$

The subscript "in" indicates contributions from the incident light. The first two terms of Eq. (56) are noise and dc current; the third gives $g^{(1)}(\tau)$ directly.

Measurements of the frequency distribution, that is, in the frequency rather than the time domain, depend on the power spectrum $P(\omega)$. Of interest is the power in the photocurrent in the range between $\omega$ and $\omega + d\omega$, $P(\omega)d\omega$. The power at any frequency is the Fourier transform of $C_i(\tau)$,

$$P(\omega) = \frac{1}{2\pi}\int_0^\infty C_i(\tau)e^{i\omega\tau}d\tau \quad (57)$$

For a homodyne determination, $P(\omega)$, like $C_i^{homo}(\tau)$, depends on shot noise, dc current, and a Lorentzian function of $\gamma$, the half-width at half-height of the scattered spectrum.

$$2\pi P_i^{homo}(\omega) = e\langle i \rangle + 2\pi\langle i \rangle^2 \delta(\omega) + \langle i \rangle^2 \frac{4\gamma}{\omega^2 + (2\gamma)^2} \quad (58)$$

The shot noise again contains a $\delta$ function whose value is 1 at center of the spectrum, $\omega = 0$, and zero at all other values of $\omega$. In the heterodyne measurement

$$\pi P_i^{het}(\tau) = e\, i_{in} + i_{in}^2 \delta'(\omega) + 2i_{in}\langle i_s \rangle \frac{\gamma}{[\omega - |\omega_0 - \omega_i|]^2 + \gamma^2} \quad (59)$$

where $\delta'(\omega) = 2$ for $\omega > 0$ or zero for $\omega < 0$.

If $k_0$ is the wave vector of the incident light and $k_s$ is the scattered wave vector, then the difference or scattering vector $K$ is defined as

$$K = |k_0 - k_s| \quad (60)$$

and this depends on the refractive index of the medium, $n$, the wavelength of the incident light *in vacuo*, $\lambda_0$, and the scattering angle, $\theta$, such that

$$|K| = \frac{4\pi n}{\lambda_0}\sin\frac{\theta}{2} \quad (61)$$

For the helium–neon laser, an aqueous medium, and 90° scattering, $K = 1.87 \times 10^5$. Particle displacement, which gives rise to the frequency distribution, is related, for isotropic scatterers, to $D$ [see Eq. (42)], and it can be shown that the terms in $g^{(1)}(\tau)$ in Eqs. (55) and (56) become

$$(C_i^{\text{homo}})' = \langle i_s \rangle^2 e^{-2K^2D\tau} \quad (62)$$

and

$$(C_i^{\text{het}})' = 2i_{\text{in}} \langle i_s \rangle e^{-K^2D\tau} \quad (63)$$

and in the power spectrum equations, $\gamma$, the half-width at half-height is

$$\gamma = K^2D \quad (64)$$

For a long, thin rodlike molecule such as collagen, where the length of the molecular long axis, $L$, is on the order of the wavelength of the incident light, rotational diffusion will also contribute to the time dependence of the scattering. The intensity, $I$, of the scattered light, averaged over all rod orientations, is related to $KL$ by Eq. (65).

$$I = \frac{2}{KL} \int_0^{1/2 \, KL} \frac{\sin^2 X}{X^2} dS = \frac{2}{KL} \int_0^{KL} \frac{\sin KL}{KL} d(KL) - \left[ \frac{\sin(KL/2)}{(KL/2)} \right]^2 \quad (65)$$

The integral, $\int_0^{KL}(\sin KL/KL) \, d(KL)$, is a standard form, $\text{Si}(X)$, tabulated in many mathematical handbooks. Pecora[26,34] has shown that Eq. (65) can be treated as the sum of two terms, one dependent only on $D$, the other on $\theta$ and $D$, that is,

$$I = I_D + I_{D,\theta} \quad (66)$$

The field autocorrelation function for a solution of long rods is an infinite series of even exponentials having nonzero values and with relaxation times $\tau_0 = (K^2D)^{-1}$, $\tau_2 = (K^2D + 6\theta)^{-1}$, $\tau_4 = (K^2D + 20\theta)^{-1}$, ..., each term weighted by a factor $B_0$, $B_2$, $B_4$ ..., respectively. The spectrum is a sum of Lorentzian's centered at $\omega_0$, weighted by $B_l$ with half-widths at half-heights $\lambda$.

$$\gamma = K^2D + \frac{l}{l+1} \theta \quad (67)$$

When $l = 0$, Eq. (66) is entirely the term $I_D$. In general the homodyne correlation function is

$$g^{(1)}(\tau) = 1 + \left( \frac{B_0}{B_0 + B_2} \right) e^{-2K^2D\tau} + \left( \frac{B_2}{B_0 + B_2} \right) e^{-2(K^2D+6\theta)\tau}$$
$$+ 2 \left( \frac{B_0 B_2}{B_0 + B_2} \right) e^{-(2K^2D+6\theta)\tau} \quad (68)$$

and the heterodyne correlation function is simpler

$$g^{(2)}(\tau) = \left( \frac{B_0}{B_0 + B_2} \right) e^{-K^2D\tau} + \left( \frac{B_2}{B_0 + B_2} \right) e^{-(K^2D+6\theta)\tau} \quad (69)$$

[34] R. Pecora, *J. Chem. Phys.* **48**, 4126 (1968).

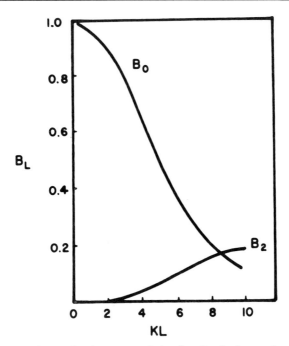

FIG. 6. Weighting factors for the autocorrelation function for long rods as a function of $KL$, adapted from Berne and Pecora.[32] $B_0$ and $B_2$ are the weighting factors for $\tau_0$ and $\tau_2$, respectively, in the discussion of Eqs. (66), (68), and (69).

The dependence of $B_l$ on $KL$ is shown in Fig. 6 for the significant terms $B_0$ and $B_2$. As indicated earlier, $|K|$ has the value $1.87 \times 10^5$ for $90°$ scattering in aqueous solution with a helium–neon laser. $KL$ is thus 5.6 for collagen monomer. According to Fig. 6, $B_2$ is near zero at scattering angle of 0, but assumes significant values at $KL > 5$. Thus, the rotational contribution, $\theta$, is important for collagen at high scattering angles, but measurements at small angles, that is where $KL < 5$, depend principally on $D$. On the other hand, Eqs. (68) and (69) clearly show that at a scattering angle equal to zero, where $|K| = 0$, $g^{(1)}(\tau)$ and $g^{(2)}(\tau)$ are entirely dependent on the value of the rotational diffusion coefficient $\theta$. It is evident that zero-angle scattering data should be most sensitive to particle length, since $\theta$ is proportional to $L^3$.

Other Methods

*Rayleigh Scattering*

Classical Rayleigh light scattering has been used effectively, as in the early work of Boedtker and Doty,[2] to assess the presence of aggregates in solution. The standard presentations of Rayleigh scattering technique and computations (see Tanford[35]) require little elaboration. The early studies all suffered from the fact that, while for rodlike particles a strong angular dependence of scattering was expected, technical considerations of instrument design and light sources made long extrapolations to zero angle a necessity. With the development of the laser light source and the availability of commercial equipment, such as the Chromatix KMX-6 in which intensity measurements can be made at angles close to the incident beam ($\sim 4°$), it is now possible to determine weight average molecular weights with minimal angular dependence and to determine the angular dependence of scattering close to scattering angle zero.

Silver and Trelstad[36] have carried out such an investigation and found $M_{r,w} = 2.82 \times 10^5$ for a monomer preparation, a very good value indeed. The problem of light scattering, however, is shown by their finding of a weight, $M_{r,w} = 8.05 \times 10^5$ for an undefined "high molecular weight" aggregate fraction. The data involved are very good but provide no insight into the nature of the weight distribution. These measurements must be coupled with other procedures to provide such data. At present I am not aware that anyone has combined molecular weight determination with gel filtration chromatography, a technique made possible by the Chromatix system. This combination, which could provide integral weight distributions, would appear to be quite promising.

Silver and Trelstad[37] have computed the values of Eq. (65), the particle scattering function for rods, as a function of $KL$ and of $L/\lambda_0$ for scattering angle 90°. These data (Fig. 7) indicate that for all cases with $KL > 4$ the particle scattering factors will be essentially independent of axial ratio. The limiting slope of the Rayleigh ratio, $[KC/R(0)]_{c=0}$ vs $\sin^2 \theta/2$, divided by the intercept, $(KC/R(0))_{c=0}$, provides a $z$-averaged value for $L$, in contrast to the weight-averaged value for $M_r$.

*Equilibrium Ultracentrifugation and Approach to Equilibrium Techniques*

In addition to the transport methods and direct measurement of $S$, equilibrium centrifugation[38,39] and the Yphantis approach to equilibrium method[40] have been used effectively to determine monomer molecular

[35] C. Tanford, "Physical Chemistry of Macromolecules," Chapters 5 and 6. Wiley, New York, 1961.
[36] F. H. Silver and R. L. Trelstad, *J. Biol. Chem.* **255**, 9427 (1980).
[37] F. H. Silver and R. L. Trelstad, *J. Theor. Biol.* **81**, 515 (1979).
[38] H. K. Schachman, this series, Vol. 4, p. 32.
[39] H. K. Schachman, "Ultracentrifugation in Biochemistry." Academic Press, New York, 1959.
[40] D. A. Yphantis, *Biochemistry* **3**, 297 (1964).

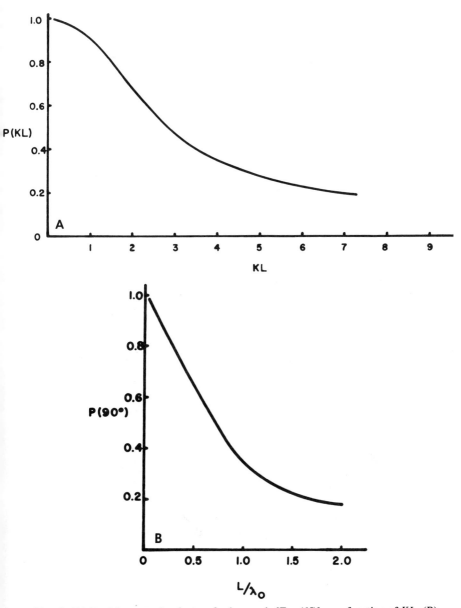

FIG. 7. (A) Particle scattering factors for long rods [Eq. (65)] as a function of $KL$. (B) Particle scattering factors for long rods at 90° as a function of $L/\lambda_0$. Note that, although $P(KL)$ and $P(90°)$ both become insensitive to increased length at high $L$, $B_2$ of Fig. 6, and hence $\theta_b$, become more important in the scattering. Data replotted, with permission, from Silver and Trelstad.[37]

weight and the presence of higher weight aggregates in collagen solutions. The photoelectric scanner system permits one to work at sufficiently low concentrations and determine the total solute concentration, $c(r)$, directly as a function of $r$, the radial distance. The integrated Svedberg equation [Eq. (70)] can then be

$$\ln c(r) = \frac{(1 - \bar{v}_2 \rho)}{2RT} M_{w,app} r^2 + \text{constant} \tag{70}$$

used to determine the weight average apparent molecular weight, $M_{w,app}$, at any point within the cell. Yphantis[40] has pointed out that for a multi-component system one can write

$$c(r) = \sum_{i=1}^{n} c_i(r) = \sum_{i=1}^{n} A_i \exp(\sigma_i r^2/2) \tag{71}$$

where

$$\sigma_i = \omega^2 S/D = \frac{M(1 - \bar{v}\rho)}{RT} \omega^2 \tag{72}$$

and

$$A_i = \frac{c_{i,0} \sigma_i (b^2 - a^2)}{2[\exp \sigma_i b^2/2 - \exp \sigma_i a^2/2]} \tag{73}$$

In these expressions $a$ and $b$ are the values of $r$ at meniscus and base, respectively, and $c_{i,0}$ is the concentration of component $i$ in the initial solution.

For the two macromolecular species we have been considering, monomer and dimer, one has

$$c(r) = A_1 e^{\sigma_1 r^2/2} + A_2 e^{\sigma_2 r^2/2} \tag{74}$$

but, as shown in Table II, $(S/D)_1 = 3.39 \times 10^{-6}$ cm$^2$ and $(S/D)_2 = 6.73 \times 10^{-6}$ cm$^2$ or $\sigma_2 = 2\sigma_1$ so that

$$c(r) = A_1 e^{\sigma_1 r^2/2} + A_2 e^{\sigma_1 r^2} \tag{75}$$

If conditions are such that the concentration of all species is zero at the meniscus, then the ratio of concentration of component 1 to component 2 at each point $r$ is given by

$$\frac{c_1(r)}{c_2(r)} = \frac{\sigma_1 f_1}{\sigma_2 f_2} \exp[(\sigma_2 - \sigma_1)(b^2 - r^2)/2] \tag{76}$$

with $f_1$ and $f_2$ being the weight fractions of the components. Equation (76) reduces to the simple form

$$\frac{c_1(r)}{c_2(r)} = \frac{f_1}{2f_2} \exp[\sigma_1(b^2 - r^2)/2] \tag{77}$$

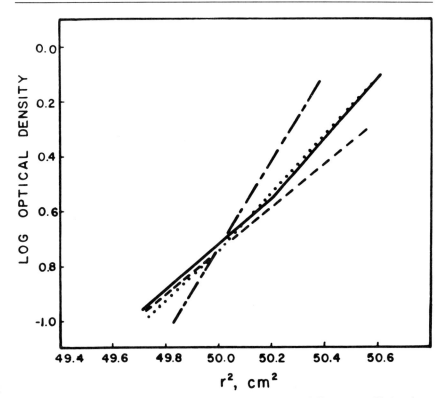

FIG. 8. Concentration distributions of collagen monomer and dimer at equilibrium in an analytical ultracentrifuge, 10,000 rpm. ----, Calculated pure monomer distribution; —-—-, calculated pure dimer distribution; ·····, calculated distribution in a mixture containing 80% monomer and 20% dimer. ——, Data of Yuan and Veis[41] for Pronase-treated acid-soluble type I collagen.

and can be used to determine the speed, $\omega$, and column length required to obtain a region of specified length containing only monomer. The plot of ln $c(r)$ vs $r^2$ will show a constant slope to this point, then a break to a new line with a higher slope. Figure 8 shows a plot of $c(r)$ calculated for an 80% monomer, 20% dimer mixture, using the $S/D$ values of Table II, at 10,000 rpm in a column length of 0.3 cm. Note that in this case, since $\sigma$ is a function of $M_r$, not of the frictional coefficient, side-by-side and end-linked dimer would distribute in the same fashion in the ideal, noninteraction situation. Yuan and Veis[41] have examined various soluble collagen preparations by this procedure under different solvent conditions. They demonstrated that nonideal behavior causes marked deviations from linearity for collagen in 0.01 $M$ acetic acid without supporting electrolyte. How-

[41] L. Yuan and A. Veis, *Biopolymers* **12**, 1437 (1973).

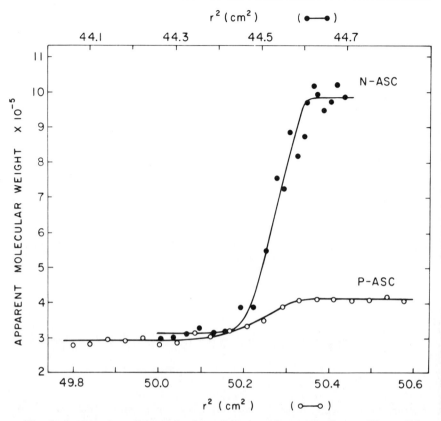

FIG. 9. A comparison of the molecular weights in various parts of a centrifuge cell for native acid-soluble collagen (N-ASC) and Pronase-treated acid-soluble collagen (P-ASC). In each case the monomer component is evident, and its weight is readily computed. The striking feature is that the aggregates in N-ASC must be much larger than dimer and that few aggregates of intermediate size are present, as shown by the steep slope in the N-ASC plot. Data of Yuan and Veis.[41]

ever, addition of neutral salt to an ionic strength of 0.17 at pH 4.6 effectively eliminated the ionic repulsive interactions, and linear plots of $\ln c(r)$ vs $r^2$ were obtained over the entire accessible column length. These data suggest that the pressure dependence of the sedimentation coefficients of monomer and aggregate species is minimal. Yuan and Veis[41] were also able to distinguish between the cases of monomer–dimer mixtures and monomer–higher aggregate size mixtures (Fig. 9). One cannot determine from these data alone whether the aggregates were present in solution initially or induced by the high pressures and concentrations near the cell bottom. It seems unlikely that the aggregates demonstrated in Fig. 9 were formed by pressure-dependent interactions because relatively uniform weight aggregates were apparent in the aggregate containing regions near the cell base.

## Strategies for Characterization Studies: Experimental Considerations

From the previous discussions it should be clear that reasonable agreement appears to exist between hydrodynamic theory, yielding translational and rotational frictional coefficients, and the values of such parameters provided by experiment. The principal question to be theoretically and experimentally resolved is the effect of molecular flexibility.[42] The existing techniques applicable to the analysis of rodlike macromolecules fall into two broad categories. The methods emphasize either the presence of aggregates in solution and their averaged molecular weights or the explicit nature of the aggregate particle. The latter data, in conjunction with the theoretical values for $f_t$ and $f_r$, allow one to construct models of specific internal arrangement; the former data can then be interpreted in terms of weight distributions within the proposed model systems.

The most rapid and least expensive way to examine a particular preparation or solution, and at the same time gain insight into the state of aggregation is the use of low-shear gradient viscometry. We have found the Zimm–Crothers rotating cylinder viscometer,[43] as produced by Beckman Instruments, to be simple to operate and reliable in terms of data reproducibility. It is readily applicable to collagen solutions and has no limitations relative to solvent composition, pH, or ionic strength. The Beckman Instruments viscometer requires about 5 ml of solution, most of which can be recovered at the conclusion of a measurement. Calibration can be made readily with glycerine–water mixtures of known viscosity. Rotation times in water, with a 0.65-cm radius stator and 0.5-cm radius rotor, set up shear gradients on the order of 1 $sec^{-1}$. No extrapolations are required, and all viscosity values can be treated in the same way. The viscosities of solutions with collagen concentrations less than 0.03 g/liter can be measured.[14] Two precautions are required: the rotor and stator must be scrupulously clean, and a time-dependent rotor slowing must be considered. One must plot rotor rotation time as a function of total elapsed time during a measurement and use the zero elapsed time extrapolated value of the rotation time for computations. The slopes of the rotation time–elapsed time plots are a function of solute concentration. Be-

[42] W. A. Wegener, *Biopolymers* **19**, 1899 (1980).
[43] B. H. Zimm and D. M. Crothers, *Proc. Natl. Acad. Sci. U.S.A.* **48**, 905 (1962).

cause of excluded volume and other interaction effects, the concentration dependence of $\eta_{\text{red}}$ becomes greater as the fraction of aggregate increases, if the aggregate is increased in length. When high aggregate content becomes evident, more data points are required in the low-concentration region. This is somewhat inconvenient because the viscometer must be cleaned between measurements, and one cannot make dilutions directly in the viscometer because of the large surface area to volume ratio. In spite of these problems, the lower shear gradients available and the ease of cleaning relative to multibulb capillary viscometers make this type of viscometry the method of choice. For rodlike particles, $[\eta]$ is essentially a weight-average value. Thus, assuming a two-component monomer–aggregate mixture and the proper choice, via computations such as those summarized in Table III, of theoretical values for $[\eta]_i$, one can readily determine the maximum weight fraction of aggregate present per model.

Collagen solutions containing paucidisperse distributions of aggregates can also be examined by equilibrium centrifugation or sedimentation velocity techniques. Use of the ultraviolet scanner system is essential in order to permit examination of solutions at sufficiently low concentrations. Unfortunately, collagen molecules have very low absorbancies in the 260 to 280 nm range, where most proteins have reasonable extinction coefficients, and one is forced to work at wavelengths near the lower limit of the Spinco scanner optics. We have routinely used 233 to 236 nm for analysis, and these wavelengths require more than routine attention to the cleanliness of the entrance and exit windows of the rotor chamber. Basement membrane collagens, with higher tyrosine content, can be examined more conveniently at higher wavelengths.[13]

Equilibrium centrifugation as practiced by Yuan and Veis,[41] with 233 nm illumination, collagen loading concentration on the order of 0.2 g/liter, and rotor speeds in the range of 10,000 to 15,000 rpm, distributed the collagen (Fig. 9) to show distinctly the paucidisperse nature of the samples examined, and permitted the calculation of the aggregate molecular weights. Figure 9 compares the ideal case of a monomer–dimer system, in the weight ratios of 80% monomer–20% dimer, with data on a Pronase-treated acid-soluble collagen in acid at pH 4.6, in 0.17 $M$ NaCl, 17° after 40 hr of centrifugation at 10,000 rpm.

One must always be aware that the partial specific volume is crucial in the interpretation of sedimentation data. Values of $\bar{v}_2$ for collagen do vary with solvent composition. Rice *et al.*[44] determined $\bar{v}_2 = 0.706$ ml/g in phosphate and 0.686 ml/g in citrate. Drake and Veis[45] found

[44] R. V. Rice, E. F. Casassa, R. E. Kerwin, and M. D. Maser, *Arch. Biochem. Biophys.* **105**, 408 (1964).
[45] M. P. Drake and A. Veis, *J. Biol. Chem.* **238**, 2003 (1963).

$\bar{v}_2 = 0.709 \pm 0.01$ for collagen in $0.05\ N$ HAc. Piez[46] recommended the use of $\bar{v}_2 = 0.700$ ml/g. Fessler and Hodge,[47] on the other hand, raised the distinct possibility that aggregated collagen might have a different density than unaggregated collagen. This point requires further study.

Rayleigh light scattering, that is, scattering intensity determinations, as a function of angle[6] or at ~zero angle,[36] provide values of $M_{r,w}$ that should correspond quite well with the weight average data of intrinsic viscosity. Estimates of particle length using the Zimm plots, as in the work of Öbrink,[6] are not particularly useful, particularly since a few very large aggregates have such prominent effect.

More detailed insight into the nature of any aggregates in solution require application of the various relaxation procedures leading to determination of $\theta_r$. The most detailed study utilizing dynamic light-scattering procedures is the beautiful work of Thomas and Fletcher,[29] to which the reader is referred. The drawback of the light scattering system is, again, the inordinate effect of a few very large aggregates and the relatively high concentrations required. Nevertheless, Thomas and Fletcher[29] were clearly able to distinguish between monomer and end-to-end aggregate forms.

Electric birefringence studies, using any of the methodologies (direct pulse decay, reversed pulse, or sine wave), is the current best alternative to light scattering, and in conjunction with laser light scattering data, will probably be the most valuable way to determine molecular parameters of both monomer and aggregate. The apparatus of Bernengo et al.[23,48,49] has the significant advantage over earlier apparatus used for collagen[12,50,51] in that heating effects have been minimized. Thus, higher ionic strengths can be tolerated and electrostatic interaction effects minimized. Unfortunately, like the highly sophisticated autocorrelation function-monitored light scattering arrangements, the instrumentation for electric birefringence measurement is complex and there are only a few scattered centers where it is possible to carry out such studies. Thus, the interplay between theoretical approaches suggested by the first discussion in this review must be matched by collaborative efforts between diverse groups if full use of these powerful techniques is to be gained.

---

[46] K. A. Piez, *in* "Treatise on Collagen" (G. N. Ramachandran, ed.), Vol. 1, p. 236. Academic Press, New York, 1967.
[47] J. H. Fessler and A. J. Hodge, *J. Mol. Biol.* **5**, 446 (1962).
[48] J.-C. Bernengo, B. Roux, and D. Herbage, *Ber. Bunsenges. Phys. Chem.* **80**, 246 (1976).
[49] J.-C. Bernengo, D. Herbage, and B. Roux, *Biochim. Biophys. Acta* **532**, 305 (1978).
[50] K. Yoshioko and C. T. O'Konski, *Biopolymers* **4**, 499 (1966).
[51] L. D. Kahn and L. P. Wittnauer, *J. Am. Leather Chem. Assoc.* **64**, 12 (1969).

## [8] Preparation and Translation of Interstitial (Calvaria) Procollagen mRNA

By JANET M. MONSON

### A Procedure for the Isolation of Biologically Active Procollagen Messenger RNAs from Connective Tissues

A prerequisite for studies of the expression of genes encoding connective tissue proteins throughout development is a procedure for the isolation of intact, biologically active messenger RNAs (mRNAs) from a variety of tissues. Detailed below is a procedure that facilitates the recovery of procollagen mRNAs from a calcified tissue—in this case chick embryo calvaria. Successful extraction requires the rapid disruption of the connective tissue accompanied by the instantaneous denaturation of tissue ribonucleases. This may be accomplished by violently homogenizing the frozen, powdered tissue in a potent denaturant, 5 M guanidinium thiocyanate, under strongly reducing conditions. The potency of guanidinium thiocyanate resides in the fact that both the anion and cation are powerful denaturants. The reducing agent permits the complete denaturation of disulfide-bonded proteins, such as ribonucleases, and aids in the dissociation of the connective tissue matrix. The RNA is separated from the clarified homogenate on the basis of its buoyant density by sedimentation through a cushion of cesium chloride.[1] A variation of this method has been described in detail for the isolation of RNA from tissues rich in ribonucleases, although degradation of RNA was reported to occur when powdered frozen tissue was used.[2] The present procedure allows tissue to be stockpiled in the frozen state. This is a significant advantage when numerous dissections are required to obtain sufficient material. Since the RNA is initially recovered by sedimentation rather than a concentration-dependent precipitation very small quantities of tissue may be utilized. In fact, 20 μg of total RNA were recovered from the extraction of *Drosophila* testes by this procedure. Since the isolation from a calcified tissue of biologically active procollagen mRNAs that are 5000 nucleotides in length represents a rather extreme case, the procedure detailed here should be generally applicable to other mRNAs and connective tissues.

[1] V. Glisin, R. Crkvenjakov, and C. Byus, *Biochemistry* **13**, 2633 (1974).
[2] J. M. Chirgwin, A. E. Przybyla, R. J. MacDonald, and W. J. Rutter, *Biochemistry* **18**, 5294 (1979).

*Reagents*

Guanidinium thiocyanate (Fluka purum grade; Tridom, Inc., Hauppage, New York)
Guanidine hydrochloride (Mann ultrapure)
$N$-Lauroylsarcosine (Sigma)
Antifoam A (Sigma)
Cesium chloride, (KBI high purity)
Diethyl pyrocarbonate (Sigma)
Disodium ethylenediaminetetraacetate (EDTA) (Mallinckrodt analytical reagent)
Ethylene glycol bis(2-aminoethyl ether)-$N,N'$-tetraacetic acid (EGTA) (Sigma analytical reagent)
2-Mercaptoethanol (Eastman)
Ethanol 100% (Gold Shield)
Chelex-100 (Bio-Rad Laboratories)

*Equipment*

Pipetman (Gilson)
Polytron tissue homogenizer (Brinkmann Industries)

*Solutions*

Solution A: 5 $M$ guanidinium thiocyanate; 25 m$M$ EDTA, pH 8.0, 25 m$M$ EGTA, pH 8.0, 2% sodium sarkosyl, 0.1% 2-mercaptoethanol; and 0.3% antifoam A

Solution B: 5.7 $M$ CsCl, 25 m$M$ EDTA, pH 8.0, and 25 m$M$ EGTA, pH 8.0

Solution C: 7.5 $M$ guanidine hydrochloride, 25 m$M$ EDTA, pH 8.0, 25 m$M$ EGTA, pH 8.0, and 0.1% 2-mercaptoethanol

*Preparations.* Solution A is prepared by dissolving guanidinium thiocyanate in $H_2O$ with heating. Sarkosyl and the metal chelators are added from concentrated stock solutions. The pH of the solution is adjusted to 8.0 with 1 $M$ NaOH and filtered through a 0.22 $\mu M$ Millipore disposable filter apparatus. Additions of 2-mercaptoethanol and antifoam A are made immediately before use. Solution C is prepared in an analogous manner.

Solution B is prepared by bringing the appropriate dry weight of cesium chloride to volume with $H_2O$ after the additions of chelators. The solution is filtered as above and treated with 0.1% diethyl pyrocarbonate (DEP) for 20 min. Finally, the solution is autoclaved to remove residual traces of DEP.

Double-glass distilled $H_2O$ is prepared for use by treating it with DEP and autoclaving.

Cellulose nitrate tubes are prerinsed with solution C, rinsed with $H_2O$, and thoroughly drained.

Glassware is rendered RNAse free by baking overnight at 180°, and Pipetman tips are autoclaved.

*Procedure for Extracting Procollagen mRNA from Chick Embryo Calvaria*

Calvaria (frontal and parietal bones) from 10 dozen day-16 Hubbard chick embryos are surgically removed and immediately frozen in liquid $N_2$ and stored at $-70°$. Prior to extraction the calvaria are ground to a powder with a mortar and pestle in the presence of liquid $N_2$.[3] Total nucleic acid is extracted by homogenizing at top speed for 30–60 sec with a Polytron in 50 ml of 37° solution A. For this it is convenient to use the 18 mm in diameter Polytron probe with a 100-ml Gibco serum bottle as a homogenizing vessel. The powdered tissue is introduced into the vessel containing 37° solution A by pouring it through a powder funnel prechilled with liquid $N_2$. The homogenate is immediately diluted with another 50 ml of 37° solution A and directly centrifuged at 27,000 $g$ for 15 min at 25° in the SS34 rotor. The supernatants are removed with a 25-ml pipette, and the large pellets are discarded. The combined supernatants are diluted to 168 ml with solution A, and 42 ml of solution B are added and mixed. The resultant homogenate is distributed into six cellulose nitrate tubes (1 × 3.5 inches) by overlaying onto 5-ml cushions of solution B. Centrifugation is performed in the SW27 rotor at 26,000 rpm for 22 hr at 20°.

After centrifugation the guanidinium thiocyanate layer and the portion of the CsCl layer containing visible DNA and polysaccharide bands are carefully removed with a pipette. The internal walls of the tubes are rinsed several times with a minimum volume of solution C; each rinse is removed with a Pasteur pipette after it has drained onto the CsCl layer. Finally, the surface of the CsCl layer is rinsed with solution C. At this point the tubes are inverted and drained thoroughly. The bottom portions of the tubes are cut off with a sterile scapel while maintaining each tube in an inverted position. The tube bottoms are then placed upright on ice, and the surface of each RNA pellet is rinsed with 200 $\mu$l of cold $H_2O$. Dissolution of the RNA is facilitated by mechanically dispersing each pellet with a sealed Pasteur pipette. The RNA pellet is dissolved in several successive additions of $H_2O$, and each solution is transferred to a 30-ml Corex tube. A total volume of 20 ml of $H_2O$ is usually sufficient for this process. The RNA solution is heated to 65° and quenched with ice $H_2O$ prior to ethanol precipitation from 0.15 $M$ lithium acetate, pH 5.5, with 2 volumes of 95% ethanol at $-20°$. The RNA is collected by centrifugation at $-10°$ for 20 min

---

[3] K. Benveniste, J. Wilczek, and R. Stern, *Nature (London)* **246**, 303 (1973).

at 10,000 rpm in Corex tubes in an HB4 rotor. The RNA is dissolved in 30 ml of 10 m$M$ EDTA, pH 8.0, and extracted with an equal volume of chloroform–butanol (4:1, v/v). The organic phase is extracted two additional times with 10 ml of 10 m$M$ EDTA, pH 8.0. The combined aqueous phases are precipitated several times from 0.15 $M$ lithium acetate, pH 5.5, to remove residual salts. A yield of 14 mg of total RNA per 10 dozen calvaria is routinely obtained by this method.

The RNA may be prepared for translation or enzymic reactions by washing the ethanol pellet several times with 95% ethanol at $-20°$ for 10 min. Subsequent to each wash the RNA is collected by centrifugation at $-10°$ for 15 min at 10,000 rpm in the HB4 rotor. The RNA pellet may be dried under $N_2$ or under vacuum and dissolved in $H_2O$. Routinely RNA is stored as an ethanol suspension at $-20°$, but smaller quantities may be stored in $H_2O$ at $-70°$.

*Adaptations*

This procedure may be readily adapted to much smaller quantities of tissue or cultured cells. In such instances it is convenient to use a Polytron probe, 11 mm in diameter, with a 10- or 30-ml Corex tube as a homogenizing vessel. A 1.2-ml cushion of solution B is recommended for each 0.5 × 2 inch cellulose nitrate tube. Centrifugation is performed in a SW50.1 rotor at 20° for 19 hr at 37,000 rpm. For extractions that do not involve calcified tissues, it is advisable to substitute EDTA for EGTA in solutions A, B, and C. Conceivably, extraction of some tissues may require a further purification of the RNA. In such cases the RNA may be precipitated from guanidine hydrochloride,[2] or sulfated polysaccharides may be eliminated by precipitating the RNA from 3 $M$ sodium acetate.[4]

*Comments*

Successful utilization of this procedure depends on several critical points. First, the rate of denaturation of tissue ribonucleases must greatly exceed the rate of RNA degradation by them. Hence, rapid dispersal of the powdered tissue within solution A by use of a Polytron is essential. Solution A is prewarmed to compensate for the chilling effect of the frozen powder. The temperature of solution A may require adjustment for different quantities of tissue, so the homogenate rapidly reaches 25°. The denaturing capacity of solution A should not be exceeded with too much tissue; 3 mg of RNA per SW27 tube is the upper limit for the procedure detailed above. In this regard it is advisable to perform the low speed

---

[4] J. M. Monson and H. M. Goodman, *Biochemistry* **17**, 5122 (1978).

centrifugation immediately after homogenization. Second, it is necessary to prevent the contamination of the RNA pellet with denatured ribonucleases that could renature at a later step. The procedures outlined for rinsing the centrifuge tubes before and after centrifugation have eliminated this problem in my hands. Third, to prevent heavy metal ion hydrolysis of RNA at 65°, solutions B and C and the $H_2O$ must be free of these ions. If necessary these solutions may be treated with Chelex-100.

Translation Products of Procollagen mRNAs

The conditions for translation of procollagen mRNAs in the mRNA-dependent reticulocyte lysate have been detailed elsewhere.[4,5] Identification of prepro-$\alpha$ chains by bacterial collagenase sensitivity and specific antibody precipitation have also been described in detail.[4,5] The purification of the translation products by ammonium sulfate precipitation followed by DEAE-cellulose and agarose column chromatography has recently been described.[6] The characterization of these purified prepro-$\alpha$ chains by tryptic peptide mapping, bacterial collagenase digestion, and radioactive amino acid sequencing by automated Edman degradation have also been reported.[6] Here I describe a simple, rapid method for the initial identification of procollagen mRNA translation products after peptic digestion.

*Rationale*

The triple helix of procollagen is resistant to digestion by pepsin under the appropriate conditions.[7,8] This property has been used as the basis of a very sensitive assay for triple-helical structures, since one can observe by sodium dodecyl sulfate–polyacrylamide gel electrophoresis (SDS-PAGE) the disappearance of pro-$\alpha$1 and pro-$\alpha$2 chains after the globular $NH_2$- and COOH-terminal domains of the molecule are digested by pepsin and the concomitant appearance of the resistant $\alpha$1 and $\alpha$2 chains.[9] The primary structure of *in vitro* synthesized prepro-$\alpha$ chains appears to contain the requisite information for the self-assembly of these chains into pepsin-

[5] D. W. Rowe, R. C. Moen, J. M. Davidson, P. H. Byers, P. Bornstein, and R. D. Palmiter, *Biochemistry* **17**, 1581 (1978).
[6] P. N. Graves, B. R. Olsen, P. D. Fietzek, D. J. Prockop, and J. M. Monson, *Eur. J. Biochem.* **118**, 363 (1981).
[7] A. L. Rubin, D. Pfahl, P. T. Speakman, P. J. Davison, and F. O. Schmitt, *Science* **139**, 37 (1963).
[8] J. Rosenbloom, M. Harsh, and S. Jimenez, *Arch. Biochem. Biophys.* **158**, 478 (1973).
[9] S. Jimenez, M. Harsh, and J. Rosenbloom, *Biochem. Biophys. Res. Commun.* **52**, 106 (1973).

resistant structures. Accordingly, peptic digestion of the entire translation reaction followed by SDS-PAGE of the resistant products results in the conversion of the prepro-$\alpha$ chains to their corresponding $\alpha$ chains, which can readily be identified by their migration relative to pepsin-resistant $\alpha$ chain standards derived from the different collagen types. In addition to simplicity and rapidity, this method also has the distinct advantage that identification is based on comigration with $\alpha$-chain standards. Thus, the ambiguity inherent in comparisons made between unglycosylated, unhydroxylated prepro-$\alpha$ chains and their corresponding glycosylated, hydroxylated pro-$\alpha$ chains is circumvented. Artifacts resulting from partial proteolysis of the prepro-$\alpha$ chains or pro-$\alpha$ chain standards are also avoided. This method is illustrated for chicken calvarial prepro-$\alpha$1(1) and prepro-$\alpha$2(1) which have been identified by antibody precipitation [4] and chemical characterization of the purified prepro-$\alpha$ chains.[6] Others have extended these studies to include chicken type II procollagen translation products as well.[10]

*Procedure for Peptic Digestion of Procollagen mRNA Translation Products*

Total chick embryo calvarial RNA is extracted as described in procedure for isolation of mRNAs and translated in a mRNA-dependent reticulocyte lysate using either [$^3$H]proline or [$^{35}$S]methionine to label the translation products.[4] Reactions are terminated by EDTA and RNase digestion at 26° for 15 min.[4] Aliquots (25 $\mu$l) of each translation reaction to be tested are preincubated at 4° for 30 min in a 1.5-ml Eppendorf polypropylene tube. An equal volume of cold (4°) 1 $M$ acetic acid with or without pepsin (final 800 $\mu$g/ml; Worthington 2 × crystallized) is added to each aliquot. Peptic digestion is performed at 4° for 12–18 hr or at 10° for 6 hr. After digestion, 3 $\mu$g of lathrytic rat skin collagen in 0.5 $M$ acetic acid are added to each sample as a carrier prior to precipitation with 5 volumes of acetone. The precipitate is immediately collected by centrifugation for 10 min at 12,000 $g$ in an Eppendorf centrifuge. The pellets are dissolved in 75 $\mu$l of loading buffer (3% SDS, 0.0625 $M$ Tris-HCl, pH 6.8, 5 $M$ urea, 10% glycerol, 5% (v/v) 2-mercaptoethanol). After heating at 100° for 5 min, 20 $\mu$l (6.7 $\mu$l of the original translation reaction) of each sample are analyzed by SDS-PAGE using a 3% stacking and a 6% separating acrylamide gel.[4] Under these conditions the sample completely penetrates the stacking gel and enters the separating gel. Gels are stained with 0.1% Coomassie Blue in 50% Cl$_3$CCOOH and destained with 7.5% acetic acid. The dried gels are autoradiographed using No-Screen Kodak medical

[10] L. A. Ouellette, L. M. Paglia, and G. R. Martin, *Collagen Rel. Res.* **1**, 327 (1981).

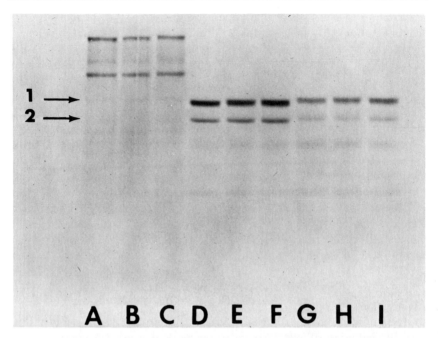

FIG. 1. Fluorogram of sodium dodecyl sulfate–polyacrylamide gel electrophoresis of [$^3$H]proline-labeled chick embryo calvarial RNA translation products. Controls incubated at 15° without pepsin after preincubation of 4° for 0, 30, or 60 min are shown in lanes A, B, and C, respectively. Samples digested with pepsin at 10° (lanes D, E, and F) or 15° (lanes G, H, and I) after the indicated preincubations are also shown.

X-ray film for $^{35}$S or for $^3$H prepared for fluorography with preflashed RP-Royal X-Omat X-ray film.[11,12]

The application of the method to [$^3$H]proline-labeled translation products is illustrated in Fig. 1. In this case, chicken calvarial translation reactions were preincubated at 4° for 0, 30, or 60 min prior to peptic digestion at 10° or 15°. Samples incubated without pepsin retained prominent prepro-$\alpha$ chains at 15° (Fig. 1, lanes A–C) as well as at 10°. Pepsin-resistant $\alpha$1(I)- and $\alpha$2(I)like chains were obtained regardless of the preincubation conditions (4° for 0, 30, or 60 min) after digestion for 6 hr at 10° (Fig. 1, lanes D–F) or 15° (Fig. 1, lanes G–I). The migration of pepsin-treated, rat $\alpha$1(I) and $\alpha$2(I) chain standards is indicated by arrows labeled 1 and 2, respectively. These internal, stained standards essentially comigrated with the pepsin-resistant translation products; a nearly negligible

[11] W. M. Bonner and R. A. Laskey, Eur. J. Biochem. 46, 83 (1974).
[12] R. A. Laskey and A. D. Mills, Eur. J. Biochem. 56, 335 (1975).

discrepancy arose from the fact that the translation products were unhydroxylated as compared to the hydroxylated standards. The appearance of the $\alpha 1$(I)- and $\alpha 2$(I)-like chains after peptic digestion allows the bands to be initially identified as type I procollagen translation products. This identification has previously been confirmed by specific antibody precipitation[4,5] and by chemical characterization of purified prepro-$\alpha$ chains.[6] The pepsin resistance of the type I procollagen translation products has been confirmed, and the observation has been extended to include type II products as well.[10]

## [9] Cell-Free Synthesis of Noninterstitial (CHL Cell) Procollagen Chains

*By* MICHAEL A. HARALSON

The class of proteins collectively referred to as collagen is composed of at least nine distinct structural gene products.[1-4] Based upon anatomic location and physicochemical function, this class of proteins can be subdivided into two major groups: the interstitial collagens (types I, II, and III) and the noninterstitial collagens (types IV and V).[5-17] Type IV colla-

[1] E. J. Miller, *Mol. Cell. Biochem.* **13**, 165 (1976).
[2] J. H. Fessler and L. I. Fessler, *Annu. Rev. Biochem.* **47**, 129 (1978).
[3] P. Bornstein and H. Sage, *Annu. Rev. Biochem.* **49**, 957 (1980).
[4] S. Gay, R. E. Gay, and E. J. Miller, *Arthritis Rheum.* **23**, 937 (1980).
[5] The following nomenclature is employed in this chapter to delineate the constituent chains that are derived from the noninterstitial collagens without reference to the molecular organization of the collagen molecules within each collagen group (type). The type IV collagen group is composed of molecules containing the $\alpha 1$(IV) chain[5a] [also designated as C-chain,[6-8] 140K,[9] P1,[10] and $\alpha''$(IV)-chain[11]] and/or molecules containing the $\alpha 2$(IV) chain[4] [also designated as D-chain,[6-8] 70K,[9] P2,[10] and $\alpha'$(IV) chain[11]]. The type V collagen group is composed of molecules containing the $\alpha 1$(V) chain[4] [also designated as B-chain,[12-14] and $\alpha B$[15,16]], the $\alpha 2$(V) chain[4] [also designated as A-chain[12-14] or $\alpha A$[15,16]], and/or the $\alpha 3$(V) chain[4] [also designated as $\alpha C$[16,17]].
[5a] N. A. Kefalides, *Biochem. Biophys. Res. Commun.* **45**, 226 (1971).
[6] S. Gay and E. J. Miller, *Arch. Biochem. Biophys.* **198**, 370 (1979).
[7] S. N. Dixit and A. H. Kang, *Biochemistry* **18**, 5686 (1979).
[8] T. F. Kresina and E. J. Miller, *Biochemistry* **18**, 3089 (1979).
[9] H. Sage, R. G. Woodbury, and P. Bornstein, *J. Biol. Chem.* **254**, 9893 (1979).
[10] R. Timple, P. Bruckner, and P. P. Fietzek, *Eur. J. Biochem.* **95**, 255 (1979).
[11] R. W. Glanville, A. Rauter, and P. P. Fietzek, *Eur. J. Biochem.* **95**, 383 (1979).
[12] E. Chung, R. K. Rhodes, and E. J. Miller, *Biochem. Biophys. Res. Commun.* **71**, 1167 (1976).
[13] R. K. Rhodes and E. J. Miller, *Biochemistry* **17**, 3442 (1978).

gen molecules are those collagens isolated from morphologically distinct basement membranes,[5a–7,18,19] where they function as cellular and tissue supportive elements within these anatomic structures. In contrast, recent immunohistochemical evidence indicates that type V collagen molecules are localized in pericellular or exocytoskeletal basement-membrane-like structures with apparent supportive functions similar to that of type IV collagen molecules.[20–26]

The intensive efforts aimed at elucidating the chemical composition of the noninterstitial collagens and their constituent chains reflect, at least in part, their presence in elements (basement membranes and exocytoskeletal matrices) that are involved in maintaining the normal function of the lung, kidney, eye, vascular tissue, and other organs and are morphologically affected in a variety of pathological states (for reviews, see footnotes 26–29). Thus an understanding of these disease processes—which include diabetes mellitus, scleroderma, lupus erythematosus, poststreptococcal glomerulonephritis, atherosclerosis, Goodpasture's syndrome, certain cancers, and others—requires not only the elucidation of the components in these tissue elements, but also precise information about the mechanisms that regulate the biosynthesis and catabolism of these components, e.g., the type IV and type V collagens and their con-

---

[14] H. Bentz, H. P. Bächinger, R. Glanville, and K. Kühn, *Eur. J. Biochem.* **92**, 563 (1978).
[15] R. E. Burgeson, F. A. El Adli, I. I. Kaitila, and D. W. Hollister, *Proc. Natl. Acad. Sci. U.S.A.* **73**, 2579 (1976).
[16] R. A. Brown, A. Shuttleworth, and J. B. Weiss, *Biochem. Biophys. Res. Commun.* **80**, 866 (1978).
[17] H. Sage and P. Bornstein, *Biochemistry* **18**, 3815 (1979).
[18] N. A. Kefalides, *Int. Rev. Connect. Tissue Res.* **6**, 63 (1973).
[19] N. A. Kefalides, R. Alper, and C. C. Clark, *Int. Rev. Cytol.* **61**, 167 (1979).
[20] S. Gay, R. K. Rhodes, R. E. Gay, and E. J. Miller, *Collagen Rel. Res.* **1**, (1981).
[21] M. A. Haralson, W. E. Mitchell, R. K. Rhodes, T. F. Kresina, R. Gay, and E. J. Miller, *Proc. Natl. Acad. Sci. U.S.A.* **77**, 5206 (1980).
[22] E. J. Miller, R. K. Rhodes, S. Gay, T. F. Kresina, and D. K. Furuto, *Front. Matrix Biol.* **7**, 3 (1979).
[23] A. J. Bailey, G. B. Shellswell, and V. C. Duance, *Nature (London)* **278**, 67 (1979).
[24] R. K. Rhodes and E. J. Miller, *J. Biol. Chem.* **254**, 12,084 (1979).
[25] R. E. Gay, R. B. Buckingham, R. K. Prince, S. Gay, G. P. Rodnan, and E. J. Miller, *Arthritis Rheum.* **23**, 190 (1980).
[26] S. Gay and E. J. Miller, "Collagen in the Physiology and Pathology of Connective Tissue." Fischer, Stuttgart, 1978.
[27] N. A. Kefalides (ed.) "Biology and Chemistry of Basement Membranes." Academic Press, New York, 1978.
[28] N. A. Kefalides, *Dermatologica* **150**, 4 (1975).
[29] C. M. Lapière and B. Nusgens, *in* "Biochemistry of Collagen" (G. N. Ramachandran and A. H. Reddi, eds.), pp. 377–447. Plenum, New York, 1976.

stituent chains. One approach to obtaining such information is to establish defined cell-free systems capable of synthesizing noninterstitial collagenous protein(s). Such systems can then be dissected at the molecular level to elucidate potential transcriptional, posttranscriptional, and/or translational regulatory elements involved in the biosynthesis of the noninterstitial collagens, as well as delineation of the posttranslational sequence of processes involved in the maturation and catabolism of the initial gene products.

The HT1 clone of cultured Chinese hamster lung (CHL)[29a] cells, a subclone of the V79 CHL line originally established by Ford and Yerganian,[30] devotes a significant proportion of its biosynthetic capacity to the synthesis of collagenous material.[31-33] This cloned line, which has been employed in a variety of both genetic and biochemical studies,[34-38] possesses several characteristics that make it advantageous for use in studying macromolecular synthesis, including ($a$) a relatively stable karotype of 21, which quantitatively, if not qualitatively, approximates the normal Chinese hamster cell complement of chromosomes (22); ($b$) the ability to grow under a variety of culture conditions; ($c$) a rapid generation time (approximately 14 hr at 37°); ($d$) growth to high cell density; and ($e$) extremely high (>90%) plating efficiency. Evidence from this laboratory has demonstrated that the HT1 clone of CHL cells produces a collagen characteristic of one member of the noninterstitial collagen group, and thus this cell line represents an appropriate model for evaluating the biosynthesis and regulation of the collagen $\alpha 1(V)$ chain.[21] In this chapter, the preparation of three systems containing CHL components that direct the cell-free synthesis of collagenous protein(s) is described, and the usefulness of each system for studying the metabolism of a noninterstitial collagen and its component chains is discussed.

[29a] The abbreviations used in this manuscript are: CHL, Chinese hamster lung; S30, supernatant fraction from 30,000 $g$ centrifugation; S250, supernatant fraction from 250,000 $g$ centrifugation; DTT, dithiothreitol; SDS, sodium dodecyl sulfate; IBOH, isobutyl alcohol; CP, creatine phosphate; CPK, creatine phosphokinase; ATA, aurin tricarboxylic acid; HEPES, 4-(2-hydroxyethyl)-1-piperazineethanesulfonic acid; PBS, phosphate-buffered saline.
[30] D. K. Ford and G. Yerganian, *J. Natl. Cancer Inst. (U.S.)* **21**, 393 (1958).
[31] M. A. Haralson, K. L. Frey, and W. M. Mitchell, *Biochemistry* **17**, 864 (1978).
[32] M. A. Haralson and W. M. Mitchell, *Fed. Proc., Fed. Am. Soc. Exp. Biol.* **37**, 1408 (1978).
[33] M. A. Haralson, J. H. Sonneborn, and W. M. Mitchell, *J. Biol. Chem.* **253**, 5536 (1978).
[34] A. L. Beaudet, D. J. Roufa, and C. T. Caskey, *Proc. Natl. Acad. Sci. U.S.A.* **70**, 320 (1973).
[35] D. J. Roufa and S. J. Reed, *Genetics* **80**, 549 (1975).
[36] D. J. Roufa, R. E. Moses, and S. J. Reed, *Arch. Biochem. Biophys.* **167**, 547 (1975).
[37] M. A. Haralson and D. J. Roufa, *J. Biol. Chem.* **250**, 8618 (1975).
[38] J. J. Wasmuth and C. T. Caskey, *Cell* **8**, 71 (1976).

General Procedures

*Reagents and Materials*

Dulbecco's modified Eagle's medium with 4.5 g of glucose per liter, fetal bovine serum, and NaHCO$_3$ (Grand Island Biological Company)
Tissue culture plasticware (Falcon)
Tris (Sigma 7-9)
Acid-washed glass beads (Sigma)
Sodium heparin (Sigma)
Sephadex G-25 and G-10 (Sigma)
Puromycin (Sigma)
DTT (Sigma)
2-Mercaptoethanol (Sigma)
HEPES, potassium creatine phosphate, and creatine phosphokinase from rabbit skeletal muscle (Calbiochem)
K$_2$ATP and NaGTP (P-L Biochemicals)
Electrophoresis reagents (Bio-Rad)
Nitrocellulose filters, HAWP, 2.5 cm (Millipore)
Raw wheat germ (General Mills, Vallejo, California)
Bacterial collagenase (CLSPA) (Worthington)
[$^3$H]Proline, [$^{14}$C]proline, [$^3$H]leucine, and guanidine-HCl (Schwarz/Mann)
Liquid scintillation reagents (New England Nuclear)

All inorganic compounds used are ACS reagent grade. The origin of the Chinese hamster lung (CHL) cell line, as well as the clone (HT1) used in these studies, has been previously reported.[30,31]

*Solutions*

Solution A: 10 m$M$ Tris-HCl, pH 7.5; 10 m$M$ KCl; 1.5 m$M$ Mg(OAc)$_2$

Solution B: 300 m$M$ Tris-HCl, pH 7.5; 1.25 $M$ KCl; 50 m$M$ Mg(OAc)$_2$; 100 m$M$ 2-mercaptoethanol

Solution C: 30 m$M$ Tris-HCl, pH 7.5; 125 m$M$ KCl; 5 m$M$ Mg(OAc)$_2$; 10 m$M$ 2-mercaptoethanol

Solution D: 25 m$M$ Tris-HCl, pH 7.5; 25 m$M$ NaCl; 5 m$M$ MgCl$_2$; heparin, 1 mg/ml; 2% (v/v) Triton X-100

Solution E: 25 m$M$ Tris-HCl, pH 7.5; 25 m$M$ NaCl; 200 m$M$ MgCl$_2$; heparin, 1 mg/ml; 2% (v/v) Triton X-100

Solution F: 8 $M$ guanidine-HCl; 10 m$M$ NaOAc, pH 5.2; 1 m$M$ DTT

Solution G: 8 $M$ guanidine-HCl; 10 m$M$ NaOAc, pH 5.2; 1 m$M$ Na$_2$EDTA; 1 m$M$ DTT

Solution H: 20 m$M$ HEPES, pH 7.6; 100 m$M$ KCl; 1 m$M$ Mg(OAc)$_2$; 2 m$M$ CaCl$_2$; 1 m$M$ DTT

Solution I: 20 m$M$ HEPES, pH 7.6; 120 m$M$ KCl; 5 m$M$ Mg(OAc)$_2$; 1 m$M$ DTT

*Growth of Cells in Culture.* CHL cells are grown in Dulbecco's modified Eagle's medium containing 4.5 g of glucose per liter, 0.15% (w/v) NaHCO$_3$, 50 μg of gentamycin sulfate per milliliter, and 10% (v/v) fetal bovine serum. Stock cultures are maintained in either 25-cm$^2$ or 75-cm$^2$ flasks and are subcultured at a 1:50 dilution by trypsinization approximately every 4 days. For the generation of large quantities of cells for the preparation of cell-free extracts, plastic dishes 150 mm in diameter are seeded with approximately 5 × 10$^5$ cells/dish and grown under 10% CO$_2$ tension at 37° to a density of approximately 2 × 10$^7$ cells per dish (85–90% confluency). The medium is removed, and the cell layer is washed with 10 ml of PBS. The cells are removed from the dishes by scraping into two 3-ml portions of cold PBS per dish, and the cells are then collected by centrifugation at 1000 $g$ for 10 min at 4°. The yield of CHL cells is approximately 150–200 mg wet weight per dish.

## Procedures for Preparing Cell-Free Protein Synthesis Components

*CHL S30.* A cell-free postnuclear extract (S30) from CHL cells is prepared by suspending the collected cells in solution A and allowing the cells to swell for 30 min at 4°. Routinely, 5 g (wet weight) of collected cells are suspended in 20 ml of solution A, and the swollen cells are disrupted in a cold Dounce homogenizer using 10 strokes of a loose-fitting pestle followed by 5 strokes of a tight-fitting pestle. The homogenate is adjusted to standard buffer and salt conditions by the addition of one-ninth the volume of solution B, followed by centrifugation at 30,000 $g$ (18,000 rpm in a Beckman JA-20 rotor) for 20 min to remove nuclei, mitochondria, and cell debris. The supernatant fluid (excluding the upper lipid layer) is withdrawn using a Pasteur pipette, and the material (~ 20 ml) is dialyzed for 8 hr at 4° against 2 × 1-liter portions of solution C containing 20% (v/v) glycerol. The dialyzed material (S30) is then quick-frozen in aliquots and stored at −70°.

*CHL Polysomes.* CHL polysomes are prepared by a modification of the Mg$^{2+}$-precipitation protocol of Palmiter.[39] CHL cells (4.5 g) are suspended in 25 ml of solution D and allowed to swell for 15 min in a chilled Dounce homogenizer. The cells are disrupted by 10 strokes of a loose-fitting glass pestle followed by 5 strokes of a tight-fitting pestle, and the homogenate is centrifuged at 4° at 15,000 rpm (16,000 $g$) in the JA-20 rotor in a Beckman J21B centrifuge. The supernatant fraction (approximately 28 ml) is withdrawn using a Pasteur pipette and adjusted to 100 m$M$ MgCl$_2$ concentration by the addition of an equal volume of solution E. The mixture is kept

[39] R. D. Palmiter, *Biochemistry* **13**, 3606 (1974).

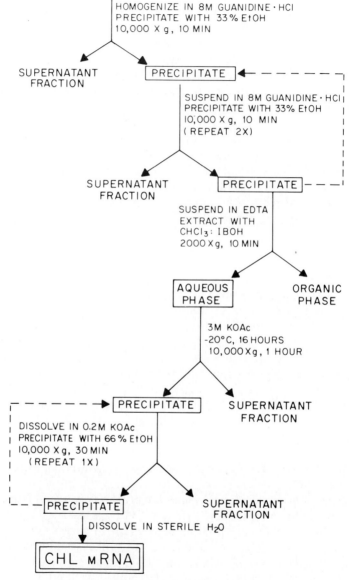

Fig. 1. Isolation scheme for CHL cell mRNA

at 4° for 1.5 hr, and the precipitated ribonucleoproteins are collected by centrifugation at 16,000 rpm for 25 min in 20-ml aliquots through a 10-ml cushion of 1 $M$ sucrose in 25 m$M$ Tris-HCl, pH 7.5, 25 m$M$ NaCl, and 5 m$M$ MgCl$_2$. The supernatant fraction is removed by aspiration, and the pelleted material is suspended in solution C by trituration using a Pasteur pipette. The suspension is then clarified by centrifugation at 10,000 rpm for 10 min, and the polysome suspension is stored at $-70°$ in aliquots at a concentration of ~40 $A_{260}$/ml. The average yield of polysomes from 4.5 g of CHL cells is ~275 $A_{260}$ units.

*CHL mRNA.* An RNA fraction, which directs the synthesis of collagenous protein in a wheat germ lysate (S30) system, is prepared from either CHL polysomes or CHL cells by a modification of the guanidine-HCl extraction procedure.[40-42] A schematic representation of this protocol is shown in Fig. 1. Routinely, a 10–20% homogenate (w/v) of either CHL cells or CHL polysomes is prepared by suspension of the cellular material in solution F using a Dounce homogenizer. The suspension is then mixed with 0.5 volume (relative to the original volume of solution F) cold absolute ETOH, the mixture is shaken and placed at $-20°$ for 2 hr. The precipitate is collected by centrifugation at 10,000 $g$ for 10 min at $-10°$. The precipitate is suspended by trituration in 0.5 volume (relative to the original volume of solution F) of solution G at 4°, 0.5 volume (relative to solution G) of cold absolute ETOH is added, the mixture is shaken, stored at $-20°$ for 2 hr, and RNA is collected by centrifugation as described above. The precipitate is extracted with solution G two additional times with subsequent reductions in the amount of solution G used. The final guanidine-extracted RNA precipitate is then dissolved by trituration at 4° in 0.2 volume (relative to solution F) of 20 m$M$ Na$_2$EDTA, pH 7.0. An equal volume of CHCl$_3$:isobutanol (4:1) is added, the suspension is shaken, and the phases are separated by centrifugation at 2000 rpm for 10 min at 4°. The aqueous (upper) phase is withdrawn and placed in a chilled graduated cylinder; the organic phase is reextracted with 20 m$M$ Na$_2$EDTA, pH 7.0 (0.3 volume relative to the original volume of Na$_2$EDTA) until the interphase material is dispersed (usually 3 to 4 additional times). The combined aqueous phases are then mixed with 2 volumes of 4.5 $M$ KOAC, pH 5.8, and the suspension is stored overnight at $-20°$. The RNA is collected by centrifugation at 15,000 $g$ for 1 hr at $-10°$,

---

[40] J. M. Chirgwin, A. E. Przybyla, R. J. MacDonald, and W. J. Rutter, *Biochemistry* **18**, 5294 (1979).

[41] S. L. Adams, M. E. Sobel, B. H. Howard, K. Olden, K. M. Yamada, B. de Crombrugghe, and I. Pastan, *Proc. Natl. Acad. Sci. U.S.A.* **74**, 3399 (1977).

[42] M. A. Haralson, S. J. Fairfield, W. E. Nicholson, R. W. Harrison, and D. N. Orth, *J. Biol. Chem.* **254**, 2172 (1979).

and the precipitate is dissolved in 0.2 $M$ KOAc, pH 5.8 (approximately 0.2 volume relative to volume of combined aqueous solution). Twice the volume of cold absolute ethanol is added, the solution is stored at $-20°$ for 2 hr, and insoluble material is collected by centrifugation at 10,000 $g$ for 30 min at $-10°$. The precipitate is dissolved in 0.2 $M$ KOAc, pH 5.8, and RNA is precipitated from the solution as described above. After drying under a stream of nitrogen, the final precipitate is dissolved in sterile deionized $H_2O$ to a concentration of $\sim 2.5$ mg of RNA per milliliter, centrifuged at 10,000 $g$ for 10 min at 4° to remove insoluble material, and the RNA solution is stored in aliquots at $-70°$. The average yield of RNA obtained by this procedure is $\sim 3$ mg of RNA per gram wet weight of cells.

*Wheat Germ S30 and S250.* Wheat germ lysate (S30) and the postribosomal supernatant fraction (S250) derived from the lysate are prepared by a modification of the procedure of Marcu and Dudock[43]; a schematic presentation of this protocol is shown in Fig. 2. All procedures are performed at 4°. Twelve grams of raw wheat germ are ground dry for 60 sec in a chilled mortar with an equal weight of acid-washed glass beads. Twenty-five milliliters of solution H are added, and the mixture is centrifuged at 16,000 rpm (15,000 $g$) rpm for 15 min in the JA-20 rotor in a Beckman J21B centrifuge. The brown supernatant fraction between the pellet and the upper lipid pellicle is withdrawn, and a portion of the material (5 ml) is applied to a 1.5 × 35 cm Sephadex G-25-80 column equilibrated in solution I. Fractions are collected at a flow rate of 1.0 ml/min, and materials eluting at the void volume of the column with a concentration greater than 100 $A_{260}$ units/ml are combined. The combined material is centrifuged at 18,000 rpm (30,000 $g$) for 20 min, and the supernatant fraction (S30) is withdrawn. If no further fractionation of the S30 is desired, the material is quick frozen in aliquots and stored at $-70°$ at a concentration of $\sim 110$ $A_{260}$/ml ($A_{260}/A_{280} \cong 1.50$). For the preparation of the postribosomal supernatant fraction (S250), fresh unfrozen wheat germ S30 is centrifuged at 50,000 rpm (250,000 $g$) for 3 hr in the 60 Ti rotor in a Beckman L5-50 ultracentrifuge. The upper three-fourths of the supernatant fluid is withdrawn and dialyzed for 12 hr against two successive 500-ml portions of solution C containing 20% (v/v) glycerol. The dialysate (S250) is quick-frozen in aliquots and stored at $-70°$ at a concentration of 19 $A_{260}$/ml ($A_{260}/A_{280} = 0.95$).

Reaction Conditions for Cell-Free Protein Synthesis Directed by CHL Cell Components

*CHL S30-Catalyzed Reactions.* Standard reaction mixtures contain in a total volume of 0.05 ml: 45 m$M$ Tris-HCl, pH 7.5; 113 m$M$ KCl; 3 m$M$

[43] K. Marcu and B. Dudock, *Nucleic Acids Res.* **1**, 1385 (1974).

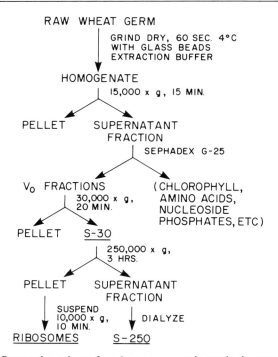

FIG. 2. Preparation scheme for wheat germ protein synthesis components.

Mg(OAc)$_2$, 10 m$M$ 2-mercaptoethanol; 1 m$M$ ATP; 0.1 m$M$ GTP; 8 m$M$ creatine phosphate; 2 µg of creatine phosphokinase; 2.04 µ$M$ [$^3$H]proline (37,900 cpm/pmol) or 2.05 µ$M$ [$^3$H]leucine (48,200 cpm/pmol); and dialyzed CHL S30 extract in the amount indicated. After incubation for 2 hr at 25°, the reaction is terminated by adding 20 µg of RNase or 0.1 ml of 1 $N$ KOH and incubating an additional 20 min at 37° to hydrolyze $^3$H-labeled aminoacyl-tRNA. Trichloroacetic acid (2 ml of a 10% (w/v) solution) is then added to the reaction, and the suspension is placed on ice for 15 min. Acid-insoluble material is collected by filtration through nitrocellulose filters (Millipore); each filter is washed with 3 × 4 ml aliquots of 5% trichloroacetic acid followed by 2 × 4 ml aliquots of deionized water. The filters are then dissolved in a toluene–dioxane based scintillation cocktail,[44] and radioactivity retained on the filter is determined in a Beckman liquid scintillation counter. Under these conditions, efficiency of counting of each isotope is approximately 36%.

*CHL Polysome-Directed Reactions.* Standard reaction mixtures contain in a total volume of 0.1 ml: 45 m$M$ Tris-HCl, pH 7.5; 125 m$M$ KCl; 4.5 m$M$ Mg(OAc)$_2$; 10 m$M$ 2-mercaptoethanol; 1 m$M$ ATP (neutralized with

[44] J. F. Chiu and S. C. Sung, *Biochim. Biophys. Acta* **209**, 34 (1970).

KOH); 0.2 m$M$ GTP; 8 m$M$ creatine phosphate; 4 $\mu$g of creatine phosphokinase; 2.08 $\mu M$ [$^3$H]proline (specific activity = 33,800 cpm/pmol); wheat germ S250 and CHL polysomes in the amounts indicated. The reaction mixture is incubated for 2 hr at 25° and terminated by the addition of 20 $\mu$g of RNase. After an additional incubation at 37° for 20 min, 2 ml of a 10% (w/v) of trichloroacetic acid are added to the reaction, and the suspension is placed on ice for 15 min. The precipitate is collected on nitrocellulose filters and washed; radioactivity is determined as described in the preceding section.

*CHL mRNA-Programmed Reactions.* Standard reaction mixtures contain in a total volume of 0.15 ml: 27 m$M$ HEPES, pH 7.6; 70 m$M$ KCl; 2.5 m$M$ Mg(OAc)$_2$; 5 m$M$ DTT; 1 m$M$ ATP; 0.2 m$M$ GTP; 8 m$M$ creatine phosphate; 6 $\mu$g of creatine phosphokinase; 10.2 $\mu M$ [$^{14}$C]proline (505 cpm/pmol) or 1.4 $\mu M$ [$^3$H]proline (31,200 cpm/pmol); wheat germ S30 and CHL mRNA in the amounts indicated. The reaction mixture is incubated for 2 hr at 25° and terminated by the addition of 20 $\mu$g of RNase. After incubation for 20 min at 37°, the incorporation of radioactivity into CHL mRNA-directed cell-free products is determined as previously detailed in this section.

### Procedures for Cell-Free Synthesized Product Characterization

*Collagenase Digestion.* Protein synthesis reaction mixtures containing either CHL S30, CHL polysomes, or CHL mRNA are prepared as previously detailed. After incubation for 2 hr at 25°, the mixtures are adjusted to 2 m$M$ in respect to CaCl$_2$ and are incubated an additional 2 hr at 25° in either the presence or the absence of 10 $\mu$g of purified bacterial collagenase, which had been previously assayed to ensure the absence of any detectable noncollagenolytic proteolytic activity.[31-33] After treatment, the decrease in acid-insoluble RNase-resistant material retained by nitrocellulose filters is determined as previously detailed, or the alteration in the molecular weight profile of the cell-free synthesized products is determined by gel electrophoresis as described in the next section.

*SDS–Polyacrylamide Gel Electrophoresis.* Molecular weight analysis of cell-free synthesized products is performed on 5.0% polyacrylamide gels using a modification of the procedure of Fairbanks *et al.*[45] Cell-free synthesized [$^3$H]proline-labeled products (with or without collagenase digestion) are prepared for electrophoresis by chromatography on a Sephadex G-10 column equilibrated in 10 m$M$ Tris-HCl, pH 8.0, 10 m$M$ EDTA; the fractions eluting at the void volume of the column are combined and lyophilized. The residue is dissolved in 20% of its original volume in H$_2$O,

[45] G. Fairbanks, T. L. Steck, and D. F. H. Wallach, *Biochemistry* **10**, 2606 (1971).

made 1% with respect to SDS and 50 m$M$ with respect to DTT, and heated for 15 min at 80°. An aliquot of the sample is then applied to a 0.6 × 10 cm polyacrylamide gel that had been electrophoresed for 2 hr prior to sample application. The gels are then electrophoresed for 4 hr at 8 mA/gel or until the tracking dye (bromophenol blue) is approximately 2 mm from the bottom of the gel.

To determine the distribution of radioactivity, the completed gels are sliced and crushed into 2-mm segments using a Gilson gel fractionator, and radioactivity is eluted from the crushed sections in 0.5 ml of $H_2O$ for 6 hr at room temperature. Each sample is then suspended in 10 ml of Aquasol, and radioactivity is measured in a Beckman liquid scintillation counter. Under these conditions, recovery of applied radioactivity is >75% and counting efficiency of the isotope is ~30%.

Molecular weight markers used for gel calibration are run on parallel gels, and their mobility as a function of size is visualized after staining with 0.05% Coomassie Brilliant Blue. For the studies reported in this chapter, the molecular weight ($M_r$) standards employed are myosin ($M_r$ = 212,000), β-galactosidase ($M_r$ = 120,000), phosphorylase α ($M_r$ = 98,500), bovine serum albumin ($M_r$ = 67,000), and human γ-globulin (heavy chain $M_r$ = 50,000 and light chain $M_r$ = 25,000).

Protein Synthesis Catalyzed by CHL Cell Components: Parameters of the Reactions and Characterization of the Cell-Free Products

*Cell-Free Procollagen Chain Synthesis Catalyzed by CHL S30 Preparations.* The dialyzed cell-free extract (S30) prepared from cultured Chinese hamster lung cells promotes the incorporation of approximately equimolar amounts of [$^3$H]proline and [$^3$H]leucine into newly synthesized product (Table I). The CHL S30-catalyzed reaction depends upon extract and incubation, is stimulated by the addition of nucleoside triphosphates and a regenerating system, and is inhibited more than 90% by the antibiotic puromycin (Table I). The CHL S30-catalyzed reaction has $Mg^{2+}$ and $K^+$ optima at 3 and 113 m$M$, respectively, and incorporation of radioactive substrate into product is linear with respect to time for 1 hr at 25° followed by a slower rate of synthesis that continues for several hours (data presented by Haralson *et al.*[31]).

Demonstration that collagenous protein(s) are among the CHL S30-catalyzed reaction products is achieved by examining the effects of bacterial collagenase on the newly synthesized material. As shown in Table II, part A, approximately 50% of the [$^3$H]proline incorporated into cell-free product is rendered acid soluble by treatment with the enzyme. Furthermore, the addition of a large excess of nonradioactive substrate for the

TABLE I
REACTION REQUIREMENTS FOR CELL-FREE PROTEIN BIOSYNTHESIS CATALYZED BY
CHINESE HAMSTER LUNG CELL S30[a,b]

| | Polypeptide synthesized | | | |
|---|---|---|---|---|
| | [³H]Proline | | [³H]Leucine | |
| Reaction components | cpm | pmol | cpm | pmol |
| Complete | 61,200 | 1.61 | 74,400 | 1.54 |
| Minus extract | 430 | 0.01 | — | — |
| Minus incubation | 600 | 0.02 | — | — |
| Minus ATP, GTP, CPK, and CP | 9,600 | 0.25 | — | — |
| Plus puromycin (0.67 m$M$) | 4,800 | 0.13 | — | — |

[a] Reprinted, with permission, from Haralson et al.[31]
[b] Cell-free incorporation of ³H-labeled amino acids into RNase-resistant acid-insoluble products was performed as described with modifications of standard reaction mixtures as indicated. Reaction mixtures were supplemented with 0.62 $A_{260}$ units of CHL S30 and were incubated for 2 hr at 25°. Values presented are corrected for radioactivity retained by the filter in the absence of extract (0.01 and 0.02 pmol, for [³H]proline and [³H]leucine, respectively).

TABLE II
COLLAGENASE EFFECTS ON CHL S30-CATALYZED REACTION PRODUCTS[a,b]

| | Acid-insoluble [³H]proline (cpm) | | Collagenase digestible product (%) |
|---|---|---|---|
| Component | Minus collagenase | Plus collagenase | |
| A. CHL S30 amount ($A_{260}$ units) | | | |
| 0.18 | 22,700 | 12,200 | 46.3 |
| 0.37 | 37,800 | 19,200 | 49.2 |
| 0.75 | 55,100 | 29,100 | 47.2 |
| B. CHL S30 + collagen | 59,200 | 60,600 | 0 |

[a] Reprinted, with permission from Haralson et al.[31]
[b] Standard reaction mixtures containing CHL S30 extract (A) in the amounts indicated or (B) 0.75 $A_{260}$ unit were incubated for 2 hr at 25°. The mixtures were then adjusted to 2 m$M$ in respect to CaCl$_2$, supplemented with bacterial collagenase when indicated, and incubated an additional 2 hr at 25°; base-resistant acid-insoluble [³H]proline-labeled products were determined as described. In B, 100 μg of rat skin collagen were added to the reaction mixture before the addition of collagenase. Values have been corrected for nonspecific binding of [³H]proline to the filters in the absence of extract (745 cpm).

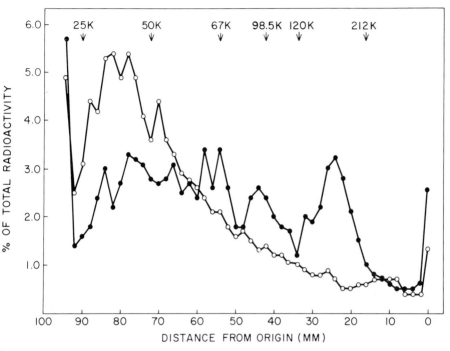

FIG. 3. SDS–polyacrylamide gel electrophoresis of CHL S30-catalyzed reaction products. Cell-free synthesized products catalyzed by 0.62 $A_{260}$ units of CHL S30 were prepared as described and analyzed on 5% polyacrylamide gels before (●——●) and after (○——○) collagenase treatment. The positions of migration of molecular weight standards are indicated by arrows at the top of the figure. Adapted, with permission, from Haralson et al.[31]

enzyme to the completed reaction mixture before collagenase digestion prevents the solubilization of the radioactive cell-free material (Table II, part B).

Further product characterization is achieved by examining the size distribution of the newly synthesized proteins both before and after collagenase digestion. As shown in Fig. 3, three distinct size classes of radioactive collagenase-digestible proteins are generated by this system. The largest of the collagenous products has a molecular mass of ~165,000 daltons and represents approximately 20% of the total incorporated radioactivity. Two additional size classes of collagenase-sensitive proteins are also among the cell-free products: (a) a species with a molecular weight of 95,000 that represents approximately 13% of the incorporated radioactivity; and (b) material with an approximate molecular weight of 65,000 that comprises approximately 12% of the cell-free product. Thus the collagenase-sensitive proteins, whose synthesis is catalyzed *in vitro* by

## TABLE III
### CELL-FREE PROTEIN SYNTHESIS DIRECTED BY CHINESE HAMSTER LUNG CELL POLYSOMES[a]

| Reaction component | [³H]Proline incorporated (cpm) | Percentage of control |
|---|---|---|
| A. Complete | 104,700 | 100 |
| Minus CHL polysomes | 5,290 | 5.1 |
| Minus wheat germ S250 | 5,160 | 4.9 |
| Minus ATP, GTP, CPK, and CP | 6,750 | 6.5 |
| Plus puromycin | 9,310 | 8.9 |
| Plus ATA | 60,200 | 57.5 |
| B. Complete | | |
| Plus $CaCl_2$ minus collagenase | 115,300 | 100 |
| Plus $CaCl_2$ and collagenase | 54,980 | 47.7 |

[a] Cell-free protein biosynthesis directed by CHL polysomes in the presence of wheat germ S250 was determined as described in reaction mixtures supplemented with 0.38 $A_{260}$ units of wheat germ S250 and 0.81 $A_{260}$ units of CHL polysomes. Reaction mixtures were modified in Part A to omit the components indicated or to include 0.1 m$M$ puromycin or 0.1 m$M$ aurin tricarboxylic acid as indicated. Complete cell-free protein synthesis mixtures were adjusted in Part B to 2 m$M$ $CaCl_2$ and incubated an additional 2 hr at 25° in the presence or the absence of bacterial collagenase as described. Values have been corrected for nonspecific binding of [³H]proline to the filters in the absence of S250 and polysomes (1534 cpm).

CHL S30 preparations, can be ascribed to three size classes with approximately 20% of the total cell-free product corresponding in size to procollagen chains (165,000 daltons).

*Cell-Free Procollagen Chain Synthesis Directed by CHL Polysomes.* Polysomes isolated from cultured Chinese hamster lung cells promote the efficient incorporation of [³H]proline into cell-free product in the presence of wheat germ postribosomal supernatant factors (S250). As shown in Table III, part A, this reaction is dependent upon both subcellular components and an energy source, is inhibited approximately 10-fold by the inclusion of puromycin, and is significantly affected by aurin tricarboxylic acid, a compound that inhibits the initiation of protein synthesis.[46] The reaction has $Mg^{2+}$ and $K^+$ optima at 4.5 m$M$ and 125 m$M$, respectively, and total incorporation of substrate into product is essentially complete by 2 hr at 25° (data not shown).

When the newly synthesized material is subjected to bacterial collagenase digestion (Table III, part B), approximately 50% of the incorpo-

[46] P. C. Tai, B. J. Wallace, and B. D. Davis, *Biochemistry* 7, 668 (1973).

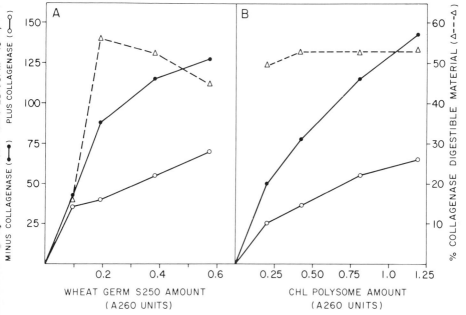

FIG. 4. Cell-free procollagen chain synthesis as a function of wheat germ S250 or CHL polysome amount. Reaction mixtures containing wheat germ S250 or CHL polysomes in the amounts indicated were prepared and incubated as described. The mixtures were then adjusted to 2 m$M$ in respect to $CaCl_2$ and incubated in the absence (●——●) or in the presence (○——○) of bacterial collagenase; the amount of [$^3$H]proline incorporated into acid-insoluble products was determined. The degree of collagenase-sensitive material produced (△---△) is expressed as the ratio of acid-insoluble material after collagenase digestion to total synthesis at each data point. In panel A, reaction mixtures contained 0.81 $A_{260}$ units of CHL polysomes and wheat germ S250 in the amounts indicated. In panel B, CHL polysomes levels were varied in the presence of 0.38 $A_{260}$ unit of wheat germ S250. Values presented have been corrected for retention of radioactivity (2250 cpm) by the filters in the absence of both subcellular fractions.

rated radioactivity is rendered acid soluble—a result similar to that obtained with CHL S30-catalyzed reaction products (Table II, part A). In contrast to the CHL S30-catalyzed reaction, in which the amount of collagenase-digestible product is independent of the amount of CHL S30 employed (Table II, part A), this value is markedly affected by the level of wheat germ S250 employed in this system. As shown in Fig. 4, panel A, the amount of collagenase-sensitive product produced directly relates to the amount of wheat germ S250 in the reaction when this value is determined in the presence of excess CHL polysomes. In contrast, the amount of collagenous material synthesized is independent of the CHL polysome

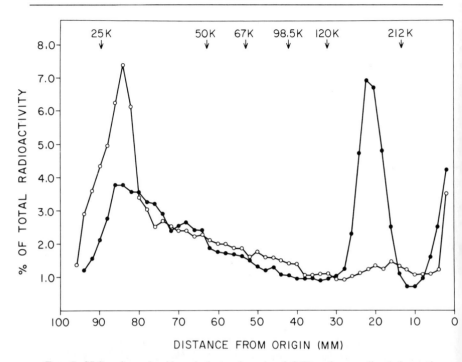

FIG. 5. SDS–polyacrylamide gel electrophoresis of CHL polysome-directed reaction products. Cell-free reaction products synthesized in the presence of 0.81 $A_{260}$ unit of CHL polysomes and 0.38 $A_{260}$ unit of wheat germ S250 were prepared as described and analyzed on 5% polyacrylamide gels before (●——●) and after (○——○) digestion with bacterial collagenase. The significance of the arrows at the top of the figure is detailed in the legend to Fig. 3.

level in the reaction when nonlimiting amounts of wheat germ S250 are employed (Fig. 4, panel B).

Characterization of the cell-free products directed by CHL polysomes in the presence of nonlimiting amounts of wheat germ S250 by polyacrylamide gel electrophoresis reveals that the collagenase-digestible material represents a single molecular weight (approximately 170,000) species of procollagen chains (Fig. 5). Furthermore, a fraction of this material (~35%) represents the products of complete (i.e., initiation, elongation, and termination) procollagen chain mRNA translation (data presented by Haralson et al.[33]). Thus CHL polysomes direct the synthesis of a single molecular weight species of procollagen chains.

*Cell-Free Procollagen Chain Synthesis Programmed by CHL mRNA.* The mRNA fraction isolated from cultured Chinese hamster lung cells promotes the efficient incorporation of [$^{14}$C]proline into cell-free product in a wheat germ lysate (S30) system. As shown in Table IV, this reaction

TABLE IV
REACTION REQUIREMENTS FOR CELL-FREE PROTEIN BIOSYNTHESIS
PROGRAMMED BY CHINESE HAMSTER LUNG CELL mRNA[a]

| Reaction component | [$^{14}$C]Proline incorporated (pmol) | Percentage of control |
|---|---|---|
| Complete | 60.4 | 100 |
| Minus CHL mRNA | 4.1 | 6.8 |
| Minus wheat germ S30 | 0.3 | 0.5 |
| Minus ATP, GTP, CPK, and CP | 3.5 | 5.8 |
| Plus puromycin | 6.8 | 11.3 |

[a] Cell-free incorporation of [$^{14}$C]proline into RNase-resistant acid-insoluble products was performed as detailed with modifications of standard reaction mixtures as indicated. Reaction mixtures were supplemented with 2.91 $A_{260}$ units of wheat germ S30 and 0.47 $A_{260}$ unit (19.6 µg) of CHL mRNA. Values presented have been corrected for radioactivity retained by the filters in the absence of CHL mRNA and wheat germ S30 (2.5 pmol).

depends upon both CHL mRNA and wheat germ S30 fractions and an energy source, and incorporation of substrate into product is inhibited approximately 90% by puromycin. The reaction has $Mg^{2+}$ and $K^+$ optima at 2.5 m$M$ and 70 m$M$, respectively, and incorporation of substrate into product is essentially complete by 2 hr at 25° (data not shown). Treatment of the cell-free reaction products programmed by CHL mRNA with bacterial collagenase reveals a significant amount of the incorporated substrate to be digested by the enzyme. When the amount of collagenase-sensitive product is assessed as a function of CHL mRNA amount in reactions containing wheat germ S30 levels that were optimized for total incorporation of substrate into product, approximately 38% of the total incorporated substrate in reactions programmed from 0 to 0.24 $A_{260}$ unit (approximately 10 µg) with CHL mRNA is digestible (Fig. 6). At higher levels of CHL mRNA, the amount of product solubilized by the enzyme decreases but still represents a significant amount (i.e., approximately 26% at 0.47 $A_{260}$ unit of CHL mRNA) of the total cell-free synthesized material.

Examination of the size distribution of the products programmed by CHL mRNA indicates that a significant fraction of the incorporated radioactivity is present in a protein with a molecular mass of approximately 170,000 daltons (Fig. 7) and that treatment of the cell-free products with bacterial collagenase results in the digestion of this species (data not shown). Thus the procollagen chains programmed by CHL mRNA appear to be similar, if not identical, to the procollagen chains directed by CHL polysomes and represent proteins with a molecular weight of 170,000.

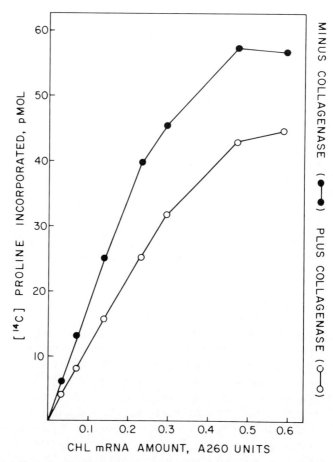

FIG. 6. Effect of collagenase on CHL mRNA-programmed reaction products. Reaction mixtures containing 2.9 $A_{260}$ units of wheat germ S30 and CHL mRNA in the amounts indicated were prepared and incubated as described. The mixtures were then adjusted to 2 mM in respect to $CaCl_2$, incubated in the absence (●——●) or in the presence (○——○) of bacterial collagenase, and the amount of [$^{14}$C]proline incorporated into RNase-resistant acid-insoluble products was determined. Values presented have been corrected for retention of radioactivity (5.1 pmol) by the filters in the absence of CHL mRNA.

General Discussion

This chapter contains a description of the methodologies employed to achieve the cell-free synthesis of procollagen chains directed by components of CHL cells, which synthesize a noninterstitial collagen.[21] In each system, the largest collagenous material synthesized corresponds in size

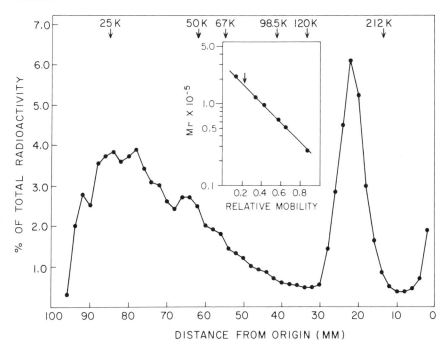

FIG. 7. SDS–polyacrylamide gel electrophoresis of CHL mRNA-programmed reaction products. Reaction mixtures containing 0.47 $A_{260}$ unit of CHL mRNA were prepared and incubated, and the size distribution of the cell-free products was determined as described. The relative mobilities of molecular weight standards are indicated by arrows at the top of the figure, and their mobilities as a function of size is shown in the inset, with the arrow indicating the position of migration of the cell-free synthesized procollagen chain cell-free product.

to a protein with a molecular mass of 170,000 daltons. As a fraction of the material directed by CHL polysomes has been demonstrated to be complete mRNA translation products,[33] this 170,000-dalton protein is thought to represent the initial gene product, i.e., procollagen chain, in the biosynthesis of CHL cell procollagen. The assignment of a molecular mass of 170,000 daltons to the CHL cell-free synthesized procollagen chain is based upon its mobility on SDS-polyacrylamide gels relative to the mobilities of noncollagen molecular weight standards. The reasons for use of these standards, as opposed to procollagen and collagen chains, have previously been detailed,[33] and although the value obtained for the molecular weight of the CHL procollagen chain ($M_r = 170,000$) may be an overestimate of its actual size, the value approximates the size reported for a subunit of native type I procollagen as determined from sedimentation equilibrium data[47] and for cell-free synthesized preprocollagen $\alpha 1(I)$ and $\alpha 2(I)$ chains.[48]

In the two systems described, which employ translational components derived from wheat germ, the synthesis of CHL procollagen chains is achieved under reaction condition that are optimal for total protein synthesis. This observation is in contrast to reports from other laboratories that $K^+$ concentrations greater than those optimal for total protein synthesis are required to achieve the synthesis of high molecular weight collagenous proteins in a wheat germ translation system.[49,50] This apparent discrepancy may reflect inherent differences in the nature of interstitial procollagen chain mRNAs[49,50] and the noninterstitial procollagen chain CHL mRNAs which influence their translational efficiency and fidelity using wheat germ protein synthesis components. On the other hand, this apparent disagreement might be resolved on the basis of the observation that the amount of collagenous protein synthesized in reactions programmed by CHL mRNA or CHL polysomes depends upon the ratio of the amount of the procollagen chain mRNA to the amount of wheat germ translational components (Figs. 4 and 6). A possible paucity of aminoacyl-tRNA synthetases, tRNAs, or elongation factors relative to the number of cell-free initiated protein chains could result in an increase in the amount of shorter polypeptides synthesized, and as procollagen chains contain large collagenase-insensitive sequences at their amino termini, an apparent lack of synthesis of collagenous material (i.e., high molecular weight procollagen chains) would be observed. Therefore, the experiences in this laboratory suggest that optimization of the amount of wheat germ translational factors relative to the amount of procollagen mRNA to be critical for achieving the cell-free synthesis of procollagen chains. As the experiments described here were performed under optimal protein synthesis conditions, it appears that the values obtained both for the size and percentage of collagenase-sensitive proteins directed by CHL polysomes or programmed by CHL mRNA in the presence of wheat germ factors are valid assessments of these parameters.

Each of the three cell-free systems described in this manuscript offers unique opportunities for investigating the biosynthesis and metabolism of noninterstitial procollagen chains. The CHL S30-catalyzed system is an homologous cell-free system, and as a consequence, any positive or negative translational element(s), i.e., specific tRNA or initiation factor, involved in noninterstitial procollagen chain mRNA expression will be pres-

[47] E. Bańkowski and W. M. Mitchell, *Biophys. Chem.* **1**, 73 (1973).
[48] L. Sandell and A. Veis, *Biochem. Biophys. Res. Commun.* **92**, 554 (1980).
[49] K. Benveniste, J. Wilczek, A. Ruggeri, and R. Stern, *Biochemistry* **15**, 830 (1976).
[50] R. Harwood, M. E. Grant, and D. S. Jackson, *FEBS Lett.* **57**, 47 (1975).

ent. Furthermore, the observation that collagenous proteins smaller than procollagen chains are among the products of cell-free reactions catalyzed by CHL S30 preparations suggest that posttranslational processing of the initial gene product may have occurred; thus this system is also appropriate for investigating these aspects of noninterstitial procollagen chain metabolism. The CHL polysome-directed system offers the advantage of allowing the discrimination between transcribed but not translated procollagen chain mRNA species from those that are actively being expressed. The heterologous cell-free protein synthesis system programmed by CHL mRNA offers the opportunity not only to discern positive translational regulatory elements but also to identify factors and enzymes involved in the posttranslational modification(s) and catabolism of a noninterstitial procollagen chain. Furthermore, one prerequisite for examining gene expression and structure is to isolate the products of the gene, i.e., mRNA, and to demonstrate that such preparations program the cell-free synthesis of the desired protein(s); thus the results presented in this manuscript indicate that the HT1 clone of cultured Chinese hamster lung cells represents an appropriate model for evaluating the expression and regulation of the $\alpha 1(V)$-procollagen chain gene.

Acknowledgments

The author expresses his appreciation to Robena Ross for her aid in the preparation of this manuscript. This work was supported by National Institutes of Health Grants RR05424-19, CA 05719, and AM 18222.

# [10] Posttranslational Enzymes in the Biosynthesis of Collagen: Intracellular Enzymes

By KARI I. KIVIRIKKO and RAILI MYLLYLÄ

Collagen biosynthesis is characterized by the presence of a large number of posttranslational modifications of the polypeptide chains, many of these modifications being unique to collagens and a few other proteins with collagen-like amino acid sequences. The posttranslational processing of this protein can be regarded as occurring in two stages. Intracellular modifications, together with the synthesis of the polypeptide chains, result in the formation of the triple-helical procollagen molecule, and extracellular processing converts this molecule into collagen and incorporates it into a stable, cross-linked fibril. The intracellular enzyme modifications consist

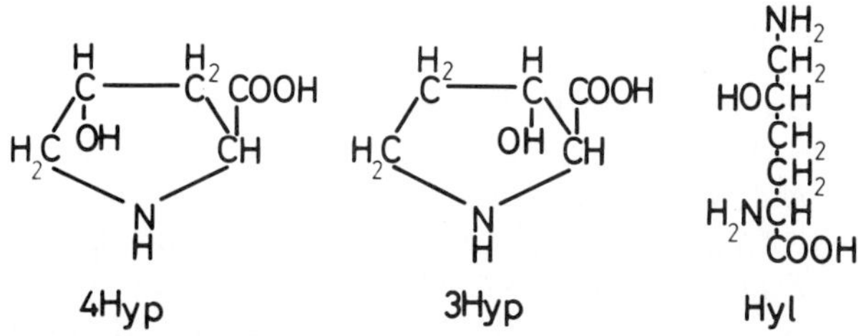

```
     H    H₂           H₂   H              NH₂
     C────C COOH       C────C  COOH        CH₂
    / |    \          / |    \            HOCH
  H₂C  OH   CH      H₂C   OH  CH           CH₂
    \  /              \  /                 CH₂
     N                 N                  H₂NCH
     H                 H                    COOH
    4Hyp              3Hyp                  Hyl
```

FIG. 1. The structures of 4-*trans*-hydroxy-L-proline, 3-*trans*-hydroxy-L-proline, and 5-hydroxy-L-lysine.

of removal of the preprotein sequences; hydroxylation of appropriate prolyl and lysyl residues to 4-hydroxyprolyl, 3-hydroxyprolyl, and hydroxylysyl residues; glycosylation of certain hydroxylysyl residues to galactosylhydroxylysyl and glucosylgalactosylhydroxylysyl residues; and glycosylation of certain asparaginyl residues in the propeptide extensions of the pro-α chains. Formation of the intrachain and interchain disulfide bonds between the propeptides may also require an enzyme, but little specific information is currently available on this process.

The purpose of this chapter is to describe the assay, purification, and properties of the enzymes catalyzing the unique intracellular modifications, i.e., the hydroxylases and hydroxylysyl glycosyltransferases. The glycosylation of the propeptides is reviewed in this volume [15]. Several reviews deal with various aspects of the posttranslational modifications of collagen biosynthesis[1] and the intracellular modifications,[2] hydroxylations,[3-6] and glycosylations.[7]

## An Overview on Hydroxylases

4-Hydroxyproline, 3-hydroxyproline, and hydroxylysine (Fig. 1) are found in vertebrate proteins almost exclusively in collagens and a few

[1] K. I. Kivirikko and R. Myllylä, in "Collagen in Health and Disease" (M. J. Jayson and J. B. Weiss, eds.). Churchill-Livingstone, Edinburgh, in press.
[2] D. J. Prockop, R. A. Berg, K. I. Kivirikko, and J. Uitto, in "Biochemistry of Collagen" (G. N. Ramachandran and A. H. Reddi, eds.), p. 163. Plenum, New York, 1976.
[3] G. J. Cardinale and S. Udenfriend, *Adv. Enzymol.* **41**, 245 (1974).
[4] O. Hayaishi, M. Nozaki, and M. T. Abbott, in "The Enzymes" (P. D. Boyer, ed.), 3rd ed., Vol. 12, p. 151. Academic Press, New York, 1976.
[5] E. Adams and L. Frank, *Annu. Rev. Biochem.* **49**, 1005 (1980).
[6] K. I. Kivirikko and R. Myllylä, in "The Enzymology of Post-Translational Modification of Proteins" (R. B. Freedman, and H. Hawkins, eds.), p. 53. Academic Press, New York, 1980.
[7] K. I. Kivirikko and R. Myllylä, *Int. Rev. Connect. Tissue Res.* **8**, 23 (1979).

other proteins with collagen-like amino acid sequences.[1-6] The 4-hydroxyprolyl residues have an important function in that their hydroxyl groups stabilize the triple helix of collagen under physiological conditions,[2,6] but the role of the 3-hydroxyprolyl residues is not known at present. The hydroxylysyl residues have two important functions: their hydroxyl groups serve as sites of attachment for carbohydrate units, and they are essential for the stability of the intermolecular collagen cross-links.[1,6]

All three amino acids are formed as posttranslational modifications by the hydroxylation of peptide-bound prolyl or lysyl residues, catalyzed by three separate enzymes, prolyl 4-hydroxylase (proline,2-oxoglutarate dioxygenase, prolyl-glycyl-peptide,2-oxoglutarate:oxygen oxidoreductase, 4-hydroxylating, EC 1.14.11.2), prolyl 3-hydroxylase (prolyl-4-hydroxyprolyl-glycyl-peptide,2-oxoglutarate:oxygen oxidoreductase, 3-hydroxylating; EC 1.14.11.?), and lysyl hydroxylase (lysine,2-oxoglutarate dioxygenase, peptidyllysine,2-oxoglutarate:oxygen 5-oxidoreductase, EC 1.14.11.4). These enzymes are widely distributed in nature, prolyl 4-hydroxylase having been identified in most tissues of higher organisms, in several lower organisms, and in several plants[8] that also contain 4-hydroxyproline in peptide linkages.

Prolyl 4-hydroxylase, prolyl 3-hydroxylase, and lysyl hydroxylase belong to the group of 2-oxoglutarate dioxygenases, the other members of which include 4-butyrobetaine hydroxylase, thymine 7-hydroxylase, pyrimidine deoxyribonucleoside 2-hydroxylase, $\epsilon$-$N$-trimethyllysine hydroxylase, and $p$-hydroxyphenylpyruvate hydroxylase.[2-4,6] These enzymes require $Fe^{2+}$, 2-oxoglutarate, $O_2$, and ascorbate, and they decarboxylate 2-oxoglutarate, one atom of the $O_2$ molecule being incorporated into the succinate while the other is incorporated into the hydroxyl group. These reactions can thus be described by the following equation:

$$\text{Substrate-H} + O_2 + \begin{array}{c} \text{COOH} \\ | \\ \text{CH}_2 \\ | \\ \text{CH}_2 \\ | \\ \text{CO} \\ | \\ \text{COOH} \end{array} \xrightarrow[\text{ascorbate}]{Fe^{2+}} \text{substrate-OH} + \begin{array}{c} \text{COOH} \\ | \\ \text{CH}_2 \\ | \\ \text{CH}_2 \\ | \\ \text{COOH} \end{array} + CO_2$$

The minimum sequence requirement for interaction with prolyl 4-hydroxylase is fulfilled by an -X-Pro-Gly- triplet, and that for interaction with lysyl hydroxylase by an -X-Lys-Gly- triplet, but both enzymes appear to hydroxylate certain other triplets in some cases.[1,2,6] Prolyl

[8] R. Kuttan and A. N. Radhakrishnan, *Adv. Enzymol.* **37**, 273 (1973).

3-hydroxylase probably requires a -Pro-4Hyp-Gly- triplet, whereas a triplet of the form -Pro-Pro-Gly- probably cannot be a substrate.[1,6] The interaction with prolyl 4-hydroxylase and lysyl hydroxylase is also influenced by the amino acid in the X position of the triplet to be hydroxylated, and interaction with all three enzymes is affected by amino acids in other parts of the peptide, by the peptide chain length, and by the peptide conformation.[1,2,6] The main effect of an increase in peptide chain length is a decrease in the $K_m$ of the peptide, and the effect of peptide conformation is seen in that triple-helical peptides are not hydroxylated by any of the three enzymes.[1,2,6]

The intracellular location of prolyl 4-hydroxylase and lysyl hydroxylase is within the cisternae of the rough endoplasmic reticulum, prolyl 4-hydroxylase being either free within the cisternae or loosely bound to the inner membrane, and lysyl hydroxylase probably being in part either free or loosely bound and in part tightly bound.[2,6,9] Several observations suggest that prolyl 3-hydroxylase is likewise located within this cell compartment.[1,6] Most of the hydroxylation takes place while the procollagen polypeptide chains are growing on the ribosomes, but these reactions continue after the release of the complete polypeptide chains from the ribosomes, until formation of the procollagen triple helix prevents any further hydroxylation.[1,2,6] In some experimental cases, however, all the hydroxylation has been found to occur after the release of the completed polypeptide chains into the cisternae of the rough endoplasmic reticulum.[1-3,6]

## Assays of the Hydroxylases

### Principles of Activity Assays and Comparison of Various Types of Method

A number of different substrates and procedures are available for the measurement of the three hydroxylase activities. As none of the three enzymes acts on the free amino acid, it is necessary to use as substrates either synthetic or biologically prepared peptides. The latter are either intact protocollagen, a protein consisting of nonhydroxylated pro-$\alpha$ chains with molecular weights of about 150,000,[1,2,6] or partially degraded forms of this protein. Any triple-helical conformation of the substrate must be thermally denatured before hydroxylation (see above), and in cases of assays of prolyl 3-hydroxylase activity, the -Pro-Pro-Gly- sequences of the protocollagen must be converted to -Pro-4Hyp-Gly- sequences (see above) by incubation with pure prolyl 4-hydroxylase *in vitro*.[10]

[9] B. Peterkofsky and R. Assad, *J. Biol. Chem.* **254**, 4714 (1979).
[10] K. Tryggvason, J. Risteli, and K. I. Kivirikko, *Biochem. Biophys. Res. Commun.* **76**, 275 (1977).

The enzyme reaction mixture must also contain $Fe^{2+}$, 2-oxoglutarate, $O_2$, and ascorbate (see above), the amount of $O_2$ being sufficient under aerobic incubation conditions. Even in the presence of optimal concentrations of all these reactants, maximal activity is not obtained with partially purified hydroxylases unless dithiothreitol, bovine serum albumin,[11,12] and catalase[13] are added to the incubation solution.[2,6] This stimulation by dithiothreitol suggests that the catalytic site of these enzymes contains free thiol groups that are essential for the activity.[2,3,6] The action of bovine serum albumin is in part explained by a "protein effect" that can also be demonstrated with other proteins, but is in part more specific, probably attributable to the presence on this protein of a number of free thiol groups.[11,12] Catalase probably acts partly by destroying peroxide, which is generated nonenzymically by solutions of $O_2$, $Fe^{2+}$, and ascorbate[13] and partly by a nonspecific "protein effect."[12] Thymol, which possesses antioxidant properties, has been reported to give a stimulation identical with that obtained with catalase.[14]

The main types of assays available for prolyl 4-hydroxylase activity are summarized in Table I. The methods involving biologically prepared protocollagen substrates[15–21] are highly sensitive, and in most instances only these procedures can be used for determination of the enzyme activity in crude tissue specimens. Standardization of these assays and preparation of the protocollagen present several problems (discussed in detail below), but with appropriate precautions it is possible to use these procedures for a reliable and reproducible measurement of the enzyme activity in a large variety of biological samples. The methods based on a radiochemical assay[22] of the labeled hydroxyproline formed in the substrate[15–19] are highly specific, whereas those based on the release of $^3H_2O$ from the 4-carbon of prolyl residues[20,21] are less specific but more rapid. The use of synthetic polypeptide substrates[13,23–26] such as (Pro-Pro-

[11] R. E. Rhoads, J. J. Hutton, and S. Udenfriend, *Arch. Biochem. Biophys.* **122**, 805 (1967).
[12] R. E. Rhoads and S. Udenfriend, *Arch. Biochem. Biophys.* **139**, 329 (1970).
[13] K. I. Kivirikko and D. J. Prockop, *J. Biol. Chem.* **242**, 4007 (1967).
[14] B. Peterkofsky, D. Kalwinsky, and R. Assad, *Arch. Biochem. Biophys.* **199**, 362 (1980).
[15] J. J. Hutton and S. Udenfriend, *Proc. Natl. Acad. Sci. U.S.A.* **56**, 198 (1966).
[16] K. I. Kivirikko and D. J. Prockop, *Arch. Biochem. Biophys.* **118**, 611 (1967).
[17] K. I. Kivirikko and D. J. Prockop, *Proc. Natl. Acad. Sci. U.S.A.* **57**, 782 (1967).
[18] R. A. Berg and D. J. Prockop, *Biochemistry* **12**, 3395 (1973).
[19] J. Risteli, L. Tuderman, and K. I. Kivirikko, *Biochem. J.* **158**, 369 (1976).
[20] J. J. Hutton, A. L. Tappel, and S. Udenfriend, *Anal. Biochem.* **16**, 384 (1966).
[21] P. H. Fleckman, J. J. Jeffrey, and A. Z. Eisen, *J. Invest. Dermatol.* **60**, 46 (1973).
[22] K. Juva and D. J. Prockop, *Anal. Biochem.* **15**, 77 (1966).
[23] R. E. Rhoads and S. Udenfriend, *Proc. Natl. Acad. Sci. U.S.A.* **60**, 1473 (1968).
[24] J. Halme, K. I. Kivirikko, and K. Simons, *Biochim. Biophys. Acta* **198**, 460 (1970).
[25] K. I. Kivirikko, Y. Kishida, S. Sakakibara, and D. J. Prockop, *Biochim. Biophys. Acta* **271**, 347 (1972).

TABLE I
ASSAYS OF PROLYL 4-HYDROXYLASE ACTIVITY[a,b]

| Substrate | Product or cosubstrate assayed | Sensitivity (in ng of enzyme)[c] | Comments |
|---|---|---|---|
| [$^{14}$C]Proline-labeled protocollagen[d] | Hydroxy[$^{14}$C]proline by chromatography or by radiochemical procedure | 2 | Highly specific, substrate has to prepared, linear over narrow range |
| [$^{3}$H]Proline-labeled protocollagen[d] | $^{3}$H$_2$O released from 4-carbon of proline | 1–10 | Less specific but more rapid than above assay; substrate has to prepared; linear over narrow r |
| (Pro-Pro-Gly)$_n$ | Hydroxyproline | 120–240 | Highly specific; substrate comm cially available; linear over b range |
| (Pro-Pro-Gly)$_n$ or Ascaris cuticle collagen | $^{14}$CO$_2$ released from 2-oxo-[1-$^{14}$C] glutarate | 10–60 | Nonspecific; substrate commerc available; linear over broad ra rapid and simple |
| (Pro-Pro-Gly)$_n$ | O$_2$ uptake | 5000 | Nonspecific; substrate commerc available; suitable for continu measurement of activity |

[a] Modified from Prockop et al.[2]
[b] For original references, see text.
[c] Sensitivity estimated as described in detail by Prockop et al.[2]
[d] Protocollagen is a biologically prepared protein consisting of nonhydroxylated pro-α chains molecular weight of about 150,000.

Gly)$_n$ or the hydroxyproline-deficient collagen from the cuticle of Ascaris worms[12,26] overcomes the difficulties in preparing protocollagen. These assays are highly specific if a chemical method for hydroxyproline[27] is used to follow the reaction,[13,23,24] but the sensitivity is quite low. These substrates become very easy to use if the reaction is followed in terms of the release of $^{14}$CO$_2$ during the hydroxylation-coupled stoichiometric decarboxylation of 2-oxo[1-$^{14}$C]glutarate,[12,23,25,26,28] but in this case the assays become less specific and can in most instances be carried out only with partially purified enzymes. There are nevertheless some examples of a reliable application of these methods for the measurement of prolyl 4-hydroxylase activity in crude fibroblast extracts.[29,30] A reported procedure that likewise involves synthetic peptide substrates is based on determination of the uptake of O$_2$.[31] This assay is suitable for the continuous

[26] R. A. Berg and D. J. Prockop, J. Biol. Chem. 248, 1175 (1973).
[27] K. I. Kivirikko, O. Laitinen, and D. J. Prockop, Anal. Biochem. 19, 249 (1967).
[28] R. E. Rhoads and S. Udenfriend, Arch. Biochem. Biophys. 133, 108 (1969).
[29] W. W.-Y. Kao, R. A. Berg, and D. J. Prockop, Biochim. Biophys. Acta 411, 202 (1975).
[30] R. A. Berg, W. W.-Y. Kao, and D. J. Prockop, Biochim. Biophys. Acta 444, 756 (1976).
[31] J. J. Nietfeld and A. Kemp, Biochim. Biophys. Acta 613, 349 (1980).

measurement of enzyme activity, and is therefore useful for studies on the reaction mechanism.

The two other enzyme activities can be determined by the same types of method as prolyl 4-hydroxylase activity. When biologically prepared protocollagen is used as a substrate,[10,17,32–34] the product of the lysyl hydroxylase reaction can be measured with a rapid and specific radiochemical procedure,[35] whereas that of the prolyl 3-hydroxylase reaction has to be assayed by chromatographic methods.[34] Another possibility is to use protocollagen labeled with $^3$H in the appropriate carbon of the lysyl or prolyl residues, and to determine the $^3H_2O$ formed during the reaction.[33,36] Several synthetic polypeptides can also be used as substrates for lysyl hydroxylase,[37] and product formation can be followed either by a specific chemical procedure[38] or by measuring the decarboxylation of 2-oxo[1-$^{14}$C]glutarate.[37] Corresponding methods are not yet available for prolyl 3-hydroxylase activity. The measurement of $O_2$ uptake also seems possible in the case of these two enzymes, but there are no reports on the use of such procedures. The limitations and advantages of these various assays are similar to those of the corresponding methods for prolyl 4-hydroxylase (above).

### Extraction of the Hydroxylase Activities from Crude Tissue or Cell Specimens

All three hydroxylases are easily inactivated during the handling of biological specimens, and their activities are also quite labile in many tissue extracts. For example, up to about 50% of prolyl 4-hydroxylase activity is lost when the supernatant of human liver homogenate is frozen and thawed once.[39] The solution used to extract the enzyme is also critical: sufficient concentrations of a detergent and sodium chloride are required for an optimal solubilization,[2,6,29,40–43] the presence of salt affecting

[32] K. I. Kivirikko and D. J. Prockop, *Biochim. Biophys. Acta* **258**, 366 (1972).
[33] R. L. Miller, *Anal. Biochem.* **45**, 202 (1972).
[34] J. Risteli, K. Tryggvason, and K. I. Kivirikko, *Eur. J. Biochem.* **73**, 485 (1977).
[35] N. Blumenkrantz and D. J. Prockop, *Anal. Biochem.* **30**, 377 (1969).
[36] J. Risteli, K. Tryggvason, and K. I. Kivirikko, *Anal. Biochem.* **84**, 423 (1978).
[37] K. I. Kivirikko, K. Shudo, S. Sakakibara, and D. J. Prockop, *Biochemistry* **11**, 122 (1972).
[38] N. Blumenkrantz and D. J. Prockop, *Anal. Biochem.* **39**, 59 (1971).
[39] E.-R. Kuutti-Savolainen, J. Risteli, T. A. Miettinen, and K. I. Kivirikko, *Eur. J. Clin. Invest.* **9**, 89 (1979).
[40] R. F. Diegelmann, L. Bernstein and B. Peterkofsky, *J. Biol. Chem.* **248**, 6514 (1973).
[41] N. A. Guzman and K. R. Cutroneo, *Biochem. Biophys. Res. Commun.* **52**, 1263 (1973).
[42] J. Risteli and K. I. Kivirikko, *Biochem. J.* **144**, 115 (1974).
[43] L. Ryhänen, *Biochim. Biophys. Acta* **438**, 71 (1976).

especially the extraction of lysyl hydroxylase.[9,32,43] To avoid these problems, our laboratory now uses the following procedure for extraction of the hydroxylase activities to be assayed in crude biological specimens.[36,39,44–46]

Tissue samples are rapidly frozen in liquid nitrogen, and pellets of cultured or freshly isolated cells at $-70°$; all these specimens are stored at $-70°$ until assayed (for up to about 2 months). The frozen samples are homogenized in a cold solution ($0°$) consisting of 0.2 $M$ NaCl, 0.1 $M$ glycine,[47] 50 $\mu M$ dithiothreitol, 0.1% (w/v) Triton X-100, 0.01% (w/v) soybean trypsin inhibitor, and 20 m$M$ Tris-HCl buffer, pH adjusted to 7.5 at $4°$. In the case of tissue specimens, the volume of this solution is usually 5–10 ml per gram of tissue, and in the case of cell pellets about 0.3–1 ml per $10^6$ cells. Soft tissues and cell pellets are homogenized with a Teflon and glass homogenizer (1200 rpm, 50 strokes), whereas hard or tough tissues, such as skin, require more vigorous treatment. In the case of skin, the samples are first minced with scissors for 5 min in the above solution and then homogenized with an Ultra-Turrax homogenizer three times for 5 sec at $0°$. The homogenate is centrifuged at 15,000 $g$ for 30 min at $4°$, and aliquots of the supernatant are used immediately for assays of the hydroxylase activities.

Procedures Using Radioactively Labeled Protocollagen Substrates

*Preparation of the Protocollagen Substrates*

Several different sources, including minces of whole chick embryos,[15,20] isolated chick embryo calvaria[48] or cartilaginous bone,[16,17,21,32] carrageenin-induced guinea pig granuloma tissue,[33] medium of cultured cells,[49] and cells isolated from chick-embryo leg tendons by enzymic digestion,[10,18,19,36] have been used to prepare the radioactively labeled intact or partially degraded protocollagen. The substrates obtained from these various sources are basically similar, but the yields of radioactive protocollagen and the linearity of the enzyme activity assays are different depending on the source. In the following we describe in detail the purification of protocollagen in freshly isolated chick embryo tendon cells,

[44] E.-R. Kuutti-Savolainen and M. Kero, *Clin. Chim. Acta* **96**, 43 (1979).
[45] L. Risteli, J. Risteli, L. Salo, and K. I. Kivirikko, *Eur. J. Biochem.* **97**, 297 (1979).
[46] R. Myllylä, K. Alitalo, A. Vaheri, and K. I. Kivirikko, *Biochem. J.* **196**, 683 (1981).
[47] J. Halme and K. I. Kivirikko, *FEBS Lett.* **1**, 223 (1968).
[48] B. Peterkofsky and R. DiBlasio, *Anal. Biochem.* **66**, 279 (1975).
[49] W. A. Gonnerman, A. B. Goral, and C. Franzblau, *Anal. Biochem.* **102**, 8 (1980).

which is now used as a standard source for these substrates in our laboratory.

*Reagents*

Modified Krebs medium: $Na_2HPO_4$, 15.7 m$M$; $KH_2PO_4$, 1.6 m$M$; NaCl, 111.2 m$M$; KCl, 5.4 m$M$; $MgCl_2$, 1.3 m$M$; $NaHCO_3$, 4.0 m$M$; glucose, 13.0 m$M$; pH adjusted to 7.4 at 37°

$CaCl_2$, 0.17 $M$ in the modified Krebs medium

Collagenase, type I from *Clostridium histolyticum* (Sigma C-0130)

Trypsin, 2.5% solution in physiological saline

Fetal calf serum

$\alpha,\alpha'$-Dipyridyl, 15 m$M$ solution in the modified Krebs medium (prepared immediately before use)

Amino acids—one of the following, depending on the substrate to be prepared: L-[$^{14}$C]proline (>200 Ci/mol); L-[$^{14}$C]lysine (>200 Ci/mol); L-[4-$^3$H]proline or L-[3,4-$^3$H]proline (>20 Ci/mmol); L-[4,5-$^3$H]lysine (>60 Ci/mmol); or L-[2,3-$^3$H]proline (>20 Ci/mmol)

Acetic acid, 0.1 $M$, containing EDTA, 1 m$M$, and soybean trypsin inhibitor, 0.1 g/liter, cooled to 4°

Acetic acid, 0.1 $M$, cooled to 4°

NaCl-Tris buffer: NaCl, 0.2 $M$, containing Tris-HCl, 50 m$M$, cooled to 4°; pH adjusted to 7.8 at 4°

Pure prolyl 4-hydroxylase, required only for preparation of substrate for prolyl 3-hydroxylase

*Procedure.* The cells are isolated from leg tendons of 50–100 17-day-old chick embryos by a slight modification of the procedure described by Dehm and Prockop.[50,51] The legs of the embryos are placed in petri dishes containing modified Krebs medium, and the tendons are isolated by pulling on the toes. Little further dissection is necessary to obtain clean tissues. The tendons from 50 chick embryos are placed in a 50-ml Erlenmeyer flask containing 15 ml of modified Krebs medium, to which is added 1.5 ml of the trypsin solution, 0.5 ml of $CaCl_2$ and 15,000 units of collagenase. The amounts of the Krebs medium and its additions are varied according to the number of chick embryos; for example, all the above values are doubled in the case of leg tendons from 100 embryos. The sample is incubated at 37° for 45 min with shaking, most of the tissue disintegrating during this time. The system is then filtered through lens paper in a small Swinnex filter (Millipore) and the filtrate is centrifuged at 600 $g$ at room temperature for 3 min. The supernatant is discarded, and the cell pellet is gently suspended in 30 ml of modified Krebs medium

[50] P. Dehm and D. J. Prockop, *Biochim. Biophys. Acta* **240**, 358 (1971).
[51] P. Dehm and D. J. Prockop, *Biochim. Biophys. Acta* **264**, 375 (1972).

containing 10% fetal calf serum. The suspension is centrifuged as above, and this washing step repeated two more times.

The isolated cells are suspended in 30 ml of modified Krebs medium (in the case of leg tendons from 50 chick embryos) containing 0.6 ml of 15 mM $\alpha,\alpha'$-dipyridyl solution, and preincubated in a siliconized Erlenmeyer flask with shaking for 30 min at 37°. The iron required as a cofactor in the hydroxylation reactions is chelated during the preincubation, and hence the collagenous protein synthesized during subsequent incubation with the radioactive amino acid consists of nonhydroxylated pro-$\alpha$ chains. [$^{14}$C]Proline (60 $\mu$Ci), [$^{14}$C]lysine (120 $\mu$Ci), or one of the $^3$H-labeled amino acids (500 $\mu$Ci) is evaporated to dryness, dissolved in 0.5 ml of modified Krebs medium and added for an additional incubation of 4 hr. The sample is then centrifuged at 1200 $g$ for 12 min at room temperature, the supernatant is removed, and the cell pellet is frozen at $-20°$. The frozen cells are homogenized with a Teflon and glass homogenizer in 5 ml of ice-cold 0.1 $M$ acetic acid containing EDTA and soybean trypsin inhibitor and dialyzed overnight against 0.1 $M$ acetic acid at 4°, with one change. The sample is centrifuged in a siliconized tube at 20,000 $g$ for 30 min, and the supernatant is dialyzed against the NaCl-Tris buffer at 4°, with several changes. The protocollagen can be purified further as described by Berg and Prockop,[18] but for normal enzyme activity assays the following procedure[10,19] is recommended. The preparation is heated to 100° for 10 min, centrifuged at 1000 $g$ to remove the precipitate formed during heating, and stored at $-20°$ or $-70°$ in suitable aliquots so that the same portion is thawed only once. [$^{14}$C]Proline, [4-$^3$H]proline, or [3,4-$^3$H]proline-labeled preparations are used for the assay of prolyl 4-hydroxylase activity, and [$^{14}$C]lysine or [4,5-$^3$H]lysine-labeled preparations are used for the assay of lysyl hydroxylase activity.

In the case of prolyl 3-hydroxylase activity assays, the substrate must contain 4-hydroxyproline.[10,34] The [$^{14}$C]proline or [2,3-$^3$H]proline-labeled protocollagen prepared as above is therefore fully hydroxylated with pure prolyl 4-hydroxylase after the substrate has been heated at 100° and centrifuged, but before it has been divided into aliquots for storage. The 4-hydroxylation is carried out for 4 hr at 37° in a final volume of 10 ml of a solution containing the protocollagen preparation, 5 $\mu$g of pure prolyl 4-hydroxylase, and sufficient amounts of the following compounds to make the final concentrations indicated: 0.8 m$M$ FeSO$_4$, 2 m$M$ ascorbic acid, 0.5 m$M$ 2-oxoglutarate, and 50 m$M$ Tris-HCl buffer, pH adjusted to 7.8 at 25°. After the incubation the solution is dialyzed against the NaCl-Tris buffer at 4°, with one change, divided into aliquots, and stored as above.

*The Hydroxylation Reactions*

*Reagents*

Radioactively labeled protocollagen substrate, prepared as described above

Enzyme standard: crude ammonium sulfate-precipitated chick embryo enzyme, obtained in the first step of the purification procedures described below, and stored in aliquots at $-20°$ to avoid repeated freezing and thawing

Ascorbic acid, 20 m$M$ in cold $H_2O$, dissolved immediately before use

2-Oxoglutaratic acid, 10 m$M$, in cold $H_2O$, dissolved immediately before use

$FeSO_4$, 1 m$M$ in cold $H_2O$, dissolved immediately before use

Dithiothreitol, 1 m$M$ in cold $H_2O$, stored as a 100 m$M$ stock solution at 4° for up to 2 weeks

Bovine serum albumin, 20 mg/ml, freshly prepared in cold $H_2O$

Catalase, 2 mg/ml in cold $H_2O$, freshly diluted from a commercially available suspension

Tris-HCl buffer, 0.5 $M$, pH adjusted to 7.8 at 25°, stored at 4°

Concentrated HCl, required in the prolyl 4-hydroxylase and prolyl 3-hydroxylase reactions when the radioactive amino acid product is to be measured

Acetone, ice-cold, required in the lysyl hydroxylase reaction when radioactive hydroxylysine is to be measured

Trichloroacetic acid, 50%, required in the hydroxylation reactions when tritiated water is to be measured

*Procedure for the Assay of Prolyl 4-Hydroxylase Activity.* The test tubes cooled in iced water each receive an appropriate addition of cold $H_2O$ to give a final volume of 2.0 ml after the subsequent additions of the substrate, enzyme, and other reactants. The following components are then added: 0.2 ml of Tris-HCl buffer, 0.2 ml of bovine serum albumin, 0.1 ml of catalase, 0.2 ml of dithiothreitol, 0.2 ml of ascorbic acid, 0.1 ml of $FeSO_4$, the enzyme to be assayed or the enzyme standard, or $H_2O$, 60,000 dpm of [$^{14}$C]proline-labeled or 200,000 dpm of [4-$^3$H]proline- or [3,4-$^3$H]proline-labeled protocollagen, and 0.1 ml of 2-oxoglutaric acid. The protocollagen substrate is heated to 100° for 5 min and rapidly cooled to 0° immediately before use. The test tubes are transferred from the iced water to a water bath of 37° or 30° and incubated with shaking for 30 min.

In the assay with [$^{14}$C]proline-labeled protocollagen the reaction is stopped by transferring the tubes to iced water and rapidly adding 2 ml of concentrated HCl. The tubes are sealed, and their contents are hydrolyzed at 120° for 16 hr. The samples are evaporated to dryness over a steam bath, dissolved in 4.0 ml of $H_2O$, and used for the assay[22] of the 4-hydroxy[$^{14}$C]proline formed, as described in this volume [17].

In the assays with the tritiated protocollagens the reaction is stopped by cooling the tubes as above and rapidly adding 0.2 ml of 50% trichloroacetic acid. The tritiated water is then measured as described below under the heading Determination of Tritiated Water.

*Procedure for the Assay of Lysyl Hydroxylase Activity.* The reaction with [$^{14}$C]lysine-labeled protocollagen is carried out in screw-capped tubes having a capacity of about 20 ml (Kimax). The enzyme reaction is similar to that described above for prolyl 4-hydroxylase, with the following differences. The final volume is 1.0 ml, and hence all volumes are one-half of those above, except for the volumes of bovine serum albumin and ascorbate, which are one-fourth (0.05 ml).[52] The substrate is either [$^{14}$C]lysine- or [4,5-$^3$H]lysine-labeled protocollagen, 200,000 dpm/tube. After the incubation at 37° or 30° for 30 min, the tubes are transferred to iced water, and in the assay with [$^{14}$C]lysine-labeled protocollagen 10 ml of cold acetone are rapidly added. The contents of the tubes are carefully mixed and the tubes are allowed to stand in the iced water for 30 min. They are then centrifuged at about 3000 $g$ for 30 min, the supernatants are removed by suction, and the pellets are dried by gently blowing $N_2$ into the tubes. The hydroxy[$^{14}$C]lysine formed is determined as described below under a separate heading. In the assay with [4,5-$^3$H]lysine-labeled protocollagen, the reaction is stopped as described for the corresponding substrates of prolyl 4-hydroxylase.

*Procedure for the Assay of Prolyl 3-Hydroxylase Activity.* The incubation mixture is the same as described for prolyl 4-hydroxylase, except that the substrate consists of 100,000 dpm of [$^{14}$C]proline- or 1,000,000 dpm of [2,3-$^3$H]proline-labeled protocollagen, fully 4-hydroxylated with prolyl 4-hydroxylase *in vitro*. The reaction with crude tissue extracts is carried out for 30 or 60 min at 20°,[34] but in the case of partially purified enzyme preparations, the incubation can also be carried out at 30°.[53] The reaction with [$^{14}$C]proline-labeled protocollagen is stopped, and the samples are hydrolyzed and evaporated as described for prolyl 4-hydroxylase. The 3-hydroxy[$^{14}$C]proline formed is then assayed using an amino acid analyzer[34] as described in this volume [17]. In the assay with [2,3-$^3$H]proline-labeled protocollagen the reaction is stopped as described for the corresponding substrates of prolyl 4-hydroxylase.

---

[52] U. Puistola, T. M. Turpeenniemi-Hujanen, R. Myllylä, and K. I. Kivirikko, *Biochim. Biophys. Acta* **611**, 40 (1980).
[53] K. Tryggvason, K. Majamaa, J. Risteli, and K. I. Kivirikko, *Biochem. J.* **183**, 303 (1979).

*Determination of Hydroxy[$^{14}C$]lysine*[35]

*Principle.* The hydroxy[$^{14}C$]lysine formed in the lysyl hydroxylase reaction is oxidized by periodate to give [$^{14}C$]formaldehyde from the $\epsilon$-carbon, and this is then caused to react with dimedon. The resulting [$^{14}C$]formaldehydemethone complex is extracted into toluene, and its radioactivity is measured.

*Reagents*

Citrate-phosphate buffer: Mix 154 ml of 0.15 $M$ citric acid and 346 ml of 0.3 $M$ $Na_2HPO_4$; adjust pH to 6.4

Hydroxylysine-HCl, mixture of DL- and DL-*allo*-, 30 mg/ml in cold $H_2O$

Sodium metaperiodate, 0.3 $M$ in $H_2O$, stored in the dark in a brown bottle with aluminum foil for up to 2 weeks

Dimedon, 2.8 g in 50 ml of 50% ethanol and water (v/v), stored in the dark in a brown bottle for several weeks

Toluene

Scintillation solution: Dissolve 6 g of 2,5-diphenyloxazole (PPO) and 20 mg of 1,4-bis-[2-(4-methyl-5-phenyloxazolyl)]benzene (POPOP) in 1000 ml of toluene; add 600 ml of ethylene glycol monomethyl ether (methyl Cellosolve) to this solution.

*Procedure.*[32,35] The pellets obtained in the lysyl hydroxylase reaction are suspended in 6.0 ml of the citrate–phosphate buffer, and the assay is continued in the original hydroxylation reaction tubes. Carrier hydroxylysine, 0.05 ml, and sodium metaperiodate, 1.0 ml, are added, and the solution is mixed rapidly. This is followed rapidly by 1.5 ml of the dimedon solution, mixed in immediately, and 7.5 ml of toluene. The tubes are sealed with their caps, shaken vigorously in a horizontal position for 30 min, and centrifuged briefly to separate the toluene and aqueous phases. Then 5.0 ml of the toluene phase is transferred into a counting vial containing 5 ml of the scintillation solution, and the radioactivity is measured in a liquid scintillation spectrometer.

*Determination of Tritiated Water*

The tritiated water is separated from the reaction mixture after the addition of the trichloroacetic acid either by vacuum distillation[20,21,33,36,49] or through a brief ion-exchange chromatography.[48] The distillation is carried out at 60–70° using the apparatus[20] shown in Fig. 2, with several such units connected to a vacuum pump. In the ion-exchange procedure[48] the samples are first centrifuged at 1500 $g$ for 5 min, the supernatant solutions are removed, the pellets are resuspended in 1.0 ml of 5% trichloroacetic

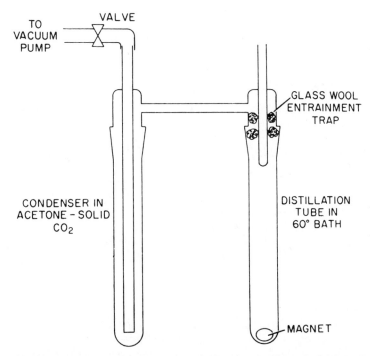

FIG. 2. Apparatus for vacuum distillation of tritiated water. Reproduced from Hutton et al.[20] with permission of the publisher.

acid, and the tubes are centrifuged as before. These supernatants are combined with the initial solutions and passed through small columns (1 × 2 cm) of Dowex 50-X8 in $H^+$ form, and the columns are washed with 2 ml of $H_2O$. A measured aliquot containing most of the tritiated water obtained by vacuum distillation or as the effluent from the ion-exchange column is dissolved in 5 or 10 ml of a counting solution (e.g., Aquasol (New England Nuclear) or Instagel (Packard), and the radioactivity is measured in a liquid scintillation spectrometer.

The hydroxylation reactions can also be carried out as micro modifications by reducing all the volumes both in the incubation with the enzyme and in the assay of tritiated water.[21,48]

## Standardization of the Hydroxylase Assays

The relationship between enzyme concentration and product formation in assays with protocollagen substrates is never entirely linear, but an almost linear relationship can be obtained within a relatively narrow range of enzyme concentrations. The shape of the standard curve differs somewhat with extracts from different tissues and even in the same tissue with separate preparations of the protocollagen substrate. A new standard curve must therefore be prepared for each new batch of protocollagen using the tissue extract in question. The volumes of the tissue extract used in the subsequent assays should then be adjusted so that the enzyme activities are measured within the relatively linear part of the curve.

Assays carried out in separate series with the same substrate preparation can be compared on the basis of the values obtained with the chick embryo ammonium sulfate enzyme standard (see reagents for the hydroxylation reactions). Nevertheless, in experiments involving samples from controls and animals or patients with certain diseases or treatments, it is advisable to divide the samples in such a way that each series of assays contains various types of specimens. Considerable additional difficulties are encountered when exact comparison of values obtained with two separate substrate preparations is necessary. This requires activity measurements from several tissue extract samples in the same assay series with both preparations and the use of separate standard curves for each substrate. The problem can be avoided in most cases, however, by preparing a large batch of protocollagen and carrying out all the assays with aliquots of the same substrate. The procedure for preparing protocollagen already described usually gives about 400–800 aliquots of 60,000 dpm of [$^{14}$C]proline-labeled substrate from the leg tendons of 100 chick embryos. If even larger amounts are required, two protocollagen preparations can be made and pooled before dividing them into aliquots for storage. The $^{14}$C-labeled substrates can be used for at least a year with no significant changes in their properties. The tritiated substrates gradually release $^3H_2O$ during storage at $-20°$ or $-70°$, resulting in elevated blank values, but the preparations are usually sufficiently stable for 2–3 months in the cases of lysyl hydroxylase and prolyl 3-hydroxylase assays, and for at least 6 months in the case of prolyl 4-hydroxylase assays.

The enzyme activities are usually expressed simply as disintegrations per minute (dpm) of product formed with the protocollagen preparation used or by defining 1 unit of enzyme activity as the amount of enzyme present in 1 mg of the corresponding ammonium sulfate standard preparation (see Tables V and VI). In some studies [e.g.,19,39,54] the prolyl 4-hydroxylase activity of crude tissue extracts is converted to micromoles of hydroxyproline formed per minute at 37° under the conditions defined below for saturating concentrations of the synthetic peptide substrates. The conversion factor can be obtained by determining the specific activity of a purified enzyme preparation with the latter substrates, and by then incubating the radioactively labeled protocollagen with this enzyme of known specific activity.

[54] L. Tuderman and K. I. Kivirikko, *Eur. J. Clin. Invest.* 7, 295 (1977).

Procedures Using Synthetic Peptide Substrates

A number of synthetic peptide substrates have been used to determine prolyl 4-hydroxylase[2,3,6,13,23-26] or lysyl hydroxylase[37] activity. Peptides synthesized by a modification of the Merrifield procedure, in which triplets of Pro-Pro-Gly are successively coupled onto a solid support,[55] are well defined and homogeneous products.[2] When these peptides of (Pro-Pro-Gly)$_n$ are used as substrates for prolyl 4-hydroxylase, the $K_m$ values decrease considerably as the chain length increases from 5 to 20 repeated tripeptide units,[25] while the $V$ remains about the same[25] and is not significantly different from that for protocollagen.[2,6,56] Our laboratory now uses peptides of this kind with $n = 5$ or 10 as standard substrates for the assay of prolyl 4-hydroxylase activity. A corresponding standard substrate for the assay of lysyl hydroxylase activity is the synthetic peptide termed L-I, which has the sequence Ala-Arg-Gly-Ile-Lys-Gly-Ile-Arg-Gly-Phe-Ser-Gly.[37]

The assays based on measurement of the stoichiometric decarboxylation of 2-oxo[1-$^{14}$C]glutarate[12,23,25,26,28,37] are rapid and easy to perform and are very useful for determining the specific activity of partially purified and pure enzyme preparations and for a number of different experiments to determine the properties of such enzymes. The methods based on measurement of the quantity of hydroxyproline[13,23,24] or hydroxylysine[37] formed are now mainly used to standardize the stoichiometric relationship between amino acid product formation and 2-oxoglutarate decarboxylation,[23,26,37,57] or for measurement of the specific activity of the enzyme in a tissue extract used as a source for enzyme purification.[57] These assays are therefore not described here in detail. The procedure based on determination of the uptake of $O_2$[31] is also of relatively limited use (see above) and hence is not given below.

*Methods Based on Decarboxylation of 2-Oxo[1-$^{14}$C]glutarate*

*Reagents*

Ascorbic acid, FeSO$_4$, dithiothreitol, bovine serum albumin, catalase, and Tris-HCl buffer—as in the reactions with the protocollagen substrates

Peptide substrate: (Pro-Pro-Gly)$_5$ · 4 H$_2$O, 5 mg/ml; (Pro-Pro-Gly)$_{10}$ · 9 H$_2$O, 1 mg/ml; or peptide L-I, Ala-Arg-Gly-Ile-Lys-Gly-Ile-Arg-Gly-Phe-Ser-Gly, 5 mg/ml in H$_2$O (all these peptides are commercially available from the Peptide Institute, Protein Research Foun-

---

[55] S. Sakakibara, Y. Kishida, Y. Kikuchi, R. Sakai, and K. Kakiuchi, *Bull. Chem. Soc. Jpn.* **41**, 1273 (1968).
[56] R. A. Berg, Y. Kishida, S. Sakakibara, and D. J. Prockop, *Biochemistry* **16**, 1615 (1977).
[57] L. Tuderman, E.-R. Kuutti, and K. I. Kivirikko, *Eur. J. Biochem.* **52**, 9 (1975).

dation, Minoh, Osaka, Japan). The peptide is heated to 100° for 10 min and rapidly cooled to 0° immediately before use.

2-Oxo[1-$^{14}$C]glutarate, mixed with the unlabeled compound to give a final concentration of 2 m$M$ with $1.2 \times 10^6$ or $2.0 \times 10^6$ dpm/ml. This solution is divided into aliquots of 1 ml, lyophilized, and stored at $-20°$. One aliquot is dissolved in 1.0 ml of H$_2$O (or two aliquots in 2.0 ml) immediately before use.

NCS, 0.6 $N$ solution in toluene, commercially available. Pieces of filter paper, $1.5 \times 3.0$ cm, are saturated with this reagent and placed on the wire hooks of the rubber stoppers used to seal the test tubes.

KH$_2$PO$_4$, 1.0 $M$, pH 5.0, stored at 4°

Scintillation solution[58]: 15 g of 2,5-diphenyloxazole (PPO) and 50 mg of 1,4-bis[2-(4-methyl-5-phenyloxazolyl)]benzene (POPOP) are dissolved in 1000 ml of toluene; 600 ml of ethylene glycol monomethyl ether (methyl Cellosolve) are then added.

*Procedure.* Test tubes cooled in iced water each receive an appropriate addition of cold H$_2$O to make a final volume of 1.0 ml after the subsequent addition of the substrate, enzyme, and other reactants. The following additions are made: 0.1 ml Tris-HCl buffer, 0.1 ml of bovine serum albumin (0.05 ml in lysyl hydroxylase assays), 0.05 ml of catalase, 0.1 ml of dithiothreitol, 0.1 ml of ascorbic acid (0.05 ml in lysyl hydroxylase assays), 0.05 ml of FeSO$_4$, the enzyme to be assayed or a partially purified enzyme standard of known specific activity, or H$_2$O, 0.1 ml of the peptide substrate, and 0.05 ml of 2-oxo[1-$^{14}$C]glutarate. The test tubes are rapidly sealed with rubber stoppers, each having a filter paper saturated with the NCS solution. The tubes are then transferred from the iced water to a water bath of 37°, incubated with gentle shaking for 30 or 40 min, and returned to iced water. The phosphate buffer, 1.0 ml, is rapidly injected through the rubber stopper with a needle and syringe, the tubes are shaken vertically for 30 min at room temperature, and the filter papers are placed into counting vials containing 10 ml of the scintillation solution. Radioactivity is then determined in a liquid scintillation spectrometer.

*Definition of Units of Enzyme Activity.* One unit of prolyl 4-hydroxylase activity is defined as the amount of enzyme required to synthesize 1 $\mu$mol of hydroxyproline per minute at 37° under conditions in which the concentrations of the cofactors and cosubstrates are those detailed above, except that the concentration of 2-oxoglutarate is corrected to a saturated value, and in which a saturating concentration of (Pro-Pro-Gly)$_n$, $n = 5$, 10, or 20, synthesized by the solid-state method, is used as the substrate.[57] One unit of lysyl hydroxylase activity is defined analogously, except that a saturating concentration of the peptide L-I, L-II,[37] or L-III[37] is used as the

[58] D. J. Prockop and P. S. Ebert, *Anal. Biochem.* **6**, 263 (1963).

substrate. It may be noted that owing to their high cost, these peptides are used only in the concentrations indicated above and product formation is corrected to a saturating concentration using values obtained from double-reciprocal plots of peptide concentration against 2-oxo[1-$^{14}$C]glutarate decarboxylation.

Assay of Prolyl 4-Hydroxylase Protein

Several methods are now available for the assay of immunoreactive prolyl 4-hydroxylase protein. The first procedures were based on the displacement of the active enzyme from its antibody by inactive enzyme protein and the determination of the unbound enzyme activity.[59,60] The use of these procedures is somewhat complicated by the problems encountered in assaying prolyl 4-hydroxylase activity in biological specimens with the protocollagen substrate (see above). An assay based on hemagglutination inhibition has also been reported.[61] More recently, radioimmunoassays have been developed that are based on the displacement of radioactively labeled prolyl 4-hydroxylase protein from its antibody by the immunoreactive enzyme protein and precipitation of the enzyme-antibody complex by a second antibody.[29,62] One of these assays[62] has been used to assay prolyl 4-hydroxylase protein in a number of animal[19,46,63] and human[39,44,46,54] cells or tissues, including serum[39,64-66] from healthy human subjects and patients with various diseases. This assay is described in detail below.

Tritiation of Prolyl 4-Hydroxylase

Reagents

Pure prolyl 4-hydroxylase, 0.5 ml, with a concentration of 0.3–0.5 mg/ml. To remove most of the free glycine present in the buffer used to purify the enzyme (see below), this preparation is dialyzed at 4° for 5 hr against 2 liters of a solution of 0.15 $M$ NaCl and 0.01 $M$ Tris-HCl, pH adjusted to 7.8 at 4°. Dialysis is continued for 2 hr

[59] J. O'D. McGee and S. Udenfriend, *Biochem. Biophys. Res. Commun.* **46**, 1646 (1972).
[60] F. L. H. Stassen, G. J. Cardinale, J. O'D. McGee, and S. Udenfriend, *Arch. Biochem. Biophys.* **160**, 340 (1974).
[61] B. R. Olsen, R. A. Berg, Y. Kishida, and D. J. Prockop, *J. Cell Biol.* **64**, 340 (1975).
[62] L. Tuderman, E.-R. Kuutti, and K. I. Kivirikko, *Eur. J. Biochem.* **60**, 399 (1975).
[63] L. Tuderman, *Eur. J. Biochem.* **66**, 615 (1976).
[64] L. Tuderman, J. Risteli, T. A. Miettinen, and K. I. Kivirikko, *Eur. J. Clin. Invest.* **7**, 537 (1977).
[65] E.-R. Kuutti-Savolainen, *Clin. Chim. Acta* **96**, 53 (1979).
[66] E.-R. Kuutti-Savolainen, K. I. Kivirikko, and O. Laitinen, *Ann. Rheum. Dis.* **39**, 217 (1980).

against 2 liters of a solution of 0.15 $M$ NaCl and 0.16 $M$ sodium borate, pH 9.0.
Sodium [$^3$H]borohydride, 0.04 $M$, 5-15 Ci/mmol
Formaldehyde, 0.04 $M$
Dialysis buffer, a solution of 0.15 $M$ NaCl, 0.1 $M$ glycine, and 0.01 $M$ Tris-HCl; pH adjusted to 7.8 at 4°

*Procedure.* The pure enzyme is labeled with $^3$H using formaldehyde and sodium [$^3$H]borohydride, the whole procedure being carried out at 0–4°. Formaldehyde solution, 30–50 μl, is added to 0.5 ml of the dialyzed enzyme preparation, with a protein concentration of 0.3–0.5 mg/ml, respectively. This is followed in 30 sec by four sequential additions of 25 μl of sodium [$^3$H]borohydride and after 2 min by an additional 50 μl of the [$^3$H]borohydride. The enzyme is then exhaustively dialyzed against the dialysis buffer at 4°, divided into aliquots, and stored at −70°. The specific activity is usually about 1 to 2 × 10$^8$ dpm per milligram of prolyl 4-hydroxylase.

*The Radioimmunoassay*

*Reagents*

Tritiated prolyl 4-hydroxylase, preferably of the species in question. One aliquot of the labeled hydroxylase is diluted with the assay buffer to an enzyme protein concentration of 2.5 μg/ml.

Nonlabeled pure prolyl 4-hydroxylase, preferably of the species in question, stored at −70° in aliquots. One aliquot is diluted with the assay buffer to an enzyme protein concentration of 1 μg/ml.

Antiserum to prolyl 4-hydroxylase, preferably of the species in question. Pure prolyl 4-hydroxylase, 0.25 mg, is dialyzed exhaustively against 0.15 $M$ NaCl, mixed with an equal volume of complete Freund's adjuvant, and injected intradermally at 10 sites in a rabbit's back. At a separate site, 0.5 ml of crude *Bordatella pertussis* vaccine is injected subcutaneously. Further injections of 0.1 mg of prolyl 4-hydroxylase mixed with incomplete Freund's adjuvant are administered, 2, 3, and 4 weeks later. The rabbit is bled at 1-week intervals by ear vein puncture beginning 2 weeks after the first injection. The antisera are stored at −20° in aliquots. One aliquot is diluted with the assay buffer to a suitable concentration, depending on the titer. The antiserum to chick prolyl 4-hydroxylase is usually diluted 1:2500 (final dilution in the first incubation 1:8330) or the antiserum to human prolyl 4-hydroxylase 1:600 (final dilution in the first incubation 1:2000).

Second antibody (DASP-antirabbit, Organon Teknika B.V., The

Netherlands). The contents of the commercial vial, 5.5 ml, are diluted with the assay buffer 1 : 10.

Assay buffer, a solution of 0.15 $M$ NaCl, 2.5 mg of bovine serum albumin per milliliter, and 0.01 $M$ Tris-HCl buffer, adjusted to pH 7.4 at 4°.

*Procedure.* The tritiated enzyme, 10 $\mu$l, is pipetted into test tubes of size 7 × 75 mm. Tubes 1 and 2 are used to indicate precipitation of the tritiated enzyme in the absence of antiserum; and tubes 3–5, the amount of tritiated enzyme bound to the first and second antibody in the absence of displacement by the nonlabeled enzyme. Tubes 6–15 receive 10–200 $\mu$l of the nonlabeled enzyme standard in duplicate, and the other tubes receive a suitable aliquot of the specimens to be examined (diluted with the assay buffer). All tubes except 1 and 2 receive 150 $\mu$l of the diluted antiserum, and all then receive assay buffer to a final volume of 0.5 ml. The tubes are incubated with agitation at room temperature for 5 hr, and the tritiated enzyme bound to the antibody is precipitated by the addition of 1.0 ml of the diluted second antibody to each tube. The tubes are incubated overnight with agitation at room temperature and centrifuged at about 600 $g$ for 1 min; the supernatants are removed by suction. The pellets are washed twice with 2 ml of the assay buffer, the centrifugations being carried out as above, and finally the pellets are suspended in 1.5 ml of H$_2$O and assayed for radioactivity in 5 ml of a counting solution (e.g., Aquasol, New England Nuclear, or Instagel, Packard) in a liquid scintillation spectrometer.

An aliquot of 10 $\mu$l of the tritiated enzyme is counted in duplicate together with the other samples in order to obtain a value for the total radioactivity added. About 2–3% of the total label is usually found in tubes 1 and 2, this value being subtracted from all the others. The amount of antiserum is adjusted so that about 50–60% of the total radioactivity is found in tubes 3–5, the ratio of bound to free label thus being about 1–1.5. The results are expressed as nanograms of pure prolyl 4-hydroxylase present, these values being obtained from the curve of bound/free radioactivity versus concentration of the nonlabeled enzyme standard (tubes 6–15).

Purification and Properties of the Hydroxylases

Purification and Molecular Properties of Prolyl 4-Hydroxylase

Prolyl 4-hydroxylase was first purified to near homogeneity from chick embryos[24,67] and the skin of newborn rats[12] by methods involving frac-

[67] M. Pänkäläinen, H. Aro, K. Simons, and K. I. Kivirikko, *Biochim. Biophys. Acta* **221**, 559 (1970).

tionating precipitations, ion-exchange chromatographies, and gel filtration. More recently, two affinity column procedures have been developed. The first is based on the affinity of the enzyme for its polypeptide substrates and consists of affinity chromatography on a column containing the hydroxyproline-deficient collagen from the cuticle of *Ascaris* worms linked to agarose, elution of the enzyme with a synthetic peptide substrate, and separation of the enzyme from this peptide by gel filtration.[26] The second procedure is based on the affinity of the enzyme for its competitive polypeptide inhibitor and involves affinity chromatography on a column containing poly(L-proline) linked to agarose, elution with poly(L-proline) of a lower molecular weight, and gel filtration.[57] Prolyl 4-hydroxylase has been isolated as a homogeneous protein from chick embryos by both procedures,[26,31,57] from cultured L-929 fibroblasts[68] and the skin of newborn rats[69] by the first procedure, and from human tissues[70] and newborn rats[19] by the second. Both procedures are rapid and easy to perform. Our laboratory favors the poly(L-proline) method, as pure and characterized preparations of this peptide are commercially available, whereas the preparation and characterization of the *Ascaris* cuticle collagen involve some difficulties. Moreover, the cost of poly(L-proline) is much lower than that of the synthetic peptide substrate used to elute the *Ascaris* column. The poly(L-proline) procedure is described in detail in the next section.

*Preparation of Poly(L-Proline)–Agarose*

Poly(L-proline) is coupled to agarose by the cyanogen bromide activation technique.[57,71] About 100 ml of 4% agarose (Sepharose 4B, Pharmacia) are washed three times with 1 liter of $H_2O$, pH adjusted to 11, and decanted to a final volume of about 200 ml. Then 25 g of carefully ground cyanogen bromide is added, and the reaction is allowed to proceed for 15–20 min in an ice bath with continuous stirring. The pH is maintained at 11 by frequent additions of 8 $M$ and 1 $M$ NaOH. The mixture is rapidly washed in a Büchner funnel with 500 ml of a solution of 0.14 $M$ NaCl, and 0.1 $M$ $NaHCO_3$, pH 9.3, and the coupling reaction is carried out by quickly stirring into the gel 1 g of poly(L-proline), molecular weight about 30,000, dissolved in 10 ml of cold $H_2O$, the washing and addition of poly(L-proline) being completed in about 40 sec. The mixture is transferred into a beaker, and the reaction is allowed to proceed with gentle stirring at 4° for 20 hr. The gel is washed with large volumes of cold $H_2O$

[68] W. W.-Y. Kao and R. A. Berg, *Biochim. Biophys. Acta* **586**, 528 (1979).
[69] S. Chen-Kiang, G. J. Cardinale, and S. Udenfriend, *Proc. Natl. Acad. Sci. U.S.A.* **74**, 4420 (1977).
[70] E.-R. Kuutti, L. Tuderman, and K. I. Kivirikko, *Eur. J. Biochem.* **57**, 181 (1975).
[71] P. Cuatrecasas and C. B. Anfinsen, this series, Vol. 22, p. 345.

followed by a cold solution of 0.1 $M$ NaCl, 0.1 $M$ glycine, 10 $\mu M$ dithiothreitol, and 0.01 $M$ Tris-HCl buffer, pH adjusted to 7.8 at 4°. The gel is poured into a column (1.5 × 30 cm) with a bed volume of about 50 ml, and the washing is continued with 3 liters of the same solution.

The efficiency of the coupling reaction is measured by eluting a small aliquot of the washed gel briefly with $H_2O$ to remove the buffer, hydrolyzing the sample in 6 $M$ HCl at 120° for 16 hr, and measuring the proline content in an amino acid analyzer. The poly(L-proline) content should be at least about 3–5 mg per milliliter of gel.

*Purification Procedure*[57,70]

All procedures are carried out at 0–4°. The term enzyme buffer is used for a solution of 0.1 $M$ NaCl, 0.1 $M$ glycine, 10 $\mu M$ dithreitol, and 0.01 $M$ Tris-HCl buffer, pH adjusted to 7.8 at 4°. The protein concentration is assayed by peptide absorbance at 225 nm or 230 nm using an absorption coefficient of $A_{225\,nm}^{1\,mg/ml} = 7.40$ with a 1-cm light path for crude enzyme preparations and $A_{230\,nm}^{1\,mg/ml} = 7.73$ for the pure enzyme.

*Step 1. Extraction.* The extract is prepared by homogenizing about 500–2000 g of tissue in enzyme buffer supplemented with 0.1% Triton X-100, using 1–4 ml of the solution per gram of tissue. The homogenization is usually carried out in a Waring blender in suitable batches at full speed for two bursts of 30 sec with a 1-min interval, but in the case of tough tissue a careful preliminary mincing with scissors is recommended. The homogenate is allowed to stand with occasional stirring for 15 min and then centrifuged at 15,000 $g$ for 30 min.

*Step 2. $(NH_4)_2SO_4$ Fractionation.* Solid $(NH_4)_2SO_4$ is slowly stirred into the supernatant to a final concentration of 20% saturation (114 g/liter) for human placenta, 25% (144 g/liter) for newborn rats, or 30% (176 g/liter) for chick embryos. After centrifugation at 15,000 $g$ for 20 min, the pellet is discarded, and solid $(NH_4)_2SO_4$ is slowly stirred into the supernatant to a final concentration of 65% saturation (267 g/liter or 235 g/liter, respectively) for newborn rats and chick embryos, or 70% (340 g/liter) for human placenta. After centrifugation at 15,000 $g$ for a further 20 min, the pellet is dissolved in enzyme buffer and dialyzed for 4 hr against 16 liters of enzyme buffer and for 12 hr against another 16 liters. The enzyme is then frozen in aliquots of up to 10 g of protein, except for that from human placenta, for which the aliquots may be of up to 50 g.

*Step 3. Affinity Chromatography.* An aliquot of the $(NH_4)_2SO_4$ fraction is thawed and centrifuged at 15,000 $g$ for 45 min, and the protein concentration of the supernatant is adjusted with enzyme buffer to 10 mg/ml (50 mg/ml for human placenta). The sample is passed through the affinity column at a flow rate of about 40–60 ml/hr, and the column is washed with enzyme buffer until the absorbance of the eluate at 230 nm is about 0.05 (usually about 1 liter). The enzyme is eluted with 20 ml of enzyme buffer containing 3 mg of poly(L-proline) per milliliter, molecular weight about 3000–6000, fractions of 3 ml being collected, and the elution is continued with about 50 ml of enzyme buffer without the peptide. The size of the poly(L-proline) is critical in that a peptide with a molecular weight of more than about 7000–9000 cannot be completely separated from the enzyme in the final gel filtration (see step 4). After the experiment the column is regenerated by washing with 6 $M$ urea in a beaker and then equilibrated with enzyme buffer.

The absorbance of the fractions eluted with poly(L-proline) is measured at 230 nm, and the fractions are pooled beginning with the first that shows any increase in absorbance and continuing after the peptide peak until the last that still has an absorbance above 2.0. The pooled fractions are concentrated to about 2 ml in an ultrafiltration cell (Amicon) with a PM-30 membrane, and the cell is washed twice with 0.5 ml of enzyme buffer, these washes being pooled with the enzyme.

*Step 4. Gel Filtration.* The sample is centrifuged at 20,000 $g$ for 10 min, and the clear supernatant is applied to an 8% agarose gel column (BioGel A-1.5m, 200–400 mesh, Bio-Rad) of size 1.5 × 90 cm, equilibrated and eluted with enzyme buffer. Fractions of 3 ml are collected and their absorbance is measured at 230 nm. The fractions containing most of the enzyme activity are pooled (as indicated in Fig. 3), concentrated to about 2 ml by ultrafiltration as above, and taken to constitute the final enzyme pool. The same gel filtration column can be used only for 3–4 enzyme preparations, as some of the poly(L-proline) becomes adsorbed to the column in spite of washing with large amounts of enzyme buffer.

*Comments on the Purification.* Examples of the purification of prolyl 4-hydroxylase from three different sources are shown in Table II. The recovery is usually about 40–80% when calculated from the total activity in the $(NH_4)_2SO_4$ fraction, and the enzyme is pure when examined by polyacrylamide gel electrophoresis either as a native protein or after reduction and denaturation in the presence of SDS.

*Molecular Properties*

The molecular weight of prolyl 4-hydroxylase from chick embryos is about 240,000 by sedimentation equilibrium,[26,57,67] whereas gel filtration gives a somewhat higher value.[24,26] The rat[19] and human[70] enzymes have a molecular weight by gel filtration identical with that of the chick enzyme. Prolyl 4-hydroxylase from all vertebrate sources studied is a tetramer

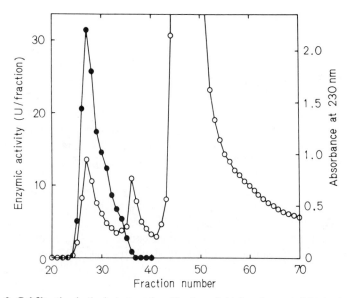

FIG. 3. Gel filtration in the last step of purification of chick embryo prolyl 4-hydroxylase. ●——●, Enzyme activity; ○——○, absorbance. The large peak corresponds to the elution position of poly(L-proline) and the first peak to that of the enzyme. Reproduced from Tuderman et al.[57] with permission of the publisher.

($\alpha_2\beta_2$) consisting of two different types of enzymically inactive monomer with molecular weights of about 64,000 ($\alpha$-subunit) and 60,000 ($\beta$-subunit) by SDS–polyacrylamide gel electrophoresis.[12,19,26,57,70] Electron microscopy indicates that both monomers are rod-shaped, and the tetramer probably consists of two V-shaped dimers that are interlocked.[72] Intrachain disulfide bonds appear to be essential for the monomers to maintain the native structure necessary for their association, whereas the presence of interchain disulfide bonds does not seem likely.[2,73]

Amino acid analyses of pure prolyl 4-hydroxylase protein from chick embryos,[26,57] rat skin,[69] and human tissues[70] demonstrate that there is a great similarity between the enzyme proteins from all these sources and that the protein is highly acidic (Table III). This finding agrees with early amino acid analyses of almost pure enzyme preparations[12,24] and with an acidic isoelectric point.[67,73]

The subunits of prolyl 4-hydroxylase can be isolated by dissociating pure enzyme tetramers with dithiothreitol in the presence of high concentrations of urea[26,73] and by separating the reduced,[26,69] or reduced and

[72] B. R. Olsen, R. A. Berg, K. I. Kivirikko, and D. J. Prockop, Eur. J. Biochem. **35**, 135 (1973).
[73] R. A. Berg, N. L. Kedersha, and N. A. Guzman, J. Biol. Chem. **254**, 3111 (1979).

## TABLE II
### PURIFICATION OF PROLYL 4-HYDROXYLASE FROM THREE SOURCES[a]

| Source and enzyme fraction | Total protein (mg) | Total activity (units) | Recovery (%) | Specific activity (units/mg protein) | Purification[b] (fold) |
|---|---|---|---|---|---|
| Chick embryos[c] | | | | | |
| $(NH_4)_2SO_4$, 30–65% satn | 7080 | 2.83 | 100 | 0.0004 | 1 |
| After affinity chromatography and gel filtration | 1.54 | 2.32 | 82 | 1.50 | 3750 |
| Fetal rats[d] | | | | | |
| $(NH_4)_2SO_4$, 25–65% satn | 6806 | 1.14 | 100 | 0.000167 | 1 |
| After affinity chromatography and gel filtration | 0.35 | 0.45 | 40 | 1.30 | 7770 |
| Human fetal tissues[e] | | | | | |
| $(NH_4)_2SO_4$, 20–70% satn | 50,000 | 1.42 | 100 | 0.000028 | 1 |
| After affinity chromatography and gel filtration | 0.45 | 0.58 | 41 | 1.30 | 45,700 |

[a] The procedure involves in all cases affinity chromatography on poly(L-proline) linked to agarose. One unit of enzyme activity is defined as the amount of enzyme required to synthesize 1 $\mu$mol of 4-hydroxyproline per minute at 37° under the conditions specified in the text.
[b] Purification calculated from the ammonium sulfate fraction.
[c] Reproduced from Tuderman et al.,[57] with permission of the publisher.
[d] Reproduced from Kuutti et al.,[70] with permission of the publisher.
[e] Reproduced from Risteli et al.,[19] with permission of the publisher.

alkylated,[73] monomers. The separation is accomplished either by affinity chromatography on concanavalin A-agarose,[69] in which only the $\alpha$-subunit is bound, or by ion-exchange chromatography on DEAE-cellulose.[73] The isoelectric point of the $\alpha$-subunit of the chick enzyme is about 5.5, and that of the $\beta$-subunit about 4.7.[73] Distinct differences are found between the two subunits in amino acid composition[69,73] (Table III) and in the peptide fragments produced by trypsin,[69] chymotrypsin,[73] or *Staphylococcus aureus* V8 protease.[73] Prolyl 4-hydroxylase, like the other two hydroxylases of collagen biosynthesis (see below), is a glycoprotein, most of its carbohydrate being located in the $\alpha$-subunit.[69,73,74] This monomer of the chick enzyme binds concanavalin A and contains 16 residues of mannose, 1 of galactose and at least 2 of N-acetylglucosamine.[73] The $\beta$-subunit of the chick enzyme does not readily bind this lectin, but contains two residues of mannose and three of galactose.[73]

[74] N. A. Guzman, R. A. Berg, and D. J. Prockop, *Biochem. Biophys. Res. Commun.* **73**, 279 (1976).

## TABLE III
### AMINO ACID COMPOSITION OF THE TETRAMER AND THE TWO SUBUNITS OF PROLYL 4-HYDROXYLASE FROM VARIOUS SOURCES

| Amino acid | Residues/1000 amino acids | | | | | | | |
|---|---|---|---|---|---|---|---|---|
| | Enzyme tetramer | | | | Isolated subunits | | | |
| | Chick[a] | Chick[b] | Rat[c] | Human[d] | Rat $\alpha^c$ | Rat $\beta^c$ | Chick $\alpha^e$ | Chick $\beta$ |
| Lysine | 93 | 88 | 93 | 93 | 86 | 96 | 90 | 96 |
| Histidine | 17 | 17 | 23 | 19 | 26 | 22 | 19 | 16 |
| Arginine | 54 | 48 | 41 | 42 | 56 | 19 | 60 | 31 |
| Aspartic acid | 99 | 117 | 118 | 122 | 100 | 113 | 105 | 112 |
| Threonine | 46 | 55 | 53 | 51 | 57 | 48 | 62 | 45 |
| Serine | 38 | 44 | 45 | 45 | 52 | 49 | 65 | 53 |
| Glutamic acid | 137 | 138 | 143 | 142 | 126 | 139 | 125 | 145 |
| Proline | 50 | 52 | 43 | 50 | 40 | 45 | 37 | 39 |
| Glycine | 82 | 70 | 72 | 72 | 75 | 62 | 81 | 80 |
| Alanine | 90 | 90 | 89 | 89 | 73 | 95 | 72 | 90 |
| Valine | 58 | 55 | 59 | 58 | 54 | 57 | 50 | 52 |
| Isoleucine | 44 | 43 | 38 | 45 | 40 | 47 | 35 | 44 |
| Leucine | 108 | 104 | 94 | 98 | 105 | 94 | 106 | 85 |
| Tyrosine | 11 | 23 | 30 | 22 | 42 | 25 | 44 | 19 |
| Phenylalanine | 57 | 56 | 53 | 52 | 34 | 70 | 37 | 83 |

[a] From Berg and Prockop.[26]
[b] From Tuderman et al.[57]
[c] From Chen-Kiang et al.[69]
[d] From Kuutti et al.[70]
[e] From Berg et al.[73]

### Purification and Molecular Properties of an Immunoreactive β-Subunit-Related Prolyl 4-Hydroxylase Protein

Prolyl 4-hydroxylase protein is present in isolated cells and intact tissues in two forms: the active enzyme tetramers and an inactive form that corresponds in size to the 60,000 molecular weight β-subunit of the enzyme.[1-3,6,75] The concentration of the active enzyme tetramers in most cells and tissues is much lower than that of the β-subunit-related protein, and the ratio of active tetramers to total enzyme protein varies markedly between tissues, and even in the same cell type or tissue, in different conditions.[1,3,6]

Studies have demonstrated that at least one function of the β-subunit-related protein is to serve as a precursor of the β-subunits in the

[75] J. O'D. McGee and S. Udenfriend, *Arch. Biochem. Biophys.* **152**, 216 (1972).

assembly of the prolyl 4-hydroxylase tetramer.[76-80] The α-subunit appears to be utilized immediately to form the active enzyme, whereas the β-subunit is synthesized in an excess and enters a precursor pool of this monomer before being incorporated into the tetramer.[78-80] In contrast to the intracellular location of the enzyme tetramers, which are found only within the cisternae of the rough endoplasmic reticulum,[1,2,6] the β-subunit-related protein is also present in association with the plasma membrane.[80] It is not known, however, whether this protein is an integral part of the plasma membrane or just associated with the membrane through some unknown mechanism; neither is it known whether the β-subunit-related protein is exclusively related to prolyl 4-hydroxylase or also has some other functions.

*Purification Procedure*

The β-subunit-related protein has been purified to near homogeneity from the skin of newborn rats by a procedure involving ammonium sulfate fractionation, affinity chromatography on a column containing specific antibodies to prolyl 4-hydroxylase linked to agarose, and gel filtration.[69] The purification can be followed by the assays for prolyl 4-hydroxylase protein described above.[29,59-62]

*Preparation of the Immunoadsorbent.* Antiserum to prolyl 4-hydroxylase is prepared in rabbits (as described above for assay of the enzyme protein) or goats,[69] and the IgG fraction is separated from the other serum components as follows.[81] Solid $(NH_4)_2SO_4$ is slowly stirred into the serum to a final concentration of 50% saturation (313 g/liter); after centrifugation at 15,000 $g$ for 20 min, the pellet is dissolved in 10 m$M$ potassium phosphate, pH 6.8, the volume being equivalent to that of the original serum sample. This solution is again precipitated with $(NH_4)_2SO_4$ at 50% saturation, centrifuged, and dissolved as before. The solution is dialyzed against three changes of 100 volumes each of 10 m$M$ potassium phosphate, pH 6.8, and is applied to a column of DEAE-cellulose (DE-52, Whatman, 10–15 ml packed material per milliliter of serum to be fractionated), equilibrated, and eluted with the same buffer. The IgG, which does not bind to the column, is collected.

[76] L. Tuderman, A. Oikarinen, and K. I. Kivirikko, *Eur. J. Biochem.* **78**, 547 (1977).
[77] C. O. Chichester, G. C. Fuller, and C.-J. Mo Cha, *Biochim. Biophys. Acta* **586**, 341 (1979).
[78] K. Majamaa, E.-R. Kuutti-Savolainen, L. Tuderman, and K. I. Kivirikko, *Biochem. J.* **178**, 313 (1979).
[79] R. A. Berg, W. W.-Y. Kao, and N. L. Kedersha, *Biochem. J.* **189**, 491 (1980).
[80] W. W.-Y. Kao and K.-L. L. Chou, *Arch. Biochem. Biophys.* **199**, 147 (1980).
[81] D. M. Livingston, this series, Vol. 34, p. 723.

The purified IgG is coupled to agarose using the cyanogen bromide activation technique,[71] as described above for the coupling of poly(L-proline) to agarose, except that the solution used for washing in the Büchner funnel is 0.1 $M$ sodium borate, pH 8.4. The coupled gel is successively washed with 0.05 $M$ sodium borate, pH 8.4, 1.0 $M$ 2-aminoethanol, pH 8.0, and a solution of 0.2 $M$ NaCl, 0.2 $M$ glycine, 10 $\mu M$ dithiothreitol, 10 $\mu M$ EDTA, and 10 m$M$ Tris-HCl, pH adjusted to 7.8 at 4°. A similar IgG fraction is prepared from preimmune serum and coupled to agarose in the same way.

*Step 1. Extraction.* This and all other steps of the purification procedure are carried out at 0-4°. Skins of 200 newborn rats are suspended in 400 ml of a solution containing 0.25 $M$ sucrose, 10 $\mu M$ dithiothreitol, 10 $\mu M$ EDTA and 10 m$M$ Tris-HCl buffer, pH adjusted to 7.8 at 4°; phenylmethanesulfonyl fluoride, freshly dissolved in 2-propanol, is added to a final concentration 50 $\mu$g/ml. The mixture is homogenized with a Sorvall Omni-Mixer at full speed for two 1-min pulses and then with a Polytron PT-10 homogenizer for two additional 1-min pulses. The homogenate is centrifuged at 15,000 $g$ for 30 min, and the pellet is discarded.

*Step 2. $(NH_4)_2SO_4$ Fractionation.* Solid $(NH_4)_2SO_4$ is slowly stirred into the supernatant to a final concentration of 25% saturation (144 g/liter); after centrifugation at 15,000 $g$ for 20 min, the pellet is discarded. The supernatant is brought to 70% saturation (307 g/liter) with $(NH_4)_2SO_4$, and the precipitate obtained by a similar centrifugation is used for purification of active prolyl 4-hydroxylase by the procedures described above. The supernatant is completely saturated by a further addition of solid $(NH_4)_2SO_4$ (237 g/liter), and the pellet is collected by centrifugation, dissolved in 50 ml of a solution containing 0.2 $M$ NaCl, 0.2 $M$ glycine, 10 $\mu M$ dithiothreitol, 10 $\mu M$ EDTA, and 10 m$M$ Tris-HCl buffer, pH adjusted to 7.8 at 4° (termed the purification buffer), and dialyzed extensively against the same solution.

*Step 3. Chromatographies on the Preimmune IgG and Immunoadsorbent Column.* The dialyzed sample is passed through the preimmune IgG-Sepharose column to remove materials that adsorb nonspecifically onto IgG. The effluent is passed through the immunoadsorbent column at a flow rate of about one column volume per hour, and the effluent is recycled twice. The column is washed with the purification buffer, and the adsorbed protein is eluted with 3 $M$ ammonium thiocyanate. The eluate is dialyzed immediately against purification buffer and concentrated to about 6 ml in an ultrafiltration cell (Amicon) with a PM-30 membrane. Control experiments have indicated that when pure prolyl 4-hydroxylase is incubated with 3 $M$ ammonium thiocyanate for up to 2 hr at 4° there is no diminution of antigenic activity, although the catalytic activity is reduced to about 1% of the original value.[69]

TABLE IV
PURIFICATION OF β-SUBUNIT-RELATED PROTEIN OF PROLYL 4-HYDROXYLASE FROM SKINS OF NEWBORN RATS[a]

| Purification step | Total protein (mg) | Total amount of β-subunit-related protein (units)[b] | Recovery (%) | Specific activity (units/mg protein) | Purification (fold) |
|---|---|---|---|---|---|
| 15,000 · g supernatant | 21,792 | 100,245 | 100 | 4.6 | 1 |
| $(NH_4)_2SO_4$, 70–100% satn. | 3087 | 41,369 | 41 | 13.4 | 2.9 |
| Effluent from the preimmune IgG column | 3007 | 40,290 | 40 | 13.4 | 2.9 |
| Eluate from the antibody column | 87.5 | 21,390 | 21 | 244.5 | 53.2 |
| Gel filtration | 4.5 | 10,770 | 11 | 2414.8 | 525.0 |

[a] From Chen-Kiang et al.[69]
[b] One unit of β-subunit-related protein is defined as that amount required to displace 1 μg of enzyme in the immunoassay.

*Step 4. Gel Filtration.* The sample is applied to an 8% agarose gel column (BioGel A-1.5m, 200–400 mesh, Bio-Rad) of size 2.5 × 90 cm equilibrated and eluted with purification buffer. The fractions containing most of the immunoreactive prolyl 4-hydroxylase protein are pooled and constitute the final purified β-subunit-related protein.

*Comments on Purification.* The purification of β-subunit-related protein from the skin of 800 newborn rats is shown in Table IV. The recovery is about 10%, and the protein is more than 90% pure when examined by polyacrylamide gel electrophoresis either under nondenaturing conditions or after reduction and denaturation in the presence of SDS.[69]

*Molecular Properties*

The migration of the β-subunit-related protein in SDS–polyacrylamide gels is identical with that of the molecular weight ($M_r$) 60,000 β-subunit of prolyl 4-hydroxylase.[68,69,76,78–80] The amino acid compositions of these proteins are likewise identical within the limits of experimental error.[69] The peptide maps produced of these two proteins by digestion with trypsin[69] or *Staphylococcus aureus* V8 protease[68,79] are very similar, but two of the 35 peptides obtained with the former protease[69] and at least one obtained with the latter[79] do not coincide. These minor differences may be due to differences in glycosylation, but it is also possible that the β-subunit-related protein must be processed in some way before being converted into the final subunit. The kinetics of assembly of the enzyme tetramer in intact cells favor the latter possibility.[79]

Purification and Molecular Properties of Lysyl Hydroxylase

Early studies demonstrated that lysyl hydroxylase and prolyl 4-hydroxylase activities are derived from separate enzyme proteins.[24,32,82-84] Lysyl hydroxylase was subsequently purified several 100- or 1000-fold from chick embryo extract by procedures consisting of conventional protein purification steps.[32,43] The enzyme was then found to have a very high affinity for columns of concanavalin A-agarose,[43] and could be eluted only with a combination of methyl $\alpha$-D-mannoside or methyl $\alpha$-D-glucoside and ethylene glycol.[85] Such a chromatography has recently been combined with either gel filtration and chromatography on hydroxyapatite (procedure A) or affinity chromatography on denatured collagen linked to agarose and gel filtration (procedure B).[86] Lysyl hydroxylase has been isolated as a homogeneous protein from chick embryos by both procedures,[86] and from human placenta by the latter.[87] Procedure B is described here in detail.

*Purification Procedure*[86]

All steps are carried out at 0–4°. The term enzyme buffer is used for a solution of 0.2 $M$ NaCl, 0.1 $M$ glycine, 10 $\mu M$ dithiothreitol, 1% glycerol, and 20 m$M$ Tris-HCl buffer, pH adjusted to 7.5 at 4°.

*Step 1. Extraction.* The procedure described is for isolating the enzyme from chick embryos, but can also be applied to other sources. A total of 200–400 14-day chick embryos are homogenized in batches of 30 in enzyme buffer supplemented with 0.1% (w/v) Triton X-100 (1 ml of solution per gram of embryos) in a Waring blender at full speed twice for 30 sec. The homogenate is allowed to stand with occasional stirring for 30 min and then centrifuged at 15,000 $g$ for 30 min.

*Step 2. $(NH_4)_2SO_4$ Fractionation.* Solid $(NH_4)_2SO_4$ is slowly stirred into the supernatant fraction to a final concentration of 17% saturation (97

---

[82] E. Weinstein, N. Blumenkrantz, and D. J. Prockop, *Biochim. Biophys. Acta* **191**, 747 (1969).
[83] R. L. Miller, *Arch. Biochem. Biophys.* **147**, 339 (1971).
[84] E. A. Popenoe and R. B. Aronson, *Biochim. Biophys. Acta* **258**, 380 (1972).
[85] T. M. Turpeenniemi, U. Puistola, H. Anttinen, and K. I. Kivirikko, *Biochim. Biophys. Acta* **483**, 215 (1977).
[86] T. M. Turpeenniemi-Hujanen, U. Puistola, and K. I. Kivirikko, *Biochem. J.* **189**, 247 (1980).
[87] T. M. Turpeenniemi-Hujanen, U. Puistola, and K. I. Kivirikko, *Collagen Rel. Res.* **1**, 355 (1981).

g/liter). The pellet obtained by centrifugation at 15,000 $g$ for 20 min is discarded, and solid $(NH_4)_2SO_4$ is slowly stirred into the supernatant to a final concentration of 55% saturation (244 g/liter). The pellet obtained by centrifugation at 15,000 $g$ for 20 min is dissolved in enzyme buffer and dialyzed for 4 and 12 hr against two separate 16-liter volumes of enzyme buffer supplemented with 3 m$M$ $MnCl_2$.

*Step 3. Chromatography on a Concanavalin A-Agarose Column.* The dialyzed $(NH_4)_2SO_4$-fractionated enzyme is centrifuged at 15,000 $g$ for 20 min, and the supernatant is diluted with enzyme buffer to a protein concentration of about 20 mg/ml. A sample of about 2000 ml is passed at a flow rate of 50–70 ml/hr through a column of concanavalin A-agarose (concanavalin A-Sepharose 4B, Pharmacia) of size 3.8 × 4.5 cm (about 50 ml), equilibrated with enzyme buffer containing 3 m$M$ $MnCl_2$; the column is washed at the same flow rate with enzyme buffer containing 3 m$M$ $MnCl_2$ and 1 $M$ methyl $\alpha$-D-glucoside until the absorbance of the eluate at 225 nm is about 0.1 (usually about 48 hr). The enzyme is eluted with 1000 ml of enzyme buffer containing 0.3 $M$ methyl $\alpha$-D-glucoside and 60% (v/v) ethylene glycol, and the eluate is dialyzed overnight against enzyme buffer.

*Step 4. Affinity Chromatography on a Collagen-Agarose Column.* The dialyzed eluate from step 3 is applied at a flow rate of 20 ml/hr to a collagen-agarose column (prepared as described in detail below for the purification of galactosylhydroxylysyl glucosyltransferase) of size 1.2 × 18 cm (about 20 ml), equilibrated with enzyme buffer. The column is washed at the same flow rate with 100 ml of enzyme buffer and then eluted with 160 ml of enzyme buffer containing 60% (v/v) ethylene glycol. The fractions are collected and pooled as indicated in Fig. 4 and concentrated to about 2 ml in an Amicon ultrafiltration cell with a PM-30 membrane. After the experiment, the column is regenerated by washing with 6 $M$ urea in a beaker and then equilibrated with enzyme buffer.

*Step 5. Gel Filtration.* The sample is centrifuged at 20,000 $g$ for 10 min, and the clear supernatant is applied to an 8% agarose gel column (BioGel A-1.5m, 200–400 mesh, Bio-Rad) of size 1 × 90 cm, equilibrated, and eluted with enzyme buffer. The fractions containing most of the enzyme activity are pooled and constitute the final purified enzyme.

*Comments on Purification.* The purification of a typical preparation from chick embryo extract is about 13,400-fold (Table V)[86] and that from human placenta extract about 63,000-fold,[87] the recovery being in both cases about 4%. The best preparations obtained from chick embryos[86] or human placenta[87] are pure when examined after reduction and denaturation by SDS–polyacrylamide gel electrophoresis under various conditions. Some preparations contain one or two minor contaminants, but

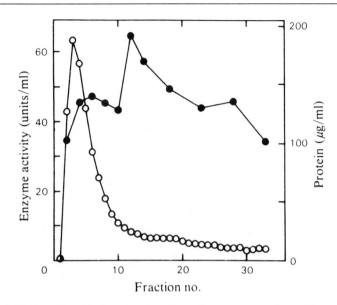

FIG. 4. Elution of lysyl hydroxylase from the collagen-agarose column. The enzyme was eluted with enzyme buffer containing 60% ethylene glycol (see text). ●——●, Enzyme activity; ○——○, protein. Reproduced from Turpeenniemi-Hujanen et al.[86] with permission of the publisher.

these represent less than 5–10% of the total protein. It is not possible to study the enzyme as a native protein by polyacrylamide gel electrophoresis, as most of the enzyme is precipitated at the top of the gel even when the experiments are carried out in the presence of 0.1% (w/v) Triton X-100.

The enzyme activity in the ethylene glycol-containing fractions from the collagen-agarose column is very stable and the fractions can be stored for several months with little, if any, inactivation at −20°, a temperature at which they do not freeze. Such preparations are therefore very useful for experiments requiring highly purified but not pure active enzyme. There are usually large losses of activity during the concentration of the enzyme for the last gel filtration, and the final purified enzyme loses about half of its activity in a few days at 4° or during one cycle of freezing and thawing, whereas this loss does not differ greatly when the enzyme is stored frozen for a few days or months.

*Molecular Properties*

The molecular weight of lysyl hydroxylase isolated by the above procedure from chick embryos,[85,86] or human placenta[87] is about 190,000 by

TABLE V
PURIFICATION OF LYSYL HYDROXYLASE FROM A CHICK EMBRYO $(NH_4)_2 SO_4$ FRACTION[a,b]

| Purification step | Total protein (mg) | Total activity (units) | Recovery (%) | Specific activity (units/mg) | Purification (fold) |
|---|---|---|---|---|---|
| $(NH_4)_2SO_4$, 17–55% satn. | 43,800 | 43,800 | 100 | 1.0 | 2 |
| Concanavalin A-agarose | 91 | 34,200 | 78 | 380 | 760 |
| Collagen agarose | 1.5 | 5010 | 11 | 3340 | 6680 |
| Gel filtration | 0.24 | 1610 | 3.7 | 6700 | 13,400 |

[a] Reproduced from Turpeenniemi-Hujanen et al.,[86] with permission of the publisher.
[b] One unit of enzyme activity is defined as the amount of enzyme present in 1 mg of $(NH_4)_2SO_4$ fraction (17–55% saturation). The purification was calculated on the basis of a specific activity of 0.5 unit/mg for the original 15,000 g supernatant.

gel filtration. The molecular weight of the enzyme monomer by SDS–polyacrylamide gel electrophoresis is about 85,000, suggesting that the active enzyme is a dimer consisting of only one type of subunit.[85–87] The activity of the enzyme partially or highly purified by conventional procedures is found both in gel filtration and in hydroxyapatite chromatography in two main forms, with apparent molecular weights of about 550,000 and 200,000.[32,43] A number of observations suggest that the larger form is not a separate enzyme protein, but represents a polymeric form of the enzyme dimer.[32,43,86] Lysyl hydroxylase is probably a glycoprotein,[43,85] but no carbohydrate analysis has yet been performed.

Purification of a lysyl hydroxylase with quite different molecular properties has been reported from porcine fetal skin.[88] As the specific activities of the porcine fetal skin and chick embryo extracts are very similar,[87] it can be estimated that the specific activity of this purified skin enzyme is about 100 times lower than that of the chick embryo enzyme. The purity and properties of the porcine enzyme thus require additional study.

Antiserum to pure chick embryo lysyl hydroxylase gives a line of identity with the enzyme from a number of chick embryo tissues in immunodiffusion, and similar amounts of the antiserum are required for a 50% inhibition of the same number of enzyme units from various tissues.[89] These observations suggest that the large differences found in the extent of lysyl hydroxylation between different collagen types and within the same collagen type from various tissues[1,6] (Chapters [1] and [2] of this volume) are not due to tissue-specific isoenzymes with significantly differ-

[88] R. L. Miller and H. H. Varner, Biochemistry 18, 5928 (1979).
[89] T. M. Turpeenniemi-Hujanen, Biochem. J. 195, 669 (1981).

ent specific activities. Additional work is nevertheless required to demonstrate either the presence or absence of such isoenzymes.

Purification and Molecular Properties of Prolyl 3-Hydroxylase

Prolyl 3-hydroxylase and prolyl 4-hydroxylase have been demonstrated to be separate enzymes. Pure preparations of prolyl 4-hydroxylase did not catalyze the formation of any 3-hydroxyproline under conditions in which crude tissue extracts would readily do so, and prolyl 3-hydroxylation in crude enzyme preparations was not inhibited by antiserum to prolyl 4-hydroxylase or by poly(L-proline), an effective inhibitor of 4-hydroxylase.[34] The two enzyme activities were also subsequently separated by gel filtration.[10] Prolyl 3-hydroxylase has now been purified up to about 5000-fold from an $(NH_4)_2SO_4$ fraction of chick embryo extract by a procedure consisting of affinity chromatography on denatured collagen linked to agarose and gel filtration,[53] but the enzyme has not yet been isolated as a homogeneous protein.

Purification Procedure[53]

All steps are carried out at 0–4°. The term enzyme buffer is used for a solution containing 0.2 $M$ NaCl, 0.1 $M$ glycine, 1 m$M$ 2-mercaptoethanol, and 50 m$M$ Tris-HCl buffer, pH adjusted to 7.5 at 4°.

*Step 1. Extraction.* The procedure is described for isolating the enzyme from chick embryos. A total of 400 14-day chick embryos are homogenized in batches of 30 in enzyme buffer supplemented with 0.1% (w/v) Triton X-100 (1 ml of solution per gram of embryos) in a Waring blender at full speed twice for 30 sec with a 1-min interval. The homogenate is left to stand with occasional stirring for about 1 hr and then centrifuged at 15,000 $g$ for 30 min.

*Step 2. $(NH_4)_2SO_4$ Fractionation.* Solid $(NH_4)_2SO_4$ is slowly stirred into the supernatant fraction to a final concentration of 55% saturation (351 g/liter). The pellet obtained by centrifugation at 15,000 $g$ for 20 min is dissolved in enzyme buffer, dialyzed for 4 and 12 hr against two separate 20-liter volumes of this solution, centrifuged at 15,000 $g$ for 20 min to remove a small amount of insoluble material, and stored in batches of 100–500 ml at −20°.

*Step 3. Affinity Chromatography on a Collagen-Agarose Column.* A portion of the $(NH_4)_2SO_4$ fraction is thawed and centrifuged at 15,000 $g$ for 20 min to remove insoluble material. The supernatant is diluted with enzyme buffer to a protein concentration of about 20 mg/ml and passed at a flow rate of about 1 column volume per hour through a collagen-agarose col-

## TABLE VI
PURIFICATION OF PROLYL 3-HYDROXYLASE FROM A CHICK EMBRYO $(NH_4)_2SO_4$ PREPARATION [a,b]

| Purification step | Total protein (mg) | Total activity (units) | Recovery (%) | Specific activity (units/mg) |
| --- | --- | --- | --- | --- |
| $(NH_4)_2SO_4$, 0–55% satn | 18,400 | 18,400 | 100 | 1 |
| Collagen-agarose | 6.58 | 8000 | 43 | 1216 |
| Gel filtration | 1.02 | 4360 | 24 | 4270 |

[a] Reproduced from Tryggvason et al.,[53] with permission of the publisher.
[b] One unit of enzyme activity is defined as the amount of enzyme present in 1 mg of $(NH_4)_2SO_4$-precipitated chick embryo extract.

umn with a bed volume of either 6 ml (1 × 8 cm, for up to 150 ml of the sample) or 40 ml (2 × 13 cm, for up to 1000 ml), equilibrated with enzyme buffer. The column is washed with enzyme buffer at the same flow rate until the absorbance of the eluate at 225 nm is less than 0.05, and then eluted with enzyme buffer containing 50% (v/v) ethylene glycol. The fractions containing most of the enzyme activity are pooled. This pool is diluted with 4 volumes of enzyme buffer and concentrated in an Amicon ultrafiltration cell with a PM-30 membrane to a volume of about 2 ml.

*Step 4. Gel Filtration.* The concentrated sample is applied to an 8% agarose gel column (BioGel A-1.5m, 200–400 mesh, Bio-Rad) of size 1.5 × 90 cm, equilibrated, and eluted with enzyme buffer. Fractions containing most of the enzyme activity are pooled and constitute the final purified enzyme.

*Comments on Purification.* The enzyme pool obtained in a typical case has a purification of about 4200-fold over the original $(NH_4)_2SO_4$-fractionated enzyme, the recovery being about 24% (Table VI). The highest specific activity observed in a single fraction from the gel filtration column corresponds to a purification of about 5200-fold. The enzyme is still not pure, however, as six bands are seen when such preparations are examined after reduction and denaturation by SDS–polyacrylamide gel electrophoresis.[53]

### Molecular Properties

The molecular weight of prolyl 3-hydroxylase from chick embryos[53] and rat kidney cortex[10] is about 160,000 by gel filtration, but as the enzyme has not yet been isolated as a homogeneous protein, the subunit structure is not known. Prolyl 3-hydroxylase is probably a glycoprotein, since its activity is inhibited by concanavalin A, and since the enzyme is bound to columns of this lectin coupled to agarose and can be eluted with a buffer containing methyl $\alpha$-D-mannoside.[53]

Catalytic Properties of the Hydroxylases

The main requirements for the peptide substrate are summarized in the overview on hydroxylases (above) and are discussed in detail elsewhere.[2,6] The $K_m$ values for protocollagen, or, in the case of prolyl 3-hydroxylase, for fully 4-hydroxylated protocollagen, are by far the lowest among those for the peptide substrates, being of the order of $10^{-8} M$ for all three enzymes.[17,18,36,90] The very low $K_m$ is probably explained in part by an effect of the chain length (see overview) and in part by a particularly favorable amino acid sequence.[2,6] Interestingly, the $V$ for protocollagen as a substrate of prolyl 4-hydroxylase is not significantly different from that for (Pro-Pro-Gly)$_n$ (see, under Assays, Procedures Using Synthetic Peptide Substrates), even though the $K_m$ values for the latter peptides are markedly higher.[2,6,25,56] The $K_m$ values for the two standard peptide substrates of prolyl 4-hydroxylase, (Pro-Pro-Gly)$_5$ and (Pro-Pro-Gly)$_{10}$, are about 150–250 $\mu M$ and 15–20 $\mu M$, respectively, when determined from initial velocity data obtained by varying two of the reaction components.[91]

The $K_m$ for peptide L-I as a substrate for lysyl hydroxylase (see above), when determined from similar plots, is about 400–500 $\mu M$.[52]

The pH optima of these hydroxylases are at about 7.4.[13,31,32,34] The kinetic constants of the three enzymes for their cosubstrates are very similar (Table VII), and the turnover numbers of about 200–400 mol/mol per minute determined for pure prolyl 4-hydroxylase[19,26,31,56,57,70] and lysyl hydroxylase[86,87] from various sources are likewise similar. Several bivalent cations, citric acid cycle intermediates, and certain other compounds inhibit the hydroxylations competitively with respect to some of the cosubstrates,[1,6] the $K_i$ values of these compounds for prolyl 4-hydroxylase and lysyl hydroxylase again being very similar.[43,92,93]

The hydroxylation mechanisms have been examined by carrying out extensive kinetic studies on the prolyl 4-hydroxylase[91,92,94] and lysyl hydroxylase[52,93] reactions. Initial velocity data were analyzed by varying the concentration of one substrate (the cosubstrates are also termed here

[90] L. Ryhänen and K. I. Kivirikko, *Biochim. Biophys. Acta* **343**, 129 (1974).
[91] R. Myllylä, L. Tuderman, and K. I. Kivirikko, *Eur. J. Biochem.* **80**, 349 (1977).
[92] L. Tuderman, R. Myllylä, and K. I. Kivirikko, *Eur. J. Biochem.* **80**, 341 (1977).
[93] U. Puistola, T. M. Turpeenniemi-Hujanen, R. Myllylä, and K. I. Kivirikko, *Biochim. Biophys. Acta* **611**, 51 (1980).
[94] R. Myllylä, E.-R. Kuutti-Savolainen, and K. I. Kivirikko, *Biochem. Biophys. Res. Commun.* **83**, 441 (1978).

TABLE VII
APPARENT $K_m$ VALUES FOR THE COSUBSTRATES OF THE THREE HYDROXYLASES FROM CHICK EMBRYOS[a]

| Cosubstrate | Prolyl 4-hydroxylase | | Lysyl hydroxylase | | Prolyl 3-hydroxylase |
| --- | --- | --- | --- | --- | --- |
| | Protocollagen substrate[b] ($\mu M$) | Synthetic substrate[c] ($\mu M$) | Protocollagen substrate[d] ($\mu M$) | Synthetic substrate[e] ($\mu M$) | Protocollagen substrate (fully 4-hydroxylated)[f] ($\mu M$) |
| $Fe^{2+}$ | 2 | 4 | 2 | 3 | 2 |
| 2-Oxoglutarate | 5 | 22 | 70 | 100 | 3 |
| $O_2$ | ND[g] | 43 | ND | 45 | 30 |
| Ascorbate | 100 | 300 | 220 | 200 | 120 |

[a] Reproduced from Kivirikko and Myllylä,[6] with permission of the publisher.
[b] From Kivirikko and Prockop.[17]
[c] From Myllylä et al.[91]
[d] From U. Puistola, Biochem. J., in press.
[e] From Puistola et al.[52]
[f] From Tryggvason et al.[53]
[g] ND, not determined.

substrates) in the presence of different fixed concentrations of the second, while the concentrations of the other substrates were held constant. Intersecting lines were obtained in double-reciprocal plots for all possible pairs involving $Fe^{2+}$, 2-oxoglutarate, $O_2$, and the peptide substrate with both enzymes, as shown for $Fe^{2+}$ and 2-oxoglutarate in Fig. 5A. In contrast, parallel lines were obtained for pairs comprising ascorbate and each of the other substrates, as shown for ascorbate and the peptide substrate in Fig. 5B. Additional kinetic data consisted of inhibition patterns with the reaction products and several other inhibitors.

The kinetic data indicate that the mechanisms of the prolyl 4-hydroxylase and lysyl hydroxylase reactions are probably identical. The kinetic [52,91-94] and certain other data[6] are consistent with a reaction scheme (Fig. 6A) involving an ordered binding of $Fe^{2+}$, 2-oxoglutarate, $O_2$, and the peptide substrate to the enzyme in this order, and an ordered release of the hydroxylated peptide, $CO_2$, succinate, and $Fe^{2+}$, in which $Fe^{2+}$ need not leave the enzyme between the catalytic cycles and in which the order of release of the hydroxylated peptide and $CO_2$ is uncertain. The oxygen must evidently be activated, probably to superoxide.[92,93,95,96] In the absence of the peptide substrate, both enzymes catalyze an uncoupled de-

[95] R. S. Bhatnagar and T. Z. Liu, FEBS Lett. 26, 32 (1972).
[96] R. Myllylä, L. M. Schubotz, U. Weser, and K. I. Kivirikko, Biochem. Biophys. Res. Commun. 89, 98 (1979).

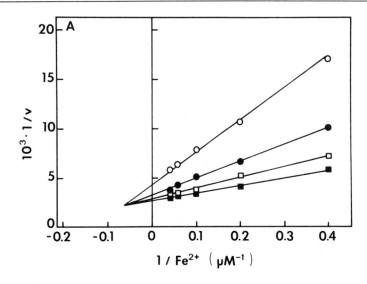

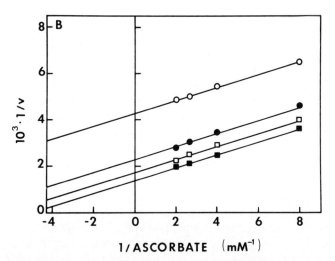

FIG. 5. Effect of the concentration of $Fe^{2+}$ (A) or ascorbate (B) on the rate of the prolyl 4-hydroxylase reaction at different fixed concentrations of 2-oxoglutarate (A) or (Pro-Pro-Gly)$_{10}$ (B) with constant concentrations of the other components. (A) The concentrations of 2-oxoglutarate were 0.02 m$M$ (○), 0.04 m$M$ (●), 0.06 m$M$ (□), 0.08 m$M$ (■). (B) The concentrations of the polypeptide substrate were: 0.01 mg/ml (○), 0.02 mg/ml (●), 0.03 mg/ml (□), 0.04 mg/ml (■). The concentrations of the other components were constant: $O_2$ 200 μ$M$, (Pro-Pro-Gly)$_{10}$ 0.1 mg/ml (A), ascorbate 1 m$M$ (A), $Fe^{2+}$ 0.05 m$M$ (B), 2-oxoglutarate 0.1 m$M$ (B). Reaction velocity ($v$) was measured in disintegrations per minute. Reproduced from Myllylä et al.,[91] with permission of the publisher.

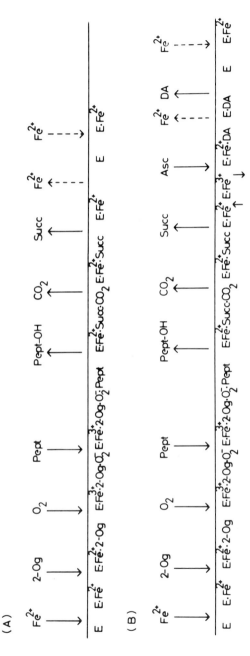

FIG. 6. Schematic representation of the mechanism for the prolyl 4-hydroxylase and lysyl hydroxylase reactions. It is suggested that the enzymes operate according to the scheme (A) for a number of catalytic cycles, but at some stage the reaction ceases, probably owing to oxidation of either the enzyme · $Fe^{2+}$ complex (as shown by the arrows below the line) or free enzyme (not shown), and ascorbate is then required to reduce this oxidized form (B). The order of release of the hydroxylated peptide and $CO_2$ in both schemes is uncertain, and it is possible that the reaction with ascorbate in scheme (B) occurs only after the release of $Fe^{2+}$. The dashed lines between the enzyme and $Fe^{2+}$ indicate that the latter need not leave the enzyme during each catalytic cycle. Under some conditions the enzymes can also form certain dead-end complexes that are not shown. E, enzyme; 2-Og, 2-oxoglutarate; Pept, peptide substrate; Succ, succinate; Asc, ascorbate; DA, dehydroascorbate. Reproduced from Kivirikko and Myllylä[6] with permission of the publisher.

carboxylation of 2-oxoglutarate at a low but significant rate.[3,31,52,92] The uncoupled reaction requires the same cosubstrates as the complete reaction,[52,92,97] and its rate is increased by the presence of competitive polypeptide inhibitors,[97,98] but the details of this reaction are unknown.

Ascorbate is a quite specific requirement for pure prolyl 4-hydroxylase and highly purified lysyl hydroxylase, but dithiothreitol, L-cysteine, and some reduced pteridines in high concentrations can in part replace this vitamin with some,[52,94] but not all,[92,99] enzyme preparations. A low hydroxylation rate has also been observed in several cultured cells in the complete absence of ascorbate,[1,6] and data have been presented suggesting that this is due to an unidentified reductant rather than cysteine or tetrahydropteridine.[14] Ascorbate is not consumed stoichiometrically during the hydroxylation,[92,99] and some,[52,94] but not all,[99] preparations of the two hydroxylases can catalyze their reactions for a number of catalytic cycles in the complete absence of this vitamin. These findings suggest that the enzyme can operate according to the scheme of Fig. 6A for a number of catalytic cycles. The kinetic data[91,93,94] suggest that the reaction with ascorbate takes place either after the release of the hydroxylated peptide, $CO_2$, and succinate or after the release of all products, including $Fe^{2+}$ (Fig. 6B), and it thus seems likely that this vitamin is required to prevent oxidation of either the enzyme–iron complex or the free enzyme during some catalytic cycles, but not the majority.[91,93,94,99] Additional studies are nevertheless required to elucidate the precise role of ascorbate in the prolyl and lysyl hydroxylase reactions.

## An Overview on Hydroxylysyl Glycosyltransferases

The only carbohydrate units in mammalian interstitial collagens are found in O-glycosidic linkage to hydroxylysyl residues. Some of the carbohydrate is present as the monosaccharide galactose and some as the disaccharide glucosylgalactose, the structure of the disaccharide unit with its peptide attachment (Fig. 7) being 2-O-α-D-glucopyranosyl-O-β-D-galactopyranosylhydroxylysine (see this volume [14]). The extent of glycosylation of the hydroxylysyl residues varies markedly between the collagen types (this volume [2] and [14]), and even in the same collagen type from different tissues and in the same tissue in many physiological and pathological states.[7]

The functions of these carbohydrate units are uncertain. Since they form the most extrusive groups of the collagen molecule, it has been

---

[97] D. F. Counts, G. J. Cardinale, and S. Udenfriend, *Proc. Natl. Acad. Sci. U.S.A.* **75**, 2145 (1978).

[98] N. V. Rao and E. Adams, *J. Biol. Chem.* **253**, 6327 (1978).

[99] J. J. Nietfeld and A. Kemp, *Biochim. Biophys. Acta* **657**, 159 (1981).

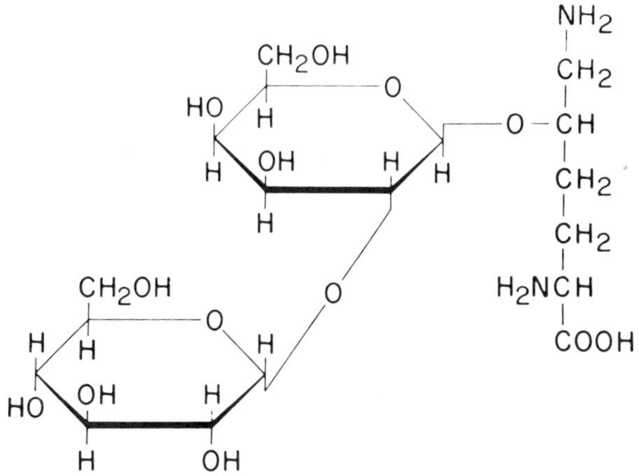

FIG. 7. The structure of 2-$O$-α-D-glucopyranosyl-$O$-β-D-galactopyranosylhydroxylysine.

suggested that they may have some role in controlling the organization of the fibrils. This is supported by data indicating that there is an inverse relationship between carbohydrate content and fibril diameter,[100] and that native fibrils of type II collagen are more swollen with water than those of type I.[101] Hydroxylysyl residues participating in intermolecular collagen cross-links can be glycosylated,[7] but it is not known whether this has any effect on cross-link formation or the properties of the cross-links.

These carbohydrate prosthetic groups are formed as posttranslational modifications catalyzed by two specific enzymes, first hydroxylysyl galactosyltransferase (UDPgalactose-collagen galactosyltransferase, UDPgalactose:5-hydroxylysine-collagen galactosyltransferase, EC 2.4.1.50), transferring galactose to hydroxylysyl residues, and then galactosylhydroxylysyl glucosyltransferase (UDPglucose-collagen glucosyltransferase, UDPglucose:5-hydroxylysine-collagen glucosyltransferase, EC 2.4.1.66), transferring glucose to galactosylhydroxylysyl residues (for terminology, see Kivirikko and Myllylä[7]). The preferential sugar donor for both transferases is the corresponding UDPglycoside, and both reactions require a bivalent cation, this requirement being best fulfilled by manganese.[7]

Hydroxylysine is formed only as a posttranslational modification (see above), and thus it is apparent that the synthesis of the carbohydrate units *in vivo* must take place on a peptide substrate. Free hydroxylysine does not serve as the carbohydrate acceptor for the galactosyltransferase, whereas free galactosylhydroxylysine does serve *in vitro* as a substrate for

[100] P. J. Morgan, H. G. Jacobs, J. P. Segrest, and L. W. Cunningham, *J. Biol. Chem.* **245**, 5042 (1970).
[101] D. R. Eyre, *Science* **207**, 1315 (1980).

the glucosyltransferase (for references on substrate requirements, see Kivirikko and Myllylä[7]). The presence of the free $\epsilon$-amino group in the hydroxylysyl residues seems to be an absolute requirement for both enzymes, as its N-acetylation or deamination entirely prevents the transfer of galactose to hydroxylysyl residues and of glucose to galactosylhydroxylysyl residues. The interaction with both transferases is further influenced by the peptide chain length of the substrate, the amino acid sequence, and the conformation. The effect of chain length is seen in that longer peptides constitute better substrates than shorter ones. The number of -X-Hyl-Gly- triplets in a polypeptide chain is an important factor for the overall glycosylation of the peptide,[102] whereas the effect of the amino acid sequence around such a hydroxylysyl residue may be only a relative factor, as the great majority of the hydroxylysyl residues have a disaccharide unit in some collagens. The reaction with both hydroxylysyl glycosyltransferases is entirely prevented by the triple-helical conformation in the substrate, indicating that these two glycosylations, like the three hydroxylations (above) must occur in the cell before triple helix formation.[103-106]

In intact cells the hydroxylysyl glycosylations are initiated while the polypeptide chains are still being assembled on the ribosomes,[107,108] there being no significant lag between the hydroxylation of lysyl residues and the glycosylation of hydroxylysyl residues.[109] These reactions, like the three hydroxylations, are probably continued, however, after the release of complete pro-$\alpha$ chains into the cisternae of the rough endoplasmic reticulum until formation of the procollagen triple helix prevents any further glycosylation.[7,109-111] Most of the two hydroxylysyl glycosyltransferase activities in subcellular fractions are associated with the rough endoplasmic reticulum, but significant amounts have also been found in the smooth endoplasmic reticulum.[108] This distribution agrees with the location of the reactions summarized above, but the significance of the activity in the smooth endoplasmic reticulum remains to be determined. Small amounts of galactosylhydroxylysyl glucosyltransferase are also

[102] H. Anttinen and A. Hulkko, *Biochim. Biophys. Acta* **632**, 417 (1980).
[103] R. Myllylä, L. Risteli, and K. I. Kivirikko, *Eur. J. Biochem.* **58**, 517 (1975).
[104] L. Risteli, R. Myllylä, and K. I. Kivirikko, *Biochem. J.* **155**, 145 (1976).
[105] S. Menashi, R. Harwood, and M. E. Grant, *Nature (London)* **264**, 670 (1976).
[106] H. Anttinen, L. Tuderman, A. Oikarinen, and K. I. Kivirikko, *Blood* **50**, 29 (1977).
[107] A. G. Brownell and A. Veis, *Biochem. Biophys. Res. Commun.* **63**, 371 (1975).
[108] R. Harwood, M. E. Grant, and D. S. Jackson, *Biochem. J.* **152**, 291 (1975).
[109] A. Oikarinen, H. Anttinen, and K. I. Kivirikko, *Biochem. J.* **156**, 545 (1976).
[110] A. Oikarinen, H. Anttinen, and K. I. Kivirikko, *Biochem. J.* **160**, 639 (1976).
[111] V.-J. Uitto, J. Uitto, W. W.-Y. Kao, and D. J. Prockop, *Arch. Biochem. Biophys.* **185**, 214 (1978).

present in human serum,[112] and assays of this serum enzyme activity may give useful information on actual hepatic collagen synthesis in patients with liver diseases[112,113] or with some other diseases affecting the collagen-producing cells.[65,112]

Assays of Hydroxylysyl Glycosyltransferase Activities

Principles and Comparison of Various Types of Method

All procedures available for determination of the hydroxylysyl glycosyltransferase activities are based on the transfer of radioactively labeled galactose or glucose from the corresponding UDPglycoside to hydroxylysyl or galactosylhydroxylysyl residues in peptide linkages or, in the case of one assay for the glucosyltransferase, to free galactosylhydroxylysine.[7] Differences exist in the acceptor substrates used and in the assay of the reaction products.

The substrates used in the early methods consisted of citrate-soluble guinea pig skin collagen from which the sugars had been removed before determination of the galactosyltransferase activity[114] or the glucose had been removed before determination of glucosyltransferase activity.[115] After the enzyme reaction, the protein of the incubation mixture was precipitated and the radioactivity of the precipitate was measured. These procedures are rapid and easy to carry out, but unfortunately very unspecific with crude enzyme samples at least, as has been demonstrated in subsequent investigations.[116-118] The specificity can be considerably increased, however, by precipitating the protein on a glass fiber disk, which is washed several times before its radioactivity is determined.[119]

Spiro and Spiro[116,117] developed procedures in which the substrates are prepared from bovine renal glomerular basement membrane, but several other sources can also be used. The carbohydrate units are removed before assay of the galactosyltransferase activity, and the glucose is removed before assay of the glucosyltransferase activity. The hydroxylysine glycoside formed in the reaction is liberated by alkaline hydroly-

[112] H. Anttinen, *Clin. Chim. Acta* **77**, 323 (1977).
[113] E.-R. Kuutti-Savolainen, H. Anttinen, T. A. Miettinen, and K. I. Kivirikko, *Eur. J. Clin. Invest.* **9**, 97 (1979).
[114] H. B. Bosmann and E. H. Eylar, *Biochem. Biophys. Res. Commun.* **33**, 340 (1968).
[115] H. B. Bosmann and E. H. Eylar, *Biochem. Biophys. Res. Commun.* **30**, 89 (1968).
[116] M. J. Spiro and R. G. Spiro, *J. Biol. Chem.* **246**, 4910 (1971).
[117] R. G. Spiro and M. J. Spiro, *J. Biol. Chem.* **246**, 4899 (1971).
[118] R. Myllylä, L. Risteli, and K. I. Kivirikko, *Eur. J. Biochem.* **52**, 401 (1975).
[119] K. E. Draeger and K. U. Weithmann, *Diabetologia* **15**, 125 (1978).

sis and purified further by a brief ion-exchange chromatography and by paper chromatography of 6 days' duration. These methods are highly specific but very slow, and for the latter reason they have been used in only a few studies.

Myllylä et al.[118] reported assays for both transferases, involving gelatinized insoluble calf skin collagen as a standard substrate, although several others can also be used.[103,104,120] The enzymic reaction is terminated by precipitation of the protein, and most of the radioactivity present in the free UDPglycoside is removed by several washings of the precipitate. The hydroxylysine glycoside is then liberated by alkaline hydrolysis and purified further by a brief ion-exchange chromatography and paper electrophoresis. The ion-exchange chromatography differs from that used by Spiro and Spiro[116,117] in that a preliminary elution of the column with citrate–phosphate buffer (pH 3) is introduced. This was found to be an important step, as large amounts of nonspecific products are removed. This procedure is entirely specific with respect to assays of the final products in an amino acid analyzer (Fig. 8).

In initial studies with the above method, precipitation of the protein and washing of the precipitate were carried out with either a phosphotungstic acid or an acetone procedure, but subsequent studies have indicated that the former procedure is considerably more effective in removing the radioactivity present in the free UDPglycoside. As a consequence, the method[118] can be shortened for a number of biological samples by omitting the final electrophoresis. Such a shortened procedure is entirely specific for both transferases after they have been purified about 10- to 20-fold from chick embryo extract (Table VIII). The glucosyltransferase activity assays have a specificity of 90% or more in all biological specimens studied, except human serum, where it is about 80% (Table VIII). Even in the latter case, however, the shortened method is preferred, as the summing of values from several paper electrophoresis fractions also introduces a significant inaccuracy. The galactosyltransferase activity assays have a specificity of about 80% with specimens from most sources, but only about 50% in human platelet extract and no specificity in human serum (Table VIII). It thus seems that although the shortened procedure can be used in most instances, the specificity should always be determined using an amino acid analyzer before the method is applied to new biological samples.

A procedure very similar to that described above uses as a substrate bovine anterior lens capsule collagen from which the appropriate carbohydrate units have been removed.[108] The enzymic reactions are termi-

[120] R. Myllylä, *Eur. J. Biochem.* **70**, 225 (1976).

[10] COLLAGEN BIOSYNTHESIS: INTRACELLULAR ENZYMES 289

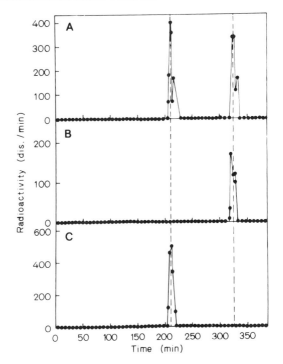

FIG. 8. Specificity of the hydroxylysyl glycosyltransferase assays of Myllylä et al.[118] when determined by studying the reaction products using an amino acid analyzer. (A) Standards of glucosylgalactosylhydroxylysine and galactosylhydroxylysine; their positions are indicated in the other parts of the figure by vertical broken lines. (B) Hydroxylysyl galactosyltransferase and (C) galactosylhydroxylysyl glucosyltransferase reaction, including purification of the product by all steps in the procedure. Reproduced with permission of the publisher.

nated by precipitation of the protein with acetone, and, after several washings, the hydroxylysine glycosides are liberated by alkaline hydrolysis. They are then purified further by ion-exchange chromatography, not involving a citrate–phosphate buffer elution step (above), and by paper electrophoresis. This method is entirely specific, but unlike the procedure of Myllylä et al.,[118] the use of paper electrophoresis is necessary with most specimens.[108,121]

Galactosylhydroxylysyl glucosyltransferase activity can also be determined by using free galactosylhydroxylysine as a substrate.[122] The en-

[121] S. Menashi and M. E. Grant, Biochem. J. **178**, 777 (1979).
[122] D. F. Smith, D. P. Kosow, C. Wu, and G. A. Jamieson, Biochim. Biophys. Acta **483**, 263 (1977).

## TABLE VIII
SPECIFICITY OF THE METHOD OF MYLLYLÄ et al.[a] WITHOUT PAPER ELECTROPHORESIS FOR THE ASSAY OF HYDROXYLYSYL GLYCOSYLTRANSFERASE ACTIVITIES IN VARIOUS BIOLOGICAL SPECIMENS

| Source of enzyme | Percentage of measured radioactivity as the specific product[b] | |
|---|---|---|
| | Galactosylhydroxylysyl glucosyltransferase | Hydroxylysyl galactosyltransferase |
| Partially purified glucosyltransferase or galactosyltransferase from chick embryos[c] | 100 | 100 |
| Chick embryo extract[d] | >95 | 90 |
| Rat embryo bone extract[e] | 95 | 80 |
| Rat liver extract[f] | >95 | 80 |
| Rat kidney extract[g] | >95 | >80 |
| Rat peritoneal macrophages[h] | 90 | 85 |
| Human cultured skin fibroblasts[i] | 90 | 85 |
| Human skin extract[j] | 95 | 75 |
| Human platelet extract[k] | 95 | 50 |
| Human serum[l] | 80 | 0 |

[a] The procedure is carried out as described by Myllylä et al.[118] using protein precipitation with phosphotungstic acid and omitting the final paper electrophoresis.
[b] The specificity of the procedure was established by assaying the final reaction product in an amino acid analyzer as shown in Fig. 8.
[c] From Myllylä et al.[125] and Risteli et al.[104]
[d] From L. Risteli, Biochim. Biophys. Acta **497**, 673 (1977).
[e] From R. Myllylä, K. Tryggvason, K. I. Kivirikko, and A. H. Reddi, Biochim. Biophys. Acta **674**, 238 (1981).
[f] From J. Risteli and K. I. Kivirikko, Biochem. J. **158**, 361 (1976).
[g] From Risteli et al.[128]
[h] From R. Myllylä and H. Seppä, Biochem. J. **182**, 311 (1979).
[i] From E.-R. Savolainen, unpublished results.
[j] From H. Antinnen, A. Oikarinen, and K. I. Kivirikko, Clin. Chim. Acta **76**, 95 (1977).
[k] From Anttinen et al.[106]
[l] From Anttinen.[112]

zyme incubation is terminated by protein precipitation, and the product, present in the supernatant, is purified by an ion-exchange chromatography. This procedure likewise has a high specificity.

Extraction of Hydroxylysyl Glycosyltransferase Activities from Crude Tissue or Cell Specimens

Part of the hydroxylysyl glycosyltransferase activity can be found in the soluble fraction of the tissue or cell homogenate without the use of any detergent, but large increases in both enzyme activities occur after incubation with these substances.[42,104,108,123-125] A relatively high salt concentration is also required.[126,127] This has been shown, for example, in a study in which the values for the hydroxylsyl glycosyltransferase activities in rat kidney were compared after extraction of the enzymes with three different solutions reported by various laboratories.[128] Galactosyltransferase activity is easily inactivated in crude tissue extracts,[127] whereas glucosyltransferase activity is quite stable. No loss of the latter enzyme activity occurs in human liver extract or serum when frozen and thawed three or five times, respectively, or in liver biopsy specimens stored at $-70°$ or the serum stored at $-20°$ for one year.[113]

Our laboratory now uses the same procedure for the extraction of hydroxylysyl glycosyltransferase activities as that described above for the extraction of the three hydroxylase activities. Thus all five enzymes can be assayed in the same extract. When the hydroxylases are not assayed, the glycine is omitted from the extraction buffer.

Procedure for the Assay of Hydroxylysyl Glycosyltransferase Activities[118]

*Preparation of Collagen Substrate*

Gelatinized insoluble calf skin collagen is prepared as follows. The hair is removed, and 50 g of the skin are first minced with scissors and then homogenized with an Ultra-Turrax homogenizer in 500 ml of a cold solution containing 0.15 $M$ NaCl and 0.05 $M$ Tris-HCl buffer, pH adjusted to 7.4 and 4°. The homogenate is extracted with shaking overnight at 4° and centrifuged at 5000 $g$ for 10 min. The residue is washed with 500 ml of $H_2O$, with 500 ml of acetone for 3 hr at room temperature, and with 500 ml of ether, the pellet being separated by centrifugation between each washing. The material is then dried at room temperature, gelatinized with 50 ml of $H_2O$ at 120° overnight, and filtered; the gelatin is stored in aliquots at $-20°$. Amino acid analyses of this gelatin agree within 5% with that for pure calf skin collagen.

Among the alternative substrates, one can use citrate-soluble rat skin collagen, obtained as described below for the preparation of the collagen-

[123] R. G. Spiro and M. J. Spiro, *J. Biol. Chem.* **246**, 4919 (1971).
[124] H. Anttinen, *Biochem. J.* **160**, 29 (1976).
[125] R. Myllylä, L. Risteli, and K. I. Kivirikko, *Eur. J. Biochem.* **61**, 59 (1976).
[126] R. Myllylä, H. Anttinen, L. Risteli, and K. I. Kivirikko, *Biochim. Biophys. Acta* **480**, 113 (1977).
[127] L. Risteli, *Biochem. J.* **169**, 189 (1978).
[128] J. Risteli, V. Koivisto, H. Åkerblom, and K. I. Kivirikko, *Diabetes* **25**, 1066 (1976).

agarose column for isolation of galactosylhydroxylysyl glucosyltransferase.

*The Glycosylation Reactions*

*Reagents*

Collagen substrate, prepared as described above. The substrate is diluted with $H_2O$ to a protein concentration of 80 mg/ml, heated to 60° for 10 min, and rapidly cooled to 0° immediately before use.[103,104]

Uridine diphosphate-D-[$^{14}$C]glucose, mixed daily with the unlabeled compound to give a final concentration of 1.5 m$M$ with a specific activity of 3.3 Ci/mol. In cases of samples with low transferase activity, the specific activity is increased 5- or 10-fold, and in cases of assays in serum samples the UDPglucose concentration is reduced to 0.75 m$M$ with a specific activity of 33 Ci/mol.

Uridine diphosphate-D-[$^{14}$C]galactose, mixed daily with the unlabeled compound to give final concentration of 1.5 m$M$ with a specific activity of 3.3 Ci/mol. This specific activity can likewise be varied (see UDPglucose, above).

$Mn^{2+}$–dithiothreitol–albumin solution, prepared daily by adding solid $MnCl_2$, final concentration 16.7 m$M$, dithiothreitol, final concentration 1 m$M$, and bovine serum albumin, final concentration 2 mg/ml, to an aliquot of 416 m$M$ Tris-HCl buffer, pH adjusted to 7.7 at room temperature.

Reaction stock solution. The above solutions are mixed in the following proportions: 44 µl of collagen substrate, 4 µl of UDP[$^{14}$C]glucose or UDP[$^{14}$C]galactose, and 12 µl of $Mn^{2+}$-dithiothreitol–albumin.

Enzyme standard, crude ammonium sulfate-precipitated chick embryo enzyme, obtained in the first step of the purification procedures described below, and stored in aliquots at −20° to avoid repeated freezing and thawing.

Phosphotungstic acid, 1% in 0.5 $M$ HCl

*Incubation with the Enzyme.* Small test tubes cooled in iced water each receive an appropriate addition of cold $H_2O$ to make a final volume of 100 µl after the subsequent additions. The following additions are then made: 60 µl of the reaction stock solution and a suitable aliquot of the enzyme to be assayed or the enzyme standard or $H_2O$. The tubes are transferred from the iced water to a water bath at 37° and incubated with shaking for 45 min. The reaction is stopped by transferring the tubes to iced water again and rapidly adding 2 ml of the phosphotungstic acid solution.

## Assay of the Reaction Product

### Reagents

Trichloroacetic acid, 10%
Ethanol–ether (1:1)
NaOH, 2 $M$
HCl, 2 $M$
Citrate–phosphate buffer, 0.1 $M$, pH 3
$NH_4OH$, 1.5 $M$
Pyridine–acetate buffer, 30 m$M$, pH 3.5
Dowex 50-X8, $H^+$ form
Scintillation solution, e.g., Instagel (Packard) or Aquasol (New England Nuclear)

*Procedure.* After addition of the phosphotungstic acid solution, the samples are centrifuged at 4000 $g$ for 10 min and the pellets are washed twice with trichloroacetic acid and once with ethanol–ether, the pellet being separated by centrifugation between each washing. The pellets are then dried and hydrolyzed with 1.5 ml of NaOH in sealed tubes at 105° for 18 hr. The hydrolyzates are diluted with 4 ml of $H_2O$ and titrated with HCl to pH 3. This pH adjustment must be carried out rapidly to avoid gel formation in the sample. The samples are then placed on small columns containing 1 g of Dowex 50-X8, and the columns are washed with 7 ml of citrate–phosphate buffer and 10 ml of $H_2O$ and eluted with 7 ml of $NH_4OH$. The eluates are evaporated to dryness on a steam bath and dissolved in 0.8 ml of pyridine–acetate buffer, and the radioactivity is determined with 5 ml of scintillation solution in a liquid scintillation spectrometer.

In cases requiring the paper electrophoresis step (see above), the evaporated column eluates are dissolved in 0.3 ml of 0.1 $M$ pyridine–acetate buffer, and aliquots of 50 $\mu$l are streaked on 4 × 30 cm strips of Whatman No. 1 paper. The electrophoresis is carried out in 0.1 $M$ pyridine–acetate buffer, pH 3.5, at 300 V for 3.5 hr. The strips are dried and cut into 5-cm fractions between 4 and 10 cm from the origin. The radioactivity of these fractions is measured in 5 ml of the scintillant used for counting paper strips in assays of the hydroxylase activities based on the decarboxylation of 2-oxo[1-$^{14}$C]glutarate (above).

### Definition of the Units of Enzyme Activity

One unit of enzyme activity is defined for both transferases as the activity required to synthesize an amount of the radioactive product (in disintegrations per minute) corresponding to 1 $\mu$mol in 1 min at 37° with the reactant concentrations indicated above and a saturating concentration of denatured purified citrate-soluble rat skin collagen as substrate.[104,125] Most of the assays are carried out with 35 mg of the gelatinized insoluble calf skin collagen substrate per milliliter, as this substrate can easily be prepared in large quantities. The product formation with this substrate is corrected to a saturating concentration using values obtained from double-reciprocal plots of substrate concentration and reaction velocity, such plots being prepared with each new batch of the substrate. The $V$ obtained is further compared with the $V$ given by denatured citrate-soluble rat skin collagen in the same experiment.

## Purification and Properties of the Hydroxylysyl Glycosyltransferases

### Purification and Molecular Properties of Galactosylhydroxylysyl Glucosyltransferase

Galactosylhydroxylysyl glucosyltransferase has been isolated from chick embryos as a homogeneous protein by two affinity column procedures. The first[126] consists of ammonium sulfate fractionation, affinity chromatographies on collagen-agarose and UDPglucose-derivative-agarose, and two gel filtrations. The second procedure[129] involves ammonium sulfate fractionation, affinity chromatographies on concanavalin A–agarose, collagen-agarose, and UDPglucose-derivative–agarose, and one gel filtration. The collagen–agarose[130] and UDPglucose-derivative–agarose[131] columns were initially eluted with small peptides prepared from collagen, but subsequently elution with ethylene glycol was introduced,[132] and this is now used in both procedures.

A high degree of purification from chick embryos has also been obtained by a procedure consisting of six conventional protein purification steps.[125] Considerably lower degrees of purification have been reported from several other sources, including rat kidney cortex,[117] chick embryo cartilage,[125] guinea pig skin,[115] bovine arterial tissue,[133] human fetal tissues,[132] and human plasma[134] and platelets.[134-136]

The procedure[129] involving three affinity columns is described in detail below.

[129] H. Anttinen, R. Myllylä, and K. I. Kivirikko, *Biochem. J.* **175**, 737 (1978).
[130] L. Risteli, R. Myllylä, and K. I. Kivirikko, *Eur. J. Biochem.* **67**, 197 (1976).
[131] H. Anttinen and K. I. Kivirikko, *Biochim. Biophys. Acta* **429**, 750 (1976).
[132] H. Anttinen, R. Myllylä, and K. I. Kivirikko, *Eur. J. Biochem.* **78**, 11 (1977).
[133] W. Henkel and E. Buddecke, *Hoppe Seyler's Z. Physiol. Chem.* **356**, 921 (1975).
[134] J. C. Leunis, D. F. Smith, N. Nwokora, B. L. Fishback, C. Wu, and G. A. Jamieson, *Biochim. Biophys. Acta* **611**, 79 (1980).
[135] A. J. Barber and G. A. Jamieson, *Biochim. Biophys. Acta* **252**, 533 (1971).
[136] H. B. Bosmann, *Biochem. Biophys. Res. Commun.* **43**, 1118 (1971).

## Preparation of Collagen–Agarose[130]

Citrate-soluble collagen is prepared from rat skin as follows.[137] The hair is removed and 100 g of the skin are first minced with scissors and then homogenized with an Ultra-Turrax homogenizer in 400 ml of cold 0.5 $M$ sodium acetate. The homogenate is mixed for about 15–18 hr at 4° and centrifuged at 5000 $g$ for 30 min; the pellet is extracted with another 400 ml of 0.5 $M$ sodium acetate as above. The pellet is washed twice with 400 ml of $H_2O$ and extracted for 15–18 hr with 400 ml of 75 m$M$ sodium citrate buffer, pH 3.7. The supernatant obtained after centrifugation at 15,000 $g$ for 30 min is exhaustively dialyzed against 20 m$M$ $Na_2HPO_4$, and the precipitate formed during dialysis is collected by centrifugation at 15,000 $g$ for 30 min. This precipitate is washed twice with $H_2O$ and lyophilized. A portion of 2 g is suspended in 100 ml of $H_2O$ and heated to 100° for 10 min. The insoluble residue is removed by centrifugation at 5000 $g$ for 10 min at room temperature, and the supernatant is cooled to 0° and used immediately for the coupling reaction.

This collagen is coupled to agarose by the cyanogen bromide activation technique,[71,130] using exactly the same procedure as described above for the coupling of poly(L-proline) to agarose. After the coupling reaction, the mixture is stirred at 4° for 20 hr, as described for poly(L-proline)–agarose. The gel is then washed first with large volumes of cold $H_2O$, then three times with 0.1 $M$ acetic acid, and finally with a solution consisting of 0.12 $M$ NaCl, 2 m$M$ $MnCl_2$, 50 $\mu M$ dithiothreitol, 1% (v/v) glycerol, and 50 m$M$ Tris-HCl buffer, pH adjusted to 7.4 at 4°. The gel is poured into a column (1.7 × 17 cm) with a bed volume of about 40 ml, and the washing is continued with the same solution.

The efficiency of the coupling reaction is measured using an amino acid analyzer, as described for poly(L-proline)–agarose (above). The collagen content should be at least about 2–4 mg per milliliter of packed gel.

## Preparation of UDPGlucose-Derivative–Agarose[131]

UDPglucuronic acid is coupled in its carboxyl group to free amino groups of aminohexyl-agarose (AH-Sepharose 4B, Pharmacia) using the carbodiimide method.[71] The AH-Sepharose 4B is allowed to swell for 6 hr in 0.5 $M$ NaCl and washed with large volumes of 0.5 $M$ NaCl and $H_2O$. Then 20 ml of the packed gel are suspended in distilled water to give a volume of 30 ml, and 450 mg of sodium UDPglucuronate, dissolved in 1.5 ml of $H_2O$, are added. The pH is adjusted to 4.7 by adding 0.2 $M$ and 0.05 $M$ HCl, and 1.5 g of 1-ethyl-3-(3-dimethylaminopropyl)carbodiimide is added in 4 ml of $H_2O$ over a period of 15 min. The pH is kept between 4.7 and 4.8 during this period, and pH control is continued for 2 hr. The

[137] P. M. Gallop and S. Seifter, this series, Vol. 6, p. 635.

reaction is allowed to proceed with gentle agitation at room temperature for 20 hr, and the gel is then washed with 1 liter of 1 $M$ NaCl at 4°. The gel is poured into a column (1 × 19 cm) with a bed volume of about 15 ml, and equilibrated with the same buffer as for the collagen-agarose (above).

*Purification Procedure*[129]

All procedures are carried out at 0–4°. The term enzyme buffer is used for a solution consisting of 0.12 $M$ NaCl, 2 m$M$ MnCl$_2$, 50 $\mu M$ dithiothreitol, 1% (v/v) glycerol, and 50 m$M$ Tris-HCl buffer, pH adjusted to 7.4 at 4°.

*Step 1. Extraction.* The procedure is described for isolating the transferase from chick embryos, but can also be applied to other sources. A total of 200–400 chick embryos are homogenized in batches of 30 in a solution of 0.2 $M$ NaCl, 0.1% Triton X-100, 50 $\mu M$ dithiothreitol, and 20 m$M$ Tris-HCl buffer, pH adjusted to 7.4 at 4° (1 ml of solution per gram of embryo) using a Waring blender at full speed, twice for 30 sec with a 1-min interval. The homogenate is allowed to stand with occasional stirring for 30 min and is then centrifuged at 15,000 $g$ for 30 min.

*Step 2. (NH$_4$)$_2$SO$_4$ Fractionation.* Solid (NH$_4$)$_2$SO$_4$ is slowly stirred into the supernatant fraction to a final concentration of 55% saturation (351 g/liter). The pellet obtained by centrifugation at 15,000 $g$ for 20 min is dissolved in enzyme buffer and dialyzed for 4 hr and 12 hr against two separate 16-liter volumes of enzyme buffer. The preparation is then frozen in aliquots at $-20°$.

*Step 3. Chromatography on a Concanavalin A–Agarose Column.* An aliquot of the (NH$_4$)$_2$SO$_4$ fraction is thawed, centrifuged at 15,000 $g$ for 30 min to remove the insoluble material, and adjusted to a protein concentration of about 20 mg/ml. A sample of about 4000 ml is then passed at a flow rate of 70–80 ml/hr through a 4 × 8 cm column (about 100 ml) of concanavalin A–agarose, equilibrated with enzyme buffer. The column is washed with enzyme buffer at the same flow rate until the absorbance of the eluate at 225 nm is about 0.1 (usually about 48 hr), and the enzyme is then eluted at the same flow rate with 1500 ml of a solution consisting of 1 $M$ methyl $\alpha$-D-glucoside and 20% (v/v) ethylene glycol in enzyme buffer. The fractions containing most of the enzyme activity are pooled and dialyzed for 4 hr and 12 hr against two separate 8-liter volumes of the enzyme buffer.

*Step 4. Chromatography on a Collagen–Agarose Column.* The enzyme pool from step 3 is passed through the collagen–agarose column at a flow rate of 40 ml/hr, and the column is washed with the enzyme buffer until the absorbance of the eluate at 225 nm is about 0.1. The enzyme is eluted with 300 ml of 50% (v/v) ethylene glycol in the enzyme buffer, and the fractions containing most of the enzyme activity are pooled and concentrated to a volume of about 5 ml in an Amicon ultrafiltration cell with a PM-10 membrane.

*Step 5. Gel Filtration.* The enzyme pool is centrifuged at 20,000 $g$ for 10 min, and the clear supernatant is applied to a 2.5 × 90 cm column of Sephadex G-150 (Pharmacia), equilibrated and eluted with enzyme buffer. The fractions containing most of the enzyme activity are pooled.

*Step 6. Chromatography on a UDPglucose-Derivative-Agarose Column.* The enzyme pool is passed through the UDPglucose-derivative-agarose column at a flow rate of 10 ml/hr, and the column is washed with 50 ml of enzyme buffer. The enzyme is eluted first with 300 ml of enzyme buffer having a NaCl concentration of 0.5 $M$ and then with 300 ml of 50% (v/v) ethylene glycol in enzyme buffer. In some cases most of the enzyme is found in the first eluate, and in other cases in the second. The fractions containing most of the enzyme activity constitute the final enzyme pool.

*Regeneration of the Affinity Columns.* The concanavalin A–agarose is regenerated by washing with a solution of 1.0 $M$ NaCl, 10 m$M$ MnCl$_2$, 10 m$M$ MgCl$_2$, 10 m$M$ CaCl$_2$, and 0.5 $M$ Tris-HCl buffer, pH adjusted to 7.5 at 4°, and then equilibrated with enzyme buffer. The two other affinity columns are washed with 6 $M$ urea and likewise equilibrated with the enzyme buffer.

*Comments on the Purification.* The specific activity of the purified enzyme is about 40,000 times that in the chick embryo extract, the recovery being about 5–10% (Table IX). Preparations obtained by either this procedure or the procedure involving two affinity columns[126] are entirely pure when studied by SDS–polyacrylamide gel electrophoresis after reduction and denaturation. It is not possible to study the purified enzyme by polyacrylamide gel electrophoresis as a native protein, since most of the protein aggregates.[126] Similarly, when purified enzyme preparations are concentrated above 0.2 mg/ml, the solution becomes cloudy with a loss of most of the activity. The final enzyme pool loses about one-third of its activity when stored in a protein concentration of about 0.1 mg/ml at 4° for 1 week or at −20° for several months, provided that the preparation is not thawed during this period.

*Molecular Properties*

The molecular weight of galactosylhydroxylysyl glucosyltransferase from chick embryos is about 72,000–78,000 by SDS–polyacrylamide gel electrophoresis, the value being dependent on the gel composition.[126] The sedimentation coefficient $s_{20,w}$ of 4.7 for the native enzyme[126] is in agree-

TABLE IX
PURIFICATION OF GALACTOSYLHYDROXYLYSYL GLUCOSYLTRANSFERASE FROM
CHICK EMBRYOS[a,b]

| Purification step | Total protein (mg) | Total activity (munits) | Recovery (%) | Specific activity (munits/mg) | Purification[c] (fold) |
|---|---|---|---|---|---|
| $(NH_4)_2SO_4$, 0–55% satn | 86,000 | 2550 | 100 | 0.003 | 2.2 |
| Concanavalin A–agarose | 514 | 1250 | 49 | 2.4 | 169 |
| After collagen–agarose and gel filtration | 11.3 | 383 | 15 | 33.8 | 2350 |
| UDPglucose derivative–agarose | 0.36 | 205 | 8 | 573.3 | 39,800 |

[a] Reproduced from Anttinen et al.[129] with permission of the publisher.

[b] One unit of enzyme activity is defined as the amount of enzyme required to synthesize 1 μmol of glucosylgalactosylhydroxylysine per minute at 37° under the assay conditions with a saturating concentration of denatured citrate-soluble rat skin collagen as the substrate.

[c] Purification was calculated on the basis of the specific activity for the original 15,000 g supernatant.

ment with this molecular weight. Gel filtration gives a lower and variable value,[125,126] probably due to partial adsorption of the transferase to the gel filtration columns.[126,132] The enzyme molecule probably consists of only one polypeptide chain.[126]

Amino acid analysis indicates that the transferase is rich in glutamic acid + glutamine, aspartic acid + asparagine, glycine, and alanine.[129] The enzyme is probably a glycoprotein, as it can be stained with the periodic acid–Schiff reagent,[125] and as the transferase has a high affinity for columns of concanavalin A–agarose, this affinity being markedly reduced in the presence of methyl α-D-mannoside.[132] An additional characteristic feature of the enzyme protein is its capacity for strong hydrophobic interaction,[132] which may explain some of its unusual properties, such as partial adsorption to gel filtration columns and a marked tendency for concentrated preparations to form aggregates.

Antiserum to pure chick embryo galactosylhydroxylysyl glucosyltransferase gives a line of identity with the enzyme from many chick embryo tissues in immunodiffusion, and quite similar amounts of the antiserum are required to inhibit by 50% the same amount of the transferase activity from different tissues.[138] These data do not support the concept that galactosylhydroxylysyl glucosyltransferase may have tissue-specific

[138] R. Myllylä, Biochim. Biophys. Acta 658, 299 (1981).

isoenzymes with markedly different specific activities. The possible existence of such isoenzymes nevertheless requires further elucidation.

Purification and Molecular Properties of Hydroxylysyl Galactosyltransferase

Hydroxylysyl galactosyltransferase has been purified up to about 1000-fold from chick embryo extract by a procedure consisting of ammonium sulfate fractionation, affinity chromatography on collagen-agarose, and gel filtration.[130] The enzyme has not yet been isolated as a homogeneous protein, however. A major problem in the purification of this transferase is its marked tendency to lose activity during all purification steps and in many cases the specific activity decreases during steps in which the purity of the enzyme protein clearly increases.[104,117,127] The transferase is also known to have a high affinity to UDPgalactose-derivative–agarose and concanavalin A–agarose,[127] but it has not been possible to use a procedure consisting of two or three successive affinity columns, as the enzyme then loses all its activity.

Procedures consisting of conventional protein purification steps have also been reported. One involves six steps of precipitations or chromatographies, and yields about a 50- to 150-fold increase in specific activity over the chick embryo extract used as a starting material.[104] As the enzyme becomes markedly inactivated, the degree of purification of the transferase protein is probably much greater than the increase in the specific activity. Low degrees of purification have also been reported from rat kidney cortex,[116] guinea pig skin,[114] and human platelets.[136,139]

The procedure involving affinity chromatography on collagen-agarose is described here in detail.

*Purification Procedure* [130]

All procedures are carried out at 0–4°. The term enzyme buffer is used for a solution consisting of 0.12 $M$ NaCl, 2 m$M$ MnCl$_2$, 50 $\mu M$ dithiothreitol, 1% (v/v) glycerol, and 50 m$M$ Tris-HCl buffer, pH adjusted to 7.4 at 4°.

*Step 1. Extraction.* This step is carried out exactly as described for galactosylhydroxylysyl glucosyltransferase.

*Step 2. $(NH_4)_2SO_4$ Fractionation.* This step is likewise performed as in the case of galactosylhydroxylysyl glucosyltransferase, except that the final $(NH_4)_2SO_4$ concentration corresponds to 60% saturation (390 g/liter).

*Step. 3. Chromatography on a Collagen–Agarose Column.* An aliquot of

[139] A. J. Barber and G. A. Jamieson, *Biochim. Biophys. Acta* **252**, 546 (1971).

the $(NH_4)_2SO_4$ fraction is thawed and centrifuged at 15,000 $g$ for 30 min, and the supernatant is adjusted to a protein concentration of 15–20 mg/ml. A sample of about 150 ml is passed at a flow rate 10 ml/hr through a 1 × 13 cm collagen–agarose column (about 10 ml), prepared and equilibrated as described for the purification of galactosylhydroxylysyl glucosyltransferase. The column is washed with enzyme buffer until the absorbance of the eluate at 225 nm is less than 0.05 (about 300–500 ml), and the enzyme is then eluted either with 80 ml 50% (v/v) ethylene glycol in enzyme buffer[127,132] or with 2 g of dialyzable collagen peptides in 2 ml of enzyme buffer, followed by 50 ml of enzyme buffer. In the elution with ethylene glycol the fractions are pooled on the basis of enzyme activity, whereas when the peptides are used they are pooled beginning with the first to show any increase in absorbance and continuing until the last one after the peptide peak that still has an absorbance above 2.0 at 225 nm. This pool is concentrated in an Amicon ultrafiltration cell with a PM-10 membrane to a volume of about 2 ml.

The dialyzable collagen peptides are prepared as follows. Commercially available bovine Achilles tendon collagen, 10 g, is suspended in 70 ml of $H_2O$ and autoclaved at 120° overnight. The volume is adjusted to 100 ml with a solution of $NH_4HCO_3$ and $CaCl_2$ having final concentrations of 20 m$M$ of the former and 0.1 $M$ of the latter and a pH of 7.5. Crude bacterial collagenase, 40 mg, is added, and the mixture is incubated for 24 hr at 37°. The sample is dialyzed against $H_2O$, and the dialyzable peptides are lyophilized.

*Step 4. Gel Filtration.* The pool from the previous step is applied to a 2.5 × 90 cm column of Sephadex G-150 (Pharmacia), equilibrated, and eluted with enzyme buffer. The fractions containing most of the enzyme activity constitute the final enzyme pool.

*Comments on Purification.* The binding of the transferase to collagen-agarose is highly variable. In some experiments most of the enzyme activity becomes bound, while in others little binding is achieved. The binding seems to vary especially when columns prepared in different coupling experiments are compared, even though the coupling efficiency is relatively similar. This suggests that minor differences in coupling are critical for the binding of this enzyme to collagen-agarose.[130] Elution with dialyzable collagen peptides probably gives a higher recovery of the enzyme activity than elution with ethylene glycol,[127] but owing to the large variation between different experiments this is only a minor factor affecting the recovery. The purification of the best preparations over the chick embryo extract is about 1000-fold, with a recovery of about 30% when calculated from the units in the $(NH_4)_2SO_4$-fractionated enzyme (Table X). These preparations are clearly not pure, but can be used for many studies on catalytic properties of the enzyme.

TABLE X
PURIFICATION OF HYDROXYLYSYL GALACTOSYLTRANSFERASE FROM CHICK EMBRYOS[a,b]

| Purification step | Total protein (mg) | Total activity (munits) | Recovery (%) | Specific activity (munits/mg) | Purification[c] (fold) |
|---|---|---|---|---|---|
| $(NH_4)_2SO_4$, 0–60% satn | 3000 | 73 | 100 | 0.024 | 1.74 |
| Collagen–agarose and gel filtration | 1.65 | 24 | 33.2 | 14.7 | 1050 |

[a] Reproduced from Risteli et al.[130] with permission of the publisher.
[b] One unit of enzyme activity is defined as the amount of enzyme required to synthesize 1 μmol of galactosylhydroxylysine per minute at 37° under the assay conditions with a saturating concentration of denatured citrate-soluble rat skin collagen as the substrate.
[c] Purification calculated on the basis of the specific activity for the original 15,000 g supernatant.

## Molecular Properties

The molecular weight of hydroxylysyl galactosyltransferase is not known, as the enzyme has not been isolated as a homogeneous protein. The activity of the partially purified transferase obtained by conventional steps is found by gel filtration in the form of two major species with apparent molecular weights of about 450,000 and 200,000 and one minor species with an apparent molecular weight of about 50,000.[104] Most of the enzyme activity obtained by affinity chromatography on collagen-agarose is found in the void volume of the Sephadex G-150 column and thus probably corresponds to one or two of the above major forms.[130] It is not known whether one or more of these forms may represent an aggregate of either the enzyme alone or the enzyme with some other proteins.

Hydroxylysyl galactosyltransferase, like all the other intracellular enzymes described above, is probably a glycoprotein, as its activity is inhibited by concanavalin A, and as the transferase becomes bound to columns of this lectin coupled to agarose and can be eluted with methyl α-D-mannoside or methyl α-D-glucoside in the presence of ethylene glycol.[127]

## Catalytic Properties of the Hydroxylysyl Glycosyltransferases

The main requirements for the glycosyl acceptors are summarized in the overview (above) and are discussed in detail elsewhere.[7] The $K_m$ value for the gelatinized insoluble calf skin collagen used as a standard substrate in the assay described above is about 7–14 g/liter in the case of the glucosyltransferase,[103,118] but is more variable in that of the galactosyltransferase. In the initial studies[118] values as high as 150 g/liter were observed, whereas more recently the values have usually ranged from about 15 to 35 g/liter.[104,124,127] Denatured citrate-soluble rat skin collagen has a considerably lower $K_m$, about 0.5–1 g/liter for the glucosyltransferase[103] and 2–4 g/liter for the galactosyltransferase.[104] When expressed in terms of the concentration of galactosylhydroxylysyl or hydroxylysyl acceptor sites, the latter values correspond to about 3–6 $\mu M$ and 0.1–0.2 m$M$, respectively. The highest $K_m$, about 4 m$M$, is found with the free galactosylhydroxylysine as a substrate for the glucosyltransferase.[122]

The specificity of both transferases for their glycosyl acceptors is very high. Galactosylhydroxylysyl glucosyltransferase fails to attach glucose to a number of carbohydrates, or to glycopeptides or glycoproteins in which the terminal galactose is linked to other sugars or to amino acids other than hydroxylysine.[7,115,117,129,133,135,136] Neither does this enzyme attach glucose to hydroxylysyl residues[117,135,136] or any additional glucose to glucosylgalactosylhydroxylysyl residues.[117] Hydroxylysyl galactosyltransferase correspondingly does not attach galactose to a number of carbohydrates, or to glycopeptides or glycoproteins in many cases in which the galactosyl residues have been specifically removed.[7,114,116,136,139] Similarly, the enzyme does not catalyze the transfer of a second galactose unit to galactosylhydroxylysyl residues or a galactose unit to glucosylgalactosylhydroxylysyl residues.[116,136,139] The only distinct exception to this high specificity is galactosylsphingosine, which acts *in vitro* as a good glucosyl acceptor for both crude[117] and pure[129] galactosylhydroxylysyl glucosyltransferase. This compound bears a close structural similarity to galactosylhydroxylysine in that both compounds have an unsubstituted amino group next to the hydroxy group to which the galactose is attached. It may be noted that the product of this reaction has not been described in nature.[117]

The pH optimum for both hydroxylysyl glycosyltransferases is about 7–7.4,[105,116–118,122] and both reactions require a bivalent cation, this requirement being best fulfilled by manganese.[7] The optimal $Mn^{2+}$ concentration for partially purified enzymes is about 0.2–2 m$M$ for the glucosyltransferase[120,140] and about 2 m$M$ for the galactosyltransferase.[127] The glucosyltransferase can bind at least two $Mn^{2+}$ ions, the first having a $K_d$ of 3–5 $\mu M$ (site I) and the second 50–70 $\mu M$ (site II).[140] The binding of the second $Mn^{2+}$ reduces the $K_m$ of UDPglucose to about one-fourth, whereas the $V$ at a saturating UDPglucose concentration is unaffected.[140] No corresponding studies have yet been carried out with respect to the galactosyltransferase.

[140] R. Myllylä, H. Anttinen, and K. I. Kivirikko, *Eur. J. Biochem.* **101**, 261 (1979).

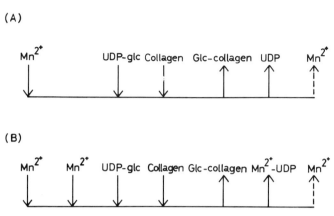

FIG. 9. Schematic presentation of the proposed mechanism for the galactosylhydroxylysyl glucosyltransferase reaction at low (A) and high (B) $Mn^{2+}$ concentrations. The dashed line between the enzyme and $Mn^{2+}$ at the end of the cycle indicates that $Mn^{2+}$ need not leave the enzyme during each catalytic cycle. Under some conditions the enzyme can also form certain dead-end complexes that are not shown. E, collagen; UDP-glc, UDPglucose; Glc-collagen, glucosylated collagen. Reproduced from Myllylä et al.[140] with permission of the publisher.

Data on the specificity of the metal requirement vary considerably, probably owing in part to the many artifacts present in these studies, as discussed elsewhere.[7,140] $Fe^{2+}$ ($K_d$ 5–7 $\mu M$) and $Co^{2+}$ ($K_d$ 30 $\mu M$) are the only other metals that can activate the glucosyltransferase at low concentrations, whereas no further activation, but instead an inhibition, is found at high concentrations.[140] The $K_d$ values for $Mn^{2+}$ at site II and for $Co^{2+}$ are probably much higher than the physiological concentrations, whereas the activation by $Mn^{2+}$ at site I and by $Fe^{2+}$ may be of physiological significance.[140] It has been demonstrated, however, that the metal requirements of both hydroxylysyl glycosyltransferases can be fulfilled in isolated chick embryo tendon cells even in the absence of $Fe^{2+}$.[102] No work has been carried out on the possible activation of hydroxylysyl galactosyltransferase at low concentrations of $Mn^{2+}$ or other cations. Several bivalent cations are inhibitors of the $Mn^{2+}$-activated hydroxylysyl glycosyltransferases.[127,140]

The preferential sugar donor for both transferases is the corresponding UDPglycoside.[7] The $K_m$ of UDPglucose for the glucosyltransferase from various sources at high $Mn^{2+}$ concentrations is about 5–30 $\mu M$,[117,118,122,125,140] and the corresponding $K_m$ of UDPgalactose for the galactosyltransferase is about 20–30 $\mu M$.[118,127] The reaction product, UDP, and several other nucleotides are inhibitors of the hydroxylysyl glycosyltransferases.[7]

The mechanism of the galactosylhydroxylysyl glucosyltransferase reaction with the enzyme from chick embryos has been studied by analyzing the initial velocity and inhibition kinetics in detail.[120,140] Initial velocity data were determined by varying the concentration of one substrate (the cosubstrates are also termed here substrates) in the presence of different fixed concentrations of the second, while the concentration of the third was held constant. Intersecting lines were obtained in double-reciprocal plots for every pair of the three substrates, $Mn^{2+}$, UDPglucose, and collagen, indicating that the reaction occurs by a sequential mechanism. These and a number of additional data are consistent with a reaction scheme (Fig. 9) involving an ordered binding of $Mn^{2+}$, UDPglucose, and the collagen substrate at low $Mn^{2+}$ concentrations, and an ordered binding of $Mn^{2+}$, $Mn^{2+}$, UDPglucose, and the collagen substrate at high concentrations.[120,140] The binding of $Mn^{2+}$, which occurs before that of UDPglucose, is in both cases at thermodynamic equilibrium.

Some kinetic studies have also been carried out on galactosylhydroxylysyl glucosyltransferase from human platelets.[122] The data are consistent with the scheme described above, although a rapid equilibrium random binding of the sugar acceptor and sugar donor was also possible. No detailed kinetic studies have been carried out on the hydroxylysyl galactosyltransferase reaction.

The hydroxylysyl glycosyltransferase reactions, unlike those involved in the biosynthesis of asparaginyl-linked carbohydrate units in the propeptides of procollagen (this volume [15]) and in other proteins,[141] do not seem to involve any lipid intermediate. Both purified enzymes can catalyze their reactions in the absence of any lipid *in vitro,* and attempts to detect a possible lipid intermediate in tissue extracts or crude enzyme preparations have given no evidence of its presence.[116,117,142]

---

[141] A. J. Parodi and L. F. Leloir, *Biochim. Biophys. Acta* **559**, 1 (1979).
[142] N. H. Behrens, A. J. Parodi, L. F. Leloir, and C. R. Krisman, *Arch. Biochem. Biophys.* **143**, 375 (1971).

## [11] Posttranslational Enzymes in the Biosynthesis of Collagen: Extracellular Enzymes

By DARWIN J. PROCKOP and LEENA TUDERMAN

### Procollagen Proteinases

Conversion of the interstitial procollagens to collagens, such as types I, II, III, $\alpha$A, and $\alpha$B, requires removal of large peptides from both the N terminus and C terminus of the proteins by specific proteinases.[1-3] A procollagen N-proteinase (EC 3.4.24.) that removes the N-propeptide from type I procollagen has been isolated from calf tendon and chick embryo tendon[4-6] and partially characterized (Table I). A procollagen C-proteinase has been identified in several sources[6-8] but has not been extensively purified or characterized (Table II).

### Assays

Two general methods are used to assay the procollagen N- and C-proteinases. One is to use polyacrylamide gel electrophoresis for examining the cleavage products obtained with the enzymes. The other is to precipitate the reaction products under conditions in which uncleaved procollagen, partially cleaved procollagen, and collagen are recovered in the pellet, but the released N- or C-propeptides are in the supernatant. The procedures of polyacrylamide gel electrophoresis are the same as those commonly employed to examine procollagen and other large proteins. The conditions for the more rapid precipitation assays are tailored to the enzyme being examined.

*Preparation of Substrates.* The substrate used to assay the N-proteinase is either procollagen, or pN-collagen, i.e., an intermediate in the conversion of procollagen to collagen that contains the N-propeptide

---

[1] J. H. Fessler and L. J. Fessler, *Annu. Rev. Biochem.* **47**, 129 (1978).
[2] D. J. Prockop, K. I. Kivirikko, L. Tuderman, and N. A. Guzman, *N. Engl. J. Med.* **301**, 77 (1978).
[3] P. Bornstein and W. Traub, in "The Proteins", (H. Neurath and R. L. Hill, eds.), 3rd ed., Vol. 4, p. 412, Academic Press, New York, 1979.
[4] L. D. Kohn, C. Isersky, J. Zupnik, A. Lenaers, G. Lee, and C. M. Lapiere, *Proc. Natl. Acad. Sci. U.S.A.* **71**, 40 (1974).
[5] L. Tuderman, K. I. Kivirikko, and D. J. Prockop, *Biochemistry* **17**, 2948 (1978).
[6] M. K. K. Leung, L. I. Fessler, D. B. Greenberg, and J. H. Fessler, *J. Biol. Chem.* **254**, 224 (1979).
[7] E. Kessler and B. Goldberg, *Anal. Biochem.* **86**, 463 (1978).
[8] F. Njieha, L. Tuderman, T. Morikawa, and D. J. Prockop, *Biochemistry*, in press.

TABLE I
PROPERTIES OF PROCOLLAGEN N-PROTEINASE

| | |
|---|---|
| Identified sources | Calf tendon; chick embryo tendons, calvaria, fibroblasts; human skin fibroblasts; fetal calf skin |
| Molecular weight | 260,000 (gel filtration) |
| Number of polypeptide chains | Uncertain |
| pH optimum | 7.4 |
| Metal cofactors | $Ca^{2+}$ and perhaps a second metal |
| Bonds cleaved | Pro-Gln in the pro$\alpha$1 chain[a] and X-Gln in the pro$\alpha$2 of native type I procollagen; also X-Gln in native type II procollagen |
| Inhibitors | Denatured type I pN-collagen and type III pN-collagen<br>Short peptides with amino acid sequences found around cleavage site in pro$\alpha$1(I)<br>Metal chelators<br>Reducing agents |

[a] D. Hörlein, P. P. Fietzek, and K. Kühn, *FEBS Lett.* **89**, 279 (1978).

TABLE II
PROPERTIES OF PROCOLLAGEN C-PROTEINASE

| | |
|---|---|
| Identified sources | Chick embryo tendons, calvaria and cultured fibroblasts; human cultured fibroblasts; mouse fibroblasts |
| Molecular weight | 80,000 (gel filtration) |
| Number of polypeptide chains | Uncertain |
| pH optimum | Neutral |
| Metal cofactors | $Ca^{2+}$ and perhaps other metals |
| Bonds cleaved | -Ala-Asp- in type I procollagen |
| Inhibitors | Metal chelators<br>Short synthetic peptides with sequence similar to that of cleavage site in pro$\alpha$1 and pro$\alpha$2 chains of type I procollagen |

but not the C-propeptide. The substrate used to assay the C-proteinase is either procollagen or pC-collagen, i.e., the intermediate that contains the C-propeptide but not the N-propeptide. These substrates can be obtained with several procedures (Table III). One approach is to incubate whole tissues, such as calvaria from chick embryos, with radioactive amino acids and then to extract the labeled procollagen.[9] Another is to isolate matrix-free cells from chick embryo tendon or sterna, incubate the cells in suspension with radioactive amino acids, and then isolate labeled procol-

[9] J. M. Monson and P. Bornstein, *Proc. Natl. Acad. Sci. U.S.A.* **70**, 3521 (1972).

TABLE III
SOURCES OF SUBSTRATES FOR PROCOLLAGEN N- AND C-PROTEINASES

| Tissue or cells | Amount of tissue | Radioactive label | Procollagen Yield (mg) | Spec. act. (cpm/μg) |
|---|---|---|---|---|
| Matrix-free tendon cells[a] | 144 chick embryos (2 × 10$^9$ cells) | 200 μCi of $^{14}$C-labeled mixture of amino acids | 1–2 | 1 × 10$^4$ |
| Cultured human fibroblasts[b] | Ten 150 cm$^2$ flasks (10$^8$ cells) | 800 μCi of $^{14}$C-labeled mixture of amino acids | 2 | 2 × 10$^3$ |
| Dermatosparactic calf or sheep skin[c] | One skin | 120 μCi of [$^{14}$C]iodoacetamide | 50[d] | 2 × 10$^3$ |

[a] Dehm and Prockop.[10,11]
[b] Peltonen et al.[12]
[c] Nusgens and Lapiere.[13]
[d] Substrate here is pN-collagen rather than procollagen and therefore can be used only for assay of N-proteinase.

lagen or pC-collagen from the medium.[10,11] A third approach is to incubate cultured fibroblasts with radioactive amino acids and isolate labeled procollagen from the medium.[12] A fourth approach, which provides substrate for the N-proteinases but not for the C-proteinases, is to extract pN-collagen from the skin of dermatosparactic calf, and label the pN-collagen with [$^{14}$C]iodoacetamide.[13] The procedure employed in our own laboratory is to use matrix-free cells from chick embryo tendons.

To prepare matrix-free cells,[10,11] legs were cut from about 12 dozen 17-day-old chick embryos. The embryos were placed in a small amount of modified Krebs medium II. The Krebs medium II was prepared by adding 5 ml of fresh 5.4% glucose solution to 110 ml of a solution containing 0.24 g of NaH$_2$PO$_4$ (15.7 m$M$), 0.02 g of KH$_2$PO$_4$ (1.6 m$M$), 0.75 g of NaCl (111.2 m$M$), 0.04 g of KCl (5.4 m$M$), 0.03 g of MgCl$_2$ (1.3 m$M$), and 0.04 g of NaHCO$_3$ (4.0 m$M$), pH 7.4. The tendons were pulled from the legs with two forceps, one grasping the upper part of a leg and the other a toe. The toes with the attached tendons were placed in a petri dish containing

[10] P. Dehm and D. J. Prockop, *Biochim. Biophys. Acta* **264**, 375 (1972).
[11] P. Dehm and D. J. Prockop, *Eur. J. Biochem.* **35**, 159 (1973).
[12] L. Peltonen, A. Palotie, T. Hayashi, and D. J. Prockop, *Proc. Natl. Acad. Sci. U.S.A.* **77**, 162 (1980).
[13] B. Nusgens and C. M. Lapiere, *Anal. Biochem.* **95**, 406 (1979).

Krebs medium II, and the tendons were cut from the toes with a scalpel. The tendons, about 10 g, were placed in 30 ml of Eagle's minimum essential medium containing 3.3 ml of 2.5% trypsin solution (Gibco, Grand Island, New York) and about 12,000 units of crude bacterial collagenase (Sigma or Worthington). The sample was incubated in a 100-ml Erlenmeyer flask at 37° for 45–60 min under 95% $O_2$ and 5% $CO_2$ with shaking. The reaction was terminated when most of the tissue was digested but a few small pieces (1–3 mm) were still present; overdigestion decreases the viability of the cells. The samples were filtered through lens paper in a small Swenex filter (Millipore Corporation, Bedford, Massachusetts), and the filtrate was centrifuged at 600 $g$ at room temperature for 3 min. The supernatant was discarded, and the cells in the pellet were washed three times with about 25 ml of Krebs medium II containing 10% fetal calf serum. The pellet of cells was suspended in about 20 ml of Krebs medium II without fetal calf serum, and the cell number was determined by examining an aliquot (diluted 1 : 100) in a hemacytometer. The yield of cells was about $2 \times 10^9$ per 12 dozen chick embryos.

To obtain procollagen, the cells were incubated at a concentration of 7.5 to $10 \times 10^6$ cells per milliliter in Krebs medium II at 37° for 4–5 hr with gentle shaking. The medium contained 100 $\mu$Ci of a $^{14}$C-labeled mixture of amino acids (New England Nuclear) per $10^9$ cells. The cells were removed by centrifuging at 600 $g$ for 5 min and discarded. A one-tenth volume of a stock solution of proteinase inhibitors was added. The stock solution contained 2.56 g of EDTA (275 m$M$); 344 mg of $N$-ethylmaleimide (125 m$M$) (Sigma); 5.5 ml of phenylmethylsulfonyl fluoride dissolved in isopropanol (11 m$M$); 47 mg of $p$-aminobenzamidine (10 m$M$) per 25 ml. Then the medium proteins were precipitated by stirring in 176 mg per milliliter of ammonium sulfate (Baker Chemical). The sample was stirred overnight and centrifuged at 15,000 $g$ for 30 min at 4°. The pellet was dissolved in about 10 ml of 0.4 $M$ NaCl in 0.1 $M$ Tris-HCl buffer, adjusted pH 7.4 at 4°, and stirred for 4–5 hr at 4°. The sample was centrifuged at 15,000 $g$ for 20 min, and the small pellet was discarded. If procollagen entirely free of medium proteinases was required, about 1 mg of carrier collagen[1] was added, and the protein was reprecipitated by adding NaCl to a final concentration of 2.5 $M$ and stirring the sample overnight at 4°. The protein was recovered by centrifuging at 15,000 $g$ for 30 min, and the pellet was dissolved in 0.4 $M$ NaCl in 1 $M$ Tris-HCl buffer, pH 7.4 at 4°. After stirring for 4–5 hr, the sample was centrifuged at 15,000 $g$ for 30 min, and the small pellet was discarded. The NaCl precipitation step is not essential; adequate substrate is obtained without it, but some degradation occurs after storage for more than 2 or 3 weeks.

The samples were then dialyzed against the start buffer for DEAE-chromatography: 2 $M$ urea in 0.1 $M$ Tris-HCl buffer, adjusted to pH 8.6 at

4°.[14,15] The chromatography was at 4° on a 1.5 × 5 cm column of DEAE-cellulose (DE-52, Whatman) which was eluted with a linear gradient of 200 ml of start buffer and 200 ml of the same buffer containing 0.12 $M$ NaCl. The flow rate was 60 ml/hr, and fractions of 5 ml were collected. The column was monitored by assay of $^{14}C$ in a liquid scintillation counter. The procollagen eluted as a peak at about 0.06 $M$ NaCl, and it usually accounted for about 70% of the total $^{14}C$-labeled protein.

The peak containing procollagen was dialyzed overnight against 0.4 $M$ NaCl in 0.1 $M$ Tris-HCl buffer, and the protein was then precipitated overnight with either 176 mg of ammonium sulfate per milliliter or 2.5 $M$ NaCl. The precipitation was occasionally less complete with the NaCl, but the precipitate tended to be more soluble. The precipitate was recovered by centrifugation at 15,000 $g$ for 30 min. The sample was dissolved in 0.4 $M$ NaCl in 0.1 $M$ Tris-HCl buffer and stored in aliquots of 0.2 ml at $-20°$. The volume of the storage buffer was adjusted to give a final concentration of about 40,000 cpm per 10 $\mu$l.

The procedure for preparing $^{14}C$-labeled pC-collagen from matrix-free cells was the same as that for procollagen (above) except the cells were incubated at a concentration of 15 to 20 × $10^6$ cells per milliliter. (Incubation of the cells at a high concentration apparently exposes the procollagen to N-proteinase secreted by the cells.) The pC-collagen was recovered from the DEAE-cellulose column in a peak that is intermediate between the void volume and the elution position of procollagen. The pC-collagen recovered accounts for 20–30% of the total $^{14}C$-labeled protein recovered from the column.

*Rapid Assay for Procollagen N-Proteinase.* Two rapid assays for the N-proteinase have been reported. They differ primarily in that ammonium sulfate is used as the precipitant in one[5] and ethanol in the other.[13] We describe here the method using ammonium sulfate that has been used in our laboratory for 3 years.[5]

The assay was carried out in a final volume of 100 $\mu$l that contained 2–4 $\mu$g of labeled procollagen, 0.02–0.40 unit of enzyme, 5 m$M$ $CaCl_2$, and 0.15 $M$ NaCl in 0.05 $M$ Tris-HCl buffer, pH 7.4 at 30°. The reaction was carried out for 90 min at 30°. It was terminated by transferring the tubes to an ice bath and immediately adding 50 $\mu$l of 500 m$M$ EDTA, 100 $\mu$l of fetal calf serum heat-inactivated at 60° for 15 min, and 750 $\mu$l of ammonium sulfate (243 mg/ml). The ammonium sulfate precipitation was allowed to proceed for 60 min at room temperature. The reaction was carried out in 1.0-ml cellulose acetate butyrate tubes (No. 00211,

[14] B. D. Smith, P. H. Byers, and G. R. Martin, *Proc. Natl. Acad. Sci. U.S.A.* **69**, 3260 (1972).
[15] H. P. Hoffmann, B. R. Olsen, H. T. Chen, and D. J. Prockop, *Proc. Natl. Acad. Sci. U.S.A.* **73**, 4304 (1976).

DuPont-Sorvall) so that the precipitate could be removed by centrifuging the samples at 15,000 g for 30 min in an SS-34 rotor with No. 00508 adaptors (DuPont-Sorvall). A 0.5-ml aliquot of the supernatant was placed in a scintillation vial, and 1.0 ml of water and 19 ml of formula L-963 (New England Nuclear) were added for liquid scintillation counting. The background value for each assay was established with control samples containing all the component of the reaction mixture except enzyme. Background values varied from 60 to 120 cpm.

One unit of enzymic activity was defined as the amount of enzyme present in 1 mg of protein in a standard extract from chick embryo tendons (see below). The reaction is linear at least over the range of 0.5–10 $\mu$l of tendon extract (0.008–0.15 unit) and with incubation times of up to 180 min. The assay is based on the principle that procollagen, pC-collagen, collagen, and the C-propeptide are precipitated by ammonium sulfate under the conditions employed here. Although no data are available, it is likely that the N-propeptides from the three chains remain associated and they are not recovered in the supernatant unless all three pro$\alpha$ chains in native procollagen are cleaved. As described here, the assay is carried out with a substrate concentration (about 0.4 $\mu$g/ml) that is below the $K_m$ value (about 0.3 $\mu M$ or 150 $\mu$g/ml); use of higher concentrations is costly, and procollagen becomes insoluble near its $K_m$.

*Assay for N-Proteinase by Gel Electrophoresis.* For assay by gel electrophoresis, the enzyme reaction was carried out as described above, but the reaction was stopped by adding 50 $\mu$l of 0.5 M EDTA and 15 $\mu$l of 20% sodium dodecyl sulfate. The sample was boiled for 3 min and then dialyzed against sample buffer for polyacrylamide gel electrophoresis in sodium dodecyl sulfate.[16] Just prior to electrophoresis the samples were reduced by adding 5% 2-mercaptoethanol and boiling for 3 min. The samples were then electrophoresed in 6% polyacrylamide gels, and fluorograms were prepared under standard conditions[17,18] (Fig. 1).

Quantitation of the assay can be obtained by scanning of the fluorogram and integrating the peaks. It should be noted that assay of the reaction in this way measures the total number of bonds cleaved, not simply the end product of a released trimer of the N-propeptides. The N-propeptides released from the pro$\alpha$1(I) and pro$\alpha$2 chains can be seen by examining the reaction products on 20% polyacrylamide gels, but the bands are weak because the N-propeptides tend to diffuse out of the gel during the processing required for fluorography.[5]

*Rapid Assay for Procollagen C-Proteinase.* Two rapid assays are available for the C-proteinase. One uses as substrate [$^3$H]tryptophan-labeled

[16] J. King and U. K. Laemmli, *J. Mol. Biol.* **61**, 465 (1972).
[17] W. M. Bonner and R. A. Laskey, *Eur. J. Biochem.* **46**, 83 (1974).
[18] R. A. Laskey and A. D. Mills, *Eur. J. Biochem.* **56**, 335 (1975).

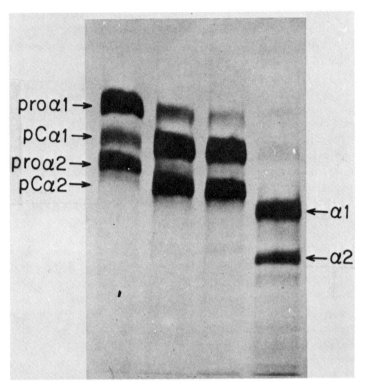

FIG. 1. Assays for procollagen N-proteinase and C-proteinase by gel electrophoresis. Lane 1: Procollagen substrate containing $^{14}$C-labeled proα1(I) and proα2 chains together with a small amount of pCα1(I) and pCα2. Lane 2: Procollagen substrate after incubation with procollagen N-proteinase. Lane 3: pC-collagen substrate consisting largely of $^{14}$C-labeled pCα1(I) and pCα2. Lane 4: pC-collagen substrate after incubation with procollagen C-proteinase; the products are α1(I) and α2 chains.

procollagen from cultured human fibroblasts or mouse 3T6 fibroblasts.[7] After incubation with enzyme, the reaction system is precipitated with 25% ethanol, and the $^3$H-labeled C-propeptide is recovered in the supernatant. The procollagen is not purified from other proteins in the medium, and the assay is based on the assumption that all the [$^3$H]tryptophan released from the substrate originates from the C-propeptide. Therefore the assay may not be specific for C-proteinase. For example, it may detect N-proteinase activity, since one tryptophan residue is present in the proα1 N-propeptide of type I procollagen[19,20] and the N-propeptide should also

[19] D. Hörlein, P. P. Fietzek, E. Wachter, C. M. Lapiere and K. Kühn, *Eur. J. Biochem.* **99**, 31 (1979).
[20] H. Rhode, E. Wachter, W. J. Richter, P. Bruckner, O. Helle, and R. Timpl, *Biochem. J.* **179**, 631 (1979).

be recovered in the ethanol supernatant. The method we describe here[8] is probably more specific but requires purification of substrate (see above).

Assay incubation conditions were the same as for the N-proteinase except that pC-collagen is used as substrate, and it was usually necessary to extend the incubation time to 4 hr in order to detect small amounts of activity. The reaction was stopped by adding 50 $\mu$l of 0.5 $M$ EDTA, 10 $\mu$l of carrier collagen, and 80 $\mu$l of absolute ethanol. The sample was incubated at 4° with shaking for 1 hr and then centrifuged at 15,000 $g$ for 30 min. A 100-$\mu$l aliquot of the supernatant was then mixed with 1.0 ml of water and assayed for $^{14}$C as described above.

The reaction is linear over a 10-fold range of amount of enzyme. With active preparations the reaction was linear with incubation times varying from 1 to 4 hr. As with rapid assay for N-protease reaction, the assay is based on release of the entire propeptide from procollagen and therefore requires cleavage of three bonds in each molecule. Hydrolysis of the C-propeptide has been shown to occur in a stepwise manner,[21,22] and therefore the assay will not measure the first steps of the process.

*Assay for C-Proteinase by Gel Electrophoresis.* Samples were processed in the same manner as for the assay of N-proteinase by gel electrophoresis. The reaction can be quantitated by densitometric scanning of the fluorograms and determining the conversion of pC$\alpha$1 and pC$\alpha$2 chains to $\alpha$1 and $\alpha$2 (Fig. 1).

A variation on this procedure is to carry out the electrophoresis without reduction. Under these conditions one detects six bands on the 6% polyacrylamide gels: (*a*) the disulfide-linked trimer of uncleaved pC-collagen; (*b*) the intermediate containing two pC$\alpha$ chains and one disulfide-linked C-propeptide; (*c*) an intermediate of one pC$\alpha$ chain and the two disulfide-linked C-propeptides; (*d*) $\alpha$1 chains; (*e*) $\alpha$2 chains; and (*f*) the disulfide-linked C-propeptides from the three chains. The number of bonds cleaved in the reaction can be assayed by densitometric scanning and integration of the peaks.[8]

*Purification of Procollagen N-Proteinase*

As of this writing, procollagen N-proteinase has been extensively purified from just one source, leg tendons of chick embryos.[5,23] Partial purification has been carried on what appears to be the same enzyme from calf tendons[4] and the medium of cultured chick embryo tendon fibroblasts.[6] We describe here the purification from chick embryo tendons.

[21] J. M. Davidson, L. S. G. McEneany, and P. Bornstein, *Eur. J. Biochem.* **81**, 349 (1977).
[22] N. P. Morris, L. J. Fessler, and J. H. Fessler, *J. Biol. Chem.* **254**, 11024 (1979).
[23] L. Tuderman and D. J. Prockop, manuscript in preparation.

Leg tendons were removed from about 150 17-day-old chick embryos (about 10 g wet weight) and minced with scissors in 1 ml of cold 2 $M$ KCl per gram and 0.1% Triton X-100 in 0.1 $M$ Tris-HCl buffer, pH 7.4 at 4°. The samples were extracted for 2.5 hr with stirring. These and all subsequent procedures were carried out at 4°. The sample was centrifuged at 15,000 $g$ for 30 min, and the supernatant was recovered. The pellet was resuspended in an equal volume of the extraction buffer diluted 1 : 1 with water. The sample was centrifuged at 15,000 $g$ for 30 min and the supernatant was combined with the first supernatant. The extracts were stored at 4° for 1–4 days, and extracts from two to four preparations were pooled. The pooled extracts were precipitated by adding an equal volume of 630 mg of ammonium sulfate per milliliter in 0.1 $M$ Tris-HCl, pH 7.4, at 4°. Then solid ammonium sulfate was added to a final concentration of 390 mg/ml. The sample was stirred for 30 min and then centrifuged at 15,000 $g$ for 30 min. The precipitate was dissolved in 1 $M$ KCl and 0.1 $M$ Tris-HCl buffer, pH 7.4, at 4° and dialyzed overnight against this same buffer.

A sample in volume of 45 ml or less (protein concentration of about 15 mg/ml) was passed through a 0.5–4.0 ml column of concanavalin A covalently linked to agarose (Con A–Sepharose; Pharmacia), which was equilibrated with 1 $M$ KCl and 0.1 $M$ Tris-HCl buffer, pH 7.4 at 4°. The flow rate for the chromatography was 1 column volume per hour. The column was washed with about 30 column-volumes of the equilibration buffer and then it was eluted with 15 column-volumes of the same buffer containing 0.3 $M$ methyl $\alpha$-D-mannoside.

The fractions eluted with methyl $\alpha$-D-mannoside were assayed for enzymic activity, and the absorbance of the fractions was measured at 230 nm. The fractions having the highest specific activities were pooled and precipitated with 390 mg of ammonium sulfate per milliliter as described above. The precipitate was dissolved in 1–2 ml of 1 $M$ KCl and 0.1 $M$ Tris-HCl, pH 7.4 at 4°, and dialyzed against this same buffer.

Then the sample was dialyzed against 0.2 $M$ KCl in 0.1 $M$ Tris-HCl, pH 7.4 at 4°, and chromatographed on a column of heparin linked to Sepharose (heparin–Sepharose 6BCL; Pharmacia).[23] The column was 1 × 2 cm, and it was equilibrated with 0.2 $M$ KCl in 0.1 $M$ Tris-HCl, pH 7.4 at 4°. The column was eluted with linear gradient prepared with 10 ml of 0.2 $M$ KCl in 0.1 $M$ Tris-HCl and 10 ml of 1 $M$ KCl in 0.1 $M$ Tris-HCl. The enzyme eluted with about 0.5 $M$ KCl. Fractions of 1.2 ml were collected and the flow rate was 3 ml/hr. The fractions of highest specific activities were pooled, and the sample (about 3 ml) was then applied to a 1.5 × 90 cm column of polyacrylamide and agarose (Ultrogel AcA-34, separation range 20,000 to 350,000 daltons; LKB). The column was

## TABLE IV
## PURIFICATION OF PROCOLLAGEN N-PROTEINASE [a]

| Step | Total protein (mg) | Total activity (units) | Specific activity (units/mg) | Recovery (% of initial) |
| --- | --- | --- | --- | --- |
| Tendon extract | 1034 | 1034 | 1 | 100 |
| Ammonium sulfate precipitation | 547 | 930 | 1.7 | 90 |
| Con A-Sepharose | 11.55 | 693 | 60 | 67 |
| Heparin-Sepharose column | 0.068 | 253 | 3720 | 24 |

[a] From Tuderman and Prockop.[23]

equilibrated and eluted with 0.5 M KCl and 0.1 M Tris-HCl buffer, pH 7.4 at 4°. Fractions of 2 ml were collected, and fractions containing enzymic activity were pooled. Because of the losses of enzyme activity, the gel filtration step was frequently omitted.

The overall yield of enzyme activity was about 25%, and the fold purification was about 3700 (Table IV). The enzyme appeared to be homogeneous; after iodination only a single radioactive band was seen when the protein was examined by gel electrophoresis in sodium dodecyl sulfate. However, the amount of protein obtained was too small for extensive characterization. By gel filtrations the protein had a molecular weight of about 260,000.

The procedure for purifying N-proteinase from calf tendons[4] involves ammonium sulfate precipitation, DEAE-cellulose chromatography, stepwise ammonium sulfate extraction, and gel filtration on Sephadex G-200. The purification was 2000-fold based on total tissue protein and 30-fold based on the protein content of the initial tissue extract. The enzyme eluted in three peaks from the G-200 column, one in the void volume and two in the included volume.

N-Proteinase activity was recovered also from the medium of cultured chick embryo fibroblasts[6] by a procedure involving ammonium sulfate precipitation and then either chromatography in DEAE-cellulose or velocity sedimentation in a sucrose gradient. The extent of purification was not determined, but the N-proteinase activity was separated from the C-proteinase.

*Purification of Procollagen C-Proteinase*

No successful procedure for purifying the C-proteinase has been developed, largely because the amount of this activity is small in tissue extracts, even those containing modest amounts of N-proteinase.

One laboratory has taken the medium from cultured fibroblasts[6] of chick embryo tendons, precipitated the proteins with 50% saturated ammonium sulfate, and chromatographed the protein in DEAE-cellulose. The C-proteinase was eluted with 0.07–0.12 $M$ NaCl whereas N-proteinase in the same preparation eluted with 0.15–0.28 $M$ NaCl. Alternatively, the dissolved ammonium sulfate precipitate was layered onto a 5 to 20% sucrose gradient and centrifuged at 56,000 rpm in a SW60 rotor (Beckman) for 25 hr. The C-proteinase was recovered in the middle of the gradient, whereas the N-proteinase sedimented more rapidly. In our own laboratory we have extracted calvaria from 17-day-old chick embryos[8] under the same conditions used to extract N-proteinase from tendons. The proteins in the extract were precipitated with 60% saturated ammonium sulfate and then passed through a Sephadex G-150 column (Pharmacia). The peak of enzyme activity, which was in the included volume, was pooled, and the sample was passed through a DEAE-cellulose column equilibrated and eluted with 0.05 $M$ NaCl, 0.2% Triton X-100 in 0.05 $M$ Tris-HCl buffer, adjusted to pH 7.4 at 25°. The protein in the void column was used as a crude source of C-proteinase activity. However, it included N-proteinase activity. Also in our hands, the total yields of C-proteinase with this procedure, as well as the alternative procedure for fibroblast medium, were small and highly variable.

Lysyl Oxidase

Lysyl oxidase (EC 1.4.3.) is the enzyme that initiates the biosynthesis of cross-links in collagen and elastin by catalyzing oxidative deamination of $\epsilon$-amino groups of specific lysyl and hydroxylysyl residues in these proteins. The aldehydes that are generated in the reaction form cross-links through either Schiff base formation or aldol condensation, frequently followed by more complex reactions (for review, see Siegel[24]). The reactions beyond formation of the aldehydes appear to be nonenzymic. The lysyl oxidase reaction differs from that of other amino oxidases because the lysyl or hydroxylysyl residues that are deaminated must be in peptide linkage and apparently in amino acid sequences similar to those of collagen and elastin. The enzyme acts more readily on aggregated collagen than on soluble, monomeric forms of the protein.

*Assays*

*Preparation of Substrates.* Either collagen of elastin substrates or synthetic peptides with primary structure similar to that of elastin[25] can be

[24] R. C. Siegel, *Int. Rev. Connect. Tissue Res.* **8**, 73 (1979).
[25] H. M. Kagan, L. Tseng, P. C. Trackman, K. Okanoto, R. S. Rapaka, and D. W. Urry, *J. Biol. Chem.* **25**, 3656 (1980).

used to assay the enzyme. For preparation of an elastin substrate,[26] 18 aortas from 17-day-old chick embryos were incubated for 24 hr in 10 ml of Eagle's minimum essential medium minus lysine and glutamine. The medium contained β-aminopropionitrile fumarate (50 µg/ml), ascorbic acid (50 µg/ml), penicillin G (2000 units/ml), and 200 µCi of either DL-[6-$^3$H]lysine or DL-[4,5-$^3$H]lysine (New England Nuclear). The aortas were rinsed with distilled water, lyophilized, and stored at 4°. On the basis of sensitivity to elastase and collagenase digestion, the substrate is about 88% elastin and 12% collagen.

A minor modification of this procedure provides both a soluble substrate that is probably a mixture of both collagen and elastin, and an insoluble substrate.[27] In this procedure, the aortas were minced before incubation, the medium was Waymouth's lysine-free medium gassed with $CO_2$–$O_2$ (5:95, v/v), and the incubation was for 72 hr. The medium was dialyzed against water at 4°, boiled 8 min to inactivate endogenous enzymes, dialyzed against 0.12 $M$ NaCl in 16 m$M$ potassium phosphate, pH 7.7, and clarified by centrifugation at 105,000 $g$ for 60 min. The supernatant served as the soluble substrate. The insoluble substrate was prepared from the tissue itself by homogenizing it in 0.15 $M$ NaCl (0.23 ml per aorta) and centrifuging the homogenate at 17,300 $g$ for 10 min. The pellet was washed twice with 0.15 $M$ NaCl, suspended in water, boiled for 8 min, recentrifuged, and suspended in 0.12 $M$ NaCl in 6 m$M$ potassium phosphate, pH 7.7, at a concentration of about 2.5 × 10$^6$ cpm/ml.

For preparation of collagen substrate,[28] calvarial parietal bones from eighteen 17-day-old chick embryos were incubated at 37° in a 50-ml Erlenmeyer flask with 10 ml of Eagle's minimum essential medium without lysine and supplemented with 4 mg of proline, 5 mg of glycine, 5 mg of β-aminopropionitrile, 5 mg of ascorbic acid, and 20,000 units of penicillin G per 100 ml of medium. After preincubation for 30 min to reduce the free lysine pool, the medium was changed, 200 µCi of DL-[6-$^3$H]lysine was added to each flask, and the flasks were gassed for 1 min with a 5% $CO_2$–air mixture. After incubation for 24 hr, the calvaria from four flasks were pooled and homogenized in 40 ml of 1 $M$ NaCl in 0.05 $M$ Tris-HCl, pH 7.4 at 4°, for 1 hr with a ground-glass homogenizer and then centrifuged at 20,000 $g$ for 10 min. The collagen in the supernatant was precipitated by the addition of solid NaCl to 20% concentration. The precipitate was collected by centrifugation, resuspended in 7–9 ml of 0.15 $M$ NaCl in 0.1 $M$ NaH$_2$PO$_4$, pH 7.8, and then dialyzed for 48–72 hr against the same buffer at 4°. The dialyzate was clarified by centrifugation at

[26] R. C. Siegel, S. R. Pinnel, and G. R. Martin, *Biochemistry* **9**, 4486 (1970).
[27] E. D. Harris, W. A. Gonnerman, J. E. Savage, and B. L. O'Dell, *Biochim. Biophys. Acta* **341**, 332 (1974).
[28] R. C. Siegel, *Proc. Natl. Acad. Sci. U.S.A.* **71**, 4826 (1974).

30,000 g for 15 min, and the supernatant was used as substrate. This substrate was stable in solution for more than 2 weeks. By amino acid analysis it was approximately 90% pure collagen with 95 residues of hydroxyproline and 300 of glycine per 1000 residues. As determined by hydroxyproline content, the collagen content varied from 300 to 400 µg/ml. The specific activity of the collagen solution was approximately $10^6$ cpm/nmol, and approximately 1 mg of collagen was obtained per flask. The collagen solution (0.3–0.5 ml) contained about 90–150 µg of collagen or 300,000–500,000 cpm $^3$H. Background values for release of tritiated water (without addition of enzyme) were about 150 cpm per million cpm of substrate, but after storage of the collagen substrate for 10 days, rose to about 500 cpm, which corresponds to 10–15% of the tritiated water released by the enzyme.[24] The background was reduced by incubating the collagen substrate in 5 m$M$ β-aminopropionitrile for 1–2 hr at 37°. During this treatment, the collagen formed reconstituted fibrils; these were recovered by centrifugation. The pellet was redissolved in 0.15 $M$ NaCl in 0.1 $M$ NaH$_2$PO$_4$, dialyzed against the same buffer, and stored frozen.[24]

*Assay of Lysyl Oxidase.* The enzymic reaction was carried out for 4 hr at 37° in a volume of 1 ml containing 200,000–500,000 cpm of substrate, varying amounts of enzyme, and 120 mM NaCl in potassium phosphate buffer, pH 7.7.[27] The tritiated water formed was isolated by vacuum distillation. Aliquots of 0.2 ml of the distillate were then assayed by liquid scintillation counting.

One modification[29] of this procedure is to use a microdistilling apparatus that makes it possible to collect the distillate directly in a liquid scintillation vial. Another is to separate the tritiated water by gel filtration.[30]

The assay is specific for lysyl oxidase as opposed to nonspecific amine oxidases. Several tests were used to examine the specificity. One was to add high concentrations of free lysine in order to rule out other enzymes not specific for peptidyllysine. A second was to determine whether the reaction, as expected for lysyl oxidase, was inhibited by β-aminopropionitrile—about 50% inhibition with 10 µ$M$ concentration. A third test was to prepare the substrate with a [$^{14}$C]lysine label, to oxidize the allysine formed by the reaction to α-aminoadipic acid, and to assay of the $^{14}$C-labeled α-adminoadipic acid by column chromatography.[26]

*Purification of Lysyl Oxidase*

Femoral and tibial epiphysial cartilages from 17-day-old chick embryos were dissected free of surrounding tissues and homogenized with a

[29] R. L. Misiorowski, J. B. Ulreich, and M. Chvapil, *Anal. Biochem.* **71**, 186 (1976).
[30] J. Melet, G. D. Vianden, and B. N. Bachra, *Anal. Biochem.* **77**, 141 (1977).

TABLE V
PURIFICATION OF LYSYL OXIDASE[a]

| | Total protein (mg) | Total activity (cpm) | Purification (fold) | Recovery (% of initial) |
|---|---|---|---|---|
| Initial extract | 200 | 718,000 | 1[b] | 100 |
| DEAE-column I | 1.62 | 356,000 | 61 | 50 |
| DEAE-column II | 0.60 | 212,000 | 982 | 29 |

[a] From Siegel and Fu.[31]
[b] The value here is for urea extract obtained after first washing the tissue with phosphate-buffered saline. The specific activity of the urea extract obtained in this way is about eightfold higher than the specific activity of a direct urea extract of the tissue.

Polytron homogenizer at 4° in 0.15 $M$ NaCl, 0.10 $M$ Na$_2$HPO$_4$, pH 7.8 (2 ml per gram of tissue). After centrifugation at 17,000 $g$ for 10 min, the pellet was again homogenized in the same buffer and recentrifuged. The supernatant from these two homogenizations was discarded. The pellet was then homogenized in 6 $M$ urea, 0.05 $M$ Tris, pH 7.6, at 25° (2 ml/per gram of original tissue), and centrifuged at 40,000 $g$ for 60 min. The pellet was reextracted with the 6 $M$ urea buffer and recentrifuged. The supernatants from both urea extracts were pooled and used for subsequent studies.[31]

The pooled urea extract was loaded directly onto a DEAE cellulose (Whatman DE-52) column (2.5 × 18 cm) previously equilibrated with 6 $M$ urea in 0.05 $M$ Tris, pH 7.6, at 25°. After loading, the column was washed with this buffer until the absorbance at 280 nm had returned to the initial value. Elution was with 6 $M$ urea in 0.05 $M$ Tris, pH 7.6, using a linear gradient of NaCl from 0 tp 0.5 $M$. The total volume of the gradient was 400 ml. Two distinct peaks of enzyme activity were routinely found. Fractions from each peak were dialyzed overnight at 4° against 0.15 $M$ NaCl in 0.1 $M$ NaH$_2$PO$_4$, pH 7.8 (PBS), and pooled. They were then stirred for 3 hr at 4° with an affinity resin that had been prepared by coupling lathyritic rat skin collagen to Sepharose 4B after cyanogen bromide activation.[28] The resin was then eluted sequentially at 25° with 0.05 $M$ Tris, pH 7.6; 1 $M$ NaCl, 0.05 $M$ Tris, pH 7.6; and 6 $M$ urea, 0.05 $M$ Tris, pH 7.6. Enzyme activity from both activity peaks was eluted from the affinity resin with 6 $M$ urea eluate. This fraction was loaded onto a DEAE-cellulose column (1.5 × 10 cm) at 25°. Elution was with 6 $M$ urea in 0.05 $M$ Tris, pH 7.6, using a linear

[31] R. C. Siegel and J. C. C. Fu, *J. Biol. Chem.* **251**, 5779 (1975).

gradient of NaCl 0 to 0.5 $M$. The total volume of the gradient was 200 ml. Fractions were dialyzed against 0.15 $M$ NaCl in 0.1 $M$ NaH$_2$PO$_4$ buffer, and the tube with highest activity was used for subsequent studies. The fraction purified from the second peak of activity on the initial DEAE-chromatogram had a protein content of approximately 2 $\mu$g/ml. For some experiments the tubes with highest activity were purified by another cycle of absorption to the affinity resin and rechromatography on DEAE. The purified protein (Table V) gave a single band with an apparent molecular weight of 62,000 by gel electrophoresis in sodium dodecyl sulfate. Antibodies to the protein were shown to react with the enzyme band on the gel by a competitive antibody assay. However, all tissues examined appeared to contain more than one form of the enzyme, since multiple peaks of activity are obtained by chromatography on DEAE-cellulose of gel filtration. Also, two forms of the enzyme with different immunobiological characteristics were isolated from bovine aorta.[32] However, it has not been established whether these various forms are isozymes or partially degraded forms of the same enzyme.

Acknowledgments

We are grateful to Dr. Robert Siegel for comments and suggestions about a preliminary form of this paper. Research work referred to here was supported in part by NIH grant AM-16516.

[32] G. P. Vidal, J. J. Shieh, and K. T. Yasunobu, *Biochem. Biophys. Res. Commun.* **64**, 989 (1975).

# [12] C1q

## By K. B. M. REID

C1q is a serum glycoprotein of 410,000 molecular weight, which in association with the two proenzymes C1r and C1s forms the C1 complex,[1-3] i.e., the first component of the classical pathway of the complement system. The proenzymes C1r and C1s, both of approximately 83,000 molecular weight, are glycoproteins present in the blood as a calcium-dependent complex with the composition C1r$_2$-Ca$^{2+}$-C1s$_2$. No enzymic ac-

[1] I. H. Lepow, G. B. Naff, E. W. Todd, J. Pensky, and C. F. Hind, *J. Exp. Med.* **117**, 983 (1963).
[2] R. J. Ziccardi and N. R. Cooper, *J. Exp. Med.* **147**, 385 (1978).
[3] R. R. Porter, and K. B. M. Reid, *Adv. Protein Chem.* **33**, 1 (1979).

tivity has been found in C1q, and its major role appears to be concerned with the recognition and binding of activators of the classical pathway of complement. This binding results in the activation of the C1r and C1s proenzymes in the C1 complex, thus allowing the sequential activation of the later components of the pathway.[3] Immune complexes containing IgM, or certain subclasses of IgG, are probably the most important activators of the C1 complex; this is consistent with the finding that C1q and C1 bind strongly to aggregated immunoglobulins but weakly to monomeric immunoglobulin.[4-6] In the electron microscope a single molecule of C1q has the appearance of a bunch of flowers.[7,8] Its six globular "heads" are considered to contain the binding sites for immunoglobulin, and the six collagen-like "stems" may be involved in the binding and activation of the C1r and C1s proenzymes,[3] the inhibition of collagen-induced platelet-aggregation,[9,10] and the interaction of C1q with C1q receptors on lymphoid cells.[11]

Serum Concentration

The concentration of C1q in human serum has been reported to be as low as 60 $\mu$g/ml and as high as 190 $\mu$g/ml in normal individuals as judged by radial immunodiffusion methods. The lower value appears to be more accurate because there are several factors that cause the overestimation of C1q levels by radial immunodiffusion,[12] and a value of 63 $\mu$g/ml was obtained by a method involving the estimation of the hydroxyproline content of the euglobulin fraction (which should contain at least 95% of the C1q present) of human serum.[13] Thus a value of approximately 70 $\mu$g/ml appears to be the best estimate of the concentration of C1q in normal adult serum.

[4] W. Augener, H. M. Grey, N. R. Cooper, and H. J. Müller-Eberhard, *Immunochemistry* **8**, 1011 (1971).
[5] V. N. Schumaker, M. A. Calcott, H. L. Spiegelberg, and H. J. Müller-Eberhard, *Biochemistry* **15**, 5175 (1976).
[6] H. Metzger, *Contemp. Top. Mol. Immunol.* **7**, 119 (1978).
[7] H. R. Knobel, W. Villiger, and H. Isliker, *Eur. J. Immunol.* **5**, 78 (1975).
[8] B. Brodsky-Doyle, K. R. Leonard, and K. B. M. Reid, *Biochem. J.* **159**, 279 (1976).
[9] J. P. Cazenave, S. N. Assimeh, R. H. Painter, M. A. Packham, and J. F. Mustard, *J. Immunol.* **116**, 162 (1976).
[10] J. L. Wautier, H. Souchon, K. B. M. Reid, A. P. Peltier, and J. P. Caen, *Immunochemistry* **14**, 763 (1977).
[11] Y. Gabay, H. Perlmann, P. Perlmann, and A. Sobel. *Eur. J. Immunol.* **9**, 787 (1979).
[12] R. J. Ziccardi and N. R. Cooper, *J. Immunol.* **118**, 2047 (1977).
[13] C. L. Rosano, N. Parhami, and C. Hurwitz, *J. Lab. Clin. Med.* **94**, 593 (1979).

### Assay

Since C1q plays both a binding role and an activation role in the C1 complex, its activity may be estimated either by the use of an agglutination assay employing IgG-coated latex particles,[14] of a binding assay employing $^{125}$I-labeled C1q and immune aggregates,[15] or of a hemolytic assay.[16,17] The hemolytic assay is probably the most sensitive monitor of biological activity, since C1q has to play both its binding role (to antibody) and activation role (of the $C1r_2$-$Ca^{2+}$-$C1s_2$ complex). Kolb et al.[17] have described the preparation from human serum of a reagent that is devoid of C1q and thus can conveniently be used in a hemolytic assay, along with sheep erythrocytes and antibody to sheep erythrocytes, which allows the detection of C1q down to the 1 ng level.

### Purification Procedure

Outdated human plasma may be used in the preparation of C1q without any apparent significant lowering of the final yield of protein or activity. Clotting of the plasma is performed by addition of $CaCl_2$ to a concentration of 20 m$M$ and incubation at 37° for 3 hr, followed by centrifugation and final separation of serum from the clot by the use of nylon gauze or muslin. The use of proteolytic inhibitors, such as diisopropyl fluorophosphate (DFP), phenylmethane sulfonyl fluoride (PMSF), phenanthroline monohydrate, and iodoacetamide, does not appear to enhance the final yield of C1q or affect the properties of the finally purified material.

Three main types of procedure have been successfully used in the preparation of human C1q: (*a*) conventional purification procedures involving euglobulin precipitation, ion-exchange chromatography, and gel filtration; (*b*) precipitation procedures involving primarily the use of DNA, or low ionic strength buffers containing the chelating agents EGTA and EDTA, to precipitate out specifically the C1q; (*c*) affinity chromatography on IgG-Sepharose 4B. In general procedures (*a*) or (*c*) might be used when relatively large amounts of serum (>500 ml) are available and quantities greater than 10 mg of C1q are routinely required, or if it is desired to purify proteins other than C1q (e.g., C1r and C1s) at the same time. If a small amount of serum (<50 ml) is available, or if only a small amount of C1q is required, procedure (*b*) employing low ionic strength buffers containing chelating agents should be considered.

[14] R. W. Ewald and A. F. Schubart, *J. Immunol.* **97**, 100 (1966).
[15] C. Heusser, M. Boesman, J. H. Nordin, and H. Isliker, *J. Immunol.* **110**, 820 (1973).
[16] M. A. Calcott and H. J. Müller-Eberhard, *Biochemistry* **11**, 3443 (1972).
[17] W. P. Kolb, L. M. Kolb, and E. R. Podack, *J. Immunol.* **122**, 2103 (1979).

*Conventional Procedure.* This procedure is adapted from several publications[1,16-19] and unpublished observations. The first step, as described by Lepow et al.,[1] involves the precipitation of the euglobulin fraction of the serum at low ionic strength using pH 5.5 acetate buffer followed by solubilization of the precipitate in 0.5 $M$ NaCl–1 m$M$ EDTA, pH 7.4, and clarification of this solution by ultracentrifugation and removal of the lipid layer. The resolubilized precipitate is then dialyzed against pH 7.4 phosphate buffer, ionic strength 0.15, containing 1 m$M$ EDTA and applied to a column of DEAE-Sephadex A-50 equilibrated with the same buffer (a 7.0 cm × 20 cm column is required for a sample derived from 2 liters of human serum). The C1q is not retained on the column, and therefore all the protein eluted with the equilibrating buffer is pooled and concentrated either in an ultrafiltration cell using a PM-10 membrane (Amicon Corporation, Lexington, Massachusetts) or by the addition of ammonium sulfate at 50% of saturation at 4°.[17] In both cases, especially during ultrafiltration, it is essential that the speed of the magnetic stirrer be kept low in order to avoid irreversible denaturation of the C1q. The concentrated sample is then dialyzed against an acetate buffer pH 5.2 for gradient elution from a CM-32-cellulose column (Whatman Ltd., Springfield Mill, Maidstone, Kent, U.K.). Most of the applied sample is not retained on the column, but gradient elution is not started until the $E_{280}^{1cm}$ of the effluent is below 0.050. C1q is one of the last proteins to be eluted in the gradient,[16,18,19] and it is useful to monitor the column fractions by SDS–polyacrylamide slab-gel electrophoresis in order accurately to detect the position of the C1q peak from the very characteristic band patterns given by C1q in reducing and nonreducing conditions.[17,20,21] This is especially necessary if C1q hemolytic activity is not being measured. The last step of the procedure involves gel filtration on BioGel A-5M (Bio-Rad Laboratories, Richmond, California) in 5 m$M$ phosphate buffer, pH 7.4, containing 0.65 $M$ NaCl and 10 m$M$ EDTA, which allows separation of C1q from IgG and IgM.[17]

*Precipitation Procedures.* The procedure involving precipitation of C1q by DNA,[15,22] followed by digestion of the DNA with DNase, appears also to require gel filtration and ion-exchange chromatography steps, which makes it similar in some respects to the conventional procedure. The other precipitation procedure is that of Yonemasu and Stroud,[21] which involves three precipitations of the C1q by means of dialysis against low ionic strength buffers containing EGTA and EDTA. This procedure pro-

[18] K. B. M. Reid, *Biochem. J.* **141**, 189 (1974).
[19] K. B. M. Reid, D. M. Lowe, and R. R. Porter, *Biochem. J.* **130**, 749 (1972).
[20] K. B. M. Reid and R. R. Porter, *Biochem. J.* **155**, 19 (1976).
[21] K. Yonemasu and R. M. Stroud, *Immunochemistry* **9**, 545 (1972).
[22] V. Agnello, R. J. Winchester, and H. G. Kunkel, *Immunology* **19**, 909 (1970).

vides a high yield of C1q, and a relatively small volume (50 ml) of serum can be used. The preparation has been reported to contain a C1q inhibitor that copurifies with the C1q and can be separated from the C1q by affinity chromatography on Con A-Sepharose[23]; the inhibitor is not retained, and the C1q can be eluted with α-methyl-glucopyranoside.

*Affinity Chromatography on IgG-Sepharose.* In this method the C1q is bound to IgG that is covalently bound to Sepharose 4B.[17,24,25] The most successful application of this technique appears to be that described by Kolb *et al.*,[17] who used a linear salt gradient to elute the bound C1q rather than the stepwise conditions, involving 1:4 diaminobutane used in previous studies.[24,25] A gel-filtration step on BioGel A-5M[17] or an ion-exchange step on DEAE-cellulose[25] is considered necessary for the final purification of C1q by this method.

*Yields.* The conventional methods yield 0.63 mg[16] to 2.5 mg[19] per 100 ml of serum. The DNA-precipitation method[15] yields approximately 1.3 mg per 100 ml of serum, and the precipitation method involving low ionic strength buffers and chelating agents[21] yields 5 mg per 100 ml of serum. The affinity methods yield 1.7 mg[17] to 5.0 mg[25] per milliliter of serum.

*Preparation of C1q from Other Species.* C1q has been prepared from the ox,[26] rabbit,[19,27,28] rat,[29] and frog.[30] The conventional procedures[19,26,27,30] or Yonemasu and Stroud's precipitation technique[28,29] have been used, and the yields and properties of the C1q samples obtained suggest that a molecule very similar to human C1q is present at a level of 5–7 mg/100 ml in these animals.

Properties

*Stability.* Repeated freezing and thawing of C1q preparations, especially at neutral pH, leads to irreversible denaturation, as does vigorous stirring, e.g., in a stirred ultrafiltration cell. Samples stored at $-70°$ in 0.25 $M$ NaCl at pH 5.3 retain full hemolytic activity for at least 6 months. Kolb *et al.*[17] recommend the precipitation of finally purified C1q by ammonium sulfate at 50% saturation and 4° followed by resuspension in 50 m$M$ NaCl containing 40% glycerol for prolonged storage at $-20°$ or $-70°$.

[23] J. D. Conradie, J. F. Volanakis, and R. M. Stroud, *Immunochemistry* **12**, 967 (1975).
[24] C. R. Sledge and D. H. Bing, *J. Immunol.* **111**, 661 (1973).
[25] S. N. Assimeh, D. H. Bing, and R. H. Painter, *J. Immunol.* **113**, 225 (1974).
[26] D. Campbell, N. A. Booth, and J. E. Fothergill, *Biochem. J.* **177**, 531 (1978).
[27] S. M. Paul and P. A. Liberti, *J. Immunol. Methods* **21**, 341 (1978).
[28] J. E. Volanakis and R. M. Stroud, *J. Immunol. Methods* **2**, 25 (1972).
[29] K. Höffken, P. J. McLaughlin, M. R. Price, V. E. Preston, and R. W. Baldwin, *Immunochemistry* **15**, 409 (1978).
[30] R. J. Alexander and L. A. Steiner, *J. Immunol.* **124**, 1418 (1980).

*Physical Properties and Composition.* The extinction coefficient of C1q ($E_{1cm}^{1\%}$) is 6.82 at 280 nm.[19] C1q has a molecular weight, in nondissociating conditions, of 410,000 and an $s_{20,w}^{\circ}$ in acetate buffer, pH 5.2, of 10.20 S as estimated by analytical ultracentrifugation.[19] It has an unusual amino acid composition for a serum protein, since it contains 17% glycine, 5% 4-hydroxyproline, and 2.1% hydroxylsine.[16,19] Approximately 80% of the hydroxylysine residues are glycosylated by glucosylgalactosyl disaccharides or galactose monosaccharides.[31] The entire molecule is composed of 18 polypeptide chains (6A, 6B, and 6C) each having a molecular weight of approximately 23,000.[19,20] Each molecule of C1q contains nine disulfide-linked dimers of these chains (six A-B dimers and three C-C dimers.[20] In each of the three types of chain after a short N-terminal section of non-collagen-like amino acid sequence of 2–8 residues, there is a region of approximately 80 residues of typical -Gly-X-Y- repeating triplet, collagen-like, sequence (which is broken at one point in each chain) followed by approximately 110 residues of globular sequence that extends to the C terminus.[3,20] Strong evidence for the view that these collagen-like sequences, found in C1q, form triple helical structure has come from studies involving limited proteolysis of the molecule with collagenase[9,32,33] or pepsin[34], from circular dichroism studies,[8] and from electron microscopy studies.[7,8] Electron micrographs of C1q show quite clearly that the molecule is divided into two distinct types of structure, i.e., six globular "heads" each connected by strands to a central fibril-like region. A model has been proposed for C1q in which the six strands and central fibril-like region are considered to be composed of six collagen-like triple helices, and each globular "head" is considered to be composed of the 110 residues of globular sequence from one A, one B, and one C chain.[3,20]

The three different types of chain found in C1q can be separated preparatively, by ion-exchange chromatography in urea solutions, after performic acid oxidation[19]; the globular heads can be isolated after collagenase digestion of the whole molecule,[32,33] and the collagen-like regions can be isolated after limited proteolysis with pepsin.[34]

Some of the purification procedures cited in this chapter are given in greater detail in another volume in this series.[35] An efficient, and convenient, conventional procedure has recently been published[36] which allows a rapid, two-step, purification of C1q.

[31] H. Shinkai and K. Yonemasu, *Biochem. J.* **177**, 847 (1979).
[32] E. P. Pâques, R. Huber, and H. Priess, *Hoppe-Seyler's Z. Physiol. Chem.* **360**, 177 (1979).
[33] N. C. Hughes-Jones and B. Gardner, *Immunochemistry* **16**, 697 (1979).
[34] K. B. M. Reid, *Biochem. J.* **155**, 5 (1976).
[35] K. B. M. Reid, this series, Vol. 80 [3].
[36] A. J. Tenner, P. H. Lesarve, and N. R. Cooper, *J. Immunol.* **127**, 648 (1981).

## [13] Acetylcholinesterase

*By* TERRONE L. ROSENBERRY, PHILIP BARNETT, and CAROL MAYS

Acetylcholinesterase (EC 3.1.1.7) is associated primarily with cells involved in cholinergic synaptic transmission, but it is also found in a variety of other neuronal, and a few nonneuronal, cells.[1,2] In extracts of most cells, more than one sedimenting acetylcholinesterase form can be identified; six forms can be resolved from mammalian muscle.[3] The predominant 18 S form found in high ionic strength extracts from the electric organs of the eel *Electrophorus electricus* can be categorized[4,5] as an $A_{12}$ molecule and is composed of 12 catalytic subunits[6] linked asymmetrically by disulfide bonds to three collagen-like tail subunit polypeptides.[7-9] The $A_{12}$ nature of the principal asymmetric form from several rat, chicken, and human muscles and from bovine superior cervical ganglion also has been confirmed,[5] but, in contrast to eel electric organ, these tissues also contain considerable amounts of globular $G_4$, $G_2$, and $G_1$ forms. The $A_{12}$ form in rat diaphragm is localized primarily in the endplate region and virtually disappears on denervation of the muscle,[10] while the globular forms are distributed nearly uniformly in this muscle and show a much smaller decrease on denervation.

Observations that bacterial collagenase degrades endplate basement membrane and releases acetylcholinesterase from the endplate[11,12] suggested that a significant fraction of the endplate $A_{12}$ enzyme was localized in the junctional basement membrane by means of its collagen-like tail. This suggestion was supported by direct demonstration of acetylcholines-

---

[1] D. Nachmansohn, "Chemical and Molecular Basis of Nerve Activity." Academic Press, New York, 1959; 2nd rev. ed., 1975.
[2] T. L. Rosenberry, *Adv. Enzymol.* **43**, 103 (1975).
[3] J. Massoulié, *Trends Biochem. Sci. (Pers. Ed.)* **5**, 160 (1980).
[4] This nomenclature[5] distinguishes two classes of acetylcholinesterase: "asymmetric" forms $A_n$ that are assemblies of $n$ catalytic subunits associated with a collagen-like tail structure, and "globular" forms $G_n$ in which the assemblies of $n$ catalytic subunits are devoid of any detectable collagen-like components.
[5] S. Bon, M. Vigny, and J. Massoulié, *Proc. Natl. Acad. Sci. U.S.A.* **76**, 2546 (1979).
[6] F. Rieger, S. Bon, J. Massoulié, J. Cartaud, B. Picard, and P. Benda, *Eur. J. Biochem.* **68**, 513 (1976).
[7] T. L. Rosenberry and J. M. Richardson, *Biochemistry* **16**, 3550 (1977).
[8] L. Anglister and I. Silman, *J. Mol. Biol.* **125**, 293 (1978).
[9] T. L. Rosenberry, P. Barnett, and C. Mays, *Neurochem. Int.* **2**, 135 (1980).
[10] Z. W. Hall, *J. Neurobiol.* **4**, 343 (1973).
[11] Z. W. Hall and R. Kelly, *Nature (London), New Biol.* **232**, 62 (1971).
[12] W. Betz and B. Sakmann, *J. Physiol. (London)* **230**, 673 (1973).

terase activity in the endplate basement membrane matrix that survived selective destruction of nerve and muscle cell elements.[13] These observations have been widely interpreted as indicating that $A_{12}$ acetylcholinesterase forms are localized through the direct interaction of their collagen-like tail structures with the junctional basement membrane. Thus these tail structures are *de facto* junctional basement membrane collagens, and the eel electric organ tail structures are the first such collagens to be specifically identified and isolated. Although muscle endplate $A_{12}$ forms have not been isolated and subjected to subunit analysis, the similarities of endplate and electric organ $A_{12}$ forms in their native molecular weights[5] and their solubility, aggregation, and collagenase digestion patterns[5,8,14-18] indicate that their collagen-like tail structures must be quite comparable.

Our approach to a structural characterization of acetylcholinesterase from eel electric organ can be summarized by reference to Fig. 1. The central structure in this figure represents an 18 S molecule, in which twelve 75,000-dalton catalytic subunits (circles) are arranged in three tetrameric groups. Within each tetramer, two catalytic subunits are directly linked by one disulfide bond and the remaining two are each covalently attached by a single disulfide bond to one 30,000-40,000 dalton tail subunit.[7-9] Pepsin degrades the catalytic subunits and the noncollagen-like domains of the tail subunits that are adjacent to the catalytic subunits to give triple-helical tail subunit fragments that correspond to 24,000 daltons.[9,19,20] Either an endogenous protease in stored eel electric organ tissue or trypsin can cleave the tail subunits at a point that releases a $G_4$ 11 S catalytic subunit tetramer and leaves an $A_8$ 14 S form that retains most of the residual tail structure.[21] The 11 S enzyme generated by the endogenous protease remains linked to an 8000-dalton residual tail subunit fragment with a noncollagenous composition,[9,22] but this fragment is degraded by trypsin.[23] In the following sections we describe isolation procedures for 18 S and 14 S eel acetylcholinesterase, for the pepsin-resistant triple-

[13] U. J. McMahan, J. R. Sanes, and L. M. Marshall, *Nature (London)* **271**, 172 (1978).
[14] S. Bon, J. Cartaud, and J. Massoulié, *Eur. J. Biochem.* **85**, 1 (1978).
[15] C. D. Johnson, S. P. Smith, and R. L. Russell, *J. Neurochem.* **28**, 617 (1977).
[16] M. S. Watkins, A. S. Hitt, and J. E. Bulger, *Biochem. Biophys. Res. Commun.* **79**, 640 (1977).
[17] G. Webb, *Can. J. Biochem.* **56**, 1124 (1978).
[18] R. L. Rotundo and D. M. Fambrough, *J. Biol. Chem.* **254**, 4790 (1979).
[19] C. Mays and T. L. Rosenberry, *Biochemistry* **20**, 2810 (1981).
[20] L. Anglister, S. J. Leibovich, and I. Silman, *Meeting of the International Society for Neurochemistry, 7th, Jerusalem, 1979*.
[21] W. F. X. McCann and T. L. Rosenberry, *Arch. Biochem. Biophys.* **183**, 347 (1977).
[22] P. Barnett and T. L. Rosenberry, *Fed. Proc., Fed. Am. Soc. Exp. Biol.* **36**, 485 (1978).
[23] T. L. Rosenberry, P. Barnett, and C. Mays, in preparation.

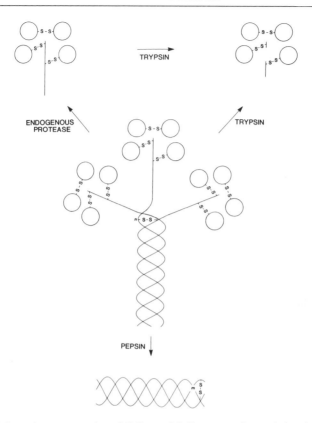

FIG. 1. Schematic representation of 18 S acetylcholinesterase from eel electric organ and the patterns of degradation shown by three different proteases.

helical tail subunit fragments, and for disulfide-reduced and denatured intact tail subunits.

## Isolation of 18 S and 14 S Acetylcholinesterase from Eel Electric Organ by Affinity Chromatography

The preparation of a methylacridinium-linked resin and its use for the purification of these acetylcholinesterase forms has been described by Dudai and Silman.[24] This synthesis of the methylacridinium ligand (10-methyl-9-[$N^\beta$-($\epsilon$-aminocaproyl)-$\beta$-aminopropylamino]acridinium)[24] is rather difficult,[7,25] and certain modifications have been proposed.[25] We

---

[24] Y. Dudai and I. Silman, this series, Vol. 34, p. 571.
[25] G. Webb and D. G. Clark, *Arch. Biochem. Biophys.* **191**, 278 (1978).

have developed an alternative acridinium-linked resin which gives similar purification values. The $pK_a$ of the protonated ring nitrogen (10 position) in the acridinium-linked resin (VIII) below is about 10.[26] Thus both the acridinium resin described here and the methylacridinium resin[24] are cationic at physiological pH values, and both show similar affinities for acetylcholinesterase.[7]

*Preparation of Acridinium-Linked Affinity Resins*

*Synthesis of 9-[$N^\gamma$-(Benzyloxycarbonyl)-$\gamma$-aminopropylamino]acridine hydrochloride (III)*. $N^1$-Benzyloxycarbonyl-1,3-propanediamine hydrochloride (II) is prepared[27] from bis(benzyloxycarbonyl)-1,3-propanediamine (I). In a typical preparation, carbobenzoxychloride (100 g, 0.58 mol) is added with rapid stirring to 1,3-propanediamine (37 g, 0.5 mol) in 800 ml of 0.75 $N$ NaOH at 0°. NaOH (1 $N$, 600 ml) is added, and the mixture is returned to 25° over 30 min. The white precipitate (I) is collected by filtration, washed twice with water (500 ml) and twice with petroleum ether (300 ml), and dried over $P_2O_5$ [96 g, 96%, mp 100–101° (lit[27]: 112–113°)]. To glacial acetic acid (550 ml) at 60° are added 12 $N$ HCl (50 ml) and (I) (96 g, 0.28 mol). This mixture is refluxed for 20 min, cooled to 4° for 3 hr, and the precipitate (propanediamine dihydrochloride, 8 g, mp 245–250°) is discarded. To the filtrate is added anhydrous ethyl ether (3.75 liters), and after 2 hr at 25° the precipitate (II) is collected and dried (30 g, mp 176–179°). The precipitate is dissolved in absolute ethanol (500 ml) at 70° and filtered. A small amount of (II) (3 g, mp 185–187°) crystallizes from the filtrate after 20 hr at 4°, but further crystallization of (II) occurs after addition of anhydrous ethyl ether (1.0 liter) for 1 hr at 25° [20 g, total 33%, mp 184–186° (lit[27]: 186–187°)].

9-Phenoxyacridine is prepared[28] from 9-chloroacridine (Eastman). Sodium hydroxide pellets (2.3 g) are dissolved in phenol (40 g) at 100° and 9-chloroacridine (10 g, 0.047 mol) is added slowly with stirring. After 1 hr at 115–120°, the mixture is poured into 2 $N$ NaOH (330 ml), stirred, and allowed to stand for 3 hr. The 9-phenoxyacridine precipitate is collected by filtration, washed with water (1 liter), and dried over first NaOH and then $P_2O_5$ [12.2 g, 96%, mp 123° (lit[28]: 127–128°)]. Compound (III) is prepared from (II) (11.0 g, 0.045 mol) and 9-phenoxyacridine (12.2 g, 0.045 mol) in phenol (40 g).[29] The mixture is maintained at 110–130° for 90 min and

---

[26] A. Albert, "The Acridines," p. 170. Arnold Press, London, 1966.
[27] W. B. Lawson, M. D. Leafer, Jr., A. Tewes, and G. J. S. Rao, *Hoppe-Seyler's Z. Physiol. Chem.* **349**, 251 (1968).
[28] A. Albert, "The Acridines," p. 282. Arnold Press, London, 1966.
[29] A. Albert, "The Acridines," p. 307. Arnold Press, London, 1966.

briefly triturated with anhydrous ethyl ether (200 ml) to form a light-brown gum. After discarding the solvent, anhydrous ethyl ether (800 ml) is added to the gum. This mixture is stirred until the gum is completely converted to a yellow precipitate (III) which is washed with additional ether (16.3 g, 86%, mp 116-118°, hygroscopic). Recrystallization is from ethanol–anhydrous ethyl ether (mp 155-157°, analysis[7]).

*Synthesis of 9-γ-Aminopropylaminoacridine Dihydrobromide (IV).* Compound (IV) is obtained after dissolving (III) (15.3 g, 0.039 mol) in glacial acetic acid (100 ml) and adding 30–32% HBr in glacial acetic acid (Fisher; 50 ml) with swirling. The yellow crystals (IV) obtained are washed repeatedly with anhydrous ethyl ether (15.9 g, 98%, mp 260–262° dec; analysis[7]).

*Synthesis of 9-[$N^\gamma$-($N^\epsilon$-Benzyloxycarbonyl-ε-aminocaproyl)-γ-aminopropylamino]acridine (VI).* Compound (VI) is synthesized from (IV) and p-nitrophenyl N-benzyloxycarbonyl-6-aminocaproate (V). To N-benzyloxycarbonyl-6-aminocaproic acid[30] (20 g, 0.075 mol) and p-nitrophenol (13 g, 0.091 mol) in ethyl acetate (300 ml) at 0° is added dicyclohexylcarbodiimide (16 g, 0.075 mol), and the mixture is stirred for 2 hr.[31] Glacial acetic acid (1 ml) is added, precipitated N,N'-dicyclohexylurea is removed by filtration, and solvent is removed from the filtrate under reduced pressure to give an oil. The oil is dissolved in isopropanol (300 ml) and acetic acid (0.3 ml), and crystals (V) are obtained after 48 hr at 4° (22 g, 78%, mp 62-65°; mass spec[32] for $C_{20}H_{22}N_2O_6$: observed pseudo-molecular ion, $(M + 1)^+$, 386.146; calculated, 386.148).

Compound (IV) (3.0 g, 7.0 mmol) is converted to the free base by dissolution in water (50 ml) and sequential addition of 1 N NaOH (8 × 1 ml) and extraction with chloroform (8 × 250 ml). The organic phases are combined, and the chloroform is reduced to 25 ml under reduced pressure. Dimethylformamide (8 ml) and (V) (3.5 g, 9.1 mmol) are added, and the mixture is stirred for 2 hr. TLC (4 : 1 : 1 n-butanol : acetic acid : water on Silica Gel G) shows the appearance of a new yellow spot (VI) at an $R_f$ of 0.5, although some free base starting material remains at an $R_f$ of 0.3. This starting material is quantitatively removed by sequential extraction of the mixture with 0.2 M sodium phosphate (2 × 250 ml) and 0.2 N NaOH (2 × 250 ml). Solvent is removed from the remaining

[30] T. L. Rosenberry, H. W. Chang, and Y. T. Chen, *J. Biol. Chem.* **247**, 1555 (1972).
[31] M. Bodanszky and V. Du Vigneaud, *Biochem. Prep.* **9**, 110 (1962).
[32] Mass spectral analyses for compounds V–VII were performed on a Kratos MS-30 mass spectrometer. Spectra were observed using electron impact ionization (70 eV) and/or chemical ionization using isobutane as reagent gas. Compounds were inserted on the direct insertion probe. $^2$H NMR spectra were also observed for compounds V–VII at 270 MHz. Spectral details are consistent with the proposed structures but will not be discussed here.

organic phase under reduced pressure, and the resultant oil partially solidifies on drying *in vacuo* overnight. Acetonitrile (10 ml) is added, and crystallization of (VI) proceeds slowly over several days at 4° (1.7 g, 50%, mp 152–156°; mass spec for $C_{30}H_{34}N_4O_3$: observed molecular ion, $(M)^+$, 498.258; calculated, 498.263).

*Synthesis of 9-[$N^\gamma$-($\epsilon$-Aminocaproyl-$\gamma$-aminopropylamino*)*acridine Dihydrobromide (VII)*. Compound (VII) is obtained after dissolving (VI) (0.4 g, 0.7 mmol) in glacial acetic acid (3 ml) and adding 30–32% HBr in glacial acetic acid (50 ml) with swirling. The yellow crystals (VII) obtained are washed repeatedly with ethyl acetate (0.4 g, 95%, mp 145–150°; mass spec for $C_{22}H_{30}Br_2N_4O$: observed molecular ion, $(M)^+$, 526; calculated, 526).

*Coupling of Acridine Ligands to Sepharose CL 4B*. Two alternative routes to the affinity resin 9-[$N^\gamma$-($N^\epsilon$-Sepharose 4B-$\epsilon$-aminocaproyl)-$\gamma$-aminopropylamino]acridinium (VIII) have been explored. Both involve initial activation of Sepharose CL 4B (Pharmacia, 200–400 ml) with cyanogen bromide.[33] In Method A, the activated resin is first linked with 6-aminocaproic acid (1–8 μmol per milliliter of packed resin) and then coupled with compound (IV) through an N-hydroxysuccinimide ester activation procedure.[7,34] In Method B, the activated resin is linked directly with compound (VII).[33] While Method A avoids the additional organic syntheses involved in extending (IV) to (VII), it suffers from a low efficiency in the second coupling step and results in maximal attached ligand concentrations of 0.5–0.6 μmol of (IV) per milliliter of packed resin, a range that is somewhat low for optimal purification of 18 S and 14 S acetylcholinesterase with this resin. In contrast, coupling of (VII) in Method B proceeds with nearly 100% efficiency and permits attached ligand concentrations of 0.4–0.8 μmol of (VII) per milliliter of packed resin, the entire range for optimal purification with this resin.[25] We evaluate the quantity of attached ligand by a direct observance measurement of the affinity resin itself in 0.1 cm pathlength cuvettes. The concentration of attached ligand is calculated assuming the spectral properties of compound III [$\lambda_{max}$ 411 ($\epsilon$ 9800), 433 ($\epsilon$ 8400), 394 ($\epsilon$ 6200)].

*Purification Procedures*

*Enzyme Assay*. The activity of acetylcholinesterase is monitored by a modification[7] of the spectrophotometric method of Ellman *et al.*[35] The

---

[33] S. C. March, I. Parikh, and P. Cuatrecasas, *Anal. Biochem.* **60**, 149 (1974).
[34] P. Cuatrecasas and I. Parikh, *Biochemistry* **11**, 2291 (1972).
[35] G. L. Ellman, K. D. Courtney, V. Andres, Jr., and R. M. Featherstone, *Biochem. Pharmacol.* **7**, 88 (1961).

assay contains 0.50 m$M$ acetylthiocholine iodide, 0.33 m$M$ 5,5'-dithiobis(2-nitrobenzoic acid), and 0.1 $M$ sodium phosphate (pH 7.0) in 3.00 ml at 25°; activity is measured as the increase in absorbance at 412 nm on addition of enzyme. Activity units ($\mu$mol/min) are defined under standard conditions (2.7 m$M$ acetylcholine bromide, 0.1 $M$ NaCl, 0.02 $M$ MgCl$_2$, pH 7.4, 25°) and here correspond to 3.94 $\Delta A_{412}$/min per unit.[7]

*Extraction of Enzyme.* An electric eel (4–5 ft, World Wide Scientific Animals, Ardsley, New York) is sacrificed, and the main electric organ (0.6–1.5 kg) is removed by dissection. The fresh tissue is immediately mixed with 1–2 volumes of sodium phosphate (pH 7.0, 10 m$M$) and homogenized at 4° in a Waring blender. The homogenate is centrifuged at 10,000 $g$ for 1 hr. The supernatant (S$_1$), containing 5–10% of the total enzyme activity, is discarded, and the residue is rehomogenized with 0.8 volume (relative to the initial tissue volume) of 1.4 $M$ NaCl, 23 m$M$ sodium phosphate, pH 7.0. After centrifugation of this homogenate at 48,000 $g$ for 30 min, the supernatant (S$_2$) is collected, and the residue is rehomogenized with 0.4 volume of 1.0 $M$ NaCl, 20 m$M$ sodium phosphate, pH 7.0 (buffer I). This homogenate is also centrifuged at 48,000 $g$ for 30 min, and the residue is discarded; the supernatant (S$_3$) is combined with S$_2$ as the crude onput to the affinity column.

*Affinity Chromatography.* All solvents are Millipore filtered. A single chromatography column (5 × 20 cm) at 4° is packed with 400 ml of affinity resin in buffer I. The packed column is washed with 3–5 volumes of buffer I, and the crude enzyme extract is introduced to the column at a flow rate of 0.1–0.2 column volume per hour. After entry of the crude extract, the column is washed with 4–6 volumes of buffer I at a flow rate of 0.1–0.2 column volume per hour. Elution of the retained enzyme is accomplished with decamethonium bromide (5 m$M$ in buffer I) at a flow rate of 0.05 column volume per hour. About 2–7% of the crude enzyme activity applied to the column is not retained; 2–10% is recovered in the buffer wash; and 55–75% is eluted sharply with decamethonium. The purified enzyme has a specific activity of 3800 units ml$^{-1}$ $A_{280}^{-1}$, about 75 times that of the crude onput. It consists of about 75% 18 S and 25% 14 S forms and is free of detectable protein contaminants as judged by sodium dodecyl sulfate gel electrophoresis.[7] The column can be regenerated with 6 $M$ guanidine hydrochloride and reused.[7]

The purified enzyme is concentrated by dialysis against 10 m$M$ sodium phosphate, pH 7 (4–16 hr, 4°) followed by centrifugation at 75,000 $g$ for 7–16 hr in an SW25.1 or an SW27 rotor. Under these low ionic-strength conditions, the 18 S and 14 S forms aggregate[3] and are recovered by suspension of the centrifugation pellet. Pellet recoveries correspond to 80% of the enzyme activity at concentrations up to 8 mg/ml.[19]

*Protein Determinations.* Concentrations of 18 S and 14 S acetylcholinesterase are estimated by absorbance measurements at 280 nm assuming $\epsilon_{280}^{1\%} = 18.0$.[7] Polypeptides containing the enzyme active site serine are monitored by prelabeling the enzyme stock (10–100 $\mu N$) with 20–200 $\mu M$ [$^3$H]diisopropylfluorophosphate (0.9 Ci/mmol, New England Nuclear) in 10 m$M$ sodium phosphate buffer for 1 hr.

Isolation of Pepsin-Resistant Fragments of 18 S and 14 S Acetylcholinesterase

Pepsin digestion[36] of tissue or tissue extracts at low pH is frequently used to solubilize collagen-like proteins or selectively to degrade noncollagen-like protein domains.

*Pepsin Digestion and Gel Exclusion Chromatography.* To suspensions of acetylcholinesterase (2–6 mg/ml in 10 m$M$ sodium phosphate, pH 7) are added glacial acetic acid and stock porcine pepsin (Worthington 2× recrystallized, 1–10 mg/ml in 50 m$M$ acetic acid). Final concentrations are 0.5 $M$ acetic acid and 0.02–0.1 mg of pepsin per milliliter. This mixture is maintained at 15° for 6 hr, after which the digestion is terminated by the addition of stock Tris-chloride (2.9 $M$, pH 8.5) to a final concentration of 1.0 $M$, pH 8.0. The pH 8 digestion mixture (2–8 ml) is applied to a 145-ml Sepharose CL 6B column (1.5 × 82 cm) equilibrated in buffered 1.0 $M$ NaCl (50 m$M$ Tris-chloride or 20 m$M$ sodium phosphate, pH 7) at 4°. A flow rate of 10 ml/hr is maintained by a hydrostatic pressure differential. Equal fractions of 2–3 ml are collected, and the absorbances of each fraction at 280, 250, 225, and 215 nm are measured. Elution of this sample results in fractions that can be grouped somewhat arbitrarily into the five pools shown in Fig. 2. Pool I corresponds to the column void volume. Protein in this pool is aggregated, as suggested by the low absorbance ratio $A_{280}/A_{250}$ of about 0.8 for the fractions in this pool. In contrast, fractions that elute after pool I are characterized by $A_{280}/A_{250}$ values of greater than 2 that are typical of acetylcholinesterase prior to digestion.

*Protein Content of Fractions Obtained from Gel Exclusion Chromatography.* The protein content of these fractions can be established by direct amino acid analysis. Analyses in our laboratory are conducted on a Beckman 119CL Amino Acid Analyzer with a Model 126 Data System. Analyzer buffers A, B, and C correspond respectively to 0.2 $M$ sodium citrate, pH 3.10; 0.2 $M$ sodium citrate, 0.2 $M$ NaCl, pH 3.95; and 0.2 $M$ sodium citrate, 0.8 $M$ NaCl, pH 5.10. Protein hydrolysis of dried samples dissolved in 6 $N$ HCl (Baker, Ultrapure) containing 80 m$M$ mercaptoethanol is conducted in tubes sealed under argon for 16–20 hr at either

[36] A. L. Rubin, D. Pfahl, P. T. Speakman, P. F. Davison, and F. O. Schmidt, *Science* **139**, 37 (1963).

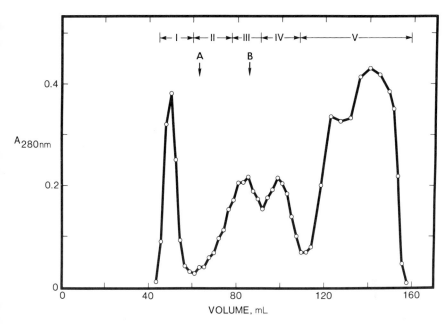

FIG. 2. Gel exclusion chromatography of a pepsin digest of 18 S and 14 S acetylcholinesterase. The column onput contained a digest of 10 mg of esterase and approximately 0.1 mg of pepsin[19] in 6.4 ml. Fractions were combined into the five pools designated by Roman numerals. Comparative elution volumes are indicated for (A) 18 S eel acetylcholinesterase (molecular weight 1,100,000) and (B) 11 S eel acetylcholinesterase (molecular weight 320,000).

110° or 115°. Hydrolyzed samples are dried, dissolved in buffer A, pH 2.2, and applied to the analyzer in 0.1–1.0 ml aliquots. Analyses involving quantitative protein determinations use nondialyzed dried samples containing up to 1.4 mmol of NaCl and 0.3 mmol of sodium phosphate in 2.0 ml of acid solution, and application aliquots contain no more than one-half of each sample in 1.0 ml. Amino acid standards for these determinations contain a similar quantity of salt and are carried through an identical hydrolysis procedure. Absorbance measurements can be related to these analyses, and approximate extinction coefficients can be estimated. For fragments in column pool I, values of $\epsilon_{280}^{1\%} = 50$ and of $(\epsilon_{215} - \epsilon_{225})^{1\%} = 150$ are observed.

A summary of the protein recoveries from the column in Fig. 2 is presented in Table I. Pool I, with nearly 30% glycine and over 5% each of hydroxyproline and hydroxylysine, is predominantly collagen-like in composition. Pool II has only one-half the mole percentages of hydroxyproline and hydroxylysine found in pool I and thus appears to consist of

TABLE I
RECOVERY OF PROTEIN AND OF AMINO ACIDS CHARACTERISTIC OF
COLLAGEN-LIKE POLYPEPTIDES FROM THE COLUMN CHROMATOGRAPHIC
FRACTIONATION OF THE PEPSIN DIGEST[a]

| Fraction | Protein (mg total amino acids) | Amino acid (mol %) | | |
|---|---|---|---|---|
| | | Hyp | Gly | Hyl |
| Column onput | 10.53 | 0.6 | 10.9 | 0.7 |
| Pool I | 0.59 | 5.2 | 27.9 | 5.6 |
| Pool II | 0.41 | 2.3 | 18.2 | 2.5 |
| Pool III | 0.76 | 0.9 | 12.8 | 1.0 |
| Pool IV | 0.73 | 0.0 | 6.1 | 0.0 |
| Pool V | 8.37 | 0.0 | 8.9 | 0.0 |
| Total recovery | 10.86 | 0.5 | 10.3 | 0.5 |

[a] Pool samples were not dialyzed to avoid the loss of small peptide fragments. Total amino acids are expressed in milligrams of mean residue weights.

about equal amounts of collagen-like and noncollagen-like fragments. Pool III seems to be at most 20% collagen-like, and pools IV and V are devoid of any fragments derived from collagen-like domains. The overall amino acid recovery is quantitative, although the total recoveries of hydroxyproline and hydroxylysine are 81% and 79%, respectively. Of the recovered hydroxyproline and hydroxylysine, 66% is obtained in pool I. The amino acid composition of the pool I fragments is given in Table III below.

*Characterization of Polypeptides Obtained from Pepsin Digestion.* Polyacrylamide gel electrophoresis in sodium dodecyl sulfate of the pepsin-digested sample followed by Coomassie Blue staining reveals only two discrete polypeptide fragment bands larger than pepsin. The molecular weights ($M_r$) corresponding to these bands can be estimated from parallel gels that contain cyanogen bromide fragments of $\alpha 1(I)$ and $\alpha 2(I)$ acid-soluble calf skin collagen as calibration standards. Both the larger band of 72,000 $M_r$, denoted $F_3$, and the smaller band of 48,000 $M_r$, $F_2$, represent polypeptides linked by disulfide bonds, as both bands are converted to a rather broad band $F_1$ of about 24,000 $M_r$ after exposure to a disulfide reducing agent. Several observations suggest that these polypeptide fragments correspond to the triple-helical collagen-like domain of the acetylcholinesterase tail structure.[19]

1. Gel bands representing these fragments are observed in only those pools in Table I that contain hydroxyproline and hydroxylysine. Pool I contains no other bands, and in pools II and III the intensity

of these bands relative to contaminant protein bands correlates well with the mole percentages of these amino acids.
2. These fragments are absent in pepsin digests of 11 S eel acetylcholinesterase, a form that is devoid of a collagen-like structure.
3. These fragments are not labeled by exposure of the 18 S and 14 S acetylcholinesterase to the catalytic site reagent [$^3$H]diisopropylfluorophosphate prior to pepsin digestion.
4. The circular dichroism spectra, including the mean residue ellipticities, of the fragments in pool I and of acid-soluble calf skin collagen are virtually identical and consistent with literature values for collagen triple helices.

Although most of the triple-helical collagen-like fragments are isolated in pool I at the column void volume, a separate peak of these fragments distinct from that in pool I can be detected in pool II.[19] Thus these collagen-like fragments can be obtained both in large aggregates and as discrete "dissociated" complexes prior to denaturation and separation into individual polypeptides in sodium dodecyl sulfate.

There are several indications that $F_2$ is derived from $F_3$ by further pepsin degradation.[19] The ratio of $F_2$ to $F_3$ is increased by higher pepsin concentrations, longer digestion times, or increased digestion temperatures. Pepsin conversion of $F_3$ to $F_2$ is particularly accelerated between 25° and 30°, suggesting that the triple-helical structure in the disulfide linkage region undergoes thermal destabilization in this temperature range. Digestion at 40° yields presumably triple-helical $F_1$ structures devoid of disulfide linkages, although further degradation to small fragments can be detected at this temperature.

Isolation of Intact Tail Subunits of 18 S and 14 S Acetylcholinesterase

Dissociation of the intact[37] tail subunits from catalytic subunits requires both reduction of the intersubunit disulfide bonds shown in Fig. 1 and exposure to a denaturing solvent. Subsequent isolation of the 30,000–40,000-dalton tail subunits is slightly complicated by the presence not only of intact 75,000-dalton catalytic subunit polypeptides, but also of 50,000-dalton and 24,000–27,000-dalton catalytic subunit fragments.[38] The 50,000-dalton fragment includes the active-site serine residue, while the 24,000–27,000-dalton fragments contain the single cysteine residue re-

[37] The word "intact" is used advisedly here. The isolated acetylcholinesterase contains some 14 S as well as 18 S enzyme, and the 14 S form itself appears to be derived from the 18 S form by the endogenous protease cleavage of the tail structure that releases one 11 S tetramer (Fig. 1).[21]
[38] T. L. Rosenberry, Y. T. Chen, and E. Bock, *Biochemistry* **13**, 3168 (1974).

sponsible for intersubunit disulfide bonds involving the catalytic subunit. The cleavage arises from an endogenous proteolytic agent, and about 20% of the catalytic subunits in the purified 18 S and 14 S acetylcholinesterase stock have undergone this fragmentation.

*Radioisotopic Labeling of Acetylcholinesterase Subunits.* The various polypeptide chains generated by disulfide reduction are conveniently monitored by double isotopic labeling. Acetylcholinesterase (5–15 mg) labeled at active sites with [$^3$H]diisopropylfluorophosphate (50–100 $\mu$Ci/$\mu$mol) as outlined above is dialyzed against water and lyophilized. The lyophilizate is dissolved (40–80 mg/ml) in 6 $M$ guanidine hydrochloride (Schwarz-Mann ultrapure) that contains 50 m$M$ dithiothreitol and 50 m$M$ Tris-chloride at pH 8.5. After 30 min at 50° the solution is cooled to room temperature and the pH is adjusted to 6–6.5 by the addition of crystalline NaH$_2$PO$_4$. A fourfold molar excess of [$^{14}$C]$N$-ethylmaleimide (0.2–0.4 $\mu$Ci/$\mu$mol) over dithiothreitol is then added, and after 45 min the protein solution is dialyzed against water and lyophilized.

Acetylcholinesterase has no free sulfhydryl groups prior to disulfide reduction. After this reduction and alkylation procedure, seven sulfhydryl groups on each catalytic subunit are $^{14}$C-labeled, five on the 50,000-dalton fragment and two on the 24,000–27,000-dalton fragments.[2,38] In addition, approximately five sulfhydryl groups per tail subunit are $^{14}$C-alkylated.

*Ion Exchange and Gel Exclusion Chromatography in Denaturants.* The double-labeled, denatured acetylcholinesterase lyophilizate is dissolved in 1 ml of 6 $M$ guanidine hydrochloride and dialyzed extensively against buffered urea (10 $M$ deionized[38] urea containing 5.0 m$M$ sodium acetate and 0.62 m$M$ acetic acid at an apparent pH of 6.5). DEAE-Cellulose (Whatman DE-52), precycled as directed by the manufacturer, is washed with 0.5 $M$ sodium acetate, pH 6.5, and equilibrated with the buffered urea in a 0.9 × 12.5 cm column. The labeled acetylcholinesterase sample is applied to the column, and eluted fractions (1.0 ml) corresponding to the column void volume (fractions 4–8) are pooled, dialyzed extensively against 10 m$M$ ammonium acetate, and lyophilized. This lyophilizate is dissolved in 0.2 ml of buffered 6 $M$ guanidine hydrochloride (20 m$M$ sodium phosphate, pH 6) and applied to a 100 ml Sepharose CL 6B column (0.9 × 157 cm) equilibrated in buffered guanidine hydrochloride. A column flow rate of 1.5–2 ml/hr is maintained hydrostatically, and 1.0 ml fractions are collected.

*Analyses of Polypeptides Recovered from the Chromatographic Procedures.* The purification procedure is evaluated by polypeptide banding patterns following polyacrylamide gel electrophoresis in sodium dodecyl sulfate and by the content of hydroxyproline and hydroxylysine, shown in

## TABLE II
RECOVERY OF PROTEIN AND OF AMINO ACIDS CHARACTERISTIC OF COLLAGEN-LIKE POLYPEPTIDES DURING PURIFICATION OF THE INTACT TAIL SUBUNITS

| Fraction | Protein[a] (mg total amino acids) | Hyp (mol %) | Hyl (mol %) | Recovery of Hyl (%) |
|---|---|---|---|---|
| Native enzyme applied to | 11.4 | 0.6 | 0.6 | 100 |
| DEAE-cellulose column | 9.8 | 0.5 | 0.6 | 91 |
| Void volume from DEAE-cellulose column | 1.18 | 3.3 | 3.4 | 63 |
| Total tail subunit pool from Sepharose CL 6B column | 0.58 | 4.0 | 4.5 | 42 |

[a] Amino acid analyses were conducted on dialyzed fractions. Total amino acids are expressed in milligrams of mean residue weights.

Table II.[9,39] The double-labeling procedure also aids in monitoring the separation of catalytic and noncatalytic subunit polypeptides throughout the fractionation procedure. All polypeptides are $^{14}$C-labeled, but only the 75,000-dalton and 50,000-dalton components contain $^3$H. These $^3$H-labeled polypeptides are quantitatively adsorbed to the DEAE-cellulose column. Only 0.3% of the $^3$H label is present in the column void volume, whereas most of the 30,000–40,000 dalton tail subunits and some of the 24,000–27,000 dalton catalytic subunit fragments are found in this fraction. These latter components are readily separated on the Sepharose CL 6B column, and fractions containing more than 40% of the initial hydroxylysine (Table II), but free of detectable catalytic subunit contamination, are obtained. The high concentrations of denaturants used in these procedures preclude direct protein determination by amino acid analysis without dialysis, and most of the loss of hydroxylysine occurs during dialysis steps. However, dialysis for the minimum necessary time (2 hr) in the presence of 10 m$M$ ammonium acetate minimizes these losses, especially at low protein concentrations. Basic amino acid hydrolyzates reveal that most of the hydroxylysine residues are linked to 2-$O$-$\alpha$-D-glucopyranosyl-$O$-$\beta$-D-galactopyranose.

The peak of the tail subunit pool is eluted from the Sepharose CL 6B column at a volume that corresponds to a 38,000 $M_r$ polypeptide relative to protein standards. Gel analyses in sodium dodecyl sulfate indicate that this pool contains a predominant band corresponding to 41,000 $M_r$ (ac-

[39] P. Barnett and T. L. Rosenberry, in preparation.

TABLE III
COMPARISON OF THE AMINO ACID MOLE
PERCENTAGES OF INTACT TAIL SUBUNITS AND
PEPSIN-RESISTANT TAIL SUBUNIT FRAGMENTS OF
14 S AND 18 S ACETYLCHOLINESTERASE[a]

| Amino acid | Intact tail subunits | Pepsin-resistant fragments |
|---|---|---|
| Hyp | 4.2 | 5.0 |
| Asp | 6.7 | 5.0 |
| Thr | 2.4 | 2.3 |
| Ser | 5.2 | 5.9 |
| Glu | 9.8 | 9.1 |
| Pro | 11.0 | 8.7 |
| Gly | 23.3 | 27.2 |
| Ala | 3.5 | 3.4 |
| Val | 4.7 | 4.3 |
| Met | 2.0 | 2.8 |
| Ile | 2.4 | 2.2 |
| Leu | 6.5 | 5.2 |
| Tyr | 1.5 | 1.7 |
| Phe | 2.0 | 1.2 |
| Hyl | 4.6 | 5.3 |
| Lys | 2.6 | 2.3 |
| His | 1.8 | 2.2 |
| Arg | 4.8 | 4.1 |
| Half-Cys | 1.4 | — |

[a] The intact subunit sample corresponded to the peak fractions from the Sepharose CL-6B column and is the average of four determinations.[9] The fragment sample corresponded to pool I fractions (Fig. 1) and is the average of nine determinations.[19]

cording to noncollagen standards) or 30,000 $M_r$ (according to the collagen standards noted above). Since the amino acid composition of this pool (Table III) is only partially collagen-like, the tail subunits contain both collagen-like and noncollagen-like domains. Thus a precise molecular weight is difficult to assign from these data. In addition to the predominant band, minor bands of both slightly larger and slightly smaller molecular weight are observed. Small amounts of components corresponding to dimers and trimers of the predominant band also are found.

*Comparison of the Amino Acid Compositions of the Intact Tail Subunits and the Pepsin-Resistant Tail Subunit Fragments.* These compositions are given in Table III. The 27% glycine content of the pepsin-resistant fragments is somewhat higher than that of the intact tail subunits, but this percentage is low enough to indicate that both preparations contain noncollagen-like domains. A major portion of the difference between the two preparations is contributed by the 8000-dalton residual tail subunit fragment generated by endogenous protease conversion of 18 S to 11 S acetylcholinesterase (Fig. 1). This fragment is not collagen-like, and its amino acid composition satisfactorily accounts for the difference between the compositions in Table III.[9]

Acknowledgments

This investigation was supported, in part, by National Institutes of Health Grants NS-16322 and NS-11766, by National Science Foundation Grant PCM77-09383, by the Muscular Dystrophy Association, and by the New York Heart Association.


[14] Hydroxylysine Glycosides

By WILLIAM T. BUTLER

Collagens and collagen-like proteins contain hexoses covalently linked to hydroxylysyl residues.[1,2] The structures of the two hydroxylysine-linked moieties are β-D-galactopyranosylhydroxylysine (Gal-Hyl) and 2-O-β-D-glucopyranosyl-O-β-D-galactopyranosylhydroxylysine (Glc-Gal-Hyl). A number of articles have appeared describing the isolation and quantitation of hydroxylysine glycosides from a variety of vertebrate and invertebrate sources. Likewise several scientific papers have documented the exact positions within collagen α chains to which the hexoses are attached. The methods described here are not exhaustive in reviewing all the techniques that are available, but are those that have been employed by the author or his colleagues.

Isolation of Glc-Gal-Hyl from Sponge

Katzman et al.[3] noted that the skeleton of a variety of sponges is a rich source of Glc-Gal-Hyl. The procedure described here is a simplification of

[1] W. T. Butler and L. W. Cunningham, J. Biol. Chem. 241, 3882 (1966).
[2] W. T. Butler, in "Glycoproteins and Glycolipids in Disease Processes" (E. F. Walborg, Jr., ed.), ACS Symp. Ser. 80, 213 (1978).
[3] R. L. Katzman, M. H. Halford, V. N. Reinhold, and R. W. Jeanloz, Biochemistry 11, 1161 (1972).

the procedure described by Katzman et al.[3] and involves digestion of the total sponge tissue in 2 N NaOH to liberate Glc-Gal-Hyl followed by separation of this glycoside from the digest by sequential chromatography on Sephadex G-25 and a sulfonated polystyrene column.[4]

A commercial ("natural") sponge purchased in a local hardware store and weighing approximately 20 g is washed thoroughly with water, dried, and cut into 1 cm$^3$ pieces. The pieces are refluxed in 150 ml of 2 N NaOH for 24 hr. The pH of the hydrolyzate is adjusted to 7.0 with glacial acetic acid, and the resulting precipitate is removed by centrifugation at 20,000 g for 1 hr. In order to concentrate the solution, the clear, brown supernatant is lyophilized and redissolved in 40 ml of water.

In order partially to separate Glc-Gal-Hyl from lower molecular weight peptides, amino acids and other components, 20 ml of this hydrolyzate is subjected to gel filtration on a 5.0 × 100 cm column of Sephadex G-25, fine (total volume 1960 ml), equilibrated and eluted with 0.2 N acetic acid. The eluent fractions are assayed for hexose with an automated orcinol procedure[5] and for ninhydrin-positive material by an automated modification[6] of the Moore and Stein[7] procedure. A sample chromatogram is illustrated in Fig. 1. The major orcinol-positive peak eluting ahead of the ninhydrin-reactive material is lyophilized to yield a brown powder.

A jacketed column of dimensions 0.9 × 150 cm constructed of glass that withstands high pressures (500–1000 psi) is packed with Technicon Chromobeads, type A resin at 37°. This column is then equilibrated with 0.2 M pyridine acetate buffer, pH 3.1[8] (starting buffer) by pumping at 50 ml/hr at pressures of 300–500 psi. The brown, hexose-containing powder (50–100 mg) from the Sephadex G-25 column, dissolved in 2–3 ml of starting buffer, is forced onto the Chromobeads A column with nitrogen pressure (30–50 psi). The column is eluted with a linear gradient formed from 750 ml of starting buffer and 750 ml of 2.0 M pyridine acetic buffer, pH 5.0[8] (limiting buffer). Ten-milliliter fractions are collected and analyzed for ninhydrin-positive material by an automated procedure.[6] A typical chromatogram is illustrated in Fig. 2 (upper panel). The Glc-Gal-Hyl is totally separated from other carbohydrate and amino acid-containing substituents by this procedure. Any alkali-catalyzed racemization (this volume [15]) is not detected on this column. Fractions containing Glc-Gal-

[4] E. J. Meehan and W. T. Butler, unpublished
[5] J. Judd, W. Clouse, J. Ford, J. van Eys, and L. W. Cunningham, Anal. Biochem. 4, 512 (1962).
[6] Technicon Instruction Manual, T-67-101 (1967).
[7] S. Moore and W. H. Stein, J. Biol. Chem. 176, 367 (1948).
[8] W. A. Schroeder, this series, vol. 11, p. 351.

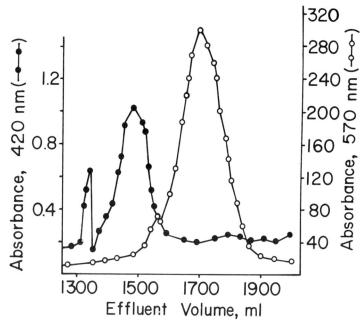

FIG. 1. Separation of the alkaline hydrolyzate of sponge skeleton on a 5.0 × 100 cm column of Sephadex G-25 (fine) eluted with 0.2 $M$ acetic acid. Fractions were assayed for hexose (-●-●-●-) by an automated orcinol procedure[5] and for peptides and amino acids (-○-○-○-) by an automated ninhydrin procedure.[6] Because of the large amount of ninhydrin-positive material in many fractions, they were diluted with water prior to analysis; the data points represent calculations taking into account these dilutions. The material eluting at 1420–1520 ml was lyophilized and used for isolation of Glc-Gal-Hyl, as illustrated in Fig. 2 (upper panel).

Hyl are lyophilized and subjected to gel filtration on a 1.5 × 100 column of BioGel P-2 eluted in 0.2 $N$ acetic acid. The Glc-Gal-Hyl, detected as above with ninhydrin, elutes from this column ahead of nonvolatile impurities contained in the pyridine acetate buffers. Lyophilized Glc-Gal-Hyl obtained in this way is a white powder and can be stored in a desiccator at $-20°$ indefinitely. The product yields only glucose and galactose when hydrolyzed in 2 $N$ HCl at 108° for 2 hr; the hydrolyzate is subjected to paper chromatography in $n$-butanol–pyridine–water (6:4:3), and the chromatogram is developed with the aniline-phthalate spray.[9] The product yields only hydroxylysine when the above hydrolyzate is subjected to amino acid analysis. For other structural proof, the reader should consult Katzman *et al.*[3]

[9] S. M. Partridge, *Nature* (*London*) **164**, 443 (1949).

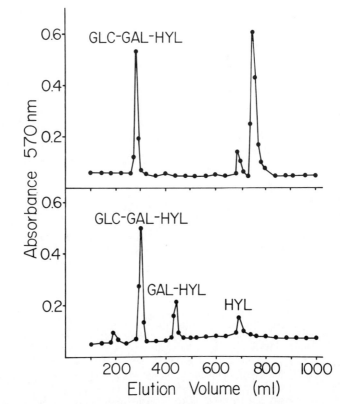

FIG. 2. Ion-exchange chromatography of hydroxylysine glycosides. A 0.9 × 150 cm column of Technicon Chromobeads, type A resin was equilibrated with 0.2 M pyridine acetate, pH 3.1[8] (starting buffer) at 37°. Samples were eluted from the column with a linear gradient formed from 750 ml each of starting buffer and 2.0 M pyridine acetate, pH 5.0.[8] Fractions were analyzed for ninhydrin-positive material[6] to detect hydroxylysine glycosides. *Upper panel:* Purification of Glc-Gal-Hyl obtained from Sephadex G-25 (see Fig. 1). *Lower panel:* Separation of the hydroxylysine glycosides and Hyl after partial acid hydrolysis of Glc-Gal-Hyl with 0.1 N HCl at 108° for 8 hr.

### Conversion of Glc-Gal-Hyl to Gal-Hyl

The Glc-Gal glycosidic bond is relatively acid labile compared to the Gal-Hyl bond. In order to produce Gal-Hyl, a sample of Glc-Gal-Hyl can be heated in 0.1 N $H_2SO_4$ of 0.1 N HCl at 108° for 8–16 hr. The reaction results in a mixture of Glc-Gal-Hyl, Gal-Hyl, and Hyl, the proportions of the three varying according to the type of acid and the length of hydrolysis.

In a typical experiment 10 mg of Glc-Gal-Hyl are placed in a Pyrex hydrolysis tube with 2 ml of 0.1 N HCl. The atmosphere of the tube is exchanged for nitrogen, and the tube is sealed. The solution is heated in an oil bath at 108° for 8 hr, after which the tube is opened and the contents are dried with a rotary evaporator.

The hydrolyzate is chromatographed on the Chromobeads A column used for separation of Glc-Gal-Hyl. As shown in Fig. 2 (lower panel), Gal-Hyl is completely separated from Glc-Gal-Hyl and Hyl by this procedure. The Gal-Hyl is lyophilized and subjected to gel filtration on BioGel P-2 (as above for Glc-Gal-Hyl). The lyophilized Gal-Hyl from BioGel P-2 is a white powder that can be desiccated at $-20°$ for indefinite periods. When a hydrolyzate (2 N HCl, 108°, 2 hr) is subjected to paper chromatography in $n$-butanol–pyridine–water (6:4:3), and sprayed with aniline phthalate, the only detectable sugar is galactose. The hydrolyzate yields only hydroxylysine on amino acid analysis.

Automatic Analysis of Hydroxylysine Glycosides in Proteins and Peptides

The quantitative determination of Glc-Gal-Hyl and Gal-Hyl in various collagens was first described by Spiro.[10] Using alkaline hydrolyzates, the amino acids and hydroxylysine glycosides were separated with a Technicon amino acid analyzer. Because of the overlap of Gal-Hyl with Tyr, the amino acids in the hydrolyzate had to be removed by paper chromatography prior to analysis of this glycoside.

An improved separation technique[11] was developed by Dr. E. J. Miller of the University of Alabama in Birmingham. Proteins or peptides containing 0.05–0.50 μmol of hydroxylysine glycosides are hydrolyzed in 2 N NaOH (0.1–1.0 ml) at 108° for 24 hr in sealed, alkali-resistant tubes in a nitrogen atmosphere. The contents of the tubes are cooled at 24°, the tubes are opened, and the pH is carefully adjusted to 5.3 with 0.5 N HCl. In order to be able to apply the sample directly to the amino acid analyzer, the neutralized hydrolysis is diluted 10-fold with water. At the same time as alkaline hydrolysis is performed, a portion of the protein or peptide is digested with constant-boiling 6 N HCl and submitted to routine amino acid analysis to determine its total amino acid content.

The hydroxylysine glycosides in the alkaline hydrolyzate are separated and quantitated on the $0.9 \times 60$ cm column of a Beckman Model 119 automatic amino acid analyzer. The column, packed with sulfonated polystyrene beads (resin M-82, Beckman Instruments, Munich Division) and eluted at a flow rate of 60 ml/hr at 51°, is equilibrated with 0.35 M

[10] R. G. Spiro, *J. Biol. Chem.* **244**, 602 (1969).
[11] W. T. Butler, J. E. Finch, Jr., and E. J. Miller, *Biochemistry* **16**, 4981 (1977).

(Na⁺) sodium citrate buffer, pH 5.3.[12] Aliquots of the diluted hydrolyzates (0.25 ml) are directly applied to the analyzer, and the column is eluted for 240 min with 0.35 $M$ (Na⁺) sodium citrate, pH 5.3, followed by a 100-min elution with 0.3 $M$ (Na⁺) sodium citrate containing 0.9 $M$ NaCl, pH 6.1. Most of the acidic and neutral amino acids are not retarded and emerge in the elution front of the column. Glc-Gal-Hyl elutes at 54 min followed by Tyr at 60 min, Phe at 68 min, Gal-Hyl at 88 min, and free Hyl at 160 min.

The amounts of Glc-Gal-Hyl and Gal-Hyl in alkaline hydrolyzate are computed relative to the amount of Hyl in the alkaline hydrolyzate, and amounts in the original protein or peptide can then be calculated from the Hyl of the total amino acid analysis. Examples of this type of analysis are given in the table. During sequence analysis of type II collagen, a CNBr peptide from bovine α1(II) chains was cleaved with trypsin. The three tryptic peptides T1, T2, and T9 were thus shown to contain Gal-Hyl, Gal-Hyl, and a mixture of Glc-Gal-Hyl and Gal-Hyl, respectively. Small amounts of unsubstituted Hyl were present in peptides T1 and T2.

Standards of Glc-Gal-Hyl and Glc-Hyl are prepared by dissolving 2-4 mg of the lyophilized samples in 10 ml of water in volumetric flasks. To standardize the amount of hydroxylysine glycosides in the solutions, 0.2-0.5 ml aliquots are transferred to hydrolysis tubes and evaporated to dryness. Two milliliters of 2 $N$ HCl are added, the contents are sealed under nitrogen, and the samples are hydrolyzed for 2 hr at 108°. The amount of hydroxylysine glycoside in the aliquot is then determined by quantitation of the hydroxylysine content by routine amino acid analysis. The color factors for Glc-Gal-Hyl and Gal-Hyl utilizing the above separation techniques are approximately 0.98 and 0.92 of that of Phe.

Sequence Analysis of Glc-Gal-Hyl and Gal-Hyl Residues

During the determination of the amino acid sequence of collagen chains and of collagen-like polypeptides, the elucidation of Glc-Gal-Hyl and Gal-Hyl residues has been problematic. When automated Edman degradation (using the liquid phase sequencer) is used for sequence analysis, the cycles where Pth-hydroxylysine glycosides should appear yield only "blanks"; that is, at that cycle of Edman degradation *no* Pth-amino acid can be detected in reasonable yields by gas-liquid chromatography, thin-layer chromatography, or high-performance liquid chromatography.[11] The derivative is probably not extracted by the $n$-butyl chloride used to recover amino acid derivatives after cyclization, but is simply retained in the spinning cup. Nevertheless, Edman degradation does result in com-

[12] E. J. Miller, *Biochemistry* **11**, 4903 (1972).

ANALYSIS OF AMINO ACIDS AND HYDROXYLYSINE GLYCOSIDES IN ACID AND
ALKALINE HYDROLYZATES OF TRYPTIC PEPTIDES FROM TYPE II COLLAGEN[a]

| Amino acid | α1(II)-CB8-T1 | α1(II)-CB8-T2 | α1(II)-CB8-T9 |
|---|---|---|---|
| *After acid hydrolysis, 6 N HCl, 108° for 24 hr* | | | |
| 4-Hyp | 1.5 | 0.8 | 3.8 |
| Asp | 0.9 | — | 1.1 |
| Thr | — | — | 1.9 |
| Hse[b] | — | — | 1.0 |
| Glu | 1.0 | 1.0 | 2.3 |
| Pro | 1.6 | 1.5 | 4.1 |
| Gly | 4.8 | 3.9 | 10.8 |
| Ala | 1.0 | 2.0 | 4.1 |
| Leu | — | 1.8 | 1.9 |
| Phe | 1.0 | — | — |
| Hyl | 1.0 | 1.1 | 1.0 |
| Lys | 1.1 | — | — |
| Arg | — | 1.1 | — |
| *After alkaline hydrolysis, 2 N NaOH, 108°, 24 hr* | | | |
| Glc-Gal-Hyl | — | — | 0.61 |
| Gal-Hyl | 0.91 | 0.94 | 0.39 |
| Hyl | 0.09 | 0.06 | — |

[a] Data are taken from Butler *et al.*[11] Data are given as residues per peptide.
[b] Homoserine.

plete cleavage of the glycosides from the polypeptide because yields of Pth-amino acids in succeeding cycles are not inordinately reduced. Thus the finding of a "blank" during a cycle of automated Edman degradation with good yields at preceding and succeeding cycles is suggestive evidence that a hydroxylysine glycoside could be present. Of course one cannot rule out the presence of Pth-Ser or Pth-Thr at this point because the recoveries of these derivatives are usually low.

In order to prove the existence of Glc-Gal-Hyl and Gal-Hyl at a particular site, relatively small peptides are isolated after digestions with trypsin or collagenase. The amino acid compositions are determined and compared to the sequence. If all the amino acids *except hydroxylysine* in the total analysis can be accounted for by the sequence analysis in which a "blank" was observed, the evidence for hydroxylysine glycosides in the blank positions is stronger. To confirm their existence and identify the type of hydroxylysine glycoside, the small peptides are subjected to alkaline hydrolysis and analysis of Glc-Gal-Hyl, Gal-Hyl, and Hyl as outlined in the preceding section.

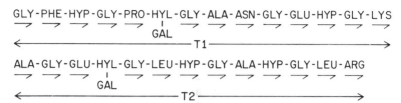

FIG. 3. The covalent structures of two tryptic peptides from α1(II)-CB8. All amino acids were placed in the sequence by identification of Pth-amino acids by automated Edman of α1(II)-CB8 (shown by the half arrows →) *except* the hydroxylysine glycosides. The *presence* of hydroxylysine was shown by amino acid analysis after acid hydrolysis and that of Gal-Hyl after alkaline hydrolysis (see the table). The exact *locations* of the hydroxylysine glycosides were then placed in the sequence by deduction.

The table gives analyses of this type used to confirm the existence of hydroxylysine glycosides in small peptides from type II collagen. In all three cases all the amino acid residues except the hydroxylysyl glycosides were placed within the sequence by automated Edman degradation (see Fig. 3). The analysis of alkaline hydrolyzates for Glc-Gal-Hyl, Gal-Hyl, and Hyl confirmed the *presence* and indicated the *nature* of the particular glycoside within the sequence. The location within the sequence is thus determined deductively. The conclusions for the sequences of peptides α1(II)-CB8-T1 and α1(II)-CB8-T2 are shown in Fig. 3.

Acknowledgment

This research was sponsored in part by Grant DE-02670 from the National Institute of Dental Research.

## [15] Asparagine-Linked Glycosides[1]

By CHARLES C. CLARK

Owing to the presence of hydroxylysine-linked glycosides (see this volume [14]), collagen (and *per force* procollagen) has been considered to be a glycoprotein. Moreover, it has been determined that the propeptide regions of procollagen contain asparagine-linked glycosides.[2-6] Many use-

[1] This work was supported in part by National Institutes of Health Grants AM-20553 and AM-00360.
[2] C. C. Clark and N. A. Kefalides, *Proc. Natl. Acad. Sci. U.S.A.* **73**, 34 (1976).
[3] C. C. Clark and N. A. Kefalides, *J. Biol. Chem.* **253**, 47 (1978).

ful procedures for characterizing such units have appeared in earlier volumes of this series,[7] but the emphasis was on quantitative analyses. In working with procollagen, however, it is often difficult to obtain sufficient quantities of materials for analysis. Thus, we have adapted many of these procedures for the analysis of radioactive samples.

In this section, the major emphasis will be on the preparation and characterization of radioactively labeled oligosaccharide units from type I procollagen and from lipid intermediates. We have utilized matrix-free chick embryo tendon cells in our experiments, but the methods should be adaptable to other systems.

### Preparation of Sugar-Labeled Procollagen

The preparation and characterization of procollagen is discussed in this volume [4]. For labeling of the propeptide oligosaccharide units, cells in suspension culture (2 to 3 × $10^7$ cells/ml) are incubated in Krebs medium II[8] with the glucose replaced by 5 m$M$ sodium pyruvate, and labeled in the presence of 10 $\mu$Ci/ml[9] of either [2-$^3$H]mannose (10–20 Ci/mmol), [6-$^3$H]glucosamine–HCl (10–25 Ci/mmol), or [6-$^3$H]glucose (5–15 Ci/mmol) (Amersham Corp.).[2,3,6] After incubation for up to 4 hr, the cells and medium are separated by centrifugation. The medium is used to prepare extracellular procollagen, and the cells are used to prepare intracellular procollagen and dolichol oligosaccharides.

### Preparation of Sugar-Labeled Glycopeptides and Oligosaccharides from Protein

*Proteolytic Digestion*

Pronase from *Streptomyces griseus* (Calbiochem-Behring) is the enzyme of choice for digestion of glycoproteins (Vol. 28 [1]).[7] Digestions are carried out in 0.5 to 1 ml of 0.1 $M$ Tris–HCl (pH 8.0) containing 2 m$M$ CaCl$_2$ in a screw-top tube. To each sample containing at least 100,000 cpm $^3$H is added 3 mg of predigested enzyme and a drop of toluene to

---

[4] B. R. Olsen, N. A. Guzman, J. Engel, C. Condit, and S. Aase, *Biochemistry* **16**, 3030 (1977).
[5] N. A. Guzman, P. N. Graves, and D. J. Prockop, *Biochem. Biophys. Res. Commun.* **84**, 691 (1978).
[6] C. C. Clark, *J. Biol. Chem.* **254**, 10,798 (1979).
[7] R. G. Spiro, this series, Vol. 8 [1] and [2]; Vol. 28 [1].
[8] P. Dehm and D. J. Prockop, *Biochim. Biophys. Acta* **240**, 358 (1971).
[9] P. D. Yurchenco, C. Ceccarini, and P. H. Atkins, this series, Vol. 50 [17].

inhibit microbial growth. Digestion is continued with stirring for 100 hr at 37° with 1.5 to 3 mg portions of Pronase added every 24 hr. Each aliquot of the enzyme is predigested for 1 hr at 37° immediately before addition to destroy endogenous glycosidases.[10]

The reaction is terminated by the addition of EDTA to 10 m$M$, any insoluble material is removed by centrifugation (1000 $g$, 20 min), and the digest is directly applied to a gel filtration column to obtain glycopeptides (see below).

*Alkaline-Borohydride Hydrolysis*

Alkaline-borohydride treatment is used to cleave the N-glycosidic linkage quantitatively between peptidylasparagine and $N$-acetylglucosamine to liberate intact oligosaccharide units. The procedure used is based upon that described by Zinn *et al.*[11] Samples are transferred to conical screw-top tubes and lyophilized. Hydrolysis is performed in 500 μl of 1 $M$ NaOH–4 $M$ NaBH$_4$ at 80° for 24 hr. After cooling and addition of an equal volume of distilled water, glacial acetic acid is added dropwise until the liberation of H$_2$ gas is complete. Using pH paper, the pH is adjusted to 5. The sample is centrifuged in the same tube (600 $g$, 10 min) to remove insoluble material, and the supernatant is directly applied to a gel filtration column to obtain oligosaccharides (see below).

Preparation and Characterization of Sugar-Labeled Oligosaccharides from Lipid Intermediates

It is well established that dolichol oligosaccharide intermediates are involved in the synthesis and assembly of asparagine-linked glycosides. A number of procedures for synthesizing, extracting, detecting, and characterizing these intermediates in cell-free systems were presented in this series.[12]

*Organic Solvent Extraction*

After the incubation of 0.25 to 1 × 10$^9$ cells with [$^3$H]sugar as described previously,[2,3,6] the cells are harvested by centrifugation (600 $g$, 10 min), washed with serum-free medium, resuspended in 1 ml of distilled water in a 10- to 15-ml siliconized conical screw-top tube, and frozen overnight at −20°.

---

[10] R. T. Schwarz. M. F. G. Schmidt, U. Anwer, and H.-D. Klenk, *J. Virol.* **23**, 217 (1977).
[11] A. B. Zinn, J. S. Marshall, and D. M. Carlson, *J. Biol. Chem.* **253**, 6761 (1978).
[12] N. H. Behrens and E. Tabora, this series, Vol. 50 [45].

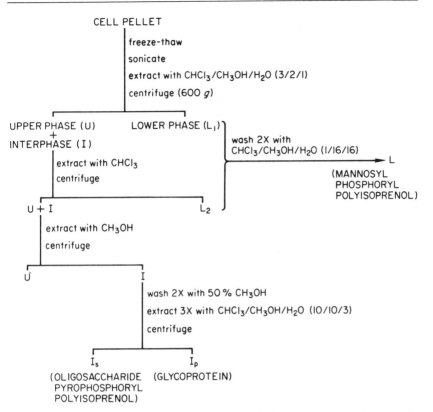

FIG. 1. Flow diagram for lipid extraction of matrix-free chick embryo tendon cells. Details are described in the text.

*Reagents and Buffers*

Chloroform ($CHCl_3$)
Methanol ($CH_3OH$)
$CHCl_3/CH_3OH/H_2O$ (1:16:16 by volume)
$CHCl_3/CH_3OH/H_2O$ (10:10:3 by volume)
50% $CH_3OH$
Tris–HCl, 0.1 $M$ pH 8.0, containing 2 m$M$ $CaCl_2$

*Procedure.* The extraction procedure is based upon those described previously (Fig. 1).[13,14] Cell pellets are thawed and dispersed by brief (0.5–1 min) immersion in an ultrasonic bath (Cole-Parmer). To the suspen-

[13] M. J. Spiro, R. G. Spiro, and V. D. Bhoyroo, *J. Biol. Chem.* **251**, 6400 (1976).
[14] J. W. Schmitt and A. D. Elbein, *J. Biol. Chem.* **254**, 12,291 (1979).

sion is added 3 ml of $CHCl_3$ and 2 ml of $CH_3OH$; the sample is extracted at room temperature for 15 min with occasional vigorous vortexing and then centrifuged (600 $g$, 10 min) to yield a lower phase (L), an interphase (I), and an upper phase (U).

$L_1$ is carefully removed with a Pasteur pipette, and 2 ml of $CHCl_3$ are added to the remaining I plus U. After vigorous vortexing to disperse the sample, centrifugation is repeated as described above. $L_2$ is removed, combined with $L_1$, and washed twice with equal volumes of $CHCl_3/CH_3OH/H_2O$ (1 : 16 : 16). After centrifugation, the upper layer is discarded and the combined lower layer represents L.

To the remaining I plus U is added 1 ml of $CH_3OH$; after vortexing and centrifuging, a supernatant and pellet are obtained. The supernatant is discarded, and the pellet is washed twice with 50% $CH_3OH$. The washes are discarded, and the pellet is extracted three times with 5-ml portions of $CHCl_3/CH_3OH/H_2O$ (10 : 10 : 3) with vigorous vortexing. After centrifugation, the supernatants ($I_s$) are combined and the pellet ($I_p$) is dried under a stream of $N_2$.

This protocol separates neutral lipids and mannosylphosphorylpolyisoprenol (L) from oligosaccharide pyrophosphorylpolyisoprenol ($I_s$) from glycoprotein ($I_p$).[15,16]

Aliquots of L and $I_s$ are evaporated to dryness in scintillation vials on a steam bath and counted for total radioactivity; $I_p$ is resuspended in 1 ml 0.1 $M$ Tris–HCl (pH 8.0) containing 2 m$M$ $CaCl_2$ and digested with Pronase as described previously.

*Anion Exchange Chromatography*

*Reagents and Buffers*

Glacial acetic acid
$CHCl_3/CH_3OH/H_2O$ (10 : 10 : 3)
$CHCl_3/CH_3OH$ (1 : 1)
$CHCl_3/CH_3OH/5$ m$M$ ammonium acetate (pH 6.0) (10 : 10 : 3)
$CHCl_3/CH_3OH/50$ m$M$ ammonium acetate (pH 6.0) (10 : 10 : 3)
$CHCl_3/CH_3OH/100$ m$M$ ammonium acetate (pH 6.0) (10 : 10 : 3)

*Procedure.* Purification of oligosaccharide pyrophosphorylpolyisoprenol by anion exchange chromatography is based upon procedures described previously.[16] DEAE–controlled-pore glass (CPG) (Electro-Nucleonics, Inc.) is converted to the acetate form by overnight equilibration in three volumes of glacial acetic acid. After degassing by water aspiration, a 1 × 50-cm solvent-resistant column (Pharmacia) is packed to a height of 20 cm and the resin is washed successively with three

[15] C. J. Waechter, J. J. Lucas, and W. J. Lennarz, *J. Biol. Chem.* **248**, 7570 (1973).
[16] B. K. Speake and D. A. White, *Biochem. J.* **170**, 273 (1978).

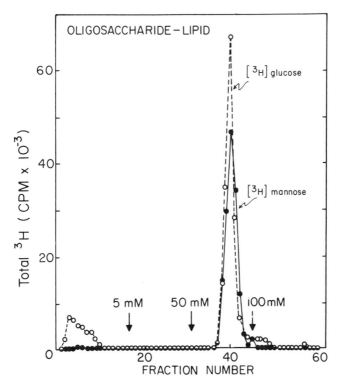

FIG. 2. Chromatography on DEAE-controlled-pore glass of [$^3$H]sugar-labeled oligosaccharide-lipid ($I_s$) extracted from matrix-free chick embryo tendon cells. The arrows show the beginning of elution with solutions containing various concentrations of ammonium acetate (pH 6.0). The procedures for extraction and chromatography are detailed in the text. ●——●, [$^3$H]mannose-labeled sample; ○---○, [$^3$H]glucose-labeled sample.

volumes of $CHCl_3/CH_3OH$ (1:1) and five volumes of $CHCl_3/CH_3OH/H_2O$ (10:10:3). Gravity flow is maintained by a Mariotte flask with an operating pressure of ~50 cm. To prevent phase separation of the eluent, the buffer reservoir is placed on a magnetic stirring motor.

The sample ($I_s$) is applied (at least 200,000 cpm) and batchwise elution is performed using 75 ml of $CHCl_3/CH_3OH/H_2O$ (10:10:3), 75 ml of $CHCl_3/CH_3OH/5$ m$M$ ammonium acetate (pH 6.0) (10:10:3), 100 ml of $CHCl_3/CH_3OH/50$ m$M$ ammonium acetate (pH 6.0) (10:10:3) and 75 ml of $CHCl_3/CH_3OH/100$ m$M$ ammonium acetate (pH 6.0) (10:10:3). The column effluent is collected in 5-ml fractions, and appropriate aliquots are evaporated to dryness and counted for total radioactivity. The results obtained from a [$^3$H]glucose-labeled sample and from a [$^3$H]mannose-labeled sample are shown in Fig. 2. In this system, neutral and cationic constituents do not bind to the column and elute with the solution containing water; most of the monosaccharide phosphoryl dolichol appears to elute with 5 m$M$ ammonium acetate, and oligosaccharide pyrophosphoryl dolichol elutes with 50 m$M$ ammonium acetate. Recovery of applied radioactivity is generally 70–80%. The eluted fractions are pooled, evaporated to dryness, resuspended in water, and lyophilized several times to remove residual ammonium acetate.

The column is regenerated by eluting with glacial acetic acid until the effluent is acidic, with $CHCl_3/CH_3OH$ (1:1) until the effluent is neutral, and finally with five volumes of $CHCl_3/CH_3OH/H_2O$ (10:10:3).

*Thin-Layer Chromatography*

Oligosaccharide pyrophosphoryl polyisoprenols can be further characterized by thin-layer chromatography as described by Chambers and Elbein.[17] Samples containing at least 10,000 cpm are suspended in a small volume of $CHCl_3/CH_3OH/H_2O$ (10:10:3) and spotted on 10 × 20-cm Avicel plates (250 $\mu$m layer) (Analtech). The plates are developed in isobutyric acid/$NH_4OH/H_2O$ (57:4:39) at room temperature until the boundary migrates to within 1 cm from the top of the plate (~7 hr). After overnight drying at room temperature, the plate is divided into 0.5-cm sections; the adsorbent is scraped onto weighing paper using a razor blade, transferred to a scintillation vial, and counted for total radioactivity (Fig. 3). The recovery of radioactivity is 40–50%.

*Mild Acid Hydrolysis*

To liberate the oligosaccharides from lipid linkage, the fraction eluting from DEAE-CPG with 50 m$M$ ammonium acetate is transferred to a 10-ml conical screw-top tube and incubated at 37° for 16 hr in 2 ml of 0.1 $M$ HCl–20% $CH_3OH$.[13] After evaporation to dryness, the samples are partitioned in 4 ml $CHCl_3/CH_3OH$ (2:1) and 1 ml $H_2O$.[16] After vortexing and centrifugation as described previously, the aqueous (upper) phase is isolated, evaporated, and characterized by gel filtration (see below).

Partial Characterization of Glycopeptides and Oligosaccharides

*Gel Filtration of Glycopeptides and Oligosaccharides*

Because of its resolving power and excellent recoveries, gel filtration in volatile buffers is the method of choice (Vol. 28 [1]).[7]

Columns (1 × 130 cm) of BioGel P-6 (200–400 mesh) (Bio-Rad Laboratories) equilibrated in 0.1 $M$ ammonium acetate (pH 6.0) containing 0.01%

---

[17] J. Chambers and A. D. Elbein, *J. Biol. Chem.* **250**, 6904 (1975).

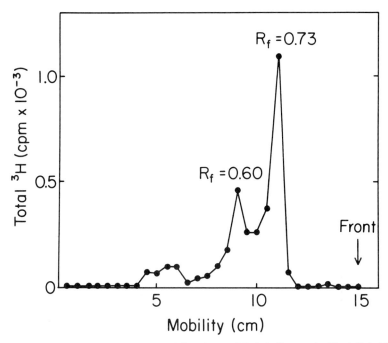

FIG. 3. Thin-layer chromatography of [$^3$H]glucose-labeled oligosaccharide-dolichol from DEAE-CPG (50 m$M$ effluent from Fig. 2). The details of the procedure are given in the text.

sodium azide have proved to be very useful.[18] To improve resolution, the sample volume is limited to 1 ml or less containing at least 20,000 cpm of sample, 5 mg of bovine serum albumin to mark the void volume, and 0.25 µCi of [$^3$H]hexose or $^3$H$_2$O to mark the totally included volume. One hundred fractions of 3 ml each are collected, and aliquots ranging from 50 µl to 1 ml are counted for total radioactivity (Fig. 4).

Fractions of interest are pooled, lyophilized, resuspended in water, and relyophilized several times to remove residual ammonium acetate. Recoveries of radioactivity are generally greater than 90%.

The relative molecular weight ($M_r$) of oligosaccharides is estimated from the linear relationship between log $M_r$ and the relative elution position of the sample ($K_d$) where

$$K_d = (V_e - V_0)/(V_t - V_0)$$

and where $V_e$ represents the elution volume of the sample, $V_0$ represents the elution volume of bovine serum albumin (determined by measuring

[18] J. R. Etchison, J. S. Robertson, and D. F. Summers, *Virology* **78**, 375 (1977).

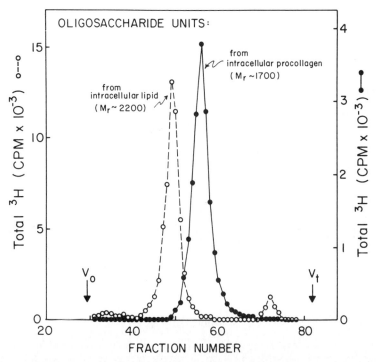

FIG. 4. Gel filtration on BioGel P-6 of [$^3$H]mannose-labeled oligosaccharide units derived either from lipid ($I_s$) by mild acid hydrolysis or from intracellular procollagen by Pronase digestion followed by alkaline hydrolysis. The details of the procedures are given in the text. Bovine serum albumin and $^3$H$_2$O were used to mark the void volume ($V_0$) and the included volume ($V_t$), respectively. The relative molecular weights ($M_r$) were estimated as described in the text; a $K_d$ of 0.390 was obtained for the lipid-derived oligosaccharide (O---O), and a $K_d$ of 0.500 was obtained for the protein-derived oligosaccharide (●——●).

$A_{280}$) and $V_t$ represents the elution volume of [$^3$H]hexose or $^3$H$_2$O.[18] Since molecular weight standards for oligosaccharide units are not routinely available, a calibration line for a P-6 column could be constructed using published $K_d$ values calculated for structures of known molecular weight.[18,19] To ascertain the validity of this relationship between log $M_r$ and $K_d$, a glycopeptide of known molecular weight ($M_r$ ~3600) and bacitracin ($M$ = 1423) (Sigma) were both tested and found to elute in the predicted volume.[6]

Probably the most popular gel filtration procedure today involves the use of relatively long columns (up to 200 cm) of BioGel P-4 (-400 mesh) at

[19] I. Tabas, S. Schlesinger, and S. Kornfeld, *J. Biol. Chem.* **253**, 716 (1978).

elevated temperatures[20] and relatively high flow rates (18 ml/hr).[21] These columns are very stable and provide excellent resolution by being able to distinguish differences of one glucose equivalent. Such resolution is not available with P-6 (200–400 mesh).

*Endoglycosidase Digestions*

A relatively easy method for determining the sequence and anomeric configuration of sugars in an oligosaccharide unit is the use of exoglycosidases. A comprehensive list of these enzymes, their source, specificity, and rationale for their use has been discussed previously (Vol. 28 [1]).[7] More recently, the use of endo-$\beta$-$N$-acetylglucosaminidases as an approach to determining the structure of asparagine-linked oligosaccharide units has been described.[22–24] The specificity of these enzymes permits designation of such oligosaccharide units as either "simple" or "complex."

Samples containing at least 20,000 cpm are incubated at 37° for 24 hr in 50 $\mu$l of the appropriate buffer, 10 mU of enzyme and a drop of toluene to prevent microbial growth. The products of digestion are monitored by gel filtration as described previously (Fig. 5).[6]

*Endo-$\beta$-$N$-acetylglucosaminidase D* (Miles Laboratories). The specificity of this enzyme has been discussed previously.[21] Ten milliunits of this enzyme from *Diplococcus pneumonia* are incubated in 50 $\mu$l of 0.15 $M$ phosphate buffer (pH 6.5) for 24 hr.

*Endo-$\beta$-$N$-acetylglucosaminidase H* (Miles Laboratories). The specificity of this enzyme has been discussed previously.[23] Ten milliunits of this enzyme from *Streptomyces plicatus* are incubated in 50 $\mu$l of 0.1 $M$ sodium citrate buffer (pH 5.5) for 24 hr.

*Monosaccharide Analysis*

The method of choice for analysis of radioactively labeled monosaccharides is ion exchange chromatography. As described previously,[7,25] neutral sugars are separated as borate complexes by anion exchange chromatography whereas hexosamines are separated by cation exchange chromatography. The emphasis in this section will be on the differences between our procedures and those already published in this series. A

[20] A. Kobata, K. Yamashita, and Y. Tachibana, this series, Vol. 50 [21].
[21] M. Natowicz and J. U. Baenziger, *Anal. Biochem.* **105**, 159 (1980).
[22] T. Muramatsu, this series, Vol. 50 [60].
[23] A. L. Tarentino, R. B. Trimble, and F. Maley, this series, Vol. 50 [63].
[24] A. Kobata, this series, Vol. 50 [62].
[25] Y. C. Lee, this series, Vol. 28 [16].

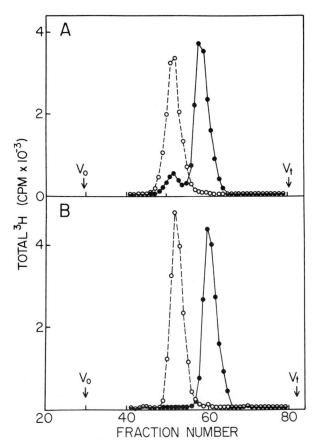

FIG. 5. Gel filtration on BioGel P-6 of [$^3$H]mannose-labeled glycopeptide from (A) pro-α1 and (B) pro-α2 after digestion with either endo-β-N-acetylglucosaminidase D (O---O), or endo-β-N-acetylglucosaminidase H (●——●) as described in the text. In the experiment shown, the samples were susceptible only to the latter enzyme ($K_d \sim 0.550$; $M_r \sim 1450$). From Clark.[6]

major advantage of this protocol is that both neutral sugars and hexosamines can be determined after a single hydrolysis.

*Hydrolysis of Glycoproteins, Glycopeptides, and Oligosaccharide Units*

This procedure is based upon that described previously by Lenhardt and Winzler.[26]

[26] W. F. Lenhardt and R. J. Winzler, *J. Chromatogr.* **34**, 471 (1968).

*Reagents*

HCl, 0.02 $M$
$CH_3OH$, 50%
$NH_4OH$, 1.5 $M$

*Procedure.* A sample containing 25,000–50,000 cpm of $^3$H-sugar-labeled material is lyophilized in a 16 × 100-mm screw-top tube. To this is added 500 µl of $H_2O$ and 500 µl of a 40% suspension (v/v) of Dowex 50-X2[H$^+$] (200–400 mesh) (Bio-Rad Laboratories) in 0.02 $M$ HCl. The sample is flushed with $N_2$, capped, and hydrolyzed for 48 hr at 100°. To facilitate subsequent recovery of radioactivity, 100 µg each of mannose, glucose, galactose, fucose, glucosamine, and galactosamine are added as carrier.

After cooling to room temperature, the hydrolyzate is passed through a column containing Dowex 1-X8[$HCO_3^-$] (200–400 mesh) and Dowex 50-X2[H$^+$]. This column is prepared in a 0.5 × 14.5-cm Pasteur pipette and consists of a small plug of glass wool, 1–1.5 cm of Dowex 1, a small plug of glass wool, and the remainder filled with Dowex 50 to 1 cm from the top. The hydrolyzate is added to the top of the column, and elution is continued with 5 ml of $H_2O$, 5 ml of 50% $CH_3OH$, and 5 ml of 1.5 $M$ $NH_4OH$. The water and methanol effluents are combined and evaporated for neutral sugar analysis; the $NH_4OH$ effluent is evaporated for hexosamine analysis.

*Neutral Sugar Analysis*

This procedure is based upon the accelerated system described by Lee et al.[27]

*Reagents and Buffers*

HCl, 1 $M$
NaOH, 1 $M$
Ethanol, 70% (w/v)
Boric acid, 0.4 $M$, pH 8.0
Boric acid, 0.4 $M$, pH 10.0
Potassium tetraborate, 10%, containing 0.1% EDTA

The borate buffers are made by titrating boric acid solutions with 50% NaOH.

*Column and Resin.* DA-X4 anion exchange resin (Durrum Chemical Corp.) is treated in a beaker as described by Mopper[28]: Wash with ~400 ml of water; wash with ~200 ml of 1 $M$ HCl; wash with water until

[27] Y. C. Lee, G. S. Johnson, B. White, and J. Scocca, *Anal. Biochem.* **43**, 640 (1971).
[28] K. Mopper, *Anal. Biochem.* **87**, 162 (1978).

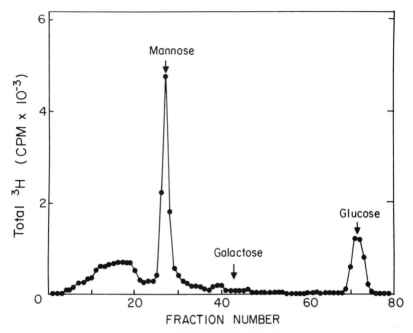

FIG. 6. Chromatography on DA-X4 of [$^3$H]glucose-labeled lipid-derived oligosaccharide after hydrolysis and elution from a mixed-bed Dowex column as described in the text. The elution positions of carrier mannose, galactose, and glucose are shown.

neutral; wash with 1 $M$ NaOH; wash with water until neutral; wash with 10% tetraborate until pH ~10; wash with water; wash with ~200 ml of 70% ethanol.

The resin is introduced into a water-jacketed column in ethanol and packed by pumping water at 65° with a flow rate of 1.4 ml/min to give a 0.6 × 25 cm-resin bed. After washing out the ethanol with distilled water, the temperature is raised to 70° and the column is equilibrated with boric acid (pH 8.0) for 30 min at a flow rate of 0.7 ml/min.

*Instruments.* A linear gradient elution system, a high-pressure pump and a pressure gauge from a Technicon sugar chromatography system, and a fraction collector are used.

*Procedure.* The lyophilized sample is resuspended in 200 µl of boric acid (pH 8.0) and applied to the column under $N_2$ pressure. A wash of 100 µl is also applied. Elution is achieved with a 140-ml gradient consisting of 70 ml of boric acid (pH 8.0) and 70 ml of boric acid (pH 10.0) at a flow rate of 0.7 ml/min. The effluent is passed directly into a fraction collector, 90 2-min fractions (1.4 ml) are collected, and aliquots of 500 µl are counted. In a typical run, mannose elutes at ~54 min, galactose at ~85 min, and

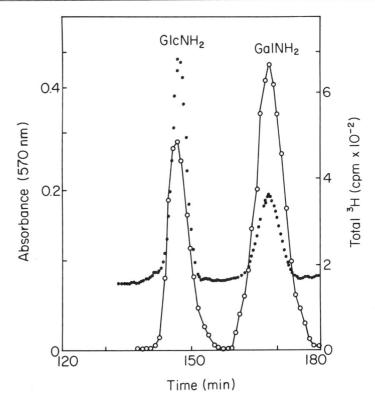

FIG. 7. Chromatography on UR-30 of [$^3$H]glucosamine-labeled medium from a chick embryo tendon cell incubation after hydrolysis and elution from a mixed-bed Dowex column as described in the text. The elution positions of carrier glucosamine (GlcNH$_2$) and galactosamine (GalNH$_2$) are shown. O——O, $^3$H; · · · · · , absorbance at 570 nm.

glucose at ~145 min (Fig. 6). The elution position of carrier sugars can be monitored by a phenol–H$_2$SO$_4$ colorimetric procedure.[29] The recovery of radioactivity is generally 50–75%.[2] The column is regenerated by elution with 10% tetraborate for 45 min followed by equilibration with boric acid (pH 8.0).

*Hexosamine Analysis*

This procedure is based upon that described by Denduchis et al.[30]

[29] M. Dubois, K. A. Gilles, J. K. Hamilton, P. A. Ribers, and F. Smith, *Anal. Chem.* **28**, 350 (1956).
[30] B. Denduchis, N. A. Kefalides, and A. Bezkorovainy, *Arch. Biochem. Biophys.* **138**, 582 (1970).

*Reagents and Buffers*

Sodium citrate, 0.35 $M$, pH 5.25
NaOH, 2.0 $M$
*Column and Resin.* The long column from an amino acid analyzer is convenient for this analysis. UR-30 cation exchange resin (Beckman Instruments, Inc.) is packed to give a 0.9 × 50-cm column. The column temperature is maintained at 55°.

*Procedure.* The sample is resuspended in 200 μl of 0.35 $M$ sodium citrate buffer (pH 5.25) and applied to the column under $N_2$ pressure. A wash of 100 μl is also applied. Elution is achieved with the same buffer at a flow rate of 1.1 ml/min. The ninhydrin-reacted effluent is passed directly into a fraction collector, and 100 2-min fractions are collected. Aliquots of 1 ml are counted in a scintillation counter. In a typical run, glucosamine elutes at ~145 min and galactosamine at ~170 min (Fig. 7). Recovery of radioactivity is greater than 90%.[2] The column is regenerated by elution with 2 $M$ NaOH for 30 min followed by equilibration in the citrate buffer.

## [16] Covalent Cross-Links in Collagen

*By* NICHOLAS D. LIGHT and ALLEN J. BAILEY

Cross-linking in the collagen molecule may be divided into two functionally separate groups. First, intramolecular cross-linking, in which two α chains within the same molecule may be covalently linked (aldol cross-links). Second, intermolecular cross-linking that involves the formation of covalent bridges between chains in different molecules (Schiff's base and more complex cross-links). Whereas in the first case there is no evidence of the involvement of more than two chains, intermolecular cross-links may link more than two α chains in more than two molecules.

Within the group of intermolecular cross-links we may add a subdivision, namely, cross-links laid down in newly formed collagen and those formed during maturation. As the cross-links of young tissue may be investigated by mild reduction utilizing tritiated borohydride, we shall refer to this group as "reducible" cross-links. The intermolecular cross-links of mature tissue cannot be reduced by borohydride and are apparently stable to high temperatures and low pH treatment. Because of their origin we shall refer to these as "mature" cross-links.

The enzyme lysyl oxidase acts upon the newly processed and precipitated collagen fibrils in the extracellular matrix to produce lysine or hydroxylysine aldehyde precursors of the aldol and Schiff's base cross-links.

The sites of action of the enzyme are limited to the N- and C-terminal nonhelical regions of the molecules. The α-aminoadipic-δ-semialdehyde residues then spontaneously react with other similar residues in the same molecule to form the intramolecular aldol or with lysine or hydroxylysine ε-amino groups in adjacent molecules to form the reducible bifunctional cross-links (Fig. 1). The latter cross-links decrease in number during aging until they attain a very low level that reflects the cross-linking of collagen laid down during normal turnover. It has been postulated that the reducible cross-links form the chemical precursors of the more complex nonreducible mature cross-links.

The mode of investigation of the reducible cross-links has been the use of tritiated borohydride. It should be remembered that cross-links exist as the unreduced form in collagen, although we shall confine ourselves to a discussion of their artificially reduced products.

Reduction Procedure

The method of borohydride reduction of isolated α chains, soluble collagen, reprecipitated fibers, or suspended tissue follows the general procedure described by Bailey et al.[1] The sample is dissolved or suspended in 0.9% (w/v) NaCl to a final concentration of 2–20 mg/ml. The solution is usually buffered at neutral pH with $0.02\,M$ phosphate, although this leads to the base-catalyzed production of an artifact, as we shall see later. Tritiated potassium borohydride, diluted with the nonradioactive reagent to give a final specific radioactivity of approximately 10 mCi/mmol, is added directly to the sample with stirring to give a substrate: reagent ratio of 30:1 (w/w) based on collagen dry weight. The reaction is allowed to proceed for 1 hr at room temperature, at which time it is terminated by the addition of acetic acid to pH 3.0. Exhaustive dialysis against water removes unreacted borohydride and reaction by-products, and the sample may then be freeze dried.

Isolation of Reduced Cross-Link Components

Reduced components are analyzed after hydrolysis of the sample. The hydrolysis may be carried out under acid (redistilled $6\,N$ HCl) or alkaline ($2\,M$ NaOH in polypropylene bottles) conditions at 110° for 24 or 48 hr, respectively. The hydrolyzate is dried *in vacuo* and, after desalting by ion-exchange chromatography (as described below) if necessary, may be analyzed using a variety of methods.

[1] A. J. Bailey, C. M. Peach, and L. J. Fowler, *Biochem. J.* **117**, 819 (1970).

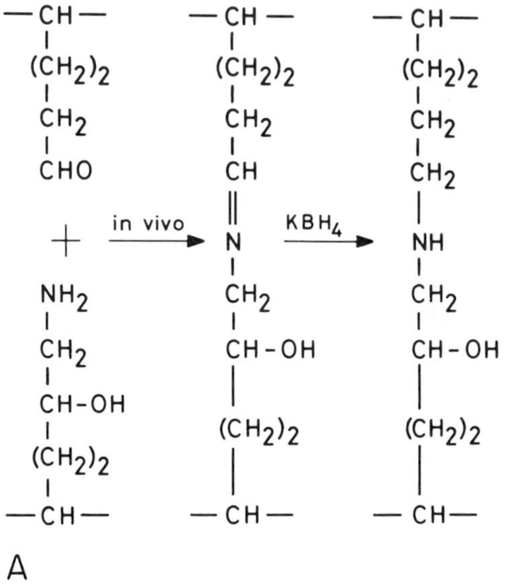

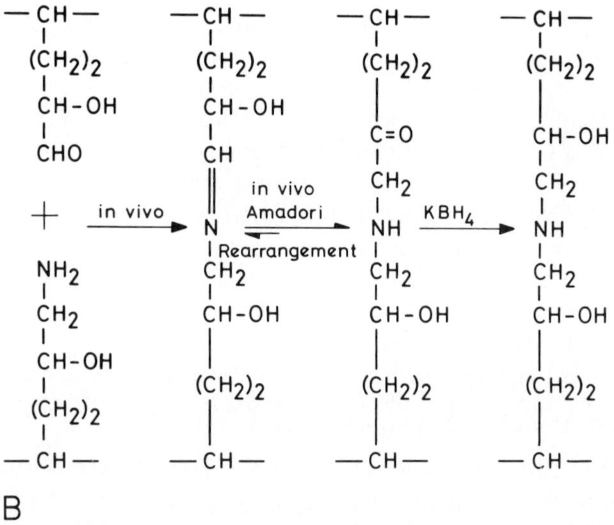

$$
\begin{array}{cccc}
-\mathrm{CH}- & -\mathrm{CH}- & -\mathrm{CH}- \\
| & | & | \\
(\mathrm{CH_2})_2 & (\mathrm{CH_2})_2 & (\mathrm{CH_2})_2 \\
| & | & | \\
\mathrm{CH_2} & \mathrm{CH_2} & \mathrm{CH_2} \\
| & | & | \\
\mathrm{CHO} & \mathrm{CH} & \mathrm{CH_2} \\
& \| & | \\
+ \xrightarrow{\text{in vivo}} & \mathrm{N} \xrightarrow{\mathrm{KBH_4}} & \mathrm{NH} \\
& | & | \\
\mathrm{NH_2} & \mathrm{CH_2} & \mathrm{CH_2} \\
| & | & | \\
\mathrm{CH_2} & \mathrm{CH-OH} & \mathrm{CH-OH} \\
| & | & | \\
\mathrm{CH-OH} & (\mathrm{CH_2})_2 & (\mathrm{CH_2})_2 \\
| & | & | \\
(\mathrm{CH_2})_2 & & \\
| & & \\
-\mathrm{CH}- & -\mathrm{CH}- & -\mathrm{CH}-
\end{array}
$$

**A**

$$
\begin{array}{cccc}
-\mathrm{CH}- & -\mathrm{CH}- & -\mathrm{CH}- & -\mathrm{CH}- \\
| & | & | & | \\
(\mathrm{CH_2})_2 & (\mathrm{CH_2})_2 & (\mathrm{CH_2})_2 & (\mathrm{CH_2})_2 \\
| & | & | & | \\
\mathrm{CH-OH} & \mathrm{CH-OH} & \mathrm{C=O} & \mathrm{CH-OH} \\
| & | & | & | \\
\mathrm{CHO} & \mathrm{CH} & \mathrm{CH_2} & \mathrm{CH_2} \\
& \| & \xrightarrow{\text{in vivo}} & | \\
+ \xrightarrow{\text{in vivo}} & \mathrm{N} \xrightleftharpoons{\text{Amadori}} & \mathrm{NH} \xrightarrow{\mathrm{KBH_4}} & \mathrm{NH} \\
& | \text{ Rearrangement} & | & | \\
\mathrm{NH_2} & \mathrm{CH_2} & \mathrm{CH_2} & \mathrm{CH_2} \\
| & | & | & | \\
\mathrm{CH_2} & \mathrm{CH-OH} & \mathrm{CH-OH} & \mathrm{CH-OH} \\
| & | & | & | \\
\mathrm{CH-OH} & (\mathrm{CH_2})_2 & (\mathrm{CH_2})_2 & (\mathrm{CH_2})_2 \\
| & | & | & | \\
(\mathrm{CH_2})_2 & & & \\
-\mathrm{CH}- & -\mathrm{CH}- & -\mathrm{CH}- & -\mathrm{CH}-
\end{array}
$$

**B**

FIG. 1. Aldimine and keto-imine cross-links from collagen. (A) Representation of the formation *in vivo* of the reducible aldimine dehydrohydroxylysinonorleucine and its reduction *in vitro* with borohydride to produce hydroxylysinonorleucine. (B) Representation of the formation *in vivo* of the reducible aldimine dehydrodihydroxylysinonorleucine, its spontaneous Amadori rearrangement to form the keto-imine hydroxylysino-5-ketonorleucine, and its reduction *in vitro* to form dihydroxylysinonorleucine.

# [16] COVALENT CROSS-LINKS IN COLLAGEN 363

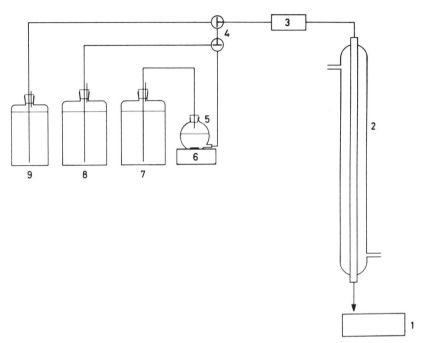

FIG. 2. Diagrammatic representation of the pyridine-formate ion exchange column system. 1, Fraction collector; 2, water-jacketed resin column; 3, positive displacement pump; 4, three-way taps; 5, mixing vessel; 6, magnetic stirrer; 7, limiting buffer; 8, starting buffer for column reequilibration; and 9, 2 $M$ pyridine reservoir for column regeneration.

During preparative procedures the hydrolyzed sample should be subjected to a preliminary chromatographic step. For example, when up to 10 g of starting material are to be used the hydrolyzate should be passed through a column (90 × 5 cm) of Sephadex G-10 or BioGel P-2 (Pharmacia Ltd., Uppsala, Sweden, and Bio-Rad Labs., Richmond, California, respectively) equilibrated in 0.1 $M$ acetic acid. The tritiated reduced cross-link compounds elute before the bulk of the other amino acids owing to their larger size. If more than 10 g of the material are to be hydrolyzed, the displacement chromatography method described by Partridge and Brimley[2] may be used (up to 500 g). However, this procedure is lengthy and complex, and for 50–100 g of material several preliminary gel filtration steps might be regarded as easier and less time-consuming.

In this laboratory, all the reduced hydrolysis products may be separated and identified using a single-column ion-exchange system (see Fig. 2). The method employs a modified Technicon autoanalyzer with a col-

[2] S. M. Partridge and R. C. Brimley, *Biochem. J.* **51**, 628 (1952).

umn (1.2 × 60 cm) of Zerolit 225 (30–40 μm bead size prepared by the method described by Hamilton[3]) maintained at 60° by a water jacket. The resin is washed with 2 $M$ HCl and converted to the pyridinium form by stirring in 2 $M$ pyridine for 24 hr prior to equilibration in 0.1 $M$ pyridine–formic acid buffer, pH 2.9. The column is eluted with a gradient formed by running 1 $M$ pyridine–formic acid buffer, pH 5.0, into a mixing vessel containing 350 ml of the starting buffer. The flow rate is 60 ml/hr, and fractions are collected every 5 min. At the end of a run the column is regenerated by elution with 2 $M$ pyridine and then starting buffer.

A stream-splitting device may be used at the base of the column to divert one-fifth of the eluate to a modified ninhydrin system as described by Padieu et al.[4] We routinely redistill pyridine from ninhydrin (2 g/liter) to avoid interference from nonspecific amine contaminants. The expense of a continuous automated ninhydrin system may be obviated by locating ninhydrin-positive peaks as follows. Aliquots from fractions are spotted on to Whatman No. 1 paper, and peaks are located by spraying with ninhydrin (0.25%, w/v, in acetone). Initial standardization of a column must be made, however, by analysis of fractions by classical chromatographic methods and the use of pure preparations of individual amino acids. Fractions are analyzed for radioactivity by counting a sample (0.2 ml) in 3 ml of Insta-Gel (Packard Inst. Co., Downers Grove, Illinois) scintillation fluid. On an analytical scale this method is found to be adequate to separate all the major components found in reduced collagen, although a second system should always be used for verification of identity. A typical chromatograph is shown in Fig. 3. On a larger scale, components isolated from the Zerolit 225 column may be further purified by elution in 0.35 $M$ sodium citrate buffer, pH 5.25, from a 60-cm modified Beckman amino acid analyzer column maintained at 56°. Isolated components are then desalted by adsorption on a column (5 × 0.5 cm) of Amberlite GC-120 washed with water and displacement with 2 $M$ $NH_3$.

Identification Procedures

In our initial studies, the reduced components eluted from the Zerolit 225 column were denoted as fractions A, B (B1 and B2), and C, as indicated in Fig. 3. Other minor radioactive compounds were identified as the two reduced cross-link precursors 5,6-dihydroxynorleucine and 6-hydroxynorleucine, the acid hydrolysis by-product of the latter, 6-chloronorleucine, the reduced intramolecular cross-link, and lysinonor-

[3] P. B. Hamilton, Anal. Chem. **30**, 914 (1958).
[4] P. Padieu, N. Malekina, and G. Schapira, Colloq. Amino Acid Anal., 3rd 1965, p. 71 (Technicon).

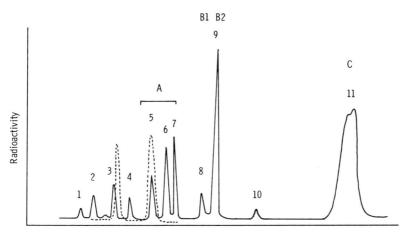

FIG. 3. Typical chromatogram obtained after elution of an amino acid hydrolyzate of reduced calf skin from the pyridine-formate ion exchange system. Solid line represents the radioactive components eluted after application of an acid hydrolyzate. The identities and structures of the various compounds are listed in the table. The dashed line represents additional peaks seen in alkali hydrolyzates and shows (in order of elution) glucosylgalactosyl derivatives of mono- and dihydroxylysinonorleucine and galactosyl derivatives of mono- and dihydroxylysinonorleucine.

leucine (see the table). Fraction A was characterized independently in two laboratories[5,6] and shown to contain the reduced condensation products of glucose or mannose with lysine or hydroxylysine—hexosyllysine and hexosylhydroxylysine, respectively (see the table). Fraction C has been named histidinohydroxymerodesmosine and fully characterized[7] but has been shown to be an artifact of the reduction reaction.[8] The compound is thought to be formed during the reaction with borohydride by a base-catalyzed Michael addition of histidine across the double bond of an intramolecular aldol.

The two reduced cross-link compounds correspond to the material chromatographing under peaks B1 and B2. The first of these peaks contains dihydroxylysinonorleucine (B1), and the second contains hydroxylysinonorleucine (see Fig. 1). The glycosylated variants of these compounds are located in positions denoted by the dashed line on the

[5] S. P. Robins and A. J. Bailey, *Biochem. Biophys. Res. Commun.* **48**, 76 (1972).
[6] M. L. Tanzer, R. Fairweather, and P. M. Gallop, *Arch. Biochem. Biophys.* **151**, 137 (1972).
[7] M. L. Tanzer, T. Housley, L. Berube, R. Fairweather, C. Franzblau, and P. M. Gallop, *J. Biol. Chem.* **248**, 393 (1973).
[8] S. P. Robins and A. J. Bailey, *Biochem. J.* **135**, 657 (1973).

## Reduced Compounds in Collagen Hydrolysates

| Compound | No. | Structure | Comments |
|---|---|---|---|
| 5:6-dihydroxynorleucine | 1 | $\text{CH(COOH)(NH}_2\text{)-(CH}_2)_2\text{-CH(OH)-CH}_2\text{OH}$ | Reduced cross-link precursors |
| 6-hydroxynorleucine | 2 | $\text{CH(COOH)(NH}_2\text{)-(CH}_2)_3\text{-CH}_2\text{OH}$ | |
| Reduced aldol | 3 | $\text{CH(COOH)(NH}_2\text{)-(CH}_2)_2\text{-CH(CH}_2\text{OH)-CH}_2\text{-(CH}_2)_3\text{-CH(NH}_2\text{)(COOH)}$ | Reduced intramolecular cross-link |
| 6-chloronorleucine | 4 | $\text{CH(COOH)(NH}_2\text{)-(CH}_2)_3\text{-CH}_2\text{Cl}$ | Acid hydrolysis derivative of 6-hydroxynorleucine |
| Hexosyl-hydroxylysine | 5 | $\text{CH(COOH)(NH}_2\text{)-(CH}_2)_2\text{-CH(OH)-CH}_2\text{-NH-CH}_2\text{-(CHOH)}_4\text{-CH}_2\text{OH}$ | Glucosyl and mannosyl derivatives of hydroxylysine and lysine |
| Hexosyl-lysine | 6 | $\text{CH(COOH)(NH}_2\text{)-(CH}_2)_4\text{-NH-CH}_2\text{-(CHOH)}_4\text{-CH}_2\text{OH}$ | |
| Anhydro derivative of hexosyl-lysine | 7 | $\text{CH(COOH)(NH}_2\text{)-(CH}_2)_4\text{-NH-CH}_2\text{-CH(OH)-CH-(CHOH)}_2\text{-CH}_2\text{-O-}$ (ring) | Artifact of acid hydrolysis |
| Hydroxylysino-hydroxynorleucine | 8 | $\text{CH(COOH)(NH}_2\text{)-(CH}_2)_2\text{-CH(OH)-CH}_2\text{-NH-CH}_2\text{-CH(OH)-(CH}_2)_2\text{-CH(NH}_2\text{)(COOH)}$ | Reduced intermolecular cross-links |
| Hydroxylysino-norleucine | 9 | $\text{CH(COOH)(NH}_2\text{)-(CH}_2)_4\text{-NH-CH}_2\text{-CH(OH)-(CH}_2)_2\text{-CH(NH}_2\text{)(COOH)}$ | |
| Lysinonorleucine | 10 | $\text{CH(COOH)(NH}_2\text{)-(CH}_2)_4\text{-NH-(CH}_2)_4\text{-CH(NH}_2\text{)(COOH)}$ | |
| Histidinohydroxy-merodesmosine | 11 | $\text{NH}_2\text{-CH(COOH)-(CH}_2)_3\text{-CH=C-CH}_2\text{-CH(COOH)(NH}_2\text{)}$ linked via imidazole ring (N—CH=N) to $\text{(CH}_2)_2\text{-CH(NH}_2\text{)(COOH)}$ and $\text{CH(COOH)(NH}_2\text{)-(CH}_2)_2\text{-CH(OH)-CH}_2\text{-NH-CH}_2$— | Artifact of borohydride reduction |

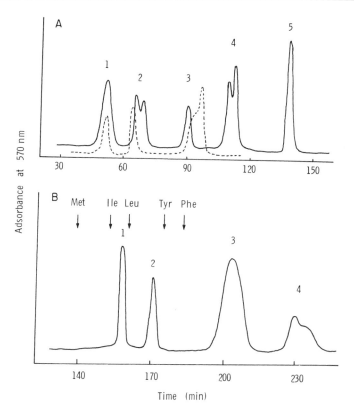

FIG. 4. Chromatograms obtained on standard amino acid analysis of reducible cross-links and their glycosylated derivatives. (A) Peaks 1, 2, 3, 4, and 5 are, respectively, dihydroxylysinonorleucine, hydroxylysinonorleucine, lysinonorleucine, hydroxylysine, and lysine. The dashed line shows the location of various hexosyllysine and hexosylhydroxylysine derivatives. (B) Peaks 1, 2, 3, and 4 represent, respectively, glucosylgalactosyldihydroxylysinonorleucine (and the reduced aldol), glucosylgalactosylhydroxylysinonorleucine, galactosyldihydroxylysinonorleucine, and galactosylhydroxylysinonorleucine.

chromatograph in Fig. 3 after alkali hydrolysis of reduced samples. Confirmation of the identity of these components must be made by analytical ion exchange chromatography in 0.35 $M$ sodium citrate buffer at pH 5.25. This procedure is essential to avoid confusion with other spurious radioactive contaminants that may confuse results obtained on the pyridine–formate column. The elution positions of dihydroxylysinonorleucine and hydroxylysinonorleucine from a 30 × 1 cm column on a Locarte analyzer are shown in Fig. 4A. The galactosyl and glucosylgalactosyl derivatives of both reduced intermolecular cross-links chromatograph in positions shown in Fig. 4B during standard amino acid analysis in 0.2 $M$ sodium citrate buffers on a Locarte amino acid analyser. The disaccharide derivative of dihydroxylysinonorleucine co-chromatographs with the reduced aldol on both pyridine–formate and citrate systems. These compounds may be resolved by gel filtration on BioGel P-2 (140 × 1.6 cm) in 0.1 $M$ acetic acid.

Further confirmation of the nature of the isolated reduced cross-link compounds can be obtained by carrying out a Smith degradation with periodate. One volume of sodium periodate (5 mg/ml in 0.1 $M$ sodium citrate buffer, pH 5.5) is added to 3 volumes of the suspected reduced cross-link in the same buffer (1.5 mg/ml). After 5 min the pH is raised to 7.5 with 5 $M$ NaOH and the products are reduced by the addition of excess $KBH_4$. After 30 min the pH is adjusted to 3.0 with HCl, and the solution can then be applied to an amino acid analyzer. Dihydroxylysinonorleucine yields predominantly proline and some norvaline, whereas hydroxylysinonorleucine yields lysine and proline.

Quantitation of Cross-Links

The relatively low resolution of the pyridine–formate system and the presence of co-chromatographing hexosyllysine derivatives on the citrate system (Fig. 4A) makes accurate one-step quantitation of cross-links in tissue hydrolyzates impossible. When samples are very small, the extremely low incidence of reducible cross-links make quantitation by ninhydrin assay equally unreliable. Similarly, the use of incorporation of radioactivity cannot be accurately used owing to the difficulty in assessing the exact extent of titration per cross-link in each reduced sample. Mindful of the need for a rapid and reliable assay for these important cross-links, particularly in small biopsy samples from patients suffering from connective tissue disorders, we are currently developing both high-performance liquid chromatography (HPLC) and radioimmunoassays for accurate quantitation of the reducible cross-links.

Isolation of Unusual Compounds Believed To Be Mature Cross-links

There are several reports in the literature of putative cross-link compounds that have been isolated from acid hydrolyzates of mature insoluble collagen. The problem with this approach is that it makes two fundamental assumptions: ($a$) that mature cross-links are stable under the conditions of acid hydrolysis; and (B) that any unusual compound may be a cross-link or cross-link derivative. Consequently, considerable biochemical and chemical evidence is required to verify the identity of any novel compound as a genuine mature cross-link. Small amounts of several fluorescent components have been observed in collagen hydrolyzates, and these have aroused some interest due to the fluorescent nature of desmosine, the elastin cross-link. Only one of these compounds, pyridinoline,[9] has received any interest from more than one laboratory, although the current dispute as to its origin[10] makes it unreasonable to discuss it in detail here. The compound has a molecular weight in excess of 400 and so can be purified relatively easily by adopting the large-scale gel filtration preparative method described in the section on isolation of reduced cross-link components. Fluorescence at 400–410 nm with excitation at 295 nm may be used to monitor for the compound. The fluorescent peak is pooled and, in this laboratory, applied to a 30 × 1 cm Jeol amino acid analyzer column run under standard conditions in citrate buffers (pH 2.9–4.65 with the last buffer extended for 40 min). The compound is pooled and desalted on a short column of Sephadex G-10 in 0.1 $M$ acetic acid. The original method described by Fujimoto et al.[9] is lengthy and, in our hands, not as successful as that described above. The compound isolated by either method shows a characteristic peak on the amino acid analyzer[10] and an acid-base shift in ultraviolet absorption from 295 nm in acid (0.1 $M$ HCl) to 325 nm in alkali (0.02 $M$ NaOH).

The only acceptable proof of the origin of these presumptive cross-link components involves their isolation from collagenous peptides prepared from the insoluble starting material. Such a procedure is discussed below and is analogous in theory to the isolation of cross-linked peptides from collagen cross-linked with reducible bonds.

Preparation and Identification of Cross-linked Peptides Containing Reducible Bonds

In general, the most successful work in this field has been carried out on peptides produced after CNBr cleavage of cross-linked collagen, since CNBr peptides can be readily characterized. Partial alkaline[11] and enzymic digestion[12,13] methods have also been used, although not with unequivocal effect. In each case the starting material was reduced with tritiated borohydride to provide a label in cross-linked peptides. Obviously, with the number of other non-cross-link reduced components de-

[9] D. Fujimoto, K. Akiba, and N. Nakamura, *Biochem. Biophys. Res. Commun.* **76**, 1124 (1977).
[10] D. F. Elsden, N. D. Light, and A. J. Bailey, *Biochem. J.* **185**, 531 (1980).
[11] K. Fujii, D. Corcoran, and M. L. Tanzer, *Biochemistry* **14**, 4409 (1975).
[12] D. R. Eyre and M. J. Glimcher, *Biochem. J.* **135**, 393 (1973).
[13] U. Becker, H. Furthmayr, and R. Timpl, *Hoppe-Seyler's Z. Physiol. Chem.* **356**, 21 (1975).

scribed above, it is essential not only to isolate and analyze putative labeled cross-linked peptides (so-called H peptides), but to back up such claims with N-terminal or partial sequence analysis as well as using periodate cleavage to break the reduced bonds linking the H peptide and obtain the two separated peptides. Studies by Miller and Robertson[14] (type II), Henkel et al.[15,16] (types I and III), and Nicholls and Bailey[17] (type III) have been the most successful, and, although all are similar in general approach, the details of the methodology used in each case are too lengthy to catalog here.

To summarize the techniques, the collagenous material is reduced as described earlier with tritiated potassium borohydride after mild washing procedures (neutral salt solution at 4° for 48–72 hr). The reduced and dialyzed material may be washed further, if required, prior to CNBr digestion. This takes place in 70% (v/v) formic acid solution at 25–30° for 4 hr with a suspension of collagen equivalent to a final concentration of 10 mg/ml. An equal weight of CNBr is added at the start of the incubation. After 4 hr the reaction solution is diluted 10-fold with water and rotary-evaporated to its original volume. After repeating this procedure twice, the solution is freeze dried.

The next task is to separate the cross-linked peptide(s) from the rest of the CNBr-digested material. This may be achieved by a combination of routine ion-exchange techniques for CNBr peptides, as described in this volume [2], on carboxymethyl cellulose in $0.02\,M$ citrate buffer at pH 3.6, on phosphocellulose in $0.001\,M$ sodium acetate pH 3.8 at 4° with either a linear gradient from 0 to $0.4\,M$ NaCl[13] or a concave gradient from 0 to $0.3\,M$ NaCl[12] and gel filtration on suitable media. Once purified, the suspected cross-linked H-peptide should be subjected to rigorous analytical procedures, such as determination of molecular weight, amino acid analysis, N-terminal analysis, or partial sequence analysis, if possible. The peptide may be treated with periodate followed by immediate reduction as detailed above. This procedure will cleave the reduced bond linking the two peptides, and they may then be separated by suitable chromatographic methods and collected for identification (see this volume [2] on the characterisation of vertebrate collagens).

Isolation and Characterization of Cross-Linked Peptides from Mature Collagen

The only work carried out in this field to date has been in this laboratory with bovine Achilles tendon from mature animals. The defatted and

[14] E. J. Miller and P. B. Robertson, *Biochem. Biophys. Res. Commun.* **54**, 432 (1973).
[15] W. Henkel, J. Rauterberg and T. Stirtz, *Eur. J. Biochem.* **69**, 223 (1976).
[16] W. Henkel, J. Rauterberg, and R. W. Glanville, *Eur. J. Biochem.* **96**, 249 (1979).
[17] A. C. Nicholls and A. J. Bailey, *Biochem. J.* **185**, 195 (1980).

finely chopped tendon is extracted with 8 $M$ urea in 0.5 $M$ acetic acid for 24–48 hr at room temperature (10 g wet weight of tendon per liter of buffer). After exhaustive washing with water, the material is digested with CNBr and freeze dried as described above. The freeze-dried material is dissolved to a final concentration of 50 mg/ml in 1 $M$ $CaCl_2$ containing 0.05 $M$ Tris-Cl, pH 7.5, by heating at 60° for 15–30 min. The small amount of undissolved material is removed by centrifugation at 40,000 $g$ for 15 min. Because of an anomaly in its behavior on A-1.5 m agarose (Bio-Rad Laboratories, Richmond, California), cross-linked peptide material may be separated from other CNBr peptides by gel filtration through this medium in the calcium chloride buffer. We use either a 140 × 5 cm column (1.5 g total useful capacity) or, for large-scale preparations (10 g), a preliminary step involving gel filtration on a column (14.5 cm × 16 cm diameter, conical base of 5 cm depth) of A-1.5 m agarose in 1 $M$ $CaCl_2$ 0.05 $M$ Tris-Cl, pH 7.5. A buffer head is maintained at the top of the column by means of a regulated drip-feed and peptide material is monitored in the effluent at 230 nm. Cross-linked material (termed poly-$\alpha$1CB6) is collected at the void volume of both types of column. The expected yield of cross-linked material, the polymeric product of both $\alpha$1CB6 and $\alpha$1CB5,[18] is about 16%, whereas the observed yield at $V_0$ is invariably between 11 and 13% of the total material.

Noncollagenous acid glycoprotein contaminants may be removed by passing the partially purified poly-$\alpha$1CB6 through DEAE-Sepharose CL 6B (Pharmacia Ltd., Uppsala, Sweden) in a buffer containing 2 $M$ urea, 0.2 $M$ NaCl, and 0.02 $M$ sodium phosphate, pH 7.5. The unbound collagenous cross-linked peptides are collected, dialyzed exhaustively against water, and freeze dried. A further purification on a second, higher-resolution A-1.5 m agarose column may be necessary. In this case, we use columns of varying size dependent on the total weight of sample to be chromatographed. The cross-linked material has been shown to be polydisperse over a molecular weight range of 50,000–350,000 by several techniques, has an amino acid analysis consistent with a content of $\alpha$1CB6 and $\alpha$1CB5, and has been shown to be covalently cross-linked by several criteria.[18] However, conventional amino acid analytical techniques in citrate or pyridine–formate buffers (as described earlier) failed to show any unusual compounds in acid hydrolyzates of this material. More recent work has indicated that the covalent bonds in poly-$\alpha$1CB6 may be labile to strong acid and alkali treatment,[19] and so we are currently utilizing enzymic digestion to isolate unusual components.

The overriding difficulty in these studies is that of selecting an efficient and specific technique for labeling the mature cross-link. As previously

[18] N. D. Light and A. J. Bailey, *Biochem. J.* **189**, 111 (1980).
[19] N. D. Light and A. J. Bailey, in preparation (1980).

stated, most reducible intermolecular cross-links are glycosylated and, as these bifunctional bonds are thought to complex with other groups in adjacent molecules to form the mature cross-links, this characteristic may be utilized as a convenient handle in locating putative mature cross-link compounds. The enzyme galactose oxidase is commercially available and may be used in concert with tritiated borohydride under mild conditions to label galactose residues. However, after initial studies using this technique we would advise considerable caution, as we have found the Sigma Chemical Co. (Poole, Dorset, U.K.) preparation, at least, to contain nonspecific protease activity. This particular enzyme preparation has also been tested in another laboratory with similar results, but with also some question as to the actual specificity of the enzyme (Dr. S. P. Robins, personal communication). Until satisfactory safeguards or further purification procedures can be outlined for this enzyme preparation, we advise the avoidance of this technique for labeling presumptive cross-link moieties in collagen.

A less convenient but very specific labeling or monitoring technique is currently being developed in this laboratory. As we now have good evidence that the mature cross-link is formed in part from the N-terminal amino acid (hydroxylysine) in $\alpha$1CB5, we can use this moiety to monitor for the presence of the cross-link. Because this hydroxylysine residue is unique in poly-$\alpha$1CB6 in containing both galactose and glucose, we can label a small proportion of any presumptive cross-link preparation (peptides of any size) with dansyl chloride and, after alkali hydrolysis, assay for the disaccharide derivative of $\alpha$-DNS-hydroxylysine. Although not yet fully tested, this technique is already proving to be extremely useful in this laboratory as DNS-derivatives can be readily resolved by HPLC.

The techniques described for the isolation and quantitation of cross-links in collagen have important applications in relation to mechanical studies, investigations of aging, and structural studies and in determination of molecular defects in pathological syndromes involving collagen.

# [17] Determination of 3- and 4-Hydroxyproline

### By RICHARD A. BERG

The determination of hydroxyproline in animal tissues has been an essential technique in the development of connective tissue biochemistry during the last 20 years.[1-4] Hydroxyproline is also present to a small

[1] R. Kuttan and A. N. Radhakrishnan, *Adv. Enzymol.* **37**, 273 (1973).

extent in some other animal proteins[1-4] and in plants,[5] but the techniques for its measurement have been primarily applied to measuring the imino acid in animal connective tissues. Although connective tissue has been found to be composed of many different proteins, the major structural protein is a family of related proteins designated as the collagens.[6,7] The presence of hydroxyproline, a unique amino acid, in collagenous molecules derived from connective tissue provides a characteristic biochemical marker that can simplify the chemical analysis of connective tissues from animals. For a recent comprehensive review of proline and hydroxyproline metabolism, see Adams and Frank.[4]

General Considerations and Purpose

Hydroxyproline in animals is synthesized by enzymes that operate on prolyl residues after they have been incorporated into peptide linkage.[2-4,6-8] At least two prolyl hydroxylases have been identified in mammals: one that synthesizes 4-hydroxy-L-proline, and a second one, which synthesizes another isomer, 3-hydroxy-L-proline.[4,8] The 4-prolyl hydroxylases that have been examined have been purified from chick embryos, newborn rats, human fetal skin, and cultured cells and are very similar in terms of their amino acid compositions, quaternary structure, and kinetic parameters.[4,8] The enzyme that synthesizes 3-hydroxyproline has been identified in chick embryos and partially characterized.[4,8] Besides collagen,[9,10] hydroxyproline has been identified in other proteins including elastin and C1q and in acetylcholin esterase.[4,8] 3-Hydroxy-L-proline is much less abundant than 4-hydroxy-L-proline and has been identified only in collagen.[4,7,8]

Since hydroxyproline is unique to collagen, the determination of hydroxyproline in various tissues and cells has been used to quantify the

---

[2] G. J. Cardinale and S. Udenfriend, *Adv. Enzymol.* **41**, 245 (1974).
[3] J. H. Fessler and L. I. Fessler, *Annu. Rev. Biochem.* **47**, 129 (78).
[4] E. Adams and L. Frank, *Annu. Rev. Biochem.* **49**, 1005 (1980).
[5] D. Ashford and A. Neuberger, *Trends Biochem. Sci.* (*Pers. Ed.*) **5**, 245 (1980).
[6] K. Piez, *in* "Biochemistry of Collagen" (G. N. Ramachandran and A. H. Reddi, eds.), p. 1. Plenum, New York, 1976.
[7] P. Bornstein and H. Sage, *Annu. Rev. Biochem.* **49**, 957 (1980).
[8] K. I. Kivirikko and R. Myllylä, *in* "The Enzymology of Post-translational Modification of Proteins" (R. B. Freedman and H. C. Hawkins, eds.), p. 53. Academic Press, New York, 1981.
[9] P. Bornstein and W. Traub, *in* "The Proteins" (H. Neurath and R. L. Hill, eds.), 3rd ed., Vol. 4, p. 411. Academic Press, New York, 1979.
[10] P. Bornstein and P. H. Byers, *in* "Metabolic Diseases and Control Metabolism" (P. K. Bondy and L. E. Rosenberg, eds.), p. 1089. Saunders, Philadelphia, Pennsylvania, 1980.

amount of collagenous protein in such tissues and cells. This has proved to be a very valuable technique in measuring the collagen content of tissues, and it has been extended to the use of measuring radioactive hydroxyproline in tissues in order to quantify collagen that is newly synthesized. Hydroxyproline is synthesized from peptide-bound proline by enzymes that operate after peptide synthesis, and it has become apparent that some collagens vary in terms of the level of hydroxylation of selected prolyl residues (4, 6–9). Although free *trans*-4-hydroxy-L-proline cannot be directly incorporated into peptide linkage, a related analog, *cis*-4-hydroxy-L-proline can be taken up and incorporated in place of prolyl residues with dramatic consequences for the conformation of collagen molecules.[11]

The measurement of hydroxyproline has been used to determine not only the quantity of collagen in a given tissue, but also the rate with which the collagen is synthesized or degraded in such tissues. A large number of specific assays for hydroxyproline have appeared in the literature over the past 20 years. The number of methods and modifications of methods indicates the lack of a simple specific method for measuring hydroxyproline. The present chapter describes two of the most successful assays for determining unlabeled or labeled *trans*-4-hydroxyproline and *cis*-4-hydroxyproline as well as two assays that can distinguish *trans*-3- from *trans*-4-hydroxyproline.

Hydrolysis

Since hydroxyproline is peptide bound, samples containing hydroxyproline must be hydrolyzed to release the hydroxyproline for analysis. A standard procedure is that the samples should be hydrolyzed in 6 $N$ HCl for 16 hr at 116°. The reaction can be carried out in a volume of between 1 and 10 ml in a culture tube (Pyrex), 15 or 25 mm in diameter, having a Teflon-lined screw cap. The volume of 6 $N$ HCl can be adjusted depending on the amount of tissue hydrolyzed, but generally 10 times the wet weight of tissue is convenient. Both the hydroxyproline and proline are fairly stable in 6 $N$ HCl, however, *trans*-4-hydroxyproline is epimerized to *cis*-4-hydroxy-D-proline[12] and *trans*-3-hydroxyproline is epimerized to *cis*-3-hydroxy-D-proline[13] in a time-dependent reaction (Fig. 1). Because of the epimerization reaction it is difficult entirely to avoid some conversion of *trans*-4-hydroxy-L-proline to *cis*-4-hydroxy-D-proline

---

[11] D. J. Prockop, R. A. Berg, K. I. Kivirikko, and J. Uitto, in "Biochemistry of Collagen" (G. N. Ramachandran and A. H. Reddi, eds.), p. 161. Plenum, New York, 1976.
[12] D. D. Dziewiatkowski, V. C. Hascall, and R. L. Riolo, *Anal. Biochem.* **49**, 550 (1972).
[13] M. Man and E. Adams, *Biochem. Biophys. Res. Commun.* **66**, 9 (1975).

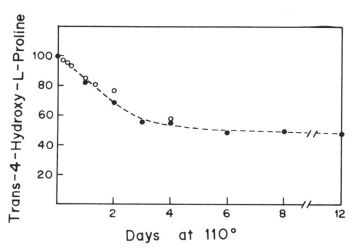

FIG. 1. The transformation of trans-4-hydroxy-L-proline to cis-4-hydroxy-D-proline with time in 12 N HCl at 110°. The data have been plotted against the time at which the reaction mixture was incubated at 110°. Adapted with permission from Dziewiatkowski et al.[12]

during hydrolysis, and up to 10% of the trans-4-hydroxy-L proline may be converted to cis-4-hydroxy-D-proline in 16 hr. In assaying hydroxyproline it is important to realize that ion-exchange chromatography and high-voltage paper electrophoresis separate trans from cis epimers of the hydroxyprolines, so both amino acids must be identified and summed to obtain the total hydroxyproline in a sample. In the colorimetric reactions, trans-4- and cis-4-hydroxyproline are determined together.

Samples to be analyzed for hydroxyproline by the procedure involving conversion to pyrrole (see below) need not be pure prior to hydrolysis, since the assay can detect hydroxyproline in as small an amount as 1 in 10,000. On the other hand, if the hydroxyproline is to be analyzed by ion-exchange chromatography (see below), the hydroxyproline should comprise at least 1% of the sample. In the latter case, where an ion exchange column is employed, interfering substances, such as salt, ammonia, acrylamide, cellulose, or strong buffers, should be removed from the sample prior to hydrolysis.

Colorimetric Reaction for 4-Hydroxyproline

The colorimetric reaction for 4-hydroxy-L-proline is based on the oxidation of hydroxyproline to pyrrole. Pyrrole can be allowed to react with p-dimethylaminobenzaldehyde to form a complex that is detected and quantified as a chromaphore having an absorbance at 560 nm. Since the procedure involves the chemical conversion of hydroxyproline (water soluble) to pyrrole (soluble in toluene), the procedure has the advantage of being able to detect very small amounts of hydroxyproline in samples containing a large amount of contaminating substances. Although a number of methods based on the oxidation of hydroxyproline have appeared in the literature (for review see Woessner[14]) the method presented here is a refinement that is particularly applicable to detecting hydroxyproline in impure samples.[15] It has proved to be highly reliable as used for 10 years in the author's laboratory. As an example of its specificity, as little as 0.5 $\mu$g of hydroxyproline can be detected in samples containing over 50 mg of other amino acids. This assay for hydroxyproline is therefore applicable to the detection of hydroxyproline in a variety of connective tissues, cells in tissue culture, and purified collagens.

*Materials*

Potassium borate buffer: The solution is prepared by adding 61.8 g of boric acid and 225 g of potassium chloride to 800 ml of distilled water. The pH is adjusted to 8.7 with concentrated potassium hydroxide, and the final volume is made up to 1 liter. The solution can be stored in a refrigerator with a few drops of toluene on its surface to inhibit bacterial growth.

Alanine solution: A 10% (w/v) alanine solution is prepared by adding 25 g of alanine to 200 ml of distilled water; the pH is adjusted to 8.7 with potassium hydroxide; and the final volume is made up to 250 ml. The solution can be stored in a refrigerator with a few drops of toluene on its surface to inhibit bacterial growth.

Sodium thiosulfate solution: The solution is prepared by gradually adding 569.20 g of anhydrous sodium thiosulfate (J. T. Baker Chemical Co.) to approximately 500 ml of warm distilled water with constant stirring. The solution is brought up to 1 liter with distilled water and stored in an opaque bottle at room temperature.

Ehrlich's reagent (p-dimethylaminobenzaldehyde): The solution is made by adding 27.4 ml of concentrated sulfuric acid to 200 ml of absolute alcohol in a beaker, and the mixture is cooled. Then 120 g of dimethylaminobenzaldehyde (J. T. Baker Chemical Co.) are added to 200 ml of absolute ethanol in another beaker, and this solution is also cooled. The acid–ethanol solution is then added with stirring to the p-dimethylaminobenzaldehyde–alcohol solution. The mixture can be

---

[14] J. F. Woessner, Jr., in "The Methodology of Connective Tissue Research" (D. A. Hall, ed.), p. 227. Joynson-Bruvvers, Oxford, England, 1976.

[15] K. I. Kivirikko, O. Laitinen, and D. J. Prockop, *Anal. Biochem.* **19**, 249 (1967).

stored for up to several months in an opaque bottle in the refrigerator. Crystals that form during storage must be redissolved on the day of use by heating the solution to approximately 60° with stirring for approximately 30 min. The final solution should have an orange color. If the color of the solution is green, $p$-dimethylaminobenzaldehyde must be recrystallized prior to making up the Ehrlich's reagent. Recrystallization is performed by dissolving the $p$-dimethylaminobenzaldehyde in absolute ethanol and allowing it to cool, then removing the purified crystals by filtration and washing them briefly with cold ethanol. This procedure should be done twice. The final crystals of $p$-dimethylaminobenzaldehyde may be stored for 6 months to a year before recrystallization should again be necessary.

Phenolphthalein solution: 1% in ethanol

Chloramine-T solution: A 0.2 $M$ solution is made up fresh daily by dissolving 2.82 g (Eastman Kodak Co.) in 50 ml of methyl Cellosolve.

Hydroxyproline for standards: A solution of 1 mg of 4-hydroxy-L-proline per milliliter in distilled water is prepared as a stock solution. The stock solution is used to prepare an appropriate set of 4-hydroxyproline standards for the assays. It is important that the hydroxyproline be dried completely before weighing so that its final concentration is not underestimated because of the presence of water in the commercial preparation. If possible, the final concentration should be assayed directly using an amino acid analyzer.

Humin precipitant: 20 g of analytical grade resin (AG1-X8, 200–400 mesh, Cl form) is mixed with 20 g of Norit A. The mixture is washed several times with 6 $N$ HCl on a coarse sintered-glass funnel and then dried to a fine powder with ethanol and ether.

Potassium chloride, reagent grade

Potassium hydroxide, reagent grade

Toluene, reagent grade

Methyl Cellosolve (2-methoxyethanol; ethylene glycol monomethyl ether, reagent grade)

Hydrochloric acid, reagent grade, concentrated solution

Culture tubes (20 mm × 150 mm, Pyrex) with screw caps having Teflon liners

*Procedure*

After hydrolysis the samples are examined to determine whether or not there is a large amount of humin in the hydrolyzate. If there is humin, a small amount of humin precipitant is added to the sample, which is then shaken. The sample is then filtered through a No. 4 filter paper (Whatman

Co.) on a funnel to remove the humin precipitant. The filter paper should be washed several times with water in order to ensure that the entire sample was transferred to the tube. To remove the HCl used for hydrolysis, the sample is then either transferred to a round-bottom flask and evaporated in a rotary evaporator or the culture tube is attached to an evaporator such as an Evapomix (Buchler Instruments) and evaporated in the culture tube itself. The evaporated sample should be dissolved in 4 ml of distilled water and transferred to a screw-cap culture tube (20 mm × 150 mm). At this point the sample may have to be diluted so that the aliquot of material to be analyzed contains between 1 and 10 $\mu$g of hydroxyproline. If the amount of hydroxyproline is unknown, a set of dilutions (usually 10-fold serial) must be prepared in order to estimate the amount of hydroxyproline present in a given sample. The samples containing up to 10 $\mu$g of hydroxyproline each are placed in a 4-ml volume of water in individual culture tubes. A drop of phenolphthalein is added to each tube, and the pH is adjusted to a pale pink color by using dilute solutions of potassium hydroxide or hydrochloric acid as necessary.

The assay begins with the addition of potassium chloride to each culture tube so that the salt is approximately half an inch deep on the bottom of each culture tube. If necessary, the pH is readjusted with potassium hydroxide so that the samples again have a pale pink color. Exactly 0.5 ml of the 10% alanine solution, pH 8.7, is added to each tube followed by the addition of 1.0 ml of potassium borate buffer, pH 8.7, to each tube. The samples are mixed well with a Vortex mixer. To ensure that the potassium chloride is saturating, the tubes are allowed to stand for 30 min with occasional shaking, and more potassium chloride is added to the tubes if necessary. Hydroxyproline is converted to pyrrole-2-carboxylic acid by oxidizing the samples with exactly 1 ml of a 0.2 $M$ chloramine-T solution added to each tube. It is most convenient to add the chloramine-T to three tubes at a time. The first three tubes are then mixed on a Vortex mixer before the chloramine-T is added to the next three tubes. A clock is set for 25 min at the end of the first addition. The tubes are incubated at room temperature for exactly 25 min. The oxidation is then stopped by adding 3 ml of the 3.6 $M$ sodium thiosulfate solution to each tube in sets of three in the same order in which the samples were oxidized.

After the oxidation step, 5 ml of toluene are added to each tube, and the tubes are capped tightly with Teflon-lined screw caps. The tubes are then vigorously shaken in a horizontal position for approximately 4 min and centrifuged at 1500 rpm in order to separate the organic from the aqueous phase. The toluene layer (the upper phase) is then removed by suction, collected, and discarded. The tubes are again tightly capped and boiled for 30 min in a rapidly boiling water bath to convert the pyrrole-2-

carboxylic acid to pyrrole. After the tubes are added to the water bath, a clock is set for 30 min when the water has just again begun to boil. The tubes are then removed from the water bath and cooled under running tap water. Exactly 5.0 ml of toluene is added to each tube, and they are again shaken for 4 min and centrifuged for 4 min.

The hydroxyproline which has been converted to pyrrole is then determined by removing 2.5 ml of the toluene extract from each tube and mixing it with 1 ml of the Ehrlich's reagent in a separate tube. Ehrlich's reagent is added quickly to each tube, and the sample is immediately mixed on a Vortex mixer. After the toluene extracts from each tube have been mixed with Ehrlich's reagent, the samples are set aside for approximately 30 min to allow the color to develop. The optical densities are then read in 1 cm path length cuvettes in a spectrophotometer at a wavelength of 560 nm. It is important that the cuvettes be completely dry and free of water before adding the samples, since the presence of even traces of water will cause Ehrlich's reagent to precipitate. The cuvettes, if necessary, can be dried by using acetone prior to the addition of the sample. The chromophore is stable, and the optical density can be read up to 2 hr after preparation. Under the conditions of the reaction described here, 10 $\mu$g of hydroxyproline give an optical density of approximately 0.5 (Fig. 2). A standard curve may be constructed in order to determine the reproducibility of the assay and to provide standards against which unknown samples containing hydroxyproline may be compared. The variation in the slope of the standard curve should be less than 5% from day to day, and the assay as described here should be sensitive to 1 $\mu$g or less of hydroxyproline in each sample.

The primary advantage in using the colorimetric reaction for 4-hydroxyproline described here is that the samples may be very impure, with hydroxyproline representing as little as 0.1% of the amino acids present in the hydrolyzate. Approximately 48 samples can be assayed in a single day by one technician. A number of improved modifications of the technique have been published that claim increased sensitivity (for example, see Inayama et al.[16]).

The colorimetric assay for trans-4-hydroxy-L-proline described here has also been shown to detect cis-4-hydroxy-L-proline.[17] Therefore cis-4-hydroxy-D-proline produced during hydrolysis due to epimerization would be expected to be measured as hydroxyproline by this method. The colorimetric reaction will not detect trans-3-hydroxyproline due to its instability during oxidation.[18]

[16] S. Inayama, T. Shibata, J. Ohtsuki, and S. Saito, *Keio J. Med.* **27,** 43 (1978).
[17] J. Uitto and D. J. Prockop, *Arch. Biochem. Biophys.* **181,** 293 (1977).
[18] G. Szymanovicz, O. Mercier, A. Randoux, and J. P. Borel, *Biochimie* **60,** 499 (1978).

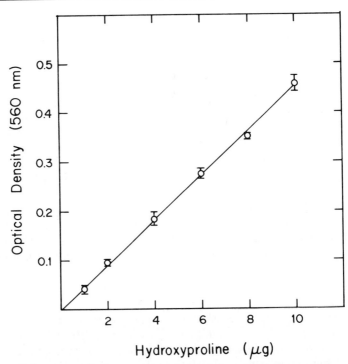

Hydroxyproline (μg)

FIG. 2. Standard curve for the assay of hydroxyproline using the colorimetric assay involving the conversion to pyrrole. The vertical lines are standard deviations for the average of three separate experiments performed on different days. Figure by courtesy of Ms. Perveen Kamani.

Radiochemical Assay for 4-Hydroxyproline

Many studies on collagen metabolism involve using radioactive proline to label newly synthesized collagen polypeptide chains, which can then be identified or characterized by determining the presence of radioactive 4-hydroxy-L-proline in hydrolyzates. Although various procedures have been published (for review see Woessner[19]) the procedure described here is based on modifications introduced by Juva and Prockop.[20] In studies on the biosynthesis of collagen using a tracer technique, minute amounts of newly synthesized collagen can be identified and subjected to preliminary chemical characterization using techniques such as polyacrylamide slab gel electrophoresis. In studies on collagen degradation, collagen that is

[19] J. F. Woessner, Jr., in "The Methodology of Connective Tissue Research" (D. A. Hall, ed.), p. 247. Joynson-Bruvvers, Oxford, England, 1976.
[20] K. Juva and D. J. Prockop, Anal. Biochem. 15, 77 (1966).

synthesized using labeled proline and then degraded can be conveniently followed by examining the amount of labeled 4-hydroxy-L-proline present in dialyzable fractions of tissues or in hydrolyzates of urine. The radiochemical assay is based on the same reactions described above for the colorimetric reaction for 4-hydroxyproline except that the hydroxyproline is converted to radioactive pyrrole, which can then be quantified using scintillation counting. An internal standard of unlabeled hydroxyproline is added to each sample prior to analysis, and its recovery is determined in order to correct the values obtained for the hydroxy[$^{14}$C]proline.

*Materials*

The materials required for the radiochemical assay of 4-hydroxyproline are similar to those required for the colorimetric determination of 4-hydroxyproline (see above). Additional materials required for the radiochemical assay for hydroxyproline are given as follows:

Sodium pyrophosphate: A 0.2 $M$ solution of sodium pyrophosphate is prepared by adding 93.22 g of sodium pyrophosphate (J. T. Baker Chemical Co.) to 800 ml of warm distilled water and adjusting the pH to 8.0 with HCl. The volume is then adjusted to 1.0 liter with distilled water.

Tris buffer: A 1.0 $M$ solution is prepared by adding 121.1 g of Tris(hydroxymethyl)aminomethane to 800 ml of distilled water. The pH is adjusted to 8.0 with HCl, and the volume adjusted to 1.0 liter.

Sodium chloride (reagent grade)

Proline solution: A solution is made by adding 10 mg/ml in distilled water.

Hydroxyproline solution: A solution of 20 mg of hydroxyproline per milliliter is made by adding the anhydrous reagent to distilled water (see note above).

Pyrrole: Commercial pyrrole (Eastman Kodak Co.) must be distilled and immediately diluted to obtain a final concentration of 1 m$M$ in toluene. It is stable at this concentration for up to 1 year when stored at $-20°$. In more concentrated solutions, polymerization occurs to form material that cannot be used as a standard, and the solution must then be redistilled.

Silicic acid: Three grams of silicic acid (Bio-Rad Laboratories, minus 325 mesh) are added per 10 ml of toluene, 5 ml of which is used to prepare a single column for each sample.

Sand suitable for use in preparing silicic acid columns

Scintillation counting solution: Formula 963 or Aquasol (New England Nuclear) or another toluene-based liquid scintillation solution)

Culture tubes: screw cap tubes (25 mm × 200 mm, Pyrex) with Teflon-lined screw caps

Sintered-glass columns (12 mm × 30 mm, Pyrex)

[$^{14}$C]Toluene: A standardized solution of [$^{14}$C]toluene is diluted from $^{14}$C-labeled toluene to give approximately 5000 dpm per vial. It is diluted to 20 ml of toluene to which is added 1 ml of scintillation solution. The [$^{14}$C]toluene solution is used as a standard to measure the efficiency of counting when $^{14}$C-labeled hydroxyproline is being quantified. If hydroxy[$^{3}$H]proline is being quantified, then an appropriate standard containing [$^{3}$H]toluene must be prepared in order to provide an appropriate standard for converting cpm to dpm in the assay. It is important that the scintillation counting solution be used in the same proportions to toluene for the standard [$^{14}$C]toluene as for the samples (see below).

*Procedure*

To each 25 × 200 mm culture tube having a Teflon-lined screw cap is added 1 mg of carrier proline (0.1 ml) and 2 mg of carrier hydroxyproline (1.0 ml). Samples for the assay that have been previously hydrolyzed in 6 $N$ HCl are evaporated and resuspended in 4 ml of deionized water in the culture tubes. If the hydrolyzed samples contain humin, they can be decolorized by the addition of humin precipitant prior to evaporation as described for the colorimetric assay (above). Exactly 0.1 ml of each 4.0-ml sample is counted in a liquid scintillation counter, and the total number of counts is determined. If the total number of counts in each 4-ml sample exceeds 100,000 cpm, then an aliquot of the sample containing 100,000 cpm or less is then diluted to 4 ml with water in a clean culture tube. One or two drops of phenolphthalein is added to each sample and the pH is adjusted to a pale pink color (approximately 8) using NaOH. If the samples have not been decolorized so that it is difficult to judge the color of the indicator dye, the pH may be determined with a pH meter. To each tube 6 ml of 0.2 $M$ sodium pyrophosphate buffer, pH 8, is added, and the tubes are mixed well on a Vortex mixer. A solution of chloramine-T is freshly prepared in methyl Cellosolve as described above, and the samples are oxidized by adding 1 ml of the freshly prepared chloramine-T solution to each tube. A clock is set for 20 min. After the first addition, chloramine-T is then added to subsequent tubes, each of which is then mixed. At the end of the 20-min incubation period at room temperature the oxidation is stopped by adding 6 ml of 3.6 $M$ sodium thiosulfate to each tube in sequence, and each tube is then mixed on a Vortex mixer. After each tube has been mixed, the pH is then readjusted with sodium hydroxide to return the solution to a pale pink color. Four milliliters of 1.0 $M$ Tris buffer, pH 8, are then added to each tube, and the tubes are mixed well. Each tube is then saturated with NaCl to a level approximately 1 inch deep. Ten milliliters of toluene are added to each tube, and the tubes are tightly capped with a Teflon-lined plastic cap. The tubes are shaken vigorously for 5 min in a horizontal position on a mechanical shaker and then centrifuged for 4 min to separate the toluene from the aqueous phase. After centrifugation the toluene layer is carefully removed from each tube and discarded. The tubes are again tightly capped and placed in a rapidly boiling water bath to convert the pyrrole-2-carboxylic acid to pyrrole. After the tubes are added to the water bath and the water returns to boiling, a timer should be set for 25 min. At the end of the 25-min boiling period the tubes are cooled to room temperature under running tap water. If more NaCl is required to maintain a half-inch depth of solid NaCl at the bottom of the tube it is added at this time. Exactly 12 ml of toluene are added to each tube, and each tube is again capped tightly with a Teflon-lined plastic cap; all the tubes are shaken for 4 min as before in a horizontal position on a mechanical shaker. The tubes are again centrifuged for 4 min to separate the toluene from the aqueous layer.

Silicic acid columns are prepared, one for each sample, by adding a thin layer of sand approximately 0.5 cm deep to each glass column. Enough dry silicic acid (30 g/100 ml) is added to toluene in a beaker with stirring using a magnetized stirring bar so that 5 ml will be available for each sample. While the suspension is stirring rapidly, 5 ml are drawn quickly into a 5-ml pipette having a wide tip and immediately added to each column. Each of the poured columns is allowed to run dry, and the toluene that runs through the columns is collected and discarded. To each column is then added exactly 10 ml of the clear toluene solution from each sample, and a large test tube is placed under each column to collect the sample. After applying the samples, each column is washed with 15 ml of fresh toluene. The total effluent from each column, 25 ml, is then pooled and mixed briefly; 20 ml of this eluate are placed in a counting vial to which is added 1 ml of the scintillation counting fluid. The vials are then counted to determine the radioactive pyrrole derived from the hydroxyproline in each sample. In order to determine the recovery of hydroxyproline in each sample, 0.1 ml of each column eluate is placed in a clean tube to which 4.9 ml of toluene are added. Standards are prepared by diluting the stock solution (1 $\mu$mol/ml) 1 part to 49 parts toluene, thus obtaining a 0.02 $\mu$mol/ml solution. Two milliliters of this are pipetted into a clean tube and then further diluted by the addition of 3.0 ml of toluene. Several tubes are prepared in this manner to obtain duplicate standards, each containing 0.04 $\mu$mol of pyrrole in a total of 5.0 ml of toluene. To each tube 2 ml of Ehrlich's reagent are quickly added, immediately mixed on a Vortex mixer, then allowed to sit for 30–45 min in the dark while the chromophore develops. The absorbance of each sample is then read in a spectrophotometer at 560 nm. By quantifying the chromaphore present, it is possible to estimate the recovery of carrier hydroxyproline added to each sample so that the radioactive pyrrole in each sample can then be corrected for losses during the assay and used to estimate the counts per minute of hydroxyproline in the original sample. By determining the counting efficiency of the [$^{14}$C]toluene, the counts per minute of radioactive hydroxyproline can be converted to disintegrations per minute.

*Calculations*

The following formula is used to calculate the amount of hydroxy[$^{14}$C]proline in the aliquot oxidized for the radiochemical assay.[20]

$$\text{dpm hydroxy[}^{14}\text{C]proline} = \text{observed cpm} \times \left[\frac{100}{E}\right]\left[\frac{25}{20}\right]\left[\frac{5}{4}\right]\left[\frac{100}{R}\right]$$

where $E$ is the percentage efficiency of the counting system, the ratio 25/20 is the correction factor for the aliquot of effluent from the silicic acid column that was counted. The ratio 5/4 is the correction factor for the loss of the carboxyl carbon in the conversion of uniformly labeled hydroxy[$^{14}$C]proline to [$^{14}$C]pyrrole, and $R$ is the percentage recovery of the carrier hydroxyproline used as an internal standard.

The percentage recovery of hydroxyproline, $R$, was calculated from the colorimetric assay as follows:

$$R = \left[\frac{OD_s}{OD_p}\right][0.04]\left[\frac{25}{0.1}\right]\left[\frac{100}{H}\right]$$

where the $OD_s$ is the absorbance of pyrrole from the sample and $OD_p$ is the absorbance of pyrrole in the standard solution, 25/0.1 is the correction factor for the aliquot of toluene taken for the colorimetric assay, and $H$ is the amount in micromoles of the carrier hydroxyproline used in the assay.[20]

The advantages of using the radiochemical assay are similar to the advantages for the colorimetric reaction for 4-hydroxyproline (above). The assay is sensitive to approximately 0.1% of the labeled hydroxyproline in a sample containing hydroxy[$^{14}$C]proline and $^{14}$C-labeled proline. Since hydroxyproline carrier is used throughout the assay, there is no lower limit on the number of moles of hydroxyproline that can be analyzed using this assay as long as there is enough radioactivity to detect. Up to 36 samples can be conveniently handled in a single day by one technician. Although the assay has not been experimentally demonstrated to apply to $^{14}$C-labeled cis-4-hydroxy-L-proline. It has been shown that the cis isomer behaves like the trans isomer in the colorimetric assay,[17] so the assay may be used also for measuring the amount of cis-4-hydroxy-L-proline isomer along with the trans isomer. In this regard, any hydroxyproline that was epimerized during the hydrolysis to cis-4-hydroxy-D-proline would be expected to be detected by this radiochemical assay for 4-hydroxyproline. The isomer 3-hydroxy-L-proline is sensitive to oxidation[18] and would be destroyed during the assay, so it could not be determined using the assay described here.

*Analysis of 3- and 4-Hydroxyproline by High-Voltage Paper Electrophoresis*

The previously described methods are convenient for determining 4-hydroxyproline, but are not applicable for the detection of 3-hydroxyproline. A rapid way to analyze radioactively labeled 3- and 4-hydroxyproline in a sample without resorting to an amino acid analyzer is the method of separation using high-voltage electrophoresis on paper. The assay was described by Gryder and Adams[21] and has more recently been described by Tseng et al.[22]

*Materials*

Electrophoresis buffer: The buffer is prepared by mixing 75 ml of glacial acidic acid with 27.5 ml of formic acid (98%) and diluting with distilled water to 1 liter. The pH of the final solution used for electrophoresis should be 1.85.

Isatin staining reagent: The reagent was made by dissolving 1 g of isatin (indole-2,3-dione, Fisher Scientific), 2.7 ml of collidine (2,4,6-trimethylpyridine), and 1.8 ml of anhydrous glacial acidic acid in 100 ml of absolute ethanol.

Extraction solution: 30% ethanol in water

Whatman No. 1 filter paper (46 cm × 57 cm)

*Procedure*

Samples of protein containing $^{14}$C-labeled or $^{3}$H-labeled hydroxyproline are hydrolyzed in 6 N HCl, evaporated to dryness, and redissolved in distilled water (above). A line is drawn approximately 8 cm from one end of the filter paper, and up to 15 samples are applied along the line. Each sample should have approximately 10 µg of proline and 20 µg of

[21] R. M. Gryder and E. Adams, *J. Bacteriol.* **97**, 292 (1969).
[22] S. C. G. Tseng, R. Stern, and D. E. Nitecki, *Anal. Biochem.* **102**, 291 (1980).

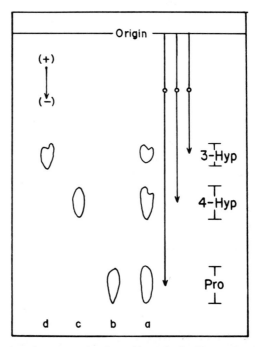

FIG. 3. Electrophoretic pattern of amino acid standards. Electrophoresis was performed for 30 min at 88 V/cm. The paper was stained, and the spots were outlined. Samples applied were as follows: d, 3-hydroxyproline; c, 4-hydroxyproline; b, proline; and a, mixture of all three; 10 μg of proline and 20 μg of hydroxyproline were used. Adapted, with permission, from Tseng et al.[22]

hydroxyproline as carrier. A hair dryer is used with gentle heating to facilitate the application of the samples to the paper. After wetting evenly in the electrophoresis buffer, the filter paper is pressed as dry as possible and is transfered to an electrophoretic chamber (Savant Instruments, Inc., or Gilson Medical Electronics, Inc.) with the origin near the anode. Electrophoresis is run for approximately 30 min at 5000 V.

The sheets are then dried and heated in an oven at 90° for 10 min and sprayed with isatin staining reagent. Bluish spots are developed on a yellowish background for the hydroxyprolines (Fig. 3). Relative mobilities are measured, and the spots containing either proline or hydroxyproline are cut from the paper and extracted in 2 ml of 30% ethanol in water. The extraction supernatant is then removed for scintillation counting. It is necessary to calculate a correction factor to account for the counting efficiency of labeled hydroxyproline, as the ethanol solution used for extraction causes some quenching even though the recovery of radioactive amino acid from the paper exceeds 97%.[22] The carrier proline and hydroxyproline were used in order to facilitate identifying the location of the spots on the paper for scintillation counting. One advantage of this method is that the sample may contain fairly high concentrations of salt, acid, or protein. A second advantage is that multiple samples can be assayed on a single sheet of paper within 2 hr. A third advantage of the method is that it can separate *trans*-3-hydroxy-L-proline from *trans*-4-hydroxy-L-proline in biological samples.[22] The procedure has been reported to separate *trans*-4-hydroxy-L-proline from *cis*-4-hydroxy-D-proline[21-23] and *trans*-3-hydroxy-L-proline from *cis*-3-hydroxy-D-proline[23,24] although the procedure does not separate the *cis*-3- from *cis*-4-hydroxyprolines from each other.[23,24]

Ion-Exchange Chromatography for 3- and 4-Hydroxyproline

Ion-exchange chromatography is perhaps the most widely used method to separate 3- and 4-hydroxyproline from proline in biological hydrolyzates. The reason for the universal application of this method for separating hydroxyprolines is that it allows the exact quantification of the hydroxyprolines and gives the analysis of other amino acids in the same chromatograph, so that an entire amino acid composition may be obtained.[25,26] It is the only reliable technique for quantifying unlabeled 3-hydroxyproline. This method can be used to separate *trans*-3-hydroxy-L-proline from *cis*-3-hydroxy-D-proline and *trans*-4-hydroxy-L-proline from *cis*-4-hydroxy-D-proline[4,27] as well as *trans*-3- from *trans*-4-hydroxy-L-proline.[4,28,29] Ion-exchange chromatography can also be used for the separation of radioactive hydroxyprolines because the eluates from the ion-exchange column can be collected as fractions and counted in a scintillation counter. A major improvement in quantifying collagen produced in biosynthetic systems involves determining the specific activity of hydroxyprolines and proline in radiolabeled collagen and using the specific activities to convert radioactivity to moles of collagen[30,31] (see below).

[23] J. C. Sheehan and J. G. Whitney, *J. Am. Chem. Soc.* **85**, 3863 (1963).
[24] T. H. Finlay and E. Adams, *J. Biol. Chem.* **245**, 5248 (1970).
[25] D. H. Spackman, this series, Vol. 11, p. 3.
[26] P. E. Hare, this series, Vol. 47, p. 3.
[27] F. Irreverre, K. Morita, A. V. Robertson, and B. Witkop, *J. Am. Chem. Soc.* **85**, 2824 (1963).
[28] J. Risteli, K. Tryggvason, and K. I. Kivirikko, *Eur. J. Biochem.* **73**, 485 (1977).
[29] K. J. Lembach, R. E. Branson, P. B. Hewgley, and L. W. Cunningham, *Eur. J. Biochem.* **72**, 379 (1977).
[30] W. W.-Y. Kao, D. J. Prockop, and R. A. Berg, *J. Biol. Chem.* **254**, 2234 (1979).
[31] S. D. Breul, K. H. Bradley, A. J. Hance, M. P. Schafer, R. A. Berg, and R. G. Crystal, *J. Biol. Chem.* **255**, 5250 (1980).

*Materials*

Ion-exchange resin: A variety of polystyrene-based ion-exchange resins are available, such as the DC-1A cation exchange resin (Dionex Chemical Corp., Sunnyvale, California) or the W-1 ion-exchange resin (Beckman Instruments, Palo Alto, California) A resin such as Dowex 50-X8 (Bio-Rad Laboratories) can also be used, but the resolution with Dowex is considerably less than that of the other resins. Generally a column of 0.9 × 30 cm is used for biological samples involving hydroxyprolines; however, microbore columns containing improved resins are also available.[32] When using the microbore columns the capacity is lower than with large columns, so that salt may be more of a problem and samples may have to be desalted prior to analysis.[33] The microbore columns also require much higher pressure to maintain flow rate and may be more amenable to clogging or compacting when relatively impure samples are applied than when using the older-type resins and larger columns.

Amino acid analyzer: A commercial amino acid analyzer is required; or, if only radioactive samples are to be analyzed, a homemade system may be constructed consisting of a high-pressure pump (Milroy), a water bath to maintain the temperature of the jacketed column, a sample injector, and a timer and switches for controlling automatic column regeneration (a kit is available from Dionex Chemical Corp., Sunnyvale, California).

Citrate buffer: Citrate buffer containing 0.2 $N$ sodium is generally used to elute the acidic and neutral amino acids including hydroxyproline and proline from an ion-exchange column. To separate the hydroxyprolines from proline, only one buffer having a pH of 3.25 is required. It may be necessary to vary the pH of the citrate buffer slightly to obtain the maximal resolution. Citrate buffer for amino acid analyzers is available in premixed quantities (Beckman, Palo Alto, California, or Pierce, Rockford, Illinois).

Detection system: Normally amino and imino acids are detected by allowing them to react with ninhydrin as they are eluted from the ion-exchange column followed by spectrophotometry,[34] but they may also be detected by allowing them to react with a fluorescent reagent followed by fluorescent spectrophotometry.[35,36] These systems are

[32] J. R. Benson, this series, Vol. 47, p. 19.
[33] R. S. Brenkowski and C. Engels, *Anal. Biochem.* in press (1981).
[34] J. R. Benson, in "Instrumentation in Amino Acid Sequence Analysis" (R. N. Perham, ed.), p. 1. Academic Press, New York, 1975.
[35] P. Bohlen and M. Millet, *Anal. Biochem.* **94**, 313 (1979).
[36] A. Szymanowicz, C. Cheron, and J. P. Borel, *Biochimie* **61**, 425 (1979).

highly sophisticated and allow the determination of as little as 10 nmol of 3- and 4-hydroxyproline using ninhydrin and 10 pmol using a fluorescent detection system.[36] If it is necessary to quantify radioactive hydroxyprolines, the eluate from an ion-exchange column is collected in scintillation vials in a fraction collector and counted directly in a scintillation counter. Alternatively, in some amino acid analyzers the stream may be split so that half of the sample is analyzed by spectrophotometry and half is collected in a fraction collector for counting.

Hydroxyproline standards: Although *trans*-4-hydroxy-L-proline is commercially available (Sigma Chemical Co.) and *cis*-4-hydroxy-L-proline is commercially available (Sigma Chemical Co., Calbiochem), *trans*-3-hydroxy-L-proline is not available and must be either synthesized or purified from a natural source (for references, see Adams and Frank[4]).

*Procedure*

Samples are prepared from protein hydrolyzates that have been evaporated to remove the HCl (see above) and then diluted in a sample application buffer consisting of 0.2 N citrate buffer at pH 2.2. The sample should normally be filtered through a 0.2-$\mu$m Millipore filter connected to a 3-ml disposable syringe in order to remove any precipitated protein, cellulose, glass, or other material that may have contaminated the sample and could potentially interfere with Teflon valves or buffer flow through the column. Because of the nature of separation of amino acids by ion-exchange chromatography, the samples for hydrolysis should not contain excessive amounts of acids, bases, or salts that are not volatile because such contaminants may alter the chromatographic properties of the amino acids and affect the separation of amino acids on the chromatogram. When it is desired to separate all amino acids in a protein hydrolyzate, it is very often necessary to adjust the pH of the citrate buffer or adjust the column temperature. *trans*-3-Hydroxyproline elutes earlier than *trans*-4-hydroxyproline and is not difficult to separate from other amino acids (Fig. 4). *cis*-4-Hydroxyl-L-proline and *cis*-4-hydroxy-D-proline are very difficult to separate from other amino acids and most often coelute with threonine on the ion-exchange column (Fig. 4).

When the hydroxyprolines are to be quantified both by using ninhydrin or a fluorescent procedure and by counting in a liquid scintillation counter, the effluent from the amino acid analyzer must be collected in a fraction collector. If the effluent has already been allowed to react with ninhydrin to allow the detection of the amino acids spectrophotometri-

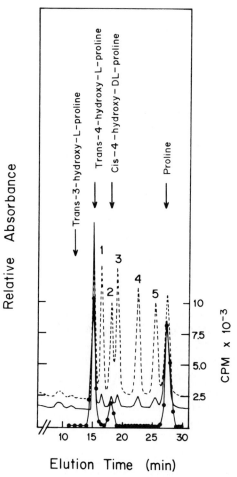

Elution Time (min)

FIG. 4. Elution positions of *trans*-3-hydroxy-L-proline, *trans*-4-hydroxy-L-proline, *cis*-4-hydroxy-D-proline, and *cis*-4-hydroxy-L-proline compared with standard amino acids by ion-exchange chromatography. The imino acids *trans*-3-hydroxy-L-proline and *cis*-4-hydroxy-L-proline were not run in this chromatogram, but their elution positions are indicated by arrows. The radioactive samples were prepared by hydrolyzing [$^{14}$C]proline-labeled collagen for 72 hr at 116°, which causes epimerization of *trans*-4-hydroxy-L-proline to *cis*-4-hydroxy-D-proline. The observed quenching of the $^{14}$C-labeled proline and hydroxyproline was 50%. The chromatography was performed on a Beckman Model 121MB amino acid analyzer using AA-10 resin (Beckman) and 0.2 $M$ citrate buffer, pH 2.93. The initial temperature was 39.2°, and it was increased to 65.5° over 71 min. The amino acids were determined by optical absorbance at 570 nm after reaction with ninhydrin, and the imino acids were similarly detected at 440 nm or, in the case of labeled imino acids, detected by liquid scintillation counting of 0.5-min fractions collected from the effluent. Dashed lines represent absorbance at 570 nm; solid lines represent absorbance at 440 nm. Connected dots indicate counts per minute of column effluent fractions. Peak 1, aspartate; peak 2, threonine; peak 3, serine; peak 4, glutamate; peak 5, glycine.

cally, a suitable standard should be run in order to obtain a correction factor for possible quenching of radioactivity. Ion-exchange columns designed only for the detection of radioactive amino acids have much simpler detection and pumping systems than do amino acid analyzers designed for quantification of all amino acids, the effluents from the column being merely collected into a fraction collector for scintillation counting. For the elution positions of trans-3-hydroxy-L-proline, trans-4-hydroxy-L-proline, and for cis-4-hydroxy-L-proline and cis-4-hydroxy-D-proline, see Fig. 4. It has been reported that trans-3-hydroxy-L-proline may give rise to a split peak due to the formation of a complex with citrate or other organic acids in the sample or the buffer.[37]

The major advantages of the ion-exchange method are, first, the high accuracy in determining the chemical amounts of both trans-3- and trans-4-hydroxy-L-prolines in a single run and, second, to be able to obtain the amount of radioactivity in each hydroxyproline peak in the same run so that the specific activities of the hydroxyprolines can be obtained. The major disadvantage of this technique is that usually at least 2 hr are required for a single analysis. The method cannot be used reliably for samples where the hydroxyprolines comprise less than 1% of the sample if the hydroxyprolines are to be detected spectrophotometrically or fluorimetrically owing to the overlap of closely eluting amino acids. Also some components of samples other than amino acids may interfere with the ion-exchange chromatography. If, however, the hydroxyprolines are to be analyzed by radioactivity only, the samples may contain much higher amounts of other amino acids as they will not interfere with the detection of the radioactive hydroxyprolines.

Specific Activity of Intracellular Proline Pool

Collagen biosynthesis has been an important aspect of connective tissue research because it has been used to answer questions about the regulation of collagen production at various transcriptional, translational, and posttranslational levels. To quantify newly synthesized collagen, cells or tissues are incubated in the presence of radioactive proline in order to label newly synthesized collagen. The method described here has been utilized for measuring collagen production under the conditions of tissue culture although it could be easily adapted to organ cultures. Cultures of fibroblasts are first labeled with radioactive [$^{14}$C]proline for a length of time determined by the experiment, for example, between 1 and 24 hr. Radioactive collagen is then obtained either by isolating collagenase digestible counts (see below) or by isolating newly synthesized collagen itself.

[37] M. Man, R. M. Gryder, and E. Adams, *Anal. Biochem.* **63**, 513 (1975).

In order to convert the radioactivity into the chemical amount of collagen, it is necessary to determine the specific activity of the proline in newly synthesized collagen.

*Procedure*

The specific activity of proline in collagen can be determined by one of two methods. The first consists of determining the specific activity of the free [$^{14}$C]proline in the intracellular pool,[31] and the second consists of determining the specific activity of the [$^{14}$C]proline and hydroxyproline in the newly synthesized, and purified, collagen.[30] To determine the specific activity of the intracellular pool of labeled proline, a culture plate is labeled and then the medium of a culture plate is removed and discarded. The cell layer attached to the plate is placed on ice and rinsed 8 times with phosphate-buffered saline at pH 7.4. The cells are then harvested by scraping with 5 ml of 1% picric acid. After the proteins have precipitated overnight, the precipitate is removed by centrifugation and the supernatant containing the free intracellular proline is passed through a disposable prefilled ion-exchange column AG 1-X8 (100–200 mesh, Bio-Rad Laboratories) to remove the picric acid. The column is then washed twice with 5 ml of water, and the combined eluates are lyophilized. The sample is solubilized in citrate buffer and chromatographed on an amino acid analyzer to which a fraction collector has been connected. By collecting fractions of the ion-exchange column eluate, the counts of [$^{14}$C]proline and the nanomoles of proline can be determined simultaneously. The specific activity of proline can be calculated as the counts divided by the nanomoles of proline.

An alternative method for determining the specific activity of proline and hydroxyproline in newly synthesized collagen[30] is to incubate the cells in fresh medium containing radioactive proline for a period of time; the medium is then separated from the cells and the collagen isolated from the medium by ammonium sulfate fractionation[30] followed by DEAE purification.[3,7] The isolated collagen (procollagen) is then hydrolyzed in 6 $N$ HCl, evaporated, and dissolved in 0.2 $M$ citrate buffer and chromatographed on an amino acid analyzer. Again, the effluent from the amino acid analyzer column is connected to a fraction collector so that the radioactivity eluted in the hydroxyproline and proline peaks can be obtained along with nanomoles of hydroxyproline and proline eluted in the same peaks containing radioactivity. From the specific activity of proline or hydroxyproline, it is possible to calculate from either the total amount of labeled hydroxyproline or the total amount of collagenase-sensitive counts (see below) synthesized in the culture the absolute amount of collagen produced in the culture.

An alternative procedure has been recently devised for determining the specific activity of proline.[38] The intracellular pool of [$^{14}$C]prolyl-acyl-transfer RNA is determined. The procedure involves first isolating the acylated transfer RNAs from cultured cells and then liberating the labeled amino acids from transfer RNAs by alkaline hydrolysis at pH 9 for 1 hr at 37°.

The released [$^{14}$C]proline (or other amino acids) is analyzed for its specific activity by reaction with [$^{3}$H]dansyl chloride of known specific activity. The amino acids are then separated by two-dimensional thin-layer chromatography. The fluorescent spots corresponding to individual dansylated amino acids are cut from the two-dimensional chromatogram, and radioactivity is determined for the amino acid and for the dansyl group. The specific activity of a given amino acid can then be determined from the counts per minute in the amino acid divided by the counts per minute in the dansyl chloride multiplied by the specific activity of the dansyl chloride times the number of dansyl chloride moieties bound per amino acid. It should be noted that the method utilizing the free intracellular pool of proline and the method utilizing the pool of activated transfer RNAs must be compared with the method involving measurement of the specific activity of proline in isolated collagen for any given biosythetic system to ensure that the specific activities of either the intracellular proline pool or the prolyl-acyl-transfer RNA pool accurately reflect the specific activity in newly synthesized collagen.[30,31]

## Use of Hydroxyproline Measurements in Quantifying Collagen in Unlabeled Cells and Tissues

Since hydroxyproline is a unique amino acid found in collagen, its measurement has often been used in determining the collagen content of various cells and tissues.

### Procedure

When quantifying the amount of collagen present in unlabeled tissues, it is most convenient simply to hydrolyze the tissues and determine the total hydroxyproline by the colorimetric assay outlined above. Since the assay is highly specific for hydroxyproline even in the presence of large amounts of other amino acids, this assay can be used to measure hydroxyproline levels in tissues when it is present even in very small amounts. After the hydroxyproline level has been determined, a correction factor must be applied in order to convert the amount of hydroxy-

[38] J. Airhardt, J. Kelley, J. E. Brayden, R. B. Low, and W. S. Stirewalt, *Anal. Biochem.* **96**, 45 (1979).

proline to the collagen content of the tissues. For most of the collagen so far examined, the average level of hydroxyproline is approximately 10% by weight of the collagen molecule, and generally this value can be used to correct the hydroxyproline values for the amount of collagen present in a given tissue.[6,7,9-11] Since underhydroxylated collagen could give a significantly altered view for the amount of collagen found in a tissue, it is necessary to be aware of the significant underrepresentation of the amount of collagen when collagen is found to be underhydroxylated. However, as discussed below much of the underhydroxylated collagen in tissues is subject to degradation—either intracellular degradation as in the case of newly synthesized collagen or extracellular degradation as is normally the case in the turnover of collagen in connective tissue. Therefore, the correction factor of approximately 10% for the amount of hydroxyproline contained in the collagen has been shown to be a fairly reliable estimate of the total collagen content of tissues, especially when total amounts of unlabeled collagen in a tissue is evaluated.

Determination of Collagenase-Sensitive Radioactively Labeled Collagen

In order to study the synthesis of collagen, tissues or cells are incubated with radioactive proline, which after incorporation into a collagen polypeptide linkage is converted to labeled hydroxyproline. Measurement of the amount of labeled hydroxyproline in tissue or cells has been extremely useful in studies on the biosynthesis of collagen and in studies on the turnover of collagen. Newly synthesized collagen, however, differs from collagen contained within the tissue in that, under a variety of circumstances, it has been shown that such newly synthesized collagen can be significantly underhydroxylated and therefore the amount of labeled collagen synthesized cannot accurately be deduced from the amount of labeled hydroxy[$^{14}$C]proline contained within a tissue.

A method described by Peterkovsky and Diegelmann[39] has been used to' circumvent this problem by using purified bacterial collagenase to digest the collagen contained in a tissue or cells in order to determine the number of radioactive counts in proline and hydroxyproline together that are collagenase sensitive. The assay does not distinguish between labeled proline or labeled hydroxyproline, but it does distinguish collagen from all other known proteins because collagen is uniquely susceptible to bacterial collagenase. The method has been applied with very good results to collagens synthesized in both tissue culture and organ cultures. The method, however, is not very well suited for determining the collagen content of

[39] B. Peterkofsky and R. Diegelmann, *Biochemistry* **10**, 988 (1971).

relatively stiff pieces of tissue where collagen may not be adequately susceptible to digestion with bacterial collagenase. However, where the method is applicable, it is possible to very accurately determine the number of counts of radioactive proline plus hydroxyproline in newly synthesized collagen. This method may be coupled with a determination of the specific activity of the intracellular proline pool or the specific activity of proline in newly synthesized collagen to estimate the absolute amount of radiolabeled collagen in cells or tissues.

*Procedure*

The amount of label actually incorporated into collagenous polypeptides can be increased by using a medium that is deficient in proline. At the end of the incubation period, the cultures are scraped using the medium to rinse the cell layer, sonicated, and then heated to 100° for 15 min to destroy all intracellular processing and degrading enzymes.[31] The samples are then dialyzed exhaustively to remove any free non-protein-bound labeled proline. Dialyzates from a single culture are then lyophilized to reduce the volume and dissolved in 2.2 ml of collagenase buffer (10 m$M$ Tris-HCl, 5 m$M$ CaCl$_2$, 25 m$M$ $N$-ethylmaleimide, pH 7.5). A 1-ml aliquot of the dissolved material is placed in a small dialysis tube, and 50 units of purified bacterial collagenase (Advanced Biofactures Corp., Lynbrook, New York) are added.[30] The bag is tied and then dialyzed against 10 ml of collagenase buffer at room temperature overnight. A blank is made containing a duplicate sample except that the collagenase is left out of the bag. After dialysis 1-ml aliquots of the solution outside of the dialysis bags are counted in a liquid scintillation counter and used to calculate the total counts that are collagenase digestible. The bags are cut open, and the entire contents of each is dissolved in counting solution (bag included) and counted, thereby determining the total nondialyzable [$^{14}$C]proline incorporated into collagen and noncollagen protein. The dialyzable counts from the control are then subtracted from the dialyzable counts from the sample containing collagenase in order to determine the net collagenase-digestible counts per sample. This value is corrected for the aliquot taken and then divided by the entire counts from the retentate of the dialysis bag plus the counts obtained with collagenase in order to determine the percentage of counts digested with collagenase.

The percentage of collagen is then calculated using a formula devised by Breul *et al.*[31] that includes correction factors for the amount of proline in collagen compared with other noncollagenous proteins. In order to determine the efficiency of collagenase digestion, aliquots of the retentates and the dialyzates of the samples are hydrolyzed in 6 $N$ HCl and analyzed for hydroxy[$^{14}$C]proline. The values for hydroxyproline can be used to determine the efficiency of the collagenase digestion, and the counts of hydroxy[$^{14}$C]proline in dialyzable peptides can be used to estimate the degree of hydroxylation of the proline in newly synthesized collagen. Since the counts liberated by collagenase digestion represent the total collagenous polypeptides present in the sample, it is possible to determine the absolute amount of collagen this represents if the specific activity of the intracellular proline pool is known (see above).

Use of Hydroxyproline Measurements in Quantifying the Degradation of Newly Synthesized Collagen

During the last few years, it has been discovered that a rather significant fraction of newly synthesized collagen is subject to intracellular degradation immediately after or during the synthesis of the collagen in tissues or in cultured fibroblasts.[40-42] Since collagen is a unique protein in that it contains the unusual amino acid, hydroxyproline, it is possible to follow the degradation of newly synthesized collagen specifically by following the appearance of dialyzable radioactive hydroxyproline in a culture.

*Procedure*

The method as described for determining the degradation of collagen in cell cultures involves scraping the entire content of each culture plate into a plastic tube, where it is heated at 100° for 15 min to destroy all proteolytic enzymes that might cause collagen degradation during the analytical procedures. The samples are sonicated to disrupt cells and precipitated proteins and then divided into two equal aliquots. One aliquot is placed onto a Centriflow filtration cone (CF25, Amicon Corp.) over a plastic centrifuge tube and centrifuged at 600 $g$ for 1 hr to separate the small peptides and free amino acids from protein. The filtrate is lyophilized to concentrate it and then hydrolyzed in 6 $N$ HCl for 16 hr at 116° to obtain the amount of $^{14}$C-labeled hydroxyproline in the culture that was present in small peptides (degraded). The second aliquot of the sonicated sample is lyophilized and then hydrolyzed directly to obtain the total amount of labeled hydroxyproline synthesized in the culture during the labeling pe-

[40] R. S. Bienkowski, B. J. Baum, and R. G. Crystal, *Nature (London)* **276**, 413 (1978).
[41] R. S. Bienkowski, M. J. Cowan, J. A. McDonald, and R. G. Crystal, *J. Biol. Chem.* **253**, 4356 (1978).
[42] R. A. Berg, M. L. Schwartz, and R. G. Crystal, *Proc. Natl. Acad. Sci. U.S.A.* **77**, 4746 (1980).

riod. The hydrolyzates are evaporated and analyzed by an ion-exchange column (see above).

The precentage of collagen degraded in each culture can be quantified by dividing the amount of labeled hydroxyproline appearing in the filtrate as small peptides by the total amount of labeled hydroxyproline synthesized in the culture. Identification of labeled hydroxyproline requires the use of an ion-exchange column, where the effluent is collected in a fraction collector and counted in a scintillation counter. An ion-exchange column is necessary in order to separate the *trans*-4-hydroxy-L-proline from any contaminating material that might be present in the cultures owing to impurities in the [$^{14}$C]proline used to label the cultures.[42] A modification of this method has been published[33] where a known amount of purified $^3$H-labeled *trans*-4-hydroxyl-L-proline is added to the culture extracts prior to hydrolysis in order to be able to quantify the recovery of hydroxy[$^{14}$C]proline in the sample.[33] The samples are then desalted by chromatography on a Dowex 50-X8 column, eluted with 0.5 N acetic acid and lyophilized. The lyophilized material is then rechromatographed on an ion-exchange column in order to separate the *trans*-4-hydroxy-L-proline from both *trans*-3-hydroxy-L-proline and *cis*-4-hydroxy-D-proline. Since hydrolysis of samples of collagen is known to cause epimerization of 4-hydroxy-L-proline (see above) to *cis*-4-hydroxy-D-proline, it is necessary to correct the values for *trans*-4-hydroxy-L-proline by accounting for that which was epimerized to *cis*-4-hydroxy-D-proline.

Summary

All the methods described for the separation of *trans*-3- and *trans*-4-hydroxyproline require that the samples containing these amino acids be hydrolyzed in order to liberate the free imino acids. There exist colorimetric assays for *trans*-4-hydroxy-L-proline that are applicable also to *cis*-4-hydroxy-L-proline and therefore presumably to *cis*-4-hydroxy-D-proline produced during hydrolysis. These assays are highly specific for hydroxyproline, and they can be used when the amount of hydroxyproline represents only 0.1% of the sample. The colorimetric assays are not useful for assaying *trans*-3-hydroxy-L-proline owing to the sensitivity of *trans*-3-hydroxy-L-proline to degradation during the oxidation step.[18] *trans*-3-Hydroxy-L-proline can be separated from *trans*-4-hydroxy-L-proline by high-voltage paper electrophoresis and by ion-exchange chromatography. High-voltage paper electrophoresis can be used when the hydroxyprolines are radiolabeled, but it is not sensitive enough to measure accurately chemical amounts of the hydroxyprolines. Ion-exchange chromatography has the advantage of being able not only to separate the labeled amino acids by collecting the effluent from the column in a fraction collector, but also (in conjunction with an amino acid analyzer) to determine the absolute amounts of these amino acids by their reaction with ninhydrin. Because ion-exchange chromatography has the advantage of being able to make both these measurements, it is possible through ion-exchange chromatography to determine the specific activities of each of these individual amino acids. The measurement of hydroxyproline and proline used in conjunction with collagenase digestion of cells or tissues can be used to estimate very accurately the quantity of labeled collagen contained in cells or tissues. The use of hydroxyproline measurements in dialyzable versus nondialyzable extracts of tissues allows the estimate of the turnover of collagen, especially the degradation of newly synthesized collagen, in various biosynthetic systems.

Acknowledgment

Supported by U.S. Public Health Service Grants AM 16516 (Project 1) and AM 21744. The author is especially grateful to Ms. Perveen Kamani for excellent technical assistance and to Ms. Nancy Kedersha for both technical and editorial assistance. The amino acid analysis was kindly provided by Dr. P. P. Fietzek, Department of Biochemistry, CMDNJ–Rutgers Medical School.

## [18] Urinary Metabolites of Collagen

By JERE P. SEGREST

The urinary excretion of hydroxyproline (free plus peptide) has been a commonly used standard for the evaluation of collagen degradation.[1] There are, however, two limitations to the use of urinary hydroxyproline excretion as a measure of collagen degradation.

1. Quantitative analysis of collagen breakdown is not possible. Weiss and Klein[2] have estimated that only 25% of the hydroxyproline of degraded collagen is excreted in the urine.
2. Hydroxyproline is nonspecific and thus cannot provide information about the rate of collagen breakdown in particular tissues (e.g., bone versus soft tissues).[3]

[1] D. J. Prockop and K. I. Kivirikko, in "Treatise on Collagen" (B. S. Gould, ed.), Vol. 2A, p. 215. Academic Press, New York, 1968.
[2] P. H. Weiss and L. Klein, J. Clin. Invest. 48, 1 (1969).
[3] J. P. Segrest and L. W. Cunningham, J. Clin. Invest. 49, 1497 (1970).

Hydroxylysine is a second amino acid, unique to collagen and collagen-like peptides, that has been used as a urinary marker for collagen degradation. Free urinary hydroxylysine does not appear to give a quantitative estimate of degraded collagen.[4] However, a variable proportion of hydroxylysine residues in collagen have been shown to be glycosylated as $\alpha$-1,2-glucosylgalactosyl-$O$-hydroxylysine (GGH) and $\beta$-1-galactosyl-$O$-hydroxylysine (GH).[5] GGH and GH have molecular weights of 485 and 323, respectively. These hydroxylysylglycosides (HLG) are also present in urine.[5] Collagen turnover equivalents calculated from urinary levels of HLG are 2–4 times that of hydroxyproline and thus account for 50–100% of degraded collagen.[3] The plasma levels of HLG, approximately 0.1 $\mu M$ of each glycoside per liter of plasma,[6] can account for all of the HLG excreted in urine (approximately 21.5 and 18.0 $\mu M$ per 24 hr of GGH and GH, respectively[3]), assuming no tubular absorption or excretion. Diet has relatively little effect on urinary HLG excretion.[3]

The ratios of the HLG are different in different collagens. For example, the ratio of GGH/GH is 1.61 and 0.15 for human adult skin and bone, respectively.[3] The ratio GGH:GH in human urine has been shown to be of some value in differentiating bone collagen turnover from turnover of collagen of the soft tissues; the GGH:GH ratio is less than 1.0 in patients with bone disorders such as Paget's disease.[7] In normal adults the urinary GGH:GH ratio is greater than 1.0, whereas in normal growing children the urinary GGH:GH ratio is less than 1.0.[3] The latter ratio presumably reflects the rapid remodeling of collagen in the bones of growing children.

Determination of Hydroxyproline

Urinary hydroxyproline can be determined by amino acid analysis or by a colorimetric assay.

*By Amino Acid Analysis*

To 1 ml of urine is added 1 ml of concentrated HCl, and the samples are hydrolyzed at 110° for 16 hr in Teflon-lined screw-cap tubes (Kimax). The samples are then dried in a desiccator containing solid NaOH under vacuum, qualitatively transferred to 5-ml volumetric flasks with 0.1 $N$ HCl, and brought to volume; 0.5 ml of final hydrolyzate is used for amino acid analysis.

[4] R. Askenasi, *J. Lab. Clin. Med.* **83** 673 (1974).
[5] L. W. Cunningham, J. D. Ford, and J. P. Segrest, *J. Biol. Chem.* **242**, 2570 (1967).
[6] J. P. Segrest, Ph.D. thesis, p. 152, Vanderbilt University Graduate School, 1969.
[7] S. M. Krane, F. G. Kantrowitz, M. Byrne, S. R. Pinnell, and F. S. Singer, *J. Clin. Invest.* **59**, 819 (1977).

Analysis is carried out by the method of Miller and Piez[8] or by use of any single-column amino acid analyzer system. As described by Miller and Piez,[8] a 51 × 0.9 cm column of resin type PA-28 (Beckman Instruments, Inc.) is used at 55° as a single-column system. Pumping rate is 80 ml/hr. Hydroxyproline elutes at about 45–50 min between methionine sulfoxide and aspartic acid.

*By Colorimetric Assay*

To 1 ml of urine is added 1 ml of concentrated HCl, and the samples are hydrolyzed at 110° for 16 hr in Teflon-lined screw-cap tubes (Kimax). The tubes are 150 × 16 mm with a capacity of 20 ml. Tubes used for hydrolysis are marked at exactly the 15 ml level with a diamond knife. Analysis is by the method of Kivirikko *et al.*,[9] whereby hydroxyproline is oxidized to pyrrole and the pyrrole is quantitated by a colorimetric reaction with Ehrlich's reagent.

*Reagents*

Potassium borate buffer (pH 8.7): Boric acid, 61.84 g, and 225 g of potassium chloride are mixed with approximately 800 ml of distilled water, the pH is adjusted to 8.7 with 10 $N$ and 1 $N$ KOH, and the final volume is made up to 1000 ml.

Alanine solution (pH 8.7): Alanine, 10 g, is dissolved in approximately 90 ml of distilled water; the pH is adjusted to 8.7 with 1 $N$ KOH, and the final volume is made up to 100 ml.

Chloramine-T solution, 0.2 $M$ (Eastman Organic Chemicals) in methyl Cellosolve, prepared daily

Sodium thiosulfate solution, 3.6 $M$ in distilled water. The solution can be stored under toluene at room temperature for several weeks.

Ehrlich's reagent: Concentrated sulfuric acid, 27.4 ml, is slowly added to 200 ml of absolute alcohol, and the mixture is cooled. *p*-Dimethylaminobenzaldehyde, 120 g, (Matheson, Coleman and Bell), is added to 200 ml of absolute alcohol in a separate beaker, and the acid-ethanol solution is slowly stirred into the second solution. The solution is stable in the refrigerator for several weeks. Crystals that precipitate when the solution is stored in a refrigerator are redissolved by warming the solution briefly.

*Procedure.* After hydrolysis, 1 drop of 1% phenolphthalein in ethanol is added, and the samples are adjusted to a faint pink color with 12 $N$, 1 $N$, and 0.1 $N$ KOH; back-titrating is by 6 $N$ HCl. The samples are then

[8] E. J. Miller and K. A. Piez, *Anal. Biochem.* **16**, 320 (1966).
[9] K. I. Kivirikko, O. Laitinen, and D. J. Prockop, *Anal. Biochem.* **19**, 249 (1967).

diluted to the 15-ml mark on the hydrolysis tubes with distilled water, mixed, and centrifuged at 1500 rpm for 10 min to remove precipitate.

An aliquot of each hydrolyzed sample containing 3–12 mg of hydroxyproline is placed in another screw-cap tube, and distilled water is used to adjust the volume to 4 ml. A water blank and 5- and 10-$\mu$g standards of hydroxyproline are assayed together with each set of samples. A second drop of 10% phenolphthalein is added, and the pH is adjusted to a pale pink color with 0.1 $N$ and 0.05 $N$ KOH, using 0.05 $N$ HCl for back-titrating. The solutions are then saturated with KCl by adding approximately 3 g of the solid salt (the exact amount is not critical but should be the same in each tube); 0.5 ml of alanine solution and 1.0 ml of potassium borate buffer are added, and the solutions are mixed vigorously. The tubes are then incubated at room temperature for 30 min with occasional mixing. The samples are then oxidized by the addition of 1.0 ml of 0.2 $M$ chloramine-T solution with rapid and complete mixing. The tubes are allowed to stand at room temperature for 25 min with occasional mixing; the reaction is then stopped with 3.0 ml of 3.6 $M$ sodium thiosulfate.

Approximately 5 ml of toluene are added to each tube, the tubes are tightly capped with Teflon-lined screw caps and then vigorously shaken for 5 min and centrifuged at 1500 rpm for 5 min. The toluene layer is then removed and discarded. The tubes are tightly capped and placed in a briskly boiling water bath for 30 min and cooled with tap water; 5.0 ml of toluene are added. The tubes are recapped, shaken for 5 min and centrifuged at 1500 rpm for 5 min. Exactly 2.5 ml of toluene extract are placed in clean test tubes, and 1.0 ml of Ehrlich's reagent is added with rapid mixing during addition. The tubes are allowed to stand at room temperature for 30 min, and the absorbance is measured at 560 nm.

Determination of Hydroxylysine

Levels of urinary hydroxylysine can be measured either by amino acid analysis or by a colorimetric assay procedure.

*By Amino Acid Analysis*

To 1.0 ml of urine is added 1.0 ml of concentrated HCl, and the samples are hydrolyzed in Teflon-lined screw-cap tubes (Kimax) at 100° for 16 hr. The samples are then dried in a desiccator containing solid NaOH under vacuum, quantitatively transferred to a 5-ml volumetric flask with 0.1 $N$ HCl, and brought to volume. Of final hydrolyzate 0.5 ml is used for amino acid analysis by the method of Miller and Piez[8] (or by any standard single column amino acid analyzer system) as described for hydroxyproline. Hydroxylysine elutes at approximately 170 min between phenylalanine and ammonia in the Miller and Piez system.

*By Colorimetric Assay*

One milliliter of urine is precipitated directly into Teflon-lined screw-cap tubes (Kimax) with 5 ml of cold acetone, and the suspension is kept on ice for 30 min. The tubes are then centrifuged at 1800 rpm for 15 min, the supernatant is discarded, and the precipitate is washed three times with 5 ml of cold acetone. The samples are then dried under a stream of nitrogen and hydrolyzed with 2 ml of 6 $N$ HCl at 110° for 16 hr. Analysis is by the method of Blumenkrantz and Asboe-Hansen[10] and is based on the periodate oxidation of hydroxylysine to glutamic semialdehyde and the conversion of the latter into its cyclic form, $\Delta'$-Pyrroline-5-carboxylic acid, which on further oxidation gives a chromogen with Ehrlich's reagent.

*Reagents*

Citrate–phosphate buffer (pH 7.0): 150 ml of 0.15 $M$ citric acid (Fisher Scientific Co.) mixed with 356 ml of 0.6 $M$ dibasic sodium phosphate (J. T. Baker Chemical Co)

Hydroxylysine standard: $\Delta$-Hydroxylysine-HCl (Sigma Chemical Corp.) is prepared at 1.0 mg/ml, and subsequent dilutions are made from this solution.

Periodic acid solution: Periodic acid (Fisher Scientific Co.), 0.0015 $M$ in distilled water is prepared and can be stored up to 4 months in the dark.

Extraction solution: 250 ml of toluene mixed with 250 ml of isobutanol and 100 ml of $n$-propanol

Ehrlich reagent: Isobutanol, 15 ml, is added to 4 g of $p$-diethylaminobenzaldehyde (Eastman Organic Chemicals), followed by 4.5 ml of perchloric acid. The solution can be stored for several weeks at 4° in the dark.

*Procedure.* Hydrolyzed samples are dried in a desiccator containing solid NaOH under vacuum and made up to 7 ml with the citrate buffer. The tubes are covered with aluminum foil, 0.5 ml of 0.0015 $M$ periodic acid is added, and the samples are stirred. Then 3 ml of extraction solution are added, and the tubes are shaken on a Vortex mixer for 15 min. The tubes are centrifuged at 1500 rpm for 10 min, and 600 $\mu$l of the organic phase are transferred to a test tube; 150 $\mu$l of Ehrlich's reagent are added, and the tubes are shaken vigorously and allowed to sit at room temperature for 15 min. The absorbance is then measured at 565 nm.

[10] N. Blumenkrantz and G. Asboe-Hansen, *Anal. Biochem.* **56**, 10 (1973).

One possible difficulty with the colorimetric procedure is a questionable quantitative precipitation of hydroxylysine and hydroxylysylglycosides by acetone. An automated version of the colorimetric procedure is available.[11]

## Hydroxylysyl Glycosides

*Principle*

A number of procedures for determination of urinary HLG levels have been published. These procedures generally consist of three basic steps: (a) sample preparation; (b) ion-exchange chromatography; and (c) quantitation by one of three methods, the ninhydrin assay for free amino groups, a colorimetric assay for hydroxylysine, or the orcinol–$H_2SO_4$ assay for hexose.

The goal of sample preparation generally has been to achieve one or more of the following: (a) concentration of urinary HLG; (b) desalting of urine to prevent interference with ion-exchange chromatography; (c) pH adjustment to prevent interference with ion-exchange chromatography; (d) removal of urinary compounds interfering with HLG quantitation; (e) addition of a radiolabeled internal standard; (f) alkaline hydrolysis of HLG peptides to GGH and GH.

Several different ion-exchange columns have been utilized for urinary HLG assay; all involve strong acid cation-exchange resins but vary in the length of time (2–12 hr) required for analysis. Often the ion-exchange column used was a component of an amino acid analyzer.

Most of the published methods for urinary HLG determination use ninhydrin for quantitation. However, as pointed out by Pearson et al.,[12] ninhydrin analysis may give erroneously high HLG levels owing to interference from amino acids eluting near to GGH or GH.

The colorimetric assay for hydroxylysine when used in conjunction with cation-exchange column chromatography, owing to its specificity, has the advantage that nearby peptides cause no interference.[13] The major disadvantage of this procedure is a requirement for acid hydrolysis to release free hydroxylysine from GGH and GH. This is a time-consuming step that makes a continuous-flow assay for hydroxylysine impossible.

The orcinol–$H_2SO_4$ procedure for hexose analysis[14] shares the advantage of specificity with the hydroxylysine assay, but has the additional

[11] N. Blumenkrantz and G. Asboe-Hansen, *Clin. Biochem.* **1**, 177 (1975).
[12] C. H. Pearson, L. Ainsworth, and A. Chovelon, *Connect Tissue Res.* **6**, 51 (1978).
[13] W. C. Bisbee and P. C. Kelleher, *Clin. Chim. Acta* **90**, 29 (1978).
[14] R. J. Winzler, *Methods Biochem. Anal.* **2**, 200 (1955).

advantages of speed and simplicity; a continuous-flow orcinol–$H_2SO_4$ procedure is available.[12]

*Procedure*

PREPARATION OF STANDARDS

GGH and GH standards can be obtained from alkaline hydrolyzates of a number of collagen or basement membrane preparations. Disadvantages of this approach are limitations in quantities of HLG attainable and racemization of the HLG by alkaline hydrolysis. Racemization is a particular problem for GH, since it leads to a double peak on ion-exchange chromatography. The most convenient source of HLG standards is urine. Large quantities of unracemized HLG can be obtained from urine by the following procedure:

*Acetone Precipitation.* Urine, 24 hr sample, is collected under 20 ml of toluene and stored frozen at −20° until use. After thawing, the toluene is decanted; the urine is filtered to remove sediment and subsequently is treated with acetone by a modification of the method of Estes and Golaszewski.[15] Four volumes of acetone are added to one volume of urine, and the solution is allowed to stand at 4° overnight. The precipitate is collected, washed twice with cold acetone, and air dried. The dried precipitate is then dissolved in a minimal quantity of distilled water (5–10 ml/g), and the pH is simultaneously adjusted to 9.5. The precipitate dissolved in distilled water is then subjected to acetone precipitation twice more. A final yield of 8–10 g of dried precipitate is obtained from a 24-hr collection of normal urine.

*Gel Filtration.* The dried precipitate, up to 10 g, is dissolved in no more than 50 ml of 0.1 $M$ pyridine acetate buffer, pH 5.0, and subjected to gel filtration on a 2000-ml column of Sephadex G-25 (Pharmacia Fine Chemicals) equilibrated with the same buffer. The temperature is approximately 25°. Fractions, 20 ml, are collected and analyzed for ninhydrin-positive material by the method of Moore and Stein[16] and for carbohydrate by the orcinol procedure.[14] The results of a typical analysis are shown in Fig. 1. At least three distinct carbohydrate-containing fractions can be distinguished. The three fractions are well separated from most of the ninhydrin-positive material. Pool III is the fraction that contains the HLG.

*Ion-Exhange Column Chromatography.* Pool III contains approximately 16% of the total hexose and amounts to approximately 200 mg of material per 24-hr sample of urine. The final step in purification of GGH

[15] F. L. Estes and T. Golaszewski, *Fed. Proc. Fed. Am. Soc. Exp. Biol.* **24**, 606 (1965).
[16] S. Moore and W. H. Stein, this series, Vol. 6, p. 819.

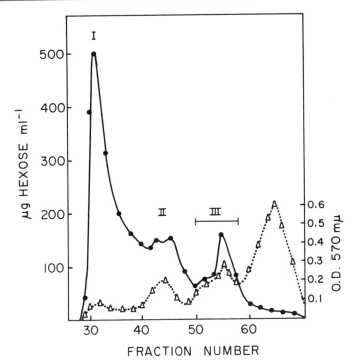

FIG. 1. Gel filtration on a 2000-ml column of Sephadex G-25 of acetone precipitation from normal urine. △---△, Ninhydrin reaction, OD 570 nm; ●——●, orcinol–$H_2SO_4$ reaction, μg hexose $ml^{-1}$.

and GH from urine is column chromatography of pool III material on a strong acid cation-exchange resin such as Dowex 50 X-12 (alternatively, AG 50 W, X-12, Bio-Rad Laboratories, can be used). The most convenient way to accomplish this step is to combine chromatography on the long column of a double-column amino acid analyzer with stream splitting; this allows dual ninhydrin analysis and collection of 2–4-ml fractions for hexose analysis by the orcinol–$H_2SO_4$ method.[14] With modern amino acid analyzers, only a small fraction of the total column effluent should be required for free amino group analysis by ninhydrin or fluorescent methods. Once the GGH and GH peaks have been identified, stream splitting may not be required, as hexose analysis of individual fractions should suffice to locate the HLG.

Figure 2 shows the results of ion exchange chromatography of pool III material on a 159 × 0.9 cm column of Dowex 50, X-12 maintained at 55°. Elution was begun with 0.2 N sodium citrate buffer, pH 3.25, with a change to pH 4.25 0.2 N sodium citrate buffer at 340 ml. The flow rate was

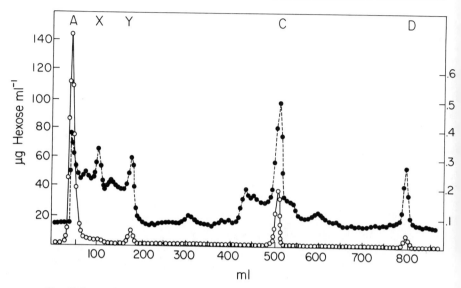

Fig. 2. Ion-exchange column chromatography on Dowex 50 X-12 of Sephadex G-25 pool III, a low molecular weight glycopeptide fraction from normal human urine. ●---●, Ninhydrin reaction, OD 570 nm; ○——○, orcinol–$H_2SO_4$ reaction, μg hexose $ml^{-1}$.

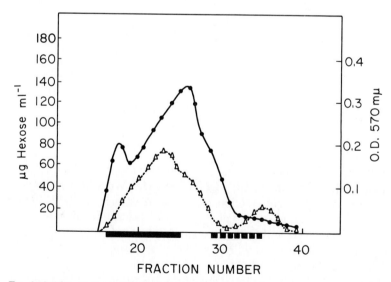

Fig. 3. Gel filtration on a 225-ml column of Sephadex G-10 of 6 ml of concentrated normal human urine. △---△, Ninhydrin reaction, OD 570 nm; ●——●, orcinol–$H_2SO_4$ reaction, μg hexose $ml^{-1}$. The solid bar represents the elution volume of purified GGH and GH; the broken bar represents the elution volume of citrate.

60 ml/hr and 4-ml fractions were collected. Fractions designated C and D correspond to GGH and GH, respectively. These two fractions together represent 22% of the total hexose from pool III or 3–4% of the total hexose in a 24-hr urine sample.

Final purification of GGH and GH requires a desalting step. Sephadex G-10, in pyridine acetate buffer, is conveniently used for this final step, as it separates HLG from citrate (Fig. 3). The fractions represented by the solid bar (Fig. 3) are pooled and lyophilized. GGH and GH can be conveniently quantitated by 20-hr hydrolysis in 6 $N$ HCl at 110°, followed by amino acid analysis of total hydroxylysine in each HLG preparation. Approximately 10 mg of GGH and 5 mg of GH can be obtained from a single 24-hr collection of urine.

SAMPLE PREPARATION

Use of the orcinol–$H_2SO_4$ reagent to quantitate urinary HLG levels, because of its specificity, simplifies the steps necessary to prepare samples of urine for HLG analysis, as one does not have to be concerned with removal of interfering compounds. The following procedure is suggested as a minimal sample preparation procedure.

Two milliliters of urine can be desalted over a 20–50-ml column of Sephadex G-10 equilibrated with the starting buffer to be used for ion-exchange column chromatography (Fig. 3). Precalibration of the column with isolated HLG and silver nitrate precipitation of NaCl allows the determination of the volume of eluting buffer required to recover HLG (specifically the smaller GH), but exclude salt; a subsequent batch preparation can then be utilized. Use of an appropriate internal radioactive standard [e.g., 5000 cpm of $^3$H-labeled thyrotropin-releasing hormone (molecular weight 362; available from New England Nuclear) per milliliter of urine] allows a rapid estimation of the percentage recovery of HLG from urine after desalting. After determination of HLG recovery, 1–2 ml of the desalted urinary sample is applied directly to the ion-exchange column without further preparation.

ION-EXCHANGE CHROMATOGRAPHY

Because of the specificity of the orcinol–$H_2SO_4$ procedure for quantitating urinary HLG using ion-exchange chromatography, column resolution is relatively unimportant; the only essential feature is that the excluded volume material be separated from the GGH peak and the GGH peak be separated from the GH peak. Given this flexibility, rapidity of elution of HLG from the ion-exchange column is the most desirable feature in designing a chromatographic system. Two previously reported ion-exchange columns, the Dowex 50 X-8 system of Segrest and Cunning-

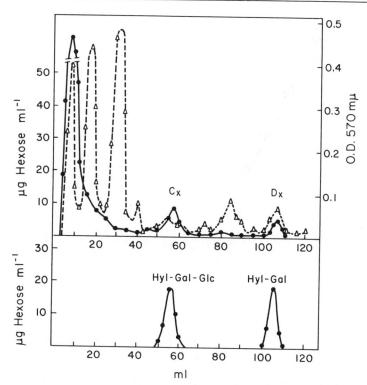

FIG. 4. Ion-exchange column chromatography on Dowex 50 X-8, of desalted, concentrated normal human urine, equivalent to approximately 1 ml of unconcentrated urine. *Upper panel:* Concentrated, desalted normal human urine: △---△, ninhydrin reaction, OD 570 nm; ●——●, orcinol–$H_2SO_4$ reaction, μg hexose ml$^{-1}$. *Lower panel:* Purified GGH and GH from an acetone precipitate of normal human urine: ●——●, orcinol–$H_2SO_4$ reaction, μg hexose ml$^{-1}$.

ham[3] and the Beckman M72 system of Askenasi,[17] seem to possess the optimal combination of rapidity of elution with adequate resolution. Both of these columns provide complete elution of HLG in less than 2 hr. The original procedure of Segrest and Cunningham[3] and a modification of the Askenasi[17] procedure are described here.

Desalted urine, 1–2 ml, in 0.24 N sodium acetate, pH 3.25, is applied to a 21 × 0.9 cm column of AG 50 W-X8 (Bio-Rad Laboratories), 50–100 mesh, maintained at 55°. (The original Dowex 50 resin used is no longer available, but AG 50 W-X8 is an equivalent resin.) The initial buffer is 0.24 N sodium acetate, pH 3.25, with a change to a pH 4.25, 0.2 N sodium acetate, at 60 ml. The flow rate is 60 ml/hr, and 1-ml fractions are collected

[17] R. Askenasi, *Biochim. Biophys. Acta* **304**, 375 (1973).

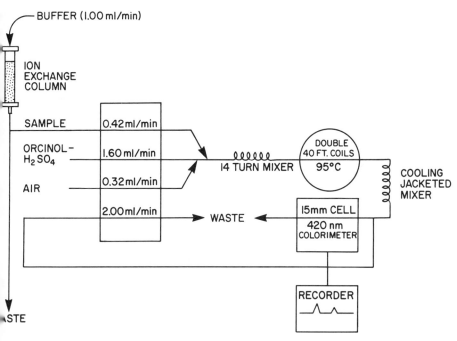

FIG. 5. Flow diagram for a continuous-flow orcinol–$H_2SO_4$ analyzer. Orcinol–$H_2SO_4$ is 0.10% orcinol (Sigma) in 70% $H_2SO_4$ (reagent grade). The pump is a Technicon Autoanalyzer II.

if a continuous-flow orcinol–$H_2SO_4$ analyzer is not used. The results of a typical chromatogram are shown in Fig. 4; GGH elutes at 55–65 min, and GH at 100–110 min.

The original Askenasi[17] procedure was a 55.5 × 0.09 cm column of Beckman resin M 72. Elution was carried out with 0.35 $M$ sodium citrate buffer, pH 5.28, at a flow rate of 50 ml/hr and 64°. The M 72 resin is no longer available; it has been replaced by Beckman resin W1. Since the Askenasi column combines rapidity (70 and 110 min of elution time for GGH and GH, respectively) with better resolution than is necessary for orcinol–$H_2SO_4$ analysis, a modification in the original procedure to obtain more rapid elution is suggested. Using the original conditions of buffer, flow rate, and temperature, a shorter column (approximately 25 × 0.9 cm) of Beckman W1 resin should provide satisfactory resolution and an elution time of approximately 60 min for GH.

QUANTITATION

There are three basic means of quantitating the urinary levels of HLG after ion-exchange chromatography: (*a*) continuous-flow assay by automated orcinol–$H_2SO_4$; (b) assay of multiple small fractions by automated or manual orcinol–$H_2SO_4$; or (c) assay of batch fractions of GGH and GH by automated or manual orcinol–$H_2SO_4$.

Figure 5 shows a flow diagram for the continuous-flow assay of hexose by the automated orcinol–$H_2SO_4$ procedure (adapted from Pearson et al.[12]). The sensitivity of the assay can be increased by increasing the flow rate of the sample line so that it takes a greater percentage of the total ion-exchange column flow than is indicated in the diagram. Use of HLG standards of known concentration in conjunction with a computing integrator allows direct readout of micromoles of GGH and GH per milliliter of urine.

In lieu of a continuous-flow assay for HLG, the ion-exchange column effluent can be collected into 1-ml fractions (0.5-ml fractions for a short Beckman resin W1 column), and the individual fractions can be analyzed for hexose by the automated orcinol–$H_2SO_4$ procedure[18] or by manual orcinol–$H_2SO_4$.[14]

The simplest procedure for quantitation of GGH and GH, if funds or equipment are limited, is to collect the effluent from the ion-exchange column into four fractions [e.g., using Fig. 4 as an example, collect, in batch, fractions 1–45 (45 ml, batch 1), 46–70 (25 ml, batch 2), 71–95 (25 ml, batch 3), and 96–100 (25 ml, batch 4)] and analyze batch 2 (GGH) and batch 4 (GH) for hexose by manual orcinol–$H_2SO_4$.[14]

[18] J. Judd, W. Clouse, J. Ford, J. van Eys, and L. W. Cunningham, Anal. Biochem. **4**, 512 (1962).

## [19] Estimation of the Size of Collagenous Proteins by Electrophoresis and Gel Chromatography[1]

By RALPH J. BUTKOWSKI, MILTON E. NOELKEN, and BILLY G. HUDSON

Practical methods for the estimation of the molecular weight of globular proteins by sodium dodecyl sulfate (SDS)-gel electrophoresis[2] and SDS-gel filtration chromatography have been outlined in numerous publications in recent years. These methods rely upon comparing the unknown protein with standard proteins. Certain classes of proteins have unique chemical properties such that a direct comparison of these with typical

[1] This work was supported by NIH Grants AM18381 and AM26178.
[2] K. Weber and M. Osborn, in "The Proteins" (H. Neurath and R. H. Hill, eds.), 3rd ed., Vol. 1, p. 179. Academic Press, New York, 1975.

globular protein standards could lead to erroneous molecular weight estimations. Collagenous proteins are an example of this, and one must be aware of the characteristic behavior of their SDS complexes in order to ensure that the correct molecular weight estimates are made. This chapter describes how SDS-gel electrophoresis can be used to obtain the best estimate of the number of residues of a collagenous polypeptide and how gel chromatography using SDS can be used to estimate its molecular weight.

## Sodium Dodecyl Sulfate-Gel Electrophoresis

The technique of SDS-gel electrophoresis is extensively used to estimate the molecular weights of proteins. This technique is described in detail elsewhere as applied to typical globular proteins.[2] In this section the unique behavior of collagens and their derivatives upon SDS-gel electrophoresis is discussed, and procedures are presented for the estimation of the molecular sizes of collagenous polypeptides and their cross-linked multimers ranging in molecular weight from 13,500 to about 1,000,000.

Collagenous proteins behave anomalously on SDS-polyacrylamide gel electrophoresis in comparison to typical globular proteins.[3] For example, the $\alpha 1$ collagen chain, molecular weight 96,000, migrates at a slightly slower rate than serum albumin dimer, molecular weight 132,000, and thus behaves as though it has a molecular weight 40% higher than the true value. We have found that this anomalous behavior is due, in large part, to interpretation of data; that is, the electrophoretic mobility of globular proteins and collagenous proteins is better correlated with the number of residues in a polypeptide chain rather than molecular weight.[4] This finding is illustrated in Fig. 1 where type I collagen $\alpha 1$ and $\alpha 2$ chains are characterized with respect to both molecular weight and residue number using globular proteins as standard markers. The collagen $\alpha$ chains have falsely high apparent molecular weights of 133,000 and 122,000 compared with the true value of 96,000 for each chain, whereas the observed number of residues is 1156 for $\alpha 1$ and 1060 for the collagen $\alpha 2$ chain compared to the true value of 1052.[5] Likewise, $\beta$ chain and the cyanogen bromide peptides in the 217–660 residue size range which are derived from type I collagen correlate well with globular proteins when the number of residues per polypeptide chain is considered, rather than the molecular weight.[6]

The reason for the better correlation between electrophoretic mobility

[3] H. Furthmayr and R. Timpl, *Anal. Biochem.* **41**, 510 (1971).
[4] J. W. Freytag, M. E. Noelken, and B. G. Hudson, *Biochemistry* **18**, 4761 (1979).
[5] P. P. Fietzek and K. Kühn, *Int. Rev. Connect. Tissue Res.* **7**, 1 (1976).
[6] M. E. Noelken, B. J. Wisdom, Jr., and B. G. Hudson, *Anal. Biochem.* **110**, 131 (1981).

and number of residues per polypeptide chain rather than the moleculjr weight is not well understood. A plausible explanation is based on the low mean residue weight of collagen.[4] Collagen contains 33 mol % glycine and about 10 mol % proline, resulting in the unusually low mean residue weight of 91.6. On the other hand, the average value for the mean residue weight of the standard proteins used in Fig. 1 is 115. Thus, collagen has 26% more residues per unit molecular weight than the standard proteins, a value quite close to the percentage by which the apparent molecular weights are too high. This explanation, however, does not reconcile the different mobilities of the $\alpha 1$ and $\alpha 2$ chains in SDS gels. In this case, the behavior could be due to localized conformation effects resulting from restricted rotation of polypeptide chains about imino acid residues as originally proposed by Furthmayr and Timpl.[3]

The relationship between electrophoretic mobility and number of residues in the polypeptide chain is thus regarded as empirical.[4] Polyacrylamide gel electrophoresis is nonetheless valuable in characterizing collagenous polypeptides, since an accurate value for the number of residues has the same importance as an accurate molecular weight and the residue number can be converted to molecular weight values by multiplying by the mean residue weight. The relationship is especially useful because it can be applied with equal validity to collagen chains and procollagen-like polypeptides, as well as to noncollagen polypeptides.

*Proteins and Polypeptides.* The following proteins are used as standards: acid-soluble calf skin collagen (Calbiochem); rabbit muscle myosin (prepared by the method of Perry[7]); other standard proteins may be obtained from Sigma Chemical Co. and include *Escherichia coli* $\beta$-galactosidase, rabbit muscle phosphorylase *a* (2 × crystallized), bovine serum albumin (fatty acid free), hen's egg ovalbumin (Grade VI), and bovine pancreatic chymotrypsinogen (6 × crystallized). Cyanogen bromide fragments of calf skin collagen are prepared as described by Butler *et al.*[8] The molecular weights and number of residues in the standard proteins and polypeptides are given in the table.[4] The number of residues and the relative locations of the cyanogen bromide fragments in the intact $\alpha$ chains are given in the report of Hunt and Dayhoff[9]; $\alpha$1-CB3, which is fragment number three obtained from the $\alpha$ collagen chain, occupies positions 419 to 567 in the sequence of the intact molecule. The molecular weights of $\alpha$1-CB3, $\alpha$1-CB6, $\alpha$1-CB7, and $\alpha$1-CB8 were calculated from

[7] S. V. Perry, this series, Vol. 2, p. 582.
[8] W. T. Butler, K. A. Piez, and P. Bornstein, *Biochemistry* **6**, 3772 (1967).
[9] L. T. Hunt and M. O. Dayhoff, *in* "Atlas of Protein Sequence and Structure" (M. O. Dayhoff, ed.), Vol. 5, Supplement 2, p. 233. Natl. Biomed. Res. Found., Washington, D.C., 1976.

## Standard Proteins Used for Gel Electrophoresis and Chromatography

| Proteins | Molecular weight[a] | Number of residues[a] |
|---|---|---|
| Myosin | 212,000 | 1690 |
| β-Galactosidase | 116,000 | 1021 |
| Phosphorylase a | 92,500 | 797 |
| Bovine serum albumin | 66,000 | 581 |
| Ovalbumin | 43,000 | 370 |
| Chymotrypsinogen | 25,700 | 245 |
| Type I collagen α chain | 96,400 | 1052 |
| Type I collagen α chain multimers | $n(96,400)$[b] | $n(1052)$ |
| Cyanogen bromide peptides of calf skin collagen α chains | | |
| α2-CB3-5 | 61,000 | 661 |
| α2-CB4 | 29,500 | 321 |
| α1-CB8 | 25,100 | 279 |
| α1-CB7 | 24,000 | 268 |
| α1-CB6 | 20,000 | 217 |
| α1-CB3 | 13,500 | 149 |

[a] References to the original data are given in the text.
[b] $n$ refers to the number of α chains present in any given naturally occurring multimer.

the amino acid compositions and the number of residues reported by Rauterberg and Kühn.[10] Similarly, the chemical composition[11] of α2-CB3-5 was used to calculate its molecular weight. The peptide α2-CB4 has 321 residues and a molecular weight of 29,500 as determined from its amino acid sequence.[12]

*Preparation of Gels*

Electrophoresis supplies, i.e., SDS, acrylamide, $N,N'$-methylenebisacrylamide, $N,N,N',N'$-tetramethylenediamine (TEMED), ammonium persulfate, and Coomassie Brilliant Blue R-250 may be purchased from Bio-Rad Laboratories. Agarose (ME) may be obtained from Marine Colloids Division, FMC Corp. Preparation of polyacrylamide gels is performed by the method of Weber and Osborn,[2] using continuous 0.1 $M$ phosphate buffer.

In the following procedure, all solutions are prepared in 0.1 $M$ sodium phosphate, pH 7.0, containing 0.1% SDS. For 5% polyacrylamide gels with a 1:37 cross-linker ratio, 10 ml of solution A (20% acrylamide, 0.54%

[10] J. Rauterberg and K. Kühn, *Eur. J. Biochem.* **19**, 398 (1971).
[11] P. P. Fietzek, M. Munch, D. Breitkreutz, and K. Kühn, *FEBS Lett.* **9**, 229 (1970).
[12] P. P. Fietzek and F. W. Rexrodt, *Eur. J. Biochem.* **59**, 113 (1975).

bisacrylamide) are mixed with 5.0 ml of solution B (0.4% TEMED), and to this is added 23.5 ml of buffer. The resulting solution is then degassed and mixed with 20 mg of ammonium persulfate in 1.5 ml of buffer (freshly prepared). This solution volume is sufficient to cast twelve 10 × 0.55 cm tube gels. After casting, the gels are overlayered with 25–30 µl of water and allowed to polymerize for 1 hr before use. For 3% polyacrylamide gels with a 1:60 cross-linker ratio, solution A is replaced with 12% acrylamide, 0.2% bisacrylamide.

Agarose gels are also prepared in 0.1 $M$ sodium phosphate, pH 7.0, containing 0.1% SDS. Generally 2.5 or 3% agarose (w/v) gels are used, although the method works well for agarose concentrations between 0.5 and 3%. For 2.5% gels, 2.0 g of agarose are mixed with 80 ml of buffer in a volumetric flask. The agarose is melted by placing the flask in a boiling water bath for 30 min with occasional shaking. Tube gels are cast by first placing about 0.5 ml of 50% sucrose in the bottom of the tube, then overlayering the hot agarose solution on top of the sucrose. After the gels have solidified, a small piece of cheesecloth is secured over the gel tube with a rubber band in order to prevent the gel from sliding out of the tube during electrophoresis. The gels are then inverted, and placed in the electrophoresis apparatus. The interface between the sucrose solution and the agarose gel thus becomes the top of the gel, which is rinsed with buffer to remove the excess sucrose. The reservoir buffer used for both the continuous polyacrylamide and agarose gel electrophoresis consists of 0.1% SDS, 0.05 $M$ sodium phosphate, pH 7.0. Ten percent polyacrylamide gels are prepared in discontinuous Tris-HCl and Tris-glycine buffers by the method of Laemmli[13] as described by Weber and Osborn.[2] This method employs a 1 cm stacking gel consisting of 3.3% polyacrylamide in 0.137 $M$ Tris-HCl, pH 6.8, on the separating gel, which is in 0.375 $M$ Tris-HCl, pH 8.8. The bisacrylamide–acrylamide ratio is 1:37 and the reservoir buffer consists of 0.05 $M$ Tris-glycine, pH 8.3. All buffers contain 0.1% SDS.

*Sample Preparation and Electrophoresis*

Lyophilized samples are dissolved at 5 mg/ml in either 0.1 $M$ sodium phosphate, pH 7.0, 1% SDS for continuous buffer gels or 0.06 $M$ Tris-HCl, pH 6.8, containing 1% SDS, for discontinuous buffer gels. Twenty percent by volume of 0.05% bromophenol blue in 50% glycerol is then added to each sample. The samples are incubated either overnight at 37° or for 5 min at 90° prior to being applied to the gels. Electrophoresis is performed at 6–8 mA per tube for continuous buffer systems or at 3 mA per tube for discontinuous buffer gels. Alternatively, gels are developed overnight at 15 V with the power supply set for voltage regulation. Gels are stained for

[13] U. K. Laemmli, *Nature* (*London*) **227**, 680 (1970).

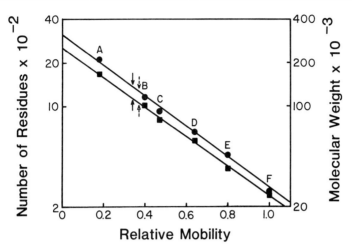

FIG. 1. Comparison of calf skin collagen $\alpha$ chains with standard globular proteins by SDS-polyacrylamide gel electrophoresis. The top curve and right-hand coordinate examine the apparent molecular weight, and the bottom curve and left-hand coordinate compare the number of residues of the $\alpha$ chains with the globular standard proteins. The standard proteins are A, myosin; B, $\beta$-galactosidase; C, phosphorylase $a$; D, bovine serum albumin; E, ovalbumin; and F, chymotrypsinogen. The solid arrow represents the collagen $\alpha 1$ chain, and the dashed arrow represents the $\alpha 2$ chain. Gels are 5% polyacrylamide developed in continuous sodium phosphate buffers in SDS. Mobilities are related to that of smallest component in the system; i.e., chymotrypsinogen equals 1.0, and the curve is based upon least squares analysis.

at least 4 hr in 7.5% acetic acid, 40% methanol, and 0.05% Coomassie Brilliant Blue. Diffusion destaining is done in 7.5% acetic acid.

*Treatment of Data*

The logarithms of the molecular weight or number of residues per polypeptide chain are plotted as a function of relative mobility, which corresponds to the migration distance of each component divided by that of the dye front. Alternatively, the mobility is determined by using a small protein component (such as chymotrypsinogen, as in Fig. 1) instead of the dye front as the point of reference. The curves shown in Figs. 1, 3, and 4 are based upon least squares analysis.

*Interpretation of Results*

Typical electrophoresis profiles are presented in Fig. 2. Gel A contains the mixture of peptides resulting from cleavage of calf skin collagen with cyanogen bromide. The peptides identified on the gel print are based upon the results of Furthmayr and Timpl[3] for calf skin collagen. Assignment of

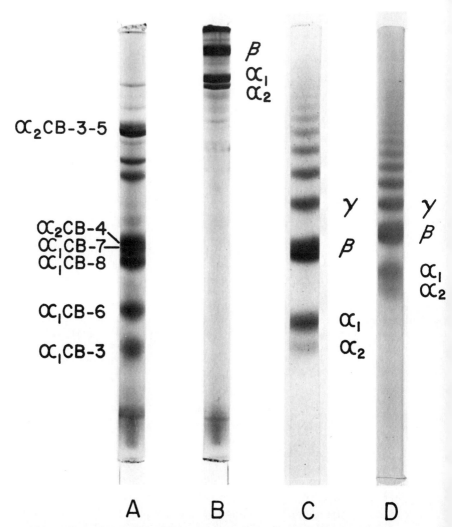

FIG. 2. Sodium dodecyl sulfate-gel electrophoretic analysis of collagenous proteins and polypeptides. Gel A represents the cyanogen bromide polypeptides derived from calf skin collagen and B represents calf skin collagen. Both gels consist of 10% polyacrylamide with a cross-linker ratio of 1:37 and were developed in discontinuous Tris buffers. Gels C and D represent analysis of calf skin collagen in 3% polyacrylamide with a cross-linker ratio of 1:60 (gel C) and in 2.5% agarose (gel D). Gels C and D were developed in continuous phosphate buffers. Calf skin collagen $\alpha$ chains are the fastest migrating components and are closest to the bottom, and bands of successively lower mobility represent $\beta$ chains ($\alpha$ dimers), $\gamma$ chains ($\alpha$ trimers), etc.

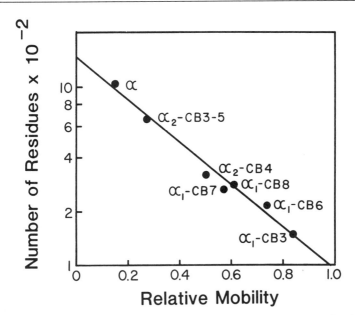

FIG. 3. Relationship between relative electrophoretic mobility and number of amino acid residues in calf skin collagen chains and their cyanogen bromide peptides. The gel consists of 10% polyacrylamide as shown in Fig. 2 and was developed in discontinuous buffers. The symbol "α" indicates the average mobility of the α1 and α2 chains. The curve is based upon least squares analysis.

the positions of peptides α2-CB4, α1-CB7, and α1-CB8 is based upon the report of Sage et al.[14] for human type I collagen. The gel consists of 10% polyacrylamide and was developed in the discontinuous buffer system in order to achieve optimal resolution of polypeptides from about 150 to 1050 amino acid residues. Gel B represents calf skin collagen developed in the same gel system. When these cyanogen bromide peptides are combined with unreacted starting material, a useful mixture is obtained for characterization of proteins in this size range. A plot of the logarithm of the number of residues as a function of electrophoretic mobility for such a mixture gives a reasonably straight line, an example of which is shown in Fig. 3.

Collagen molecules contain covalent cross-links near the ends of the chains resulting in dimers, trimers . . . $n$(monomers) of the α-size component.[15] It is convenient to take advantage of this phenomenon for the

[14] H. Sage, R. G. Woodbury, and P. Bornstein, *J. Biol. Chem.* **254**, 9893 (1979).
[15] M. L. Tanzer, *Science* **180**, 561 (1973).

purpose of developing standard curves to characterize proteins that are larger than 1000 residues. Gel C of Fig. 2 shows a 3% polyacrylamide gel of calf skin collagen. The gel shown was cast using a low ratio of bisacrylamide : acrylamide and developed in continuous phosphate buffers in order to permit resolution of a maximum number of the $\alpha$-chain multimers. By this method it is typically possible to resolve molecules containing from 8 to 11 of the cross-linked $\alpha$-size chains covering a size range from about 100,000 to $>10^6$ molecular weight. These are designated as $\beta(\alpha$ dimers), etc. in gel C (Fig. 2).

Often collagenous complexes, such as in basement membranes, are too large to penetrate polyacrylamide gels. When this occurs, agarose can be effectively substituted for polyacrylamide as the gel medium.[16,17] The size range of components that can be studied in agarose is extended to about $5 \times 10^6$, however, with some loss of resolving power. A standard developed from calf skin collagen multimers is shown in Fig. 2, gel D.

Plots of the number of residues as a function of mobility for a 3% agarose gel and a 3% polyacrylamide gel are shown in Fig. 4. To determine the mobility of the $\alpha$ size components, an average of the migration distance of the $\alpha 1$ and $\alpha 2$ chains is used. Also, the farthest migrating component in the system (average migration distance of $\alpha 1$ and $\alpha 2$) is used as the reference point in calculating mobilities.

The agarose gel system yields a straight line between 1050 residues and 5250 residues, whereas in the polyacrylamide gel system a linear relationship is observed between 2100 and 11,500 residues. The line from the polyacrylamide gel appears to have a slight upward curvature for components of 9 or more monomeric units. Upon extrapolation of the straight portion of these lines to the abscissa, it is evident that agarose gels should permit resolution of larger molecules.

The gel systems that have been described were selected to illustrate the feasibility of resolving components over a maximum size range, but the gel concentrations used do not necessarily provide optimal resolution within a more select size range. For example, 12.5% polyacrylamide gels developed in discontinuous buffers provide superior resolution of the cyanogen bromide fragments of type I collagen.[6] Also, gels intermediate in polyacrylamide concentration and cross-linker ratio may be selected to give optimal resolution of components from 100,000 to 500,000 molecular weight (about 1000–5000 residues). It is technically difficult to increase the agarose concentration beyond 3% in order to resolve relatively smaller components by this method. However, it is possible to decrease the

[16] R. J. Butkowski, P. Todd, J. J. Grantham, and B. G. Hudson, *J. Biol. Chem.* **254**, 10503 (1979).
[17] T. W. West, J. W. Fox, M. Jodlowski, J. W. Freytag, and B. G. Hudson, *J. Biol. Chem.* **255**, 10451 (1980).

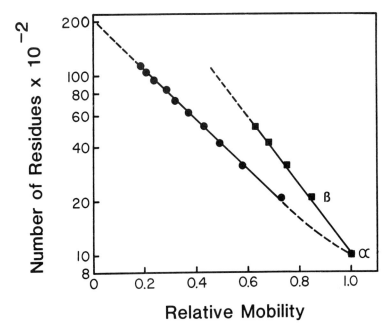

FIG. 4. Plots of number of residues as a function of electrophoretic mobility of calf skin collagen chains and their naturally occurring multimers. The symbols $\alpha$ and $\beta$ represent the average mobility of $\alpha1$ and $\alpha2$ and $\beta11$ and $\beta12$ chains, respectively. The top curve is from analysis of calf skin collagen in 3% agarose gels, and the bottom curve represents a similar analysis using 3% polyacrylamide gels with a cross-link ratio of 1:60. Both gel systems employed continuous 0.1 $M$ sodium phosphate buffer in SDS. The average mobility of the $\alpha$ chains was taken as the reference in determining the relative mobility of the remaining components. Solid lines represent the portion of the curves drawn from least squares analysis, and dashed lines are extrapolations of these.

agarose concentration for characterization of still larger components. Gels with as little as 0.5% agarose are easy to handle, and components several times larger than those discussed here will penetrate them. However, this is accomplished at the expense of resolving power. Gels that contain a mixture of agarose and polyacrylamide have also been used to characterize high molecular weight components of tubular basement membrane.[16] These gels offer an advantage in that components in the molecular weight range of 0.5 to 1 million are seen as sharper, more distinct bands than on either agarose or 3% polyacrylamide gels. With a gel composition of 2.5% polyacrylamide–1% agarose, the upper size limit of components that penetrate the gel is the same as it is for the 3% polyacrylamide gels discussed above, and the standard curve generated using calf skin collagen is virtually identical to that for the 3% gel shown in Fig. 4.

## Gel Chromatography

The principles of gel chromatography and applications to studies of proteins have been reviewed by Ackers.[18] The theoretical basis for using aqueous SDS[19-22] as well as aqueous guanidine hydrochloride[22,23] in gel chromatography to determine the molecular weights of simple globular proteins with their disulfide bonds reduced, is well established. Mann and Fish[24] have reviewed the experimental aspects of gel chromatography in 6 $M$ guanidine hydrochloride, and many of the details are applicable to the aqueous SDS system.

Gel chromatography, using aqueous SDS as solvent, is frequently used to estimate the molecular weights of collagenous polypeptides. This solvent is particularly useful because of the solubility characteristics of collagen and the convenience of detection of polypeptides in column fractions by SDS gel electrophoresis.

### Gel Chromatography Material

The choice of chromatography gel to use in analyzing SDS complexes of collagenous polypeptides is critical because they have expanded or extended structures characterized by high Stokes' radii. Nozaki et al.[25] found evidence which suggested that end-on insertion of large asymmetric particles (Stokes' radius >60 A) is a factor in determining their elution volume. The relative importance of end-on insertion increases with increasing particle size and leads to a curved calibration plot for large asymmetric complexes whereas that for large native proteins is linear, when Sepharose 4B (Pharmacia) is used. We recommend Sepharose CL-4B (Pharmacia) for SDS-collagenous chain complexes, since its use results in a linear calibration curve in the molecular weight range 43,000 to 289,000[4] (Fig. 5).

### Preparation of the Gel and Column

Sepharose CL-4B, which is supplied in a fully swollen suspension in distilled water containing 0.01% Merthiolate, is equilibrated at room tem-

---

[18] G. K. Ackers, in "The Proteins" (H. Neurath and R. L. Hill, eds.), 3rd ed., Vol. 1, p. 1. Academic Press, New York, 1975.
[19] C. Tanford, Adv. Protein Chem. 23, 121 (1968).
[20] W. W. Fish, J. A. Reynolds, and C. Tanford, J. Biol. Chem. 245, 5161 (1970).
[21] T. B. Nielsen and J. A. Reynolds, this series, Vol. 48, p. 3.
[22] W. W. Fish, K. G. Mann, and C. Tanford, J. Biol. Chem. 244, 4989 (1969).
[23] J. A. Reynolds and C. Tanford, J. Biol. Chem. 245, 5161 (1970).
[24] K. G. Mann and W. W. Fish, this series, Vol. 26, p. 28.
[25] Y. Nozaki, N. M. Schechter, J. A. Reynolds, and C. Tanford, Biochemistry 15, 3884 (1976).

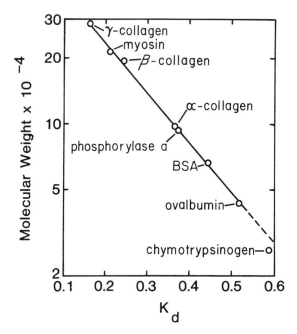

FIG. 5. Gel chromatograph of SDS–protein complexes. BSA, bovine serum albumin. Adapted, with permission, from Freytag et al.[4]

perature with solvent of the composition 0.01% 2-mercaptoethanol, 0.1 $M$ sodium phosphate, pH 7.0, and poured carefully into a 1.2 cm × 90 cm glass column to a total packed bed height of 80 cm. The column is then equilibrated further by passing two column volumes of solvent through it at 4 ml/hr; a Mariotte flask is used to control the flow rate.

*Calibration of Column*

The standard proteins and internal volume marker, $N$-dinitrophenylalanine (DNP-alanine) are applied as a mixture. For the molecular weight range 26,000 to 289,000 a solution is made that contains 2 mg of rabbit muscle myosin, 2 mg of calf skin collagen, 1 mg of rabbit muscle phosphorylase $a$, 1 mg of bovine serum albumin, 1 mg of hen's egg ovalbumin, 1 mg of bovine pancreatic $\alpha$-chymotrypsinogen, and 1 mg of DNP-alanine, 1 ml of solvent composed of 5% SDS, 5% 2-mercaptoethanol, 0.1 $M$ sodium phosphate, pH 7.0. The calibration solution is incubated at 100° for 5 min and then at 37° for 3 hr. It is then applied to the column and eluted at a rate of 4 ml/hr; 0.6-ml fractions are collected.

The elution profile of DNP-alanine is determined by absorbance measurements at 410 nm. The elution profile of each of the standard proteins is determined by SDS-polyacrylamide gel electrophoresis on 0.1-ml aliquots of each fraction. The continuous phosphate buffer system described in the first part of this chapter is used with cylindrical 5% gels (5.5 × 90 mm). After staining with Coomassie Brilliant Blue R-250, the gels are scanned at 565 nm in a Gilford Model 240 spectrophotometer equipped with a Model 42108 linear transport gel scanner. The elution volume of a protein–SDS complex can be determined with an accuracy of ±0.3 ml by locating the peak in a plot of maximum absorbance vs elution volume where the maximum absorbance is obtained from gel scans. The electrophoresis method is advantageous because it eliminates the error involved in making several column runs for the calibration.

The void volume, $V_o$, of the column is determined by measuring the elution volumn of Blue Dextran 2000 (Pharmacia). A 1-ml sample of a 0.6% solution is applied to the column, and the elution profile of Blue Dextran is determined by 630 nm absorbance measurements on the eluent fractions. When a column is used for the first time, the elution profile of Blue Dextran should be measured prior to running the calibration proteins, so that it can be used to check the column for uniform packing. In subsequent runs it can be incorporated into the calibration sample.

*Treatment of Data*

The partition coefficient, $K_d$, of each protein is calculated from the following equation:

$$K_d = \frac{V_e - V_o}{V_i - V_o}$$

where $V_e$ is the elution volume of the protein; $V_o$, the void volume, is the elution volume of Blue Dextran; and $V_i$, the internal volume, is the elution volume of DNP-alanine.

A calibration curve is obtained by plotting the logarithm of molecular weight ($M_r$) vs $K_d$. The molecular weights of the standard proteins used are listed in the table. A calibration curve is shown in Fig. 5.

*Applications*

The calibration curve is linear in the 43,000 to 289,000 $M_r$ range, indicating the usefulness of the SDS-gel chromatography system in this size range. Especially significant is the fact that the $\alpha$, $\beta$, and $\gamma$ chains of collagen fit on the curve as well as protein chains prepared from reduced globular proteins. This indicates that the SDS-gel chromatography method can be used for the determination of the molecular weight of polypeptide chains that contain both collagenous and noncollagenous regions, such as procollagen chains and collagenous chains derived from basement membranes. It should be noted, however, that some basement membrane proteins may contain nonreducible cross-links in unknown locations and relatively high amounts of carbohydrate. For these proteins, molecular weights estimated by gel chromatography may be somewhat inaccurate.[18,22-25]

Summary

Collagenous polypeptides have apparent low electrophoretic mobilities compared to those of typical globular proteins when their molecular weights are examined as a function of mobility. If, however, the number of amino acid residues per polypeptide chain is examined as a function of electrophoretic mobility, both collagenous polypeptides and globular proteins obey the same empirical linear relationship. The relationship suggests that the number of residues in collagenous polypeptides from sources such as basement membranes, which may contain both collagenous and noncollagenous regions, may be estimated accurately from SDS-gel electrophoresis. Calf skin collagen and its naturally occurring cross-linked multimers, as well as polypeptides derived from it, provide useful molecular standards in a range from 150 residues to 11,500 residues per polypeptide chain (corresponding to a molecular weight range from 13,500 to about 1,000,000 for collagen). The number of residues can in turn be translated into molecular weight values with a knowledge of the mean residue weight of the polypeptide of interest. In contrast, the molecular weight of collagenous chains can be directly estimated by gel chromatography in Sepharose CL-4B, since collagen chains give the correct molecular weight when compared to globular protein standards.

## [20] Vertebrate Collagenases

By EDWARD D. HARRIS, JR. and CAROL A. VATER

Vertebrate collagenases have been identified in many cells and tissues since the original demonstration and partial purification of collagenase from tadpole tails by Gross and his colleagues in 1962 (reviewed by Gross[1]). The term collagenase, as used in this chapter, refers to enzyme

[1] J. Gross, in "Biochemistry of Collagen" (G. N. Ramachandran and A. H. Reddi, eds.), p. 275. Plenum, New York, 1976.

TABLE I
STEPS IN COLLAGENASE PRODUCTION AND CATALYSIS

| Step | Comments[a] |
|---|---|
| 1. Synthesis | Estimated to occur in 15 min in monolayer cultures of skin fibroblasts[5] |
| 2. Secretion | Found by immunoprecipitation within 30 min after labeling skin fibroblasts in monolayer culture,[5] and can represent up to 6% of proteins released into culture medium by cells |
| 3. Activation | Accomplished by many different proteinases including trypsin, plasmin, kallikrein, and "specific" activator enzymes in certain cells and tissues. *In vitro,* many latent collagenases can be activated by organic mercurial compounds. |
| 4. Catalysis | Specific cleavage through all three polypeptide chains in helical conformation at a Gly—Ile bond in types I, II, and III collagens[3,4] |
| 5. Inhibition | $\alpha_2$-Macroglobulin and $\beta_1$-anticollagenase circulate in plasma and complex with active forms of collagenase. Many tissues and cells produce specific collagenase inhibitors[6,7] |

[a] Superscript numbers refer to text footnotes.

with the capacity to cleave native collagen in triple helical conformation. In general, the substrates cleaved are types I, II, and III. Type IV collagen is cleaved by leukocyte elastase[2] and perhaps by other proteinases, but not by enzymes that degrade types I, II, or III. The cleavage site of $\alpha$ chains by vertebrate collagenases is known and is specific[3,4]; other identical peptide bonds in the $\alpha$ chain are not cleaved.

Table I summarizes the steps in collagenase production and activation by most mammalian cells.[3-7]

Almost without exception, vertebrate collagenases are metallobiomolecules dependent upon calcium ions for activity and probably containing zinc as a constitutive component of active and latent molecules. Consistent with their function in extracellular tissues, activity is maximal at neutral pH (e.g., 7–8). Neutral proteinases with specificity for denatured (gelatin) chains are frequently associated with collagenases produced by many diverse cells.[8,9]

[2] C. L. Mainardi, S. N. Dixit, and A. H. Kang, *J. Biol. Chem.* **255**, 5435 (1980).
[3] J. Gross, E. Harper, E. D. Harris, Jr., P. A. McCroskery, J. H. Highberger, C. Corbett, and A. H. Kang, *Biochem. Biophys. Res. Commun.* **61**, 555 (1974).
[4] E. J. Miller, E. D. Harris, Jr., E. Chung, J. E. Finch, P. A. McCroskery, and W. T. Butler, *Biochemistry* **15**, 787 (1976).
[5] K-J. Valle and E. A. Bauer, *J. Biol. Chem.* **254**, 10115 (1979).
[6] C. A. Vater, C. L. Mainardi, and E. D. Harris, Jr., *J. Biol. Chem.* **254**, 3045 (1979).
[7] J. C. Nolan, S. Ridge, A. L. Oronsky, L. L. Slakey, and S. S. Kerwar, *Biochem. Biophys. Res. Commun.* **83**, 1183 (1978).
[8] E. D. Harris, Jr. and S. M. Krane, *Biochim. Biophys. Acta.* **258**, 566 (1972).
[9] J. L. Seltzer, S. A. Adams, G. A. Grant, and A. Z. Eisen, *J. Biol. Chem.* **256**, 4662 (1981).

## Substrates for Collagenase Assays

### Purification of Collagen

Details of purification of soluble forms of collagen are given in Chapter [2] of this volume. It is crucial that collagen prepared for fibril assays be purified at 4° to prevent denaturation and/or increased susceptibility to nonspecific proteinases. Pepsin-solubilized collagen does not form good fibrils. The capacity for tight fibril formation diminishes as a function of shelf life of collagen. Freeze-dried collagens are best stored in small aliquot portions at $-70°$ in a desiccated environment. Pepsin-solubilized collagen is well suited to assays of collagenase using soluble substrates, since fibrillar material or aggregates form less readily.

Use of lathyrogens (e.g., $\beta$-aminopropionitrile) in the diet of animals several weeks before sacrifice results in larger amounts of extractable collagen. However, background or "blank" radioactivity in fibril assays using collagen made from lathyritic animals is often unacceptably high.

### Isotopic Labeling of Collagen

Initial studies using labeled collagens utilized *in vivo* procedures. Isotope was given by intraperitoneal injection over a 1- or 2-day period before killing animals and harvesting the skins. This technique is expensive because most of the isotope is taken up by tissues other than skin. There are no chemical alterations, however, in this naturally produced substrate. In general, collagen soluble in neutral salt solutions should not be used because endogenous latent and active proteinases are found bound even to purified specimens.[10]

### Labeling of Collagen in Vivo

This method is modified from the original methods of Lapiere and Gross.[11]

#### Animals

Guinea pigs (six to twelve animals). To assure maximal incorporation of labeled amino acids the animals must be actively growing (each between 175 and 300 g) at time of injection.

#### Materials

$^{14}$C-Labeled (or 1 mCi $^3$H) L-glycine and L-Proline (100–200 $\mu$Ci per animal) in a glycine : proline mixture of 3 : 1

---

[10] R. Perez-Tamayo, *Am. J. Pathol.* **92**, 508 (1978).
[11] C. M. Lapiere and J. Gross, in "Mechanisms of Hard Tissue Destruction" (R. F. Sognnes, ed.), Vol. 75, p. 663. Am. Assoc. Adv. Sci., Washington D.C., 1963.

*Procedure.* Animals are injected intraperitoneally with one-half the isotope at 48 hr before killing. The remaining one-half is given 24 hr before killing. At sacrifice by barbiturate overdose, the shaved skins are quickly removed and placed on ice. After mincing and weighing, a 2-hr wash with 0.15 $M$ NaCl is performed. This is discarded. Subsequently, two 24 hr extractions of dermal collagen in 0.5 $M$ acetic acid are performed, and the extracts are pooled for further purification by methods detailed in Chapter [2]. Collagen with specific activity >5000 dpm/mg can be prepared using this procedure.

## In Vitro Labeling of Collagen

The standard method for *in vitro* labeling is that of Gisslow and McBride,[12] in which collagen in solution is labeled with [1-$^{14}$C]acetic anhydride.

*Principle.* This method involves acetylation of $\epsilon$-NH$_2$ groups of lysine residues on collagen. There is minimal risk of denaturing the substrate.

*Reagents*

Purified collagen, 250 mg in 0.01% acetic acid at a concentration of 2 mg/ml

K$_2$HPO$_4$, 1 $M$

[1-$^{14}$C]Acetic anhydride (100 $\mu$Ci/mg) in 1.5 ml of benzene. Use 250 $\mu$Ci for every 200–250 mg collagen.

NaOH, 1 $N$

Glacial acetic acid

*Procedure.* This reaction is carried out at $\leq$10°. Immediately prior to acetylation, the pH of the collagen is brought to 8.9 by addition of K$_2$HPO$_4$. The [$^{14}$C]acetic anhydride is added dropwise over a period of 2 hr to the collagen in a conical flask with a magnetic stirring bar. The pH is maintained at 8.0 by addition of NaOH. After stirring for an additional hour, the pH is adjusted to 4.0 with acetic acid. Exhaustive dialysis against 0.01% acetic acid removes residual [$^{14}$C]acetic acid, a by-product of the reaction. Labeled collagen is lyophilized and stored at $-20$°.

Cawston and Barrett[13] have suggested the following modifications in this procedure. Collagen dissolved in 0.2 $M$ acetic acid is dialyzed against 0.01 $M$ disodium tetraborate, 0.2 $M$ CaCl$_2$ adjusted to pH 9.0 with NaOH. [1-$^{14}$C]Acetic anhydride (condensed by using Dry Ice in one end of the container before opening) is solubilized in 1 ml of dry dioxane. This mixture is added to the collagen being stirred at 4°. After 30 min of stirring,

---

[12] M. T. Gisslow and B. C. McBride, *Anal. Biochem.* **68**, 70 (1975).
[13] T. E. Cawston and A. J. Barrett, *Anal. Biochem.* **99**, 340 (1979).

the acetylated collagen is dialyzed exhaustively against 0.05 $M$ Tris-HCl, pH 7.6, 0.2 $M$ NaCl, and 0.005 $M$ calcium acetate before freezing or lyophilization. $CaCl_2$ in the borate buffer prevents collagen precipitation at the elevated pH during acetylation.

*Comments*

In vitro labeling of soluble collagen produces substrate with high specific activity. The drawbacks to these labeling procedures include the possibility of excessive isotope being inserted into nonhelical portions of the molecule, which are susceptible to lysis by nonspecific proteinases without true collagenolytic capabilities, and that the *in vitro* radiolabeling process, which includes a high pH, may partially denature the collagen. In general, however, these methods are effective and are much less wasteful of radioisotope.

Collagen labeled at more than 250,000 dpm/mg can be obtained in this fashion. Two of the seven lysine residues labeled by acetylation in collagen $\alpha$ chains are in the nonhelical portion of the molecule and, therefore, susceptible to noncollagenolytic proteinases. The nonhelical portions may be removed by addition of pepsin to collagen dissolved in 0.2 $M$ acetic acid (pepsin : collagen = 1 : 10) for a 2- to 4-day period. The residual pepsin is inactivated at neutral pH but persists as a significant contaminant (10%) in the $^{14}$C-labeled collagen preparation. Salt and/or ethanol precipitation (see Chapter [2]), is sufficient to remove the pepsin.

*Storage of Labeled Collagen.* Deterioration of radiolabeled collagen during storage is manifested by gradually rising "blank" radioactivity in assays for collagenase. It can be minimized by meticulous attention to attaining complete lyophilization with storage of the freeze-dried collagen in airtight aliquot portions at temperatures of $\leq -20°$.

Synthetic Substrates for Vertebrate Collagenase

The specificity of vertebrate collagenases for a single Gly-Ile locus within collagen $\alpha$ chains has led to development of synthetic substrates for assay of the enzymes. 2,4-Dinitrophenyl derivatives of hexa- to octapeptides based on the amino acid sequence around the cleavage site have been synthesized.[14] Seven of these have been studied in detail; the structures of these are as follows:

  I. DNP-Pro-Leu-Gly-Ile-Ala-Gly-Arg-$NH_2$
 II. DNP-Leu-Gly-Ile-Ala-Gly-Arg-$NH_2$

[14] Y. Masui, T. Takemoto, S. Sakakibara, H. Hori, and Y. Nagai, *Biochem. Med.* **17**, 215 (1977).

III. DNP-Pro-Gln-Gly-Ile-Ala-Gly-Gln-D-Arg-OH
IV. DNP-Gln-Gly-Ile-Ala-Gly-Gln-D-Arg-OH
V. DNP-Pro-Gln-Gly-Ile-Ala-Gly-Arg-NH$_2$
VI. DNP-Pro-Leu-Gly-Ile-Ala-Gly-Gln-D-Arg-OH
VII. DNP-Pro-Gln-Gly-Ile-Ala-Gly-Gln-Arg-NH$_2$

Peptide III has been found to have a high degree of specificity for vertebrate collagenases[14] and is sufficiently soluble for general use. The melting point and analysis of purity of the compounds are described by Masui et al.[14] Sensitivity for measuring collagenase using these synthetic substrates is not equal to that achieved using collagen substrates. Substrates can be obtained from the Protein Research Foundation (Minoh, Osaka, Japan).

Collagenase Assays

Assays for vertebrate collagenase active against types I, II, and III collagens must be based on the specificity of the enzyme cleavage of collagen. Collagenases cleave $\alpha$1(I) collagen chains between residues 772 and 773 (or 774 and 775, by some calculations).[3,4] The cleavage site is found in $\alpha$1(I)CB7 peptides and appears to be very near the attachment region for collagen to fibronectin. Residues on either side of the cleavage site in chick skin $\alpha$1(I) are as follows: Gly-Ala-Hyp-Gly-Thr-Pro-Gly-Pro-Gln-Gly-(cleavage site) Ile-Ala-Gly-Gln-Arg-Gly-Val-Val-Gly-Leu-Hyp.

Although there are other Gly-Ile residues in $\alpha$1(I) collagen chains, these are not cleaved readily by highly purified mammalian collagenases. $\alpha$1(II) and $\alpha$1(III) collagens are cleaved at a Gly-Ile site at a similar place in the molecule (~75% from the N terminus of the $\alpha$ chains) although the primary sequence in the adjacent portions of the molecule differs, particularly when comparing type I with type III collagen. A discussion of possible structural determinants of the unique cleavage site is given in a review by Gross et al.[15]

Type I collagen is used exclusively for most collagenase assays. Type II collagen is cleaved at a much slower rate than type I, and at least one form of vertebrate collagenase (e.g., that from PMN leukocytes) is inactive against type III collagen. Although the triple helical conformation of collagen provides protection for the molecule from hydrolysis by nonspecific proteinases, highly purified rabbit tumor collagenase cleaved a helical collagen substrate at a faster rate than denatured substrate.[16] It is possible that with prolonged incubation, sites other than the specific one

[15] J. Gross, J. H. Highberger, B. Johnson-Wint, and C. Biswis, in "Collagenase in Normal and Pathological Connective Tissues." (D. E. Woolley and J. M. Evanson, eds.), p. 11. Wiley, New York, 1980.
[16] P. A. McCroskery, S. Wood, and E. D. Harris, Jr., Science 182, 70, (1973).

TABLE II
MAXIMUM RATES OF DEGRADATION FOR
TYPES I AND II COLLAGENS[a]

| Collagen (type and species) | $V_{max}$ (molecules collagen degraded per molecule collagenase/hr) |
|---|---|
| Type I | |
| Human | 102 |
| Calf | 34 |
| Guinea Pig | 22 |
| Rat | 19 |
| Type II | |
| Human | 0.59 |
| Calf | 2.7 |
| Rat | 4.1 |

[a] From Welgus et al.[17]

described above will be cleaved, but it is also possible that minute quantities of proteinases copurify with or are bound to the collagen substrates to account for the apparent broader specificity. Type III collagen in helical form is susceptible to cleavage by some serine proteinases at a site near, but not identical to, that cleaved by collagenases.

Welgus et al.[17] have utilized collagen in solution as substrate to determine the $K_m$ and $V_{max}$ for many different types of collagen substrates using purified human skin collagenase. The $K_m$ for all collagen types of all species was nearly identical to human type I: $K_m \sim 1 \times 10^{-6} M$. This is similar to values reported for rheumatoid synovial collagenase and mouse bone collagenase using type I substrates. In contrast, measurements of maximum rates of degradation for types I and II collagens showed marked species variation (Table II).

Types IV and placental AB collagen(s) were not cleaved by the human skin collagenase.

## Assays for Vertebrate Collagenase Using Soluble Substrate

Assays using soluble substrate can measure the specific cleavage rate, and analysis of products can be adapted to identify the cleavage site.

### Viscometry

Viscometric analysis of collagenolysis provides direct monitoring of cleavage. The reaction mixtures can be analyzed by polyacrylamide gel

[17] H. G. Welgus, J. J. Jeffrey, and A. Z. Eisen, J. Biol. Chem. 256, 9511 (1981).

electrophoresis or by electron microscopy after forming segment-long-spacing (SLS) aggregates to identify the amino-terminal 75% fragment (TC[A]) and/or the carboxy-terminal 25% piece (TC[B]). A method using collagen in solution as substrate follows.

*Principle.* The high Stokes' radius of collagen molecules in solution is directly responsible for the high intrinsic viscosity of collagen. A fall in viscosity can be measured over time of incubation with collagenase.

*Equipment and Reagents*

Circulatory water bath
Thermoregulator (Haake) to keep temperature constant ($\pm 0.1°$)
Stopwatch, accurate to 0.01 sec
Cannon-Manning semimicro viscometers (1 ml capacity) (150 series, Cannon Instruments, State College, Pennsylvania) with "water time" at 25°, ~30 sec. Flow time with collagen–collagenase mixture at zero time should be 50–60 sec. Viscometers are cleaned with chromic acid, rinsed extensively with distilled water, and dried with acetone/air before each use.
Acid-soluble collagen substrate, 0.2%, free of particulate aggregates in 0.05 $M$ Tris-HCl, pH 7.6, 0.2 $M$ NaCl. Pepsin-treated collagen is most appropriate although not essential.
Tris-HCl, 0.05 $M$, pH 7.6, 0.2 $M$ NaCl, 0.01 $M$ CaCl$_2$
Test sample clarified by centrifugation

*Procedure.* Four to six readings using 1 ml of water at the set temperature in each viscometer are made to establish the "water time." At temperatures $\geq 27°$, the test sample is added at time zero to a mixture of collagen (250 $\mu$l) and Tris–NaCl–CaCl$_2$ in a total volume of 1200 $\mu$l. The sample is mixed, and 1 ml is removed immediately to the viscometer for an initial reading. A substrate control is included using buffer solution in place of the collagenase sample. During long incubations, inclusion of 0.02% NaN$_3$ will ensure bacteriostasis. The following observations are recorded at rapid intervals: temperature, time elapsed, and flow time.

*Data Calculation.* From the data collected from viscometry, the following calculations and expression of data can be made:

$$\text{Relative viscosity} = \eta_{\text{rel}} = \frac{\text{flow time (sample)}}{\text{flow time (water)}}$$

$$\text{Specific viscosity} = \eta_{\text{sp}} = \eta_{\text{rel}} - 1.0$$

Data are frequently presented graphically by plotting time on the abscissa and percentage of initial specific viscosity on the ordinate. Specific viscosity falls to 40–50% of initial values if all collagen substrate

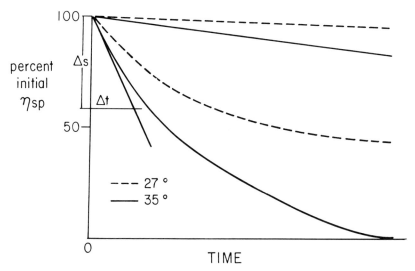

FIG. 1. Use of viscosity assays to determine initial velocity of collagenase action. This schematic diagram illustrates typical viscosity curves at 27° and 35°. At 27° the reaction products $TC^A$ and $TC^B$ formed by specific collagenolytic action have significant residual viscosity because they still are in helical conformation. At 35°, using 0.05 $M$ arginine to inhibit fibril formation [J. Gross and D. Kirk, *J. Biol. Chem.* **233**, 355 (1958)], the reaction products denature immediately and contribute a negligible viscosity to the final reaction mixture. Constitution of the ratio $\Delta S/\Delta T$ permits calculation of micrograms of collagen degraded per unit time.

in the viscometer has been cleaved to $TC^A$ and $TC^B$ and all reaction products have remained helical.

If collagenase preparations are used that have been purified free of nonspecific proteinases, a viscometer bath temperature of 35° can be used. At this temperature, in solution at pH 7–8, native mammalian collagens undergo relatively little denaturation (or loss of initial $\eta_{sp}$) over a period of several hours. Conversely, the specific reaction products ($TC^A$ and $TC^B$) produced by collagenolysis are denatured almost immediately and contribute almost nothing to the viscosity of the reaction mixture. Whereas collagen molecules have an intrinsic viscosity of ~10 dl/g, gelatin chains and/or fragments have an intrinsic viscosity of <0.4 dl/g. Thus, by assuming that the viscosity of cleaved products is zero at this temperature, one may calculate the initial velocity of specific cleavage by collagenases (Fig. 1).

In order to prevent fibril formation within the viscometer at 35°, arginine (0.05 $M$) or glucose (0.25 $M$) should be added as a component of the final 1.0-ml reaction mixture in both experimental and control samples.

The reaction within a viscometer can be stopped at any time by adding EDTA sufficient to chelate all calcium in the reaction mixture. The products can then be analyzed on SDS-PAGE.[18,19] Use of 7.5% acrylamide gels will show intact $\alpha$ chains as well as $TC^A$ and $TC^B$ products. Reaction products obtained at 27° or 24° will remain helical and can be transferred directly to dialysis membranes for dialysis against a weak solution of acetic acid preparatory to SLS aggregate precipitation and visualization using electron microscopy.[20]

*Assay for Collagenase Using Soluble, Radiolabeled Substrate*

The advantages to this assay are that (*a*) no time is necessary for fibrils to form; (*b*) background radioactivity is less than in assays based on fibril formation; (*c*) the assay is approximately twice as sensitive at 35° than is the fibril assay at 37°; and (*d*) enzyme kinetics can be determined more accurately than with fibrillar substrates.

The principal disadvantage is that, unlike assays using preformed collagen fibrils, the investigator cannot follow the assay visually, making the assay termination time for unknown samples difficult to predict.

An assay method using [$^{14}$C]collagen as substrate follows.

*Principle.* This assay[21] is based on the observation that native collagen is precipitated in 50% v/v dioxane at room temperature, whereas denatured collagen and denatured products of collagenolysis remain soluble.

*Materials*

Microcentrifuge system (e.g., Beckman Microfuge) using polypropylene disposable centrifuge tubes

[$^{14}$C]collagen (4 mg/ml) in 0.005 $M$ acetic acid

Solution A: 0.1 $M$ Tris-HCl, pH 7.8, 0.01 $M$ $CaCl_2$, 0.4 $M$ NaCl, in 1 $M$ glucose

$o$-Phenanthroline, 0.08 $M$, in 50% dioxane

*Procedure.* One hundred microliters each of collagen solution and solution A are added to each centrifuge tube; 200-$\mu$l aliquot portions of samples to be assayed are added. The tubes are capped, mixed, and incubated in a water bath at 35°. The assay is terminated by addition of 20 $\mu$l of $o$-phenanthroline in dioxane. Tubes are incubated for 60 min at 35° to denature reaction products and then are cooled to room temperature. Dioxane is added to a final volume of 50%, and the tubes are capped,

[18] D. M. Neville, Jr., *J. Biol. Chem.* **246**, 6328 (1971).
[19] U. K. Laemmli, *Nature (London)* **227**, 680 (1970).
[20] J. Gross and Y. Nagai, *Proc. Natl. Acad. Sci. U.S.A.* **54**, 1197 (1965).
[21] K. Terato, Y. Nagai, K. Kawaniski, and Y. Shinro, *Biochim. Biophys. Acta* **445**, 753 (1976).

shaken, and centrifuged. An aliquot portion of each supernatant is counted in Bray's solution or an equivalent scintillant in a liquid scintillation spectrometer. Results are expressed as a percentage of the total radioactivity in each tube and can be translated easily into micrograms of collagen degraded per unit time.

Assay for Collagenase Using Fibrillar Substrates

Fibrils formed by warming solutions of collagen at pH 7-8 serve as excellent substrates for collagenase. It has been determined that simultaneous with cleavage of molecules in fibrils, the reaction products $TC^A$ and $TC^B$ become soluble even if no further degradation to smaller fragments results.[16] Even so, the assay is dependent on both cleavage and solubilization of collagen molecules, in contrast to assay of collagen in solution, which is dependent only on cleavage of single molecules. Reconstituted fibrils that are cross-linked before assay by formaldehyde[22] or by lysyl oxidase[23] are much less sensitive substrates for collagenase than are non-cross-linked fibrils.

The fibril assay is, perhaps, the easiest to use as a screening test because it can be followed visually and the assay terminated after short or long (up to 24 hr) periods, depending on the time needed for the enzyme to degrade up to 70% of the fibrils, beyond which the reaction is no longer linear.

*Assay for Collagenase Using [$^{14}C$]Collagen Fibrils*

*Principle.* Cleavage by vertebrate collagenase of collagen molecules incubated at 37° produces soluble fragments that can be separated from intact fibrils by centrifugation and/or filtration.

*Materials*

$^{14}C$-Labeled acid-soluble collagen (3 mg/ml) in 0.2 $M$ NaCl, 0.01 $M$ Tris-HCl (pH 7.6)

Stock buffer: 0.1 $M$ Tris-HCl, pH 7.6, 0.2 $M$ NaCl, 0.03 $M$ $CaCl_2$, in 0.02% $NaN_3$

Microcentrifuge system (e.g., Beckman Microfuge) utilizing polypropylene disposable tubes

Trypsin (TPCK) 100 $\mu$g/ml in 0.1 $M$ Tris-HCl, pH 7.6

Bacterial collagenase: 1.0 mg/ml in 0.1 $M$ Tris-HCl, pH 7.6, 5 m$M$ $CaCl_2$

---

[22] E. D. Harris, Jr. and M. E. Farrell, *Biochim. Biophys. Acta* **278**, 133 (1972).
[23] C. A. Vater, E. D. Harris, Jr., and R. C. Siegel, *Biochem. J.* **151**, 639 (1979).

*Procedure.* Collagen, 50–100 µl, is pipetted into the bottom of each centrifuge tube. The tubes are capped and incubated at 35° overnight, although shorter periods (3–6 hr) may be sufficient for adequate fibril formation. Three sets of assays other than the test samples are used.

1. "Blank" tubes containing [$^{14}$C]collagen fibrils and buffer solutions. Radioactivity in these supernatants indicates the presence of collagen that was denatured or for other reasons did not form fibrils.
2. Tubes to which bacterial collagenase (10 µl) is added to degrade all the collagen to provide a number for total radioactivity in the aliquot portion used as substrate.
3. Tubes to which trypsin (10 µl) is added. Trypsin "blanks" have 5–20% radioactivity in solution after centrifugation and serve as a control for the presence of nonhelical collagen substrate.

For assay, buffer solution and test samples are added to give a final volume of 200–300 µl with a final calcium concentration of 5–10 m$M$. Fibrils can be "rimmed" away from the tube walls either with the stylet of a No. 21 gauge lumbar puncture needle or by vigorous shaking or agitation using a Vortex. The assay should be carried out at 35–37°. Progress can be monitored visually and the assay terminated when one or more test samples appears to have degraded 70–80% of the collagen fibrils.

The assay is most simply terminated by centrifugation of the tubes, although some investigators filter the reaction products through filters with pore size ~0.45 µm. Aliquot portions (all but 100 µl) of the reaction mixture are counted in an appropriate liquid scintillant. Assays should be done in duplicate or triplicate. From the radioactivity in each test sample is subtracted the radioactivity in tubes incubated with trypsin or enzyme-free buffer. Results are expressed as micrograms of collagen degraded per minute at 37°.

*The "Diffuse Fibril" Assay for Vertebrate Collagenase*

This simplified fibril assay, published by Cawston and Barrett,[13] is reported to be quicker, more sensitive, and more reproducible than the traditional fibril assay.

*Principle.* The test solutions are mixed with substrate collagen *before* fibril formation to ensure uniform distribution of enzyme.

*Materials*

Buffer: 0.1 $M$ Tris-HCl, pH 7.6, 0.015 $M$ CaCl$_2$
$^{14}$C-Labeled collagen (1 mg/ml) in 0.1 $M$ Tris-HCl, pH 7.6, 0.2 $M$ NaCl
Microcentrifuge system
Liquid scintillant: 4 g of 2,5-diphenyloxazole in 480 ml of 2-ethoxyethanol and 560 ml of toluene

*Procedure.* Test sample (100 µl), buffer solution, and [$^{14}$C]collagen (100 µl of a solution of 1 mg/ml) in 50 m$M$ Tris-HCl, pH 7.6, 0.2 $M$ NaCl are immediately mixed and incubated at 37°. At the termination of the assay, tubes are centrifuged at 10,000 g. Samples (200 µl) are counted for $^{14}$C in a liquid-scintillation counter.

*Assessment.* In this assay there is fibril formation occurring at the same time as fibrils are being degraded. Since high salt concentration (e.g., 0.4 $M$ NaCl or 30 m$M$ CaCl$_2$) interferes with fibril formation, blanks are unacceptably high in the presence of high salt. Radioactivity remaining soluble after centrifugation is 1.5- to 2-fold higher than in the standard fibril assay, and this assay is linear only to 50% of total substrate, compared with linearity to >75% with the conventional fibril assay. An assay of collagenase at 25° using $^{14}$C-labeled substrate in solution followed by addition of EDTA to stop the reaction and subsequent reincubation at 37° to allow fibrils to form from remaining intact molecules[23a] is a rational one, but is more time consuming than the use of dioxane to precipitate intact molecules (see above).

[$^{14}$C]*Collagen Film Collagenase Assay*

This screening assay for large numbers of samples has been introduced recently.[24]

*Principle.* Thermally reconstituted radiolabeled collagen fibrils are dried to a film that remains attached to a microwell while soluble degradation products are liberated into the supernatant.

*Materials*

[$^{14}$C]Acetylated rat tail tendon collagen (2 mg/ml; 200,000 cpm/ml) in 0.5 $M$ acetic acid
Phosphate buffer, 0.15 $M$, pH 7.4
Linbro flat-bottom 96-well tissue culture plates
Buffer A: 0.2 $M$ NaCl, 50 m$M$ Tris-HCl, 1 m$M$ CaCl$_2$, 0.02% NaN$_3$, pH 7.4
Trypsin (0.01%) in buffer A
Bacterial collagenase (Worthington; 4 mg/ml) in buffer A

*Procedure.* [$^{14}$C]Collagen solution is dialyzed at 4° against 0.15 $M$ phosphate buffer, pH 7.4; 25-µl aliquots are added to each well of the tissue culture plates maintained at a 45 degree angle, and incubated at 37° for 1 hr to polymerize the collagen. The gels are washed free of phosphate buffer by immersing the plates in distilled water for 1 hr at room temperature and then are allowed to dry overnight in air. Collagenase samples are added (200 µl per well) and incubated at 37° for 0.5–3 hr. The entire

[23a] G. Vaes, *Biochem. J.* **126**, 275 (1972).
[24] B. Johnson-Wint and J. Gross, *Anal. Biochem.* **104**, 175 (1980).

supernatant is collected from the non-collagen-coated site of each wekk and counted in a scintillation spectrometer. Three sets of control wells are included: (a) buffer A; (b) trypsin control (0.01%) to measure denatured substrate; (c) bacterial collagenase to determine total counts per minute per well. Buffer control counts per minute are subtracted from experimental counts per minute.

The collagen film method is a rapid, simple collagenase assay designed for screening large numbers of samples. It is reported to be of comparable sensitivity to the fibril assay. For large-scale screening assays, unlabeled collagen can be used and the degree of lysis estimated visually using indirect illumination under unstained gels or after staining residual substrate with Coomassie Brilliant Blue R-250.

### Assays for Collagenase Adapted for Special Situations

*Microassay for Collagenase Using Endogenous Substrate*

This assay can be useful for study of specific biological events involving collagenolysis in localized anatomic sites. It is based on methods suggested by Morales et al.[25]

*Principle.* Breakdown of endogenous collagen by active collagenolytic systems is determined by measuring the amount of hydroxyproline released from explants in organ culture.

*Procedure.* The tissue to be examined (e.g., Graffian follicles, synovium) is dissected in assay buffer (0.05 $M$ Tris-HCl, pH 7.6, 0.01 $M$ $CaCl_2$, 0.145 $M$ NaCl, 0.25% Triton X-100). Samples are homogenized at 4° in assay buffer and centrifuged (6000 $g$ for 20 min at 4°) and washed twice. Latent enzymes can be activated at this stage by adding TPCK trypsin (0.5 mg/ml, 3 $\mu$l per 100 $\mu$l of homogenate), incubating for 3–10 min at 37°, and then cooling in an ice bath after addition of soybean trypsin inhibitor (0.5 mg/ml, 12 $\mu$l per 100 $\mu$l of homogenate). EDTA (final concentration 0.02 $M$) is added to control samples. Assays are done in Kimax capillary tubes using 25 $\mu$l of homogenate per assay. Sealed capillary tubes are placed in 37° water baths and incubated for 18 hr.

At the conclusion of the assay, samples are centrifuged using a microcapillary centrifuge. Supernatants are removed using a Hamilton syringe. Pellet and supernatant are assayed for hydroxyproline[25a] adapted to detect 25–600 ng of the amino acid in a starting volume of 80 $\mu$l.

It is impossible to determine which collagen type is being degraded in this assay, a factor of importance considering that the different collagen

[25] T. I. Morales, J. F. Woessner, D. S. Howell, J. M. Marsh, and W. J. LeMaire, *Biochim. Biophys. Acta* **524**, 428 (1978).
[25a] J. F. Woessner, *Arch. Biochem. Biophys.* **93**, 440 (1961).

types have different inherent susceptibilities to collagenases and other enzymes.

This assay system applied to uterine tissue was linear with time to at least 20 hr, at which time 20% of the collagen substrate was solubilized.

*Plaque Assay for Collagenase Using an Agarose-Collagen Gel*

Visible lysis of collagen gelled in agarose[26] can be adapted for single cell plaque assays or for demonstration of functional antibody specific for collagenase (Fig. 2).

*Principle.* Degradation of fibrillar collagen incorporated in agarose layers by radially diffusing collagenase is accompanied by visual clearing of opaque nondegraded substrate.

*Procedure.* A low melting temperature agarose or agarose derivative (e.g., Sea Plaque Agarose, Marine Colloids) is dissolved at 2% in 0.05 $M$ Tris-HCl, pH 7.6, 0.17 $M$ NaCl, 0.02 $M$ CaCl$_2$, 0.02% NaN$_3$ and cooled to 37°. Acid-soluble collagen at 4 mg/ml in 0.05 $M$ Tris-HCl, pH 7.6, 0.2 $M$ NaCl, 0.02% NaN$_3$ is warmed to 37° immediately prior to mixing (1 part collagen to 1 part agarose). The thoroughly mixed solution is pipetted onto a glass slide and left on a level plate at room temperature to solidify for 10 min. Wells are punched in the gel, and 20-$\mu$l active enzyme is added to the wells. The plates are allowed to diffuse in a moist chamber at 37° for 24 hr or until a zone of clearing is apparent. Visibility may be enhanced by washing in 1% NaCl, staining with 0.1% Coomassie Brilliant Blue R250 in 45% methanol–10% acetic acid, and destaining in methanol–acetic acid.

This method may be adapted to correlate immunological and functional data by substituting collagen-agarose for agarose in the traditional double diffusion of antigen and antibody. Anti-collagenase IgG antibody is placed in one well and active enzyme in another. The zone of clearing halts abruptly in a straight zone precisely where collagenase antigen and specific antibody meet and precipitate as Ag/Ab complex (see Fig. 2).

*Use of Synthetic Substrates for Assay of Collagenase*

As mentioned previously, the only appropriate synthetic substrates for collagenase are ones that contain the site of specific cleavage in $\alpha$ chains of collagen and appropriate amino acid residues known to surround that site.

The use of DNP-peptides for collagenase assay[14] is described below.

[26] J. A. Yankeelov, W. B. Wacker, and M. M. Schweri, *Biochim. Biophys. Acta* **482**, 159 (1977).

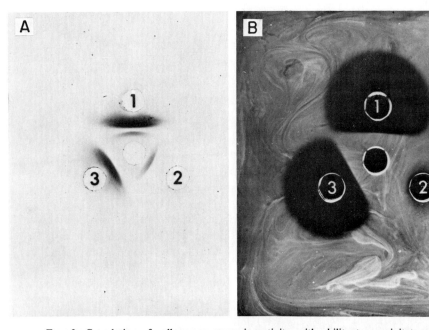

FIG. 2. Correlation of collagenase enzymic activity with ability to precipitate anticollagenase antibody. (A) A standard double diffusion assay in 1% agarose, 0.02 M Tris-HCl, pH 7.6, 0.17 M NaCl, 0.01 M CaCl$_2$, 0.02% NaN$_2$. The center well contained 20 μl of purified sheep IgG antibody to rabbit synovial fibroblast collagenase. The antigen wells contained 20 μl of concentrated (1) crude collagenase-containing culture medium from rabbit synovial fibroblasts, (2) crude culture medium from which collagenase had been removed by affinity chromatography, and (3) collagenase purified from crude culture medium. The major antigen has been intentionally overloaded to permit visualization of the contaminating antigen. (B) Result of diffusing the same antibody and antigen samples in collagen-agarose after activation of latent collagenases with 1.7 mM aminophenylmercuric acetate. In this case, an agarose (SeaPlaque) gelling at low temperature was mixed with guinea pig skin collagen (final concentration 0.2%) at 37–40° and allowed to set for 10 min at 20° before addition of samples and subsequent diffusion for 24 hr at 37°. The zone of radially diffusing collagen degradation indicates the presence of functional collagenase. It is abruptly inhibited upon formation of complex and precipitation with antibody to collagenase. Comparison of plates A and B indicates that the major antigenic protein in the crude culture medium represents collagenase. Plate A, but not B, has been stained with Coomassie Brilliant Blue R-250.

*Materials*

DNP-peptide III (DNP-Pro-Gly-Gly-Ile-Ala-Gly-Gly-D-Arg-OH). This may be obtained from the Protein Research Foundation, Minoh, Osaka, Japan.

Buffer solution: 0.05 M Tris-HCl, pH 7.6, 0.15 M NaCl, and 5 mM CaCl$_2$ in 0.02% bovine serum albumin

Solvents: ethylacetate-$n$-butanol (1 : 0.15, v/v)

*Procedure.* DNP-peptide III is dissolved in buffer at a concentration of $5 \times 10^{-4}$ M. Test enzyme solution (0.1 ml) is added to 0.1 ml of peptide solution at time zero and incubated at 37°. Reactions are stopped by adding 0.5 ml of 1 N HCl. DNP-peptide fragments released are extracted by vigorous shaking with 1.0 ml of ethylacetate-*n*-butanol followed by centrifugation at 1000–1500 g at room temperature. Absorbance in the organic layer is measured at 365 nm; the molar extinction coefficient is 1.49.

A possibly unique use for the synthetic substrate of collagenase is for assay of collagenase complexed to $\alpha_2$-macroglobulin ($\alpha_2$M). Using collagen as substrate, vertebrate collagenases are inhibited by $\alpha_2$M. Hori and Nagai[27] have demonstrated that collagenase complexed with $\alpha_2$M still has an unmasked active site to which the small synthetic DNP-peptide III has access without any significant blunting of activity at the specific Gly-Ile sequence.

## Vertebrate Collagenase: Isolation and Purification

Study and characterization of collagenases from different cells and tissues have been facilitated by enhanced insight into the factors regulating the activity of these enzymes once they are released from cells.

With the exception of polymorphonuclear leukocyte collagenase, these enzymes are not stored intracellularly to any significant degree before release from cells. Using immunoprecipitation techniques, Valle and Bauer[5] calculated that less than 9% of total collagenase produced by skin fibroblasts in culture was located intracellularly at any given time. Secretion from the cells is linear in culture. The collagenases of almost all tissues are released in a latent form. Many proteinases (e.g., plasmin, trypsin, kallikrein) can activate latent collagenase, as can organic mercurial compounds (e.g., *p*-chloromercuribenzoate, aminophenylmercuric acetate, mersalyl). Conversion from latent to active forms usually, but not necessarily, is associated with a decrease in apparent molecular weight of the enzyme protein.

Latent collagenase binds less well than active enzyme to substrate. The biologic half-life of collagenase bound to substrate has not been determined in any system. Activity of collagenase once it has been activated is a function of nonspecific circulating inhibitors of collagenase (e.g., $\alpha_2$-macroglobulin), specific tissue and circulating collagenase inhibitors, and other factors such as temperature and pH of the microenvironment, the

[27] H. Hori and Y. Nagi, *Biochim. Biophys. Acta* **566**, 211 (1979).

type of collagen serving as substrate, and the degree to which the substrate is cross-linked.

Once thought to be limited to very few tissues, collagenase has been found to be produced in virtually all mesenchymal tissues, some epithelial cells, and by macrophages and polymorphonuclear phagocytes. Induction of enzyme synthesis results from many different types of stimuli, including phagocytosis, membrane fusion, microcrystalline sodium urate, the presence of small concentrations of proteinases, soluble factors produced by mononuclear cells, and agents that appear to perturb cell membranes and alter cellular function, such as phorbol esters, colchicine, and cytochalasin B.

Isolation of collagenases, with this understanding of the biology of these enzymes, may be achieved through extraction from tissues or by accummulation of enzyme in culture medium from explants or monolayer cultures of cells.

Collection of Enzyme

*Direct Extraction of Collagenase from Tissues*

This method has been reported for the resorbing uterus and inflamed gingiva, but should be effective in any tissue containing large amounts of collagen undergoing rapid remodeling.

*Principle.* This is based on the likelihood that collagenase binds to substrate collagen fibers and that heating in the presence of calcium ions in high concentration may dissociate this complex.[28]

*Materials*

Homogenizing buffer: 0.01 $M$ $CaCl_2$, 0.25% Triton X-100
Extraction buffer: 0.02 $M$ Tris-HCl, pH 7.5, in 0.1 $M$ $CaCl_2$

*Procedure.* The tissue is minced and washed free of serum proteins in 0.1 $M$ Tris-HCl, pH 7.6. Tissue is homogenized at 4° in the homogenizing buffer and is centrifuged at 5000 $g$. The pellet is resuspended in extraction buffer and heated to 60° for 4 min. After immediate cooling to 4°, the mixture is centrifuged at 10,000 $g$. Collagenase in the supernatant solution is dialyzed and concentrated by pressure dialysis and can be activated or assayed directly without prior activation. Estimated recoveries of collagenase from two sequential extractions is ~80%.

Wirl has used similar techniques to extract collagenase from skin and tumors in skin.[29] Homogenization is carried out in 0.01 $M$ Tris-HCl (pH 7.5) containing 0.4 $M$ sucrose and 5 m$M$ $CaCl_2$ at 4°. The sediment is

[28] J. G. Weeks, J. Halme, and J. F. Woessner, Jr., *Biochim. Biophys. Acta* **445**, 205 (1976).
[29] G. Wirl, *Connect. Tissue Res.* **5**, 171 (1977).

washed and extracted for 2 hr with 5 $M$ urea in 0.05 $M$ Tris-HCl (pH 7.5), 0.2 $M$ NaCl, and 5 m$M$ CaCl$_2$.

*Extraction of Collagenase from Cells*

As mentioned above, only small amounts of immunoreactive collagenase can be found in cells in monolayer cultures relative to the amount of enzyme released into culture medium surrounding these cells. Recovery of intracellular enzyme can be achieved by exploring the properties of a nonionic detergent such as Triton X-100.

A functional assay of intracellular collagenase is described below.

*Materials*

Hanks' balanced salt solution (HBSS)
TNC buffer: 0.02 $M$ Tris-HCl, pH 7.6, 0.17 $M$ NaCl, 0.01 $M$ CaCl$_2$, 0.02% NaN$_3$
Disruption buffer: 0.1% Triton X-100 in TNC
Cultures of cells in confluent monolayers using serum-free culture medium. Cells must be releasing easily measurable amounts of collagenase into culture medium.

*Procedure.* The culture medium is removed and the cell layer is rinsed with HBSS. Fresh HBSS is added; the cells are gently scraped off the dish with a rubber policeman and transferred to a centrifuge tube. To ensure that intracellular enzyme, not merely associated extracellular enzyme, is to be detected, the cells must be washed extensively by repeated centrifugation and resuspension in HBSS. After at least five washes, the cell pellet is resuspended in a minimal volume of 0.1% Triton X-100 in TNC and triturated to disrupt cell membranes. The preparation is then dialyzed extensively against TNC before being assayed for collagenase.

Care should be taken to perform the cell disruption at 4° as quickly as possible, as Triton X-100 appears to decrease recoveries of active collagenase and may cause some activation of latent enzyme.

*Recovery of Collagenase from Explant Cultures*

This method can be adapted to any tissue. Mesenchymal tissues survive for long periods in culture, even without the presence of serum.

Large amounts of tissue (e.g., tendons pooled from many animals or taken from large animals) may be minced, thoroughly washed in buffered saline, and placed in minimum essential or Dulbecco's modified Eagle's medium in a 3–5 (volume of medium in ml) to 1 (weight of tissue in grams) ratio in sterile bottles or large plastic culture dishes and incubated at 37° in 5% $CO_2$, 95% air. Culture medium should be changed when phenol red indicators show pH change to acid conditions.

Smaller quantities of tissue can be minced into fragments of less than 3 mm³ and held in place on culture dishes by a sterile glass coverslip fixed to the dish with vacuum grease sterilized in an autoclave. This technique is identical to that used to grow cells out from explants.

*Recovery of Collagenase from Conditioned Media of Monolayer Cultures*

This technique has the merit of providing a homogeneous denominator (e.g., cell number) for quantitating the amount of collagenase released by cells as well as effecting a partial purification by producing enzyme free of contamination by serum proteins.

Cells can either be grown out from explants and passaged by trypsinization or mechanical agitation, or else dissociated from matrix proteins immediately after removal from the patient or animal. The following method is effective for dissociating rheumatoid synovial cells[30] and can be adapted to many other tissues.

*Principle.* This method[30] produces monolayer cultures more quickly than the technique of growing cells out from explants and provides cell types more typical of the *in vivo* situation at time of surgery or sacrifice of animals.

*Materials*

Phosphate-buffered saline (PBS), calcium- and magnesium-free
Dulbecco's modification of Eagle's medium (DMEM), with penicillin and streptomycin (50 µg/ml each)
Clostridiopeptidase A (Worthington Biochemical CLS, 125–200 units/g)
Trypsin 0.05% and EDTA 0.02% in saline

*Procedure.* Fresh tissue is washed in PBS and minced. The mince is added to DMEM containing 4 mg of clostridiopeptidase per milliliter and allowed to incubate at 37° in 5% $CO_2$, 95% air for 1–2 hr. After dispersion repeatedly by aspiration and expulsion through a sterile Pasteur pipette, an equal volume of 0.25% trypsin and 0.02% EDTA in a balanced salt solution is added and incubated for 30–60 min in a trypsinization flask with a magnetic stirring bar. The dissociated cells are filtered from debris through sterile gauze pads, washed three times with Hanks' balanced salt solution, and plated at $5 \times 10^5$ cells per milliliter in DMEM with 10% heat-inactivated fetal calf serum and cultured at 35–37° in 5% $CO_2$ 95% air. After 24 hr of culture, serum-containing medium can be replaced in most cultures with DMEM supplemented with 0.2% lactalbumin hydroly-

---

[30] J-M. Dayer, S. M. Krane, R. G. G. Russell, and D. R. Robinson, *Proc. Natl. Acad. Sci. U.S.A.* **73**, 945 (1976).

zate (DMEM-LH). Culture medium removed from these cultures during the next 6–9 days may be frozen at $-20°$ before assay.

*Comments.* Several comparisons of the quantity of latent collagenase released by cells under various conditions are appropriate:

1. Dissociated cells and explants prepared from identical amounts (wet weights) of dissected rheumatoid synovial tissue have been shown to produce the same amounts of collagenase if cultured over a 20-day period in DMEM-LH.[31]
2. Culture in serum-containing medium (1% fetal calf serum) results in 3- to 5-fold more collagenase production by monolayer cultures of fibroblasts. Preliminary inactivation of proteinase inhibitors in serum to be used for cell culture by acid treatment (shock to pH 3.0 for 30 min followed by adjustment of to pH 7.4) enables one to detect collagenase in the culture medium.
3. Fibroblasts in culture produce maximal quantities of collagenase per cell at early confluence.
4. Collagenase accumulation in culture medium of skin fibroblasts is 2- to 10-fold greater at pH 7.6–8.2 than at pH 6.8–7.2.[32]

### Activation of Latent Collagenase to Active Forms

Collagenase is often found in tissues or cell cultures as a latent enzyme. These latent forms were first identified in tadpole tissue[33] and mouse bone cultures.[23a]

Depending on species and the organ or tissue sources of collagenase, variable percentages of total enzyme will be in active form in the crude extraction or culture medium. In general, collagenase tends to become activated during purification procedures or during repeated freezing-thawing. Complete activation is the goal for determination of total enzyme activity, however, and can be accomplished by limited proteolysis or use of organic mercurial compounds.

*Activation by Limited Proteolysis*

Optimal activation conditions must be determined experimentally for each source of collagenase to ensure detection of maximal activity. In general, purer preparations of collagenase require milder activation procedures (e.g., less activating proteinase) for less time or at lower temperature. Although several neutral proteases have been used successfully to

[31] E. D. Harris, Jr. and C. A. Vater, *in* "Collagenase in Normal and Pathologic Connective Tissues" (D. E. Woolley and J. M. Evanson, eds.), p. 37. Wiley, New York, 1980.

[32] D. F. Busiek and E. A. Bauer, *Biochim. Biophys. Acta* **585**, 389 (1979).

[33] E. Harper, K. J. Bloch, and J. Gross, *Biochemistry* **10**, 3035 (1971).

activate latent collagenases *in vitro,* trypsin has been chosen most commonly. Crude, serum-free culture medium is activated routinely with a final concentration of 10 µg of TPCK-trypsin per milliliter for 30 min at room temperature. Activation at 37° requires a shorter incubation than at room temperature. When higher trypsin concentrations, up to 100 µg/ml, are used, the incubation is shortened accordingly. For activating serum-containing culture medium, 125–150 µg of trypsin per milliliter for 30 min at room temperature are used, the excess presumably being needed to saturate $\alpha_2$-macroglobulin. All trypsin-activation reactions are terminated by the addition of a 4- to 5-fold excess of soybean trypsin inhibitor.

Most latent collagenases appear to be sensitive to an overactivation phenomenon, as a loss of activity is apparent if activation is continued excessively. This occurs more commonly with proteinase activation than with organic mercurial activation, presumably owing to a direct degradative effect of proteinase on the activated collagenase molecule. Care must be taken to avoid the "overkill" phenomenon, especially when precise quantitative determinations are being made.

*Activation by Mercurial Compounds*

Although various organic mercurial compounds are now used routinely for activation of latent collagenases, the mechanism of activation is by no means clear. There are no data to suggest that the mercurial compounds cleave peptide bonds. Comparable levels of activation may be seen using either proteinases or mercurials. Mercurial compounds, such as phenylmercuric chloride and 4-aminophenylmercuric acetate (APMA) are dissolved in 0.2 M NaOH and the pH is adjusted to 10.0.[34] They may be preincubated with latent enzyme in concentrations from 1 to 10 mM or added directly to the fibril assay. APMA is more soluble than is phenylmercuric chloride and, therefore, easier to use.

Purification of Vertebrate Collagenases

Units for the expression of collagenase activity have evolved as one unit equaling 1 µg of collagen degraded (specify fibril form or solution) per unit time (minute/hour) at 35–37°. Assays using collagen in solution in viscometers will produce units degrading soluble collagen at 35°. Fibril assays are reported at 37°. Specific activities of enzymes purified to homogeneity have included those listed in Table III.[27,35–40]

[34] T. E. Cawston, E. Mercer, and J. A. Tyler, *Biochim. Biophys. Acta* **657**, 73 (1981).
[35] D. E. Woolley, R. W. Glanville, M. J. Crossley, and J. M. Evanson, *Eur. J. Biochem.* **54**, 611 (1975).
[36] P. A. McCroskery, J. F. Richards, and E. D. Harris, Jr., *Biochem. J.* **152**, 131 (1975).
[37] G. P. Stricklin, E. A. Bauer, J. J. Jeffrey, and A. Z. Eisen, *Biochemistry* **16**, 1607 (1977).

TABLE III
SPECIFIC ACTIVITY OF COLLAGENASE FROM VARIOUS SOURCES

| Source of collagenase | Specific activity (μg collagen fibrils degraded/min at 37°/mg protein) | Reference[a] |
|---|---|---|
| Human rheumatoid synovium | 312 | 35 |
| Rabbit tumor $V_2$ ascities cell in muscle | 3,583 | 36 |
| Human skin | 958 | 37 |
| Human skin | 32 | 38 |
| Mouse bone | 2,850 | 39 |
| Pig synovium | 26,300 | 40 |
| Tadpole back skin | 8,440 | 27 |

[a] Numbers refer to text footnotes.

Amino acid analyses available for two enzymes, human skin fibroblast collagenase[41] and tadpole skin collagenase[27] offer striking similarities (Table IV). It is interesting that the molecular weights of the active tadpole enzyme (43,000) and human skin enzyme (47,500) are similar. The amino acid compositions of collagenases reported to have significantly lower or higher molecular weights are not known. Amino acid analyses for the zymogen of human skin collagenase are shown in column 4 of Table IV. The active enzyme is less basic than the zymogen owing to a significant reduction in lysine content. It has been estimated that proteolytic activation of the skin collagenase zymogen is accompanied by loss of a peptide of approximately 85 amino acids with the composition estimated in column 5. The human skin enzyme has been shown to have no hexose or amino sugars.

Purifications of collagenase have involved ion-exchange chromatography, gel-filtration chromatography, and various types of specialized affinity chromatography.

*Concentration of Pooled Culture Medium*

Human skin collagenase has been precipitated quantitatively by ammonium sulfate added to a final concentration of 60% at 4°.[42] This is effec-

---

[38] D. E. Woolley, R. W. Glanville, and J. M. Evanson, *Biochem. Biophys. Res. Commun.* **51**, 729 (1973).
[39] S. Sakamoto, M. Sakamoto, P. Goldhaber, and M. J. Glimcher, *Arch. Biochem. Biophys.* **188**, 438, 1978.
[40] T. E. Cawston and J. A. Tyler, *Biochem. J.* **183**, 647 (1979).
[41] G. P. Stricklin, A. Z. Eisen, E. A. Bauer, and J. J. Jeffrey, *Biochemistry* **17**, 2331 (1978).
[42] D. E. Woolley, *Protein Degradation Health Dis. Ciba Found. Symp. 1980*, p. 73 (1980).

TABLE IV
COMPARATIVE AMINO ACID ANALYSES OF TADPOLE AND HUMAN SKIN COLLAGENASES

| Amino acid | Residues/1000 | | | Residues/molecule of putative peptide piece removed from human enzyme during activation |
|---|---|---|---|---|
| | Tadpole (active)[a] | Human (active)[b] | Human (latent)[b] | |
| Lysine | 49 | 48 | 63 | 12 |
| Histidine | 26 | 34 | 36 | 4 |
| Arginine | 31 | 51 | 54 | 6 |
| Asparatic acid | 143 | 122 | 117 | 8 |
| Glutamic acid | 97 | 99 | 105 | 11 |
| Threonine | 57 | 57 | 55 | 4 |
| Serine | 60 | 60 | 54 | 2 |
| Proline | 62 | 64 | 61 | 4 |
| Glycine | 81 | 85 | 78 | 3 |
| Alanine | 67 | 66 | 66 | 5 |
| ½-Cystine | 10 | 12 | 13 | 2 |
| Valine | 52 | 51 | 56 | 6 |
| Methionine | 10 | 15 | 16 | 2 |
| Isoleucine | 44 | 44 | 39 | 2 |
| Leucine | 79 | 62 | 61 | 5 |
| Tyrosine | 49 | 43 | 39 | 3 |
| Phenylalanine | 70 | 69 | 71 | 7 |
| Tryptophan | 13 | 18 | 16 | 0 |

[a] Hori and Nagai.[27]
[b] Stricklin et al.[41]

tive for both serum-free and serum-containing culture medium. After centrifugation at 9000 $g$, the precipitate is redissolved in 0.05 $M$ Tris-HCl, pH 7.5, 0.01 $M$ CaCl$_2$ and dialyzed against this buffer before storage.

Pressure filtration through Amicon Diaflo UM-10 membranes effectively concentrates vertebrate collagenases. The presence of 0.01% polyethylene glycol 1000 in the starting crude medium prevents leakage of enzyme.

Lyophilization of crude culture medium is useful for large volumes but is accompanied by greater losses of activity.

*Stability of Collagenase during Purification*

The presence of a detergent (e.g., Brij 35) may stabilize collagenase through purification steps. In purification of pig synovial collagenase, the final yield of enzyme present in the starting material was 70%.[40] All buffer solutions contained Brij 35 at a final concentration of 0.05%. Calcium-free preparations of collagenase are inactive and less stable than preparations containing calcium[36] at concentrations between 5 and 10 m$M$. In the absence of calcium ions, proteinases such as trypsin rapidly destroy latent and active collagenases. Sodium azide (0.02%) can be added to all buffer solutions to inhibit bacterial growth.

## Cation-Exchange Chromatography

Mammalian collagenases bind tightly to phosphocellulose. Recovery of collagenase from a heparin-Sepharose column may be greater than that from CM-cellulose but the nature of the operations is very similar.[37]

### Phosphocellulose

Buffer solution A: 0.05 $M$ Tris-HCl, pH 7.5

Sample preparation: Crude culture medium containing ±100 mg protein is prepared by dialysis against buffer solution A for 24 hr at 4°. The precipitate formed is removed by filtration through Whatman No. 1 filter paper.

Column preparation and sample run: Phosphocellulose (Whatman) equilibrated in buffer A at 4° is packed in a jacketed column (2.5 × 5 cm) under 8 psi at 4°. After sample application buffer A is pumped through at 60 ml/hr until optical density at 280 nm returns to baseline. Bound proteins are eluted in a linear gradient from 0.0 to 0.4 $M$ $(NH_4)_2SO_4$ in 0.05 $M$ Tris-HCl, pH 7.5. If storage of fractions after elution before assay is planned, addition of an equal volume of 0.1 $M$ Tris-HCl (pH 7.6), 0.01 $M$ $CaCl_2$ is appropriate for enzyme stabilization.

### Carboxymethyl Cellulose

Buffer solution B: 0.01 $M$ Tris-HCl, pH 7.5, 1 m$M$ $CaCl_2$

Sample preparation: Bovine serum albumin is added to culture medium (concentrated or crude) in a final concentration of 0.5 mg/ml in order to reduce tendencies of medium proteins to precipitate when dialyzed at 4° against 40 volumes of buffer B. Precipitate that does form is removed by filtration.

Column preparation and sample run: CM-cellulose (CM-52, Whatman) is equilibrated in 0.01 $M$ Tris-HCl, pH 7.5, and packed into a 2.5 × 20-cm jacketed column at 8 psi run at 4°. The sample (<110 mg of protein, <500 ml volume) is applied and after washing the column, a linear gradient of 0 to 0.3 $M$ NaCl in buffer B is applied to elute bound protein.

## Anion Exchange Columns

*Diethylaminoethyl (DEAE).* DEAE functional groups linked to cellulose or to dextran (DEAE-Sephadex), when used in moderately low

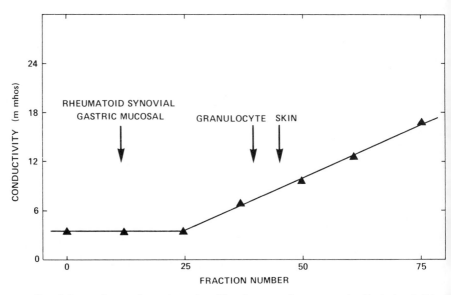

FIG. 3. Ion-exchange chromatography of four human collagenases using Sephadex QAE A-50. Partially purified collagenase preparations derived from rheumatoid synovium, gastric mucosa, skin, and granulocytes were separately eluted from columns of Sephadex QAE A-50 (20 cm × 1.6 cm) using a linear NaCl gradient in a starting buffer of 50 m$M$ Tris-HCl, 5 m$M$ CaCl$_2$, pH 8.1. Arrows indicate each peak of eluted activity determined by the [$^{14}$C]-collagen fibril assay. Reprinted, with permission of author and publisher, from D. E. Woolley.[42]

ionic strength buffers, provide excellent recovery and moderate purification for vertebrate collagenases. The collagenase does not bind to the column (similar to IgG), whereas many other proteins do bind.

Buffer solution A: 0.05 $M$ Tris-HCl, pH 7.5, 0.05 $M$ NaCl, 0.005 $M$ CaCl$_2$.

Sample preparation: Crude culture medium (total protein 500 mg → 2.0 g) is dialyzed against buffer A exhaustively at 4° after concentration.

Column preparation and sample run: Whatman DE-52 equilibrated with the same buffer A is packed in a column (2.5 × 30 cm) at 4°. Enzyme activity is not retarded by this column. In such crude samples of starting material, IgG is the principal contaminant in the enzyme peak. DEAE-Sephadex A-50 may be used instead of DE-52. Cawston and Tyler[40] have substituted cacodylate (0.25 $M$) for Tris-HCl using DEAE-cellulose and obtained similar elution characteristics for pig synovial collagenase.

*Sephadex QAE A-50.* Woolley[42] has demonstrated differential elution characteristics of collagenases from different organs of the same species from the resin. A graph of this is shown in Fig. 3.

Buffer solution: $0.02\ M$ Tris-HCl, pH 8.1, $0.01\ M$ $CaCl_2$. All procedures are carried out at 2–4°.

Sample preparation: The sample, partially purified, and containing 30–100 mg of protein, is dialyzed extensively against the buffer.

Procedure: Sephadex [(QAE A-50 (Pharmacia)] equilibrated with buffer is packed in a column (1.5 × 32 cm). After application of sample, the flow rate is adjusted to 15 ml/hr. After elution of several bed volumes of starting buffer, a gradient 0 to $0.3\ M$ NaCl in buffer is applied.

## Gel Filtration Chromatography

This technique is effective in separating latent from active enzymes that differ by more than 10,000 in molecular weight and in separating other proteinases from collagenases. It is helpful to use buffer solutions of high ionic strength (e.g., $\sim 0.5$–$1\ M$ NaCl). Jacketed columns, reverse pumping, and low flow rates all increase precision and help achieve narrow elution peaks. Dextran (Sephadex G-100, G-150, G-200), agarose (BioGel A-0.5m, A-1.5m), and acrylamide/agarose (Ultrogel AcA 54 and AcA 44) have been used effectively to purify collagenases.

## Special Purification Techniques

### Affinity Chromatography

Substrate collagen bound to activated agarose has been used in a purification step for collagenase.[43] Specificity for collagenase has been increased by using the cyanogen bromide cleavage peptide $\alpha$1-CB7 (268 residues) from $\alpha$1 chains, which contains the specific cleavage site for collagenase.[36]

### Materials

$\alpha$1-CB7 peptides prepared as described in Chapter [2]

Affi-Gel 10 (Bio-Rad). This is an $N$-hydroxysuccinimide ester of cross-linked succinylated aminoalkyl BioGel A-5m.

Procedure. $\alpha$1-CB7 linkage to Affi-Gel 10 is performed at 4° in 25 ml of $0.1\ M$ sodium phosphate, pH 7.0, using a 1 : 200 (w/w) ratio of $\alpha$1-CB7 to Affi-Gel 10. The mixture is shaken for 12 hr, and the reaction is stopped with $1\ M$ ethanolamine-HCl (pH 8.0). The material is packed into a small (0.9 × 10 cm) column and equilbrated with $0.05\ M$ Tris HCl, pH 7.6, 5

---

[43] C. Gillet, Y. Eeckhout, and G. Vaes, *FEBS Lett.* **74**, 126 (1977).

m$M$ CaCl$_2$ at 4°. The sample of partially purified collagenase (less than 20 mg) is equilbrated with the same buffer and added to the column. Collagenase is collected in a batch elution with 0.1 $M$ Tris-HCl, pH 7.6, 0.75 $M$ NaCl, 0.01 $M$ CaCl$_2$. Recovery ranges from 50 to 60%. It is useful for final purification steps.

### Heparin-Substituted Agarose

Heparin, when added to mouse bone explants, enhances production and/or release of collagenase. Studies by Sakamoto et al. suggested that only bone collagenase produced in tissue culture in the presence of heparin binds to heparin-agarose.[44] Subsequently, it has been shown that pig synovial collagenase and rabbit bone and uterine collagenases bind to heparin-agarose, and the heparin-agarose presumably acts as a cation exchanger.[40]

*Preparation of Heparin-Agarose.* Heparin, 400 mg (sodium heparin, 170 U.S.P. units/mg, Sigma Chemical Co.) is mixed and allowed to react with 50 ml of CNBr-activated Sepharose 4B (Pharmacia Fine Chemicals) according to the manufacturer's instructions. Heparin-Sepharose CL-6B may be purchased commercially from Pharmacia Fine Chemicals.

*Preparation of Sample.* Partially-purified samples of collagenase are equilbrated against 0.05 $M$ Tris HCl, pH 7.6, 0.05 $M$ NaCl, 0.05 $M$ calcium acetate and applied to a column (2.5 × 10 cm) of heparin-substituted Sepharose 4B gel. Fractions are eluted by a linear gradient of NaCl in calcium acetate from 0.05 to 0.2 $M$ each in 0.05 $M$ Tris-HCl, pH 7.6. Mouse bone collagenase is eluted under these circumstances at about 0.15 $M$ buffer. Cawston and Tyler[40] have used the following buffer with heparin-Sepharose: 0.025 $M$ sodium cacodylate, pH 7.6, 0.01 $M$ CaCl$_2$, in 0.05% Brij 35. The elution buffer was a linear gradient from 0 to 0.4 $M$ NaCl in the cacodylate buffer. Purification of 7-fold with 100% recovery was found.

### Blue Dextran Sepharose

Stricklin et al.[41] reported separation of two forms of active collagenase (one of 50,000 daltons, the other of 45,000 daltons, and both with very similar amino acid composition) by chromatography on Blue Dextran Sepharose. Blue Dextran, 0.1 g in 50 ml of 0.1 $M$ NaHCO$_3$, pH 9.5, was added to Sepharose activated with CNBr and stirred gently for 18 hr at 4°.

---

[44] S. Sakamoto, M. Sakamoto, P. Goldhaber, and M. J. Glimcher, *Biochim. Biophys. Acta* **385**, 41 (1975).

The slurry was washed with 2 liters of $0.1 M$ $NaHCO_3$, pH 9.5, 1 $M$ NaCl and equilibrated with 0.05 $M$ Tris-HCl, pH 7.5, 0.01 $M$ $CaCl_2$. Three milligrams of enzyme were applied in this buffer to a $1.6 \times 30$ cm column. A 600-ml gradient of $0.05 M$ to $0.75 M$ NaCl in the buffer produced separation of the two forms of enzyme.

*Zinc-Chelate Affinity Chromatography*

Porath et al.[45] designed this matrix and proposed that exposed histidine and/or thiol groups of proteins bind to the chelated zinc attached to the column. Cawston and Tyler[40] reported 2-fold purification of already highly purified pig synovial collagenase with 90% recovery using the zinc-chelate column; nickel and copper-chelate columns were not effective.[39]

The sample and zinc-chelate column ($0.6 \times 12$ cm) are equilibrated in 0.025 $M$ sodium borate, pH 8.0, 0.15 $M$ NaCl. The sample is applied. Sequential elution is by the following buffers: 0.025 $M$ sodium cacodylate, pH 6.5; 0.025 $M$ sodium cacodylate, pH 6.5, 0.8 $M$ NaCl; 0.05 $M$ sodium acetate, pH 4.7, 0.8 $M$ NaCl. All buffers contained Brij 35 (0.05%) and 0.001 $M$ $CaCl_2$. In one application of the method, pig synovial collagenase was eluted with the third buffer.[40]

Degradation of Type IV Collagen

Type IV and type V (AB) collagens have been isolated from tissues rich in basement membranes. Type IV collagen has properties that distinguish it from types I, II, and III collagen; it has a high content of glycosylated hydroxylysine and disulfide bonds. Vertebrate collagenases described above do not degrade type IV collagen.

Human polymorphonuclear leukocyte elastase degrades type IV collagen prepared from bovine anterior lens capsule,[2] and an enzyme harvested from culture medium from murine tumor explants has been shown to degrade type IV collagen purified from basement membranes.[46]

*Preparation of Type IV Collagen from Bovine Lens Capsule*

Anterior lens capsules were separated from bovine lenses and washed and lyophilized. Freeze-dried material was pulverized in liquid $N_2$ and resuspended at 2 mg/ml in 0.1 $M$ acetic acid and extracted for 24 hr at 4°. A

---

[45] J. Porath, J. Carlsson, I. Olsson, and G. Belfrage, *Nature (London)* **258**, 598 (1975).
[46] L. A. Liotta, S. Abe, P. G. Robey, and G. R. Martin, *Proc. Natl. Acad. Sci. U.S.A.* **76**, 2268 (1979).

second extraction in 0.05 $M$ citrate, pH 3.7, 0.5 $M$ NaCl, and a third in 0.1 $M$ acetic acid was followed by suspension in Tris-HCl, pH 7.6.

Soluble collagen was obtained by incubating lens capsule in 0.5 $M$ acetic acid, pH 2.5, in pepsin (100 mg/ml). Solubilized collagens were precipitated in 10% (w/v) NaCl, and after several reprecipitations were solubilized in 0.5 $M$ Tris-HCl, pH 7.6, 0.15 $M$ NaCl.

*Preparation of Enzyme*

Granules from polymorphonuclear leukocytes were prepared by lysing these leukocytes in 0.2 $M$ sucrose plus 100 µg/ml heparin for 16 hr at 4° followed by forcing the suspension through a 25-gauge needle. After differential centrifugation in 0.34 $M$ sucrose, granules were suspended in 0.01 $M$ phosphate, pH 7.2, 1.0 $M$ NaCl, and sonicated. Soluble enzyme activity was purified by gel filtration on Ultrogel AcA 44 and affinity chromatography using 4-phenylbutylamine-Affi-Gel.

*Assay for Type IV Collgenase*

Insoluble substrate (200 µg of collagen) was incubated with 100 µl of enzyme in 0.03 $M$ Tris-HCl, pH 7.6, 5 m$M$ CaCl$_2$ at 37°. After 2 hr, the reaction mixtures were centrifuged in a Beckman Microfuge, and the supernatants were decanted, hydrolyzed, and assayed for hydroxyproline.

Soluble collagen was used in viscometric assays.

Elastin agarose plates were prepared as follows: 1% solution of Seakem agarose was boiled in 0.01 $M$ phosphate, pH 7.2, and allowed to cool slowly; when the temperature of the solution reached 37°, pulverized lens capsule was added at 1 mg/ml, stirred rapidly, and pipetted onto petri dishes. For enzyme assays, holes were punched in the agarose, and 20 µl of enzyme solution was added to each well; the plates were incubated in a moist atmosphere at 37° for 24 hr.

Elastase purified free of cathepsin G activity had the capacity to degrade type IV collagen. Five reaction products of low molecular weight were produced. This enzyme has proteolytic activity against proteoglycan, fibrinogen, and the nonhelical portion of collagen. Type V collagen is unaffected by this multifunctional enzyme, named for only one of its substrates.

## [21] Bacterial Collagenase

*By* BEVERLY PETERKOFSKY

The subject of proteolysis by *Clostridium histolyticum* collagenase (EC 3.4.24.3) has been discussed in two previous volumes of this series.[1,2] These chapters covered historical aspects, purification of the enzyme from crude culture media, and specificity and physical properties of the purified enzyme. There also have been several more recent reviews that discuss these same topics.[3-5] Therefore, this chapter only briefly discusses these aspects of the enzyme and instead focuses on the use of the enzyme for the analysis and identification of collagen species. Purification of the enzyme and sequence specificity is discussed mainly from the standpoint of obtaining an enzyme preparation free of other proteolytic enzyme activity and capable of quantitatively and specifically digesting collagen in a mixture of proteins.

### Properties of Bacterial Collagenases

The generally accepted definition of a collagenase has been an enzyme that degrades collagen only in the undenatured, triple helical form. It has become clear, however, that some bacterial enzymes that do not adhere to this definition because they degrade denatured collagen or gelatin, nevertheless have amino acid sequence requirements for cleavage that are similar or identical to "true collagenases." The minimum sequence requirement is R-Pro-X-Gly-Pro, where Pro can be substituted by Hyp and the first imino group must be blocked.[6,7] Cleavage occurs between X and Gly. As pointed out by Keil,[5] the original definition is rather restrictive and, according to the purposes for which the enzyme is used at present, it seems appropriate to call any enzyme a collagenase that specifically cleaves the typical collagenous sequence of repeating glycine tripeptides, whether the molecule is in the native or denatured form, and this broad

---

[1] S. Seifter and P. Gallop, this series, Vol. 5, p. 659.
[2] S. Seifter and E. Harper, this series, Vol. 19, p. 613.
[3] I. Mandl, *in* "The Methodology of Connective Tissue Research" (D. A. Hall, ed.), p. 167. Joynson-Bruvvers, Oxford, 1976.
[4] S. Seifter and E. Harper, *in* "The Enzymes" (P. D. Boyer, ed.), 3rd ed., Vol. 3, p. 649. Academic Press, New York, 1971.
[5] B. Keil, *Mol. Cell. Biochem.* **23**, 87 (1979).
[6] Y. Nagai and H. Noda, *Biochim. Biophys. Acta* **34**, 298 (1959).
[7] Y. Nagai, S. Sakakibara, H. Noda, and S. Akabori, *Biochim. Biophys. Acta* **37**, 567 (1960).

definition is used here. In fact, such an enzyme activity may be more desirable, since collagen in biological samples often is denatured in the process of analysis.

The most commonly used microbial collagenase is that of *C. histolyticum*, but enzymes from several other organisms have been purified. *Vibrio B-30*, a marine bacterium, produces a collagenase with properties similar to those of the clostridial enzyme but that differs from it immunologically.[8] The enzyme has a molecular weight of 105,000, is composed of subunits with molecular weights of 24,000 and 28,000, and it has essentially the same sequence specificity for cleavage as the clostridial enzyme. In spite of these similarities, the degradation of native *Nereis* cuticle collagen by the clostridial enzyme was barely detectable while the *Vibrio* enzyme extensively degraded this substrate.[9] The *Vibrio* enzyme also degraded denatured cuticle collagen twice as extensively as did the clostridial enzyme. Native calf skin collagen also was degraded to a greater degree by the vibrio compared to the clostridial enzyme, but the reverse was true with denatured calf skin collagen. The *Vibrio* enzyme may prove to be useful in probing the structure of native collagenous molecules, which are relatively resistant to clostridial collagenase.

*Achromobacter iophagus*, a nonpathogenic, aerobic bacterium, produces a collagenase that has been highly purified.[5] Its properties also are similar to those of the clostridial enzyme, but its specific activity against either native collagen or a synthetic peptide is higher than that of any purified preparation of the clostridial enzyme yet reported.

Heterogeneity of Clostridial Collagenase with Regard to Substrate

Although the term collagenase implies that there is a single enzyme produced by *C. histolyticum*, this may not be the case. Purification of collagenase activity by various techniques has resulted in identification of two to six different collagenase fractions that can be differentiated on the basis of substrate preference. Activity has been measured against native or denatured collagen (gelatin), Azocoll, a denatured collagen with bound dye, and synthetic penta- or hexapeptides containing the specific collagenase cleavage site. Grant and Alburn first obtained three fractions after DEAE-cellulose chromatography of crude clostridial collagenase and designated them as A, B, and C, in order of elution from the column.[10] Mandl and co-workers also found three peaks of activity after DEAE-Sephadex chromatography (I-III),[11] but Yoshida and Noda used DEAE-

[8] J. R. Merkel and J. H. Dreisbach, *Biochemistry* **17**, 2857 (1978).
[9] J. H. Waite, M. L. Tanzer, and J. R. Merkel, *J. Biol. Chem.* **255**, 3596 (1980).
[10] M. H. Grant and H. E. Alburn, *Arch. Biochem. Biophys.* **82**, 245 (1959).
[11] I. Mandl, S. Keller, and J. Manahan, *Biochemistry* **3**, 1737 (1964).

cellulose and obtained two well separated peaks (II and I).[12] Seifter and Harper obtained similar results using DEAE-cellulose and called the two fractions A and B,[2] and Kono obtained three distinct collagenase activities after ion-exchange chromatography.[13] At least six peaks of activity were obtained by gel filtration of a partially purified commercial preparation when column fractions were assayed against Azocoll and denatured collagen.[14] Fractions called II by Yoshida and Noda, I by Mandl and coworkers, and A-α by Kono, preferentially exhibited activity against native collagen and synthetic penta- or hexapeptides containing the specific cleavage site, but had relatively little activity against Azocoll and denatured collagen. In contrast, fractions called I, II, and B-α by these same groups, respectively, had much greater activity against Azocoll and denatured collagen. Harper's fraction A from DEAE-cellulose resembled the first group of enzymes in that it had a greater specific activity against native collagen and a synthetic peptide than did fraction B.[4] Miyoshi and Rosenbloom,[15] as well as Lwebuga-Mukasa et al.,[16] used electrofocusing to obtain additional collagenase activities from the DEAE-cellulose fractions A and B. These exhibited a spectrum of relative activities against native collagen, Azocoll, and synthetic peptide. The molecular weights of three of the enzymes obtained by the latter group ranged from 86,000 to 92,000 and one was 65,000. All cross-reacted immunologically. The existence of these multiple forms with different substrate preferences has raised the question of whether they represent individual gene products or result from proteolytic cleavage by other proteases in the culture filtrate. The immunological data suggest that they arise from a single progenitor molecule or at least share a subunit.

The presence of several of these enzymes appears to be required for optimal digestion of collagen to acid-soluble peptides,[13,14] and this may be due to slight differences in their sequence specificity. Some of the enzymes may be less stringent than others with regard to the Pro-X-Gly-Pro sequence, thus producing smaller fragments of the collagen polypeptide than more stringent species of enzyme. Harper and Kang[17] found that only one out of seven sites in the chick skin collagen peptide α1-CB2 which were cleaved by collagenase B was also cleaved by collagenase A. Some of the sequences cleaved by collagenase B contained an alanine or leucine in place of one or both of the proline residues in the generalized sequence.

[12] E. Yoshida and H. Noda, *Biochim. Biophys. Acta* **105**, 562 (1965).
[13] T. Kono, *Biochemistry* **7**, 1106 (1968).
[14] B. Peterkofsky and R. Diegelmann, *Biochemistry* **10**, 988 (1971).
[15] M. Miyoshi and J. Rosenbloom, *Connect. Tissue Res.* **2**, 77 (1974).
[16] J. S. Lwebuga-Mukasa, E. Harper, and P. Taylor, *Biochemistry* **15**, 4736 (1976).
[17] E. Harper and A. H. Kang, *Biochem. Biophys. Res. Commun.* **41**, 482 (1970).

Cleavage of Ser-Leu bonds in β-casein,[18] and a similar bond in an immunoglobin light chain,[19] have been reported but the former experiment was carried out with Sigma type III collagenase and the latter with Worthington CLSPA collagenase, both of which have been reported to contain considerable contamination with nonspecific proteolytic activities.[14,20]

For the purposes of analytical estimation or identification of collagen in a mixture of other proteins, it is essential that the collagenase preparation does not contain nonspecific proteolytic activity that can solubilize noncollagenous proteins. There are several proteases present in crude collagenase preparations, some of which persist even in purified preparations. Clostripain, the most abundant enzyme in crude preparations,[5] contains an essential SH group, is activated by cysteine, and can be inhibited by sulfhydryl binding reagents.[21,22] It has a trypsin-like specificity but is not inhibited by soybean trypsin inhibitor.[21] Clostripain activity can be assayed using synthetic peptides, such as $N$-benzoyl-L-arginine β-naphthylamide (BANA). Another protease present can cleave casein, and the activity is not significantly affected by the sulfhydryl reagent $N$-ethylmaleimide (NEM).[15]

Assays for Collagenase Activity

Because of the heterogeneity of collagenase activity with regard to substrate preference, as discussed above, using a single substrate one may incorrectly estimate the presence of a particular species in a preparation. Assays using Azocoll, insoluble tendon collagen, radioactively labeled acid-soluble collagen, and synthetic peptides are described. In all assays, phosphate buffer should be avoided, since it chelates calcium ion, which is essential for activity and stability.

*Azocoll as Substrate*

*Principle.* Azocoll (Calbiochem-Behring Corp.) is insoluble, powdered cowhide to which a red dye is attached; it consists mainly of denatured collagen. Collagenase activity against this substrate closely parallels activity against denatured collagen and collagen that has been labeled with a radioactive amino acid in biological systems.[13,14,15] It is therefore very useful for assaying collagenase preparations intended for analyzing collagen synthesis in such systems. When peptide bonds are cleaved, the

---

[18] A. M. Gilles and B. Keil, *FEBS Lett.* **65**, 369 (1976).
[19] M. C. Coletti-Previero and J. C. Cavadore, *Immunochemistry* **12**, 93 (1975).
[20] V. Lee-Owen and J. C. Anderson, *Prep. Biochem.* **5**, 229 (1975).
[21] A. Nordwig and L. Strauch, *Hoppe-Seyler's Z. Physiol. Chem.* **330**, 10 (1963).
[22] W. M. Mitchell and W. F. Harrington, *J. Biol. Chem.* **243**, 4683 (1968).

bound dye becomes soluble while the undigested protein remains insoluble and can be separated by centrifugation. The absorbance of the solubilized dye at 530 nm is used as a measure of collagenolytic activity. Since the collagen is denatured, other proteolytic enzymes also can degrade this substrate. The assay is carried out in the presence of NEM to suppress clostripain activity, but proteases not susceptible to inhibition by NEM, such as the caseinolytic enzyme, may contribute to solubilization if present in the preparation.

*Reagents*

Ethylenediamine tetraacetate, disodium salt (EDTA), 10 m$M$
Tris-HCl, pH 7.6, 50 m$M$/CaCl$_2$, 5 m$M$ (elution buffer)
Purified collagenase in elution buffer, 600–800 µg/ml
Azocoll
$N$-Ethylmaleimide (NEM), 62.5 m$M$

*Assay Procedure.* Reactions are carried out in 10 × 75 mm test tubes containing 3 mg of Azocoll, 0.25 ml of elution buffer with or without 1–3 µg of purified collagenase, and 20 µl of NEM in a final volume of 0.5 ml. The tubes are incubated at 37° with shaking for 15–30 min or any time interval that is in the linear portion of a plot of digestion versus time. The reaction is stopped by adding 0.5 ml of 10 m$M$ EDTA, the tubes are mixed thoroughly and centrifuged at 1000 $g$ for 5 min. The absorption of the supernatant solution is measured at 530 nm. Results are calculated as milligrams of substrate digested, using an absorbance of 0.75 for 3 mg of completely solubilized Azocoll. The extinction may vary for different batches.

*Insoluble Tendon Collagen as Substrate*

*Principle.* The appearance of peptides soluble in acetic acid as a result of collagenolysis of the insoluble substrate is measured by the Hartree protein assay.[23] The procedure is similar to that of Mandl et al.[24]

*Reagents.* The reagents used are essentially the same as for the assay with Azocoll except for those required for the protein assay.

*Assay Procedure.* Reactions are carried out in 10 × 75 mm tubes containing 2 mg of insoluble tendon collagen (Sigma) and 0.10 ml of elution buffer without or with collagenase in a final volume of 0.20 ml. Control samples without collagen are included to correct for the reaction given by collagenase in the protein assay. The tubes are incubated for 20 min at 37° with shaking, and the reaction is stopped by the addition of 0.2 ml of cold 0.1 $M$ acetic acid. After the contents of the tubes are mixed, they are

[23] E. F. Hartree, *Anal. Biochem.* **48**, 422 (1972).
[24] I. Mandl, J. D. MacLennan, and E. L. Howes, *J. Clin. Invest.* **32**, 1323 (1953).

centrifuged at 13,000 g for 5 min, the supernatant solution is removed, and a portion is assayed by the Hartree procedure using a gelatin solution as a standard to calculate the amount of degraded collagen present. Activity is expressed as micrograms of collagen degraded per minute. Collagenolysis does not follow the usual kinetic pattern for enzyme reactions, in that linearity with respect to enzyme and substrate occurs only at very low enzyme concentrations. This is probably due to the nature of the substrate, since it has been reported that with polymeric or fibrous collagen as substrate there appeared to be an excess of enzyme when it could be calculated that there was, in fact, a vast excess of substrate in the reaction.[25]

*Radioactive Acid-Soluble Collagen as Substrate*

*Principle.* Undigested noncollagenous protein is precipitated by trichloroacetic/tannic acid, and the radioactivity of the solubilized collagen peptides is measured. This procedure can be used with limiting collagenase incubated for short time intervals to measure enzyme activity or with excess enzyme to measure collagen quantitatively.

*Reagents.* These are the same as for the assays above except for the following:

HEPES buffer, 1 $M$, pH 7.2

Trichloroacetic acid (TCA), 10%, and tannic acid, 0.5%

Radioactive acid soluble collagen. The preparation of acid-soluble collagen labeled biologically is described here. Methods also have been described for labeling commercially available acid-soluble collagen chemically with [$^{14}$C]acetic anhydride or tritium from $^3H_2S$.[26,27] Methyl-$^{14}$C-labeled and tritiated soluble collagens also are available from New England Nuclear Corporation.

A small hole is made in the egg shell above the air space of eggs containing 10-day-old chick embryos, and 50 μl of a 10 m$M$ β-aminopropionitrile solution in $H_2O$ are placed on the shell membrane of each. The hole is covered with transparent tape, and the eggs are incubated, with the air space at the top, at 37.5° for 30 min. The tape is removed, and 100 μl of radioactive proline (10 μCi of $^{14}$C or 30 μCi of tritium-labeled proline) are placed on the membrane of each. Radioisotope solutions in 0.01 $N$ HCl should be neutralized with a small volume of 0.2 $N$ NaOH before addition. The hole is retaped, and incubation is continued for 5 hr. The embryos are removed from the eggs, the brain tissue is

---

[25] F. S. Steven, *Biochim. Biophys. Acta* **452**, 151 (1976).
[26] M. T. Gisslow and B. C. McBride, *Anal. Biochem.* **68**, 70 (1975).
[27] K. R. Labrosse, I. E. Liener, and P. A. Hargrave, *Anal. Biochem.* **70**, 218 (1976).

dissected out, and the remainder is placed in 0.15 M NaCl (0°). All further procedures should be carried out at 0-4°. The embryos are rinsed with saline, blotted dry, weighed, and minced. The minced tissue is homogenized with 0.5 N acetic acid (5 ml/g wet weight) in a blender for 30 sec, and the collagen is extracted for 16 hr. The homogenate is centrifuged at 20,000 g for 20 min, the supernatant solution is removed, and the pellet is resuspended in one-tenth the volume of acetic acid used originally. The suspension is centrifuged as before and the resultant supernatant solution pooled with the first. The solution is dialyzed to equilibrium against 10 volumes of 0.4 M NaCl—0.1 M Tris-HCl, pH 7.6; the buffer is replaced with a fresh solution and again dialyzed to equilibrium (6-8 hr). The collagen is precipitated with solid ammonium sulfate (176 mg per milliliter of solution) as described previously,[28] and the precipitate is dissolved in a volume of 0.2 M NaCl-0.05 M Tris-HCl, pH 7.6, equal to one-half of the original tissue weight. A portion of the solution (up to 10 ml) is placed on an agarose A-5m (Bio-Rad) column that has been equilibrated with 0.2 N NaCl-0.05 M Tris-HCl, pH 7.6, and maintained at 4°. Protein is eluted with the same buffer, and 1.5-ml fractions are collected. Protein content of the fractions is measured at 280 nm, and 20-μl protions of each fraction are used to measure radioactivity. Those fractions containing the major portion of the radioactivity and/or protein, are assayed for collagen content using purified collagenase as described below. Collagen elutes close to the void volume of the column (50-60 ml), and the small amount of noncollagen protein still present appears spread out in the low molecular weight region (Fig. 1). The collagen-containing fractions are pooled and stored in small portions in liquid nitrogen.

*Assay Procedure.* Components of the reaction are added to 2-ml conical tubes to give the following concentrations: HEPES buffer, pH 7.2, 0.1 M; NEM, 2.5 mM; CaCl$_2$, 0.5 mM. Elution buffer or purified collagenase in elution buffer equal to 1/50th of the final volume of the reaction mixture and containing sufficient enzyme to give a final concentration of 1-3 μg/ml is added, and the tubes are incubated at 37° for 10-15 min. The reaction is terminated by the addition of 20 μl of a 25 mg/ml solution of crystalline bovine serum albumin and one volume of 10% TCA-0.5% tannic acid. The tubes are centrifuged at 1000 g for 5 min, and the supernatant solutions are transferred to counting vials. The precipitates are resuspended with 0.5 ml of 5% TCA/0.25% tannic acid, and the tubes are centrifuged as before. The resultant supernatant solutions are combined with the first ones, scintillation fluid is added and the samples are counted to measure radioactivity in the collagen-derived peptides. The radioactivity in the

---

[28] S. A. Jiminez, P. Dehm, B. R. Olsen, and D. J. Prockop, *J. Biol. Chem.* **248**, 720 (1973).

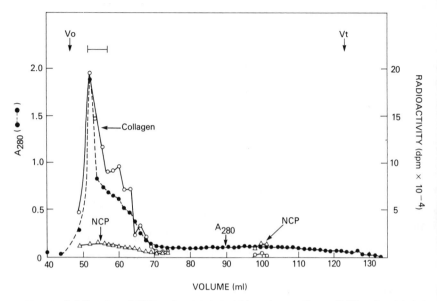

FIG. 1. Gel filtration of [$^{14}$C]proline-labeled chick embryo collagen. [$^{14}$C]Proline-labeled collagen was extracted with acetic acid and precipitated with ammonium sulfate as described in the text. The precipitate was dissolved in 0.2 $M$ NaCl–0.05 $M$ Tris-HCl, pH 7.6, and 4 ml were applied to a 1.5 × 66 cm BioGel A-5m column that had been equilibrated with the same buffered saline solution at 4°. The column was eluted with this solution at a flow rate of 0.9 ml/min. Fractions of 1.5 ml were collected, and their absorbance at 280 nm was used to measure protein content. Thirty microliters of each fraction in the major protein-containing region (49–73) and in several of the fractions in the lower molecular weight region were incubated without or with purified *Clostridium histolyticum* collagenase as described in the text to measure collagen content. The results are expressed as total disintegrations per minute (dpm) per fraction of either collagenase-digestible (○——○) or nondigestible (△——△) radioactivity, or absorbance at 280 nm (●---●). The fractions indicated were pooled. The labeled collagen was judged to be in native form based on its resistance to pepsin digestion. NCP, noncollagen protein.

sample that had been incubated without enzyme is subtracted from the radioactivity in the peptide sample to determine collagen radioactivity. Enzyme activity is expressed as disintegrations per minute of collagen degraded per unit of time.

This procedure also may be used to quantitate the collagen content of samples that have been purified to the extent that most of the radioactivity in free amino acids has been removed in prior steps. For quantitation, however, collagenase must be present in excess (8–16 $\mu$g of purified collagenase per milliliter) and incubation should be carried out for 90 min. The amount of enzyme required to be in excess depends on the amount of collagen in the reaction. With a constant amount of [$^{14}$C]proline-labeled chick embryo collagen (4000 dpm, 10 μg of protein) in a 0.5-ml reaction, there was no competition for digestion of the radioactive substrate by 8 μg of Sephacryl S-200-purified collagenase when up to 3.0 mg of unlabeled, acid-soluble calfskin collagen was added.[29]

*Synthetic Peptides as Substrates*

Cleavage of the peptides Cbz-Gly-Pro-Leu-Gly-Pro,[7] Cbz-Gly-Pro-Gly-Gly-Pro-Ala,[30] and 4-phenylazobenzyloxycarbonyl-L-Pro-L-Leu-Gly-L-Pro-D-Arg[31] between the Leu-Gly or Gly-Gly residues has been used as a measure of collagenase activity. In the case of the first two peptides, cleavage is proportional to the appearance of ninhydrin reactivity,[12] and in the case of the latter peptide, cleavage is followed by measuring the appearance of a dipeptide containing a chromogen that can be extracted into ethyl acetate.[31] This peptide substrate and the chromogenic cleavage product are available from Fluka Chemical Company.

Purification of *C. histolyticum* Collagenase

Several methods have been described for purification starting with an ammonium sulfate-precipitated fraction from crude culture filtrate[2,11] or a similar commercial preparation.[12,13,16,20,32,33] Gel filtration of a commercially available, partially purified preparation (Worthington collagenase CLSPA) on Sephadex G-200 has been used to remove nonspecific protease contaminants from collagenase activity.[14] Sephacryl S-200 has been used by several laboratories,[33,34] including our own,[35] for this purpose, since this gel has a faster flow rate than G-200. Figure 2 shows gel filtration of Worthington CLSPA enzyme on S-200 using Azocoll as substrate to detect collagenase activity. This enzyme preparation contained considerable impurities, including nonspecific protease activity, which were present as a shoulder on the main protein peak and a broad peak of protein eluting in the lower molecular weight region. Resolution into several collagenase peaks was not obtained with this preparation, and this may be due either to a change in the initial purification steps or to a difference in resolving power of S-200 compared to G-200. The three most active frac-

[29] B. Peterkofsky and W. Prather, unpublished results.
[30] W. Grassman and A. Nordwig, *Hoppe-Seyler's Z. Physiol. Chem.* **322**, 267 (1960).
[31] E. Wünsch and H. G. Heidrich, *Hoppe-Seyler's Z. Physiol. Chem.* **333**, 149 (1963).
[32] I. Emöd and B. Keil, *FEBS Lett.* **77**, 51 (1977).
[33] F. Oppenheim and C. Franzblau, *Prep. Biochem.* **8**, 387 (1978).
[34] M. L. Mailman, *J. Dent. Res.* **58**, 1424 (1979).
[35] B. Peterkofsky and R. Oneson, unpublished results.

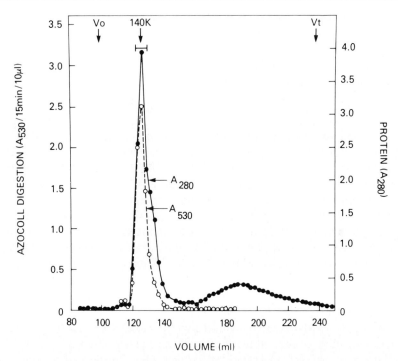

FIG. 2. Gel filtration of commercially purified *Clostridium histolyticum* collagenase. A 1.8 × 131 cm column of Sephacryl S-200 gel (Pharmacia) was equilibrated with elution buffer (Tris-HCl, pH 7.6, 50 m$M$–CaCl$_2$, 5 m$M$) at 24°, and 56 mg of Worthington CLSPA collagenase was applied to the column. Elution buffer was passed through the column at a flow rate of 14 ml/hr, and 2.7-ml fractions were collected. Absorption of fractions at 280 nm was measured as an index of protein content, and 10-$\mu$l portions of each fraction were assayed for collagenase activity for 15 min with Azocoll as substrate in the presence of 2.5 m$M$ $N$-ethylmaleimide (NEM), as described in the text, except for the three fractions with the highest protein content. These were diluted for assay, and the results were corrected to 10 $\mu$l of undiluted solution; ●——●, absorption at 280 nm; ○---○, collagenase activity. Fractions indicated by the bar were pooled to give a solution with a protein concentration of 3.33 mg/ml measured by the Bio-Rad protein assay with an immunoglobin standard. The solution was lyophilized as described in the text.

tions were pooled, and the purity was tested by the method described below. One-milliliter portions of the pooled enzyme were distributed into tubes, lyophilized, and stored desiccated at 4°. When needed, the material was reconstituted with 1 ml of cold H$_2$O, diluted 1:4 with elution buffer, and stored in small portions at −20°. Lyophilized enzyme is stable for at least 5 years, and solutions are stable for several months if they are not thawed and frozen more than 3 or 4 times.

*Assay for Contamination of Collagenase with Nonspecific Proteases*

*Principle.* [$^{14}$C]Tryptophan-labeled proteins are prepared from cultured cells to serve as substrate for proteases not specific for collagen. A duplicate sample of cells is labeled with radioactive proline in order to determine the percentage of collagen in the cell protein being assayed. Radioactive proteins are precipitated with TCA, unincorporated radioactive amino acid is removed by rinsing the precipitate with TCA, and the proteins are redissolved in NaOH. Duplicate portions of the solution are incubated without or with the enzyme preparation to be tested under conditions identical to those used to measure collagen content. Undigested proteins are reprecipitated by TCA–tannic acid, and proteolysis is measured by the appearance of acid-soluble, radioactive peptides. In the case of radioactive tryptophan-labeled proteins, the appearance of radioactive peptides is indicative of contamination by nonspecific proteolytic activity. This assay is based on the fact that the portion of collagen that contains the typical glycine tripeptide repeat, and is the substrate for collagenase, does not contain tryptophan, so that radioactive peptides can be derived only from noncollagenous proteins.

*Reagents.* Reagents are the same as those listed for assay of collagenase activity against radioactive acid-soluble collagen with the exception of 0.2 $N$ NaOH and 0.15 $N$ HCl solutions. These should be prepared from Fisher certified 1 $N$ stock solutions or titrated to precise concentrations and stored at $-20°$.

*Preparation of Radioactive Cell Protein.* The preparation of protein from cell cultures has been found to be more convenient than using chick embryos as previously described,[14] but any type of cultured cells or tissue may be used. Growth medium is removed from eight dishes (100 × 20 mm) of cells in late logarithmic phase of growth. The cell layers are rinsed twice with serum-free medium (MEM-O) and 3 ml of this medium containing 5 $\mu M$ colchicine are added to each dish. Incubation is carried out in the absence of ascorbate but in the presence of colchicine, so as to prevent the secretion of collagen and other proteins[36] and thus assure a complete mixture of cell proteins in the isolated material. To four dishes are added 0.15 ml of 10 $\mu$Ci/ml L-[$^{14}$C]proline (2 m$M$), and to the other four are added 0.15 ml of 90 $\mu$Ci/ml L-[$^{14}$C]tryptophan. The cultures are incubated at 37° under 95% air–5% $CO_2$ for 3 hr. Shortly before this time, growth medium is removed from eight additional dishes of cells, and the cell layers are rinsed as described above. The MEM-O containing colchicine and radioactive amino acids is removed from the previously incubated cultures and transferred to the fresh cultures, which are then incubated for

[36] R. F. Diegelmann and B. Peterkofsky, *Proc. Natl. Acad. Sci. U.S.A.* **69**, 892 (1972).

3 hr as described above. After the radioactive medium has been removed, the cells are harvested at 0–4°.

The cells are scraped off the dish with 2 ml of 0.11 $M$ NaCl–0.05 $M$ Tris-HCl, pH 7.4, containing either 1 m$M$ L-proline or L-tryptophan, depending on which isotope was used for labeling. The suspension is transferred to a plastic conical tube and centrifuged at 1000 $g$ for 5 min. The supernatant medium is discarded, the cell pellet is resuspended in 2 ml of buffered saline, and the cells are again pelleted by centrifugation. The supernatant solution is discarded and after the addition of 1 ml of 0.05 $M$ Tris-HCl, pH 7.6, the cells are broken by sonication. The sonicates from the sequentially labeled batches of cells are pooled, and an equal volume of 20% TCA containing 2 m$M$ L-proline or tryptophan is added with mixing. After 5 min, the precipitated proteins are isolated by centrifugation at 1000 $g$ for 5 min and resuspended in 1 ml of 5% TCA containing 1 m$M$ proline or tryptophan, and the suspension again is centrifuged; the process is repeated twice more. Each precipitate is dissolved in 9 ml of 0.20 $N$ NaOH. This solution may be stored in liquid nitrogen.

*Assay Procedure.* Portions (0.15–0.20 ml) of each radioactively labeled solution are adjusted to the appropriate pH range (6.8–7.4) by adding 0.1 ml of 1 $M$ HEPES buffer, pH 7.2, and an amount of 0.15 $N$ HCl equivalent to 60% of the equivalents of NaOH present. Two solutions are prepared, both containing one volume of 25 m$M$ CaCl$_2$, two volumes of 62.5 m$M$ NEM, and one volume of either elution buffer (minus) or enzyme in elution buffer (plus). Forty microliters of the "minus" solution are added to one and 40 $\mu$l of the "plus" solution are added to the other of the duplicate tubes containing the neutralized solutions of radioactive proteins. The final volume is adjusted to 0.50 ml with H$_2$O. The tubes are incubated at 37° for 90 min, and the reaction is stopped by the addition of 0.50 ml of 10% TCA–0.5% tannic acid. The tubes are allowed to remain at 0° for 5 min and then are centrifuged at 1000 $g$ for 5 min. The supernatant solutions are transferred to counting vials, the precipitates are resuspended in 0.5 ml of 5% TCA–0.25% tannic acid, and the tubes are again centrifuged. The resultant supernatant solution of each is added to the appropriate vial, and radioactivity in the combined supernatant solutions is measured as described in the section on quantitation of radioactive, acid-soluble collagen. To measure radioactivity in the undigested proteins, the pellets are resuspended in 1.0 ml of 6 $N$ HCl and hydrolyzed either at 120° under pressure for 3 hr,[37] or at 100° for 16 hr. A portion of the hydrolyzate is mixed with scintillation fluid, and radioactivity is measured in a liquid scintillation spectrometer. Radioactivity in the supernatant fractions and the acid hydrolyzates are converted to distintegrations per minute by

[37] D. J. Prockop, B. Peterkofsky, and S. Udenfriend, *J. Biol. Chem.* **237**, 1581 (1962).

comparison with radioactive standards counted in the appropriate solutions. The percentage of the substrate degraded is calculated from the radioactivity in the TCA–tannic acid-soluble peptides (corrected for the value obtained in the control sample without collagenase) compared to the sum of radioactivity in the peptides plus the undigested protein or to radioactivity of the undigested protein in the minus enzyme control sample.

*Purity of Commercial Collagenase Preparations.* Several enzyme preparations of varying degrees of purity are available from commercial sources. Of these, Worthington CLSPA enzyme repeatedly has been reported to contain nonspecific protease contaminants in significant amounts.[14,15,20,38] The main contaminant appears to be clostripain, but some preparations contain other contaminants, since the nonspecific activity could not be inhibited totally by sulfhydryl-reactive reagents,[14] as discussed above. Sigma type III, Grade A collagenase also has been reported to contain significant contaminating protease.[20] One preparation reported to contain virtually no contaminants is that of Advance Biofactures Corporation.[39] We have retested CLSPA collagenase before and after gel filtration on Sephacryl S-200 for protease contamination, as described above, and also tested Advance Biofactures purified collagenase and an analytical grade collagenase from Boehringer Mannheim. Specific enzyme activity was determined against insoluble tendon collagen and Azocoll. The extent of nonspecific protease contamination and the effect of NEM on this activity was determined with [$^{14}$C]tryptophan-labeled cell proteins and compared to the solubilization of the identical proteins labeled with [$^{14}$C]proline. The results of these tests are shown in Table I.

The collagenase from Boehringer Mannheim appeared to be a crude enzyme preparation, since it contained brown pigment and was the most contaminated with nonspecific proteolytic activity. Almost one-half of the [$^{14}$C]tryptophan-labeled substrate was degraded, and very little of this activity was inhibited by NEM. Worthington CLSPA collagenase contained about the same degree of contamination as reported previously, and again only a small fraction was inhibited by NEM.[14] In the previous study, 21% of a [$^{14}$C]tryptophan-labeled substrate was degraded by 35 $\mu$g of enzyme, but only 10 $\mu$g were used in this study. Gel filtration removed essentially all the contaminant activity based on the observations that the amount of [$^{14}$C]tryptophan-derived radioactivity appearing in the TCA–tannic acid fraction was insignificant (0.1%) compared to the minus enzyme blank (2.9% of the substrate) and was essentially the same with or without NEM. The same was true of the enzyme from Advance Biofac-

[38] W. M. Mitchell, *Johns Hopkins Med. J.* **127**, 192 (1970).
[39] R. Miller and S. Udenfriend, *Arch. Biochem. Biophys.* **139**, 104 (1970).

TABLE I
ASSAY OF COMMERCIAL COLLAGENASE PREPARATIONS FOR NONSPECIFIC
PROTEASE ACTIVITY[a]

| Source of enzyme[b] | Enzyme protein (μg) | NEM | Radioactive protein digested | | | |
|---|---|---|---|---|---|---|
| | | | [$^{14}$C]Pro (dpm) | Percent of substrate | [$^{14}$C]Try (dpm) | Percent of substrate |
| Worthington CLSPA | 10.0 | − | 5261 | 20.8 | 3939 | 6.5 |
| | 10.0 | + | 4174 | 16.5 | 3082 | 5.1 |
| S-200 purified CLSPA | 8.3 | − | 2708 | 10.7 | 62 | 0.1 |
| | 8.3 | + | 2568 | 10.2 | 92 | 0.1 |
| Advance Biofactures | 7.0 | − | 2907 | 11.5 | 62 | 0.1 |
| | 7.0 | + | 2941 | 11.6 | 0 | 0 |
| Boehringer Mannheim | 14.2 | − | 15,789 | 62.4 | 28,383 | 46.7 |
| | 14.2 | + | 13,331 | 52.7 | 23,672 | 39.0 |

[a] The assay was carried out in a total volume of 0.5 ml as described in the text except that N-ethylmaleimide (NEM) (2.5 mM) was either absent (−) or present (+), as indicated in column 3. Radioactive proteins were prepared from Balb 3T3-P13 cells as described in the text. The trichloroacetic acid-precipitated protein was dissolved in 0.20 N NaOH, and portions containing 660 μg of cellular protein (approximately 10–15 μg of collagen) and either 25,290 dpm of [$^{14}$C]proline or 60,752 dpm of [$^{14}$C]tryptophan-labeled, TCA-precipitable protein were added to each assay. Radioactivity has been corrected for control samples to which no enzyme was added, and these values were as follows: 658 dpm or 2.6% of the [$^{14}$C]proline-labeled substrate and 1745 dpm or 2.9% of the [$^{14}$C]tryptophan-labeled substrate.

[b] Worthington CLSPA collagenase was further purified on Sephacryl S-200 as described in the legend to Fig. 2. Two vials of Advance Biofactures collagenase (3000 units/vial) were dissolved with 1 ml of elution buffer to give a protein concentration of 139 μg/ml. Boehringer Mannheim enzyme also was dissolved in elution buffer. The specific activity of these preparations against Azocoll (milligrams of Azocoll digested per minute per milligram of enzyme protein) were as follows: Worthington CLSPA, 8.1; S-200 purified CLSPA, 36.7; Advance Biofactures, 21.1; Boehringer Mannheim, 15.9.

tures. The amount of [$^{14}$C]proline-labeled protein (collagen) released by these two preparations also was similar. The amount of [$^{14}$C]proline-labeled substrate degraded by the other two enzymes was greater than degradation of the [$^{14}$C]tryptophan-labeled substrate, since it included the radioactivity from collagen in the substrate.

The S-200 purified and Advance Biofactures enzyme also were tested for completeness of digestion of purified, acid-soluble collagen (Table II). Unlabeled acid soluble collagen (100 μg) was added to approximate a collagen concentration that might result during analysis of a connective tissue sample. The higher concentration of S-200-purified enzyme (8.33 μg) had been found previously to be in excess even when 3 mg of unlabeled collagen were added,[29] and here approximately 94% of the sub-

TABLE II
DIGESTION OF PURIFIED [$^{14}$C]PROLINE-LABELED ACID-SOLUBLE COLLAGEN BY COLLAGENASE PREPARATIONS[a]

| Enzyme source[b] | Enzyme protein (μg) | Solubilized substrate (dpm) | Undigested substrate (dpm) | Percentage degraded |
|---|---|---|---|---|
| S-200 purified CLSPA | 1.67 | 1331 | 80 | 94.3 |
|  | 8.33 | 1304 | 73 | 94.7 |
| Advance Biofactures | 1.39 | 1286 | 153 | 89.4 |
|  | 6.95 | 1360 | 66 | 95.3 |

[a] [$^{14}$C]Proline-labeled acid-soluble collagen from chick embryo was purified as described in the text. A portion of the solution was mixed with unlabeled calf skin acid-soluble collagen and chick embryo carrier protein (prepared as described elsewhere[54]). Protein was precipitated with trichloroacetic acid (final concentration of 10%), and the isolated precipitate was dissolved in 0.2 N NaOH. Portions of 0.15 ml containing 100 μg of acid-soluble collagen and 0.5 mg of carrier protein were neutralized and incubated with either 50 μl of elution buffer or an equivalent volume of enzyme dissolved in elution buffer in a final volume of 0.5 ml containing, in addition, 0.5 mM CaCl$_2$ and 2.5 mM NEM. Samples were incubated at 37° for 90 min and were processed as described in the section on assay of collagenase with radioactive acid-soluble collagen as substrate.

[b] The enzyme preparations were the same as those used in the experiment described in Table I.

strate was degraded even with one-fifth this amount of enzyme. Similar results were obtained with the Advance Biofactures enzyme at 6.95 μg/ml, and there was only slightly lower degradation at one-fifth the amount of enzyme.

The results of these two tests indicate that either of these purified enzymes are suitable for specifically quantitating the amount of collagen in biological samples. We found, however, that two separate batches of Advance Biofactures collagenase differed greatly in their activity against Azocoll. The batch tested in the experiments shown in Tables I and II had a specific activity against Azocoll that was only slightly lower than that of the S-200 purified enzyme. We found that a previous batch, however, had only about 2% the specific activity of this batch or of the S-200 enzyme. In contrast, the specific activity of both Advance Biofactures batches against insoluble tendon collagen were similar and severalfold higher than that of the S-200 enzyme. Since the activity of purified collagenases against Azocoll appears to reflect their activity against denatured collagen derived from biological samples during analytical procedures (see above and reference 14), it is advisable to check the activity of each batch of commercial preparation against this substrate if the enzyme is intended for such use.

Uses of Protease-Free Bacterial Collagenase

*Structural Probe.* Limited digestion with bacterial collagenases has been used to obtain information on the helical region of collagen. The site of cleavage has been determined by electron microscopy of segment long-spacing (SLS) fragments or Edman degradation.[40,41] At a low enzyme-to-substrate ratio, cleavage occurred mainly at two sites—interbands 26–27 and 33–34 in calf skin collagen, but at bands 18 and 22 in rat skin collagen. Highly purified *A. iophagus* collagenase produced cleavages mainly at bands 33–34 and 41 in both types of collagen.

*Clostridium histolyticum* collagenase has been used to isolate the nonhelical regions of procollagen. The general structure of the N- and C-terminal extension regions of procollagen has been reviewed recently.[42] Most of the amino acids in the procollagen extension peptides are not in the characteristic sequence found in the region that forms the triple helix in the collagen molecule. The exception is a short region of 30–40 amino acid residues in the N-terminal propeptides. Prolonged collagenase digestion of either pro-$\alpha$1 type I and II chains leaves the C-terminal propeptide, which has a molecular weight of about 30,000, and two peptides, designated Col 1 and 2, from the N-propeptides.[43–45] The helical region of the N-terminal propeptide is designated Col 3. In the unreduced N-propeptide of procollagen III, Col 3 is relatively resistant to collagenase unless the molecule is reduced and alkylated.[46] Digestion of chick and sheep pro-$\alpha$2 yields only Col. 2.[44,47] The Col 1 peptides from the various pro-$\alpha$1 chains have molecular weights in the range of 10,000–11,000. The Col 2 peptides, which span the C-terminal region of the N-propeptide extension and telopeptide end of the processed $\alpha$ chains, have molecular weights of about 3000–6000. Col 1 and 2 also have been obtained by collagenase digestion of a peptide secreted into the medium of chick embryo tendon cells.[48] The N- and C-terminal propeptides are isolated after collagenase digestion either by gel filtration or DEAE-cellulose chromatography or by a combination of both techniques.[43,45–50]

[40] M. Stark and K. Kühn, *Eur. J. Biochem.* **6**, 534 (1968).
[41] A. LeCroisey and B. Keil, *Biochem. J.* **179**, 53 (1979).
[42] J. H. Fessler and L. I. Fessler, *Annu. Rev. Biochem.* **47**, 129 (1978).
[43] D. Hörlein, P. P. Fietzek, and K. Kühn, *FEBS Lett.* **89**, 279 (1978).
[44] N. P. Morris, L. I. Fessler, and J. H. Fessler, *J. Biol. Chem.* **254**, 11024 (1979).
[45] M. Wiestner, T. Krieg, D. Hörlein, R. W. Glanville, P. Fietzek, and P. K. Müller, *J. Biol. Chem.* **254**, 7016 (1979).
[46] H. Nowack, B. R. Olsen, and R. Timpl, *Eur. J. Biochem.* **70**, 205 (1976).
[47] U. Becker, R. Timpl, O. Helle, and D. J. Prockop, *Biochemistry* **15**, 2853 (1976).
[48] D. M. Pesciotta, M. H. Silkowitz, P. P. Fietzek, P. N. Graves, R. A. Berg, and B. R. Olsen, *Biochemistry* **19**, 2447 (1980).
[49] W. H. Murphy, K. von der Mark, L. S. G. McEneany, and P. Bornstein, *Biochemistry* **14**, 3243 (1975).

A species of basement membrane collagen, 7 S collagen, has been isolated on the basis of its resistance to digestion by clostridial collagenase.[51]

*Identification of Collagenous Components after Separation by Column or Gel Electrophoresis Procedures.* Two different techniques may be used to identify radioactively labeled collagenous components separated by column chromatography. Each fraction can be assayed for collagen content as described above (Fig. 1) or the sample can be digested with collagenase before applying it to the column. In the latter case, two samples must be chromatographed, each of which has been incubated under identical conditions except for the addition of purified collagenase. An internal standard of a radioactive protein with a different isotope or a substance that can be measured specrophotometrically is added to each sample in order to align the elution patterns from the two runs. Subtracting the radioactivity of the fractions derived from the sample incubated with collagenase from the corresponding fractions of the sample without enzyme will give the elution pattern of the collagenous components.

An example of this procedure is illustrated by the gel filtration patterns in Fig. 3. Media of chick embryo bone cells containing [$^{14}$C]proline-labeled procollagen at varying levels of proline hydroxylation were heated to inactivate proleolytic enzymes and then incubated with or without purified collagenase. The enzyme will digest the collagen polypeptide regardless of whether proline is hydroxylated or not. Procollagen was denatured by heating at 100° for 5 min after the addition of SDS and dithiothreitol to give final concentrations of 0.5% and 1 m$M$, respectively. Samples were applied to a gel filtration column of agarose A-5m (Bio-Rad), and radioactivity in the fractions was measured. Phenol red, which was present in the medium and has an absorption maximum at 550 nm, served as an internal marker. Radioactivity appearing in the lower molecular weight region after collagenase digestion represents peptide extensions of the procollagen. This procedure also has been used with CM-cellulose chromatography.[52]

The same principle of predigestion is followed for analysis by gel electrophoresis, except that the incubation is carried out on a microscale. Solutions containing procollagen or collagen derived by pepsin treatment are usually dialyzed against 0.5 $N$ acetic acid and lyophilized. The residue is dissolved in 20–25 $\mu$l of buffer containing 0.5 m$M$ CaCl$_2$ and 2.5 m$M$ NEM and incubated with or without purified collagenase. Media from

---

[50] B. R. Olsen, N. A. Guzman, J. Engel, C. Condit, and S. Aase, *Biochemistry* **16**, 3030 (1977).
[51] J. Risteli, H. P. Bächinger, J. Engel, H. Furthmayr, and R. Timpl, *Eur. J. Biochem.* **108**, 239 (1980).
[52] R. Hata and B. Peterkofsky, *Proc. Natl. Acad. Sci. U.S.A.* **74**, 2933 (1977).

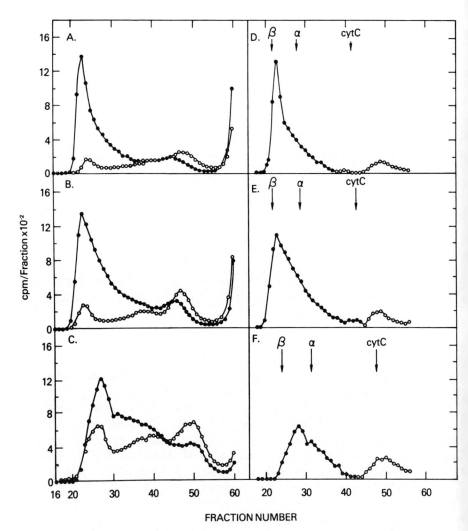

FIG. 3. Gel filtration of radioactive collagenous components in medium from cultured chick embryo bone cells. Growth medium was removed from chick embryo bone cell cultures, and the cells were incubated with [$^{14}$C]proline in serum-free medium for 2 hr as described elsewhere [T. J. J. Blanck and B. Peterkofsky, *Arch. Biochem. Biophys.* **171**, 259 (1975)]. Either no additions were made (B) or 0.1 m$M$ ascorbate (A) or 0.5 m$M$ $\alpha,\alpha'$-dipyridyl (C) were present. The medium was removed from cultures and heated 5 min at 100°. Duplicate 1.5 ml (A,B) or 5 ml (C) portions were incubated without or with purified collagenase (10 µg/ml) for 90 min at 37°, under conditions described in the text and then sodium dodecyl sulfate (SDS) and dithiothreitol (DTT) were added to give final concentrations of 0.5% and 1.0 m$M$, respectively. The solutions were heated at 100°, cooled and applied to a 1.5 × 50 cm column of A-5m gel (Bio-Rad) equilibrated with 50 m$M$ Tris-HCl, pH 7.6–0.1% SDS–0.1 m$M$ DTT. Elution was carried out with a solution containing the same concentrations of these reagents, 1-ml fractions were collected, and radioactivity in the entire fraction was measured. Panels A–C show the patterns obtained from samples preincubated without (●——●) or with (○——○) collagenase. Panels D–F show the difference

radioactively labeled cell cultures usually have very small amounts of collagenous protein so that digestion can be accomplished very rapidly at the collagenase concentrations used routinely (8–16 μg/ml). This is advantageous when working with procollagen samples, since cleavage of the propeptides will be minimized. A preliminary experiment should be carried out to determine the minimum time required for digestion. After incubation of the samples, an equal volume of a denaturing solution containing 2% SDS and 10% sucrose is added and the samples are heated for 5 min at 65°. If the effect of reduction is to be tested, dithiothreitol is added to the denaturing solution at a concentration of 20 m$M$. The samples are cooled and applied to disk or slab gels. Radioactivity can be measured by gel slicing,[52] or by scanning fluorograms. The difference between the patterns obtained with and without collagenase will give the specific collagen pattern. Preflashing of the X-ray film under reproducible conditions is required in order to obtain linearity with respect to varying density of bands on the fluorogram.[53]

*Measurement of Collagen Biosynthesis and Proline Hydroxylation.* The use of *C. histolyticum* collagenase rather than hydroxyproline formation to measure collagen synthesis has the advantage of dissociating the process of protein synthesis from proline hydroxylation. If cofactors required for the hydroxylation, such as ascorbate or ferrous ion, become rate limiting, the rate of collagen synthesis will be underestimated if hydroxyproline formation is used. The procedures for carrying out routine assays to measure collagen biosynthesis in tissue and cell cultures are reviewed in detail elsewhere,[54] and these include labeling of tissues or cells with radioactive amino acids, preparation of samples for digestion by collagenase, and formulas for calculating the relative rate of collagen or procollagen synthesis. A method has been devised to measure the extent of proline hydroxylation concomitantly with collagen synthesis.[55] This is accomplished by using a mixture of [4-$^3$H]- and [$^{14}$C]proline to label proteins and measuring the $^3$H : $^{14}$C ratio in collagenase digests. When proline in collagen is completely hydroxylated, the ratio will decrease to almost one-half of the initial ratio, whereas complete inhibition of hydroxylation will result in an unaltered ratio.

[53] R. A. Laskey and A. D. Mills, *Eur. J. Biochem.* **56**, 335 (1975).
[54] B. Peterkofsky, M. Chojkier, and J. Bateman, *in* "Immunochemistry of Extracellular Matrix: Methods" (H. Furthmayr, ed.), Vol. 1 CRC Press, Boca Raton, in press, 1981.
[55] M. Chojkier, B. Peterkofsky, and J. Bateman, *Anal. Biochem.* **108**, 385 (1980).

---

curves which represent collagen (radioactivity that disappeared from fractions after digestion) (●——●), or collagenase-resistant polypeptides appearing in fractions (○——○). Acid-soluble calf skin collagen and cytochrome *c* (cytC) dissolved in serum-free medium and treated exactly like the labeled samples, served as molecular weight markers.

## [22] Antibodies to Collagens and Procollagens

By RUPERT TIMPL

### Preparation and Characterization

During the last decade antibodies have become important tools for characterizing structural, biological, and pathological properties of collagens. The most striking achievements were the development of reagents capable of distinguishing among various types of the protein as well as precursor-specific portions of procollagens. It was also possible to identify the structure of individual antigenic determinants, thus allowing the use of such reagents at the submolecular level. This chapter will mainly discuss methodological and applied aspects. The reader is referred to general reviews on the immunology of collagens and procollagens,[1-4] to special treatments on the immunohistology of collagens,[5-8] and to a recent book on immunochemical methods[9] for a more comprehensive discussion.

### Choice of Immunogens and Preparation of Antisera

Even though it is an almost trivial claim that immunogens should be as pure as possible, it was found to be of particular importance for collagen antigens. This is because the relatively poor immunogenicity of collagen and a low contamination (0.1% or less) by serum or other noncollagenous proteins[1] or procollagens[10] may provoke antibody titer against these impurities, which are as high as those against the collagenous proteins. Collagenous proteins should be used for immunization in soluble form, since insoluble, cross-linked material lacks substantial immunogenicity.[1] Materials obtained by extraction of tissues with neutral or acidic buffers have

---

[1] R. Timpl, in "Biochemistry of Collagen" (G. N. Ramachandran and A. H. Reddi, eds.), p. 319. Plenum, New York, 1976.
[2] H. Furthmayr and R. Timpl, *Int. Rev. Connect. Tissue Res.* **7**, 61 (1976).
[3] H. K. Beard, W. P. Faulk, L. B. Conochie, and L. E. Glynn, *Prog. Allergy* **22**, 45 (1977).
[4] D. Michaeli, in "Immunochemistry of Proteins" (M. Z. Atassi, ed.), Vol. 1, p. 317. Plenum, New York, 1977.
[5] R. Timpl, G. Wick, and S. Gay, *J. Immunol. Methods.* **18**, 165 (1977).
[6] S. Gay and E. J. Miller "Collagen in the Physiology and Pathology of Connective Tissue." Fischer, Stuttgart, 1978.
[7] K. von der Mark, *Curr. Top. Dev. Biol.* **14**, 199 (1980).
[8] K. von der Mark, *Int. Rev. Connect. Tissue Res.* **9**, in press (1981).
[9] H. Furthmayr (ed.), "Immunochemistry of the Extracellular Matrix." CRC Press, Boca Raton, Florida, in press, 1982.
[10] H. Rohde, H. Nowack, U. Becker, and R. Timpl, *J. Immunol. Methods.* **11**, 135 (1976).

been found to be useful particularly when care was taken to prevent loss of terminal telopeptides due to endogenous proteases. However, successful production of antibodies has also been described against collagens solubilized by limited pepsin digestion[11-14] that may have lost most of their telopeptide segments. These materials should be purified extensively by precipitation with NaCl at acidic and neutral pH and by ion-exchange chromatography (see Chapter [2]) prior to their administration to animals.

The use of denatured collagens as immunogens, particularly in the form of their constituent polypeptide chains or smaller fragments, is attractive because of the possibility of subjecting these materials to more extensive purification. Compared to native, triple-helical collagens, the denatured, unfolded products are very often weaker immunogens[1] or may even be inactive in animal species that mainly respond against helical antigenic determinants.[15] The production of antibodies against small collagenous structures such as the N-terminal region of type III collagen[16] or a synthetic peptide[17] has been reported when these materials were coupled to potent carriers such as serum albumin. Collagen-like synthetic polypeptides were frequently found to be reasonably good immunogens,[1,3] but the antibodies with a few exceptions[18] did not cross-react, or did so only weakly, with authentic collagenous proteins.

Procollagens were consistently found to be better immunogens than collagens, and the antisera obtained showed only weak antibody titers against collagen.[19-21] The immunogens used were intact procollagens obtained from cell and organ cultures,[20,21] pN-collagens from dermatosparactic animals (type I)[10,19] or fetal skin (type III),[22] or isolated amino- and carboxypropeptides.[10,23] In spite of their good immunogenicity, sub-

[11] E. Hahn, R. Timpl, and E. J. Miller, *J. Immunol.* **113,** 421 (1974).
[12] H. von der Mark, K. von der Mark, and S. Gay, *Dev. Biol.* **48,** 237 (1976).
[13] J. A. Madri and H. Furthmayr, *Human Pathol.* **11,** 353 (1980).
[14] J. A. Grimaud, M. Druguet, S. Peyrol, O. Chevalier, D. Herbage, and N. el Badrawy, *J. Histochem. Cytochem.* **28,** 1145 (1980).
[15] H. Nowack, E. Hahn, and R. Timpl, *Immunology* **29,** 621 (1975).
[16] U. Becker, H. Nowack, S. Gay, and R. Timpl, *Immunology* **31,** 57 (1976).
[17] E. Benjamini, D. Michaeli, C. Y. Leung, K. Wong, and H. Scheuenstuhl, *Immunochemistry* **10,** 629 (1973).
[18] A. Maoz, S. Fuchs, and M. Sela, *Biochemistry* **12,** 4246 (1973).
[19] R. Timpl, G. Wick, H. Furthmayr, C. M. Lapiere, and K. Kühn, *Eur. J. Biochem.* **32,** 584 (1973).
[20] K. von der Mark, E. M. Click, and P. Bornstein, *Arch. Biochem. Biophys.* **156,** 356 (1973).
[21] C. J. Sherr, M. B. Taubman, and B. Goldberg, *J. Biol. Chem.* **248,** 7033 (1973).
[22] H. Nowack, S. Gay, G. Wick, U. Becker, and R. Timpl, *J. Immunol. Methods.* **12,** 117 (1976).
[23] P. Dehm, B. R. Olsen, and D. J. Prockop, *Eur. J. Biochem.* **46,** 107 (1974).

stantial purification of the material seems to be still required, since antibodies against unrelated contaminants have been detected in antisera against procollagen obtained from cell cultures.[21] The strength and specificity of the antibody response against procollagen structures is dependent on intact disulfide bridges.[10,19] Although it is apparently not a problem to produce antibodies against individual procollagen peptides, studies in inbred mice showed that procollagens are still better immunogens. This is due to carrier determinants in the collagenous segments of the molecules that augment the antibody response against antigenic determinants localized in the noncollagenous domains.[24] Aminopropeptides that lack a substantial noncollagenous domain, such as that from pro-$\alpha$2 chain, lacked, however, immunogenicity.[25]

Only a few data are available on the affinity of antibodies produced against collagens or procollagens. Particularly high apparent affinity constants in the order of $10^{11}$ liters mol$^{-1}$ were reported in studies with the aminopropeptides of types I and III collagens.[26] Anti-collagen antibodies were considered to be of low affinity on the basis of radioimmunoassay studies,[26,27] but precise data are lacking except for an affinity constant of $10^7$ liters mol$^{-1}$ for antibodies against a synthetic, collagenous peptide.[17] These parameters are obviously of practical importance for designing radioimmunoassays or related assays, since affinity of antibodies determines sensitivity.[28,29]

A variety of animal species and immunogens have been used to produce antibodies against collagens and procollagens (Table I). The amounts of protein injected were usually adjusted empirically to the different immunogenicity of the material and to the size of the animals used. Production of antibodies to collagen in rabbits[30] and mice[15,31] required the presence of complete Freund's adjuvant at least for the first injection. Guinea pigs showed a better antibody response when collagen was administered in incomplete Freund's adjuvants.[27] Using a good emulsion between the antigen solution and the adjuvant (ratio 1 : 1, v/v) is often of crucial importance and can be achieved by extensive mixing of the components within a

---

[24] H. Nowack, H. Rohde, D. Götze, and R. Timpl, *Immunogenetics* **4**, 117 (1977).
[25] H. Rohde, U. Becker, H. Nowack, and R. Timpl, *Immunochemistry* **13**, 967 (1976).
[26] R. Timpl and L. Risteli, in "Immunochemistry of the Extracellular Matrix" (H. Furthmayr, ed.), CRC Press, Boca Raton, Florida, in press, 1982.
[27] B. C. Adelmann, G. J. Gentner, and K. Hopper, *J. Immunol. Methods* **3**, 319 (1973).
[28] W. H. C. Walker and P. M. Keane, in "Handbook of Radioimmunoassay" (G. E. Abraham, ed.), Chap. 3. Dekker, New York, 1977.
[29] A. Zettner and P. E. Duly, *Clin. Chem.* **20**, 5 (1974).
[30] M. Stoltz, R. Timpl, H. Furthmayr, and K. Kühn, *Eur. J. Biochem.* **37**, 287 (1973).
[31] E. Hahn, H. Nowack, D. Götze, and R. Timpl, *Eur. J. Immunol.* **5**, 288 (1975).

TABLE I
IMMUNIZATION WITH COLLAGENS AND PROCOLLAGENS

| Animal species | Amounts of antigen injected[a] | References[b] |
|---|---|---|
| Rabbit | 5–10 mg: interstitial collagens (I, II, III, V) | 1–4 |
| | 0.2–1 mg: type IV collagen, procollagen, 7 S collagen | 5–9 |
| | 0.1–0.5 mg: procollagen peptides | 10, 11 |
| Guinea pig | 0.75 mg: types I and II collagens | 4, 12 |
| | 0.2 mg: type IV collagen | 5 |
| Rat | 0.5 mg: interstitial collagens (I, II) | 13–15 |
| Chick | 5 mg: type I collagen | 16 |
| Mouse | 5–50 µg: interstitial collagens, procollagens (high responder) | 17–19 |
| | 50–500 µg: interstitial collagens, procollagens (low responder) | |
| | 0.2–0.5 µg: procollagen peptides (high responder) | 20 |
| Goat, sheep | 10–20 mg: interstitial collagens (I, II, III) | 21, 22 |

[a] The amounts indicated were usually administered 2 or 3 times together with complete Freund's adjuvant. Intervals between the injections were 3–6 weeks.

[b] Key to references: (1) C. Steffen, R. Timpl, and I. Wolff, Immunology **15**, 135 (1968); (2) M. Stoltz, R. Timpl, H. Furthmayr, and K. Kühn, Eur. J. Biochem. **37**, 287 (1973); (3) U. Becker, H. Nowack, S. Gay, and R. Timpl, Immunology **31**, 57 (1976); (4) H. von der Mark, K. von der Mark, and S. Gay, Dev. Biol. **48**, 237 (1976); (5) R. Timpl, R. W. Glanville, G. Wick, and G. R. Martin, Immunology **38**, 109 (1979); (6) R. Timpl, G. Wick, H. Furthmayr, C. M. Lapiere, and K. Kühn, Eur. J. Biochem. **32**, 584 (1973); (7) H. Nowack, S. Gay, G. Wick, U. Becker, and R. Timpl, J. Immunol. Methods **12**, 117 (1976); (8) K. von der Mark, E. M. Click, and P. Bornstein, Arch. Biochem. Biophys. **156**, 356 (1973); (9) J. Risteli, H. P. Bächinger, J. Engel, H. Furthmayr, and R. Timpl, Eur. J. Biochem. **108**, 239 (1980); (10) H. Rohde, H. Nowack, U. Becker, and R. Timpl, J. Immunol. Methods **11**, 135 (1976); (11) D. M. Pesciotta, M. H. Silkowitz, P. P. Fietzek, P. N. Graves, R. A. Berg, and B. R. Olsen, Biochemistry **19**, 2447 (1980); (12) B. C. Adelmann, G. J. Gentner, and K. Hopper, J. Immunol. Methods **3**, 319 (1973); (13) W. Beil, R. Timpl, and H. Furthmayr, Immunology **24**, 13 (1973); (14) E. Hahn, R. Timpl, and E. J. Miller, J. Immunol. **113**, 421 (1974); (15) D. E. Trentham, A. S. Townes, and A. H. Kang, J. Exp. Med. **146**, 857 (1977); (16) W. Beil, H. Furthmayr, and R. Timpl, Immunochemistry **9**, 779 (1972); (17) E. Hahn, H. Nowack, D. Götze, and R. Timpl, Eur. J. Immunol. **5**, 288 (1975); (18) H. Nowack, E. Hahn, and R. Timpl, Immunology **29**, 621 (1975); (19) H. Nowack, E. Hahn, and R. Timpl, Immunology **30**, 29 (1976); (20) H. Nowack, H. Rohde, D. Götze, and R. Timpl, Immunogenetics **4**, 117 (1977); (21) J. W. Chidlow, F. J. Bourne, and A. J. Bailey, Immunology **27**, 665 (1974); (22) H. K. Beard, R. Ryvar, R. Brown, and H. Muir, Immunology **41**, 491 (1980)

syringe. As a quality control for a good emulsion, a drop should be put on a water surface and should not spread out within the next 3–5 mins.

Raising antisera against collagens or procollagens normally does not require more than 2 or 3 injections of the antigen. Immunization over a longer period of time usually does not increase antibody titers, but increases the risk of producing antibodies against impurities. Hyperimmunization, however, may also change the response pattern against different antigenic determinants, as found for rat type I collagen.[32]

Many other biological parameters not yet completely understood may contribute to the successful production of antibodies. Phenomenologically, they may cause a considerable variation in the strength and specificity of the antibody response among individual animals (see, e.g., Furthmayr and Timpl.[33]). Studies in inbred strains of mice have related some of these parameters to the control by immune response genes linked or not linked to the major histocompatibility locus $H$-$2$.[15,24,31] Apparently, different immune response genes are governing the optimal production of antibodies against types I and II collagens and against procollagen structures. These controls are presumably exerted at the level of T cell activation in agreement with other observations that collagens[34] and procollagens[24] belong to the class of T cell-dependent antigens. A selective stimulation of T cells by collagens (cell-mediated immunity) without strong antibody responses could be obtained in guinea pigs using particular immunization schedules.[35,36]

Hybridoma models have been used to produce mouse antibodies on a large scale against chick types I and II collagens.[37,38] These antibodies are usually of monoclonal origin and hence directed against a single antigenic determinant,[39] although this has not been proved in the above studies. Monoclonal antibodies may have an extremely high specificity and distinguish among various forms of type I collagens[37] that by other criteria are considered as identical proteins. These possibilities have to be kept in mind for devising the use of conventional, polyclonal, or monoclonal antibodies in a particular experiment.

*Purification of Antibodies*

Immunological studies with raw antisera may be sufficient for many purposes, particularly when the antigens used in the reactions are well characterized and purified and when the stoichiometry of the reaction can

---

[32] R. Timpl, H. Furthmayr, and W. Beil, *J. Immunol.* **108**, 119 (1972).
[33] H. Furthmayr and R. Timpl, *Biochem. Biophys. Res. Commun.* **47**, 944 (1972).
[34] H. Nowack, E. Hahn, and R. Timpl, *Immunology* **30**, 29 (1976).
[35] B. C. Adelmann, J. A. Kirrane, and L. E. Glynn, *Immunology* **23**, 723 (1972).
[36] G. Senyk and D. Michaeli, *J. Immunol.* **111**, 1381 (1973).
[37] T. F. Linsenmayer, M. J. C. Hendrix, and C. D. Little, *Proc. Natl. Acad. Sci. U.S.A.* **76**, 3703 (1979).
[38] T. F. Linsenmayer and M. J. C. Hendrix, *Biochem. Biophys. Res. Commun.* **92**, 440 (1980).
[39] C. Milstein and E. Lennox, *Curr. Top. Dev. Biol.* **14**, 1 (1980).

be assessed properly. Studies of complex antigen mixtures, such as tissues or cell cultures, with the aim of identifying a single component usually requires antibody reagents with a low level of reaction against unrelated or cross-reacting, related antigens. This could be achieved by using purified antibodies as has been illustrated in one of the first, carefully controlled immunohistological studies on type I collagen.[40] Here, raw antisera stained interstitial connective tissue and kidney glomeruli whereas antibodies purified on a type I collagen adsorbent reacted only with the former tissue. This observation agreed with later studies[41,42] demonstrating lack of cross-reaction between interstitial and basement membrane collagens.

The rationale for using affinity chromatography on insolubilized antigens for improving the specificity of antibodies is based on the general experience that low levels of impurities (1% or less) may be responsible for a disproportionally strong antibody response. Since binding of antibody is determined by stoichiometry, adsorption of the antiserum on a column of the same antigen should decrease the content of antibodies against the impurity owing to a low capacity for this contaminant. However, higher levels of contamination (1% or more) may already make this approach unsuccessful. Collagens coupled to diazotized $p$-aminobenzyl cellulose have been used initially for antibody purification[43] and were used in many subsequent studies.[22,44] Other methods of insolubilizing collagens or procollagen peptides included cross-linking to other proteins[23] or coupling to agarose that has been activated by CNBr treatment.[45] The latter procedure has now become a standard method in collagen immunology.[12,16,46]

*Coupling of Collagen to CNBr-Activated Sepharose.*[46] Collagens are dissolved in 0.1 $M$ acetic acid (1 mg/ml) and dialyzed at 4° against 0.1 $M$ phosphate pH 8.0, 0.4 $M$ NaCl. Other antigens with a low ability to precipitate (type IV collagen, procollagen) may be directly dissolved in 0.1 $M$ phosphate, pH 8.0. Commercially obtained CNBr-activated Sepharose (e.g., Pharmacia, Uppsala) is prewashed prior to use on a funnel with 0.001 $M$ HCl. Alternatively, agarose can be activated with CNBr following an established procedure.[45] About 20 ml gel is then incubated with 50

---

[40] G. Wick, H. Furthmayr, and R. Timpl, *Int. Arch. Allergy Appl. Immunol.* **48**, 664 (1975).
[41] D. E. Gunson and N. A. Kefalides, *Immunology* **31**, 563 (1976).
[42] R. Timpl, R. W. Glanville, G. Wick, and G. R. Martin, *Immunology* **38**, 109 (1979).
[43] R. Timpl, H. Furthmayr, C. Steffen, and W. Doleschel, *Z. Immunitaetsforsch. Allerg. Klin. Immunol.* **134**, 391 (1967).
[44] W. Beil, R. Timpl, and H. Furthmayr, *Immunology* **24**, 13 (1973).
[45] R. Axén, J. Porath, and S. Ernback, *Nature (London)* **214**, 1302 (1967).
[46] G. Wick, R. W. Glanville, and R. Timpl, *Immunobiology* **156**, 372 (1979).

ml of antigen solution for 2–3 days at 4° under slow stirring to avoid mechanical damage of the gel beads. The reaction is stopped by filtering the gel followed by incubation in 50 ml 0.1 $M$ Tris base for 2 hr at room temperature. The gel is then successively washed in the cold on a funnel with 200–400 ml of 0.1 $M$ phosphate, pH 8.0, phosphate-buffered saline, pH 7.2, containing 0.02% sodium azide (PBS), 1 $M$ acetic acid, 0.05 $M$ HCl containing 0.15 $M$ NaCl (or alternatively 3 $M$ KSCN, pH 7), and finally again with PBS. The yield of coupling is usually 50–90%.

*Purification of Antibodies by Immunoadsorption.* All operations should be carried out in the cold room. Columns with a volume of about 20 ml of adsorbent gel were prewashed with PBS and, if they are new, followed by 30–50 ml of normal rabbit serum and the elution procedure outlined below. This helps to clean the column from last amounts of noncovalently attached antigen and unreacted protein binding sites. The capacity of the column is usually sufficient to bind most of the antibodies in 20–50 ml of antiserum. Antibody concentrations have been previously determined to be in the range 100–500 µg/ml for anti-collagen sera but may be 1–2 mg/ml for stronger immunogens, such as procollagens.[1] After passage of the antiserum the column is washed with PBS until the optical density at 280 nm of the effluent drops below 0.02–0.03. Antibodies are then displaced by elution with 50 ml of 1 $M$ acetic acid followed by 30 ml of 0.05 $M$ HCl, 0.15 $M$ NaCl.[44] The acidic eluates are immediately neutralized by adding 4 $M$ Tris base and dialyzed against PBS. After removal of precipitated, presumably denatured protein (normally 10–20% of total), the supernatants are concentrated by ultrafiltration (Diaflo, filter UM-20) to a volume of 5–10 ml. Column-bound antibodies can also be eluted by 3 $M$ KSCN, pH 6–7,[12,23] which is somewhat more efficient but has the risk of denaturing native collagen adsorbents.

The concentration of purified antibodies can be estimated from the optical density at 280 nm, which is about 1.4 for 1 mg/ml. If stored at $-20°$ to $-70°$ at a concentration of 0.5–3 mg/ml the antibodies usually retained full activity over 3–5 years. Diluted solutions as required in many studies become more readily inactivated. Immunoelectrophoretic analysis of such antibodies in most cases showed a single band corresponding to IgG. About 80–90% of the antibodies could be bound again to an appropriate adsorbent in a second passage.[40] The procedure described was also useful to purify antibody Fab fragments.[14,47] In our hands, a particular antigen column could be used several dozens of times over a period of several years without large loss of antibody-binding capacity. Between experiments, collagen-adsorbents should be stored at 4° in 0.02% sodium azide or be kept frozen.

[47] L. Balleisen, H. Nowack, S. Gay, and R. Timpl, *Biochem. J.* **184**, 683 (1979).

For the isolation of antibodies specific for certain types of collagens the antisera were usually passed over columns containing cross-reacting collagens prior to a final passage over a column containing the collagen used for immunization.[22] This procedure allows the elimination of antibodies against determinants shared by several collagen types and may be done as a precaution even for antisera without detectable cross-reactivity. Similar methods can be designed for the purification of antibodies against antigenic determinants located in distinct regions of a single protein, e.g., in the aminopropeptide of pN-collagens.

Antisera or purified antibodies have also been used after coupling to an insoluble carrier for the specific binding of antigens such as cross-linked peptides of type I collagen.[48] Here elution of bound material was accomplished by 1 $M$ aqueous ammonia. Our own studies[49] on antibody adsorbents specific for aminopropeptides showed a good capacity to remove the antigens from complex biological samples (fetal serum, cell culture medium). The capacity of such adsorbents showed a distinct decrease after several uses presumably owing to the degradation of antibody by endogenous proteases.

*Immunochemical Assays*

Commonly used and recommended methods for characterizing antisera to collagens and procollagens include radioimmunoassays,[26,27] enzyme immunoassays,[50] and passive hemagglutination.[44,51] Complement fixation has been used in several previous studies (see Timpl[1]), but is usually a very complex reaction and not easy to interpret. Gel diffusion suffers from a low sensitivity and has the additional complication that triple helical collagen antigens may not readily diffuse into such gels.[23,51]

*Labeling of Collagen with Iodine-125*.[26,42] Collagens are dissolved in 0.1 $M$ acetic acid (1 mg/ml) and dialyzed at 4° against 0.1 $M$ phosphate buffer, pH 7.2, 0.2 $M$ NaCl. A 25-$\mu$l aliquot of the solution is mixed with 0.5–1 mCi of Na$^{125}$I in 50 $\mu$l of 0.1 $M$ phosphate buffer, pH 7.2, and 100 $\mu$l of chloramine-T solution (3 mg per 5 ml of 0.1 $M$ phosphate, pH 7.2). After incubation for 5–10 min at room temperature the reaction is stopped by adding 120 $\mu$l of sodium metabisulfite (3 mg per 5 ml of H$_2$O) and 500 $\mu$l of 0.1 $M$ sodium phosphate, pH 7.2. Nonbound iodine may then be removed on a BioGel P-2 column,[26] by dialysis for 2–3 days against phosphate-buffered saline, pH 7.2 (PBS), containing 0.04% Tween 20 or by dialysis

[48] J. W. Chidlow, F. J. Bourne, and A. J. Bailey, *Immunology* **27**, 665 (1974).
[49] H. Rohde and R. Timpl, unpublished observation, 1980.
[50] S. I. Rennard, R. Berg, G. R. Martin, J. M. Foidart, and P. Gehron Robey, *Anal. Biochem.* **104**, 205 (1980).
[51] W. Beil, H. Furthmayr, and R. Timpl, *Immunochemistry* **9**, 779 (1972).

against 0.1 M acetic acid. The labeled product is then stored as a stock solution (10 μg/ml) at 4°. With a variety of connective tissue proteins, the range of incorporated radioactivity was found to be 4000–15,000 cpm/ng. About 70–95% of the radioactivity could be precipitated with 10% trichloroacetic acid in the presence of serum protein (test for protein-bound radioactivity), and nonspecific binding to 0.5% normal serum was usually in the range 1–12%. The labeled proteins usually showed a good stability and could be used in radioimmunoassays for at least 8 weeks.

The procedure described could also be used for labeling procollagens and procollagen peptides.[10,22,25] Other procedures applied to collagens or collagen fragments with equivalent efficiency included labeling with iodine in the presence of lactoperoxidase[52] or with the Bolton–Hunter reagent[53] (3-[$^{125}$I]iodo-1-p-hydroxyphenyl-propionic acid N-hydroxysuccinimide ester). Labeling of collagens with [$^{14}$C]acetic acid anhydride has been described.[54] Alternatively, proteins labeled with $^3$H or $^{14}$C amino acids in cell or organ culture systems may be used. Both these procedures yield material with a relatively low specific radioactivity not allowing assays of the same sensitivity as achieved with iodinated proteins.

*Binding of $^{125}$I-Labeled Proteins to Antibody.* Rabbit antisera or purified antibodies are diluted in 0.5% nonimmune rabbit serum serving as a carrier; 0.1-ml aliquots are then mixed with 0.2 ml of 0.04% Tween 20 in PBS and 0.1 ml of labeled protein diluted in Tween–PBS to a radioactivity of about 2000–10,000 cpm (0.2–1 ng/0.1 ml). The mixture is incubated for 16 hr at 4° and then mixed with a pretested amount of goat antiserum against rabbit immunoglobulin G (second antibody) allowing its complete precipitation. After another incubation for 16–24 hr at 4°, the precipitates are collected by centrifugation and are washed in the cold twice with 0.5 ml of Tween/PBS prior to measuring their radioactivity in a gamma counter. Other ways to separate free labeled antigen from antibody-bound antigen may use second antibodies or protein A (from *Staphylococcus aureus*) bound to an insoluble support. Binding data can be plotted in various ways, for example, as percentage of the total fraction added, as shown in Fig. 1a, which is calculated by Eq. (1).

$$\% \text{ antigen bound} = (a - c)/(b - c) \times 100 \qquad (1)$$

where $a$ = cpm in precipitate obtained with antibody, $b$ = total protein-bound radioactivity added, and $c$ = cpm in precipitate obtained with nonimmune serum carrier only (background binding).

[52] R. B. Clague, R. A. Brown, J. A. Weiss, and P. J. L. Holt, *J. Immunol. Methods.* **27**, 31 (1979).
[53] F. J. Roll, J. A. Madri, and H. Furthmayr, *Anal. Biochem.* **96**, 489 (1979).
[54] J. Menzel, *J. Immunol. Methods.* **15**, 77 (1977).

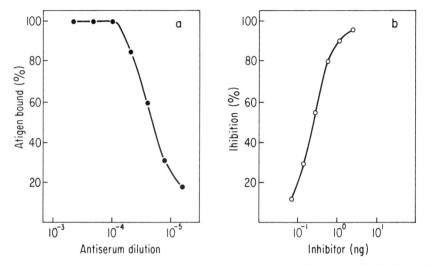

FIG. 1. Radioimmunoassay for the aminopropeptide of type III collagen. (a) Binding of [125]I-labeled aminopropeptide by rabbit antiserum to bovine pN-collagen. (b) Inhibition of the reaction by nonlabeled aminopropeptide. Here, 0.1 ml of antiserum (diluted 1 : 40,000) was preincubated with inhibitor dissolved in 0.2 ml of buffer prior to the addition of 1 ng of labeled aminopropeptide in 0.1 ml of buffer (sequential saturation). Binding and inhibition were calculated according to Eqs. (1) and (2) (see text) (H. Rohde and R. Timpl, unpublished data).

*Inhibition Assay.* The most universally employed variant of radioimmunoassays exploits the competitive reaction of labeled protein (antigen) and nonlabeled protein (inhibitor) for the same antibody binding sites. The tests can be carried out under equilibrium conditions (all three reactants are mixed simultaneously) or by preincubation of inhibitors with a fixed amount of antibody prior to the addition of labeled antigen. The latter variant is referred to as sequential saturation analysis,[29] which is usually more sensitive than an equilibrium analysis. Here, all the tubes receive 0.1 ml of antiserum dilution (in 0.5% normal serum) capable of binding about 50% of the labeled antigen that will be added later. Two to three tubes receive only nonimmune serum carrier for determining background binding. Antibody-containing tubes are then mixed with inhibitor solutions prepared in Tween–PBS except for two or three tubes that are mixed with plain buffer for determining binding in the absence of inhibitor. After incubation of 16 hr at 4°, all tubes are mixed with 0.1 ml of labeled antigen (0.2–1 ng/ml, see above) and incubation is continued for 4–6 hr at 4°. Separation of bound from free antigen (e.g., with second antibody) follows then the protocol described above.

Inhibition data may be calculated and plotted in the following way (see Fig. 1b):

$$\% \text{ inhibition} = 1 - [(a - c)/(b - c)] \times 100 \qquad (2)$$

where $a$ = cpm in the precipitate obtained with antibody in the presence of inhibitor; $b$ = cpm in the precipitate obtained with antibody in the absence of inhibitor; and $c$ = cpm in precipitate obtained with nonimmune serum carrier only (background binding).

Many other ways to express inhibition data are known[28,29,55,56] and are usually based on calculating the ratio between bound and free antigen. These variants are particularly useful for evaluating data with assays carried out under equilibrium conditions.

For the immunochemical interpretation of data obtained by radioimmunoassays, the reader is referred to several reviews.[26,28,29] Such assays are also widely used for the quantitative determination of minute amounts of antigens in complex biological samples. For these purposes, assays have to be optimized for a high sensitivity and precision.[28,55,56] Complications in these quantitative studies may arise by observing a nonparallel shift in the inhibition curve produced by the material in the unknown sample. This may be due to incomplete cross-reaction or to several unknown factors and may be overcome by varying a number of technical parameters (e.g., batch of antibody, incubation time, background control) of the assay.[57]

Enzyme immunoassays have become popular during the last 10 years and are in many features similar to radioimmunoassays. Normally, the antigen is used in insoluble form, i.e., after adsorption to plastic tubes under slightly alkaline conditions, and the enzyme label (alkaline phosphatase, peroxidase) is introduced by the second antibody (anti-immunoglobulin). The uptake of the label by the insoluble phase is proportional to the reaction of the first antibody with antigen and measured by an appropriate colorimetric reaction. This assay has been applied to a variety of collagens,[50,58] using it both for determination of antibody titers and for inhibition assays with soluble antigens. The development of special spectrophotometers that allow to measure the colored product quantitatively in a rapid and precise manner (e.g., in wells of a microtiter plate) have now made enzyme immunoassays as versatile as radioimmunoassays.

Passive hemagglutination has been applied in many of the earlier stud-

[55] A. Zettner, *Clin. Chem.* **19**, 699 (1973).
[56] D. Rodbard and G. R. Frazier, this series, Vol. 37, p. 3.
[57] J. P. Felber, *Methods Biochem. Anal.* **22**, 1 (1974).
[58] B. Gosslau and H. J. Barrach, *J. Immunol. Methods* **29**, 71 (1979).

ies on the immunology of collagen.[44,59] Here antigens are bound to red blood cells and then exposed to antibodies. The reaction is evaluated by scoring the agglutination pattern, e.g., by analysis of settling patterns of the red cells in the well[5] of microtiter plates. The sensitivity of such assays is largely determined by optimal coating of the cells.

*Coupling of Collagen onto Glutaraldehyde-Fixed Erythrocytes.*[45,51] Ten milliliters of a 10% suspension of washed human erythrocytes in 0.15 $M$ NaCl are mixed with 3 ml of 2.5% aqueous glutaraldehyde and stirred for 15 min at room temperature. Cells are collected by centrifugation (10 min, 2000 $g$), washed twice with 0.15 $M$ NaCl and then resuspended in 10 ml of 0.1 $M$ calcium acetate. This suspension is mixed with 10 ml of collagen (3 mg/ml) dissolved in 0.1 $M$ calcium acetate and stirred for 90 min at room temperature. For preparing the antigen solution collagen is first dissolved in 0.1 $M$ acetic acid and extensively dialyzed at 4° against 0.1 $M$ calcium acetate until a transient precipitate has disappeared. For preparing denatured collagen the solution is heated for 30 min at 50° just before addition to the erythrocytes. Cells are finally harvested by centrifugation and washed twice with each 0.1 $M$ calcium acetate followed by 0.15 $M$ NaCl. For storage at 4° they are then suspended in 20 ml of PBS, pH 7.2, containing 0.1% serum albumin or normal serum. Usually, such cells could be used without loss of activity for a period of 4–12 months.

Erythrocytes could also be coated with collagens after treatment with a 0.25% solution of tannic acid.[59] Such cells have to be prepared freshly each day and show a low sensitivity for coating with native collagen.[44,51] They require, however, smaller amounts of collagen (0.01–1 $\mu$g/ml) for optimal coating, which is usually determined by chessboard titration.

*Titration and Inhibition Assay.* Serial dilutions of antisera in a volume of 25 $\mu$l were prepared in PBS containing 0.5% serum albumin, most conveniently in U-shaped plastic titration plates using microtiter equipment. Each well is then mixed with 50 $\mu$l of diluent and 25 $\mu$l of a 0.5% suspension of coated erythrocytes. The settling pattern is evaluated after incubation at 4° overnight. For assessing the specificity of the reaction, coated erythrocytes are allowed to react in the same assay against several dilutions of normal serum. A further negative control includes reaction of the antiserum against uncoated erythrocytes.

In the inhibition assay, wells are filled with 25 $\mu$l of a single antiserum dilution. An optimal dilution (8 agglutinating units) is the eight-fold higher concentration of that antiserum dilution just still agglutinating the coated erythrocytes. The antiserum is then mixed with 50 $\mu$l of antigen (range 0.1–100 $\mu$g/ml) dissolved in PBS and incubated for 1–2 hr at 4°. After

[59] C. Steffen, R. Timpl, and I. Wolff, *Z. Immunitaetsforsch. Allorg. Klin. Immunol.* **134**, 91 (1967).

addition of 25 µl of erythrocyte suspension, incubation is continued at 4° for 6–16 hr. Finally, the smallest amount of antigen still blocking agglutination is then determined by reading the settling pattern. This inhibition assay allows a semiquantitative determination of antigens, but with a large error of at least ± 50%.

Radioimmunoassays and enzyme-immunoassays are the most sensitive and reliable assays so far known for determining antibody titers against collagens and to determine collagenous material in biological samples in the nanogram range. Disadvantages of radioimmunoassays are general problems encountered in the work with radioactive substances and the limited stability of labeled antigens. Enzyme immunoassays may suffer from the use of adsorbed antigens in the reaction, a step that is not easy to control in terms of its reproducibility and the potential reaction with antigens that are contaminants of the major protein. Passive hemagglutination assays have the disadvantage of being semiquantitative and of having a 10- to 100-fold lower sensitivity when compared to the other assays.[22,58,60] They allow, however, a rapid screening for antibodies and/or antigens if they are available in sufficiently high concentrations.

### Specificity of Antibody Reagents

Collagens and procollagens owing to their large size are complex antigens possessing a variety of different antigenic determinants. This has allowed the development of various antibody reagents specific for distinct regions within an individual protein or capable of distinguishing related proteins (Table II). Interstitial collagens type I, II, and III possess potent antigenic determinants in the terminal telopeptide regions which lack the regular Gly-X-Y structure. These determinants are particularly recognized when rabbits are used for immunization, presumably owing to certain differences in the amino acid sequence of the telopeptide regions in rabbit collagen.[61] Reaction of these sites with antibodies is usually not dependent on an intact triple helix. The structure of some of these antigenic determinants has been elucidated.[1,2]

Other animal species, such as guinea pigs,[27] rats,[11,44] mice,[15] and chicken,[51] responded mainly against conformation-dependent antigenic determinants located in the triple helix of collagens. Reaction with these sites required a correct chain composition and allowed, for example, to distinguish molecules of the composition $[\alpha 1(I)]_2 \alpha 2$, $[\alpha 1(I)]_3$, and $[\alpha 1(II)]_3$[11,62] as well as identical types of collagens obtained from different animal species.[62] Similarly, rabbit antibodies to type V collagen have also

---

[60] G. Wick, H. Nowack, E. Hahn, R. Timpl, and E. J. Miller, *J. Immunol.* **117**, 298 (1976).
[61] U. Becker, P. P. Fietzek, H. Furthmayr, and R. Timpl, *Eur. J. Biochem.* **54**, 359 (1975).
[62] E. Hahn and R. Timpl, *Eur. J. Immunol.* **3**, 442 (1973).

TABLE II
IMMUNOLOGICAL RECOGNITION OF DIFFERENT STRUCTURES IN
COLLAGENS AND PROCOLLAGENS

| Distinction by antibodies of | Antigenic structures involved | References[a] |
|---|---|---|
| Structures within collagen $\alpha$ chains | Telopeptide regions of $\alpha 1(I)$ and $\alpha 2$ chain | 1, 2 |
| | Terminal and central regions in type I collagen | 3, 4 |
| | Triple helix vs unfolded chains | 5, 6 |
| Collagens and procollagens | Procollagen vs collagen | 7-9 |
| | Aminopropeptides vs carboxypropeptides | 10-12 |
| Types of collagen | Types I, II, and III collagens | 13-19 |
| | Triple helices of the compositions $[\alpha 1(I)]_2 \alpha 2$ and $[\alpha 1(I)]_3$ | 13 |
| | Type IV vs interstitial collagens | 19-21 |
| | $\alpha 1(IV)$ and $\alpha 2(IV)$ chains | 22 |
| | Type V collagen | 23 |
| Types of procollagens | Aminopropeptides types I and III | 24, 25 |
| Collagens from different animal species | Mammalian types I and II collagens | 6, 26, 27 |
| | Mammalian type IV collagens | 21, 22, 28 |
| | Vertebrate vs invertebrate collagens | 29, 30 |

[a] *Key to references:* (1) D. Michaeli, G. R. Martin, J. Kettman, E. Benjamini, D. Y. K. Leung, and B. A. Blatt, *Science* **166**, 1522 (1969); (2) H. Furthmayr and R. Timpl, *Int. Rev. Connect. Tissue Res.* **7**, 61 (1976); (3) R. Timpl, W. Beil, H. Furthmayr, W. Meigel, and B. Pontz, *Immunology* **21**, 1017 (1971); (4) H. Furthmayr, M. Stoltz, U. Becker, W. Beil, and R. Timpl, *Immunochemistry* **9**, 789 (1972); (5) W. Beil, R. Timpl, and H. Furthmayr, *Immunology* **24**, 13 (1973); (6) E. Hahn, and R. Timpl, *Eur. J. Immunol.* **3**, 442 (1973); (7) R. Timpl, G. Wick, H. Furthmayr, C. M. Lapiere, and K. Kühn, *Eur. J. Biochem.* **32**, 584 (1973); (8) K. von der Mark, E. M. Click, and P. Bornstein, *Arch. Biochem. Biophys.* **156**, 356 (1973); (9) H. Rohde, H. Nowack, U. Becker, and R. Timpl, *J. Immunol. Methods* **11**, 135 (1976); (10) C. Nist, K. von der Mark, E. D. Hay, B. R. Olsen, P. Bornstein, R. Ross, and P. Dehm, *J. Cell Biol.* **65**, 75 (1975); (11) D. M. Pesciotta, M. H. Silkowitz, P. P. Fietzek, P. N. Graves, R. A. Berg, and B. R. Olsen, *Biochemistry* **19**, 2447 (1980); (12) L. I. Fessler, R. Timpl, and J. H. Fessler, *J. Biol. Chem.* **256**, 2531 (1981); (13) E. Hahn, R. Timpl, and E. J. Miller, *J. Immunol.* **113**, 421 (1974); (14) H. Nowack, S. Gay, G. Wick, U. Becker, and R. Timpl, *J. Immunol. Methods* **12**, 117 (1976); (15) G. Wick, H. Nowack, E. Hahn, R. Timpl, and E. J. Miller, *J. Immunol.* **117**, 298 (1976); (16) H. von der Mark, K. von der Mark, and S. Gay, *Dev. Biol.* **48**, 237 (1976); (17) U. Becker, H. Nowack, S. Gay, and R. Timpl, *Immunology* **31**, 57 (1976); (18) F. J. Roll, J. A. Madri, and H. Furthmayr, *Anal. Biochem.* **96**, 489 (1979); (19) S. I. Rennard, R. Berg, G. R. Martin, J. M. Foidart, and P. Gehron Robey, *Anal. Biochem.* **104**, 205 (1980); (20) D. E. Gunson and N. A. Kefalides, *Immunology* **31**, 563 (1976); (21) R. Timpl, R. W. Glanville, G. Wick, and G. R. Martin, *Immunology* **38**, 109 (1979); (22) J. Risteli, D. Schuppan, R. W. Glanville, and R. Timpl, *Biochem. J.* **191**, 517 (1980); (23) J. A. Madri and H. Furthmayr, *Am. J. Pathol.* **94**, 323 (1979); (24) H. Rohde, U. Becker, H. Nowack, and R. Timpl, *Immunochemistry* **13**, 967 (1976); (25) H. Rohde, P. Bruckner, and R. Timpl, in preparation; (26) E. Hahn, R. Timpl, and E. J. Miller, *Immunology* **28**, 561 (1975); (27) B. Sykes and E. Solomon, *Nature (London)* **272**, 548 (1978); (28) J. Risteli, H. P. Bächinger, J. Engel, H. Furthmayr, and R. Timpl, *Eur. J. Biochem.* **108**, 239 (1980); (29) D. Michaeli, G. Senyk, A. Moaz, and S. Fuchs, *J. Immunol.* **109**, 103 (1972); (30) W. Meigel, B. Pontz, R. Timpl, E. Hieber, A. Nordwig, C. Steffen, and K. Kühn, *J. Immunol.* **107**, 1146 (1971).

been found to react mainly with triple helical determinants.[63] A third class of antigenic structures, named central antigenic determinants, could be identified in the central regions of unfolded $\alpha$ chains. Antibodies against these sites may be specific for unfolded chain constituents and showed a generally broad interspecies cross-reaction.[64]

The various types of collagens are sufficiently different in their covalent structure to allow the production of type-specific antibodies. Crude antisera, however, normally show some cross-reactions between related collagens, such as types I, II, and III.[12,14,22,60] This makes it necessary to restrict the specificity by appropriate immunoadsorption procedures.[12-15] Monoclonal antibodies[37,38] however, may exhibit already a high specificity for a single type of collagen. Types IV and V collagens are apparently more unique in their structure and usually do not share substantial antigenicity with other collagenous proteins.[41,42,63] It was also possible to distinguish between fragments of the $\alpha 1(IV)$ and $\alpha 2(IV)$ chains of type IV collagen.[65] A similar low level of cross-reaction was found for aminopropeptides of types I and III collagens.[25] So far cross-reactions between antibodies to collagens or procollagens and unrelated proteins have not been observed or could be attributed to contaminating antibodies raised by impurities in the material used for immunization. Exceptions may include proteins that possess a triple helical collagenous structure such as found in acetylcholinesterase.[66]

Quite considerable cross-reactions have been found by comparing the same types of collagens or procollagens obtained from different animal species. As a rule such cross-reactions are as strong, as most of the antigenic structures involved are independent of a particular conformation. Antibodies showing a broad interspecies cross-reaction were particularly useful in studies of antigens that cannot easily be obtained, e.g., material from human sources. Typical examples are antisera to the aminopropeptide of type III collagen, which showed almost complete cross-reaction with corresponding human and chick antigens even in radioimmunoassays.[67,68] In most cases cross-reactions will be incomplete, i.e., involve only a particular fraction of the total antibodies, but such antisera may still be useful for qualitative analyses, such as by immunofluorescence or by immunoprecipitation of labeled products. An-

---

[63] F. J. Roll, J. A. Madri, J. Albert, and H. Furthmayr, *J. Cell Biol.* **85**, 597 (1980).
[64] H. Furthmayr, M. Stoltz, U. Becker, W. Beil, and R. Timpl, *Immunochemistry* **9**, 789 (1972).
[65] J. Risteli, D. Schuppan, R. W. Glanville, and R. Timpl, *Biochem. J.* **191**, 517 (1980).
[66] L. Anglister, R. Tarrab-Hazdai, S. Fuchs, and I. Silman, *Eur. J. Biochem.* **94**, 25 (1979).
[67] H. Rohde, L. Vargas, E. Hahn, H. Kalbfleisch, M. Bruguera, and R. Timpl, *Eur. J. Clin. Invest.* **9**, 451 (1979).
[68] L. I. Fessler, R. Timpl, and J. H. Fessler, *J. Biol. Chem.* **256**, 2531 (1981).

tibodies of high species-specificity may be required in some studies, i.e., in identifying products synthesized by hybrid cells used in the chromosomal localization of collagen genes.[69,70] Since with a few exceptions[1] antisera do not cross-react with collagens or procollagens obtained from the same animal species used for immunization, an appropriate design of the immunization experiments will be helpful to obtain antisera of the required specificity.

Applications

The comprehensive immunochemical characterization of various antibodies against collagens and procollagens has provided the basis for their successful use in numerous other studies. Major applications of such reagents included the visualization of various types of collagens in tissue sections and cultured cells, the identification of metabolically labeled products by immunoprecipitation, and the development of sensitive quantitative assays for experimental and clinical studies.

*Tissue Distribution of Collagens*

The indirect immunofluorescence technique has been most widely used for localizing collagens in tissues and cells. As discussed above, the use of affinity-purified antibodies appears to be mandatory for obtaining reliable results. The specificity of antibody reagents may also be evaluated in a limited way at the histological level, e.g., using tissues containing mainly a single type of collagen, such as bone (type I), hyaline cartilage (type II), or selected basement membranes (type IV). Alternatively, fibrils prepared from purified collagens may be used as targets for antibodies, which are then visualized by the immunoperoxidase technique.[47]

*Indirect Immunofluorescent Tests on Tissues and Cells.*[40,71] Analyses are routinely done on 4–6 μm thick air-dried unfixed cryostat sections that can easily be obtained from most tissues. Certain soft tissues such as early embryos may require fixation in ethanol–acetic acid at 4° and embedding in paraffin prior to sectioning.[72] Further pretreatments that were found to improve the accessibility of collagenous antigens may include removal of minerals from bone or teeth with 0.3 $M$ EDTA[12,73,74] and of proteoglycans

[69] B. Sykes and E. Solomon, *Nature (London)* **272**, 548 (1978).
[70] C. V. Sundar Raj, R. L. Church, L. A. Klobutcher, and F. H. Ruddle, *Proc. Natl. Acad. Sci. U.S.A.* **74**, 4444 (1977).
[71] S. Gay, G. R. Martin, P. K. Müller, R. Timpl, and K. Kühn, *Proc. Natl. Acad. Sci. U.S.A.* **73**, 4037 (1976).
[72] I. Leivo, A. Vaheri, R. Timpl, and J. Wartiovaara, *Dev. Biol.* **76**, 100 (1980).
[73] S. Gay, P. K. Müller, C. Lemmen, K. Remberger, K. Matzen, and K. Kühn, *Klin. Wochenschr.* **54**, 969 (1976).
[74] I. Thesleff, S. Stenman, A. Vaheri, and R. Timpl, *Dev. Biol.* **70**, 116 (1979).

from hyaline cartilage by hyaluronidase digestion.[12] Cultured cells are grown on microscope slides or in small petri dishes and fixed by drying at room temperature or by treating with acetone at $-20°$ or with 70% ethanol at room temperature. These procedures fracture most cellular membranes, allowing the visualization of intra- and extracellular antigens. Exclusive reaction of antibodies with extracellular material may be achieved by fixing the cells with 3.5% formaldehyde[72,75] or by staining living cells. These variants require careful controls on antibody penetration and possible damage to antigenic structures.[14]

Properly exposed biological samples are incubated with 10–25 μl of antibodies against collagen or procollagen for 20 min at room temperature followed by two or three washes with an excessive volume of PBS. Negative controls include similar sections treated with unrelated antibodies or normal immunoglobulin. Bound antibodies are then visualized by treatment with antiserum against the corresponding immunoglobulin containing an appropriate fluorescing label (e.g., fluorescein, tetramethylrhodamine) followed by two or three washes with PBS. The optimal dilution of the second antibody conjugate, which can be obtained from a variety of commercial sources, is usually determined by chessboard titration. For a precise description of this and other technical details (appropriate illumination, selection of filters, double-staining technique, etc.) the reader is referred to established procedure, e.g., those reviewed by Wick et al.[76]

Optimal staining is usually achieved with antibody concentrations in the range 20–50 μg/ml. Concentrations as low as 0.5–1 μg/ml may, however, give a positive reaction.[5] Thus indirect immunofluorescence is less sensitive by one to three orders of magnitude when compared to radioimmunoassays or passive hemagglutination.[22,40,60] Density of antigens in tissues and also accessibility for antibodies contribute to the sensitivity of the assays. This possibility was illustrated by immunofluorescence staining for type III collagen, which was intense in the upper portion of skin but virtually absent in the lower portion.[16,77] Biochemical analysis[78] and studies with cultured cells.[79] however, failed to demonstrate such a gradient of production and deposition of type III collagen in skin. The characterization and standardization of antibody reagents and variations in the

---

[75] A. Vaheri, M. Kurkinen, V. P. Lehto, E. Linder, and R. Timpl, *Proc. Natl. Acad. Sci. U.S.A.* **75**, 4944 (1978).

[76] G. Wick, S. Baudner, and F. Herzog, "Immunofluorescence," 2nd ed. Med. Verlagsgesellschaft, Marburg, 1978.

[77] W. N. Meigel, S. Gay and L. Weber, *Arch. Dermatol. Res.* **259**, 11 (1977).

[78] E. H. Epstein and N. H. Munderloh, *J. Biol. Chem.* **253**, 1336 (1978).

[79] R. Fleischmajer, J. S. Perlish, T. Krieg, and R. Timpl, *J. Invest. Dermatol.* **76**, 400 (1981).

recognition of antigenic determinants[40] also determines the quality of immunofluorescence data and may explain controversial results, such as those on the presence or the absence of type III collagen in the vessel subendothelium[80,81] or in the early embryo,[72,82] and the failure to detect type I collagen in the liver.[83] All these data emphasize the qualitative nature of immunofluorescence tests even though developments in the quantitative determination of fluorescence intensity[84] may allow a more critical interpretation of staining patterns.

In spite of a number of pitfalls in the evaluation of immunofluorescence data, the technique is quite useful for localizing collagens at the anatomical level, as shown in numerous studies (Table III). It was possible to detect collagens types I, II, III, IV, and V and the aminopropeptides of types I and III collagen in a variety of tissues. Very often the antigens showed a restricted distribution and allowed, for example, to distinguish between cartilage matrix (type II collagen) and surrounding perichondrium (type I collagen)[12,60] or between interstitial connective tissue (types I and III collagens) and basement membrane zones (types IV and V collagens).[13,14,46,85] Other data, particularly when obtained in double-staining experiments with two different fluorescing markers, indicated extensive codistribution of types I and III collagens, types IV and V collagens, and between collagens and fibronectin or between type IV collagen and laminin.[13,14,63,75,86,87] Because of the limited resolution at the light microscopical level (200–500 nm), this does not necessarily indicate molecular interaction between such components. High-resolution epifluorescence as used in a study on kidney collagens[85] may be a helpful improvement. Type I procollagen showed a more restricted tissue distribution when compared with type I collagen, suggesting that it is localized to metabolically active regions of tissues.[19,40,77] The aminopropeptide of type III collagen showed a more abundant distribution, which was interpreted as indicating a major structural role in the formation of fibrils.[22] The distribution of the various collagens and procollagens was found to change considerably during embryonic development and to some extent under pathological conditions (Table III).

[80] S. Gay, L. Balleisen, K. Remberger, P. P. Fietzek, B. C. Adelmann, and K. Kühn, *Klin. Wochenschr.* **53**, 889 (1975).
[81] J. A. Madri, B. Dryer, F. A. Pitlick, and H. Furthmayr, *Lab. Invest.* **43**, 303 (1980).
[82] M. I. Sherman, R. Gay, S. Gay, and E. J. Miller, *Dev. Biol.* **74**, 470 (1980).
[83] K. Remberger, S. Gay, and P. P. Fietzek, *Virchows Arch. A: Pathol. Anat. Histol.* **367**, 231 (1975).
[84] K. Schauenstein and G. Wick, *Acta Histochem. Suppl.* **22**, 101 (1980).
[85] J. I. Scheinman, J. M. Foidart, and A. F. Michael, *Lab. Invest.* **43**, 373 (1980).
[86] H. Rohde, G. Wick, and R. Timpl, *Eur. J. Biochem.* **102**, 195 (1979).
[87] K. Alitalo, M. Kurkinen, A. Vaheri, T. Krieg, and R. Timpl, *Cell* **19**, 1053 (1980).

## TABLE III
### IMMUNOHISTOLOGICAL LOCALIZATION OF COLLAGENS AND PROCOLLAGENS IN TISSUES

| Tissue or organ | Collagens detected[a] | Technique[b] | References[c] |
|---|---|---|---|
| *A. Normal mature tissues* | | | |
| Skin | I, III, IV, pI, pIII | IIF | 1–3 |
|  | III | EM | 4 |
|  | IV | IP | 5 |
| Tendon | I, III, V | IIF | 6–8 |
| Bone | I, II | IIF | 9–12 |
| Hyaline cartilage | I, II, III | IIF | 9, 10, 12, 13 |
|  | V | IIF | 14 |
| Intervertebral disk | I, II | IIF, IP | 13, 15 |
| Blood vessels | I, III | IIF, IP | 16–18 |
|  | IV, V | IIF, IP | 18, 19 |
| Trachea | I, II, III, IV | IIF | 20 |
| Lung | IV, V | IIF | 19, 21 |
| Kidney | I, IV, pI, pIII | IIF | 19, 22, 23 |
|  | IV, V | IIF, EM | 23, 24 |
| Liver | I, III, IV, V, pI, pIII | IIF | 25–29 |
|  | I, III, IV, V | IIF, IP | 30 |
| Spleen | I, III | IIF | 6 |
| Muscle | I, III, V | IIF | 31 |
|  | pI, pIII, IV | IIF | 32 |
| Intestine | III, IV, V | IIF | 33 |
| Nerve | IV, V | IIF | 34 |
| Mummified skin | I, III | IIF | 35 |
| *B. Developing tissues* | | | |
| Early embryo, mouse | I, III, IV, V, pIII | IIF | 36, 37 |
|  | I, IV | IP | 38 |
| Early embryo, chick | I, II | IIF | 39, 40 |
| Eye, chick | I, II, pIII | IIF | 41 |
|  | V | IIF | 42 |
|  | IV | IIF | 43 |
| Bone formation | | | |
|   Mouse | I, II | IIF | 44 |
|   Chick | I, II | IIF | 12, 40 |
|   Implant | I, II, III, V | IIF | 45, 45a |
| Kidney tubulogenesis, mouse | I, IV, pIII | IIF | 46 |
| Tooth formation | I, III, IV, pIII | IIF | 47, 48 |
| Muscle formation | I, III, V | IIF | 49 |
| Amniotic membrane | I, IV, pIII | IIF | 50 |

TABLE III (continued)

| Tissue or organ | Collagens detected[a] | Technique[b] | References[c] |
|---|---|---|---|
| C. Pathological tissues | | | |
| Skin | | | |
| Scleroderma | I, III, IV, pI, pIII | IIF | 51, 52 |
| Dermatosparaxis | I, pI | IIF, EM | 53 |
| Various disorders | I, III, IV, pI, pIII | IIF | 1, 54, 55 |
| Liver fibrosis | | | |
| Human | I, III, V, pI, pIII | IIF | 25, 27, 29, 56 |
| | IV | IIF | 28 |
| | I, III, IV, V | IIF, IP | 30 |
| Rat | I, III, pIII | IIF | 57 |
| Lung fibrosis | I, II, III, IV, V, pI, pIII | IIF | 58, 58a |
| Scar + granulation tissue | I, III, pIII | IIF | 59, 60 |
| Joint + bone lesions | I, II, III | IIF | 9, 11, 61 |
| Artherosclerotic plaques | I, II, III | IIF | 16, 33 |

[a] Collagens types I, II, III, IV, or V and procollagens (pN-collagens) types I (pI) and III (pIII).

[b] IIF, indirect immunofluorescence; IP, immunoperoxidase reaction (light and/or electron microscope); EM, electron microscopy after ferritin labeling.

[c] *Key to references:* G. Wick, D. Kraft, E. M. Kokoschka, and R. Timpl, *J. Clin. Immunol. Immunpathol.* **6**, 182 (1976); (2) W. Meigel, S. Gay, and L. Weber, *Arch. Derm. Res.* **259**, 1 (1977); (3) R. Timpl, G. R. Martin, P. Bruckner, G. Wick, and H. Wiedemann, *Eur. J. Biochem.* **84**, 43 (1978); (4) R. Fleischmajer, S. Gay, J. S. Perlish, and J. P. Cesarini, *J. Invest. Dermatol.* **75**, 189 (1980); (5) H. Yaoita, J. M. Foidart, and S. I. Katz, *J. Invest. Dermatol.* **70**, 191 (1978); (6) U. Becker, H. Nowack, S. Gay, and R. Timpl, *Immunology* **31**, 57 (1976); (7) V. C. Duance, D. J. Restall, H. Beard, F. J. Bourne, and A. J. Bailey, *FEBS Lett.* **79**, 248 (1977); (8) H. Herrmann, W. Dessau, L. I. Fessler, and K. von der Mark, *Eur. J. Biochem.* **105**, 63 (1980); (9) S. Gay, P. K. Müller, C. Lemmen, K. Remberger, K. Matzen, and K. Kühn, *Klin. Wochenschr.* **54**, 969 (1976); (10) K. von der Mark, and H. von der Mark, *J. Bone Joint Surg.* **59**, 458 (1977); (11) K. Remberger and S. Gay, *Z. Krebsforschg.* **90**, 95 (1977); (12) K. von der Mark, H. von der Mark, and S. Gay, *Dev. Biol.* **53**, 153 (1976); (13) G. Wick, H. Nowack, E. Hahn, R. Timpl, and E. J. Miller, *J. Immunol.* **117**, 298 (1976); (14) S. Gay, R. K. Rhodes, R. E. Gay, and E. J. Miller, *Collagen Rel. Res.* **1**, 53 (1981); (15) H. K. Beard, R. Ryvar, R. Brown, and H. Muir, *Immunology* **41**, 491 (1980); (16) S. Gay, L. Balleisen, K. Remberger, P. P. Fietzek, B. C. Adelmann, and K. Kühn, *Klin. Wochenschr.* **53**, 889 (1975); (17) S. Gay, P. Walter, and K. Kühn, *Klin. Wochenschr.* **54**, 889 (1976); (18) J. A. Madri, B. Dreyer, F. A. Pitlick, and H. Furthmayr, *Lab. Invest.* **43**, 303 (1980); (19) G. Wick, R. W. Glanville, and R. Timpl, *Immunobiology,* **156**, 372 (1979); (20) R. Timpl, K. von der Mark, and H. von der Mark, in "Biology of Collagen" (A. Viidik and J. Vuust, eds.), p. 211. Academic Press, New York, 1980; (21) J. A. Madri and

(*continued*)

H. Furthmayr, *Am. J. Pathol.* **94**, 323 (1979); (22) G. Wick, H. Furthmayr, and R. Timpl, *Int. Arch. Allergy Appl. Immunol.* **48**, 664 (1975); (23) J. I. Scheinman, J. M. Foidart, and A. F. Michael, *Lab. Invest.* **43**, 373 (1980); (24) F. J. Roll, J. A. Madri, J. Albert, and H. Furthmayr, *J. Cell Biol.* **85**, 597 (1980); (25) U. Ott, E. Hahn, E. Moshudis, J. C. Bode, and G. A. Martini, *Verh. Dtsch. Ges. Inn. Med.* **83**, 537 (1977); (26) H. Nowack, S. Gay, G. Wick, U. Becker, and R. Timpl, *J. Immunol. Methods* **12**, 117 (1976); (27) G. Wick, H. Brunner, E. Penner, and R. Timpl, *Int. Arch. Allergy Appl. Immunol.* **56**, 316 (1978); (28) E. Hahn, G. Wick, D. Pencev, and R. Timpl, *Gut* **21**, 63 (1980); (29) L. Biempica, R. Morecki, C. H. Wu, M.-A. Giambrone, and M. Rojkind, *Am. J. Pathol.* **98**, 591 (1980); (30) J. A. Grimaud, M. Druguet, S. Peyrol, O. Chevalier, D. Herbage, and N. el Badrawy, *J. Histochem. Cytochem.* **28**, 1145 (1980); (31) A. J. Bailey, G. B. Shellswell, and V. C. Duance, *Nature (London)* **278**, 67 (1979); (32) T. Krieg, R. Timpl, K. Alitalo, M. Kurkinen, and A. Vaheri, *FEBS Lett.* **104**, 405 (1979); (33) S. Gay and E. J. Miller, "Collagen in the Physiology and Pathology of Connective Tissue." Fischer, Stuttgart, 1978; (34) G. B. Shellswell, D. J. Restall, V. C. Duance, and A. J. Bailey, *FEBS Lett.* **106**, 305 (1979); (35) G. Wick, M. Haller, R. Timpl, H. Cleve, and G. Ziegelmayer, *Int. Arch. Allergy Appl. Immunol.* **62**, 76 (1980); (36) M. I. Sherman, R. Gay, S. Gay, and E. J. Miller, *Dev. Biol.* **74**, 470 (1980); (37) I. Leivo, A. Vaheri, R. Timpl, and J. Wartiovaara, *Dev. Biol.* **76**, 100 (1980); (38) E. D. Adamson and S. E. Ayers, *Cell* **16**, 953 (1979); (39) K. von der Mark, H. von der Mark, and S. Gay, *Dev. Biol.* **48**, 237 (1976); (40) W. Dessau, H. von der Mark, K. von der Mark, and S. Fischer, *J. Embryol. Exp. Morphol.* **57**, 51 (1980); (41) K. von der Mark, H. von der Mark, R. Timpl, and R. L. Trelstad, *Dev. Biol.* **59**, 75 (1977); (42) A. Pöschl and K. von der Mark, *FEBS Lett.* **115**, 100 (1980); (43) E. D. Hay, T. F. Linsenmayer, R. L. Trelstad, and K. von der Mark, *Curr. Top. Eye Res.* **1**, 1 (1979); (44) H. J. Barrach and K. Angermann, in "Methods in Prenatal Toxicology" (D. Neubert, H. J. Merker, and E. Kwasigrock, eds.), p. 332. Thieme, Stuttgart, 1977; (45) A. H. Reddi, R. Gay, S. Gay, and E. J. Miller, *Proc. Natl. Acad. Sci. U.S.A.* **74**, 5589 (1977); (45a) J. M. Foidart, and A. H. Reddi, *Dev. Biol.* **75**, 130 (1980); (46) P. Ekblom, E. Lehtonen, L. Saxén, and R. Timpl, *J. Cell. Biol.* **89**, 276 (1981); (47) H. Lesot, K. von der Mark, and J. V. Ruch, *C. R. Hebd. Seances Acad. Sci., Ser. D* **286**, 765 (1978); (48) I. Thesleff, S. Stenman, A. Vaheri, and R. Timpl, *Dev. Biol.* **70**, 116 (1979); (49) G. B. Shellswell, A. J. Bailey, V. C. Duance, and D. J. Restall, *J. Embryol. Exp. Morphol.* **60**, 245 (1980); (50) K. Alitalo, M. Kurkinen, A. Vaheri, T. Krieg, and R. Timpl, *Cell* **19**, 1053 (1980); (51) R. Fleischmajer, S. Gay, W. N. Meigel, and J. S. Perlish, *Arthritis Rheum.* **21**, 418 (1978); (52) R. Fleischmajer, W. Dessau, R. Timpl, T. Krieg, C. Luderschmidt, and M. Wiestner, *J. Invest. Dermatol.* **75**, 270 (1980); (53) G. Wick, B. R. Olsen, and R. Timpl, *Lab. Invest.* **39**, 151 (1978); (54) S. Gay, T. T. Kresina, R. Gay, E. J. Miller, and L. F. Montes, *J. Cutaneous Pathol.* **6**, 91 (1979); (55) G. Wick, H. Hönigsmann, and R. Timpl, *J. Invest. Dermatol.* **73**, 335 (1979); (56) S. Gay, P. P. Fietzek, K. Remberger, M. Eder, and K. Kühn, *Klin. Wochenschr.* **53**, 205 (1975); (57) G. Kent, S. Gay, T. Inonye, R. Bahu, O. T. Minick, and H. Popper, *Proc. Natl. Acad. Sci. U.S.A.* **73**, 3719 (1976); (58) G. Wick, R. Timpl, and R. G. Crystal, in preparation; (58a) J. A. Madri and H. Furthmayr, *Human Pathol.* **11**, 353 (1980); (59) S. Gay, J. Viljanto, R. Penttinen, and J. Raekallio, *Acta Chir. Scand.* **144**, 205 (1978); (60) M. Kurkinen, A. Vaheri, P. J. Roberts, and S. Stenman, *Lab. Invest.* **43**, 47 (1980); (61) S. Gay and K. Remberger, *Verh. Dtsch. Ges. Pathol.* **60**, 290 (1976).

Various types of collagens or procollagens could also be detected by immunofluorescent tests in the pericellular matrix and/or the intracellular compartment of cultured fibroblasts, chondrocytes, muscle cells, endothelial and epithelial cells, and tumor cells.[71,75,81,87-96] Extracellular material often showed fibrillar patterns, whereas intracellular material usually appeared as granular deposits around the nucleus. Since this approach allows the identification of single cells, it was possible to demonstrate simultaneous production of types I and III procollagens by individual fibroblasts[71,75,92] and of types I and II collagens by certain cultivated chondrocytes.[89] Decreased production and secretion of collagenous proteins by transformed cells or after chemical treatment could be visualized by loss of extracellular material while intracellular staining appeared to be unimpaired.[75,96] Immunoperoxidase staining of endothelial cells showed type V collagen close to the cell membrane and between cell junctions, indicating that this protein resembles a "membranous" form of collagen.[81]

The ultrastructural localization of collagenous proteins has been attempted in only a few studies. This approach is feasible, as shown in Fig. 2 for type III collagen fibrils.[97] Other studies have demonstrated aminopropeptides in the close vicinity of dermatosparactic collagen fibrils,[98] types IV and V collagen being restricted to certain regions of renal[63] and epidermal basement membranes[99] and procollagen antigens within the rough endoplasmic reticulum of cultured cells.[100,101] Further examples are listed in Table III. Staining of collagen fibrils frequently showed a periodicity resembling the quarter-stagger of the molecules,[47,97] indicating

[88] P. Bornstein and J. F. Ash, *Proc. Natl. Acad. Sci. U.S.A.* **74**, 2480 (1977).
[89] K. von der Mark, V. Gauss, H. von der Mark, and P. Müller, *Nature (London)* **267**, 531 (1977).
[90] G. W. Conrad, W. Dessau, and K. von der Mark, *J. Cell Biol.* **84**, 501 (1980).
[91] W. Dessau, J. Sasse, R. Timpl, F. Jilek, and K. von der Mark, *J. Cell Biol.* **79**, 342 (1978).
[92] H. Herrmann, W. Dessau, L. I. Fessler, and K. von der Mark, *Eur. J. Biochem.* **105**, 63 (1980).
[93] K. Alitalo, A. Vaheri, T. Krieg, and R. Timpl, *Eur. J. Biochem.* **109**, 247 (1980).
[94] J. M. Foidart, J. J. Berman, L. Paglia, S. Rennard, S. Abe, A. Perantoni, and G. R. Martin, *Lab. Invest.* **42**, 525 (1980).
[95] J. Sasse, H. von der Mark, U. Kühl, W. Dessau, and K. von der Mark, *Dev. Biol.* **83**, 79 (1981).
[96] T. Krieg, M. Aumailley, W. Dessau, M. Wiestner, and P. Müller, *Exp. Cell Res.* **125**, 23 (1980).
[97] R. Fleischmajer, S. Gay, J. S. Perlish, and J. P. Cesarini, *J. Invest. Dermatol.* **75**, 189 (1980).
[98] G. Wick, B. R. Olsen, and R. Timpl, *Lab. Invest.* **39**, 151 (1978).
[99] H. Yaoita, J. M. Foidart, and S. I. Katz, *J. Invest. Dermatol.* **70**, 191 (1978).
[100] B. R. Olsen and D. J. Prockop, *Proc. Natl. Acad. Sci. U.S.A.* **71**, 2033 (1974).
[101] C. Nist, K. von der Mark, E. D. Hay, B. R. Olsen, P. Bornstein, R. Ross, and P. Dehm, *J. Cell Biol.* **65**, 75 (1975).

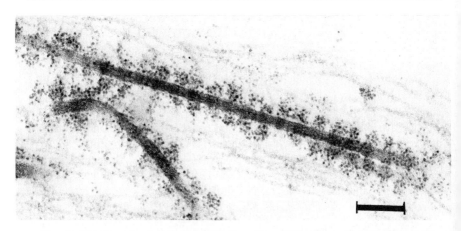

FIG. 2. Electron microscopical visualization of type III collagen fibrils in human skin. The specimen was treated with rabbit antibodies to type III collagen followed by a ferritin conjugate of anti-rabbit immunoglobulin and was poststained with uranyl acetate for visualizing cross-striated collagen. Note the heavy decoration of the two thick fibrils by ferritin molecules (small dots) with some regular spacing. The bar indicates a length of 100 nm. Taken from Fleischmajer et al.,[97] with permission.

that only a restricted number of antigenic determinants are involved in the reaction. The antibodies used in these studies were either directly conjugated with ferritin[98,100,101] or peroxidase[14] or were used in a double-antibody assay or similar variants.[63,97,100] A number of technical problems, such as appropriate penetration of antibody, exposure of antigenic determinants, and high levels of unspecific staining, may determine the quality and reliability of immunoelectron microscopic studies. Some of these difficulties were overcome by staining ultrathin sections[63] or by exposing antibodies over a long period of time to larger tissue fragments or cells that were partially disrupted to enhance penetration and prefixed to maintain the morphological structure.[98,100] Double-staining experiments have not yet been carried out at the ultrastructural level but may become feasible by using antibodies adsorbed to colloidal gold particles of different diameter.[102]

*Use of Antibodies in Biosynthetic Studies*

The selective precipitation by antibodies of metabolically labeled proteins from complex culture media or cell extracts has become a convenient method for identifying newly synthesized collagens and procollagens. The precipitated material is usually dissolved under reducing condi-

[102] M. Horisberger and J. Rosset, *J. Histochem. Cytochem.* **25**, 295 (1977).

tions and subjected to gel electrophoresis, which is followed by fluorography[103] of the gel in order to visualize the radioactive polypeptide bands. If the antibodies are sufficiently specific almost complete separation of collagen types could be achieved, as shown for type III and IV procollagens in Fig. 3. This method has been successfully applied in several studies on types I, II, III, IV, and V collagens[68,81,87,91,92,95,104,105] using both antibodies to collagens and to aminopropeptides. It was possible, for example, to determine the order of release of procollagen peptides from type III procollagen,[68] to identify two constituent chains in type IV procollagen,[93,106-108] and to isolate precursor material in a single step by using antibody adsorbents.[68] Since the amounts of antigen in the biological samples are usually below 1 $\mu$g/ml, sufficient precipitation requires addition of a second antibody to immunoglobulin or of staphylococcal protein A, which are often used in insolubilized form, e.g., after coupling to agarose. The addition of nonionic detergents (e.g., 0.1% Triton X-100) is recommended in order to reduce background binding.

The reaction of antibodies has been also exploited for the quantitative analysis of biosynthetic products.[101,105,108,109] Here, constant aliquots of the sample are incubated with increasing amounts of antibody until further additions will not increase the amount of radioactivity in the precipitate. The failure to observe a plateau region in the antibody excess zone normally indicates heterogeneity of antigen or a poor quality of the antibody. Background binding, which is determined in the presence of normal immunoglobulin, may be as high as 20%[81]; this will seriously interfere with the precise determination of minor components. The relative proportions of major components, such as types I and II collagens produced by cultured chondrocytes,[109] may be determined with sufficient accuracy.

Antibodies were also applied to enrich polysomes producing type I collagen[110] and may become useful for identifying products obtained in cell-free translation systems by immunoprecipitation. Even though the latter approach has not yet been attempted, it can be predicted that antibodies against conformation-independent antigenic determinants in the N-terminal regions of procollagens and collagens may be particularly important for this purpose.

[103] W. M. Bonner and R. A. Laskey, *Eur. J. Biochem.* **46**, 83 (1974).
[104] T. Krieg, R. Timpl, K. Alitalo, M. Kurkinen, and A. Vaheri, *FEBS Lett.* **104**, 405 (1979).
[105] H. Sage, E. Crouch, and P. Bornstein, *Biochemistry* **18**, 5433 (1979).
[106] I. Leivo, K. Alitalo, L. Risteli, A. Vaheri, R. Timpl, and J. Wartiovaara, *Exp. Cell. Res.* in press (1981).
[107] K. Tryggvason, P. Gehron Robey, and G. R. Martin, *Biochemistry* **19**, 1284 (1980).
[108] E. Crouch, H. Sage, and P. Bornstein, *Proc. Natl. Acad. Sci. U.S.A.* **77**, 745 (1980).
[109] K. von der Mark and H. von der Mark, *J. Cell Biol.* **73**, 736 (1977).
[110] P. J. Pawlowski, M. T. Gilette, J. Martinell, L. N. Lukens, and H. Furthmayr, *J. Biol. Chem.* **250**, 2135 (1974).

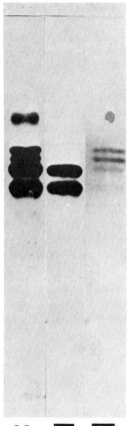

FIG. 3. Identification of type III procollagen (III) and type IV collagen (IV) in the culture medium (M) of rhabdomysarcoma cells by immunoprecipitation with specific antibodies. Proteins in the medium were labeled with [$^3$H]proline and concentrated by precipitation with 40% saturated ammonium sulfate. Immunoprecipitates were dissolved in the presence of 2-mercaptoethanol, subjected to sodium dodecyl sulfate slab gel electrophoresis followed by the fluorographic visualization of radioactive polypeptide bands. Antibodies to the aminopropeptide of type III collagen precipitated the intact [pro-α1 (III)] and a truncated [pN-α1 (III)] form of the chains of type III procollagen. Antibodies against type IV collagen precipitated both the α1(IV) (upper band) and α2(IV) chain (lower band) and minute amounts of smaller fragments. The slowest migrating band of the medium proteins could be identified as fibronectin and was not precipitated by any of the anti-collagen antibodies. Reproduced from Krieg et al.,[104] with permission.

## Quantitative Determination of Antigens

Radioimmunoassays[26] and enzyme immunoassays[50] in the variant of an inhibition test were the major methods used for the immunochemical quantitation of collagenous proteins and procollagen peptides. Both assays are apparently of similar sensitivity, although they have not yet been strictly compared in the analysis of connective tissue proteins. Optimized radioimmunoassays allowed the determination of procollagen peptides[10,67,111] and of 7 S collagen[112] which comprises a cross-linking domain of type IV collagen in the range of 0.1–10 ng ($10^{-13}$ to $10^{-15}$ mol). Assays for other collagenous proteins were 10- to 100-fold less sensitive, presumably owing to a relatively low affinity of antibodies.

Soluble forms of aminopropeptides of type I[10] and type III[67] collagen and of the carboxypropeptide of type I collagen[111] could be detected in serum, urine, ascites and amniotic fluid. The concentrations in normal human serum were in the range 2–12 ng/ml (aminopropeptide type III) or 50–150 ng/ml (carboxypropeptide type I) and were distinctly increased in certain liver diseases associated with fibrosis[67] or in bone disease.[113] The urinary form of the aminopropeptide of type III collagen was smaller than that found in serum, indicating further processing of this structure after release from procollagen. Studies have also demonstrated circulating forms of 7 S collagen antigen that changed in concentration during aging and during tumor growth.[112] The whole range of diagnostic and prognostic applications of such assays and their value for studying experimental models and biological development remain to be established.

Enzyme immunoassays[50] and radioimmunoassays[87,104,111] were applied to study collagen synthesis by cultured fibroblasts or other cells. When compared to normal fibroblasts, increased production of type III procollagen was found for cells from scleroderma patients[79] whereas synthesis was 10-fold decreased in a patient with Ehlers–Danlos syndrome type IV.[114] Production of both types I and III procollagens was augmented in cells obtained from dermatosparactic sheep. Here, the separation of procollagen from procollagen peptides by precipitation allowed one to measure the conversion of procollagen into collagen in a quantitative manner and hence offered an immunological assay for determining the activity of procollagen N-protease.[115] A potential further application would be the

[111] M. B. Taubman, B. Goldberg, and C. J. Sherr, *Science* **186**, 1115, (1974).
[112] J. Risteli, H. Rohde, and R. Timpl, *Anal. Biochem.* **113**, 372 (1981).
[113] M. B. Taubman, S. Kammerman and B. Goldberg, *Proc. Soc. Exp. Biol. Med.* **152**, 284 (1976).
[114] M. Aumailley, T. Krieg, W. Dessau, P. K. Müller, R. Timpl, and M. Bricaud, *Arch. Dermatol. Res.* **269**, 169 (1980).
[115] M. Wiestner, H. Rohde, O. Helle, T. Krieg, R. Timpl, and P. K. Müller, *in* "New Trends in Basement Membrane Research" (K. Kühn, R. Timpl, and H. H. Schöne, eds.). Raven, New York, 1982, in press.

development of quantitative assays suitable for the analysis of small tissue samples. This would require complete solubilization of the material under conditions preserving antigenic structure and thus selection of assays for particular antigenic determinants. This approach seems feasible as shown for 7 S collagen in kidneys.[112]

*Other Applications*

Several previous studies have shown that antibodies to collagens interfere with fibril formation, are cytotoxic for fibroblasts,[116] and can detect membrane-bound forms of collagen.[117] Most of these studies (see review[1]) were done with antisera not characterized with respect to their specificity for distinct types of collagen. Antibodies have also been used to isolate[48] or identify[118] cross-linked peptides obtained from insoluble collagen or to monitor purification of aminopropeptides.[119] Coating of collagen fibrils by antibodies blocked aggregation of platelets,[47] demonstrating that various types of collagen possess this activity. It was also reported that antisera specifically enhanced uptake of collagen by macrophages.[120]

Serological methods developed for the characterization of experimentally produced antibodies were used to detect autoantibodies against collagens in several human diseases. Such studies included the analysis of antibodies against denatured collagens in rheumatoid arthritis[54,121] and in chronic liver disease[122] and of antibodies against type II collagen in polychondritis.[123] Other radioimmunoassay studies failed to demonstrate significant reactions with type III collagen and procollagen in the serum of patients with coeliac disease[124] and with type IV collagen in patients with Goodpasture syndrome.[125]

---

[116] D. Duksin, A. Maoz, and S. Fuchs, *Cell* **5**, 83 (1975).
[117] W. P. Faulk, L. B. Conochie, A. Temple, and M. Papamichail, *Nature (London)* **256**, 123 (1975).
[118] U. Becker, H. Furthmayr, and R. Timpl, *Hoppe-Seyler's Z. Physiol. Chem.* **356**, 21 (1975).
[119] H. Nowack, H. Rohde, and R. Timpl, *Hoppe-Seyler's Z. Physiol. Chem.* **357**, 601 (1976).
[120] K. E. Hopper, B. C. Adelmann, G. Gentner, and S. Gay, *Immunology* **30**, 249 (1976).
[121] N. A. Adriopolous, J. Mestecky, G. P. Wright, and E. J. Miller, *Immunochemistry* **13**, 709 (1976).
[122] E. J. Menzel, J. S. Smolen, F. Renner, C. Steffen, and W. Horak, *Int. Arch. Allergy Appl. Immunol.* **63**, 424 (1980).
[123] J. M. Foidart, A. Shigeto, G. R. Martin, T. M. Zizic, E. V. Barnett, T. J. Lawley, and S. I. Katz, *N. Engl. J. Med.* **299**, 1203 (1978).
[124] R. Timpl, G. Wick, and G. Granditsch, *Clin. Exp. Immunol.* **28**, 546 (1977).
[125] G. Wick and R. Timpl, *Clin. Exp. Immunol.* **39**, 733 (1980).

## [23] Cell–Collagen Interactions: Overview

By FREDERICK GRINNELL

Collagen is a ubiquitous component of the connective tissue, and it forms the essential framework of the tissues and organs. Many cells are found resting on a collagen-containing basal lamina (e.g., muscle, nerve, epithelial, and endothelial cells) or within a collagenous matrix (e.g., many connective tissue cells). Cell–collagen interactions are an essential feature of cell movements during development in the embryo and under a variety of circumstances in the adult. The latter may occur as a result of normal processes (e.g., inflammation, wound healing, trophoblast implantation) or as a result of malignancy.

The potential usefulness of studying cell–collagen interactions is to develop simulated tissue and organ structures that can be used to study cell behavior in a well defined *in vitro* system.[1] Steps toward this goal have been made with the development of techniques that permit cell growth and differentiation within collagen matrices.[2,3] A similar model system has also been used to study tumor cell invasiveness.[4]

Despite the importance of cell–collagen interactions, the majority of studies on cell behavior *in vitro* have been carried out with cells cultured on material surfaces (e.g., plastic or glass). Nevertheless, there have been some studies on different aspects of cell–collagen interactions. These are briefly reviewed in this overview, and many of the relevant techniques are described in detail in subsequent chapters of this volume.

### Cell–Collagen Interactions in Cell Growth and Differentiation

Systematic investigations on the use of collagen gels for supporting cell growth were reported more than 20 years ago.[5] A variety of cell strains and tissue explants of both normal and transformed cells were tested, and in most cases cell growth on hydrated collagen gels was found to be better than growth on glass substrata. Once the usefulness of collagen gels for improving cell growth was established, the ability of collagen to promote cell development *in vitro* was also analyzed. Development of

---

[1] T. Elsdale and J. Bard, *J. Cell Biol.* **54**, 626 (1972).
[2] J. Yang, J. Richards, P. Bowman, R. Guzman, J. Enami, K. McCormick, S. Hamamoto, D. Pitelka, and S. Nandi, *Proc. Natl. Acad. Sci. U.S.A.* **76**, 3401 (1979).
[3] E. Bell, B. Ivarsson, and C. Merrill, *Proc. Natl. Acad. Sci. U.S.A.* **76**, 1274 (1979).
[4] S. L. Schor, *J. Cell Sci.* **41**, 159 (1980).
[5] R. L. Ehrmann and G. O. Gay, *J. Natl. Cancer Inst.* **16**, 1375 (1956).

embryonic epidermis[6] and muscle[7] was found to occur on hydrated collagen gels, but not on glass substrata. It was suggested that collagen might play a permissive[8] role or even a general inductive role[9] in epithelial morphogenesis.

Maintenance of the differentiated function of hepatocytes[10] and mammary epithelium[11] in cell culture has also been shown to require the use of collagen gels. Otherwise, the cells rapidly degenerate. Similarly, responsiveness of epithelial[12] and endothelial[13] cells to some mitogenic factors occurs when the cells are cultured on collagen but not on glass. Differences have also been noted in the extracellular matrix components secreted by cells cultured on collagen gels compared to cells cultured on some other substrata.[14,15]

### Adhesion and Motility of Cells on Collagen Substrata

The adhesion of cells on collagen as on other substrata is exceedingly complex and involves both cell surface and cytoskeletal components. Fibroblasts, epithelial cells, endothelial cells, macrophages, monocytes, and neutrophils have all been shown to attach to substrata and undergo characteristic shape changes. Platelets not only attach to substrata and change shape, but also, they then release the contents of cytoplasmic vesicles and form platelet–platelet aggregates. Lymphocytes also attach to substrata but usually more tenuously than other cells. A complete review on cellular adhesiveness and extracellular substrata has been published elsewhere.[16] Several of the different adhesion parameters that have been useful in analyzing cell interactions with collagen as well as with material surfaces are described below.

*Initial Cell Attachment*

It is possible to measure the percentage of total cells that attach to the substratum as a function of incubation time. The final extent of attachment may vary considerably depending upon the kind of cells and their physio-

---

[6] J. W. Dodson, *Exp. Cell. Res.* **31**, 233 (1963).
[7] S. D. Hauschka and I. R. Konigsberg, *Proc. Natl. Acad. Sci. U.S.A.* **55**, 119 (1966).
[8] N. K. Wessells and J. H. Cohen, *Dev. Biol.* **18**, 294 (1968).
[9] C. Grobstein, *Natl. Cancer Inst. Mongr.* **26**, 279 (1967).
[10] G. Michalopoulos, F. Russell, and C. Biles, *In Vitro* **15**, 796 (1979).
[11] J. T. Emerman, S. J. Burwen, and D. R. Pitelka, *Tissue Cell* **11**, 109 (1979).
[12] D. Gospodarowicz, G. Greenburg, and C. R. Birdwell, *Cancer Res.* **38**, 4155 (1978).
[13] A. M. Schor, S. L. Schor, and S. Kumar, *Int. J. Cancer* **24**, 225 (1979).
[14] S. Meier and E. D. Hay, *Dev. Biol.* **38**, 249 (1974).
[15] G. David and M. R. Bernfield, *Proc. Natl. Acad. Sci. U.S.A.* **76**, 786 (1979).
[16] F. Grinnell, *Int. Rev. Cytol.* **53**, 65 (1978).

logical condition. Passive cell attachment (without spreading) can sometimes occur and is generally thought to be a nonphysiological cell-substratum adsorption phenomenon.[16] Therefore, in any given system, it is worth while to demonstrate the dependence of cell attachment on metabolic energy. To inhibit active attachment, it is usually necessary to inhibit glycolysis (e.g., with 2-deoxyglucose) as well as electron transport. Inhibitors of electron transport alone are generally less effective, especially in the presence of glucose in the incubation medium.[17] Sulfhydryl binding reagents are also general inhibitors of active cell adhesion.[16,18]

*Cell Spreading (Morphology)*

Cells initially attach in a rounded morphology, but can then change shape. Reorganization of the cytoskeleton occurs and leads to the protrusion of cell extensions. Subsequently, the cell cytoplasm becomes redistributed along and in between the extensions. The final morphology of spread cells varies depending upon the cell type, the substratum, and the incubation conditions. In general, cell spreading is inhibited by low temperature, absence of divalent cations, or cytochalasins, but not by colchicine. The influence of the physical organization of the substratum on cell morphology is discussed below.

*Cell Motility*

Once the initial sequence of attachment and spreading is complete, fibroblastic cells often become polarized. On material surfaces (e.g., glass or plastic) there is usually a major ruffling lamellipodium, and cell motility involves the advance of this lamellar region of spread cytoplasm over the substratum, which is accompanied by retraction of the cell's trailing end.[19] The motility of fibroblasts on and within hydrated collagen gels is quite different from that which occurs on material surface. On hydrated collagen, cells have large, pseudopodial-like processes that extend into the collagen matrix, and motility involves cytoplasmic flow into these leading protrusions.[20,21] This behavior is more characteristic of what is observed *in vivo*. Significantly, fibroblasts can become aligned and move along the collagen fibrils of hydrated collagen gels, but after the fibrils are collapsed by drying, cells move as though they were on a material surface.[22]

[17] R. J. Klebe, *J. Cell Physiol.* **86**, 231 (1975).
[18] F. Grinnell, M. Milam, and P. A. Srere, *Arch. Biochem. Biophys.* **153**, 193 (1972).
[19] J. P. Trinkaus, in "The Cell Surface in Animal Embryogenesis and Development" (G. Poste and G. L. Nicolson, eds.), p. 225. Elsevier, Amsterdam, 1976.
[20] J. B. L. Bard and E. D. Hay, *J. Cell Biol.* **67**, 400 (1975).
[21] E. M. Davis, *J. Embryol. Exp. Morphol.* **55**, 17 (1980).
[22] G. A. Dunn and T. Ebendal, *Exp. Cell Res.* **111**, 475 (1978).

*Cell Separation*

One other way to study cell adhesion is to determine the conditions under which cells can be separated from the substratum. The removal of cells from the substratum to which they are attached is probably not a reversal of cell adhesion per se. Rather, it is likely to involve physical shearing of attenuated portions of the cells from the substratum.[23] The reagents most frequently used for cell separation (e.g., trypsin, EDTA, lidocaine) probably have their primary effect by disrupting the cell cytoskeleton, thereby causing cell rounding.[16]

Influence of Collagen Type on Cell–Collagen Interactions

As discussed in detail in Chapter [2] of this volume, there are different collagens with distinct chemical compositions that also vary with regard to their anatomical locations. Therefore, one might expect that certain cells would have preferences for certain types of collagen. Only a few experiments along these lines have been carried out. Certain epithelial cells[24] and transformed skin fibroblasts[25] have been shown to prefer type IV collagen, and chondrocytes have been shown to prefer type II collagen.[26]

Influence of Collagen Organization on Cell–Collagen Interactions

The ability of cells to change shape in response to the organization of the matrix in which they reside is a well characterized phenomenon.[27] This is a point that may be critical in order to understand cell interactions with collagen. The significance of the hydrated collagen substrate may reside in the ability of cells to penetrate into the collagen matrix.[28] Most of the studies on cell growth and differentiation described above were carried out with collagen gels that are composed of a lattice of fine collagen fibrils.[1] In addition, the studies on cell morphology and cell motility—in which cell behavior resembling that which occurs *in vivo* were observed—were carried out with collagen gels. On the other hand, many cell adhesion experiments have been carried out with dried collagen sub-

---

[23] L. Weiss, "The Cell Periphery, Metastasis and Other Contact Phenomena." Wiley, New York, 1967.
[24] M. S. Wicha, L. A. Liotta, S. Garbisa, and W. R. Kidwell, *Exp. Cell Res.* **124**, 81 (1979).
[25] J. C. Murray, L. Liotta, S. I. Rennard, and G. R. Martin, *Cancer Res.* **40**, 347 (1980).
[26] A. T. Hewitt, H. K. Kleinman, J. P. Pennypacker, and G. R. Martin, *Proc. Natl. Acad. Sci. U.S.A.* **77**, 385 (1980).
[27] P. Weiss, *Int. Res. Cytol.* **7**, 391 (1958).
[28] F. Grinnell and M. H. Bennett, *J. Cell Sci.* **48**, 19 (1981).

strata, but on these substrata cells spread with lamellipodia similarly as observed on material surfaces. These points are discussed more thoroughly in Chapter [28].

Requirement for Adhesion Proteins in Cell–Collagen Interactions

Finally, the interaction of cells with collagen may be mediated by specific adhesion proteins. For instance, on dried collagen substrata, adhesion of fibroblasts requires fibronectin[29,30] and adhesion of chondrocytes requires chondronectin.[26] In some cases the addition of an adhesion protein may not be required for cell adhesion to dried collagen if the cells can synthesize enough of their own factor. For instance, human fibroblasts can bind to dried collagen in protein-free medium by secreting endogenously synthesized fibronectin.[31] Although an absolute specificity for fibronectin has been reported for fibroblasts to attach on dried collagen substrata, these cells can interact with hydrated collagen gels by a direct mechanism that does not require fibronectin.[31,32] This observation emphasizes the possible importance of the interplay between the physical organization of the substratum and other factors.

[29] H. K. Kleinman, E. B. McGoodwin and R. J. Klebe, *Biochem. Biophys. Res. Commun.* **72**, 426 (1976).
[30] E. Pearlstein, *Int. J. Cancer* **22**, 32 (1978).
[31] F. Grinnell and D. Minter, *Proc. Natl. Acad. Sci. U.S.A.* **75**, 4408 (1978).
[32] S. L. Schor and J. Court, *J. Cell Sci.* **38**, 267 (1979).

## [24] Fibroblast Adhesion to Collagen Substrates

*By* HYNDA K. KLEINMAN

It has been shown that fibroblasts utilize fibronectin to attach to plastic and to collagen substrates.[1–3] Presumably because of their unique adsorptive properties, the plastic dishes used for tissue culture bind fibronectin or other attachment proteins that are in serum or are produced by the cells. Collagen has a specific binding site for fibronectin.[4,5] It has been

[1] R. J. Klebe, *Nature (London)* **250**, 248 (1974).
[2] E. Pearlstein, *Nature (London)* **262**, 497 (1976).
[3] F. Grinnell, *Int. Rev. Cytol.* **53**, 65 (1978).
[4] W. Dessau, B. C. Adelmann, R. Timpl, and G. R. Martin, *Biochem. J.* **169**, 55 (1978).
[5] H. K. Kleinman, E. B. McGoodwin, G. R. Martin, R. J. Klebe, P. P. Fietzek, and D. W. Woolley, *J. Biol. Chem.* **253**, 5642 (1978).

shown that fibronectin binds to the substrate and then is able to attach to the surface of the cells.[1] Sustained attachment requires divalent cations and cellular metabolism.[6] There is evidence that even on a substrate of tissue culture plastic fibroblasts must synthesize and deposit a collagenous substratum to survive.[7] Besides promoting the adhesion of cells, collagen substrates also influence their growth,[8] differentiation,[9] and migration.[10] These substrates can be prepared as a film dried onto the culture dish or as a hydrated gel that either adheres to the dish or floats in the culture medium.[8,11-13] In addition to purified collagen, less defined collagen-containing substrates have been successfully used for cell attachment and long-term culture of cells. These include extracts of tissues,[14] devitalized tissues such as freeze-thawed lens[15] or bone powders,[16] and the extracellular matrix deposited on the surface of the culture vessel by cultured cells.[17,18] Described herein are various ways in which collagenous substrates can be prepared and their use in adhesion assays.

Preparation of Collagen Films

Purified and lyophilized collagen is dissolved in 0.5 $N$ acetic acid at 1–2 mg/ml by stirring in the cold overnight. Commercially available collagen solutions, such as "vitrogen" (Collagen Corporation, Palo Alto, California,), are satisfactory. The surface of the dish is coated by adding a small amount of the solution to cover the surface and removing the excess. It is equally effective to add the collagen solution to the culture dish and allow it to air-dry. As little as 10 $\mu$g of collagen per 35-mm culture dish forms an even coating sufficient for cell attachment[11] and differentiation.[9] A more uniform coat of collagen is obtained on bacteriological plastic than on tissue culture plastic. The collagen-coated plastic dishes can be

[6] R. J. Klebe, *J. Cell. Physiol.* **86**, 231 (1975).
[7] L. A. Liotta, D. Vembu, H. K. Kleinman, G. R. Martin, and C. Boone, *Nature (London)* **272**, 622 (1978).
[8] R. L. Erhmann and G. O. Gey, *J. Natl. Cancer Inst.* **16**, 1375 (1956).
[9] S. D. Hauschka and I. R. Konigsberg, *Proc. Natl. Acad. Sci. U.S.A.* **55**, 119 (1966).
[10] A. E. Postlethwaite, R. Snyderman, and A. H. Kang, *J. Exp. Med.* **144**, 1188 (1976).
[11] H. K. Kleinman, E. B. McGoodwin, S. I. Rennard, and G. R. Martin, *Anal. Biochem.* **94**, 308, (1979).
[12] T. Elsdale and J. Bard, *Nature (London)* **236**, 152 (1972).
[13] C. A. Sattler, G. Michalopoulos, G. A. Sattler, and H. C. Pitot, *Cancer Res.* **38**, 1539 (1978).
[14] L. G. M. Reid and M. Rojkind, this series, Vol. 58, p. 263.
[15] J. W. Dodson and E. D. Hay, *J. Exp. Zool.* **189**, 51 (1974).
[16] M. Nakagawa and M. R. Urist, *Proc. Soc. Exp. Biol. Med.* **154**, 568 (1977).
[17] S. B. Mizel and J. R. Bamburg, *Dev. Biol.* **49**, 20 (1976).
[18] D. Gospodarowicz and C. R. Ill, *Proc. Natl. Acad. Sci. U.S.A.* **77**, 2726 (1980).

sterilized by exposure to ultraviolet light with a lamp at a distance of 1-2 feet for 6-20 hr. Glass dishes coated with collagen retain attachment activity after autoclaving.

Preparation of Reconstituted Collagen

Collagen dissolved in 0.1% acetic acid at 1-2 mg/ml forms a gel at room temperature when brought to neutral pH with 0.1 volume of 2 $M$ buffer (phosphate, Tris, or HEPES) plus some 0.5 $N$ NaOH.[12] Alternatively, the acidic collagen solution may be dialyzed in the cold against a buffered salt solution or culture medium, and the pH subsequently can be adjusted to 7.0. The collagen can be sterilized by dialysis against buffer saturated with chloroform (5 ml/liter) followed by dialysis against buffer alone. Antibiotics, such as penicillin-streptomycin or gentamicin, may also be included in the buffer. After the sample is neutralized, 1 ml of the collagen solution is pipetted onto a 35-mm dish and the temperature is raised to 37° for 1 hr to allow the gel to form. These gels can be used directly or can be gently displaced from the culture dish to serve as floating supports.[13]

Other Collagenous Matrices

Aqueous salt or acid extracts of tissues can be used to coat culture dishes.[14] Such extracts contain collagen as well as other substances, such as fibronectin and proteoglycans. Several investigators have also used as a substrate the material remaining on the culture dish after removal of the cells with EDTA or detergent.[17,18] In this case, cells are removed under sterile conditions by incubation for 30 min with 0.5% Triton X-100 in phosphate-buffered saline or by brief treatment with 0.05% EDTA in $Ca^{2+}$-$Mg^{2+}$-free balanced salt solution. Under these conditions, the dishes appear to be free of cells and cellular fragments but are coated with a complex substrate of collagen, procollagen, fibronectin, and proteoglycans. Tissues such as denuded endothelium,[19] freeze-thawed lens,[15] and demineralized bone[16] have also been used as substrates for cell culture.

Cell Attachment Assay

Since most cells that are freshly dissociated from culture dishes or from tissues adhere to collagen substrates within 1-3 hr, attachment assays need not be carried out under sterile conditions.[1] Cells (1 to 10 × 10$^6$/ml) are suspended in minimal essential medium containing

[19] D. Gospodarowicz, J. Greenburg, M. Bialecki, and B. Zetter, *In Vitro* **14**, 85 (1978).

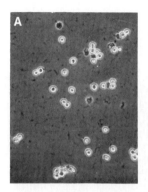  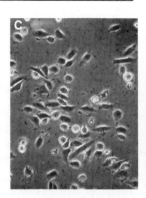

FIG. 1. Appearance of Chinese hamster ovary cells on collagen-coated dishes after 1 hr in the presence and in the absence of serum. Thirty-five-millimeter dishes coated with 10 μg of dried collagen were preincubated at 37° in 95% air, 5% $CO_2$ for 1 hr with 1.0 ml of minimal essential medium containing 200 μg of bovine serum albumin and 10% bovine serum where indicated. Cells were trypsinized, washed, and resuspended at $1 \times 10^6$ cells per milliliter in minimal essential medium containing 200 μg of bovine serum albumin. Then 0.1 ml of the suspension was added to each dish and allowed to incubate for 1 hr. (A) Cells after 1 hr in the absence of serum; unwashed plate. (B) Cells after 1 hr in the absence of serum; plate washed three times with phosphate-buffered saline. (C) Cells after 1 hr in the presence of 10% serum; plate washed three times with phosphate-buffered saline. Unwashed plates were identical.

bovine serum albumin (200 μg/ml). Factors such as serum, fibronectin, or other attachment proteins are dissolved in the same buffer. In most assays, an aliquot (~0.1 ml) of the cells is added to a 35-mm dish containing 0.9 ml of medium and attachment factors, and the cells are incubated at 37° in an atmosphere of 95% air, 5% $CO_2$. Unattached cells are removed by decanting the medium and washing the dish three times with 1-ml portions of 0.02 $M$ phosphate-buffered saline, pH 7.4, or with other physiological solutions. The attached cells are then released with 1 ml of 0.1% trypsin–0.1% EDTA in phosphate-buffered saline. Detachment usually occurs within 5 min but should be monitored in each case by microscopic examination. The number of unattached and attached cells are measured with an electronic particle counter or a hemacytometer. More than 80% of Chinese hamster ovary cells and other fibroblasts attach in the presence of 1–5% serum or 1–5 μg of purified fibronectin per milliliter (Fig. 1).

Characteristics of Fibroblast Attachment

The interstitial collagens from various sources and species appear to be equally active in binding fibronectin and promoting adhesion of fibroblasts. Likewise, fibronectin appears to be equally active in cell adhesion regardless of the animal species or tissue from which it is derived.[20] In the absence of fibronectin, fibroblasts synthesize their own attachment factor.[21] In the presence of fibronectin, however, protein synthesis is not required and cycloheximide does not inhibit cell adhesion. Adhesion is temperature-dependent. At 37° adhesion is maximal, whereas at 4°, little or no cell adhesion occurs, perhaps owing to the reduced rate of metabolism.[6] Attachment is inhibited by compounds that affect cell membranes,[22] (i.e., local anesthetics, nonionic detergents, and aliphatic alcohols) and by substances affecting cytoskeletal organization[23] (i.e., cytochalasin B, vinblastine sulfate, colchicine, and Colcemid).

The role of proteoglycans in cell adhesion is not understood. Although no direct evidence for their biological activity in the attachment assay has been described, certain proteoglycans are known to bind to collagen[24] and to fibronectin[25] and appear to strengthen and accelerate the binding of fibronectin to collagen.[26,27] In addition, certain proteoglycans are present in the material left on the dish after removal of the cells with detergent,[28] further suggesting they may have a role in cell adhesion. High concentrations of proteoglycans from cartilage, however, inhibit the attachment of fibroblasts, but since these macromolecules are not produced by fibroblasts, the biological significance of this observation is unclear.[29] Certain gangliosides have been shown to bind to fibronectin and inhibit adhesion, suggesting that the cell surface receptor for fibronectin may be a ganglioside or a glycoconjugate with similar oligosaccharide components.[30]

Approaches for Determining Attachment Requirements of Cells

The requirement for specific attachment factors is tested by their addition to the assay. Serum is a good source of some attachment factors, and most cells appear to be capable of synthesizing their own factors. For

[20] K. M. Yamada and D. W. Kennedy, *J. Cell Biol.* **80**, 492 (1979).
[21] F. Grinnell and M. K. Field, *Cell* **18**, 117 (1979).
[22] R. L. Juliano and E. Gagalang, *J. Cell. Physiol.* **98**, 483 (1979).
[23] K. Nath and P. A. Srere, *J. Cell. Physiol.* **92**, 33 (1977).
[24] T. R. Oegema, J. Laidlaw, V. C. Hascall, and D. D. Dziewiakowski, *Arch. Biochem. Biophys.* **170**, 698 (1975).
[25] K. M. Yamada, D. W. Kennedy, K. Kimata, and R. M. Pratt, *J. Biol. Chem.* **255**, 6055, (1980).
[26] S. Johansson and M. Höök, *Biochem. J.* **187**, 521 (1980).
[27] E. Ruoslahti and E. Engvall, *Biochim. Biophys. Acta* **631**, 350 (1980).
[28] B. J. Rollins and L. A. Culp, *Biochemistry* **18**, 5621 (1980).
[29] P. Knox and P. Wells, *J. Cell Sci.* **40**, 77 (1979).
[30] H. K. Kleinman, G. R. Martin, and P. H. Fishman, *Proc. Natl. Acad. Sci. U.S.A.* **76**, 3367 (1979).

example, fibroblasts adhere rapidly in the presence of fibronectin, but not in serum from which the fibronectin has been removed by collagen affinity chromatography.[31] Fibronectin, however, is not present in all tissues, and cells in general interact with extracellular matrix components present in the tissues from which they are derived. For example, chondrocytes adhere preferentially to type II collagen,[32] which is found only in cartilage, whereas epithelial and endothelial cells specifically adhere to type IV (basement membrane) collagen.[33-35] It is therefore not surprising that attachment factors other than fibronectin exist. For example, chondrocytes, in contrast to fibroblasts, utilize both complete and fibronectin-free serum.[32] It has been shown that chondrocytes utilize chondronectin, a serum component distinct from fibronectin, which is also present in salt extracts of cartilage and in the vitreous of the eye.[36] Epithelial cells attach slowly and serum and fibronectin do not stimulate their attachment. Cycloheximide (25 µg/ml) inhibits their attachment. Addition of crude salt extracts of basement membrane-rich tissues or purified laminin, a basement membrane glycoprotein, increases the rate and extent of adhesion of these cells.[34]

The identification of cell-specific attachment factors could lead to improved methods for culturing cells with better maintenance of differentiated functions. Standard culture conditions alter some cellular functions. For example, specific attachment factors that are not normally seen by the cells change the phenotypic expression of cells such as chondrocytes, which dedifferentiate in the presence of fibronectin,[37] and myoblasts, which do not fuse in the presence of exogenous fibronectin.[38] Thus, fibronectin and other attachment factors can replace serum in many defined media.

[31] E. Engvall and E. Ruoslahti, *Int. J. Cancer* **17**, 1 (1977).
[32] A. T. Hewitt, H. K. Kleinman, J. P. Pennypacker, and G. R. Martin, *Proc. Natl. Acad. Sci. U.S.A.* **77**, 385 (1980).
[33] J. C. Murray, G. Stingl, H. K. Kleinman, G. R. Martin, and S. I. Katz, *J. Cell Biol.* **80**, 197 (1979).
[34] V. P. Terranova, D. H. Rohrbach, and G. R. Martin, *Cell* **22**, 719 (1980).
[35] R. C. Hughes, G. Mills, Y. Courtois, and J. Tassin, *Biol. Cell* **36**, 321 (1979).
[36] A. T. Hewitt, H. H. Varner, and G. R. Martin, in "Immunochemistry of Collagen" (H. Furthmayr, ed.), Vol. I. CRC Press, Boca Raton, Florida, 1981.
[37] J. P. Pennypacker, J. R. Hassell, K. M. Yamada, and R. M. Pratt, *Exp. Cell Res.* **121**, 411 (1979).
[38] T. R. Podleski, I. Greenberg, J. Schessinger, and K. M. Yamada, *Exp. Cell Res.* **122**, 317 (1979).

## [25] Platelet–Collagen Adhesion

*By* SAMUEL A. SANTORO *and* LEON W. CUNNINGHAM

When the integrity of the vascular endothelium is disrupted, circulating blood platelets rapidly adhere to exposed fibrillar collagen and are subsequently stimulated to aggregate. The platelet–collagen system appears to be a good model system with which to explore the nature of the interactions of collagen with cell surfaces. The end result of the interaction of platelets with collagen, platelet aggregation, may be easily monitored by the decreased turbidity of the platelet suspension as aggregation occurs.[1]

The initial event, adhesion of platelets to collagen, has been more difficult to quantitate. Desirable characteristics of the selected method include ease and rapidity of performance, capacity for the multiple determinations usually required in biochemical work, and reproducibility. The adhesion observed should reflect the physiological interaction of interest. For example, conditions that decrease the observed adhesion should also decrease the physiological consequence of adhesion, that is, platelet aggregation. An assay that appears to be useful in exploring the biochemical basis of the platelet–collagen interaction has been described.[2]

### The Membrane-Filtration Method

*Principle.* The assay measures the adhesion of washed, radiolabeled platelets to fibrillar collagen. The platelet suspension is incubated with a homogenate of reconstituted collagen fibrils in the presence of EDTA to inhibit platelet aggregation. Platelets adherent to the collagen fibers are separated from single, nonadherent platelets by filtration, using polycarbonate membranes with 5-$\mu$m pores. The collagen fibers and adherent platelets are retained whereas single platelets pass through. The extent of adhesion is determined by counting the radioactivity retained on the membrane filters.

*Special Equipment*

Gamma counter
Reciprocal shaker-water bath
Multiwell membrane filter manifold (Millipore)
Unipore 2.5-cm polycarbonate membrane filters with 5-$\mu$m pores (Bio-Rad Laboratories)

[1] G. V. R. Born, *Nature* (*London*) **194**, 927 (1962).
[2] S. A. Santoro and L. W. Cunningham, *Proc. Natl. Acad. Sci. U.S.A.* **76**, 2644 (1979).

*Reagents and Solutions*

Anticoagulant: 1% EDTA (w/v), 0.9% NaCl (w/v) adjusted to pH 7.4
$^{51}$Cr, as sodium chromate
Platelet washing solution (solution A): NaCl, 0.137 $M$; KCl, 2.7 m$M$; Na$_2$HPO$_4$, 8.0 m$M$; KH$_2$PO$_4$, 1.5 m$M$; glucose, 5.0 m$M$; pH 7.4
Adhesion assay solution (solution B): NaCl, 0.137 $M$; KCl, 2.7 m$M$; Na$_2$HPO$_4$, 8.0 m$M$; KH$_2$PO$_4$, 1.5 m$M$; glucose, 5.0 m$M$; EDTA, 2.0 m$M$; bovine serum albumin, 0.3% (w/v); pH 7.4
Sodium dodecyl sulfate (SDS), 1% (w/v)
Na$_2$HPO$_4$, 0.02 $M$
Collagen: Rat skin acid soluble collagen is dissolved in 3% (v/v) acetic acid at a concentration of 2 mg/ml. See Chapter [2] of this volume for details of collagen preparation.

*Preparation of Collagen Fibers.* Native-type collagen fibrils are prepared by dialysis of small volumes of collagen dissolved in 3% (v/v) acetic acid against 0.02 $M$ Na$_2$HPO$_4$ in the cold.[3] Further details regarding the preparation of collagen fibrils are given in Chapter [5] of this volume. The fibers are gently homogenized to a uniform suspension with 1 or 2 strokes of a Potter–Elvehjem tissue grinder.

*Preparation of Platelet Suspension.* Venous blood (30 ml) is obtained from a healthy subject by careful venipuncture. The needle is removed from the syringe, and the contents are gently expressed into a 50-ml polypropylene centrifuge tube containing 3 ml of anticoagulant solution. The tube is capped, and the contents are mixed by gentle inversion several times.

Next, platelet-rich plasma is prepared. The blood is centrifuged at 180 $g$ for 15 min at room temperature. The platelet-rich plasma (supernatant) is carefully removed with a siliconized or plastic Pasteur pipette. Contaminating red cells are removed by two 1-min centrifugations (1000 $g$).

The platelet-rich plasma is then subjected to centrifugation (1240 $g$) for 10 min. The supernatant is discarded unless it is desired to resuspend the platelets later in autologous plasma. The platelet pellet is resuspended in 2.5 ml of solution A by gentle aspiration with a Pasteur pipette; 20–50 $\mu$Ci of $^{51}$Cr are added, and the suspension is incubated at room temperature for 20 min. The suspension is then diluted to 30 ml with solution A; 3 ml of anticoagulant solution are added, and the suspension is centrifuged for 10 min at 1240 $g$. The supernatant is discarded, the platelet pellet is resuspended, and the platelets are washed one additional time as above. The washed platelets are then resuspended in solution B, and the platelet count is adjusted to $2.5 \times 10^8$ per milliliter (250,000/$\mu$l). Thirty milliliters of blood will yield approximately 15 ml of washed platelets.

[3] R. R. Bruns and J. Gross, *Biopolymers* 13, 931 (1974).

*Performance of Adhesion Assay.* The adhesion assay is initiated by the addition of 0.1 ml of a suspension containing the desired quantity of reconstituted collagen fibers (usually 100–200 µg) or buffer to 0.4 ml of the labeled platelet suspension in 12 × 75 mm polypropylene tubes. The mixture is incubated for 20 min at 37° in a reciprocal shaker operating at 100 cycles per minute. At the conclusion of the incubation, the suspension is decanted onto a membrane filter, which has been rinsed with solution B, mounted in the multiwell manifold connected to light suction. The assay tubes are rinsed twice with 1-ml portions of solution B, which are then decanted onto the membrane filter. The manifold is then disassembled, and the filter is carefully removed with a forceps and placed in a 12 × 75 mm tube in a manner such that contact between the walls of the tube and the platelet and collagen-coated surface of the filter is minimized. The walls are then rinsed with 1 ml of 1% SDS. The filter is compressed so that it is entirely below the surface of the detergent solution, and the tube is counted in the gamma counter.

*Analysis of Data.* Controls should always be run in which collagen is omitted from the assay. This baseline adhesion is typically about 1%. All data should be corrected for background radiation.

It is convenient to express the data in terms of the percentage of platelets in the incubation mixture that have adhered to collagen. This requires knowledge of the total counts per minute used per tube and the counts retained on the filter.

We have found it convenient to adjust the collagen concentration to obtain approximately 40% adhesion under the routine conditions described above. Several characteristics of this assay were described in the original report.[2] The extent of adhesion is essentially a linear function of the collagen fiber concentration. The adhesion is not due to trapping of platelets within the collagen mat on the filter, because application of platelets to a filter on which collagen has already been deposited results in baseline values of adhesion. Maximum adhesion will have occurred in 15–20 min, although significant adhesion can be detected at early times. Scanning electron microscopy reveals that aggregation does not occur, as only single platelets are seen attached to the collagen fibers.

Alternative Methods

*Column Filtration Methods*

The "Sepharose test" devised by Fauvel *et al.*[4] is similar in principle to the method described above. Columns of Sepharose 4B are used to

---

[4] F. Fauvel, Y. J. Legrand, and J. P. Caen, *Thromb. Res.* **12**, 273 (1978).

separate single platelets from large platelet-collagen complexes that cannot pass through the column. These procedures are somewhat tedious and require separate columns for each assay point. A potential complication is alteration of the flow properties of the column by the insoluble collagen fibers or derivatives so that nonadherent platelets may be retained.

*Turbidimetric Method*

This technique, a modification of that used to measure platelet aggregation, was originally described by Spaet and Lejnieks.[5] Platelet-rich plasma anticoagulated with EDTA to prevent platelet aggregation is allowed to react in the aggregometer with connective tissue fragments. The resulting small increase in transmission that occurs as the platelets adhere to the collagen is recorded and considered to be a measure of adhesion. This assay is simple to perform, can be performed directly on platelet-rich plasma, and does not require the use of radioisotopes. The result is a continuous tracing of adhesion as a function of time. Unfortunately, the changes in transmission are small and require the use of an expanded scale. Lack of reproducibility from run to run is a potential problem. The assay is highly dependent on the nature of the collagen preparation, especially the turbidity. In our hands homogenates of insoluble collagen work well whereas homogenates of fibers reconstituted from soluble collagen are ineffective.

*Binding of Radiolabeled Collagen*

Gordon and Dingle[6] have described the binding of radiolabeled collagen to platelets. Platelets were separated from the collagen by centrifugation through Ficoll. The data of Gordon and Dingle suggested that collagen did not bind to the platelet until it became fibrillar. Specific binding of native collagen to the platelet has not been demonstrated in this or other similar systems. Obviously the binding of multimolecular collagen fibers to the platelet surface complicates the analysis of the binding data.

*Affinity Chromatography*

Affinity chromatography is an initially attractive method by which to measure the adhesion of platelets to collagen. Brass *et al.*[7] used as a measure of adhesion the difference in radioactivity, in the form of radiolabeled platelets, applied to and eluted from collagen-Sepharose. Un-

---

[5] T. H. Spaet and I. Lejnieks, *Proc. Soc. Exp. Biol. Med.* **132**, 1038 (1969).
[6] J. L. Gordon and J. T. Dingle, *J. Cell Sci.* **16**, 157 (1974).
[7] L. F. Brass, D. Faile, and H. B. Bensusan, *J. Lab. Clin. Med.* **87**, 525 (1976).

fortunately this technique requires that a separate column be prepared for each assay point. The polymeric state of the collagen actually covalently attached to the Sepharose is not easily defined. Alteration of the flow properties of the column by fibrillar collagen is a potential complication. Nonspecific adhesion of negatively charged cells to positively charged products of side reactions that occur during the coupling of substrate to Sepharose has been a problem in another cell adhesion system.[8]

*Adhesion to Collagen-Coated Surfaces*

The adhesion of platelets to several collagen-coated surfaces has been described. For example, Meyer and Weisman[9] determined by microscopy the number of platelets adherent to coverslips that had been exposed to collagen solutions. Cazenave and co-workers[10] measured the number of platelets adherent to collagen-coated test tubes. The state of collagen polymerization is not well defined in these assays. Surfaces uniformly coated with collagen, especially fibrillar collagen are difficult to prepare reproducibly. The analysis of this type of experiment may be further complicated by leaching of collagen from the glass into the platelet suspension during the course of the assay.

[8] R. L. Schnaar, T. F. Sparks, and S. Roseman, *Anal. Biochem.* **79**, 513 (1977).
[9] F. A. Meyer and Z. Weisman, *Thromb. Res.* **12**, 431 (1978).
[10] J. P. Cazenave, M. A. Packham, and J. F. Mustard, *J. Lab. Clin. Med.* **82**, 978 (1973).

## [26] Hepatocyte– Collagen Adhesion

*By* BJÖRN ÖBRINK

Anchorage of cells to collagenous structures seems to be of fundamental importance for cellular physiology and pathology in vertebrate organisms.[1] The phenomenon as such also draws the attention to an important property, other than pure structural functions, of collagen and of other matrix components. Although vertebrate cells are able to adhere to other components than collagen or fibronectin *in vitro,* it has been found from work with hepatocytes that substrates made from these proteins are unique in supporting survival and important physiological responses, such

[1] B. Öbrink, C. Ocklind, K. Rubin, M. Höök, and S. Johansson, *in* "Biochemistry of Normal and Pathological Connective Tissues" (A. M. Robert and L. Robert, eds.) (*Colloq. Int. C.N.R.S.* 287), p. 239 (1980).

as spreading, of the attached cells.[2-4] Some data point to the existence of specific cell surface glycoproteins that are involved in attachment of hepatocytes to collagen.[5] The analysis of such collagen receptors and their influence on cellular behavior is a most important task in modern cell biology.

The cells in vertebrate tissues interact not only with their collagenous matrices, but also with each other.[1] There is now convincing evidence that intercellular interactions also involve specific cell surface glycoproteins.[6-10] However, most studies on cell–cell interactions *in vitro* have been performed with cells in suspension, in spite of the fact that most cells *in vivo* are anchored to collagenous structures. As already mentioned, the attachment to a collagen or a fibronectin substrate profoundly changes cellular behavior. Thus intercellular interactions must be studied on matrix-attached cells as well as on suspended cells in order to clarify the cell biological implications of cell–cell interactions.[1,2]

Rat hepatocytes have proved to be a useful model system for analyses of cell–cell and cell–matrix adhesion phenomena at the molecular level.[1-5,9] *In vivo* these cells have well developed intercellular junctional complexes, and they are associated with collagen-and fibronectin-containing reticular fibers at their sinusoidal surfaces.[4] *In vitro* the hepatocytes attach efficiently to both collagen and fibronectin, but, unlike fibroblasts, they do not need fibronectin, exogenously added or endogenously produced, for attachment to or spreading on collagen.[4,11-13] Data have been presented which indicate that different molecular mechanisms are involved in the attachment of rat hepatocytes to collagen and to fibronectin, respectively.[11] Furthermore it has been demonstrated that fibronectin is not involved in intercellular adhesion of rat hepatocytes[9] and that different cell surface glycoproteins seem to be involved in cell–cell and cell–collagen adhesion of rat hepatocytes.[5]

[2] K. Rubin, L. Kjellén, and B. Öbrink, *Exp. Cell Res.* **109**, 413 (1977).
[3] M. Höök, K. Rubin, Å. Oldberg, B. Öbrink, and A. Vaheri, *Biochem. Biophys. Res. Commun.* **79**, 726 (1977).
[4] K. Rubin, Å. Oldberg, M. Höök, and B. Öbrink, *Exp. Cell Res.* **117**, 165 (1978).
[5] C. Ocklind, K. Rubin, and B. Öbrink, *FEBS Lett.* **121**, 47 (1980).
[6] W. Frazier and L. Glaser, *Annu. Rev. Biochem.* **48**, 491 (1979).
[7] J. P. Thiery, R. Brackenbury, U. Rutishauser, and G. M. Edelman, *J. Biol. Chem.* **252**, 6841 (1977).
[8] K. Müller and G. Gerisch, *Nature (London)* **274**, 445 (1978).
[9] B. Öbrink and C. Ocklind, *Biochem. Biophys. Res. Commun.* **85**, 837 (1978).
[10] M. Takeichi, *J. Cell Biol.* **75**, 464 (1977).
[11] K. Rubin, S. Johansson, I. Pettersson, C. Ocklind, B. Öbrink, and M. Höök, *Biochem. Biophys. Res. Commun.* **91**, 86 (1979).
[12] K. Rubin, M. Höök, B. Öbrink, and R. Timpl, *Cell* **24**, 463 (1981).
[13] K. Rubin, S. Johansson, M. Höök, and B. Öbrink, *Exp. Cell Res.* **135**, 127 (1981).

This chapter describes methods that currently are used in the study of the molecular mechanisms of adhesion of hepatocytes to collagen and to fibronectin and of intercellular adhesion between hepatocytes while they are anchored to a collagen matrix. Techniques are emphasized rather than rationale or methodological validation.

Collagenase

Bacterial collagenase is used both for hepatocyte preparation and in some of the cell–collagen attachment techniques. Several companies provide collagenase preparations that they claim to be particularly well suited for cell preparation from specific tissues, such as liver and adipose tissue. However, it is our experience that batches of collagenase claimed to be effective on liver do not work better than batches that have not been selected for any special type of tissue. We have obtained the best and most reproducible results with collagenase type I obtained from Sigma Chemical Co., although occasionally one may get a batch of this type that does not disintegrate the tissue adequately. When this happens, it is better to purchase another batch of the same type of collagenase than to switch to a different type of collgenase.

Animals

Male, Sprague-Dawley rats are used throughout in our laboratory, but other strains can be used as well. For collagen preparation rats weighing 50–100 g are used, and for hepatocyte preparation it is most convenient to use animals weighing 200–300 g. The animals are fed ad libitum and are housed at 20° with alternating 12-hr cycles of light and darkness.

Collagen

*Solutions*

NaCl, 0.45 $M$: 26.3 g of NaCl, distilled $H_2O$ to 1 liter
NaCl, 5 $M$: 292.2 g of NaCl, distilled $H_2O$ to 1 liter
Phosphate buffer (NaCl) (0.01 $M$ phosphate buffer, pH 7.6, ionic strength adjusted to 0.4 with NaCl): 21.3 g of NaCl, 46.6 ml of 0.2 $M$ $Na_2HPO_4$, 3.4 ml of 0.2 $M$ $NaH_2PO_4$, distilled $H_2O$ to 1 liter
$Na_2HPO_4$, 0.02 $M$: 3.56 g of $Na_2HPO_4 \cdot 2 H_2O$, distilled $H_2O$ to 1 liter
Acetic acid, 0.1 $M$
Ethanol, 96%

Procedures for preparing soluble collagen are given in Chapter [2] of this volume. However, many of these procedures result in collagen preparations that are difficult to keep in solution at physiological pH and ionic strength at 4°. Therefore, a procedure is presented here that yields a collagen preparation completely soluble at pH 7.4 and ionic strength 0.15 at 4°. As a source for collagen, rat skins are used. After sacrificing the animals, the skins are dehaired and defatted mechanically and then ground in a meatgrinder. The ground skins are extracted three times with three volumes of 0.45 $M$ NaCl at 4° (the whole procedure is carried out at 4°). The extracts are precipitated with an equal volume of 5 $M$ NaCl. The precipitate is centrifuged and dissolved in cold phosphate buffer (NaCl). The precipitation step with 5 $M$ NaCl is repeated twice; after the third precipitation, the dissolved precipitate is dialyzed against 0.02 $M$ $Na_2HPO_4$ until a fibrous precipitate appears in the dialysis bag. This precipitate is centrifuged and dissolved in 0.1 $M$ acetic acid and dialyzed against phosphate buffer (NaCl). Then cold (4°) 96% ethanol is added to a final concentration of 14% (v/v), and the precipitate is centrifuged and dissolved in 0.1 $M$ acetic acid. The acid solution is centrifuged at 100,000 $g$ for 3–5 hr; this procedure yields a clear supernatant. The clear supernatant contains pure collagen, mostly type I, and is not contaminated with fibronectin. The collagen can be stored at 4° in 0.1 $M$ acetic acid for more than a year, or it can be lyophilized from acetic acid. The acid collagen solution is sterile, and no microbial growth usually occurs in gels made by heat gelation of neutral salt solutions prepared from such collagen solutions. However, if the collagen solution needs to be sterilized, this can be done by adding a few drops of chloroform per 1000 ml of the acid collagen solution. The chloroform will disappear during the dialysis steps preceding gel formation and will not interfere with cell cultivation on the collagen gels. The collagen concentration can be determined by the biuret method[14] with human or bovine serum albumin as a standard. The color yield of collagen is 0.94 that of albumin.

Collagen Substrates

Several procedures have been published for the preparation of collagen substrates that are suited for cell attachment.[4,15,16] In our laboratory we have tried a variety of procedures with both native and denatured collagen and including both drying of the substrates and treatment of culture dishes with collagen solutions without drying. Since all the various

[14] M. Dittebrandt, *Am. J. Clin. Pathol.* **18**, 439 (1948).
[15] T. Elsdale and J. Bard, *J. Cell Biol.* **54**, 626 (1972).
[16] R. J. Klebe, *Nature (London)* **250**, 248 (1974).

methods have resulted in substrates with essentially the same properties with respect to attachment of hepatocytes only three procedures currently used in our laboratory are listed here. These procedures yield substrates consisting of denatured collagen, of native collagen monomers and of native collagen fibers, respectively.

*Denatured Collagen*

*Solution:* Buffer 3: see Hepatocyte Preparation
Petri dishes made of bacteriological plastic are used (e.g., dishes 35 mm in diameter from Falcon Plastics, Cat. No. 1008). Immediately prior to coating of the dishes, an aliquot of the collagen stock solution (containing 1–2 mg of collagen per milliliter in 0.1 $M$ acetic acid) is denatured by warming at 50° for 30 min. One hundred micrograms of collagen from the denatured collagen stock solution are then added to each 35-mm petri dish together with 1 ml of buffer 3. The dishes are incubated at 37° for 30 min and are then decanted and washed with buffer 3. They are then ready to use and should be used immediately without storage. Care is taken to prevent the dishes from drying during any step of the coating procedure.

*Native Collagen Monomers*

Bacteriological plastic petri dishes are coated exactly in the same way as described for denatured collagen except that the collagen in the collagen stock solution is not denatured prior to addition to the dishes. These dishes should also be used immediately and should not be allowed to dry.

*Native Collagen Fibers*

*Solutions*

Buffer 3: see Hepatocyte Preparation
Phosphate-buffered saline (PBS): 8.00 g of NaCl, 0.20 g of KCl, 1.44 g of $Na_2HPO_4 \cdot 2 H_2O$, 0.20 g of $KH_2PO_4$, distilled $H_2O$ to 1 liter, pH 7.3

Reconstituted native collagen fibers that can be used for cell attachment are prepared by heat gelation of soluble collagen in petri dishes. However, for this purpose petri dishes made of bacteriological plastic cannot be used, since the gels do not adhere to them. Instead, tissue culture plastic dishes are used (e.g., dishes 35 mm in diameter from Falcon Plastics, Cat. No. 3001). An aliquot of the collagen stock solution is dialyzed at 4° under sterile conditions against a neutral salt solution of pH 7.4 and physiological ionic strength (e.g., buffer 3 or PBS; the presence of phosphate does not influence the attachment or spreading behavior of rat hepatocytes on collagen, as has been suggested to be the case for fibroblasts[17]). The neutral collagen solution is adjusted with cold dialysis medium to a collagen concentration of 1 mg/ml, and 0.8 ml of this collagen solution is added to each 35-mm petri dish. The dishes are incubated overnight at 37° in humid air. Under these conditions a transparent, firm gel adhering to the bottom of the dish is formed.

Fibronectin

*Solutions*

Citrate-protease inhibitor solution (citrate-PIS): 0.15 $M$ sodium citrate, $10^{-3}$ $M$ phenylmethylsulfonyl fluoride, 1 $M$ $\epsilon$-aminocaproic acid, 0.1 $M$ benzamidine, and $10^3$ units of Trasylol per milliliter

Phosphate buffer: 0.05 $M$ sodium phosphate, 0.1 $M$ NaCl, 0.01 $M$ sodium citrate, 0.01 $M$ benzamidine, and 0.02% (w/v) NaN$_3$, pH 6.5

Urea, 4 $M$ in phosphate buffer

Tris buffer: 0.01 $M$ Tris-HCl, 0.01 $M$ benzamidine, and 0.02% (w/v) NaN$_3$, pH 8.0

NaCl, 0.4 $M$, in Tris buffer

PBS: see Native Collagen Fibers

Fibronectin can be prepared from human, bovine, or rat plasma.[11] It is essential that protease inhibitors be present from the beginning of the blood drawing because fibronectin is sensitive to a variety of proteases, including plasmin.

To 90 ml of freshly drawn blood, 10 ml of citrate-PIS are added and the blood is centrifuged at 1000 $g$ for 20 min. The supernatant is centrifuged at 4000 $g$ for 20 min, and the new supernatant is collected. If the plasma (supernatant) cannot be used immediately for fibronectin isolation, it can be frozen at this stage by rapid freezing in ethanol–Dry Ice and stored at $-70°$. The remaining procedure for fibronectin preparation will be described for 1000 ml of plasma.

One liter of plasma is applied to a 250-ml column of gelatin-Sepharose (prepared from gelatin type I, Sigma Chemical Co., as described by Engvall and Ruoslahti[18]). The column is eluted at room temperature with phosphate buffer at a flow rate of approximately 125 ml/hr until $A_{280}$ approaches zero (approximately 400 ml). The column is then eluted with 400 ml of 4 $M$ urea in phosphate buffer, and fractions of 10 ml are collected. The absorbancy at 280 nm is determined, and the absorbancy peak is

[17] H. K. Kleinman, E. B. McGoodwin, S. I. Rennard, and G. R. Martin, *Anal. Biochem.* **94**, 308 (1979).
[18] E. Engvall and E. Ruoslahti, *Int. J. Cancer* **20**, 1 (1977).

pooled and dialyzed at 4° against 100 volumes of Tris buffer. Tris buffer should be changed once during dialysis. The dialyzed material is applied to a 2.5 × 25 cm column of DEAE-cellulose (Whatman DE-52) equilibrated with Tris buffer. This column is operated at 4°. The column is first eluted with 200 ml of Tris buffer and then with 1 liter of a linear gradient from 0 to 0.4 $M$ NaCl in Tris buffer. The absorbancy at 280 nm is monitored. Fibronectin is eluted at about 0.2 $M$ NaCl. The pooled material is concentrated by precipitation from 50% saturated ammonium sulfate. The protein is stored as an ammonium sulfate precipitate at 4° in the presence of protease inhibitors and 0.02% $NaN_3$. When the protein is to be used, it is first dialyzed against PBS.

Fibronectin Substrates

*Solution:* Buffer 3: see Hepatocyte Preparation

Petri dishes made of bacteriological plastic are used (see section entitled Collagen Substrates). The dishes can be coated with fibronectin, either by incubating the dishes with fibronectin solutions as described for collagen substrates or by drying solutions of fibronectin in the dishes. Dishes that are dried will contain more fibronectin than incubated dishes and can be stored, but hepatocytes attach efficiently to both types of fibronectin-coated dishes.

Coating by incubation is performed as described for native collagen monomers by addition of 20 µg of fibronectin from the fibronectin stock solution to 1 ml of buffer 3 in a 35-mm petri dish.

Coating by drying is performed by dilution of 20 µg of fibronectin from the stock solution with 2 ml of deionized, distilled water in a 35-mm petri dish. The dishes are then allowed to air-dry at 40°. The dried dishes are rinsed extensively with deionized water and are air-dried again. Dishes coated in this way can be stored at room temperature for several months without loss in attachment efficiency.

Hepatocyte Preparation

*Solutions*

Buffer 1: 8.3 g of NaCl, 0.5 g of KCl, 2.4 g of $N$-2-hydroxyethylpiperazine-$N'$-2-ethanesulfonic acid (HEPES), distilled $H_2O$ to 1 liter, adjusted to pH 7.4 with 1 $M$ NaOH

Buffer 2: 3.9 g of NaCl, 0.5 g of KCl, 0.7 g of $CaCl_2 \cdot 2 H_2O$, 24.0 g of HEPES, 15 g of bovine serum albumin (type V, Sigma Chemical Co.), distilled $H_2O$ to 1 liter, adjusted to pH 7.6 with 1 $M$ NaOH

Buffer 3: 8.0 g of NaCl, 0.35 g of KCl, 0.16 g of MgSO$_4$ · 7 H$_2$O, 0.18 g of CaCl$_2$ · 2 H$_2$O, 2.4 g of HEPES, distilled H$_2$O to 1 liter, adjusted to pH 7.4 with 1 $M$ NaOH

Ethanol, 70% (v/v)

Rats to be used for hepatocyte preparation should be starved overnight, since an empty stomach makes the operation much easier to perform from a technical point of view. Cells prepared from nonstarved or starved rats have identical adhesion properties. For a detailed description of the perfusion equipment and of the handling of the liver after it has been excised, see Seglen.[19]

A rat is anesthetized with ether by first putting the animal to sleep in a beaker and then placing it on its back with a mask, infiltrated with ether, placed over the nose. The abdomen is washed with 70% ethanol and cut open, a small incision is made in the portal vein, and a polyethylene catheter is inserted into the portal vein *in situ* and ligated. The lower caval vein is cut to allow free draining of the liver, and the perfusion is started. The whole perfusion procedure is carried out at a flow rate of 50 ml/min by aid of a peristaltic pump and with oxygenated media warmed to 37°.

First, 500 ml of the calcium- and magnesium-free buffer 1 is perfused through the liver for 10 min. This step is essential for a successful perfusion; without effective removal of calcium ions from the liver, the yield of viable hepatocytes would be very low. During this step the liver is gently excised and placed on a nylon screen fitted to a funnel, thus allowing the perfusate to be collected. The liver should be placed upside down; care should be taken so that no torsion of any liver lobe occurs, which would impair the flow through the liver. After perfusion with buffer 1 (the perfusate is discarded), perfusion is continued with 50 ml of buffer 2 containing collagenase, which is recirculated through the liver for 15 min by connecting the outlet of the funnel to the medium reservoir. Collagenase type I from Sigma Chemical Co. is used at a concentration of 0.6 mg/ml in buffer 2. Before use, the collagenase solution should be filtered through a 0.22-$\mu$m Millipore filter in order to remove small particles that otherwise might clog the liver sinusoids. Serum albumin is present in buffer 2 to serve as a competitive substrate for contaminating proteases in the collagenase preparation. Finally, the liver is perfused with buffer 1 for 3 min to remove the collagenase.

The liver is placed in a sterile petri dish, the liver capsule is peeled off, and the liver cells are gently dispersed in buffer 1 supplemented with bovine serum albumin (15 mg/ml). It is extremely important that the liver cells be handled with great care and that the dispersion be carried out by light shaking of the tissue in the medium, assisted by very gentle manipu-

[19] P. O. Seglen, *Methods Cell Biol.* **13**, 29 (1976).

lations of a forceps to allow entrance of the medium to all parts of the soft tissue. Rough scraping to detach cells from the connective tissue residue should be avoided, since it results in damaged cells. The cell suspension is filtered through a 132-$\mu$m nylon filter (obtained from Nitex or Monyl, Züricher Beutelfabrik AG), diluted to 75 ml with buffer 1 containing 15 mg of bovine serum albumin per milliliter and is then filtered sequentially through nylon filters (Nitex or Monyl filters) of pore sizes 50 and 25 $\mu$m, respectively. The cells are further purified at room temperature by four centrifugations in polycarbonate tubes at 70 $g$ for 2 min each, in a swinging-bucket rotor. These centrifugations enrich for hepatocytes in the pellets, since the hepatocytes are considerably larger than other types of liver cells. The cell pellets obtained in each centrifugation (except the last one) are gently resuspended in 75 ml of buffer 3 supplemented with bovine serum albumin (15 mg/ml). After the last centrifugation the pellet is resuspended in 10–20 ml of buffer 3 (without albumin) or any other desired medium in which the experiment is to be performed. The viability is checked by Trypan Blue exclusion.

Prepared in this way the cell preparation consists of more than 90% hepatocytes, mostly in the form of single cells with a viability of 80–95%. If the viability is less than 80%, the cells should not be used. The cells can be stored on ice for at least 5 hr before the adhesion experiments are performed, without loss of viability or adhesion properties.

Seeding, Incubation, and Determination of Attachment

*Solution:* Buffer 3: see Hepatocyte Preparation

If only the initial attachment of the hepatocytes is to be determined the cells can be seeded in plain buffer 3. No serum should be added. The attachment behavior of the cells is not changed by the addition of glucose, amino acids, vitamins, or other components that usually are present in cell culture media. However, the presence of $Mg^{2+}$ ions is an absolute requirement for hepatocyte attachment to collagen and fibronectin.[4]

The cells in the suspension are counted in a hemacytometer, and the cell number is adjusted to $0.5 \times 10^6$ cells/ml. Two milliliters of this cell suspension are pipetted into each of the desired number of coated 35-mm dishes, each dish thus receiving $1 \times 10^6$ cells. The dishes are incubated at 37° in humid air. If the kinetics of attachment is to be analyzed, the incubation is interrupted at various times and the number of attached cells are determined as described below. On dishes coated with the amount of collagen described above, the maximal number of cells (i.e., 80–100% depending on the viability, since only viable cells attach under these conditions) attach in approximately 30 min (Fig. 1). If the dishes are coated with less than 5 $\mu$g of collagen per dish, the rate of attachment will be

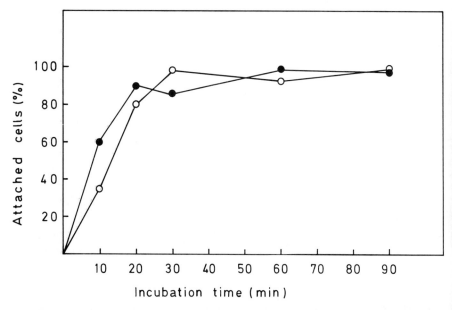

FIG. 1. Attachment kinetics of rat hepatocytes seeded in collagen-coated dishes. Cells ($1 \times 10^6$) in 2 ml of buffer 3 were seeded in 35-mm dishes coated with native (●) or denatured (○) collagen as described in the text. The amount of attached cells (shown on the ordinate) is given as percentage of the number of cells seeded per dish.

slower, but maximal attachment will always have occurred within 90 min of incubation.[12] On dishes not coated with collagen or fibronectin (or with inadequate amounts of these proteins), no attachment will occur under the described conditions.

The amount of cells that have attached to the dishes might be determined in several ways. One procedure is to measure the turbidity of the initial cell suspension and of the medium removed from the dishes after the appropriate incubation periods.[20] The amount of attached cells is given by the difference in turbidity between these two suspensions. However, it is more satisfactory to measure directly the amount of cells that have attached. The most common way to do that is to detach the cells from the dishes, e.g., by light trypsinization, and count them. However, rat hepatocytes are extremely sensitive to trypsin, and many cells are destroyed by such a procedure; hence, it is not applicable to these cells. An alternative procedure, which is used in our laboratory, is to lyse the cells and determine the amount of lactate dehydrogenase (LDH), which is proportional to the cell number.

[20] P. O. Seglen and J. Fosså, *Exp. Cell Res.* **116**, 199 (1978).

*Determination of LDH activity*

*Solutions*

Phosphate buffer (0.1 $M$ phosphate, pH 7.0): 5.36 g of $KH_2PO_4$, 13.9 g of $K_2HPO_4 \cdot 3 H_2O$, distilled $H_2O$ to 1 liter
Lysis buffer: 1% (v/v) Triton X-100, 1% (w/v) bovine serum albumin (type V, Sigma Chemical Co.) in phosphate buffer.
Assay solution: 8 mg of NADH (grade III, Sigma Chemical Co.), 25 mg of pyruvate (type II, Sigma Chemical Co.), phosphate buffer to 100 ml
Buffer 3: see Hepatocyte Preparation

At the end of the incubation the medium in the dishes is swirled and aspirated off, and the dishes are washed once with 1 ml of buffer 3. Further washings do not significantly reduce the number of attached cells. To dishes coated with fibronectin, denatured collagen, or native collagen monomers, lysis buffer can be added directly. If native collagen gels are used, the gels must first be degraded by collagenase. Thus to each dish containing collagen gel, 1 ml of buffer 3 containing 0.5 mg/ml of collagenase (type I, Sigma Chemical Co.) and 15 mg/ml of bovine serum albumin (type V, Sigma Chemical Co.) is added; the dishes are incubated at 37° for 10 min. Then 200 μl of the collagenase digest is mixed with 800 μl of lysis buffer. To dishes coated with fibronectin or denatured or native collagen, 1 ml of lysis buffer is added per dish. The contents of the dishes are thoroughly mixed, and 100 μl of the respective lysate are diluted with lysis buffer to a suitable concentration (generally varying from 1 : 1 to 1 : 4) depending on the amount of attached cells.

To measure the LDH activity of the lysates, 50 μl of the diluted lysate are added to 1 ml of assay solution in a cuvette and mixed thoroughly; the decrease in absorbancy at 340 nm is measured in a recording spectrophotometer. If the lysates have been diluted correctly, the first part of the decrease is linear with time and corresponds to the initial rate of the enzyme reaction. The initial rate of the reaction is a measure of the amount (in arbitrary units) of LDH present.

In order to translate the amount of LDH to cell number, a standard is needed. The standard is prepared from the initial cell suspension used in the attachment experiment. An aliquot of this cell suspension is added to lysis buffer so that a final mixture corresponding to $1 \times 10^6$ cells per milliliter of lysis buffer is obtained. This standard solution is then diluted to various concentrations with lysis buffer. The various dilutions are assayed for LDH activity as described above in order to construct a standard curve relating LDH activity (initial reaction rate) to cell number. This method to determine the amount of hepatocytes is very sensitive, and as little as 200 cells can be detected.

Seeding, Incubation, and Determination of Spreading

*Solutions*

Buffer 3: see Hepatocyte Preparation
Ham's F-10 medium, obtainable as dry powder medium from, e.g., Gibco, Inc.

Under appropriate conditions, cells that have adhered to a surface go through a spreading reaction; that is, they change from being initially spherical to a more flattened configuration, and, depending on the cell type and the geometry of the supporting substrate, they gain specific morphological features. The spreading reaction is thought to be an active process involving the cytoskeletal system.[21] Spreading seems to be essential for fundamental cell biological properties such as motility, regulation of RNA and DNA synthesis and of proliferation, and for survival.[1,21,22] Thus it is essential to study the mechanisms of cell spreading.

Although hepatocytes can attach to a variety of substrates, they have so far been found to spread only on collagen and on fibronectin substrates.[3,4] This emphasizes the opinion that the spreading reaction is biologically important and that collagen and fibronectin are essential components of the cell-associated matrices *in vivo*. Hepatocytes, being of epithelial origin, gain a polygonal, epithelium-like morphology upon spreading. This is a less drastic change in shape than that occurring in spreading of, for example, fibroblasts. It may thus be difficult to decide whether a certain hepatocyte is spread or unspread, as can easily be done with fibroblasts. In quantification of fibroblast spreading, the proportion of spread versus unspread cells is usually determined.[23] However, this is difficult to do with hepatocytes, and instead the dimensions of a number of cells have to be measured in a somewhat different way.

In experiments where the spreading is to be analyzed, the hepatocytes must be incubated longer than when the initial attachment is measured. Hence the cells should be seeded and incubated in a complete culture medium rather than in the balanced salt solution buffer 3. After the last centrifugation the hepatocytes are thus brought up in Ham's F-10 medium supplemented with penicillin (100 units/ml) and streptomycin (50 $\mu$g/ml), counted, and adjusted to a concentration of 0.25 to 0.50 $\times$ 10$^6$ cells/ml. Of this suspension, 2 ml are pipetted into each 35-mm dish. The dishes are incubated at 37° in humid air containing 5% $CO_2$ for 30–60 min to allow attachment. Then the medium is aspirated off, the attached cells are washed with fresh medium, and the incubation is continued at 37° and 5%

[21] F. Grinnell, *Int. Rev. Cytol.* **53**, 65 (1978).
[22] A. Ben-Ze'ev, S. R. Farmer, and S. Penman, *Cell* **21**, 365 (1980).
[23] F. Grinnell and D. G. Hays, *Exp. Cell Res.* **116**, 275 (1978).

FIG. 2. Schematic illustration of hepatocyte spreading. The contours of hepatocytes attached to collagen-coated dishes are illustrated. The group of cells to the left represents unspread cells that have just attached. The a's represent the projections of the diameters of these unspread cells. The group of cells to the right represents spread hepatocytes, both single and in contact with other cells. The b's represent the projections of the longest dimensions of these spread cells. For further explanations, see the text.

$CO_2$ in 1 ml of F-10 medium per dish. At various times the cells on different dishes are fixed with 2.5% glutaraldehyde in buffer 3. Thus the kinetics of cell spreading can be analyzed. Hepatocyte spreading is usually completed in 5–10 hr. For a qualitative analysis of the spreading, the cells are watched and photographed in an inverted phase contrast microscope. To quantify the spreading, an ocular with a precision micrometer is used in the inverted microscope. The projections on the substrate of randomly selected cells are measured with the micrometer. Unspread cells are spherical, and the projections of the diameters of these cells are measured. Spread cells have a more irregular shape, and on these cells the projections of the longest dimensions are measured, as illustrated in Fig. 2. If a sufficient number of cells are measured, the data can be analyzed statistically, so that spreading under various experimental conditions can be investigated.[13]

Effects of Antibodies on Cell Attachment

Solution: Buffer 3: see Hepatocyte Preparation

In the work to identify cell surface molecules that are involved in various cell adhesion phenomena, antibodies directed against cell surface components have proved to be quite useful.[5,7–9,11,24] Thus multispecific antibodies against plasma membranes have been found to inhibit both intercellular adhesion and attachment to substrates, including collagen. Solubilized cell surface components are able to neutralize this inhibition,[7–9] and it is thus possible to use this neutralizing ability to follow the purification of components involved in, for example, hepatocyte attachment to collagen.

Antibodies against isolated liver plasma membranes are raised in rabbits. Standard procedures are used for membrane isolation,[4,9] immunization,[9] IgG and Fab preparation[9,11]; the reader is referred to the original references cited[9,11] for these methods. Both intact IgG molecules and Fab fragments against hepatocyte plasma membranes inhibit attachment of rat hepatocytes to collagen[11] (Fig. 3).

Hepatocytes are prepared and suspended in buffer 3 as described above. The cells are then incubated together with antibodies (IgG molecules or Fab fragments) at 4° for 30 min. The cell concentration during this incubation is adjusted to $1 \times 10^6$ cells/ml, and the amount of antibodies should vary from 50 to 1000 $\mu$g/$10^6$ cells in different samples. After this preincubation period, 1 ml of the incubation mixture is seeded in 35-mm collagen-coated petri dishes, and incubation at 37° and attachment determination are carried out as described above (see also Fig. 3). The antibodies should be present both during the preincubation period and during the incubation at 37°. To test neutralizing activity of solubilized membrane components, the antibodies are preincubated with the solubilized components at 20° for 30 min prior to preincubation together with cells.

Determination of Cell–Cell Adhesion of Collagen-Attached Cells

*Solutions*

Ham's F-10 medium: see Seeding, Incubation, and Determination of Spreading
Buffer 3: see Hepatocyte Preparation
Lysis buffer: see Determination of LDH Activity
Assay solution: see Determination of LDH Activity

For this purpose dishes containing gels consisting of native collagen fibers are used. The principle of the assay is to seed single-cell suspensions of hepatocytes densely on the collagen gels so that the cells that attach to the collagen will be in contact with each other from the beginning, without the need of cell locomotion.[2] The attached cells are then incubated for the desired time period, and the fraction of the cells that

[24] D. E. Wylie, C. H. Damsky, and C. A. Buck, *J. Cell Biol.* **80**, 385 (1979).

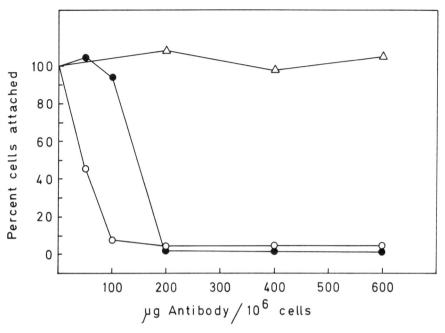

FIG. 3. Effects of antibodies against hepatocyte plasma membranes on attachment of hepatocytes to denatured collagen. Hepatocytes ($1 \times 10^6$ cells per milliliter of buffer 3) were preincubated for 30 min at 4° with the indicated amounts of antibodies or preimmune IgG; $1 \times 10^6$ cells (still in the presence of antibodies) were added to each dish and incubated at 37° for 30 min. The medium was aspirated off, and the dishes were washed with 1 ml of buffer 3. Attached cells were determined as described in the text. Attachment is given as the percentage of cells attached, with 100% set to equal the number of cells attached in the absence of antibodies. ○, Incubation with anti-plasma membrane IgG; ●, incubation with anti-plasma membrane Fab; △, incubation with IgG from preimmune serum.

remain as single cells is determined. The decrease in the percentage of single cells is due to the formation of stable lateral contacts between adjacent cells. To determine the fraction of single cells, the collagen substrate is destroyed by collagenase so that all the cells will appear in suspension again (see Fig. 4).

Hepatocytes are prepared and suspended in F-10 medium at a concentration of $1.6 \times 10^6$ cells/ml, and collagen gels are prepared in 35-mm petri dishes as described above. Of the cell suspension, 2 ml are seeded in each dish. The dishes are then incubated at 37° in humid air containing 5% $CO_2$. After 20 min the cells that have not attached to the gels are aspirated off, the dishes are washed with 1 ml of medium, and then 0.8 ml of F-10 medium is added. The collagen gels should now be covered with a dense

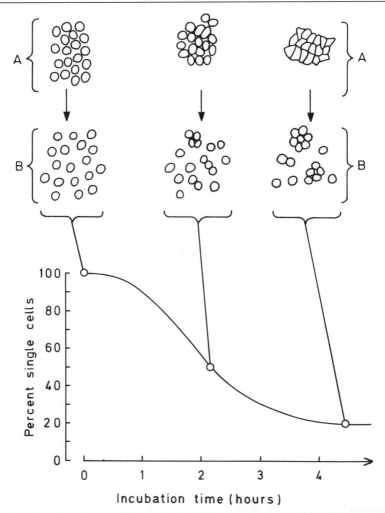

FIG. 4. Schematic illustration of the determination of cell–cell contacts formed between hepatocytes attached to collagen fibers. The upper portion of the diagram shows the appearance of cells (at different times) that are attached to the collagen gels (A). At various times (indicated in the aggregation curve in the lower portion of the diagram) the gels are degraded by collagenase, thus leaving the cells in suspension again (B). With increasing incubation time, more and more of the cells have formed stable lateral bonds with each other, resulting in cellular aggregates present in the suspension. The lower portion of the diagram demonstrates the aggregation kinetics as measured by the decrease in percentage of single cells versus incubation time.

layer of cells lying close to each other. Incubation is continued at 37° and 5% $CO_2$ for varying times. At the end of the incubation time the medium is removed from the dishes, which are then washed twice with 1 ml of buffer 3 containing bovine serum albumin (15 mg/ml). One milliliter of buffer 3, containing per milliliter 0.5 mg of collagenase (type I, Sigma Chemical Co.) and 15 mg of bovine serum albumin, is added to each dish; the dishes are incubated at 37° for 10 min. The collagen gel is completely lysed by this treatment, and the cells, which now are in suspension again, are gently transferred to a clean petri dish with a large-bore Pasteur pipette. At this stage both the number of single cells and the amount of total cells per unit volume should be determined to give the percentage of single cells. The concentration of single cells is determined with an electronic particle counter, e.g., a Coulter Counter (Coulter Electronics, Limited) or a Celloscope 302 (Lars Ljungberg & Co., Stockholm, Sweden). The aperture of the capillary in the counting piece should be minimum 100 $\mu$m and maximum 190 $\mu$m. The samples have to be diluted with buffer 3 in order to avoid coincidence error in the cell counting. How much the samples have to be diluted depends on the size of the aperture. The larger the aperture, the more the cells should be diluted to avoid coincidence error (follow the instructions of the manufacturer). The amount of total cells per unit volume is determined by determination of the LDH activity as described above.

By this method the percentage of single cells decreases with time owing to the formation of stable lateral intercellular contacts while the cells are attached to collagen fibers (see Fig. 4).[2] The collagenase treatment does not destroy the cell contacts, or, stated differently, this method measures the formation of cell contacts resistant to collagenase. It is, however, important that the collagenase digestion of the gels be performed in a balanced salt solution, such as buffer 3, which contains calcium, since calcium is vital for the formation and integrity of hepatocyte intercellular contacts and, further, is necessary for optimal activity of the collagenase. Serum albumin is added during the collagenase digestion to serve as a competitive substrate for other proteases present in the collagenase preparation.

## [27] Biosynthetic Matrices from Cells in Culture

By CARL G. HELLERQVIST

The molecular mechanisms involved in cell adhesion have prompted considerable interest over the years. The study of interactions between cells and matrices *in vitro* is aimed at shedding some light on the extremely complicated system of interactions that must occur *in vivo*.

*Maintenance of Cells*

The experimental approaches to be discussed below utilize human diploid fibroblasts derived from an explant of newborn foreskin.[1] The cells are routinely maintained in Dulbecco's medium (Gibco, Inc.) supplemented with 10% (v/v) fetal calf serum (medium B). Cells between passages 10 and 20 are used for experimental purposes, and stock cultures are stored in liquid nitrogen.

*Media*

Phosphate-buffered saline (PBS), pH 7.3, containing (in grams per liter) NaCl, 8.00; KCl, 0.20; $Na_2HPO_4$, 1.16; $KH_2PO_4$, 0.20. This PBS is used for washing cell layers and microexudate carpets (see below).

Incubation medium (medium A): Incubations performed with cells under serum-free conditions utilize a HEPES-buffered Hanks' balanced salt solution, pH 7.3, containing (in grams per liter) HEPES (Sigma Chemical Co.) 2.38; NaCl, 8.00; KCl 0.40; $MgS_4 \cdot 7\ H_2O$, 0.20; $Na_2HPO_4 \cdot 2\ H_2O$, 0.06; $KH_2PO_4$, 0.06; $CaCl_2$, 0.14; $NaHCO_3$, 0.35, supplemented with D-glucose 1.00 g/liter; ascorbic acid, 50 mg/liter; and essential and nonessential amino acids at the concentrations given for Minimal Essential Media (Gibco Inc). When a radioactive amino acid is used for labeling purposes the concentration of that cold amino acid in the medium is reduced by 90%.

Medium B: Dulbecco's medium supplemented with 10% (v/v) fetal calf serum.

Microexudate Carpets

*Preparation of Urea Carpet.* The urea microexudate carpet is prepared in the following manner. Confluent 7-day cultures are washed twice with

[1] K. J. Lembach, R. E. Branson, P. B. Hewgley, and L. W. Cunningham, *Eur. J. Biochem.* **72**, 369 (1977).

PBS and incubated with 0.25% trypsin (Gibco Solution B) for a minimum length of time (5–15 min). Serum containing medium B is added, and the dispersed cells are transferred to centrifuge tubes and spun at 200 $g$ for 4 min. The pellet is resuspended in medium A and recentrifuged as above twice. The cells are then resuspended in medium A, plated onto tissue culture dishes so as to give a density of $5 \times 10^4$ cells/cm$^2$, and incubated for 4 hr at 37° in 10% $CO_2$/moisture-saturated air.

These human fibroblasts will attach over the first hour and by the end of a 4-hr incubation period they will have formed a monolayer with a morphology quite similar to that which would be observed under incubation conditions employing serum.

The monolayers are washed twice with PBS; 1 $M$ urea buffered to pH 8 with 0.1 $M$ Tris-HCl is then added, and the dishes are incubated for 30 min at 37°.[1] The cell suspension is removed by aspiration, and the residue (urea carpet) is washed twice with the buffered urea solution followed by five washes with PBS. The urea microexudate carpet so obtained is overlayered with medium A until it is to be inoculated with a cell suspension and/or chemically or enzymically modified.

*Preparation of SDS Carpet.* To prepare SDS carpet, monolayers obtained during a 4-hr incubation as above, are overlaid with 0.05% SDS. The instantly lysed cells are removed, and the SDS carpet is washed five times with PBS. Aliquots of medium A are then added to the dish, pending further experiments.

Preparation of Enzyme-Digested Carpets

*Preparation of Trypsin-Digested Carpets.* Urea carpets, prepared from incubations of $5 \times 10^4$ cells/cm$^2$ in medium A as above, are incubated for 15 min with a 0.1% (w/v) crystalline trypsin solution. Crystalline trypsin (~250 units/mg) (Worthington Biochemical Corp.) is dissolved in PBS, and the carpets are incubated with 125 $\mu$l of the solution per square centimeter of surface area. After 15 min at 37°, the protease-digested carpet is washed three times with PBS and overlaid with medium A pending further use.

*Preparation of Collagenase-Digested Carpets.* Urea carpets are incubated with collagenase solutions prepared by dissolving the enzyme (Worthington CLSPA) (~400 units/mg) in medium A so as to obtain a concentration of 160 units/ml. Urea carpets are incubated with 125 $\mu$l of this solution per square centimeter of dish surface area for 15 min at 37°; they are then washed three times with PBS, and medium A is added to prevent drying prior to further use.

EFFECT OF METHOD OF CELLULAR RELEASE ON
L-[U-¹⁴C]PROLINE-LABELED RESIDUE[a]

| Residue | Total counts per minute | Hydroxyproline : proline ratio |
|---|---|---|
| Urea | 3420 | 0.118 |
| Collagenase | 3450 | 0.024 |
| Trypsin | 3255 | 0.114 |

[a] Cells were incubated in 10-cm dishes with 10 µCi of [¹⁴C]proline for 4 hr in HEPES–Hanks' buffer. Cells were removed by 1 $M$ urea, 0.1 $M$ in Tris-HCl (pH 8.0; 30 min at 37°), 0.1% Worthington CLSPA collagenase in HEPES-Hanks' buffer (30 min at 37°), or 0.1% crystalline trypsin in phosphate-buffered saline (5 min at 37°). The residue was solubilized as described.

## Characterization of Carpets

*Solubilization of Carpets.* Isotopically labeled carpets are obtained by introducing an appropriate labeled precursor during the 4-hr incubation. The microexudate carpet obtained by either urea or SDS removal of the cells or by enzymic digestion can then be removed in 80–90% yield, as judged by recovery of radioactivity, by incubating the dish with 1% SDS. The remaining radioactivity can be released from the dish by incubation with 2 $N$ NaOH for 4 hr at 37°.

For a 10-cm tissue culture dish the procedure is as follows. Five ml of 1% SDS are added, and the plate is sonicated at 40° in the Ultramet II (Buchler LTD) by removing the lid, placing it, floating, in the water bath, then placing the plate in the lid. The dish is then placed overnight in an incubator at 37°. After this incubation the sonication procedure is repeated.

The solubilized residue is transferred to a test tube, and the dish is washed with an additional 5 ml of 1% SDS, which is combined with the initial solution and heated at 100° for 2 min.

The bulk of the SDS is removed by dialysis against 3% aqueous $n$-butanol. At the onset of the dialysis, approximately 1 ml of $n$-butanol per 10 ml solution is added into the contents of the dialysis bag. After dialysis the material can be lyophilized and subjected to analysis.

An example of such an analysis is given in the table. The cells were incubated for 4 hr in medium A containing 50 µCi [1-¹⁴C]proline, and the residues were collected as above and subjected to amino acid analysis.[2]

[2] C. E. Schwartz, C. G. Hellerqvist, and L. W. Cunningham, *Ann. N. Y. Acad. Sci.* **312**, 450 (1978).

## Adhesion of Single Cells to Urea Carpet

*Characteristics of the Urea Carpet.* Scanning electron microscopy (SEM) demonstrates that the microexudate carpet does not contain any visible structures such as "foot pads."[3] When a freshly trypsin- or collagenase-released cell suspension obtained from a 4-hr serum free incubation is allowed to adhere to the carpet, the cells will attach and spread completely within 5 min.[3]

*Quantitation of Cell Substrate Adhesion.* The microexudate carpet as evidenced by SEM observations promotes cell attachment and spreading of the cells. Thus biochemical studies of the mechanism of this phenomenon are feasible utilizing a modification of a quantitative adhesion assay.[4]

The rate at which single cells adhere to the microexudate carpet is determined as follows.

A labeled single cell suspension is prepared by incubating a 6-day culture for 24 hr with 100 $\mu$Ci of L-[4,5-$^3$H]leucine in 20 ml of medium B in a 15-cm tissue culture dish. These cells are dispersed with trypsin as described earlier, and a 10-cm tissue culture dish is inoculated with $5 \times 10^4$ cells per square centimeter in medium A and incubated for 4 hr at 37°.

The cells are subsequently released by incubating the dish with 2.5 ml of either medium A containing 160 units of collagenase per milliliter for 30 min, or PBS containing 250 units of crystalline trypsin per milliliter for 5 min on a rocking platform at 37°. The collagenase-dispersed cells are diluted with 10 ml of ice cold medium A and centrifuged at 200 $g$ for 4 min. The pellet is washed once more in ice-cold medium A, and the pellet is resuspended in ice-cold medium A to a final concentration of 100,000 cells/ml. Bovine serum albumin (BSA) is added to the cell suspension to a concentration of 3 mg/ml to prevent adhesion of cells to plastic. The trypsin-dispersed cells are diluted with medium B (10 ml) to inhibit the trypsin and 10 ml of medium A to maintain the pH near 7.3. This mixture is centrifuged at 200 $g$ for 4 min. The pellet is resuspended in medium A, repelleted, and washed once more with medium A prior to preparing a final concentration at 100,000 cells/ml in medium A containing 3 mg of BSA per milliliter.

Urea carpets are prepared by inoculating each well in a multiple-well dish (such as a Linbro Chemical Co. Model Fb 16-24TC) with 1 ml of a cell suspension containing $10^5$ cells in medium A. After a 4-hr incubation at 37°, carpets are prepared as described previously. Every effort should be

---

[3] C. E. Schwartz, L. Hoffman, C. G. Hellerqvist, and L. W. Cunningham, *Exp. Cell. Res.* **118**, 427 (1979).

[4] B. T. Walter, P. Ohman, and S. Roseman, *Proc. Natl. Acad. Sci. U.S.A.* **70**, 1469 (1973).

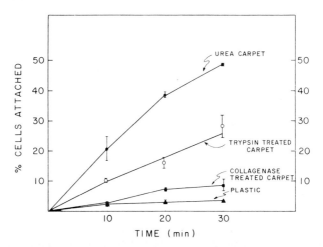

FIG. 1. Adhesion of collagenase-released single cells. Collagenase-released [$^3$H]leucine-labeled single cells were added to unlabeled urea carpets at 100,000 cells per Linbro well. Collagenase-treated carpets were subjected to 100 units of collagenase solution per milliliter for 15 min at 37°. Trypsin carpets were treated with 0.1% crystalline trypsin for 5 min at 37°. All assay media were 0.3% in bovine serum albumin. From Schwartz et al.[2]

made to coordinate the preparation of single-cell suspension and carpet to minimize the time required before the assay can be initiated.

The assay is started by adding 1 ml of the [$^3$H]leucine-labeled single-cell suspension to each well of the multiple-well dish. The dish is then placed in a reciprocal shaker water bath at 37° and 50 strokes per minute.

Adhesion kinetics are measured by removing by aspiration the nonadherent cells from wells at various times. The well is subjected to two washes with medium A immediately after the removal of the cell suspension at any given time point.

When the assay is completed, the dish is placed in a freezer for 10 min to facilitate cell lysis. The adhered cells are transferred to scintillation vials with three washes of 0.5 ml of $NH_4OH$ each. Each time $NH_4OH$ is added, the dish is placed in the sonicator bath for 5 min. In order to obtain good duplicates, each well is transferred with a separate pipette, which should be prewetted with 0.5 $N$ $NH_4OH$.

The rate of adhesion is expressed as percentage of cells adhered by comparing the radioactivity in each well with that of a 1-ml aliquot of the original cell suspension, to which 0.5 ml of 1.5 $N$ $NH_4OH$ is added.

Figure 1 shows typical adhesion data obtained in this system. Active metabolism is required for adhesion, as evidenced by the inhibition by KCN and iodoacetate combined.

Comments

The methodology described herein can be utilized to investigate the biochemistry of cell substrate adhesion. The adhesive quality of the urea carpet can be altered either chemically of enzymically, and in addition metabolic inhibitors, such as $\alpha,\alpha'$-dipyridyl or cyclohexamide, can be administered to the cells employed in urea carpet synthesis.

Furthermore the single cell suspension can be modified in similar ways, making possible many permutations that can be investigated to cast some light on the important phenomenon of cell substrate adhesion and on the possible role for the various components of the carpet as well as of the cell membrane in adhesion.

## [28] Ultrastructural Studies of Cell–Collagen Interactions

*By* FREDERICK GRINNELL and MARYLYN HOY BENNETT

One of the major goals of studies on cell–substratum interactions is to develop *in vitro* models of normal tissue and organ structure that can be used to study such phenomena as cell growth and differentiation.[1,2] There are a variety of cell behaviors expressed on hydrated collagen gel substrata that are absent when cells are cultured on material surfaces (e.g., glass, plastic). Some of these were reviewed in the Overview on cell–collagen interactions" (this volume [23]). It is particularly noteworthy that cell shape[3,4] and cell motility[5,6] are more typical of *in vivo* behavior when cells are cultured on hydrated collagen gels compared to when they are cultured on material surfaces. The crucial difference may be that, on hydrated collagen gels, cells can penetrate into the three-dimensional matrix of the substratum.[3,4,7] In support of this notion is the finding that cells interact with dried collagen gels similarly to the way in which they interact with material surfaces.[8] Dried collagen gels lack a three-dimensional matrix because they collapse as a result of the drying procedure (see below).

[1] J. Yang, J. Richards, P. Bowman, R. Guzman, J. Enami, K. McCormick, S. Hamamoto, D. Pitelka, and S. Nandi, *Proc. Natl. Acad. Sci. U.S.A.* **76**, 3401 (1979).
[2] E. Bell, B. Ivarsson, and C. Merrill, *Proc. Natl. Acad. Sci. U.S.A.* **76**, 1274 (1979).
[3] T. Elsdale and J. Bard, *J. Cell Biol.* **54**, 626 (1972).
[4] T. Ebendal, *Zoon* **2**, 99 (1974).
[5] J. B. L. Bard and E. D. Hay, *J. Cell Biol.* **67**, 400 (1975).
[6] E. M. Davis, *J. Embryol. Exp. Morphol.* **55**, 17 (1980).
[7] F. Grinnell and M. H. Bennett, *J. Cell Sci.* **48**, 19 (1981).
[8] G. A. Dunn and T. Ebendal, *Exp. Cell Res.* **111**, 475 (1978).

In order to analyze cell morphology, the organization of the substratum, and physical relationship of cells to the substratum, it is necessary to use ultrastructural methods. In the following sections, techniques are described for using scanning and transmission electron microscopy to study cell interactions with dried and hydrated collagen gels.

Preparation of Collagen

All steps are carried out using sterile technique. Collagen is prepared essentially as described by Price[9] by extracting the minced tendons from two rat tails in 200 ml of 0.1% acetic acid for 48 hr at 4°. Insoluble material is removed from the preparation by centrifugation at 5500 rpm for 30 min at 4° (Sorvall GSA rotor) and then 7800 rpm for 30 min at 4° (Sorvall HB4 rotor). Subsequently, the collagen is precipitated from solution by the addition of NaCl (10% w/v). The precipitate is collected by centrifugation and resuspended in 50 ml of 0.1% acetic acid and dialyzed overnight against the same solution. The purified preparations (about 2 mg/ml) are stored at 4° and are stable for several weeks. Collagen type I (Vitrogen 100) of reasonable purity is also prepared commercially by the Collagen Corporation, Palo Alto, California.

Preparation of Substrata

All steps are carried out using sterile technique. Collagen substrata are prepared on glass coverslips 12 mm in diameter (Scientific Products) or on 15-mm Thermanox plastic coverslips (Lux Scientific Corporation) for scanning electron microscopy (SEM) and on Permanox dishes 60 mm in diameter (Lux Scientific Corporation) for transmission electron microscopy (TEM). The use of the Permanox dishes is essential because these dishes are resistant to the organic reagents used during dehydration of the specimens. In addition, the embedded specimens can be removed more easily from these dishes than any others that we have tested. The coverslips for SEM are placed in 60-mm tissue culture dishes for handling purposes.

Dried collagen gels are prepared by the method of Klebe[10] except that the urea treatment step is omitted. Aliquots of the above collagen solution (adjusted to 0.5 mg/ml with 0.1% acetic acid) are placed in the center of the coverslips (0.1 ml) or Permanox dishes (0.2 ml), gelled by exposure to ammonia vapors for 30 min at 22°, and then air-dried for 48 hr under a

[9] P. J. Price, *Tissue Cult. Assoc. Man.* **1**, 43 (1975).
[10] R. J. Klebe, *Nature* (*London*) **250**, 248 (1974).

tissue culture hood. If higher collagen concentration are used, the dried gels sometimes crack.

Hydrated collagen gels are prepared as described by Elsdale and Bard.[3] Aliquots (1.0 ml) of the collagen solution (about 1.5–2.0 mg/ml) are maintained at 4° and brought to physiological ionic strength by the addition of 10 times concentrated phosphate buffer saline to a final concentration of 0.01 $M$ sodium phosphate, 0.15 $M$ NaCl. The solutions are then neutralized to pH 7 by the addition of 0.8 $M$ NaOH (approximately 0.015 ml). Subsequently, samples of these solutions are placed on coverslips (0.1 ml) or Permanox dishes (0.2 ml) and incubated for 60 min at 37° in a humidified incubator. The gels that form are approximately 1 mm thick. These substrata are used immediately. With lower concentrations of collagen, the hydrated gels sometimes float off the coverslips and the Permanox dishes.

Incubation with Cells

All steps are carried out using sterile technique. The collagen substrata should be preincubated briefly (about 15 min) with 5 ml of the medium to be used in the experiments. This can be a basic salts solution or complete medium with or without serum or other proteins. Subsequently, cells in 3–6 ml of medium are plated out using routine tissue culture techniques. Care must be taken not to physically disrupt the hydrated collagen gels, which are easily collapsed and/or torn by harsh treatment. Cells can be cultured on the collagen gels for short or extended times. Cells that secrete collagenases, however, may eventually destroy the collagen substratum.[11] Using an inverted phase contrast microscope, it is possible to see through the collagen gels (hydrated or dried) and determine whether cells are attached and spread.

SEM Analysis

Samples for SEM analysis are fixed for 45 min at 22° with 4 ml of 3% glutaraldehyde (Electron Microscopy Sciences) in 0.1 $M$ sodium phosphate, pH 7. Cells are then washed twice for 15 min at 22° in phosphate buffer containing 0.2 $M$ sucrose and stored in phosphate sucrose overnight at 4°. The samples are postfixed with 1% $OsO_4$ (Polysciences) in phosphate buffer (just enough to cover the samples) for 30 min at 22° and then washed for 15 min in 4 ml of phosphate buffer and for 15 min in 4 ml of

[11] J. P. Revel, G. Darr, E. B. Griepp, R. Johnson, and M. M. Miller, in "Molecular Basis of Cell–Cell Interaction" (R. A. Lerner and D. Bergsma, eds.), p. 57. Liss, New York, 1978.

glass-distilled $H_2O$ at 22°. The specimens are then dehydrated through a graded series of ethanol at 22° with two times 5 min each in 4 ml of 30%, 50%, 70%, 95%, and three times in 100%. The dehydrated samples are critically point dried (e.g., Sorvall) with liquid $CO_2$ using six cycles of 5 min soak and 2 min flush. Finally, the samples are placed in an argon atmosphere at $10^{-6}$ to $10^{-7}$ torr in a vacuum evaporator (e.g., Denton DV5022) and coated with approximately 100 Å of metal from a 60% gold/ 40% palladium target using a sputter coater (e.g., Denton DSM-5). When viewing the coated specimens with the scanning electron microscope, accelerating voltages ranging from 10 to 25 kV are used depending upon the stability of the samples and the magnification desired.

TEM Analysis

Samples for TEM analysis are fixed and washed similarly to samples for SEM except that 2% glutaraldehyde/1% tannic acid is used instead of 3% glutaraldehyde to enhance the contrast of membranes. Also, postfixation with $OsO_4$ is carried out for 60 min instead of 30 min. The samples are dehydrated through ethanol as for SEM except that 15-min periods are used. This is particularly important with the hydrated gels, which are often difficult to embed properly. The dehydration is completed with two 10-min treatments with propylene oxide (which evaporates very rapidly), and then infiltration is carried out overnight beginning with 50% propylene oxide/50% Epon, allowing the propylene oxide to evaporate. The Epon mixture used (Epon 812, Ladd Scientific) is prepared by mixing 4.5 ml of solution A, 5.5 ml of solution B, and 2% DMP-30. Finally, the samples are placed in 4 ml of fresh 100% Epon and polymerized in a 60° oven for 24 hr. Immediately after the samples are removed from the oven, the Epon-embedded specimens can easily be peeled away from the Lux Permanox dishes. The samples are trimmed to 1-mm × 5-mm block faces, and thick sections (1 μm) are cut using glass or diamond knives with an ultramicrotome (e.g., Sorvall MT-2).

An alternative method for preparing samples is useful when it is necessary to view a large number of cells in each section (e.g., when studying the distribution of receptors), but the precise geometric arrangements of the cells to each other and to the substratum are less important. This is accomplished by using propylene oxide to dissolve the surface of the culture dish and then preparing a compact pellet containing all the cells and the substratum. In this case, the collagen substrata are prepared on routine 60-mm tissue culture dishes, instead of the Permanox dishes. Everything else is done the same through the ethanol dehydration. At that point in the procedure, 2 ml of propylene oxide are added to the dish, and

## [28] ULTRASTRUCTURE OF CELL–COLLAGEN INTERACTIONS 539

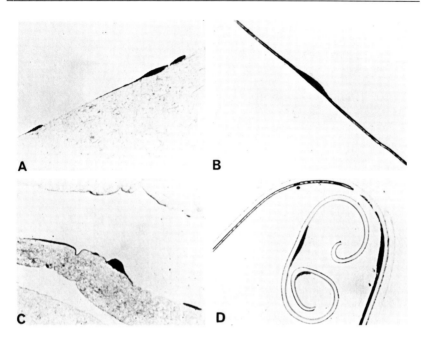

FIG. 1. Thick sections of human skin fibroblast cells on dried and hydrated collagen substrata. The specimens were prepared by embedding in dishes (A, B) or by embedding pellets (C, D). (A) and (C) are hydrated collagen substrata; (B) and (D) are dried collagen substrata. With the pellet preparations, a layer of plastic can be observed underlying the layer of dried collagen (D). Other details are given in the text. × 550.

the cells and substratum are immediately scraped off using a Teflon policeman. This sample is placed in a microfuge tube and microfuged (e.g., Beckman Model 152) for 1 min at 22°. The tip of the tube is then cut off with a razor blade, and the pellet is mechanically dislodged and transferred to a 12-ml glass snap-cap vial (Wheaton). Two additional washes with 4 ml of propylene oxide are carried out to complete dehydration and to remove plastic from the culture dish that is trapped in the pellet. Embedding and polymerization are carried out as above except that the last step is performed with the pellets in 1 ml of fresh Epon in Beem capsules (Ernest F. Fullam). For thick sections, excess plastic is trimmed away and the entire pellet is sectioned.

The thick sections from samples prepared by either technique are picked up on glass slides and stained with toluidine blue (0.75 g of toluidine blue and 0.75 g of sodium borate in 50 ml of distilled $H_2O$, freshly filtered) at 60°. The sections can then be viewed by light microscopy. Figure 1 illustrates the appearance of cells on native and dried collagen

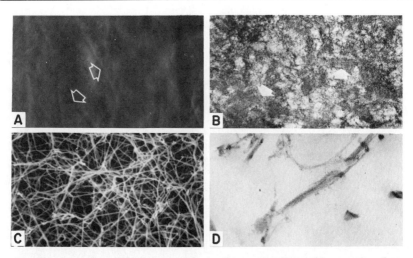

FIG. 2. Appearance of dried and hydrated collagen gels observed by scanning electron microscopy and by transmission electron microscopy. (A) Dried collagen, 45° tilt, × 6000; (B) dried collagen, × 42,000; (C) hydrated collagen, 25° tilt, × 6000; (D) hydrated collagen, × 42,000. Other details are given in the text. From Grinnell and Bennett.[7]

substrata prepared by the two different methods. Once the cells are located, the block faces of the dish-polymerized specimens are further trimmed to 1 mm × 2 mm (1 mm × 1 mm for the pellet specimens).

Thin (gold) sections are cut with a diamond knife and picked up on 200-mesh copper grids (Pelco) coated with Formvar (Ladd, 0.5% in dichloroethylene). The specimens are stained for 10 min with uranyl acetate (saturated solution in 50% ethanol) followed by 7 min with Reynold's lead citrate.[12]

Appearance of Collagen Gels by SEM and TEM

The differences in physical organization of dried and hydrated collagen gels are readily visualized as shown in Fig. 2. Observed by SEM, the hydrated collagen gels are composed of an interlaced network of fine collagen fibrils approximated 50–100 nm in diameter (Fig. 2C), and cross sections and longitudinal sections of the fibrils are seen by TEM (Fig. 2D). The dried collagen gels are markedly different and appear to have collapsed. They are characterized by a relatively smooth surface with only

[12] E. S. Reynolds, *J. Cell Biol.* **17**, 208 (1963).

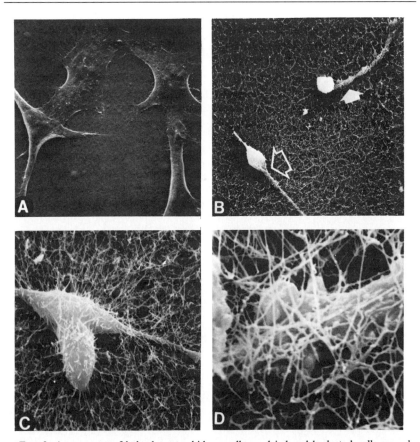

FIG. 3. Appearance of baby hamster kidney cells on dried and hydrated collagen substrata observed by scanning electron microscopy. The incubations contained $0.75 \times 10^6$ cells per dish and were carried out for 2 hr. In the experiments on dried collagen, plasma fibronectin (10 units/ml) was added; otherwise, the cells did not attach. In the experiments on hydrated collagen, there were no proteins added to the incubation medium. The preparation of fibronectin and the incubation medium used have been described by F. Grinnell and D. Minter [*Proc. Natl. Acad. Sci. U.S.A.* **75**, 4408 (1978)]. (A) Cells on dried collagen; × 700. (B) cells on hydrated collagen with monopolar (filled arrow) or bipolar (open arrow) morphology; × 700. (C) cells on hydrated collagen that have completely penetrated in the matrix; × 2125. (D) monopolar cell on hydrated collagen showing edge of cell body and pseudopodia within the collagen matrix; × 7100. Other details are given in the text. From Grinnell and Bennett.[7]

occasional raised fibrils (Fig. 2A, arrows). Observed by TEM, the dried collagen gels appear to consist predominantly of amorphous, densely packed material, although a few collagen fibrils are visible (Fig. 2B, arrows).

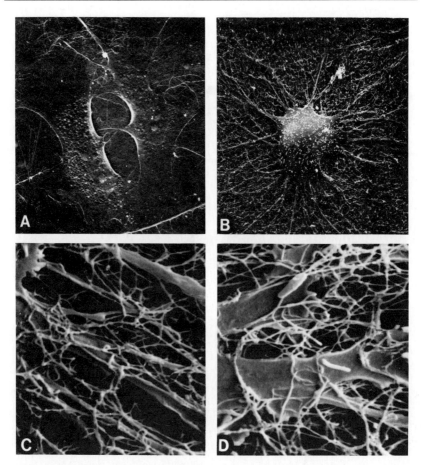

FIG. 4. Appearance of human skin fibroblast (HSF) cells on dried and hydrated collagen substrata observed by scanning electron microscopy. The incubations contained $0.4 \times 10^6$ HSF cells per dish and were carried out for 2 hr. Protein-free medium was used in all the experiments. These cells, however, can secrete their own fibronectin and use it as an adhesion factor [F. Grinnell and M. Feld, Cell **17**, 117 (1979)]. (A) Cells on dried collagen, × 700. (B) Cells on hydrated collagen; × 900. (C) Detail of thin cell extensions just beneath surface of collagen matrix; × 7100. (D) Detail of broad cell extension just beneath surface of collagen; × 7100. Other details are given in the text. From Grinnell and Bennett.[7]

Appearance of Fibroblasts on Dried and Hydrated Collagen Gels as Observed by SEM

The morphologies of baby hamster kidney (BHK) cells and early passage human skin fibroblast (HSF) cells after incubation for 2 hr on the collagen substrata are observed by SEM as shown in Figs. 3 and 4, respectively. Both cell types have broad, flat lamellipodia and fine retraction

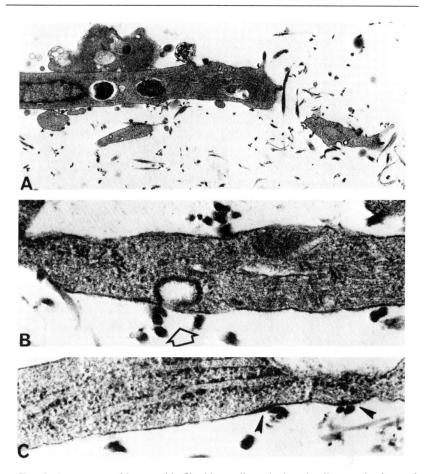

FIG. 5. Appearance of human skin fibroblast cells on hydrated collagen gels observed by TEM. The incubation conditions are the same as in Fig. 4. (A) Cell body on the surface of the collagen matrix with extensions within the matrix. × 11,200. (B) Detail of cell extension within collagen matrix showing collagen fibrils nearby what may be a coated vesicle (arrow). × 55,000. (C) Detail of cell extension within collagen matrix showing what appear to be adhesion plaques (arrow heads). × 55,000. The specimens were observed and photographed using a Philips 300 electron microscope at 80 kV. Other details are given in the text.

fibrils on dried collagen (Figs. 3A, 4A). On the other hand, cell spreading on hydrated collagen is characterized by the protrusion of large cell pseudopodia. The BHK cells are generally monopolar or bipolar (Fig. 3B), whereas the HSF cells are stellate (Fig. 4B). With longer incubation times, the HSF cells become bipolar. It also can be seen that the BHK cells sometimes burrow within the collagen (Fig. 3C) as well as send extensions deep within the collagen (Fig. 3D), whereas the extensions of the HSF cells stay just beneath the surface of the collagen and become intertwined with the collagen fibrils (Figs. 4C, 4D).

### Appearance of HSF Cells on Hydrated Collagen Gels as Observed by TEM

The presence of the HSF cell bodies on top of the collagen gel with cell extensions projected into the collagen can be observed by TEM (Fig. 5A). When observed at higher magnification, special morphological features can be observed. For instance, collagen fibrils sometimes appear to be associated with structures that look like coated vesicles[13] (Fig. 5B, arrow) or adhesion plaques[14] (Fig. 5C, arrows).

[13] R. G. W. Anderson, J. L. Goldstein, and M. S. Brown, *Proc. Natl. Acad. Sci. U.S.A.* **73**, 2434 (1976).
[14] M. Abercrombie, G. A. Dunn, and J. P. Heath, *in* "Cell and Tissue Interactions" (J. W. Lash and M. M. Burger, eds.), p. 57. Raven, New York, 1977.

## [29] Collagen as a Substrate for Cell Growth and Differentiation

*By* STEPHEN C. STROM and GEORGE MICHALOPOULOS

It has been long recognized that established cell culture lines can maintain a stable differentiated phenotype and replicate on plastic surfaces whereas primary cultures occasionally require specialized substrates. This is more applicable to primary cultures of epithelial cells, such as hepatocytes,[1] mammary epithelial cells,[2] transitional epithelium,[3] squamous epithelium,[4] and thyroid cells.[5] The particular type of collagen substrate used may also affect cellular attachment and differentiation in culture. There is evidence that epithelial cells in primary culture attach better and maintain differentiated properties for longer periods of time when maintained on collagen gels as compared to collagen-coated plates.[1,2] These effects of collagen are not surprising in view of the fact that collagen in its various types is a ubiquitous component of intercellular matrix.

[1] G. Michalopoulos and H. C. Pitot, *Exp. Cell Res.* **94**, 70 (1975).
[2] J. T. Emerman and D. R. Pitelka, *In Vitro* **13**, 316 (1977).
[3] F. J. Chlapowski and L. Haynes, *J. Cell Biol.* **83**, 605 (1979).
[4] M. A. Karasek and M. E. Charlton, *J. Invest. Dermatol.* **56**, 205 (1970).
[5] M. Chambard, J. Gabrion, and J. Mauchamp, *C. R. Hebd. Seances Acad. Sci. Ser. D* **291**, 79 (1980).

Other components of intercellular matrix, such as fibronectin and mucopolysaccharides, may also be of importance in promoting attachment and differentiation of epithelial cells in primary culture. Rojkind et al.[6] indicated that when a crude preparation of intercellular matrix was extracted from whole liver and was used to coat plastic culture surfaces, the hepatocytes that attached on the strands of this "biomatrix" survived longer and maintained properties characteristic of hepatic differentiation for long periods of time. We describe here the methods that in our hands give the best results in terms of coating plain plastic plates and making collagen gels as substrates for cell culture. We also present evidence on the effects of collagen, using examples from normal hepatocytes and hepatoma cells.

Methods

*Preparation of Collagen Solutions.* Collagen is available through several commercial sources. Of the various types of dry collagen that are marketed, we found that only the ones declared to be acid soluble could be used to prepare collagen solutions for tissue culture purposes. A preferable source of collagen for our purposes has been the collagen fibers obtained from the tendons that attach to the vertebrae of the rat tail.[7] Rat tails are removed when rats are sacrificed for experimental purposes. Care should always be taken to avoid using tails of animals that have been exposed to toxic substances. The rat tails can be frozen at $-20°$ for more than a year without appreciable loss of solubility of collagen fibers. Approximately 2 hr prior to the preparation of fibers, the rat tails are brought to room temperature and allowed to thaw slowly. When the rat tails are thawed, the base of the tail is held with forceps, and with sharp scissors a longitudinal incision is made along the length of the tail all the way from the base to the tip of the tail. After the incision is made, the skin is held with a hemostat and pulled off the tail. The exposed surface of the tail reveals the long collagen fibers and vessels that run along the length of the tail. The vessels that are on the surface are removed. Using a forceps and a hemostat, the distal three to four small vertebrae at the tip of the tail are broken off and pulled away slowly. The muscle tendons (collagen fibers) that are attached to them are removed along with the vertebrae. This procedure is repeated every two or three vertebrae, going from the tip of the tail toward the base. After removal of the collagen fibers and vertebrae, the collagen fibers are severed with scissors from the points of at-

[6] M. Rojkind, Z. Gatmaitan, S. Mackensen, M. Giambrone, P. Ponce, L. M. Reid, *J. Cell Biol.* **87**, 255 (1980).
[7] R. L. Ehrmann, G. O. Gey, *J. Natl. Cancer Inst.* **16**, 1375 (1956).

tachment on the vertebrae. The collagen fibers are left to dry at room temperature for approximately 3 hr; then the fibers are weighed.

The yield of fibers from the tail depends on the size of the animals from which the tails have been removed. The dried fibers are placed on shallow plastic containers and exposed for 48 hr in UV light generated by a germicidal UV lamp. A solution of acetic acid in water is prepared by adding 1 ml of glacial acetic acid to 1000 ml of double-distilled, filter-sterilized water. The UV-sterilized collagen fibers are transferred with a sterile forceps into a sterile container that contains the acetic acid solution, and a presterilized magnetic stirring bar. One gram of collagen fibers is added to 300 ml of the acetic acid solution. The container is placed on top of a magnetic stirrer and is left to stir at low speed for approximately 48 hr. Within 1 hr after the collagen fibers have been placed in the acid solution, the fibers become more flexible and clear. After stirring for 48 hr the resultant collagen solution (solution A) contains crude fibers in suspension and dissolved collagen. A solution free of undissolved fibers can be obtained by filtering through a sterile triple gauze filter. In our hands the presence of undissolved fibers has not been found to interfere with either the coating of plastic plates or the preparation of collagen gels. Filtration of the collagen solution does, however, allow easier pipetting.

*Coating of Plastic Tissue Culture Plates.* A collagen solution prepared as described above is diluted 1 : 10 with sterile double-distilled water. From this diluted solution (solution B), 2 ml are added into tissue culture plates 60 mm in diameter. The plates are shaken by hand to assure uniform distribution of the solution. The plates with the solution are then placed in a dry incubator at 37° for 1–2 days until the plates appear dry. It is important to keep the temperature of the incubator at 37° or below to minimize denaturation of the collagen fibers. After the plates are dry, they can be stored in a humidified incubator indefinitely and used for tissue culture.

*Preparation of the Collagen Gels.* The principles for the preparation of collagen gels were originally described by Elsdale and Bard.[8] Briefly, when solutions of collagen at the proper concentrations are treated in such a way that there is a simultaneous rise in the ionic strength and the pH of the solution, the collagen fibers precipitate *in situ* to form a "hydrated collagen lattice." In this form, which we shall call collagen gel, the collagen is in bundles of native collagen that, as described by the above authors, vary in diameter from 500 to 5000 Å. When the acidic collagen solutions are exposed to raised ionic strength and pH close to the normal physiological level, then they gel fairly fast. This necessitates simultaneous rise of the ionic strength and the pH. The method that worked best

[8] T. Elsdale and J. Bard, *J. Cell Biol.* **54**, 626 (1972).

in our hands is a modification[2] of a method that we originally described.[1] The solutions required for making collagen gels are the following:

Collagen solution (solution A)

Medium Waymouth MB 752/1, 10-fold concentrated and supplemented with bovine serum albumin (20 g/liter). This solution can be prepared from commercially available medium powder, sterilized by filtration, and stored in the refrigerator.

Sodium hydroxide solution, 0.34 $N$

All the solutions are kept at refrigerator temperature and are mixed in a sterile flask kept on ice at volume relations of 17:2.66:1.10, respectively. The mixture is kept on ice; 2 ml of it are added into tissue culture plates 60 mm in diameter. When left at room temperature the solution forms a gel within approximately 5–10 min. The advantages of this method are that the three solutions are evenly mixed by shaking the flask (instead of individual plates), and the pH of the final collagen gel can be visually monitored by the color of the indicator dye (phenol red) that is present in the medium. After the gels form, they can be stored in a humidified incubator for at least 1 month.

The gels attached to the tissue culture plates can be used as substrates as they are. The experience of most laboratories indicates that if the cells added cover at least 60% of the surface of the collagen gel or, alternatively, if the added cells grow to cover most of the surface of the collagen gel, then the collagen gel contracts from the periphery of the plate and occasionally lifts and floats in the medium. Observations with hepatocytes and mammary epithelial cells indicate that the attached epithelial cells remain longer in culture and maintain differentiated properties longer on floating collagen gels compared to attached gels.[1,2] To prepare floating collagen gels, 4 hr after the cells are added a sterile Pasteur pipette with a flame-fused tip is used to cut the gels around the internal periphery of the tissue culture plate and slowly lift the gels off the plate. Some of the collagen gels may fragment during the cutting procedure. The presence of bovine serum albumin was found in our hands to add to the resilience of the gels and prevent fragmentation. Special care should also be taken during the change of the media in the first 3–4 days after the floating collagen gels are made because, owing to the large size of the gel, it can be accidentally destroyed by the vacuum aspirator.

Results and Discussion

Studies with hepatocytes have provided an example of the effect of collagen on cell longevity and differentiation in culture.[1,9] The effect of the

[9] C. A. Sattler, G. Michalopoulos, G. L. Sattler, and H. C. Pitot, *Cancer Res.* **38**, 1539 (1978).

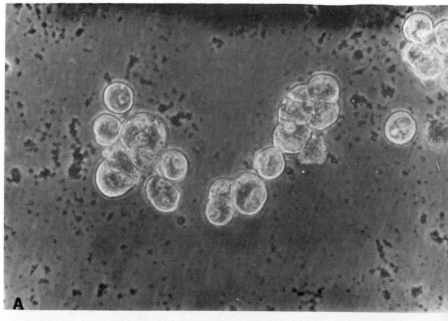
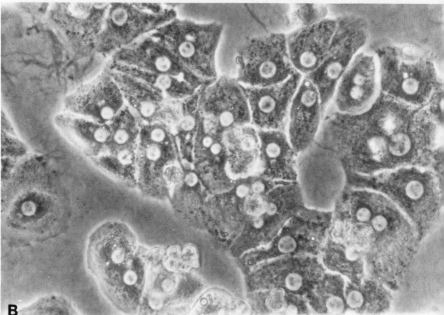

Fig. 1. Morphology of hepatocytes maintained on plain plastic (A) compared to hepatocytes on collagen-coated (B) plates. Hepatocytes were in culture for 24 hr.

substrate on the morphology of the haptocytes is shown in Fig. 1 (A, B). Hepatocytes were prepared by perfusion of rat liver with a collagenase solution.[1] Cells were inoculated in L-15 media with 5% fetal bovine serum in plastic tissue culture plates as well as plates coated with collagen as described above. The morphology of the cells at 24 hr in culture is shown in Fig. 1. Very few cells attach on the plain plastic plate in the absence of collagen. The attached cells did not spread out and remain rounded as shown in Fig. 1A. In contrast, 60% of the inoculated cells attach on collagen-coated plates. The attached cells within 4 hr adhere and spread over the collagen substratum. The morphology of the cells attached to collagen-coated plates is shown in Fig. 1B. Longer follow-up of the cultures shows that the cells attached on plain plastic die and are removed with changes of the media within 3 days in culture. The cells in the collagen-coated plates remain attached and morphologically identifiable as hepatocytes for periods of 6-7 days in culture. Analogous phenomena have been described for a variety of other cell types.[1-5]

In all the above instances, the effects of collagen were seen on nonneoplastic cells. Similar phenomena on the cell morphology and differentiation, however, can be shown on neoplastic cell types. A hepatoma cell line (JMI) produced in our laboratory shows different morphology when grown on plastic versus collagen substrates. The morphology of the JMI cells grown on plain plastic plates is shown in Fig. 2A. The cells grow rapidly on the plain plastic surface to reach a saturation density of at least $3 \times 10^6$ cells per plain plastic plate 60 mm in diameter. The JMI cells, however, do not establish a continuous monolayer, do not acquire an epithelial phenotype, and have the morphological characteristics of transformed cells. Open spaces can be seen between the cells, and cells growing on top of other cells are noted. In contrast when the JMI cells are grown in collagen-coated plates, after reaching saturation density, they lose the morphology of the transformed phenotype and acquire an epithelial appearance with minimal multilayer formation, as shown in Fig. 2B. Prior to reaching saturation density the morphology of the cells growing on plain plastic cannot be distinguished from that of cells on collagen-coated plates. The role of the cell density and the substrate in producing this drastic morphological change in the JMI hepatoma cell line is currently under investigation.

More subtle changes can be seen when one compares cells grown on collagen-coated plates with cells grown on floating collagen gels. Again with hepatocytes, it was found that cells growing on floating collagen gels remained viable longer and also maintained inducibility of characteristic hepatocellular enzymes for longer times in culture.[1,9] When hepatocytes are grown on collagen-coated plates they rapidly lose the levels of cyto-

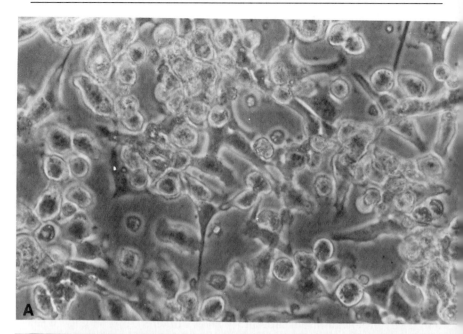
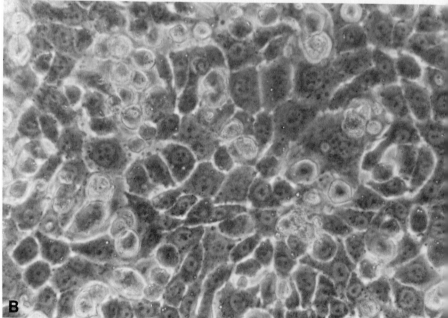

FIG. 2. Morphology of JMI hepatoma cells at saturation density on plain plastic (A) compared to that on collagen-coated (B) plates.

chrome P-450 within 72 hr. The cytochrome P-450 complex cannot be further induced by the addition of characteristic chemical inducers such as phenobarbital and 3-methylcholanthrene. In contrast, cells maintained on floating collagen gels maintain measurable levels of cytochrome P-450 complex for longer periods of time. When the inducers phenobarbital and 3-methylcholanthrene are added, higher levels of cytochrome P-450 and cytochrome P-448 are induced, in a manner analogous to the liver *in vivo*.[10] There are several possible reasons for the differences in behavior of the cells on different types of collagen substrates. Floating collagen gels are in close proximity to the surface of the tissue culture media and the location of the cells on top of the collagen gel may facilitate the exchange of $O_2$ and $CO_2$. Alternatively, it is possible that it is the composition of the collagen gels (higher concentration of collagen fibers, lesser degree of collagen denaturation, presence of other contaminating proteins of extracellular matrix) that is of importance in explaining the differences in behavior of cells rather than the proximity of the cells to the surface of the media.

To investigate this question we compared hepatocytes grown on collagen-coated plates and collagen gels with hepatocytes grown on confluent monolayers of human fibroblasts.[11] Hepatocytes grown on confluent fibroblasts attach on the extracellular matrix produced by the fibroblasts. This matrix contains different types of collagen as well as fibronectin. The thickness of the fibroblast monolayer is very small compared to the thickness of the overlying nutrient medium. Thus the hepatocytes on top of the human fibroblasts are maintained at a distance from the surface of the medium almost identical to that of hepatocytes maintained on collagen-coated plates. The levels and inducibility of cytochrome P-450 complex in hepatocytes maintained on the three substrates were examined; the P-450 levels are shown in Fig. 3. Hepatocytes maintained on top of human fibroblasts behaved in a manner almost identical to hepatocytes maintained on floating collagen gels. These results indicate that the proximity of the cells to the surface of the medium is not a determining factor. It rather appears that subtle effects due to the chemical composition of the substrate may be the important parameters in affecting cell differentiation. The mechanisms by which cell differentiation in culture is affected by collagen are not apparent. This experiment also indicates that several other factors that were thought to be of overriding importance in the design of the floating collagen gels may not be as important. For example, the transfer of metabolites from the medium into the hepatocytes from both sides of the hepatocytes (the one exposed to the medium and the one

[10] G. Michalopoulos, G. L. Sattler, and H. C. Pitot, *Science* **139**, 907 (1976).
[11] G. Michalopoulos, F. Russell, and C. Biles, *In Vitro* **15**, 796 (1979).

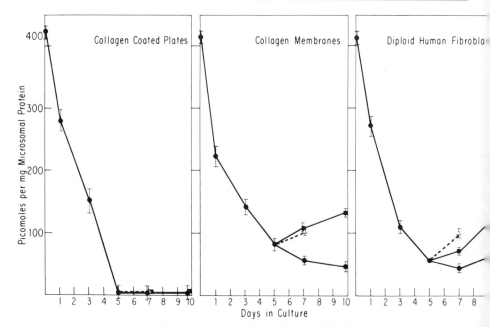

FIG. 3. Levels of cytochrome P-450 in hepatocytes maintained on different substrates as a function of the age of the cultures. ●, control cells; ■, cells treated with phenobarbital (2 × $10^{-3}$ M) from day 5 to day 10 in culture; △, cells treated with methylcholanthrene (10 μm) from day 5 to day 7 in culture. Reprinted from Michalopoulos et al.[10]

attached to the collagen gel) was considered to be important. This mechanism however, probably should not operate in the case of cells attached on top of fibroblasts. Hepatocytes grown on collagen-coated plates are usually, in the presence of serum, overgrown by contaminant cells (fibroblasts, clear epithelial cells, etc.). This phenomenon does not occur on cells on floating collagen gels because of the contraction of the gel (see below). Again, however, this phenomenon does not appear to be of overriding importance, since hepatocytes entirely surrounded by human fibroblasts appear to behave identically to the hepatocytes on floating collagen cells. It is possible that the interplay of mechanical factors between the substrate and the epithelial cells may also be important. Emerman and Pitelka[2] observed that, when mammary epithelial cells were grown on attached collagen cells, the rate of protein synthesis was slow, but the rate of protein synthesis was dramatically increased when the collagen gels were cut and detached from the tissue culture plate.

When hepatocytes are placed on top of floating collagen gels, the collagen gels contract slowly to a smaller size within 4–6 days. A floating gel that is carrying hepatocytes at day 4 in culture is shown in Fig. 4. The

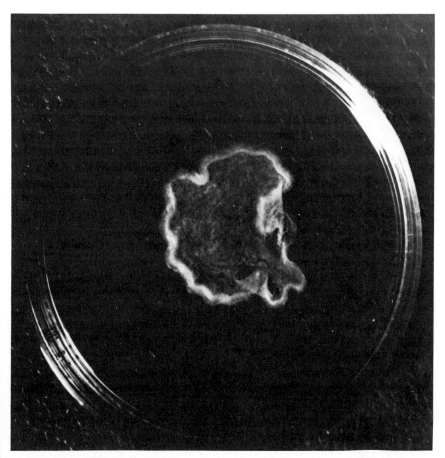

FIG. 4. The appearance of a floating collagen gel at day 4 in culture. The diameter of the collagen gel was originally 60 mm (identical to that of the culture plate shown in this figure). At time zero, $2.5 \times 10^6$ hepatocytes were added, and at 4 hr the gel was released as described in the text.

original size of the gel was the same as the size of the culture plate. The rate of contraction depends on the number of the hepatocytes added. The contraction of the gel is also dependent on the actual physical presence of cells on top of the gel surface. When only half the surface of the collagen gel is overlayed with hepatocytes, the part of the collagen gel that is not overlayed with hepatocytes does not contract. When cell culture medium that was exposed to hepatocytes is added on to collagen gels that do not carry hepatocytes, no contraction of the collagen gels is seen. Analogous phenomena are seen with mammary epithelial cells.[2] The mechanism of the contraction of the collagen gels is not entirely clear. It appears that breakdown of collagen is not involved.[12] When cytochalasin is added to the cultures, the contraction of the collagen gel is prevented. In studies with human fibroblasts, it was also found that addition of colchicine to the cultures prevents contraction of the collagen gel induced by human fibroblasts.[13] Electron microscopy of cultures of hepatocytes indicates that they accumulate microfilaments with increasing age in culture.[9,14] The abovementioned observations with contractile protein inhibitors indicate that microfilaments and microtubules may be involved in the process of the contraction of the collagen gel. Cross-linking of collagen fibers may also be involved. We have observed in our laboratory that the contracted gels of cultures older than 7 days are more resistant to solubilization by acid or alkaline solutions and are less amenable to digestion by collagenase. The mechanism of collagen contraction deserves further investigation especially in view of the possible implications to the pathogenesis of cirrhosis of the liver.

Several modifications of the collagen gel technique have recently appeared in the literature. Of particular interest are techniques which involve embedding of the epithelial cells within the collagen cell matrix. This approach was used for primary culture of human mammary epithelial cells by S. Nandi et al.[15] In the presence of hormones, growth and organization to histotypic ductlike structures was noted. The same approach was followed for culture of thyroid follicular cells.[5] When the thyroid cells are maintained on top of the collagen gel they maintain a basal–apical differentiation without forming follicles. When these cultures are embedded into collagen gels, however, they form follicular structures. Another interesting variation of the collagen cell technique is the use of a nylon mesh support system for the collagen cell.[16] The presence of the nylon mesh prevents the collagen gel from contraction. The advantage of this technique is that it requires very small amounts of collagen material, the main support being provided by the nylon mesh. Because of the small amount of collagen required the cells can be easily removed intact after a short period of treatment with collagenase in the presence of soybean trypsin inhibitor. A disadvantage of this technique might be that in the absence of contraction of the gel there is surface area of the gel available for overgrowth of contaminating nonparenchymal elements. This, how-

[12] L. J. Cuprak and W. F. Lever, *Proc. Soc. Exp. Biol. Med.* **146**, 309 (1974).
[13] E. Bell, B. I. Varsson, and C. Merill, *Proc. Natl. Acad. Sci. U.S.A.* **76**, 1274 (1979).
[14] W. W. Mak, C. A. Sattler, and H. C. Pitot, *Cancer Res.* **40**, 4552 (1980).
[15] J. Yang, R. Guzman, J. Richards, V. Jentoft, M. R. Devault, S. R. Willings, and S. Nandi, *J. Natl. Cancer Inst.* **65**, 337 (1980).
[16] A. E. Sirica, C. G. Hwang, G. L. Sattler, and H. C. Pitot, *Cancer Res.* **40**, 3259 (1980).

ever, can be overcome by increasing the initial seeding density of the hepatocytes.

Regardless of the particular type of collagen gel technique, all collagen gel preparations can be processed by the usual histological techniques for embedding and sectioning. The collagen gels can be fixed in formaldehyde or in ethyl alcohol. Alternatively, the cells can be removed from the gel by treatment with collagenase and processed separately.

The use of collagn substrates in tissue culture has allowed maintenance of structure and characteristic differentiation for a variety of epithelial types (hepatocytes, mammary epithelial cells, thyroid follicular cells, etc.) whose long-term cultures were not possible before. Despite its wide applicability the basic mechanisms that mediate the effects of collagen on epithelial cells in culture are not precisely understood. Future research into the components of the intercellular substance of different organs may allow design of collagen cells in which elements of the intercellular matrix specific to the various organs can be incorporated. Such developments may allow better understanding of the nature of the stimuli that are important for maintaining the epithelial phenotype.

# Section II

# Elastin

A. Elastin Structure and Biosynthesis
*Articles 30 through 41*

B. Immunology of Elastin
*Articles 42 through 44*

## [30] Elastin Structure and Biosynthesis: An Overview

*By* JUDITH ANN FOSTER

The biological distribution of elastic fibers in the mammalian system is such that the highest concentration is found in those tissues that are subjected to continual physical deformation, tension, and high pressure differentials. Because of the intrinsic "elastic" nature of elastin, its presence in a tissue imparts a stretchability and subsequent recoil that is dependent only on the application of some physical force. Morphologically, elastic fibers can be distinguished into two intimate, but separate, components characterized by either a fibrillar or amorphous microscopic appearance.[1] Ontogenetically, the fibrillar component appears first in the extracellular matrix and is proposed to act as framework upon which the amorphous component is deposited or, perhaps more accurately, insolubilized.[2] Chemically, elastic fibers can also be divided into at least two classes of proteins, which correspond to the morphological entities. The amorphous component is referred to as elastin. It is characterized by an amino acid composition possessing 95% nonpolar amino acid residues and unique lysine-derived cross-links.[3] The fibrillar component is chemically distinguished from elastin by containing a high polar amino acid content and a significant amount of disulfide cross-links.[4] The amino acid compositions of the elastic fiber components are given in the table.

Although investigations into the synthesis, degradation, and structure of elastic fibers progressed significantly during the 1970s, much still remains unknown concerning the metabolism and interactions of the elastin and microfibrillar components. The principal difficulty in studies related to elastic fibers rests in the extreme insolubility of its components, especially elastin. The first obstacle faced when trying to characterize elastin is the purification of an insoluble protein to homogeneity without disrupting the integrity of the polypeptide chains. This would not be such a formidable task if elastin were not so closely associated with the microfibrillar protein and collagen, both of which are also insoluble. Consequently, the classical approach to isolating elastin involves the removal of all other connective tissue protein by some harsh and rather unorthodox protein treatments, such as boiling in 0.1 $N$ NaOH for 45 min.[5] As harsh as this latter proce-

[1] R. Ross, *J. Cell Biol.* **50,** 172 (1971).
[2] R. Ross, P. J. Fialkow, and L. K. Altman, *Adv. Exp. Med. Biol.* **79,** 7 (1977).
[3] C. Franzblau, *Compr. Biochem.* **26C,** 659 (1971).
[4] R. Ross and P. Bornstein, *J. Cell Biol.* **4D,** 366 (1969).
[5] A. I. Lansing, T. Rosenthal, M. Alex, and E. Dempsey, *Anat. Rec.* **144,** 555 (1952).

AMINO ACID COMPOSITIONS OF ELASTIC FIBER
COMPONENTS[a]

| Amino acid | Elastin[b] | Microfibril[c] |
|---|---|---|
| Cysteic acid | 1 | 12 |
| Lysine | 4 | 50 |
| Histidine | 1 | 22 |
| Arginine | 5 | 47 |
| Aspartic acid | 2 | 82 |
| Threonine | 3 | 38 |
| Serine | 5 | 72 |
| Glutamic acid | 12 | 119 |
| Proline | 128 | 72 |
| Glycine | 352 | 157 |
| Alanine | 176 | 108 |
| Valine | 175 | 70 |
| Isoleucine | 19 | 42 |
| Leucine | 47 | 70 |
| Tyrosine | 12 | 7 |
| Phenylalanine | 23 | 31 |
| Allysine[d] | 6 | — |
| ACP[e] | 12 | — |
| Lysinonorleucine | 1 | — |
| Isodesmosine | 3 | — |
| Desmosine | 3 | — |
| Merodesmosine | 2 | — |

[a] Compositions are expressed as residues per 1000 amino acid residues.
[b] From Foster et al.[35]
[c] Unpublished data.
[d] Allysine was determined as $\epsilon$-hydroxynorleucine, the reduced derivative.
[e] ACP is the aldol condensation product of two allysine residues determined as the reduced derivative. Cross-links are expressed as lysine equivalents.

dure is, after base treatment a residue remains that possesses an amino acid composition identical to soluble elastin (tropoelastin) highlighted by the fact that it contains no methionine, cysteine, or histidine residues.[6] These latter amino acid residues are many times taken as a measure of the degree of contamination present in elastin preparations.

A major difficulty in purification of elastin arises when the connective tissue under investigation contains a small amount of elastin relative to other proteins, such as in lung tissue, or when the tissue under investiga-

[6] L. G. Sandberg, N. Weissman, and D. W. Smith, *Biochemistry* **8**, 2940 (1969).

tion is obtained from a very young, an aged, or a diseased animal. In all the above cases, the elastin purified by the hot base treatment contains more polar amino acids and is slightly different in composition from soluble elastin. It is interesting to note that the apparent difference in the amino acid composition of elastic isolated from tissues of different ages is paralleled in smooth muscle cell cultures. Working with rabbit aortic smooth muscle cells, Faris et al.[7] have reported on the amino acid composition of the extracellular matrix protein remaining after hot alkali treatment. Their findings demonstrate that the composition of elastin recovered from second-passage cultures of different ages contains varying amounts of polar amino acids atypical of soluble elastin. The explanation for this discrepancy in elastin composition is commonly thought to reflect a very close association of an acidic protein or glycoprotein (microfibril) with elastin during various stages of maturation as well as in disease states.[8] This association is apparently not disrupted by hot alkali treatment. The chemical nature, mechanism, or significance of this association is presently unknown.

It is apparent from the latter discussion that the isolation and purification of insoluble elastin is an initial problem in structural and comparative studies. Within the last 4 years there has been a concerted effort to develop "nondegradative" methodologies to isolate elastin. As a result of these studies, several mild methods have been published in which insoluble elastin can be isolated using a combination of denaturants,[4,9-11] enzymes,[4,9] and hydrolytic agents.[12] The published amino acid compositions of elastin isolated from a variety of organs and species closely resembles hot alkali-prepared elastin with the exception that more polar amino acids and less detectable $NH_2$-terminal amino acids are present when milder isolation procedures are used.[9-12] However, these latter procedures do not eliminate the problem of contamination of elastin in different-aged tissue or diseased tissue but, in fact, may result in a magnification of the problem, since closely associated proteins are less likely to be removed.

Elastin Cross-links

Many of the early studies performed on elastin in the 1950s and early 1960s were directed at understanding the extremely insoluble nature of

[7] B. Faris, L. L. Salcedo, V. Cook, L. Johnson, J. A. Foster, and C. Franzblau, *Biochim. Biophys. Acta* **418**, 93 (1976).
[8] J. W. Downie, F. S. LaBella, and S. Whitaker, *Connect. Tissue Res.* **2**, 37 (1973).
[9] V. Richmond, *Biochim. Biophys. Acta* **351**, 173 (1974).
[10] D. S. Jackson and E. G. Cleary, *Methods Biochem. Anal.* **15**, 25 (1967).
[11] M. A. Paz, D. A. Keith, and P. M. Gallop, *Adv. Exp. Med. Biol.* **79**, 343 (1977).
[12] B. L. Rasmussen, E. Bruenager, and L. B. Sandberg, *Anal. Biochem.* **64**, 255 (1975).

this protein. Extensive extraction of elastic tissue with a wide spectrum of polar and nonpolar solvents, with and without reducing agents, failed to solubilize elastin. Since elastin contains very few cysteine residues, it was concluded that insolubilization resulted from some unique cross-link formation distinct from the disulfide bond. Largely through the efforts of Partridge,[13] Miller,[14] Gallop,[15] Martin,[16] and Franzblau,[17] a lysine-derived cross-linking system was elucidated. In summary, it was demonstrated that the mechanism of elastin cross-link formation involved an initial oxidative deamination of key lysyl residues in the then postulated soluble precursor to insoluble elastin. The lysyl residues are thus converted to aldehydes ($\alpha$-aminoadipic acid $\delta$-semialdehyde) that represent very reactive chemical groups. The fate of the newly formed aldehydes, referred to as allysine for brevity, are then to undergo Schiff base formation with unmodified lysyl residues or to react with another allysine residue resulting in an aldol condensation. Ultimately, through an unknown mechanism, three allysine residues and one lysine residue react to form a pyridinium ring alkylated in four positions. This latter cyclic compound, originally described by Thomas et al.,[13] exists in elastin as two isomers referred to as desmosine and isodesmosine. The isomers differ only in the position of ring alkylation. Gallop and Paz[15] and Pereyra et al.[18] have shown that cross-link formation is very complex because of presence of numerous isomeric compounds, which can result from the various aldehyde and Schiff base compounds present in elastin. These authors have also demonstrated the presence of proposed precursor ring forms to the desmosines, which differ in their oxidation levels.

Currently, at least two theories for the mechanism of desmosine formation have been proposed. These include the reaction of the aldol condensation product with dehydrolysinonorleucine[19,20] or the reaction of dehydromerodesmosine with an allysine residue.[21] Which of these pathways is correct, or if both may be functional, has not yet been definitely established. Each of the above-described lysine-derived cross-links can theo-

[13] J. Thomas, D. F. Elsden, and S. M. Partridge, *Nature* (*London*) **200**, 651 (1963).
[14] E. J. Miller and H. M. Fullmer, *J. Exp. Med.* **123**, 1097 (1966).
[15] P. M. Gallop and M. A. Paz, *Physiol. Rev.* **55**, 418 (1975).
[16] E. J. Miller, G. R. Martin, C. E. Mecca, and K. A. Piez, *J. Biol. Chem.* **240**, 3623 (1965).
[17] C. Franzblau and R. Lent, *in* "Structure, Function and Evolution in Proteins" (Brookhaven Symp. Biol.), p. 358. Upton, New York, 1968.
[18] B. Pereyra, O. O. Blumenfeld, M. A. Paz, E. Henson, and P. M. Gallop, *J. Biol. Chem.* **249**, 2212 (1974).
[19] W. R. Gray, *Adv. Exp. Med. Biol.* **79**, 285 (1977).
[20] J. A. Foster, L. Rubin, H. M. Kagen, C. Franzblau, E. Bruenger, and L. B. Sandberg, *J. Biol. Chem.* **249**, 6191 (1974).
[21] G. Francis, R. John, and J. Thomas, *Biochem. J.* **136**, 45 (1973).

retically serve as either intra- or intermolecular cross-links (with the exception of allysine), depending upon whether or not the original lysyl residues are located within the same or different polypeptide chains. Potentially, the structure of these cross-links allow for the linkage of 1 (intra), 2, 3, or 4 polypeptide chains.

The key step in cross-link formation is the initial oxidation of lysyl residues. The enzyme responsible for this reaction is lysyl oxidase, which requires $Cu^{2+}$ for activity.[22] It has also been demonstrated that lathyrogens, such as β-aminopropionitrile and α-aminoacetronitrile, can effectively inhibit lysyl oxidase both *in vitro* and *in vivo*.[23,24] Since cross-link formation results in the complete insolubilization of soluble precursor molecules of elastin, inhibition of lysyl oxidase by either eliminating copper or administering lathyrogens results in an accumulation of soluble elastin.[6,24] These approaches form the basis for the isolation of soluble elastin, as discussed in more detail in chapters [36], [37], and [38] of this volume.

In early studies by Partridge[25] and Franzblau et al.,[26] the areas of elastin cross-links were reported to be enriched in alanine residues. Cross-linked peptides containing the desmosine[25,26] and lysinonorleucine[26] cross-links were shown to contain high levels of alanine residues relative to the overall composition of elastin and suggested a restricted distribution of alanine in the elastin primary structure. Foster et al.[27] later showed that alanine enrichment was also found in the area immediate to the aldol condensation product. This latter study also suggested that the aldol cross-link may function as an intramolecular bridge, since one of the peptides examined showed one $NH_2$-terminal amino acid and one peptide chain after periodate–permanganate oxidation followed by $NaBH_4$ reduction.[27] The significance of this finding relative to the mechanism and structure of desmosine cross-links will be discussed below.

Gerber and Anwar[28,29] have reported on the isolation and partial amino acid sequence of desmosine peptides isolated from bovine ligamentum nuchae and porcine aortas. Using the Edman degradation procedure as a cleavage reagent, single-chain peptides were released from the COOH-terminal of the desmosine cross-links. The single-chain released peptides

---

[22] R. C. Siegel, S. R. Pinnel, and G. R. Martin, *Biochemistry* **9**, 4486 (1970).
[23] R. C. Page and E. P. Benditt, *Lab. Invest.* **26**, 22 (1972).
[24] J. E. Savage, D. W. Bird, G. Reynolds, and O'Dell, *J. Nutr.* **88**, 15 (1966).
[25] S. M. Partridge, *Adv. Protein Chem.* **17**, 227 (1962).
[26] C. Franzblau, F. M. Sinex, and B. Faris, *Nature (London)* **204**, 802 (1965).
[27] J. A. Foster, W. R. Gray, and C. Franzblau, *Biochim. Biophys. Acta* **303**, 363 (1973).
[28] G. E. Gerber and R. A. Anwar, *J. Biol. Chem.* **249**, 5200 (1974).
[29] G. E. Gerber and R. A. Anwar, *Biochem. J.* **149**, 685 (1975).

were purified, and the primary sequence was determined. A comparison of the sequences obtained, with those reported by Foster et al.[30] from porcine tropoelastin, permitted the identification of 12 possible sequences adjacent to lysine residues that may be involved in desmosine formation. Further, these investigators found that peptides adjacent to the desmosine cross-link fall into two broad categories based on their $NH_2$-terminal residue. One of these groups possesses a hydrophobic group (phenylalanine, tyrosine, or leucine), and the other possesses alanine as $NH_2$-terminal. Because of this distinction, it was proposed that the hydrophobic amino acid residues prevent the oxidative deamination of the adjacent lysine $\epsilon$-amino group, which contributes its nitrogen to the pyridinium ring of the desmosine.

Foster et al.[20,30] published data on the primary sequence and conformation of desmosine cross-linked peptides isolated from bovine ligamentum nuchae. The purified desmosine peptides revealed several unique features in their amino acid composition. On a molar basis each of the peptides contained one serine, one glutamic acid, one tyrosine, and two or three residues of phenylalanine per peptide. The combined amount of the above-mentioned amino acid residues in the total ligament elastin is only 6–7%; this would suggest some clustering of these residues within the area of the desmosine cross-link. The purified desmosine peptides were sequenced on an automated Beckman Sequencer directly. Results from this study indicated that the desmosine cross-link two chains based on (a) micromoles of $NH_2$-terminal amino acids recovered per micromole of desmosine contained in the peptide; and (b) the finding of two peptide sequences comparable to those obtained from the soluble elastin precursor. Of special interest in this study was the finding of the two isomers, desmosine and isodesmosine, in the same primary sequence approximately equally substituted. This led to a hypothesis that the prime determinant in the formation of either desmosine or isodesmosine would be the direction of the condensation of the two allysine residues. Specifically, the aldol condensation can result in two products, chemically indistinguishable, depending upon which of the allysine residues retains its aldehyde group. The remaining half of the desmosine moiety, whether it be in the form of dehydrolysinonorleucine or as an unreacted allysine and lysine residue, is limited to one reactive direction. Implicit in this model is the fact that two polypeptide chains form the desmosine cross-link by each donating two lysine residues. Examination of the circular dichroism spectrum of the desmosine peptides suggested the presence of an extended helical conformation based on its general shape and behavior toward temperature changes.[30]

[30] J. A. Foster, L. Rubin, H. M. Kagan, and C. Franzblau, *Polymers* **249**, 6191 (1974).

## Soluble Elastin

The difficulties inherent in isolating and characterizing insoluble elastin were greatly alleviated with the discovery of a soluble form of elastin. Primarily owing to the efforts of Smith and Carnes[31] and Sandberg et al.,[6] a soluble elastin was isolated from the aortas of piglets maintained on a copper-deficient diet. Since copper is a cofactor for the enzyme lysyl oxidase, restriction of this metal from the diets of young, developing animals results in an inhibition of cross-link formation.[22] Although not 100% effective, copper deficiency does allow for the accumulation of significant amounts of soluble elastin. Rucker et al.[32] have also reported on the isolation of soluble elastin from the aortas of copper-deficient chicks, and Whiting et al.[33] have isolated soluble elastin from the ligament of copper-deficient calves.

As might be anticipated, another experimental situation in which elastin cross-linking can be inhibited is administration of an inhibitor of lysyl oxidase such as the lathyrogen $\beta$-aminopropionitrile (BAPN). Sykes and Partridge[34] and Foster et al.[35] have reported on the isolation of soluble elastin from the aortas of lathyritic chicks. More recently, Foster et al.[36] have reported on the isolation of tropoelastin from aortas and ear cartilage of lathyritic piglets. In the latter study, the yield of tropoelastin from aortic tissue was comparable to that previously reported for copper-deficient piglets.

There are currently two widely used methodologies for the isolation of soluble elastin, both utilizing the distinctive solubility properties of the uncross-linked elastin. Smith et al.[37] have developed a neutral salt procedure that uses aqueous buffers at neutral pH and a series of acid and salt precipitations. Sandberg et al.[38] have published another popular technique taking advantage of the nonpolar nature of elastin. In this procedure the tissue is first extracted at low pH, and the solubilized protein is extracted and is mixed with a butanol–propanol mixture aimed at retaining only soluble elastin in the organic solvent.

The soluble elastin isolated by either of the above procedures possesses a molecular weight of approximately 72,000 and contains 850 amino

[31] D. W. Smith and W. H. Carnes, J. Biol. Chem. **248**, 8157 (1973).
[32] R. B. Rucker, W. Goettlich-Riemann, K. Tom, M. Chen, J. Poaster, and S. Koerner, J. Nutr. **105**, 46 (1967).
[33] A. H. Whiting, B. C. Sykes, and S. M. Partridge, Biochem. J. **141**, 573 (1974).
[34] B. C. Sykes and S. M. Partridge, Biochem. J. **141**, 567 (1974).
[35] J. A. Foster, R. Shapiro, P. Voynow, G. Crombie, B. Faris, and C. Franzblau, Biochemistry **74**, 5343 (1975).
[36] J. A. Foster, C. B. Rich, and M. D. DeSa, Biochim. Biophys. Acta **626**, 383 (1980).
[37] D. W. Smith, D. Brown, and W. H. Carnes, J. Biol. Chem. **247**, 2427 (1972).
[38] L. G. Sandberg, R. D. Zeikus, and S. M. Coltrain, Biochim. Biophys. Acta **236**, 542 (1971).

acid residues.[39] Soluble elastin, as well as the mature insoluble elastin, is rich in glycine (33%) and the nonpolar amino acids alanine, valine, and proline (see the table). It differs from the insoluble protein in its high content of lysine residues and lack of any significant cross-links. Sandberg et al.[39] have reported on the sequences of short tryptic peptides from tropoelastin. Their data reveal that two peptides, Lys-Ala-Ala-Ala-Lys and Lys-Ala-Ala-Lys, each repeat six times in the tropoelastin molecule. This clustering of lysine and alanine residues is in direct agreement with data obtained on elastin cross-linked peptides. It also adds credibility to the postulate that desmosines cross-link two polypeptide chains, each chain donating two lysyl residues. Space-filling models of the alanine and lysine sequences reveal that either an $\alpha$-helix[40] or an extended helix[30] will accommodate the formation of the pyridinium ring of the desmosine cross-link.

Through the characterization of large tryptic peptides, Foster et al.[41] have been able to sequence over 400 residues that correspond to approximately one-half of the residues present in the tropoelastin molecule. To date, the sequences obtained reveal a tropoelastin primary structure quite distinct from that of tropocollagen. Tropoelastin possesses repeat units of a tetrapeptide, Gly-Gly-Val-Pro; a pentapeptide, Pro-Gly-Val-Gly-Val; and a hexapeptide, Pro-Gly-Val-Gly-Val-Ala. In addition, alanine residues in the tryptic peptides are distributed primarily at the amino and carboxyl terminals, suggesting that these residues are close to the lysyl residues of elastin. Several of the peptides contain a partial substitution of hydroxyproline or proline, especially in the sequences Gly-Leu-Pro-Gly and Gly-Ile-Pro-Gly. Since the sequences of these peptides resemble those of other tropoelastin peptides and also lack glycine in every third position (contrary to collagen), this is definitive evidence for the existence of hydroxyproline in elastin.

Biosynthesis

Thus far, several cell types including smooth muscle cells, fibroblasts,[42] chondrocytes,[43] and endothelial cells[44] have been both histochemically and chemically identified as synthesizing elastin. A major problem in

[39] L. G. Sandberg, N. Weissmann, and W. R. Gray, *Biochemistry* **10**, 52 (1971).
[40] W. R. Gray, L. B. Sandberg, and J. A. Foster, *Nature (London)* **246**, 461 (1973).
[41] J. A. Foster, E. Bruenger, W. Gray, and L. B. Sandberg, *JBC* **248**, 2879 (1973).
[42] R. P. Mecham and G. Lange, *Connect. Tissue Res.* **7**, 247 (1980).
[43] G. Quintarelli, B. C. Starcher, A. Vocaturo, F. Di Gianfilippo, L. Gotte, and R. P. Mecham, *Connect. Tissue Res.* **6**, 1 (1979).
[44] W. H. Carnes, P. A. Abraham, and V. Buonassisi, *Biochem. Biophys. Res. Commun.* **90**, 1393 (1979).

trying conclusively to identify elastin synthesis by a particular cell type rests in the absolute reliance of determining insoluble, cross-linked elastin as the sole index of elastin synthesis. In order to identify elastin synthesis in cell cultures, either insoluble elastin is isolated and identified by amino acid analysis[7] or desmosine is chromatographically identified by pulsing the cells with [$^{14}$C]lysine[7] or by a radioimmune assay.[45] This is not as straightforward a task as might be expected. Faris et al.[7] showed that the actual deposition of insoluble elastin in cultures of rabbit smooth muscle cells is very dependent on the age of the culture. Further, these authors demonstrated that the level of the desmosine cross-links in the cell cultures was very low relative to elastin isolated directly from aortic tissue. These observations suggest that either the actual synthesis of elastin or the insolubilization of soluble elastin is somehow regulated in cultured smooth muscle cells. Whether this reflects a failure to optimize conditions for elastin synthesis or is a problem intrinsic to homogeneous smooth muscle cell cultures is not known. Since measurements of elastin rely almost entirely on cross-linked elastin, the question of the actual synthesis and secretion of soluble elastin cannot be answered.

Within the last 6 years many studies have been directed at the synthesis of elastin both *in vivo* and *in vitro*. The basic methodology utilized in these studies is the use of radioactive amino acids (usually valine) to label and follow the fate of newly synthesized elastin. A number of investigators have examined elastin synthesis in organ cultures of aortic tissue from chick and pig sources. In summary, through the use of either [$^3$H]valine, [$^{14}$C]proline, or [$^3$H]glycine under a variety of pulse-chase conditions, the following conclusions were drawn: (*a*) elastin is first synthesized as a 72,000–76,000 molecular weight ($M_r$) species[46-55]; (*b*) the newly synthesized elastin is very susceptible to proteolytic cleavage[46]; (*c*) the disappearance of the 72,000–76,000 $M_r$ isotopically labeled elastin is accompanied by increased incorporation of radioactivity into either insoluble elastin[50,51] or the desmosine cross-links[52]; (*d*) the process of trans-

[45] G. S. King, D. S. Moylan, and B. C. Starcher, *Connect. Tissue Res.* **7**, 263 (1980).
[46] A. S. Narayanan and R. C. Page, *FEBS Lett.* **44**, 59 (1974).
[47] L. Murphy, M. Harsch, T. Mori, and J. Rosenbloom, *FEBS Lett.* **21**, 113 (1972).
[48] M. Miyoshi, M. Kanamori, and J. Rosenbloom, *J. Biochem.* **79**, 1235 (1976).
[49] H. Ito, *Kumamoto Med. J.* **26**, 143 (1973).
[50] D. W. Smith, P. A. Abraham, and W. H. Carnes, *Biochem. Biophys. Res. Commun.* **66**, 893 (1975).
[51] D. W. Smith and W. H. Carnes, *J. Biol. Chem.* **248**, 8157 (1973).
[52] D. W. Smith, P. A. Abraham, and W. H. Carnes, *Adv. Exp. Med. Biol.* **79**, 385 (1977).
[53] R. B. Rucker, M. C. Yau, C. Keen, M. Mohanram, and W. Rieman, *Artery* **3**, 1 (1977).
[54] R. B. Rucker, I. Lee, M. Lefevre, and K. Tom, *Adv. Exp. Med. Biol.* **79**, 461 (1977).
[55] R. B. Rucker, K. Tom, M. Tanaka, M. Haniu, and K. Yasunobu, *Biochem. Biophys. Res. Commun.* **66**, 287 (1975).

ference of radioactivity from the soluble elastin fraction to insoluble elastin is inhibited by BAPN[56,57]; and (e) newly synthesized soluble elastin appears to act as substrate for lysyl oxidase.[57]

More recently, several papers have been published concerning the synthesis of soluble elastin in cell suspensions from chick embryonic aortae. It should be pointed out that the cell suspensions under investigation were derived from whole aortic tissue and, therefore, do not represent a homogeneous cell population. Uitto et al.[58] have demonstrated the synthesis of an 80,000-dalton protein in the media and intracellular fraction of the suspended aortic cells. This protein was indirectly identified as elastin by valine incorporation, molecular size, and cyanogen bromide treatment.

Rosenbloom and his co-workers have also reported on the synthesis of elastin in aortic cell suspensions.[59] In contrast to the study described above, Rosenbloom reported that the major [$^3$H]valine-labeled protein found in the matrix-free system was 70,000 daltons. In addition, a significant amount of low molecular weight [$^3$H]valine-labeled protein was found in the cell media. It was therefore necessary to add bovine trypsin inhibitor to the cell suspensions in order to recover sufficient amounts of the 70,000 $M_r$ component. Through the use of a double immunoprecipitation technique, using antisera developed against chick insoluble elastin, the 70,000-dalton protein was identified as elastin.[60]

Several laboratories have reported on the synthesis of soluble elastin by homogeneous cultures of aortic smooth muscle cells. Abraham et al.[61] have examined the synthesis of soluble elastin and lysine-derived cross-links in long-term cultures of pig aortic smooth muscle cells. Utilizing either [$^2$H]proline or [$^3$H]valine labels, the identification of soluble elastin of 72,000 $M_r$ was accomplished by copurification with added purified tropoelastin isolated from copper-deficient piglets.[31] In addition, [$^3$H]lysine-labeled tropoelastin was isolated from pig aorta organ cultures and added to the medium of confluent smooth muscle cell cultures. The subsequent incorporation of tritium into the insoluble elastin cross-links of the cell matrix was then measured to determine uptake of the added tropoelastin into the insoluble elastin protein. The data obtained were interpreted as demonstrating a direct precursor relationship between tropoelastin and insoluble elastin.[61]

[56] A. S. Narayanan and R. C. Page, J. Biol. Chem. 251, 1125 (1976).
[57] F. W. Keeley, Connect. Tissue Res. 4, 195 (1976).
[58] J. Uitto, H. P. Hoffmann, and D. J. Prockop, Arch. Biochem. Biophys. 173, 187 (1976).
[59] J. Rosenbloom and A. Cyivinski, Biochem. Biophys. Res. Commun. 69, 613 (1976).
[60] P. Christner, M. Doxon, A. Cyivinski, and J. Rosenbloom, Biochem. J. 157, 525 (1976).
[61] P. A. Abraham, D. W. Smith, and W. H. Carnes, Biochem. Biophys. Res. Commun. 58, 597 (1974).

Narayanan et al.[62] have also reported on the synthesis of soluble and insoluble elastin in aortic smooth muscle cell cultures. In addition to identifying tropoelastin indirectly via molecular weight and coelectrophoresis with purified pig tropoelastin, the secreted 70,000 $M_r$ protein was shown to be a substrate for lysyl oxidase.

More recently, Foster et al.[63] have reported on the isolation of a high molecular weight (120,000–140,000) form of soluble elastin from lathyritic chick aortas. The high molecular weight soluble elastin was also found in cultures of rabbit aortic smooth muscle cells.[64] These findings prompted the hypothesis that the high molecular weight protein was a proelastin molecule and therefore represented the primary elastin gene product.[64] Studies by Rucker et al.[65] and Heng-Khoo et al.[66] in chick embryonic aorta cultures and copper-deficient chick aortas confirmed the existence of the putative proelastin molecule.

The concept of a proelastin species has remained controversial. For this reason several laboratories have been studying the synthesis of elastin via the cell-free translation of elastin mRNA. This system offers the advantage of viewing the initial gene product in a system devoid of potential processing enzymes.

Ryhanen et al.[67] have reported on the synthesis of tropoelastin using polysomes isolated from chick embryonic aortas and translated with either rabbit reticulocyte or Ehrlich ascites cell lysates. Burnett and Rosenbloom[68] have reported on the cell-free synthesis of chick aortic mRNAs. They find a major 70,000-dalton protein that was immunoprecipitable with antibody directed against chick insoluble elastin.

Foster et al.[69,70] examined the cell-free translation of elastin mRNAs isolated from chick and pig aortic and lung tissues. These results show that the mRNA populations from all tissues examined contain elastin mRNAs coding for the synthesis of two distinct species of tropoelastin

---

[62] A. S. Narayanan, L. B. Sandberg, R. Ross, and D. L. Layman, *J. Cell Biol.* **68**, 411 (1976).
[63] J. A. Foster, R. P. Mecham, and C. Franzblau, *Biochem. Biophys. Res. Commun.* **72**, 1399 (1976a).
[64] J. A. Foster, R. P. Mecham, C. B. Rich, M. Cronin, A. Levine, M. Imberman, and L. L. Salcedo, *J. Biol. Chem.* **253**, 2797 (1978).
[65] R. B. Rucker, J. Murray, M. Lefevre, and J. Lee, *Biochem. Biophys. Res. Commun.* **75**, 358 (1977).
[66] C. S. Heng-Khoo, R. B. Rucker, and K. W. Buckingham, *Biochem. J.* **177**, 559 (1979).
[67] L. Ryhanen, P. N. Graves, G. M. Bressan, and D. J. Prockop, *Arch. Biochem. Biophys.* **185**, 344 (1978).
[68] W. Burnett and J. Rosenbloom, *Biochem. Biophys. Res. Commun.* **86**, 478 (1979).
[69] J. A. Foster, C. B. Rich, S. Fletcher, S. R. Karr, and A. Przybyla, *Biochemistry* **19**, 857 (1980).
[70] L. L. Barrineau, C. B. Rich, and J. A. Foster, *Connect. Tissue Res.* **8**, 189 (1981).

possessing 70,000 and 73,000 molecular weights. The two elastin proteins, referred to as tropoelastins a and b, differ in molecular weight, isoelectric point, cysteine content, and primary structure.[69,70] Both tropoelastins contain leader sequences of approximately 24 amino acid residues.[71]

The finding of the two tropoelastins presents the possibility that these two proteins may associate posttranslationally to form a dimer prior to secretion. This might explain the origin of the putative proelastin molecule seen in other systems. The fact that both tropoelastins a and b possess cysteine residues offers the potential for an initial disulfide bond between the two polypeptide chains to stabilize the dimer.[69]

Characterization of tropoelastin b by peptide mapping and amino acid sequencing of the $NH_2$ terminus has shown that this molecule is identical with the tropoelastin described by a number of investigators.[71,72] Further, within an animal species tropoelastin b is identical in different tissues, such as aorta, lung, and cartilage.[72,73] Sufficient amounts of tropoelastin a have not been isolated to perform a thorough chemical characterization, although initial data described above clearly demonstrate that it is a protein distinct from tropoelastin b.

As stated previously, Barrineau et al.[70] have found mRNAs directing the synthesis of both tropoelastins a and b in all tissues thus far examined. Interestingly, the ratio of the a and b proteins is variable in different tissues,[70] although between different animals, the ratio of tropoelastin a and b is essentially the same for a given tissue.[70] This finding, coupled with the fact that tropoelastin b appears to be identical in different tissues,[36,72] raises the possibility that the proportion of the two tropoelastins may dictate the unique molecular organization of elastic fibers seen among different tissues.

Acknowledgment

The author acknowledges the assistance of Beverly Marrano in typing and organizing these chapters.

---

[71] J. A. Foster, C. B. Rich, S. B. Karr, and M. D. DeSa, *Connect. Tissue Res.* **8**, 259 (1981).
[72] J. A. Foster, C. B. Rich, S. Fletcher, S. R. Karr, M. D. DeSa, T. Oliver, and A. Przybyla, *Biochemistry* **20**, 3528 (1981).
[73] J. A. Foster, C. B. Rich, and M. D. DeSa, *Biochim. Biophys. Acta* **626**, 383 (1980).

## [31] Elastin Isolation and Cross-Linking

*By* MERCEDES A. PAZ, DAVID A. KEITH, and PAUL M. GALLOP

Elastin is an insoluble, highly cross-linked hydrophobic protein, rich in nonpolar amino acid residues, such as valine, leucine, isoleucine, and phenylalanine. On the average, about every third residue is glycine and about every ninth residue is a prolyl residue. There are only modest amounts of hydroxyproline in contradistinction to collagen, which is relatively rich in hydroxyproline.[1]

Lysyl residues serve as the precursors of the cross-links in elastin, and all the known cross-links have structures consistent with a posttranslational derivation from either lysyl or allysyl residues. There is no evidence for any hydroxylysine in elastin preparations free of collagen. Elastic fibers in tissues generally contain a significant number of reactive aldehydes in the form of allysyl and other cross-linking precursor compounds. These aldehydes can be detected histologically by sulfurous acid–Schiff reagent procedures or by use of dansyl hydrazine,[2] which binds to the aldehydes and converts the elastic fibers to yellow fluorescent fibers easily visualized in the fluorescent microscope. In tissues, elastic fibers can also be recognized by characteristic staining patterns with relatively hydrophobic dyes.

Microscopically and biochemically at least two types of elastin are now distinguishable. The familiar elastin (type I) derived from either ligamentum nuchae, aorta, lung pleura, lung parenchyma, or skin, appears to be quite similar, if not identical, in amino acid composition although the posttranslational cross-linking patterns are variable in a qualitative sense. Most of the methods employed to prepare elastin have been quite destructive of certain of the cross-links and their intermediates. A second type of elastin (type II) can be isolated from elastic cartilage and is related to type I elastin. It may be a different genetic variety, much in the way that type II collagen found in cartilage is a different variety from the familiar type I collagen found in other tissues.

### Isolation of Elastin from Lung and Other Tissues

Elastin may be isolated from tissues after other more soluble connective tissue elements have been removed, most often by the use of harsh

[1] P. M. Gallop and M. A. Paz, *Physiol. Rev.* **55**, 418 (1975).
[2] D. A. Keith, M. A. Paz, and P. M. Gallop, *Adv. Exp. Med. Biol.* **79**, 57 (1977).

methods involving autoclaving and exposure to hot alkali. The Lansing procedure[3] involves the use of 0.1 $N$ NaOH at 98° for 45 min and has been used by many investigators[4-7] to prepare elastin from various tissues. Such procedures first remove collagen and other components, leaving behind "intact," but undoubtedly "degraded," elastin. In these elastin preparations, a number of N-terminal amino acid residues have been found,[8,9] demonstrating the probable hydrolysis of peptide bonds in elastin. Because of these problems, the use of less degradative procedures based on collagenase digestion and subsequent removal of other proteins by reduction in 5 $M$ guanidine HCl and dithioerythritol had been proposed by Ross and Bornstein.[10] Richmond[11] employed similar procedures to prepare elastin from guinea pig lung parenchyma, but found that lung tissue required further treatment for a satisfactory purification. She added an extraction (three times) with 6 $M$ urea, 1% sodium dodecyl sulfate, 0.05 $M$ dithioerythritol, 0.05 $M$ Tris (pH 7.7) at room temperature in a stoppered flask under $N_2$. She found that this procedure resulted in better recovery of the cross-linking amino acids desmosine and isodesmosine when compared to hot alkali-prepared elastin. An interesting variation of this procedure was introduced by Rasmussen et al.,[12] who followed the guanidine–HCl–mercaptoethanol extraction with a treatment for 24 hr at room temperature with 97% formic acid and an excess of cyanogen bromide. Cyanogen bromide in formic acid removed the collagen and produced a less degraded elastin than the Lansing procedure. Serafini-Fracassini et al.[13] proposed an affinity chromatographic procedure for the purification of the collagenase used in the elastin preparation. The crude collagenase is purified by batchwise treatment with DEAE-cellulose followed by affinity chromatography on a column prepared with finely milled bovine ligamentum nuchae elastin purified by hot alkali treatment. They found that this purified collagenase had no elastolytic activity and, when

[3] A. I. Lansing, R. B. Rosenthal, M. Alex, and E. W. Dempsey, *Anat. Rec.* **114**, 555 (1952).
[4] S. M. Partridge, H. F. Davis, and G. S. Adair, *Biochem. J.* **61**, 21 (1955).
[5] M. Fitzpatrick and V. D. Hospelhorn, *J. Lab. Clin. Med.* **60**, 799 (1962).
[6] R. John and J. Thomas, *Biochem. J.* **127**, 261 (1972).
[7] D. S. Jackson and E. G. Cleary, *Methods Biochem. Anal.* **15**, 25 (1967).
[8] L. Gotte, P. Stern, D. R. Elsden, and S. M. Partridge, *Biochem. J.* **87**, 344 (1963).
[9] A. Serafini-Fracassini and J. W. Smith, in "The Structure and Biochemistry of Cartilage" (A. Serafini-Fracassini and J. W. Smith, eds.), p. 7. Churchill-Livingstone, Edinburgh, 1974.
[10] R. Ross and P. Bornstein, *J. Cell Biol.* **40**, 366 (1969).
[11] V. Richmond, *Biochim. Biophys. Acta* **351**, 173 (1974).
[12] B. L. Rasmussen, E. Bruenger, and L. B. Sandberg, *Anal. Biochem.* **64**, 255 (1975).
[13] A. Serafini-Fracassini, J. M. Field, G. W. Rodger, and M. Spina, *Biochim. Biophys. Acta* **386**, 80 (1975).

used in the elastin preparation from bovine ligamentum nuchae, resulted in a considerable reduction in N-terminal residues compared to the amount when using collagenase purified by the Yoshida and Noda procedure.[14] Spina et al.[15] applied a similarly purified collagenase to the isolation of insoluble elastin from bovine aorta and compared the final product to that obtained by the Lansing procedure. They found that serine, threonine, and aspartic and glutamic acids are reduced and N-terminal amino acids are increased in hot alkali-treated elastin. Starcher and Galione[16] prepared elastin from different animal species using a procedure that combined some of the features mentioned above. An extraction with $0.05 M$ $Na_2HPO_4$, pH 7.6, containing 1% NaCl and 0.1% EDTA is followed by autoclaving for 45 min in $H_2O$ which they claimed to be essential for consistent results. Unfortunately, this step may cause some elastin degradation, although Starcher and Galione claimed that they were unable to detect any N-terminal amino acids from elastin purified by their procedure. Malanik and Ledvina[17] compared elastin prepared by hot alkali treatment ($0.05 N$ NaOH at 100° for 15 min, four times) and the Rasmussen et al.[12] procedure from fetal and adult rabbit aorta. In agreement with Richmond's[11] results and our own,[18] they found that the hot alkali treatment reduced the percentage of iso(desmosines). Dihydrodesmosines are converted in part to desmosines and other products artifactually by alkali treatment and acid hydrolysis.

The procedure to be described set out to isolate and purify elastin from histologically defined regions of calf lung, other bovine tissues, and dog aorta and to measure the nature and distribution of cross-links. The tissues were separated as far as technically possible into anatomically defined regions.

Trachea: Dissection was performed from below the larynx to the carina.

Bronchial tree: From the carina, the bronchial tree was dissected peripherally and cleaned of parenchyma and blood vessels. Bronchioles with a diameter of about 2 mm could be dissected free of surrounding tissue.

Pleura: Visceral pleura could be stripped off the parenchyma. This was more easily accomplished when the lung was distended with air.

Parenchyma: Parenchyma was more difficult to isolate as a defined entity. Peripheral lung, 3–4 mm thick, freed of pleura, major airways, and major blood vessels was used.

[14] E. Yoshida and H. Noda, *Biochim. Biophys. Acta* **105**, 562 (1965).
[15] M. Spina, G. Garbin, J. M. Field, and A. Serafini-Fracassini, *Biochim. Biophys. Acta* **400**, 162 (1975).
[16] B. C. Starcher and M. J. Galione, *Anal. Biochem.* **74**, 441 (1976).
[17] V. Malanik and M. Ledvina, *Connect. Tissue Res.* **6**, 139 (1978).
[18] M. A. Paz, D. A. Keith, H. P. Traverso, and P. M. Gallop, *Biochemistry* **15**, 4912, 1976.

Thoracic bovine and dog aortas were dissected free from loose connective tissue adherent to the adventitia.

Representative samples were taken for histological examination to confirm that the tissues were normal and that adequate isolation of the defined areas had occurred. A modified Verhoeff's iron hematoxylin stain[19] was used for this purpose and, in addition, standard electronmicroscopy methods were employed to examine the elastin preparations.

The anatomically defined regions were minced finely in 50% aqueous pyridine at 4°. The suspension was centrifuged, and the residue was washed thoroughly by centrifugation with water until free of pyridine and dried under vacuum. This material was finely powdered in a nitrogen mill for 3 min and then extracted for two 24-hr periods with $2\,M$ $CaCl_2$ at 4° followed by four water washes. The treatment with 50% pyridine and $2\,M$ $CaCl_2$ removed cell membranes and other tissue components. The remaining tissue was further defatted and dehydrated with ethanol for 20 min, filtered through a sintered-glass filter, treated with chloroform–methanol (2:1, v/v) for 20 min, washed with ether, and lyophilized. This defatted tissue was incubated with purified collagenase (ratio of substrate to enzyme approximately 50:1) in $0.025\,M$ Tris at pH 7.5 in the presence of 0.5 $mM$ $CaCl_2$ at 37° overnight to remove collagen. The supernatant was separated by centrifugation, and the residue was again digested with collagenase. The collagenase-insoluble residue was washed with water and was further extracted with $5\,M$ guanidine HCl, $0.05\,M$ dithioerythritol, 2.6 $mM$ EDTA in $0.1\,M$ Tris buffer at pH 8.5 at 37° overnight under a $N_2$ atmosphere in a stoppered bottle to remove microfibrillar and other remaining proteins. The residue obtained after centrifugation was twice washed with $H_2O$ and extracted with $6\,M$ urea, 34.7 $mM$ sodium dodecyl sulfate, $0.05\,M$ dithioerythritol in $0.05\,M$ Tris buffer at pH 7.7 at room temperature under $N_2$ in a stoppered bottle. This last extraction was repeated twice. After centrifugation, the residue was dialyzed vs distilled $H_2O$ until sodium dodecyl sulfate was no longer detectable and was finally lyophilized. Elastin was converted to a fine powder in a liquid nitrogen Spex freezer mill.

After completion of the procedure, an aliquot of the preparation was hydrolyzed following the procedure described below and the amino acid composition was determined. If the amino acid composition differed significantly from that shown in Table I for type I elastin, the collagenase digestion and the subsequent steps were repeated. Elastin from visceral pleura and aorta are more easily purified than elastin from lung paren-

---

[19] F. H. Verhoeff, *J. Am. Med. Assoc.* **50**, 876 (1908).

TABLE I
AMINO ACID COMPOSITION OF ELASTIN FROM VARIOUS BOVINE TISSUES

| Amino acid | Ligamentum nuchae | Aorta | Visceral pleura | Lung parenchyma | Ear cartilage |
|---|---|---|---|---|---|
| Hydroxyproline | 9 | 16 | 12 | 8 | 19 |
| Aspartic acid | 7 | 7 | 7 | 10 | 37 |
| Threonine | 9 | 10 | 8 | 10 | 12 |
| Serine | 9 | 9 | 8 | 10 | 17 |
| Glutamic acid | 17 | 17 | 22 | 24 | 59 |
| Proline | 115 | 106 | 117 | 108 | 115 |
| Glycine | 312 | 310 | 332 | 302 | 304 |
| Alanine | 241 | 244 | 212 | 213 | 157 |
| Cystine | — | — | — | — | — |
| Valine | 140 | 131 | 133 | 148 | 74 |
| Methionine | — | — | — | 1 | 1 |
| Isoleucine | 26 | 28 | 29 | 17 | 17 |
| Leucine | 54 | 58 | 57 | 66 | 68 |
| Tyrosine | 8 | 8 | 7 | 7 | 20 |
| Phenylalanine | 30 | 30 | 29 | 38 | 28 |
| Isodesmosine[a] | 3.7 | 3.8 | 3.1 | 3.2 | 2.9 |
| Desmosine[a] | 4.6 | 5.3 | 8.0 | 7.6 | 4.1 |
| Lysinonorleucine | 0.9 | 1.3 | 1.9 | ND[b] | 1.0 |
| Histidine | 1 | 2 | 1 | 3 | 4 |
| Lysine | 8 | 8 | 8 | 13 | 9 |
| Arginine | 6 | 6 | 6 | 10 | 15 |

[a] The iso(desmosine) peaks represent only a portion of the total ring cross-links present in elastin (see text).
[b] Not determined.

chyma. Treatments with 50% pyridine and 2 $M$ $CaCl_2$ can be omitted in these cases.

Previously we included in our procedure an extraction with 5% trichloroacetic acid at 90° for 30 min to denature and remove much of the collagen.[18] We found that this step is unnecessary to obtain pure elastin, and it may be undesirable as it could break aldimine cross-links.

## Isolation of Elastin from Elastic Cartilage

Bovine cartilages from young animals were obtained from a local slaughterhouse. The cartilages were dissected free of pericondrium and attached soft tissue and finely minced in 50% aqueous pyridine at 4°. The procedure is identical to that described above and includes the 2 $M$ $CaCl_2$ step.

Characterization

*Amino Acid Composition*

Amino acid analyses, including the separation of the desmosine and isodesmosine cross-link components, were carried out on aliquot samples, which had been hydrolyzed for 44 hr in 6 $N$ HCl under vacuum at 105°, on a Beckman Model 121-M automatic amino acid analyzer with an elution gradient programmed for elastin. To estimate possible collagen contamination, aliquot samples were analyzed using an extended chromatogram of the basic amino acids to measure the hydroxylysine content accurately. Samples for hexosamine analysis were hydrolyzed in 8 $N$ HCl for 3 hr at 95° and similarly chromatographed on the amino acid analyzer with the same conditions used for hydroxylysine. Galactosamine and glucosamine peaks were resolved, and their contents were calculated. Consistent with common, although erroneous, practice, desmosine and isodesmosine as found after acid hydrolysis are listed as such. It should be noted that they may account for less than half of the total ring-containing cross-links, which are partially converted to the iso(desmosines) during extraction and acid hydrolysis.

The amino acid composition of elastin prepared as above from bovine ear cartilage, ligamentum nuchae, visceral pleura, lung parenchyma, and aorta, is shown in Table I. It should be noted that there are some consistent differences between the composition of cartilage elastin and that derived from other tissues. Specifically, when compared to the more familiar ligament elastin, aspartic acid, serine, glutamic acid, tyrosine, and arginine residues are increased whereas valine and alanine residues are decreased in elastin derived from cartilage. In this type of elastin, two or three residues of methionine are also present, and the sum of isodesmosine and desmosine residues is 80–90% of that found in the more familiar type of elastin. Further analysis of the purified samples demonstrated that collagen contamination amounted to no more than 1.5% and that neither glucosamine nor galactosamine were present at a greater concentration than 1.0 residue per 1000 amino acid residues. Similar differences in amino acid composition between cartilage elastin and other elastins were also obtained using the harsher hot alkali purification technique (D. A. Keith, unpublished results).

These results suggest that cartilage elastin has a quite different amino acid composition when compared with the other elastin(s). In previous reports elastin has been isolated by hot alkali treatment from rabbit ear cartilage[20] and by hot alkali and autoclaving from bovine ear cartilage.[8]

[20] R. A. Anwar, *Can. J. Biochem.* **44**, 725 (1966).

Although impure and degraded elastin was obtained, the same characteristic trends in the amino acid analysis can be seen. A more recent study[21] also demonstrates the tissue-related compositional variability of elastin. Using an enzymic purification technique, the authors demonstrated an overall twofold increase in polar residues and a 20% reduction in valine when compared to elastin from aorta and ligamentum nuchae. Although the increase in the amount of aspartic acid, glutamic acid, and tyrosine, and the decrease in alanine, valine, and isoleucine were not as dramatic as in our preparations, the overall trends were similar.

*Valylprolyl Content*

A further way of characterizing elastin is afforded by an analyses of the Val-Pro sequence. Hauschka and Gallop,[22] following Cannon's[23] original observation, have shown that after base hydrolysis valylproline is preserved to a large extent. If labeled valine is used, elastin biosynthesis can be measured as described by Hauschka and Gallop[22] for mouse lung.

*Procedure.* Elastin samples (10 mg) are hydrolyzed in alkali-resistant tubes with 1 ml of 2 $N$ KOH under vacuum at 100° for 22 hrs. Cooled hydrolyzates are neutralized with concentrated $HClO_4$ using a trace of phenol red as an indicator. After cooling on ice for 1 hr, the samples are centrifuged (2000 $g$, 10 min) to remove $KClO_4$. The clear yellow supernatant is chromatographed on a 60 × 0.9 cm column of cation exchange resin MR-205 at 55° using a citrate gradient (pH 2.8, 0.2 $M$ $Na^+$ to pH 7.0, 0.75 $M$ $Na^+$). Valylproline elutes after and close to phenylalanine, while the valylprolyl anhydride elutes at the front.[22] Three separate preparations are hydrolyzed for each type of elastin.

Such investigations[22,23] have shown that in the elastins studied, the sequence Val-Pro is preserved about 50% as the dipeptide and about 50% in the cyclized diketopiperazine under alkaline hydrolysis conditions. At present,[24] valylproline recovery is calculated taking into consideration the ninhydrin peak of valylproline eluting after phenylalanine. The calculation is based on a color factor obtained with an authentic sample of valylproline, which is 62% of the color factor of phenylalanine. The phenylalanine peak is also used to calculate Val-Pro sequences per 1000 amino acid residues as the recovery of phenylalanine is the same after acid and alkaline hydrolysis.

[21] J. M. Field, G. W. Rodger, J. C. Hunter, A. Serafini-Fracassini, and M. Spina, *Arch. Biochem. Biophys.* **191**, 705 (1978).
[22] P. V. Hauschka and P. M. Gallop, *Anal. Biochem.* **92**, 61 (1979).
[23] D. J. Cannon, Ph.D. Dissertation, Boston University, Boston, Massachusetts, 1968.
[24] P. V. Hauschka, *Anal. Biochem.* **80**, 212 (1977).

Five reference proteins containing known Val-Pro sequences were base-hydrolyzed, and the recovery of valylproline was obtained. The recoveries ranged from 24.0 to 55.6% with an average of 38.4%. When the two different elastins were compared by this procedure, 19 Val-Pro sequences (uncorrected for recovery) and 5 Val-Pro sequences (uncorrected) were obtained from ligamentum nuchae and ear cartilage elastins, respectively. Correcting for recovery, employing the range obtained from the five reference proteins, indicated that ligamentum nuchae had 56 ± 22 Val-Pro sequences per 1000 residues whereas ear cartilage elastin had 15 ± 6 Val-Pro sequences per 1000 amino acid residues. Clearly, this reinforces the thesis that the amino acid sequence in cartilage elastin is quite different from the more familiar elastin(s) and may thus represent a different genetic type.

On the basis of amino acid composition, there are 115 proline and 141 valine residues in ligamentum nuchae and 120 proline and 78 valine residues in cartilage elastin. On the basis of a random distribution, this calculates to 18 and 11 Val-Pro sequences for the two elastins. From sequence data, ligamentum nuchae is estimated to have about 41 Val-Pro sequences.[25] The sequence data for ear cartilage elastin is unknown as yet. Clearly, the Val-Pro sequence range (56 ± 22) for ligamentum nuchae obtained by alkaline hydrolysis is consistent with the sequence data. The range (15 ± 6) for the ear cartilage is much lower than that of the ligament elastin, in fact closer to the calculated random range. It is apparent that very significant sequence differences occur in these two elastins even in such characteristic residues as valine and proline. Even more dramatically, if one assumes the same efficiency of recovery of valylproline after base hydrolysis for the two elastins, there would be about four times as many of these sequences in ligamentum nuchae as in elastic cartilage elastin.

*Histochemical and Electron Microscopic Characterization*

Histochemical and electron microscopic studies have pointed to distinct morphological differences between the elastic fibers found in bovine ligamentum nuchae and those found in the bovine elastic cartilage of the external ear, epiglottis, and larynx. Using Verhoeff's[19] stain and standard electron microscopy methods,[26] it was shown (D. A. Keith, unpublished observations) that elastic fibers from ligamentum nuchae have a generally smooth appearance, run parallel to one another, appear not to branch, and have a mean diameter of 1.7 $\mu$m.[27] They are similar to those observed in

[25] L. B. Sandberg, *Int. Rev. Connect. Tissue Res.* **7**, 160 (1976).
[26] M. E. Holtrop, L. G. Raisz, and H. A. Simmons, *J. Cell Biol.* **60**, 346, 1974.
[27] D. A. Keith, *Arch. Oral Biol.* **24**, 211 (1979).

the aorta and lung. In contrast, elastic fibers in ear cartilage of animals of the same age have an irregular, somewhat beaded appearance, form a distinct network that envelops the cells, and branch extensively. They are similar to those found in epiglottic and laryngeal cartilage and have a mean diameter of 0.7 μm.[27]

*Analysis of Elastin Cross-links—Labeled NaBH$_4$ Reduction*

Finely powdered elastin (10 mg) is suspended in 10 ml of 0.1 $M$ sodium phosphate buffer, pH 7.8. To the suspension is added 2.5 mg of $^3$H-labeled sodium borohydride (specific activity approximately 5 to 10 × 10$^6$ dpm/ μmol, calibrated according to Paz *et al*.[28] The mixture is kept for 30 min at room temperature with continuous stirring. The reaction is stopped by addition of glacial acetic acid to pH 5, and the suspension is dialyzed for 36 hr against several changes of H$_2$O. The lyophilized, reduced elastin is divided into two aliquots: one is hydrolyzed with 3 $N$ HCl in a sealed evacuated tube for 44 hr at 105°; the other is hydrolyzed with 2 $N$ KOH at 100° for 22 hr. After evaporation of the HCl from the acid hydrolyzate under vacuum in a rotary evaporator, the residue is dissolved in 2 ml of H$_2$O. To the alkaline hydrolyzates, perchloric acid is added to pH 3.0 at 4°, and the precipitate of potassium perchlorate is separated by centrifugation. The cross-linking profiles are obtained by chromatography of the hydrolyzates on a single column automatic amino acid analyzer using a gradient of sodium citrate buffers from pH 2.8, 0.2 $M$ Na$^+$ to pH 7.0, 0.75 $M$ Na$^+$.

*Labeled NaBD$_4$, D$_2$O Reduction*

Finely powdered elastin (10 mg) is suspended in 2 or 3 ml of D$_2$O and is evaporated under vacuum to remove residual H$_2$O; the process is repeated several times. The elastin then is allowed to stand overnight at room temperature in D$_2$O adjusted to pH 8–9 by addition of a small amount of anhydrous K$_2$CO$_3$.[29] To the suspension is added 2.5 mg of tritiated sodium borodeuteride (NaBD$_4$) calibrated as described previously,[28] and the mixture is kept for 30 min at room temperature with continuous stirring. The procedure is continued as described for the NaBH$_4$ reduction and the reduced elastin is then hydrolyzed with 3 $N$ HCl in a sealed evacuated tube for 44 hr at 105°. The HCl is removed under vacuum and the hydrolyzate is fractionated, then further purified by col-

[28] M. A. Paz, E. Henson, R. Rombauer, L. Abrash, O. O. Blumenfeld, and P. M. Gallop, *Biochemistry* **9**, 2123 (1970).
[29] M. A. Paz, B. Pereyra, P. M. Gallop, and S. Seifter, *J. Mechanochem. Cell Motil.* **2**, 231 (1974).

umn chromatography using the procedure of Paz et al.[30] The purified reduced iso(desmosine) (tetrahydrodesmosines) are next converted to the N-trifluoroacetylmethyl esters, which are then examined in the Hitachi RMU6E mass spectrometer.

*Reaction of Elastin with [$^{14}C$]NaCN and $NH_3$*

[$^{14}C$]NaCN of specific activity 1 mCi/mmol was purchased from New England Nuclear Corporation (Boston, Massachusetts) and stored at −20°. The exact specific activity was determined as [$^{14}C$]alanine using a Strecker reaction with acetaldehyde as described by Pereyra et al.[31] Finely powdered elastin (10 mg) is suspended in 2 ml of $H_2O$. To the suspension are added 2.0 mg of [$^{14}C$]NaCN[31,32] and 2 ml of 30% $NH_4OH$. The pH of the reaction mixture is approximately 11.5 and is maintained at 11.5 for 1 hr. Then the pH is progressively decreased, one unit at a time, over a 5-hr period to a value of pH 7. After acidification to pH 1 under the hood and dialysis, the protein, now containing $^{14}C$-labeled aminonitriles, is treated with 6 N HCl at 105° for 48 hr. The HCl is removed under vacuum in a rotary evaporator; the hydrolyzate is taken up five times in distilled $H_2O$ and reevaporated. The hydrolyzate is chromatographed on a single-column amino acid analyzer using the gradient described above.

*Effect of [$^3H$]NaBH$_4$ Reduction on Elastin Cross-linking Compounds*

Upon reduction, allysine residues in elastin are reduced to ε-hydroxynorleucine residues that upon acid hydrolysis are partially converted into chloronorleucine[33]; the extent of this conversion to chloronorleucine is much less with 3 N HCl hydrolysis.[34] The reduced aldol compound that results from the chemical reduction of the aldol condensation product of two alysine residues is unstable to acid hydrolysis. After acid hydrolysis, two or more radioactive peaks appear in the tyrosine region of the eluate. The reduced aldol and the hydroxynorleucine are best measured after alkaline hydrolysis. The cross-linkages, lysinonorleucine and dehydrolysinonorleucine, formed by the reactions between allysine and lysine, are present in elastin as both the naturally reduced compound lysinonorleucine (75%) and the aldimine, dehydrolysinonorleucine (25%);

[30] M. A. Paz, P. M. Gallop, O. O. Blumenfeld, E. Henson, and S. Seifter, *Biochem. Biophys. Res. Commun.* **43**, 289 (1971).
[31] B. Pereyra, O. O. Blumenfeld, M. A. Paz, E. Henson, and P. M. Gallop, *J. Biol. Chem.* **249**, 2212 (1974).
[32] B. Pereyra, M. A. Paz, P. M. Gallop, and O. O. Blumenfeld, *Biochem. Biophys. Res. Commun.* **55**, 96 (1973).
[33] C. Franzblau and R. Lent, *Brookhaven Symp. Biol.* **21**, 358 (1969).
[34] M. L. Tanzer, G. Mechanic, and P. M. Gallop, *Biochim. Biophys. Acta* **207**, 548 (1970).

the latter compound will incorporate tritium upon chemical reduction with [$^3$H]NaBH$_4$ to form labeled lysinonorleucine, which is stable to acid hydrolysis. The lysinonorleucine peak appears in the chromatogram of reduced elastin after the reduced iso(desmosine) (tetrahydrodesmosine) peak.

It is likely that intact elastin when present in tissues contains crosslinks derived from three allysyl residues and one lysyl residue, which are condensed into dihydropyridine compounds (I).

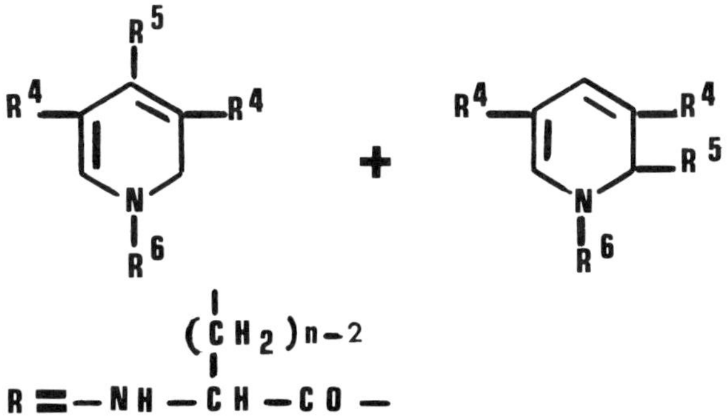

$$R = -NH-CH-CO- \quad \text{with} \quad (CH_2)_{n-2}$$

I

It is known that related synthetic compounds are easily oxidized to the pyridinium salts by oxygen, halogens, silver nitrate and other oxidizing agents[35] (II).

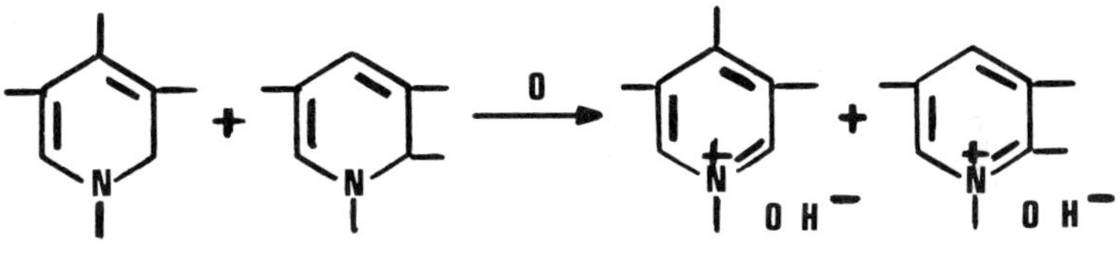

II

Pyridinium salts in the presence of alkali are easily hydrated and may be further oxidized to pyridones (III).

Thus, it would seem likely that the isolation and preparation of elastin with hot alkali would promote the destruction of both the dihydropyridine and pyridinium forms of the cross-links and lead to the artifactual production of further oxidized compounds. With elastin prepared by the Lansing

---

[35] J. A. Joule and G. F. Smith, "Heterocyclic Chemistry," 2nd ed. Van Nostrand-Reinhold, New York, 1978.

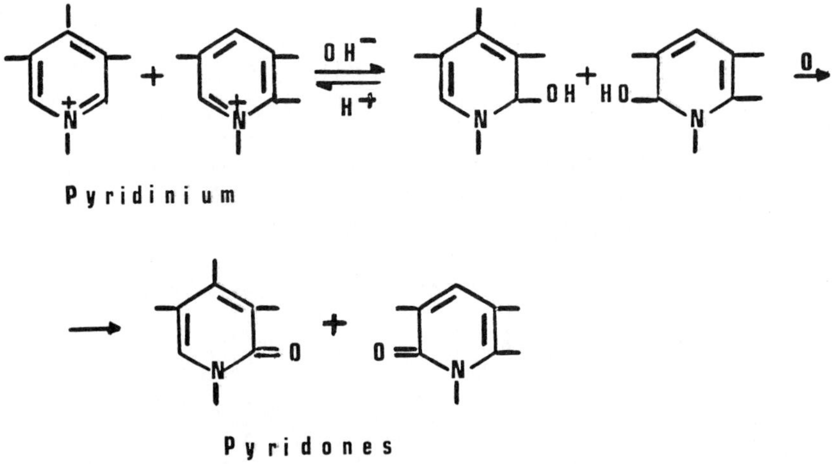

**Pyridinium**

**Pyridones**

**III**

procedure,[3] such compounds may account for as much as 20% of the cross-links[30]; other laboratories have also observed such compounds,[36] which have not been adequately characterized as yet.

*Reduction of the Ring Cross-links in Elastin*

In analogy with the reduction of pyridinium and dihydropyridine compounds,[35] reduction of the ring cross-links of elastin should result in the formation of 1,2,5,6-tetrahydropyridines and from mass spectral studies in our laboratory[29] this seems to be the major pathway (IV).

The first step in the reduction of the pyridinium compounds by borohydride is the hydride transfer to $C_2$ followed by a second reduction of the 5,6 enamine double bond (the olefinic 3,4 double bond is not reduced). In the reduction of the dihydropyridine cross-link, reduction of the enamine double bond occurs also without reduction of the olefinic double bond. Accordingly, if one employs $NaBD_4/D_2O$ reduction, the deuterated reduction products resulting from the pyridinium and dihydropyridines can be distinguished by mass spectrometry of appropriate derivatives.[29] Furthermore, with the pyridinium compounds, the methylene hydrogens adjacent to the quaternary nitrogen in the $R^5$-$CH_2$-$N\overset{+}{\leq}$ side chain are exchangeable with $D_2O$ before, but not after, reduction (V).

[36] R. Baurain, J. F. Larochelle, and F. Lamy, *Eur. J. Biochem.* **67**, 155 (1976).

## IV

## V

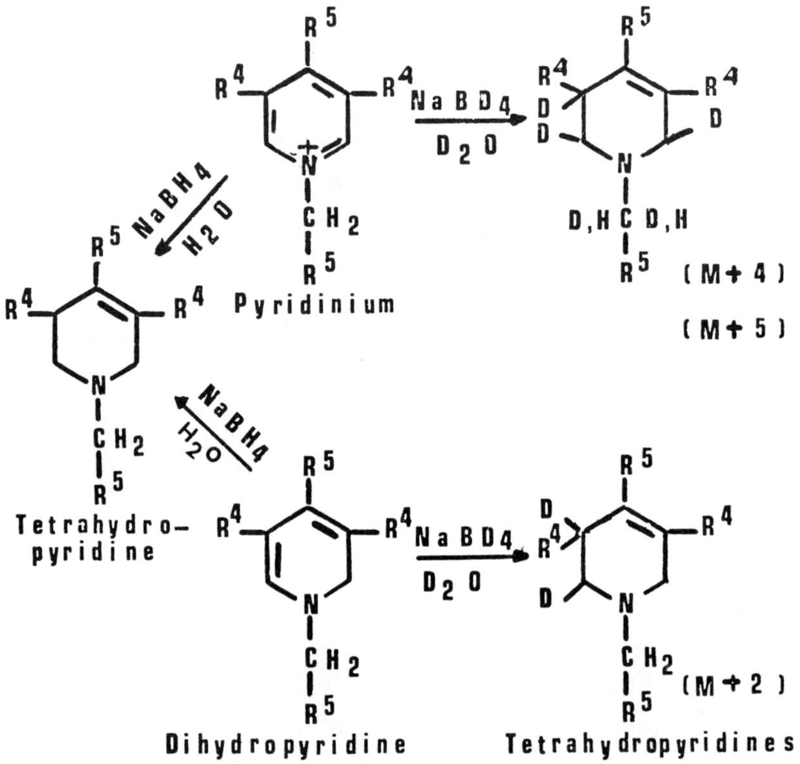

[Figure VI: Reduction scheme showing Pyridinium, Tetrahydropyridine, Dihydropyridine, and Tetrahydropyridines with NaBH₄/H₂O and NaBD₄/D₂O reactions, yielding (M+4), (M+5), and (M+2) species]

**VI**

Thus, if the mass of the tetrahydrodesmosines obtained by the reduction of the desmosines or dihydrodesmosines with $NaBH_4$ is $M$ then the reduction with $NaBD_4/D_2O$ will give masses of $M + 4$ and $M + 5$ with the desmosines and a mass of $M + 2$ with the dihydrodesmosines[29,30] (VI).

Suffice it to say that mass spectrometric analysis of the products of [³H]$NaBH_4$ and [³H]$NABD_4/D_2O$ reduction have led to the following conclusions:

1. In native elastin dihydrodesmosine and dihydroisodesmosine are the major ring cross-links with lesser levels of desmosines and isodesmosines. A small level of the naturally reduced tetrahydrodesmosines are also present.
2. If elastin is isolated after base treatment, there is significant destruction of both desmosines and dihydrodesmosines and the appearance of more highly oxidized forms, which are formed artifactually.
3. Acid hydrolysis of elastin without prior $NaBD_4/D_2O$ reduction

cannot give an accurate view of the cyclic cross-links, since acid hydrolysis is highly destructive of the dihydrodesmosine forms of the cross-links converting a portion to the iso(desmosines).
4. There is evidence also from mass spectral analysis that in addition to the pyridinium and dihydropyridine forms of the cross-links there are cyclic forms of the cross-links in which a double bond is exocyclic and present on the $R^5$ sidechain.
5. It is possible, indeed even likely, that in native elastin, certain of the enamine and olefin double bonds present in the ring cross-links may be hydrated.

*Effect of [$^{14}C$]NaCN and NH$_3$ Treatment on Elastin Cross-linking Compounds*

The reaction of the elastin with [$^{14}$C]NaCN and NH$_3$ leads to the stabilization and labeling of certain cross-links and their intermediates, which can be isolated after acid hydrolysis. Four $^{14}$C-labeled amino acids have been previously identified: $\alpha,\epsilon$-diaminopimelic acid (DAP) and $\epsilon$-hydroxy-$\alpha$-aminopimelic acid (OHAP) (derived from $\alpha$-amino adipic acid $\delta$-semialdehyde); 1,9-diamino-1,9-dicarboxy-glycino-non-4-ene [carboxyamino aldol (CAA), derived from the aldol condensate of two residues of $\alpha$-aminoadipic acid $\delta$-semialdehyde]; and 2-amino-6-lysinopimelic acid [carboxylysinonorleucine (CLNL) derived from dehydrolysinonorleucine[31,32]]. Table II shows the analyses of elastin cross-links from various tissues. Desmosine, isodesmosine, and lysinonorleucine are calculated from the ninhydrin reaction in the amino acid analyzer, and OHAP, DAP, CLNL, and CAA are calculated from the recovered radioactivity and the specific activity of the cyanide reagent. The results are expressed as lysine equivalents. In Table II, the sum of OHAP and DAP measures the allysine content of the elastin, CAA measures aldol condensation product content, and CLNL measures dehydrolysinonorleucine. There are various other, as yet unidentified, peaks, and one lysine equivalent is assumed for each peak. In the elastin from cartilage the recovery of lysine-derived cross-links and intermediates varies between a total of 9.5 and 11.3 residues, which is lower than the corresponding figures for visceral pleura and lung parenchyma elastin. In addition, the total lysine residues accounted for is lower in elastin derived from cartilage.

*Photolysis of Desmosines in Ultraviolet Light*

Baurain, Larochelle, Guay, and Lamy[36-38] developed a method to cleave specifically the (iso)desmosines. Photolysis of (iso)desmosines in

[37] F. Lamy, R. Baurain, M. Guay, and J. F. Larochelle, *Adv. Exp. Med. Biol.* **79**, 177 (1977).
[38] M. Guay and F. Lamy, *Trends Biochem. Sci.* **4**, 160 (1979).

## TABLE II
### ANALYSIS OF ELASTIN CROSS-LINKS[a]

| | Bovine tissue source | | |
|---|---|---|---|
| | Ear | Lung parenchyma | Visceral pleura |
| Isodesmosine[b] | 3.2 | 3.2 | 3.1 |
| Desmosine[b] | 4.6 | 7.6 | 8.0 |
| Lysinonorleucine[b] | 0.20 | 0.50 | 1.9 |
| Pre-DAP[c] | 0.40 | 0.54 | 0.15 |
| DAP + OHAP[c] | 0.17 | 0.43 | 0.44 |
| CAA[c] | 0.48 | 0.57 | 0.95 |
| CLNL[c] | 0.32 | 0.17 | 0.65 |
| Others[c] | 0.09 | 0.09 | 0.39 |
| Lysine derived cross-links and intermediates | 9.5 | 13.1 | 15.6 |
| Lysine residues | 6.2 | 13.0 | 8.5 |
| Total lysine residues accounted for[d] | 15.7 | 26.1 | 24.1 |

[a] The average residue molecular weight of elastin is 83.8; the number of micromoles per 83.8 mg of elastin is equivalent to the number of residues per 1000 residues; all the data are expressed as lysine equivalents per 1000 residues.

[b] Calculated from absorbance at 570 nm (ninhydrin) using the color value of lysine.

[c] Calculated from radioactivity.

[d] Pig aortic tropoelastin contains about 46 residues of lysine, assuming the same value for tropoelastin of other species; about one-half are accounted for by lysine-remaining and lysine-derived cross-links and intermediates. There are other unidentified radioactive peaks not considered in this calculation. The iso(desmosine) represents only a portion of the lysine-derived ring compounds in elastin (see text).

ultraviolet light (280 nm) results in the formation of an azabicyclo compound, which at low pH further cleaves to give an aminoketone derivative. When the desmosine is in a peptide linkage, the aminoketone derivative is hydrolyzed, resulting in the formation of a new lysine residue and a substituted ketoaldehyde. In the case of free desmosine, the aminoketone could rearomatize to give a new isomer of desmosine, photodesmosine (VII).

This procedure allows a specific cleavage of iso(desmosine) and may be useful in sequence studies of cross-linked (pyridinium)-containing peptide. However, to reduce the size of the peptide, an additional double-bond cleavage will be necessary if it is assumed that two peptides are cross-linked by the pyridinium compound.

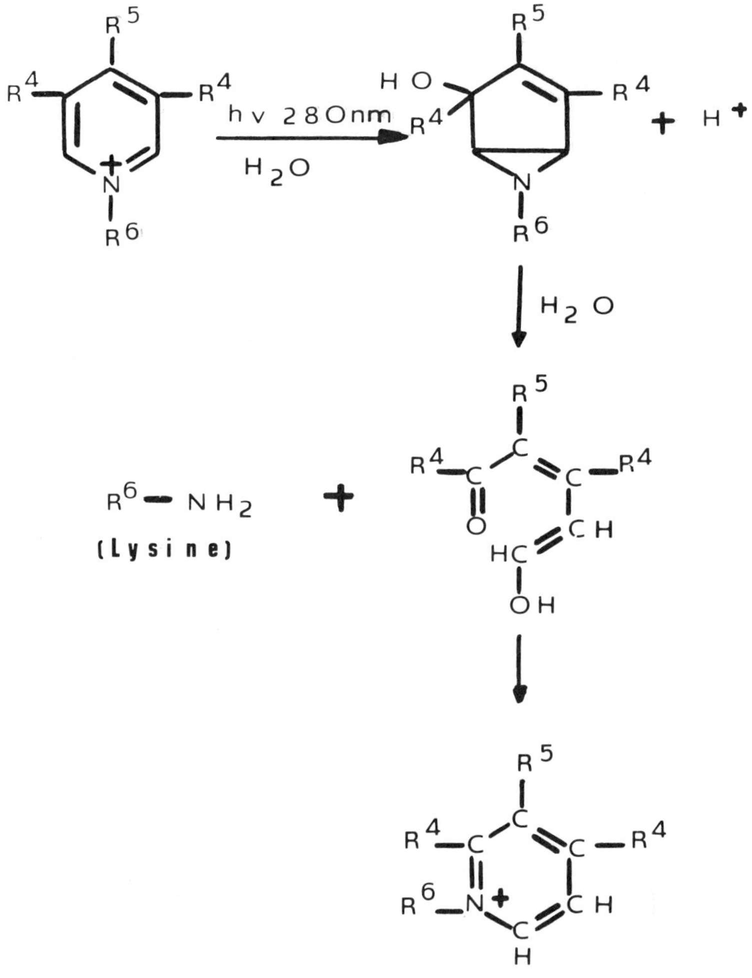

**Photodesmosine**

**VII**

Accordingly, further methodology based on photolysis and specific methods of cleavage of the remaining double bonds should be developed in order to facilitate sequence analysis of cross-linked elastin peptides. Guy and Lamy[38] have reviewed some of the analytical difficulties involved in analysis of the ring compounds in elastin.

## [32] Proteolysis of Insoluble Elastin

*By* P. J. STONE, C. FRANZBLAU, and H. M. KAGAN

### Elastases

Although insoluble elastin is unusually stable in mature, healthy elastic tissue, there is evidence that proteolytic degradation of elastin is involved in certain diseases including emphysema, pancreatitis, and advanced atherosclerosis.[1-3] The unusual nature of the relationship between this insoluble protein and soluble proteases that degrade it has prompted considerable investigative effort on the nature and sources of elastolytic enzymes as well as on the mechanism of the elastolytic process. The physiological balance between elastases and their inhibitors and the role of such imbalances in certain disease states has also been the subject of considerable study.[1]

Elastolytic proteases were found sufficiently rarely in the early years of research on elastolysis that an enzyme demonstrated to have such potential was commonly named an "elastase." The evolution of this terminology has led to the occasional misconception that elastin is the only protein that such elastases degrade, a conclusion now clearly shown not to be correct. For example, digestion of lung tissue with porcine pancreatic elastase results in complete solubilization of the tissue except for the collagen component.[4] Elastase can also catalyze the cleavage of soluble proteins such as ribonuclease and insulin.[5-7]

Enzymes capable of digesting insoluble elastin have now been identified in several sources including pancreas and human polymorphonuclear leukocytes. A number of enzymes from plants and microorganisms have been found to digest elastin, as well; many of these are metalloproteases.[8,9] Pancreatic elastase and polymorphonuclear leukocyte (PMN) elastase, both serine proteases, are the most abundant elastases in mam-

---

[1] M. Galdston, A. Janoff, and A. L. Davis, *Am. Rev. Respir. Dis.* **107**, 718 (1973).
[2] M. C. Geokas, *Arch. Pathol.* **86**, 127 (1968).
[3] W. A. Loeven, *J. Atheroscler. Res.* **10**, 379 (1969).
[4] R. H. Goldstein, B. Faris, C.-L. Hu, G. L. Snider, and C. Franzblau, *Am. Rev. Respir. Dis.* **117**, 281 (1978).
[5] A. S. Narayanan and R. A. Anwar, *Biochem. J.* **114**, 11 (1969).
[6] H. Levy and G. Feinstein, *Biochim. Biophys. Acta* **567**, 35 (1979).
[7] D. Atlas, *J. Mol. Biol.* **93**, 39 (1975).
[8] N. Nishino and J. C. Powers, *J. Biol. Chem.* **255**, 3482 (1980).
[9] G. L. Snider, J. A. Hayes, C. Franzblau, H. M. Kagan, P. J. Stone, and A. L. Korthy, *Am. Rev. Respir. Dis.* **110**, 254 (1974).

mals and are the most thoroughly characterized.[10,11] Certain of their properties are listed in Table I. The possibility that elastase activity derived from neutrophils may contaminate tissue samples, thus introducing a rich source of elastase in such samples, must be considered. Thus, elastases found in the alveolar macrophage,[12] sputum,[13] and spleen[14] appear to represent PMN elastase. Cathepsin G, a serine protease found with PMN elastase in the azurophilic granules of PMNs, exhibits weak elastolytic activity.[15] An elastase isolated from murine peritoneal macrophages exhibits the inhibitor profile of a metalloenzyme (Table I). Elastase-like activities have been obtained from alveolar macrophages,[12,16] human platelets,[17] and human aorta[18] and from human fibroblasts and rat smooth muscle cells in culture,[19] although additional work is required to establish their properties.

Characterization and quantification of elastases depends critically upon the selection of appropriate enzyme assay procedures. Although a variety of excellent synthetic ester and amide substrates exist that the neutrophil and pancreatic enzymes will hydrolyze, designation of an enzyme as an "elastase" requires the clear demonstration that such an enzyme digests insoluble elastin. A number of sensitive assay methods now exist that can be relied upon to establish and quantify the ability of a protease to degrade elastin. This chapter describes assay procedures found to be most useful in our laboratories for quantifying elastolytic activity.

Preparation of Elastin for Use as a Substrate

Insoluble Elastin

It is known that the susceptibility of insoluble elastin to elastolytic enzymes can be markedly influenced by the method of purification of

[10] D. M. Shotton, this series, Vol. 19, p. 113.
[11] R. J. Baugh and J. Travis, *Biochemistry* **15**, 836 (1976).
[12] L. M. Hinman, C. A. Stevens, R. A. Matthay, and J. B. L. Gee, *Am. Rev. Respir. Dis.* **121**, 263 (1980).
[13] D. Y. Twumasi and I. E. Liener, *J. Biol. Chem.* **252**, 1917 (1977).
[14] P. M. Starkey and A. J. Barrett, *Biochem. J.* **155**, 255 (1976).
[15] C. F. Reilly and J. Travis, *Biochim. Biophys. Acta* **621**, 147 (1980).
[16] H. DeCremoux, W. Hornebeck, M.-C. Jaurand, J. Bignon, and L. Robert, *J. Pathol.* **125**, 171 (1978).
[17] Y. Legrand, J. Caen, F. M. Booyse, M. E. Rafelson, B. Robert, and L. Robert, *Biochim. Biophys. Acta* **305**, 406 (1973).
[18] W. Hornebeck and L. Robert, *Adv. Exp. Med. Biol.* **79**, 145 (1977).
[19] M. C. Bourdillon, D. Brechemier, N. Blaes, J. C. Derouette, W. Hornebeck, and L. Robert, *Cell Biol. Int. Rep.* **4**, 313 (1980).

TABLE I
PROPERTIES OF THREE ELASTASES[a-c]

| Property | Porcine pancreatic | Human leukocytic | Mouse peritoneal macrophage |
|---|---|---|---|
| Rate of solubilization | | | |
| [$^3$H]Elastin[a,e] | 1.7 μg/μg enzyme/hr | 2.4 μg/μg/hr | Elastolysis stimulated by SDS pretreatment of elastin |
| [$^3$H]Elastin-SDS[d-f] | 40 μg/μg enzyme/hr | 12 μg/μg/hr | |
| Specificity ($K_{cat}/K_m$) against | | | |
| Ac-Ala-Ala-Pro-X-$p$-nitroanilide[g] | X = Ala ≫ Leu > Val | X = Val ≫ Ile > Ala | Not available |
| Inhibitors | | | |
| MeO-Suc-Ala$_2$-Pro-Val-CH$_2$Cl | + | + | − |
| EDTA | − | ± | + |
| Phenylmethylsulfonylfluoride | + | + | − |
| Soybean trypsin inhibitor | − | + | − |
| $\alpha_1$-Protease inhibitor (human) | + | + | − |
| $\alpha_2$-Macroglobulin (human) | + | − | + |
| Serum albumin | − | | |
| Charge at pH 7 | Cationic | Cationic | Cationic |

[a] D. M. Shotton, this series, vol. 19, p. 113.
[b] R. J. Baugh and J. Travis, *Biochemistry* **15**, 836 (1976).
[c] R. R. White, D. Norby, A. Janoff, and R. Dearing, *Biochim. Biophys. Acta* **612**, 233 (1980).
[d] P. J. Stone, G. Crombie, and C. Franzblau, *Anal. Biochem.* **80**, 572 (1977).
[e] P. J. Stone, J. D. Calore, and C. Franzblau, unpublished data, 1979.
[f] SDS, sodium dodecyl sulfate.
[g] M. Zimmerman and B. M. Ashe, *Biochim. Biophys. Acta* **480**, 241 (1977).

insoluble elastin from connective tissue. Elastin occurs as an insoluble protein together with a microfibrillar glycoprotein in the elastic fiber. This elastic fiber is found in close association with other structural elements of connective tissue, particularly collagen and proteoglycans. Insoluble elastin as isolated by currently used procedures is considered to be that protein material which remains insoluble after all other connective tissue components have been removed. Since such procedures of necessity resort to relatively harsh extraction conditions to solubilize the nonelastin components, a degree of uncertainty may exist about the correspondence of the physical-chemical properties of the isolated elastin to those properties of the unperturbed protein in living tissue. For the purposes of examining the characteristics of insoluble elastin as a substrate for elastase, the description of the methods of purification of insoluble elastin can be considered under three categories: hot alkali or acid procedures, autoclaving procedures, and milder enzymic and detergent procedures.

*Extraction with Hot Alkali*

For many years, the most commonly used method for the purification of insoluble elastin was that described by Lansing et al.[20] According to this method, a sample of connective tissue is incubated in 0.1 N NaOH at 95° for 45 min. The residue is then thoroughly washed with water and dried with a series of washes in alcohol–ether mixtures and finally in ether. The insoluble residue is characterized at least by amino acid analysis. Preferably, analyses should also describe the content of lysine-derived cross-links. It should be noted that the number of peptide bonds cleaved in insoluble elastin by the Lansing method may be variable.[21] This would result in varying degrees of polarity of this protein, which is an important factor in its susceptibility to elastases.[22]

*Autoclaving*

Insoluble elastin can also be prepared by repeated autoclaving of crude connective tissues.[23] The tissue is suspended in water and autoclaved for successive periods of 45 min to 1 hr. The supernatant after each autoclaving period is tested for the presence of soluble protein by an appropriate procedure. Autoclaving is continued until protein is no longer solubilized.

[20] A. I. Lansing, T. B. Rosenthal, M. Alex, and E. W. Dempsey, *Anat. Rec.* **114**, 555 (1952).
[21] S. M. Partridge, *Adv. Protein Chem.* **17**, 227 (1962).
[22] P. J. Stone, H. M. Kagan, B. Faris, and C. Franzblau, *Adv. Exp. Med. Biol.* **79**, 597 (1977).
[23] S. M. Partridge, H. F. Davis, and G. S. Adair, *Biochem. J.* **61**, 11 (1955).

## Newer Purification Procedures

Denaturants such as guanidine-HCl have also been used to purify elastin from tissues. Miller and Fullmer[24] employed successive treatment of tissues with $Na_2HPO_4$, KCl, and guanidine-HCl for 48 hr at 5°. The product was treated with purified bacterial collagenase at 37° for an additional 48 hr. Ross and Bornstein[25] modified this procedure further employing $0.05\,M$ dithioerythritol buffered with $0.1\,M$ Tris, pH 8.5, and 0.1% EDTA together with the collagenase and $5\,M$ guanidine-HCl treatment. Richmond[26] further modified the treatment by using $6\,M$ urea and 1% sodium dodecyl sulfate.

A more recently described procedure for the preparation of insoluble elastin has employed CNBr to remove all of the nonelastin components, an approach that exploits the fact that insoluble elastin does not contain methionine. Rasmussen et al.[27] treated tissue with $5\,M$ guanidine-HCl containing 1% mercaptoethanol, following this with 5 hr of autoclaving, and finally treating with CNBr in 97% formic acid. These procedures have been reviewed in detail.[28] It should be noted that it is essential to characterize the purity of the preparation of collagenase used in any of these procedures, since enzymic impurities may exist in preparations of collagenase that might catalyze the hydrolysis of elastin.

### Soluble Elastin

Relatively large polypeptides can be solubilized from insoluble elastin by partial hydrolysis, and such soluble elastin fragments may serve as substrates for various elastolytic enzymes. Successive treatments of insoluble elastin with $0.25\,M$ oxalic acid at 90° will solubilize two major polydispersed fractions that can be separated and are identified as $\alpha$- and $\beta$-elastin.[23,29] $\alpha$-Elastin has an average molecular weight of approximately 80,000, and $\beta$-elastin has an $M_r$ of 5000. Pancreatic elastase will digest both fractions.[29] Other similar procedures for the partial solubilization of elastin have been described. For example, treatment of insoluble elastin with $1\,N$ NaOH in 80% ethanol at room temperature yields soluble $\kappa$-elastin.[30] The use of such soluble fragments as substrates in elastolytic assays must be considered with caution, since there are several kinds of proteases that can catalyze their hydrolysis, not all of which may be able to digest insoluble elastin.

[24] E. J. Miller and H. M. Fullmer, *J. Exp. Med.* **123**, 1097 (1966).
[25] R. Ross and P. Bornstein, *J. Cell Biol.* **40**, 366 (1969).
[26] V. Richmond, *Biochim. Biophys. Acta* **351**, 173 (1974).
[27] B. L. Rasmussen, E. Bruenger, and L. B. Sandberg, *Anal. Biochem.* **64**, 225 (1975).
[28] L. B. Sandberg, *Int. Rev. Connect. Tissue* **7**, 159 (1976).
[29] S. M. Partridge and H. F. Davis, *Biochem. J.* **61**, 21 (1955).
[30] N. Kornfeld-Poullain and L. Robert, *Bull. Soc. Chim. Biol.* **50**, 759 (1968).

## Modified Elastins

Several covalently or noncovalently modified derivatives of soluble and insoluble elastin preparations have been tested as substrates for elastolytic assays. These include both chromophorically labeled and isotopically labeled forms of elastin.

### Chromophoric Ligands

Many elastolytic assay procedures employ insoluble elastin substrates that have been complexed with chromophoric substances. Congo red elastin[10] has been widely used, for example. Fluorescently labeled derivatives of elastin substrates include rhodamine-elastin,[31] fluorescein-elastin,[32] and Remazolbrilliant blue-elastin.[32] Elastolysis is monitored by spectrophotometrically quantifying the appearance of peptide-bound dye molecules solubilized from insoluble elastin by proteolysis.

### Radiolabeled Elastin

Elastin can be covalently derivatized with isotopically labeled reagents by, for example, iodination[33] with $^{125}$I and $^{131}$I, sodium borohydride reduction of the protein in the presence of [$^{14}$C]formaldehyde leading to the incorporation of labeled methyl groups,[34] or direct radioactive reduction of insoluble elastin with $^3$H-labeled sodium borohydride.[35,36] In each case, elastolysis of such substrates releases radioactive soluble fragments from the insoluble protein. The measurement of solubilized radioactivity provides a convenient, sensitive, and accurate measurement of rates of elastolysis. The preparation and use of $^3$H-labeled sodium borohydride-reduced insoluble elastin is described in detail in the next section.

## Assays for Elastolytic Activity

As discussed at the outset of this chapter, the identification of an enzyme as having elastase activity must include the demonstrated ability of such an enzyme to digest an insoluble elastin substrate. In this section we will describe that method, utilizing a tritiated reduced derivative of insoluble elastin as a substrate; we believe this to be a reliable and ex-

---

[31] P. F. Huebner, *Anal. Biochem.* **74**, 419 (1976).
[32] H. Rinderknecht, M. C. Geokas, D. Silverman, Y. Lillard, and B. J. Haverback, *Clin. Chim. Acta* **19**, 327 (1968).
[33] B. Robert and L. Robert, *Eur. J. Biochem.* **11**, 62 (1969).
[34] D. R. Bielefeld, R. M. Senior, and S. Y. Yu, *Biochem. Biophys. Res. Commun.* **67**, 1553 (1975).
[35] S. Takahashi, S. Seifter and F. C. Yang, *Biochim. Biophys. Acta* **327**, 138 (1973).
[36] R. W. Lent, B. Smith, L. L. Salcedo, B. Faris, and C. Franzblau, *Biochemistry* **8**, 2837 (1969).

tremely sensitive method for demonstrating and quantifying elastolytic activity.

Insoluble Elastin Substrates

*Solubilization Assays*

Many elastolytic assay procedures are based on the solubilization of elastin. Such procedures may monitor the appearance of soluble elastin as measured by the physical or chemical properties of the solubilized elastin, or they may monitor the appearance of a chromophoric or isotopic label covalently or noncovalently bound to insoluble elastin substrate. Elastin substrate may be prepared as elastin–sodium dodecyl sulfate (SDS) complexes to further enhance the sensitivity of assays, since elastin-SDS is hydrolyzed much more readily by pancreatic elastase than is the free insoluble protein.[37]

[$^3$H]ELASTIN SOLUBILIZATION ASSAY

*Principle.* Measurement of enzyme-solubilized radiolabeled elastin from an insoluble radiolabeled form of elastin. As discussed previously, it is essential to characterize elastolytic activity using an insoluble elastin substrate. We have found the use of purified elastin radiolabeled with [$^3$H]NaBH$_4$ to be a convenient, sensitive, and accurate method of identifying and quantifying elastase activity by this criterion.[38]

*Reagents*

Insoluble bovine ligamentum nuchae elastin, prepared by the autoclaving method[23]
Sodium borate, 0.1 $M$; Na EDTA, 0.001 $M$; pH 9.7
NaOH, 0.01 $N$
[$^3$H]NaBH$_4$, about 300 mCi/mmol
Acetic acid, 10%
Sodium dodecyl sulfate, 0.5%
NaHCO$_3$, 0.1 $M$; NaN$_3$, 0.05%; pH 8.9
Porcine pancreatic elastase (200 µg/ml, $A_{280\,nm}$ = 0.404) in 0.5 m$M$ CaCl$_2$, 0.05% NaN$_3$. The solution should be prepared and clarified by centrifugation at 2°. Aliquots of this elastase solution may be stored at $-20°$. Depending upon the purposes of the assay, it may be desirable to determine the concentration of elastase by active site titration.[39]

[37] H. M. Kagan, G. D. Crombie, R. E. Jordan, W. Lewis, and C. Franzblau, *Biochemistry* **11**, 3412 (1972).
[38] P. J. Stone, G. Crombie, and C. Franzblau, *Anal. Biochem.* **80**, 572 (1977).
[39] J. C. Powers and D. L. Carroll, *Biochem. Biophys. Res. Commun.* **67**, 639 (1975).

*Preparation of Substrate.* The insoluble elastin powder (15 g) is suspended at 2° in 200 ml of 0.1 $M$ sodium borate buffer, pH 9.7, 0.001 $M$ EDTA by stirring for several hours. The solubility of the sodium borate is limited at 2°. One Curie of [$^3$H]NaBH$_4$ is dissolved in 2 ml of 0.01 $N$ NaOH at 2°. The borohydride solution is added dropwise to the well stirred elastin suspension in a fume hood. After 2 hr of stirring, the suspension is allowed to settle at room temperature. The supernatant is decanted in the fume hood and discarded. The elastin is washed twice with water by centrifugation in tightly capped centrifuge bottles followed by several washes with 10% acetic acid and several washes with water. The reduced elastin product is dried by lyophilization. Typically, such preparations have a specific activity of $1.5 \times 10^7$ dpm/mg, determined by scintillation counting of acid-hydrolyzed aliquots.

[$^3$H]*Elastin Solubilization Assay Procedure.* Tritiated powdered elastin (5 mg per assay tube) is homogenized as one batch in 8 ml of 0.5% sodium dodecyl sulfate using a ground-glass homogenizer and is centrifuged at room temperature. The supernatant is discarded, and the pellet is resuspended in 0.5% sodium dodecyl sulfate and centrifuged as before. This procedure is repeated with 0.1 $M$ sodium bicarbonate buffer, pH 8.9, containing 0.05% sodium azide. The pellet is resuspended in this bicarbonate assay buffer at 1.25 mg of tritiated elastin per milliliter. Aliquots (4 ml) are removed from the vigorously stirred suspension using a pipette with a wide-bore opening and placed in screw-cap culture tubes. The tubes are centrifuged (1000 $g$) and the supernatant is discarded. Assay buffer (4 ml) or assay buffer containing the enzyme sample to be assayed is added to each tube containing the elastin pellet.

Aliquots of pancreatic elastase in a solution stabilized with 0.5 m$M$ CaCl$_2$ are used directly or are freshly diluted in plastic tubes and are added to several assay tubes containing [$^3$H]elastin prepared as described. The tubes are capped, and the contents are gently mixed and then incubated at 37° without agitation. The incubation period ranges from 2 hr to 3 weeks depending upon the amount of elastolytic activity present and the sensitivity desired. After incubation, the contents of the tubes are filtered through 25-mm conical funnels containing 5.5 cm medium-porosity filter paper. An aliquot of each filtrate is counted by liquid scintillation spectrometry.

*Interpretation of Results.* The total amount of soluble tritium in each assay tube is corrected for the tritium release observed in the blank samples. The standard deviation of the tritium released within a typical experiment is ± 5%.[38] Elastolytic activity is expressed in microgram equivalents of pancreatic elastase based upon the results obtained with the standard elastase solution. The rate of solubilization of elastin in micrograms of elastin digested per hour can also be calculated by dividing the net soluble tritium by the product of the specific radioactivity of the [³H]elastin preparation (cpm/μg) and the assay incubation time (hr).

*Precautions.* Dilution of pancreatic elastase solutions in glassware often results in loss of enzyme activity due to adsorption to the glass. Dilution of the assay buffer with relatively large volumes of the sample to be assayed should be avoided, since the enzyme activity of elastases is often sensitive to ionic strength and pH. In such cases an assay buffer, such as $0.05\,M$ Tris, $0.14\,M$ NaCl, $0.05\%$ $NaN_3$, may be employed that is similar in pH and ionic strength to that of the sample. Since elastin-bound SDS is in equilibrium with SDS in solution, the effects of even small amounts of dissolved SDS on the material to be assayed must be considered. For example, weak complexes containing elastase may be disrupted and unstable elastase inhibitor molecules such as $\alpha_1$-protease inhibitor may be denatured during the assay procedure.

*Sensitivity and Advantages.* Less than 0.001 μg of porcine pancreatic elastase or its equivalent activity can be detected using this procedure. When elastin is reduced with relatively small amounts of sodium borohydride, its susceptibility to elastase is similar to autoclaved ligamentum nuchae elastin that has not been reduced. The rate of digestion of reduced tritiated elastin, when pretreated with SDS, is enhanced to approximately the same degree as is the SDS complex of bovine ligamentum nuchae elastin. Enzyme-solubilized radiolabeled peptides can also be further characterized by gel filtration and amino acid analysis to gain information on the endopeptidyl specificity of the particular elastase being studied. The reduced and labeled cross-linking amino acids of the solubilized elastin can also be separated and quantified by amino acid analysis.[36]

OTHER SOLUBILIZATION ASSAYS

Underivatized insoluble elastin is also useful for solubilization assays, since the intrinsic protein nature of the enzyme-solubilized material can be quantified by spectrofluorometric,[40] spectrophotometric,[41] conductimetric,[42] or chemical means.[43] Visual measurement of the area of clearing around sample wells in a suspension of finely divided elastin in agar is also widely used.[44] However, some elastase-like enzymes, such as that from rabbit alveolar macrophages, produce only partial clearing of the elastin suspension, making quantification difficult.[45] This suggests that this enzyme activity cleaves only a limited number of peptide bonds in

[40] R. S. Quinn and E. R. Blout, *Biochem. Biophys. Res. Commun.* **40**, 328 (1970).
[41] W. Ardelt, S. Ksiezny, and I. Niedzwiecka-Namyslowska, *Anal. Biochem.* **34**, 180 (1970).
[42] H. Bakala, J. Wallach, and M. Hanss, *Biochimie* **60**, 1205 (1978).
[43] I. Banga, J. Balo, and M. Horvath, *Biochem. J.* **71**, 544 (1959).
[44] R. M. Senior, P. F. Huebner, and J. A. Pierce, *J. Lab. Clin. Med.* **77**, 510 (1971).
[45] Z. Werb and S. Gordon, *J. Exp. Med.* **142**, 361 (1975).

elastin. Immunological procedures can be used to quantify solubilized elastin.[46,47] The elastin cross-linking amino acid, desmosine, has been similarly assessed.[48]

A number of assay procedures employ elastin that has been pretreated with a chromophoric dye, as noted above in the subsection on modified elastins. Orcein-elastin has been used to assess elastolytic activity and also to visualize the position of elastolytic enzyme activity in polyacrylamide gels.[49,50] When the dye is not covalently attached to the elastin, care must be used in associating the appearance of dye in solution with the solubilization of elastin. The presence of dye-absorbing materials, such as albumin, in the incubation mixture might facilitate the nonenzymic release of dye from the elastin substrate. It is also likely that the dye may affect the substrate properties of the elastin, either enhancing or inhibiting the rate of elastolysis (see subsection on modulation of elastolysis). There are reports that the presence of chymotrypsin and trypsin apparently enhance the rate of elastin solubilization as measured by the appearance of dye in solution.[51] This phenomenon is not observed when the [$^3$H]elastin solubilization procedure is used.[38]

*Peptide Bond Hydrolysis—Assay in a Recording pH Stat*

*Principle.* Although proteolysis of the peptide bonds of elastin eventually leads to solubilization, characteristically there is a significant lag in the time of appearance of soluble peptides, although initial proteolysis has occurred. This lag reflects the extensive cross-linking of elastin polypeptides that prevents solubilization until several bonds have been hydrolyzed. Therefore, accurate measurement of initial rates of elastolysis must rely upon assessment of peptide bond hydrolysis. pH-stat titration of the proton release accompanying peptide bond hydrolysis provides a sensitive measure of this process. Proton release consequent to peptide bond hydrolysis can be followed only at pH values where the ionizable proton of the newly formed $\alpha$-amino group is released into solution. We have found that the pH-stat assay employed at pH 8.9 to be a sensitive and quantitative measure of enzymic peptide bond hydrolysis in insoluble elastin.[37]

[46] U. Kucich, P. Christner, G. Weinbaum, and J. Rosenbloom, *Am. Rev. Respir. Dis.* **122**, 461 (1980).
[47] R. P. Mecham and G. Lange, *Connect. Tissue Res.* **7**, 247 (1980).
[48] G. S. King, V. S. Mohan, and B. C. Starcher, *Connect. Tissue Res.* **7**, 263 (1980).
[49] L. A. Sachar, K. K. Winter, N. Sicher, and S. Frankel, *Proc. Soc. Exp. Biol. Med.* **90**, 323 (1955).
[50] J. Dijkhof and C. Poort, *Anal. Biochem.* **83**, 319 (1977).
[51] G. Gnosspelius, *Anal. Biochem.* **81**, 315 (1977).

*Reagents*

Insoluble ligamentum nuchae elastin prepared by the autoclaving method.[21] The final product is a finely ground free-flowing powder, passing through a fine-mesh sieve of 16 divisions per centimeter. The elastin is uniformly suspended in $CO_2$-free water at 20 mg ml$^{-1}$ by homogenizing in a ground glass, hand-held conical homogenizer.
$CO_2$-free distilled, deionized water
NaCl, 1 $M$, in $CO_2$-free water
NaOH, 0.005 $M$, standardized against potassium acid phthalate

*Procedure.* The assay is performed in an automatic recording pH stat, e.g., a Radiometer Autotritor TTT-11 with a Titrigraph SBR-2, a 0.5 ml Autoburette filled with 0.005 $N$ NaOH, and a Titration Assembly TTA-3. The double-walled reaction vessel is maintained at 37°, and the reaction mixture is continuously stirred by a disk magnet supplied with the reaction vessel. A stream of pure nitrogen gas, saturated by first bubbling through a vessel containing $CO_2$-free water, is continuously passed over the surface of the reaction mixture. The reaction vessel contains 20 mg of elastin added as 1 ml of the elastin stock suspension. It is essential to thoroughly stir the elastin stock suspension just prior to removal of aliquots to prevent uneven distribution of the insoluble protein. The specific amount of the NaCl solution to be added may vary depending upon the purpose of the assay and the specific elastase to be assayed. Thus, activity of pancreatic elastase against elastin is optimal in 0.03 $M$ NaCl,[52,52a] whereas PMN elastase is stimulated at higher salt concentrations.[53] The volume of the reaction mixture is brought to 1.9 ml with water and the pH adjusted to 8.9 with NaOH. The pH-stat is set at this end point, and a 5-min baseline recording of base uptake is established. Enzyme (0.1 ml) is added to start the reaction and followed for 3–10 min to establish the initial reaction rate (recorder speed 1 cm/2 min; proportional band, 0.3).

*Units.* One unit of elastase is defined as the amount of enzyme releasing 1 $\mu$mol of protons per minute under the conditions of the assay. The corrected quantity of protons released can be obtained by substituting the p$K_{app}$ value of 8.15 ($\alpha$-amino functions) into the Henderson–Hasselbach equation, yielding a correction factor of 1.18 by which proton release measured at pH 8.9 is multiplied.

[52] R. E. Jordan, N. Hewitt, W. Lewis, H. Kagan, and C. Franzblau, *Biochemistry* **13**, 3497 (1974).
[52a] F. Lamy, C. P. Craig, and S. Tauber, *J. Biol. Chem.* **236**, 86 (1961).
[53] C. Boudier, K. K. Andersson, C. Balny, and J. G. Bieth, *Biochem. Med.* **23**, 219 (1980).

Soluble Substrates for Elastase-like Activity

Since porcine pancreatic elastase cleaves preferentially at alanine peptide bonds and human PMN elastase cleaves preferentially at valine peptide bonds, and since each enzyme appears to have an extended substrate binding site, a number of peptidyl and nonpeptidyl synthetic oligopeptides, amides, and esters have been prepared and used as substrates for elastase-like activities. Such assays are convenient and quite sensitive, but, as discussed, final designation of any uncharacterized enzyme activity as elastolytic must require that it digest insoluble elastin. For example, human pulmonary alveolar macrophages secrete an enzyme that hydrolyzes the commercially available substrate $N$-succinyl-L-alanyl-L-alanyl-L-alanine-$p$-nitroanilide (SAPNA) but does not solubilize [$^3$H]elastin.[12] Similarly, serum hydrolyzes SAPNA, but not insoluble elastin.[54] Efforts to overcome this drawback include the use of chloromethyl ketone peptides, which appear to be specific inhibitors for pancreatic and/or PMN elastase.[55]

Hydrolysis of $N$-acetyl-(L-Ala)$_3$-methyl ester can be followed by pH-stat titration, described earlier. The ester bond of this substrate is only slowly hydrolyzed by trypsin and $\alpha$-chymotrypsin but is readily susceptible to pancreatic elastase.[56] Similarly, elastase hydrolyzes the ester linkage of $N$-acetyl-(L-Ala)$_3$-$p$-nitroanilide releasing $p$-nitroaniline, which is highly chromophoric.[57] This substrate is also hydrolyzed only very slowly by chymotrypsin or trypsin but may be susceptible to other proteases or esterases. Fluorogenic peptidyl synthetic substrates with the sequence MeO-Suc-Ala-Ala-Pro-Val-X provide increased sensitivity for assays of elastase-like esterase activity.[58] Thus, the hydrolysis of the 1-methoxy-3-naphthylamide derivative was followed spectrofluorometrically, allowing the detection of as little as 0.33 ng of human PMN elastase and 0.47 ng of porcine pancreatic elastase per milliliter.[58]

Histochemical substrates are available for the localization of elastase-like activity. One of these permits the localization of elastase-like activity at the ultrastructural level. Thus, fixed tissue or cell specimens are incu-

---

[54] K. Katagiri, K. Ito, M. Miyaji, T. Takeuchi, K. Yoshikane, and M. Sasaki, *Clin. Chim. Acta* **95**, 401 (1979).
[55] J. C. Powers, B. F. Gupton, A. D. Harley, N. Nishino, and R. J. Whitley, *Biochim. Biophys. Acta* **485**, 156 (1977).
[56] A. Gertler and T. Hofman, *Can. J. Biochem.* **48**, 384 (1970).
[57] G. Feinstein, A. Kupfer, and M. Sokolovsky, *Biochem. Biophys. Res. Commun.* **50**, 1020 (1973).
[58] M. J. Castillo, K. Nakajima, M. Zimmerman, and J. C. Powers, *Anal. Biochem.* **99**, 53 (1979).

bated with $N$-$t$-Boc-L-alanine-$p$-nitrothiophenyl ester in the presence of gold ions.[59] Hydrolysis of this thiol ester results in the release of the $p$-nitrothiophenyl ion, which precipitates as the electron dense gold-mercaptide at the site of hydrolysis. Another technique suitable for the light microscope level utilizes the $N$-acetylalanine ester of 1-naphthol, which is hydrolyzed by elastase-like esterases and simultaneously coupled with hexazonium pararosaniline.[60] Ac-Ala-ONP which releases $p$-nitroaniline on hydrolysis has been used to visualize elastolytic activity in polyacrylamide gels.[61]

Soluble elastin[62] as well as tropoelastin[63] have been used in a number of assay procedures. These soluble elastin substrates may be susceptible to a broader range of proteases than insoluble elastin. The physical-chemical properties, and particularly the polarity of soluble fragments of elastin, are markedly different from those of insoluble elastin. Oxalic acid-solubilized elastin peptides have been tritiated by reductive methylation using formaldehyde followed by reduction of the resulting Schiff base with [$^3$H]NaBH$_4$.[62] Proteolysis of this substrate is assessed by quantification of the radioactivity in soluble tritiated fragments after trichloroacetic acid precipitation of the reaction mixtures. It is noteworthy that the presence of sodium dodecyl sulfate in a concentration that stimulates the elastolysis of insoluble elastin suppresses the rate of degradation of this soluble elastin substrate produced by elastase, consistent with earlier studies.[37]

## Modulation of Elastolysis

### Physical–Chemical Environment

The optimal pH for solubilization of elastin by pancreatic elastase is approximately 8.9.[10] The optimal temperature is between 37° and 45°, depending upon a number of other variables.[64] The optimal concentration of sodium chloride is 0.03 $M$ for assays of pancreatic elastase and approximately 0.7 $M$ for PMN elastase.[52,52a,53]

---

[59] J. M. Clark, D. W. Vaughan, B. M. Aiken, and H. M. Kagan, *J. Cell Biol.* **84**, 102 (1980).
[60] L. Ornstein, A. Janoff, F. R. Sweetman, and H. R. Ansley, *J. Histochem. Cytochem.* **21**, 411 (1973).
[61] G. Feinstein and A. Janoff, *Biochim. Biophys. Acta* **403**, 493 (1975).
[62] W. Bieger and G. Scheele, *Anal. Biochem.* **104**, 239 (1980).
[63] P. Christner, G. Weinbaum, B. Sloan, and J. Rosenbloom, *Anal. Biochem.* **88**, 682 (1978).
[64] R. E. Jordan, *Diss. Abs. Int. B* **35**, 87 (1974).

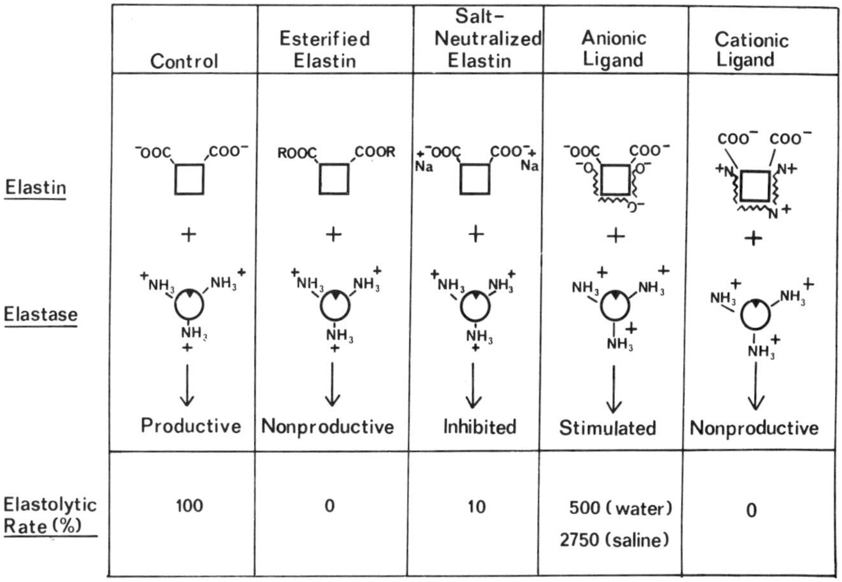

| | Control | Esterified Elastin | Salt-Neutralized Elastin | Anionic Ligand | Cationic Ligand |
|---|---|---|---|---|---|
| Elastin | $^-OOC\ \ COO^-$ | $ROOC\ \ COOR$ | $^-OOC\ \ COO^-$ / Na Na | $^-OOC\ \ COO^-$ with anionic ligand | $COO^-\ COO^-$ with $+N\ \ N+$ / $N+$ |
| | + | + | + | + | + |
| Elastase | $^+NH_3\ \ NH_3^+$ / $NH_3^+$ | $^+NH_3\ \ NH_3^+$ / $NH_3^+$ | $^+NH_3\ \ NH_3^+$ / $NH_3^+$ | $^+NH_3\ \ NH_3^+$ / $NH_3^+$ | $NH_3\ \ NH_3$ / $NH_3^+$ |
| | ↓ | ↓ | ↓ | ↓ | ↓ |
| | Productive | Nonproductive | Inhibited | Stimulated | Nonproductive |
| Elastolytic Rate (%) | 100 | 0 | 10 | 500 (water) 2750 (saline) | 0 |

FIG. 1. Surface lysyl residues of pancreatic elastase are individually indicated as $-NH_3^+$. The enzyme-active site is represented as a triangular wedge. Carboxylate functions of elastin are represented as $-COO^-$. All rates were established by pH-stat titration.

## Elastin Ligands

Digestion of insoluble elastin by pancreatic elastase may be inhibited or totally prevented if sources of electrostatic charge on the enzyme or on the substrate are specifically altered.[52,65-67] As summarized in Fig. 1, elastolysis is partially inhibited by neutralization of elastin carboxylate groups with sodium ions or is completely inhibited by esterification of these anionic sites in the elastin substrate.[52,65,66] Alternatively, elimination of surface positive charge of pancreatic elastase by maleylation of its ε-amino groups produces an enzyme derivative that is inert toward elastin, but hydrolyzes synthetic esters at nearly normal rates.[52,67] Thus, nonspecific electrostatic forces distributed between anionic sites in elastin and the cationic enzyme are essential for the formation of productive enzyme–substrate complexes. Elastolysis may also be affected by elastin ligands. Amphiphilic agents of appropriate charge can generate elastin–

[65] H. M. Kagan and R. M. Lerch, *Biochim. Biophys. Acta* **434**, 223 (1976).
[66] D. A. Hall and J. W. Czerkowski, *Biochem. J.* **80**, 128 (1961).
[67] A. Gertler, *Eur. J. Biochem.* **20**, 541 (1971).

ligand complexes with selectively altered charge species and/or charge density. Anionic ligands, including a variety of fatty acids, bile salts, anionic detergents, and anionic dyes stimulate elastolysis, whereas a complex of elastin with dodecyltrimethylammonium bromide (DTAB), a cationic detergent, or crystal violet, a cationic dye, is inert as a substrate for pancreatic elastase (Table II). However, DTAB-elastin is a substrate for maleylated elastase, a situation in which the normal charge relationships have been completely reversed by rendering cationic sites in the enzyme anionic by chemical modification and by binding a cationic ligand to elastin,[52] emphasizing the nonspecificity of this charge relationship. The optimal conditions for the action of the anionic ligands differ as suggested by the data presented in Table II; however, what is clear is the dramatic levels of stimulation of elastolysis achievable with anionic agents that are similar with respect to their hydrophobicities and the presence of highly

TABLE II
RELATIVE RATES OF DIGESTION OF INSOLUBLE ELASTIN BY PORCINE PANCREATIC ELASTASE

| | Assay procedure[a] | | | |
|---|---|---|---|---|
| | pH stat | | Solubilization of [$^3$H]Elastin | |
| Ligand | Relative rate | Ligand conc. (mM) | Relative rate | Ligand conc. (mM) |
| None | 1[a,b] | | 1[a,c] | |
| Anionic | | | | |
| Sodium dodecyl sulfate | 25 | 15[b,d] | 25 | 15[c,d] |
| Sodium oleate | 25 | 6[b,d] | 32 | 15[d,e] |
| Congo red | 11 | 2[e] | 22 | 0.3[e] |
| Cationic | | | | |
| Crystal violet | 0.2 | 0.8[e] | 0.4 | 0.2[e] |
| Dodecyltrimethyl-ammonium bromide | 0 | 5[b] | 0 | 5[e] |

[a] See text for a discussion of the assay procedures. Except where otherwise noted, the solubilization assay was carried out at 37° with 0.05 M Tris buffer, pH 7.6, 0.14 M NaCl, and 0.05% NaN$_3$; the pH-stat procedure was carried out at 37° and pH 8.9 in the presence of 0.15 M NaCl.
[b] R. E. Jordan, N. Hewitt, W. Lewis, H. Kagan, and C. Franzblau, *Biochemistry* **13**, 3497 (1974).
[c] P. J. Stone, G. Crombie, and C. Franzblau, *Anal. Biochem.* **80**, 572 (1977).
[d] Samples were pretreated with the concentrations indicated, washed, and resuspended.
[e] P. J. Stone, J. D. Calore, and C. Franzblau, unpublished data, 1979.

anionic functional groups. It is of further interest that elastin contains very few intrinsic sites of negative charge, since 70% of its very few dicarboxylic amino acids exist as amides.[65]

The elastase activity in crude preparations of stimulated mouse peritoneal macrophages, like pancreatic elastase, is dramatically enhanced using SDS-elastin as a substrate, suggesting that elastolysis by this enzyme also may be charge dependent.[37,45] Although human neutrophil elastase is also a cationic protein, elastolysis by this enzyme is considerably more resistant to high salt concentrations and is elevated only four- to fivefold by addition of SDS to the elastin substrate (Table I). Other agents, including a series of aliphatic alcohols,[68] divalent metal ions such as $Ca^{2+}$ and $Mg^{2+}$ [52,68a] and bile salt complexes[52,68a] also elevate elastolytic rates, presumably by acting as elastin ligands. Taken together these effects raise the possibility that elastolysis may be influenced by naturally occurring agents, such as free fatty acids, that may bind to elastin *in vivo*. Further, such findings have permitted the development of more sensitive elastolytic assays, as illustrated by the use of SDS-elastin as a substrate for elastase.[35]

Influence of the Method of Preparation of Elastin

It is not surprising that the method of purification of elastin may affect its susceptibility to proteolysis by elastase. Thus, elastin purified with hot alkali appears to be more susceptible to elastase digestion than elastin prepared by autoclaving.[21,22] Conversely, significant stimulation of elastolysis is brought about by anionic ligands such as sodium dodecyl sulfate, using elastin prepared by autoclaving.[37,66] The stimulation is significantly less, however, with elastin purified with hot alkali.[22] Furthermore, the substrate that is partially purified from calf aorta by extraction only with 1 $M$ NaCl is more similar to autoclaved aorta elastin than to hot alkali-purified aorta elastin.[22] Elastins similarly prepared from aorta, ligament, or lung are similarly susceptible to pancreatic elastase. Thus, the substrate properties of elastin of different tissues vary primarily in a manner related to the method of purification.[22] Elastin treated with hot alkali contains more N termini and C termini than elastin that has been purified by autoclaving.[21] The increased number of carboxyl groups might serve as binding sites for the cationic elastase molecule and thus could facilitate elastolysis.[65,66]

[68] B. M. Ashe and M. Zimmerman, *Biochem. Biophys. Res. Commun.* **75**, 194 (1977).
[68a] W. Hornebeck, *in* "Biochemistry of Normal and Pathological Connective Tissues" (A. M. Robert and L. Robert, eds.), Colloq. Internat. CNRS, 287, Vol. 2, p. 115. CNRS, Paris, 1980.

## Measurement of Functional Anti-Elastase Activity

Naturally occurring inhibitors of elastolytic enzymes are found in the circulation of mammals and other species. Such inhibitors must play extremely important roles in preventing unrestricted proteolytic degradation in blood vessels, lung, and other organs and systems. Alteration of the levels of such inhibitors is associated with certain states in which there is an imbalance in the elastase–antielastase system.[69] The principal inhibitors of elastase in mammals are $\alpha_1$-protease inhibitor (also known as $\alpha_1$-antitrypsin) and $\alpha_2$-macroglobulin. In this section we will outline procedures for quantifying functional antielastase inhibitor levels.

$\alpha_1$-Protease inhibitor ($\alpha$-1-PI) and $\alpha_2$-macroglobulin ($\alpha$-2-M) inhibit elastase by different mechanisms. Proteolytic attack of serine proteases, such as elastase or trypsin, on an $\alpha$-1-PI molecule results in one of two possible events[70]: (a) inactivation of the $\alpha$-1-PI as an inhibitor molecule with the elastase molecule retaining its enzyme activity; (b) alternatively, the elastase molecule may cleave and then form a covalent bond with a portion of the $\alpha$-1-PI molecule, yielding a complex that exhibits no proteolytic or elastolytic activity and characteristically has a molecular weight of about 70,000. The much larger inhibitor molecule, $\alpha$-2-M, exhibits a molecular weight of about 750,000 and inhibits a much wider variety of proteolytic enzymes than does $\alpha$-1-PI.[71] Proteolytic interaction of porcine pancreatic elastase with $\alpha$-2-M produces in the inhibitor molecule a dramatic conformational change that physically traps the elastase.[72] The enzyme is functional in this complex toward small polypeptides or synthetic substrates, but the enzyme is sterically prevented from attacking larger protein substrates, such as elastin.[72] Furthermore, enzymatically active elastase molecules are slowly released from such complexes, presumably by proteolysis of the $\alpha$-2-M.[73]

Those inhibitor molecules that no longer inhibit elastase owing to proteolysis may still retain many of their antigenic properties. Therefore, functional assessment of $\alpha$-1-PI and $\alpha$-2-M can best determine the concentration of each inhibitor class that is capable of inactivating elastase. This assessment is often carried out by mixing a given amount of elastase with increasing volumes of the serum, plasma, or other inhibitor sample.[74] The residual elastolytic or proteolytic activity of each mixture is assessed, and

---

[69] C. B. Laurell and F. Ericksson, *Scand. J. Clin. Lab. Invest.* **15**, 132 (1963).
[70] H. L. James and A. B. Cohen, *J. Clin. Invest.* **62**, 1344 (1978).
[71] P. C. Harpel, *J. Exp. Med.* **138**, 508 (1973).
[72] P. M. Starkey and A. J. Barrett, in "Proteinase in Mammalian Cells and Tissues" (A. J. Barrett, ed.), p. 663. North-Holland Publ., Amsterdam, 1977.
[73] J. S. Baumstark, *Biochim. Biophys. Acta* **207**, 318 (1970).
[74] T. Klumpp and J. G. Bieth, *Clin. Chem.* **25**, 969 (1979).

a plot of percentage of residual elastase activity vs volume of inhibitor sample is determined by extrapolation to zero enzyme activity. There are several difficulties that can be encountered in this procedure. Often a synthetic substrate is used that can be hydrolyzed by elastase which is complexed with $\alpha$-2-M. Extrapolation will yield a value for the inhibitor activity attributable mainly to $\alpha$-1-PI. In addition, the substrate may also be susceptible to hydrolysis by uninhibited nonspecific esterases in the sample (see earlier discussion of SAPNA, for example). The possible effects of competitive substrates for elastase that may be present in the sample are also difficult to assess. However, the use of insoluble elastin substrates generally requires longer incubations for measurement of elastolytic activity. Treatment of the elastin with sodium dodecyl sulfate to enhance elastolysis, as previously noted, may destabilize the $\alpha$-2-M–elastase complex[71] and may also denature $\alpha$-1-PI molecules. Even with the omission of such anionic agents, free elastase will eventually be released from the $\alpha$-2-M–elastase complex, resulting in elevated residual elastase activity.[73]

An alternative procedure for assessing the amount and type of inhibitor activity requires treatment of the inhibitor sample with saturating amounts of elastase followed by rapid inactivation of free elastase and $\alpha$-2-M-bound elastase with a synthetic elastase inhibitor, such as a chloromethyl ketone peptide.[55] The amount of each elastase-inhibitor complex in the sample may be quantified by immunoelectrophoresis[75] or by use of a radiolabel on the elastase followed by separation of the $\alpha$-1-PI–elastase and $\alpha$-2-M–elastase by gel chromatography. In the latter case, the amount of elastase bound to each inhibitor class is calculated from the amount of radioactivity detected in the appropriate molecular weight region divided by the specific radioactivity of the original radiolabeled elastase preparation. It is important to compare the amount of radioactivity recovered from the molecular sieve column with that placed on the column. Furthermore, the radiolabeling procedure should, ideally, produce a uniform population of elastase molecules that exhibit properties unaltered from the original elastase preparations.[76]

Although this method is more time consuming than the others previously mentioned, there are several advantages to this chromatographic procedure. Small amounts of inhibitors that complex with elastase can be quantified. Nonspecific competitive substrates present in the sample are not measured inadvertently. Furthermore the remaining elastase complexes of $\alpha$-1-PI and $\alpha$-2-M are separated and available for other studies.

[75] A. R. Bradwell and D. Burnett, *Clin. Chim. Acta* **58**, 283 (1975).
[76] P. J. Stone, J. D. Calore, G. L. Snider, and C. Franzblau, *Am. Rev. Respir. Dis.* **120**, 577 (1979).

## [33] Primary Structure of Insoluble Elastin

*By* RASHID A. ANWAR

The problems encountered in the study of the primary structure of mature elastin are attributable to its extreme insolubility and high degree of cross-linking. The protein is insoluble in all nonhydrolytic solvents. For this reason, it is practically impossible to establish homogeneity (or lack of it) of an elastin preparation. The solubilization of elastin by chemical or enzymic means usually generates a complex mixture of peptides. Fortunately, in some cases a soluble precursor of elastin (tropoelastin) can be isolated and its primary structure studied essentially by classical sequencing techniques (see Chapters [36]-[40] on tropoelastin). Nevertheless, the study of mature elastin, especially around the cross-links, was clearly required to understand the structure–function relationship of elastin and the mechanism of formation of the cross-links. These considerations led to the structural studies of solubilized cross-linked peptides of elastin.[1-5] In order to determine the amino acid sequences of the cross-linked peptides, the cross-linked peptide chains must be resolved into single-chain peptides, so that the amino acid sequence assignments can be made unambiguously. With the use of Edman degradation, single-chain peptides can be released from the carboxyl groups of the cross-links present in the elastolytic peptides of elastin. These peptides can then be purified and their amino acid sequences can be determined.[1]

*The Rationale behind the Use of the Preparative Edman Degradation for the Release of Carboxyl-Terminal Peptides from the Elastolytic Cross-linked Peptides.* In earlier work,[6] the release of single-chain peptides from the desmosine-containing elastolytic peptides of elastin was attempted by means of the chemical cleavage of the pyridine rings (periodate-permangnate oxidation[7]). The characterization of peptides thus released from the desmosine cross-links indicated that elastase was cleaving at and very close to the $NH_2$ terminals of the cross-links. In retrospect, this was

[1] G. E. Gerber and R. A. Anwar, *J. Biol. Chem.* **249**, 5200 (1974).
[2] G. E. Gerber and R. A. Anwar, *Biochem. J.* **149**, 685 (1975).
[3] K. M. Baig, M. Vlaovic, and R. A. Anwar, *Biochem. J.* **185**, 611 (1980).
[4] J. A. Foster, L. Rubin, H. M. Kagan, C. Franzblau, E. Bruenger, and L. B. Sandberg, *J. Biol. Chem.* **249**, 6191 (1974).
[5] R. P. Mecham and J. A. Foster, *Biochem J.* **173**, 617 (1978).
[6] W. Shimada, A. Bowman, N. R. Davis, and R. A. Anwar, *Biochem. Biophys. Res. Commun.* **37**, 191 (1969).
[7] R. U. Lemieux and E. Von Rudloff, *Can. J. Chem.* **33**, 1701 (1956).

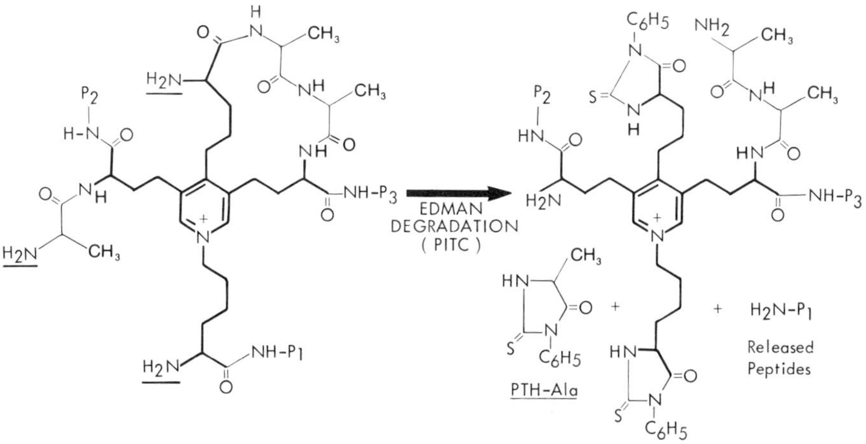

FIG. 1. The use of the Edman degradation for the release of carboxyl-terminal peptides from the elastolytic, desmosine-containing, cross-linked peptides of elastin. PITC, phenylisothiocyanate; PTH, phenylthiohydantoin.

expected because the purified elastase has a strong specificity for alanyl peptide bonds and the cross-linked regions of elastin are very rich in alanine content. The amino-terminal analysis of the elastolytic cross-linked peptides of elastin confirmed the presence of free $\alpha$-amino groups of desmosines and isodesmosines in these peptides. It was, therefore, thought possible to release carboxyl-terminal peptides from the elastolytic, desmosine- and isodesmosine-containing peptides by Edman degradation.[1] Figure 1 shows the Edman degradation of one of the many possible structures present in the cross-linked peptide mixtures and illustrates the rationale of this approach. The thick dark lines represent the desmosine cross-links. At those positions, where the cross-link itself is the $NH_2$ terminal, the Edman degradation will release the peptides attached to the $\alpha$-carboxyl groups of the cross-link ($P_1$ in this case). The released single-chain peptides can then be separated from the cross-linked ones on the basis of size. At position $P_2$, after the first cycle of Edman degradation, the amino-terminal alanine will be released as PTH-Ala, with the concomitant liberation of the $\alpha$-amino group of the cross-link. Therefore, the second cycle of Edman degradation will release $P_2$ from the carboxyl terminal of the cross-link. The peptide $P_3$ will not be released until the cross-link $NH_2$ terminal immediately preceding $P_3$ becomes available. Thus, $P_3$ will be released after four cycles of Edman degradation. It is apparent that after each cycle of Edman degradation the released single-chain peptides must be separated from the cross-linked peptides.

Procedures

*Solubilization and Exhaustive Digestion of Elastin with Porcine Pancreatic Elastase.* It is important to use elastase purified to homogeneity.[8] Partially purified elastase preparations contain carboxypeptidase A and other proteolytic activities. Therefore, the purity of commercially obtained elastase should be ascertained before use (see Chapter [32]). All buffers are saturated with chloroform to minimize bacterial contamination.

Elastin powder (10 g) is suspended in 1 liter of 0.01 $N$ $NH_4HCO_3$ buffer, pH 8.8, and stirred overnight at room temperature. The next morning, 20 mg of elastase are added, and the mixture is allowed to incubate at room temperature (22–24°). After 24 hr of incubation another 20 mg of elastase are added, and the incubation is allowed to continue for additional 24 hr. The pH of the incubation mixture is maintained between 8.7 and 8.9 with the addition of concentrated $NH_4OH$. The incubation mixture is then concentrated to 150 ml, under nitrogen pressure (80–90 psi) in a Diaflo apparatus (Amicon) equipped with a UM-10 membrane. The retentate is diluted with 350 ml of 0.01 $N$ $NH_4HCO_3$ buffer, pH 8.8, and reconcentrated to 150 ml; this procedure is repeated two more times. The desmosine-containing cross-linked peptides do not pass through a UM-10 membrane and are thus retained. This permits the removal of small elastolytic peptides that inhibit elastase. The retentate is then diluted to 500 ml with the same buffer and further digested with two more additions of elastase (20 mg each). The mixture is allowed to incubate for 24 hr after each addition. Finally, the solution is boiled for 5 min and filtered through a Whatman No. 5 paper.

*Isolation of Cross-linked Peptides.* The pH of the elastase digest containing the cross-linked peptides is adjusted to 4.0 with glacial acetic acid. The digest is then applied to a column of cellulose phosphate (Cellex-P; 2.0 × 30 cm), preequilibrated with 0.01 $N$ sodium acetate buffer, pH 4.5. The column is washed with 500 ml of the preequilibration buffer and then eluted with 0.1 $N$ NaCl in the same buffer. Fractions of 7.5 ml are collected, and their absorbance at 280 nm is measured after suitable dilution when necessary. A typical elution profile is shown in Fig. 2. Under these conditions, over 90% of the desmosine-containing peptides present in an elastase digest of elastin absorb onto a cellulose phosphate (CP) column and are eluted in fractions labeled CPB in Fig. 2.

It is useful and desirable first to study the desmosine-containing cross-linked peptides, which are essentially free of other cross-links. This provides unambiguous results about the environment of the desmosines. The desmosine-containing cross-linked peptides essentially free of other

[8] A. S. Narayanan and R. A. Anwar, *Biochem. J.* **114**, 11 (1969).

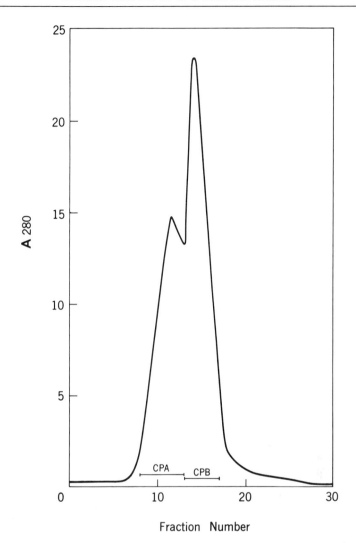

FIG. 2. Chromatographic separation of elastolytic peptides of elastin on cellulose phosphate.

cross-links can be obtained by further fractionation of CPB (Fig. 2) on Sephadex G-50 and Sephadex G-25. Fractions labeled CPB are pooled, dried under reduced pressure, and dissolved in 3–5 ml of 0.01 $M$ $NH_4HCO_3$ buffer, pH 8.8. This solution is applied on a Sephadex G-50 (fine) column (4 × 80 cm) previously equilibrated with 0.01 $M$ $NH_4HCO_3$ buffer, pH 8.8. The column is then eluted with the same buffer at a flow

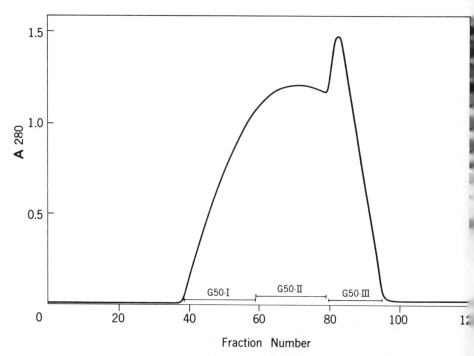

FIG. 3. Chromatography of fraction CPB (Fig. 2) on Sephadex G-50.

rate of 100 ml/hr. Fractions of 10 ml are collected, and their absorbance is measured at 280 nm. A typical elution profile of a Sephadex G-50 chromatography is shown in Fig. 3. The fractions are pooled as indicated in Fig. 3.

The pooled fractions appearing under the peak marked G-50-III are dried under reduced pressure on a rotary evaporator and dissolved in about 1 ml of 0.01 $M$ NH$_4$HCO$_3$ buffer, pH 8.8. This solution is applied on to a Sephadex G-25 (fine) column (2.0 × 35 cm), previously equilibrated with 0.01 $M$ NH$_4$HCO$_3$ buffer, pH 8.8, and the column is eluted with the same buffer. Fractions of 5.0 ml are collected, and absorbance is measured at 275 nm after suitable dilution, when necessary. A typical elution profile of Sephadex G-25 chromatography is shown in Fig. 4. The desmosine-containing cross-linked peptides present in fractions labeled G-25.2 are free of other cross-links and contain 25–30% of the total desmosine and isodesmosine present in elastin. Before pooling the fractions, it is prudent to analyze (amino acid analysis) suitable aliquots from individual fractions on either end of the peak marked G-25.2 to ascertain that the fractions are essentially free of cross-links other than the desmosines (see Chapters [16] and [31]).

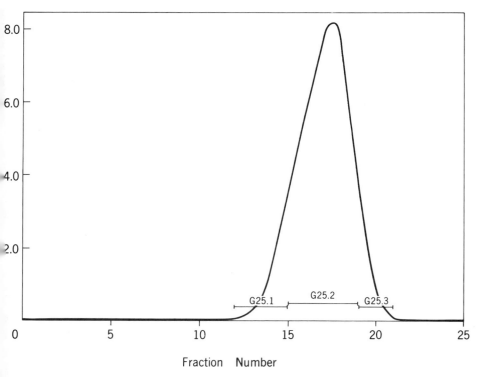

FIG. 4. Chromatography of fraction G-50-III (Fig. 3) on Sephadex G-25.

The pooled fractions (G-25.2) are dried under reduced pressure on a rotary evaporator and the dried material is dissolved in 1 ml of pyridine-water-acetic acid (4:4:1) mixture and applied on to a Sephadex LH-20 column (1.9 × 60 cm) previously equilibrated with 50% pyridine solution. The column is eluted with 50% pyridine at a flow rate of 28 ml/hr. Fractions of 3.5 ml are collected, absorbance is measured at 330 nm, and aliquots (10 µl) are assayed with ninhydrin after alkaline hydrolysis. Owing to the presence of 50% pyridine in elution solvent, absorbance at 280 nm or 275 nm cannot be measured. The excluded material elutes in fractions labeled A in Fig. 5. These fractions, which contain all the desmosine-containing cross-linked peptides applied to the column, are pooled and dried under reduced pressure on a rotary evaporator. This dried material is used for the release of single-chain peptides from the carboxyl terminals of the desmosine cross-links. As an example of the degree of purification expected at each step, amino acid compositions of cross-linked bovine elastin peptides at various stages of purification are shown in the table.

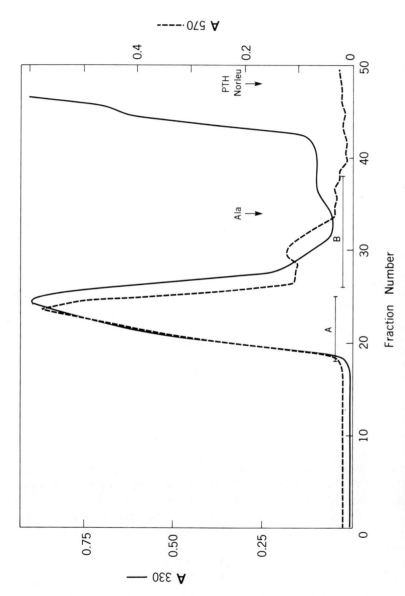

FIG. 5. Chromatography of peptides on Sephadex LH-20. The elution volumes of alanine (Ala) and phenylthiohydantoin of norleucine (PTH-norleu) are indicated by markers.

AMINO ACID COMPOSITION OF CROSS-LINKED BOVINE ELASTIN PEPTIDES AT
VARIOUS STAGES OF PURIFICATION[a]

| Amino acid | CPB | G-50-I | G-50-II | G-50-III | G-25.1 | G-25.2 | G-25.3 | A |
|---|---|---|---|---|---|---|---|---|
| Aspartic acid | 0.2 | 0.4 | 0.1 | 0.2 | 0.1 | 0.1 | 0.5 | 0.1 |
| Threonine | 0.3 | 0.4 | 0.3 | 0.3 | 0.3 | 0.2 | 0.4 | 0.2 |
| Serine | 0.6 | 0.6 | 0.5 | 0.4 | 0.4 | 0.4 | 0.9 | 0.3 |
| Glutamic acid | 0.5 | 0.7 | 0.5 | 0.4 | 0.3 | 0.3 | 1.0 | 0.3 |
| Proline | 4.8 | 6.6 | 4.4 | 1.5 | 2.0 | 1.4 | 3.1 | 1.6 |
| Glycine | 10.4 | 15.3 | 14.2 | 4.5 | 5.7 | 4.4 | 9.2 | 4.1 |
| Alanine | 16.8 | 19.6 | 17.6 | 11.5 | 13.0 | 11.4 | 16.9 | 11.6 |
| Valine | 1.7 | 3.7 | 2.3 | 0.8 | 0.9 | 0.7 | 1.5 | 0.6 |
| Isoleucine | 0.6 | 0.9 | 0.5 | 0.3 | 0.2 | 0.2 | 0.7 | 0.2 |
| Leucine | 1.6 | 2.7 | 1.8 | 0.7 | 0.9 | 0.7 | 1.4 | 0.7 |
| Tyrosine | 0.4 | 0.7 | 0.6 | 0.3 | 0.3 | 0.3 | 0.4 | 0.3 |
| Phenylalanine | 1.2 | 2.0 | 1.7 | 1.1 | 1.0 | 1.0 | 2.9 | 1.0 |
| Isodesmosine | 0.41 | 0.42 | 0.35 | 0.31 | 0.34 | 0.33 | 0.32 | 0.32 |
| Desmosine | 0.59 | 0.58 | 0.65 | 0.69 | 0.66 | 0.67 | 0.68 | 0.68 |
| Lysinonorleucine | 0.1 | 0.1 | 0.1 | 0.1 | 0.1 | — | 1.4 | — |
| Lysine | 0.9 | 0.4 | 0.3 | 0.2 | 0.1 | 0.1 | 0.3 | 0.1 |
| Arginine | 1.1 | 0.5 | 0.4 | 0.3 | 0.3 | 0.3 | 1.4 | 0.1 |

[a] Values are expressed as residues per residue of cross-link (i.e., desmosine plus isodesmosine).

*Release of Carboxyl-Terminal Peptides and Separation of the Released Single-Chain Peptides from the Cross-linked Ones.* The dried material (20–30 μmol of desmosine plus isodesmosine) is dissolved in 4 ml of coupling buffer (50% pyridine, 2% triethylamine) in an acid-washed Pyrex test tube, and the tube and the contents are flushed with nitrogen. At least 50-fold excess of phenylisothiocyanate is added to the solution and the mixture is incubated under nitrogen at 50° for 45 min. The reaction mixture is then dried on a rotary Evapo-Mix at 60° under reduced pressure. After drying, the vacuum is released under nitrogen. This is accomplished with the use of a three-way stopcock; one end leads to the test tube, through rotary Evapo-Mix connections; the other end to a vacuum pump; and the third end to a nitrogen tank. The dried reaction mixture is incubated with 1 ml of trifluoroacetic acid at 50° for 15 min, and trifluoroacetic acid is removed under reduced pressure.

The residue is dissolved in 1.0 ml of pyridine–water–acetic acid (4:4:1) mixture and applied to a Sephadex LH-20 column (1.9 × 60 cm) previously equilibrated with 50% pyridine. The column is eluted with 50% pyridine at a flow rate of 28 ml/hr. Fractions of 3.5 ml are collected, absorbance is measured at 330 nm, and aliquots (10 μl) are assayed with ninhydrin after alkaline hydrolysis. The cross-linked peptides elute in fractions labeled A, and the released single-chain peptides appear in fractions labeled B (Fig. 5). Fractions under A are pooled and dried, and the Edman degradation and Sephadex LH-20 chromatography are repeated. It may be emphasized that after each cycle of Edman degradation the released single-chain peptides (fraction B, Fig. 5) are separated from the cross-linked peptides (fraction A). Five cycles of Edman degradation can be carried out for this purpose without any problem. It may be mentioned that most of the information becomes available after three cycles of degradation.

*Purification of Released Single-Chain Peptides.* The released single-chain peptides obtained after each cycle of Edman degradation may be purified separately and then sequenced, or these peptides from several cycles may be pooled and then purified. The pooling of released peptides from several cycles allows better recovery of low-yield peptides. Several methods of peptide purification have been described.[9,10] In author's laboratory[2,3] initial separation is carried out on a column (0.9 × 55 cm) of Beckman AA-15 resin (a substitute for AA-15 resin is W-2 resin) with pyridine acetate buffers (I: 0.05 $N$ pyridine acetate, pH 2.6; II: 0.10 $N$ pyridine acetate, pH 3.15; III: 0.5 $N$ pyridine acetate, pH 3.7; IV: 2.0 $M$ pyridine acetate, pH 5.0). The column is first developed with a gradient consisting of 300 ml of each of buffers I, II, and III connected in sequence in the order of increasing pH and molarity. This is followed by a gradient consisting of 150 ml of buffers III and IV. The column is eluted at 55° and a flow rate of 42 ml/hr. The peptide peaks are detected with ninhydrin after alkaline hydrolysis. Further purification of peptides (when required) is achieved by paper chromatography (solvent: $n$-butanol–$n$-butyl acetate–acetic acid–water, 135 : 6 : 30 : 50) and/or high-voltage paper electrophoresis, in 0.25 $M$ pyridine acetate buffer, pH 3.5.

*Sequencing of Purified Released Peptides.* Methods for sequence analysis of purified single chain peptides have been described.[11,12] In the author's laboratory, the purified single-chain peptides, after amino acid analysis, are sequenced manually with the use of Edman degradation followed by dansylation.[13] It may be emphasized that coupling and cyclization are carried out under nitrogen. Dansyl amino acids are identified by chromatography on polyamide sheets.[14]

Other elastolytic, cross-linked peptides of elastin (G-50-I and G-50-II) may be examined by a procedure identical to that described above for

[9] This series, Vol. 25, Section VI.
[10] This series, Vol. 47, Section VII.
[11] This series, Vol. 25, Section VII.
[12] This series, Vol. 47, Section VIII.
[13] W. R. Gray, this series, Vol. 25, p. 333.
[14] W. R. Gray, this series, Vol. 25, p. 135.

fraction G-25.2.[3] The above approach has been successfully used in a comparative study of elastins from different species and tissues.[2]

*Simultaneous Sequencing of Two Cross-linked Peptide Chains.* Foster and co-workers[4,5] have studied desmosine-containing cross-linked peptides isolated from subtilisin and thermolysin digests of elastin. Peptides were purified by chromatography on ion-exchange resin (Technicon peptide resin, 4% cross-linked, or Aminex 50W-X4, or type-P chromobead) with pyridine acetate buffers and by gel filtration (e.g., Sephadex G-50 or Sephadex G-25). In this approach, cross-linked peptide, judged to be homogeneous, is sequenced on a Beckman sequencer. As two or more chains are sequenced simultaneously, the cross-link positions are assigned on the basis of amino acid residue yield obtained at each cycle of degradation. For example, the sequence data (first cycle, Ala, 127 nmol; second cycle, Ala, 85 nmol; third cycle, Ala 61 nmol; fourth cycle, Ala, 81 nmol; fifth cycle, Ala, 34 nmol) are interpreted to show the sequencing of two chains. At first cycle, both chains contain Ala. At second cycle the yield of Ala is about half that of the first cycle; therefore, one chain has Ala at this position, and in the other chain this position is occupied by the cross-link, and the same is true of the third cycle. At the fourth cycle both positions are assigned to Ala, and in the fifth cycle one position is assigned to Ala and the other to cross-link. As the peptides carboxyl terminal to the cross-links are released, the yield drops sharply and the assignments become difficult. Therefore, in a way this approach is complementary to the one described above.

It is apparent that the cross-linked peptide under examination must be essentially homogeneous. It may also be pointed out that amino-terminal analysis alone is not a sufficient proof of homogeneity of desmosine cross-linked peptides, because all desmosine cross-links have alanine residues at the amino terminals and thus will show two alanine residues per residue of desmosine (i.e., desmosine plus isodesmosine) whether the peptide is pure or is a mixture of several peptides.


# [34] Biosynthesis of Insoluble Elastin in Cell and Organ Cultures

*By* CARL FRANZBLAU and BARBARA FARIS

The formation of insoluble elastin in *in vitro* systems such as cell and organ cultures has been the focus of several laboratories in recent years. As noted in the overall introduction (Chapter [30]) to the chapters on elastin (Section II of this volume), elastic fibers play a vital role in the structural integrity of many different organ systems. In nature, the disruption or lack of synthesis, or both, of the elastin fiber component leads to devastating diseases, such as pulmonary emphysema and possibly atherosclerosis. Thus, studies on the biosynthesis of the insoluble elastin fiber in cell and organ cultures provide valuable tools for our understanding of the formation and turnover of this important connective tissue component.

Elastin fibers are composed of two discrete proteins, amorphous elastin and a thin surrounding and intrapenetrating network of glycoprotein, the latter known as the microfibrillar protein. This section will focus only on the amorphous elastin component. It is thought that the microfibrillar protein serves as a framework into which elastin is secreted and is eventually organized into a functional insoluble elastic fiber. As described in Chapter [31], one obtains pure elastin by removing all the other components normally associated with the insoluble elastin. This is possible because the elastin that has been processed *in vivo* into an elastic fiber is insoluble in all solvent systems. This insolubility is most likely due to a combination of a high proportion of nonpolar amino acid residues and the formation of covalent cross-links.

Since elastic fibers are extracellular connective tissue components, the biosynthesis of the insoluble protein component has to have its origin in the intracellular protein synthetic machinery, followed by secretion from the cell and, finally, incorporation into the fiber. By examining the end product, which is the insoluble fiber, one realizes that such studies are measuring the sum of all these processes. Thus scientific questions posed in cell or organ cultures may not be readily answerable by studying only the insoluble elastin component. Indeed, much of the focus in the elastin biosynthesis field has been on the soluble precursor component(s) of insoluble elastin. Both cell-free and cell-culture methods for producing the soluble elastin precursors are described in this volume [40] and [41]) and are not within the scope of this chapter.

The processes required for the elastic fiber formation are many. By measuring the end product, the insoluble elastin, we are able to learn something of a cell's ability to ($a$) synthesize protein; ($b$) hydroxylate prolyl residues; ($c$) transport macromolecules to the extracellular milieu; ($d$) assemble larger protein aggregates; ($e$) form connective tissue cross-links. Studies of this type will yield some insight into mechanisms of elastin turnover.

Studies in cell cultures have been limited since very few culture systems are capable of producing insoluble elastin. As of this writing, the only cells that have been reported to produce insoluble elastin in culture are vascular smooth muscle cells, human and rat endothelial cells from umbilical cord veins, chondroblasts, and, indirectly, fibroblasts from bovine ligamentum nuchae. One must point out that vascular smooth muscle cells from all species do not produce elastin under the same conditions in culture. The formation of insoluble elastin can also be evaluated in organ culture systems, although it is more difficult to do so. A major obstacle encountered in organ cultures is the significant amounts of previously formed *in vivo* elastin in the explant. Nevertheless, radioactive precursor amino acids can be utilized in assessing insoluble elastin formation. The organ culture systems will be described separately.

## Biosynthesis of Elastin by Cell Culture

### Smooth Muscle Cell Cultures

The prominent cell of the medial layer of large arteries is the smooth muscle cell. Smooth muscle cells form several layers within the media and are embedded in an extracellular matrix of proteoglycans, elastic fibers, and collagen, in an orderly array. Depending on the type of artery, the exact content of extracellular components will vary, as will the number of cell layers. The cells are responsible for the synthesis of the extracellular matrix and for the contractile elements of the blood vessels. The medial layer is separated from the other cell layers (the intimal and adventitial) by elastic membranes, the internal and external elastic laminae, respectively.

Smooth muscle cells are isolated and grown from the medial layer of the aortic arch of weanling rabbits according to the method of Ross[1] with some modifications.[2] The arch of the aorta is rapidly excised and immediately placed in a petri dish containing Dulbecco's modified Eagle's medium containing 2.2 g of sodium bicarbonate per liter, 10% fetal bovine serum, penicillin (1000 units/ml), Aureomycin (50 $\mu$g/ml), and fungizone (0.25 mg/ml). To each 100 ml of medium is added 1.0 ml each of minimum essential medium (MEM) nonessential amino acids (100 ×) and sodium pyruvate solution (100 ×).

The aortas are grossly cleaned of extraneous tissue, blood, and fat and transferred to fresh medium. They are then cut open, cleaned again, and placed in another petri dish with fresh medium; the intimal/inner medial layer is stripped off according to the method of Wolinsky and Daly.[3] These layers are cut into 1 mm$^3$ pieces and transferred to 75-cm$^2$ tissue culture

---
[1] R. Ross, *J. Cell Biol.* **50,** 172 (1971).
[2] B. Faris, L. L. Salcedo, V. Cook, L. Johnson, J. A. Foster, and C. Franzblau, *Biochim. Biophys. Acta* **418,** 93 (1976).
[3] H. Wolinsky and M. M. Daly, *Proc. Soc. Exp. Biol. Med.* **106,** 364 (1970).

flasks. The flasks are allowed to stand with the cap end up to drain most of the medium in order to promote tissue adherence. Medium as above, but which contains 100 units of penicillin and 100 μg of streptomycin per milliliter and 3.7 g of sodium bicarbonate per liter is added to cover the explants. The flasks are loosely capped and allowed to remain undisturbed in a 5% $CO_2$/95% air humidified incubator at 37°. The medium is changed weekly. In 2 weeks, the cells that grew out from the explants are detached with a 0.05% trypsin/0.02% EDTA solution in modified Puck's saline A solution, centrifuged at 200 g for 10 min, resuspended in complete medium, counted, and seeded into new flasks. Routinely, 75-cm$^2$ plastic tissue culture flasks are seeded with $1.5 \times 10^6$ cells. Cell culture medium (20 ml) is replaced twice weekly. After 1 week, the cells are again detached by trypsinization and subcultivated as above (second passage). These cells are maintained for the specified times with 20 ml of medium per flask, which is changed twice weekly. Phase micrographs of typical rabbit smooth muscle cells at various times after the second subcultivation are shown in Fig. 1. All experiments described in this section use cells in the second passage. In those instances in which the pulmonary artery smooth muscle cells are used, the preparation of the cell cultures from pulmonary arteries of weanling rabbits are prepared as described above.

*Assessment of Elastin Formation in Smooth Muscle Cell Cultures*

There are relatively few criteria that are relied upon to substantiate the formation of insoluble elastin in a cell culture. It is suggested that the cell cultures to be studied should be derived, if possible, from a primary culture that has undergone at least two subcultivations as described above. Thus one is relatively sure that no preexisting insoluble elastin is present in the culture system. In this manner one can truly assess the appearance, accumulation, and turnover of the insoluble elastin. There are three criteria that can be used in assessing the presence of elastin in cell cultures. These include (a) evaluation of the culture by ultrastructural analyses; (b) evaluation of the insoluble elastin formed by amino acid composition of the product obtained after purification of the elastin; and (c) detection and possible quantification of the formation of the unique lysine-derived cross-links. Each of these will be described below. We suggest that no one criterion is sufficient for providing evidence that a cell culture is producing insoluble elastin.

ULTRASTRUCTURAL ANALYSIS

As noted, elastic fibers consist of two structurally and chemically distinct components, the amorphous elastin and the acidic glycoproteins known as the microfibrillar component. Their presence has been shown

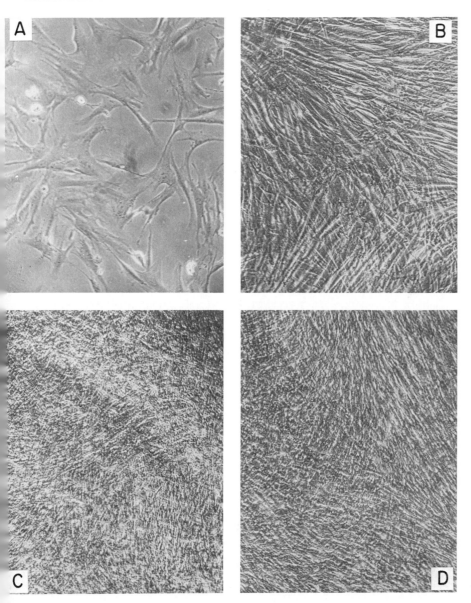

Fig. 1. Phase-contrast micrographs of aortic smooth muscle cells in culture at different periods of time after the second subcultivation. × 300. (A) 1 day; (B) 1 week; (C) 6 weeks; (D) 14 weeks.

morphologically in cultures derived from guinea pig, rat, swine, rabbit, and monkey aortas.[4,5] Reports in the literature suggest that human endothelial cells from umbilical cord veins,[6] rat lung endothelial cells,[6a] and rabbit chondroblasts[7] are also capable of producing insoluble elastin. It appears that fibroblasts derived from bovine ligamentum nuchae can produce insoluble elastin if a previously formed insoluble elastin matrix is provided to the cell culture.[8]

The sequence of appearance of the two components of the elastic tissue in primary 5-day-old rat aorta smooth muscle cells grown in culture for 28 days was reported by Hinek and Thyberg[9] to be initially comprised of small bundles of microfibrils that become associated with small conglomerates of a dense amorphous material. In the more mature elastic fibers (older cultures), such material appears to form confluent amorphous areas in which it is difficult to detect microfibrils. Studies with the rabbit smooth muscle cell cultures reveal similar findings and are presented in more detail to serve as a model for all cell culture systems.

*Preparation of Cell Culture for Ultrastructural Analysis.* In the typical vascular smooth muscle cell, the culture medium is first removed and the cell layer in each flask is rinsed twice with 4.3% glutaraldehyde in 0.1 $M$ Veronal buffer, pH 7.4, then kept in the same buffer for 1 hr at room temperature. The cultures are again rinsed with the buffer, and postfixed with 1% $OsO_4$ in 0.075 $M$ Veronal buffer for 30 min at room temperature. This is followed by dehydration in a graded series of ethanol solutions and embedding in a 1 : 1 mixture of Araldite 502 and dodecyl succinic anhydride. After polymerization, thin sections are cut, stained with uranyl acetate and lead citrate, and examined in the electron microscope. Some modifications of the procedure can be used to accommodate individual cell culture needs. These include the use of different fixatives and embedding agents.

*Pattern of Elastin Fiber Formation in Rabbit Smooth Muscle Cell Culture.* Studies of the ultrastructure of smooth muscle cells in culture can be carried out for several weeks from the time of seeding. Figure 2A shows a typical micrograph of a pulmonary artery smooth muscle cell culture that

---

[4] R. B. Rucker and D. Tinker, *Int. Rev. Exp. Pathol.* **17**, 1 (1977).
[5] J. M. Burke and R. Ross, *Int. Rev. Connect. Tissue Res.* **8**, 119 (1979).
[6] E. A. Jaffe, C. R. Minick, B. Adelman, C. G. Becker, and R. Nachman, *in* "Biology and Chemistry of Basement Membranes" (N. A. Kefalides, ed.), p. 355. Academic Press, New York, 1978.
[6a] J. O. Cantor, S. Keller, M. S. Parshley, T. V. Darnule, A. T. Darnule, J. M. Cerreta, G. M. Turino, and I. Mandle, *Biochem. Biophys. Res. Commun.* **95**, 1381 (1980).
[7] G. Quintarelli, B. C. Starcher, A. Vocaturo, F. DiGianfilippo, L. Gotte, and R. P. Mecham, *Connect. Tissue Res.* **7**, 1 (1979).
[8] R. P. Mecham, *Connect. Tissue Res.* **8**, 241 (1981).
[9] A. Hinek and J. Thyberg, *J. Ultrastruct. Res.* **60**, 12 (1977).

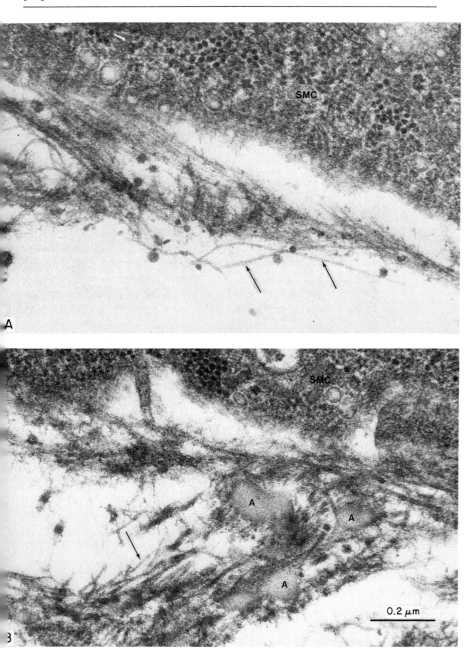

FIG. 2. Electron micrographs of pulmonary smooth muscle cells in culture for (A) 14 days and (B) 53 days after the second subcultivation. × 115,000. SMC, smooth muscle cell; arrows, microfibrillar component; A, insoluble amorphous elastin.

is 14 days in second passage. Little amorphous elastin is detected, but there are significant quantities of the microfibrillar component. As the cell cultures become older, there is an increasing amount of insoluble elastin fibers (Fig. 2B). As noted above, we suggest that if ultrastructural analysis indicates the presence of insoluble elastin in a cell culture, the results should be corroborated, if possible, by chemical studies. The exact time in culture at which insoluble amorphous elastin can be detected in the electron microscope may vary in different culture preparations.

INSOLUBLE ELASTIN PREPARATION FROM CELL CULTURES

The spent medium is aspirated from the cell cultures and the cell layers are washed with 0.9% NaCl. They are then harvested by scraping with the aid of a rubber policeman and lyophilized. The lyophilized samples are suspended in 0.1 $N$ NaOH at 98° for 45 min with occasional shaking according to the method of Lansing et al.,[10] and the resulting insoluble material is centrifuged. These residues are washed with $H_2O$ and hydrolyzed under vacuum, in 6 $N$ HCl at 110° for 20 hr.

The residues from cell cultures after this hot NaOH treatment have an amino acid analysis compatible with elastin only after the cell cultures have reached confluency. In early experiments[2] in which the cells were routinely maintained with 10 ml of medium, confluence was not achieved until at least 7 days after seeding. In later experiments that utilized 20 ml of maintenance medium, the cell layers reached confluence in approximately 4 days after seeding. Therefore, as is well known in tissue culture systems, the time course of events may vary somewhat depending on the conditions used, but the trend should remain consistent. The amino acid compositions of the hot NaOH residues from cells cultured for various numbers of days in second passage and maintained with 10 ml of medium are given in Table I. Also included in the table for comparison purposes is elastin prepared in the same manner from the medial tissue of the donor rabbit aorta. It is interesting to note the differences between the earliest and latest time points when compared to the 13- and 20-day-old cell cultures, which yield definitive elastin amino acid analyses. To estimate with some accuracy the small amount of desmosines present in the insoluble elastin, two separate concentrations of the acid hydrolyzate from the 20-day-old cells were applied to the amino acid analyzer. The higher concentration of hydrolyzate used was 15 times greater than used for the other samples. In a separate set of experiments, the accumulation of insoluble elastin from cells cultured with 20 ml of medium for various periods of time was monitored by ninhydrin (Fig. 3). As can be seen, rabbit smooth muscle cells provide an excellent culture for studying long-term accumulation of insoluble elastin. However, the desmosine content

[10] A. I. Lansing, T. B. Rosenthal, M. Alex, and E. W. Dempsey, *Anat. Rec.* **114**, 555 (1952).

TABLE I
AMINO ACID ANALYSES OF ELASTIN FROM SMOOTH MUSCLE CELLS IN CULTURE[a]

| Amino acid | Days | | | | Medical tissue from rabbit aorta |
| --- | --- | --- | --- | --- | --- |
| | 9 | 13 | 20[b] | 43 | |
| Hydroxyproline | NC[c] | 4.8 | 3.8 | 3.7 | 15.3 |
| Aspartic acid | 81.2 | 14.9 | 13.3 | 32.2 | 3.1 |
| Threonine | 53.0 | 15.8 | 15.1 | 26.7 | 9.9 |
| Serine | 56.0 | 18.7 | 17.3 | 25.8 | 11.7 |
| Glutamic acid | 106.7 | 28.3 | 26.2 | 46.0 | 16.7 |
| Proline | 84.9 | 121.8 | 124.6 | 116.0 | 119.2 |
| Glycine | 164.1 | 333.2 | 319.1 | 293.7 | 344.6 |
| Alanine | 105.7 | 232.0 | 222.9 | 186.4 | 243.3 |
| Valine | 73.9 | 99.1 | 103.9 | 94.2 | 98.6 |
| Isoleucine | 38.8 | 22.9 | 23.1 | 28.3 | 20.9 |
| Leucine | 75.6 | 53.5 | 60.1 | 62.5 | 54.0 |
| Tyrosine | 25.5 | 13.7 | 27.1 | 26.3 | 23.2 |
| Phenylalanine | 32.5 | 20.7 | 25.0 | 23.8 | 19.1 |
| Lysine | 33.5 | 7.7 | 5.9 | 10.3 | 5.2 |
| Histidine | 22.2 | 3.5 | 4.6 | 8.7 | 0.5 |
| Arginine | 45.4 | 9.2 | 11.7 | 18.5 | 5.8 |
| Isodesmosine | — | — | 0.5 | — | 2.8 |
| Desmosine | — | — | 0.4 | — | 4.2 |
| Lysinonorleucine | — | — | — | — | 1.2 |

[a] Values are expressed as residues per 1000 residues.
[b] Values were obtained from two different amino acid analyses (see text).
[c] NC, not calculated.

in the elastin, while significant, is always considerably less than that found in the elastin obtained directly from aorta or ligament. On the other hand, as described in detail below, the cells in culture appear to accumulate large quantities of the aldol condensation product of two residues of allysine.

The amino acid composition of the insoluble elastin is quite consistent throughout the age of the culture, except at the very early and late time periods in second passage. Early in the passage the amino acid composition of the insoluble elastin preparation from the cultures contains more acidic amino acids than normal. This most likely reflects a higher proportion of the microfibrillar component in the early developing elastic fiber. Older cultures also contain higher contents of polar amino acids suggesting accumulation or association with other glycoprotein components, similar to the observation made by Lansing et al. in aged human tissues.[11]

[11] A. I. Lansing, E. Roberts, G. B. Ramasarma, T. B. Rosenthal, and M. Alex, Proc. Soc. Exp. Biol. Med. **76**, 714 (1951).

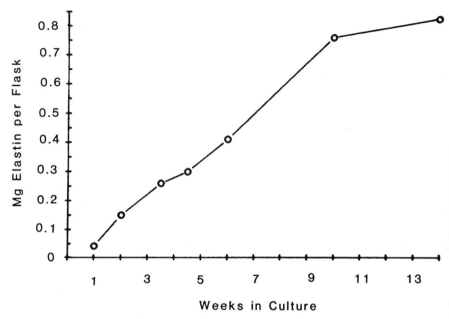

Fig. 3. The accumulation of insoluble elastin from aortic smooth muscle cells in culture at different periods of time after the second subcultivation.

*Pulse-Chase Procedures and Cross-link Formation*

To prepare the cultures for pulsing with precursor amino acids, the spent medium is aspirated off and the cell layers are washed twice with calcium- and magnesium-free Puck's saline G. The cells are then incubated for 1 hr at 37° in medium free of either proline or lysine and fetal bovine serum, but containing penicillin and streptomycin as in the maintenance period described above, as well as 50 $\mu$g of sodium ascorbate per milliliter. The medium in each flask is then replaced with 10 ml of medium that contains either 1 $\mu$Ci of [$^{14}$C]proline or [$^{14}$C]lysine per milliliter, but is otherwise identical to that used in the 1 hr period. Normally, the pulse period is for 24 hr. One could, however, vary the pulse time and the amount of radioactivity used to meet the needs of a particular experiment.

For the "chase" experiments, the spent radioactive medium is aspirated off after the pulse period (24 hr) and replaced with the routine maintenance medium containing fetal bovine serum. Each flask is changed twice weekly with 20 ml of medium for the specified "chase" period.

At various times, the medium is removed from the cells, and each cell layer is washed twice with 2.0 ml of 0.9% NaCl. Pooled cell layers from two flasks of [$^{14}$C]proline-pulsed cells are harvested as described above, dialyzed versus H$_2$O, and lyophilized. In the case of the [$^{14}$C]lysine-pulsed cells, groups of five flasks of cells are pooled, dialyzed, and lyophilized. The lyophilized [$^{14}$C]lysine-pulsed cell layers are homogenized in 5.0 ml of H$_2$O in a motorized glass homogenizer and separated into aliquots corresponding to two and three flasks of cells. That portion of the homogenate representing two flasks of cells is lyophilized directly, whereas the aliquot of homogenate corresponding to three flasks of cells is reduced with NaBH$_4$ in the following manner before lyophilization.

NaBH$_4$ REDUCTION PROCEDURE

In order to determine the distribution of lysine-derived cross-links and their precursors formed in the cultures, one is required to stabilize these aldehyde and Schiff-base components. The use of sodium borohydride (NaBH$_4$), or sodium borotritide (NaB$^3$H$_4$), or, more recently, the sodium cyanide–ammonia method[12] have served this purpose well. Not only do these reductants reduce and stabilize the lysine-derived cross-links, but when tritium is used, as in NaB$^3$H$_4$, the result is the production of radiolabeled reduced cross-links. A complete description of the methods for determination of the cross-links is given in this volume [31].

To each cell homogenate (2–3 ml) to be reduced, 0.5 ml of a solution containing 0.5 mg of NaBH$_4$ in 0.001 $N$ NaOH is added, and the reaction is allowed to proceed at 4° with occasional shaking for 90 min. The reaction mixture is maintained at pH 7.0–9.0 with dilute HCl for the 90-min duration, then acidified with 50% acetic acid and lyophilized. Some flasks of cells that were never pulsed are harvested and homogenized directly, then reduced with NaB$^3$H$_4$, acidified, and lyophilized. Those cell layers that have been reduced with NaBH$_4$ (or NaB$^3$H$_4$) are hydrolyzed with 2 $N$ NaOH at 110° for 20 hr and analyzed on a Technicon amino acid analyzer equipped with a split stream arrangement. The reason for the alkaline hydrolysis is that the reduced aldol condensation product is not stable to 6 $N$ HCl hydrolysis. In addition, the ε-hydroxynorleucine is partially converted to ε-chloronorleucine in 6 $N$ HCl. A gradient system described previously for separating the reduced lysine-derived aldehydes in elastin[13] is employed.

The presence of radioactive hydroxyproline (when [$^{14}$C]proline is employed as a precursor), desmosine, and lysine in the cell fraction is deter-

[12] B. Pereyra, O. O. Blumenfeld, M. A. Paz, E. Henson, and P. M. Gallop, *J. Biol. Chem.* **249**, 2212 (1974).
[13] R. W. Lent, B. Smith, L. L. Salcedo, B. Faris, and C. Franzblau, *Biochemistry* **8**, 2237 (1969).

## TABLE II
### RADIOACTIVE PROLINE AND HYDROXYPROLINE DISTRIBUTION IN [$^{14}$C]PROLINE-PULSED AORTIC SMOOTH MUSCLE CELLS IN CULTURE

|  | Proline (cpm/flask) | Hydroxyproline (cpm/flask) | Hydroxyproline in elastin fraction (cpm/flask) |
|---|---|---|---|
| 24-Hour pulse | 1,311,500 | 180,900 | 4,520 |
| 1-Week chase | 939,050 | 205,750 | 18,930 |
| 2-Week chase | 648,500 | 195,450 | 39,285 |
| 3-Week chase | 778,550 | 307,550 | 55,670 |

mined by hydrolyzing the samples, without prior reduction, in 6 $N$ HCl at 110° for 20 hr. These hydrolyzates are placed on an automatic amino acid analyzer also equipped with a split-stream arrangement.

As seen in Table II, the protein-bound [$^{14}$C]proline associated with the cell layer decreases with time. On the other hand, the [$^{14}$C]hydroxyproline in this same fraction remains constant throughout the first 2 weeks of the chase period. However, the radioactive hydroxyproline that remains insoluble after the hot alkali treatment (elastin fraction) increases during the chase period. When measured by ninhydrin, the hydroxyproline and proline content of these fractions are 7 and 110 residues per 1000 residues, respectively. These values are compatible with those of insoluble elastin.

LYSINE-PULSED CELLS—TYPICAL RESULTS

*Not Reduced.* Trace amounts of radioactive desmosine and isodesmosine are detected in the insoluble elastin fraction of the cell layer obtained after the 24-hr pulse. The radioactivity obtained in these crosslinking amino acids increases significantly during the 3-week chase period. Between 3 and 5 weeks after pulsing, the radioactivity in the demosines (Table III) appears to have leveled off.

*Reduced.* The radioactivity content in the ϵ-hydroxynorleucine (HNL) and reduced aldol condensate (ALD) fractions appear to be of the same order of magnitude after the 24-hr pulse (Table III). One should of course be aware of the fact that two lysyl residues are required for each ALD cross-link formed whereas only one lysine is incorporated into a HNC residue. Thus, the ALD residue contains twice the number of lysine equivalents per molecule when compared to a molecule of HNL. During the entire chase period, the HNL increases less than 2-fold above the pulse period value whereas the increase in the ALD is 15-fold. As in the case of the desmosines, there is a leveling off of the radioactivity in the

TABLE III
LYSINE-DERIVED CROSS-LINKS IN INSOLUBLE ELASTIN FROM
CELL CULTURES AND AORTIC TISSUE

|  | DES[a] (cpm/flask) | HNL[b] (cpm/flask) | ALD[c] (cpm/flask) | ALD/HNL |
|---|---|---|---|---|
| Cells |  |  |  |  |
| [$^{14}$C]Lysine (NaBH$_4$ reduced) |  |  |  |  |
| 24-Hour pulse | Trace | 725 | 655 | 0.9 |
| 1-Week chase | 515 | 860 | 4135 | 4.8 |
| 3-Week chase | 1885 | 1250 | 8960 | 7.2 |
| 5-Week chase | 1940 | 1235 | 9795 | 7.9 |
| NaB$^3$H$_4$ reduced |  |  |  |  |
| 3-Week culture | — | — | — | 4.0 |
| Tissue (aorta) |  |  |  |  |
| [$^{14}$C]Lysine-pulsed | — | — | — | 2.3 |
| NaB$^3$H$_4$ reduced | — | — | — | 7.0 |

[a] DES, isodesmosine plus desmosine.
[b] HNL, $\epsilon$-hydroxynorleucine.
[c] ALD, reduced aldol condensate.

ALD and HNL between 3 and 5 weeks after the pulse. There are approximately 7–8 times as much radioactive ALD as HNL in the elastin synthesized 3–5 weeks after the lysine pulse. If, however, the cell layer is not pulsed, but is reduced directly with NaB$^3$H$_4$, there is 4 times as much radioactivity associated with the ALD when compared to the HNL. These latter values represent the reduced elastin that has accumulated during the entire culture, not just the pulse period. Since only one tritium atom is incorporated into either the HNL or the ALD cross-link upon reduction, these data are consistent with the lysine pulse studies described above. Table III also presents the radioactive ratio of the donor tissue reduced with NaB$^3$H$_4$ as well as the donor rabbit tissue pulsed with [$^{14}$C]lysine and reduced with NaBH$_4$.

Several important points need to be mentioned since two distinct sets of data are presented. If one reduces a cell culture directly with NaB$^3$H$_4$, one obtains a measure of the accumulation and turnover of the entire insoluble elastin fraction the cells have produced from the inception of that particular passage. On the other hand, if one pulses for a short period with radioactive lysine and then reduces with NaBH$_4$, one measures the accumulation of only the newly synthesized elastin during that particular pulse period. By taking a large number of flasks of cells derived from the same explants, it is possible to obtain information as to the cell's ability to accumulate elastin over long periods of time versus the cell culture's

TABLE IV
FORMATION OF THE LYSINE-DERIVED CROSS-LINKS IN INSOLUBLE ELASTIN FROM
[$^{14}$C]LYSINE-PULSED AORTIC SMOOTH MUSCLE CELLS IN CULTURE[a]

|  | Total cpm/flask[b] | cpm HNL/nmol Leu | cpm ALD/nmol ALD | nmol ALD/flask |
|---|---|---|---|---|
| 24-Hour pulse | 4,500 | 17,300 | 101,400 | 6.5 |
| 1-Week chase | 6,900 | 10,300 | 286,100 | 14.5 |
| 3-Week chase | 15,700 | 6,500 | 398,100 | 22.5 |
| 5-Week chase | 15,700 | 5,700 | 287,300 | 34.1 |

[a] HNL, ε-hydroxynorleucine; ALD, reduced aldol condensate.
[b] Total cpm: cpm Lys + cpm HNL + cpm ALD.

ability to produce elastin over a short period at any given time in the culture.

An interesting observation occurs when one examines ALD:HNL ratios. The intact rabbit aorta displays a ratio of approximately 7:1 while the 24-hr lysine-pulsed aortic tissue shows a ratio of 2.3:1. This is strikingly similar to that which is observed in the cell cultures. The data reveal that the ratio of ALD to HNL increases with time in culture. However, even 20-day-old cultures of rabbit cells, when reduced with NaB$^3$H$_4$, do not yield the values obtained from the intact rabbit aorta. Yet, the absolute quantity of the aldol is significantly greater (twofold). Similar findings are obtained when one does a pulse and chase experiment. It takes approximately 3–5 weeks after the 24-hr pulse with [$^{14}$C]lysine to approach values of ALD:HNL that are obtained in the NaB$^3$H$_4$-reduced intact tissue. Thus, as stated many times, the complete process of elastic fibrogenesis is a slow process even in cell culture. This allows one to study the process over long periods of time.

While the total radioactivity in the insoluble elastin fraction from these cell cultures (lysine + HNL + ALD) increases to a maximum level between 3 and 5 weeks after pulsing, the decrease in the specific activity of the lysine-derived cross-links seen during the same time period suggests a continuous accumulation of insoluble elastin in the culture (Table IV) during the chase period. This observation is corroborated by noting the continual increase in the ALD concentration present at the various times studied, again suggesting an increase in total elastin. Because of difficulties in accurately calculating the concentration of HNL with ninhydrin with the chromatographic conditions employed in these experiments, the radioactivity of HNL is calculated on a per nanomole of leucine basis.

The distribution of the radioactivity between lysine and the various lysine-derived cross-links with time after the pulse is given in Table V. The

TABLE V
DISTRIBUTION OF RADIOACTIVITY BETWEEN LYSINE AND
LYSINE-DERIVED CROSS-LINKS IN INSOLUBLE ELASTIN FROM
LYSINE-PULSED AORTIC SMOOTH MUSCLE CELLS IN
CULTURE[a]

|  | LYS (%) | DES (%) | HNL (%) | ALD (%) |
|---|---|---|---|---|
| 24-Hour pulse | 75% | Trace | 13% | 12% |
| 1-Week chase | 22% | 7% | 12% | 58% |
| 3-Week chase | 28% | 11.3% | 7.5% | 54% |
| 5-Week chase | 24% | 11.1% | 7.3% | 58% |

[a] LYS, lysine; DES, isodesmosine + desmosine; HNL, $\epsilon$-hydroxynorleucine; ALD, reduced aldol condensate.

percentage of the total radioactivity in lysine sharply decreases whereas the ALD increases within 1 week after the pulse. The radioactivity in HNL is shown to decrease gradually, whereas that in the desmosines, which is barely measurable after the pulse, reaches a maximum of approximately 11%. All the data indicate that the appearance of desmosine in the elastin in these cell cultures with time is a relatively slow process, whereas the lysine-derived intermediate cross-links (aldehydes) appear quite rapidly. The data also suggest that in culture the insoluble elastin contains significantly more ALD, but less desmosine and isodesmosine.

If the formation of the desmosines occurs by the condensation of the ALD and a dehydrolysinonorleucine moiety, the presence of the latter may be rate limiting in the formation of the desmosines, since the cells display little or no dehydrolysinorleucine. The explanation for the low content of desmosine and isodesmosine in the cultures is unclear as of this writing. One should be aware that if desmosine and isodesmosine are the true cross-linking units in the insoluble elastin, not some reduced form of the same as suggested by Paz et al.,[14] then an additional oxidation, possibly catalyzed by an enzyme system different from the lysyl oxidase, could be required. It may well be that this system is not functional in the cell cultures.

Other Cell Cultures

*Endothelial Cells.* Jaffe et al.[6] have reported that human endothelial cells from umbilical cord veins produce insoluble elastin. The data pre-

[14] M. A. Paz, B. Pereyra, P. M. Gallop, and S. Seifter, *J. Mechanochem. Cell Motil.* **2**, 231 (1974).

sented clearly suggest the presence of insoluble elastin by ultrastructural analyses. However, no data have been presented on the chemical nature of the elastin from such cultures. A culture of rat lung endothelial cells has been shown to produce desmosine and isodesmosine.[6a]

*Fibroblasts.* No evidence for the formation of insoluble elastin has been noted from cultures of fibroblasts except for those grown on an elastin matrix as noted above.[8]

*Chondroblasts.* Quintarelli *et al.*[7] have reported that chondroblasts from rabbit ear cartilage grown *in vitro* are capable of synthesizing insoluble elastin. By both ultrastructural and amino acid analyses, the insoluble elastin obtained from the *in vitro* system compared well with *in vivo* cartilage elastin.

### Addition of Soluble Precursor Elastin (Tropoelastin) to Cell Cultures

By adding appropriate radioactive precursors of elastin, such as tropoelastin, to the tissue culture medium, Abraham *et al.*[15] have demonstrated that such soluble precursors have been incorporated into insoluble elastin. The source of their cells was newborn pig aorta. The synthesis of cross-linked elastin was shown by the isolation and identification of the lysine-derived cross-links after incubation of the cultures with $^3$H-labeled soluble elastin isolated from copper-deficient pig aorta, and then isolating labeled cross-links from the insoluble residue. The incubation was carried out for periods of up to 60 min, and the insoluble residue was defined as that portion of the cell layer that remains after extraction with $1.0\,M$ NaCl (pH 7.2). The residue was reduced with NaBH$_4$ and hydrolyzed for 72 hr. The formation of desmosine was not demonstrated, but radioactivity was detected from the amino acid analyzer in fractions where lysinonorleucine and desmosine elute. This approach is an interesting one, and it remains to be seen how useful such a system will be in probing the mechanisms of insoluble elastin formation.

### Effects of Various Reagents on the Formation of Insoluble Elastin in Cell Culture

Addition of different agents at various times during the life of a culture has been evaluated for their effect on insoluble elastin formation. The most prominent chemicals to be employed in this regard are β-aminopropionitrile (15–50 µg/ml of culture medium), $\alpha,\alpha'$-dipyridyl, DL-penicillamine, ascorbic acid, and azetidine-2-carboxylic acid.

[15] P. A. Abraham, M. L. Hart, A. R. Winge, and W. H. Carnes, *Adv. Exp. Med. Biol.* **79**, 397 (1977).

TABLE VI
AMINO ACID COMPOSITION OF NaOH RESIDUES (INSOLUBLE ELASTIN) FROM PULMONARY ARTERY SMOOTH MUSCLE CELLS GROWN WITH AND WITHOUT ASCORBATE[a]

| Amino acid | Day 7 | | Day 14 | | Day 21 | |
|---|---|---|---|---|---|---|
| | −Asc | +Asc | −Asc | +Asc | −Asc | +Asc |
| Hydroxyproline | 0 | 0 | 2.2 | 0 | 3.6 | 12.3 |
| Aspartic acid | 35.2 | 66.5 | 12.7 | 79.3 | 10.6 | 81.9 |
| Threonine | 22.2 | 33.1 | 10.0 | 32.3 | 10.0 | 45.1 |
| Serine | 67.7 | 50.0 | 14.1 | 52.7 | 12.7 | 47.3 |
| Glutamic acid | 40.6 | 92.7 | 26.4 | 109.1 | 24.1 | 113.3 |
| Proline | 77.9 | 70.0 | 110.9 | 65.1 | 104.7 | 65.3 |
| Glycine | 275.1 | 182.5 | 316.2 | 171.1 | 320.2 | 154.9 |
| Alanine | 185.6 | 133.0 | 232.4 | 124.3 | 238.0 | 108.5 |
| Valine | 88.6 | 80.7 | 107.4 | 77.7 | 111.4 | 71.6 |
| Isoleucine | 19.8 | 23.4 | 22.8 | 31.0 | 20.7 | 42.4 |
| Leucine | 54.2 | 69.2 | 58.6 | 74.6 | 57.3 | 76.7 |
| Tyrosine | 25.1 | 23.6 | 28.7 | 10.9 | 27.0 | 26.3 |
| Phenylalanine | 23.3 | 32.8 | 22.6 | 34.5 | 21.7 | 38.4 |
| Lysine | 24.5 | 56.5 | 8.8 | 64.9 | 5.7 | 47.6 |
| Histidine | 14.7 | 19.1 | 3.5 | 19.3 | 3.0 | 19.2 |
| Arginine | 17.4 | 41.2 | 11.3 | 41.9 | 9.2 | 44.6 |
| Isodesmosine | 0 | 0 | 0.48 | 0 | 0.32 | 0 |
| Desmosine | 0 | 0 | 0.38 | 0 | 0.37 | 0 |

[a] Values are expressed as residues per 1000 residues.

β-Aminopropionitrile inhibits lysyl oxidase, and thus cross-linking is impaired and no accumulation of insoluble elastin should occur. Accordingly, no evidence for insoluble elastin accumulation, either ultrastructurally or biochemically, is seen when the cells are grown in the presence of BAPN.

Ascorbate treatment of the cells in culture is of interest. One report in the literature[16] suggests that ascorbate feeding of rat smooth muscle cells stimulates the formation of insoluble collagen fibers and at the same time inhibits the accumulation of insoluble elastin. Our laboratory has observed similar results studying cultures of rabbit pulmonary artery smooth muscle cells. The insoluble material obtained after the Lansing treatment (hot alkali) does not resemble elastin when the cells are grown in the presence of ascorbate, whereas the elastin amino acid composition is normal from those cells grown in the absence of ascorbate (Table VI). The ascorbate effect in calf aortic smooth muscle cells is also worth noting. Calf smooth muscle cells, in our hands, do not grow well unless they are

[16] Y. A. DeClerk and P. A. Jones, *Biochem. J.* **186**, 217 (1980).

fed ascorbate. Consequently, we do not observe the formation of insoluble elastin in our calf cell cultures. Schwartz et al.[17] have also shown that calf smooth muscle cells do not produce insoluble elastin in the presence or in the absence of ascorbate. Thus the ascorbate phenomenon is an important one to evaluate in each culture system studied, since it may be quite influential in determining the nature and extent of the insoluble elastin produced. Additionally, its proposed effects on collagen synthesis are substantial (see Chapter [10]).

It has been shown that $\alpha,\alpha'$-dipyridyl has little effect on the synthesis of soluble elastin (tropoelastin) from aortic cells suspended in culture.[18] Thus one would conclude from such an observation that insoluble elastin accumulation would be minimally affected, and this appears to be the case. Not many studies have been forthcoming on the effects of other perturbants on insoluble elastin formation. Azetidine-2-carboxylic acid may have some effect on the conformation of the soluble precursor elastins that are produced by the cells and thus impair insoluble elastin accumulation. Hormone studies are under way in several laboratories, and the results from such studies should be appearing in the literature.

### Biosynthesis of Insoluble Elastin in Organ Cultures

It is more difficult to assess the *in vitro* formation of insoluble elastin when one employs organ culture systems, or more correctly stated, tissue explants. Such tissues by their very nature have already accumulated significant quantities of insoluble elastin. Thus, amino acid analyses after isolation or ultrastructural studies will have little significance unless one can reasonably define the portion of the insoluble elastin fiber that was synthesized in culture versus the portion that had accumulated in the donor tissue before sacrifice. Studies of elastin synthesis in such organ culture systems have therefore been limited to the use of radioactive precursors, such as valine or lysine. Lysine incorporation experiments focus on the formation of cross-links in much the same manner described in the cell culture section. Radiolabeled valine is also commonly used in elastin studies, since this protein has an unusually high content of valine.

The earliest studies on *in vitro* insoluble elastin formation were carried out in embryonic aortic tissues. They focused exclusively on the formation of the desmosine cross-links described by Partridge and his collaborators and the lysinonorleucine and aldol condensate described by Franzblau and his collaborators.

[17] E. Schwartz, A. M. Ademany, and O. O. Blumenfeld, *Biochim. Biophys. Acta* **624**, 531 (1980).
[18] J. Rosenbloom and A. Cywinski, *FEBS Lett.* **65**, 246 (1976).

## Description of Organ Culture System (Typical)

Independent studies by Miller et al.[19] in chick embryo aorta and by Partridge et al.[20] on rat aorta confirmed the original suggestion by Partridge et al.[21] that lysine is the precursor of both desmosine and isodesmosine. Miller et al. removed approximately 3 mm of the ascending aorta from 16-day-old chick embryos, which were then attached to a Millipore filter by means of a fibrin clot and placed individually in test tubes. For each time period studied, 25–50 aortas were used. The culture medium employed was medium 199 (with Hanks' salts) containing penicillin and streptomycin, as well as 0.1–0.5 $\mu$Ci of uniformly labeled [$^{14}$C]lysine per milliliter.

### Studies on Demosine Formation

Those aortas that had been treated with radioactive lysine for 1 day and then grown in medium containing no lysine for an additional 11 days (the medium was changed every other day throughout the chase period) gave the following results after isolation and hydrolysis of the elastin. Lysine had a specific activity of 73,700 cpm/$\mu$mol and the quarter-desmosines (desmosine and isodesmosine/4) had a specific activity of 78,100 cpm/$\mu$mol. Similar data were obtained by injecting [$^{14}$C]lysine into Sprague-Dawley rats.[20] Additional evidence that four lysines were incorporated into one desmosine was obtained by Anwar and Oda.[22] Radioactive desmosine was obtained from the ascending aortas of 13- to 14-day-old chick embryos that had been grown in tissue culture in the presence of uniformly labeled [$^{14}$C]lysine. Both desmosine and isodesmosine were isolated by column chromatography and subjected to alkaline degradation. The alkaline degradation had been previously shown to yield 41% of the theoretical lysine from desmosine and 73% of the theoretical lysine from isodesmosine (see Chapter [31] for structures). The ratio of the specific activity of a quarter-desmosine or quarter-isodesmosine was the same as the specific activity of the [$^{14}$C]lysine derived from them. In addition, the isolated radioactive desmosines were treated with ninhydrin, and the ratio of the total radioactivity of the molecule to the total radioactivity of the $CO_2$ recovered after such a procedure was found to be approximately 6.0. In a later communication[23] the authors used [1-$^{14}$C]lysine

---

[19] E. J. Miller, G. R. Martin, and K. A. Piez, *Biochem. Biophys. Res. Commun.* **17**, 248 (1964).
[20] S. M. Partridge, D. F. Elsden, J. Thomas, A. Dorfman, A. Tesler, and P.-L. Ho, *Nature (London)* **209**, 399 (1966).
[21] S. M. Partridge, D. F. Elsden, and J. Thomas, *Nature (London)* **197**, 1297 (1963).
[22] R. A. Anwar and G. Oda, *J. Biol. Chem.* **241**, 4638 (1966).
[23] R. A. Anwar and G. Oda, *Biochim. Biophys. Acta* **133**, 151 (1967).

and [6-$^{14}$C]lysine in place of the uniformly labeled amino acid. The data obtained after alkaline degradation and ninhydrin–$CO_2$ studies were entirely compatible with the concept that the α-amino and α-carboxyl groups of the lysine remain intact in peptide linkage. Thus, formation of desmosine and isodesmosine must occur at carbons 5 and 6 of the lysyl side chain.

That lysine is also the precursor of the lysinonorleucine was shown by Franzblau et al.[24] In extending those studies of Miller et al., 12-day chick embryo aortas were pulsed for 1 day and chased for an additional 8 days. During this time the medium was changed every other day. Elastin was isolated from the organ culture by the Lansing procedure, described above, and hydrolyzed. The hydrolyzates were placed on an amino acid analyzer equipped with a stream-splitting device. The specific activities obtained for lysine, the desmosines, and lysinonorleucine are given in Table VII. The absolute values obtained in this type of experimental approach will depend upon the quantity and quality of the radioactive lysine employed in the studies. As with the cell cultures, one should be quick to note that the data presented are only representative of a phenomenon. In any one culture dish, more or less lysine may be incorporated and the time course of desmosine formation may be slower or faster depending upon the conditions of the culture and the inherent individuality of cultures in general.

Rapid Method for Organ Culture

To perform rapid organ culture experiments on insoluble elastin accumulation one can simply employ methods that are quite similar to the preparation of lysyl oxidase or prolyl hydroxylase substrates from aortic explants (see Chapter [35]).

As an example of such a system one can remove the aortic arches from chick embryos and place them in groups of 25–50 in small petri dishes that are designated for tissue culture. A preincubation for 1 hr in Dulbecco's MEM containing no serum but supplemented with 3.7 g of $NaHCO_3$ per liter and 100 units of penicillin and 100 µg of streptomycin per milliliter of medium is carried out. Considerations on the addition of antibiotics as well as ascorbate, BAPN, and other reagents can be made for the appropriate experiment. Each petri dish of 25 aortas can then represent a single time point in a pulse-chase experiment. The number of dishes required for each time point will vary according to the experiment, but one should be cautious in making sweeping generalities from the data obtained from a

[24] C. Franzblau, F. M. Sinex, B. Faris, and R. Lampidis, Biochem. Biophys. Res. Commun. 21, 575 (1965).

TABLE VII
SPECIFIC ACTIVITIES OF LYSINE, DESMOSINES,
AND LYSINONORLEUCINE OBTAINED FROM
[$^{14}$C]LYSINE-LABELED CHICK EMBRYONIC
AORTIC ELASTIN

| Amino acid | cpm/$\mu$mol[a] |
|---|---|
| Lysine | 34,000 |
| Isodesmosine | 31,000 |
| Desmosine | 27,000 |
| Lysinonorleucine | 33,000 |

[a] Based on lysine equivalents. Isodesmosine and desmosine are each assumed to equal four lysine equivalents. Lysinonorleucine is assumed to equal two lysine equivalents.

single petri dish. The variation from dish to dish can be substantial, and appropriate controls must be evaluated at each time point in much the same manner as the cell cultures described above.

*Lysine-Derived Cross-links and Cross-link Intermediates*

Original studies on the formation of lysine aldehyde cross-links and intermediates were also performed in organ culture systems. To quantify the distribution of lysine-derived cross-links in cell or organ culture systems, one must essentially stabilize the aldehyde and Schiff-base components. This can be readily accomplished by reduction with NaBH$_4$ or NaB$^3$H$_4$ or by treatment with sodium cyanide and ammonia as described above and in Chapter [31]. If the content of the lysine-derived cross-link profiles are to be related to insoluble elastin formation, one must first reduce the tissue and thereby stabilize the cross-links before proceeding with the purification of the insoluble elastin component. If purification of elastin is attempted before reduction, the methods involved in such purification tend to destroy the cross-links, significantly altering the results. It should also be pointed out that this reduction and stabilization treatment may give slightly different results with respect to the yield of insoluble elastin than if the cultures were not reduced. As noted in the cell culture section above, the aldol condensate is not stable to acid hydrolysis, but it is stable to alkaline hydrolysis. Depending upon the information required, it may be necessary to carry out both a 2 $N$ NaOH hydrolysis as well as a 6 $N$ HCl hydrolysis of the purified reduced insoluble elastin before analyzing for lysine-derived cross-links. That radioactive lysine is incorporated into the aldol condensation product in the chick embryo aortas in a fashion similar to the cell culture studies described in detail above was demonstrated by Salcedo et al.[25]

The importance of the organ cultures at this time in elastin research may be of value in assessing the precursor molecules required for insoluble elastin biogenesis. Experiments are being conducted in which chick embryo aortas can serve as matrices to evaluate the incorporation and assimilation of previously isolated soluble elastin (tropoelastin) into an insoluble fiber. In such a system, the precursor molecule may be isolated from a different culture system or directly from a donor animal, such as a copper-deficient piglet. The radiolabeled soluble precursor can then be added to the insoluble chick embryo aortas that have been placed in appropriate flasks with medium as described above. Data are much too premature to evaluate at this time. Another importance of the organ culture system relates to evaluation of the possible enzyme systems involved in insoluble elastin formation (see Chapter [35]).

Other Sources of Organ Cultures

Organ culture systems other than chick embryo aorta have been studied in much the same way. For example, one can examine weanling rabbit aorta by removing a section of the mid-thoracic aorta, cleaning it of adhering tissue, stripping the adventitia, and carefully cutting it into 1 mm$^3$ explants by employing a McIlwain tissue chopper. Approximately 25–50 pieces (explants) are placed in a petri dish 60 mm in diameter in the presence of Dulbecco's medium without serum. Incubation with appropriate radioactive precursors as described above can then be carried out. One can also employ the same medium containing fetal bovine serum. The presence of serum in the culture medium appears to cause significantly more [$^{14}$C]lysine incorporation into the insoluble elastin. Inhibitors and other reagents can be used in the same manner as described for the cell cultures. Required concentrations of the reagents may vary from tissue source to tissue source.

*Definition of Insoluble Elastin*

One final word of precaution concerning the evaluation of the insoluble elastin matrix in these organ and cell cultures should be included at this time. As pointed out, the purification of elastin in most tissues requires rather harsh treatment. Much of the elastin data presented to date used the hot alkali or autoclaving[26] procedures. Such treatments, including the

[25] L. L. Salcedo, B. Faris, and C. Franzblau, *Biochim. Biophys. Acta* **188**, 324 (1969).
[26] S. M. Partridge, H. F. Davis, and G. S. Adair, *Biochem. J.* **61**, 191 (1961).

more mild treatments involving collagenase and denaturants,[27] may be removing what might be considered as partially insoluble elastin. Thus newly incorporated tropoelastin molecules, not yet fully cross-linked, may be released by such treatments. Therefore one must fully describe the system being studied and define quite clearly the insoluble elastin obtained under specified conditions. At best, all definitions of insoluble elastin are strictly operational.

An interesting approach in this regard, which should be useful in the cell culture systems as well, is that described by Sykes and Hawber.[28] They studied the formation of elastin biosynthesis using an immunoprecipitant procedure. After obtaining a sheep anti-chick elastin serum they examined the incorporation of tritiated valine and/or proline into both the soluble and the insoluble elastin produced by organ cultures of chick embryo aortas. After appropriate incubation times of up to 3 hr they extracted the aortas with 0.5 $M$ acetic acid or neutral salt solutions with and without protease inhibitors. Significant amounts of immunoprecipitated radioactivity were detected in the extracts, suggesting partial solubilization of previously insolubilized elastin. It would be of interest to evaluate the cross-link content in these solubilized materials. Such observations tend to reinforce the suggestion that well defined operational procedures must be employed in all experimental approaches. As one becomes more sophisticated in the definition of insoluble elastin, the experimental protocols and interpretation of data from organ and cell cultures will have to be reevaluated.

Acknowledgments

The authors wish to thank Diane Dunn for the pulmonary artery smooth muscle cell data; Dr. Paul Toselli, who performed the electron microscopy; and Dr. Lily L. Salcedo for her many contributions in the preparation of this manuscript.

[27] L. B. Sandberg, *Int. Rev. Connect. Tissue Res.* **7**, 159 (1976).
[28] B. Sykes and S. Hawber, *Adv. Exp. Med. Biol.* **79**, 453 (1977).


# [35] Lysyl Oxidase: Preparation and Role in Elastin Biosynthesis

*By* HERBERT M. KAGAN and KATHLEEN A. SULLIVAN

Lysyl oxidase is the enzyme that initiates the formation of covalent cross-linkages in elastin and collagen by oxidatively deaminating certain endopeptidyl lysine residues in these proteins to peptidyl $\alpha$-aminoadipic-

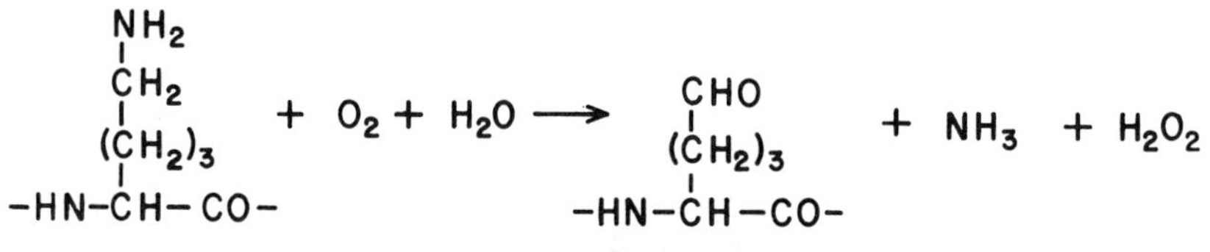

$$\underset{\substack{\text{PEPTIDYL}\\\text{LYSINE}}}{\begin{array}{c}NH_2\\|\\CH_2\\|\\(CH_2)_3\\|\\-HN-CH-CO-\end{array}} + O_2 + H_2O \longrightarrow \underset{\substack{\text{PEPTIDYL}\\\text{AMINOADIPIC SEMIALDEHYDE}}}{\begin{array}{c}CHO\\|\\(CH_2)_3\\|\\-HN-CH-CO-\end{array}} + NH_3 + H_2O_2$$

Fig. 1. Reaction catalyzed by lysyl oxidase.

δ-semialdehyde. Once generated, these peptidyl aldehydes can undergo a series of spontaneous condensation reactions with ε-amino groups or other aldehyde residues to yield the inter- and intramolecular cross-linkages in these connective tissue proteins. The chemistry of the reaction catalyzed by lysyl oxidase is presented in Fig. 1. The chemistry of the cross-linkages and proposed biosynthetic routes for their formation are reviewed in this volume.[1] A review of the properties and biological role of lysyl oxidase has appeared.[2] Clearly, lysyl oxidase plays an essential role in elastin biosynthesis by converting soluble elastin precursors to the insoluble cross-linked elastic fibers, which are essential to the normal function of the connective tissue matrix.

Lysyl oxidase has been demonstrated in several connective tissues that vary in their relative contents of elastin and collagen. Further, substrates utilized for its assay have included various forms of elastin, collagen, mixtures of both, and chemically undefined protein preparations which presumably contain forms of collagen and/or elastin. Studies of this enzyme are rendered still more complex by the finding in each of several connective tissues of multiple, catalytically functional enzyme forms. In all cases, however, enzyme activity oxidizing peptidyl lysine in elastin or collagen substrates should be 50% inhibited by less than 10 $\mu M$ β-aminopropionitrile (BAPN), an irreversible naturally occurring inhibitor of this enzyme,[3-5] to confirm such amine oxidase activity as deriving from lysyl oxidase.

The present review describes purification and assay methods utilized in our laboratory for the study of aortic lysyl oxidase and its activity toward elastin substrates.

[1] M. A. Paz, D. A. Keith, and P. M. Gallop, this volume [31].
[2] R. C. Siegel, *Int. Rev. Connect. Tissue Res.* **8**, 73 (1979).
[3] S. R. Pinnell and G. R. Martin, *Proc. Natl. Acad. Sci. U.S.A.* **61**, 708 (1968).
[4] A. S. Narayanan, R. C. Siegel, and G. R. Martin, *Biochem. Biophys. Res. Commun.* **46**, 745 (1972).
[5] P. C. Trackman and H. M. Kagan, *J. Biol. Chem.* **254**, 7831 (1979).

## Assay Methods

### Tritium Release Assay

*Principle.* The original report of Pinnell and Martin[3] identifying lysyl oxidase activity in an extract of chick cartilage also first defined the tritium release assay, which remains a principal method of assay for lysyl oxidase, with some modification. Chick embryo aortas are incubated in organ culture with [6-$^3$H]lysine in the presence of BAPN. The saline-insoluble, elastin-rich fraction is isolated from the pulsed aortas after incubation in organ culture. Oxidative deamination of the $\epsilon$-carbon of radioactive peptidyl lysine in this insoluble substrate releases a tritium ion that forms tritiated water by exchange during the assay. Tritiated water is distilled *in vacuo* from assay mixtures and quantified by liquid scintillation spectrometry.

Substrates can also be prepared by pulsing aortic tissue with [4,5-$^3$H]lysine, which is available in higher specific activity than is [6-$^3$H]lysine. Release of tritium from C-5 accompanies keto-enol tautomerism of the enzyme-produced aldehyde.[6]

### Reagents

L-[4,5-$^3$H]Lysine, 60–80 Ci-mmol
BAPN fumarate (Aldrich Chemical Company)
Sodium ascorbate
Sodium bicarbonate
Penicillin (10,000 units/ml)–streptomycin (10,000 µg/ml)
L-Valine
Chick embryos, 16 days old
Dulbecco's modified Eagle medium (dry concentrate) without lysine, containing glucose (4.5 mg/liter) and L-glutamine
NaCl, 0.15 $M$
Sodium borate, 0.1 $M$, pH 8.0, containing NaCl, 0.15 $M$
Purified bacterial collagenase (Biofactures; 2500 units/ml)
Tris, 0.05 $M$, pH 7.4, containing CaCl$_2$, 1 m$M$, and BAPN, 0.1 m$M$

*Pulsing of Tissue.* NaHCO$_3$ (3.7 g) and L-valine (97 mg) are added to a liter equivalent of medium concentrate dissolved in 500 ml of distilled, deionized water. The pH is adjusted to 7.35, the volume brought to 1 liter with water, and the solution is sterilized by filtration through a 0.2 µm filter assembly (Millipore). The medium is warmed to 37° and supplemented with sterile precautions with BAPN (50 µg/ml), sodium ascorbate (50 µg/ml), and with the antibiotic solution (10 ml/liter). The supplemented

---

[6] R. C. Siegel, S. R. Pinnell, and G. R. Martin, *Biochemistry* **9**, 4486 (1970).

medium is dispensed in 15-ml aliquots into sterile, plastic petri dishes (100 × 15 mm) to serve as a preincubation medium into which freshly excised aortas are placed (60 aortas/dish).

Chick embryos, 16 days postfertilization, are removed from the eggs and approximately 0.5 to 1 cm lengths of the ascending thoracic aortas are excised. In practice, the aortic segments are removed along with corresponding lengths of the pulmonary artery and pulmonary vein. The vascular tissue is blotted free of excess blood and placed into the preincubation medium for 30 min at 37° to deplete the issue of endogenous lysine. The aortas are then transferred to sterile 125-ml Erlenmeyer flasks containing fresh 25-ml aliquots of medium further supplemented with [4,5-$^3$H]lysine (20 $\mu$Ci per milliliter of medium). Each flask contains 60 aortas in 25 ml of radioactive medium. The flasks are incubated with gentle shaking at 37° for 22 hr in a 95% $O_2$–5% $CO_3$ atmosphere. The aortas are then removed and rinsed several times in distilled water, lyophilized, and stored at $-20°$. It should be noted that the decanted medium contains tritium-labeled soluble proteins that may be prepared for use as a soluble substrate for lysyl oxidase, as described by Harris et al.[7,8]

Substrate may also be prepared using aortic tissue of 1-day-old hatchling chicks, thus obtaining more tissue from each animal. The procedure is the same as outlined above, with the exceptions that the aortas are cut with a razor blade to 1-mm$^3$ pieces and the tissue is incubated at the equivalent of 20 aortas/50 ml of radioactive medium.

*Extraction of Pulsed Tissue.* Pulsed, lyophilized aortas are extracted in cold 0.15 $M$ NaCl (5 aortas/ml) in a conical, ground-glass homogenizer until an even suspension is obtained. The homogenate is sedimented at 10,000 $g$ for 15 min, the supernatant is removed, and the pellet is extracted and sedimented again in this manner. The resulting pellet is homogenized in 1 $N$ HCl to inactivate endogenous lysyl oxidase[9] and centrifuged as above. Two additional extractions in 0.1 $M$ sodium borate, 0.15 $M$ NaCl, pH 8.0, neutralize the remaining acid. The pellet is finally suspended in this borate–NaCl buffer at 5 aortas per milliliter, and the radioactivity in an aliquot of the evenly stirred suspension is quantified by liquid scintillation spectrometry. Typically, one aorta equivalent contains 6 × 10$^5$ cpm of tritium, counted at 30% efficiency.

The insoluble substrate has been found to contain approximately 88% elastin and 12% collagen.[10] To remove the collagen contaminant, the in-

[7] E. D. Harris, W. A. Gonnerman, J. E. Savage, and B. L. O'Dell, *Biochim. Biophys. Acta* **341**, 332 (1974).
[8] E. D. Harris and M. C. Garcia-de-Quevedo, *Arch. Biochem. Biophys.* **190**, 227 (1978).
[9] H. M. Kagan, N. A. Hewitt, L. L. Salcedo, and C. Franzblau, *Biochim. Biophys. Acta* **365**, 223 (1974).
[10] A. S. Narayanan, R. C. Page, and G. R. Martin, *Biochim. Biophys. Acta* **351**, 126 (1974).

soluble substrate pellet equivalent to 20 mg of lyophilized aortas is incubated with 600 units of bacterial collagenase in 50 m$M$ Tris, 1 m$M$ $CaCl_2$, 0.1 m$M$ BAPN, pH 7.4, for 3 hr at 37°. The suspension is centrifuged; the isolated pellet is extracted three times with 0.15 $M$ NaCl and finally suspended in borate–NaCl buffer, and the radioactivity of the pellet is quantified as above.

*Assay Procedure.* Aliquots of the evenly suspended substrate containing 125,000 cpm are transferred to 13 × 100 mm glass tubes on ice containing borate–NaCl buffer. Lysyl oxidase is added to the assays in 0.1-ml aliquots or less to bring the final volume of each assay to 0.75 ml. Control assays are incubated without added enzyme or with enzyme and BAPN (50 μg/ml). The assay tubes are incubated at 37° with shaking for 2 hr. The reaction is terminated by chilling the tubes to 0° on ice. Tritiated water is distilled from the reaction mixtures under vacuum; the distillates are collected at −50°. Aliquots (0.5 ml) of the distillates are counted by liquid scintillation spectrometry. A microdistillation apparatus suitable for this procedure has been described.[11]

*Assay Precautions.* The assay is linear over a range of 600 cpm of tritium released in the 2-hr incubation. Lysyl oxidase is inhibited by urea carried over from stock solutions of enzyme, although inhibition plots can be used to correct for this effect.[12] Collagen, elastin, or nonspecific proteins contaminating enzyme extracts interfere with the assay. Multiple dilutions of enzyme samples should be assayed to ensure linearity in each case.

*Specific Activity.* A convenient enzyme unit is defined as 1 cpm released in 2 hr per milliliter of assay mixture under the assay conditions described. There is no uniformly accepted enzyme unit for lysyl oxidase owing to the different substrates and assay conditions employed by investigators thus far. Specific activity is expressed as units per milligram of enzyme protein, the latter determined by the Lowry procedure.[13]

*Chemical Identification of Enzyme Product.* It is advisable to confirm the generation of aldehyde products of lysyl oxidase activity by direct chemical analysis of forms of substrates not previously analyzed. The radioactive aldehyde produced from tritium-labeled or [14]C-labeled peptidyllysine by the action of lysyl oxidase can be reduced with sodium borohydride to peptidyl-ε-hydroxynorleucine, the alkali-stable alcohol derivative. The substrate is then base hydrolyzed, and the ε-hydroxynorleucine product is identified by split-stream amino acid anal-

[11] R. Misiorowski, J. B. Ulreich, and M. Chvapil, *Anal. Biochem.* **71**, 186 (1976).
[12] H. M. Kagan, K. A. Sullivan, T. A. Olsson, and A. L. Cronlund, *Biochem. J.* **177**, 203 (1979).
[13] O. H. Lowry, N. J. Rosebrough, A. L. Farr, and R. J. Randall, *J. Biol. Chem.* **193**, 265 (1951).

ysis.[9,14] Alternatively, radioactive aldehyde deriving from [14]C-labeled peptidyllysine can be oxidized to peptidylaminoadipic acid by treatment of the substrate with performic acid after its incubation with lysyl oxidase. The [14]C-labeled aminoadipic acid is released by acid hydrolysis and identified by split-stream amino acid analysis.[3]

*Enzyme-Coupled Spectrophotometric Assays*

*Principle.* Bovine aortic lysyl oxidase has been shown to oxidize a variety of nonpeptidyl amines producing the corresponding aldehydes and $H_2O_2$.[5] The oxidation of $n$-butylamine, for example, can be followed fluorometrically by coupling this assay to the $H_2O_2$-dependent oxidation of homovanillic acid by horseradish peroxidase. The accumulation of the fluorescent product of homovanillate oxidation is followed at 37° at an excitation wavelength of 320 nm and an emission wavelength of 420 nm. Alternatively, the production of $n$-butyraldehyde by lysyl oxidase can be coupled to the reduction of $NAD^+$ by aldehyde dehydrogenase, measuring the change in absorption at 340 nm.[5] Although the oxidation of nonpeptidyl amines is limited to 300 turnovers of lysyl oxidase, since the enzyme becomes inactivated in the process, initial rates of oxidation are proportional to enzyme concentration.[5] This assay is reliable for partially or highly purified lysyl oxidase of bovine aorta, although its application to enzyme of other tissues or crude extracts remains to be established.

Purification

Lysyl oxidase is mostly insoluble in neutral salt buffers in a variety of connective tissues, although enzyme activity is present in the growth medium of fibroblast cell cultures as a soluble protein.[15] The enzyme can be solubilized from connective tissue sources by buffers containing 4–6 $M$ urea.[16] Most purification schemes that have evolved begin with the urea-solubilized fraction and generally share other purification steps as well. We shall describe the method used in our laboratory for the purification of lysyl oxidase from bovine aorta.[12]

*Reagents*

Thoracic aortas of 2- to 6-week-old calves, obtained at slaughter
Potassium phosphate, 0.016 $M$, pH 7.7

[14] R. C. Siegel, *J. Biol. Chem.* **251**, 5786 (1976).
[15] D. L. Layman, A. S. Narayanan, and G. R. Martin, *Arch. Biochem. Biophys.* **149**, 97 (1972).
[16] A. S. Narayanan, R. C. Siegel, and G. R. Martin, *Arch. Biochem. Biophys.* **162**, 231 (1974).

Potassium phosphate, 0.016 $M$, pH 7.7, containing 1 $M$ NaCl or containing urea at 1 $M$, 2 $M$, 4 $M$, or 6 $M$
Urea, 8 $M$, freshly deionized by passage through a mixed-bed ion-exchange resin (Barnstead, Ultrapure). This stock solution is used to prepare each of the buffered urea solutions.
(Diethylaminoethyl) cellulose (DE-52, Whatman)
Sephacryl S-200 (Pharmacia)
Sepharose CL-4B (Pharmacia)
Citrate-soluble calf skin collagen
Cyanogen bromide

*Extraction.* The entire purification procedure is carried out at 4°. Each liter of extracting buffer receives 0.5 ml of a 2 $M$ solution of phenylmethanesulfonyl fluoride (PMSF; Sigma) in dimethyl sulfoxide just prior to each extraction. The calf aortas are cleaned of adhering muscle and fat and are finely ground by passage through a chilled meat grinder. The ground tissue is homogenized at high speed in a Waring blender for 90 sec in 0.016 $M$ potassium phosphate, pH 7.7, containing 0.15 $M$ NaCl, and 1 m$M$ PMSF at a ratio of 2.5 ml of buffer per gram of ground tissue. This ratio is utilized for all extractions. The homogenate is centrifuged at 11,000 $g$ for 20 min, the supernatant is decanted, and the pellet is extracted once more with this buffer. The salt extracts contain negligible lysyl oxidase activity and may be discarded. The pellet is homogenized in 0.016 $M$ potassium phosphate, pH 7.7, containing PMSF. The supernatant is discarded after centrifugation, and the pellet is homogenized in 0.016 $M$ potassium phosphate, pH 7.7, containing 1 $M$ urea and 1 m$M$ PMSF, and stirred slowly at 4° for 1 hr. The pellet is isolated by centrifugation and homogenized in 4 $M$ urea in phosphate buffer and PMSF. The homogenate is stirred slowly for 18 hr and centrifuged; the supernatant is decanted and saved. The 4 $M$ urea extraction is repeated twice more, and the three 4 $M$ urea supernatants are pooled and saved.

*Affinity Chromatography.* The collagen affinity column is prepared by coupling citrate-soluble calf skin collagen, prepared as described,[17] to Sepharose CL-4B, activated with CNBr.[18] Lyophilized calf skin collagen is denatured and dissolved in water at 60° just prior to coupling, utilizing a ratio of 1 g of collagen per 200 g of Sepharose. One 200-g affinity column retains enzyme extracted from 160 g of aortic tissue.

The pooled 4 $M$ urea extracts are diluted with an equal volume of 0.016 $M$ potassium phosphate, pH 7.7, and passed into the affinity column. Unbound material is removed by washing the column with 2 $M$ urea, 0.016 $M$ potassium phosphate, until the $A_{280}$ falls to zero. The column is succes-

[17] P. M. Gallop and S. Seifter, this series, Vol. 6, p. 635.
[18] P. Cuatrecasas, *J. Biol. Chem.* **245**, 3059 (1970).

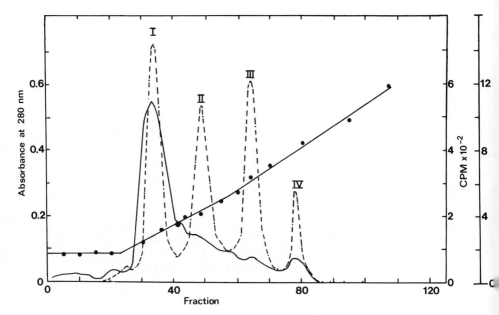

FIG. 2. Chromatography of 6 $M$ urea effluent of affinity column on DEAE-cellulose. Extraction was performed in the presence of phenylmethanesulfonyl fluoride (0.001 $M$). -----, Enzyme activity; ———, $A_{280}$; ●, salt gradient.

sively eluted with 600-ml quantities per 200 g of collagen–Sepharose of 0.016 $M$ potassium phosphate; 1 $M$ NaCl, 0.016 $M$ potassium phosphate; 0.016 potassium phosphate; and, finally, 6 $M$ urea, 0.016 $M$ potassium phosphate, each adjusted to pH 7.7. The 1 $M$ NaCl wash elutes considerable bound protein as measured at 280 nm but does not remove enzyme. Lysyl oxidase elutes with the 6 $M$ urea buffer.

*DE-52 Chromatography.* The conductivity of the 6 $M$ urea effluent of the affinity column is measured to ensure that its ionic strength is not greater than that of 0.016 $M$ potassium phosphate, 6 $M$ urea, pH 7.7. A maximum volume of 1 liter of this effluent is then loaded onto a 2.5 × 84 cm column of DE-52 equilibrated in this buffer at 4°. The column is eluted with a linear gradient of NaCl (0 to 0.4 $M$) in 6 $M$ urea, 0.016 $M$ potassium phosphate, pH 7.7. The total gradient volume is 2680 ml. Fractions (8.5 ml) are collected and analyzed for $A_{280}$, enzyme activity, and conductivity. A typical elution pattern is shown in Fig. 2. Although the same multiplicity of peaks of enzyme activity results regardless of the presence or the absence of PMSF in the extraction buffers, the addition of PMSF as described yields purified enzyme species with somewhat greater stability to storage, suggestive of the effects of serine protease activity.

*Gel Exclusion Chromatography.* The pools of the four species resolved in DE-52 (Fig. 2) are individually concentrated to approximately 15-ml volumes by ultrafiltration (Amicon type YM-10 filters at 30 psi). Each enzyme concentrate is passed through a 120 × 3 cm column of Sephacryl S-200 equilibrated and eluted with 6 $M$ urea, 0.016 $M$ potassium phosphate, pH 7.7. Each enzyme species elutes at or just prior to the elution position of carbonic anhydrase ($M_r$ = 29,000) (Fig. 3). Typical purification results are summarized in Table I.

*Purity.* Polyacrylamide gel electrophoresis in SDS resolves each purified enzyme species into a major band at a molecular weight of 32,000 ± 1000[12] but reveals the presence of a minor band at 24,000 molecular weight constituting <5 to 15% of the stainable protein on the gel. However, only the major band of each preparation of each enzyme species becomes labeled when the preparations are incubated with [$^{14}$C]BAPN at 37° and then analyzed by SDS gel electrophoresis. Further, the peptide maps of the minor band and of the major band isolated from the gels are very similar in each case.[19] We conclude that the band at 32,000 ± 1000 is the protomer of catalytically functional lysyl oxidase. The smaller band appears to be a degradation product of the enzyme.

Properties

*Stability.* The enzyme in the final purification step as described above elutes from the Sephacryl column in 6 $M$ urea, 0.016 $M$ potassium phosphate, pH 7.7. Storage of the enzyme at 4° as the pooled fractions pooled from the Sephacryl column in the 6 $M$ urea buffer results in 85-90% recovery of activity for as long as 6 weeks. Dialysis of enzyme into urea-free buffers or into water decreases activity in shorter time periods. As much as 90% of the activity is lost upon lyophilization of enzyme previously dialyzed into water. Enzyme activity is not restored by addition of copper salts, pyridoxal phosphate, or flavins to assays. Enzyme can be dialyzed out of urea into 0.1 $M$ sodium borate, 0.15 $M$ NaCl, pH 8.0, assay buffer and stored at 4° with nearly full retention of activity for 1 week. Freezing of such solutions, however, causes marked losses of activity.

*Assay Optima.* The enzyme has a fairly broad pH optimum between 7.8 and 8.3, and is routinely assayed at pH 8.0. The plot of the temperature dependency of the assay using the aortic elastin substrate exhibits a sharp maximum at 50-52°.[9] The enzyme is irreversibly inactivated at 85° and is, therefore, quite thermostable. The optimal assay temperature may reflect temperature-induced changes in the conformational properties of the substrate. Assays are usually conducted at 37°, however, because of the

[19] K. A. Sullivan and H. M. Kagan, in preparation, 1981.

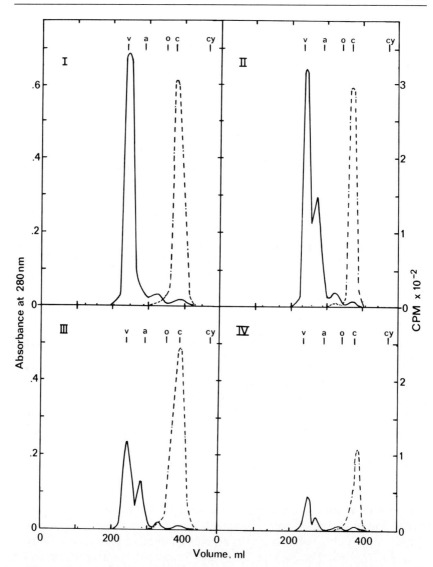

FIG. 3. Chromatography of DEAE-cellulose peaks I–IV on Sephacryl S-200 in 6 $M$ urea–0.016 $M$ potassium phosphate, pH 7.7. Abbreviations used to identify elution positions of standards are as follows: v, void; a, bovine serum albumin; o, ovalbumin; c, carbonic anhydrase; cy, cytochrome $c$. Roman numerals refer to peaks of activity resolved on DEAE-cellulose, as labeled in Fig. 2. -----, Enzyme activity; ———, $A_{280}$.

TABLE I
PURIFICATION OF FOUR SPECIES OF LYSYL OXIDASE FROM BOVINE AORTA[a,b]

| Purification step | Activity (total units × $10^{-6}$) | Total protein (mg) | Specific activity (× $10^{-5}$) | Yield (%) | Purification (fold) |
|---|---|---|---|---|---|
| Urea extract, 4 M | 11.7 | 1554 | 0.075 | 100 | 1 |
| Affinity chromatography | 8.75 | 160 | 0.547 | 74.7 | 7.3 |
| DEAE-cellulose | | | | | |
| Peak I | 3.02 | 30.4 | 0.994 | 25.8 | 13.2 |
| Peak II | 1.59 | 35.8 | 0.443 | 13.5 | 5.9 |
| Peak III | 1.03 | 16.4 | 0.627 | 8.8 | 8.3 |
| Peak IV | 0.41 | 7.4 | 0.557 | 3.5 | 7.4 |
| (Sum of four peaks) | (6.05) | (90.0) | | (51.6) | |
| Sephacryl S-200 | | | | | |
| Peak I | 0.49 | 0.54 | 11.0 | 4.2 | 146 |
| Peak II | 0.30 | 0.39 | 6.9 | 2.3 | 92 |
| Peak III | 0.48 | 0.67 | 12.2 | 4.1 | 162 |
| Peak IV | 0.26 | 0.37 | 6.0 | 2.2 | 80 |
| (Sum of four peaks) | (1.54) | (1.97) | | (12.8) | |

[a] Reproduced, with permission from the publisher, from Kagan et al.[12]
[b] Purification began with 200 g of bovine aorta.

increased nonenzymic release of tritium from the substrate, which occurs at elevated temperatures. The aortic elastin substrate is not markedly affected by ionic strength above that of the borate-NaCl buffer. However, significant inhibition results at unusually high (e.g., 2 M NaCl) ionic strength.

*Amino Acid Composition.* Amino acid analyses have been reported for the four species of bovine aortic enzyme,[4] and one species of the chick cartilage enzyme[20] (Table II) with similarities but not identities in composition noted. The aortic enzymes are each apparently acidic species, although the amide contents have not been established. Preliminary isoelectric focusing experiments on the bovine aortic enzyme species reveal p*I* values between 4.5 and 5.

*Molecular Weight.* The molecular weights of the bovine aortic lysyl oxidase species, determined by the SDS–gel electrophoresis technique of Laemmli,[21] each are 32,000 ± 1000,[12] consistent with their elution just

[20] F. L. H. Stassen, *Biochim. Biophys. Acta* **438**, 49 (1976).
[21] U. K. Laemmli, *Nature (London)* **72**, 248 (1976).

TABLE II
AMINO ACID ANALYSES OF PURIFIED LYSYL OXIDASE SPECIES[a,b]

Content (residues/1000 residues)

| Residue | Peak I | Peak II | Peak III | Peak IV | Chick-embryo cartilage enzyme[c] |
|---|---|---|---|---|---|
| Aspartic acid | 123 | 125 | 122 | 94 | 136 |
| Threonine | 57 | 55 | 56 | 48 | 53 |
| Serine | 104 | 86 | 104 | 122 | 82 |
| Glutamic acid | 113 | 136 | 136 | 124 | 106 |
| Proline | 60 | 51 | 50 | 51 | 58 |
| Glycine | 120 | 87 | 108 | 177 | 97 |
| Alanine | 71 | 82 | 83 | 85 | 66 |
| Valine | 39 | 42 | 35 | 32 | 39 |
| Cysteine | 27 | 24 | 18 | 14 | 30 |
| Methionine | 15 | 16 | 15 | 17 | 15 |
| Isoleucine | 30 | 27 | 31 | 29 | 40 |
| Leucine | 64 | 86 | 78 | 65 | 67 |
| Tyrosine | 25 | 31 | 21 | 18 | 65 |
| Phenylalanine | 27 | 30 | 26 | 25 | 27 |
| Lysine | 31 | 36 | 36 | 29 | 31 |
| Histidine | 39 | 25 | 27 | 23 | 29 |
| Arginine | 56 | 61 | 56 | 49 | 59 |

[a] Reproduced, with permission from the publisher, from Kagan et al.[12]
[b] Tryptophan was not determined. Cysteine was analyzed as cysteic acid in separate samples oxidized with performic acid.
[c] Data were taken from Stassen.[20]

prior to a 29,000 molecular weight marker from Sephacryl S-200 in 6 $M$ urea. In the absence of denaturants, however, lysyl oxidase polymerizes into a series of multimeric states, ranging up to 1,000,000 $M_r$.[12] Such enzyme polymers may exist under assay conditions, although the functional molecular weight of the enzyme has not been established.

*Cofactors.* A number of investigations support the conclusion that lysyl oxidase is a copper metalloprotein. Nutritional deprivation of copper lowers cross-linkage content and biosynthesis of elastin and collagen.[22,23] Enzyme activity is markedly reduced in copper-deficient chick aorta but can be restored by incubating the copper-deficient aortas with $Cu^{2+}$ in organ culture.[24] However, the stoichiometric relationship between the metal atom and enzyme protein is uncertain owing to the states of purity and/or

[22] E. J. Miller, G. R. Martin, C. E. Mecca, and K. A. Piez, *J. Biol. Chem.* **240**, 3623 (1965).
[23] E. D. Harris and B. L. O'Dell, *Adv. Exp. Biol. Med.* **48**, 267 (1974).
[24] E. D. Harris, *Proc. Natl. Acad. Sci. U.S.A.* **73**, 371 (1976).

to uncertainties about the subunit molecular weight of preparations in which $Cu^{2+}$ has been directly measured. The bovine aortic enzymes are each inhibited by $\alpha,\alpha'$-dipyridyl and other copper chelators,[4] but the content of enzyme-bound copper has not been measured with this enzyme.

There is indirect evidence for the presence of a second organic cofactor in lysyl oxidase. Nutritional studies[25,26] and binding studies with tritiated pyridoxine[26] suggest a role for pyridoxal phosphate. The purified bovine aortic enzymes as well as enzymes of other sources are fully inhibitable by carbonyl reagents, such as isoniazid, semicarbizide, or NaCN,[7,12] consistent with such a prosthetic group. However, fluorescence and absorption spectrophotometric analysis of the purified bovine aortic enzyme species pursued in the author's laboratory have not revealed the presence of such an identifiable organic cofactor, either as an enzyme complex or as material solubilized from purified enzyme.

*Substrate Specificity.* Insoluble elastin[9] and soluble tropoelastin[27] as well as fibrillar forms of collagen[28] are oxidized by lysyl oxidase. Although activity is expressed toward tropoelastin in solution, a coacervated aggregate of this protein is much more readily oxidized and becomes cross-linked to a preexistent insoluble elastin matrix[27] upon the addition of purified cartilage lysyl oxidase. Bovine aortic lysyl oxidase oxidizes elastin and collagen substrates as well as synthetic polypeptides modeled after valine- and proline-rich repeating sequences found in elastin.[29] The repeat polypentapeptide HCO-(Val-Pro-Gly-X-Gly)$_n$-Val-OMe, where X = Val or Lys at a 4:1 ratio and where $n \geq 40$, is much more readily oxidized and cross-linked by the enzyme if it is coacervated prior to and during exposure to the enzyme.[29] Thus, intermolecular alignment, intramolecular conformational changes, and the increased hydrophobicity resulting from coacervation of natural or synthetic polypeptide substrates all may contribute to the susceptibility of potential substrates to oxidation by lysyl oxidase. This possibility seems to be supported upon examination of the sequences surrounding lysine residues in elastin, collagen, and synthetic polypeptide substrates known to be oxidized by the enzyme (Table III).[29-32] The variety of permissible sequences strongly suggests that fac-

[25] B. C. Starcher, *Proc. Soc. Exp. Biol. Med.* **132**, 379 (1969).
[26] J. C. Murray and C. I. Levene, *Biochem. J.* **167**, 463 (1978).
[27] A. S. Narayanan, R. C. Page, F. Kuzan, and C. G. Cooper, *Biochem. J.* **173**, 857 (1978).
[28] R. C. Siegel, *Proc. Natl. Acad. Sci. U.S.A.* **71**, 4826 (1974).
[29] H. M. Kagan, L. Tseng, P. C. Trackman, R. Okamoto, R. S. Rapaka, and D. W. Urry, *J. Biol. Chem.* **255**, 3656 (1980).
[30] J. A. Foster, R. Shapiro, P. Voynow, G. Crombie, B. Faris, and C. Franzblau, *Biochemistry* **14**, 5343 (1975).
[31] P. P. Fietzeck and K. Kuhn, *Int. Rev. Connect. Tissue Res.* **7**, 1 (1976).
[32] U. Becker, H. Furthmayr, and R. Timpl, *Hoppe-Seyler's Z. Physiol. Chem.* **356**, 21 (1975).

TABLE III
SPECIFICITY OF LYSYL OXIDASE FOR PEPTIDYLLYSINE[a]

| Peptide | Source | Reference |
|---|---|---|
| -Ala-Ala-Lys-Ala-Ala-Ala-Lys-Ala- | Elastin and tropoelastin | 30 |
| -Gly-Tyr-Asp-Glu-Lys-Ser-Ala-Gly- | $NH_2$-terminal, collagen $\alpha1(I)$ chain | 31 |
| -Gln-Glu-Glx-Lys-Ala-His-Asp-Gly- | COOH-terminal, collagen $\alpha1(I)$ chain | 32 |
| HCO-(X-Pro-Gly-Gly)$_n$-Val-OMe[b] | Synthetic repeat polytetrapeptide | 29 |
| HCO-(Val-Pro-Gly-X-Gly)$_n$-Val-OMe[b] | Synthetic repeat polypentapeptide | 29 |
| HCO-Val(Ala-Pro-Gly-X-Gly-Val)$_n$-OMe[b] | Synthetic repeat polyhexapeptide | 29 |

[a] Lys or X (as lysine) residues susceptible to oxidation by lysyl oxidase.
[b] X = Val or Lys at a 4:1 ratio; $n \geq 40$.

tors in addition to sequence are important in the expression of enzyme activity. It has been suggested, however, that tyrosine or phenylalanine peptide linked through $\alpha$-amino groups to endopeptidyl lysine in elastin prevents oxidation of lysine by lysyl oxidase.[33,34]

Purified bovine aortic lysyl oxidase can oxidize nonpeptidyl amines in addition to polypeptide substrates, and, as noted, such compounds can serve as substrates in continuously monitored spectrophotometric assays.[5] Such substrates include free lysine and its $\alpha$-N-acetylated and/or esterified derivatives; monoamines, including propyl-, butyl-, and octylamine; and 3-, 4-, 5-, and 6-carbon primary diamines. The biological significance of the oxidation of such compounds by lysyl oxidase has not been established.

[33] J. A. Foster, L. Rubin, H. M. Kagan, C. Franzblau, E. Bruenger, and L. Sandberg, *J. Biol. Chem.* **249**, 6191 (1974).
[34] K. M. Baig, M. Vlaovic, and R. A. Anwar, *Biochem. J.* **185**, 611 (1980).

## [36] Isolation of Soluble Elastin from Copper-Deficient Chick Aorta

*By* ROBERT B. RUCKER

Lysyl oxidase activity and the rate of cross-linking must be reduced effectively to cause an increase in the net pool of soluble elastins prior to isolation. This may be achieved by rendering animals nutritionally copper-deficient or lathrytic.[1,2] The choice of experimental animals is also

[1] R. B. Rucker and D. Tinker, *Int. Rev. Exp. Pathol.* **17**, 1 (1977).
[2] L. B. Sandberg, *Int. Rev. Connect. Tissue Res.* **7**, 159 (1976).

important. In this regard the chick system has been of fundamental importance, since the isolation of soluble elastin from other common experimental animals, such as rats and mice, has proved to be difficult. If a nutritional or pharmacological treatment is used to precondition animals for subsequent isolation of soluble elastin, it is often necessary to initiate the treatment during periods of neonatal development. The chick is easily manipulated nutritionally during neonatal development, and with a sufficient number of chicks the yield of thoracic aorta, an excellent source of soluble elastin, is adequate for soluble elastin isolation.

The predominant product, tropoelastin is a protein with a molecular weight of 72,000. Although higher molecular weight forms of soluble elastin have been isolated and described,[3,4] it is now the general feeling that these products may represent cross-linked soluble elastin peptides or complexes of soluble elastin with structural glycoproteins, lysyl oxidase, or elastin-related proteinases.[5,6] Soluble elastins with molecular weights less than 72,000 appear to be products derived from the proteolysis of tropoelastin.[5,7]

Nutritional Conditioning

The nutritional conditioning of chicks for the isolation of tropoelastin from aorta requires that attention be given to dietary formulation and husbandry. A diet based on bovine spray-dried skim milk powder is used to achieve a level of copper that is less than 1 mg per kilogram (<1 ppm) of diet. Many commercially available skim milk powders contain less than 0.4 mg of copper per kilogram. With the exception of iron and manganese, skim milk powder also contains sufficient concentrations of other known essential elements, such that they need not be added to the ration. This is advantageous, since it precludes the addition of copper as a contaminant from a separate mineral source.

The composition of the diet that we have used successfully contains the following ingredients expressed as grams per 100 g of diet: spray-dried skim milk, 60 g; DL-methionine, 0.3 g; L-arginine-HCl, 0.5 g; glycine, 0.6 g; corn oil, 5 g; vitamin premix, 1 g; NaCl, 0.5 g; $FeSO_4 \cdot 7 H_2O$, 0.065 g; $MnSO_4 \cdot H_2O$, 0.035 g; and glucose, 32 g. The vitamin premix is made to

---

[3] J. A. Foster, R. P. Mecham, C. B. Rich, M. F. Cronin, A. Levine, M. Imberman, and L. L. Salcedo, *J. Biol. Chem.* **253**, 2797 (1978).
[4] R. B. Rucker, C. S. Heng-Khoo, M. Dubick, M. Lefevre, and C. E. Cross, *Biochemistry* **18**, 3854 (1979).
[5] J. A. Foster, this volume [30].
[6] S. M. Partridge and A. H. Whiting, *Biochim. Biophys. Acta* **576**, 71 (1979).
[7] R. P. Mecham and J. A. Foster, *Biochemistry* **16**, 3825 (1977).

supply the following vitamins expressed as grams per 100 g of premix: thiamine-HCl, 1.6 g; riboflavin, 1 g; nicotinic acid, 5 g; calcium D-pantothenate, 2 g; pyridoxine-HCl, 0.6 g; biotin, 0.06 g; cyanocobalamin, 0.003 g; inositol, 25 g; menadione sodium bisulfite, 0.152 g; retinol palmitate, 0.03 g; DL-α-tocopherylacetate, 10 g; cholecalciferol, 0.0025 g; and choline chloride, 10 g. The vitamins are mixed with glucose (to 100 g), so that their subsequent addition to a diet based on skim milk at 1 g per 100 g assures their presence at or over the optimal requirements for early growth. It has been found that the addition of 80 mg of zinc (as zinc carbonate) or 0.1 g of ascorbic acid per 100 g of diet intensifies the signs of nutritional copper deficiency.[8] We often add zinc as an antagonist to ensure a severe deficiency. However, the use of antagonists does not alleviate the need for careful analysis of dietary components for copper prior to formulation. The presence of copper in amounts greater than 200 $\mu$g per 100 g (2 ppm) of diet (equivalent to an intake of 20–40 $\mu$g of copper per day per bird) will markedly reduce the yield of aorta tropoelastin, even in the presence of excessive zinc or ascorbic acid. Only distilled or deionized water should be used. Likewise, conventional galvanized feed and water trays can be a source of copper.

Careful husbandry is also required throughout all phases of the conditioning period. When day-old chicks are fed the diet, gross signs of copper deficiency are observed after 7–10 days. Many of the chicks have enlarged hocks and lack mobility. Appropriate measures are required to ensure that these birds receive food and water when housed in large groups in typical brooders. Our highest yields of tropoelastin have been obtained from birds that are maintained and fed the copper-deficient diet for 14–20 days from the day of hatching. This is approximately the time when the relative net content of elastin in aorta increases significantly (1.5- to 2.0-fold).[9] With severe copper deficiency, this is also the period when mortality sharply increases and clear signs of dissecting aneurysms may be observed at autopsy (10–20% of the birds).[10]

Tissue Homogenization and Extraction of Tropoelastin

For the isolations of tropoelastin in which yields of at least 30 mg are needed, 500–700 copper-deficient chicks are required. The birds are killed by cervical dislocation or by $CO_2$ inhalation. The thoracic aortas are rapidly removed, cleansed of adhering adventitial tissue, and immersed in liquid nitrogen. A Wiley mill with a 20-mesh screen is then used to powder

[8] R. B. Rucker and J. Murry, *Am. J. Clin. Nutr.* **31**, 1221 (1978).
[9] I. Lee, M. Chen Yau, and R. B. Rucker, *Biochim. Biophys. Acta* **442**, 432 (1976).
[10] R. B. Rucker and W. Goehlich-Riemann, *J. Nutr.* **102**, 563 (1972).

the tissue. The mill is first chilled with the liquid nitrogen, and the aortas are slowly milled in the presence of liquid nitrogen. Once the liquid nitrogen has evaporated from the collecting vessel, the powder is weighed and then suspended in the extracting solvent.

The choice of extracting solvent depends largely upon isolation protocol preferred. Elsewhere in this volume are procedures in which neutral salt solutions are used as extractants.[11] Routinely, we use 0.5 $M$ acetic acid containing 20 $\mu$g of pepstatin per milliliter as the initial extracting solvents. At low pH, tropoelastin is not subject to excessive proteolysis. Further, the amounts of structural glycoprotein extracted are reduced at low pH.

Typically, three serial 4-hr extractions are carried with the following ratios of solvent to initial wet weight or volume of tissue: 9:1, 4:1, 2:1. It is essential to keep the protein concentrations in the extract as high as possible, since many of the subsequent purification steps are dependent upon tropoelastin concentration. After each of the extractions and separation of the residues by centrifugation (10,000 $g$, 60 min), the supernatant fractions are filtered through a fast-flow paper to remove solid lipid particles. The combined supernatant fractions are used for the isolation.

Ammonium Sulfate Precipitation and Alcohol Fractionation

*Buffers*

AAI: 0.5 $M$ ammonium acetate (adjusted to pH 5.0 with glacial acetic acid), 0.005 $M$ EDTA, 10 $\mu$g/ml pepstatin, 0.001 $M$ phenylmethyl sulfonyl fluoride, and 0.01 $M$ $\beta$-aminopropionitrile fumarate

PA: 0.2 $M$ pyridine acetate (adjusted to pH 5.0 with glacial acetic acid).

*Procedure.* Except where noted, all of the following steps are carried out at 4° and with continued slow stirring of solutions. First, solid ammonium sulfate is added (over a 2-hr period) to the combined supernatant fractions to 50% saturation. This solution is stirred for 6–10 hr, and the precipitated protein is collected by centrifugation at 20,000 $g$ for 30 min. The precipitate is then suspended in 5–8 volumes of AAI and immediately dialyzed overnight against a 30-volume excess of AAI. No attempt is made at this point to remove flocculated material. The dialyzed suspension of protein is next fractionated directly with alcohol.

The use of $n$-propanol and $n$-butanol as described by Sandberg[2] has given excellent recoveries of tropoelastin. First, a volume of chilled

[11] C. B. Rich and J. A. Foster, this volume [38].

$n$-propanol 1.5 times that of the dialyzed protein suspension is added slowly and dropwise to the suspension. This is followed by the dropwise addition of a volume of $n$-butanol 2.5 times that of the original starting suspension. Tropoelastin is soluble in the aqueous alcoholic mixture, and any precipitated material is separated by means of a filtering flask under vacuum.

To recover the tropoelastin the aqueous alcoholic mixture is then subjected to rotary flash evaporation at 40°. When the volume of solution has been reduced by 90–95%, the apparatus is disconnected and the residue is collected. This is usually done by the addition of PA to form a thick slurry. This slurry and subsequent washes are transferred directly to dialysis tubing of appropriate size for dialysis against PA. Dialysis is continued with multiple changes until ammonium is not easily detected by reaction with ninhydrin.

The lyophilized product after dialysis should possess a composition similar to elastin. Its electrophoresis in a sodium dodecyl sulfate–polyacrylamide gel system similar to that described by Furthmayer and Timpl[12] will usually yield a major band (following staining with Coomassie Brilliant Blue) at 72,000 daltons (approximately 60–70% of the total), higher molecular weight components (10–20%), and a pattern of smaller molecular weight peptides similar to those described by Mecham and Foster (10–30%).[7]

Ion Exchange and Gel Filtration Isolation Steps

*Buffers*

UTI: 4 $M$ Urea, 0.05 $M$ Tris, 0.005 $M$ $\beta$-aminoproprionitrile fumarate, 0.05 $M$ $\epsilon$-aminocaproic acid, and 0.001 $M$ phenylmethylsulfonyl fluoride adjusted to pH 8.0 with glacial acetic acid.

PA: 0.2 $M$ pyridine acetate adjusted to pH 5.0 with glacial acetic acid

*Procedure.* An ion-exchange chromatography step followed by gel filtration is usually sufficient to assure a high degree of homogeneity. For the ion-exchange step, DEAE-cellulose (Whatman DE-52, preswollen) is equilibrated with UTI and poured as a slurry into a 2 × 12 cm siliconized column. Soaking the column in 1% Siliclad (Clay Adams, Parsippany, New Jersey) followed by drying in an oven is usually sufficient for siliconization. Prior to use of the column for tropoelastin isolation, it is also advisable to condition the column. For conditioning we usually apply 20 mg of cytochrome $c$ and elute by successive additions of UTI. We have found that such conditioning substantially improves the recovery of soluble elastin.

[12] H. Furthmayer and R. Timpl, *Anal. Biochem.* **41**, 510 (1971).

TABLE I
RECOVERY OF TROPOELASTIN FOR ACETIC ACID EXTRACTS OF
COPPER-DEFICIENT CHICK AORTA

| Purification step | Recovery of soluble elastin per gram of aorta ($\mu$g) |
|---|---|
| 1. Acetic acid extraction | 2500–4200 [a] |
| 2. Ammonium sulfate precipitation | 1900–3800 [b] |
| 3. Alcohol fractionation and dialysis | 1650–2650 [a] |
| 4. DEAE-cellulose | 1200–2000 [b] |
| 5. Sephadex G-150 | 1000–1900 [a,b] |

[a] Range of values from four experiments based on radioimmunoassay. Birds were approximately 20 days old.
[b] Range of values based on crude estimates following amino acid analysis using criteria suggested by Sandberg.[2]

A stepwise gradient of successive 150-ml volumes of UTI containing 0.1 $M$, 0.5 $M$, or 1 $M$ NaCl (flow rate 0.5 ml/min) is used for elution of protein. With respect to the products soluble in $n$-propanol and $n$-butanol, soluble elastin and elastin peptides elute with the first major peak of protein (between 0 and 0.1 $M$ concentrations of NaCl). This protein fraction is collected, dialyzed free of urea against PA, and concentrated. After DEAE-cellulose chromatography, a gel filtration step is necessary for removal of the elastin-derived peptides. Again, care must be taken to precondition the column. For example, we have observed that recovery of tropoelastin is often poor from Sephadex or Sephacryl columns that have not been used previously.

For routine gel filtration, Sephadex G-150 or G-200 is used. Recovery of tropoelastin is near 100% if the gel is treated in the following manner. First, 1 g of NaBH$_4$ is added per 50 ml of swollen gel (in deionized H$_2$O) in an equal volume of 0.01 $M$ NaOH. The gel–NaBH$_4$ suspension is then heated to 80° for 2 hr. After cooling to room temperature, the gel is suspended in PA and washed free of NaBH$_4$. A column is then poured of sufficient length to resolve 72,000-dalton protein subunits from 58,000-dalton protein subunits (size of the largest tropoelastin-derived degradation product); PA is used as eluent. The column is standardized in the usual manner with globular proteins of known molecular weight before application of soluble elastin.

After gel filtration it is usually possible to recover tropoelastin with homogeneity suitable for sequencing and structural studies. In many instances we have omitted the DEAE-cellulose chromatography step without compromising purity. Table I gives the approximate recoveries of

## TABLE II
### AMINO ACID COMPOSITION OF TROPOELASTIN FROM COPPER-DEFICIENT CHICK AORTA

| Amino acid | Tropoelastin[a] |
|---|---|
| Lys | 42 |
| His | 0 |
| Arg | 6 |
| Hyp | 8 |
| Asp | 5 |
| Thr | 10 |
| Ser | 8 |
| Glu | 13 |
| Pro | 127 |
| Gly | 335 |
| Ala | 175 |
| ½-Cys | 0 |
| Val | 177 |
| Met | 0 |
| Ile | 18 |
| Leu | 54 |
| Tyr | 11 |
| Phe | 30 |

[a] Expressed as moles per 1000 mol of amino acid residues.

tropoelastin at the various steps outlined above. These estimates are based in part on assessment of gel electrophoretic patterns or estimates using radioimmunoassay techniques.[13] The amino acid analysis of the final product (Table II) should not contain in excess of 5 residues per 1000 total amino acid residues of aspartic acid. No histidine, cystine, or methionine or carbohydrate-derived residues should be detected. As noted above in sodium dodecyl sulfate–polyacrylamide gels similar in composition to those described by Furthmayer and Timpl,[12] tropoelastin migrates more slowly than bovine serum albumin with a molecular weight of 72,000; however, in systems similar to that described by Laemmli,[14] tropoelastin migrates nearer to 62,000–64,000 daltons.

### Other Comments

The best yields are obtained when the above steps are carried out rapidly. Isolations that have been carried out over a 7- to 8-day period

[13] J. A. Foster, C. B. Rich, S. Chen Wu, and M.-T. Marzullo, this volume [44].
[14] U. K. Laemmli, *Nature (London)* **227**, 680 (1970).

have resulted in increased recovery of the smaller tropoelastin-derived peptides. It is clear that activity from a contamining proteinase or proteinases persists throughout the early steps of isolation (cf. Mecham and Foster[7]). For example, if the DEAE-cellulose chromatography step is omitted, incubation of tropoelastin in physiological buffers without inhibitors will result in proteolysis (fraction obtained after gel filtration). The final product is always best stored in the lyophilized state. The addition of 1 mg of α-proteinase inhibitor from bovine or human serum per 200 mg of tropoelastin prior to lyophilization has been used to protect against proteolysis during storage and subsequent dissolution. This inhibitor may be removed by precipitating tropoelastin by addition of NaCl to 15% (w/v).[11]

## [37] Production and Isolation of Soluble Elastin from Copper-Deficient Swine

*By* L. B. SANDBERG and T. B. WOLT

Introduction and Rationale

Soluble elastin or tropoelastin has been a known entity since 1969. It was first described as a protein present in the arteries of copper-deficient swine.[1,2] Copper deficiency apparently renders the enzyme system lysyl oxidase ineffectual so that the normal cross-linking process that leads to the formation of the elastic fiber does not take place. The protein thus accumulates as an extracellular soluble material in the media of the blood vessels of these animals. They have weak arteries and eventually will succumb to a ruptured aortic aneurysm or a myocardial infarction. In order to induce a sufficient degree of copper deficiency, the experimental animal must grow relatively rapidly and be able to subsist on a diet composed largely of milk products, milk being a low-copper food. The pig ideally fits these criteria. The yields of tropoelastin from these animals is 0.1% of the aortic wet weight, that is a 15-kg pig will possess a 15-g aorta, which will yield 150 mg of tropoelatin. By far, the most difficult aspect of producing this protein is the raising of the copper-deficient swine. Therefore, we describe this in careful detail below; then follows the technique of protein isolation. The isolation scheme takes advantage of the extremely hydrophobic nature of the protein, which renders it soluble in low molecular weight alcohols. Thus, the technique centers on an alcohol

[1] D. W. Smith, N. Weissman, and W. H. Carnes, *Biochem. Biophys. Res. Commun.* **31**, 309 (1968).
[2] L. B. Sandberg, N. Weissman, and D. W. Smith, *Biochemistry* **8**, 2940 (1969).

fractionation utilizing a mixture of n-propanol and n-butanol. Extensive use is made of enzyme inhibitors during the isolation, because of a proteolytic component that isolates with tropoelastin and appears to be tightly bound to it. The repeated addition of inhibitors is necessary to inactivate the protease as it is slowly released from the substrate.

Procedures

*Raising of Copper-Deficient Swine*

Piglets 3–7 days old are obtained from local farmers. Usually, whole litters of animals are purchased (7–10 animals). Pigs can be weaned very early, so that with care the mortality rate of even a 3-day liter is very low. It is important that the age of the litter be well documented, because animals more than 14 days old are difficult to render sufficiently copper deficient. The piglets are sensitive to temperature change, therefore a transfer box is utilized to bring them from the farm into our swine facility. The facility consists of 1 by 1.5 meter pens with either galvanized steel or concrete walls approximately 1.2 meters high. Floors consist of 2.5-cm galvanized steel mesh positioned 20–30 cm above a concrete floor. Temperature control is critical during the entire experiment (20 ± 2°). Two piglets are placed together in a pen when they are very young, but later, at 1–2 months of age, they are placed in individual pens. The pens are washed three times daily to prevent coprophagia, and pigs are sprayed with deionized water twice a day. For the first 30 days, the piglets sleep on a blanket with a heat lamp and are protected underfoot by a fine expanded stainless steel mesh screen placed over the galvanized steel floor. The entire swine care facility is disinfected with a soap and water spray followed by steam cleaning prior to the arrival of a new brood. The most important aspect of care is that the animal handler be a responsible person who recognizes the need for keeping the facility completely free of copper. Copper is easily introduced by vermin, by manure or dirt tracked in on shoes, and by dust, as might result from handling of ground feeds. The animal handler must become the surrogate mother for the piglets during the first few weeks while they are adapting to their new facility. We have observed that the more intimate this relationship, the better the degree of copper deficiency as well as the general health of the animal. This aspect of care cannot be overemphasized. It is the single most important factor in completing a successful experiment.

Copper-deficient swine are fed entirely on reconstituted tinned evaporated milk. Diets are mixed in the proportions of 300 ml of evaporated milk to 400 ml of deionized water containing 7% dextrose. Diets for the young animals are not controlled in amount. The animals are fed ad libitum five times a day during the first month; during the last week of this period, sulfide water (0.234 m$M$ sodium sulfide in deionized water) gradually replaces the 7% dextrose solution. The entire feeding procedure is described in detail in an earlier publication.[3] To assist the prevention of respiratory and bowel infection, 5 mg of gentamicin sulfate in solution are mixed with each milk feeding for the first 30 days of life. Also during the first month, the pigs are started on intramuscular iron injections given as the iron dextran complex (Pigdex, American Cyanamid Company). Fifty milligrams of elemental iron (0.5 ml) is given weekly for the first 2 weeks in alternate hams. After that, 100 mg of iron are given weekly until a total of 1.0 g has been administered to each animal.

After 1 month of age, the pigs receive a strict sulfide milk diet at the level of 230 ml/kg per day divided into two equal feedings. This diet provides all the nourishment and fluid that the animal requires. No minerals or vitamins are added to the milk. Should the older animals develop diarrhea, this is adequately controlled by again adding 1 ml of gentamicin to each of the milk feedings.

*Monitoring Progress of the Copper Deficiency*

The second most essential item, in addition to a conscientious animal handler, is adequate monitoring of the clinical status of the animals with respect both to their general health and to the level of their copper deficiency. We monitor plasma copper levels by drawing carotid artery blood into heparinized tubes and evaluating the resultant plasma by atomic absorption spectrometry. The method of copper determination is according to Parker *et al.*[4] with the exception that the final trichloroacetic acid (TCA) concentration is 5% instead of 4% and the samples are boiled for 15 min prior to centrifugation. Copper standards are made in concentrations to bracket the expected sample concentration and are diluted with 5% TCA. Although some laboratories have advocated whole-blood copper evaluation, we find little advantage in this. Results are essentially identical to the plasma levels and involve considerably greater work because of the necessity of nitric acid digestion. Copper determinations are best carried out on fresh plasma. We find that results vary considerably if the separated plasma is allowed to stand even for several hours in the cold. Copper determinations are begun at approximately 8 weeks of age (see Fig. 1) and are carried out weekly along with weight evaluations and blood hematocrit levels. Weights of the copper-deficient animals generally plateau at 16

[3] L. B. Sandberg, *Int. Rev. Connect. Tissue Res.* 7, 159 (1976).
[4] M. M. Parker, F. L. Humoller, and D. J. Mahler, *Clin. Chem.* 13, 40 (1967).

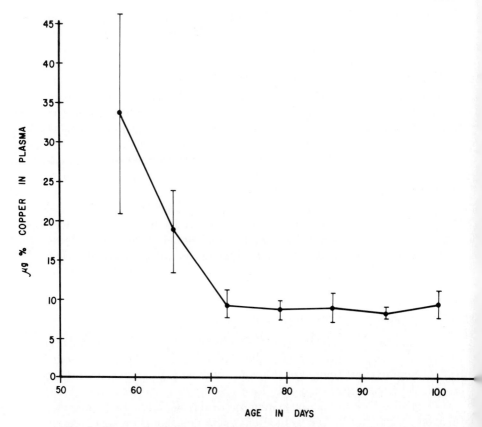

FIG. 1. Changes in plasma copper levels over a period of 6 weeks from pigs made deficient by the low-copper diet. The data show mean values (plus or minus one standard deviation) for a litter of eight animals. Copper levels were evaluated using atomic absorption spectrometry of trichloroacetic acid-treated plasma.

weeks of age. Those of control animals continue to rise and are approximately double those of the copper-deficient pigs at this age. Control diets are made by adding 0.5 mg/kg per day of elemental copper in the form of copper sulfate to the sulfide milk just prior to feeding. Hematocrit levels generally parallel the fall in copper levels and are a good check on the accuracy of the copper determinations and the degree of copper deficiency. When an animal's hematocrit drops below 20% packed cell volume, it should be considered critical. With hematocrits less than 10% the animals usually survive only a few days. Therefore, weekly records and graphs should be kept of these three parameters: weight, hematocrit, and plasma copper levels. Copper levels of less than 10 μg/100 ml are essential for obtaining reasonable yields of tropoelastin. This point cannot be overemphasized. There seems to be a threshold at the 10–12 µg/100 ml level for tropoelastin production. Also, below this level the animals appear to be susceptible to a number of other abnormalities, such as myocardial infarction, kidney infarction, lung hemorrhage, and respiratory infection. Therefore, the clinical status must be monitored closely.

*Harvesting of Aortas*

The swine are killed at 15–17 weeks of age by the intraarterial injection of pentobarbital. This is best given immediately after blood is drawn for the copper and hematocrit determination, leaving the needle in the artery and simply switching syringes. At the time of sacrifice, animals weigh 12–15 kg and tolerate the carotid punctures with relative ease because of the weekly blood samplings. The thoracic and abdominal cavities are immediately opened by a midline incision, and the abdominal vena cava is incised just below the liver to exsanguinate the animal. The viscera are removed en bloc, and the entire aorta is dissected free from the accompanying organs. Adventitial tissue is removed, and, after a dorsal longitudinal incision, the aorta is briefly rinsed in cold tap water. The aortas are stored frozen at $-70°$ until needed for tropoelastin extraction. They can be kept in this fashion, if air is excluded, for a number of years.

*Isolation of Tropoelastin*

The aorta is thawed and cut transversely into 0.5-cm strips. The strips are frozen in liquid nitrogen and milled through a 20-mesh screen using a Wiley mill equipped with a Dry Ice chamber surrounding the blade housing. All of the following steps are carried out at 0–4° with the exception of the flash evaporation. The milled frozen tissue is homogenized for 3 min at high speed in a conventional Waring blender using 150 ml of buffer containing 0.5 m$M$ ammonium acetate, 25 m$M$ EDTA, 5 m$M$ $N$-ethylmaleimide, and 1 m$M$ diisopropylfluorophosphate (DFP), pH 7.0. The homogenate is centrifuged at 30,000 $g$ for 1 hr; 1 m$M$ DFP is added to the resultant supernatant at hourly intervals, while the pulp is reextracted. The pulp is rehomogenized as before and shaken for 2 hr on a wrist-action shaker. After centrifugation, the supernatants are combined; 1 m$M$ DFP is added, and the preparation is filtered. Ammonium sulfate is added to 45% saturation (28 g/100 ml), and 1 m$M$ DFP is again added. After standing overnight, the ammonium sulfate precipitate is resuspended in 0.1 M ammonium formate containing 1 m$M$ DFP and dialyzed against 0.1 $M$ ammonium formate overnight. In the morning, the tenate is again treated with DFP, and the molarity is brought to 0.5 $M$ ammonium formate. One

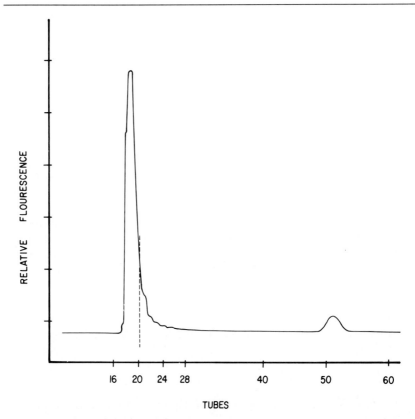

FIG. 2. Elution pattern of tropoelastin, previously isolated according to the procedure described in the text and eluted from a Sephadex G-75 column, 60 cm by 0.9 cm, using a 0.05 $M$ pyridine acetate buffer, pH 4.9, with a flow rate of 11 ml/hr, collecting 0.7 ml per tube. The dashed line represents the cut made to remove the smaller molecular weight material from the major peak. With a good preparation, these components represent less than 5% of the total material.

and one-half volumes of $n$-propanol are added dropwise while stirring. This is followed by 2.5 volumes of $n$-butanol similarly added. A precipitate results that is filtered out with a large-pore paper (S S No. 588), and the alcoholic protein solution is flash evaporated to near dryness on a rotary flash evaporator at 30°. The evaporator is connected to a full vacuum, and the condensing coil is chilled with a $-20°$ cooling bath. The condensate is frozen with Dry Ice. Lipids are removed from the damp protein residue with two 30–40 ml extractions by chloroform, again utilizing the large-pore paper for filtration. The residue is dried by a gentle stream of nitrogen, redissolved in 0.02 $N$ formic acid containing 1 m$M$ DFP, and dialyzed against 4 liters of 0.02 $N$ formic acid overnight. Before

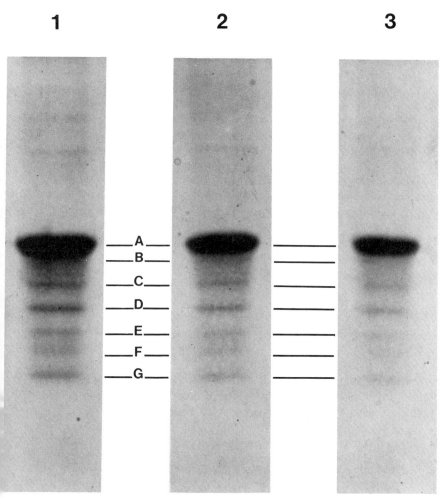

FIG. 3. Evaluation for purity and homogeneity of tropoelastin using 12.5% sodium dodecyl sulfate acrylamide gels according to the method of O'Farrell and Gold.[5] The loading for wells 1, 2, and 3 were 25, 15, and 5 µg, respectively. Band A is the homogeneous protein of molecular weight 68,000. Bands B through G are the proteolytic degradation products derived from the whole molecule. They range in molecular weight from 60,000 to 30,000.

the solution is lyophilized, any remaining particulate material is removed by centrifugation at 30,000 g for 1 hr. The soluble elastin may be further purified by gel filtration over Sephadex G-75, eluting with 0.01 M pyridine acetate, pH 4.9, as shown in Fig. 2. This serves to rid the preparation completely of salt as well as some of the lower molecular weight breakdown products of soluble elastin that are carried through with this isola-

RESULTS OF AMINO ACID ANALYSES IN RESIDUES PER 1000 RESIDUES
OF 11 PORCINE TROPOELASTIN SAMPLES PREPARED ACCORDING TO THE
METHOD DESCRIBED IN THE TEXT

| Amino acid | Mean value | Standard deviation | Coefficient of variation | Residues per molecule[a] |
|---|---|---|---|---|
| Lys | 46.6 | 1.5 | 3.1 | 37 |
| Arg | 4.8 | 0.4 | 7.4 | 4 |
| Hyp | 9.3 | 0.3 | 3.2 | 7 |
| Asp | 3.5 | 0.7 | 18.9 | 3 |
| Thr | 13.1 | 0.5 | 4.0 | 11 |
| Ser | 9.6 | 0.5 | 5.4 | 8 |
| Glu | 14.5 | 1.1 | 7.7 | 12 |
| Pro | 113.9 | 2.6 | 2.3 | 91 |
| Gly | 306.2 | 6.6 | 2.2 | 245 |
| Ala | 233.1 | 3.9 | 1.7 | 187 |
| Val | 129.2 | 3.9 | 3.0 | 103 |
| Ile | 17.3 | 0.7 | 4.0 | 14 |
| Leu | 51.0 | 1.4 | 2.8 | 41 |
| Tyr | 16.8 | 0.5 | 3.1 | 14 |
| Phe | 30.4 | 0.9 | 3.0 | 24 |

[a] Based on an estimated size of 800 residues.

tion procedure. The constant addition of the DFP assists in minimizing these fragments, but a small amount of them is unavoidable, as shown in the polyacrylamide electrophoresis patterns of Fig. 3.

*Evaluation of Tropoelastin Purity*

We utilize two criteria, that of amino acid analysis and acrylamide gel electrophoresis. The latter method gives us the most information and is routinely applied to all samples. The gel patterns shown in Fig. 3 demonstrate a product of high purity by amino acid analysis[5] and reasonable homogeneity with respect to molecular weight. The low molecular weight breakdown products can be further removed by gel permeation, on Sephadex G-75 as described above (Fig. 2). The table shows an acceptable amino acid analysis derived from 11 samples purified by the above isolation techniques. Previous analyses of ours[6] showed a considerably higher level of glycine and hydroxyproline, which we now feel were due to collagen contamination. Important considerations for a pure tropoelas-

[5] P. Z. O'Farrell and L. M. Gold, *J. Biol. Chem.* **248**, 5499 (1973).
[6] L. B. Sandberg, R. D. Zeikus, and I. M. Coltrain, *Biochim. Biophys. Acta* **236**, 542 (1971).

tin sample are as follows: absence of cystine, methionine, histidine, and desmosines; low hydroxyproline, aspartate, and glutamate levels; threonine greater than serine; a valine level approaching 130 residues per 1000 residues; and a isoleucine : leucine ratio of 1 : 3.

Acknowledgments

This research was supported by NHLBI Grants HL11963 and HL22446 as well as funds from the Department of Pathology, University of Utah Medical Center. The expert technical assistance and care of Mr. Lewis Corlett of the Department of Vivarial Science is deeply appreciated. Ms. Helen Ashenbrucker of the Department of Internal Medicine faithfully performs our copper analyses.

## [38] Isolation of Soluble Elastin-Lathyrism

By CELESTE B. RICH and JUDITH ANN FOSTER

Studies aimed at examining elastin structure and biosynthesis necessitate sufficient amounts of purified and chemically characterized tropoelastin to use as a standard. Clearly, an efficient and relatively inexpensive method for obtaining tropoelastin is desirable. The present report describes the development and use of experimentally induced lathyrism[1] as a means of obtaining tropoelastin from a variety of different elastic tissues. In the following paragraphs, the background and rationale for inducing lathyrism in chicks and pigs are described.

Lysyl oxidase is the enzyme responsible for catalyzing the oxidative deamination of lysyl residues,[2] which is a key step in the insolubilization of elastin via the formation of cross-links, such as desmosine and isodesmosine.[3] Inhibition of lysyl oxidase allows for the accumulation and subsequent isolation of appreciable quantities of soluble elastin from elastin-rich tissues. Lysyl oxidase can be inhibited by rendering animals copper-deficient, since lysyl oxidase requires $Cu^{2+}$ for activity.[4] However, copper deficiency requires special cages, water, and diet.[5–7] Since

[1] B. J. Geiger, H. Steinback, and H. T. Parsons, *J. Nutr.* **6**, 427 (1933).
[2] S. R. Pinnell and G. R. Martin, *Proc. Natl. Acad. Sci. U.S.A.* **61**, 708 (1968).
[3] S. M. Partridge, *Adv. Protein Chem.* **17**, 227 (1962).
[4] E. D. Harris, W. A. Gonnerman, J. E. Savage, and B. L. O'Dell, *Biochim. Biophys. Acta* **341**, 332 (1974).
[5] L. B. Sandberg, N. Weissman, and D. W. Smith, *Biochemistry* **8**, 2940 (1969).
[6] D. W. Smith, D. M. Brown, and W. H. Carnes, *J. Biol. Chem.* **247**, 2427 (1972).
[7] R. B. Rucker and W. Goettlich-Rieman, *J. Nutr.* **102**, 563 (1972).

lathyrogens are also known to inhibit lysyl oxidase,[8,9] an alternative to copper-deficiency is the administration of a lathyrogen directly to the normal diet of an animal.[10,11]

One commonly used lathyrogen is $\beta$-aminopropionitrile.[11,12] Sykes and Partridge[13] and Foster et al.[14] have reported on the isolation of non-crosslinked, soluble elastin from the aortas of chicks made lathyritic by administration of $\beta$-aminopropionitrile. However, the yield, based on starting wet weight of soluble elastin from lathyritic chick aortas[14] was considerably less than the yield from copper-deficient chick aortas.[15] Therefore, copper deficiency is either a more efficient inhibitor of lysyl oxidase or $\beta$-aminopropionitrile–fumarate may have other biological actions beside inhibition of lysyl oxidase. This possibility was pursued.

In summary, we have made the following observations in our studies dealing with lathyritic chicks:

1. The amount of fumarate present in commercially available $\beta$-aminopropionitrile–fumarate preparations varies, and this variability in different preparations significantly affects the yield of tropoelastin[16]; i.e., the less fumarate, the higher the yield of tropoelastin and the lower the rate of mortality.
2. The lathyrogens $\beta$-aminopropionitrile and $\alpha$-aminoacetonitrile are both competitive inhibitors of trypsin[16] (see Fig. 1) and also both inhibit the trypsin-like protease[17] associated with tropoelastin preparations[16] (see Fig. 2).
3. Administration of both $\alpha$-aminoacetonitrile-HCl (the more potent competitive inhibitor of trypsin-like enzymes) and $\epsilon$-aminocaproic acid (also a competitive inhibitor of trypsin-like enzymes)[17] to the diet of chicks beginning on day 8 after hatching results in maximum yields of tropoelastin and minimum mortality rates.[18]

[8] R. C. Siegal, S. R. Pinnell, and G. R. Martin, *Biochemistry* **9**, 4486 (1970).
[9] A. S. Narayanan, R. C. Siegel, and G. R. Martin, *Biochem. Biophys. Res. Commun.* **46**, 745 (1972).
[10] I. V. Ponseti and W. A. Baird, *Am. J. Pathol.* **28**, 1059 (1952).
[11] R. C. Page and E. P. Benditt, *Lab. Invest.* **26**, 22 (1972).
[12] E. E. Peacock and J. W. Madden, *Surgery* **66**, 215 (1969).
[13] B. C. Sykes and S. M. Partridge, *Biochem. J.* **130**, 1171 (1972).
[14] J. A. Foster, R. Shapiro, P. Voynow, G. Crombie, B. Faris, and C. Franzblau, *Biochemistry* **14**, 5343 (1975).
[15] R. B. Rucker, W. Goettlich-Riemann, K. Tom, M. Chen, J. Poaster, and S. Kouner, *J. Nutr.* **105**, 46 (1967).
[16] J. A. Foster, C. B. Rich, N. Berglund, S. Huber, R. P. Mecham, and G. Lange, *Biochim. Biophys. Acta* **587**, 477 (1979).
[17] R. P. Mecham and J. A. Foster, *Biochemistry* **16**, 3825 (1977).
[18] J. A. Foster, C. B. Rich, M. D. DeSa, A. J. Jackson, and S. Fletcher, *Anal. Biochem.* **108**, 233 (1980).

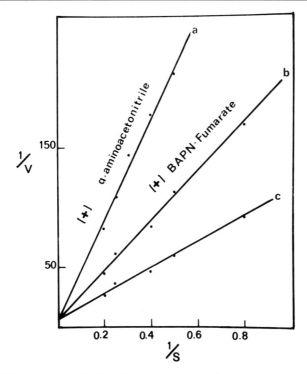

FIG. 1. Lineweaver–Burk plot of trypsin vs azocasein in the presence of 100 m$M$ β-aminopropionitrile–fumarate (b) and α-aminoacetonitrile-HCl (a). c, Normal curve without addition of lathyrogen. From Foster et al.[16]

Induction of Lathyrism

Chickens. The methodology described herein represents optimum conditions for isolating tropoelastin from chick tissues.[18] Two hundred newly hatched chicks are fed normal diets of commercial starting feed for 7–8 days. The chicks are then placed on a diet supplemented with 0.1% (w/w) α-aminoacetonitrile-HCl and 0.05% ε-aminocaproic acid. Both the feed and water are changed twice a day. In our experience the chicks generally show definite symptoms of lathyrism (the most obvious being the inability to move about the cages) within 5–8 days after initiation of the lathyrogen feeding. At this point the chicks are sacrificed using prolonged exposure to an anesthetic, i.e., ether. This is most conveniently done in large desiccators placed in a well ventilated hood wherein a large number of chicks can be handled at once.

Piglets. Piglets, just weaned (about 6 weeks of age), are fed ad libitum diets of commercial starting mash with 0.1% (w/w) α-aminoaceto-

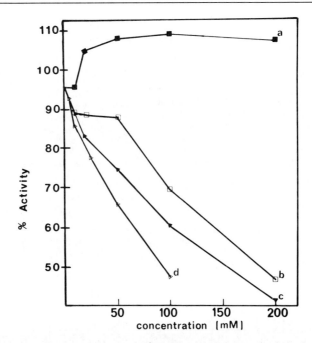

FIG. 2. The activity of tropoelastase vs [$^3$H]valine-labeled tropoelastin in the presence of (a) fumarate; (b) β-aminopropionitrile–fumarate; (c) ε-aminocaproic acid; and (d) β-aminopropionitrile. The concentration of fumarate used was half of that given for the other three compounds. From Foster et al.[16]

nitrile-HCl and 0.05% (w/w) ε-aminocaproic acid.[19] After 9 days, when the animals are definitely showing signs of lathyrism, they are sacrificed. Normally, two piglets can be used for the isolation of sufficient amounts of aortic and ear cartilage tropoelastin.[19]

Isolation of Tropoelastin

The isolation procedure is essentially the same as described by Foster et al.[14] Briefly, aortas are excised and immersed immediately in an ice cold solution of 0.1 M iodoacetamide, 0.01 M chloroquine, 0.002 M disodium ethylenediaminetetraacetate, and 0.001 M phenylmethylsulfonyl fluoride. After the removal of all organs, the tissue is homogenized in 10 volumes (w/v) of neutral salt buffer solution consisting of 0.5 M sodium chloride, 0.02 M sodium phosphate (pH 7.4), 0.1 M N-ethylmaleimide,

[19] J. A. Foster, C. B. Rich, and M. D. DeSa, *Biochim. Biophys. Acta* **626**, 383 (1980).

0.01 $M$ chloroquine, and 0.002 $M$ disodium ethylenediaminetetraacetate. After homogenization, 0.001 $M$ diisopropylfluorophosphate is added. The mixture is extracted overnight and then centrifuged at 16,000 $g$ for 1 hr. The supernatant is filtered through Whatman No. 1 filter paper to remove lipid material, and the salt-soluble collagen is acid-precipitated by adjusting the pH to 4.0 using 4 $N$ acetic acid. The suspension is allowed to sit for 2 hr and then centrifuged at 10,000 $g$ for 1 hr. The resulting supernatant is neutralized with 2 $N$ sodium hydroxide, and sodium chloride is added to 15% (w/v) to precipitate the soluble elastin. The resulting precipitate is redissolved in the neutral salt buffer, and DFP is added as above. A second salt (15%) precipitation is performed. This pellet is then dissolved in a water solution of 0.1 $M$ $N$-ethylmaleimide and 0.001 $M$ DFP. Insoluble material is removed by centrifugation (10,000 $g$), and the supernatant is made 3% in sodium chloride (w/v). A dropwise addition of 4 $N$ acetic acid is then done with very gentle agitation until the pH (by pH indicator paper) is 3-4 and no more precipitation occurs. The precipitated material is removed by centrifugation at 50,000 $g$ for 1 hr, and the supernatant is dialyzed extensively against water and then lyophilized.

This procedure employed on chick aortic tissue results in a homogeneous protein of 70,000 daltons, which corresponds to the tropoelastin b species of soluble elastin.[20,21] An SDS–polyacrylamide gel of the isolated tropoelastin is shown in Fig. 3, and the amino acid composition is given in the table. The average yield of tropoelastin from aortic tissue is 40 mg per 200 chicks.

However, when this procedure is applied to chick lung tissue, another purification step is required.[22] Taking advantage of the reversible property of coacervation exhibited by soluble elastin, a lyophilized sample of lathyritic lung tropoelastin b plus contaminants is redissolved in the neutral salt buffer at a concentration of 1 mg/ml. Diisopropylfluorophosphate (0.001 $M$) is added, and the solution is incubated at 37° for 20 min. The coacervate is pelleted at 16,000 $g$ for 30 min at room temperature. The pellet is redissolved in water and dialyzed against water extensively at 4° and lyophilized. The resulting lung tropoelastin b is homogeneous with an apparent molecular weight of 70,000. The amino acid composition is given in the table. The yield of tropoelastin from lung tissue is 16 $\mu$g per gram wet weight.

When dealing with pig tissue, the aortas are excised and maintained at

[20] J. A. Foster, C. B. Rich, S. Fletcher, S. R. Karr, and A. Przybyla, *Biochemistry* **19**, 857 (1980).
[21] J. A. Foster, this volume [30].
[22] J. A. Foster, C. B. Rich, S. Fletcher, S. R. Karr, M. D. DeSa, T. Oliver, and A. Przybyla, *Biochemistry* **20**, 3528 (1981).

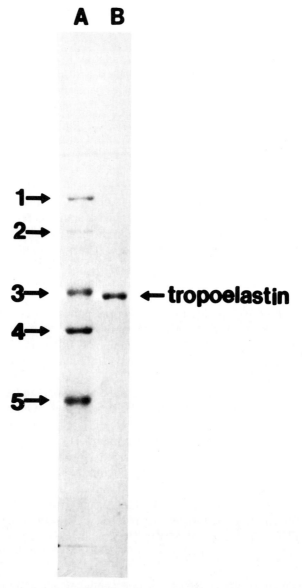

FIG. 3. Sodium dodecyl sulfate–slab gel electrophoresis of (A) molecular weight standards including β-glactosidase (130,000) (1), phosphorylase $a$ (100,000) (2), human albumin (66,000) (3), pyruvate kinase (57,000) (4), and ovalbumin (43,000) (5); (B) chick aortic tropoelastin b. From Foster et al.[18]

AMINO ACID COMPOSITIONS OF CHICK AND PIG TROPOELASTIN b[a]

| Amino acid | Chick lung[b] | Chick aorta[b] | Pig aorta[c] | Pig cartilage[c] |
|---|---|---|---|---|
| Lysine | 33 | 29 | 38 | 30 |
| Histidine | 3 | 2 | 2 | 2 |
| Arginine | 8 | 7 | 7 | 7 |
| Hydroxyproline | 10 | 11 | 10 | 20 |
| Aspartic acid | 7 | 6 | 10 | 6 |
| Threonine | 13 | 12 | 16 | 15 |
| Serine | 11 | 9 | 15 | 13 |
| Glutamic acid | 16 | 16 | 18 | 21 |
| Proline | 135 | 135 | 107 | 101 |
| Glycine | 310 | 314 | 320 | 327 |
| Alanine | 179 | 186 | 230 | 222 |
| Valine | 172 | 179 | 122 | 123 |
| Isoleucine | 18 | 19 | 15 | 18 |
| Leucine | 55 | 50 | 49 | 45 |
| Tyrosine | 11 | 9 | 16 | 13 |
| Phenylalanine | 19 | 16 | 29 | 29 |

[a] Compositions are expressed as residues per 1000 amino acid residues.
[b] From J. A. Foster, C. B. Rich, S. Fletcher, S. Karr, M. D. DeSa, T. Oliver, and A. Przybyla, *Biochemistry* **20**, 3528 (1981).
[c] J. A. Foster, C. B. Rich, and M. D. DeSa, *Biochim. Biophys. Acta* **626**, 383 (1980).

4°. The tissue must be minced finely with scissors and then homogenized and extracted in neutral salt buffer as described for lathyritic chicks. The amino acid composition of the pig aortic tropoelastin is indicated in the table.

Also in the table is the amino acid composition of tropoelastin b derived from pig auricular cartilage tissue. In order to isolate ear tropoelastin, the ears are cleaned of hair and skin. The tissue is then minced with scissors and homogenized and extracted in neutral salt buffer as described for lathyritic chick aortic tropoelastin b. However, because the sample remains contaminated with collagen after this procedure, the sample is coacervated as described for the chick lung tropoelastin b. Both the pig aortic and ear cartilage tropoelastins possess apparent molecular weights of 70,000 as determined by SDS–polyacrylamide gel electrophoresis.[19]

The yields of tropoelastin (expressed as milligrams per gram wet weight) obtained from lathyritic pig aortic and cartilage tissue were 1.8 mg and 0.6 mg, respectively. Hence, the yield of tropoelastin is comparable to that for copper-deficient piglets.[23] The induction of lathyrism is more rapid than that of obtaining a copper-deficient state, and the feeding of the lathyrogen is simple because no special diets or environmental conditions are needed. With the limited facilities that we and most laboratories have available, the lathyritic model is less expensive and not as time-consuming as the copper-deficient model.

A further purification of soluble elastin is often necessary to remove minor contaminants. Chromatography on CM-cellulose is useful for this purpose.[18] The contaminated soluble elastin preparation is chromatographed on a CM-cellulose column (1.5 × 16 cm, 0.5 ml/min) using a 0.02 $M$ sodium citrate buffer (pH 3.6) with a linear gradient from 0 to 1 $M$ sodium chloride as described by Foster et al.[18] This column is more advantageous than the usual DEAE-cellulose column in that the soluble elastin is actually fractionated on the column instead of coming through in the void volume.[17,24,25] Therefore, it can be separated from both neutral and basic proteins including collagen. On a chromatogram of the CM-cellulose column two major protein peaks were seen. Peak 1, which comes right through in the void volume, is a serine-enriched, acidic protein that is a common contaminant of soluble elastin preparations.[18] Peak 2 is tropoelastin b, as judged by amino acid analysis and SDS-polyacrylamide gels.[18]

The previous procedures are all designed for the isolation of tropoelastin b, the 70,000 dalton species.[20,21] To obtain any tropoelastin a, the 73,000 dalton species of soluble elastin, the following procedures is necessary.

The lathyritic chick aortas are excised and placed in an ice cold wash solution consisting of 0.002 $M$ disodium ethylenediaminetetraacetate, 0.01 $M$ chloroquine, and 0.001 $M$ PMSF, but no iodoacetamide. The aortas are then homogenized in 10 volumes of a 2 $M$ urea solution with 0.01 $M$ Tris (pH 8.0), 0.01 $M$ chloroquine, and 0.002 $M$ disodium ethylenediaminetetraacetate. Nitrogen is then bubbled through the solution, and 10 mg of dithiothreitol per milliliter are added for reduction. Then 0.001 $M$ DFP is added, and the mixture is extracted for 2 hr at 4°. The solution is alkylated with 0.1 $M$ iodoacetamide and stirred overnight at 4°. The mixture is then centrifuged at 16,000 $g$ for 1 hr. The supernatant is filtered to remove lipid material. It is dialyzed extensively against water adjusted to pH 3.2 with glacial acetic acid. The resulting solution is centrifuged at 16,000 $g$ for 1 hr, and the supernatant is made 0.02 $M$ sodium phosphate and 0.5 $M$ sodium chloride. The pH is adjusted to 7.5 with 2 $N$ sodium hydroxide.

[23] L. B. Sandberg, E. Bruenger, and E. G. Cleary, *Anal. Biochem.* **64**, 249 (1975).
[24] B. C. Sykes and S. M. Partridge, *Biochem. J.* **141**, 567 (1975).
[25] P. A. Abraham and W. H. Carnes, *Anal. Biochem.* **91**, 115 (1978).

The solution is then made 15% in sodium chloride. The rest of the procedure is the same as described above for the lathyritic chick aortas isolation of tropoelastin b. The resulting product contains both tropoelastins a and b. However, in order to detect tropoelastin a, the sample must be reduced with tritiated sodium borohydride and run on SDS–polyacrylamide gels together with rabbit reticulocyte lysate and exposed to X-ray plates as described by Foster et al.[20] A sample just extracted with neutral salt buffer and reduced with tritiated sodium borohydride does not contain tropoelastin a when examined on SDS–polyacrylamide gels, only tropoelastin b, whereas the urea-extracted and reduced sample of lathyritic aortas shows both a and b tropoelastins. The amount of tropoelastin a is a very small fraction of the soluble elastin preparations from adult lathyritic chick aortas. We have not as yet developed a simple preparative procedure for purifying significant amounts of tropoelastin a.

Acknowledgments

The authors acknowledge the financial support of the National Institutes of Health Grant HL-22208 and Council for Tobacco 1179.

## [39] Characterization of Soluble Peptides of Elastin by Physical Techniques

*By* DAN W. URRY

Tropoelastin,[1,2] the single precursor protein of fibrous elastin, and chemical fragmentation products of fibrous elastin, such as α-elastin,[3,4] are soluble in water at low temperatures. With the solution pH near the isoelectric point of the polypeptide fragment or protein, raising the temperature toward the physiological range causes the solution to become cloudy. On standing of an hour or so, the suspension clears with the formation of two phases. The upper phase is called the equilibrium solution, and the lower transparent phase is called the coacervate. The process is reversible and is referred to as coacervation.

[1] D. W. Smith, N. Weissman, and W. H. Carnes, *Biochem. Biophys. Res. Commun.* **31**, 309 (1968).
[2] L. B. Sandberg, N. Weissman, and D. W. Smith, *Biochemistry* **8**, 2940 (1969).
[3] S. M. Partridge, H. F. Davis, and G. S. Adair, *Biochem. J.* **61**, 11 (1955).
[4] S. M. Partridge and H. F. Davis, *Biochem. J.* **61**, 21 (1955).

As the coacervate is the stable state in aqueous solution at physiological temperatures, as it has the approximate water content[5] of purified fibrous elastin[6] and has other similarities with fibrous elastin discussed below, coacervation of tropoelastin has been considered to be the concentrating process required for fiber formation.[5] The process results in a particular aggregated state that is expected to be altered during stretching and to be re-formed on retraction or relaxation of the fiber. Accordingly, elements of the coacervating process should be relevant to elastomeric function. It becomes fundamental, therefore, to characterize the structural elements central to this uncommon property of coacervation.

Efforts to disentangle the structure–function problem of fibrous elastin have been immeasurably assisted by the working of Sandberg and his colleagues,[7-10] who have found repeating peptide sequences in tropoelastin, and by the interesting properties that these sequential polypeptides exhibit.[5] There have been observed three repeating sequences: a hexapeptide, $Ala_1$-$Pro_2$-$Gly_3$-$Val_4$-$Gly_5$-$Val_6$ (APGVGV), which repeats 5 plus times in a single run; a pentapeptide, $Val_1$-$Pro_2$-$Gly_3$-$Val_4$-$Gly_5$ (VPGVG), which repeats 11 plus times in a continuous sequence[10]; and a tetrapeptide, $Val_1$-$Pro_2$-$Gly_3$-$Gly_4$ (VPGG), which is less striking with about 4 repeats at the amino terminus. Additionally these repeats occur elsewhere in the sequence, and related repeats with one or more conformationally isomorphous replacements are also present.

These sequential polypeptides effectively allow for a dissection of the coacervation process into composite quaternary structural elements. For related reasons these repeat peptides are particularly favorable subjects for the study of secondary and tertiary structure by nuclear magnetic resonance (NMR) methods and have provided the first direct evidence for processes previously implied by complex arguments. For example, the polytetrapeptide allowed for the definition of the specific hydrophobic associations attending an inverse temperature transition[11] and for the characterization of that transition in terms of mobility changes.[12] A temperature-elicited crystallization of a cyclic analog of the polyhexapep-

[5] D. W. Urry, *Perspect. Biol. Med.* **21**, 265 (1978).
[6] S. M. Partridge, *Biochim. Biophys. Acta* **140**, 132 (1967).
[7] W. R. Gray, L. B. Sandberg, and J. A. Foster, *Nature (London)* **246**, 461 (1973).
[8] J. A. Foster, E. Bruenger, W. R. Gray, and L. B. Sandberg, *J. Biol. Chem.* **248**, 2876 (1973).
[9] L. B. Sandberg, W. R. Gray, J. A. Foster, A. R. Torres, V. L. Alvarez, and J. Janata, *Adv. Exp. Med. Biol.* **79**, 277 (1977).
[10] L. B. Sandberg, N. T. Soskel, and J. G. Leslie, *N. Engl. J. Med.* **304**, 566 (1981).
[11] D. W. Urry, M. A. Khaled, R. S. Rapaka, and K. Okamoto, *Biochem. Biophys. Res. Commun.* **79**, 700 (1977).
[12] D. W. Urry, T. L. Trapane, and M. A. Khaled, *J. Am. Chem. Soc.* **100**, 7744 (1978).

tide, in the most unquestionable manner, demonstrates that these peptides have the structural capacity in aqueous solution to increase order on increasing temperature.[13] The cyclopentadecapeptide analog of the linear polypentapeptide has shown intermolecular hydrophobic association features and intramolecular hydrogen bonding in the aqueous crystal structure,[14] which had been previously deduced in aqueous solution for its linear conformational correlate, the elastomeric[15] polypentapeptide.[5,16-18]

As proposed in 1973, the dominant secondary structural feature occurring in all three repeat peptides of elastin[19,20] is the $Pro_2$-$Gly_3$ $\beta$-turn, which is a 10-atom hydrogen-bonded ring utilizing the C-O of residue 1 and the NH of residue 4. A subsequent survey of 29 crystal structures of proteins and peptides demonstrated that of all sequences the Pro-Gly sequence is at the highest probability for $\beta$-turn formation.[21] The $\alpha$-helix with a 38% frequency, the $\beta$-pleated sheet with a 20% frequency, and the $\beta$-turn with a 32% frequency are now recognized as the three most common conformational features of proteins.[21] With the frequency of occurrence of the Pro-Gly sequence in tropoelastin,[10] with its regular recurrence in the repeat peptides of elastin, and particularly with the polypentapeptide being elastomeric,[15] structure–function descriptions of fibrous elastin are necessarily incomplete without explicit consideration of the regular recurrence of the Pro-Gly $\beta$-turn (see Fig. 11B).

With this brief introductory perspective of soluble elastin peptides, this review of the characterization of soluble elastins by physical techniques initially will consider morphological and implied quaternary structural aspects of coacervation, then will treat more detailed secondary and tertiary structural aspects of the sequential polypeptides of elastin, to arrive in the final section at a working model for a fundamental structural unit of elastin that includes considerations of how the detailed structural elements might relate to elastomeric function.

The objective of the use of physical techniques to characterize soluble

[13] D. W. Urry, M. M. Long, and H. Sugano, *J. Biol. Chem.* **253**, 6301 (1978).
[14] W. J. Cook, H. M. Einspahr, T. L. Trapane, D. W. Urry, and C. E. Bugg, *J. Am. Chem. Soc.* **102**, 5502 (1980).
[15] D. W. Urry, K. Okamoto, R. D. Harris, C. F. Hendrix, and M. M. Long, *Biochemistry* **15**, 4083 (1976).
[16] D. W. Urry, W. D. Cunningham, and T. Ohnishi, *Biochemistry* **13**, 609 (1974).
[17] D. W. Urry and M. M. Long, *CRC Crit. Rev. Biochem.* **4**, 1 (1976).
[18] D. W. Urry and M. M. Long, *in* "Elastin and Elastic Tissue" (L. B. Sandberg, W. R. Gray, and C. Franzblau, eds.), p. 685. Plenum, New York; *Adv. Exp. Med. Biol.* **79**, 685 (1977).
[19] D. W. Urry, *Res/Dev.* **24**, 30 (1973).
[20] D. W. Urry, *in* "Arterial Mesenchyme and Arteriosclerosis" (W. D. Wagner and T. B. Clarkson, eds.), p. 211. Plenum, New York; *Adv. Exp. Med. Biol.* **43**, 211 (1974).
[21] P. Y. Chou and G. D. Fasman, *J. Mol. Biol.* **115**, 135 (1977).

elastin peptides is to arrive at a molecular description for the structure and function of fibrous elastin. The particular problem here is that these methodologies indicate the presence of preferred and describable conformations, whereas characterizations of fibrous elastin by certain physical techniques and theoretical arguments have been taken as indicating a random structure. Because of this, in the present review of methodologies the focus includes achieving a molecular description that can then be analyzed in terms of the physical properties previously considered to be indicative of randomness. It will be argued that the new types of molecular conformations derived from the physical characterizations of soluble elastins do, in fact, give rise to certain properties previously considered to be unique to random structures. This approach is deemed necessary in order that the methodologies employed can be considered without the unresolved concern of apparent contradictions.

Characterization of Temperature-Dependent Elastin Peptide Aggregation

For the heating of a protein solution to result in the formation of two phases, a solution and a more dense protein-rich phase, e.g., the precipitate, is not surprising. This occurs with denaturation, and since the product of a temperature-driven aggregation, fibrous elastin, is viewed as a random construct due to the approximate adherence of its elastomeric properties to the approximate classical theory of rubber elasticity,[22-24] it is comfortable to view coacervation of tropoelastin in a similar light. But there are important differences.

The white of an egg gets its name from becoming opaque as the result of heating, and the white of an egg does not return to a transparent solution on cooling. This denaturation, this randomization, is irreversible. To the eye the coacervation of tropoelastin or of $\alpha$-elastin is entirely and rather quickly reversible, and the coacervate, the product of heating, is transparent. Interestingly there is substantial mobility in the coacervate[18] and in the elastic fiber,[25-27] whereas temperature-denatured precipitates do not show such mobility. Clearly the coacervation process and the fibrous product are unique in some way, and the process and structures justify careful examination.

[22] C. A. J. Hoeve and P. J. Flory, *J. Am. Chem. Soc.* **80**, 6523 (1958).
[23] C. A. J. Hoeve and P. J. Flory, *Biopolymers* **13**, 677 (1974).
[24] C. A. J. Hoeve, *Adv. Exp. Med. Biol.* **79**, 607 (1977).
[25] D. A. Torchia and K. A. Piez, *J. Mol. Biol.* **76**, 419 (1973).
[26] J. R. Lyerla and D. A. Torchia, *Biochemistry* **14**, 5175 (1975).
[27] W. W. Fleming, C. E. Sullivan, and D. A. Torchia, *Biopolymers* **19**, 597 (1980).

## Temperature Profiles of Coacervation

The process of aggregation can be characterized by following the change in turbidity with increasing temperature. This is called a temperature profile for turbidity formation, or a TPτ. When the process is readily reversible by reversing the temperature scan direction after achieving maximal turbidity, the curve is referred to as a temperature profile for coacervation or a TPC. Common forward and backward scan rates are 30°/hr, and the sample cuvette is sonicated at high frequency to delay settling and to facilitate the attaining of equilibrium.

## Tropoelastin and α-Elastin

Using a 20 mg/ml solution of tropoelastin (courtesy of R. B. Rucker), the TPC is seen in Fig. 1A. The TPC for α-elastin is given in Fig. 1B, where in addition the concentration is varied. With increasing concentration, the process initiates at lower temperatures and the steepness of the transition increases. The aggregational process exhibits a positive cooperativity. While this is not unexpected, the effect of high concentration could have been one where initial aggregates stabilized the original structure and delayed the onset of the temperature-elicited process. Instead, the intermolecular interactions facilitate the structural change in this binary coacervation involving only peptide and water. Our concern is to characterize the product and to identify the intermolecular forces.

Bungenberg de Jong[28] has defined simple and complex coacervation. The latter involves "salt bonds" whereas the former is "concerned with non-ionized groups in the molecule." As tropoelastin and α-elastin contain ionized groups, the question as to whether these are essential to this coacervation was largely resolved by having formylated the free amino groups and methylated the free carboxyl groups of α-elastin, but the identification of the required groups can be best achieved by considering synthetic high polymers of the repeat peptides of elastin. Indeed all three high polymers exhibit concentration-dependent TPτs as shown in Fig. 1C, D, and E.

## The Polypentapeptide[5] and the Polytetrapeptide[29]

The TPC of the N-formyl-O-methyl polypentapeptide (PPP) is of particular interest because it is entirely soluble below 25°, and for concentra-

---

[28] H. G. Bungenberg de Jong, in "Colloid Science" (H. R. Kruyt, ed.), Vol. 2, p. 232. Elsevier, Amsterdam, 1949.
[29] D. W. Urry, in "Frontiers of Matrix Biology" (L. Robert, ed.), Vol. 8, p. 78. Karger, Basel, 1980.

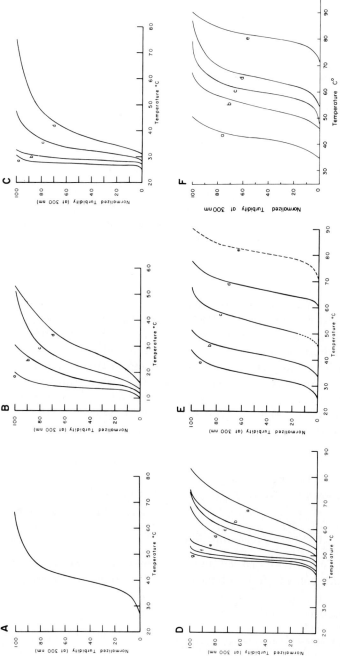

FIG. 1. Temperature profiles for turbidity formation and concentration dependence of coacervation. (A) Chick aortic tropoelastin coacervation (20 mg/ml). (B) $N$-Formyl-$O$-methyl-$\alpha$-elastin coacervation; concentration (mg/ml): a, 40; b, 10; c, 3.9; d, 1.0. (C) Polypentapeptide HCO-(Val$_1$-Pro$_2$-Gly$_3$-Val$_4$-Gly$_5$)$_n$-Val-OMe coacervation; concentration (mg/ml): a, 40–10; b, 4.9; c, 0.98; d, 0.25. (D) Polytetrapeptide H-(Val$_1$-Pro$_2$-Gly$_3$-Gly$_4$)$_n$-Val-OMe coacervation; concentration (mg/ml): a, 0.7; b, 1.3; c, 2.7; d, 5.4; e, 10.9; f, 21.5; g, 43.0. (E) Polyhexapeptide HCO-Val-(Ala$_1$-Pro$_2$-Gly$_3$-$\phi_4$-Gly$_5$-Val$_6$)$_n$-OMe ($\phi_4$ = Val; Cbz-Lys$_4$, 4:1; $n \approx 9$) turbidity formation; concentration (mg/ml): a, 39; b, 19.5; c, 9.8; d, 4.9; e, 60 ($\phi$ = Val; $n \geq 7$). (F) Turbidity formation: a, polyhexapeptide with lysine, 20 mg/ml; b, polyhexapeptide with lysine, 10 mg/ml; c, cyclodecapeptide, 15 mg/ml; d, cyclodecapeptide, 10 mg/ml; e, polyhexapeptide, 60 mg/ml.

The turbidity was followed at 300 nm in a Cary 14 spectrophotometer. The plots are given with respect to maximum turbidity as 100.

tions of 1 mg/ml and higher the transition is essentially complete by 40°, as seen in Fig. 1C. Also the effect of increasing the concentration is primarily to sharpen the transition with only limited translation to lower temperatures. The process is entirely reversible and occurs just below physiological temperatures. Interest in the PPP is further enhanced because, on cross-linking, it spontaneously forms anisotropic fibers that are elastomeric and the cross-linked product can, depending on water content, exhibit the same elastic modulus as fibrous elastin.[15]

The polytetrapeptide (PTP) also exhibits concentration-dependent coacervation, but this occurs at a higher temperature with initiation, depending on concentration, varying from 45° to 55°.

Since the coacervation process of PPP and PTP occurs without charged groups and with only hydrophobic side chains the intermolecular interactions can only involve hydrophobic bonding and possibly backbone hydrogen bonding. To date these are the simplest known coacervatable polypeptides. Since the addition of a Val residue in going from $(VPGG)_n$ to $(VPGVG)_n$ lowers the temperature at which coacervation occurs, hydrophobic interactions are implicated. Also, since it has been demonstrated that $(APGG)_n$ and $(APGVG)_n$ do not coacervate[30] (the former remains in solution and the latter precipitates), this suggests a key role for the Val-Pro sequence.

*The Polyhexapeptide*

The polyhexapeptide (PHP) exhibits a temperature profile for turbidity formation (TP$\tau$), but the polymer precipitates irreversibly and the temperature for the transition depends on the polymer length in a complex way for low values of $n$, between 3 and 7. The temperature-driven irreversible precipitation appears to be a kinetic problem, however, since the sample can be dissolved by addition of trifluoroethanol, lyophilized, and redissolved in water for a duplicate TP$\tau$. Furthermore, if the PHP sequence is disrupted by the random introduction of an $\epsilon$-carbobenzoxy-Lys residue in place of Val$_4$ for one out of every five repeats, the process becomes slowly reversible, though the dense phase is still not a transparent coacervate.[5,31]

The TP$\tau$ of the PHP containing Cbz-Lys exhibits an interesting concentration dependence (see Fig. 1E). On increasing the temperature the aggregational process shifts to lower temperatures with little or no increase in the steepness of the transition. As seen in Fig. 1F, the cyclododecapeptide analog of PHP exhibits similar TP$\tau$s. The concentration-dependent interactions do not predispose the interacted

[30] R. S. Rapaka, K. Okamoto, and D. W. Urry, *Int. J. Pept. Protein Res.* **12**, 81 (1978).
[31] R. S. Rapaka, K. Okamoto, and D. W. Urry, *Int. J. Pept. Protein Res.* **11**, 109 (1978).

polymers to a more facile subsequent interaction with additional polypeptide as might occur with an aggregation-induced conformational change. One possible explanation of these results is that the PHP forms a relatively rigid structure with a substantial barrier for the aggregational process and a very stable interacted state. This becomes relevant to the results of the secondary structural studies on the PHP.

The physical properties of the PHP are very different from those of PPP or PTP. Rather than being elastomeric, the PHP can be formed into cellophane-like sheets that can be torn and can be seen to have sharp edges in the light microscope.[31] This is quite distinct from the sticky elastomeric PPP and suggests different roles for the PHP and PPP in fibrous elastin structure and function.

Structure of the Elastin Peptide Aggregates

In the preceding section heating was seen to cause elastin peptides to aggregate with interesting and distinguishable characteristics. In this section our concern is with the structure of the aggregates. Are the aggregates isotropic as would occur by the coalescence of amorphous spheres (much as oil droplets coalesce), or as would be the case for a completely random tangle of chains (as fibrous elastin has been described[23]), or as would occur with particular quaternary structures such as the three-dimensional packing of spheres of the corpuscular model of elastin[32]? Alternatively, are the aggregates anisotropic as has also been reported for the quaternary structure of fibrous elastin[33,34]? Additionally in this section the question of the dominant intermolecular interactions will be largely answered.

*Electron Microscopy of Negatively Stained Aggregates*

A droplet of a turbid solution of Fig. 1 can be placed on a carbon-coated grid, negatively stained with uranyl acetate/oxalic acid (pH 6.2), dried, and examined in a high-resolution electron microscope. The uranyl oxalate solution wets the polypeptide subunits, filling the spaces between and around the subunits, and on drying leaves an electron-dense cast that outlines the polypeptide and protein subunits. In the case of tropoelastin coacervates (Fig. 2A), the protein subunits are seen to have aligned in the formation of parallel filaments.[35] In Fig. 2B, parallel aligned filaments are seen for α-elastin coacervates.[36] Also included as insets in Figs. 2A and B

[32] S. M. Partridge, *Symp. Fibrous Proteins* [*Pop.*] *1967*, p. 246 (1968).
[33] L. Gotte, M. G. Giro, D. Volpin, and R. W. Horne, *J. Ultrastruct. Res.* **46**, 23 (1974).
[34] L. Gotte, D. Volpin, R. W. Horne, and M. Mammi, *Micron* **7**, 95 (1976).
[35] B. A. Cox, B. C. Starcher, and D. W. Urry, *J. Biol. Chem.* **249**, 997 (1974).
[36] B. A. Cox, B. C. Starcher, and D. W. Urry, *Biochim. Biophys. Acta* **317**, 209 (1973).

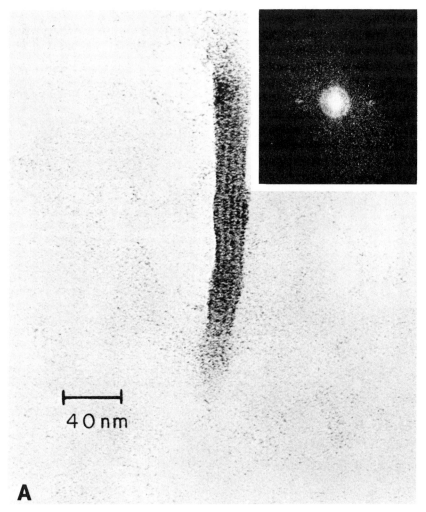

FIG. 2. Electron micrographs of negatively stained elastin coacervates and another heat-derived aggregate, and a light micrograph of heat-driven crystallization. Also included are optical diffraction patterns of the electron micrographs. (A) Tropoelastin coacervate (From reference 35) and optical diffraction pattern (from Volpin et al.[37]). (B) α-Elastin coacervate (from Cox et al.[36]) and optical diffraction pattern (from Volpin et al.[37]). (C) Polypentapeptide coacervate (from Urry[5]) and optical diffraction pattern (from Volpin et al.[40]). (D) Polytetrapeptide coacervate and optical diffraction pattern (from Long et al.[41]). (E) Polyhexapeptide aggregate and optical diffraction pattern (from Volpin et al.[40]). (F) Crystals of cyclododeca analog of the polyhexapeptide; light micrograph of cuvette after heating for temperature profile of turbidity formation in Fig. 1F (from Urry et al.[13]).

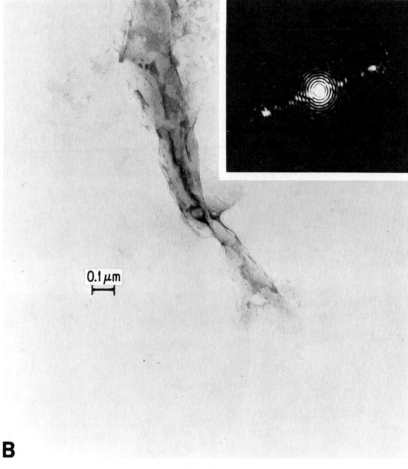

B

Fig. 2B.

are the optical diffraction patterns derived from the electron micrographs.[37] Clearly the result is one of the casting of anisotropic coacervates, and the major diffraction spots are similar to those found for fibrous elastin by Gotte and colleagues.[34]

The protein subunits, which begin in random relative orientation in solution prior to coacervation, appear to be transformed by coacervation into ordered filamentous arrays with lateral packing distances between filaments of between 50 and 55 Å. This ordering, coacervation process is

[37] D. Volpin, D. W. Urry, B. A. Cox, and L. Gotte, Biochim. Biophys. Acta **439**, 253 (1976).

# CHARACTERIZATION OF PEPTIDES OF ELASTIN

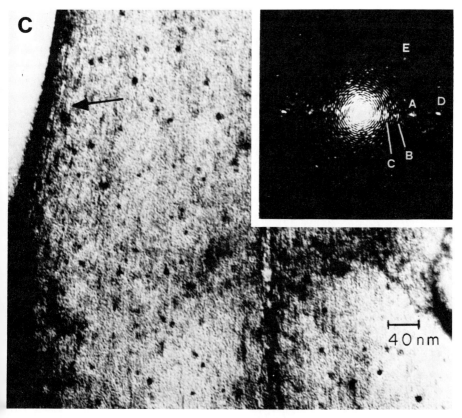

FIG. 2C.

to be considered an inverse temperature transition. Instead of disordering the protein subunits as increasing temperature commonly does, the inverse occurs and an ordering of protein subunits is brought about by increasing temperature. As the second law of thermodynamics requires that the disorder (entropy) of the total system increase with increasing temperature, it is necessary to consider a solvent involvement in which the solvent disorders sufficiently through the transition to more than offset the increase in order in the peptide. In aqueous systems this is referred to as the hydrophobic effect.[38,39] Hydrophobic groups exposed to water occur with a clathrate-like structuring of water surrounding them. On

[38] W. Kauzmann, Adv. Protein Chem. **14**, 1 (1959).
[39] C. Tanford, "The Hydrophobic Effect: Formation of Micelles and Biological Membranes." Wiley, New York, 1973.

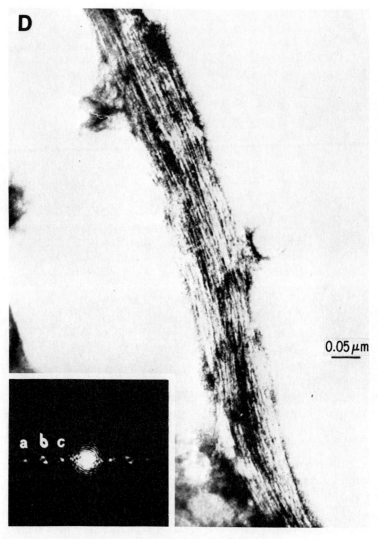

FIG. 2D.

raising the temperature, the increase in entropy on loss of the clathrate-like water is accompanied by a decrease in entropy resulting from the specific juxtaposition of hydrophobic groups. Accordingly, it is argued that the increase in order of the protein subunits is due to hydrophobic association such that the increase in entropy due to the loss of clathrate-

# [39] CHARACTERIZATION OF PEPTIDES OF ELASTIN 685

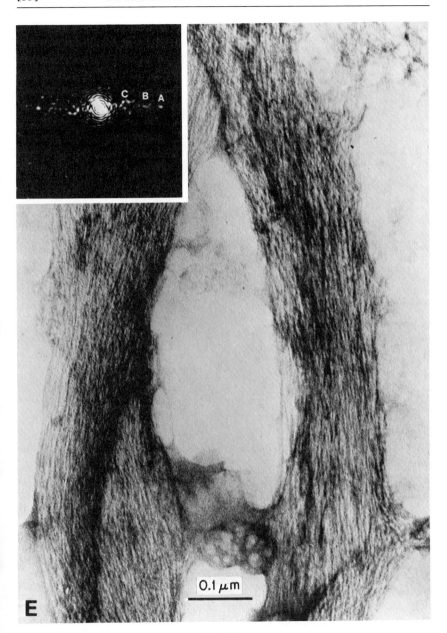

FIG. 2E.

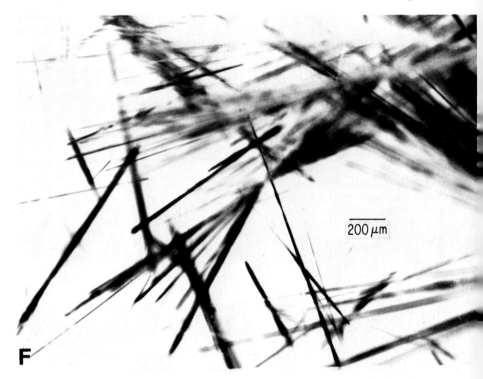

Fig. 2F.

like water, originally surrounding the hydrophobic groups, more than offsets the entropy decrease represented by the formation of regular filamentous arrays. Thus, indirectly by means of the second law of thermodynamics, one concludes that the dominant intermolecular association is hydrophobic.

Owing to the amino acid composition of the polypentapeptide (PPP) and of the polytetrapeptide (PTP) of elastin, it was found necessary when discussing the temperature profiles of coacervation to consider hydrophobic association as being responsible for coacervation. This perspective is confirmed by the electron micrographs of negatively stained PPP and PTP coacervates,[40,41] which are seen in Figs. 2C and D to contain filaments, and even to contain filaments with substructure. Throughout

[40] D. Volpin, D. W. Urry, I. Pasquali-Ronchetti, and L. Gotte, *Micron* **7**, 193 (1976).
[41] M. M. Long, R. S. Rapaka, D. Volpin, I. Pasquali-Ronchetti, and D. W. Urry, *Arch. Biochem. Biophys.* **201**, 445 (1980).

Fig. 2C, and particularly at the arrow, left-handed twisted filaments are seen. It can be argued with complete consistency, therefore, that the dominant intermolecular interaction giving rise to the twisted filament is hydrophobic and that the hydrophobic association occurs between polypentapeptide chains in some sort of loosely helical or spiral conformation.

Interestingly the electron micrographs of the polyhexapeptide (PHP) also exhibit filamentous arrays with some appearance of a crisscrossing of the filaments[40] (see Fig. 2E). The products of heating are not amorphous random masses, but rather structured material. Accordingly, the implications of structure, based on the concentration dependence of the TP$\tau$s and on the properties of the PHP aggregates, are supported by the electron microscope studies. Even so, up to this point the arguments for structure in the heat-derived aggregates have derived substantially from the negative staining technique, and the question of preparation artifacts must be considered.[42] While the artifact question can be answered directly by listing the numerous and different preparative techniques that can be used to demonstrate the filaments,[43] perhaps an effective way to address this concern here is to look for additional demonstrations of order using other methods and the synthetic repeat peptides of elastin.

*Additional Evidence for Increasing Order on Heating*

A most striking demonstration of the capacity of these types of polypeptides to order with increasing temperature is seen with the cyclododecapeptide analog of PHP, i.e., cyclo-Ala$_1$-Pro$_2$-Gly$_3$-Val$_4$-Gly$_5$-Val$_6$-Ala$_7$-Pro$_8$-Gly$_9$-Val$_{10}$-Gly$_{11}$-Val$_{12}$. In Fig. 1F, it is seen that the cyclododecapeptide exhibits temperature profiles for turbidity formation (TP$\tau$s) with profiles and with concentration dependences that are analogous to those of the PHP. The interesting thing about the turbid suspension of the cyclododecapeptide is that the turbidity, the light scattering, is due to crystals, as seen in the light micrograph of the cuvette in Fig. 2F. On raising the temperature, the cyclic polyhexapeptide crystallizes with TP$\tau$s that are analogous to those of the linear polyhexapeptide. On standing the crystals slowly redissolve, and the process can be repeated interminably. The scanning electron micrograph of a pair of crystals with adhering fragments is seen in Fig. 3A. The TP$\tau$s, of curves c and d of Fig. 1F, are actually temperature profiles of crystallization. Crystal structure data on another synthetic elastin analog will be presented in the section Detailed Structural Characterization of Elastin Peptides, which demon-

[42] B. B. Aaron and J. M. Gosline, *Nature (London)* **287**, 865 (1980).
[43] E. G. Cleary and W. J. Cliff, *Exp. Mol. Pathol.* **28**, 227 (1978).

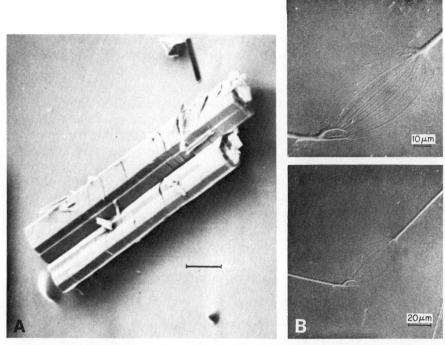

FIG. 3. Scanning electron micrographs showing ordered structures of repeat elastin peptides. (A) Heat-derived crystal of cyclododecapeptide seen as a twinned crystal with adhering crystalline fragments. (B) Polypentapeptide, cross-linked during coacervation, exhibits self-assembly into fibers that splay out into fibrils and recoalesce back into the fiber again. This fibrillar nature is anisotropic and requires that there be underlying filamentous structures of the sort seen in the electron micrographs of negatively stained coacervates, e.g., in Fig. 2C. (A) From Urry et al.[13]; (B) from Urry et al.[15]

strates both dominant intermolecular hydrophobic interactions and the $Pro_2$-$Gly_3$ β-turn. Both structural features were deduced years earlier in the solution studies.

Additional independent evidence that the coacervates are filamentous ordered structures comes from the synthetic cross-linked polypentapeptide. On cross-linking the PPP under coacervate conditions as described elsewhere,[15] the synthetic, cross-linked PPP orders itself into fibers that can be seen in the light microscope with no fixative or coating. The product is easily observed in the scanning electron microscope, as seen in Fig. 3B. In this figure, a synthetic fiber 2–3 μm in diameter is seen to splay out into many fine fibrils and to recoalesce laterally into the fiber again. This cross-linked PPP is anisotropic, and it is elastomeric, capable of the same elastic modulus as native fibrous elastin.[15] The fibrillar anisotropy must arise owing to yet smaller parallel aligned filaments, filaments of the sort observed in the electron micrographs of negatively stained coacervate. All in all, the evidence appears rather incontrovertible that coacervation is an intermolecular ordering process and that elasticity can arise from molecular systems having preferred, albeit dynamic, conformations, and the filaments seen by the EM negative staining technique should not be dismissed as artifacts of preparation.

*Circular Dichroism of the Coacervates*

The preceding studies provide evidence for intermolecular order in the coacervates. It is also possible to obtain general information on the question of intramolecular order or disorder using circular dichroism (CD). The CD spectra of $\alpha$-elastin in 0.01 $M$ sodium acetate solution and in the coacervate state are seen as curves a and b, respectively, in Fig. 4A, where a large conformational change is observed on coacervation.[44] When corrections are made for distortions due to light scattering from the coacervate film, the differences are even larger. The CD spectrum of the coacervate is characteristic of proteins containing a small amount of $\alpha$-helix, e.g., about 20–25%. It is reasonable to consider that the CD spectrum of the coacervate is due to the alanine-rich cross-linking regions, which have been proposed to be $\alpha$-helical by Gray et al.[7] In any case the change in the CD spectrum on coacervation indicates a significant increase in intramolecular order on coacervation of $\alpha$-elastin.

The CD spectra of the polypentapeptide in aqueous solution and in the coacervate state are seen in Fig. 4B, where again coacervation results in a large change in the CD pattern.[45] A similar change is seen for the polytetrapeptide in Fig. 4C and D where the transition from the solution spectrum to that of the coacervate can be partially achieved in solution by heating.[41] The coacervate CD spectra of Figs. 4B and C are characteristic of a $Pro_2$-$Gly_3$ type of $\beta$-turn[46] and are clearly not the spectra of random polypeptides. Owing to the low ellipticity of the $\beta$-turn CD spectra, the presence of this conformation is obscured in $\alpha$-elastin by the order of magnitude larger ellipticities exhibited by the $\alpha$-helical conformation. The CD spectra of Fig. 4 indicate the presence of preferred conformations in the coacervate state of elastin peptides. These conformations can be de-

---

[44] B. C. Starcher, G. Saccomani, and D. W. Urry, *Biochim. Biophys. Acta* **310**, 481 (1973).
[45] D. W. Urry, M. M. Long, B. A. Cox, T. Ohnishi, L. W. Mitchell, and M. Jacobs, *Biochim. Biophys. Acta* **371**, 597 (1974).
[46] D. W. Urry, M. M. Long, T. Ohnishi, and M. Jacobs, *Biochem. Biophys. Res. Commun.* **61**, 1427 (1974).

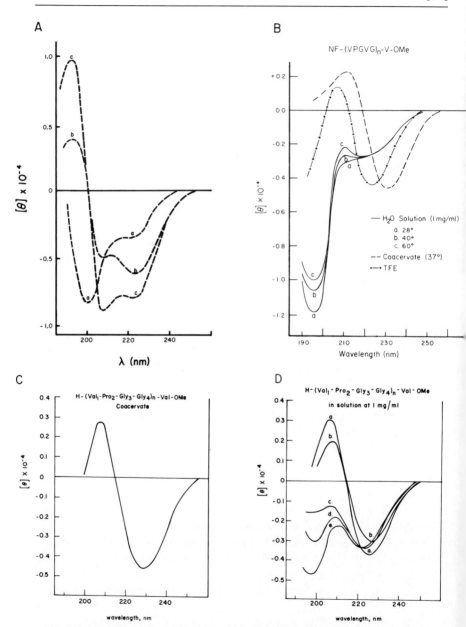

FIG. 4. Circular dichroism patterns of soluble elastin peptides in solution and coacervate states. (A) α-Elastin in aqueous solution (curve a), as the coacervate (curve b), and as the coacervate, corrected spectrum (curve c). (B) Polypentapeptide NF-(VPGVG)$_n$-V-OMe in aqueous solutions (——, 1 mg/ml) at different temperatures (a, 28°; b, 40°; c, 60°; equilibrium solutions above 40°); as the coacervate (- -, 37°); and in trifluoroethanol (TFE) solution

scribed in considerable detail for the repeat peptides of elastin using the nuclear magnetic methods described below.

By way of summarizing the preceding results, if the coacervate state has any relevance to the fibrous state of elastin and if the separately studied component sequential polypeptides of elastin have relevance, then on the basis of the physical techniques considered above it becomes exceedingly difficult to view fibrous elastin as a random construct.

Detailed Structural Characterization of Elastin Peptides

The physical techniques of light scattering to follow temperature and concentration dependences for aggregation and crystallization, of transmission electron microscopy with negative staining and optical diffraction, and of circular dichroism of the coacervates have been used above to provide general structural information indicating unambiguously the presence of both intermolecular and intramolecular order in the coacervates and other heat-derived states of elastin sequential polypeptides. In this section numerous approaches of nuclear magnetic resonance (NMR) together with conformational energy analyses are used to provide detailed information on the preferred conformations of the sequential polypeptides of elastin. These approaches are coupled with the concept of cyclic conformations with linear conformational correlates,[20,47] which allows for greater conformational detailing of the linear sequential polypeptides and for the utilization of X-ray diffraction analysis of single crystals with the demonstration of relevance to the elastin structure.

Owing to (a) the relatively recent introduction of the methods used to develop solution conformations (some of which were first shown on the repeat peptides of elastin), (b) to limited work on the conformation of these peptides in other laboratories (which constitutes two publications[48,49]), and (c) to continual reports in the literature arguing that fibrous elastin is random,[23-27,42] there is much impetus to continue testing the approaches and interpretations. Furthermore, in spite of the derived

[47] D. W. Urry, *Proc. Natl. Acad. Sci. U.S.A.* **69**, 1610 (1972).
[48] K. D. Kopple and A. Go, *Biopolymers* **15**, 1701 (1976).
[49] E. Ralston and R. L. Somorjai, *Biopolymers* **18**, 359 (1979).

---

(·—·—) (from Urry et al.[45]). (C) Polytetrapeptide H-(Val$_1$-Pro$_2$-Gly$_3$-Gly$_4$)$_n$-Val-OMe as the coacervate (from Long et al.[41]). (D) Polytetrapeptide in aqueous solution (1 mg/ml) at different temperatures (c, 60°; d, 45°; e, 25°; showing the conversion toward a coacervate spectrum on heating) and in organic solvents (a, methanol; b, trifluoroethanol) (from Urry et al.[45]). These difference absorbance measurements between left and right circularly polarized light indicate the presence of preferred nonrandom conformations in the coacervate states.

conformational detail of the repeating units as monomers[50-52] and polymers,[17,53,54] there is yet a lack of information about the details of the β-spirals, e.g., the number of repeating units per turn of spiral, the helix axis translation per turn of spiral, and the interactions between turns. It is apparent that additional analytical approaches are required. One such approach would be fiber diffraction studies. If single chains of the polypentapeptide, for example, could be dried at 60°, the helical relationships between the large repeating units might be characterized by means of fiber diffraction studies. The problem is that on elevating the temperature of aqueous solutions to 37° interchain hydrophobic interactions dominate to form twisted filaments, which grossly distort the intrachain relationships between pentamers and cause the mean chain axis to vary widely from the filament axis. Accordingly, some other approach must be found to achieve a more detailed description of the sequential polypeptides of elastin. An additional approach is to use cyclic analogs.

*The Concept of Cyclic Conformations with Linear Conformational Correlates*

This concept states that a relatively strain-free cyclic conformation with a sufficient number of residues in the cycle can be conformationally related to a helical structure with low pitch and approximately the cyclic number of residues in one turn of the helix. This concept was initially proposed to conceive of new types of helices.[55] When the repeating conformational unit is a β-turn, the resulting helix is termed a β-spiral. In the case of the sequential polypeptides of elastin, the repeating β-turns have been proposed to repeat on a helix axis to produce β-spirals with different properties.[17,20] If the β-spiral is of the type proposed as a working model for the polypentapeptide,[17] then it is expected that there would be a cyclic polypentapeptide with conformational properties similar to those of the linear polypentapeptide. Accordingly, a cyclopentapeptide series, $\lfloor(VPGVG)_n\rfloor$ with $n = 1, 2, 3, 4, 5$, and 6, has been synthesized and partially characterized.[56] As part of the demonstration of the conforma-

[50] D. W. Urry, M. A. Khaled, V. Renugopalakrishnan, and R. S. Rapaka, *J. Am. Chem. Soc.* **100**, 696 (1978).
[51] V. Renugopalakrishnan, M. A. Khaled and D. W. Urry, *J. Chem. Soc., Perkin Trans 2*, p. 111 (1978).
[52] M. A. Khaled, V. Renugopalakrishnan, and D. W. Urry, *J. Am. Chem. Soc.* **98**, 7547 (1976).
[53] D. W. Urry, T. Ohnishi, M. M. Long, and L. W. Mitchell, *Int. J. Pept. Protein Res.* **7**, 367 (1975).
[54] D. W. Urry, L. W. Mitchell, T. Ohnishi, and M. M. Long, *J. Mol. Biol.* **96**, 101 (1975).
[55] D. W. Urry, *Proc. Natl. Acad. Sci. U.S.A.* **69**, 1610 (1972).
[56] D. W. Urry, T. L. Trapane, H. Sugano, and K. U. Prasad, *J. Am. Chem. Soc.* **103**, 2080 (1981).

tional methods, data on $n = 1, 2$, and 3 will be considered in the search for a cyclic conformational correlate of the linear polypentapeptide. The nomenclature used will be $C_5^5$ for the cyclopentapeptide, $C_5^{10}$ for the cyclodecapeptide, $C_5^{15}$ for the cyclopentadecapeptide, and PPP for the linear polypentapeptide. The subscript indicates the number of residues in the repeating conformational unit.

NMR Characterization of Secondary Structure

In the use of nuclear magnetic resonance (NMR) to determine secondary structure, approaches that do not depend on exchange rates will be used, because in exchange processes a conformation occurring only a few percent of time but with, for any number of causes, a very high exchange rate for a particular NH proton could be misrepresented as a major conformation. The methods that will be used here relate to the degree of exposure of N$\underline{H}$ and $\underline{C}$-O moieties to the solvent with the perspective that a moiety shielded from the solvent becomes a candidate for intramolecular hydrogen bonding.

*Assignment of Resonances*

Owing to space limitations the means of achieving assignments will not be discussed. The procedures for achieving assignments can be obtained in the original articles.[50–57] Also for the purposes here it is necessary only to have the assignments for the PPP. These are given in Fig. 5 for all the proton and carbon-13 resonances. The resonances of the cyclic peptide with a conformation similar to that of the PPP will have similar chemical shifts in all solvents and at all temperatures. In the delineation of secondary structure, the chemical shifts of the peptide N$\underline{H}$ and $\underline{C}$-O resonances will be followed as a function of temperature and solvent.

*Temperature Dependence of Peptide N$\underline{H}$ Chemical Shift*

It has been demonstrated, for example, on the model systems of gramicidin S and valinomycin that the temperature dependence of peptide N$\underline{H}$ chemical shifts can be used to identify hydrogen-bonded peptide NH moieties.[58] In both these cases, as well as numerous others, the complete and confirming crystal structure occurred subsequently. Peptide NH protons with steep slopes of about $-9$ to $-10 \times 10^{-3}$ ppm/deg are characteristic of complete solvent exposure, whereas values of $-4 \times 10^{-3}$ ppm/deg are characteristic of peptide N$\underline{H}$ protons, which are shielded from the

[57] D. W. Urry, L. W. Mitchell, and T. Ohnishi, *Biochemistry* **13**, 4083 (1974).
[58] M. Ohnishi and D. W. Urry, *Biochem. Biophys. Res. Commun.* **36**, 194 (1969).

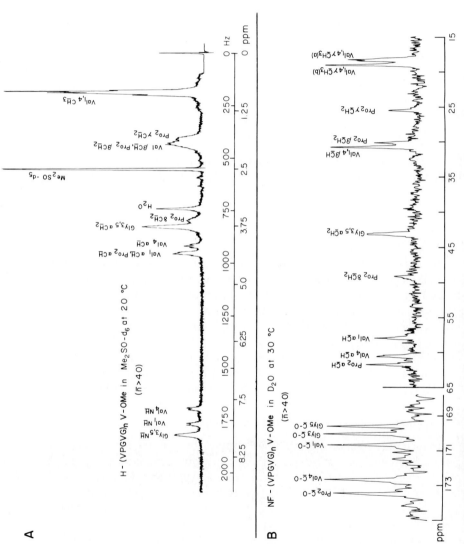

FIG. 5. Assigned nuclear magnetic resonance spectra of the polypentapeptide (VPGVG)$_n$-V-OMe of elastin.

solvent by hydrogen bonding. Care must be exercised when interpreting the data, and complicating factors such as temperature-dependent interconversion between conformations have been considered elsewhere.[17] The repeat elastin peptides are some of the most favorable peptides to study by these methods because of the presence of only aliphatic side chains, since all other side chains add significant, sometimes severe, complications.

The temperature dependences of the peptide N$\underline{\text{H}}$ chemical shifts ($\Delta\delta/\Delta T$) for the $C_5^n$ series in water are given in Fig. 6 with the data for the PPP included on each plot as dotted lines. The patterns for the chemical shift changes with temperature are very different for $C_5^5$ and $C_5^{10}$ than for the PPP, whereas the pattern for $C_5^{15}$ is strikingly similar to that of the PPP. Since these molecules all have the same amino acid sequence, the differences must be due to different conformations. In addition to showing that $C_5^{15}$ and PPP have similar conformations, Fig. 6 shows the high degree of sensitivity of the approach to differing conformation. The change in slope near 50° will be discussed below in the section NMR Characterization of Hydropholic Tertiary Structure and an Inverse Temperature Transition. Interestingly, the very low slope for the Val$_1$N$\underline{\text{H}}$ of $C_5^5$ is probably due to supershielding by two C-O moieties in addition to the hydrogen-bonded C-O, as shown in Fig. 7. This conformation of $C_5^5$ was derived for solution using the NMR methods presented here and conformational energy calculations,[59] and it is very closely that found in independent crystal structure determination.[60]

The $\Delta\delta/\Delta T$ for peptide N$\underline{\text{H}}$s in Me$_2$SO are given in Fig. 8, where again the same results are apparent. $C_5^{15}$ and PPP have very similar conformations in Me$_2$SO, whereas the conformations of $C_5^5$ and $C_5^{10}$ are very different. Interestingly, in the comparison of the patterns for $C_5^{15}$ and PPP, the Val$_1$N$\underline{\text{H}}$ in both polypeptides exhibits the same slope but is shifted upfield in $C_5^{15}$. This suggests a similar degree of solvent exposure, but a slightly different orientation of the N$\underline{\text{H}}$ with respect to the plane of a nearby peptide moiety. The chemical shift difference arises from the magnetic anisotropy of the peptide moiety. This shift has been analyzed to suggest that, if this were the only change in torsion angle, the difference would occur on going from a cyclic structure to a linear right-hand $\beta$-spiral.[56]

*Solvent Dependence of Peptide C-$\underline{O}$ Chemical Shift*

The carbonyl carbon chemical shift changes with changes in solvent can also be used to evaluate secondary structure and to identify the C-O

[59] M. A. Khaled, C. M. Venkatachalam, H. Sugano, and D. W. Urry, *Int. J. Pept. Protein Res.* **17**, 23 (1981).
[60] H. Einspahr, W. J. Cook, and C. E. Bugg, American Crystallographic Association, Winter Meeting, Eufaula (Alabama), March 17–21, Abstract PA15 (1980).

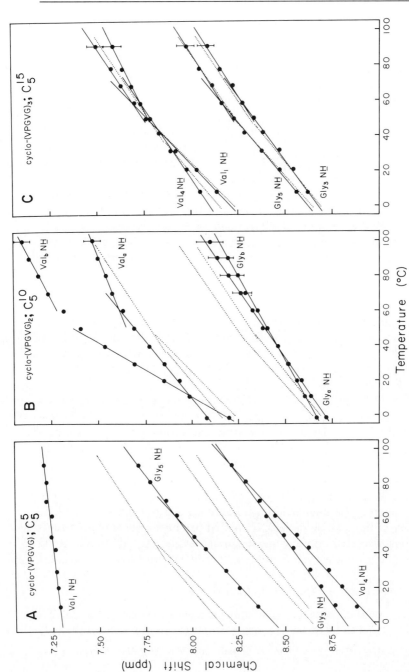

FIG. 6. Temperature dependence of peptide NH chemical shift in water. Superimposed on each curve is the dotted curve for the data on the linear polypentapeptide. (A) The cyclopentapeptide, $C_5^5$; (B) the cyclodecapeptide, $C_5^{10}$; and (C) the cyclopentadecapeptide, $C_5^{15}$. Since these molecules all have the same amino acid sequence (VPGVG), the differences are necessarily due to difference in conformation. Note how dissimilar are $C_5^5$ and $C_5^{10}$ are from each other and from the linear polypentapeptide. Note how similar are $C_5^{15}$ and

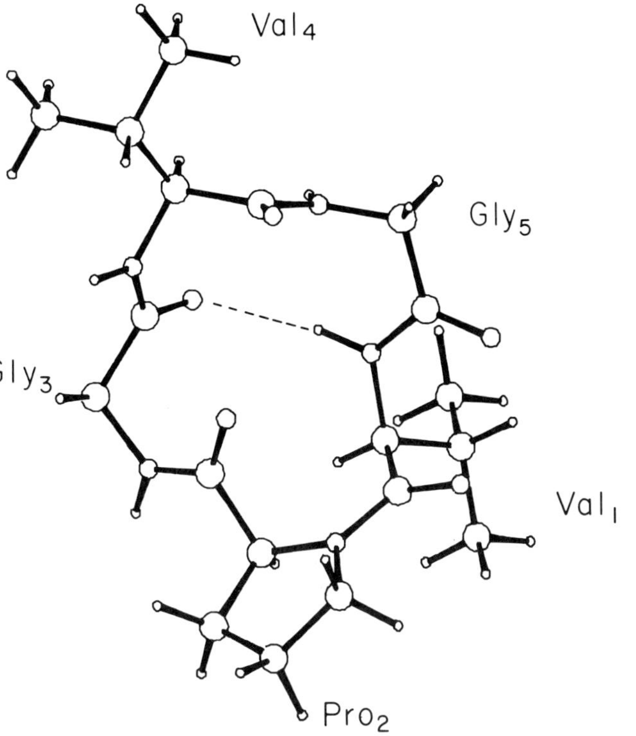

FIG. 7. Conformation of the cyclopentapeptide derived from the solution nuclear magnetic resonance studies and confirmed by crystal structure analysis. From Khaled et al.[59]

component of a hydrogen bond. In fact, this approach was first worked out on the elastin peptides[61,62] and checked on the gramicidin S model by predicting the carbonyl carbon assignments. In this approach, the C-O that shifts downfield least on adding a solvent such as $D_2O$ or trifluoroethanol to $Me_2SO$ is considered to be the most solvent shielded, and the chemical shifts are then plotted with respect to this carbonyl carbon resonance. Plots for an $Me_2SO-D_2O$ titration are given in Fig. 9 for the $C^{5n}$ series. Again different patterns are observed for $C_5^5$ and $C_5^{10}$, but the patterns for PPP and $C_5^{15}$ are virtually superimposable. The lesser downfield shift of the $Val_1\underline{C}$-O resonance makes this a candidate for intramolecular H-bonding. On the basis of numerous studies,[17] it has been shown that this C-O pairs in hydrogen bonding with the $Val_4$N-H, which has the lowest slope in Fig. 6C. This is the $Pro_2$-$Gly_5$ β-turn involving the

[61] D. W. Urry, L. W. Mitchell, and T. Ohnishi, Proc. Natl. Acad. Sci. U.S.A. 71, 3265 (1974).
[62] D. W. Urry, L. W. Mitchell, and T. Ohnishi, Biochem. Biophys. Res. Commun. 59, 62 (1974).

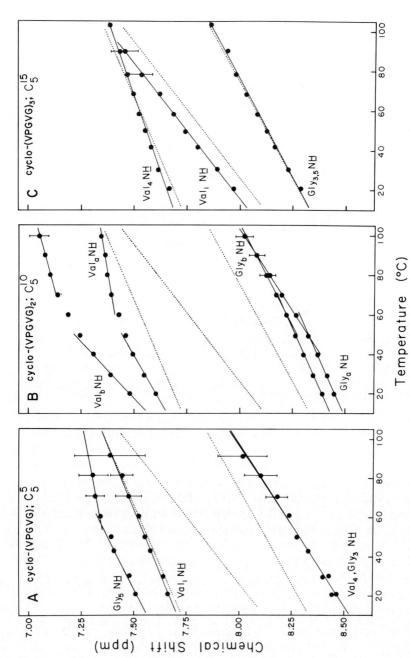

FIG. 8. Temperature dependence of peptide NH chemical shift in Me$_2$SO-$d_6$. Included for comparison are the dotted curves for the linear polypeptide. (A) The cyclopentapeptide, cyclo-(VPGVG), C$_5^5$; (B) the cyclodecapeptide, cyclo-(VPGVG)$_2$, C$_5^{10}$; and (C) the cyclopentadecapeptide, cyclo-(VPGVG)$_3$, C$_5^{15}$. See text for discussion. From Urry et al.[56]

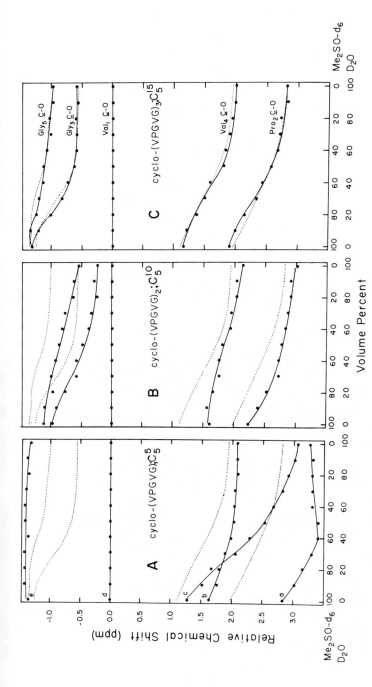

FIG. 9. Solvent dependence of the peptide $\underline{C}$-O chemical shift for the $Me_2SO$-$d_6$ to water ($D_2O$) solvent titration for (A) the cyclopentapeptide, $C_5^5$; (B) the cyclodecapeptide, $C_5^{10}$; and (C) the cyclopentadecapeptide, $C_5^{15}$. Data for the linear polypentapeptide are included as the dotted curves. The three cyclic molecules are seen to be conformationally different, and $C_5^{15}$ is seen to be a cyclic conformational correlate of the linear polypentapeptide. From Urry et al.[56]

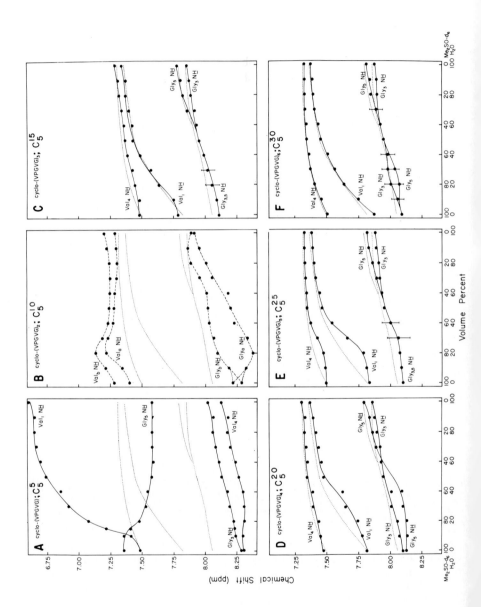

Val$_1$C-O and the Val$_4$NH in a 10-atom hydrogen-bonded ring. The Gly$_3$C-O is also substantially solvent shielded, and in previous studies it has been paired with the Gly$_3$NH in Me$_2$SO to form a $\gamma$-turn, an 11-atom hydrogen-bonded ring. Although this hydrogen bond has a high probability (approximately 0.6) in Me$_2$SO, it is a much more transient structure in water, with an estimated probability of 0.3 below 50° and increasing to 0.6 above 50° (see Table 11 of Urry and Long[17]).

*Solvent Dependence of Peptide NH Chemical Shift*

The change in peptide NH chemical shift on going from Me$_2$SO to water is shown in Fig. 10 for the entire cyclopentapeptide series. Interest here can focus on the comparison of the solvent titration patterns for the $C^{5n}$ series with the superimposed dotted curves for the linear polypentapeptide. As with the studies in Figs. 6, 8, and 9, $C_5^3$ and $C_5^{10}$ are conformationally distinct molecules. For $n = 3$ the comparison with the PPP is very close. The comparison is not as close for $n = 4$ and 5, but then the curves become extraordinarily similar for $n = 6$, the cyclotricosapeptide ($C_5^{30}$). The close conformational similarity of $C_5^{15}$ and PPP, becoming even more striking for $C_5^{30}$, raises the question of whether using $C_5^{15}$ as the cyclic conformational correlate of PPP can be related to $C_5^{30}$ by the introduction of a helix sense, i.e., a pitch. This will be considered in detail below. What is particularly noteworthy about the studies in Figs. 6, 8, 9, and 10 is that the cyclopentadecapeptide is identified as a cyclic correlate of the linear polypentapeptide. What is equally gratifying is that $C_5^{15}$ crystallizes and its crystal structure has been determined.

Crystal Structure of the Cyclic Correlate of the Polypentapeptide

The crystal structure of the cyclopentadecapeptide has been determined by single crystal X-ray diffraction.[63] The salient aspects of the molecular structure are seen in Fig. 11A and B. The Pro$_2$-Gly$_3$ $\beta$-turn, which had been deduced from proton magnetic resonance studies on the pentamer and its linear oligomers 7 years earlier,[19,20] is observed in the crystal as seen in Fig. 11A. Even the orientation of the end peptide moiety of the $\beta$-turn, i.e., the peptide moiety joining the Pro and Gly residues, has

[63] W. J. Cook, H. M. Einspahr, T. L. Trapane, D. W. Urry, and C. E. Bugg, *J. Am. Chem. Soc.* **102**, 5502 (1980).

FIG. 10. Solvent dependence at 20° of peptide NH chemical shift for the Me$_2$SO-$d_6$ to water solvent titration for the cyclopentapeptide series with $n = 1, 2, 3, 4, 5,$ and 6: (A) $C_5^5$; (B) $C_5^{10}$; (C) $C_5^{15}$; (D) $C_5^{20}$; (E) $C_5^{25}$; (F) $C_5^{30}$. Included on each of the six plots are the dotted curves for the linear polypentapeptide to simplify comparisons. $C_5^{15}$ and $C_5^{30}$ exhibit dependencies, and hence conformations, that are very nearly identical to those of the linear polypentapeptide. See text for discussion. From Urry *et al.*[56]

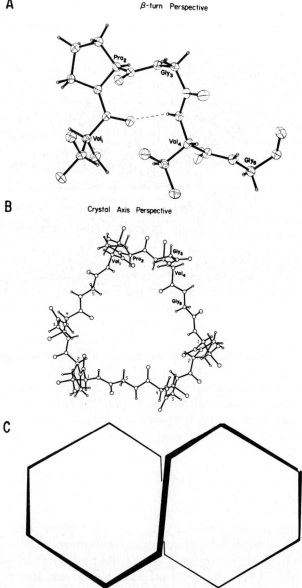

FIG. 11. Molecular structure of crystalline cyclopentadecapeptide, which, as shown in Figs. 6, 8, 9, and 10, is a cyclic conformational correlate of the linear polypentapeptide. (A) β-turn perspective showing the $Pro_2$-$Gly_3$ β-turn previously derived from solution studies; (B) molecular structure of $C_5^{15}$ showing the threefold symmetry with the β-turn of (A) being seen on end; and (C) schematic representation of two $C_5^{15}$ molecules combined in such a way as to introduce a helix sense with maintenance of the basic conformation. Only the $Gly_5$ of the crossing segments differ in absolute torsion angles from the other segments, but since this is a glycine, the angles can still be essentially equivalent. This explains how in Fig. 10F the cyclotricosapeptide, $C_5^{30}$, can be such an excellent cyclic conformational correlate of the linear polypentapeptide. (A) and (B) from Cook et al.[14]; (C) from Urry et al.[56]

FIG. 12. Crystal structure of the cyclic correlate, $C_5^{15}$, of the linear polypentapeptide. Note that the dots inside the cyclic structures are water molecules and that there is no water separating molecules. At the locations where three molecules come together are hydrophobic associations resulting in rods of hydrophobic side chains running parallel to the crystal axis. As the molecules stack exactly on top of each other, it becomes possible to go conceptually from a stack of cyclic molecules to a helix (β-spiral) containing all the elements considered in solution as resulting in the twisted filaments of the coacervate. See text for discussion. From Cook et al.[14]

the type II orientation[64] as originally deduced[19,20] and as subsequently experimentally verified in solution using the nuclear Overhauser effect.[65] The complete molecule is seen in Fig. 11B with the β-turn being seen on edge. In this perspective, it is apparent that the $Val_1NH \cdots OC\ Val_4$ hydrogen bond, which has been proposed to occur in chloroform[52] and to occur with increased probability at elevated temperature in water,[17] is poised to form with only slight rotations about single bonds. Also the $Gly_3NH \cdots OC\ Gly_5$ hydrogen bond, which is thought to occur in $Me_2SO$, can be formed with two or three rotations of about 30°. The crystal-derived molecular structure of the cyclic correlate provides a satisfying confirmation of the solution conformation studies.

In the crystal the molecules stack exactly on top of each other, the threefold symmetric crystal axis coinciding with the threefold molecular symmetry. As seen in Fig. 12, the water, which is about one-third of the crystal by weight, occurs inside the stacked cyclic molecules with no solvent separating the molecules. Furthermore, the contacts between the molecules are dominantly hydrophobic, with hydrophobic rods running parallel to the molecular symmetry axis at positions where three molecules meet. Now if one conceptually breaks the cyclic molecules at a common site (e.g., at the $Gly_5-Val_1$ bond), tilts one end up and the other down, then joins the lower end to the raised end of the similarly treated molecule below and the raised end to the lowered end of the molecule above, the stack of cyclic molecules are converted into a linear helical structure consisting of repeating β-turns, as has been proposed for the β-spiral of the polypentapeptide.[17] The previously proposed working

[64] C. M. Venkatachalam, *Biopolymers* **6**, 1425 (1968).
[65] M. A. Khaled and D. W. Urry, *Biochem. Biophys. Res. Commun.* **70**, 485 (1976).

model of the polypentapeptide $\beta$-spiral at elevated temperature was constructed of space-filling molecular models with about 2.8 repeats per turn and about 3.2 Å for the helix axis translation per repeat.[17] Now if the number of repeats per turn of $\beta$-spiral is nonintegral, then the hydrophobic rods become spiraling lipophilic ridges also, as previously considered.[17] On maximizing the hydrophobic interactions between two (or perhaps three, as implied by the crystal structure) $\beta$-spirals, as occurs on coacervation, the spirals will twist or coil around each other to give rise to a twisted filament of the sort seen in Fig. 2C, also as previously proposed.[5]

Accordingly, the crystal structure of the cyclopentapeptide supplies a verification of the recurring $\beta$-turns. By means of the concept of cyclic conformations with linear conformational correlates and the experimental data in Figs. 6, 8, 9, and 10 showing the nearly identical conformations of $C_5^{15}$ and PPP, the crystal structure furnishes an insight to the $\beta$-spiral of the polypentapeptide; and, by the observation of hydrophobic rods resulting from the intermolecular interactions, the crystal structure provides for arguments leading directly to the twisted filaments of the coacervate. The identification of a cyclic correlate of the linear polypentapeptide and its crystallization with subsequent crystal structure determination furnishes an independent derivation of the twisted filaments seen in the electron micrographs of negatively stained linear polypentapeptide. As similar structures are seen in electron micrographs of similarly stained natural fibrous elastin[33,34] and as the repeating pentamer is the most striking feature of the protein sequence, repeating eleven times in a single continuous sequence,[10] the present work also suggests a structural interpretation of the supercoiling or twisting of $\beta$-spirals in native elastin.

One additional point, which can now be clarified, is an answer to the question of how the cyclotricosapeptide can be such a good cyclic correlate in the solution studies. This is schematically demonstrated in Fig. 11C. In this representation the shorter line segment is for the $\beta$-turn seen on end (as in Fig. 11B) and the longer line segment represents the largely extended $Val_4$-$Gly_5$-$Val_1$ chain segment, again as seen in Fig. 11B. By introducing a pitch or helix sense, two $C_5^{15}$ structures can be joined together using the $Gly_5$ residues as the hinge for the reversals of chain direction. The structure has a twofold symmetry axis giving both halves the same helix sense. Thus the crystal structure also allows for an explanation as to how $C_5^{30}$ can be perhaps an even better cyclic correlate than the cyclopentapeptide and reinforces the perspective of the linear conformational correlate, i.e., of the $\beta$-spiral of the linear polypentapeptide.

NMR Characterization of Torsion Angles Coupled with Conformational Energy Analyses

A complete description of the conformation of a polypeptide can be achieved by specifying all the torsion angles, that is the $\phi$, $\psi$, $\omega$, and $\chi$

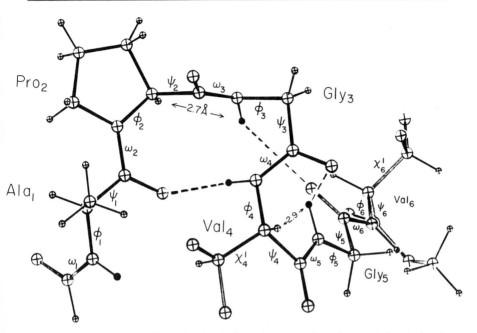

FIG. 13. A preferred conformation for the linear hexamer, shown here to define the bonds about which rotations can occur, i.e., to define the $\phi$, $\psi$, $\omega$, and $\chi$ torsion angles.

angles. These angles are defined in Fig. 13 in a derived conformation of the hexamer monomer.[50] Possible $\phi$ angles can be evaluated from fine structure of the N$\underline{H}$ and C$\underline{H}$ resonances, which gives the three-bond (vicinal) $\alpha$CH-NH coupling constant, $^3J_{\alpha\text{CH-NH}}$. The fine structure is apparent as the doublets of the NH resonance of Fig. 5A. The splitting in Hertz of the two lines of the doublet is the value of $^3J$. The equation of Bystrov et al.[66] may be used to evaluate the possible torsion angles, i.e.,

$$^3J_{\alpha\text{CH-NH}} = 9.4 \cos^2 \theta - 1.1 \cos \theta + 0.4 \quad (1)$$

The relationship between $\phi$ and $\theta$ is plotted in Fig. 14, where the ranges of possible $\phi$ values for the cyclododecapeptide, $C_2^{12}$, are plotted.[67] The cyclododecapeptide analog of the linear polyhexapeptide, it may be recalled, crystallizes on raising the temperature (see Figs. 1F, 2F, and 3A). For the special case of glycine residues, the expression becomes[66]

$$^3J_{\alpha\text{CH-NH}} + {}^3J_{\alpha\text{CH}'-\text{NH}} = -9.8 \cos^2 \phi - 1.3 \cos \phi + 15 \quad (2)$$

[66] V. F. Bystrov, S. L. Portnova, T. A. Balashova, S. A. Kozmin, Y. D. Gavrilov, and V. A. Afanasev, *Pure Appl. Chem.* **36**, 19 (1973).
[67] M. A. Khaled, C. M. Venkatachalam, T. L. Trapane, H. Sugano, and D. W. Urry, *J. Chem. Soc., Perkin Trans. 2*, p. 1119 (1980).

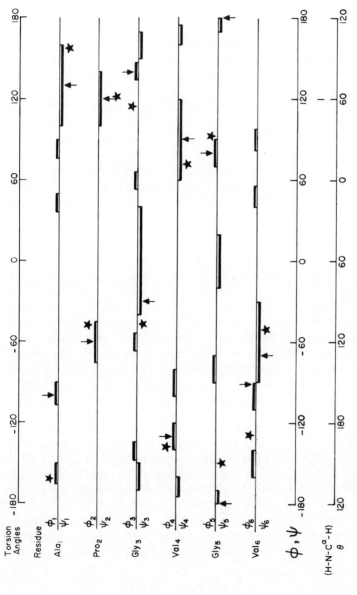

FIG. 14. Plot of the restricted $\phi$ and $\psi$ torsion angle ranges (━) as indicated from the nuclear magnetic resonance data of the cyclododecapeptide, $C_2^{12}$. In a relatively rigid molecule like $C_2^{12}$, the final structure can be expected to have a given dihedral angle in one of the indicated ranges. The arrows indicate the values derived from a constrained hydrogen-bonded wire model, and the stars are for the *in vacuo* calculated structure. Values of $\phi$ were obtained using Bystrov equations. Values of $\psi$ for $Gly_3$ and $Gly_5$ are derived from the $^2J$ analysis; for $Pro_2$ and $Val_4$, from the nuclear Overhauser effect; and for $Ala_1$ and $Val_6$, from the Dreiding model. The dihedral angle $\theta$ (see bottom line) is related to the torsion angle $\phi$ by the equation $\theta = |\phi - 60|$ for L-amino acids. The terminology is in accordance with IUPAC—IUB nomenclature [*J. Mol. Biol.* **52**, 1 (1970)]. From Khaled *et al.*[67]

With the restriction by the Pro side chain of $\phi_2$ to values near $-60°$, this gives ranges for all the $\phi$ torsion angles.

Given the value of $\phi$ for a glycine, the two-bond (geminal) coupling between the $\alpha$-protons of a glycine gives rise to a coupling constant, $^2J_{\alpha CH-\alpha CH'}$, that can place limits on the range of $\psi$ for the glycine residue by means of the expression of Barfield et al.[68]

$$^2J_{\alpha CH-\alpha CH'}(\phi,\psi) = -13.91 - 1.55 \cos^2 \psi - 2.8 \cos^4 \psi + 4.65 \cos^2 \phi \quad (3)$$

The original papers should be consulted for the somewhat complex analysis required to obtain $^2J$ from the spectral data.

Additional values of $\psi$ may be obtained for a polypeptide by means of the nuclear Overhauser effect. When a proton is irradiated that is within 3 Å or less of a second proton, the intensity of the resonance of the second proton will change. For rapidly moving molecules, the intensity will increase and the percentage of enhancement (% NOE) is proportional to the inverse sixth power of the distance between the two protons,[69] $i$ and $j$, i.e.,

$$\% \text{ NOE} = 100/kr_{ij}^6 \quad (4)$$

where $k = 1.8 \times 10^{-2}$ Å$^{-6}$. When the pair of protons are the $\alpha C\underline{H}_i$ and the $NH_{i-1}$, then the calculated distance $r_{i,i-1}$ gives an estimate of the value of $\psi$, assuming the peptide moiety to be planar.

The peptide torsion angle is given by $\omega$. When $\omega = 180$ the peptide is trans and when $\omega = 0$ the peptide is cis. Whether the peptide is cis or trans can often be deduced from the carbon-13 magnetic resonance data. Since cis most commonly occurs for X-Pro peptide bonds, the chemical shifts of the $\gamma$ and $\beta$ carbons of Pro have been used to identify the correct isomer.[70,71] These arguments have been considered for the cyclopentapeptide[59] shown in Fig. 7. Additionally, with $^{15}N$ enrichment, the $^1J_{^{15}N-^1H}$ coupling constant can be obtained and the value of this one-bond coupling constant has been shown to be sensitive to small excursions from planarity.[72]

Finally the side-chain torsion angles can be obtained from the $^3J_{\alpha CH-\beta CH}$ coupling constant by use of the expression of Abraham and McLauchlan,[73] i.e.,

[68] M. Barfield, V. J. Hruby, and J.-P. Meraldi, *J. Am. Chem. Soc.* **98**, 1308 (1976).
[69] J. H. Noggle and R. E. Schirmer, "The Nuclear Overhauser Effect." Academic Press, New York, 1971.
[70] D. G. Dorman and F. A. Bovey, *J. Org. Chem.* **38**, 2379 (1973).
[71] R. Deslauriers and I. C. P. Smith, *Top. Carbon-13 NMR Spectros.* **2**, 1 (1976).
[72] V. Renugopalakrishnan, M. A. Khaled, K. Okamoto, and D. W. Urry, *Int. J. Quantum Chem., Quantum Biol. Symp.* **4**, 97 (1977).
[73] R. J. Abraham and K. A. McLauchlan, *Mol. Phys.* **5**, 513 (1963).

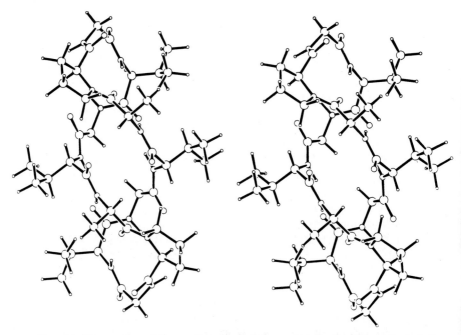

FIG. 15. Stereo view of the conformation of the cyclododecapeptide analog of the polyhexapeptide derived from solution nuclear magnetic resonance studies and conformational energy calculations. If the reader finds it difficult using distance focus to see the three-dimensional structure, the figure can be Xeroxed, the left and right sides changed, and three-dimensional structure seen by focusing near, as on a pencil point placed between the figure and the eyes. From Khaled et al.[67]

$$J_{\chi i} = \begin{matrix} 10.5 \text{ Hz cos}^2|\chi_i| - 0.28 \text{ Hz} & \text{for } |\chi_i| = 0 \text{ to } 90 \\ 13.7 \text{ Hz cos}^2|\chi_i| - 0.28 \text{ Hz} & \text{for } |\chi_i| = 90 \text{ to } 180 \end{matrix} \quad (5)$$

Using the secondary structural constraints derived as above in the section NMR characterization of Secondary Structure and the possible values of torsion angles derived from the NMR studies (as shown in Fig. 14 for the cyclododecapeptide), conformational energy calculations are carried out to find an energetically favorable structure that satisfies all these constraints. This has been carried out for the cyclododecapeptide, and the resulting structure is shown as the stereopair in Fig. 15. Interestingly, the hydrophobic side chains are all pointing outward as required for the temperature-elicited crystallization seen in Figs. 1F, 2F, and 3A.

This is the methodology used to derive the detailed conformation of repeat peptides of elastin and of many interesting and related analogs.

## NMR Characterization of Hydrophobic Tertiary Structure and an Inverse Temperature Transition

*Nuclear Overhauser Effect Demonstration of Hydrophobic Interactions*[11]

In the section Structure of the Elastin Peptide Aggregates, above, the second law of thermodynamics was used to argue for hydrophobic intermolecular interactions as being responsible for the increase in intermolecular order with an increase in temperature. Nuclear magnetic resonance can be used to demonstrate directly the hydrophobic interaction responsible for the inverse temperature transition. On raising the temperature of the polytetrapeptide (PTP),[17] the polypentapeptide (PPP) and the cyclopentadecapeptide in aqueous solutions (e.g., see Fig. 6C), the $Val_1$ NH becomes more shielded from the solvent (as evidenced by the decrease in slope of the $\Delta\delta(N\underline{H})$ temperature dependence). The carbonyl of residue 4 also becomes more shielded on raising the temperature.[17] This suggests an increase in the occurrence of a $Val_1$ NH $\cdots$ O-C residue 4, 14-atom hydrogen-bonded ring. Adding this hydrogen bond to the $Pro_2$-$Gly_3$ β-turn places the $Val_1$ and $Pro_2$ side chains in juxtaposition and suggests that this tertiary structural feature might be the hydrophobic association responsible for the increase in intramolecular order. As noted above, the nuclear Overhauser effect can provide information on the proximity of protons. Since the hydrophobic association would place the Pro $\delta CH_2$ protons adjacent to the $Val_1$ $\gamma CH_3$ protons, the $Pro_2$ $\delta CH_2$ protons of the PTP were irradiated and the area of the Val $\gamma CH_3$ protons was integrated at different temperatures. At 20° it was found that the enhancement was about 10%, but at 40° the enhancement became 43%. With 50% being the maximum possible enhancement, this result demonstrates the two hydrophobic side chains to become fixed in a close contact on elevating the temperature. To our knowledge this study on the polytetrapeptide of elastin provided the first direct observation of the hydrophobic interactions responsible for an inverse temperature transition.[11]

When the $Val_1$ is replaced by an $Ala_1$ the increase in secondary structure with elevated temperature does not occur. In fact a partial proximity of the $Ala_1$ $CH_3$ protons to the $Pro_2$ $\delta CH_2$ protons is lost on raising the temperature. Also when $Val_1$ is replaced by $Ala_1$ in both the polytetrapeptide and the polypentapeptide, as noted above in the section Temperature Profiles of Coacervation, the polymers no longer coacervate.[30] This temperature-dependent $Val_1$-$Pro_2$ side-chain juxtapositioning may be a key element in the coacervation process. Again we have seen, when examining these sequential polypeptides of elastin, that an increase in temperature brings about an increase in order.

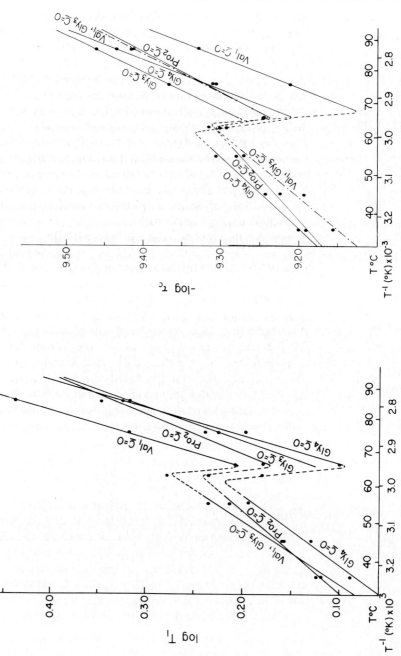

FIG. 16. Temperature dependence of the longitudinal relaxation times, $T_1$ (left) and calculated rotational correlation times, $\tau_c$ (right) of each of the polytetrapeptide carbonyl carbons (150 mg/ml $H_2O$). A sharp transition is observed on raising the temperature above 60°. The sharp transition is to a state with decreased mobility and with a barrier to mobility that is greater than prior to the transition. Even though the calculated rotational correlation times are 1 nsec or less, the sharp transition indicates a cooperative

*Carbon-13 Relaxation Studies*

The polytetrapeptide of elastin also provides an opportunity for a fundamental characterization of an inverse temperature transition. As temperature means motion, an increase in temperature means an increase in motion. Similarly an inverse temperature transition implies a decrease in mobility through the transition. NMR relaxation studies, specifically longitudinal or spin lattice ($T_1$) relaxation experiments, can be used to calculate relative mobility. When such $T_1$ studies are carried out on the PTP,[12] below the transition the mobility of the backbone is seen to increase with temperature until the transition, through which the mobility drops precipitously, and then above the transition the mobility steadily increases with increasing temperature as shown in Fig. 16. Consistent with an increased order above the transition, the energy barriers to motion are greater than below the transition. An important point to be made here is that rotational correlation times calculated from the experimental longitudinal relaxation times, i.e., from the $T_1$'s, are very near 1 nsec, which is very fast. Even so, there is a sharp transition to a state of increased order. Devoid of any interpretations, what is observed is a sharp transition that requires that there be a change in order. What this means is that these $T_1$-derived rotational correlation times of $10^{-9}$ sec, and particularly similarly derived longer rotational correlation times (which indicate slower relative motion), should not be used to argue for the absence of order.

A Nonrandom Anisotropic Twisted-Filament Model[5] and the Physical Properties Characterizing Randomness in Fibrous Elastin

As a result of the conformational studies,[17,50-54] of the coacervation properties,[5] of the electron micrographs of negatively stained coacervate[35-37,40] and their relationship to those of native elastin,[33,34] of properties of the cross-linking region,[7] of the demonstrated anisotropic and elastomeric properties of the synthetic cross-linked polypentapeptide,[15] and of the cellophane-like properties of the polyhexapeptide,[31] a specific fibrillar structure, a multichain twisted-filament model with three components in the repeating unit has been proposed[5] and is shown in Fig. 17.

The major component of the three-component twisted filament is based on the polypentapeptide model because the cross-linked polypentapeptide is an anisotropic elastomer with an elastic modulus which, depending on the water concentration, can be the same as that of aortic fibrous elastin,[15] because large peptide fragments have been isolated with

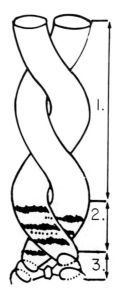

FIG. 17. Twisted filament model, composed serially of three major structural components and laterally of two chains, which has been proposed as a working model for a fundamental structural unit of fibrous elastin. The proposed elastomeric portion (region 1) is modeled after the dynamic β-spiral of the linear polypentapeptide, which is shown in Fig. 19 as the single chain of β-spiral before intertwining with additional chains. 2, Interlocked β-spiral region (polyhexapeptide model); 3, cross-linking region (α-helical model). From Urry.[5]

the pentamer composition,[7-10,74] because in the carbon-13 magnetic resonance spectra the five major carbonyl peaks of α-elastin (a 70,000-dalton chemical fragmentation product of fibrous elastin) coincide exactly with the five carbonyl peaks of the polypentapeptide,[75] because the mobilities of the polypentapeptide coacervate state[18] are at least as high as those of the elastic fiber,[25-27] and because of the above-mentioned similarity of the twisted-filament structures seen in electron micrographs of negatively stained polypentapeptide coacervate and of similarly twisted filaments in fibrous elastin.

Working models of the preferred single-chain conformations for the polypentapeptide and polyhexapeptide, which were derived from secondary structural studies, are given in Fig. 18. The β-spiral of the polypentapeptide has been derived above as a dynamic structure with spiraling lipophilic bands (see Fig. 18A), which, on maximizing intermolecular hy-

[74] S. Keller, I. Mandl, S. Birken, and R. Canfield, *Biochem. Biophys. Res. Commun.* **70**, 174 (1976).
[75] D. W. Urry and L. W. Mitchell, *Biochem. Biophys. Res. Commun.* **68**, 1153 (1976).

[39] CHARACTERIZATION OF PEPTIDES OF ELASTIN 713

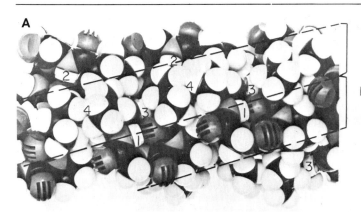

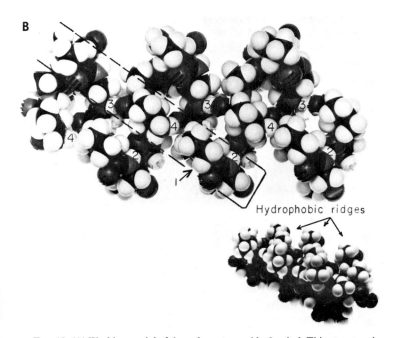

FIG. 18. (A) Working model of the polypentapeptide β-spiral. This structure is considered for the conformation that would result from drying a single chain at 60°. The structure has about 2.8 repeats per turn and a 3.2 Å helix translation per repeat. Note the long spiraling lipophilic band. 1, $Val_4$ NH · · · O—C $Val_1$; 2, $Gly_3$ NH · · · O—C $Gly_5$; 3, $Val_1$ NH · · · O—C $Val_4$; 4, paired valyl side chains. From Urry and Long.[17] (B) Proposed conformation for the polyhexapeptide β-spiral. Hydrophobic ridges result that could align and interlock during a temperature-elicited hydrophobic intermolecular interaction. 1, $Val_4$ NH · · · O—C $Ala_1$; 2, $Gly_3$ NH · · · O—C $Gly_5$; 3, $Val_6$ NH · · · O—C $Val_6$; 4, Pro-Val side chain contact. From Urry et al.[53]

drophobic interactions during coacervation, results in the electron micrograph-observed twisted filaments. The β-spiral of the polyhexapeptide is described as less flexible with spaced hydrophobic ridges (see Fig. 18B), which on hydrophobic intermolecular association could function in an aligning and interlocking manner. The role of the second, the polyhexapeptide, component would be to assist in the alignment of the two chains such that the four lysines that form a single desmosine cross-link can be in proper juxtaposition. With the repeating hexameric sequence being contiguous in the sequence data with the cross-linking sequence,[8,10] the third component, utilizing a perspective of Gray et al.,[7] is an alanine-rich α-helical segment with lysines in positions $i$ and $i + 4$ in one chain and $i$ and $i + 3$ in the second chain. Components 2 and 3 combine with the polypentapeptide model to continue the twisted filament appearance. While the structure of elastin is clearly going to be more complex, this working model is presented to contain in a single figure much of what the physical studies on the soluble elastin peptides appear to require.

At first sight, this model appears to be at odds with deductions of the structure of fibrous elastin (a) based on approximate adherence to the kinetic theory of rubber elasticity[22-24]; (b) based on mobility characterization[25-27]; and (c) based on the absence of significant optical birefringence.[42] What the kinetic theory of elasticity requires is that there be a large entropy decrease on stretching, i.e., a structure with much entropy in the relaxed state. The mobility studies also require a structure with large entropy, expressed as mobility, in the relaxed state. What the birefringence studies require is that there not be a parallel stacking of atomic densities as would occur with parallel aligned α-helices or collagen molecules.

While a dynamic random structure has been viewed as the way to achieve these three requirements, there is nothing that states that a dynamic random structure is the only way to satisfy these properties. There is good reason to believe that the β-spiral of the polypentapeptide can satisfy these requirements, particularly viewed in the light of the crystal structure of the cyclopentadecapeptide, $C_5^{15}$ (see Figs. 11 and 12), and the demonstration that the cyclopentadecapeptide is the cyclic correlate of the linear polypentapeptide (PPP). The mobility is an easy requirement to satisfy since the coacervate of PPP has a mobility fully as great as that reported for fibrous elastin, but the reason why the β-spiral of the PPP can have such mobility is more complex to answer. This answer also is the answer to the entropy requirement of the kinetic theory of elasticity.

As seen in Fig. 19, the PPP β-spiral has no hydrogen bonding between turns, and the interturn interactions are not structurally restricting, allowing for a large number of states to have a similar topological appearance.

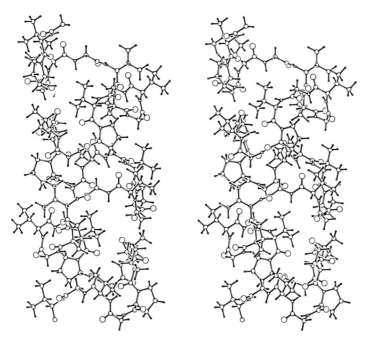

FIG. 19. Stereo view of the polypentapeptide β-spiral that can be derived from the crystal structure of the cyclopentadecapeptide using the argument of cyclic conformations with linear conformational correlates. Note the space within the spiral that would be filled with water molecules as in the cyclic correlate (see Fig. 12) and the orientation of the β-turns, which act as spacers between turns of the β-spiral. Importantly, there is no hydrogen bonding between the β-turn spacers and there are open areas that allow continuity of bulk water and water within the spiral. This low-energy conformation is one of a class of conformations with similar topological features. This structure has 2.7 repeats per turn and a 3.5 Å helix axis translation per repeat. From Venkatachalam and Urry.[76]

There is water within the β-spiral which, when not periodically blocked by the β-turns functioning as interturn spacers, is continuous with bulk water. This loose structure, with the $Gly_5$ residue strategically placed, allows for many librational states that would provide an exponential multiplying effect to give a large number of a priori equally probable states accessible to the β-spiral. Computations are underway to provide an enumeration of the states and a comparison with an extended stretched chain (Venkatachalam and Urry, in preparation).

With respect to the birefringence requirements, the presence of a large amount of water within the β-spiral which is periodically contiguous with the surrounding water is the structural aspect of the polypentapeptide β-spiral that would result in little or no birefringence even with the parallel

[76] C. M. Venkatachalam and D. W. Urry, *Macromolecules* **14**, 1225 (1981).

alignment of twisted filaments comprised of the β-spirals. Thus there is good reason to believe that physical studies on the soluble elastin peptides and selected analogs are resulting in the description of component conformations that can provide an understanding of the structure and function of elastin and the result is a proposed librational entropy mechanism for elasticity (Urry, Venkatachalam, Long, Trapane, and Prasad, in preparation).

Acknowledgments

The author gratefully acknowledges the contributions of many past and present members of the Laboratory of Molecular Biophysics to these studies on elastin and particularly wishes to thank Dr. C. M. Venkatachalam for calculating and plotting the stereo view of the polypentapeptide β-spiral used in Fig. 19. This work was suppported in part by the National Institutes of Health Grant No. HL-11310.

# [40] Biosynthesis of Soluble Elastin in Organ and Cell Culture

By JOEL ROSENBLOOM

A soluble form of elastin was first isolated by nondegradative procedures by Smith et al.[1] from the aortas of copper-deficient swine. Subsequent studies, which included amino acid composition and fingerprint analysis of elastase digests, clearly established a chemical relationship between this soluble form, which was named tropoelastin, and the amorphous component of the elastic fiber.[2] A similar protein was isolated from the aortas of copper-deficient or lathyritic chicks.[3,4] The identification of tropoelastin led to studies designed to demonstrate that tropoelastin was a legitimate biosynthetic intermediate, not an abnormal product induced in the copper-deficient or lathyritic animals. Many of these experiments utilized tissues derived from embryonic or young animals, since such tissues presumably synthesized elastin at a much greater rate relative to older animals. When embryonic chick aortas or cells derived from the aortas were incubated *in vitro*, they synthesized a protein suggestive of tropoelastin in its properties.[5-9] These properties included molecular

[1] D. W. Smith, N. Weissman, and W. H. Carnes, *Biochem. Biophys. Res. Commun.* **31**, 309 (1968).
[2] L. B. Sandberg, N. Weissman, and D. W. Smith, *Biochemistry* **8**, 2940 (1969).
[3] R. B. Rucker, W. Gottlich-Riemann, and K. Tom, *Biochim. Biophys. Acta* **317**, 193 (1973).
[4] J. A. Foster, R. Shapiro, P. Voynow, G. Crombie, B. Faris, and C. Franzblau, *Biochemistry* **14**, 5343 (1975).
[5] L. Murphy, M. Harsch, T. Mori, and J. Rosenbloom, *FEBS Lett.* **21**, 113 (1972).

weight as determined by polyacrylamide gel electrophoresis, labeling pattern with various radioactive amino acids, solubility in water–alcohol mixtures, and cross reactivity with elastin-specific antibodies. During pulse-chase experiments with the aortas, this soluble protein was incorporated into an insoluble residue. Aortas and cells from normal newborn pigs also synthesized a labeled protein that manifested many of the properties of added carrier tropoelastin.[10,11] A higher molecular weight form of soluble elastin has been isolated from the lungs and aortas of lathyritic chicks[12] and appears to be synthesized by rabbit aortic smooth muscle cells in culture.[13] However, based on experiments in which the mRNA for elastin has been translated in a cell-free system, it does not appear to be a primary translation product and its relation to tropoelastin is not clear at present.[14,15]

The following narrative describes procedures for establishing biological systems that synthesize soluble elastin intermediates, largely tropoelastin, and techniques for isolation, identification, and quantitation of the tropoelastin.

Incubation in Organ Culture

*Chick Aorta.* A section of the thoracic aorta extending from the root to a point just proximal to the entrance of the ductus arteriosus is dissected with sterile precautions from 10- to 18-day-old chick embryos.[5,9,16] In the chicken, the arteries equivalent to the innominate in mammals branch from the aorta very near the base of the heart. Their histological structure is essentially identical to the aorta itself, so they are included up to the point of their bifurcation into the subclavian and carotid. Adherent loose connective tissue is carefully stripped from the external surface, and the

---

[6] A. S. Narayanan and R. C. Page, *J. Biol. Chem.* **251**, 1125 (1976).
[7] G. M. Bressan and D. J. Prockop, *Biochemistry* **16**, 1406 (1977).
[8] J. Rosenbloom and A. Cywinski, *Biochem. Biophys. Res. Commun.* **69**, 613 (1976).
[9] J. Uitto, H.-P. Hoffmann, and D. J. Prockop, *Arch. Biochem. Biophys.* **173**, 187 (1976).
[10] D. W. Smith and W. H. Carnes, *J. Biol. Chem.* **248**, 8157 (1973).
[11] P. A. Abraham, D. W. Smith, and W. H. Carnes, *Biochem. Biophys. Res. Commun.* **58**, 597 (1974).
[12] J. A. Foster, R. P. Mecham, and C. Franzblau, *Biochem. Biophys. Res. Commun.* **72**, 1399 (1976).
[13] J. A. Foster, R. P. Mecham, C. B. Rich, M. F. Cronin, A. Levine, M. Imberman, and L. L. Salcedo, *J. Biol. Chem.* **253**, 2797 (1978).
[14] W. Burnett, R. Eichner, and J. Rosenbloom, *Biochemistry* **19**, 1106 (1980).
[15] J. A. Foster, C. B. Rich, S. Fletcher, S. R. Karr, and A. Przybyla, *Biochemistry* **19**, 857 (1980).
[16] R. Eichner and J. Rosenbloom, *Arch. Biochem. Biophys.* **198**, 414 (1979).

vessels are slit along their long axis. They are washed in modified Krebs medium[17] (15.7 mM $NaH_2PO_4$, 1.6 mM $KH_2PO_4$, 111.2 mM NaCl, 5.4 mM KCl, 1.3 mM $MgCl_2$, 4.0 mM $NaHCO_3$, and 13.0 mM glucose at pH 7.4) without fetal calf serum and incubated in the same medium supplemented, per milliliter, with 25 µg of sodium ascorbate and 100 µg of β-aminopropionitrile for 1–4 hr at 37° in air. From 3 to 6 aortas are incubated per milliliter of medium, to which are added 1–5 µCi of radioactive amino acids per milliliter, depending upon the amount of incorporation desired. Radioactive glycine, proline, or valine are the amino acids preferred for radioactive labeling, since they predominate in elastin. Labeled lysine may be used if it is desired to follow incorporation into cross-links. At the end of the incubation period, the medium is removed, and the aortas are homogenized and extracted by one of the procedures described below. Similar procedures may be used for aortas from young chicks, but the tissue is cut into small fragments (1–2 mm$^3$) before incubation.[18]

*Pig Aorta Tissue.* Newborn piglets (aged 1–2 days) are killed by exsanguination under pentobarbital anesthesia.[10] The thoracic aortas are removed with aseptic technique, stripped of adventitia, weighed, and then minced with scalpels in sterile Krebs–Ringer solution and used immediately. Each minced aorta (about 1 g wet weight) is incubated for 2 hr at 37° with gentle shaking in 10 volumes of sterile Krebs–Ringer containing radioactive amino acids (5 µCi of [$^{14}$C]valine per milliliter, for example). At the end of the incubation, the suspension is centrifuged at 2000 g for 10 min, and the aortic tissue is washed twice by centrifugation with phosphate-buffered saline (0.05 M sodium phosphate, pH 7.4, 0.15 M NaCl). The tissue is extracted as described below.

*Cartilage.* Hamsters 7–12 days old are killed by decapitation, and their ears are removed.[19] Using a dissecting microscope, the skin and adherent tissue are removed and the clean translucent cartilage is washed in warm modified Krebs medium. Samples of the cartilage are fixed in 2% glutaraldehyde in Millonig's phosphate buffer, embedded in paraffin, and sectioned by conventional histological techniques in order to evaluate the adequacy of removal of skin and any adjacent blood vessels. A convenient stain for insoluble elastin is methylene blue, azure II, basic fuchsin.[20] The cartilage fragments are incubated in the modified Krebs medium supplemented with 25 µg of ascorbate and 100 µg of β-aminopropionitrile per milliliter in air at 37° for 20 min. Labeled amino acids are then added, and the incubation is continued for up to 2 hr. The tissue is extracted as described below.

[17] H. A. Krebs, *Biochim. Biophys. Acta* **4**, 249 (1950).
[18] F. W. Keeley, *Can. J. Biochem.* **57**, 1273 (1979).
[19] P. Heeger and J. Rosenbloom, *Connect. Tissue Res.* **8**, 21 (1980).
[20] C. D. Humphrey, F. E. Pittman, *Stain Technol.* **49**, 9 (1974).

## Synthesis in Short-Term Cell Cultures

Cells freshly isolated from differentiated tissues and free of extracellular matrix provide a convenient system for studying the synthesis and secretion of soluble elastin.[8,9] The cells are obtained from the matrix by digestion of the tissue with a trypsin:collagenase solution, then washed free of the enzymes and incubated in suspension culture for up to 4 hr. The tropoelastin synthesized by the cells is secreted into the incubation medium. Tropoelastin can be isolated from the extracellular medium as well as from the cells and then characterized by various techniques (see below).

*Isolation of Cells.* Thoracic aortas from 10- to 18-day-old embryonic chicks are isolated and cleaned of adherent connective tissue as described above. The aortas are suspended in Eagle's minimal essential medium (Gibco or Microbiological Associates) containing per milliliter 1 mg of purified collagenase (Worthington CLSPA or Sigma Chemical Co.) and 2.5 mg of trypsin (Gibco). The blood vessels from 10–15 aortas (about 300 mg wet weight) are incubated in 1 ml of medium equilibrated with 5% $CO_2$, 95% air, and the aortas are shaken in a water bath for 60–120 min. The exact time of digestion in each experiment is monitored by visual inspection of the incubate; and digestion is stopped as soon as it becomes apparent that most of the tissue has been dispersed. The digest is filtered through lens paper placed in a small Swinnex filter (Millipore Corporation, Bedford, Massachusetts); the filtrate is centrifuged at 600 g at room temperature for 5 min. The cells are washed three times by resuspension and centrifugation in modified Krebs medium containing 10% fetal calf serum. The yield of cells may be monitored by counting samples with a hemacytometer or by assaying for DNA by the Schmidt–Thannhauser procedure.[21] Usually 3–6 million cells are obtained from the vessels of a single 17-day-old embryo. Proportionately fewer cells are obtained from younger embryos (approximately $5.5 \times 10^5$ cells from 11-day embryo). Figure 1 illustrates the progressive recovery of cells from aortas of 17-day-old embryos as a function of digestion time with the collagenase–trypsin mixture.

*Characterization of the Cells.* The isolated cells should be tested for viability by a method such as Trypan Blue exclusion.[22] It is advisable to discard the preparation if fewer than 85% are viable by this criterion. More extensive morphological characterization can be carried out by conventional electron microscope examination of the cells.[9]

*Incubation of Cells.* The cells are incubated at a concentration of 2 to $7.5 \times 10^6$ cells/ml in modified Krebs medium containing 2% fetal calf

---

[21] S. Lindy, H. Turto, J. Uitto, P. Helm, and I. Lorenzen, *Circ. Res.* **30**, 123 (1972).
[22] P. O. Seglen, *Methods Cell Biol.* **13**, 29 (1976).

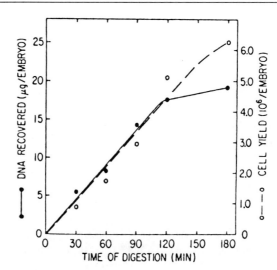

FIG. 1. Time course for the release of cells and DNA from blood vessels digested with collagenase and trypsin. Thoracic aortas and associated large blood vessels were removed from 17-day-old chick embryos, and the tissues were digested with bacterial collagenase and trypsin as described in the text. Each sample contained aortas and large blood vessels from 12 chick embryos. At the times indicated, the samples were passed through lens paper and the cells were recovered by centrifugation. The cells were washed and counted in a hemacytometer, and DNA was assayed by the Schmidt–Thannhauser procedure.[21] Each value represents the mean of duplicate assays. Reproduced, by permission of Academic Press, Inc., from Uitto et al.[9]

serum in air at 37° in a water bath with moderate shaking. The cells are preincubated for 15 min, and then radioactive amino acids are added (New England Nuclear or Amersham). Valine, proline, or glycine are the amino acids of choice because of their relatively high abundance in elastin. The cells may be incubated for up to 4 hr. At the end of the incubation, the medium is separated from the cells by centrifuging the cell suspension at 1500 g for 10 min at room temperature. Labeled proteins secreted into the medium and those found in the cells may be characterized as described below. If labeled tropoelastin secreted into the medium appears to be degraded, 200 µg of pancreatic trypsin inhibitor per milliliter are included in the incubation medium.[8]

*Comment.* Although only chick embryo cells have been used to date in these short-term cultures, the aortas of small laboratory mammals should produce similar results.

Synthesis in Long-Term Cell Culture

Long-term cell cultures offer several advantages over short-term cultures, including longer periods of incubation with radioactive amino acids and the ability to test various reagents and conditions that may modify the rate of elastin synthesis. Disadvantages are that cells must be in culture for at least 2–3 weeks after plating before they begin to synthesize elastin, and the rate of elastin synthesis is generally considerably less than that of freshly isolated aortic cells.

*Isolation of Smooth Muscle Cells from Rabbit or Pig Aorta.* Thoracic aortas are rapidly excised from weanling rabbits or newborn pigs using sterile technique and are immediately placed in 100-mm petri dishes containing Dulbecco's modified Eagle's medium with 2.2 g per liter of sodium bicarbonate, 10% fetal calf serum, and 1000 units of penicillin, 50 µg of Aureomycin, 250 µg of fungizone per milliliter.[11,23] Under a dissecting microscope, the aortas are grossly cleaned of extraneous tissue, blood, and fat and are transferred to fresh medium. They are then cut open, cleaned again, and placed in fresh medium, and the intimal/inner medial layer is pinched off and stripped according to the method of Wolinsky and Daly.[24] These layers are cut into approximately 1-mm$^3$ pieces, which are attached to the bottom of 75-cm$^2$ plastic tissue culture flasks. The explants are covered with Dulbecco's medium containing 3.7 g of sodium bicarbonate per liter and 100 units of penicillin and 100 µg of streptomycin per milliliter, and 10% fetal calf serum. The flasks are loosely capped and allowed to remain undisturbed for a week in a moist atmosphere of 5% $CO_2$, 95% air at 37°. The medium is then changed weekly. In 2–3 weeks, the cells that grow out from the explants are detached with a 0.05% trypsin–0.02% EDTA solution in modified Puck's saline A solution,[25] centrifuged at 200 g for 10 min, and resuspended in complete medium; an aliquot is counted in a hemacytometer. The cells are then reseeded at a density of $1.5 \times 10^6$ cells per 75-cm$^2$ tissue culture flask. They may be passed at confluency at intervals of 1–2 weeks using a 1:2 split for 4–6 passages.

*Pulse Labeling of Cell Cultures.* Three to four weeks after seeding, the cell cultures are washed twice with Dulbecco's modified Eagle's medium deficient in the amino acid used for radioactive labeling and containing 50 µg of β-aminopropionitrile per milliliter to inhibit cross-linking. The same medium containing 1–5 µCi of tritium labeled amino acids per milliliter is added, and the incubation is continued for 1–24 hr depending on the objectives of the experiment. [$^3$H]Valine and/or [$^3$H]proline are the amino acids of choice in this system.

*Synthesis by Endothelial Cells.* A cloned line from an established culture of rabbit aortic endothelial cells has been reported to synthesize

---

[23] B. Faris, L. L. Salcedo, V. Cook, L. Johnson, J. A. Foster, and C. Franzblau, *Biochim. Biophys. Acta* **418**, 93 (1976).
[24] H. Wolinsky and M. M. Daly, *Proc. Soc. Exp. Biol. Med.* **135**, 364 (1970).
[25] T. T. Puck, S. J. Cieciura, and H. W. Fisher, *J. Exp. Med.* **106**, 145 (1957).

soluble elastin, largely tropoelastin.[26] The cultures are grown to confluence in F-12 medium (Gibco) supplemented with 10% fetal calf serum in 5% $CO_2$ in standard plastic tissue culture ware. The culture dishes containing a total of 4 to $5 \times 10^7$ cells are rinsed three times with serum-free F-12 medium lacking the amino acid to be used for radioactive labeling. The cells are then incubated in serum-free F-12 medium containing per milliliter 50 $\mu$g of ascorbic acid, 50 $\mu$g of $\beta$-aminopropionitrile, and 1–5 $\mu$Ci of radioactive amino acids, usually valine, proline, or lysine, for 1–24 hr. At the end of the incubation, the medium is removed and the cells are washed and extracted. The labeled proteins in the medium and those in the cell extract are analyzed by the procedures described below.

Because this cell line may not be widely available, the following directions describe a general procedure for the isolation of endothelial cells. Specific directions are given for isolation from calf aortas, but the method can be readily scaled down to handle other animals.[27] The thoracic aortas (15–20 cm) are removed sterilely within 1 hr after death and immediately placed in chilled (4°) Dulbecco's phosphate-buffered saline and supplemented with glucose (2 g/liter), 2.5 $\mu$g/ml amphotericin B (2.5 $\mu$g/ml), penicillin (50 units/ml), streptomycin (50 $\mu$g/ml), and mycostatin (100 units/ml). Routinely, 8–10 aortas are collected, and the vessels are processed for culture within 3 hr. Individual aortas are washed with Dulbecco's phosphate-buffered saline, and the fat is carefully dissected away to free the intercostal arteries. The intercostal vessels are tied or clamped, and the lower end of the vessel is closed with an intestinal clamp. The vessel is filled with 10–15 ml of collagenase, 1 mg/ml (Worthington Biochemical Corporation, CLS II, catalog No. 4176 obtained from *Clostridium histolyticum* containing 125–250 units per milligram dry weight in medium 199). The upper end of the vessel is clamped, and the collagenase is distributed throughout the lumen by inverting the clamped vessel several times. The vessel is then allowed to incubate for 35–40 min at room temperature. After this incubation, the vessel contents are discarded, and the aorta is washed four times with approximately 10 ml of medium. Each wash consists of the following steps: filling of the aorta with medium, gentle agitation of the contents back and forth approximately four times, and subsequent collection of the effluent in a beaker. When wash effluents from all aortas have been collected, they are evenly distributed in petri dishes (75 cm², approximately 10 ml per dish). Routinely, each calf aorta (approximately 20 cm in length) yields 5 to $7 \times 10^6$ cells after enzyme treatment.

[26] W. H. Carnes, P. A. Abraham, and V. Buonassisi, *Biochem. Biophys. Res. Commun.* **90**, 1393 (1979).
[27] E. J. Macarak, B. V. Howard, and N. K. Kefalides, *Lab. Invest.* **36**, 62 (1977).

Cells are cultured at 37° in a humidified atmosphere of 5% $CO_2$ in air. The cells are grown in medium 199 as modified by Lewis et al.[28] supplemented with 20% fetal bovine serum (FBS Rehatuin, Armour Chemical Company) and antibiotics (amphotericin B, 2.5 µg/ml, and gentamicin, 50 µg/ml). Culture medium is changed on day 1 or 2 after the initial culture and biweekly thereafter. For subculture, cells are treated briefly with 0.25% trypsin and 0.05% ethylenediaminetetraacetic acid in a calcium and magnesium-free saline. Released cells (3 to 4 × $10^6$) are split in a 1:2 or 1:4 ratio. Subcultures and primary cultures are treated similarly with regard to medium changes and incubation conditions. The cells are examined by routine light microscopy to verify their epithelial appearance. Figure 2 illustrates the morphological appearance of various cells in culture.[28a] Among other criteria that should be used to prove the endothelial nature of the cells are the reaction of the cells with anti-factor VIII antibody using immunofluorescence techniques and the presence of angiotensin converting enzyme and Weibel–Palade bodies.[29] The cells may be incubated with radioactive amino acids as described above.

Extraction of Soluble Elastin

Because of the unusual solubility properties and great susceptibility to proteolysis of soluble elastin, the extraction of the protein in an undegraded state is a challenging problem. Generally, representatives of all the major classes of protease inhibitors should be present in the extraction medium, and efforts should be made to inactivate rapidly any proteases that may be present. The following procedures are useful in obtaining soluble elastin from cells and tissues that have been incubated with radioactive amino acids as described above.

*Extraction with Hot Sodium Dodecyl Sulfate Solutions.* At the end of the incubation, the medium is separated from cells growing in monolayer culture by decanting it or with cells or tissues in suspension culture, by centrifugation at 1500 g for 10 min. That medium is immediately placed in an ice bath, and protease inhibitors are added to the following final concentrations: 5 mM EDTA, 0.2 mM phenylmethanesulfonyl fluoride, 20 mM N-ethylmaleimide, 1 mM p-aminobenzamidine, and 70 mM ε-aminocaproic acid. An effective means of solubilizing uncross-linked elastin in cells and tissues is by extraction with a sodium dodecyl sulfate solution at elevated temperatures.[7–9] The method is particularly useful if the primary objective is to characterize the molecular weight of the newly

[28] L. J. Lewis, J. C. Hoak, and R. D. Maca, *Science* **181**, 453 (1973).
[28a] E. J. Macarek, B. V. Howard, and N. A. Kefalides, *Ann. N.Y. Acad. Sci.* **275**, 104 (1976).
[29] H. von der Mark, K. von der Mark, and S. Gay, *Dev. Biol.* **48**, 237 (1976).

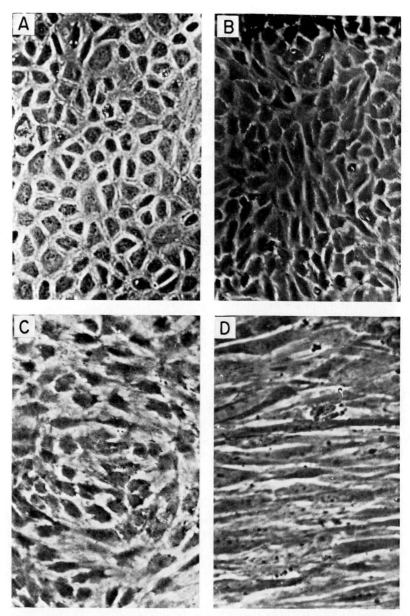

FIG. 2. Phase-contrast photographs of various cell cultures. (A) Calf endothelial cell cultures prepared from aortas as described in the text. (B) Endothelial cell cultures from human umbilical vein. (C) Fibroblasts from human foreskin. (D) Smooth muscle cells from calf aorta. All × 170. Reproduced, by permission of the New York Academy of Sciences, from Macarak et al.[28a]

synthesized elastin and to inhibit proteolysis rapidly. The tissue or cells are suspended in two volumes of 0.01 $M$ sodium phosphate buffer, pH 7.4, containing the protease inhibitors listed above, and homogenized using either a glass–Teflon homogenizer or a Polytron (Brinkman). Sodium dodecyl sulfate and 2-mercaptoethanol are added to final concentrations of 1%, and the homogenate is immediately placed in a boiling water bath for 5 min. The sample is centrifuged at room temperature at 20,000 $g$ for 10 min to remove insoluble matter. The supernatant is dialyzed against 0.01 $M$ sodium phosphate, pH 7.4, containing 0.1% sodium dodecyl sulfate and 2-mercaptoethanol to remove any unincorporated radioactive amino acids. The media from incubations may be similarly treated with sodium dodecyl sulfate and mercaptoethanol and dialyzed before characterizing the labeled proteins as described below.

*Extraction at Neutral or Acidic pH.* If it is desirable not to have sodium dodecyl sulfate in the preparation, the tissue may be extracted with 0.5 $M$ acetic acid or 0.05 $M$ Tris, pH 7.6, containing the protease inhibitors listed above (add 1 $\mu$g of pepstatin per milliliter to the mixture when extracting with acid). After homogenization, the homogenate is extracted using a magnetic stirrer for 2–24 hr at 4°. It is then centrifuged at 20,000 $g$ for 10 min. This method extracts between 40 and 90% of the elastin that can be solubilized by the hot sodium dodecyl sulfate method. An increase in the yield can sometimes be achieved by the addition to the extraction medium of Triton X-100 and/or sodium deoxycholate to final concentrations of 5.0%[16] or 2 $M$ deionized urea.[15]

## Techniques for Identification of Soluble Elastin

The following procedures provide methods for identification of tropoelastin and other soluble forms of elastin in media and extracts of cells and tissue that have been incubated with labeled amino acids.

### Polyacrylamide Gel Electrophoresis and Gel Filtration in Sodium Dodecyl Sulfate

Disc gel electrophoresis is performed as described by Weber and Osborn[5,30] using a concentration of acrylamide which may vary from 5 to 10%. Calibration of molecular weight as a function of position in the gel is made by using standard proteins of known molecular weight and staining with Coomassie Blue. The gels are cut into 1.5-mm fractions; these are placed into counting vials, and 0.25 ml of 30% hydrogen peroxide solution is added. The vials are capped and incubated at 55° for 8 hr. Then, 0.3 ml

[30] K. Weber and M. Osborn, *J. Biol. Chem.* **244**, 4406 (1969).

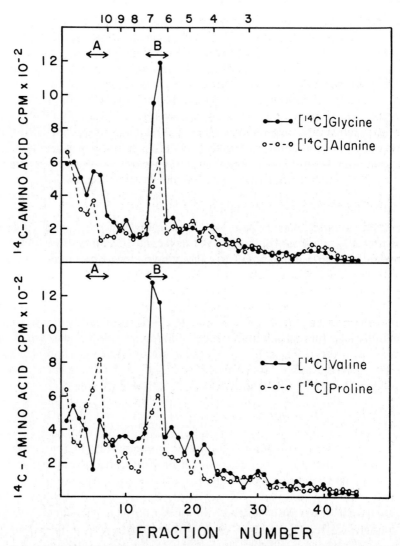

FIG. 3. Disc gel electrophoresis of sodium dodecyl sulfate (SDS)-soluble proteins from aortas labeled with $^{14}$C-labeled amino acids. Four sets of four aortas from 12-day-old embryos were incubated for 1.5 hr in 1.0 ml of modified Krebs medium[17] to which 2 μCi of either [$^{14}$C]glycine, [$^{14}$C]proline, [$^{14}$C]alanine, or [$^{14}$C]valine had been added. They were then homogenized in 0.45 ml of 0.01 $M$ sodium phosphate, pH 7.0, in a Dounce homogenizer. The homogenate was transferred to another tube; 50 μl of 10% SDS solution and 5 μl of mercaptoethanol were added, and the homogenate was placed in a boiling water bath for 5 min. The homogenate was dialyzed overnight at room temperature against 0.01 $M$ sodium phosphate, pH 7.0, 0.1% SDS, and 0.1% mercaptoethanol. After dialysis, the homogenate was centrifuged at room temperature for 10 min at 2000 $g$. Aliquots (50 μl) of the supernatant were

of water and 4 ml of a Triton X-100-based counting solution (New England Nuclear 950A or similar solution) are added, and the samples are counted in a scintillation counter. Figure 3 illustrates results obtained with labeled extracts of embryonic chick aortas.

Slab gel electrophoresis is performed according to the method of Laemmli.[7,15,31] Gels are fixed in 10% trichloroacetic acid and 50% ethanol for 30 min and stained in 0.2% Coomassie Blue in 10% trichloroacetic acid and 25% ethanol for 30 min. Destaining is accomplished by washing overnight in 10% acetic acid. Fluorography of the gel is performed according to Bonner and Laskey.[32] The dried gel is exposed to Kodak X-Omat R film at $-80°$.

Gel filtration is carried out at room temperature on a $1.5 \times 90$ cm column of BioGel A-5m, 200–400 mesh (Bio-Rad) equilibrated and eluted with 0.1% sodium dodecyl sulfate and 0.1 $M$ sodium phosphate buffer, pH 7.4.[9] Samples containing 10,000–50,000 cpm are chromatographed; 2.0-ml fractions are collected, and aliquots of 0.2–0.4 ml are assayed in a liquid scintillation counter.

*Resistance to Cleavage with CNBr*

Because elastin does not contain any methionine residues, it is resistant to cleavage with CNBr.[9,33] Samples are dissolved in 1 ml of 70% formic acid (Baker Chemical Co.) and flushed with $N_2$ for 10 min; 10 mg of CNBr added, and the sample is incubated at 37° for 4 hr. After the incubation, the sample is diluted with 10 ml of distilled water, and most of the CNBr is removed by aspiration with a water pump for 1 hr. The sample is then lyophilized, dissolved in 1 ml of 1% sodium dodecyl sulfate in 0.01 $M$ sodium phosphate buffer, and analyzed as described above.

*Solubility in Aqueous Alcohol Mixture*

Because of its unusually high content of hydrophobic amino acids, soluble elastin has the unusual property of being soluble in mixtures of

---

[31] U. K. Laemmli, *Nature (London)* **277**, 680 (1970).
[32] W. M. Bonner and R. A. Laskey, *Eur. J. Biochem.* **46**, 83 (1974).
[33] E. H. Epstein, R. D. Scott, E. J. Miller, and K. A. Piez, *J. Biol. Chem.* **246**, 1718 (1971).

---

then electrophoresed on polyacrylamide gels as described by Weber and Osborn.[30] The gels were fractionated and counted as described in the text. Calibration of molecular weight as a function of position in the gel was made by electrophoresing the following standard proteins and staining with Coomassie Blue: lactate dehydrogenase (36,000), creatine kinase (40,000), pyruvate kinase (57,000), serum albumin (68,000), rat tail α-collagen chains (95,000). Region A contains procollagen and region B contains tropoelastin. Reproduced, by permission of North-Holland Publishing Co., Amsterdam, from Murphy et al.[5]

water, propanol, and butanol.[34] Samples containing labeled soluble elastin are dialyzed against 0.1 $M$ ammonium formate, pH 4.5, and then the concentration of ammonium formate is adjusted to 0.5 $M$ by adding the solid salt. The sample pH is adjusted to 5.5 and then 1.5 volumes of $n$-propanol are added dropwise at 4° with stirring. This addition is followed by 2.5 volumes of $n$-butanol. Any precipitate that forms is removed by centrifugation at 10,000 $g$ for 10 min. The supernatant is evaporated in a rotary evaporator at 30°, and then the sample is extracted five times with 0.5 ml of 1% sodium dodecyl sulfate in 0.01 $M$ sodium phosphate buffer, pH 7.4. The combined extracts are then dialyzed against the same buffer and analyzed by gel filtration or electrophoresis in sodium dodecyl sulfate.

### [$^{14}C$]Hydroxyproline Analysis

Quantitation of [$^{14}C$]hydroxyproline in samples labeled with [$^{14}C$]-proline may be made using standard ion exchange techniques or by a specific chemical procedure.[35]

### Reaction with Elastin-Specific Antibody

Antibodies are prepared in sheep against chick elastin.[36,37] Fifty aortas from 20-day-old chick embryos are extracted four times with 100 ml of 0.1 $M$ NaOH at 100° for 10 min.[38] The insoluble elastin residue is washed successively with water, ethanol, and then acetone. After the material is air-dried, it is pulverized in liquid nitrogen or using a Wiley mill. Ten milligrams of the product are suspended in 50% Freund's complete adjuvant and injected intradermally into the back of a sheep every 2 weeks. Each animal receives a total of five injections and is bled 2 weeks after each injection. Sheep are utilized for immunization because polymeric elastins from many species appear to elicit a poor immunogenic response in rabbits. (However, see below for antibody preparation to ligamentum nuchae elastin.)

Rabbits are immunized to tropoelastin isolated from lathyritic chicks by injection into the hind toe pads of 50 $\mu$g of tropoelastin in 0.1 ml of 0.02 $M$ sodium phosphate, pH 7.5, emulsified in an equal volume of Freund's complete adjuvant.[15,39] Injections are given weekly for a month. After a waiting period of 3 weeks, the rabbits are given another injection and then

[34] L. B. Sandberg, E. Bruenger, and E. G. Cleary, *Anal. Biochem.* **54**, 249 (1975).
[35] K. Juva and D. J. Prockop, *Anal. Biochem.* **15**, 77 (1966).
[36] P. Christner, M. Dixon, A. Cywinski, and J. Rosenbloom, *Biochem. J.* **157**, 525 (1976).
[37] B. C. Sykes and J. W. Chidlow, *FEBS Lett.* **47**, 722 (1974).
[38] A. I. Lansin, T. B. Rosenthal, M. Alex, and E. W. Dempsey, *Anat. Rec.* **114**, 555 (1952).
[39] J. A. Foster, D. Knaack, B. Faris, P. Moscaritolo, L. Salcedo, M. Skinner, and C. Franzblau, *Biochim. Biophys. Acta* **446**, 51 (1976).

bled after 8 days. Antibodies are prepared from the sera by ammonium sulfate precipitation and DEAE chromatography.

The radioactively labeled samples containing tropoelastin are dialyzed at 4° against 0.05 M sodium phosphate, pH 7.4, 0.15 M NaCl, and 0.1% Triton X-100 just prior to testing. From 1 to 5 µl (1–3 mg of antibody per milliliter) of the antibody solution are mixed with 6000–10,000 cpm of the labeled proteins in a 200-µl final volume of 0.05 M sodium phosphate, pH 7.4, 0.15 M NaCl, 0.1% Triton X-100 and incubated at 16° for 2 hr. If sheep antibody is being used, then the appropriate quantity of rabbit anti-(sheep globulin) is added to precipitate the sheep globulin quantitatively. If rabbit antibody is used, the antigen–antibody complex may be precipitated using heat-killed formalin-fixed *Staphylococcus aureus*. Forty microliters of *S. aureus* (30% suspension in the phosphate incubation buffer) are added, and the incubation is continued for 10 min. The immunoprecipitates are isolated by centrifugation through layered sucrose solutions in plastic microcentrifuge tubes. The tubes are frozen by immersion in an acetone–Dry Ice bath, the tips are cut off, and the immunoprecipitates are dissolved in 120 µl of 0.01 N HCl containing phenol red indicator. The solutions are carefully neutralized with 0.1 N NaOH. The samples are then counted or analyzed further by acrylamide electrophoresis or other techniques.[14,15]

*Radioimmune Displacement*[40]

*Immunizations.* Insoluble elastin is prepared from minced bovine ligamentum nuchae by the hot alkali method of Lansing.[38] The finely milled (>400 mesh) elastin is suspended at 1 mg/ml in 0.1 M sodium phosphate, 0.15 M NaCl, pH 7.4, and emulsified in an equal volume of complete Freund's adjuvant. New Zealand white rabbits are immunized by multiple-site subcutaneous injections on the back, each animal receiving a total of 2 ml. The animals are boosted monthly with 2.0 mg of elastin in complete Freund's adjuvant and are bled 8–12 days after immunization.

*Radiolabeling with $^{128}I$.* Purified α-elastin is prepared from the insoluble elastin by five successive 1-hr extractions with 0.25 M oxalic acid at 100°. The fourth and fifth extractions are combined and dialyzed against water and lyophilized. One hundred micrograms of the α-elastin dissolved in 10 µl of 0.1 M sodium phosphate buffer, pH 8.0, are iodinated by conjugation with 0.5 mCi of $^{128}I$-labeled Bolton–Hunter reagent (Amersham). After desalting on Sephadex G-25, iodinated antigens are diluted with assay buffer (0.01 M sodium phosphate buffer pH 7.4, 0.15 M NaCl, 0.5%, w/v, bovine serum albumin, 0.03%, v/v, NP-40) such that 0.1 ml contains 10,000–12,000 cpm. Bolton–Hunter-labeled probes are most stable when diluted in immunoassay sample buffer and frozen at 70°. Diluted antigen

[40] R. P. Mecham and G. Lange, *Connect. Tissue Res.* **7**, 247 (1980).

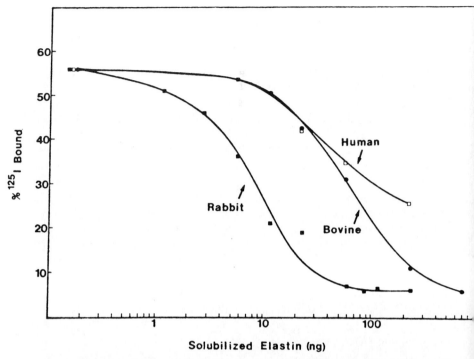

FIG. 4. Interspecies reactivity of antiserum against bovine ligamentum nuchae. Antiserum against ligamentum nuchae was prepared as described in the text and tested in the radioimmune displacement assay against various concentrations of $\alpha$-elastins prepared from rabbit lung, bovine ligament, and human aorta. $^{125}$I-labeled rabbit lung $\alpha$-elastin was used as the probe with an antiserum dilution of 1:1000. Reproduced, by permission of Gordon & Breach Science Publishers, from Mecham and Lange.[40]

should be frozen in small volumes, and unused portions should not be subjected to repeated freeze-thawing.

*Competitive Protein Binding Assay.* Antiserum dilution curves for each antigen are determined by incubating constant antigen concentrations and various dilutions of elastin antiserum for 1 hr at room temperature at 4° overnight. Antigen–antibody complexes are precipitated with 0.04 ml of a 10% suspension of IgGSORB (protein A fixed to inactivated *S. aureus* cells, The Enzyme Center, Inc. Boston, Massachusetts) by incubating for 60 min at 37°. Precipitates are pelleted by centrifugation (1500 g for 20 min), washed with assay buffer, recentrifuged, and counted for radioactivity in a gamma counter. Antiserum dilutions resulting in 50–60% labeled antigen binding are selected for competition protein binding assays.

The assay is carried out in duplicate in a final volume of 0.22 ml.

Unknown samples or standards in a volume of 0.02 ml are added to 0.1 ml of appropriately diluted antiserum. After incubation for 30 min at 37°, 0.1 ml of radiolabeld antigen in assay buffer (10,000–20,000 cpm) is added and the tubes are placed at 4° overnight. Total binding is determined by incubating antiserum and 0.1 ml of radiolabeled antigen in 0.05 ml assay buffer, and nonspecific binding is determined using normal rabbit serum. After overnight incubation at 4°, antigen–antibody complexes are precipitated with IgGSORB as described above, and the pellets are counted for radioactivity. Standard curves are prepared from stock solutions of oxalic acid-solubilized elastins, in which the protein concentration has been determined by amino acid analysis. Percentage of binding is calculated as

(CPM bound-nonspecific binding/total CPM-nonspecific binding) × 100

Nonspecific binding should average 150–200 cpm above background. The amount of elastin in experimental samples is determined from the standard curves constructed using the $\alpha$-elastin. Solubilized elastins from species other than bovine compete poorly with radiolabeled bovine ligament $\alpha$-elastin, whereas solubilized elastins from bovine ligament and aorta are equally efficient at displacing the radioactive probe. However, when solubilized elastins from nonbovine species are used as radiolabeled antigens, broad interspecies cross-reactivity using antiserum to ligamentum nuchae may be observed. Shown in Fig. 4 are competitive binding curves for rabbit lung, bovine ligament, and human aorta solubilized elastins using radiolabeled, solubilized rabbit lung elastin as a specific probe.

Acknowledgments

This work was supported by National Institutes of Health Grants AM-20863, AM-20553, HL-20994, and DE-02623.

## [41] Cell-Free Translation of Elastin mRNAs

*By* JUDITH ANN FOSTER, CELESTE B. RICH, STEVEN B. KARR, and ALAN PRZYBYLA

In order to cover fully the system for effective cell-free translations of elastin mRNAs, the methodology is divided into four sections dealing with: (a) isolation of RNA and preparation of mRNA; (b) preparation of the mRNA-dependent reticulocyte lysate and optimal conditions for the cell-free translation of elastin mRNAs; (c) identification of tropoelastins in a cell-free translation assay; and (d) quantitation of tropoelastins in the cell-free translation system. It is important to note at this point that our discussion of the cell-free synthesis of elastin involves the synthesis of two proteins designated tropoelastins a and b.[1] All our evidence to date suggests strongly that these two tropoelastins are separate gene products.[1-4] Consequently, the methodologies described are geared not only for the production of soluble elastin, but also for the separate identification of tropoelastins a and b.

Isolation of Elastin mRNAs

Our laboratory routinely uses two techniques for the isolation of total RNA from tissues. Which of the two techniques is used is predicated on what tissue as well as what animal source is being investigated. In both techniques, extreme care is taken to prevent RNase contamination. All glassware is heated to 190° for 16 hr where appropriate or heat-labile materials are treated with a freshly prepared solution of diethylpyrocarbonate (0.5% in sterile $H_2O$). Solutions are sterilized through a 0.2-$\mu$m membrane filter, or autoclaved. Sterile techniques are employed throughout RNA isolation and translation procedures.

Two techniques are described for the isolation of total RNA from various connective tissues. The first technique is based on that described by Strohman et al.[5] and is best applied to chick lung and pig lung parenchymal tissues. These tissues are placed immediately into 6 $M$ guanidine-HCl (1–3 g of tissue per 19 ml). Potassium acetate (2 $M$, pH 5.0) is added to make a 5% solution. The suspension is homogenized with a Brinkmann Polytron at high speed for 2 min at room temperature and then precipitated by the addition of 0.5 volume of 95% ethanol at $-20°$ overnight. The precipitate is centrifuged for 1 hr at 16,000 $g$ at $-20°$, and the resulting pellet is resuspended at room temperature in 6 $M$ guanidine-HCl (0.5 volume of the original extracting volume of 6 $M$ guanidine-HCl), and 5% 0.5 $M$ EDTA (pH 7). The mixture is vortexed thoroughly, followed by precipitation with addition of 0.25 volume of 2 $M$ potassium acetate (pH 5.0) and 0.5 volume of 95% ethanol at $-20°$. After storage overnight at

[1] J. A. Foster, C. B. Rich, S. Fletcher, S. R. Karr, and A. Przybyla, *Biochemistry* **19**, 857 (1980).
[2] J. A. Foster, C. B. Rich, S. Fletcher, S. R. Karr, M. D. DeSa, T. Oliver, and A. Przybyla, *Biochemistry* **20**, 3528 (1981).
[3] L. L. Barrineau, C. B. Rich, and J. A. Foster, *Connect. Tissue Res.* **8**, 189 (1981).
[4] L. L. Barrineau, Master's Thesis, Department of Biochemistry, University of Georgia, Athens, Georgia, 1980.
[5] R. C. Strohman, P. S. Moss, J. Micou-Eastwood, D. Spector, A. E. Przybyla, and B. Peterson, *Cell* **10**, 265 (1977).

−20°, the precipitate is centrifuged for 10 min at 16,000 $g$ (−20°), and the pellet is dried under a stream of $N_2$ to ensure complete removal of ethanol. The pellet is then resuspended in 20 m$M$ EDTA, pH 7 (0.15 volume of the original extraction volume of 6 $M$ guanidine-HCl), and is extracted with 2 volumes of chloroform–butanol (4:1).

The aqueous and organic phases are separated by centrifugation at 10,000 $g$ for 10 min, and the aqueous phase is carefully removed. The organic phase is then extracted four times with 20 m$M$ EDTA, pH 7 (same volume as described above). The aqueous layers are pooled and precipitated by the addition of 0.1 volume of 2 $M$ potassium acetate (pH 5.0) and 2.5 volumes of 95% ethanol at −20°. After storage overnight at −20°, the pellet is collected by centrifugation at 16,000 $g$ for 1 hr (−20°), dried under a stream of $N_2$, and resuspended in sterile $H_2O$ (0.15 volume of the initial extraction volume). The RNA is precipitated by the addition of 2 volumes of 4.5 $M$ sodium acetate (pH 6) and stored overnight at −10°. The precipitate is centrifuged for 1 hr at 16,000 $g$ (−10°), and the pellet is washed with 95% ethanol at −20°. The washed RNA pellet is dried under a stream of $N_2$, dissolved in sterile $H_2O$, and stored precipitated at −20° in 0.1 volume 2 $M$ potassium acetate and 2.5 volumes of 95% ethanol.

A second technique for the isolation of total RNA is based on the methodology of Przybyla et al.[6] This technique is especially applicable to chick aortic and pig aortic, cartilage, and lung airway tissues.[1,3] The excised tissue is suspended (3 g/20 ml) in 5 $M$ guanidinium thiocyanate, 50 m$M$ EDTA, 50 m$M$ Tris (pH 7.5), 2% Sarkosyl, and 5% 2-mercaptoethanol and is homogenized with a Brinkmann Polytron at high speed for 2 min. The homogenate is layered onto 5 ml of 5.7 $M$ cesium chloride, 100 m$M$ EDTA and is centrifuged at 180,000 $g$ for 17 hr at 20°. The pellet is redissolved in sterile $H_2O$, and the RNA is precipitated overnight by the addition of 0.1 volume of 2 $M$ potassium acetate (pH 5.0) and 2.5 volumes of 95% ethanol at −20°. The RNA is collected by centrifugation at 10,000 $g$ for 1 hr at −20°, redissolved in $H_2O$, and precipitated by the addition of 2 volumes of 4.5 $M$ sodium acetate (pH 6.0) overnight at −10°. The final precipitate is recovered by centrifugation at 10,000 $g$ for 1 hr at −10°, redissolved in $H_2O$, and stored precipitated at −20° in 0.1 volume of 2 $M$ potassium acetate and 2.5 volumes of 95% ethanol.

Isolation of poly(adenylic acid)-rich mRNA is accomplished by a modification of the procedure of Aviv and Leder[7] using a column of oligothymidylic acid-cellulose (oligo(dT)-cellulose, type 2, Collaborative Research Inc.).[8] The column consists of 2 g of oligo(dT)-cellulose in a

---

[6] A. E. Przybyla, R. J. MacDonald, J. D. Harding, R. L. Pictet, and W. J. Rutter, *J. Biol. Chem.* **254**, 2154.
[7] H. Aviv and P. Leder, *Proc. Natl. Acad. Sci. U.S.A.* **69**, 1408 (1972).
[8] R. J. MacDonald, A. E. Przybyla, and W. J. Rutter, *J. Biol. Chem.* **252**, 5522 (1977).

50-ml sterile, disposable plastic syringe. Prior to the application of RNA, the column and resin are washed with a freshly prepared solution of diethyl pyrocarbonate followed by equilibration in binding buffer (0.5 $M$ KCl, 10 m$M$ Tris-HCl, pH 7.4). Total RNA, which has been precipitated with 95% ethanol and dried under a stream of $N_2$, is dissolved in sterile $H_2O$ (10 mg/ml maximum), heated at 80° for 2 min, and quick cooled; 1 volume of 2 × binding buffer is added. The oligo(dT) column is drained dry, and the RNA solution is applied slowly with the bottom of the column open. The column effluent is collected in one tube and reapplied two more times, running the column dry each time. The column is then washed with binding buffer until the $OD_{260}$ is less than 0.010. The bound, poly(adenylic acid)-rich mRNA is eluted with 10 m$M$ Tris-HCl, pH 7.4, the effluent is collected in 2-ml fractions, and the optical density is monitored at 260 nm. Those tubes possessing significant absorbance are pooled, and the RNA is precipitated by the addition of 0.1 volume of 2 $M$ potassium acetate (pH 5.0) and 2.5 volumes of 95% ethanol overnight at $-20°$. The precipitated RNA is reapplied to the oligo(dT)-cellulose column one more time to ensure adequate removal of ribosomal RNA.

Preparation of Lysate

Rabbit reticulocyte lysates are prepared according to the procedure of Schimke et al.[9] with the exception that blood is collected by cardiac puncture into 50-ml polycarbonate tubes containing 0.1 ml of sodium heparin (10,000 U/ml). Lysates are dripped into liquid nitrogen and stored as frozen pellets at $-80°$.

The prepared lysate is rendered mRNA-dependent by a modification of the method of Pelham and Jackson.[10] To 90 $\mu$l of thawed lysate is added 1 $\mu$l of 75 m$M$ $CaCl_2$, 5 $\mu$l of creatine kinase (5 mg/ml in 50% aqueous glycerol), 2.5 $\mu$l of 1 m$M$ hemin (in 90% ethylene glycol, 20 m$M$ Tris-HCl, pH 8.2, and 50 m$M$ KCl), and 1 $\mu$l of micrococcal nuclease (mg/ml $H_2O$). Incubation is for 10 min at 20°. The digestion is terminated by the addition of 1 $\mu$l of 200 m$M$ EDTA. Undigested lysate preparations are treated in an identical manner with the exception that nuclease is not added.

Translation Conditions[1,11]

A regular translation assay consists of 25 $\mu$l of reticulocyte lysate or nuclease-treated lysate, 19 $\mu$l of master mix, and 18 $\mu$l for the addition of variable amounts of RNA, $H_2O$, and $\gamma$-globulins. The final concentrations

[9] R. T. Schimke, R. E. Rhoades, and G. S. McKnight, this series, Vol. 30, p. 694.
[10] H. R. Pelham and R. J. Jackson, Eur. J. Biochem. 67, 247 (1976).
[11] S. R. Karr, C. B. Rich, A. Przybyla, and J. A. Foster, Coll. Res. 1, 1 (1980).

of compounds in the translation assay contributed by the master mix are 75 $\mu M$ glycine, valine, proline, and alanine and 25 $\mu M$ all other amino acids, 20 m$M$ HEPES (pH 7.6), 30 m$M$ KCl, 0.65 m$M$ Mg(C$_2$H$_3$O$_2$)$_2$, 1 m$M$ ATP, 0.2 m$M$ GTP, 11 m$M$ creatine phosphate, 10 $\mu$g/ml creatine phosphokinase (115 U/mg), 0.4 m$M$ spermidine and 0.065 $\mu M$ DFP. The two most effective radioisotopes for labeling newly translated elastin polypeptides are valine and proline. Proline is especially helpful in detecting procollagen synthesis. The concentrations of L-[2,3-$^3$H]valine (Schwarz–Mann, 23 Ci/mmol) or L-[5-$^3$H]proline (Schwarz–Mann, 38 Ci/mmol) are 130 $\mu$Ci per assay (62 $\mu$l total volume). In all cases where a radioactive amino acid is used, that particular unlabeled amino acid is omitted from the master mix.

Prior to translation, the RNA (either total RNA or mRNA) is centrifuged at 10,000 $g$ for 1 hr at $-20°$, dried under a stream of N$_2$, and dissolved in H$_2$O (50 $\mu$g of total RNA/18 $\mu$l or 2–4 $\mu$g of mRNA/18 $\mu$l). The RNA solution is then heated at 70° for 1 min and quickly immersed in an ice bath. The importance of the heating step is best illustrated in Fig. 1, wherein the effect of heating the RNA prior to translation on the translation products directed by chick aortic RNA is shown.

For the actual translation of the RNA, 50 $\mu$g of total RNA or 2–4 $\mu$g of mRNAs in 18 $\mu$l of H$_2$O are mixed together with the nuclease-treated lysate and master mix. Translations are carried out for 90 min at 26°. Incorporation of radiolabeled amino acids into proteins is monitored by the procedure described by Pelham and Jackson.[10] Two-microliter aliquots are removed from the assay at 15-min intervals and spotted onto 2 × 2 cm Whatman 3-mm filter squares. The filters are allowed to air dry and are then soaked for 10 min in cold 10% TCA on ice in the presence of excess unlabeled amino acid corresponding to the radioactive amino acid used. The filter squares are then washed twice in ice cold 5% TCA and heated to 95° for 10 min in 5% TCA. After cooling, the filters are washed twice in 5% TCA, twice in 95% ethanol, once in ethanol– ether, 1 : 1, once in ether, and then placed under a heat lamp for 15 min to dry. The dried TCA precipitates are then soaked overnight in 0.4% Omnifluor (w/v), 2.6% tissue solubilizer (v/v), and 0.4% water in toluene. Radioactivity is then determined in a scintillation counter.

Examination of the molecular weight distribution of the radiolabeled proteins directed by the translation of mRNAs is conveniently performed by SDS–polyacrylamide slab gel electrophoresis followed by fluorography. Samples from translation assays are prepared by precipitating 5–10 $\mu$l of the reaction mixture with 8 volumes of acetone and drying under a stream of N$_2$. The precipitate is then dissolved in 20–30 $\mu$l of gel sample mix (1 ml of glycerol, 0.5 ml of 2-mercaptoethanol, 2.3 ml of 10% SDS,

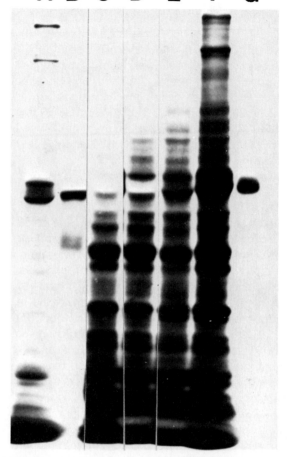

FIG. 1. Sodium dodecyl sulfate–polyacrylamide gel electrophoresis of translation products in which the RNA was heated at various temperatures prior to translation in an mRNA-dependent reticulocyte lysate, with added aortic RNA. Lane A: [$^3$H]proline label; lanes B and G, tropoelastin reduced with [$^3$H]NaBH$_4$; lanes C–F: [$^3$H]valine label. RNA was incubated at room temperature (lane C); at 60° (lane D); at 70° (lane E); and at 80° (lane F). Exposure time was 1 week. From S. R. Karr et al.[11]

1.25 ml of 0.5 M Tris-HCl, pH 6.8, brought to 10 ml with H$_2$O) and heated for 2 min in a boiling H$_2$O bath. Vertical slab gel electrophoresis is performed according to Laemmli[12] using an 8% running gel and 3% stacking gel. At completion of electrophoresis, the gels are fixed for 30 min in 50% ethanol, 10% TCA; stained in 25% ethanol, 10% TCA, 0.2% Coomassie

[12] U. K. Laemmli, Nature (London) **227**, 680 (1970).

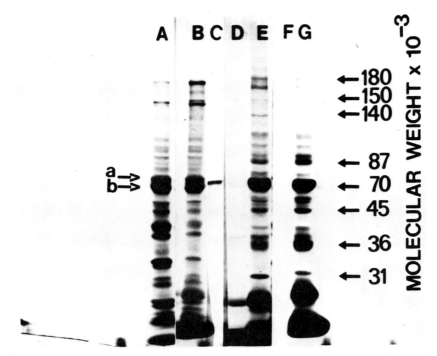

FIG. 2. Sodium dodecyl sulfate–polyacrylamide gel electrophoresis of translation products from mRNA-dependent reticulocyte lysate with added aortic RNA labeled with [$^3$H]valine (lane A) and [$^3$H]proline (lane B). Lane C: Authentic tropoelastin reduced with NaB[$^3$H]$_4$; lanes D and F: digested reticulocyte lysate with no added RNA; incubated with [$^3$H]valine and [$^3$H]proline, respectively; lanes E and G: untreated reticulocyte lysate with no added RNA; incubated with [$^3$H]valine and [$^3$H]proline, respectively. Translation time 90 min. Arrows a and b indicate tropoelastins a and b. Lanes A and B were exposed for 2 weeks; lanes C–G for 1 week. From Foster et al.[1]

Brilliant Blue; and destained overnight in 10% acetic acid. Fluorography is performed according to the procedure of Bonner and Laskey.[13] A fluorogram of an SDS-gel showing the molecular weight distribution of [$^3$H]valine- and [$^3$H]proline-labeled proteins directed by chick aortic mRNA is given in Fig. 2. Of special importance in the translation and eventual identification of newly translated tropoelastins is the necessity of generating a mRNA-dependent reticulocyte lysate system. As can be seen in Fig. 2, among the proteins directed by endogenous lysate, mRNA is a major protein possessing an apparent molecular weight of 70,000. This protein is labeled significantly with both valine and proline as shown. Hence, it is apparent that an mRNA-dependent lysate generated by treat-

[13] W. M. Bonner and R. A. Laskey, *Eur. J. Biochem.* **46**, 83 (1974).

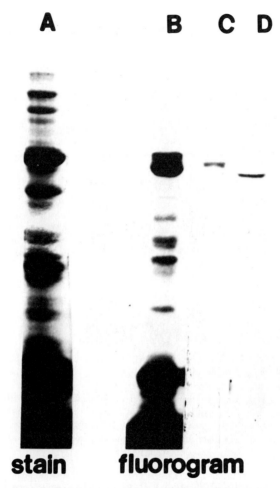

FIG. 3. Sodium dodecyl sulfate–polyacrylamide gel electrophoresis. Lane A: Coomassie Blue stain of reticulocyte lysate proteins. Fluorography of translation products from digested reticulocyte lysate plus aortic RNA incubated with [$^3$H]proline (lane B); authentic [$^3$H]tropoelastin (lane C); authentic [$^3$H]tropoelastin plus unlabeled reticulocyte lysate proteins (lane D). Translation time was 90 min. Exposure of lanes B–D was for 1 week. From Foster et al.[1]

ment with micrococcal nuclease is necessary to visualize the two tropoelastin proteins migrating in the same area of the gel. However, it should be noted that the normal presence of the unlabeled reticulocyte lysate protein of molecular weight 70,000 in the translation assay serves to accentuate the electrophoretic separation of tropoelastins a and b. This phenomenon is best illustrated in Fig. 3, where the effect of adding unlabeled reticulocyte lysate to the electrophoretic mobility of tropoelastin b is seen.

Detection of Translated Tropoelastins

One-dimensional electrophoresis of the radiolabeled translation products directed by connective tissue mRNAs is suggestive in assigning the presence of tropoelastin. Another technique that is useful for identifying tropoelastin by charge in addition to molecular size is a two-dimensional system utilizing nonequilibrium pH gradient electrophoresis (NEPHGE) in the first dimension and SDS-slab gel electrophoresis in the second dimension according to the procedure of O'Farrell et al.[14] The first-dimension cylindrical gels are run at 400 V for 8 hr, six gels being the normal number processed at one time. The second dimension slab gel is performed as described by Laemmli using an 8% running gel. The advantage of the NEPHGE system over conventional isoelectric focusing is that the nonequilibrium situation allows definition in the migration of very basic proteins. The results of two-dimensional electrophoresis of the cell-free translation of chick aortic mRNA are given in Fig. 4. By this technique not only can tropoelastin b be separated from other proteins, but tropoelastin a can be separated from tropoelastin b on the basis of overall charge as well as molecular weight.[1]

Although the two-dimensional system described above is good indirect evidence for the presence of tropoelastins in a cell-free translation of mRNAs, the best criterion for direct evidence of tropoelastin is immunoprecipitation with specific antibody.[1] Optimum precipitation of tropoelastin is obtained by translation of elastin mRNAs in the presence of tropoelastin antibody.[1] Antibody to purified tropoelastin is raised in rabbits as previously reported.[15] The γ-globulins are partially purified from both immune and nonimmune sera by ammonium sulfate according to the precipitation procedure of Harboe and Ingild.[16] The γ-globulins are then made RNase-free by passage over a combined DEAE- and CM-cellulose column exactly as described by Palacios et al.[17] Under these chromatography conditions, the γ-globulins elute first, and their concentration is determined assuming 1.46 $A_{280}$ per milligram of γ-globulin. Prior to the addition of γ-globulins (immune and nonimmune) to the translation assay,

---

[14] P. Z. O'Farrell, H. M. Goodman, and P. H. O'Farrell, Cell **12**, 1133 (1977).
[15] J. A. Foster, D. Knaak, B. Faris, R. Moscaritolo, L. Salcedo, M. Skinner, and C. Franzblau, Biochim. Biophys. Acta **446**, 51 (1976).
[16] N. Harboe and A. Ingild, in "A Manual of Quantitative Immunoelectrophoresis" (N. H. Axelson, J. Kroel, and B. Weeke, eds), p. 161. Universitetsforlagel, Oslo, Norway, 1973.
[17] R. Palacios, R. D. Palmiter, and R. T. Schimke, J. Biol. Chem. **247**, 2316 (1972).

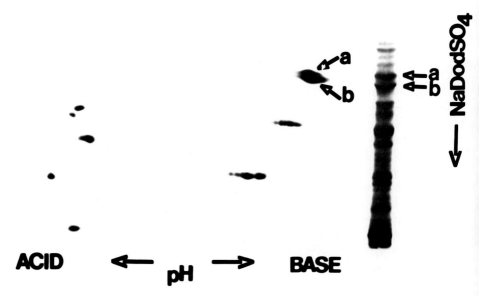

FIG. 4. Two-dimensional gel electrophoresis of aortic RNA translation products labeled with [$^3$H]valine (90 min). First-dimension electrophoresis in 3.5–10 ampholines was performed for 3200 V hr. Exposure time was 1 week. Arrows a and b indicate tropoelastin a and b. From Foster et al.[1]

they are dialyzed against H$_2$O for 8 hr, lyophilized, and redissolved in sufficient H$_2$O (3.5 µg/µl) to allow the addition of 35 µg of γ-globulins per translation assay (62 µl total). Immunoprecipitation of tropoelastins from among the cell-free translation products is very easily and efficiently accomplished using heat-killed, formalin-fixed *Staphylococcus* as an immunoadsorbent prepared as described by Kessler.[18] After the translation performed in the presence of γ-globulin, 0.2 volume of 5 × immunoprecipitation buffer (1 × = 0.5% Nonidet P-40, 150 m$M$ NaCl, 50 m$M$ Na$_2$HPO$_4$, pH 7.6, 1% sodium deoxycholate, 5 m$M$ EDTA, and 2 m$M$ unlabeled amino acid corresponding to the radioactive amino acid used for labeling) and 0.2 volume of 6 $M$ urea in 0.01 $M$ Na$_2$HPO$_4$ (pH 7.6) are added. The reaction mixture is vortexed briefly and is then centrifuged at 10,000 $g$ for 5 min at 4° to remove any particulate material; 10 µl of additional γ-globulin (3.5 µg/µl) are added, and the mixture is kept at room temperature for 15 min, after which time 40 µl of *S. aureus* (30% suspension in 1 × immunoprecipitation buffer) is added. After 5 min at room temperature, the suspension is layered on to 300 µl of 1.5 $M$ sucrose

[18] S. W. Kessler, *J. Immunol.* **115**, 1617 (1975).

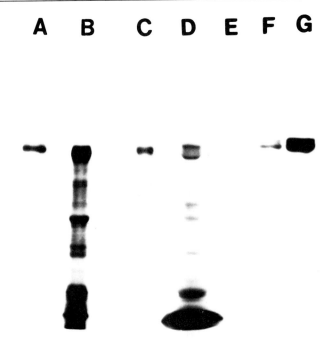

FIG. 5. Sodium dodecyl sulfate–polyacrylamide gel electrophoresis of immunoprecipitates from aortic organ culture and aortic organ culture and aortic RNA translation products. Lane A: Immunoprecipitate of NSB extract of chick aortas; lane B: NSB extract of chick aortas labeled with [³H]valine (5 µl); lane C: Authentic [³H]tropoelastin; lane D: translation products of aortic RNA labeled with [³H]proline; lane E: control immunoprecipitate of cell-free translation of aortic RNA with nonimmune IgG; lane F: immunoprecipitation of cell-free translation of aortic RNA by the addition of 70 µg of IgG after translation was complete; lane G: immunoprecipitate of cell-free translation of aortic RNA by the presence of 35 µg of IgG during translation and an additional 35 µg of IgG after translation was complete. In all immunoprecipitations *Staphylococcus aureus* was used as an immunoabsorbent, as described in the text. Exposure time was 1 week. From Foster *et al.*[1]

in 1 × buffer and centrifuged at 4° for 5 min at 10,000 g; the pellet is resuspended in 100 µl of 1 × buffer. The suspension is layered onto another 300 µl of 1.5 M sucrose and centrifuged as described above. The resulting pellet is resuspended and centrifuged 3 times in 300 µl of 1 × buffer. The radiolabeled antigen is then eluted from the *S. aureus* by incubation at 37° for 20 min in 50 µl of 3% SDS, 5 M urea and centrifuged at 10,000 g for 5 min. Prior to examination on SDS-slab gels, 2 µl of 2-mercaptoethanol are added, and the sample is placed in a boiling $H_2O$ bath for 3 min. Figure 5 pictures the application of the above-described immunoprecipitation technique to the cell-free translation products directed by chick aortic mRNAs. As can be seen, immunoprecipitation of translation products performed in the presence of added antibody (lane G) resulted in a threefold increased recovery over antibody added after translation (lane F).

Quantitation of Translated Tropoelastins[1,19]

A very sensitive technique for the direct quantitation of tropoelastins synthesized in the cell-free system is rocket immunoelectrophoresis.[20] The immunoelectrophoretic plate is prepared by heating 12 ml of 1% agarose in 0.0125 $M$ $Na_2HPO_4$ buffer (pH 7.4) in a boiling $H_2O$ bath. After the agarose cools to 55°, 200 µl of tropoelastin antiserum (the amount must be experimentally determined since antibody titers vary) is added, rapidly mixed, and poured onto a glass plate (75 × 85 mm). When the agarose has congealed, 5 µl-volume sample application wells are punched, allowing sufficient wells to accommodate duplicate application of all standards and samples to be quantitated.

Electrophoresis is performed on an LKB Multiphor Unit with the cooling plate maintained at 6° with a Haake constant-temperature circulator. The tray buffer is 0.05 $M$ $Na_2HPO_4$, pH 7.4. Electrophoresis is performed at 300 V for 4 hr. Tropoelastin migrates toward the cathode under these conditions. For detection of radiolabeled rockets, fluorography is performed. After electrophoresis, the gel is washed overnight in 0.1 $N$ NaCl with several changes. During this washing the gel becomes detached from the glass and must be handled carefully. The gel is then washed in deionized $H_2O$ for 1 hr. Fluorography of the gel is performed essentially as described by Bonner and Laskey[13] with the exceptions that methanol was substituted for dimethyl sulfoxide and the percentage of 2,5-diphenyloxazole was 18% (w/v) in methanol.[1] The impregnated gel is dried onto filter paper using a Bio-Rad slab dryer and exposed to Kodak X-Omat R film at −80°. Standard radiolabeled tropoelastin is obtained by incubating chick embryonic aortas in medium containing [$^3$H]valine (500 µCi/ml) for 1 hr.[19] After the labeling period, the medium is decanted and the aortas are homogenized in 0.01 $M$ phosphate, 0.15 $N$ NaCl containing 0.1 $M$ $N$-ethylmaleimide, 0.01 $M$ chloroquine, 0.002 $M$ EDTA, and 0.01 $M$ DFP. The aortas are extracted for 2 hr at 4° and centrifuged (15,000 $g$) for 1 hr; the supernatant is sampled directly by rocket immunoelectrophoresis. The amount of tropoelastin in the extract is determined by running different aliquots of the extract together with known amounts of chemically

---

[19] J. A. Foster, S. C. Wu, M. T. Marzullo, and C. B. Rich, *Anal. Biochem.* **101**, 310 (1980).
[20] C. B. Laurell, *Anal. Biochem.* **15**, 42 (1966).

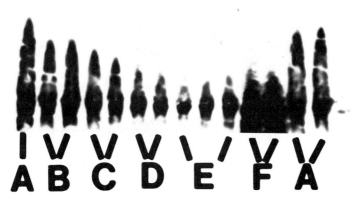

FIG. 6. Rocket gel electrophoresis of aortic RNA translation and aortic organ culture products. The concentrations of soluble elastin in the NSB-extract samples are as follows: lane A, 2.8; B, 2.0; C, 1.56; D, 1.27; E, 0.67. The linear correlation ($r$) was 0.9985. Five microliters of the translation assay (lane F) revealed 1 µg/ml of soluble elastin. Exposure time was 2 days. From Foster et al.[1]

purified tropoelastin.[21] Detection of the resulting rockets is performed by Coomassie Blue staining of the gel. Determination of the amount c tropoelastin in the aortic extract is made by measuring peak heights and calculating a line through the points using least-square analysis.[19] Once the amount of radiolabeled tropoelastin is established, this sample is di luted so that fluorography of the extract samples allows a significant linear correlation between rocket height and antigen concentrations of 1–3 µg/ml. Then 5-µl samples of the cell-free translation assay can be run, and the amount of tropoelastin can be quantitated using different dilutions of the aortic extract as standards. The results from such an analysis are shown in Fig. 6.

Acknowledgments

This work was supported by National Institutes of Health Grant HL 22208 and the Council for Tobacco Research 1179.

[21] J. A. Foster, R. Shapiro, P. Voynow, G. Crombie, B. Faris, and C. Franzblau, *Biochemistry* **14**, 5343 (1975).

## [42] Antibodies to Insoluble and Solubilized Elastin

By ROBERT P. MECHAM and GARY LANGE

A major problem in elastin biochemistry has been the development of methods for quantifying elastin synthesis and degradation. Significant progress has been made with the development of immunoassays using antisera prepared against tropoelastin (see Chapter [44] of this volume) and various forms of elastin.[1]

Tropoelastin and elastin are multideterminant antigens, and their antisera are inevitably heterogeneous. In addition, structural dissimilarities between the two proteins result in serological nonidentity. Notwithstanding these differences, insoluble elastin and tropoelastin share enough common determinants so that antiserum to elastin cross-reacts with tropoelastin (see Fig. 1) and can be a valuable tool in studying tropoelastin biosynthesis. In addition, anti-elastin antibodies recognize elastin peptides released from the intact protein by proteolysis and can be useful to investigate elastin degradation. The use of elastin as an antigen, instead of tropoelastin, offers the advantages that elastin is easy to purify, is available in larger quantities than tropoelastin, and is not readily susceptible to proteolysis and degradation. This chapter will describe the preparation of antisera to insoluble or solubilized elastin and the use of the antisera in immunoassays.

Solubilization of Elastin

Although the insolubility of elastin precludes its use directly in immunoassays, solubilized forms of elastin retain immunological activity and are applicable to immunological techniques. Four methods for solubilizing elastin are described below. The reaction products in each case are different, and the original papers should be consulted for a more thorough characterization of the solubilized peptides.

*Preparation of α-Elastin*[2]

Finely milled elastin is suspended in 0.25 M oxalic acid (1 part elastin to 8 parts acid, w/v). The mixture is heated on a steam bath or in a boiling

---

[1] To avoid confusion in subsequent paragraphs, tropoelastin refers to the salt-soluble, uncross-linked elastin precursor. Elastin is the mature, fully cross-linked, insoluble protein, and solubilized elastin refers to cross-linked insoluble elastin that has been solubilized either enzymatically or chemically.

[2] S. M. Partridge, H. F. Davis, and G. S. Adair, *Biochem. J.* **61**, 11 (1955).

[42] ANTIBODIES TO INSOLUBLE ELASTIN 745

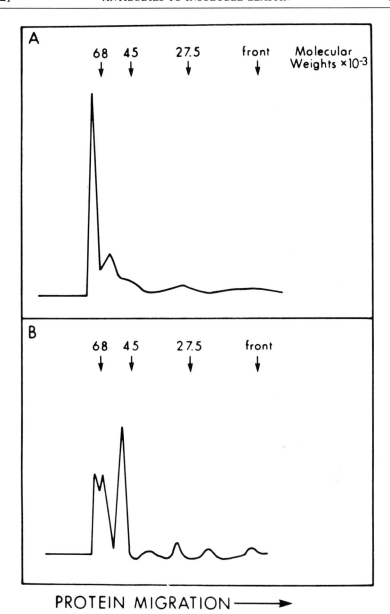

FIG. 1. (A) The specificity of anti-insoluble elastin serum for tropoelastin was shown by immunoprecipitation of ligamentum nuchae fibroblast culture medium after 16 hr of continuous labeling with [$^3$H]valine. Culture medium was boiled for 5 min after the addition of proteolytic inhibitors and dialyzed extensively against 0.1 $N$ acetic acid. One part Hanks' balanced salt solution (10 × concentrated) was added to 10 parts dialyzed sample, and the pH was adjusted to 7.5 with 1.0 $N$ NaOH. Rabbit anti-bovine insoluble elastin serum was

water bath for 1 hr and then rapidly cooled and centrifuged (16,000 $g$, 20 min). After the extract is removed, the precipitate is washed with 0.25 $M$ oxalic acid, and the washings are discarded. The residue is extracted again with 0.25 $M$ oxalic acid for 1 hr at 100°, and the extract and washings are collected as before. The cycle of extraction and washing is repeated five times in all. Bovine ligamentum nuchae elastin is completely dissolved by the fifth extraction, giving a clear, yellow solution. The solution of solubilized elastin is dialyzed against distilled water until free of oxalate and then lyophilized. The dried powder is dissolved in 0.01 $M$ acetate or phosphate buffer, pH 5.5, containing 0.1 $M$ NaCl and heated to 37°. The solution is centrifuged at the same temperature, and the coacervate, representing $\alpha$-elastin, is collected at the bottom of the tube as a viscous syrup. Coacervation is reversible so that $\alpha$-elastin quickly redissolves upon cooling.

*Preparation of $\kappa_2$-Elastin* [3,4]

Elastin powder (10 g) is suspended with stirring in 250 ml of 1 $N$ KOH–ethanol (80:20, v/v) at 37° for 60 min. The mixture is centrifuged (16,000 $g$ for 15 min), and the supernatant is discarded. The pellet is resuspended in the KOH–ethanol mixture and heated at 37° for 60 min. The soluble fraction is neutralized with perchloric acid, and any resulting precipitate is discarded. The supernatant is lyophilized (yielding $\kappa$-elastin) and chromatographed on Sephadex G-200 with 0.02 $M$ acetic acid. The initial high molecular weight peaks ($\kappa_2$-elastin) are pooled and lyophilized.

[3] L. Robert and N. Poullain, *Bull. Soc. Chim. Biol.* **45**, 1317 (1963).
[4] D. Houle and F. LaBella, *Connect. Tissue Res.* **5**, 83 (1977).

---

added to a dilution of 1:000. After an overnight incubation at 4°, goat anti-rabbit IgG was added to a final dilution of 1:100 and the sample was incubated at 4° overnight. Antigen–antibody complexes were precipitated by centrifugation, and the pellet was washed five times with HEPES buffer, pH 7.4, containing 0.15 $M$ NaCl, 1% Tween-20, 0.5% deoxycholate, 10 m$M$ EDTA, and 2.5 m$M$ phenylmethylsulfonyl fluoride, The washed pellet was dissolved by boiling for 5 min in sodium dodecyl sulfate (SDS)-gel electrophoresis sample buffer containing 10 m$M$ dithiothreitol and electrophoresed on a 10% polyacrylamide gel according to U. K. Laemmli [*Nature (London)* **227**, 680 (1970)]. The gel was fixed in trichloroacetic acid–methanol, impregnated with EN³HANCE (New England Nuclear, Boston, Massachusetts), and dried. The dried gel was exposed to Kodak X-Omat R film at −70°. Shown is the densitometric tracing of the autoradiogram. (B) Immunoprecipitation (as above) of degraded tropoelastin showing that anti-insoluble elastin serum recognizes fragments as well as the intact molecule. The fragmentation pattern of tropoelastin and fragment molecular weights have been described previously [R. P. Mecham and J. A. Foster, *Biochemistry* **16**, 3825 (1977)].

Enzymic Digestion of Elastin

*Pancreatic and Leukocyte Elastase.* Porcine pancreatic elastase, readily obtainable commercially, is added to insoluble elastin suspended in 0.05 $M$ Tris buffer, pH 8.8, at an enzyme:substrate ratio of 1:100. The mixture is incubated at 37° with constant stirring until the elastin is completely solubilized, usually between 4 and 12 hr, depending upon the specific activity of the elastase. Enzymic activity can be inhibited by heating to 100° for 15 min or by the addition of phenylmethylsulfonyl fluoride to 2 m$M$.

Optimal digestion of elastin with leukocyte elastase occurs at pH 8 in buffers containing 0.1 $M$ NaCl.[5] Leukocyte and pancreatic elastases have different peptide bond specificities and produce different peptides from elastin.[6]

*Pepsin.*[4] Unlike pancreatic elastase, which cleaves elastin at amino acid residues with small hydrophobic side chains, pepsin cleaves peptide bonds primarily at aromatic residues. Houle and LaBella[4] have reported that pepsin completely solubilizes elastin, although more slowly than porcine pancreatic elastase. Elastin is suspended in 5% acetic acid, and pepsin is added to an enzyme:substrate ratio of 1:100. The solution is allowed to react at 37° with constant stirring. The elastin is solubilized after 24 hr.

*Other Enzymes.* Other enzymes that digest elastin include subtilisin,[7] thermolysin,[8] papain,[9,10] ficin,[9,10] bromelain,[10,11] and nagarase.[12] The original papers should be consulted for specific reaction conditions.

Preparation of Antiserum

Antisera have been successfully prepared to insoluble elastin and to solubilized peptides from elastin prepared as described above. The quantity of elastin antigen necessary for optimal antibody production, however, is not known. As little as 0.67 mg and as much as 100 mg of elastin have been used to immunize rabbits (Tables I and II). Unfortunately,

[5] A. Janoff and J. Scherer, *J. Exp. Med.* **128**, 1137 (1968).
[6] R. M. Senior, D. R. Bielefeld, and B. C. Starcher, *Biochem. Biophys. Res. Commun.* **72**, 1327 (1976).
[7] J. A. Foster, L. Rubin, H. M. Kagan, C. Franzblau, L. E. Bruenger, and L. B. Sandberg, *J. Biol. Chem.* **249**, 6191 (1974).
[8] R. P. Mecham and J. A. Foster, *Biochem. J.* **173**, 617 (1978).
[9] D. S. Miyada and A. L. Tappel, *Food Res.* **21**, 217 (1956).
[10] J. Thomas and S. M. Partridge, *Biochem. J.* **74**, 600 (1960).
[11] H. Wang, C. E. Weir, M. L. Birkner, and B. Ginger, *Food Res.* **23**, 423 (1958).
[12] G. Crombie, J. A. Foster, and C. Franzblau, *Biochem. Biophys. Res. Commun.* **52**, 1228 (1973).

TABLE I
ANTISERA TO INSOLUBLE ELASTINS

| Elastin antigen | Host animal | Immunization amount (mg) | Boosting amount (mg) | Boosting interval | References |
|---|---|---|---|---|---|
| Chicken aorta | Sheep | 10–20 | 10–20 | 2 Weeks | a, b |
| Dog or human lung | Sheep[c] | 10 | 10 | 2 Weeks | d |
| Bovine ligament | Rabbits | 1 and 10 | 1 and 10 | 2–3 Weeks | e, f |
| Bovine ligament | Guinea pigs | 1.7 | 1.7 | Weekly | g |
| Human lung | Rabbits | 20 | 20 | Twice at 10-day intervals | h |

[a] Sykes and Chidlow.[18]
[b] P. Christner, M. Dixon, A. Cywinski, and J. Rosenbloom, Biochem. J. 157, 525 (1976).
[c] No antibodies were detected when rabbits were immunized with dog or human lung insoluble elastin.
[d] Kucich et al.[27]
[e] Mecham and Lange.[21]
[f] Mecham and Lange.[12]
[g] Houle and LaBella.[4]
[h] T. V. Darnule, V. Likhite, G. M. Turino, and I. Mandl, Connect. Tissue Res. 5, 67 (1977).

insufficient data are available to allow comparisons of antibody responses at each dose.

Usable antiserum can be obtained and maintained in rabbits immunized with 1–10 mg of elastin in complete Freund's adjuvant. The immunization is best given as multiple, subcutaneous, or intradermal injections in the back and in the toe pads. Booster injections are administered every 2–3 weeks, beginning 4–6 weeks after immunization, in areas near the original injection sites. The animals are bled 4–10 days after each boosting. Generally, the best antisera are obtained 3–8 months after immunization. Figures 2A and 2B show binding of radiolabeled bovine α-elastin by rabbit antiserum raised to bovine α-elastin or insoluble elastin. In both cases, maximum binding occurred approximately 6 months after the initial immunization. It is not always possible to predict when maximum antibody titers will be present, especially since antibody production can be quite variable from animal to animal. Although animals that produce good antisera (in terms of titer and specificity) early in the immune response generally provide the best antisera in later bleedings,[13] it may be worth while to continue immunizing animals producing early antisera having marginally acceptable antibody levels. Moreover, because the

[13] C. W. Parker, "Radioimmunoassay of Biologically Active Compounds." Prentice-Hall, Princeton, New Jersey, 1976.

TABLE II
ANTISERA TO SOLUBILIZED ELASTINS

| Elastin antigen | Host animal | Immunization amount (mg) | Boosting amount (mg) | Boosting interval | References |
|---|---|---|---|---|---|
| Dog lung α-elastin | Rabbits | 1.0 | 0.5 | Weekly for 2 months | a |
| | | | 0.25 | Weekly | |
| Bovine ligament α-elastin | Rabbits | 10 | 10 | Weekly for 7 weeks | b |
| Bovine ligament α-elastin | Rabbits | 1 and 10 | 2 | 2–3 Weeks | c, d |
| Porcine lung and aortic α-elastins[a] | Rabbits | 0.67 | 0.67 | 3 Times/week for 5 weeks | e |
| Bovine aortic κ$_2$ | Rabbits | 0.5, 1.0, 2.0 | 0.5, 1.0, 2.0 | 3 Times/week for 5 weeks | f, g |
| Bovine ligament elastase digest | Rabbits | 100 | 100 | Weekly for 3 weeks | h |
| Human lung elastase digest | Rabbits | 20, 50, 100 | 20, 50, 100 | Twice at 10-day intervals | i |
| Dog or human lung leukocyte elastase digest | Rabbit | 1.0 | 0.5 | Weekly for 2 months | j |
| | | | 0.25 | Weekly | |
| Sheep ligament α-elastin[k] | Rabbits | 1.0 | 1.0 | 2 Weeks | l |

[a] U. Kucich, P. Christner, G. Weinbaum, and J. Rosenbloom, *Biochem. J.* **157**, 525 (1976).
[b] Jackson, *et al.*[17]
[c] Mecham and Lange.[21] *Connect. Tissue Res.* **7**, 247 (1980).
[d] Mecham and Lange.[12] Unpublished data, 1981.
[e] R. A. Daynes, M. Thomas, V. L. Alvarez, and L. B. Sandberg, *Connect. Tissue Res.* **5**, 75 (1977).
[f] I. Tosonev, P. Hadjiisky, J. Renais, and L. Scebat, *Pathol. Biol.* **20**, 383 (1972).
[g] Houle and LaBella.[4]
[h] S. Keller, M. M. Levi, and I. Mandl, *Arch. Biochem. Biophys.* **13**, 567 (1969).
[i] T. V. Darnule, V. Likhite, G. M. Turino, and I. Mandl, *Connect. Tissue Res.* **5**, 67 (1977).
[j] Kucich *et al.*[27]
[k] Conjugated with keyhole limpet hemocyanin.
[l] Davidson *et al.*[26]

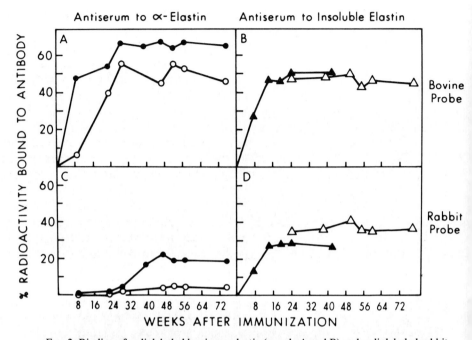

FIG. 2. Binding of radiolabeled bovine α-elastin (panels A and B) and radiolabeled rabbit α-elastin (C and D) by antisera from two rabbits immunized with bovine ligamentum nuchae α-elastin (○, ●) and by antisera from two rabbits immunized with insoluble elastin (△, ▲). Serum samples from each animal bleeding were stored frozen and assayed simultaneously by incubating with 10,000–12,000 cpm of radiolabeled antigen for 1 hr at 37° then overnight at 4°. Antigen–antibody complexes were precipitated with *Staphylococcus aureus* as described in the text. The washed pellets were counted for bound radioactivity.

specificity of late and early antisera from the same animal can differ, serum from each bleeding must be examined separately.

Antisera ordinarily can be stored at $-20°$ for several years with little loss of immunological reactivity.[13] Antisera should be stored in small volumes and not subjected to repeated freezing and thawing.

When antisera are to be used in competitive protein binding radioimmunoassays and enzyme-linked immunoabsorbent assays (see below), there is no need to purify specific antibodies. In fact, affinity purification techniques may be detrimental owing to an inability completely to recover high affinity antibodies. However, affinity-purified antibodies can be useful for other immunological techniques, e.g., immunoprecipitation, and the reader should consult Volume 34 of this series for immunoaffinity methodology.

The IgG fraction of antisera to elastin can be recovered by standard procedures of ammonium sulfate precipitation and by chromatography on

DEAE-cellulose using low ionic strength buffers at or near neutral pH.[14] Under these conditions, IgG passes through the column while other serum proteins are retained by the resin. Some caution is needed when DEAE is used to purify elastin antisera, however, since rabbit antiserum to bovine ligamentum nuchae insoluble elastin contains at least two antibody fractions, one that passes through DEAE with the void volume in low ionic strength buffers and a second retained by the resin until eluted with buffers of higher ionic strength.[15] The retained antibody fraction has not been well characterized, but similar properties have been noted for other antibody classes, such as IgM, and for IgG antibodies to positively charged proteins, which adhere to DEAE more avidly than do most IgG antibodies.[16]

Elastin Immunoassays

Sensitive hemagglutination or competitive protein binding assay techniques are useful for characterizing antisera, quantifying elastin production, and monitoring turnover of elastin *in vitro* and *in vivo*. Passive hemagglutination assays do not require expensive equipment, such as gamma counters or spectrophotometers, or the use of radioisotopes. Competitive protein binding assays using radiolabeled solubilized elastin antigen (radioimmunoassay) or enzyme-linked immunoabsorbent assays (ELISA) are more sensitive (both assays can detect as little as 10 ng of solubilized elastin per milliliter) and better suited for antigen quantification, but equipment and reagents for the radioimmunoassay are expensive. ELISA, on the other hand, is more cost effective and safer, since the assay does not require radioactive materials.

It is important to note that antisera to insoluble or solubilized elastins may not give precipitating antibodies against either solubilized elastin or tropoelastin antigens. Immunodiffusion or immunoelectrophoresis, therefore, may not be the most appropriate method to monitor antisera titers or measure antigen concentrations. Antisera to tropoelastin and to insoluble elastin from lathyritic animals[17] do have precipitating antibodies.

*Passive Hemagglutination Assay*[18,19]

Red blood cells are tanned with solubilized elastin as follows: One ml of fresh sheep red blood cells (SRBC) is washed three times with 40

[14] C. W. Parker, S. M. Godt, and M. C. Johnson, *Biochemistry* **6**, 3417 (1967).
[15] R. P. Mecham and G. Lange, unpublished data, 1981.
[16] M. Sela and E. Mozes, *Proc. Natl. Acad. Sci. U.S.A.* **55**, 445 (1966).
[17] D. S. Jackson, L. B. Sandberg, and E. G. Cleary, *Nature (London)* **210**, 195 (1966).
[18] B. C. Sykes and J. W. Chidlow, *FEBS Lett.* **47**, 222 (1974).
[19] L. Hudson and F. C. Hay, "Practical Immunology." Blackwell, Oxford, 1976.

volumes of NaCl (0.9% w/v) by centrifugation (300 g, 10 min) and suspended at 5%, v/v, in phosphate-buffered saline (PBS) (0.1 $M$ sodium phosphate, 0.15 $M$ NaCl), pH 7.5. Tannic acid (2.5 mg in 50 ml of PBS) is added to 50 ml of the 5% SRBC suspension and incubated at 37° for 15 min. The cells are collected by gentle centrifugation (100 g, 20 min) and divided into two batches. One batch, to be used as control cells, is resuspended in 100 ml of borate–succinate buffer (0.05 $M$ sodium tetraborate containing 0.15 $M$ NaCl and 1%, v/v, heat-inactivated horse serum titrated to pH 7.5 with 0.05 $M$ succinic acid). The remaining cells are resuspended in PBS containing solubilized elastin (1 mg of bovine ligamentum nuchae $\alpha$-elastin per milliliter has been used successfully) and incubated at 37° for 30 min. After washing by gentle centrifugation in PBS, the cells are resuspended in 100 ml of borate–succinate buffer. Ten milliliters of 40% formalin are added dropwise to the control and elastin-coated cells over a period of 30 min. The suspensions are left overnight at 4°, and an additional 10 ml of formalin are added. After the cells have settled overnight, the supernatants are replaced with 75 ml of fresh borate–succinate buffer. The cells are resuspended by vigorous shaking and allowed to settle overnight in the cold. The wash with borate–succinate buffer is repeated, and the cell suspensions are adjusted to 1%, v/v, in borate–succinate buffer containing 0.2% formalin as a preservative.

Hemagglutination titer is determined in microtiter plates (Flow Laboratories, Inc., McLean, Virginia) with conical wells by incubating 50 $\mu$l of serially diluted antiserum with 50 $\mu$l of elastin-tanned SRBC suspension and 100 $\mu$l of borate–succinate buffer. The contents of each well are thoroughly mixed with a micropipette, and the plates are left overnight at 4°. Agglutination is evident as a diffuse layer of cells at the base of the well, whereas a dense button indicates no agglutination. Hemagglutination scoring is as described by Stavitsky.[20] As controls, 50 $\mu$l of the lowest antiserum dilution are incubated with 100 $\mu$l of buffer and 50 $\mu$l of control cells, and rabbit serum is incubated with elastin-coated cells.

The quantity of soluble elastin in experimental samples is determined by the capacity of the sample to inhibit agglutination. Samples to be tested are incubated with undiluted antiserum for 1 hr at 37° and overnight at 4°. The residual agglutination titer is compared to the control established above and the decrease in titer is a measure of added antigen.

*Competitive Protein-Binding Radioimmunoassay*[21]

*Preparation of* $^{125}I$-*Labeled Solubilized Elastin.* The specific activity of the radiolabeled elastin probe is an important determinant of assay sen-

[20] A. B. Stavitsky, *J. Immunol.* **72**, 360 (1954).
[21] R. P. Mecham and G. Lange, *Connect. Tissue Res.* **7**, 247 (1980).

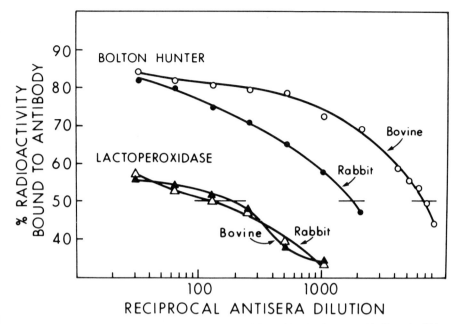

FIG. 3. Antiserum dilution curves showing anti-insoluble elastin serum binding of rabbit and bovine α-elastins radiolabeled with $^{125}$I using lactoperoxidase or Bolton–Hunter reagent. Reproduced, by permission of Gordon & Breach Science Publishers, from Mecham and Lange.[21]

sitivity. Figure 3 shows the difference in effectiveness of the same antiserum for binding bovine ligamentum nuchae α-elastin labeled with $^{125}$I by a solid-phase lactoperoxidase technique and by conjugation with iodinated Bolton–Hunter reagent (p-hydroxyphenylpropionic acid N-hydroxysuccinimide ester).[22] The antiserum dilution resulting in 50% binding of labeled antigen was 1:128 for lactoperoxidase-catalyzed labeled elastin, whereas a dilution of 1:8000 was sufficient for Bolton–Hunter $^{125}$I-labeled α-elastin. Thus, competitive protein binding assays with Bolton–Hunter-labeled α-elastin are approximately 20 times more sensitive than assays using α-elastin labeled with lactoperoxidase. This occurs because $^{125}$I-labeled Bolton–Hunter reagent forms an amide bond with free amino groups on proteins and peptides, whereas lactoperoxidase iodinates principally tyrosine residues, which comprise only 0.3% of the total amino acid residues in bovine ligament α-elastin. Since α-elastin has such a small amount of tyrosine and is a polydisperse mixture of peptides, it is likely that some antigenic peptides lack tyrosine and, hence, will not

[22] A. E. Bolton and W. M. Hunter, *Biochem. J.* **133**, 529 (1973).

be labeled using oxidative labeling techniques. Unlabeled antigenic peptides compete for antibody, thereby reducing binding of radiolabeled antigen and assay sensitivity. With Bolton–Hunter reagent, every antigenic peptide with free $\epsilon$- or $\alpha$-amino groups will be radiolabeled.

Iodinated Bolton–Hunter reagent is routinely supplied as 1 mCi in 100 $\mu$l of dry benzene (New England Nuclear, Boston, Massachusetts; Amersham, Arlington Heights, Illinois). Aliquoting the reagent is possible, but extreme care must be exercised to avoid introduction of moisture. We have obtained best results in using fresh labeling reagent and by labeling directly in the isotope vial. The benzene is evaporated immediately before use with a stream of dry nitrogen or air.

To prepare $^{125}$I-labeled elastin, solubilized elastin (100 $\mu$g) in 10 $\mu$l of ice cold 0.01 $M$ phosphate buffer, pH 8.0, is added to the dried iodinated ester and agitated to mix the protein solution thoroughly with the labeling reagent. The reaction temperature should be as close to 0° as possible. Although the half-life of Bolton–Hunter reagent in aqueous solution is on the order of minutes,[22] we routinely allow the labeling reaction to proceed overnight at 4°, ensuring complete hydrolysis of the active ester and obviating the need for addition of glycine to destroy excess reagent.

Labeled solubilized elastin is separated from the other products in the mixture by gel filtration chromatography on Sephadex G-25 (fine). The Sephadex resin is first washed in assay buffer [0.01 $M$ phosphate buffer, pH 7.4, 0.15 $M$ NaCl, 0.5% (w/v) bovine serum albumin, 0.03% NP-40], and the sample is eluted with phosphate-buffered saline. The glass column and all collection tubes are siliconized. Ten-drop fractions are collected, and those fractions having more than 95% trichloroacetic acid-precipitable counts are pooled and diluted to approximately $6 \times 10^6$ cpm/ml with assay buffer containing 0.02% azide. Aliquots (300 $\mu$l) are stored at $-70°$.

*Determining the Proper Antiserum Dilution.* The antiserum dilution to be used in the assay is determined empirically by placing 10,000–12,000 cpm of radiolabeled antigen in assay tubes and determining the minimal concentration of antibody needed to give 30–50% binding of the radioactivity. After a 1-hr incubation at 37°, antigen and antibody are incubated overnight at 4°. Antigen–antibody complexes can be separated by precipitation with a second antibody[14] or, more conveniently, using heat-killed, formalinized *Staphylococcus aureus* cells (IgGSORB, The Enzyme Center, Boston, Massachusetts). Various strains of *S. aureus* contain protein A, a cell wall constituent that reacts with the $F_c$ portion of IgG. *Staphylococcus aureus* as an immunoabsorbent offers advantages compared to the double-antibody system: the reaction of *S. aureus* with IgG is extremely rapid, going to completion in less than 30 sec; and nonspecific binding is slightly but consistently lower than with double-antibody techniques.[23,24] Protein A binds rabbit, guinea pig, human, and monkey IgG, but sheep and goat IgG is poorly bound.[24] Thus, with the exception of sheep and goat, *S. aureus* can be an effective immunoabsorbent for most species in which antibodies are produced.

Forty microliters of a 10% suspension of *S. aureus* are added to the antigen–antibody mixture, incubated for 60 min at 37°, and centrifuged at 1500 $g$ for 20 min. The pellet is washed with assay buffer, recentrifuged, and counted for bound radioactivity. Even though antiserum dilutions resulting in 30–50% radioactive antigen binding are selected to maximize assay sensitivity, it is important to establish that at least 80–90% antigen binding is achieved when larger quantities of antiserum are used.[13] This can be a useful indicator of losses in antigenic reactivity as a result of denaturation or iodination. We have found that the optimal antiserum dilution will change as the radiolabeled probe ages, so that optimal binding must be revalidated periodically.

Competitive binding assays are conducted in a final volume of 250 $\mu$l to 500 $\mu$l. Unknown samples or standards in a volume of 20–50 $\mu$l are incubated for 60 min at 37° in 100 $\mu$l of assay buffer containing 100 $\mu$l of appropriately diluted antiserum. Radiolabeled antigen is added, and the tubes are placed at 4° overnight. Immune complexes are precipitated with *S. aureus* as described above, and the washed pellets are counted for bound radioactivity. A standard curve is included with each assay and should extend from minimal to complete inhibition of radioactive antigen binding.[13] Total binding is determined by incubating labeled antigen with antiserum in the absence of competing antigen. Nonspecific binding is determined using preimmune or normal rabbit serum. Percentage of binding is calculated as

$$\left(\frac{\text{cpm bound} - \text{nonspecific binding}}{\text{total cpm} - \text{nonspecific binding}}\right) \times 100$$

The standard curve is plotted as absolute or percentage counts precipitated (ordinate) versus the logarithm of the unlabeled antigen concentration (abscissa) (see Fig. 4). The antigen content in experimental samples is determined from the standard curve. If solubilized elastin is used to generate the standard inhibition curve, results are expressed as solubilized elastin ($\alpha$-elastin) equivalents. Because the affinity of anti-insoluble elastin antiserum for solubilized elastin may be different from affinity for tropoelastin, the calculated value for tropoelastin in the sample is inaccu-

[23] S. W. Kessler, *J. Immunol.* **115**, 1617 (1975).
[24] M. A. Frohman, L. A. Frohman, M. B. Goldman, and J. N. Goldman, *J. Lab. Clin. Med.* **93**, 614 (1979).

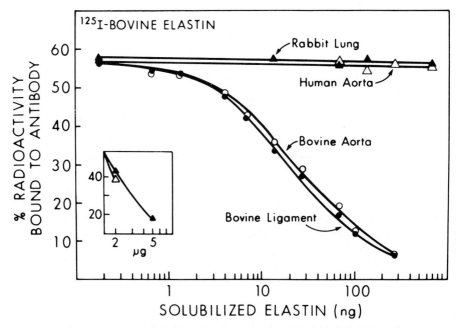

FIG. 4. Inhibition of radiolabeled bovine ligament α-elastin marker binding to anti-bovine ligament elastin antibody by bovine α-elastins from ligament or aorta. Rabbit lung and human aortic elastins compete for marker only at high concentrations. Antiserum dilution was 1:6000. Reproduced, by permission of Gordon & Breach Science Publishers, from Mecham and Lange.[21]

rate. Nevertheless, relative differences in tropoelastin content between samples are indicated. The same is true when elastin fragments are being measured and compared to an α-elastin inhibition curve. More precise values can be obtained if tropoelastin or elastin fragments are used as standards.

*Enzyme-Linked Immunoabsorbent Assay (ELISA)*

An alternative to the radioimmunoassay is the enzyme-linked immunoabsorbent assay (ELISA).[25] ELISA is a sensitive solid-phase immunoassay utilizing antigen or antibody bound to polystyrene or polyvinyl surfaces and offers the advantage of not requiring radioactive materials. In ELISAs that have been developed for elastin[26,27] and other connective

[25] E. Engvall and P. Perlmann, *Immunochemistry* **8**, 857 (1971).
[26] J. M. Davidson, S. I. Rennard, and R. Crystal, unpublished data, 1981.
[27] U. Kucich, G. Weinbaum, and J. Rosenbloom, unpublished data, 1981.

tissue proteins,[28] the solid phase (usually a microtiter plate or test tube) is coated with antigen. Bound antigen is incubated with antigen-specific serum, washed, and incubated with enzyme-linked second antibody. Unbound antibody is washed from the tube, and enzyme substrate is added. The optical density of the enzyme reaction product is determined with a spectrophotometer and is proportional to the amount of bound antibody. Antigen concentrations in samples are measured by incubating sample with antiserum in a nonabsorbent tube. Sample antigen, if present, will bind to the antibody and preclude subsequent binding of the antibody to the solid-phase antigen.

The optimal antigen concentration and antiserum dilution for an ELISA are determined empirically by washing wells in each row of a microtiter plate with serial dilutions of antigen. To each column is added serial dilutions of antiserum. Antigen concentrations and antiserum dilutions giving between 0.8 and 1.2 absorbance unit are selected for subsequent assays. An example of this technique is given by Rennard et al.[28] Sheep ligament $\alpha$-elastin (150–300 ng)[26] and 2.0 $\mu$g of elastase-digested human lung elastin[27] per milliliter have been used to coat plastic substrates for ELISAs.

*Microtiter ELISA for Elastin.* Solubilized elastin, 200 $\mu$l in 0.02 $M$ carbonate buffer, pH 9.6, is added to wells of a microtiter plate (Cook Laboratory Products, Alexandria, Virginia) and allowed to absorb to the plastic overnight at 4°. The wells are washed with phosphate-buffered saline (PBS), pH 7.4, containing 0.025% Tween-20 and stored at 4°. Coated plates can be stored for as long as 1 year without loss of the absorbed antigen.[28] Anti-elastin serum, diluted with PBS–Tween, is incubated with antigen-containing sample in Linbro Titertek microtiter plates (Flow Laboratories, Inc., McLean, Virginia) in a volume of 0.22 ml. The mixture is incubated for 1 hr at 37°, then overnight at 4°; 200 $\mu$l are transferred to the washed antigen-coated plate and incubated for 1–2 hr at room temperature (a nonequilibrium assay).[26] The wells are washed three times with PBS–Tween, and 200 $\mu$l/well of a 1:1000 dilution of horseradish peroxidase–goat antirabbit antibody (Miles Laboratories, Elkhart, Indiana) are added. The plate is incubated for 1 hr at room temperature, the wells are washed with PBS–Tween, and 200 $\mu$l/well of $o$-phenylenediamine dissolved in methanol (10 mg/ml diluted 1:100 into 0.05% $H_2O_2$ in distilled water, made fresh daily) are added. After 1 hr at room temperature, or a longer reaction time if more color development is needed, the reaction is stopped with 50 $\mu$l of 4 $N$ $H_2SO_4$. The optical density of each sample is read at 480 nm, and the antigen concentration is

[28] S. I. Rennard, R. Berg, G. R. Martin, J. M. Foidart, and P. G. Robey, *Anal. Biochem.* **104**, 205 (1980).

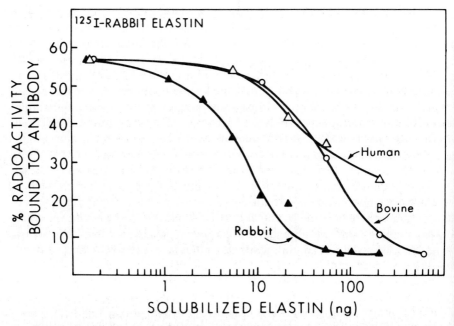

FIG. 5. Interspecies cross-reactivities using anti-bovine ligament elastin serum and radiolabeled rabbit lung α-elastin marker in a competitive binding assay. Antiserum dilution was 1:1000. Reproduced, by permission of Gordon & Breach Science Publishers, from Mecham and Lange.[21]

related to the standard curve, which must be included on each microtiter plate to compensate for plate-to-plate variation. Spectrophotometers specifically designed for ELISA are now commercially available.

Cross-Reactivity of Antisera for Elastins of Different Species

A useful property of antiserum to insoluble or solubilized elastin is the ability to cross-react with elastins from other species.[21,26] Figure 2, C and D, shows binding of radiolabeled rabbit lung α-elastin by sera to bovine elastins. It is notable, however, that the appearance of cross-reacting antibodies after immunization, and the degree of cross-reactivity, vary unpredictably between rabbits, even in antisera raised to the same batch of antigen. For example, bovine and rabbit α-elastin binding activities appeared concomitantly in antiserum to bovine insoluble elastin (Fig. 2D), but cross-reactivity was delayed 12–18 weeks in one rabbit immunized with bovine α-elastin and did not develop to a useful level in a second α-elastin-immunized rabbit (Fig. 2C). In general, our experience has shown that late sera have more cross-reactivity than early sera. Also, cross-reactivity appears earlier and occurs at higher levels in antisera when insoluble elastin is used to immunize rather than solubilized elastin. Because cross-reacting titers vary among immunized animals, it is possible to immunize many animals and select the most or least specific antisera, depending upon the intended use.

The presence of cross-reacting antibodies is convenient, enabling use of a single antiserum to study elastins from a wide range of animal species. As shown in Fig. 5, cross-reactivity can be measured in a competitive binding radioimmunoassay when radiolabeled solubilized elastins from species other than bovine are used as a specific probe.[21] Using radiolabeled rabbit lung α-elastin as a marker and anti-bovine elastin serum, we have measured soluble elastin synthesis in cultured rabbit smooth muscle cells and in tissue slices from rabbit and hamster lung and rat aorta. Chicken aortic elastin does not react appreciably with anti-bovine elastin serum.

### Acknowledgments

This work was done during the tenure of R. P. M. of an Established Investigatorship from the American Heart Association and with funds contributed in part by the St. Louis Heart Association, Missouri Affiliate and by National Institutes of Health Grant HL-16118.

## [43] Antibodies to Desmosine

### By BARRY C. STARCHER

With the exception of eggshell membrane,[1] elastin is the only known protein to contain desmosine. Since desmosine levels in elastin are constant within species, this cross-linking amino acid can be used to quantitate mature, insoluble elastin. The following procedure describes a radioimmunoassay specific for desmosine.[2,3]

*Reagents and Materials*

Desmosine (Elastin Products, St. Louis, Missouri)
Bovine serum albumin (BSA), 2 × crystallized

---

[1] S. Baumgartner, D. J. Brown, E. Slaevsky, and R. M. Leach, *J. Nutr.* **108**, 804 (1978).
[2] G. S. King, V. S. Mohan, and B. C. Starcher, *Connect. Tissue Res.* **7**, 263 (1980).
[3] S. Harel, A. Janoff, S. Y. Yu, A. Hurewitz, and E. H. Bergofsky, *Am. Rev. Respir. Dis.* **122**, 769 (1980).

1-Ethyl-3-(3-dimethylaminopropyl)carbodiimide-HCl
BioGel P2, 200–400 mesh (Bio-Rad, Richmond, California)
$^{125}$I-labeled Bolton-Hunter reagent, 1 mCi in 100 $\mu$l of benzene (New England Nuclear, Boston, Massachusetts)
*Staphylococcus aureus*-IgGsorb (The Enzyme Center, Boston, Massachusetts)
Potassium phosphate buffer, 0.05 $M$, pH 7.4, 0.1 $M$ NaCl, 0.05 m$M$ EDTA, and 0.02% sodium azide (30 g of $K_2HPO_4$, 20 g of NaCl, 60 mg of EDTA · 4 Na, 759 mg of sodium azide, 1.0 ml of $H_3PO_4$ in 3500 ml of distilled $H_2O$)
Sodium acetate buffer, 0.4 $M$, pH 6.0, 0.02% sodium azide (54.4 g of sodium acetate · 3 $H_2O$, 200 mg of sodium azide, 1.6 ml of 12 $N$ HCl in 1000 ml of distilled $H_2O$)
Glycine, 0.2 $M$, in potassium phosphate buffer
Cellulose nitrate filters: Gelman GN-6, 0.45 $\mu$m, or Millipore HAWP 02500, 0.45 $\mu$m

Assay Methods

*Antigen Preparation.* To couple desmosine to bovine serum albumin, 2 mg of desmosine and 10 mg of albumin are combined with 100 mg of 1-ethyl-3-(3-dimethylaminopropyl)carbodiimide-HCl in 1 ml of distilled water. The reaction is allowed to proceed at room temperature for 4 hr, and the product is dialyzed against distilled water and freeze-dried. The level of substitution should be 10–12 desmosine per BSA molecule. This can be confirmed by amino acid analysis.

*Antibody Production.* Rabbits are immunized with 1 mg of protein conjugate emulsified in Freuend's complete adjuvant by injection into each foot pad. Booster injections are maintained at 4-week intervals, and the rabbits are bled 10 days later. Useful titers of antibody develop in approximately 6 months. The serum is divided into 200-$\mu$l aliquots and stored frozen at $-70°$.

*$^{125}$I-Labeled Bolton–Hunter Desmosine Probe.* To prepare the radioactive probe, the benzene is removed from the 1 mCi $^{125}$I-labeled Bolton–Hunter reagent with a stream of dry nitrogen or air, and 20 $\mu$l of a solution containing 1 mg of desmosine per milliliter in distilled water are added. The mixture is constantly agitated while cooling over ice for 5 min. To the reaction, 100 $\mu$l of 0.2 $M$ glycine in phosphate buffer are added and the products are applied to a 1 × 28 cm column of BioGel P2 equilibrated with 1.0 $M$ NaCl. The column is eluted with 1.0 $M$ NaCl, and 1-ml fractions are collected. $^{125}$I-labeled Bolton–hunter desmosine elutes between 10 and 13 ml, and free $^{125}$I at 23 ml. The $^{125}$I-labeled Bolton–Hunter desmosine

fractions are pooled, and small aliquots are stored frozen at $-20°$ in *siliconized* glass tubes. The iodinated probe is stable for at least 2 months.

*Standards.* Standard desmosine solutions are prepared in potassium phosphate buffer in siliconized tubes ranging in concentration from 5 to 1000 ng/ml. The dilutions must be made from an initial desmosine concentrate, calibrated with an amino acid analyzer. An internal standard prepared from an elastin hydrolyzate should be carried through each assay.

*Sample Preparation.* Peptide-bound desmosine does not react with the antibody, so samples must first be hydrolyzed in 6 N HCl for 40 hr at 105°. The hydrolyzate is evaporated and dissolved in acetate buffer. The dilution depends on the amount of elastin hydrolyzed. For close approximation, most elastin contains about 1% desmosine.

*Assay Procedure.* The antiserum is diluted with phosphate buffer so that 100 μl will bind 50% of the total radioactive probe in a final volume of 350 μl. Usually this means a 1 : 100 to 1 : 250 dilution of the serum. For the radioimmunoassay, 100 μl of diluted antiserum are added to 200 μl of a solution containing desmosines. For the standards, this means 100 μl of each standard plus 100 μl of acetate buffer. This is mixed and allowed to incubate at room temperature for 1 hr. Fifty microliters of $^{125}$I-labeled Bolton–Hunter desmosine probe (approximately 20,000 cpm) are added, mixed, and incubated at 4° for 18 hr. All assays are done in duplicate.

Immunoglobulin-bound probe can be determined by either of two procedures with identical results.

1. The mixture is filtered with vacuum through prewashed (phosphate buffer) cellulose nitrate filters and rinsed twice with 1 ml of cold phosphate buffer; the radioactivity of the filters is measured.
2. To each mixture is added 50 μl of a 1% suspension of *S. aureus*. After vortexing, the suspension is diluted with 2 ml of phosphate buffer and centrifuged at 2000 rpm for 10 min. The supernatant is removed by aspiration, and the radioactivity is measured in the pellet.

Discussion

The desmosine radioimmunoassay performs well in all hydrolyzates analyzed to date. There are, however, a few pitfalls that should be pointed out.

Ideally one would carry a blank of the unknown containing everything that was in the sample except desmosine. This is very difficult, if not impossible, to do. The relative desmosine values obtained within an experiment are fine, but caution should be exercised when accepting absolute values without an independent means of confirmation, i.e., amino acid analysis. This is particularly true for very crude hydrolyzates, such as urine.

The pH optimum for the RIA is above pH 6. However, above pH 7 the antibody–probe interaction is inhibited in the presence of citrate ions and perhaps other polyanions. For this reason the assay is performed in an acetate buffer that maintains the pH at less than 7.

Reduced desmosine, isodesmosine, lysinonorleucine, and lysine do not compete for the probe in this assay. Merodesmosine may compete to a very limited extent.

## [44] Preparation of Antiserum to Tropoelastin

*By* JUDITH ANN FOSTER, CELESTE B. RICH, SU CHEN WU, and MARIE-THERESE MARZULLO

The availability of antibody to soluble elastin is invaluable for detection, quantitation, and isolation of tropoelastin in a variety of *in vivo* and *in vitro* systems. The advantage to eliciting antiserum to tropoelastin, as opposed to insoluble, cross-linked elastin,[1,2] is that the resulting antiserum contains precipitating antibody. The precipitin reaction between antibody and antigen is a prerequisite for performing various immunodiffusion[3] and immunoelectrophoretic[4,5] techniques available.

*Preparation of Antiserum*[6]

Normally three rabbits weighing approximately 3 kg are used to ensure adequate antiserum titers. Purified tropoelastin[7] (200 μg) is dissolved in 200 μl of 0.01 $M$ $Na_2HPO_4$, 0.15 $M$ NaCl, pH 7.4 and emulsified in an equal volume of Freund's complete adjuvant.[8] Each rabbit is injected intracutaneously with 0.1 ml (50 μg of tropoelastin) each week for a month. The actual injection is given so that a small bolus can be felt under

[1] P. Christner, M. Dixon, A. Cywinski, and J. Rosenbloom, *Biochem. J.* **157**, 525 (1976).
[2] R. P. Mecham and G. Lange, *Connect. Tissue Res.* **7**, 247 (1980).
[3] O. Ouchterlony, *Acta Pathol. Microbiol. Scand.* **26**, 507 (1949).
[4] P. Grabar and C. A. Williams, *Biochim. Biophys. Acta* **10**, 193 (1953).
[5] J. A. Foster, S. C. Wu, M. T. Marzullo, and C. B. Rich, *Anal. Biochem.* **101**, 310 (1980).
[6] J. A. Foster, D. Knaack, B. Faris, R. Moscaritolo, L. L. Salcedo, M. Skinner, and C. Franzblau, *Biochim. Biophys. Acta* **446**, 51 (1976).
[7] J. A. Foster, R. Shapiro, P. Voynow, G. Crombie, B. Faris, and C. Franzblau, *Biochemistry* **14**, 5343.
[8] J. Freund, *Annu. Rev. Microbiol.* **1**, 291 (1947).

FIG. 1. Immunodiffusion plate showing the reaction of samples in wells 1–6: 1, chick aortic tropoelastin; 2, chick gizzard tropoelastin; 3, pig aortic tropoelastin; 4, chick aortic tropoelastin; 5, chick lung tropoelastin; 6, chick aortic extract.

the skin. After a rest period of 3 weeks, another injection is given, and the rabbits are bled 6–8 days later. Booster injections can be administered at 2–3 month intervals in Freund's incomplete adjuvant. The presence of precipitating antibody can easily be tested by immunodiffusion as shown in Fig. 1. For most immunological procedures, with the exception of immunoprecipitation in cell-free translations of elastin mRNAs,[9] antiserum can be used directly without prior fractionation.

### Preparation of Tropoelastin Antibody

Specific antibody to tropoelastin can be prepared by affinity chromatography. The ideal ligand to attach onto a resin would be tropoelastin itself. However, because tropoelastin is very difficult to isolate in sufficient amounts, tryptic peptides released from lathyritic, insoluble elastin[10] are an excellent alternative.

[9] J. A. Foster, C. B. Rich, S. Fletcher, S. R. Karr, and A. Przybyla, *Biochemistry* **19**, 857 (1980).
[10] R. P. Mecham and J. A. Foster, *Biochim. Biophys. Acta* **577**, 147 (1979).

Seven-day chicks are raised on diets of 0.1% (w/w) α-aminoacetonitrile · HCl and 0.05% (w/w) ε-aminocaproic acid as previously described.[11] After extraction of tropoelastin with phosphate buffer,[7] the insoluble residue is extracted twice with chloroform : methanol (3 : 1, v/v) using 100 ml/g dry weight. The suspension is filtered through Whatman No. 1 filter paper, rinsed with acetone, and allowed to air dry overnight at room temperature. The dried material is suspended in 0.1 $N$ NaOH (50 ml/g dry weight) and placed in a boiling $H_2O$ bath for 50 min.[12] After cooling, the suspension is centrifuged at 15,000 $g$ for 30 min. The pellet is washed 4 times with deionized $H_2O$ and lyophilized. Amino acid analysis is performed at this point to check the purity of the insoluble elastin (see this volume [30] and [31]).

In order to isolate tryptic peptides from the insoluble elastin, two different temperatures are used consecutively for trypsin digestion.[10] One gram of the purified elastin is suspended in 50 ml of 0.046 $M$ Tris buffer, pH 8.1, containing 0.0115 $M$ $CaCl_2$ and dispersed briefly with a Polytron (3 sec, high speed). The mixture is equilibrated at 4°, and then 20 mg of TPCK-trypsin (217 units/mg) are added. Digestion proceeds for 17 hr at 4°. The digestion is then brought to 37°, and another 20 mg of trypsin are added. Digestion is allowed to proceed for 17 hr at 37°. The suspension is centrifuged at 10,000 $g$ for 15 min. The supernatant, containing the tryptic peptides, is desalted on a Sephadex G-10 column (1.7 × 50 cm, flow rate 0.32 ml/min) using deionized $H_2O$ and lyophilized.

The tryptic peptides are immobilized on Bio-Rad Affi-Gel 10 which contains $N$-hydroxysuccinimide activated esters attached to 4% cross-linked agarose beads. Six milliliters of the Affi-Gel 10 resin (directly from the vial) are transferred to a small glass column (0.7 × 4 cm), and the solvent is allowed to drain. The resin is then washed with three bed volumes of cold (4°) deionized $H_2O$, and the excess moisture is removed by applying a gentle suction to the bottom of the column. The resin is transferred to a small test tube, and 6 ml of 0.1 $M$ $Na_2HPO_4$, pH 6.8, containing 69 mg of the elastin tryptic peptides are added; the tube is agitated gently overnight at 4°. In order to block excess reactive groups, 1 ml of ethanolamine is added, and the pH is adjusted to 8 with concentrated HCl. The reaction is allowed to proceed for 1 hr at 4°, and the reaction mixture is transferred to a small glass column. The resin is washed with the phosphate buffer until the $OD_{230}$ reading is insignificant. In order to bind specific tropoelastin antibody, 0.5 ml of chick tropoelastin antiserum is put onto the column and allowed to penetrate fully into the gel. The

---

[11] J. A. Foster, C. B. Rich, M. D. DeSa, A. Jackson, and S. Fletcher, *Anal. Biochem.* **108**, 233 (1980).

[12] A. I. Lansing, T. B. Rosenthal, M. Alex, and E. W. Dempsey, *Anat. Rec.* **114**, 555 (1952).

column is shut off for 15 min, then elution of serum proteins and nonelastin immunoglobulins are accomplished with 0.01 $M$ $Na_2HPO_4$, 0.14 $M$ NaCl (pH 7.4). The effluent is monitored at $OD_{230}$, and elution is stopped when the baseline stabilizes at an $OD_{230}$ reading of 0.005. Elution of the specific tropoelastin antibody is accomplished with 0.01 $M$ sodium citrate, pH 2.8. In our experience the antibody elutes immediately with this buffer and can easily be tested by immunodiffusion[6] or immunoprecipitation.[9]

Acknowledgments

The authors acknowledge the financial support of the National Institutes of Health Grant HL 22208 and Council for Tobacco Research 1179.

# Section III

# Proteoglycans

## [45] Proteoglycans: Isolation and Characterization

By VINCENT C. HASCALL and JAMES H. KIMURA

### Background

#### Definition of Glycosaminoglycan

*Glycosaminoglycans* are polyanionic chains of variable length constructed from disaccharide repeating units that contain a hexosamine residue. In mammalian tissues there are four major classes of glycosaminoglycans: hyaluronic acid, the chondroitin sulfates, keratan sulfate, and the heparan sulfate–heparin class. An example of a disaccharide repeat unit of each of these classes as well as variations in structure are shown in Fig. 1. The properties of glycosaminoglycans and the techniques used to study them have been presented in this series[1] and will not be presented in detail in this chapter.

#### Definition of Proteoglycan

With the possible exception of hyaluronic acid, glycosaminoglycans are covalently linked to core proteins to form a class of macromolecules referred to as *proteoglycans*. The chondroitin sulfate and heparan sulfate–heparin glycosaminoglycans are attached to protein by a characteristic oligosaccharide linkage region that contains[1] glucuronic acid, galactose, and xylose in the structure shown in Fig. 2.[2] The elucidation of this linkage structure and details of its biosynthesis are described by Rodén.[3] The xylose is bound to a hydroxyl group on a serine residue in the protein. With the possible exception of the xylose, the biosynthesis of the linkage structure and the elongation of the glycosaminoglycan chains on the appropriate core protein occurs rapidly in the Golgi region of the cell.[4–7]

[1] L. Rodén, J. R. Baker, J. A. Cifonelli, and M. B. Mathews, this series, Vol. 28 [7].
[2] L. Rodén and R. Smith, *J. Biol. Chem.* **241**, 5949 (1966).
[3] L. Rodén, in "Biochemistry of Glycoproteins and Proteoglycans" (E. Lennarz, ed.), p. 267. Plenum, New York, 1980.
[4] K. Kimata, M. Okayama, S. Suzuki, I. Suzuki, and M. Hoshino, *Biochim. Biophys. Acta* **237**, 606 (1971).
[5] L. S. Freilich, R. G. Lewis, A. V. Reppucci, and J. E. Silbert, *J. Cell Biol.* **72**, 655 (1977).
[6] A. L. Horowitz and A. Dorfman, *J. Cell Biol.* **38**, 358 (1968).
[7] J. R. Silbert and L. S. Freilich, *Biochem. J.* **190**, 307 (1980).

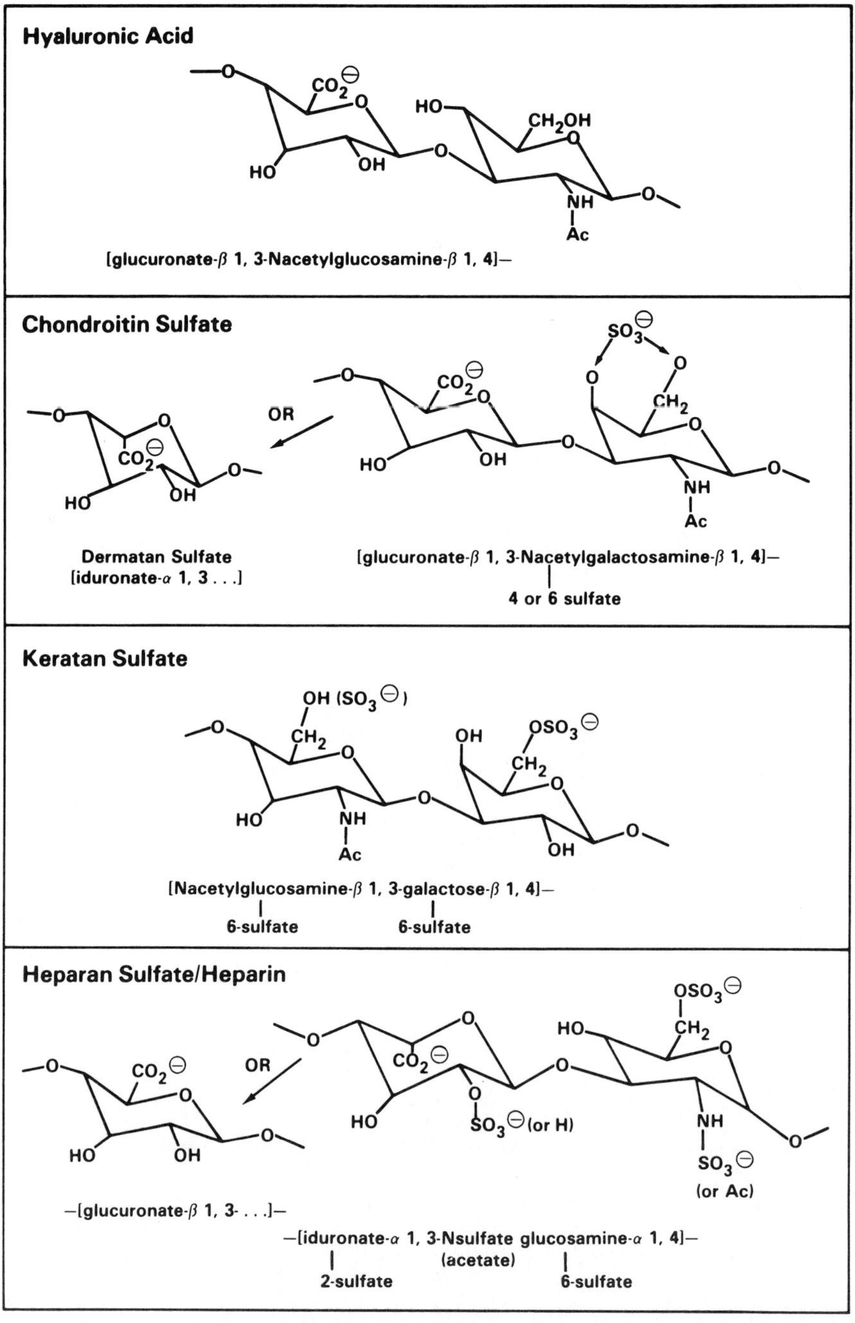

**Hyaluronic Acid**

[glucuronate-$\beta$ 1, 3-Nacetylglucosamine-$\beta$ 1, 4]–

**Chondroitin Sulfate**

Dermatan Sulfate
[iduronate-$\alpha$ 1, 3 . . .]

[glucuronate-$\beta$ 1, 3-Nacetylgalactosamine-$\beta$ 1, 4]–
            |
       4 or 6 sulfate

**Keratan Sulfate**

[Nacetylglucosamine-$\beta$ 1, 3-galactose-$\beta$ 1, 4]–
        |                           |
     6-sulfate                  6-sulfate

**Heparan Sulfate/Heparin**

–[glucuronate-$\beta$ 1, 3- . . .]–

–[iduronate-$\alpha$ 1, 3-Nsulfate glucosamine-$\alpha$ 1, 4]–
        |              (acetate)            |
     2-sulfate                          6-sulfate

FIG. 1. The basic repeating disaccharide units for the four major classes of mammalian glycosaminoglycans are shown. Positions where structural variability is found are indicated. Positions with $O$-sulfate groups do not always contain sulfate groups, depending upon the extent of sulfation. Heparin in general contains more iduronate 2-sulfate and $N$-sulfated glucosamine residues in the polymer than the closely related heparan sulfate. When iduronate residues are present in chondroitin sulfate, the glycosaminoglycan is referred to as dermatan sulfate.

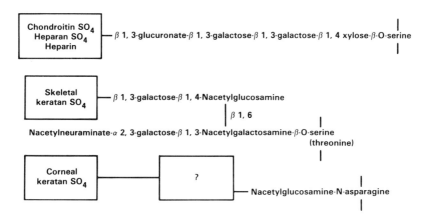

FIG. 2. The specialized linkage region structures for attaching glycosaminoglycan chains to protein are shown. The complete structure for the linkage region of corneal keratan sulfate is not yet known, but it does contain mannose. See text for discussion.

Keratan sulfate is attached to protein by two different oligosaccharide structures (Fig. 2). In cartilage (skeletal keratan sulfate), the linkage involves an $N$-acetylgalactosamine glycosidically bound to a hydroxyl group on either a serine or threonine in the protein backbone.[8,9] The rest of the linkage structure is closely related to the O-glycosidically linked oligosaccharides found on a large number of macromolecules, and it is probable that keratan sulfate chains are elongated on "primer" oligosaccharides.[10–12] In the corneal stroma, keratan sulfate is attached to protein by a different linkage in which $N$-acetylglucosamine is bound to asparagine by an $N$-glycosylamine bond (Fig. 2).[8,13] This linkage is the same as that by which the complex oligosaccharides found in glycoproteins are attached to protein.[14] While details of the rest of the linkage structure remain to be determined, mannose is present; further, the synthesis of keratan sulfate in corneal stroma explants in culture is inhibited by tunicamycin,[15] an antibiotic that selectively blocks the synthesis of

[8] N. Seno, K. Meyer, B. Anderson, and P. Hoffman, *J. Biol. Chem.* **240**, 1005 (1965).
[9] B. A. Bray, R. Lieberman, and K. Meyer, *J. Biol. Chem.* **242**, 3373 (1967).
[10] S. DeLuca, L. S. Lohmander, B. Nilsson, V. C. Hascall, and A. I. Caplan, *J. Biol. Chem.* **255**, 6077 (1980).
[11] L. S. Lohmander, S. DeLuca, B. Nilsson, V. C. Hascall, C. B. Caputo, J. H. Kimura, and D. Heinegard, *J. Biol. Chem.* **255**, 6084 (1980).
[12] V. C. Hascall, in "Biology of Carbohydrates" (V. Ginsberg, ed.), Vol. 1, p. 1. Wiley, New York, 1981.
[13] J. R. Baker, J. A. Cifonelli, and L. Rodén, *Biochem. J.* **115**, 11 (1969).
[14] R. Kornfeld and S. Kornfeld, in "Biochemistry of Glycoproteins and Proteoglycans" (E. Lennarz, ed.), p. 1. Plenum, New York, 1980.
[15] G. W. Hart and W. J. Lennarz, *J. Biol. Chem.* **253**, 5795 (1978).

$N$-glycosylamine-linked oligosaccharides by inhibiting the synthesis of dolichol diphospho-$N$-acetylglucosamine, a required intermediate.[16] It is likely, then, that in the corneal stroma, keratan sulfate is assembled on a "primer" oligosaccharide closely related to the N-linked oligosaccharides found on glycoproteins.

Examples of Proteoglycan Structures and Functions

A proteoglycan is defined as any macromolecule that has a core protein containing at least one covalently bound glycosaminoglycan chain. Within this general definition, there are a very large number of structures that serve a variety of functions in different tissues. This section briefly describes the characteristics of some of the proteoglycans that have been studied.

*Cartilage*

Cartilages are particularly rich in proteoglycans, which constitute up to 50% of the dry weight. For this reason and because tissue is readily available, cartilages have been used extensively as a source for proteoglycans, and most of the methods currently used to extract, purify, and characterize proteoglycans have been developed using these tissues.

The chemical structure of a typical cartilage proteoglycan is exceedingly complex. The core protein is very large, with estimated molecular weight of over 200,000[17-19] and possibly more than 300,000.[20-21] Nevertheless, the core protein generally represents only 5–15% of the total macromolecule, the remainder consisting of glycosaminoglycan chains and oligosaccharides. For most hyaline cartilages, chondroitin sulfate is the predominant component of the proteoglycans, and there are an average of almost 100 chains per core protein with an average chain size of about 20,000 (Fig. 3). These are concentrated along about 60% of the polypeptide at one end of the core protein referred to as the *chondroitin sulfate-rich region*.[22] Keratan sulfate chains and the related O-glycosidically linked oligosaccharides[10-12] are concentrated in a region of the core of about 30,000 molecular weight referred to as the *keratan sulfate-rich region*.[22]

[16] J. S. Tkacz and J. O. Lampden, *Biochem. Biophys. Res. Commun.* **65**, 248 (1975).
[17] V. C. Hascall and S. W. Sajdera, *J. Biol. Chem.* **245**, 4920 (1970).
[18] V. C. Hascall and R. L. Riolo, *J. Biol. Chem.* **247**, 4529 (1972).
[19] S. G. Pasternack, A. Veis, and M. Breen, *J. Biol. Chem.* **249**, 2206 (1974).
[20] W. B. Upholt, B. M. Vertel, and A. Dorfman, *Proc. Natl. Acad. Sci. U.S.A.* **59**, 1144 (1979).
[21] J. H. Kimura, E. M.-J. Thonar, V. C. Hascall, A. Reiner, and A. R. Poole, *J. Biol. Chem.* **256**, 7890 (1981).
[22] D. K. Heinegård and I. Axelsson, *J. Biol. Chem.* **252**, 1971 (1977).

## (a) Cartilage

- chondroitin SO$_4$ rich region
- keratan SO$_4$ rich region
- HA-binding region
- core protein
- link protein

**Aggregate Structure**

## (b) Corneal stroma

keratan SO$_4$ proteoglycan

dermatan SO$_4$ proteoglycan

## (c) Follicular fluid

## (d) Hepatocyte plasma membrane

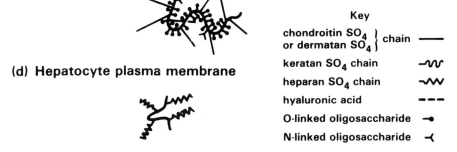

**Key**

- chondroitin SO$_4$ or dermatan SO$_4$ chain ——
- keratan SO$_4$ chain ~w~
- heparan SO$_4$ chain ~W~
- hyaluronic acid - - -
- O-linked oligosaccharide —•
- N-linked oligosaccharide —⊰

FIG. 3. Schematic diagrams of proteoglycan structures.

Most of the cartilage proteoglycans contain a specialized region at one end of the core referred to as the *hyaluronic acid-binding region*. This region, with a molecular weight of about 65,000,[23-25] contains few or no glycosaminoglycan chains but does contain oligosaccharides linked by *N*-glycosylamine bonds between *N*-acetylglucosamine and asparagine.[10,11,26] The structures of the N-linked oligosaccharides on a proteoglycan isolated from a transplantable rat chondrosarcoma were determined to be of the triantennary and biantennary complex type typical for many glycoproteins.[26] The overall average molecular weight of a monomer proteoglycan would be 2 to 3 × $10^6$.

The hyaluronic acid-binding region contains an active site that binds through noncovalent, but highly specific, interactions with hyaluronic acid.[27-30] The binding site interacts strongly with a decasaccharide of hyaluronic acid consisting of five of the repeat disaccharide units shown in Fig. 1; smaller hyaluronic acid oligomers interact only weakly.[28-30] The interaction of proteoglycans with hyaluronic acid is critical for the formation of the large aggregates that are present in the extracellular matrix of cartilages.[23,31-33] The interaction of the proteoglycan with hyaluronic acid is stabilized further by the presence of link proteins[34-38] that interact with a portion of the hyaluronic acid-binding region in the core of the proteoglycan as well as with a length of hyaluronic acid near that bound to the interactive site in the proteoglycan (Fig. 3).[29,39-43] The aggregates are

[23] V. C. Hascall and D. K. Heinegård, *J. Biol. Chem.* **249**, 4232 (1974).
[24] D. K. Heinegård and V. C. Hascall, *J. Biol. Chem.* **249**, 4250 (1974).
[25] C. B. Caputo, E. K. MacCallum, J. H. Kimura, J. Schrode, and V. C. Hascall, *Arch. Biochem. Biophys.* **204**, 220 (1980).
[26] S. Lohmander, B. Nilsson, S. DeLuca, and V. Hascall, *Semin. Arthritis Rheum.* **11**, Suppl. 1, 12 (1981).
[27] T. E. Hardingham and H. Muir, *Biochim. Biophys. Acta* **279**, 401 (1972).
[28] T. E. Hardingham and H. Muir, *Biochem. J.* **135**, 905 (1973).
[29] V. C. Hascall and D. K. Heinegård, *J. Biol. Chem.* **249**, 4242 (1974).
[30] J. E. Christner, M. L. Brown, and D. D. Dziewiatkowski, *J. Biol. Chem.* **254**, 4624 (1979).
[31] L. L. Faltz, A. H. Reddi, G. K. Hascall, D. Martin, J. C. Pita, and V. C. Hascall, *J. Biol. Chem.* **254**, 1375 (1979).
[32] J. C. Pita, F. J. Muller, S. M. Morales, and E. J. Alarcon, *J. Biol. Chem.* **254**, 10313 (1979).
[33] B. Caterson and J. R. Baker, *J. Biol. Chem.* **254**, 2394 (1979).
[34] V. C. Hascall and S. W. Sajdera, *J. Biol. Chem.* **244**, 2384 (1969).
[35] H. Keiser, H. J. Shulman, and J. I. Sandson, *Biochem. J.* **126**, 163 (1972).
[36] T. R. Oegema, M. Brown, and D. D. Dziewiatkowski, *J. Biol. Chem.* **252**, 6470 (1977).
[37] J. R. Baker and B. Caterson, *J. Biol. Chem.* **254**, 2387 (1979).
[38] J.-P. Périn, F. Bonnet, and P. Jollès, *FEBS Lett.* **94**, 257 (1978).
[39] J. D. Gregory, *Biochem. J.* **133**, 382 (1973).
[40] T. E. Hardingham, *Biochem. J.* **177**, 237 (1979).

assembled after newly synthesized proteoglycans and link proteins are secreted from chondrocytes,[43,44] and aggregation is probably important for immobilizing proteoglycans within the collagen network of the tissue.

Cartilage proteoglycans, unlike their isolated glycosaminoglycans, occupy large solvent volumes in solution with radii of gyration about 70 nm,[17,19,45–49] and they are reversibly compressible.[46] In the cartilage matrix, the collagen fibrils provide an interlaced meshwork, and the interspersed proteoglycan aggregates provide within this meshwork a hydrated, viscous gel that absorbs compressive load. In most hyaline cartilages, the concentration of proteoglycans is 3–5 times more than would be possible if they were fully extended, restricting individual molecules to domains as small as 20% of their maximum.[46] The increased charge density of this molecular configuration, like a partially compressed spring, can resist equivalent load on the tissue with less total deformation. While aggregation immobilizes proteoglycans in the matrix, $^{13}$C nuclear magnetic resonance studies indicate that the glycosaminoglycan chains even at the high concentrations *in situ* retain a high degree of segmental mobility[50,51] as does the polypeptide backbone of purified monomer proteoglycans in solution.[52] Thus the tensile strength and elasticity of cartilages are determined by the composite properties of the collagen fibrils and the surrounding proteoglycan phase.

*Tissues with Proteoglycans Similar to Cartilage Proteoglycans*

It is now apparent that other tissues contain proteoglycans with similar properties to those described above for the predominant proteoglycans in cartilages. Aortic tissue contains a population of proteoglycans that are of

---

[41] D. K. Heinegård and V. C. Hascall, *J. Biol. Chem.* **254**, 921 (1979).
[42] L.-H. Tang, L. Rosenberg, A. Reiner, and A. R. Poole, *J. Biol. Chem.* **254**, 10523 (1979).
[43] J. H. Kimura, T. E. Hardingham, and V. C. Hascall, *J. Biol. Chem.* **255**, 7134 (1980).
[44] J. H. Kimura, T. E. Hardingham, V. C. Hascall, and M. Solursh, *J. Biol. Chem.* **254**, 2600 (1979).
[45] I. A. Nieduzynski, J. K. Sheehan, C. F. Phelps, T. E. Hardingham, and H. Muir, *Biochem. J.* **185**, 107 (1980).
[46] V. C. Hascall, *J. Supramol. Struct.* **7**, 101 (1977).
[47] A. R. Foweraker, M. Isles, B. R. Jennings, T. E. Hardingham, and H. Muir, *Biopolymers* **16**, 1367 (1977).
[48] H. Reihanian, A. M. Jamieson, L.-H. Tang, and L. Rosenberg, *Biopolymers* **18**, 1727 (1979).
[49] W. D. Comper and T. C. Laurent, *Physiol. Rev.* **58**, 255 (1978).
[50] C. F. Brewer and H. Keiser, *Proc. Natl. Acad. Sci. U.S.A.* **72**, 3421 (1975).
[51] D. A. Torchia, M. A. Hasson, and V. C. Hascall, *J. Biol. Chem.* **252**, 3617 (1977).
[52] D. A. Torchia, M. A. Hasson, and V. C. Hascall, *J. Biol. Chem.* **256**, 7129 (1981).

large size and can interact with hyaluronic acid[53,54] and link protein[55] to form aggregates. Additionally, smooth muscle cells isolated from aorta and grown in culture also synthesize a population of proteoglycans, about 50% of the total, which are able to aggregate.[56] Further, glial cells in culture synthesize proteoglycans that interact with hyaluronic acid.[57] It is probable that these proteoglycans serve similar structural functions in these tissues to those of cartilage proteoglycans, and it is possible that the primary structure of their core proteins are from related genes similar to the different $\alpha$ chains of collagen.[58] However, the core protein of many other classes of proteoglycan are very different and probably unrelated to those described above.

*Corneal Stroma*

There are two separate classes of proteoglycans in the corneal stroma, one that contains keratan sulfate and one that contains chondroitin sulfate, as dermatan sulfate (Fig. 3).[53,59-64] Both types of proteoglycan are small. The keratan sulfate proteoglycan has an estimated molecular weight of 40 to 70 $\times 10^3$ and contains 1–3 keratan sulfate chains of molecular weight about 7 $\times$ 10$^3$ that are attached to the core protein through $N$-glycosylamine bonds (Fig. 2). The dermatan sulfate proteoglycan is 100 to 150 $\times$ 10$^3$ in molecular weight ($M_r$) and contains 1–2 glycosaminoglycan chains of about 55,000 $M_r$. Both proteoglycans have oligosaccharides, probably of the complex glycoprotein type, in addition to the glycosaminoglycan chains.[62,63]

The collagen lattice in the corneal stroma is very regular with uniform diameter collagen fibrils closely spaced.[65,65a] The proteoglycans are lo-

---

[53] C. A. Antonopoulos, I. Axelsson, D. Heinegård, and S. Gardell, *Biochim. Biophys. Acta* **338**, 108 (1974).
[54] T. R. Oegema, V. C. Hascall, and R. Eisenstein, *J. Biol. Chem.* **254**, 1312 (1979).
[55] S. Gardell, J. R. Baker, B. Caterson, D. Heinegård, and L. Rodén, *Biochem. Biophys. Res. Commun.* **95**, 1823 (1980).
[56] T. Wight, R. Ross, and V. C. Hascall, *J. Cell Biol.* **87**, 119A (1980).
[57] B. Norling, B. Glimelius, B. Westermark, and Å. Wastesson, *Biochem. Biophys. Res. Commun.* **84**, 914 (1978).
[58] E. J. Miller and R. K. Rhodes, this volume [2].
[59] I. Axelsson and D. K. Heinegård, *Biochem. J.* **145**, 491 (1975).
[60] P. Muthiah, H. W. Stuhlsatz, and H. Greiling, *Hoppe-Seyler's Z. Physiol. Chem.* **355**, 924 (1974).
[61] P. Speziale, M. Speziale, L. Galligani, and C. Balduini, *Biochem. J.* **173**, 935 (1978).
[62] I. Axelsson and D. K. Heinegård, *Biochem. J.* **169**, 517 (1978).
[63] J. R. Hassell, D. A. Newsome, and V. C. Hascall, *J. Biol. Chem.* **254**, 12346 (1979).
[64] J. R. Hassell, D. A. Newsome, J. H. Krachmer, and M. M. Rodriguez, *Proc. Natl. Acad. Sci. U.S.A.* **77**, 3705 (1980).
[65] R. L. Trelstad and A. J. Coulombre, *J. Cell Biol.* **50**, 850 (1971).
[65a] D. A. D. Parry and A. S. Craig, *Nature (London)* **282**, 213 (1979).

cated in the interfibrillar spaces and probably interact specifically with the collagen fibrils and with each other to maintain the highly ordered matrix structure required for optical transparency. Proteoglycans of larger hydrodynamic size, such as those from cartilage, would undoubtedly distort the collagen lattice.

The presence of the corneal proteoglycans is critical for normal function. A corneal explant culture system[63] was used to study proteoglycan synthesis in corneas isolated during corneal replacement transplantation from patients with the genetic anomaly corneal macular dystrophy.[64] While the corneal stromacytes synthesized normal dermatan sulfate proteoglycans, they synthesized an abnormal keratan sulfate proteoglycan that lacked keratan sulfate chains. It is likely that this defect in structure leads to the precipitates in the corneal stroma of these patients which cause cloudiness and eventual blindness.

*Follicular Fluid*

About 0.3% of porcine follicular fluid consists of a large proteoglycan (Fig. 3)[66] with a similar average hydrodynamic size to that for cartilage proteoglycans, but with a very different composition. The macromolecules each contain about 20 dermatan sulfate chains with approximate molecular weights of 55,000. The core protein is large, probably greater than 300,000 $M_r$. In addition to the glycosaminoglycan chains, the protein contains 300 or more oligosaccharides covalently linked by $O$-glycoside bonds between $N$-acetylgalactosamine and serine and threonine hydroxyl side chains; $N$-glycosylamine-linked oligosaccharides appear to be present as well.

The follicular fluid proteoglycans do not interact with hyaluronic acid, and it is not known whether they interact with other macromolecules in the follicle. In solution, they would have extended, open structures; and they are probably responsible for the viscosity of the follicular fluid and the expansion of the follicles during maturation. Their synthesis by granulosa cells *in vivo* is probably stimulated by hormones, such as follicle-stimulating hormone, which are involved in follicular growth and oocyte maturation.[67] Selective degradation of the proteoglycan at ovulation may also occur in response to hormone stimulation.[66,68,69]

In addition to the population of large proteoglycans, follicular fluid contains a smaller dermatan sulfate proteoglycan.[69] This population has general features which are very similar to those described for the derma-

[66] M. Yanagishita, D. Rodbard, and V. C. Hascall, *J. Biol. Chem.* **254**, 911 (1979).
[67] P. L. Mueller, J. R. Schreiber, A. W. Lucky, J. D. Schulman, D. Rodbard, and G. T. Ross, *Endocrinology* **102**, 824 (1978).
[68] W. J. Beers, S. Strickland, and E. Reich, *Cell* **6**, 387 (1975).
[69] M. Yanagishita and V. C. Hascall, *J. Biol. Chem.* **254**, 12355 (1979).

tan sulfate proteoglycan from corneal stroma. The function of this proteoglycan in follicular fluid is not known.

*Heparan Sulfate Proteoglycans*

A heparan sulfate proteoglycan population has been isolated from liver plasma membranes.[70,70a] The macromolecules have an average molecular weight of about 80,000 with four bound heparan sulfate chains of 14,000 average molecular weight. In subsequent studies, it was shown that a large proportion of these cell surface-associated proteoglycans could be displaced by adding heparin to hepatocytes in culture; this suggests that the cell surface contains receptors that interact with heparan sulfate.[71] Trypsin was required to remove the remaining heparan sulfate, and it was suggested that part of the core protein, possibly a hydrophobic region, might interact with the cell membrane.[71] The ability of the multichain heparan sulfate proteoglycan to interact with different sites on the membrane and with macromolecules, such as collagen and fibronectin, in the matrix may be important for cell-matrix organization.

A similar class of heparan sulfate proteoglycans has been identified in the lamina rarae of basement membranes of glomeruli.[72-74a] These proteoglycans are organized in a two-dimensional lattice associated with the basement membrane and provide an anionic barrier through which the plasma ultrafiltrate must pass.[72,75]

Organ perfusion of rat kidney has been used to study biosynthesis of the basement membrane proteoglycans. Two proteoglycans, a heparan sulfate proteoglycan, about 85% of the total, and a chondroitin sulfate proteoglycan were found.[74] Both populations had similar properties; an overall molecular weight of about 130,000 with about four glycosaminoglycan chains of 26,000 average molecular weight attached. These proteoglycans, because of their strategic location in the basement membrane, probably play important roles in organization of basement membrane structure and in selective permeability.[75a]

---

[70] Å. Oldberg, L. Kjellén, and M. Höök, *J. Biol. Chem.* **254,** 8505 (1979).
[70a] S. Mutoh, I. Funakoshi, N. Ui, and I. Yamashina, *Arch. Biochem. Biophys.* **202,** 137 (1980).
[71] L. Kjellén, Å. Oldberg, and M. Höök, *J. Biol. Chem.* **255,** 10407 (1980).
[72] Y. S. Kanwar and M. G. Farquhar, *J. Cell Biol.* **81,** 137 (1979).
[73] Y. S. Kanwar and M. G. Farquhar, *Proc. Natl. Acad. Sci. U.S.A.* **76,** 1303 (1979).
[74] Y. S. Kanwar, V. C. Hascall, and M. G. Farquhar, *J. Cell Biol.* **90,** 527 (1981).
[74a] M. C. Lemkin and M. G. Farquhar, *Proc. Natl. Acad. Sci. U.S.A.* **78,** 1726 (1981).
[75] Y. S. Kanwar, A. Linker, and M. G. Farquhar, *J. Cell Biol.* **86,** 688 (1980).
[75a] T. N. Wight, *Prog. Hemostasis Thromb.* **5,** 1 (1980).

General Properties

From the few examples given above, it is clear that the variety of structures and properties of proteoglycans in different tissues will be large. For this reason, there is no prescribed set of protocols that can be applied uniformly to isolate and characterize proteoglycans from a new tissue or culture system. There are, however, several properties of proteoglycans that the investigator can use to advantage in designing procedures appropriate for the system under study. Because the macromolecules contain glycosaminoglycans, they (a) are highly polyanionic; (b) usually occupy large hydrodynamic volumes relative to globular proteins or glycoproteins of equivalent molecular weight; and (c) usually have high buoyant densities. Once an effective extraction procedure is developed, techniques such as ion-exchange chromatography, molecular-sieve chromatography, and density gradient centrifugation can be used to isolate and characterize the proteoglycans on the basis of differences in these parameters. The rest of the chapter will be devoted to descriptions of such techniques and examples of their applicability to specific systems.

Extraction, Purification, and Characterization

Development of Extraction Procedures

The first objective in studying proteoglycans is to extract them from the tissue in as high a yield as possible and under conditions that minimize the chances for degradation. Proteoglycans were first identified as separate entities from glycosaminoglycans by Shatton and Schubert[76] in 1954. In their study, hyaline cartilage was dispersed by high speed homogenization in water, or low concentrations of salt.[77] The high speed homogenization procedures, referred to as *disruptive extraction* procedures, were initially used extensively as the primary method for solubilizing proteoglycans until the introduction of dissociative extraction methods.

In 1969, Sajdera and Hascall[78] reported that solvents with high ionic strengths were effective in solubilizing proteoglycans from cartilage slices without requiring homogenization. The effectiveness of extraction correlated with the ability of the solvent to dissociate proteoglycan aggregates in the tissue. The procedures, therefore, were referred to as *dissociative extractions*.[34] The very effective denaturing solvent, guanidine-HCl at

[76] J. Shatton and M. Schubert, *J. Biol. Chem.* **211**, 565 (1954).
[77] S. Pal, P. T. Doganges, and M. Schubert, *J. Biol. Chem.* **241**, 4261 (1966).
[78] S. W. Sajdera and V. C. Hascall, *J. Biol. Chem.* **244**, 77 (1969).

concentrations of 4–6 $M$ was found to be the most effective extractant and is now used as the basis for most extraction procedures (see next section). After dialysis of the dissociative extract to lower ionic strength, *associative* conditions, proteoglycan aggregates re-formed. Equilibrium CsCl density gradients[79] in both associative and dissociative solvents were then developed to purify the proteoglycan fractions.[34]

In a subsequent study of proteoglycans from a transplantable rat chondrosarcoma, it was found that the proteoglycans in dialyzed extracts were often degraded by proteolysis from renatured endogenous proteases.[80] The use of a combination of protease inhibitors and low temperature throughout minimized this problem, and currently most extraction procedures used by investigators include protease inhibitors (see next section).

While 4 $M$ guanidine-HCl is effective in dissociating ionic and hydrophilic interactions, it is less so for hydrophobic interactions. In some cases, such as for studies of membrane-associated proteoglycans[70,71] or of intracellular proteoglycans,[81] it may be necessary to use extraction solvents that include detergents. The charged detergent deoxycholate or a combination of Triton X-100 and NaCl was used to solubilize the heparan sulfate proteoglycans from plasma membrane,[70] and a newly introduced zwitterionic detergent in combination with guanidine-HCl proved to be effective for solubilizing intracellular, [$^{35}$S]sulfate-labeled proteoglycans in chondrocyte cultures,[81] as described below.

Proteoglycan Extraction Procedures

*Preparation of Tissue*

In general, tissue or cultures should be processed for extraction quickly and using procedures that will minimize endogenous degradation. Fresh tissue should be chilled on ice, cleaned, and minced quickly. Extraction efficiency is often improved when finer tissue pieces are used to expose more surface area to the solvent.[81a] Whenever possible the fresh tissue should be extracted immediately, as described in the next section. If necessary, the tissue can be rapidly frozen and stored for later use. In this case, however, the tissue should be thawed directly in the extracting solvent if possible rather than thawed and processed further before extraction. This helps avoid degradation from enzymes released by the freeze–thaw cycle. Once the small tissue pieces are in the extracting solvent,

[79] M. D. Franek and J. R. Dunstone, *J. Biol. Chem.* **242**, 3460 (1967).
[80] T. R. Oegema, V. C. Hascall, and D. D. Dziewiatkowski, *J. Biol. Chem.* **250**, 6151 (1975).
[81] J. H. Kimura, C. B. Caputo, and V. C. Hascall, *J. Biol. Chem.* **256**, 4368 (1981).
[81a] M. T. Bayliss and S. Y. Ali, *Biochem. J.* **169**, 123 (1978).

homogenization at low temperature (0-4°) with a Polytron homogenizer operated at low speeds for short times can be used to disperse the tissue further, if necessary.

*Extraction with 4 M Guanidine-HCl in the Presence of Protease Inhibitors*

A variety of chaotropic agents, $6 M$ LiCl, $2.0 M$ CaCl$_2$, $3.0 M$ MgCl$_2$,[78] and $0.5 M$ LaCl$_2$[82] in addition to 4–6 $M$ guanidine-HCl, have been used to extract proteoglycans from cartilage. Unlike guanidine-HCl, however, the other electrolytes have narrow concentration ranges for optimal extraction.[78] Guanidine-HCl has proved to be the most convenient and widely used extractant. A variety of protease inhibitors with a broad range of specificities have been used in extracting solvents: 6-aminohexanoic acid (for cathepsin D activity), benzamidine-HCl (for trypsin-like activity) and NaEDTA (for metalloproteases)[80]; phenylmethylsulfonyl fluoride (for serine-dependent proteases),[42] iodoacetamide (for thiol-dependent proteases)[42] and soybean trypsin inhibitor;[83] and pepstatin (for cathepsin D) and $N$-ethylmaleimide (for thiol proteases as well as to prevent nonspecific disulfide exchange, which can occur in denaturing solvents).[84] Additionally, the pH of the extracting solvent, usually about 6.0, is above the optimum for acid pH proteases and below the optimum for neutral pH proteases.

An effective solvent that can be used is 4 $M$ guanidine-HCl, 0.1 $M$ 6-aminohexanoic acid, 0.01 $M$ Na$_2$EDTA, 0.05 $M$ sodium acetate, pH 6.0 with 1 m$M$ benzamidine-HCl, 1 m$M$ phenylmethylsulfonyl fluoride, and 10 m$M$ $N$-ethylmaleimide added just prior to use. The phenylmethylsulfonyl fluoride (and pepstatin, 0.4 m$M$ if used) can be first dissolved at 100 × concentration in methanol, since they are difficult to dissolve in H$_2$O.[84] Tissue is added to chilled extractant (10–15 volumes per gram wet weight for cartilages, 5–10 volumes per gram wet weight for other tissues). Extraction is continued at 4° with stirring usually for 12–20 hr. Extraction efficiency can be determined by measuring the amount of proteoglycan released into the solvent with time. For cartilages this can be done by monitoring solubilization of hexuronic acid.[78] For other tissues with less proteoglycan, glycoproteins and nucleic acids will interfere with the hexuronic acid assay, and it may be necessary first to isolate the glycosaminoglycans from papain or alkali digests of portions of extracts to quantitate amounts of proteoglycans solubilized.[1] After extraction, insoluble residue is removed by coarse filtration and/or centrifugation (45,000 g

---

[82] R. M. Mason and R. W. Mayes, *Biochem. J.* **131**, 535 (1973).
[83] J. P. Pearson and R. M. Mason, *Biochim. Biophys. Acta* **498**, 176 (1977).
[84] Y. Oike, K. Kimata, T. Shinomura, K. Nakazawa, and S. Suzuki, *Biochem. J.* **191**, 193 (1980).

for 20 min at 4°). The clarified extract is then processed further as described below. The amounts of proteoglycans remaining in the residue can be determined after solubilizing glycosaminoglycans with papain or alkali.

*Extraction with Detergents*

Oldberg et al.[70] reported that extraction at 4° with 1% deoxycholate in 0.05 $M$ Tris-HCl, pH 8, effectively solubilized 90% of the $^{35}$S-labeled heparan sulfate proteoglycan from liver plasma membranes isolated from rats after labeling *in vivo* with [$^{35}$S]sulfate. A combination of 1% Triton X-100 and 2 $M$ NaCl was as effective, but either the Triton X-100 or the NaCl alone extracted much less. Thus, a combination of electrolyte and detergent can be used for effective extraction in some cases. The use of deoxycholate, however, introduces some problems if density gradients are to be used in subsequent purification steps, and the authors used molecular-sieve chromatography and ion-exchange procedures to achieve partial purification of the proteoglycan and removal of the detergent before density-gradient centrifugation.

Kimura et al.[81] have developed an extraction procedure using a zwitterionic detergent that circumvents this problem. Previously, it was found that 4 $M$ guanidine-HCl was a very poor extractant (solubilizing less than 10% of total) for intracellular, [$^{35}$S]sulfate-labeled proteoglycans in chondrocyte cultures.[44] However, when the cell layer was first exposed to one-half volume of 8% (w/v) $N$-dodecyl-$N,N$-dimethyl-3-amino-1-propanesulfonate (Zwittergent 3-12) for 10 min and then one-half volume of 8 $M$ guanidine-HCl added directly to make the final solution 4% detergent and 4 $M$ guanidine-HCl, more than 95% of the intracellular proteoglycans were solubilized. Both solutions contained protease inhibitors, and the extraction was done at 0°. The advantages of this zwitterionic detergent are its solubility in high concentrations of electrolytes and at low temperatures. Thus, its presence in extracts does not interfere with subsequent dissociative density-gradient or molecular-sieve chromatography procedures.[81] The zwitterionic detergent by itself failed to solubilize any of the intracellular proteoglycans, and a solvent of 4% detergent in 4 $M$ guanidine-HCl added directly to the cell layer was not as effective as the sequential procedure described above.[81]

*Extraction with Associative Solvents*

For most hyaline cartilages as well as other connective tissues, dissociative solvents are required for effective extraction. In a few cases, however, high extraction yields can be obtained using lower ionic strength, associative solvents. Faltz et al.[31] compared a proteoglycan preparation isolated from a transplantable rat chondrosarcoma by dissociative procedures with preparations extracted for short times (4 hr) with 0.5 $M$ guanidine-HCl in the presence of protease inhibitors at 0°. Although somewhat lower yields were obtained (60% versus 90% of total tissue proteoglycans were extracted), the proteoglycans purified from the associative extraction showed no detectable signs of degradation. Such a procedure has the advantage that the proteoglycan aggregate structures in the preparation represent those present in the tissue at the time of extraction and are not reconstituted after dialysis of a dissociative solvent to associative conditions.[31,33] Such associative extraction procedures may be useful for extracting proteoglycans from tissues, such as the chondrosarcoma, which are highly hydrated and have a loose, more open, collagen network.

Micropuncture techniques[32] have been developed to aspirate minute amounts of fluid (10–30 nl) from rat growth plate cartilages, and microanalytical techniques have been used to investigate the properties of the proteoglycans in such aspirates. These studies have clearly shown that a large proportion of the proteoglycans in such associatively extracted samples are aggregated. The ingenious, but specialized techniques used in these studies, which involve miniaturization of standard techniques, are beyond the scope of this chapter. See Pita and Howell[85] for further discussion.

*Proteoglycans in the Residue*

The 4 $M$ guanidine-HCl dissociative solvent effectively denatures protein and dissociates most noncovalent interactions between macromolecules. A proportion of proteoglycans in the tissues studied (10–50% of the total in general) are not solubilized by the solvent, and may represent proteoglycans that are covalently bound in the matrix, perhaps to collagen. If the covalent interactions are disulfide bridges, as has been suggested recently for some of the heparan sulfate proteoglycans in glomerular basement membranes,[86] then the inclusion of a reducing reagent, such as 10 m$M$ dithiothreitol, in the extracting solution while leaving out the alkylating agents may facilitate the solubilization of additional proteoglycans from extracted tissue residue. Alternative procedures such as the use of highly purified collagenase to solubilize the matrix have been attempted in some instances in order to solubilize proteoglycans in the residue.[87] Because the proteoglycans are very susceptible to proteolysis, the effects of the enzymic procedures should be tested on soluble, well

[85] J. C. Pita and D. S. Howell, *in* "The Joints and Synovial Fluid" (L. Sokoloff, ed.), Vol. 1, p. 273. Academic Press, New York, 1978.
[86] N. Parthasarathy and R. G. Spiro, *J. Biol. Chem.* **256**, 507 (1981).
[87] B. Radhakrishnamurthy, H. A. Ruiz, and G. S. Berenson, *J. Biol. Chem.* **252**, 4831 (1977).

characterized proteoglycan samples to determine the extent to which the digestion conditions may degrade the proteoglycans during the digestion of the residue. At present, there are no totally satisfactory procedures for isolating intact proteoglycans from the residue of extracted tissues.

Purification and Separation

Once the proteoglycans are in solution, different fractionation procedures, which take advantage of their macromolecular properties, are used to purify them from other macromolecules in the extract and to begin to separate them from each other. For many tissues where proteoglycan contents are small, it is usually necessary to keep the proteoglycans in dissociative conditions until some partial purification is achieved; dialysis of extracts of such tissues into associative solvents, for example, often leads to intractable coacervates. For extracts of cartilage, however, dialysis to associative conditions is required in order to allow the proteoglycans to re-form aggregates. The following sections describe various procedures that can be used to purify and separate the proteoglycans in tissue extracts.

*Isopycnic CsCl Density Gradients*

*Associative Conditions.* Franek and Dunstone[79,88] were the first to use isopycnic density gradient centrifugation methods to purify proteoglycans. The cartilage proteoglycans have a high ratio of glycosaminoglycan to protein and hence a high buoyant density; therefore they migrate to the bottom of such gradients. Proteins and glycoproteins, which have much lower buoyant densities, migrate to the top of the gradient unless they are strongly associated with the proteoglycan fraction.

This gradient procedure was subsequently applied to dialyzed dissociative extracts of cartilage.[34] Extracts in 4 $M$ guanidine-HCl were dialyzed at 4° for 20 hr against 7 volumes of the extracting solvent minus the guanidine-HCl to bring the final solvent conditions to approximately 0.5 $M$ guanidine-HCl. During the dialysis step a large proportion (usually 50–90%) of the proteoglycans reaggregate. CsCl (1.19 g/g of solution) was added to give an initial density of 1.69 g/ml, and gradients were established either in a swinging-bucket rotor or a fixed-angle rotor by centrifugation at 40,000 rpm for 44 hr at 20°. The proteoglycan fraction was recovered from the bottom two-fifths of the gradient in a fraction now referred to as the associative-1 (A1) fraction.[89]

[88] M. D. Franek and J. R. Dunstone, *Biochim. Biophys. Acta* **127**, 213 (1966).
[89] D. Heinegård, *Biochim. Biophys. Acta* **285**, 181 (1972).

Currently, associative gradients are usually established using slightly different conditions. The dissociative extract is dialyzed at 4° in the presence of protease inhibitors to 0.5 $M$ guanidine-HCl, or in the case of associative extracts the extracts are used directly.[31] Proteoglycan concentrations up to about 6 mg/ml are acceptable. Any insoluble material that forms during dialysis is removed by brief centrifugation (45,000 $g$, 20 min, 4°). CsCl (1.1 g/g of solution) is added to give an initial density of 1.60 g/ml. In many cases, a precipitate forms when the CsCl dissolves. This precipitate should be removed by a brief centrifugation (45,000 $g$, 20 min, 4°) during which it will form a floating pellicle. The clarified solution is then centrifuged in either a Beckman SW 50.1 swinging-bucket rotor, a 50 Ti or a 50.2 Ti angle rotor (or their equivalent), depending upon the volume used, at about 35,000 rpm (~150,000 $g_{max}$) 10°, 48–72 hr. The lower speed and lower initial density than those originally used prevent the formation of CsCl pellets in the bottom of the gradients. The lower temperature is recommended to decrease the chances for endogenous proteolysis. If CsCl pellets are detected in the gradients, slightly lower initial densities and/or centrifugal speeds should be used. The proteoglycan A1 fraction is often recovered from the bottom one-fourth to two-fifths of the gradient by using a tube slicer or by puncturing the bottom of the tube and collecting fractions from the bottom. In some cases, such as when polycarbonate tubes are used for large preparations, fractions are recovered by aspiration either from the top, which may cause some top to bottom mixing, or from the bottom by inserting a long needle to the bottom of the gradient before aspirating. For hyaline cartilages, a large proportion of the proteoglycans in the A1 fraction (40–90%) will be aggregated. Procedures described below can be used to separate aggregated from nonaggregated proteoglycans such that the two fractions can be analyzed separately.

Dialysis of dissociative extracts of tissues other than cartilage to associative conditions frequently leads to precipitation of proteoglycans in a coacervate with other macromolecules in the matrix, and proteoglycan recovery in subsequent associative density gradients is poor.[54] This may be circumvented by dialyzing the extract to a somewhat higher guanidine-HCl concentration of 0.75 $M$ rather than the 0.5 $M$ concentration normally used and by using less CsCl in the gradient such that the initial density (1.40–1.45 g/ml) is lower than usual. Proteoglycan aggregation can still occur in these conditions, but the electrolyte concentrations are sufficient to prevent the extensive precipitation that occurs when the normal conditions are used.

*Dissociative Conditions.* Studies on the mechanism of proteoglycan aggregation indicated that 4 $M$ guanidine-HCl was an effective solvent for dissociation.[34] When CsCl was added to aggregate fractions dissolved in this solvent and a density gradient was established by centrifugation, more than 90% of the proteoglycans were recovered in the bottom two-fifths of the gradient in a monomeric form, which by itself was unable to aggregate when restored to associative conditions. Low buoyant density components required for the aggregation process were recovered in the top fractions. Subsequent investigations identified the link proteins[23,24,35,39] and hyaluronic acid[27,39] as the factors in the upper portions of the dissociative density gradient that were necessary for re-forming aggregates. Dissociative density gradients have since become widely used to separate those proteoglycans that have sufficiently high buoyant densities from other macromolecules in tissue extracts.

Conditions suitable for preparing isopycnic CsCl dissociative density gradients for samples from cartilaginous tissues are as follows: The solutions are made approximately 4 $M$ in guanidine-HCl. For an A1 sample isolated directly from the bottom of an associative density gradient as described above, the preparation can be diluted with an equal volume of 7.5 $M$ guanidine-HCl in the same buffer-protease inhibitor solution used initially. The final proteoglycan concentration can be up to 6 mg/ml. CsCl is added to give an initial density of about 1.47 g/ml (0.55 g CsCl per gram of solution if there is no CsCl already present, or 0.15 g CsCl per gram of solution for every 0.1 g/ml difference between 1.47 g/ml and the density of the solution if some CsCl is already present). The gradient is established by centrifugation at 10° for 48–72 hr at 35,000 rpm in one of the rotors described above for the associative density gradients. The gradient is fractionated as described for the associative gradients. If an A1 sample was used, the bottom one-fourth to two-fifths of the dissociative gradient will contain most of the monomer proteoglycans, with smaller amounts of lower buoyant density proteoglycans distributing into the upper fractions.[90-92] This fraction is equivalent to the A1-D1 (associative gradient fraction 1–dissociative gradient fraction 1) as originally defined by Heinegård.[89] The link proteins are recovered in the top one-fourth of the gradient along with small amounts of low buoyant density proteoglycans and hyaluronic acid in a fraction equivalent to the A1-D4 fraction originally described.[89] The link proteins in this fraction can subsequently be purified by molecular-sieve chromatography in 4 $M$ guanidine-HCl on Sephadex G-200,[24] Sephacryl S-200,[42] or Sepharose CL-6B.[25] Preparative sodium dodecyl sulfate–polyacrylamide gel electrophoresis has also been

[90] L. Rosenberg, C. Wolfenstein-Todel, R. Margolis, S. Pal, and W. Strider, *J. Biol. Chem.* **251**, 6439 (1976).
[91] T. E. Hardingham, R. J. F. Ewins, and H. Muir, *Biochem. J.* **157**, 127 (1976).
[92] D. K. Heinegård, *J. Biol. Chem.* **252**, 1980 (1977).

used to purify the link proteins from A1-D4.[37,93] The middle fractions (A1-D2,3) contain most of the hyaluronic acid in the original A1 samples,[39,94] although small amounts of hyaluronic acid are usually present in both the A1-D1 and A1-D4 fractions as well. The same precautions should be taken for the dissociative gradients as for the associative gradients to prevent the formation of a CsCl pellet at the bottom.

*Direct dissociative* gradients can be used to isolate proteoglycans of high buoyant density directly from dissociative extracts of tissues. In this case, CsCl is added directly to the clarified extract to the desired initial density. For cartilage extracts, the same conditions as described in the preceding paragraph will yield a monomer (D1) preparation.[34] For other tissues, such as aorta, the proteoglycans in the extract often have lower average buoyant densities than the majority of the population of cartilage proteoglycans, and lower initial densities in the dissociative density gradient (1.35–1.45 g/ml) can be used.[53,54,63,95] Fractions recovered from the gradients can be analyzed for glycosaminoglycan contents to determine the distribution of proteoglycans in the gradients. For example, after dialysis to remove CsCl the proteoglycans (along with a large proportion of other macromolecules) in aliquots can be precipitated at 4° with 3 volumes of 95% ethanol containing 1.3% (w/v) potassium acetate.[96] The precipitates, recovered by centrifugation, can then be analyzed for glycosaminoglycan contents.[1]

The major contaminants which may be present in high buoyant density fractions with proteoglycans are glycogen and nucleic acids, especially if tissues with high cellularity and low proteoglycan content are being studied. A second purification step may then be required. For example, Eisenstein *et al.*[95] precipitated the proteoglycans in a D1 fraction derived from bovine aorta using the organic cation cetylpyridinium chloride, which selectively binds to the anionic groups in the glycosaminoglycans. The precipitation was done with a final concentration of 1% (w/v) cetylpyridinium chloride in the presence of 0.5 $M$ KCl, which permits the precipitation of sulfated proteoglycans but not of the nucleic acids.[97] Glycogen is not precipitated by the procedure. The proteoglycans were redissolved in 4 $M$ guanidine-HCl and reisolated in a second dissociative density gradient. Alternatively, an ion-exchange procedure such as that discussed in the next section can be used to free the proteoglycan fraction from contaminants.

[93] F. Bonnet, J.-P. Périn, and P. Jollès, *Biochim. Biophys. Acta* **532**, 242 (1978).
[94] T. E. Hardingham and H. Muir, *Biochem. Soc. Trans.* **1**, 282 (1973).
[95] R. Eisenstein, S.-E. Larsson, K. E. Kuettner, N. Sorgente, and V. C. Hascall, *Atherosclerosis* **22**, 1 (1975).
[96] K. Kimata, M. Okayama, A. Oohira, and S. Suzuki, *J. Biol. Chem.* **249**, 1646 (1974).
[97] J. E. Scott, *Method Biochem. Anal.* **8**, 145 (1960).

*Ion Exchange in 7 M Urea*

Antonopoulos et al.[53] devised an ion-exchange procedure for purifying proteoglycans from 4 M guanidine-HCl extracts as an alternative to the use of density gradients as a first purification step. Because the guanidine-HCl is an electrolyte, these investigators replaced it with 7 M urea to maintain denaturing conditions in a solvent that at the same time permits the use of ion exchange columns. The denaturing conditions prevent the formation of the coacervates which occur when such extracts from many tissues are dialyzed to associative conditions.

The procedure is as follows, with all steps at 4°. Extracts of bovine sclera, aorta, and cornea were made in 15 volumes of 4 M guanidine-HCl per gram of tissue. The extracts were ultrafiltered to concentrate to about $1/20$ to $1/30$ of the original volume, and then the guanidine-HCl was exchanged with urea either by dialysis against several changes of 7 M urea, 0.05 M Tris, pH 6.5–6.8, or by continuous ultrafiltration with the same solution. The urea solution was passed through a mixed-bed ion-exchange resin of the type used to deionize $H_2O$ prior to use. [While not used in the originally described procedure, the addition of some of the protease inhibitors described above (under Proteoglycan Extraction) to the urea solution, especially those present in low concentrations, such as phenylmethylsulfonyl fluoride, benzamidine-HCl, and $N$-ethylmaleimide, should be considered.] The concentrated extracts in 7 M urea were often slightly turbid and were clarified by a brief centrifugation. The extracts were then applied to DEAE-cellulose columns (in chloride form) which had been equilibrated with the 7 M urea solution. A 4 × 30 cm column was found satisfactory for up to ~200 mg of proteoglycan. Columns were eluted with 2–3 bed volumes each of (*a*) the 7 M urea solution; (*b*) 0.15 M NaCl in the 7 M urea solution; and (*c*) 2 M NaCl in the 7 M urea solution. In each tissue 85–90% of the proteoglycans were recovered in the 2 M NaCl, 7 M urea elution fraction whereas the large majority of the other macromolecules in the extracts were eluted in the first two steps. The proteoglycan fraction, then, can be recovered for further studies. A modification of this procedure that can be useful is to elute the proteoglycans with a linear, continuous NaCl gradient in the urea solvent. This will often help resolve from each other different classes of proteoglycans that may be present in the extracts.[98] Another ion-exchange procedure using DEAE-Sephacel and a NaCl gradient in the presence of 0.1% Tween 80 has been recently described and used to isolate $^{35}S$-labeled heparan sulfate proteoglycans from deoxycholate-solubilized hepatocyte membrane fractions.[70]

[98] J. R. Hassell, P. G. Robey, H.-J. Barrach, J. Wilczek, S. I. Rennard, and G. Martin, *Proc. Natl. Acad. Sci. U.S.A.* **77**, 4494 (1980).

## Rate Zonal-Velocity Centrifugation

*Associative.* Franek and Dunstone[79] first utilized zonal-velocity centrifugation on preformed gradients to fractionate proteoglycans. They layered a proteoglycan preparation from cartilage on 5 to 30% (w/v) NaCl gradients for subsequent fractionation by centrifugation in a swinging-bucket rotor. Unlike the isopycnic gradients, which separate proteoglycans on the basis of their buoyant density, which is related to their glycosaminoglycan to protein ratio, the band velocity method partitions proteoglycans on the basis of their sedimentation velocity, which is related in part to their molecular weight.

Hoffman and co-workers have subsequently used similar gradients under associative conditions to subfractionate proteoglycan fractions with aggregate from those without aggregate for further analysis.[99,100] In these studies, 26-ml linear gradients from 5% to saturated NaCl were formed in 36-ml capacity polyallomer tubes for a Beckman angle Ti 60 rotor. The sample (about 2 mg of proteoglycan per milliliter) in 5 ml of 2% NaCl is layered on top of the gradient and then centrifuged at 45,000 rpm, 20° for 40 min. The gradients are then fractionated with a tube slicer. Aggregate migrates to the bottom while nonaggregated proteoglycans remain in the top fractions.

Kimata *et al.* have developed a different associative rate zonal procedure.[101] Linear gradients of either 13 ml or 32 ml between $0.15\,M$ and $0.5\,M$ $Cs_2SO_4$ are formed on cushions of 1 ml or 3 ml of $2\,M$ $Cs_2SO_4$ in centrifuge tubes. The $Cs_2SO_4$ solutions contain $0.2\,M$ sodium acetate, pH 7 throughout. Samples in $0.2\,M$ sodium acetate (1 ml or 3 ml) with up to 10 mg of proteoglycan per milliliter are layered on top. Centrifugation is done for 6 hr at 10° or 4° in a swinging-bucket Beckman SW27 rotor at 25,000 rpm. The gradients are then fractionated with a tube slicer or a gradient fractionator using $2.5\,M$ $Cs_2SO_4$ to push fractions out of the top into a fraction collector. For cartilage proteoglycans, aggregate is recovered at the boundary with the $2\,M$ $Cs_2SO_4$ cushion in the bottom of the tube while nonaggregated proteoglycans migrate to a peak region approximately two-fifths of the distance into the gradient. The advantages of the use of $Cs_2SO_4$ for the preformed gradient are that (*a*) hexuronic acid analyses can be done directly on fractions from the gradient using automated procedures that are not possible when high concentrations of chloride ions are

---

[99] P. Hoffman, *J. Biol. Chem.* **254**, 11854 (1979).
[100] P. Hoffman, T. A. Mashburn, D. Hsu, D. Trivedi, and J. Diep, *J. Biol. Chem.* **250**, 7251 (1975).
[101] K. Kimata, J. H. Kimura, E. J.-M. A. Thonar, H.-J. Barrach, S. I. Rennard, and V. C. Hascall, *J. Biol. Chem.* in press (1982).

present; (b) the shallow $Cs_2SO_4$ concentration gradient still achieves an effective stabilizing density gradient, hence there is less dependence of sedimentation rates on viscosity changes or temperature; and (c) the gradients have a high capacity for proteoglycans making then suitable for both preparative and analytical separations. Gradients formed in swinging-bucket rotors also have the advantage that they can be fractionated with a density-gradient fractionator. Such $Cs_2SO_4$ gradients can also be formed in the presence of 4 $M$ guanidine-HCl to provide a dissociative solvent and have been used at low solvent pH (pH 2.7–3.6) to study dissociation of cartilage proteoglycan aggregates.[102]

The use of rate-zonal centrifugation under associative conditions permits the recovery of purified aggregates from samples of Al[99] or aAl[31,101] (an Al preparation isolated after an associative extraction step). These purified aggregates can then be dissociated in 4 $M$ guanidine-HCl and monomers prepared with a dissociative isopycnic gradient.[101] In this case a population of monomers, each of which contains a functional hyaluronic acid-binding region, is recovered, and the properties of this subpopulation can be evaluated in the absence of the nonaggregated proteoglycans that contain population(s) with different chemical properties.[99,101,103,104]

*Dissociative Conditions.* Kimata *et al.*[96] have described a zonal centrifugation procedure for separation of proteoglycans on preformed glycerol (or sucrose) gradients formed in the presence of 4 $M$ guanidine-HCl. This procedure has been used to compare the hydrodynamic sizes of proteoglycans synthesized in cultures derived from a variety of tissues.[96,105–110] Linear gradients from 10% to 35% (w/v) glycerol are formed in 4 $M$ guanidine-HCl, 0.02 $M$ Tris HCl, 0.01 $M$ sodium EDTA, pH 8.0, on a cushion of 50% glycerol in the same solvent using the volumes for the small- or large-capacity tubes of the SW 27.1 rotor as described in the preceding section. The sample, 1 ml in the same solvent without glycerol, is layered on the preformed gradient and centrifuged for 25 hr at 25,000 rpm and 20°. Fractions are obtained with a density gradient fractionator using 60% glycerol to push the gradients out of the tubes. For biosynthetic experiments, aliquots of the fractions are counted. The glycerol and

[102] K. Kimata, V. C. Hascall, and J. H. Kimura, *J. Biol. Chem.* in press (1982).
[103] D. K. Heinegård and V. C. Hascall, *J. Biol. Chem.* **254**, 927 (1979).
[104] M. B. E. Sweet, E.J.-M. A. Thonar, and A. R. Immelman, *Arch. Biochem. Biophys.* **189**, 28 (1978).
[105] M. Okayama, M. Pacifici, and H. Holtzer, *Proc. Natl. Acad. Sci. U.S.A.* **73**, 3224 (1976).
[106] K. Kitamura and T. Yamagata, *FEBS Lett.* **71**, 337 (1976).
[107] M. Pacifici, S. A. Fellini, H. Holtzer, and S. DeLuca, *J. Biol. Chem.* **256**, 1029 (1981).
[108] S. A. Fellini, M. Pacifici, and H. Holtzer, *J. Biol. Chem.* **256**, 1038 (1981).
[109] S. A. Fellini, J. H. Kimura, and V. C. Hascall, *J. Biol. Chem.* **256**, 7883 (1981).
[110] P. D. Royal, J. J. Sparks, and P. F. Goetnick, *J. Biol. Chem.* **255**, 19870 (1980).

guanidine-HCl in the fractions precludes direct chemical analyses for sugars or protein; thus, the macromolecules in the fractions must first be recovered for such analyses by either dialysis or ethanol precipitation.[96] In general, monomer proteoglycans from cartilages migrate to a peak position about three-fourths of the distance into the gradient, whereas smaller proteoglycans found in small proportions in cartilage or chondrocyte cultures, but in larger proportions in cultures of prechondrogenic cells or other cell types,[96,105,110–114] are recovered in intermediate or upper fractions.

Variations in the conditions for the dissociative rate-zonal centrifugation can be used depending upon the rotor, the nature of sample being analyzed, analytical or preparative gradients, etc. For example, Royal et al.[110] used 5 to 20% linear sucrose gradients in $4\,M$ guanidine-HCl, $0.05\,M$ sodium acetate, pH 5.8, containing protease inhibitors in tubes of smaller capacity (SW 40 rotor). Satisfactory separations were obtained for runs at 38,000 rpm, 18.5 hr, and 20°. In general the glycerol and sucrose give a high viscosity gradient, and viscosities increase markedly as the temperature is lowered, so exact centrifugal conditions for a given experiment will have to be determined empirically.

*Molecular-Sieve Chromatography*

Molecular-sieve chromatography is one of the most widely applied analytical procedures for identifying and characterizing proteoglycans. Because of the relatively low capacity of this technique for intact proteoglycans, it is seldom used as a large-scale preparative technique, although Heinegård,[92] for example, utilized preparative Sepharose 2B chromatography to isolate proteoglycan subfractions of different average hydrodynamic size from a cartilage monomer (A1-D1) preparation. In biosynthetic studies and other cases where the amounts of material are relatively small, however, molecular sieve chromatography has become a primary preparative method for resolution of proteoglycan classes for further study (see Hassell et al.,[63,64] for example). It is beyond the scope of this chapter to provide details for all the various molecular-sieve procedures that have been developed. Rather, some general considerations and examples will be presented.

Three molecular-sieve supports are routinely used—Sepharose, agarose (BioGel A), and controlled-pore glass beads—using a variety of

[111] M. J. Palmoski and P. F. Goetinck *Proc. Natl. Acad. Sci. U.S.A.* **69**, 3385 (1972).
[112] D. Levitt and A. Dorfman *Proc. Natl. Acad. Sci. U.S.A.* **69**, 1253 (1972).
[113] S. DeLuca, D. Heinegård, V. C. Hascall, J. H. Kimura, and A. I. Caplan, *J. Biol. Chem.* **252**, 6600 (1977).
[114] L. Cöster, I. Carlstedt, and A. Malmström, *Biochem. J.* **183**, 669 (1979).

both associative and dissociative solvent conditions. The general considerations for choice of an appropriate procedure are the hydrodynamic range and resolution of the support, the capacity of the support, time for elution, recovery of solute from column, and the kinds of analytical procedures to be done on the fractions.

Cartilage proteoglycans have very large hydrodynamic volumes and the most porous supports, Sepharose 2B (or more recently the cross-linked CL-2B), agarose (BioGel A-50 m), and controlled-pore glass CPG-1400 (mean pore diameter 125 nm) or CPG-2500 (mean pore diameter 257 nm) must be used so that the macromolecules will partition within the column volume. For the Sepharose and agarose supports, the samples should be applied in volumes of less than about 2% of the total column volume for good resolution. For Sepharose 2B using associative conditions such as $0.5 M$ sodium acetate, pH 7 (a solvent routinely used when colorimetric analyses of fractions are to be done[115]), the concentration of monomer proteoglycan applied should be less than about 10 mg/ml.[92] For samples with aggregate, such as A1 preparations from cartilage, the viscosity of the solution will be much higher and lower loading concentrations must be used (less than 2–3 mg/ml). When small amounts of solute are applied, as when analyzing radiolabeled proteoglycan samples, recovery in the eluted fractions may be poor, especially for large proteoglycans, and it may be necessary to add carrier, unlabeled proteoglycan to give sufficient concentration (greater than 1–2 mg/ml) to yield better recovery. When using associative solvents, precautions must be made to prevent bacterial or fungal growth. Boiling the buffer briefly before use and ultrafiltration are recommended. The inclusion of a crystal of thymol in the buffer has been used as a preservative,[116] and, unlike sodium azide which interferes with analyses for hexuronic acid, the thymol does not interfere with the colorimetric procedures (anthrone reaction for neutral sugars,[115] carbazole reaction for hexuronic acid,[115] or resorcinol reaction for sialic acid[11]) commonly used for sugar analyses. It does interfere with the Folin reaction for protein, however. Elution times for Sepharose or agarose columns should be from about 15 hr up to 2–3 days depending upon the size of column, the solvent, and temperature. Columns run in the cold require longer time than those eluted at room temperature owing to increased solvent viscosity. Resolution is not as good if shorter times are used.

For the controlled-pore glass columns, elution times can be shorter while still obtaining satisfactory resolution.[110,117] This support has some advantages over the Sepharose and agarose supports. The glass beads are

[115] D. K. Heinegård, *Chem. Scr.* **4**, 199 (1973).
[116] E. J.-M. A. Thonar, personal communication.
[117] P. L. Lever and P. F. Goetinck, *Anal. Biochem.* **75**, 67 (1976).

incompressible, permitting higher flow rates and identical flow rates at lower temperature, and the columns can be cleaned with strong acid solutions. Some disadvantages are the smaller capacity per bed volume and somewhat reduced hydrodynamic range. The controlled pore glass supports must be periodically treated with polyethylene glycol to improve recovery of solute in the eluate. Details for preparing these columns are provided elsewhere.[117]

Such dissociative solvents as 4 $M$ guanidine-HCl and 0.1% sodium dodecyl sulfate have been used with the cross-linked Sepharose and agarose supports to analyze proteoglycans.[e.g.,37,43,63,69,84] For sodium dodecyl sulfate, columns must be eluted at room temperature or above because of the poor solubility of the sodium salt of dodecyl sulfate. The inclusion of a lithium phosphate buffer[43,118] helps avoid this problem because of the increased solubility of the lithium salt. These columns have proved to be effective in separating components such as the link proteins from proteoglycans.[37,43]

The study by Oike et al.[84] utilized 0.2% (w/v) Triton X-100 in 4 $M$ guanidine-HCl with a variety of protease inhibitors as an eluting solvent for Sepharose CL-4B and BioGel A-15m chromatography. The inclusion of the detergent was required to obtain good recoveries of solute from the column when small amounts were applied. The inclusion of a detergent, then, where practical, will minimize losses of proteoglycan on the columns. It should be noted, however, that removal of detergents from the samples at later stages of purification is often difficult. The presence of the detergent may also interfere with assay procedures and prevent recovery of solute functionality such as the ability of cartilage proteoglycans to interact with hyaluronic acid.

*Electrophoresis in Composite Polyacrylamide-Agarose Gels*

McDevitt and Muir[119] have described a procedure for resolving intact proteoglycans by electrophoresis in large-pore composite polyacrylamide-agarose gels. This procedure has been used to demonstrate the presence of two, or sometimes three, closely spaced but distinct proteoglycan bands for purified monomer preparations from different hyaline cartilages.[119a-d] The method appears to separate proteoglycans on the basis of discrete differences in keratan sulfate to chondroitin sulfate levels,[119e] and

[118] B. Caterson, personal communication.
[119] C. A. McDevitt and H. Muir, *Anal. Biochem.* **44**, 612 (1971).
[119a] P. J. Roughley and R. M. Mason, *Biochem. J.* **157**, 357 (1976).
[119b] J. P. Pearson and R. M. Mason, *Biochim. Biophys. Acta* **498**, 187 (1977).
[119c] V. Stanescu, P. Maroteaux, and E. Sobczak, *Biochem. J.* **163**, 193 (1977).
[119d] V. Stanescu, P. Maroteaux, and E. Sobczak, *Biochim. Biophys. Acta* **629**, 371 (1980).
[119e] J. P. Pearson and R. M. Mason, *Biochem. Soc. Trans.* **6**, 244 (1978).

it provides a convenient analytical method (4–10 µg of proteoglycan required) for comparing proteoglycans from different cartilages or from cartilages at different stages of development. Tube or slab gels can be prepared, run, and stained by the detailed procedures described by McDevitt and Muir[119] with the modification suggested by Stanescu et al.[119d] (1.2% acrylamide and 0.7% agarose).

Further Characterization

Application of the various techniques described above will have identified the different classes of proteoglycans in a starting extract based upon differences in charge density, hydrodynamic size, and buoyant density. Such analyses not only provide information about the physical properties of the proteoglycans but help design effective strategies for purifying at least some of the subpopulations of proteoglycans identified. Because of the inherent polydisperity of proteoglycans and the critical role of the glycosaminoglycan chains in determining many of the macromolecular properties, different classes of proteoglycans may easily copurify, and further analyses are required to determine purity as well as additional properties of the macromolecules. These analyses can include chemical, physicochemical, and immunological methods. Some general discussion of these is given below, but it is beyond the scope of this chapter to provide detailed descriptions.

*Chemical*

*Glycosaminoglycans.* The hyaline cartilage proteoglycans have two glycosaminoglycans, chondroitin sulfate and keratan sulfate, attached to the same core protein. In general, however, it appears that most other proteoglycans will contain a single glycosaminoglycan, and proteoglycan samples that contain more than one glycosaminoglycan probably contain more than one population of proteoglycan. For example, the D1 fraction of $^{35}$S-labeled proteoglycans isolated from rat granulosa cells contained about 8% heparan sulfate, which was shown to reside on a different proteoglycan of similar size to the predominant dermatan sulfate proteoglycan,[69] and the proteoglycans synthesized in organ perfusion of kidney contained 15% chondroitin sulfate, which appears to be present on a proteoglycan that differed only slightly in physical properties from the predominant heparan sulfate proteoglycan.[74]

The glycosaminoglycans can be released from the protein of the proteoglycan by treatment with alkali[120] or alkali borohydride[10,121] or with papain. In the latter case, however, peptides with oligosaccharides or

[120] B. Anderson, P. Hoffman, and K. Meyer, *J. Biol. Chem.* **240**, 156 (1965).
[121] D. M. Carlson, *J. Biol. Chem.* **243**, 616 (1968).

more than one glycosaminoglycan chain may remain attached.[11] The isolated glycosaminoglycans can be analyzed for types by (a) selective enzyme digestion (such as treatment with chondroitinases, endogalactosidases which degrade keratan sulfate,[84] *Streptomyces* hyaluronidase selective for hyaluronic acid, or enzymes that degrade heparan sulfate or heparin); (b) selective chemical degradation of heparan sulfate–heparin with nitrous acid; or (c) various cellulose acetate electrophoretic procedures for separating glycosaminoglycans before and after selectively degrading certain classes. (Details for many of these procedures are described elsewhere in this series.[1]) The average chain sizes of the different types can be determined by elution on Sepharose 6B or Sephadex G-200 to compare with the chondroitin sulfate molecular weight standards analyzed by Wasteson.[122]

Some idea of the arrangement of glycosaminoglycan chains on the core protein can be obtained by more restrictive cleavage of the protein than described above. For example, treatment of cartilage proteoglycan with trypsin produces peptides that contain several chondroitin sulfate chains. Further digestion with chymotrypsin creates peptides with fewer chains,[113,123] and, finally, papain treatment releases primarily single chains attached to small peptides. Digestion of cartilage proteoglycans with other proteases, such as clostripain,[25] or the cathepsins B, D, or G,[124,125] produce similar but not identically sized fragments that can be examined by column chromatography[25,113,123,125] or by composite agarose-acrylamide large pore gel electrophoresis.[119,125] These procedures allow comparisons to be made as to the arrangement of glycosaminoglycan chains on the core protein between cartilage proteoglycans and those isolated from other tissues (see Yanagishita *et al.*,[66] for example), and also between cartilage proteoglycans isolated from different cartilages or from the same cartilage at different stages of development.

*Oligosaccharides.* Studies have shown that most, if not all, proteoglycans contain either O-linked oligosaccharides, N-linked oligosaccharides, or both.[11,12,62,63,66,69] Analyses of the structures of these oligosaccharides and their distribution on subfractions of proteoglycans have provided additional information about their localization within the core protein and the probable relationship with keratan sulfate.[10–12] The O-linked oligosaccharides can be released from the core by treatment with alkali in the presence of borohydride[10,11]; this converts the released reducing termini to *N*-acetylgalactosaminitol residues. The same solvent conditions extensively hydrolyze the core protein and release the N-linked oligosaccharides still attached to small peptides. The different oligosaccharides

---

[122] Å. Wasteson, *J. Chromatogr.* **59**, 87 (1971).
[123] D. Heinegård and V. C. Hascall, *Arch. Biochem. Biophys.* **165**, 427 (1974).
[124] P. J. Roughley and A. J. Barrett, *Biochem. J.* **167**, 629 (1977).
[125] P. J. Roughley, *Biochem. J.* **167**, 639 (1977).

can then be separated by molecular-sieve chromatography on BioGel P-10 for subsequent structural analyses by standard sugar and mass fragmentation procedures.[10,11,26]

*Core Protein.* Eventually, peptide mapping and primary sequence information will be required to compare the core protein of different kinds of proteoglycans to establish, for example, the degree of similarity between aggregating proteoglycans from different tissues. Presently, such information is difficult to obtain because of the generally large size of the core proteins and the difficulty in removing the carbohydrate elements while leaving the polypeptide intact. For example, treatment of bovine hyaline cartilage proteoglycan with chondroitinase to remove chondroitin sulfate, the major constituent, leaves a core preparation that is still more than 50% carbohydrate, derived from keratan sulfate chains, oligosaccharides, and the linkage structures for chondroitin sulfate chains.[18] Oike et al.[84] have described procedures using chondroitinase and keratanase in the presence of protease inhibitors to digest the glycosaminoglycans from the core while leaving the polypeptide intact. The hydrodynamic sizes of such core preparations can be compared by molecular-sieve chromatography or by sodium dodecyl sulfate–polyacrylamide gel electrophoretic methods.[21,54,102]

Procedures for purifying the hyaluronic acid-binding region of the aggregating proteoglycans from cartilage have been developed.[22,24,25] The presence of the link protein in aggregates protects the hyaluronic acid-binding region from digestion with trypsin[24] or clostripain.[25] A complex of link protein, hyaluronic acid-binding region and hyaluronic acid can then be purified from the protease digest by (*a*) centrifugation in an associative isopynic gradient, where the complex migrates to the top and the chondroitin sulfate-peptides migrate to the bottom; and (*b*) molecular sieve chromatography in associative conditions on Sepharose 6B, where the complex is excluded and smaller peptides and the enzyme are included. The hyaluronic acid-binding region is then separated from the hyaluronic acid and the link protein by molecular-sieve chromatography on Sepharose CL-6B in a dissociative, 4 $M$ guanidine-HCl solvent. The purified hyaluronic acid-binding region isolated from the rat chondrosarcoma proteoglycan is a polypeptide of about 67,000 molecular weight,[25] which contains some N-linked oligosaccharides.[11,26] The protein structure of this fragment, then, can probably be determined by conventional protein chemistry techniques. Specialized methods, such as the recently developed anhydrous hydrogen fluoride treatment,[126,127] may eventually be

[126] A. J. Mort and D. T. A. Lamport, *Anal. Biochem.* **82**, 289 (1977).
[127] L. Philipson, C. Coudron, K. Ellis, and N. B. Schwartz, *in* "Glycoconjugates, Proceedings of the 5th International Symposium" (R. Schauer, P. Boer, E. Buddecke, M. F. Kramer, J. F. G. Vliegenthart and H. Wiegandt, eds.), p. 151. Thieme, Stuttgart, 1979.

useful for removing the carbohydrate structures from the core proteins (or peptides derived from core proteins) without breaking the peptide bonds, thereby allowing studies of regions which are highly substituted with carbohydrate.

## Physical Characterization of Proteoglycans

The physicochemical properties of proteoglycans have been studied by a large number of techniques which are beyond the scope of this chapter to describe in detail (but see this series, Vols. 48, 49, and 61). These include analytical centrifugation, light scattering, viscometry, nuclear magnetic resonance analysis, and electron microscopy.

*Analytical Centrifugation.* Analytical sedimentation velocity centrifugation has been used routinely by many investigators to determine sedimentation coefficients for proteoglycans. There are large effects of solute concentration and electrolyte on the sedimentation characteristics of proteoglycans, however, and care must be taken to include sufficient electrolyte, typically 0.5 $M$ salt to minimize charge effects and to use low enough solute concentrations for a reliable extrapolation to zero concentration. The concentration dependence of sedimentation coefficients of proteoglycan aggregates and monomers is best fit by the equation $s = s_0 \exp^{-kc}$, based upon values of correlation coefficients after regression analysis.[128] For proteoglycan aggregates, the value of $k$ (the slope of a plot of ln $s$ against $c$) is usually constant at concentrations less than 3 mg/ml. Values of $s_0$ for proteoglycan aggregates and monomers are therefore usually obtained by extrapolation of plots in ln $s$ against $c$ to obtain $s_0$.

Sedimentation velocity centrifugal techniques have also been used to determine the distribution of sedimentation coefficients, $g(s)$, which is represented by the whole population of proteoglycans in a preparation.[17,78,129] For any given solute concentration, this procedure determines the proportion of proteoglycans in the total population that sediment with sedimentation coefficients between given values. When distributions are determined at different solute concentrations, the curves can be extrapolated to give an ideal distribution, $g(s_0)$ representing the range of sedimentation coefficients and proportions of solute for each value under the ideal conditions of infinite dilution.[129] This provides an assessment of the extent of polydispersity in sedimentation coefficients of the proteoglycans in a preparation.

Sedimentation equilibrium centrifugation has been used to estimate weight average molecular weights for proteoglycans or proteoglycan

---

[128] L. C. Rosenberg, S. Pal, and R. J. Beale, *J. Biol. Chem.* **218**, 3681 (1973).
[129] J. C. Pita, R. J. Muller, T. R. Oegema, and V. C. Hascall, *Arch. Biochem. Biophys.* **186**, 66 (1978).

fragments. However, the large size of the intact proteoglycans from cartilage and their inherent macromolecular polydisperity pose difficult technical problems for subsequent analyses and interpretations. For smaller proteoglycans, such as the heparan sulfate proteoglycans from plasma membranes,[70] this procedure can provide an effective way to estimate molecular weight.

*Light Scattering.* Mathews and Lozaityte[130] were the first to use conventional light scattering to study proteoglycans. Their studies, which used samples isolated from cartilage by disruptive extraction procedures, provided the first hydrodynamic model for cartilage proteoglycan structure, suggesting a basic proteoglycan unit of $4 \times 10^6$ molecular weight and of 1000 Å radius of gyration $R_g$. Pasternack et al.[19] first applied conventional light scattering to cartilage proteoglycans isolated by dissociative techniques, providing estimates of $2.4 \times 10^6$ molecular weight and about 600 Å for the $R_g$ of monomer proteoglycans from bovine nasal cartilage. Most recently, Reihanian et al.[48] used quasielastic light scattering to study a high molecular weight subfraction of bovine nasal cartilage monomers, obtaining values of $3.9 \times 10^6$ for the molecular weight and 700 Å for $R_g$.

*Viscometry.* Viscosity measurements in combination with sedimentation coefficient analyses were first used to estimate molecular weight parameters and to propose a hydrodynamic model for monomer proteoglycans isolated by dissociative methods.[17,78] In most studies, capillary flow viscosimeters have been used. The effects of electrolyte concentration on viscosity of monomer preparations have been described,[17,19,48,78] indicating that, in guanidine-HCl, electrolyte concentrations above 0.5 $M$ are required to minimize solute charge effects. Viscosimetric analyses have also been used to study dissociation of aggregates when solvent pH is lowered or guanidine-HCl concentration increased[34] and for studies of monomer interactions with hyaluronic acid and HA-oligomers.[28,29,40] This procedure is relatively straightforward and does not require extensive instrumentation, making it useful for certain types of investigation.

*Nuclear Magnetic Resonance.* Natural abundance $^{13}$C nuclear magnetic resonance (NMR) has been used to investigate the molecular motion of the glycosaminoglycan chains in proteoglycan solutions as well as in intact cartilage.[50,51] This technique is unusual in that it can be used to study some characteristics of the macromolecules directly in the tissue. Even at the high concentrations in the tissue, the glycosaminoglycan chains on the proteoglycans still exhibit a high degree of segmental mobility, which is indicated from the $^{13}$C NMR spectra obtained. Such mobility would con-

[130] M. B. Mathews and I. Lozaityte, *Arch. Biochem. Biophys.* **74**, 158 (1958).

tribute to the viscoelastic properties of the tissue. Cultures can be used to introduce specific $^{13}$C-labeled precursors into proteoglycans. For example, the use of [$^{13}$C]serine and [$^{13}$C]glycine has permitted analyses of the segmental mobility of the core protein of monomer proteoglycans, and distinguished local differences in mobilities introduced by nearby carbohydrate substituents.[52]

*Electron Microscopy.* Rosenberg et al.[131] first used the Kleinschmidt[132] cytochrome c monolayer spreading technique, originally used for nucleic acids, to visualize cartilage proteoglycans. This procedure has since been used by a number of investigators to compare aggregates and monomers.[31,133-137] Measurements have included monomer lengths and degree of substitution of the core with side branches representing groups of glycosaminoglycan chains,[131,134] distributions of lengths of monomers bound to aggregate structures[31,133-136] and to assess relationships of spread aggregate structures with thin sections of cartilage matrix.[137]

*Immunology*

The core protein of proteoglycans is antigenic, and immunological techniques are now being developed to quantitate proteoglycans and to study their structure and interactions with other macromolecules. Radioimmunoassays have been developed by a number of workers[138-140] to quantitate the amount of proteoglycan core protein[138,139] as well as link protein.[139,140] Enzyme-linked immunoassay (ELISA) procedures that do not rely on radiolabeled material and are in general nearly as sensitive as radioimmunoassays, have also been developed for cartilage proteoglycan[141] and for purified link protein and hyaluronic acid-binding region.[142]

Immunoelectrophoresis has also been used to quantitate proteoglycans[143] and to compare antigenic determinants on different regions of the

[131] L. Rosenberg, W. Hellmann, and A. K. Kleinschmidt, *J. Biol. Chem.* **245**, 4123 (1970).
[132] A. K. Kleinschmidt, this series, Vol. 12, p. 361.
[133] L. Rosenberg, W. Hellmann, and A. K. Kleinschmidt, *J. Biol. Chem.* **250**, 1877 (1975).
[134] J. Thyberg, L. S. Lohmander, and D. K. Heinegård, *Biochem. J.* **151**, 157 (1975).
[135] J. H. Kimura, P. Osdoby, A. I. Caplan, and V. C. Hascall, *J. Biol. Chem.* **253**, 4721 (1978).
[136] D. K. Heinegård, L. S. Lohmander, and J. Thyberg, *Biochem. J.* **175**, 913 (1978).
[137] G. K. Hascall, *J. Ultrastruc. Res.* **70**, 369 (1980).
[138] P.-L. Ho, D. Levitt, and A. Dorfman, *Dev. Biol.* **55**, 233 (1977).
[139] J. Weislander and D. K. Heinegård, *Biochem. J.* **187**, 687 (1980).
[140] B. Caterson, J. R. Baker, D. Levitt, and J. W. Paslay, *J. Biol. Chem.* **254**, 9369 (1979).
[141] S. I. Rennard, K. Kimata, B. Dusemunde, H.-J. Barrach, J. Wilczek, J. H. Kimura, and V. C. Hascall, *Arch. Biochem. Biophys.* **207**, 399 (1981).
[142] E. J.-M. Thonar, J. H. Kimura, V. C. Hascall, and A. R. Poole, submitted (1982).
[143] J. Weislander and D. K. Heinegård, *Biochem. J.* **179**, 35 (1979).

proteoglycan core protein[143] or between core protein and link protein.[42,143] Immunofluorescence studies have been used to examine extracellular[144] as well as intracellular proteoglycan[145] and link protein.[146] Antibodies have been used to immunoprecipitate core protein synthesized by chick cartilage mRNA *in vitro*[20] or by cultures of chondrosarcoma chondrocytes.[21] Details of immunological techniques as applied to studies of proteoglycans are given elsewhere in this series.[147]

Concluding Comments

The techniques described above have been developed primarily in studies for cartilage proteoglycans. It is apparent, however, that a direct application of protocols that are suitable for cartilage proteoglycans to other tissues will often prove inadequate to the task. Thus, the investigator should approach a study of proteoglycans with flexibility, exploring several ways to take advantage of the different properties of these macromolecules to devise procedures for extraction, purification, and characterization. The experience gained in the many investigations described above, then, can provide guidance and helpful clues as to how to proceed with such studies, but cannot a priori provide the best route. Since the knowledge of proteoglycans in tissues other than cartilage is still quite limited, it is probable that many new techniques and approaches will have to be devised for successfully isolating and characterizing noncartilage proteoglycans.

---

[144] B. M. Vertel and A. Dorfman, *Dev. Biol.* **62**, 1 (1978).
[145] B. M. Vertel and A. Dorfman, *Proc. Natl. Acad. Sci. U.S.A.* **76**, 1261 (1979).
[146] A. R. Poole, I. Pidoux, A. Reiner, L.-H. Tang, H. Choi, and L. Rosenberg, *J. Histochem. Cytochem.* **28**, 621 (1980).
[147] J. R. Baker, B. Caterson, and J. E. Christner, this series, Vol. 83 (1982).

# Section IV

# Other Connective Tissue Proteins

## [46] Fibronectin: Purification, Immunochemical Properties, and Biological Activities

*By* ERKKI RUOSLAHTI, EDWARD G. HAYMAN, MICHAEL PIERSCHBACHER, and EVA ENGVALL

Fibronectin is a cell-surface and blood glycoprotein that apparently mediates adhesion of cells to the extracellular matrix. Malignant cells tend to lack cell surface fibronectin, and this may contribute to their capacity for invasive and metastatic growth.

A striking characteristic of fibronectin is that it is present in an insoluble form at the cell surface and in the connective tissue, and it is found also in a soluble form in plasma and other body fluids. The different forms of fibronectin have been found to be involved in a variety of seemingly unrelated events ranging from cell surface phenomena to blood coagulation.

The protein has been given many names, but uniformity of nomenclature has been established and most laboratories now use the name fibronectin [*fibra* (Lat.) = fiber; *nectere* (Lat.) = connect, link] indicating, when appropriate, the source of the protein (plasma fibronectin, fibroblast fibronectin, etc.). The properties of fibronectin have been reviewed by several authors.[1-5] The characteristics of fibronectin relevant to this chapter on the methods used in fibronectin research will be briefly discussed here.

### General Characteristics of Fibronectin

*Chemical Characteristics of Fibronectin*

Fibronectin was first identified in plasma. The characteristic that drew attention to plasma fibronectin was its presence in the cryoprecipitate, the precipitate that forms when plasma is stored at 4°, hence the earlier name for fibronectin, "cold insoluble globulin." The plasma level of fibronectin is relatively high, about 300 $\mu$g/ml, and plasma has subsequently proved to be a convenient source for its isolation.

[1] A. Vaheri and D. F. Mosher, *Biochim. Biophys. Acta* **516**, 1 (1978).
[2] K. M. Yamada and K. Olden, *Nature (London)* **275**, 179 (1978).
[3] R. O. Hynes *in* "Cytoskeletal Elements and Plasma Membrane Organization" (G. Poste and G. L. Nicolson, eds.), *Cell Surface Rev.* **7**, in press.
[4] D. F. Mosher, *Prog. Haemostasis Thromb. Res.* **5**, 111 (1980).
[5] E. Ruoslahti, E. Engvall, and E. G. Hayman, *Collagen Res.* **1**, 95 (1981).

Fibronectin is a dimer of two unusually large polypeptide chains, both with a molecular weight close to 220,000.[6,7] In gel electrophoresis in the presence of sodium dodecyl sulfate (SDS), reduced plasma fibronectin gives two closely spaced bands corresponding to a molecular weight ($M_r$) of about 220,000. The unreduced molecule appears to be 450,000 $M_r$.

Peptide mapping has revealed no differences between the two polypeptides of plasma fibronectin separated on SDS gels, which suggests that the two subunits are closely similar.[8] This conclusion is supported by the fact that plasma fibronectin has a single amino-terminal sequence. The amino-terminal residue is blocked because of the presence of pyroglutamic acid. Removal of this residue has allowed the identification of the next few amino acids, giving an amino terminal sequence of pyroGlu-Ala-Glu-Glu-Met-Val.[6,9]

Fibronectin is a glycoprotein. Its carbohydrate content varies between 5 and 9%, depending on the source of fibronectin (see below). Phosphate in the form of phosphoserine has also been found in fibronectin.[10]

*Are Fibronectins from Different Sources Chemically Different?*

The mobility of fibronectins from different sources varies in SDS-gel electrophoresis. Fibronectins derived from some cultured cells have been found to give a single band, which appears to coincide with the upper band of the closely spaced doublet in plasma fibronectin.[11] When fibronectin from embryonal chicken or human fibroblasts is compared to plasma fibronectin, the difference in mobility on SDS gels seems to be more marked.[12,13] The same is true when fibronectins from human amniotic fluid or from amniotic epithelial cells are compared to plasma fibronectin.[14,15] Amniotic fluid fibronectin gives a diffuse band with a mobility slightly lower than that of the sharp doublet of plasma fibronectin (Fig. 1). The differences observed in SDS polyacrylamide gel electrophoresis among fibronectins from different sources may be largely due to variations in the

[6] M. W. Mosesson, A. B. Chen, and R. M. Huseby, *Biochim. Biophys. Acta* **386**, 509 (1975).
[7] K. M. Yamada, D. H. Schlesinger, D. W. Kennedy, and I. Pastan, *Biochemistry* **16**, 5552 (1977).
[8] M. Kurkinen, T. Vartio, and A. Vaheri, *Biochim. Biophys. Acta* **624**, 490 (1980).
[9] M. B. Furie and D. B. Rifkin, *J. Biol. Chem.* **255**, 3134 (1980).
[10] M.-H. Teng and D. B. Rifkin, *J. Cell Biol.* **80**, 784 (1979).
[11] A. Quaroni, K. J. Isselbacher, and E. Ruoslahti, *Proc. Natl. Acad. Sci. U.S.A.* **76**, 5548 (1978).
[12] E. Engvall, E. Ruoslahti, and E. J. Miller, *J. Exp. Med.* **147**, 1584 (1978).
[13] K. M. Yamada and D. W. Kennedy, *J. Cell Biol.* **80**, 492 (1979).
[14] E. Crouch, G. Balian, K. Holbrook, D. Duksin, and P. Bornstein, *J. Cell Biol.* **78**, 701 (1980).
[15] E. Ruoslahti, E. Engvall, E. G. Hayman, and R. Spiro, *Biochem. J.*, **193**, 295 (1981).

FIG. 1. Polyacrylamide gel electrophoresis in the presence of sodium dodecyl sulfate. Fibronectins from plasma (P) and amniotic fluid (A) analyzed under reducing conditions in a 5% gel.

glycosylation of the fibronectin molecule. Fibronectin in amniotic fluid, which differs most markedly from plasma fibronectin in its electrophoretic mobility, has a carbohydrate content of 9.5%,[15] whereas fibronectins from plasma and from fibroblasts have only about 5% carbohydrate.[6,7,16] The patterns of peptides obtained after limited proteolysis of different fibronectins are similar.[15–17] Furthermore, fibronectins from different sources are immunologically indistinguishable,[18] at least when conventional antisera are used (see also the subsection on monoclonal an-

[16] M. Vuento, M. Wrann, and E. Ruoslahti, *FEBS Lett.* **82**, 227 (1977).
[17] G. Balian, E. Crouch, E. M. Clock, W. G. Carter, and P. Bornstein, *J. Supramol. Struct.* **12**, 505 (1979).
[18] E. Ruoslahti and E. Engvall, *Ann. N. Y. Acad. Sci.* **312**, 178 (1978).

tibodies). These results led us to conclude that the differences observed among fibronectins from different sources are mainly due to variations in the glycosylation of the molecule and that the polypeptide part is likely to be the same.

Fibronectin appears to exist as disulfide-bonded multimeric aggregates at the cell surface,[19,20] whereas soluble fibronectin mainly or exclusively consists of dimers. The presence of variable amounts of aggregates in fibronectins isolated from different sources may also contribute to what appears to be differences between cell surface derived and soluble fibronectins.

*Functional Properties of Fibronectin*

Fibronectin interacts with a number of macromolecules and with cell surfaces. The interaction with collagen is of considerable interest, and its specificity has been studied in detail.[12,21,22]

Denatured collagens are more efficient binders of fibronectin than their native forms. Of the native forms of the different genetic types of collagen, type III is the most active. Studies with cyanogen bromide fragments of the individual collagen chains show that binding sites for fibronectin can be found in several different fragments,[12,21] but one particular site in the polypeptide chain, represented by the $\alpha_1$CB7 fragment, may be more active than the other sites.[22] This site contains the only region in collagen that is cleaved by mammalian collagenase. For the purposes of this presentation, the most significant aspect of the fibronectin–collagen interaction is that it forms the basis of isolation for fibronectin by affinity chromatography on gelatin–Sepharose,[23] which is described in detail below.

It has been known for a long time that the plasma form of fibronectin, "cold insoluble globulin," tends to coprecipitate with fibrinogen in the cold. This depends on a direct binding of fibronectin to fibrinogen, demonstrable by chromatography of fibronectin on Sepharose to which fibrinogen or fibrin monomer is coupled.[24,25] The binding is not readily demonstrable unless the temperature is lowered. This probably reflects the weakness of the interaction.

[19] J. Keski-Oja, D. F. Mosher, and A. Vaheri, *Biochem. Biophys. Res. Commun.* **74,** 699 (1977).
[20] R. O. Hynes and A. Destree, *Proc. Natl. Acad. Sci. U.S.A.* **74,** 2855 (1977).
[21] W. Dessau, B. C. Adelmann, R. Timpl, and G. Martin, *Biochem. J.* **169,** 55 (1978).
[22] H. Kleinman, E. B. McGoodwin, G. R. Martin, R. J. Klebe, P. P. Fietzek, and D. E. Woolley, *J. Biol. Chem.* **253,** 5642 (1978).
[23] E. Engvall and E. Ruoslahti, *Int. J. Cancer* **20,** 1 (1977).
[24] E. Ruoslahti and A. Vaheri, *J. Exp. Med.* **141,** 497 (1975).
[25] A. Stemberger and H. Hörmann, *Hoppe-Seyler's Z. Physiol. Chem.* **357,** 1003 (1976).

Fibronectin interacts with a number of polyanionic substances. Heparin, hyaluronic acid, dextran sulfate, and DNA show binding to fibronectin.[26-30] The binding of heparin, heparan sulfate, or hyaluronic acid to fibronectin–gelatin complexes conveys upon them greater stability than fibronectin–gelatin complexes have in the absence of heparin.[27-29]

Other activities of fibronectin include its binding to staphylococci,[31] and actin,[32] and its interaction with cell surfaces.

That fibronectin has affinity for the surfaces of many kinds of eukaryotic cells is implied by the fact that cells in suspension attach preferentially to surfaces coated with fibronectin and that the adhesion of cells to collagen matrices is mediated by fibronectin.[33-35] The cell surface receptor for fibronectin may be gangliosides.[36]

The sites in the fibronectin molecule for the various interactions of fibronectin with collagen, glycosaminoglycans, staphylococci, and cells reside in different parts of the fibronectin polypeptide chains. Fragments containing one or more of the binding sites have been isolated by limited proteolysis of the fibronectin molecule (for a recent review of this aspect of fibronectin, see Ruoslahti et al.[5]). The precise significance of the various interactions of fibronectin is not known, but it seems that the collagen and glycosaminoglycan interactions are involved in the formation of extracellular matrix, which may be based on insolubilization of a complex of all these components.[28,29] Such a matrix then provides a substratum for cells that attach through the fibronectin–cell interaction. Fibronectin bound to a fibrin clot could provide an adhesive temporary matrix similar to the fibronectin–collagen matrix for cells to grow into a wound. Fibronectin is known to become incorporated in a blood clot,[24] where it is cross-linked to fibrin by the transglutaminase that cross-links fibrin.[37] The cross-linking could compensate for the weakness of the interaction between fibronectin and fibrin. Another function for fibronectin in wound

[26] N. E. Stathakis and M. W. Mosesson, *J. Clin. Invest.* **60**, 855 (1977).
[27] E. Ruoslahti, A. Pekkala, and E. Engvall, *FEBS Lett.* **107**, 51 (1979).
[28] E. Ruoslahti and E. Engvall, *Biochim. Biophys. Acta* **631**, 350 (1980).
[29] F. Jilek and H. Hörmann, *Hoppe-Seyler's Z. Physiol. Chem.* **360**, 597 (1979).
[30] L. Zardi, C. Cecconi, O. Barbieri, B. Carnemolla, P. Marina, and L. Santi, *Cancer Res.* **39**, 3774 (1979).
[31] P. Kuusela, *Nature (London)* **276**, 718 (1978).
[32] J. Keski-Oja, A. Sen, and G. J. Todaro, *J. Cell Biol.* **85**, 527 (1980).
[33] R. J. Klebe, *Nature (London)* **250**, 248 (1974).
[34] F. Grinnell, D. G. Hays, and D. Minter, *Exp. Cell Res.* **110**, 175 (1977).
[35] M. Höök, K. Rubin, Å. Oldberg, B. Öbrink, and A. Vaheri, *Biochem. Biophys. Res. Commun.* **79**, 726 (1977).
[36] H. Kleinman, G. R. Martin, and P. H. Fishman, *Proc. Natl. Acad. Sci. U.S.A.* **76**, 3367 (1979).
[37] D. F. Mosher, *J. Biol. Chem.* **250**, 6614 (1975).

healing involves the proposal[38] that circulating fibronectin serves as a nonspecific opsonin promoting removal of insoluble tissue debris containing collagen, fibrin, and possibly also actin, by the reticuloendothelial system.

Malignantly transformed cells tend to lack cell-associated fibronectin, but there are many exceptions to this generalization.[1,2] The significance, if any, of the lack of surface fibronectin in transformed cells is a matter of some controversy, but if fibronectin is, indeed, the main mechanism that anchors cells to the extracellular matrix, disturbances of this mechanism could play an important role in malignancy and many other diseases.

*Are Cellular and Plasma Fibronectins Functionally Different?*

The plasma and cell culture-derived fibronectins share the molecular interactions described above and are equally active in promoting cell attachment.[4,5] A difference has been found in their capacity to normalize the morphology of transformed fibroblasts in culture and to agglutinate formalinized sheep red blood cells. Plasma fibronectin was less active in these tests.[13] However, it has been reported that fibronectin preparations containing proteoglycans are more active in correcting the morphology of transformed cells than preparations devoid of such contamination.[39] While the subject remains controversial, it seems that it will be necessary carefully to exclude the possibility that the fibronectins compared in various functional assays contain different contaminants. The degree of aggregation could also be important in determining the activity of a given fibronectin preparation in such assays. If, indeed, the capacity of fibronectin preparations to correct the morphology of transformed cells depends on a component other than fibronectin, it would be important to isolate and characterize this substance(s).

Isolation of Fibronectin

Early isolation methods for plasma fibronectin made use of its propensity to precipitate with fibrinogen and some other proteins when plasma is allowed to stand in the cold. A large fraction of the plasma cryoprecipitate is fibronectin, and it can be purified from such precipitate by using a combination of differential precipitation steps and ion exchange chromatography.[6,37] Fibronectin from the surface of cultured fibroblasts has been isolated by extraction with low concentrations of urea.[40] Antifibronectin columns have also been used to isolate fibronectin from both

[38] T. M. Saba and E. Jaffe, *Am. J. Med.* **68**, 577 (1980).
[39] M. E. Perkins, I. Summerhayes, P. Hsieh, C. Lee, and L. B. Chen, *J. Supramol. Struct. Suppl.* **4**, 179 (1980).
[40] K. M. Yamada and J. A. Weston, *Proc. Natl. Acad. Sci. U.S.A.* **71**, 3492 (1974).

plasma and cell cultures.[16] These methods have now largely been replaced by affinity chromatography on gelatin–Sepharose.[23]

*Gelatin-Sepharose Affinity Chromatography*

This method makes use of the affinity of fibronectin for collagen in a procedure that allows isolation of electrophoretically pure fibronectin from human plasma in one step, with excellent yields. The method has now been applied to the isolation of fibronectins from the plasma, cell culture media, and extracts of cells of a number of mammalian species, chicken, and even the electric eel, *Torpedo californica*.[12] The nature of the binding of fibronectin to gelatin–Sepharose has been explored by studying the reversal of the binding by various treatments.[18]

In our initial work,[23] we used 8 $M$ urea in 50 m$M$ Tris-HCl, pH 7.5, to elute fibronectin from gelatin–Sepharose. The use of gradients has revealed that elution of fibronectin begins at urea concentrations of about 2 $M$ and is complete at 4 $M$ urea.[18] Consequently, we now use 4 $M$ urea for the elution. At higher salt concentrations, the concentration of urea can be lowered (see below). Urea decomposes to cyanate and ammonia in solution, and cyanate can react with the free amino groups of proteins. We deionize urea prior to use to eliminate cyanate. In addition, we use Tris buffer in the urea solutions to provide amino groups that will scavenge any cyanate forming during the isolation process. Fibronectin eluted from gelatin–Sepharose with urea is active in all the biological tests available.

Fibronectin can also be eluted with a glycine-HCl buffer, pH 2.6, and with high concentrations of chaotropic ions, such as sodium iodide, sodium thiocyanate, and magnesium chloride. Ethylene glycol, which is thought to disrupt hydrophobic interactions, is also effective in eluting fibronectin from gelatin columns.[18] Treatment of gelatin columns, to which fibronectin is bound, with bacterial collagenase also releases fibronectin. This elution method may be useful when it becomes necessary to use the mildest possible elution conditions. Antibodies to fibronectin prevent binding of fibronectin to gelatin-Sepharose and cause elution of already bound fibronectin.[41]

Arginine has been found to elute fibronectin from gelatin-Sepharose, and this has been claimed to provide a gentler way of eluting fibronectin than is achieved with other eluting agents.[42] However, guanidine-HCl is also quite efficient in eluting fibronectin from gelatin–Sepharose (Fig. 2). Also, when 1 $M$ NaCl is added to urea solutions (to achieve an ionic strength comparable to that of 1 $M$ guanidine or arginine), urea is nearly as

[41] E. Ruoslahti, M. Vuento, and E. Engvall, *Biochim. Biophys. Acta* **534**, 210 (1978).
[42] M. Vuento and A. Vaheri, *Biochem. J.* **183**, 331 (1979).

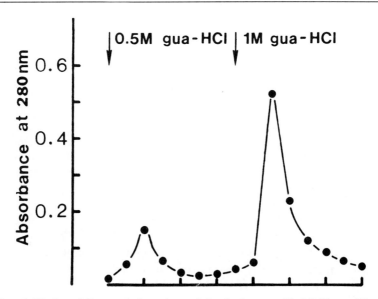

FIG. 2. Elution of fibronectin bound to gelatin–Sepharose with 0.5 $M$ guanidine-HCl followed by 1 $M$ guanidine-HCl. Subsequent elution with 8 $M$ urea did not cause elution of additional amounts of protein, indicating that the elution of fibronectin with 1 $M$ solution was complete.

efficient as arginine. A complete elution of fibronectin can be achieved with 2 $M$ urea in the presence of 1 $M$ NaCl (Fig. 3). In view of the similarity of the structures of arginine, guanidine, and urea, it is doubtful that the mechanism of the arginine elution would be any different from that of the other eluting agents in use. Arginine is more expensive than urea, and commercial arginine contains impurities that absorb light at 280 nm, making it difficult to use UV absorbance to monitor the isolation of fibronectin.

The purity of the fibronectin obtained by gelatin–Sepharose chromatography can be further improved by modifying the isolation procedure to include pretreatment of the sample from which fibronectin is to be isolated with plain Sepharose.[18,41] Plasma of several species seems to contain antibodies that bind to Sepharose. These are possibly antibodies reacting with carbohydrate determinants present in Sepharose. Absorption of plasma with Sepharose followed by elution of bound material with urea gives a preparation containing IgM and IgG immunoglobulins (Fig. 4). If not previously removed, these antibodies will bind to gelatin-Sepharose and coelute with fibronectin.

While fibronectin isolated by affinity chromatography on gelatin-Sepharose usually appears homogeneous in gel electrophoresis, under

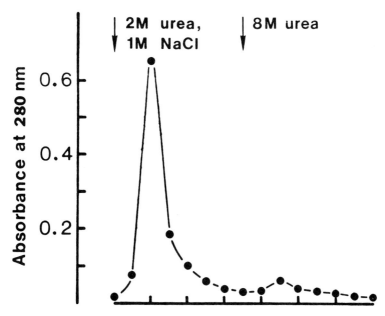

FIG. 3. Elution of fibronectin bound to gelatin–Sepharose with a solution containing 2 $M$ urea and 1 $M$ NaCl. Subsequent elution with 8 $M$ urea did not cause elution of protein.

some circumstances it may be advisable to add another purification step. Fibronectin interacts with glycosaminoglycans, and it has been shown that heparin binds to gelatin–Sepharose and coelutes with fibronectin.[27,28] Glycosaminoglycans and proteoglycans present in the original sample could behave similarly. These substances carry a strong negative net charge and bind tightly to positively charged ion exchange materials. We use chromatography on DEAE-cellulose as an additional step after gelatin–Sepharose in cases where it is important to ensure that fibronectin is free of glycosaminoglycans.[15] Fibronectin elutes from DE-52 columns as a sharp peak at the approximate salt concentration of 0.1 $M$ NaCl (Fig. 5), whereas glycosaminoglycans and proteoglycans elute at higher salt concentrations from such columns. If significant quantities of glycosaminoglycans are present in the sample fractionated on gelatin–Sepharose, elution of the bound fibronectin requires higher concentrations of urea than is the case otherwise (Fig. 6). High concentrations of NaCl (0.3 $M$ or higher) elute heparin from fibronectin-loaded gelatin–Sepharose columns without causing elution of fibronectin.[28] It may be advisable, therefore, to wash a gelatin–Sepharose column to which fibronectin is bound with salt prior to elution of fibronectin. Another potential contaminant of affinity-purified fibronectin is a collagenase inhibitor that ap-

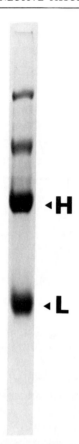

FIG. 4. Polyacrylamide gel electrophoresis in the presence of sodium dodecyl sulfate. The material that binds from normal plasma to underivatized Sepharose was analyzed under reducing conditions in an 8% gel. The positions of standard IgG heavy (H) and light (L) chains are shown. The second band from the top of the gel represents the IgM heavy chain.

pears to be a protein with a low molecular weight. This can also be separated from fibronectin on chromatography on DEAE-cellulose.[42a]

Fibronectin is quite sensitive to proteolysis. The first proteolytic cleavages result in fragments almost as large as the original subunit, but these are no longer held together in dimer form.[6] It seems that the disulfide bond(s) that link the monomers are located near one end of the molecule (believed to be the C-terminal end of the polypeptide chain) and that this is a site particularly susceptible to proteolysis. To obtain intact fibronectin from plasma we use, as starting material, freshly drawn citrated plasma (20 ml of 5% trisodium citrate for every 80 ml of blood) with

[42a] J. Aggeler, E. Engvall, and Z. Werb, *Biochem. Biophys. Res. Commun.* **100**, 1195 (1981).

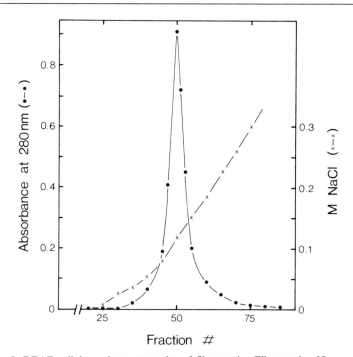

FIG. 5. DEAE-cellulose chromatography of fibronectin. Fibronectin, 27 mg, isolated from plasma by affinity chromatography on gelatin–Sepharose, was applied to a 2.5 × 9 cm column of Whatman DE-52, equilibrated with 4.5 $M$ urea, 0.01 $M$ Tris, pH 7.2. An NaCl gradient was applied, and the eluate was collected in 4-ml fractions.

phenylmethylsulfonyl fluoride (PMSF) as protease inhibitor. We also add PMSF to all solutions used in the isolation procedure. Others have used more than one protease inhibitor.[9] EDTA-plasma can also be used as starting material. Heparinated plasma is less suitable for this purpose, since heparin tends to precipitate fibronectin,[26] resulting in a lowered concentration of fibronectin in heparin plasma. As discussed above, heparin also strengthens the binding of fibronectin to gelatin–Sepharose columns.

Serum contains less fibronectin than plasma because part of the fibronectin becomes incorporated in the blood clot.[24] If blood is allowed to coagulate in the cold, almost all fibronectin is bound to the clot with little left in the serum. Another disadvantage with serum is that the proteases involved in blood coagulation are activated, and proteolysis of fibronectin may occur.

When fibronectin is isolated from media of cultured cells, it is necessary to take into account the presence of fibronectin in the serum used to

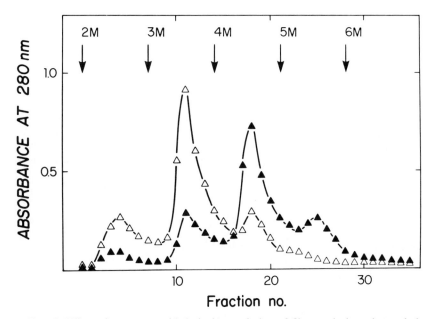

FIG. 6. Effect of treatment with heparin on elution of fibronectin bound to gelatin–Sepharose. Fibronectin from human plasma was bound to two identical columns of gelatin–Sepharose. Heparin (0.5 mg/ml in phosphate-buffered saline) was passed through one of the columns, and buffer through the other. After washing, the columns were eluted by applying progressively higher concentrations of urea in 50 m$M$ Tris-HCl (pH 7.5) as indicated, and the amount of fibronectin eluted was measured by absorbance. △, Buffer-treated column; ▲, heparin-treated column.

culture the cells. We have eliminated fibronectin from fetal bovine serum using gelatin–Sepharose followed by treatment with anti-bovine fibronectin coupled to Sepharose.[43] The fibronectin in commercial fetal bovine serum is partly proteolyzed, and although most of it binds to gelatin–Sepharose, some immunoreactive fragments remain and can be eliminated only with antibody columns. Preparations consisting solely of cell-derived fibronectin can be obtained from culture media previously depleted of bovine fibronectin.

To isolate cell layer-derived fibronectin, the washed cell layer may be extracted with 8 $M$ urea and the extract diluted 20-fold or more to allow isolation on gelatin-Sepharose.[12] Lower concentrations of urea will also release a large fraction of the cell surface fibronectin.[7,40]

Storage of isolated fibronectin may sometimes be a problem. Fibronectin in concentrations above 1 mg/ml tends to become insoluble. The insol-

[43] E. G. Hayman and E. Ruoslahti, *J. Cell Biol.* **83**, 255 (1979).

ubilization involves disulfide bonding. When analyzed by sodium dodecyl sulfate-polyacrylamide gel electrophoresis, the insoluble aggregates do not enter the gel without reduction. In the presence of reducing agent, the aggregated fibronectin is indistinguishable from nonaggregated fibronectin (unpublished results). We store isolated fibronectin frozen at $-20°$ in the urea solution that was used to elute it from gelatin-Sepharose. When necessary, the urea can be removed by passing the fibronectin through a desalting column (e.g., Sephadex G-15) equilibrated with the desired buffer. By using a volatile salt, such as ammonium bicarbonate, for the removal of urea, fibronectin can be lyophilized (unpublished results). The resulting dry fibronectin has poor solubility, but can be dissolved using a cyclohexylaminopropanesulfonic acid (CAPS) buffer, pH 11.[44] Diluted from this buffer to neutral pH, fibronectin can be shown to be active immunologically and in promoting cell attachment.

*Isolation of Fibronectin by Gelatin-Sepharose Affinity Chromatography—Detailed Procedure*

*Preparation of Gelatin-Sepharose.* Fifty milligrams of gelatin (e.g., Sigma, type I from swine skin) are dissolved in 25 ml of coupling buffer ($0.5 M$ NaCl, $0.1 M$ NaHCO$_3$) by heating the solution. The gelatin solution is cooled to 20–30° and then mixed with 5 g of cyanogen bromide-activated Sepharose (Pharmacia, Uppsala, Sweden), previously swelled and washed according to the manufacturer's instructions. The gel is stirred on a shaker in the gelatin solution overnight at $+4°$ or for 4 hr at room temperature. The unbound protein is removed by washing the gel with the coupling buffer on a sintered-glass filter. The Sepharose is then incubated for 1 hr at room temperature in $1 M$ ethanolamine-HCl, pH 8.0, to block remaining active groups in the gel. After washing with PBS, several gel volumes of $8 M$ urea, and finally again with PBS, the adsorbent is ready for use. It is stored at 4° in PBS containing 0.02% NaN$_3$.

*Isolation Procedure.* The sample is freed of particulate material by centrifugation at $30,000 g$. Phenylmethylsulfonyl fluoride (PMSF) ($10^{-4} M$) is added as protease inhibitor. If plasma is used as starting material, it is first passed through a column, twice the size of the plasma sample, of underivatized Sepharose 4B. The Sepharose can be reused after regeneration by washing with $8 M$ urea in $0.05 M$ Tris-HCl buffer, pH 7.5.) The plasma sample is then applied to a column of gelatin-Sepharose half the size of the original plasma sample. The column is washed with several column volumes of PBS, two column volumes of $1 M$ NaCl, and again with PBS. Elution is carried out using $4 M$ urea in $0.05 M$ Tris-HCl buffer,

---

[44] K. M. Yamada, *J. Cell Biol.* **78**, 520 (1978).

pH 7.5; PMSF, $10^{-4}$ $M$, is added to all buffers. Elution of fibronectin is monitored by measuring 1 cm optical density at 280 nm. The $E_{280\,nm}$ for a 1% solution of fibronectin is 13. The isolated fibronectin is stored frozen at $-20°$ in the urea solution or purified further, e.g., by DEAE-cellulose chromatography. Details for this procedure are given in the legend of Fig. 5.

Immunochemical Properties of Fibronectin

*Antisera*

Fibronectin is a good immunogen; i.e., it is relatively easy to produce antisera against it. In the early work, antibodies used to study cell surface fibronectin were prepared against material solubilized by protease treatment of the surface of cultured fibroblasts.[45-47] Using such antisera, it was found that an immunologically indistinguishable protein, subsequently identified as cold insoluble globulin,[48] was present in plasma. Fibronectin from plasma, because it is readily available, is currently used by most laboratories to raise antibodies to fibronectin.

The antigenic determinants of fibronectin appear to be quite stable. Preparations treated with sodium dodecyl sulfate and reduced and alkylated fibronectin react almost the same way as native fibronectin, whether tested against antibodies to SDS-treated or native fibronectin.[49] Antisera against fibronectin isolated by sodium dodecyl sulfate–polyacrylamide gel electrophoresis react with the intact molecule.[47]

Antibodies to fibronectin, whether prepared by immunization of rabbits or goats with fibroblast-derived or plasma fibronectin, show identical reactivity against the plasma and cellular forms of a particular species.[12,45,46] The immunological identity of the fibroblast and plasma forms of fibronectin is additional evidence in favor of the contention that they are the same protein. The value of this argument is somewhat diminished by the fact that some areas of the fibronectin molecule seem to be immunologically "silent." A 30,000-dalton collagen-binding peptide binds only about 0.1% of all antibody in anti-fibronectin sera, and gives an equally poor response when used to immunize rabbits.[50] However, for the

[45] E. Ruoslahti, A. Vaheri, P. Kuusela, and E. Linder, *Biochim. Biophys. Acta* **322**, 352 (1973).
[46] E. Ruoslahti and A. Vaheri, *Nature (London)* **248**, 789 (1974).
[47] R. O. Hynes, *Biochim. Biophys. Acta* **458**, 73 (1976).
[48] E. Ruoslahti and A. Vaheri, *J. Exp. Med.* **141**, 497 (1975).
[49] P. Kuusela, E. Ruoslahti, E. Engvall, and A. Vaheri, *Immunochemistry* **13**, 639 (1976).
[50] E. Ruoslahti, E. G. Hayman, P. Kuusela, J. E. Shively, and E. Engvall, *J. Biol. Chem.* **254**, 6054 (1979).

purposes of immunochemical detection and quantitation of fibronectin, the source of the immunogen used to prepare the antibodies is not important. Plasma fibronectin seems preferable because it can be obtained in high yields and is less likely to be contaminated with other cell surface components.

In spite of the relatively high purity of fibronectin isolated from plasma by using gelatin–Sepharose affinity chromatography, antisera to plasma fibronectin tend to have contaminating antibodies to major plasma proteins. These can be easily removed by absorption of the antiserum with the plasma proteins that do not bind to gelatin–Sepharose.[41]

It is possible to prepare antibodies against different domains of the fibronectin molecule. Antibodies that react with a collagen-binding 70,000-dalton fragment of fibronectin have been raised in rabbits[50] and, more recently, monoclonal antibodies reactive with different domains of fibronectin have been prepared against all the functional domains. Such antibodies may have proven to be useful tools in the elucidation of the structure–function relationships in fibronectin (see below).

*Interspecies Cross-Reactions and Species-Specific Antibodies*

Antisera to fibronectin show extensive interspecies cross-reactivity.[49] Most of the antibody in a given anti-fibronectin serum is cross-reactive with fibronectins from one species or another, making the serum reactive with fibronectin from most species.

In immunodiffusion, anti-human fibronectin can be shown to precipitate all mammalian fibronectins, except for that of the species of origin of the antiserum (Fig. 7). Antisera to mammalian fibronectins also precipitate chicken fibronectin, but not fibronectin from the electric ray *Torpedo californica*.[12,47] It is unusual for homologous proteins from species as widely separated as the chicken and the mammals to show such strong cross-reactivity, and this indicates that fibronectin has been highly conserved in evolution. The extent that interspecies cross-reactions are observed depends on the type of assay used. Immunodiffusion, while rather insensitive, easily reveals cross-reactions. The same is true of the more sensitive immunofluorescent technique. Interspecies cross-reactivity has been widely utilized in the detection of fibronectin by immunofluorescence in cell cultures. A single reagent, such as rabbit anti-human fibronectin, has been used to stain cell surface fibronectin in cell lines from a variety of species. Although this can be done, it is not advisable to try to draw any conclusions as to the quantity of fibronectin because the antiserum will react best with the fibronectin used in the immunization, and its reactivity with fibronectins from other species is proportional to the phylogenetic distance from the immunizing species (Fig. 7). In most cell

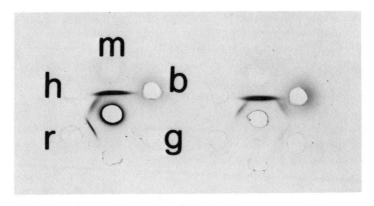

FIG. 7. Immunodiffusion analysis of the reactivity of anti-mouse fibronectin made in goat (center well, left) and in rabbit (center well, right) against goat (g), bovine (b), mouse (m), human (h), and rabbit (r) fibronectins. Note the strong reactivity against the fibronectin (mouse) used in immunization and the cross-reactivities with other species. The goat antiserum does not react with goat fibronectin and reacts only weakly with fibronectin from the closely related bovine species.

cultures, fetal bovine serum is used to support the growth of the cells. Fetal bovine serum contains fibronectin, and this fibronectin can become incorporated into the extracellular matrix of cultured cells.[43] Immunofluorescence with species-nonspecific anti-fibronectin, therefore, does not prove that the fibronectin detected originates from the cells. Such limitations have been widely ignored in the past, and this may sometimes have led to erroneous conclusions.

Radioimmunoassay, when performed utilizing the principle of a competitive assay, is much less likely to reveal interspecies cross-reactivity than other assays. The interactions of antibodies with cross-reactive antigens tend to occur with lower affinity than the interaction with the immunizing antigen. Such interactions, while of sufficiently high affinity to give a positive reaction in other assays, may not be detectable in competitive radioimmunoassay, which measures only interactions with the highest affinities.

Anti-fibronectin reagents that are operationally species-specific can be prepared by absorbing the antiserum with the fibronectin against which reactivity is not wanted. We routinely absorb our antisera with bovine fibronectin to make them suitable for detection of fibronectin in cell cultures,[41] and with other fibronectins as necessary. The absorption is performed using the relevant fibronectin insolubilized on Sepharose. Affinity-purified antibodies are similarly prepared on insolubilized fibronectin (see Ruoslahti et al.[41] and below). Species-specific reagents have been useful in the detection of the incorporation of bovine fibronectin into the matrix of cultured rat cells[43] and have also been used to clarify controversies in the chromosomal localization of the structural gene for human fibronectin.[51]

*Absorption of Anti-Fibronectin Sera and Affinity Purification of Antibodies—Procedure*

Initial absorption of the antiserum is carried out using Sepharose-coupled proteins of plasma from which fibronectin has been removed with gelatin–Sepharose. These proteins are coupled to cyanogen bromide-activated Sepharose using 20 mg of plasma protein per gram of activated Sepharose (see preparation of gelatin–Sepharose for details). Antiserum is absorbed by passing it through a column made of an equal amount of Sepharose-coupled plasma proteins. The nonbound fraction containing antibodies to fibronectin, but not to impurities, is applied to a column of fibronectin–Sepharose. For coupling to Sepharose, fibronectin eluted from a gelatin–Sepharose column is fractionated on a Sephadex G-15 column equilibrated with the coupling buffer to remove the urea and Tris. The volume of the fibronectin sample applied to the G-15 columns can be up to one-fourth of the volume of the column. Fibronectin is eluted from the column with the coupling buffer and coupled using 10 mg of fibronectin per gram of activated Sepharose. One milliliter of fibronectin–Sepharose usually binds the antibodies from 2 ml of antiserum. The bound antibodies are eluted from the column using a $0.1\ M$ glycine-HCl buffer, pH 2.6. When a species-specific reagent is desired, the antiserum is first absorbed as described above with fibronectin from the species against which reactivity is not wanted, and the remaining antibodies are isolated on the homologous fibronectin. After neutralization, the eluate is ready for use.

*Monoclonal Antibodies to Fibronectin*

Using a modification of the methods developed by Köhler and Milstein[52] for the construction of antibody-producing somatic cell hybrids, it has been possible to obtain a large number of hybridomas producing different antibodies specific for fibronectin. These antibodies may become valuable in fibronectin research. Homogeneous antibodies with well defined specificities can be used for localization and quantitation of fibronectin and as probes for the elucidation of the molecular structure of fibronectin.

[51] H. P. Klinger and E. Ruoslahti, *Cytogenet. Cell Genet.* **28**, 271 (1980).
[52] G. Köhler and C. Milstein, *Nature (Nature)* **256**, 495 (1975).

*Procedure.* Mice were immunized with fibronectin isolated from human amniotic fluid. The priming dose of antigen was 100 µg emulsified in complete Freund's adjuvant and injected subcutaneously. This was followed in 4 weeks by 50-µg booster doses given intravenously on 4 consecutive days. The spleen was taken from the animal on the day after the final boost, and the cells were fused with the 8-azaguanine-resistant myeloma cell line X-63-Ag8.6531.[53] Spleen cells ($2 \times 10^8$) were mixed with $2.5 \times 10^7$ myeloma cells in the presence 0.2 ml of 30% polyethylene glycol (average $M_r$ 1500; BDH Chemicals Ltd., Poole, England) in serum-free medium for 4 min at room temperature and then centrifuged at 800 rpm for 4 min at room temperature in a clinical centrifuge (Beckman Model TJ-6). The supernatant above the cell pellet was diluted by dropwise addition of 5 ml of medium at room temperature. The cells were again centrifuged for 5 min at 800 rpm. Finally, the cells were washed twice, resuspended in medium containing 4500 mg of glucose per liter, 10% fetal calf serum, and standard concentrations of hypoxanthine, aminopterin, and thymidine,[52] and then distributed into three 96-well plates. These cultures were maintained in this medium for 2 weeks with medium changes every third day. After 2 weeks, the aminopterin was removed, leaving the hypoxanthine and thymidine in the medium for an additional 2 weeks, after which they too were deleted from the medium.

The growing cultures were assayed using an ELISA (for details see Engvall[54] and below) in which human fibronectin was adsorbed to untreated polystyrene microtiter wells. The culture supernatants were incubated in the microwells and the nonbound protein was washed away. Bound antibody was then detected using rabbit anti-mouse immunoglobulin to which alkaline phosphatase or peroxidase had been cross-linked by glutaraldehyde.

*Reactivities of Monoclonal Antibodies to Fibronectin.* Significant immunological cross-reactivity exists among fibronectins obtained from widely varying animal species (see above). It was found that about one-third of 100 cultures that were producing antibody to fibronectin made antibody that reacted with both human and bovine fibronectin, while the remaining two-thirds reacted only with human fibronectin. This suggests that certain antigenic determinants are conserved (perhaps by function) while others are species specific.

To further define the antibodies from the various cultures, ELISAs were performed using microtiter plates that had been coated with different fragments of fibronectin (Ruoslahti *et al.*[50]). In each case, several cultures were identified which contained antibody to one and only one fragment.[54a]

[53] J. F. Kearny, A. Radbruch, B. Liesegang, and K. Rajewsky, *J. Immunol.* **124**, 1548 (1979).
[54] E. Engvall, this series, Vol. 70, p. 419.
[54a] M. Pierschbacher, E. G. Hayman, and E. Ruoslahti, *Cell,* in press.

Two antibodies were found that react with an antigenic site(s) that was destroyed by trypsin. Clones from these cultures, unique in some aspect of their reactivity with the fibronectin molecule, will contribute to a library of monoclonal antibodies, each of which will define a single site on the molecule. These can be used to define the various properties of fibronectin.

## Quantitation of Fibronectin

The methods used in immunochemical quantitation of fibronectin include radioimmunoassay (RIA),[32,41,49,55] enzyme immunoassay (ELISA)[23,54,56,57] and rocket immunoelectrophoresis.[58] Of these methods, ELISA is probably the most useful one. It is as sensitive as RIA and some of the problems associated with RIA are avoided in ELISA. The main problem with RIA is the lability of iodinated fibronectin. The proportion of radioactivity in labeled fibronectin that no longer reacts with antibody increases upon storage far more rapidly than is the case with most other proteins. Even with frequent reisolation of immunoreactive label by gelatin–Sepharose chromatography[41] this unstability limits the usefulness of the assay.

A major problem in all current fibronectin assays is the different dose-response obtained with intact and partially proteolyzed fibronectin. As discussed above, fibronectin is quite sensitive to proteolysis and the resultant monomeric fragments are found in various biological samples. The large fragments react with all antibodies to fibronectin and are indistinguishable from intact fibronectin in immunodiffusion while short fragments react to a lesser degree (Ruoslahti *et al.*[50] and unpublished data). In the rocket immunoelectrophoresis technique, molecular weight difference such as the one caused by the aggregate dimer–monomer transition in fibronectin is known to affect the results. We have found this to be the case also to some extent in RIA and in ELISA. It may be possible to find monoclonal antibodies that allow determination of intact fibronectin without interference by fragments.

A further source of interference in fibronectin assays is the presence in the samples of substances with which fibronectin can interact. These include collagen, glycosaminoglycans, and proteoglycans. How the latter two might affect fibronectin assays is not known. The presence of dena-

---

[55] D. F. Mosher and A. Vaheri, *Exp. Cell Res.* **112**, 323 (1978).
[56] S. I. Rennard, R. Berg, G. R. Martin, J.-M. Foidart, and P. G. Robey, *Anal. Biochem.* **104**, 205 (1980).
[57] R. Timpl, H. Rohde, P. G. Robey, S. I. Rennard, J.-M. Foidart and G. R. Martin, *J. Biol. Chem.* **254**, 9933 (1979).
[58] F. Blumenstock, P. Weber, and T. M. Saba, *J. Biol. Chem.* **252**, 7156 (1977).

tured collagen does not affect radioimmunoassay,[41] but does reduce the height of peaks obtained in rocket immunoelectrophoresis (our unpublished results).

*Radioimmunoassay in Quantitation of Fibronectin—Detailed Procedure (Modified from Ruoslahti et al.[41])*

*Iodination.* Fibronectin obtained from the gelatin–Sepharose column is used for iodination. Typically, the peak eluted fractions contain 1 mg or more of fibronectin per milliliter. Urea is not removed, since fibronectin tends to precipitate in the absence of urea. To 10 $\mu$l of the dialyzed solution containing 1 mg of fibronectin per milliliter, 10 $\mu$l of 1 $M$ phosphate buffer, pH 7.0, are added. This is followed by 1 mCi of $^{125}$I (New England Nuclear, Boston, Massachusetts) and 10 $\mu$l of chloramine-T (4 mg/ml). After 1 min, the reaction is stopped with 10 $\mu$l of a 2.4 mg/ml solution of sodium metabisulfite followed by 0.5 ml of a 1 mg/ml solution of bovine serum albumin in phosphate-buffered saline. The radioactively labeled fibronectin is adsorbed to a small column of gelatin–Sepharose and eluted with 4 $M$ urea, 0.05 Tris-HCl, pH 7.5, containing 0.05% bovine serum albumin. The eluate is stored refrigerated and used in the radioimmunoassay diluted to give approximately 30,000 cpm per tube.

*Radioimmunoassay Procedure.* The following procedure is a modification of that described by Ruoslahti *et al.*[41] The main difference is that we now use 10% bovine serum and 0.05% Tween-20 in phosphate-buffered saline as diluent. Addition of bovine serum to the buffer mimics the composition of culture media which may be the samples to be assayed. Addition of Tween-20 lowers the otherwise relatively high background of fibronectin radioimmunoassay.

Purified fibronectin and unknown samples are serially diluted. To each tube containing a 1-ml sample is then added about 30,000 cpm $^{125}$I-labeled fibronectin in 0.1 ml and 10 $\mu$l nonimmune rabbit (or goat) serum. An amount of rabbit (or goat) anti-fibronectin (0.1 ml) capable of binding 50% of added $^{125}$I-labeled fibronectin in the absence of unlabeled fibronectin is finally added. The tubes are incubated for 20 hr at room temperature. Antibody-bound and free fibronectin are separated by precipitation with goat antiserum to rabbit immunoglobulin (or rabbit antiserum to goat immunoglobulin). The optimal amount of second antibody is added, and precipitation is allowed to proceed for 90 min. The immunoprecipitates are collected by centrifugation, the supernatants are removed by suction, and the radioactivity in precipitates is counted in an automatic gamma counter. The sensitivity limit of this radioimmunoassay is about 3 ng/ml. A representative standard curve is shown in Fig. 8.

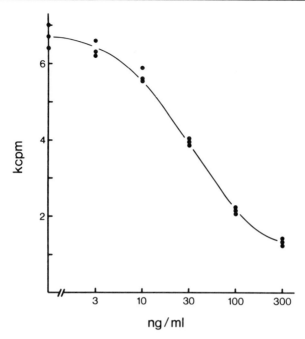

FIG. 8. A radioimmunoassay standard curve obtained by inhibiting the binding of $^{125}$I-labeled human fibronectin to anti-fibronectin by increasing amounts of unlabeled human fibronectin.

We have found that the quality of the anti-immunoglobulin serum used to precipitate immune complexes is of great importance for the quality of the fibronectin radioimmunoassay. A specific anti-IgG serum gives assays with higher sensitivity and precision and lower background than an anti-immunoglobulin serum that also precipitates serum proteins other than IgG.

*Enzyme Immunoassay*

There are many different principles that can be used in assaying for fibronectin by ELISA.[59] Space does not allow description of all. Instead, we will describe here one assay that we frequently use in our laboratory. For other types of enzyme immunoassay, including competitive and sandwich ELISA, the reader is referred to Engvall's review.[54]

The assay we will describe here is based on the interaction of fibronectin with collagen. We originally demonstrated binding of fibronectin to

[59] E. Engvall and P. Perlmann, *Immunochemistry* **8**, 871 (1971).

collagen using this kind of assay. We showed that fibronectin bound to collagen-coated microtiter plate wells. The binding was detected with enzyme-labeled antifibronectin.

This assay can be used as a direct quantitation of functionally (collagen-binding) active fibronectin. It can also be used to quantitate soluble collagen by the capacity of collagen to inhibit binding of fibronectin to collagen on the solid phase. Finally, it can be used to quantitate fibronectin from a species that does not cross-react with the enzyme-labeled antibodies used.

We use 96-well polystyrene microtiter plates as a solid phase in our assays. These provide a compact assay system that is easy to handle and allows semi- or fully automated pipetting and readout of results (e.g., using multichannel pipettes and Multiscan, Flow Laboratories). The assay can also be done using coated tubes or beads, or with proteins coupled covalently to Sepharose or paper.

*Coating of Microtiter Plates.* The microtiter plate should be of plain polystyrene and not have been treated for cell culture. The wells in the plate are coated with gelatin (Sigma type I). A stock solution of gelatin (e.g., 1 mg/ml) is diluted in $0.1\,M$ sodium carbonate buffer, pH 9.5, to give $0.1-3\,\mu g/ml$; 100 $\mu$l are added to each well, and the plate is incubated at room temperature overnight. The plate is now "coated" and can be stored as such for up to a month in the cold.

The optimal concentration of gelatin used for coating is different for different assays; $1-3\,\mu g/ml$ is best for binding assay, while $0.1-0.3\,\mu g/ml$ gives high sensitivity for competitive assays. A buffer of high pH is required for efficient and reproducible coating with gelatin. Coating at room temperature for longer times minimizes "edge effects" and gives more reproducible assays than coating for shorter times at 37°, which can also be used.

*Enzyme Labeling of Anti-fibronectin.* An IgG fraction is prepared from a specific anti-fibronectin serum (see above). This is then labeled with alkaline phosphatase or peroxidase using glutaraldehyde.

*Labeling of Antibodies with Alkaline Phosphatase.* Alkaline phosphatase from calf intestinal mucosa (Sigma type VII), 0.3 ml of a 5 mg/ml suspension in $3.4\,M$ $(NH_4)_2SO_4$, is added to 0.1 ml of a solution of antibody, 5 mg/ml in PBS. If the antibody solution is available only in concentrations lower than 5 mg/ml, the enzyme suspension is first centrifuged and the supernatant is replaced by a suitable amount of the antibody solution. The mixture of enzyme and antibody is dialyzed overnight against 1 liter of PBS. Glutaraldehyde in PBS is then added to give a final concentration of 0.2%. The conjugation is allowed to proceed for 2–3 hr at room temperature; during this time the initially colorless solution becomes pale yellow. The conjugate can then be diluted to any desired volume and dialyzed free of excess glutaraldehyde. The conjugate is stored in PBS or Tris buffer, pH 8, at +4° (conjugates must not be frozen). An unrelated protein (e.g., serum albumin) is added for stabilization, and a preservative ($NaN_3$) to prevent microbial growth.

*Measurement of Alkaline Phosphatase (ALP) Activity.* The reaction utilizes 1 mg of *p*-nitrophenyl phosphate (Sigma) per milliliter in 1 $M$ diethanolamine buffer, pH 9.8. The reaction is stopped by addition of one-fourth volume of 2 $M$ NaOH. The yellow *p*-nitrophenol is measured at 405 nm.

*Coupling of Antibodies to Horseradish Peroxidase (HRP).* A two-step glutaraldehyde procedure is used. The HRP (10 mg, Sigma type VI, RZ = 3) is dissolved in 0.2 ml of 1.25% glutaraldehyde (technical grade) in PBS, pH 7.2, and left at room temperature overnight. The reaction mixture is then diluted to 1 ml and dialyzed against 0.1 $M$ carbonate buffer, pH 9.2 (two changes, 1 liter, 4 hr each). IgG or purified antibodies (5 mg) in 0.25 ml of the carbonate buffer is added, and the mixture is incubated again overnight at room temperature. Possible remaining reactive groups are blocked by the addition of 0.1 ml of 0.2 $M$ lysine. The conjugate can be stored in 50% glycerol at 4°.

*Detection of HRP Activity.* The substrate for HRP used in assays is $H_2O_2$ or sometimes other peroxides. The cleavage of $H_2O_2$ is coupled to the oxidation of a hydrogen donor (chromogen) and goes through several intermediary steps that have different rate constants.

A variety of chromogens that yield attractive colors are available. The one we prefer is *o*-phenylenediamine. The substrate mixture contains 0.4 mg of *o*-phenylenediamine per milliliter and 0.01% $H_2O_2$ in 0.1 $M$ citrate phosphate buffer, pH 5.0. The enzymic reaction is stopped by addition of one-half volume of 4 $N$ $H_2SO_4$. The tangerine-colored product is measured at 492 nm.

*Assay; Quantitation of Fibronectin by Binding to Gelatin.* Wells in a microtiter plate are coated with 1 $\mu$g of gelatin per milliliter. Immediately before the assay, the plate is washed twice with 0.9% NaCl containing 0.15% Tween-20. Fibronectin standard and unknown samples are serially diluted, and 100 $\mu$l are added to each well. The plate is incubated at room temperature for 3 hr or longer and then washed three times with NaCl-Tween. A constant amount of enzyme-labeled anti-fibronectin is added to each well, and the plate is again incubated. Finally, after washing, the enzymic activity is measured. Figure 9 shows a standard titration curve of fibronectin obtained with this assay.

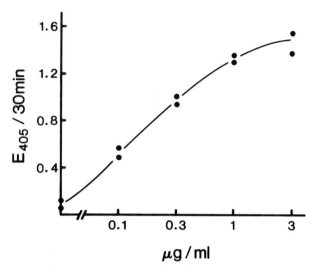

FIG. 9. Enzyme affinity assay of fibronectin. A standard binding curve is obtained with increasing amounts of fibronectin in gelatin-coated polystyrene wells. Anti-fibronectin labeled with alkaline phosphatase was used to detect the bound fibronectin.

Localization of Fibronectin by Immunofluorescence

Fibronectin can readily be visualized by immunofluorescent staining. In fibroblasts, fibronectin appears as fibrillar structures surrounding the cell. Cultured epithelial cells tend to have their fibronectin as short fibrils connecting the individual cells, and in some cells all fibronectin is found between the cells and the substratum.

Antisera to fibronectin show extensive interspecies cross-reactivity. This cross-reactivity may be helpful when one is merely interested in detecting the presence of fibronectin. However, specificity of the antisera must be taken into consideration when the purpose is to prove that a cell is synthesizing fibronectin. For instance, cultured cells are capable of utilizing the fibronectin present in fetal bovine serum along with their own fibronectin to make a matrix.[43] The use of species-specific antisera will safeguard against reaching a false conclusion about the ability of cultured cells to synthesize fibronectin.

*Staining for Fibronectin in Vitro—Detailed Procedure*

Cells are cultured on glass (coverslips or slides). Regular tissue culture plastic is not compatible with fluorescence. Coverslips are marked and sterilized by autoclaving. Each sterile slide is placed in a 60-mm tissue culture petri dish, and the cells are plated and cultured under optimal conditions for 3–4 days.

The following steps are all done for 30 min at 25° or 37°, depending on how adhesive the cells are to the glass. The coverslips are placed in a small rack and washed in a large staining dish with 500 ml of PBS with stirring. The buffer is replaced twice. The cells are then fixed in 3% paraformaldehyde in PBS. Paraformaldehyde fixes the cells in such a manner as to maintain the integrity of the cell membrane. Fibronectin visualized in such cells is primarily extracellular. To detect intracellular fibronectin, other fixatives, such as acetone (−20°, 2 min), can be used. Alternatively, the cells can be fixed with paraformaldehyde and made permeable with Triton X-100 (0.02% in PBS for 2 min).[60] After fixation, the cells are again washed in PBS as before and placed cell side up in moist chambers (a petri dish with a piece of filter paper). The antiserum in a suitable dilution (usually 1 : 50) is then applied in a quantity sufficient to cover the cells on the coverslip and incubated. The coverslips are washed as before. The bound antibodies are visualized by a similar incubation with a fluorescent conjugate of the appropriate anti-immunoglobulin (e.g., goat anti-rabbit IgG if the first antibody was prepared in rabbits). After a further wash, the coverslips are mounted on glass slides in phosphate-buffered glycerol (1 ml of PBS, 9 ml of glycerol). With nonadherent cells, the same procedures can be followed, but the washing steps are carried out by centrifugation and the slides are mounted in 50% glycerol PBS. It is important that the slides do not dry at any time during this procedure, for this will introduce artifacts. Once the slide is mounted, the cells are examined under a fluorescent microscope. The slides can be counterstained with dyes such as Evans Blue (0.01% in buffered saline) to reveal cellular details.

The usual negative controls include the use of normal rabbit serum in the place on antiserum and inhibition of fluorescence by addition of the antigen to the antiserum. Animal sera often contain antibodies that will give positive staining (e.g., antibodies to intermediate filaments are common in rabbits). Interference by such antibodies can be eliminated by using affinity-purified antibodies. Our experience has been that purified antibodies generally give crisper staining patterns. This may be due to the fact that much less IgG needs to be added in the first step and this reduces background staining.

[60] A. I. Gotlieb, M. H. Heggeness, J. F. Ash, and S. J. Singer, *J. Cell. Physiol.* **100**, 563 (1979).

*Staining for Fibronectin in Tissues*

Fibronectin can be demonstrated by immunofluorescence in frozen sections or in sections fixed according to Sainte-Marie.[61] Tissues are snap frozen in liquid nitrogen. Frozen sections are cut to a thickness of 5 μm, mounted on microscope slides, and then fixed with acetone at room temperature for 10 min. The subsequent steps are the same as those described above for the staining of cell cultures.

Fibronectin in Cell Attachment

Most types of cells, when added to plates coated with fibronectin or fibronectin–collagen complexes, attach and spread rapidly.[33–36,62,63] This suggests that fibronectin plays a role in the attachment of cells to the surrounding extracellular matrix *in vivo*.

*Attachment Assay*

We have used a simple assay for testing the role of fibronectin cellular adhesion modeled after published methods.[34,63] The assay consists of coating 96-well microtiter plates with various dilutions of fibronectin, incubating single-cell suspensions in the wells for 1 hr at 37°, removing unattached cells, and counting those that remain. We have sought optimal conditions for cell attachment, making the fibronectin coating the only variable. The assay will be described in detail, covering the plate, the cells, and the attachment results.

*The Plate.* We use 96-well polystyrene plates (Flow No. 76-203-05) that have not been treated for tissue culture. Most cells will not adhere to untreated polystyrene. Cells will attach after the plastic is coated with an adhesive protein, such as fibronectin or collagen. The untreated polystyrene has excellent coating characteristics (see above), but the plates treated for tissue culture do not.

We have elected not to coat the plate with a layer of collagen and with or without protein as described by Klebe.[33] The addition of collagen complicates the assay because some cells may attach directly to collagen. Another advantage of coating the substrate directly with the attachment protein is that it has allowed us to study fragments of fibronectin that, while lacking a collagen binding site, are still capable of promoting cell attachment.[63] We have also used the same approach to show that laminin

---

[61] G. Sainte-Marie, *J. Histochem. Cytochem.* **10**, 250 (1962).
[62] E. Pearlstein, *Nature (London)* **262**, 497 (1976).
[63] E. Ruoslahti and E. Hayman, *FEBS Lett.* **107**, 51 (1979).

can promote the attachment of liver cells, especially those from regenerating liver.[64]

*The Coating.* The fibronectin we have used for coating has come from plasma, from culture media of both normal and transformed cells, and from amniotic fluid. In our hands, the fibronectin from all these sources appears to be equally active in cell attachment. We have noted that the "freshness" of the preparations influences the results more than the source of fibronectin.

The coating procedures were modeled after those developed for the ELISA assay[54] (see above). The coating is done with fibronectin or other proteins diluted from concentrated solutions in urea to 0.1 $M$ $NaHCO_3$. The fibronectin solution, 0.1 ml containing 0.3–10 $\mu$g of protein per milliliter, is incubated in the wells for 2 hr at room temperature. The unattached protein is removed by washing three times with PBS. Once coated, the plates are not allowed to dry. The procedures used in the adhesive assay are done cleanly; sterile technique is not required because the assay takes only a couple of hours.

*The Cells.* We have used normal rat kidney cells to perform the attachment assay. However, most cells, both normal and transformed, seem to work in a similar fashion.[65] Actively growing cells have the best attachment properties for the assay. Therefore, 1 day before the assay, a confluent culture of cells is split 1:2 using standard tissue culture trypsin for the split.

For the assay, the cells are washed three times with PBS, pH 7.2, then 10 ml of 2 × crystallized trypsin (Sigma, type III), 0.1 mg/ml in PBS, is added, and the cells are incubated at 37° until they are detached. This incubation also removes all cell surface fibronectin. The cells are then collected by centrifugation and washed with a solution of 0.5 mg of soybean trypsin inhibitor per milliliter in PBS three times to ensure neutralization of the trypsin. Before the last centrifugation, a sample is taken, and cell numbers and viability by trypan blue exclusion are determined. The cells are now suspended in minimal essential medium (MEM) at a concentration of $10^6$ cells/ml and dispersed by pipetting until a single-cell suspension is obtained.

*The Assay.* To ensure even dispersal of the cells in the microtiter well, 0.1 ml of MEM is added to each well. This is followed by the addition of 0.1 ml of the cell suspension. The plate is incubated for 1 hr at 37° with or without $CO_2$, depending on the requirements of the cells used.

[64] R. Carlsson, E. Engvall, A. Freeman, and E. Ruoslahti, *Proc. Natl. Acad. Sci. U.S.A.* **78**, 2403 (1981).
[65] E. G. Hayman, E. Engvall, and E. Ruoslahti, *J. Cell Biol.* **88**, 352 (1981).

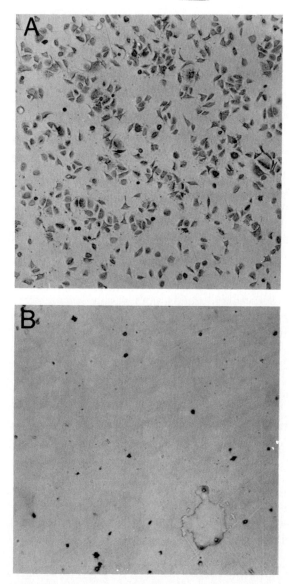

FIG. 10. Cell attachment assay showing the cell attachment promoting activity of fibronectin bound to a surface (A) as compared with a surface coated with albumin (B).

After 1 hr at 37° the unattached cells are removed simply by tossing out the medium and washing. The plate is flooded with PBS, and the washing solution is tossed out. The cells are then fixed with 3% paraformaldehyde in PBS and stained with 1% toluidine blue, 3% formaldehyde in PBS, and attached cells are counted. Their number can be best determined using a cell counter that is capable of counting cells attached to a surface (e.g., Dynatech Laboratories, Alexandria, Virginia). A typical result is shown in Fig. 10. Under optimal conditions, 70–90% of the cells remain attached to the dish after washing.

Acknowledgments

We thank Ms. Nancy Beddingfield for excellent secretarial assistance. The authors' work was supported by Grants CA 27455 and CA 28101 from the National Cancer Institute and AM 26693 from the National Institute of Arthritis and Metabolism, DHHS.

## [47] Laminin

*By* RUPERT TIMPL, HEILWIG ROHDE, LEILA RISTELI, URSULA OTT, PAMELA GEHRON ROBEY, and GEORGE R. MARTIN

Basement membranes are extracellular matrices that are widely distributed in the body and consist mainly of collagenous and noncollagenous proteins. The latter are still incompletely characterized[1,2] but include laminin,[3] a glycoprotein of high molecular weight. The isolation of laminin in amounts sufficient for comprehensive studies was accomplished by use of a mouse tumor, the EHS sarcoma, which was originally classified as a poorly differentiated chondrosarcoma, but later identified as a tumor producing large amounts of an extracellular basement membrane.[4] Other suitable sources for isolating laminin may include endodermal cells,[5-7] tumor cell cultures,[8-11] and a variety of other cultivated cells (see below).

[1] N. A. Kefalides, R. Alper, and C. C. Clark, *Int. Rev. Cytol.* **61**, 167 (1979).
[2] G. R. Martin and R. Timpl, *in* "Immunochemistry of the Extracellular Matrix" (H. Furthmayr, ed.). CRC Press, Boca Raton, Florida, 1982. In press.
[3] R. Timpl, H. Rohde, P. Gehron Robey, S. I. Rennard, J. M. Foidart, and G. R. Martin, *J. Biol. Chem.* **254**, 9933 (1979).
[4] R. W. Orkin, P. Gehron, E. B. McGoodwin, G. R. Martin, T. Valentine, and R. H. Swarm, *J. Exp. Med.* **145**, 204 (1977).
[5] B. L. M. Hogan, A. R. Cooper, and M. Kurkinen, *Dev. Biol.* **80**, 289 (1980).
[6] S. Sakashita and E. Ruoslahti, *Arch. Biochem. Biophys.* **205**, 283 (1980).
[7] C. C. Howe and D. Solter, *Dev. Biol.* **77**, 480 (1980).
[8] A. E. Chung, R. Jaffe, I. L. Freeman, J. P. Vergues, J. E. Braginski, and B. Carlin, *Cell* **16**, 277 (1979).

Purification and Properties

*Purification of Laminin from EHS Sarcoma.*[3] The tumor is grown subcutaneously in C57BL mice[4] and harvested after reaching a weight of 5–15 g. Approximately 1 kg of wet tumor tissue is homogenized three times with a total of 6–8 liters of cold 3.4 $M$ NaCl, 0.05 $M$ Tris-HCl, pH 7.4, containing protease inhibitors (0.01 $M$ EDTA, 50 $\mu$g/ml $p$-hydroxymercuribenzoate, 50 $\mu$g of phenylmethanesulfonyl fluoride per milliliter). Each homogenization is immediately followed by centrifugation (4°, 30 min, 9000 $g$) to remove soluble protein. The residue is then extracted three times overnight at 4° by stirring in 1–1.5 liters of 0.5 $M$ NaCl, 0.05 $M$ Tris-HCl, pH 7.4, containing the same protease inhibitors as above. The first extract contains most of the laminin, but also serum and various other tumor proteins. Subsequent extracts contain a greater proportion of laminin, but the laminin in these extracts has a stronger tendency to aggregate and to precipitate. Type IV collagen is usually a minor contaminant in these extracts but can be removed if necessary by precipitation at a concentration of 1.7 $M$ NaCl. The extracts are dialyzed against large volumes of 0.05 $M$ Tris-HCl, pH 8.6, 2 $M$ urea, and aliquots are subjected to DEAE-cellulose chromatography (Fig. 1). The material that does not bind to the column is concentrated by ultrafiltration (Amicon, Diaflo, filter XM 100A) to a volume of about 50 ml and passed over an agarose A-1.5m column (3.5 × 140 cm) equilibrated in and eluted with 1 $M$ CaCl$_2$, 0.05 $M$ Tris-HCl, pH 7.4, Electrophoretically pure laminin appears in the void volume of the column and is stored at 4° after dialysis against 0.4 $M$ NaCl, 0.05 $M$ Tris-HCl, pH 7.4, followed (if required) by dialysis against 0.2 $M$ ammonium bicarbonate, pH 7.9. Lyophilization of the protein should be avoided, since it is insoluble even in 8 $M$ urea under these conditions. The yield of purified laminin is about 0.5–1 g per kilogram of wet tumor. Milligram amounts of laminin could be prepared from cell cultures prior to or after treatment with bacterial collagenase.[6,10]

Although a substantial portion of the laminin fails to bind to DEAE-cellulose, some of it binds to the column and is eluted by a NaCl gradient (Fig. 1). The laminin in the various fractions is similar in composition, including sialic acid content, subunit composition, and antigenicity,[12] and the binding upon rechromatography on DEAE-cellulose is reproducible except for the front peak material. Laminin obtained from certain

---

[9] L. Risteli, H. Rohde, T. Krieg, P. K. Müller, R. Timpl, and J. C. Murray, in preparation.

[10] I. Leivo, K. Alitalo, L. Risteli, A. Vaheri, R. Timpl, and J. Wartiovaara, *Exp. Cell. Res.*, in press (1981).

[11] S. Strickland, K. K. Smith, and K. R. Marotti, *Cell* **21**, 347 (1980).

[12] H. Rohde and R. Timpl, unpublished observation.

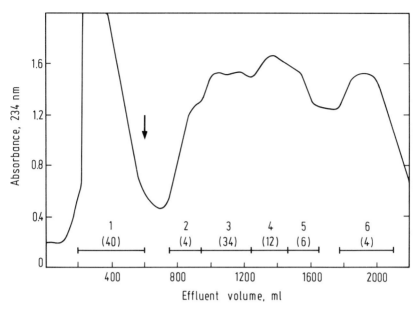

FIG. 1. Purification of laminin from a neutral salt extract of EHS sarcoma on DEAE-cellulose. The distribution of laminin in the different pools (as percentage) was determined by radioimmunoassay and is shown by the numbers in parentheses. The column (2.5 × 25 cm) was equilibrated at room temperature in 0.05 $M$ Tris-HCl, pH 8.6, 2 $M$ urea, and loaded with 200–250 ml of 0.5 $M$ NaCl extract normally obtained from 150–200 g of tumor. Elution was performed with a linear NaCl gradient from 0 to 0.4 $M$ NaCl (800/800 ml). The arrow denotes the start of the gradient.

cell cultures[9] appears to be less heterogeneous, and more than 95% of the material binds to and is eluted in the position of pools 3–4 (0.1–0.18 $M$ NaCl) from DEAE-cellulose. This position corresponds to the elution position of fibronectin which can be separated from laminin by binding to gelatin.[13,14] Laminin,[15,16] like fibronectin, binds to heparin–agarose which may be useful as another step in its purification. Here, the column (2 × 12 cm, heparin–Sepharose from Pharmacia) is equilibrated in 0.2 $M$ ammonium bicarbonate, pH 7.9, containing 0.02% sodium azide. This column binds 35–80% of 10–20 mg of laminin applied to it in the same solvent. Elution is achieved by linear gradient of NaCl from 0 to 0.5 $M$ or by the higher concentration alone. Incomplete binding to heparin is not due to a low capacity of the column and is not yet understood.[16]

[13] K. Alitalo, M. Kurkinen, A. Vaheri, T. Krieg, and R. Timpl, *Cell* **19**, 1053 (1980).
[14] K. Alitalo, A. Vaheri, T. Krieg, and R. Timpl, *Eur. J. Biochem.* **109**, 247 (1980).
[15] S. Sakashita, E. Engvall, and E. Ruoslahti, *FEBS Lett.* **116**, 243 (1980).
[16] H. Rohde and R. Timpl, unpublished observation.

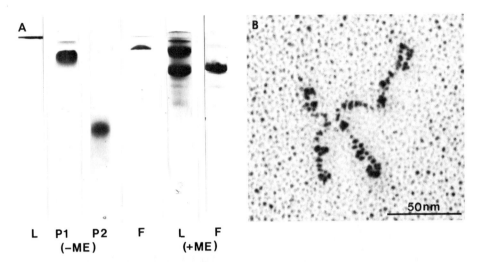

L  P1  P2   F   L   F
 (−ME)       (+ME)

FIG. 2. Characterization of laminin by sodium dodecyl sulfate (SDS) gel electrophoresis (A) and electron microscopy (B). The samples were analyzed in 3.5% polyacrylamide gels containing 2% SDS[3] prior to (−ME) or after (+ME) reduction with 5% mercaptoethanol. The samples were mouse laminin (L), its pepsin fragments P1 and P2, and, for comparison, human plasma fibronectin (F). The electron micrograph was obtained by rotary shadowing with platinum–carbon after spraying a solution of laminin in 70% glycerol onto mica disks.[18] Electron micrograph by courtesy of Dr. H. Furthmayr.

Laminin shows a characteristic pattern in SDS gel electrophoresis (Fig. 2A). The intact, nonreduced protein migrates as a single sharp band barely penetrating the gel, while after reduction, two broad bands with faster mobility are observed.[3] They migrate slightly faster than nonreduced and reduced fibronectin, indicating apparent molecular weights of 200,000–220,000 and 400,000–440,000, respectively. Occasionally, splitting of these bands into closely spaced pairs is observed with laminin prepared from the EHS sarcoma, and more regularly with proteins obtained from cell cultures.[5,9,13,14] The major constituent chains of laminin could be isolated after SDS electrophoresis,[6,8] but their nature is still controversial. While the sizes of these chains suggest either a precursor product or a monomer–dimer relationship, preliminary data[8,16] indicated that they are related but not identical. Based on electron microscopic and physical data,[17,18] it is likely that both chains are constituents of the same molecule.

[17] R. Timpl, H. Rohde, U. Ott-Ulbricht, L. Risteli, and H. P. Bächinger, in "Glycoconjugates. Proceedings of the Fifth International Symposium" (R. Schauer, P. Boer, E. Buddecke, M. F. Kramer, J. F. G. Vliegenthart, and H. Wiegandt, eds.), p. 145. Thieme, Stuttgart, 1979.

[18] J. Engel, E. Odermatt, A. Engel, J. A. Madri, H. Furthmayr, H. Rohde, and R. Timpl, J. Mol. Biol. **150**, 97 (1981).

Ultracentrifugal analyses[17,18] indicated that laminin is a large protein with a molecular weight of about 900,000. However, the protein has a tendency to aggregate, and this estimate should be considered tentative. Electron microscopy[18] of the molecule demonstrated a unique asymmetric structure (Fig. 2B). The elongated shape is characterized by a long arm (length about 75 nm) and three short arms (length about 35 nm) connected to each other in the form of a cross. This structure showed a low flexibility and contained compact domains, particularly at the ends of each arm. Circular dichroism studies[17] showed a spectrum indicating about 30% α-helix in laminin and a distinctly different spectrum from that observed with fibronectin. The α-helix elements show a sharp thermal transition around 58°.[19]

The amino acid composition of laminin is not sufficiently unusual to be distinctive even though it is different from that of fibronectin.[3] Laminin contains about 12–15% carbohydrate,[8,17] including a relatively high level of sialic acid (4–6%). The intact protein as well as the subunits stain with the periodic acid–Schiff reagent[20] and bind to a variable degree to lectins such as concanavalin A or wheat germ agglutinin.[17]

To date, laminin has not been isolated and characterized in intact form from authentic basement membranes. It may not be fully soluble, since even in the EHS sarcoma matrix about 20% of laminin requires digestion with pepsin[3] or extraction with 8 $M$ urea[12] for solubilization. Previously, large proteins were obtained from glomerular basement membranes[1,21,22] under proteolytic or denaturing conditions. They may have included laminin, but this was not demonstrated.

*Solubilization by Proteases.* Incubation of laminin with various proteases[19,23] indicated that the protein contains several distinct structural domains that are even more resistant to proteases than those in fibronectin. Two unique fragments, P1 and P2 ($M_r$ 290,000 and 45,000) were isolated from laminin after digestion with pepsin or a variety of neutral proteases. Little or no digestion was obtained with plasmin with the native protein, but some occurred after denaturation.[19] P1 and P2 differ in electrophoretic mobility (Fig. 2A) and have a high content of cysteine and

---

[19] U. Ott, E. Odermatt, J. Engel, H. Furthmayr, and R. Timpl, *Eur. J. Biochem.*, submitted.

[20] R. Timpl, H. Rohde, G. Wick, P. Gehron Robey, S. I. Rennard, J. M. Foidart, and G. R. Martin, in "Biochemistry of Normal and Pathological Connective Tissues" (A. M. Robert and L. Robert, eds.), Vol. II, p. 225. C.N.R.S., Paris, 1980.

[21] J. W. Freytag, P. N. Dalrymple, M. H. Maguire, D. K. Strickland, K. L. Carraway, and B. G. Hudson, *J. Biol. Chem.* **253**, 9069 (1978).

[22] P. Bardos, J. P. Muh, B. Luthier, B. Devulder, and A. Tacquet, *Comp. Biochem. Physiol. B* **53B**, 49 (1976).

[23] H. Rohde, H. P. Bächinger, and R. Timpl, *Hoppe-Seyler's Z. Physiol. Chem.* **361**, 1651 (1980).

disulfide bonds. The larger fragment probably arises from the region in the native protein that connects the four arms.[18]

*Isolation of Laminin Fragment P1 from a Pepsin Digest of Human Placenta.*[24] The high degree of protease resistance of certain laminin structures was exploited for isolating characteristic fragments directly from basement membrane-containing tissues. Placental tissue is homogenized in 3% Triton X-100, 0.02 $M$ EDTA, pH 7.4 (3.3 l/kg) and extracted for 2–3 hr at 4° to remove serum and cellular proteins. The insoluble residue is collected by filtration through a nylon net, washed with copious amounts of water and 0.5 $M$ acetic acid, and finally homogenized in 0.5 $M$ acetic acid (5 liters/kg). After addition of pepsin (0.5 g/kg) the mixture is incubated with stirring for 20 hr at 6–8°, the supernatant fluid is collected by centrifugation (30 min, 15,000 $g$), and solubilized protein is precipitated by adding 1.2 $M$ NaCl. The precipitate is dissolved in 0.2 $M$ NaCl, 0.05 $M$ Tris-HCl, pH 7.4, dialyzed against the same buffer (1 day, 6–8°), and subjected to fractional precipitation by raising the NaCl concentration first to 2 $M$ and then to 4 $M$.[25] Material precipitating at 2 $M$ NaCl was found to consist mainly of collagenous protein and laminin fragments, which can be separated on DEAE-cellulose under the conditions shown in Fig. 1. Here, collagens elute in the front while the laminin fragment, P1, is recovered in a broad peak within the gradient (0.1 to 0.18 $M$ NaCl). The material precipitated at 4 $M$ NaCl still contains fragment P1 and several other proteins, but is poor in collagen. If required, collagenous protein can be removed from fragment P1 by brief digestion with bacterial collagenase (1 hr, 37°; 1 mg of enzyme/200 mg of protein) in 0.2 $M$ ammonium bicarbonate, pH 7.9. The material is then passed over Sepharose CL-4B (column 2.5 × 85 cm) equilibrated in 0.005 $M$ Tris-HCl, pH 8.6, 8 $M$ urea, from which fragment P1 may emerge as a distinct peak at $V_E/V_T = 0.58$. As an additional step in purification, fragment P1 may be bound to a column (1 × 5 cm) of concanavalin A–Sepharose 4B equilibrated with 0.15 $M$ NaCl, 0.05 $M$ Tris-HCl, pH 7.4, 1 m$M$ CaCl$_2$. Usually more than 90% of fragment P1 binds to the lectin and can be eluted with 0.4 $M$ α-methyl mannoside. Purified material is finally dialyzed against water at pH 8–9 and lyophilized. The yield of fragment P1 from 1 kg of human placenta is about 10–15 mg.

As outlined below, antibodies to native laminin can be used to detect and quantitate the P1 fragment during its isolation from various digests. Other characteristic properties of P1 are the high half-cystine content (120–150 residues/1000) and the appearance on electrophoresis of a com-

---

[24] L. Risteli and R. Timpl, *Biochem. J.* **193**, 749 (1981).

[25] J. Risteli, H. P. Bächinger, J. Engel, H. Furthmayr, and R. Timpl, *Eur. J. Biochem.* **108**, 239 (1980).

ponent with a molecular weight of about 290,000 (Fig. 2A), which changes into a complex and faster moving peptide mixture after reduction.[23,24] Pepsin also produced small amounts of cystine-rich peptides, which are smaller than P1 but are related to it by chemical and immunological properties.[23,24] The origin of these peptides is not known, but they may be produced from laminin-like proteins.

*Immunological Properties.* Injection of as little as 0.1–0.25 mg of laminin or fragment P1 induces antisera of high titer in rabbits and guinea pigs. The specificity of such antibodies can be assessed by gel diffusion, radioimmunoassay, and enzyme immunoassay.[3,6,26,27] The major antigenic determinants are localized on fragment P1 and require intact disulfide bonds. Minor antigenic determinants are found on pepsin fragment P2 and on laminin reduced under denaturing conditions.[23,26] The latter material is also a good immunogen in rabbits.[6,8,12] Inhibition assays are useful to quantitate laminin or its fragments in the range of 2–200 ng/ml[26,27] and are of sufficient sensitivity to detect cross-reacting material in blood.[28] Usually, antisera and affinity-purified antibodies to laminin showed little or no reaction with fibronectin, type IV collagen, or basement membrane proteoglycan.[3,27,29] Interspecies cross-reactions were found in the range 10–30%.[24,26] The data also indicate that tumor laminin is not different from authentic laminin.

*Biosynthesis and Tissue Distribution.* A variety of cultured cells including various epithelial cells, endothelial cells, smooth muscle cells, myotubes, and several tumor cell lines were found to produce and secrete laminin.[3,30] Fibroblasts, chondrocytes, and myoblasts usually fail to synthesize significant amounts of laminin. Immunoprecipitation and electrophoresis of labeled material from cell culture showed after reduction two bands with apparent molecular weights of 210,000–220,000 and 400,000 and occasionally higher aggregates.[6,9,10,13,14] Usually, these bands were doublets, indicating that there were two polypeptide chains of each size. Similar bands produced by teratocarcinoma or endodermal cells (named GP1 and GP2[8] or PYSA and PYSB[31]) are likely to resemble the two basic subunits of laminin.[3,5] Preliminary data also indicate that laminin was not modified further after deposition into a matrix.[6,10]

[26] H. Rohde, G. Wick, and R. Timpl, *Eur. J. Biochem.* **102**, 195 (1979).
[27] S. I. Rennard, R. Berg, G. R. Martin, J. M. Foidart, and P. Gehron Robey, *Anal. Biochem.* **104**, 205 (1980).
[28] J. Risteli, H. Rohde, and R. Timpl, *Anal. Biochem.* **113**, 372 (1981).
[29] J. R. Hassell, P. Gehron Robey, H. J. Barrach, J. Wilczek, S. I. Rennard, and G. R. Martin, *Proc. Natl. Acad. Sci. U.S.A.* **77**, 4494 (1980).
[30] J. M. Foidart, E. W. Bere, M. Yaar, S. I. Rennard, M. Gullino, G. R. Martin, and S. I. Katz, *Lab. Invest.* **42**, 336 (1980).
[31] B. L. M. Hogan, *Dev. Biol.* **76**, 275 (1980).

Immunofluorescence studies have suggested that laminin is an abundant component in almost every basement membrane of the body.[3,26,30] Ultrastructural studies have localized laminin to the lamina rara of epidermal[30] and glomerular[32] basement membranes. Immunohistological studies have also detected a number of qualitative and quantitative changes in the distribution of laminin under pathological and experimental conditions.[33-37]

*Biological Activity.* Laminin shows distinct binding of heparin[15,16] indicating that similar interactions with basement membrane heparan sulfate-proteoglycan[29,38] may occur *in situ*. Laminin also promotes attachment of cells to plastic dishes[39] and to substrates of type IV collagen.[40,41] The potential interaction of laminin with collagenous proteins, fibronectin, or proteoglycans has not yet been studied *in vitro*. A crucial role of laminin in cell aggregation and formation of early extracellular matrices was suggested by studies indicating that laminin is produced and secreted as early as the morula (16–32 cells) stage of development[42] and that its synthesis precedes tubule formation in the embryonic kidney.[43] Retinoic acid and cyclic AMP, which induce differentiation of carcinoembryonic cells, also stimulated laminin synthesis.[7,11] Some cultured cells produced only laminin,[9-11] whereas other cells produced both laminin and fibronectin.[13,14] Together the data indicate that laminin functions as an adhesive protein.

---

[32] J. E. Madri, S. J. Roll, H. Furthmayr, and J. M. Foidart, *J. Cell Biol.* **86**, 682 (1980).
[33] G. Wick, H. Hönigsmann, and R. Timpl, *J. Invest. Dermatol.* **73**, 335 (1979).
[34] E. Hahn, G. Wick, D. Pencev, and R. Timpl, *Gut* **21**, 63 (1980).
[35] J. M. Foidart, and A. H. Reddi, *Dev. Biol.* **75**, 130 (1980).
[36] J. I. Scheinman, J. M. Foidart, P. Gehron Robey, A. J. Fish, and A. F. Michael, *Clin. Immunol. Immunopathol.* **15**, 175 (1980).
[37] R. Fleischmajer, W. Dessau, R. Timpl, T. Krieg, C. Luderschmidt, and M. Wiestner, *J. Invest. Dermatol.* **75**, 270 (1980).
[38] Y. S. Kanwar and M. G. Farquhar, *Proc. Natl. Acad. Sci. U.S.A.*, **76**, 4493, 1979.
[39] S. Johansson, L. Kjellén, M. Höök, and R. Timpl, *J. Cell Biol.* **90**, 260 (1981).
[40] V. P. Terranova, D. H. Rohrbach, J. C. Murray, G. R. Martin, and S. H. Yuspa, *in* "Biology of the Vascular Endothelial Cell," p. 37. Cold Spring Harbor Laboratory, Cold Spring Harbor, New York, 1980.
[41] V. P. Terranova, D. H. Rohrbach, and G. R. Martin, *Cell* **22**, 719 (1980).
[42] I. Leivo, A. Vaheri, R. Timpl, and J. Wartiovaara, *Dev. Biol.* **70**, 100 (1980).
[43] P. Ekblom, K. Alitalo, A. Vaheri, R. Timpl, and L. Saxén, *Proc. Natl. Acad. Sci. U.S.A.* **77**, 485 (1980).

## [48] Structural Glycoproteins

By LADISLAS ROBERT and MADELEINE MOCZAR

The term structural glycoproteins designates glycoproteins synthesized by mesenchymal cells, present in the intercellular matrix (some are also present in or near the cell membrane), where they associate with other matrix macromolecules, such as collagen, elastin, or proteoglycans. This definition and designation was proposed as a result of the realization that formerly isolated "noncollagenous proteins" and "neutral heteropolysaccharides" of connective tissues may well be derived from the same macromolecular components, namely, connective tissue glycoproteins.[1-8] Several other laboratories have isolated similar substances from a variety of connective tissues (see reviews[9-14]).

It soon became evident that structural glycoproteins represent a rather large, heterogeneous class of glycoproteins. We proposed therefore to consider this group of glycoproteins as representatives of the fourth family of macromolecules of the intercellular matrix (the three others being the collagens, elastins, and proteoglycans).[12-16]

---

[1] Z. Dische, A. Danilczenko, and G. Zelmenis, in "Chemistry and Biology of Mucopolysaccharides" (G. E. W. Wolstenholme and M. O'Connor, eds.), p. 116. Churchill, London, 1958.
[2] L. Robert and Z. Dische, *Biochem. Biophys. Res. Commun.* **10**, 209 (1963).
[3] B. Robert, J. Parlebas, and L. Robert, *C. R. Hebd. Seances Acad. Sci.* **256**, 323 (1963).
[4] L. Robert, J. Parlebas, N. Poullain, and B. Robert, *Protides Biol. Fluids* **11**, 109 (1963).
[5] L. Robert, J. Parlebas, P. Oudea, A. Zweibaum, and B. Robert, in "Structure and Function of Connective and Skeletal Tissue" (G. R. Tristram, ed.), p. 406. Butterworth, London, 1965.
[6] L. Robert and B. Robert, *Protides Biol. Fluids* **15**, 143 (1967).
[7] L. Robert, B. Robert, and E. Moczar, in "Fourth International Colloquium on Cystic Fibrosis of the Pancreas," Part II, p. 309. Karger, Basel, 1968.
[8] A. M. Robert, B. Robert, and L. Robert, in "Chemistry and Molecular Biology of the Intercellular Matrix (E. A. Balazs, ed.), Vol. 1, p. 237. Academic Press, New York, 1970.
[9] L. Robert and B. Robert, in "Connective Tissues. Biochemistry and Pathophysiology" (R. Fricke and F. Hartmann, eds.), p. 240. Springer-Verlag, Berlin and New York, 1974.
[10] J. C. Anderson, *Int. Rev. Connect. Tissue Res.* **7**, 251 (1976).
[11] J. C. Anderson, in "Biochimie des Tissus Conjonctifs Normaux et Pathologiques" (A. M. Robert and L. Robert, eds.), Vol. II, p. 195. C.N.R.S., Paris, 1980.
[12] L. Robert, S. Junqua, and M. Moczar, *Frontiers Matrix Biol.* **3**, 113 (1976).
[13] M. Moczar and E. Moczar, in "Biochimie des Tissus Conjonctifs Normaux et Pathologiques" (A. M. Robert and L. Robert, eds.), Vol. 2, p. 201. C.N.R.S., Paris, 1980.
[14] M. Moczar and E. Moczar, *Pathol. Biol.* **26**, 64 (1978).
[15] B. Robert and L. Robert, *Frontiers Matrix Biol.* **1**, 1 (1973).

TABLE I
APPARENT MOLECULAR WEIGHTS ($M_r$) OF SOME
STRUCTURAL GLYCOPROTEIN PREPARATIONS, DETERMINED
MOSTLY BY GEL ELECTROPHORESIS

| Source | $M_r \times 10^3$ | Reference[a] |
|---|---|---|
| Dermis, rabbit | 16 | 19 |
| Aorta, pig | 35 | 14 |
| Ligamentum nuchae, beef | 35 | 63 |
| Nasal cartilage, calf | 47 | b |
| Proteoglycan link-proteins | 51 | b |
| Achilles tendon, bovine | 60 | c |
| Cornea, calf | 75 | 8 |
| Fibroblast cultures from bovine | 150 | 58, 59 |
| Ligamentum nuchae | 300 | 58, 59 |
| Lung, monkey | 35 | 31, 31 |
|  | 75 | 31, 32 |
|  | 110 | 31, 32 |
|  | 140 | 31, 32 |
| Fibronectin, human plasma | 450 | d |
| Laminin, mouse tumor | 850 | 20 |

[a] Numbers refer to text footnotes.
[b] J. Baker and B. Caterson, *Biochem. Biophys. Res. Commun.* **77**, 1 (1977).
[c] J. Anderson, *Biochim. Biophys. Acta* **379**, 444 (1975).
[d] A. B. Chen, D. L. Amrani, and M. W. Mosesson, *Biochim. Biophys. Acta* **493**, 310 (1977).

Although their apparent molecular weight varies between large limits (see Table I), several properties are common to most representatives of this family of macromolecules, such as a relatively low solubility or insolubility in aqueous buffers, a rather similar amino acid composition (high dicarboxylic and aliphatic amino acid content, and the presence of variable but relatively high amounts of cysteine and sometimes of tryptophan[8,10–12,14,17] (see Table II). This amino acid composition is quite different from that of the collagens, elastin, or proteoglycan peptides (except link-glycoproteins, which may well be considered as belonging to the structural glycoprotein family of matrix macromolecules). It should be noticed, however, that similar amino acid compositions can be found in a number of other proteins, such as blood serum glycoproteins.

[16] L. Robert, *in* "Biochimie des Tissus Conjonctifs Normaux et Pathologiques" (A. M. Robert and L. Robert, eds.), Vol. 2, p. 189. C.N.R.S., Paris, 1980.
[17] L. Robert and P. Comte, *Life Sci.* **7**, 493 (1968).

TABLE II
AMINO ACID COMPOSITION OF SEVERAL CONNECTIVE TISSUE STRUCTURAL GLYCOPROTEIN PREPARATIONS[a]

| Amino acid | Calf cornea | | Spongia officinalis Junqua et al.[d,e] | Metridium dianthus sea anemone Katzman[f] | Rabbit dermis | | Puppy cartilage Shipp and Bowness[i] | Bovine Achilles tendon Anderson and Jackson[j] | Pig aorta 35-kilodalton glycoprotein Moczar et al.[k] |
|---|---|---|---|---|---|---|---|---|---|
| | Robert and Comte[b] | Moczar and Robert[c] | | | Timpl et al.[g] | Stoupy et al.[h] | | | |
| Asp | 102 | 90 | 98 | 109 | 96.7 | 94 | 123 | 107 | 96.2 |
| Thr | 51 | 54 | 60 | 56 | 59.6 | 59 | 52 | 65 | 52 |
| Ser | 59 | 57 | 47 | 56 | 60.8 | 54 | 56 | 63 | 60 |
| Glu | 113 | 100 | 101 | 124 | 131.3 | 161 | 116 | 122 | 139 |
| Pro | 57 | 52 | 52 | 51 | 60.0 | 45 | 63 | 67 | 58 |
| Gly | 105 | 88 | 90 | 73 | 63.9 | 71 | 98 | 57 | 106 |
| Ala | 70 | 81 | 87 | 75 | 73.8 | 91 | 67 | 75 | 82 |
| 1/2-Cys | 15 | 27 | 9 | 21 | 7.1 | — | 25 | 41 | — |
| Val | 86 | 74 | 76 | 64 | 58.8 | 59 | 68 | 72 | 55 |
| Met | 11 | 1 | 27 | 22 | 21.6 | 26 | 10 | 9 | 2.6 |
| Ile | 47 | 43 | 64 | 47 | 58.0 | 66 | 44 | 33 | 51.0 |
| Leu | 47 | 77 | 83 | 78 | 91.9 | 101 | 94 | 83 | 97 |
| Tyr | 17 | 28 | 32 | 32 | 27.9 | 34 | 26 | 32 | 24 |
| Phe | 38 | 31 | 46 | 35 | 39.3 | 41 | 36 | 38 | 41 |
| Trp | * | ND | ** | ND | ND | 16 | ND | ND | ND |
| Lys | 44 | 57 | 48 | 74 | 70.5 | 42 | 50 | 71 | 55 |
| His | 15 | 22 | 20 | 17 | 21.1 | 10 | 21 | 21 | 15 |
| Arg | 60 | 46 | 60 | 58 | 57.3 | 29 | 54 | 44 | 49 |

[a] Values are expressed as residues per 1000 residues. *, 2% of total protein; **, 0.8% of total protein; ND, not done.   [b] Robert and Comte.[7]   [c] Moczar and Robert[36], Moczar et al.[37]   [d] Junqua et al.[53]   [e] S. Junqua, J. Fayolle, and L. Robert, Protides Biol. Fluids 22, 167 (1975).   [f] R. L. Katzman, Life Sci. 11, 131 (1972).   [g] R. Timpl, I. Wolff, and M. Weiser, Biochim. Biophys. Acta 168, 168 (1968).   [h] Stoupy et al.[18]   [i] D. W. Shipp and M. Bowness, Biochim. Biophys. Acta 379, 282 (1975).   [j] J. C. Anderson and D. S. Jackson, Biochem. J. 127, 179 (1972).   [k] Moczar et al.[52a]

Structural glycoproteins have a sugar content varying from trace amounts[18,19] up to more than 15%.[20] They usually exhibit a pronounced antigenicity with variable tissue and/or species specificity.[6,21,22] It was claimed that, as they strongly associate with the other matrix macromolecules as well as the cell membrane, they play an important role in the ordered biosynthesis of tissue- and organ-specific matrix patterns, that is, in cell–matrix interaction during normal and pathological histogenesis and morphogenesis.[6,7,9,12,16]

The isolation and characterization of two major representatives of this family of glycoproteins, namely, fibronectin and laminin, largely confirmed these predictions. These two major structural glycoproteins are dealt with in Chapters [46] and [47] of this volume.

We shall describe here some of the methods of isolation and characterization of structural glycoproteins other than fibronectin and laminin.

General Principles of the Isolation Procedures

There are several important steps involved in the isolation procedures of structural glycoproteins. The first is their solubilization and separation from cells or cell debris, then from the other intercellular matrix components, such as the collagens, elastin, and proteoglycans. Another difficulty is their easy association with lipids.[23] It was claimed that structural glycoproteins can behave as apolipoproteins and associate with lipids extracted from plasma lipoproteins.[24] Therefore a delipidation step is of importance when dealing with organs such as blood vessels. A third problem of importance can be the demonstration that the extracted glycoproteins are locally synthesized by the differentiated mesenchymal cells of the organ under study. This difficulty can be overcome by the incubation in tissue culture condition of the organ with radioactive precursors. Radioactive glycine, proline, lysine, glucosamine, galactose, and mannose were used with cornea, aorta, sclera, tendon, or skin organ cultures for

---

[18] J. Stoupy, A. Vieillard, M. Desanti, J. P. Borel, and A. Randoux, *C. R. Hebd. Seances Acad. Sci. Ser. D* **275**, 2997 (1972).

[19] A. Randoux, J. Cornillet-Stoupy, M. Desanti, and J. P. Borel, *Biochim. Biophys. Acta* **446**, 77 (1976).

[20] R. Timpl, H. Rohde, P. Gehron Robey, S. I. Rennard, J. M. Foidart, and G. R. Martin, *J. Biol. Chem.* **254**, 9933 (1979).

[21] L. Robert, R. W. Darrell, and B. Robert, in "Chemistry and Molecular Biology of the Intercellular Matrix" (E. A. Balazs, ed.), Vol. 3, p. 1591. Academic Press, New York, 1970.

[22] L. Robert, P. Payrau, Y. Pouliquen, J. Parlebas, and B. Robert, *Nature (London)* **207**, 383 (1965).

[23] M. Claire, B. Jacotot, and L. Robert, *Connect. Tissue Res.* **4**, 61 (1976).

[24] J. M. Bowness, *Atherosclerosis* **31**, 403 (1978).

such demonstrations.[12,25-27] Structural glycoproteins incorporate these precursors quite actively. Even a 4-hr incubation period is sufficient to obtain labeled products for isolation and characterization.

Most of the available methods have been used for the fractionation, purification, and characterization of such tissue glycoproteins: gel filtration, ion-exchange chromatography, gel electrophoresis with PAS-staining and determination of [$^{14}$C]glucosamine incorporation pattern,[28-32] density gradient centrifugation, and isoelectric focusing in sucrose or polyacrylamide gels.[32a,33] A final problem is the determination of the precise localization of the isolated components in the native tissues, normal or pathological. This can be achieved by the indirect immunofluorescent method and by immunoelectron microscopy using peroxidase- or ferritin-labeled monospecific antibodies. As structural glycoproteins were shown to be quite strong antigens, these and other immunochemical procedures can usually be adopted for their study.[4-6,21,22]

In the next section, a few typical examples are given to illustrate the extraction and characterization procedures. More details can be found in a monograph entirely devoted to this topic.[34]

*Extraction Procedures for Structural Glycoproteins*

The first steps are usually disintegration of the tissues and defatting. This can be achieved either by cutting the tissue at cold-room temperature into small pieces (not larger than a few millimeters each side) or alternatively by reducing it to a powder after freezing the tissue in liquid nitrogen or in acetone–Dry Ice mixtures and using a prechilled stainless-steel hammer mill. Defatting can be achieved by the Folch procedure (chloroform–methanol 3:1, v/v) or by successive ethanol and chloroform–methanol (2:1, v/v) extractions. Acetone or alternate acetone-$n$-butanol extractions or ethanol–ether were also used for aorta and ligamentum nuchae.[34]

[25] L. Robert and J. Parlebas, *Bull. Soc. Chim. Biol.* **47**, 1853 (1965).
[26] L. Robert and J. Parlebas, *C. R. Hebd. Seances Acad. Sci.* **261**, 842 (1965).
[27] L. Robert, S. Junqua, A. M. Robert, M. Moczar, and B. Robert, in "Biology of the Fibroblast" (E. Kulonen and J. Pikkarainen, eds.), p. 637. Academic Press, New York, 1973.
[28] M. Moczar, J. Ouzilou, Y. Courtois, and L. Robert, *Protides Biol. Fluids* **22**, 157 (1975).
[29] M. Moczar, J. Ouzilou, Y. Courtois, and L. Robert, *Gerontology* **22**, 461 (1976).
[30] M. Moczar and L. Robert, *Paroi Artérielle* **3**, 105 (1976).
[31] C. Lafuma, M. Moczar, and L. Robert, *Bull. Eur. Physiopathol. Respir.* **15**, 38 (1979).
[32] C. Lafuma, M. Moczar, and L. Robert, in "Le Lavage Broncho-Alvéolaire," Vol. 84, p. 497. INSERM, Paris, 1979.
[32a] C. Lafuma, M. Moczar, and L. Robert, *Biochem. J.*, in press (1982).
[33] M. Moczar, E. Moczar, and L. Robert, *Biochimie* **59**, 141 (1977).
[34] M. Moczar and E. Moczar, *Frontiers Matrix Biol.* **10**, in press.

These first steps are sometimes preceded or followed, or both, by the extraction of proteins supposedly of blood plasma origin, always present in any organ. This is usually done with 0.15 $M$ NaCl or phosphate-buffered saline (PBS). Salt solutions (NaCl, $CaCl_2$, $MgCl_2$) or buffers (Tris-HCl) of higher ionic strength (0.5–2 $M$) are also frequently used for this purpose.

Although all these extractions are carried out at cold-room temperature (+4°), the addition of antiseptic agents and of protease inhibitors (a few drops of toluene, 0.1% $NaN_3$) is also required. Usually three to five extractions of 24 hr, following homogenization (Waring blender, Virtis or Omnimixer type of homogenizers) are used, followed by centrifugation at low temperature (4°) at 5000 to 10,000 $g$.

It has to be remembered that these procedures will solubilize and extract not only plasma proteins, but also diffusible and salt-soluble proteins and glycoproteins, of local tissue origin, as well as part of the proteoglycans. It could be shown that antibodies raised against the insoluble, stroma-bound corneal structural glycoproteins reacted with a soluble component present in the 1 $M$ $CaCl_2$ extract of calf or rabbit corneas.[25] When the corneas were previously incubated with radioactive precursors (such as $^{14}C$-labeled amino acids or sugars), the immune precipitate contained radioactive material.[6,7,25,26] For this reason the "structural glycoprotein" designation should not be restricted to salt-insoluble tissue glycoproteins. A similar situation prevails with the most-studied structural glycoprotein, fibronectin, which is known to be present both in soluble and insoluble form in plasma, amniotic fluid, and solid tissues.[35]

*Extraction of Glycoproteins Present in the Salt-Insoluble Stroma*

The tissue residue that remains after the above-mentioned salt extractions will be designated as the "insoluble stroma." This residue can be treated in several ways. In the early procedures, this insoluble stroma was submitted to a partial hydrolysis in 0.1–0.5 $M$ trichloroacetic acid at 90° for 5–30 min.[1–8] This method was shown to "solubilize" rather selectively and extensively the insoluble collagen fibers[2,6,9,36–40] Cold dilute NaOH extraction was also used.[41] These methods were superseded during the 1970s by the use of highly purified bacterial collagenase. Several methods

---

[35] A. Vaheri, K. Alitalo, K. Hedman, M. Kurkinen, O. Saksela, and T. Vartio, *in* "Biochimie des Tissus Conjonctifs Normaux et Pathologiques" (A. M. Robert and L. Robert, eds.), Vol. 2, p. 249. C.N.R.S., Paris, 1980.
[36] M. Moczar and L. Robert, *Atherosclerosis* **11**, 7 (1970).
[37] M. Moczar, E. Moczar, and L. Robert, *Atherosclerosis* **12**, 31 (1970).
[38] S. M. Fitch, M. L. R. Harkness, and R. D. Harkness, *Nature (London)* **176**, 163 (1955).
[39] R. L. Langner and R. A. Neumann, *Anal. Biochem.* **48**, 73 (1972).
[40] J. Laszt, J. B. Baer, and C. Milson, *Connect. Tissue Res.* **4**, 149 (1976).
[41] M. J. Barnes and S. M. Partridge, *Biochem. J.* **109**, 883 (1968).

have been recommended for the final purification of the best available commercial products and for the control of their purity.[42,43] This is often done by Peterkovsky and Digelman's procedure[42] using radioactive tryptophan-labeled chick embryo proteins. Usually the insoluble stroma is suspended in a 0.05 $M$ Tris-HCl buffer, pH 7.2, containing 0.01 $M$ $CaCl_2$; purified collagenase is added at a concentration of 2–10 $\mu$g/ml for 24–48 hr at room temperature (20–22°) or at 37° under toluene, $NaN_3$, or antibiotics as bacteriostatic agents. This treatment may be repeated several times if necessary (e.g., in collagen-rich organs, such as aorta, lung, or skin).

It should be mentioned that none of the above methods is entirely satisfactory. Hot TCA was shown to hydrolyze peptide bonds. Purified collagenase was also shown to attack immunoglobulin light chains possessing the required amino acid sequence[44] as well as possibly other noncollagenous protein components. Extraction of collagen with collagenase is never complete and was sometimes used in conjunction with the TCA treatment.[45]

For these reasons several authors adopted the use of dissociating, hydrogen bond, and hydrophobic interaction disrupting solvents prior to (or instead of) the collagenase treatment.

Structural glycoproteins were extracted from rabbit skin with 1% sodium dodecyl sulfate solution[18,19] and from aorta[33] and lung[31,32] by 5 $M$ guanidine-HCl prior to collagenase treatment.[32a] It could be shown with these two tissues that different families of glycoproteins are extracted with 5 $M$ guanidinium chloride prior to and following collagenase treatment.[32a]

An important precaution during all these treatments and extractions was shown to be the addition of inhibitors of tissue proteases. The most frequently used "cocktail" of inhibitors contained 0.1 $M$ EDTA, phenylmethylsulfonyl fluoride (50 $\mu$g/ml), $N$-ethylmaleimide or $p$-hydroxymercuribenzoate (50 $\mu$g/ml), and chloromethylketones specific for some tissue proteases.[48]

Most authors also used reducing agents such as 0.1 $M$ 2-mercaptoethanol[8,36] or 0.01–0.05 $M$ dithiothreitol[46] together with 8 $M$ urea or 5–6 $M$ guanidine-HCl for the extraction of glycoproteins remaining after collagenase or TCA treatment.[47]

[42] B. Perkofsky and S. Udenfriend, *J. Biol. Chem.* **238**, 3966 (1963).
[43] V. Lee-Own and J. C. Anderson, *Biochem. Soc. Trans.* **3**, 145 (1975).
[44] B. Vallet, C. Tannelle, and M. A. Coletti-Previero, *Pathol. Biol.* **25**, 369 (1977).
[45] M. A. Paz, D. A. Keith, H. P. Traverso, and P. M. Gallop, *Biochemistry* **15**, 4912 (1976).
[46] R. Ross and P. Bornstein, *J. Cell Biol.* **40**, 366 (1969).
[47] B. Robert, M. Szigeti, J. C. Derouette, L. Robert, H. Buissou, and M. T. Fabre, *Eur. J. Biochem.* **21**, 507 (1971).
[48] The concentrations in parentheses are only indicative and should be adapted to the protease activity of the tissue studied.

*Improving Solubility and Extractability by Reversible Chemical Modifications*

Several authors used reversible chemical modifications of some reactive sites in proteins in order to enhance solubility in aqueous buffers. Free amino groups were made to react with maleic anhydride, citraconic anhydride, tetrafluorosuccinic anhydride, or dicetene.[34] Squid skin structural glycoproteins could be extracted after maleylation.[48]

*Purification and Fractionation of the Tissue Extracts*

The first step is often the elimination of the extracting agents, urea, guanidine-HCl, or SDS from the glycoprotein solution. This is a delicate problem because of the easy and rapid reprecipitation of most structural glycoproteins from aqueous solutions. Dialysis against running tap water and distilled water in the cold is the most convenient procedure, followed by the further fractionation of the glycoprotein mixture by gel filtration or other methods.[34] Dialysis against SDS-containing buffers or just reducing the concentration of urea and guanidine by dilution or dialysis against, for instance, a $0.1-0.2\ M$ Tris-buffered (pH 7.5–8.0) $2\ M$ urea or guanidine-HCl solution may also be used. The elimination of detergents, such as SDS, usually cannot be achieved by dialysis alone. SDS was shown to be efficiently removed by passing the extract over a Dowex-1 column.[49] Dialysis against distilled water containing a suspension of Dowex-1 also can be used.

It should be mentioned that some tissue glycoproteins are extractable at low ionic strength or even in distilled water, as was shown to be the case for some of the amyloid proteins and for a family of marine sponge-derived glycoproteins (see below).[50]

The combined extracts can be concentrated by pressure dialysis or by ultrafiltration under $N_2$ in order to achieve the desired concentration for further fraction. Lyophilization is not recommended because of the low solubility of the dried preparations.

*Fractionation and Separation Techniques Used*

Ample evidence is now available to show that structural glycoproteins extracted as described above are heterogeneous and contain several components with widely variable molecular weights (see Table I), even when all possible precautions were taken to avoid degradation by tissue proteases. This was also shown by the distinctly different specific radioactivities found in glycoproteins isolated from tissues (lung,[31,32,32a]

---

[48] M. Moczar and E. Moczar, *Comp. Biochem. Physiol. B* **53B**, 255 (1976).
[49] K. Weber and D. J. Kuter, *J. Biol. Chem.* **246**, 4504 (1971).
[50] S. Junqua, M. Lemmonier, and L. Robert, *Comp. Biochem. Physiol.* **69B**, 445 (1981).

aorta,[28-30,33]) incubated prior to extraction with radioactive precursors. It is therefore reasonable to assume that any tissue contains several distinct structural glycoproteins. Lung (baboon, rat) was shown to contain at least four or five distinct *in vitro* labeled [$^{14}$C]glucosamine glycoproteins as shown by PAGE in SDS–urea with molecular weights in the range of 34,000, 75,000, 140,000, and 400,000.[31,32,51] Fibronectin could be detected in the extracts, but appeared to be a minor component.

The 400-kilodalton glycoprotein of lung tissue could not be reduced in the presence of guanidine and dithiothreitol to smaller subunits.[51] Aorta (rabbit, pig, calf) was shown to contain glycoproteins in the 5 $M$ guanidine extracts obtained before collagenase treatment as well as (apparently different) glycoproteins extractable from the residue left after collagenase treatment by 5 $M$ guanidine–0.1 $M$ 2-mercaptoethanol or 0.05 $M$ dithiothreitol.[14,32a,33,34,52] Both guanidine extracts (obtained before and after collagenase treatment) are heterogeneous and contain several glycoproteins that can be separated on SDS–PAGE or by isoelectric focusing.[32a,33,52,52a] An example for the preparation of a 35-kilodalton glycoprotein from the guanidine extract obtained prior to collagenase will be given below. As proteolytic or glycolytic enzymes may still be present at this stage of the purification, it is advisable to continue adding inhibitors to the extracts and to maintain a low temperature during further fractionation. This is usually achieved by gel filtration, ion-exchange chromatography, density gradient centrifugation in $CsCl_2$ or sucrose gradients, isoelectric focusing in sucrose solution or in polyacrylamide gels. Affinity chromatography on lectin columns is also used.

The solvent systems used for these procedures usually contain urea, guanidine, SDS, or EDTA as well as buffer (Tris-HCl or other). The examples of isolation procedures given in the next section contain more detailed information on this topic.

Examples of Extraction and Purification Procedures

*Isolation of Structural Glycoproteins from Marine Sponges*[50,53,54]

This example is given because of the simplicity of the procedure, which avoids the use of any rough treatment.[50,53,54] The thoroughly washed

---

[51] C. Lafuma, M. Moczar, and L. Robert, *Bull. Physiopathol. Respir.* **16** (Suppl.), 91 (1980).

[52] M. Moczar, B. Phan Dinh Tuy, and E. Moczar, *in* "Glycoconjugates" (R. Schauer, P. Boer, E. Buddecke, M. F. Kramer, and J. G. Vliegenhart, and H. Wiegandt, eds.), p. 557. Thieme, Stuttgart, 1979.

[52a] M. Moczar, B. Phan Dinh Tuy, E. Moczar, and L. Robert, in preparation.

[53] S. Junqua, J. Fayolle, and L. Robert, *Comp. Biochem. Physiol. B* **50B**, 305 (1975).

[54] S. Junqua and L. Robert, *in* "Glycoconjugate Research" (J. D. Gregory and R. Jeanloz, eds.), Vol. 1, p. 177. Academic Press, New York, 1979.

sponge tissue (*Spongia officinalis*) cut in small pieces is homogenized in a Sorvall Omnimixer at 4° in a solution containing 1% SDS, 0.01 $M$ EDTA adjusted to pH 8.0 with 1 $N$ NaOH, and 0.01 $M$ Na tetrathionate and 0.02% NaN$_3$, 1 m$M$ DFP. After low speed sedimentation (1000 $g$) of the debris (spongine fibers, cell residues), the supernatant is acidified by adding acetic acid to a final concentration of 4%. The heavy precipitate containing collagen (the "intercellular" type[55]) and associated glycoproteins is sedimented at low speed (1000 $g$) and resuspended in the above extraction buffer, the pH being readjusted to 8.0 with 1 $N$ NaOH. Methanol was added to a final concentration of 70% (v/v). The centrifuged (1000 $g$) sediment is again resuspended in the extraction buffer, and the collagen fibrils are eliminated by ultracentrifugation at 40,000 $g$ for 30 min. The supernatant contains the glycoproteins. It was concentrated by pressure-filtration on Amicon PM-10 Diaflo membranes and further fractionated by gel filtration after reduction with an excess of 2-mercaptoethanol and carboxamidomethylation with a 2 $M$ excess (with respect to mercaptoethanol) of $N$-ethylmaleimide.

The glycoprotein solution was applied to a Sepharose 4B column using 0.01 $M$ EDTA and 0.1% SDS at pH 8.0 as an eluent. The SDS was removed by an anion exchange resin (Dowex AGIX2, 200–400 mesh).

Three major fractions could be separated by this procedure, the first minor (excluded) fraction containing some hydroxyproline. The second, major, peak contained most glycoconjugates, and a third, minor, peak contained low molecular weight substances. The second major peak was subfractionated on lectin columns (Con A-Sepharose), yielding again three distinct fractions; a first nonretained fraction (fraction A, 65% of total protein applied) contained sulfated heteropolysaccharide protein complexes as well as the fibronectin-like protein of sponges[56,56a]; a second peak (fraction B, 29% of total proteins) eluted with 0.25 $M$ $\alpha$-methylglucoside or $\alpha$-methylmannoside; and a third fraction (fraction C, 6% of total proteins) eluted with borate buffer (0.02 $M$ sodium borate, pH 8.0, containing 1 m$M$ CaCl$_2$, 1 m$M$ MgCl$_2$, 1 m$M$ MnCl$_2$, and 0.001% phenylmercuric acetate).

The first fraction (A) could be further subfractionated on WGA Sepharose (fraction AI, ~57% of proteins, not retained, and fraction AII retained and eluted with 0.1 $M$ $N$-acetyl-D-glucosamine). The second Con

[55] R. Garrone, in "Frontiers of Matrix Biology" (L. Robert, ed.), Vol. 5. Karger, Basel, 1978.
[56] J. Labat-Robert, M. Pavans de Ceccatty, L. Robert, C. Auger, C. Lethias, and R. Garrone, in "Glycoconjugates" (R. Schauer, P. Boer, E. Buddecke, M. F. Kramer, J. F. G. Vliegenthart, and H. Wiegandt, eds.), p. 431. Thieme, Stuttgart, 1979.
[56a] J. Labat-Robert, L. Robert, C. Auger, and R. Garrone, *Proc. Natl. Acad. Sci. U.S.A.* **78**, 6261 (1981).

A-fraction (B) was entirely retained on WGA-Sepharose and desorbed as a single peak. The third Con A-fraction (C) yielded a nonretained (CI, 76%) and a WGA-retained fraction (CII). All subfractions exhibited a rather similar and typical amino acid composition (see Table II) although, as expected, their carbohydrate composition was different.[50,54] Molecular weights determined by SDS-PAGE gave two bands in fractions B and AII of 16 and 32 kilodaltons and one band at 21 kilodaltons in fraction C. On crossed immunoelectrophoresis and immunodiffusion, AI, AII, B, and C fractions showed a continuous line (using anti-fraction B immunserum). This indicates that all the fractions share at least one major, common antigenic determinant. Using the indirect Coon's technique, these glycoproteins appeared to be localized pericellularly and/or on sponge cell membranes. Fluorescent lectins (Con-A, WGA) showed a similar pattern of localization.[57]

*Extraction of Microfibrillar Glycoproteins of Elastic Tissues*

Microfibrillar glycoproteins have been extracted from human and porcine aorta,[36,37,47] bovine ligamentum nuchae,[46] and from cell cultures obtained from these tissues.[58-60]

Two glycoproteins ($M_r$ 150,000 and 300,000) were recently purified from fibroblast cultures obtained from bovine ligamentum nuchae explants by Sear *et al.*[59] One of them ($M_r$ 150,000) was shown to contain collagen-like sequences.[59] Human skin fibroblasts were also shown by these authors to synthesize and release a major glycoprotein.[58] Confluent cultures were incubated with labeled amino acids and/or L-[1-$^3$H]fucose (10 $\mu$Ci/ml) for 24–48 hr. The culture medium was dialyzed and freeze-dried. Alternatively, the glycoproteins were precipitated (in the presence of protease inhibitors) at 4° with ammonium sulfate at 75% saturation. This precipitate when reduced with dithiothreitol in SDS and alkylated with iodoacetamide, dialyzed against 0.02 $M$ Tris-HCl buffer (pH 7.4) containing 0.1% SDS and 0.03% NaN$_3$ gave on gel filtration (BioGel A-5m) three major and three minor peaks in the molecular weight range of 130,000, 85,000, 45,000, 30,000, and 15,000. SDS-polyacrylamide gel electrophoresis of the nonreduced culture medium showed several glycoprotein components with molecular weights in the range of 260,000 to 75,000. The major component had an apparent molecular weight of 230,000. The second most prominent component had a molecular weight of approxi-

[57] S. Junqua, R. Garrone, and L. Robert, in preparation.
[58] C. H. J. Sear, M. E. Grant, and D. S. Jackson, *Biochem. J.* **168**, 91 (1977).
[59] C. H. J. Sear, M. A. Kewley, C. J. P. Jones, M. E. Grant, and D. S. Jackson, *Biochem. J.* **170**, 715 (1978).
[60] L. W. Muir, P. Bornstein, and R. Ross, *Eur. J. Biochem.* **64**, 105 (1976).

mately 140,000. It was shown conclusively that the $M_r$ 230,000 component was not collagenous in nature and was distinctly different from fibronectin.

Muir et al.[60] isolated from monkey aorta smooth muscle cell cultures an $M_r$ 270,000 glycoprotein by a method derived from that of Ross and Bornstein.[46] The culture medium (preincubated with 2 μCi/ml [$^3$H]cysteine) was removed and dialyzed against several changes of 0.05 $M$ Tris-HCl, pH 7.5, containing 25 m$M$ EDTA, 1 m$M$ phenylmethylsulfonyl fluoride, and 1 m$M$ $N$-ethylmaleimide, then against distilled water, and lyophilized. The cells (pellet) were extracted separately by a slight modification of the previously described procedure[46] using 5 $M$ guanidine–1 $M$ Tris-HCl, 10 m$M$ EDTA, 2 m$M$ phenylmethylsulfonyl fluoride, 1 m$M$ $N$-ethylmaleimide at pH 8.5. A second extraction was carried out with the same solvent in the presence of 0.25 m$M$ dithiothreitol. The major glycoprotein band in all three extracts (culture supernatant, guanidine-pellet extract, and reduced guanidine pellet extract) was a 280 kilodalton band (or 220 kilodalton on a methylene bisacrylamide cross-linked gel). This band represented about 25% of the incorporated cysteine in the culture medium, was unaffected by collagenase, and contained no hydroxyproline or hydroxylysine. Further purification of this fraction was achieved by DEAE-cellulose chromatography of the carboxymethylated derivative.

*Extraction of Microfibrillar Glycoprotein from Ligamentum Nuchae Exhibiting Lysyl Oxidase Activity* [61-63]

Bovine ligamentum nuchae was defatted in chloroform–methanol, extracted with 1% NaCl for 24 hr at 4°, and lyophilized; 100 g of the lyophilized material was extracted in 1000 ml of 5 $M$ guanidine HCl–0.1 $M$ Tris at pH 7.4 with continuous stirring for 24 hr at 4°. This extraction was repeated for a total of six 24-hr periods, the residue was suspended in 500 ml of 5 $M$ guanidinium-HCl–0.1 $M$ Tris (pH 7.4) containing 0.4% EDTA and placed under $N_2$. Dithiothreitol (DTT) was introduced to a final concentration of 0.05 $M$, and the suspension was stirred at 37° for 48 hr, then centrifuged at 23,000 $g$ for 30 min, still under $N_2$. Iodoacetic acid was added to the supernatant in a fourfold molar excess over DTT, and the pH was adjusted to 8.6; the preparation was stirred for 30 min. Then a fivefold molar excess (over iodoacetate) of 2-mercaptoethanol was added and dialyzed against distilled water for 48 hr, when a white precipitate formed.

[61] E. D. Harris, J. K. Rayson, and J. E. DeGroot, *Adv. Exp. Med. Biol.* **79**, 543 (1976).
[62] A. Serafini-Fracassini, J. M. Field, and C. Armitt, *Biochem. Biophys. Res. Commun.* **65**, 1146 (1975).
[63] A. Serafini-Fracassini, G. Ventrella, R. Griffiths, and J. Hinnie, *Eur. Symp. Connect. Tissue Res., 7th*, p. 296 (1980).

Three volumes of ethanol were added to the retentate, and the precipitate was collected. Collagen was removed from the precipitate by collagenase treatment. Further purification was achieved by preparative gel electrophoresis.[63] Preparations obtained by this procedure were shown to contain a 35 kilodalton glycoprotein endowed with lysyl oxidase activity. The amino acid and carbohydrate composition and solubility properties of other lysyl oxidase preparations were shown to be similar to that found for some other aortic structural glycoproteins. This enzyme may well be part of the glycoprotein forming the microfibrils of elastic tissue.[64,65]

A somewhat similar procedure of Sear et al.[59] was used for the isolation of the microfibrillar glycoproteins from fibroblast cultures obtained from bovine ligamentum nuchae explants. Two components were obtained, and the 300 kilodalton protein had the composition and solubility properties of structural glycoproteins. The other, at 150 kilodaltons, contains some hydroxyproline and was shown to contain collagenase-sensitive sequences.

*Isolation of Glycoproteins from Pig, Rabbit, and Calf Aortic Media*[28-30,33,52]

Several distinct glycoproteins can be isolated from defatted and salt-extracted aortic media.[36,37,41,47,64,65] The usual procedure is to carry out two to five 8 $M$ urea or 5 $M$ guanidine-HCl extractions followed by collagenase treatment and by two to five extractions with 8 $M$ urea–0.1 $M$ 2-mercaptoethanol or 5 $M$ guanidinium chloride–0.05 $M$ dithiothreitol. The final residue contains elastin, still contaminated with some glycoproteins, which can be completely removed by trypsin treatment,[66] by EDTA treatment,[67] or by heating in 0.1 $N$ NaOH for 45 min.[47,68]

Molecular weights and incorporation patterns (obtained by preincubating the aortic explants in tissue culture conditions with [$^{14}$C]glucosamine or [$^{14}$C]mannose) on SDS-PAGE indicate the presence of three major components in the guanidine extract obtained before collagenase treatment and of two major components in the guanidine–DTT extract obtained after collagenase treatment. In the following, we shall describe the isolation of a 35-kilodalton glycoprotein from the guanidine extract of rabbit aorta before collagenase treatment.[52,52a]

The fragments of aortic media delipidated with $CHCl_3$–methanol (2:1) (1:10, w/v) are homogenized and extracted with 0.2 m$M$ $CaCl_2$–0.5 m$M$ DTT–0.5 m$M$ ATP (1:5, w/v) in order to remove actin.[69] Two extractions

[64] K. G. McCullagh, S. Derouette, and L. Robert, *J. Exp. Mol. Pathol.* **18**, 202 (1973).
[65] A. Kadar, B. Robert, and L. Robert, *Pathol. Biol.* **21**, 80 (1973).
[66] R. John and J. Thomas, *Biochem. J.* **127**, 261 (1972).
[67] F. W. Keeley and S. M. Partridge, *Atherosclerosis* **19**, 287 (1974).
[68] L. Robert and W. Hornebeck, in "The Methodology of Connective Tissue Research" (D. A. Hall, ed.), p. 81. Joynson-Bruvvers, Oxford, 1976.
[69] P. M. Sender, *FEBS Lett.* **17**, 106 (1971).

of 6 and 12 hr are carried out at 4°. The glycoproteins are further solubilized and reduced with 5 $M$ guanidinium chloride–0.1% EDTA–0.05 M DTT (1:5, w/v) at pH 7.4 and room temperature with 2 m$M$ phenylmethylsulfonyl fluoride added as protease inhibitor. Two extractions of 16 hr following homogenization in a Waring blender are carried out with continuous mechanical shaking. The supernatants were centrifuged at 30,000 $g$ for 30 min.

The components of the guanidinium chloride extract were $S$-carboxamidomethylated according to Haeberli et al.[70] The mixture was concentrated under $N_2$ using a Diaflo PM-10 membrane, and fractionated on a Sepharose 4B column with 6 $M$ urea–0.2% SDS, 20 m$M$ EDTA, pH 7.4 buffer. The peak material (eluted in the range of 30–40 kilodaltons was further fractionated by consecutive isoelectric focusing in sucrose gradient in 6 $M$ urea in the pH range 2–11 and 4–6. Several glycoproteins can be separated with this technique; the major fraction has an apparent molecular weight of 35,000.

This 35-kilodalton glycoprotein isolated from pig, calf, and rabbit aorta has an amino acid composition similar to that described for structural glycoproteins (Table II), and was devoid of hydroxyproline and of 3-methylhistidine. The glycan part (ca. 2.5%) contains $N$-acetylglucosamine, galactose, mannose, and sialic acid. The trypsinates of pig, calf, and rabbit glycoproteins gave similar patterns on fingerprinting. This 35-kilodalton glycoprotein was shown to associate strongly with soluble collagen forming aggregates during isoelectric focusing.[33]

*Note added in proof.* Recently several reports appeared on newly isolated and characterized structural glycoproteins, such as chondronectin,[71] hyaluronectin,[72,73] and entactin,[74,75] which is a basement membrane associated, sulfated glycoprotein. As their names indicate, the authors assumed or demonstrated an interaction between these glycoproteins, cells, and matrix components. It was also shown that in embryonic chick cornea and sclera, fibronectin is present but replaced during morphogenesis by other glycoprotein(s). The name keratonectine was proposed for the glycoprotein appearing in adult chick cornea and replacing fibronectin.[76] This glycoprotein may well be identical with the one isolated from calf cornea using the TCA-urea procedure and which was also shown to be a sulfated glycoprotein.[2–9]

[70] A. Haeberli, J. Bilstad, H. Edelhoch, and E. Rall, *J. Biol. Chem.* **25**, 7294 (1975).
[71] A. T. Hewitt, H. K. Kleinmann, J. P. Pennypacker, and G. R. Martin, *Proc. Natl. Acad. Sci. U.S.A.* **77**, 385 (1980).
[72] B. Delpech, *C.R. Hebd. Seances Acad. Sci.* **290**, 1067 (1980).
[73] B. Delpech and C. Halavent, *J. Neurochem.* **36**, 855 (1981).
[74] B. Carlin, R. Jaffe, B. L. Bender, and A. E. Chung, *J. Biol. Chem.* **256**, 5209 (1981).
[75] B. L. Bender, R. Jaffe, B. Carlin, and A. E. Chung, *Am. J. Pathol.* **103**, 419 (1981).
[76] J. Labat-Robert, M. Menasche, G. Godeau, and L. Robert, *Proc. Int. Soc. Eye Res.* **1**, 97 (1980).

# Author Index

Numbers in parentheses are reference numbers and indicate that an author's work is referred to although the name is not cited in the text. Bold face numbers indicate references in tables.

## A

Aaron, B. B., 687, 691(42), 714(42)
Aase, A., 22, 24(100), 99, 346(4), 347, 468(50, 469)
Abbott, M. T., 246, 247(4)
Abe, S., 451, 493
Abercombie, M., 544
Abraham, P. A., 566, 567, 568, 630, 672, 717, 721(11), 722
Abraham, R. J., 707
Abrash, L., 579
Ackers, G. K., 420, 423(18)
Adair, G. S., 572, 591, 592(23), 594(23), 636, 673, 744
Adams, E., 65, 88, 89, 93(4), 94, 95(b), 96, 246, 247(5), 284, 372(4), 373, 374, 385, 387(4, 21), 391, 393(37)
Adams, S. A., 424
Adams, S. L., 231
Adamson, E. D., 490(**38**), 492
Adelmann, B. C., 474, 475, 476, 479(27), 484(27), 489, 490(**16**), 491, 498, 503, 806
Ademany, A. M., 632
Adriopolous, N. A., 498
Afanasev, V. A., 705
Agnello, V., 322
Aiken, B. M., 600
Ainsworth, L., 403, 404(12), 410(12)
Airhardt, J., 393
Åkerblom, H., 290(128), 291
Akabori, S., 453
Akiba, K., 369
Alarcon, E. J., 774, 783(32)
Albert, A., 328, 329
Albert, J., 486, 489(63), 490(**24**), 492, 493(63), 494(63)
Alburn, H. E., 454
Alexa, M., 572, 582(3), 559, 591, 622, 623, 728, 729(38), 764
Alexander, R. J., 323
Alitalo, K., 99, 100(12, 16), 105(16), 127(12), 252, 262(46), 489, 490(**32, 50**), 492, 493(87), 495(87, 93), 496(104), 497(87, 104), 831(10), 832, 833, 834(13, 14), 837(10, 13, 14), 838(10, 13, 14), 844
Allan, R. E., 17, 29(82)
Allerhand, A., 175
Alper, R., 31, 226, 831, 835(1)
Altman, L. K., 559
Alvarez, V. L., 674, 712(9), 749
Amrani, D. L., 840
Ananthanarayanan, S., 196, 217(12)
Anderson, B., 771, 794
Anderson, J. C., 456, 461(20), 465(20), 839, 840(10, 11), 845
Anderson, R. G. W., 544
Andersson, K. K., 598, 600(53)
Andres, V., Jr., 330
Anfinsen, C. B., 265, 272(71), 295(71)
Angermann, K., 17, 106, 107(35), 109(35), 117, 490(**44**), 492
Anglister, L., 325, 326(7), 486
Angus, J. C., 204
Ansley, H. R., 600
Antonopoulos, C. A., 776, 787(53), 788
Anttinen, H., 274, 276(85), 277(85), 286, 287, 290, 291(113), 294(126), 295(131), 296(129), 297(126), 298(126, 132), 300(132), 302(124, 129), 303(102, 140), 304(140)
Anwar, R. A., 563, 576, 588, 606, 607(1), 608, 614(2, 3), 615(2, 3), 633, 650
Anwer, U., 348
Ardelt, W., 596
Armitt, C., 850
Armstrong, R., 196(15), 197
Aro, H., 264, 267(67), 268(67)
Aronson, R. B., 274
Asboe-Hansen, G., 402, 403
Ash, J. F., 493, 827
Ashe, B. M., 590, 603
Ashford, D., 373
Ashhurst, D., 65, 95(1)

Askenasi, R., 399, 408, 409
Assad, R., 248, 249, 252(9), 284(14)
Assimeh, S. N., 320, 323, 324(9)
Atkins, P. H., 347, 348(9)
Atlas, D., 588
Augener, W., 320
Auger, C., 848
Aumailley, M., 493, 497
Axelsson, I., 772, 776, 787(53), 788(53), 795(62), 796(22)
Axén, R., 477
Aviv, H., 733
Ayers, S. E., 490(**38**), 492

## B

Bacetti, B., 80
Bächinger, H. P., 8, 12(35), 15(35), 18(35), 25, 30(117), 42, 44(36), 49, 225(14), 226, 834, 835(17), 836, 837(23)
Bachinger, M. P., 196(14), 197, 215(14)
Bachra, B. N. 317
Baenziger, J. U., 355
Baer, J. B., 844
Bahu, R., 491(**57**), 492
Baig, K. M., 606, 614(3), 615(3), 650
Bailey, A. J., 7, 11, 12(23), 13(51), 18(23), 169, 226, 361, 365, 369, 370, 371, 475, 479, 490(**7, 31, 34, 49**), 491, 492, 498(48)
Baird, W. A., 666
Bakala, H., 596
Baker, J. R., 769, 771, 774, 776, 783(33), 787(1, 37), 793(37), 795(1), 799, 800
Balashova, T. A., 705
Balduini, C., 776
Baldwin, R. W., 323
Balian, G., 12, 101, 120(34), 121, 804, 805
Balleisen, L., 478, 487(47), 489, 490(**16**), 491, 493(47), 498(47)
Balny, C., 598, 600(53)
Balo, J., 596
Bamburg, J. R., 504, 505(17)
Banga, I., 596
Bańkowski, E., 244
Barber, A. J., 294, 299, 302(135, 139)
Barbieri, O., 807
Bard, J., 499, 502(1), 504, 505(12), 516, 535, 537, 546
Bardos, P., 835

Barfield, M., 707
Barnes, G. R. C., 154, 163(74), 167(74)
Barnes, M. J., 157, 167(78), 844, 851(41)
Barnett, E. V., 498
Barnett, P., 325, 326(9), 337(9), 338(9), 339(9)
Barrach, H.-J., 17, 106, 107(35), 109(35), 482, 484(58), 490(**44**), 492, 788, 789, 790(101), 799, 837, 838(29)
Barrett, A. J., 37, 426, 434, 589, 604, 795
Barrineau, L. L., 569, 570(70), 732, 733(3)
Bartlett, M. W., 145, 149(59)
Barzin, S., 77
Bashey, R. I., 167
Batchelder, L. S., 177, 181(25), 183(25)
Bateman, J., 471
Baudner, S., 488
Bauer, E. A., 16, 24(78), 424, 443, 444, 445(36), 446(40), 447(36), 450(40)
Baugh, R. J., 589, 590
Baum, B. J., 396
Baumgartner, S., 759
Baumstark, J. S., 604, 605(73)
Baurain, R., 582, 585
Beale, R. J., 797
Bear, R., 152
Beard, H. K., 472, 473(3), 475
Beaudet, A. L., 227
Becker, C. G., 620, 629(6)
Becker, E. D., 174
Becker, U., 13, 17, 22(80), 23(80), 24(80), 30, 31(145), 133, 369, 370(13), 468, 472, 473(10), 474(10), 475, 477(16, 22), 479(22), 480(10, 22, 25), 484(22), 485, 486(22, 25), 488(16, 22), 489(22), 490(**6, 26**), 491, 492, 497(10), 498, 649, 650(32)
Beers, W. J., 777
Behrens, N. H., 304, 348
Beighton, P., 169
Beil, W., 475, 476, 477, 478(44), 479(44), 483(44, 51), 484(44, 51), 485, 486
Belfrage, G., 451
Bell, E., 499, 535, 554
Bellamy, G., 3
Bell-Booth, P. G., 154
Benazzo, F., 136, 147(44), 164(44), 165(44)
Benda, P., 325
Benditt, E. P., 563
Benditt, M., 167
Benedetti, E., 183

Benjamini, E., 473, 474(17), 485
Bennett, M. H., 502, 535, 540, 541, 542
Benoit, H., 199, 200, 201(19)
Benson, J. R., 388
Bensusan, H. B., 512
Bentz, H., 8, 12(35), 15(35), 18(35), 49, 95(g), 96, 225(14), 226
Benveniste, K., 220, 221(3), 244
Benya, P., 100
Ben-Ze'er, A., 524
Bere, E. W., 837, 838(30)
Berenson, G. S. 783
Berg, R., 479, 482(50), 485, 497(50), 757, 821, 837
Berg, R. A., 22, 27, 137, 246, 247(2), 248(2), 249(2, 26), 250(2), 251(2, 29), 252(18), 254(18), 260(2, 26), 262(29), 265(26), 267(26), 268(2, 26), 269(73), 270(2), 271(2, 29, 61), 273(68, 79), 274(79), 280(2, 18, 26, 56), 374, 387, 392(30, 31), 393(31), 394(11), 395(30, 31), 396, 397(42), 468, 475, 485
Berglund, N., 34, 666, 667(16), 668(16)
Bergofsky, E. H., 759
Berman, J. J., 493
Berne, B. J., 205, 209
Bernengo, J.-C., 200, 201, 217
Bernfield, M. R., 500
Bernstein, L., 251
Berthet-Colominas, C., 147, 148, 151, 152, 153(65), 157(73), 158(77)
Berube, L., 365
Betz, W., 325
Bezkorovainy, A., 359
Bhatnagar, R. S., 281
Bhoyroo, V. D., 78, 79, 95(18), 349, 352(13)
Bialecki, M., 505
Bieger, W., 600
Bielefeld, D. R., 593, 747
Biempica, L., 490(**29**), 491(**29**), 492
Bienkowski, R. S., 396
Bieth, J. G., 598, 600(53), 604
Bignon, J., 589
Biles, G., 500, 551
Bilstrad, J., 852
Bing, D. H., 323
Bird, D. W., 563
Bird, R., 196, 197
Birdwell, C. R., 500
Birken, S., 712

Birkner, M. L., 747
Bisbee, W. C., 403
Black, C. M., 169
Blaes, N., 589
Blanck, T. J. J., 470
Blatt, B. A., 485
Bloch, F., 175
Bloch, K. J., 443
Bloom, M., 176
Blout, E. R., 128, 596
Blumenfeld, O. O., 562, 579, 580, 582(30), 584(30), 585(31, 32), 625, 632
Blumenkrantz, N., 251, 257(35), 274, 402, 403
Blumenstock, F., 821
Bock, E., 335, 336(37)
Bode, J. C., 490(**25**), 491(**25**), 492
Boedtker, H., 97, 128, 189, 191(2), 203, 209
Boesman, M., 321, 322(15), 323(15)
Bohlen, P., 388
Bolton, A. E., 753, 754(22)
Bon, S., 325, 326(5)
Bonner, W. M., 102, 224, 310, 495, 727, 737, 742
Bonnet, F., 774, 787
Booth, B. A., 99, 100(14), 104, 114(30), 120(30), 124(30)
Booyse, F. M., 589
Bordas, J., 149, 161
Borel, J. P., 379, 385(18), 388, 389(36), 397(18), 840(19), 841(18), 842, 845(18, 19)
Born, G. V. R., 509
Bornstein, P., 3, 5, 7(34), 8(5), 12(10, 24, 34), 13, 16, 18(24, 34), 20, 21(10, 24), 22, 23(99), 24(10, 99), 26(79), 27, 28, 30(97), 33, 52, 55, 56(53), 57(53), 58(1), 96, 97, 98, 99, 100(13), 101, 105(7), 108, 111, 113(7), 117(7, 43), 118(7, 43), 120(17, 24), 121, 122(13, 48), 123(1, 48, 52), 124(7), 128, 129(8), 131, 154(8), 160(17), 168, 169(109), 222, 225(5), 305, 306, 312, 332, 373, 392(7), 394(7, 9, 10), 412, 417, 468, 473, 475, 485, 493, 494(101), 495(101), 559, 561(4), 572, 592, 804, 805, 841(60), 845, 849(46), 850(60)
Bosmann, H. B., 287, 294(115), 299(114, 136), 302(114, 115, 136)
Boudier, C., 598, 600(53)
Bourdillon, M. C., 589

Bourne, F. J., 475, 479, 490(**7**), 491, 498(48)
Bovey, F. A., 707
Bowling, D. I., 202
Bowman, A., 606
Bowman, P., 499, 535
Bowness, M., 841, 842, 844(24)
Brackenbury, R., 514, 526(7)
Bradley, K. H., 387, 392(31), 393(31), 395(31)
Bradwell, A. R., 605
Braginski, J. E., 831, 834(8), 835(8), 837(8)
Branson, R. E., 387, 530
Brass, L. F., 512
Bray, B. A., 771
Brayden, J. E., 393
Brechemier, D., 589
Breen, M., 772, 775(19), 798(19)
Breitkreutz, D., 413
Brendel, K., 35
Brenkowski, R. S., 388, 397(33)
Brenner, S., 90
Brent, R. L., 26, 30(120)
Bressan, G. M., 569, 716(6), 717, 723(7), 727(7)
Breul, S. D., 387, 392(31), 393(31), 395(31)
Brewer, C. F., 775, 798(50)
Bricaud, M., 497
Brimley, R. C., 363
Brodsky, B., 129, 138, 149, 154, 155(58, 76), 157(76), 158, 159(53), 160, 163, 164(100), 165(100), 166(101), 167, 170, 172
Brodsky-Doyle, B., 145, 148(58), 159(58), 160(58), 320, 324(8)
Broek, D., 158
Broersma, S., 195
Brown, D. J., 759
Brown, D. M., 665
Brown, M. L., 774
Brown, M. S., 544
Brown, R. A., 7, 19(33), 225(16), 226, 480
Brownell, A. G., 286
Bruckner, P., 7, 12, 15(59), 16(59), 18(26), 23, 25, 30(117), 38, 44, 83, 110, 111(42), 225, 311, 485, 490(**3**), 491
Bruenger, E., 572, 573(12), 592, 606, 615(4), 650, 672, 674, 712(8), 714(8), 728, 747
Bruguera, M., 486, 497(67)
Brunner, H., 490(**27**), 491(**27**), 492
Bruns, R. R., 83, 132, 133, 135, 162(39), 510

Buck, C. A., 526
Buckingham, K. W., 569
Buckingham, R. B., 226
Buddecke, E., 294, 302(133)
Bugg, C. E., 675, 695, 700, 702(14), 703(14)
Buissou, H., 845, 849(47), 851(47)
Bulger, J. E., 326
Bump, S., 31
Bungenberg de Jong, H. G., 677
Buonassisi, V., 566, 722
Burgeson, R. E., 7, 8(31), 12(31), 19(31), 42(42), 43, 160, 225(15), 226
Burgers, J. M., 195
Burke, J. M., 101, 120, 121, 620
Burnett, D., 605
Burnett, W., 569, 717, 729(14)
Burwen, S. J., 500
Busiek, D. F., 443
Butkowski, R. J., 65, 66(3), 96(3), 418
Butler, W. J., 6, 55, 56(53), 57(53), 128, 339, 340, 343, 344(11), 345, 412, 424, 428(4)
Byers, P. H., 11, 16(46), 22, 23(99), 24(99), 30(46), 106, 117(37), 118(37), 120, 168, 169(109), 172, 222, 225(5), 309, 373, 394(10)
Byrne, M., 399
Bystrov, V. F., 705
Byus, C., 218

# C

Caen, J. P., 320, 511
Calcott, M. A., 320, 321, 322(16), 323(16), 324(16)
Calore, J. D., 590, 602, 605
Cameron, D., 157
Campbell, D., 323
Campbell, J. H., 32
Campo, R. D., 161
Canfield, R., 712
Cannon, D. J., 18, 577
Cantor, J. O., 620, 630(6a)
Caplan, A. I., 771, 772(10), 774(10), 791, 794(10), 795(10, 13), 796(10), 799
Caputo, C. B., 771, 772(11), 774(11), 780, 782(81), 786(25), 792(11), 794(11), 795(11, 25), 796(11, 25)
Cardinale, G. J., 246, 247(3), 248(3), 260(3), 262, 265, 268(69), 269(69), 270(3, 69),

271(60, 69), 272(69), 273(69), 284(3), 372(2)
Carlin, B., 831, 834(8), 835(8), 837(8)
Carlson, D. M., 348, 794
Carlson, E. C., 35
Carlsson, J., 451
Carlsson, R., 829
Carlstedt, I., 791
Carmichael, D. J., 11
Carnemolla, B., 807
Carnes, W. H., 565, 566, 567, 568(31), 630, 657, 665, 672, (73, 716, 717, 718(10), 721(11), 722
Carraway, K. L., 835
Carroll, D. L., 594
Cartaud, J., 325, 326
Carter, W. G., 805
Carvalho, A. C. A., 14
Carver, J. P., 128
Casassa, E. F., 216
Caskey, C. T., 227
Cassidy, K., 149(101), 164, 166(101), 170, 172
Castillo, M. J., 599
Caterson, B., 774, 776, 783(33), 787(37), 793(37), 799, 840
Cavadore, J. C., 456
Cawston, T. E., 426, 434, 444, 445, 448, 450(39), 451
Cazenave, J. P., 320, 324(9), 513
Ceccarini, C., 347, 348(9)
Cecconi, C., 807
Cerreta, J. M., 620, 630(6a)
Cesarini, J. P., 490(4), 491, 493, 494(97)
Chambard, M., 544, 549(5), 554(5)
Chambers, J., 352
Chandrakasan, G., 12, 40, 133
Chandrarajan, J., 44
Chapman, J. A., 132, 133, 146(21)
Charlton, M. E., 544, 549(4)
Chen, A. B., 804, 805(6), 808(6), 812(6), 840
Chen, H.-T., 163, 164(100), 165(100), 309
Chen, L. B., 808
Chen, M., 565, 666
Chen, Y. T., 335, 336(37)
Chen-Kiang, S., 265, 268(69), 269(69), 270, 271(69), 272(69), 273
Chen Wu, S., 656
Chen Yau, M., 652
Cheron, C., 388, 389(36)

Chevalier, O., 473, 478(14), 486(14), 488(14), 489(14), 490(**30**), 490(**30**), 492
Chiba, T., 181
Chichester, C. O., 271
Chidlow, J. W., 475, 479, 498(48), 728, 748, 751
Childs, B., 161
Chirgwin, J. M., 218, 231
Chiu, J. F., 233
Chlapowski, F. J., 544, 549(3)
Choi, H., 800
Chojkier, M., 471
Chou, K.-L. L., 271, 273(80)
Chou, P. Y., 675
Chovelon, A., 403, 404(12), 410(12)
Christner, J. E., 774
Christner, P., 568, 597, 600, 728, 748, 749, 762
Chu, G. H., 7, 34
Chung, E., 5, 7, 11(12), 20, 41, 42(30), 50, 57(48), 95($e$), 96, 133, 166(32), 225, 424, 428(4)
Church, R. L., 16, 22, 134, 487
Cieciura, S. J., 721
Cifonelli, J. A., 769, 771, 787(1), 795(1)
Cimmino, C. C., 80
Clague, R. B., 480
Claire, M., 842
Clark, C. C., 24, 26, 30(120), 31, 99, 100(15), 105(15), 226, 346(6), 347(2, 3), 348(2, 3, 6), 354(6), 355(6), 356, 359(2), 831, 835(1)
Clark, D. G., 329
Clark, J. M., 600
Cleary, E. G., 561, 572, 672, 687, 728, 749(17), 751
Cleve, H., 490(**35**), 492
Click, E. M., 16, 22(75), 23(99), 24(99), 111, 117(43), 118(43), 473, 475, 485
Cliff, W. J., 687
Clock, E. M., 805
Clouse, W., 340, 410
Coggeshall, R., 80, 136
Cohen, A. B., 604
Cohen, J. H., 500
Cole, R. H., 197
Coletti-Previero, V., 456, 845
Coltrain, I. M., 664
Coltrain, S. M., 565
Comper, W. D., 775

Comte, P., 840, 841
Condit, C., 22, 24(100), 99, 346(4), 347, 468(50), 469
Conochie, L. B., 472, 473(3), 498
Conrad, G. W., 493
Conradie, J. D., 323
Constable, B. J., 157, 167(78)
Cook, V., 561, 567(7), 617, 622(2), 721
Cook, W. J., 675, 695, 700, 702, 703
Cooper, A. R., 831, 834(5), 837(5)
Cooper, C. G., 649
Cooper, N. R., 319, 320
Corbett, C., 424, 428(3), 442(3)
Corcoran, D., 369
Corey, R. B., 159
Cornillet-Stoupy, J., 840(19), 842, 845(19)
Cöster, L., 791
Coudron, C., 796
Coulombre, A. J., 776
Counts, D. F., 284
Court, J., 503
Courtney, K. D., 330
Courtois, Y., 508, 843, 846(28, 29), 851(28, 29)
Cowan, M. J., 396
Cox, B. A., 680, 681(37), 682, 689, 691(45), 711(35, 36, 37)
Cox, G. N., 90
Craig, A. S., 137, 146, 147(51, 62), 154, 163(74), 167(74), 776
Craig, C. P., 598, 600(52a)
Crist, B., 144
Crkrenjakov, R., 218
Crombie, G., 560(35), 565, 590, 594, 595(38), 597(38), 602, 649, 650(30), 666, 668(14), 716, 743, 747, 762, 764(7)
Cronin, M. F., 651, 717
Cronlund, A. L., 641, 642(12), 645(12), 647(12), 648(12), 649(12)
Crossley, M. J., 444, 445(35)
Crothers, D. M., 215
Crouch, E., 7, 16, 22, 26(79), 30(97), 96, 98, 99, 100(13), 105(7), 108, 113(34), 117(7), 118(7), 120(7), 122(13, 48), 123(48), 124(7), 495, 804, 805
Crystal, R. G., 387, 392(31), 393(31), 395(31), 396, 397(42), 491(**58**), 492
Cuatrecasas, P., 265, 272(71), 295(71), 329, 643
Culp, L. A., 507

Cummins, H. Z., 205
Cunningham, L. W., 285, 339, 340, 387, 398, 399(3), 408, 410, 509, 511(2), 530, 532, 533
Cunningham, W. D., 675
Cuprak, L. J., 554
Curran-Patel, S., 128
Curtiss, C., 196(15), 197
Cutroneo, K. R., 251
Cywinski, A., 632, 716(8), 717, 719(8), 720(8), 723(8), 728, 748, 762
Czerkowski, J. W., 601, 603(66)

# D

Dalrymple, P. N., 835
Daly, M. M., 617, 721
Damsky, C. H., 526
Daniels, J. R., 7, 34
Danilczenko, A., 839, 844(1)
Darnule, A. T., 620, 630(6a)
Darnule, T. V., 620, 630(6a), 748, 749
Darr, G., 537
Darrell, R. W., 842, 843(21)
David, G., 500
Davidson, J. M., 27, 28, 99, 108, 222, 225(5), 312, 749, 756, 757(26), 758(26)
Davis, A. L., 588
Davis, B. D., 238
Davis, E. M., 501, 535
Davis, H. F., 572, 591, 592(23), 594(23), 636, 673, 744
Davis, J. H., 176
Davis, N. R., 606
Davison, P. F., 13, 18, 332
Dayer, J.-M., 442
Dayhoff, M. O., 412
Daynes, R. A., 749
Dearing, R., 590
Debye, P., 198
DeClerk, Y. A., 631
DeCremoux, H., 589
DeCrombrugghe, B., 24, 231
DeGroot, J. E., 850
Dehm, P., 37, 96, 99, 104, 253, 259(51), 307, 347, 459, 473, 477(23), 478(23), 479(23), 485, 493, 494(101), 495(101)
deLeon, L. D., 28
DeLuca, S., 771, 772(10, 11), 774(10, 11),

790, 791, 792(11), 794(10, 11), 795(10, 11, 113), 796(10, 11, 26)
Dempsey, E. W., 572, 582(3), 591, 622, 728, 729(38), 764
Denduchis, B., 359
Derouette, J. C., 589, 845, 849(47), 851(47)
DeSa, M. D., 565, 570(36), 666, 667(18), 668, 669, 670(18), 671(19), 672(18), 732, 764
Desanti, M., 840(19), 841(18), 842, 845(18, 19)
Deshmukh, K., 34
Deslauriers, R., 707
Dessau, W., 490(**8**, **40**), 491(**52**), 492, 493, 495(91, 92, 95), 497, 503, 806, 838
Destree, A., 806
Devault, M. R., 554
Devulder, B., 835
DiBlasio, R., 252, 257(48), 258(48)
Diegelmann, R., 394, 455, 456(14), 461(14), 463(14), 464(14)
Diep, J., 789
DiGianfilippo, F., 566, 620, 630(7)
Dijkhof, J., 597
Dingle, J. T., 512
Dische, Z., 839, 844(1, 2)
Dittebrandt, M., 516
Dixit, S. N., 7, 18(22), 21, 96, 225, 226(7), 424
Dixon, M., 728, 748, 762
Djaczenko, W., 80
Doddrell, D., 175
Dodson, J. W., 500, 504, 505(15)
Doering, T., 161
Dogange, P. T., 779
Doleschel, W., 477
Dorfman, A., 11, 21(43), 109, 633, 769, 772, 791, 799, 800(20)
Dorman, D. G., 707
Doty, P., 128, 189, 191(2), 203, 209
Douglas, D. M., 22
Downie, J. W., 561
Doxon, M., 568
Doyle, B. B., 132, 133(20), 134, 146(20), 162(20)
Draeger, K. E., 287
Drake, M. P., 13, 191, 216
Dreisbach, J. H., 454
Druguet, M., 473, 478(14), 486(14), 488(14), 489(14), 490(**30**), 491(**30**), 492
Dryer, B., 489, 490(**18**), 491, 493(81), 495(81)

Duance, V. G., 7, 12(23), 18(23), 226, 490(**7**), **31**, **34**, **49**), 491, 492
Dubick, M., 651
Dubois, M., 359
Dudai, Y., 328, 329(24)
Dudock, B., 232
Duksin, D., 498, 804
Duly, P. E., 474, 481
Dunn, G. A., 501, 535, 544
Dunstone, J. R., 780, 784, 789
Dupont, Y., 139
Dusemunde, B., 799
Dziewiakowski, D. D., 507, 774, 780, 781(80)

E

Eanes, E. D., 152
Ebendal, T., 501, 535
Ebert, P. S., 261
Edelhoch, H., 852
Edelman, G. M., 514, 526(7)
Eder, M., 162, 491(**56**), 492
Edgar, R. S., 90
Eeckhout, Y., 449
Ehrlich, H. P., 15
Ehrmann, R. L., 499, 504, 545
Eichner, R., 717, 725(16), 729(14)
Eiger, A., 71
Eigner, E. A., 52
Eikenberry, E. F., 24, 139, 149(101), 154, 155(76), 157(76), 158, 159(53), 160, 161, 163, 164(100), 165(100), 166(101), 167, 170, 172
Einspahr, H. M., 675, 700, 702(14), 703(14)
Eisen, A. Z., 249, 252(21), 257(21), 258(21), 424, 429, 444, 445(36), 446(40), 447(36), 450(40)
Eisenstein, R., 776, 785(54), 787(54), 796(54)
Ekblom, P., 490(**46**), 492, 838
El Adli, F. A., 7, 8(31), 12(31), 19(31), 42(42), 43, 95(*d*), 96, 225(15), 226
el Badrawy, N., 473, 478(14), 486(14), 488(14), 489(14), 490(**30**), 491(**30**), 492
Elbein, A. D., 349, 352
Elder, J. H., 124
Elgin, S. C. R., 32
Ellis, K., 796
Ellman, G. L., 330

Elsdale, T., 499, 502(1), 504, 505(12), 516, 535, 537, 546
Elsden, D. F., 369, 562, 633
Emerman, J. T., 500, 544, 547(2), 549(2), 553(2)
Emöd, I., 461
Enami, J., 499, 535
Engel, A., 834, 835(18), 836(18)
Engle, J., 8, 22, 24(100), 25, 30(117), 42, 44(36), 99, 346(4), 347, 468(50), 469, 475, 485, 834, 835(18), 836(18)
Engels, C., 388, 397(33)
Engstrom, A., 158
Engvall, E., 507, 508, 518, 756, 803, 804, 805(15), 806(12), 807(5), 808(5), 809(12, 18, 23), 810(18, 41), 811(15, 27, 28), 814(12), 816(12), 817(12, 41, 49, 50), 818(41), 820(50), 821(23, 41, 49, 50, 54), 822(41), 823, 829(54), 833, 838(15)
Epstein, E. H., 146, 163(60), 488, 727
Ericksson, F., 604
Ernbäch, S., 477
Estes, F. L., 404
Ethison, J. R., 353, 354(18)
Evans, H. J., 87, 88, 90, 91(43), 92, 93(43), 94(43), 95(43)
Evanson, J. M., 444, 445(35), 446(37)
Everitt, M., 122
Ewald, R. W., 321
Ewins, R. J. F., 786
Eyre, D. R., 152(96), 161, 285, 369, 370(12)
Eylar, E. H., 287, 294(115), 299(114), 302(114, 115)

## F

Fabre, M. T., 845, 849(47), 851(47)
Faile, D., 512
Fairbanks, G., 234
Fairfield, S. J., 231
Fairweather, R., 365
Faltz, L. L., 774, 782, 783(31), 785(31), 790(31), 799(31)
Fambrough, D. M., 326
Faris, B., 11, 560(35), 561, 563, 565, 567(7), 588, 591, 593, 596(36), 603(22), 617, 622(2), 625, 634, 636, 649, 650(30), 666, 668(14), 716, 721, 728, 739, 743, 762, 764(7), 765(6)
Farquhar, M. G., 778, 794 (74), 838
Farmer, S. R., 524
Farr, A. L., 641
Farrar, T. C., 174
Farrell, M. E., 433, 435(22)
Fasman, G. D., 675
Faulk, W. P., 472, 473(3), 498
Fauvel, F., 511
Fayolle, J., 841(53), 847
Featherstone, R. M., 330
Feinstein, G., 588, 599, 600
Felber, J. P., 482
Feld, M., 542
Fellini, S. A., 790
Ferris, C. D., 200
Fessler, J. H., 17, 26(81), 27, 28, 30(81), 97, 99, 100(17), 105, 106, 114(33), 115(17, 33), 120(17), 127(17), 217, 225, 305, 312(1, 6), 314(6), 315(6), 372(3), 373, 392(3), 468, 485, 486, 495(68)
Fessler, L. I., 17, 26(81), 27, 28, 30(81), 97, 105, 114(33), 115(33), 225, 305, 312(6), 314(6), 315(6), 372(3), 373, 392(3), 468, 485, 486, 490(8), 491, 493, 495(68, 92)
Fialkow, P. J., 559
Field, J. M., 572, 573, 577, 850
Field, M. K., 507
Fietzek, P. P., 5, 7, 11, 12(25), 13, 15(25), 16(47), 18(25), 21(9), 22(25), 47), 23, 24(9), 30, 42, 44(37), 45(37), 49(44), 97, 99(4), 106, 107(36), 108(36), 117(36), 118(36), 128, 129(4), 132, 154(4), 162, 163(4), 225, 311, 411, 413, 468, 475, 484, 485, 489, 490(16), 491(56), 492, 503, 649, 650(31), 806
Finch, J. E., Jr., 6, 128, 343, 344(11), 345(11), 424, 428(4)
Finlay, T. H., 387
Fish, A. J., 838
Fish, W. W., 420, 423 (22, 24)
Fisher, H. W., 721
Fischer, S., 490(40), 492
Fishback, B. L., 294
Fishman, P. H., 507, 807, 828(36)
Fitch, S. M., 844
Fitton Jackson, S., 167
Fitzpatrick, M., 572
Fjolstad, M., 169
Fleckman, P. H., 249, 252(21), 257(21), 258(21)

Fleischmajer, R., 488, 490(4), 491(**51**, **52**), 492, 493, 494, 497(79), 838
Fleming, W. W., 676, 691(27), 712(27), 714(27)
Fletcher, G. C., 178, 204, 217
Fletcher, S., 569, 570(69), 666, 667(18), 669, 670(18), 671, 672(18, 20), 673(20), 717, 725(15), 727(15), 728(15), 729(15), 732, 733(1), 734(1), 737(1), 738(1), 739(1), 740(1), 741(1), 742(1), 743(1), 763, 764, 765(9)
Flint, M. H., 137, 154
Flory, P. J., 676, 680(23), 691(23), 714(22, 23)
Foidart, J. M., 479, 482(50), 485, 489, 490(**5**, **23**, **45a**), 491, 492, 493, 497(50), 498, 757, 821, 831, 832(3), 834(3), 835(3), 837(3), 838(3, 30), 840(20), 842
Ford, D. K., 227, 228(30)
Ford, J., 340, 399, 410
Fosså, J., 522
Foster, J. A., 34, 560, 561, 562, 563, 564, 565, 566(30), 567(7), 569, 570(36, 69, 70), 606, 615(4, 5), 617, 622(2), 649, 650(30), 651, 653, 654(7), 657(11), 666, 667(18), 668, 669, 670, 671(19), 672(17, 18, 20, 21), 673, 674, 689(7), 711(7), 712(7, 8, 9), 714(7, 8), 717, 721, 725(15), 727(15), 728(15), 729(15), 732, 733(1, 3), 734(1), 737, 738, 739(1), 740, 741, 742(1), 743(19), 746, 762, 763, 764(7, 10), 765(6, 9)
Fothergill, J. E., 323
Foweraker, A. R., 775
Fowler, L. J., 361
Fox, J. W., 418
Francis, G., 562
Francis, M., 19, 64
Franek, M. D., 780, 784, 789
Frank, L., 246, 247(5), 372(4), 373, 387(4)
Frankel, S., 597
Franzblau, C., 11, 252, 257(49), 365, 461, 559, 560(35), 561, 562, 563, 564(20), 565, 566(30), 567(7), 569, 580, 588, 590, 591, 593, 594, 594(38), 596(36), 597(37, 38), 598, 600(37, 52), 601(52), 602(52), 603(22, 37, 52), 605, 606, 615(4), 617, 622(2), 625, 634, 636, 640, 642(9), 643(9), 649(9), 650(30), 666, 668(14),
716, 717, 721, 728, 739, 743, 747, 762, 764(7), 765(6)
Fraser, R. D. B., 128, 141(9), 144
Frazier, G. R., 482
Frazier, W., 514
Freeman, A., 829
Freeman, I. L., 831, 834(8), 835(8), 837(8)
Freilich, L. S., 769, 784
French, M. J., 204
Freund, J., 762
Frey, K. L., 227, 228(31), 234(31), 235(31), 236(31), 237(31)
Freytag, J. W., 411, 412(4), 418, 421, 835
Frohman, L. A., 755
Fu, J. C. C., 318
Fuchs, S., 473, 485, 486, 498
Fujii, K., 11, 369
Fujii, T., 41, 129, 133, 166(32)
Fujimoto, D., 93, 369
Fuller, F., 97
Fuller, G. C., 271
Fullmer, H. M., 562, 592
Funakoshi, I., 778, 780(71)
Furie, M. B., 804, 813(9)
Furthmayr, H., 8, 18, 23, 29, 42, 44(36), 59, 369, 370(13), 411, 412, 415, 469, 472, 473, 474(19), 475, 476, 477, 478(40, 44), 479(44), 480, 483(44, 51), 484(2, 44, 51), 485, 486(13), 487(40), 488(40), 489(13, 19, 40, 63), 490(**18**, **21**, **22**, **24**), 491(**58a**), 492, 493(63, 81), 494(63), 495(81), 498, 649, 650(32), 654, 656, 834, 835(18), 836(18), 838
Furuto, D. K., 8, 42, 44(38), 226

## G

Gabay, Y., 320
Gabriel, A., 139
Gabrion, J., 544, 549(5), 554(5)
Gagalang, E., 507
Gagnon, J., 27, 99
Galdston, M., 588
Galione, M. J., 573
Galligani, L., 776
Gallop, P. M., 295, 365, 561, 562, 571, 573, 575(18), 577, 579, 580, 582(29, 30), 584(29, 30), 585(31, 32), 625, 629, 638, 643, 845
Garbisa, S., 164, 165(103), 502

Garcia-de-Quevedo, M. C., 640
Gardell, S., 776, 787(53), 788(53)
Gardner, B., 324
Garrone, R., 65, 848, 849
Gathercole, L. J., 169
Gatmaitan, Z., 545
Gauss, V., 493
Gavrilov, Y. D., 705
Gay, G. O., 499
Gay, R., 8, 12(38), 15(38), 137, 225, 226, 227(21), 242(21), 489, 490(**36, 45**), 491(**54**), 492
Gay, S., 7, 8, 12(36), 14(21), 15(21, 36), 18(21), 21(21), 30, 31(145), 32, 38, 42(40), 43, 47(40), 49(40), 56, 127, 137, 138, 160(2), 162(52), 225, 226(6), 472, 473, 475, 477(12, 16, 22), 478(12), 479(22), 480(22), 484(22), 485, 486(22), 487(12, 47), 488(5, 12, 16, 22), 489(12, 22, 77), 490(**2, 4, 6, 9, 11, 12, 14, 16, 17, 26, 33, 36, 39, 45**), 491(**33, 51, 54, 56, 57, 59, 61**), 492, 493(47, 71), 494(97), 498(47)
Gee, J. B. L., 589
Gehron, P., 831, 832(4)
Gehron Robey, P., 831, 832(3), 834(3), 835(3), 837(3), 838(3, 29), 840(20), 842
Geiger, B. J., 665
Gentner, G. J., 474, 475, 479(27), 484(27)
Geokas, M. G., 588, 593
Gerard, S., 17
Gerber, G. E., 563, 606, 607(1), 614(2), 615(2)
Gerisch, G., 514, 526(8)
Gerth, N., 36
Gertler, A., 599, 601
Gey, G. O., 504, 545
Giambrone, M.-A., 490(**29**), 491(**29**), 492
Gibby, M. G., 175, 176(9)
Gieseking, R., 147
Gilbert, M.-A., 88, 90, 93(45), 95(45)
Gilette, M. T., 495
Gillard, G. C., 154
Gilles, A. M., 456
Gilles, K. A., 359
Gillet, C., 449
Ginger, B., 747
Giro, M. G., 680, 704(33), 711(33)
Gisslow, M. T., 426, 458
Glanville, R. W., 7, 11, 12(25), 15(25), 16(47), 18(25), 22(25, 47), 28, 30, 42, 44(37), 45(37), 49(44), 98, 106, 107(36), 108(36), 117(36), 118(36), 123(8), 225, 370, 444, 445(35), 446(37), 468, 475, 477, 479(42), 485, 486(42), 489(46), 490(**19**), 491
Glaser, L., 514
Glimcher, M. J., 158, 369, 370(12), 445, 450
Glimelius, B., 776
Glisin, V., 218
Glushko, V., 175
Glynn, L. E., 472, 472(3), 476
Gnosspelius, G., 597
Go, A., 691
Godt, S. M., 751, 754(14)
Goetinck, P. F., 161, 790, 791(110), 792(110)
Goettlich-Riemann, W., 565, 652, 665, 666, 716
Goetze, D., 474, 475, 476(24, 31)
Golaszewski, T., 404
Gold, L. M., 664
Goldberg, B., 30, 305, 311(7), 473, 474(21), 497
Goldhaber, P., 445, 450
Goldman, J. N., 755
Goldman, M. B., 755
Goldstein, A., 89, 95(*b*), 96
Goldstein, J. L., 544
Goldstein, R. H., 588
Gonnerman, W. A., 316, 317(27), 640, 649(7), 665
Goodman, D., 80
Goodman, H. M., 221, 222(4), 223(4), 225(4), 739
Goral, A. B., 252, 257(49)
Gordon, J. L., 512
Gordon, S., 596, 603(45)
Gosline, J. M., 687, 691(42), 714(42)
Gospodarowicz, D., 500, 504, 505
Gosslau, B., 482, 484(58)
Gotlieb, A. I., 827
Gotte, L., 566, 572, 576(8), 620, 630(7), 680, 681(40), 682(34), 686, 687(40), 704(33, 34), 711(33, 34, 37, 40)
Grabar, P., 762
Granditsch, G., 498
Grant, G. A., 424
Grant, M. E., 25, 26, 30(118, 119, 121), 99, 244, 286, 288(108), 289(108), 291(108), 302(105), 840(59), 849, 851(59)

Grant, M. H., 454
Grantham, J. J., 418
Grassmann, W., 89, 461
Graves, P. N., 22, 25, 222, 223(6), 225(6), 346(5), 347, 468, 475, 485, 569
Gray, W. R., 562, 563, 566, 614, 674, 689, 711(7), 712(7, 8, 9), 714(8)
Greenberg, D. B., 28, 305, 312(6), 314(6), 315(6)
Greenberg, I., 508
Greenburg, G., 500
Greenburg, J., 505
Gregory, J. D., 774, 786(39), 787(39)
Greiling, H., 776
Grey, H. M., 320
Griepp, E. B., 537
Griffiths, R., 840(63), 850, 851(63)
Grimaud, J. A., 473, 478(14), 486(14), 488(14), 489(14), 490(**30**), 491(**30**), 492
Grinnell, F., 500, 501(16), 502(16), 503, 507, 524, 535, 540, 541, 542, 807, 828(34)
Grobstein, C., 500
Gross, J., 5, 11(7), 14, 42, 80, 83, 128, 132, 133(6), 134, 135, 162(39), 332, 423, 424, 425, 428(3), 431, 432, 435, 436(24), 439(20), 442(3), 443, 510
Gryder, R. M., 385, 387(21), 391, 393(37)
Grynpas, M., 152, 158, 161
Guay, M., 585, 587
Gullino, M., 837, 838(30)
Gunson, D. E., 477, 485, 486(41)
Gupton, B. F., 599, 605(55)
Guzman, N. A., 22, 24(100), 25, 99, 251, 268, 269(73), 270(73), 305, 346,(4, 5), 347, 468(50), 469
Guzman, R., 499, 535, 554

# H

Hadjiisky, P., 749
Haeberli, A., 852
Hahn, E., 29, 473, 474(15), 475, 476(15, 31), 484(11, 15), 485, 486(15, 60), 488(60), 489(60), 490(**13**, **25**, **28**), 491(**25**, **28**), 492, 497(67), 838
Halford, M. H., 339, 340(3), 341(3)
Hall, D. A., 601, 603(66)
Hall, Z. W., 325
Haller, M., 490(**35**), 492

Halme, J., 249, 250(24), 252, 260(24), 264(24), 267(24), 268(24), 274(24), 440
Haltner, A. J., 201
Hamamoto, S., 499, 535
Hamilton, J. K., 359
Hamilton, P. B., 364
Hampton, J., 124
Hance, A. J., 387, 392(31), 393(31), 395(31)
Haniu, M., 567
Hanss, M., 200, 596
Haralson, M. A., 8, 12(38), 15(38), 56, 226, 227(21), 228(31), 231, 234(31, 32, 33), 235, 236, 237, 242(21), 243(33)
Harboe, N., 739
Hardcastle, R. A., 132, 146(21)
Harding, J. D., 733
Hardingham, T. E., 774(43), 775, 782(44), 786(27), 787, 793(43), 798(28, 40)
Hare, P. E., 387
Harel, S., 759
Hargrave, P. A., 458
Harkness, M. L. R., 844
Harkness, R. D., 844
Harley, A. D., 599, 605(55)
Harpel, P. C., 604, 605(71)
Harper, E., 22, 23(99), 24(99), 424, 428(3), 442(3), 443, 453, 455(2), 461(2, 17)
Harrington, W. F., 85, 86(35), 87, 88, 89, 90(35), 91(48), 92(35, 48), 93(35, 48, 49), 456
Harris, E. D., Jr., 424, 428(3, 4), 433, 435(22), 442(3), 447, 449(41)
Harris, R. D., 675, 679(15), 688(15), 689(15), 705(15), 711(15)
Harrison, R. W., 231
Harsch, M., 567, 716, 717(5), 725(5), 727(5)
Hart, G. W., 771
Hart, M. L., 630
Hartmann, S. R., 175
Hartree, E. F., 457
Harwood, R., 25, 26, 27, 30(119), 99, 244, 286, 288(108), 289(108), 291(108), 302(105)
Hascall, G. K., 774, 782(31), 783(31), 785(35), 790(31), 799(31)
Hascall, V. C., 36, 135, 374, 375(12), 507, 771, 772(10, 11, 12), 774(10, 11, 41, 43), 775(17), 776, 777(63), 778, 779, 780(34), 781(78, 80), 782(31, 44, 81), 783(31), 784(34), 785(31, 54), 786(23, 24, 25, 34),

787(34, 54, 63), 789, 790(31), 101), 791(63), 792(11), 793(43, 63, 69), 794(10, 11, 69, 74), 795(10, 11, 12, 25, 63, 66, 69, 113), 796(10, 18, 21, 24, 25, 26, 54, 102), 797(17, 78), 798(29, 51, 78), 799(31, 52), 800(21)
Hassager, O., 196(15), 197
Hassell, J. R., 508, 776, 777(63, 64), 787(63), 788, 791, 793(63), 795(63), 837, 838(29)
Hasson, M. A., 775, 798(51), 799(52)
Hata, R., 469, 471(52)
Hauschka, P. V., 87, 89, 90, 91(48), 92(48), 93(48), 577
Hauschka, S. D., 500, 504
Haverback, B. J., 593
Hawber, S., 637
Hay, E. D., 485, 490(**43**), 492, 493, 494(101), 495(101), 500, 501, 504, 505(15), 535
Hay, F. C., 751
Hayaishi, O., 246, 247(4)
Hayashi, T., 18, 19(89), 99, 100(11), 128, 307
Hayat, M. A., 135
Hayduk, U., 65
Hayes, J. A., 588
Hayman, E. G., 803, 804, 805(15), 807(5), 808(5), 811(15), 814, 816, 817(50), 818(43), 819(43), 820(50), 821(50), 826(43), 828, 829
Haynes, L., 544, 549(3)
Hays, D. G., 524, 807, 828(34)
Heath, J. P., 544
Heathcote, J. G., 26, 30(121)
Hedman, K., 844
Heeger, P., 718
Heggeness, M. H., 827
Heidrich, H. G., 461
Heinegård, D. K., 774(41), 775, 776, 786(23, 24), 790, 791, 792(92), 793(117), 795(62), 796(22, 24), 798(29), 799, 800(143)
Heinrickson, R. L., 92
Helle, O., 23, 169, 311, 368, 497
Hellerqvist, C. G., 532, 533
Hellmann, W., 799
Helm, P., 719
Helseth, D. L., 129
Hendricks, R. W., 145, 149(59)
Hendrix, C. F., 675, 679(15), 688(15), 689(15), 705(15), 711(15)
Hendrix, M. J. C., 476, 486(37, 38)
Heng-Khoo, C. S., 569, 651

Henkel, W., 11, 13, 294, 302(133), 370
Henson, E., 562, 579, 580, 582(30), 584(30), 585(31), 625
Herbage, D., 201, 217(23), 473, 478(14), 486(14), 488(14), 489(14), 490(**30**), 491(**30**), 492
Hermans, O., 80
Herring, G. M., 157
Herrmann, H., 41, 490(**8**), 491, 493, 495(92)
Herzog, F., 488
Heusser, C., 321, 322(15), 323(15)
Hewgley, P. B., 387, 531
Hewitt, A. T., 502, 503(26), 508
Hewitt, N., 598, 600(52), 601(52), 602(52), 603(52)
Hieber, E., 485
Higgs, T. P., 176
Highberger, J. H., 128, 133(6), 134, 424, 428(3), 442(3)
Hind, C. F., 319, 322(1)
Hinek, A., 620
Hinman, L. M., 589
Hinnie, J., 840(63), 850, 851(63)
Hitt, A. S., 326
Hjelle, J. T., 35
Ho, P.-L., 633, 799
Hoak, J. C., 723
Hodge, A. J., 217
Hoeve, C. A. J., 676, 680(23), 691(23, 24), 714(22, 23, 24)
Höffken, K., 323
Hoffman, L., 533
Hoffmann, H.-P., 25, 117, 118(47), 309, 568, 716(9), 717, 719(9), 720(9), 723(9), 727(9)
Hofman, T., 599
Hogan, B. L. M., 831, 834(5), 837(5)
Holbrook, K., 172, 804
Hollister, D. W., 7, 8(31), 12(31), 19(31), 42(42), 43, 95(d), 96, 160, 225(15), 226
Holt, P. J. L., 480
Holtrop, M. E., 578
Holtzer, H., 790, 791(105)
Hong, B.-S., 18
Hönigsmann, H., 491(**55**), 492, 838
Hood, L., 32
Höök, M., 507, 513, 514(1), 516(4), 518(11), 521(4), 522(12), 524(1, 3, 4), 525(13), 526(4, 11), 778, 780(70), 782(70), 788(70), 798(70), 807, 828(35), 838

Hopper, K., 474, 475, 479(27), 484(27)
Horak, W., 498
Hori, H., 427, 428(14), 437(14), 439, 445(27), 446
Horisberger, M., 494
Hörlein, D., 23, 28, 97, 99(4), 311, 468
Hörmann, H., 806, 807
Horne, R. W., 680, 682(34), 704(33, 34), 711(33, 34)
Hornebeck, W., 589, 603, 851
Horowitz, A. L., 769
Horvath, M., 596
Hoshino, M., 769
Hospelhorn, V. D., 455, 572
Houle, D., 746, 747(4), 748, 749
Housley, T., 77, 95(c), 96, 365
Howard, B. V., 722, 723, 724 (28a)
Howe, C. C., 831, 838(7)
Howell, D. S., 436, 783
Howes, E. L., 457
Hruby, V. J., 707
Hsieh, P., 808
Hsu, D., 789
Hu, C.-L., 588
Huber, R., 324
Huber, S., 34, 666, 667(16), 668(16)
Hudson, B. G., 35, 65, 66(3), 96(3), 411, 412(4), 418(6), 421(4), 835
Hudson, L., 751
Huebner, P. F., 593, 596
Hughes, R. C., 508
Hughes-Jones, N. C., 324
Hukins, D. W. L., 134, 149(93), 161
Hulkko, A., 286, 303(102)
Hulmes, D. J. S., 132, 133(20), 134, 145, 146(20), 147, 148, 152(95), 153(65), 155(58), 158, 159, 160, 161, 162(20)
Humoller, F. L., 659
Humphrey, C. D., 718
Hung, C., 65, 66(3), 96(3)
Hunt, L. T., 412
Hunter, J. C., 577
Hunter, W. M., 753, 754(22)
Hurewitz, A., 759
Hurwitz, C., 320
Huseby, R. M., 804, 805(6), 808(6), 812(6)
Hutcheson, E. T., 11
Hutton, J. J., 249, 252(15, 20), 257(20), 258
Hynes, R. O., 803, 806, 816, 817(47)

## I

Igarashi, S., 5, 11(7)
Ikehara, K., 87(42), 89
Ill, C. R., 504
Imberman, M., 569, 651, 717
Immelman, A. R., 790
Inayama, S., 379
Ingild, A., 739
Inonye, T., 491(**57**), 492
Irreverre, F., 387
Isersky, C., 305, 312(4), 314(4)
Isles, M., 775
Isliker, H., 320, 321, 322(15), 323(15), 324(7)
Iso, N., 69, 75(11), 76(11), 95(11)
Isselbacher, K. J., 804
Ito, H., 567
Ito, K., 599
Ivarsson, B., 499, 535

## J

Jackson, A. J., 666, 667(18), 670(18), 672(18)
Jackson, D. S., 25, 244, 286, 288(108), 289(108), 291(108), 561, 572, 749, 751, 840(59), 849, 851(59)
Jackson, R. J., 734, 735
Jacobs, H. G., 285
Jacobs, M., 689, 691(45)
Jacotot, B., 842
Jaffe, E. A., 620, 629
Jaffe, R., 831, 834(8), 835(8), 837(8)
James, H. L., 604
Jamhawi, O., 104, 110(31)
Jamieson, A. M., 775, 798(48)
Jamieson, G. A., 289, 294, 299, 302(122, 135, 139), 303(122), 304(122)
Janata, J., 674, 712(9)
Janin, J., 183
Janoff, A., 588, 590, 600, 747, 759
Jaurand, M.-C., 589
Jeanloz, R. W., 339, 340(3), 341(3)
Jeffrey, J. J., 249, 252(21), 257(21), 258(21), 429, 444, 445(36), 446(40), 447(36), 450(40)
Jeffrey, K. R., 176
Jelinski, L. W., 176, 177, 178(22), 179(22), 180(22), 181(24, 25), 183(23, 25)
Jennings, B. R., 775

Jentoff, V., 554
Jesoir, J.-G., 147, 153(65), 157
Jilek, F., 493, 495(91), 807
Jimenez, S. A., 167, 459
Jodlowski, M., 418
Johansson, S., 507, 513, 514(1), 518(11), 524(1), 525(13), 526(11), 838
John, R., 562, 572, 851
Johnson, C. D., 326
Johnson, G. S., 357
Johnson, L., 561, 567(7), 617, 622(2), 721
Johnson, M. C., 751, 754(14)
Johnson, R., 537
Johnson-Wint, B., 435, 436(24)
Johnson, V. P., 192
Jollès, P., 774, 787
Jones, P. A., 631
Jordan, R. E., 594, 597(37), 598, 600(37, 52), 601(52), 602(52), 603(37, 52)
Josse, J., 85, 86(35), 87, 88, 90(35), 92(35), 93(35)
Joule, J. A., 581, 582(35)
Judd, J., 340, 410
Juliano, R. L., 507
Junqua, S., 839, 840(12), 841, 842(12), 843(12), 846, 847(50), 849(54)
Juva, K., 249, 380, 384(20), 728

## K

Kadar, A., 851
Kagan, H. M., 34, 315, 564, 566(30), 588, 591, 594, 597(37), 600(37), 601, 603(22, 37, 65), 606, 615(4), 638, 640, 641, 642(5, 9, 12), 645(9, 12), 647, 648, 649(9, 12), 650(5, 29), 747
Kahn, L. D., 217
Kaitila, I. I., 7, 8(31), 12(31), 19(31), 42(42), 43, 225(15), 226
Kaitils, I. I., 95(d), 96
Kakiuchi, K., 260
Kalbfleisch, H., 486, 497(67)
Kalwinsky, D., 249, 284(14)
Kammerman, S., 497
Kanamori, M., 567
Kang, A. H., 5, 11(7), 13, 14, 21, 42, 133, 225, 226(7), 332, 424, 428(3), 442(3), 475, 504
Kantrowitz, F. G., 399

Kanwar, Y. S., 778, 794(74), 838
Kao, W. W.-Y., 250, 251(29), 262(29), 265, 271(29), 273(68, 79, 80), 274(79), 286, 387, 392(30), 395(30)
Karasek, M. A., 544, 549(4)
Karr, S. R., 569, 570(69), 669, 671, 672(20), 673(20), 717, 725(15), 727(15), 728(15), 729(15), 732, 733(1), 734(1), 737(1), 738(1), 739(1), 740(1), 741(1), 742(1), 743(1), 763, 765(9)
Kasai, M., 197
Katunuma, N., 122
Katz, E. P., 152, 158
Katz, S. I., 490(5), 491, 493, 498, 508, 837, 838(30)
Katzman, R. L., 339, 340, 341, 841
Kauzmann, W., 683
Kawaniski, K., 432
Keane, P. M., 474, 482(28)
Kearny, J. F., 820
Kedersha, N. L., 268, 269(73), 270(73), 271, 273(79), 274(79)
Keech, M. K., 135
Keele, E. M., 50, 57(48)
Keeley, F. W., 718, 851
Keen, C., 567
Kefalides, N. A., 6, 7, 12(17), 26, 30(118, 120), 31, 37, 96, 99, 100(15), 105(15), 225, 226(59), 346, 347(2, 3), 348(2, 3), 359(2), 477, 485, 486(41), 723, 724(28a), 831, 835(1)
Keil, B., 453, 454(5), 456(5), 461, 468
Keiser, H., 774, 775, 786(35), 798(50)
Keith, D. A., 561, 571, 573, 575(18), 578, 579(27), 638, 845
Kelleher, P. C., 403
Keller, S., 454, 461(11), 620, 630(6a), 712, 749
Kelley, J., 393
Kelly, R., 325
Kemp, A., 250, 260(31), 265(31), 280(31), 284(31)
Kennedy, D. W., 507, 804, 805(7), 808(13), 814(7)
Kent, G., 491(57), 492
Kero, M., 252, 262(44)
Kerwar, S. S., 424
Kerwin, R. E., 216
Keski-Oja, J., 806, 807, 821(32)
Kessler, E., 305, 311(7)

Kessler, S. W., 740, 755
Kettman, J., 485
Khaled, M. A., 674, 692, 693(50, 51, 52), 695, 697, 700, 703(52), 705(50), 706, 707(59), 708, 709(11), 710(12), 711(12, 50, 51, 52)
Kidwell, W. R., 502
Kikuchi, Y., 260
Kimata, K., 507, 769, 781, 787, 789, 790(101), 791(96), 793(84), 795(84), 796(84, 102), 799
Kimura, J. H., 771, 772(11), 774(11, 43), 775, 780, 782(44), 786(25), 789, 790(101), 791, 792(11), 793(43), 794(11), 795(11, 25, 113), 796(11, 21, 25, 102), 799, 800(21)
Kimura, S., 65, 67, 69(9), 72, 74(15), 75(9), 76(9, 10), 77(15), 78, 79, 95(9, 10)
King, G. S., 597, 759
King, J., 310
Kirk, D., 431
Kirrane, J. A., 476
Kirschner, D., 152(96), 161
Kishida, Y., 137, 249, 250(25), 260(25), 262, 271(61), 280(25, 56)
Kitamura, K., 790
Kivirikko, K. I., 27, 29, 246, 247(1, 2, 6), 248(1, 2, 6), 249(2, 6), 250(2, 13, 24, 25), 251(2, 6, 10, 17), 252(10, 16, 17, 19, 32, 36, 39), 254(10, 19, 34), 256(34), 257(32, 36), 259(19, 39), 260(2, 6, 13, 24, 25, 37), 261(37, 57), 262(19, 39, 46, 54), 264(24), 265(19, 57), 266(57, 70), 267(19, 24, 57, 67, 70), 268(2, 19, 24, 57, 67, 70), 269(19, 57, 70), 270(1, 2, 6, 57, 70), 271(1, 2, 6, 62), 273(76, 78), 274(24, 32), 275(86, 87), 276(85, 86, 87), 277(1, 6, 32, 85, 86, 87), 278(10, 34, 53), 279(10, 53), 280(1, 2, 6, 13, 17, 19, 25, 32, 34, 36, 52, 53, 57, 70, 86, 87), 281(53, 91, 92, 93, 94), 282(91), 283, 284(1, 6, 7, 52, 91, 92, 93, 94), 285(7), 286, 287(7), 288(103, 104, 118), 289(118), 290(104, 106, 118, 125, 128), 291(42, 104, 113, 118), 292(103, 104), 294(104, 125, 126), 295(130, 131), 296(129), 297(126), 298(125, 126, 129, 132), 299(104, 130), 300(130, 132), 301(7, 103, 104, 118, 130), 302(7, 103, 104, 118, 129), 303(7, 118, 125, 140), 304(140), 305, 309(5), 310(5),
312(5), 373, 374, 376, 387, 394(11), 398, 400
Kjellén, L., 514, 526(2), 529(2), 778, 780(70), 782(70), 788(70), 798(70), 838
Klebe, R. J., 501, 503, 504(1), 505(1), 507(6), 516, 536, 806, 807, 828(33)
Klein, L., 398
Kleinman, H. K., 502, 503(26), 504, 507, 508, 518
Kleinschmidt, A. K., 799
Klenk, H.-D., 348
Klinger, H. P., 819
Klobutcher, L. A., 487
Klumpp, T., 604
Knaack, D., 728, 739, 762, 765(6)
Knobel, H. R., 320, 324(7)
Knox, P., 507
Kobata, A., 355
Koehler, J. K., 136
Koerner, S., 565
Kohler, G., 819, 820(52)
Kohn, L. D., 305, 312(4), 314(4)
Koivisto, V., 290(128), 291
Kokoschka, E. M., 490(1), 491
Kolb, L. M., 321, 322(17), 323(17)
Kolb, W. P., 321, 322(17), 323(17)
Konigsberg, I. R., 500, 504
Kono, T., 455, 456(13), 461(13)
Kopple, K. D., 691
Kornfeld, R., 771
Kornfeld, S., 354, 771
Kornfeld-Poullain, N., 592
Korthy, A. L., 588
Kosow, D. P., 289, 302(122), 303(122), 304(122)
Koszalka, T. R., 26, 30(120), 99, 100(15), 105(15)
Kotin, R., 167
Kounter, S., 666
Kozmin, S. A., 705
Krachmer, J. H., 776, 777(64), 791(64)
Kraft, D., 490(1), 491
Krane, S. M., 159, 399, 424, 442
Krebs, H. A., 718
Kresina, T. F., 7, 8, 12(20, 38), 14(20), 15(20, 38), 18(20), 20, 38, 42(24), 45(24), 56, 225, 226, 227(21), 242(21)
Krieg, T., 28, 99, 100(12, 16), 106(16), 127(12), 468, 488, 489, 490(**32, 50**), 491(**52**), 492, 493(87), 495(93), 496,

497(79, 87, 104), 831(9), 832, 833(9), 834(9, 13, 14), 837(9, 13, 14), 838(9, 13, 14)
Krigbaum, W. R., 188
Krisman, C. R., 304
Ksiezny, S., 596
Kuchich, U., 748, 749, 756, 757(27)
Kucich, U., 597, 749
Kuettner, K. E., 787
Kühl, U., 493, 495(95)
Kühn, K., 5, 8, 11, 12(35), 13, 15(35), 16(47), 18(35), 21(9), 22(47), 23, 24(9), 41, 44, 49(44), 89, 97, 99(4), 106, 107(36), 108(36), 117(36), 118(36), 129, 132, 133, 134, 162, 166(32), 225(14), 226, 311, 411, 413, 468, 473, 474(19), 475, 485, 487, 489(19), 490(**9, 16, 17**), 491(**56**), 492, 493(71), 649, 650(31)
Kumamoto, C., 99, 100(17), 105, 106, 115(17), 120(17), 127(17)
Kumar, S., 500
Kunkel, H. G., 322
Kupfer, A., 599
Kurata, M., 87(42), 89
Kurg, T., 145, 149(59)
Kurkinen, M., 99, 100(11), 127(12), 488, 489(75), 490(**32, 50**), 491(**60**), 492, 493(75, 87), 495(87), 496(104), 497(87, 104), 831, 833, 834(5, 13), 837(5, 13), 838(13), 844
Kusch, M., 90
Kuter, D. J., 846
Kuttan, R., 247, 372, 373(1)
Kuusela, P., 807, 816, 817(49, 50), 820(50), 821(49, 50)
Kuutti, E.-R., 260, 261(57), 262, 265(57), 266(57, 70), 267(57, 70), 268(57, 70), 269(57), 270(57), 271(62), 280(57, 70)
Kuzan, F., 649

## L

Labat-Robert, J., 848
LaBella, F., 746, 747(4), 748, 749
Labrosse, K. R., 458
Laemmli, U. K., 57(58), 59, 102, 117(27), 120(27), 310, 414, 432, 647, 656, 727, 736
Lafuma, C., 840(31, 32), 843, 845(31, 32), 846(31, 32), 847(31, 32)

Lagunoff, D., 122
Laidlaw, J., 135, 507
Laitinen, O., 250, 262, 376, 400
Lampden, J. O., 772
Lampidis, R., 634
Lamport, D. T. A., 796
Lamy, F., 582, 585, 587, 598, 600(52a)
Lane, J. M., 58
Lange, G., 34, 566, 597, 620, 666, 667(16), 668(16), 729, 730, 748, 749, 751, 752, 753, 756, 758, 759(21), 762
Langley, K. H., 204
Langner, R. L., 844
Lansin, A. I., 728, 729
Lansing, A. I., 559, 572, 582(3), 591, 622, 623, 746
Lapiere, C. M., 11, 23, 29, 30, 31(136), 97, 99(4), 166, 226, 305, 307, 309(13), 311, 312(4), 314(4), 425, 473, 747(19), 475, 485, 489(19)
Larochelle, J. F., 582, 585
Larsson, S.-E., 787
Laskey, R. A., 102, 224, 310, 314(4), 471, 495, 727, 737, 742
Laszt, J., 844
Laufer, J. S., 90
Laurell, C. B., 604, 742
Laurent, T. C., 775
Lawley, K. R., 14
Lawley, T. J., 498
Lawson, W. B., 328
Layman, D. L., 3, 569, 642
Leach, R. M., 759
Leafer, M. D., Jr., 328
Lechner, J. H., 129
LeCroisey, A., 468
Leder, P., 733
Ledvina, M., 573
Lee, G., 305, 312(4), 314(4)
Lee, I., 567, 652
Lee, J., 569
Lee, Y. C., 89, 93(38), 95(38), 355, 357
Lee-Owen, V., 456, 461(20), 465(20), 845
Lefevre, M., 567, 569, 651
Legrand, Y. J., 511
Lehto, V. P., 488, 489(75), 493(75)
Lehtonen, E., 490(**46**), 492
Leïbovich, S. J., 326
Leivo, I., 487, 488(72), 489(72), 490(**37**), 492, 495, 831(10), 832, 837(10), 838(10)

Lejnieks, I., 512
Leloir, L. F., 304
LeMaire, W. J., 436
Lembach, K. J., 387, 530
Lemieux, R. U., 606
Lemkin, M. C., 778, 795(74)
Lemmen, C., 487, 490(**9**), 491
Lemmonier, M., 846, 847(50)
Lenaers, A., 30, 305, 312(4)
Lenhardt, W. F., 356
Lennarz, W. J., 350, 771
Lennox, E., 476
Lent, R., 562, 580, 593, 596(36)
Leonard, K. R., 320, 324(8)
Lepow, I. H., 319, 322(1)
Lerch, R. M., 601, 603(65)
Lerner, R. A., 124
Leslie, J. G., 674, 675(10), 704(10), 712(10), 714(10)
Lesot, H., 490(**47**), 492
Lethias, C., 848
Leung, C. Y., 473, 474(17)
Leung, D. Y. K., 485
Leung, M. K. K., 28, 305, 312(6), 314(6), 315(6)
Leunis, J. C., 294
Leushner, J., 87, 90, 93(44), 94(44)
Levene, C. I., 649
Lever, P. L., 792
Lever, W. F., 554
Levi, M. M., 749
Levine, A., 11, 569, 651, 717
Levitt, D., 791, 799
Levitt, M., 183
Levy, G.-C., 174
Levy, H., 588
Lewis, L. J., 723
Lewis, M. S., 5, 52, 71
Lewis, R. G., 769
Lewis, W., 594, 596(37), 598, 600(37, 52), 601(52), 602(52), 603(37, 52)
Li, S., 152, 158
Liberti, P. A., 323
Lichtenstein, J. R., 11, 16(46), 24(78), 30(46), 106, 117(37), 118(37), 120
Lieberman, R., 771
Liener, I. E., 458, 589
Liesegang, B., 820
Light, N. D., 7, 12(23), 18(23), 369, 371
Likhite, V., 747, 749

Lillard, Y., 593
Limeback, H. F., 30
Linder, E., 488, 489(75), 493(75), 816
Lindy, S., 719
Linker, A., 778
Linsenmayer, T. F., 476, 486(37, 38), 490(**43**), 492
Liotta, L. A., 451, 502, 504
Little, C. D., 476
Liu, T. Z., 281
Livingston, D. M., 271
Loeven, W. A., 588
Lohmander, L. S., 771, 772(10, 11), 774(10, 11), 792(11), 794(10, 11), 795(10, 11), 796(10, 11), 799
Long, M. M., 675, 678(13), 679(15), 681(13), 686, 688(13, 15), 689(15, 41), 691(45), 692(17), 693(53, 54), 695(17), 697(17), 700, 702(14), 703(17), 704(17), 705(15), 709(17), 711(15, 17, 53, 54), 712(18), 713(53)
Lorenzen, I., 719
Low, R. B., 393
Lowe, D. M., 322, 323(19), 324(19)
Lowry, O. H., 641
Lozaityte, I., 798
Lucas, J. J., 350
Lucky, A. W., 777
Luderschmidt, C., 491(**52**), 492, 838
Lukens, L. N., 495
Lunde, L. G., 11, 21(45), 36
Lundy, D. R., 152
Lustberg, T. J., 24, 106, 107(34), 108(34), 112(34)
Luthier, B., 835
Lwebuga-Mukasa, T. S., 455, 461(17)
Lyrela, J. R., 676, 691(26), 712(26), 714(26)

# M

McBride, B. C., 426, 458
McBride, O. W., 87, 91, 93(49)
MacCallum, E. K., 774, 786(25), 795(25), 796(25)
McCann, W. F. X., 326, 335(21)
McCormick, K., 499, 535
McCroskery, P. A., 424, 428(3, 4), 442(3), 447, 449(41)

McCullagh, K. G., 851
McDonald, J. A., 396
MacDonald, R. J., 218, 231, 733
McDevitt, C. A., 793, 794, 795(119)
McEneany, L. S. G., 28, 108, 312, 468
McGee, J. O'D., 262, 270, 271(59, 60)
McGoodwin, E. B., 3, 12, 34, 42(3), 44(3), 99, 105(21), 106(21), 107(21), 110(21), 503, 504, 518, 806, 831, 832(4)
McKenney, K. H., 11, 16(46), 24, 30(46), 106, 107(34), 108(34), 112(34), 117(37), 118(37)
Mackensen, S., 545
McKnight, G. S., 734
McKusick, V. A., 29
McLauchlan, K. A., 707
McLaughlin, P. J., 323
McLees, B. D., 11
MacLennan, J. D., 457
McMahan, U. J., 326
MacRae, T. P., 128, 141(9), 144
Maca, R. D., 723
Macarak, E. J., 722, 723, 724
Madden, J. W., 666
Madri, J. A., 480, 485, 486(13), 489(13), 489(63), 490(**18, 21, 24**), 491(**58a**), 492, 493(63, 81), 494(63), 495(81), 834, 835(18), 836(18)
Maguire, M. H., 835
Mahler, D. J., 659
Maigret, B., 183
Mailman, M. L., 461
Mainardi, C. L., 424
Majamaa, K., 256, 271, 273(78), 278(53), 279(53), 280(53), 281(53)
Mak, W. W., 554
Malanik, V., 573
Malekina, N., 364
Maley, F., 355
Malmström, A., 791
Mammi, M., 680, 682(34), 704(34), 711(34)
Man, M., 374, 391, 393(37)
Manahan, J., 454, 461(11)
Mandel, L., 206
Mandl, I., 453, 454, 457, 461(11), 712, 748, 749
Mann, K. G., 420, 423(22, 24)
Mantsch, H. H., 176, 180(17)
Maox, A., 473
March, S. C., 329

Marcu, K., 232
Margolis, R., 786
Marina, P., 807
Maroteaux, P., 793
Marotti, K. R., 831(11), 832, 838(11)
Marsh, J. M., 436
Marshall, J. S., 348
Marshall, L. M., 326
Martin, D., 774, 782(31), 783(31), 785(31), 788, 790(31), 799(31)
Martin, G. N., 152
Martin, G. R., 3, 5, 7, 10, 11, 12, 15(59), 16(28, 46, 59), 21(42, 43), 22(30), 26(28, 30), 28, 29, 30(46), 31(30), 33, 34, 36(2), 42(3), 44(3), 83, 98, 99, 105(6, 21), 106(21), 107(21), 109, 110(6, 21), 111(6, 42), 117(37), 118(37), 120, 223, 225(10), 309, 316, 317(26), 451, 475, 477, 479(42), 482(50), 485, 486(42), 487, 490(**3**), 493(71), 495, 497(50), 498, 502, 503(26), 504, 507, 508, 518, 562, 563, 565(22), 633, 638, 639, 640, 642(3), 647(4), 648, 649(4), 665, 666, 757, 806, 807, 821, 828(36), 831, 832(3, 4), 834(3), 835(3), 837(3), 838(3, 29, 30), 840(20), 842
Martinelli, J., 495
Martini, G. A., 490(**25**), 490(**25**), 492
Marzullo, M.-T., 656, 742, 743(19), 762
Maser, M. D., 80, 83(28), 85, 86(34), 89, 96(34), 216
Mashburn, T. A., 789
Mason, R. M., 781, 793
Massoulié, J., 325, 326(5), 331(3)
Masui, Y., 427, 428(14), 437(14)
Mathews, M. B., 80, 130, 769, 787(1), 795(1), 798
Matsumura, T., 65, 94, 95(8)
Matthay, R. A., 589
Matukas, V. J., 3, 5(1), 11(1)
Matzen, K., 487, 490(**9**), 491
Mauchamp, J., 544, 549(5), 554(5)
Mayes, R. W., 781
Mayne, R., 6, 101
Mayes, C., 325, 326(9), 331(19), 334(19), 335(19), 337(9), 338(9, 19), 339(9)
Mecca, C. E., 562, 648
Mecham, R. P., 34, 566, 567, 569, 597, 606, 615(5), 620, 630(7), 651, 654(7), 657, 666, 667(16), 668(16), 672(17), 717, 729,

730, 746, 747, 748, 749, 751, 752, 753, 756, 758, 759(21), 762, 763, 764(10)
Mechanic, G., 580
Meehan, E. J., 340
Meek, K. M., 132, 146(21)
Meezan, E., 35
Mehring, M., 175
Meier, S., 500
Miegel, W. N., 488, 489(77), 491(**51**), 492
Melet, J., 317
Menashi, S., 286, 289, 302(105)
Menzel, J., 480, 498(54)
Meraldi, J.-P., 707
Mercer, E., 444
Mercier, O., 379, 385(18), 397(18)
Merkel, J. R., 73, 77(14), 454
Merill, G., 554
Merrilees, M. J., 137, 154
Merrill, C., 499, 535
Merry, A. H., 25
Mestecky, J., 498
Metzger, H., 320
Meyer, F. A., 513
Meyer, K., 771, 794
Michael, A. F., 489, 490(**23**), 492, 838
Michael, D., 472, 473, 474(17), 476, 485
Michalopoulos, G., 500, 504, 505(13), 544, 547(1), 549(1, 9), 551, 552, 554(9)
Micou-Eastwood, J., 732
Miettinen, T., 251, 252(39), 259(39), 262(39), 287, 291(113)
Milam, M., 501
Miller, A., 132, 133(20), 134, 144, 145, 146(20), 147, 148(58), 151(77), 152(77), 153(64, 65), 155(58), 157(73), 158(77), 159(58), 160(58), 160(20), 162, 165
Miller, E. J., 3, 5(1), 6(4), 7, 8, 10(4), 11(1, 8, 12), 12(20, 36, 38), 13, 14(20, 21), 15(8, 20, 21, 36, 38), 18(4, 20, 21, 37), 19(37), 20(32), 21(4, 21, 42, 43, 45), 29, 32(4), 33, 36(2), 38, 41, 42(24, 30, 40), 43, 44(38), 45(24), 47(40), 49(40), 50, 52(33), 55, 56(52), 57(48), 58(52), 64(54), 95(*e, f*), 96, 101, 108, 109, 127, 128, 129, 132, 133, 160(2), 166(32), 225, 226(6), 227(21), 242(21), 562, 592, 633, 648, 727, 776, 343, 344(11), 345(11), 370, 400, 401, 424, 428(4), 472, 473, 475, 484(11), 485, 486(60), 488(60), 489(60), 490(**13, 14, 33, 36, 45**), 491(**33, 54**), 492, 498, 804,
806(12), 809(12), 814(12), 816(12), 817(12)
Miller, M. M., 537
Miller, R. L., 251, 252(33), 257(33), 274, 277
Millet, M., 388
Mills, A. D., 102, 224, 310, 471
Mills, G., 508
Milson, C., 844
Milstein, C., 476, 819, 820(52)
Minick, C. R., 620, 629(6)
Minick, O. T., 491(**57**), 492
Minor, R. R., 26, 30(120), 99, 100(15), 105(15), 168, 172
Minter, D., 503, 541, 807, 828(34)
Misiorowski, R. L., 317
Mitchell, L. W., 689, 691(45), 692, 693(53, 54), 697, 711(53, 54), 712, 713(53)
Mitchell, W. M., 8, 12(38), 15(38), 17, 56, 226, 227(21), 228(31), 234(31, 32, 33), 235(31), 236(31), 237(31), 242(21), 243(33), 244, 456, 465
Miyada, D. S., 747
Miyaji, M., 599
Miyoshi, M., 455, 456(16), 465(16), 567
Mizel, S. B., 504, 505(17)
Mizuno, H., 69, 75, 76(11), 95(11)
Mo Cha, C.-J., 271
Moczar, E., 839, 840(14), 841(37, 29), 842(7), 843, 844(7), 845(33), 846(33, 34), 847(33), 849(37), 851(33, 37, 52), 852(33)
Moczar, M., 839, 840(12, 14, 31, 32), 841(37), 842(12), 843(12), 844, 845(31, 32, 33, 36), 846(28–33, 34), 847(31, 32, 33), 849(36, 37), 851(28–30, 33, 36, 37, 52), 852(33)
Moen, R. C., 222, 225(5)
Mohan, V. S., 597, 759
Mohanram, M., 567
Monson, J. M., 16, 22(75), 111, 117(43), 118(43), 221, 222(4), 223(4, 6), 225(4, 6), 306
Montes, L. F., 491(**54**), 492
Moore, S., 340, 404
Mopper, K., 357
Morales, S. M., 774, 783(32)
Morales, T. I., 436
Morecki, R., 490(**29**), 491(**29**), 492
Morgan, P. J., 285
Mori, T., 567, 716, 717(5), 725(5), 727(5)
Morita, K., 387

Morikawa, T., 305, 315(8)
Moro, L., 11, 42(41), 43
Morris, N. P., 28, 312, 468
Mort, A. J., 796
Morton, L. F., 157, 167(78)
Moscaritolo, R., 11, 739, 762, 765(6)
Moseley, V. M., 157, 158(80)
Moses, R. E., 227
Mosesson, M. W., 804, 805(6), 807, 808(6), 812(6), 813(26), 840
Moshudis, E., 490(**25**), 491(**25**), 492
Mosher, D. F., 803, 806, 807, 808(1, 4, 37), 821
Moss, P. S., 732
Mozes, E., 751
Muh, J. P., 835
Muir, H., 475, 490(**15**), 491, 774, 775, 786(27), 787, 793, 794, 795(119), 798(28)
Muir, L., 89, 93(38), 95(38)
Müller, K., 514, 526(8)
Müller, P. K., 28, 104, 110(31), 468, 487, 490(9), 491, 493(71), 497, 831(9), 832, 833(9), 834(9), 837(9), 838(9)
Mueller, P. L., 777
Muller, F. J., 774, 783(32)
Muller, R. J., 797
Müller-Eberhard, H. J., 320, 321, 322(16), 323(16), 324(16)
Munch, M., 413
Munderloh, N. H., 11, 146, 163(60), 488
Muramatsu, T., 355
Murphy, L., 567, 716, 717(5), 725(5), 727
Murphy, W. H., 468
Murray, J. C., 502, 508, 649, 831(9), 832, 833(9), 834(9), 837(9), 838(9)
Murray, L. W., 76(33), 83, 95(c), 96
Mustard, J. F., 320, 324(9), 513
Muthiah, P., 776
Mutoh, S., 778, 780(71)
Myllylä, R., 246, 247(1, 6), 248(1, 6), 249(6), 251(6), 252, 256, 260(6), 262(46), 270(1, 6), 271(1, 6), 277(1, 6), 280(1, 6, 52), 281(52, 92, 93, 94), 282, 283, 284(1, 6, 7, 52, 91, 92, 94), 285(7), 286, 287(7), 288(103, 104), 289, 290(104), 291(104, 118), 292(103, 104), 294(104, 125, 126), 295(130), 296(129), 297(126), 298(125, 126, 129, 132), 299(104, 130), 300(130, 132), 301(7, 103, 104, 118, 130), 302(7, 103, 104, 118, 120, 129), 303(7, 118, 125), 304(120, 140), 373

## N

Nachman, R., 620, 629(6)
Nachmansohn, D., 325
Naff, G. B., 319, 322(1)
Nagai, Y., 18, 19(88), 427, 428(14), 432, 437(14), 439(20), 445(27), 446, 453
Nakagawa, M., 504, 505(16)
Nakajima, K., 599
Nakamura, N., 369
Nakazawa, K., 781, 793(84), 795(84), 796(84)
Nandi, S., 499, 535, 554
Narayanan, A. S., 567, 568, 569, 588, 608, 638, 640, 642, 647(4), 649(4), 666, 716(6), 717
Nath, K., 507
Natowicz, M., 355
Nelson, G. L., 174
Neuberger, A., 373
Neumann, R. A., 844
Neurath, H., 122
Neville, D. M., Jr., 432
Newsome, D. A., 776, 777(63, 64), 787(63), 791, 793(63), 795(63)
Nicholls, A. C., 370
Nicholson, W. E., 231
Niedzwiecka-Namyslowska, I., 596
Nielsen, T. B., 420
Nietfeld, J. J., 250, 260(31), 265(31), 280(31), 284(31)
Nilsson, B., 771, 772(10, 11), 774(10, 11), 792(11), 794(10, 11), 795(10, 11), 796(10, 11, 26)
Nimni, M. E., 34
Nishino, N., 588, 599, 605(55)
Nist, C., 485, 493, 494(101), 495(101)
Nitecki, D. E., 385, 386(22), 387(22)
Njieha, F., 305, 315(8)
Noda, H., 453, 455, 461(12), 573
Noelken, M. E., 411, 412(4), 418(6), 421(4)
Noggle, J. H., 707
Nolan, J. C., 424
Norby, D., 590
Nordin, J. H., 321, 322(15), 323(15)
Nordwig, A., 65, 456, 461

Norling, B., 776
Nowack, H., 11, 16(47), 22(47), 24, 29, 30(116), 31(145), 44, 49(44), 106, 107(36), 108(36), 117(36), 118(36), 138, 162(52), 468, 472, 473(10), 474(10, 15), 475, 476(15, 24, 31), 477(16, 22), 478, 479(22), 480(10, 22, 25), 484(15, 22), 485, 486(15, 22, 25, 60), 487(47), 488(16, 22, 60), 489(22, 60), 490(**6, 13, 26**), 491, 492, 493(47), 497(10), 498(47)
Nozaki, M., 246, 247(4)
Nozaki, Y., 420, 423(25)
Nusgens, B., 166, 226, 307, 309(13)
Nwokora, N., 294
Nylen, M. U., 157, 158(80)

## O

Öbrink, B., 191, 192(6), 203, 217(6), 513, 514(1), 516(4), 518(11), 521(4), 522(12), 524(1, 3, 4), 525(13), 526(2, 4, 5, 9, 11), 529(2), 807, 828(35)
Ocklind, C., 513, 514(1), 518(11), 524(1), 526(5, 9, 11)
Oda, G., 633
O'Dell, B. L., 316, 317(27), 640, 648, 649(7), 665
Odermatt, E., 834, 835(18), 836(18)
Oegema, T. R., 135, 507, 774, 776, 780, 781(80), 785(54), 787(54), 796(54), 797
O'Farrell, P. Z., 664, 739
Ogawa, H., 69, 75(11), 76(11), 95(11)
Ogston, A. G., 192
Ohman, P., 533
Ohnishi, T., 675, 689, 691(45), 692, 693(53, 54), 697, 711(53, 54), 713(53)
Ohno, M., 35
Ohtsuki, J., 379
Oikarinen, A., 271, 273(76), 286, 290(106)
Oike, Y., 781, 793(84), 795(84), 796
Okamoto, K., 649, 650(29), 674, 675, 679(15), 680(31), 688(15), 689(15), 705(15), 707, 709(11, 30), 711(15, 31)
Okayama, M., 769, 787, 790(96), 791(96, 105)
Okanoto, K., 315
O'Konski, C. T., 201, 217
Oldberg, A., 778, 780(70), 782, 788(70), 798(70), 807, 828(35)

Oldberg, B., 514, 516(4), 521(4), 524(3, 4), 526(4)
Olden, K., 231, 803, 808(2)
Oliver, T., 570, 669, 671, 732
Olsen, B. R., 22, 24(100), 25, 30(116), 99, 134, 136, 137, 146, 147(61), 169(47), 222, 223(6), 225(6), 262, 268, 271(61), 309, 346(4), 347, 459, 468(50), 469, 473, 475, 477(23), 478(23), 479(23), 485, 491(**53**), 492, 493, 494(98, 100, 101), 495(101)
Olsson, I., 451
Olsson, T. A., 641, 642(12), 645(12), 647(12), 648(12)
Onda, N., 69, 75(11), 76(11), 95(11)
Oneson, R., 461
Oohira, A., 787, 790(96), 791(96)
Oosawa, F., 197
Oppenheim, F., 461
Orkin, R. W., 12, 34, 42(3), 44(3), 99, 105(21), 106(21), 107(21), 110(21), 831, 832(4)
Ornstein, L., 600
Orodsky, B., 161
Oronsky, A. L., 424
Ortho, D. N., 231
Osborn, M., 410, 411(2), 413, 414, 725, 727
Osdoby, P., 799
Ott, U., 490(**25**), 491(**25**), 492, 835
Ouazana, R., 88, 90, 93(45), 95(45)
Ouchterlony, O., 762
Oudea, P., 839, 843(5), 844(5)
Ouellette, L. A., 223, 225(10)
Ouzilou, J., 843, 846(28, 29), 851(28, 29)

## P

Pacifici, M., 790, 791(105)
Packham, M. A., 320, 324(9), 513
Padieu, P., 364
Padilla, S., 100
Page, R. C., 563, 567, 568, 640, 649, 716(6), 717
Paglia, L., 28, 223, 225(10), 493
Painter, R. H., 320, 323, 324(9)
Pal, S., 779, 786, 797
Palacios, R., 739
Palmiter, R. D., 27, 99, 222, 225(5), 229, 739
Palmoski, M. J., 791
Palotie, A., 99, 100(11), 307

Pänkäläinen, M., 264, 267(67), 268(67)
Papamichail, M., 498
Pâques, E. P., 324
Parhami, N., 320
Parikh, I., 329
Park, E.-D., 22, 134
Parker, C. W., 748, 750(13), 751, 754(14), 755(13)
Parlebas, J., 839, 842, 843(4, 5, 22), 844(4, 5, 25, 26)
Parodi, A. J., 304
Parrish, W. B., 80
Parry, D. A. D., 132, 133(20), 137, 146(20), 147(51, 62), 152, 154, 162(20), 163(74), 167(74), 776
Parshley, M. S., 620, 630(6a)
Parsons, H. T., 665
Parthasarathy, N., 783
Partridge, S. M., 341, 363, 562, 563, 565, 572, 576(8), 591, 592(23), 594(23), 598(21), 603(21), 633, 636, 651, 665, 666, 672, 673, 674, 680, 744, 747, 844, 851(41)
Paslay, J. W., 799
Pasquali-Ronchetti, I., 681(40, 41), 686, 687(40), 689(41), 691(41), 711(40)
Pastan, I., 24, 231, 804, 805(7), 814(7)
Pasternak, J., 87, 90, 93(44), 94(44)
Pasternack, S. G., 772, 775(19), 798
Paul, S. M., 323
Parans de Ceccatty, M., 848
Pawlowski, P. J., 495
Payrau, P., 842, 843(22)
Paz, M. A., 561, 562, 571, 573, 575(18), 579, 580, 582(29, 30), 584(29, 30), 585(31, 32), 625, 629, 638, 845
Peach, C. M., 361
Peacock, E. E., 666
Pearlstein, E., 503, 828
Pearson, C. H., 403, 404(12), 410
Pearson, J. P., 781, 793
Pecora, R., 203, 205, 208, 209
Pekkala, A., 807, 811(27)
Pelham, H. R., 734, 735
Peltier, A. P., 320
Peltonen, L., 29, 99, 100(11), 307
Pencev, D., 490(**28**), 491(**28**), 492, 838
Penman, S., 524
Penner, E., 490(**27**), 491(**27**), 492
Pennypacker, J. P., 161, 502, 503(26), 508

Pensky, J., 319, 322(1)
Penttinen, R., 491(59), 492
Perantoni, A., 493
Pereyra, B., 562, 579, 580, 582(29), 584(29), 585(31, 32), 625, 629
Perez-Tamayo, R., 425
Périn, J.-P., 774, 787
Perkins, M. E., 808
Perkofsky, B., 845
Perlish, J. S., 488, 490(**4**), 491(**51**), 492, 493, 494(97), 497(79)
Perlmann, H., 320
Perlmann, P., 320, 756
Perrin, F., 179, 190, 195
Perry, A., 36
Perry, S. V., 412
Pesciotta, D. M., 22, 24, 468, 475, 485
Peterkotsky, B., 116, 248, 249, 251, 252(9), 257(48), 258(48), 284(14), 394, 455, 456(14), 461(14), 463(14), 464, 465(14), 466(29), 469, 470, 471(52)
Pettersson, I., 514, 518(11), 526(11)
Peyrol, S., 473, 478(14), 486(14), 488(14), 489(14), 490(**30**), 491(**30**), 492
Pfahl, D., 222, 332
Pfister, R., 8, 12(36), 15(36), 42(40), 43, 47(40), 49(40)
Phan Dinh Tuy, B., 841(52), 847, 851(52)
Phelps, C. F., 775
Philipson, L., 796
Phillips, S. J., 161
Picard, B., 325
Pickett, R. A., 124
Pictet, R. L., 733
Pidoux, I., 800
Pierard, G. E., 166
Pierce, J. A., 596
Piez, K. A., 5, 10, 12, 13, 21(42), 33, 36(2), 40, 50, 52, 55, 56(53), 57(53), 58(1, 47), 71, 87(43), 88(43), 90, 91(43), 92(43), 93(43), 94(43), 95(43), 120, 132, 133(20), 146(20), 153, 163(20), 217, 400, 401, 412, 562, 633, 648, 676, 691(25), 712(25), 714(25), 727
Pines, A., 175, 176(9)
Pinnel, S. R., 316, 317(26), 562, 565(22), 638, 639, 642(3), 665, 666
Pita, J. C., 774, 782(31), 783(31, 32), 785(31), 790(31), 797, 799(31)

Pitelka, D. R., 500, 544, 547(2), 549(2), 553(2)
Pitlick, F. A., 489, 490(**18**), 491, 493(81), 495(81)
Pitot, H. C., 504, 505(13), 544, 547(1), 549(1, 9), 551, 552(10), 554(9)
Pittman, F. E., 718
Poaster, J., 565, 666
Podack, E. R., 321, 322(17), 323(17)
Podleski, T. R., 508
Polak, K. L., 17, 29(82), 99, 100(14), 104(14), 114(30), 120(30), 124(30)
Ponce, P., 545
Ponseti, I. V., 666
Pontz, B., 485
Poole, A. R., 772, 774(42), 775, 781(42), 786(42), 796(21), 799, 800(21, 42)
Poort, C., 597
Popenoe, E. A., 274
Popper, H., 491(**57**), 492
Porath, J., 451, 477
Porter, R. R., 319, 320(3), 322, 323(19), 324(3, 19, 20)
Portnova, S. L., 705
Pöschl, A., 490(**42**), 492
Postlethwaite, A. E., 504
Pouliquen, Y., 842, 843(22)
Poullain, N., 746, 839, 843(4), 844(4)
Powers, J. C., 588, 594, 599, 605(55)
Powers, M. J., 10, 21(42), 33, 36(2)
Prasad, K. U., 692, 693(56), 694(56), 695(56), 696(56), 698(56), 699(56), 701(56), 702(56)
Prather, W., 461, 466(29)
Pratt, R. M., 507, 508
Preston, V. E., 323
Price, M. R., 323
Price, P. J., 536
Priess, H., 324
Prince, R. K., 226
Pritzl, P., 124
Prockop, D. J., 17, 22(80), 23(80), 24(80), 25, 26, 27, 29, 30(118), 93, 99, 100(11), 104, 117, 118(47), 128, 137, 222, 223(6), 225(6), 246, 247(2), 248(2), 249(2, 26), 250(13, 25), 251(2, 17, 29), 252(16, 17, 18, 32), 253, 254(18), 257(32, 35), 259(51), 260(2, 13, 25, 26, 37), 261(37), 262(29), 265(26), 267(26), 268(2, 26), 269, 270(2), 271(2, 29, 61), 274(32), 277(32), 280(2, 13, 17, 18, 25, 26, 32, 36, 56), 281, 286, 305, 307, 309(5), 310(5), 312(5), 313(23), 314, 315(8), 346(5), 347, 374, 376, 379, 380, 384(20), 385(17), 387, 392(30), 394(11), 395(30), 398, 400, 459, 464, 468, 473, 477(23), 478(23), 479(23), 493, 494(100), 568, 569, 716(7, 9), 717, 719(9), 720(9), 723(7, 9), 727(7, 9), 728
Przybyla, A., 569, 570(69), 669, 672(20), 673(20), 717, 725(15), 727(15), 728(15), 729(15), 732, 733(1), 734(1), 737(1), 738(1), 739(1), 740(1), 741(1), 742(1), 743(1), 763, 765(9)
Puck, T. T., 721
Puddle, B., 19, 64
Puistola, U., 256, 274, 275(86, 87), 276(85, 86, 87), 277(85, 86, 87), 280(52, 86, 87), 281(52), 284(52), 284(52, 93)

## Q

Quaroni, A., 804
Quinn, R. S., 590
Quintarelli, G., 566, 620, 630

## R

Radbruch, A., 820
Radhakrishnamurthy, B., 783
Radhakrishnan, A. N., 247, 372, 373(1)
Raekallio, J., 491(**59**), 492
Rafelson, M. E., 589
Raisz, L. G., 578
Rajewsky, K., 820
Ralston, E., 691
Ramachandran, G. N., 128
Ramasarma, G. B., 623
Randall, R. J., 641
Randoux, A., 379, 385(18), 397(18), 840(19), 841(18), 842, 845(18, 19)
Rao, G. J. S., 328
Rao, N. V., 89, 93, 284
Rapaka, R. S., 315, 649, 650(29), 674, 679, 680(31), 681(41), 686, 689(41), 691(41), 692, 705(50), 709(11, 30), 711(31, 50)
Rasmussen, B. L., 561, 572, 573, 592
Rauter, A., 7, 12(25), 15(25), 18(25), 22(25), 42, 44(37), 45(37), 225
Rauterberg, J., 11, 13, 23, 24, 133, 370, 413
Rayns, D. G., 136, 164(45)

Rayson, J. K., 850
Reale, E., 136, 147(44), 164(44), 165(44)
Reddi, A. H., 18(90), 19, 290, 490(**45, 45a**), 492, 774, 782(31), 783(31), 785(31), 790(31), 799(31), 838
Reed, S. J., 227
Reich, E., 777
Reid, K. B. M., 319, 320(3), 322, 323(19), 324(3, 8, 19, 20)
Reid, L. G. M., 504, 505(14)
Reihanian, H., 775, 798
Reilly, C. F., 589
Reilly, H. C., 154
Reiner, A., 772, 774(42), 775, 781(42), 786(42), 796(21), 800(21, 42)
Reinhold, V. N., 339, 340(3), 341(3)
Remberger, K., 162, 487, 489, 490(**9, 11, 16**), 491(**56, 61**), 492
Renais, J., 749
Rennard, S. I., 479, 482(50), 485, 497(50), 502, 504, 518, 749(26), 756, 757(26), 758(26), 788, 799, 821, 831, 832(3), 834(3), 835(3), 837(3), 838(3, 29, 30), 840(20), 842
Renner, F., 498
Renugopalakrishnan, V., 692, 693(50, 51, 52), 703(52), 705(50), 707, 711(50, 51, 52)
Reppucci, A. V., 769
Restall, D. J., 490(**7, 34, 49**), 491, 492
Revel, J. P., 537
Rexrodt, F. W., 24, 413
Reynolds, E. S., 540
Reynolds, G., 563
Reynolds, J. A., 420, 423(23, 25)
Rhoads, R. E., 249, 250(12, 23), 251(23), 260(12, 23, 28), 264(12), 268(12), 734
Rhode, H., 311
Rhodes, R. K., 7, 8, 12(36, 38), 15(36, 38), 18(37), 19(37), 20(32), 21, 42(40), 43, 47(40), 49(40), 56, 64(54), 95(*e, f*), 96, 225, 226, 227(21), 242(21), 490(**14**), 491, 776
Ribers, P. A., 359
Rice, R. V., 80, 83(28), 85, 86(34), 89, 96(34), 216
Rich, C. B., 34, 565, 569, 570(36, 69, 70), 651, 653, 656, 657(11), 666, 667(16, 18), 668(16), 669, 670(18), 671(19), 672(18, 20), 673(20), 717, 725(15), 727(15), 728(15), 729(15), 732, 733(1, 3), 734(1), 737(1), 738(1), 739(1), 740(1), 741(1), 742(1), 743(1, 19), 762, 763, 764, 765(9)
Richards, J. F., 447, 449(41)
Richardson, J. M., 325, 326(7), 328(7), 329(7), 330(7), 331(7)
Richmond, V., 561, 572, 573, 592
Richter, W. J., 23, 311
Ridge, S., 424
Rieger, F., 325
Rieman, W., 567
Rifkin, D. B., 804, 813(9)
Rinderknecht, H., 593
Riolo, R. L., 374, 375(12), 772, 796(18)
Ripamonti, A., 164, 165(103)
Risteli, J., 8, 42, 44(36), 196(14), 197, 215(14), 248, 249, 251(10), 252(10, 36, 39), 254(10, 19, 34), 256(34), 257(36), 259(19, 39), 262(19, 39), 265(19), 267(19), 268(19), 269, 278(10, 34, 53), 279(10, 53), 280(19, 34, 36, 53), 281(53), 290, 291(42), 387, 469, 475, 485, 486, 497, 498(112)
Risteli, L., 252, 286, 287, 288(103, 104, 118), 289(118), 290(118, 125), 291(104, 118), 292(103, 104), 294(104, 125, 126), 295(130), 297(126), 298(125, 126), 299(104, 127, 130), 300(127, 130), 301(103, 104, 118, 127), 302(103, 104, 118, 127), 303(118, 125, 127), 474, 479(26), 482(26), 495(26), 831(9, 10), 832, 833(9), 834(9), 835(17), 836, 837(9, 10, 24), 838(9, 10)
Robert, A. M., 839, 840(8), 843, 844(8), 845(8)
Robert, B., 589, 593, 839, 840(8), 842(6, 7, 9), 843(4, 5, 6, 21, 22), 844(3, 4, 5, 6, 7, 8, 9), 845(8), 849(47), 851(47)
Robert, L., 589, 592, 593, 746, 839(16), 840(8, 12, 31, 32), 841(37, 52a, 53), 842(6, 7, 9, 12), 843(4, 5, 6, 12, 21, 22), 844(2, 3, 4, 5, 6, 7, 8, 9, 25, 26), 845(8, 31, 32, 33, 36), 846(28–33), 847(31, 32, 33, 50), 848, 849(36, 37, 47, 54), 851(28–30, 33, 36, 37, 47), 852(33)
Roberts, E., 623
Roberts, P. J., 491(60), 492
Robertson, A. V., 387
Robertson, J. S., 353, 354(18)
Robertson, P. B., 370

Robey, P. G., 7, 16(28), 22(30), 26(28, 30), 31(30), 42, 98, 105(6), 110(6), 111(6), 451, 757, 788, 821
Robins, S. P., 12, 365
Robinson, D. R., 442
Rodbard, D., 777, 795(66)
Rodèn, L., 769, 771, 776, 787(1), 795(1)
Rodger, G. W., 572, 577
Rodnan, G. P., 226
Rodriquez, M. M., 776, 777(64), 791(64)
Rohde, H., 23, 472, 473(10), 474(10), 475, 476(24), 479, 480(10, 25), 485, 486(25), 489, 497(10, 67), 498(112), 821, 831(9), 832(3), 833(9), 834(3, 9, 16), 835(3, 12, 17, 18), 836(18), 837(3, 9, 12), 838(3, 9, 16, 26), 840(20), 842
Rohrbach, D. H., 508, 838
Rojkind, M., 490(**29**), 491(**29**), 492, 504, 505(14), 545
Roll, F. J., 480, 486, 489(63), 490(**24**), 492, 493(63), 494(63)
Rollins, B. J., 507
Rombauer, R., 579
Rosano, C. L., 320
Rosebrough, N. J., 641
Roseman, S., 513, 533
Rosenberg, L., 774(42), 775, 781(42), 786(42), 798(48), 799, 800(42)
Rosenberry, T. L., 325, 326(7, 9), 328(7), 329(7), 330(7), 331(7, 19), 334(19), 335(19, 21), 336(2, 37), 337(9), 338(9, 19), 339(9)
Rosenbloom, J., 222, 455, 456(16), 465(16), 567, 568, 569, 597, 600, 632, 716(8), 717(5), 718, 719(8), 720(8), 723(8), 725(5, 16), 727(5), 728, 729(14), 748(27), 749(27), 756, 757(27), 762
Rosenthal, R. B., 572, 582(3)
Rosenthal, T. B., 591, 622, 623, 728, 729(38), 764
Ross, G. T., 777
Ross, R., 101, 121, 485, 493, 494(101), 495(101), 559, 561(4), 569, 572, 592, 617, 620, 776, 841(60), 845, 849(46), 850(60)
Rosset, J., 494
Rotundo, R. L., 326
Roufa, D. J., 227
Roughley, P. J., 793, 795
Rous, B., 201, 217(23)
Roveri, N., 164, 165(103)

Rowe, D. W., 27, 99, 222, 225(5)
Rowlands, R. J., 144
Royal, P. D., 790, 791(110), 792(110)
Royce, P. M., 157, 167(78)
Rubin, A. L., 222, 332
Rubin, K., 513, 514(1), 516(4), 518(11), 521(4), 522(12), 524(1, 3, 4), 525(13), 526(2, 4, 5, 11), 529(2), 807, 828(35)
Rubin, L., 562, 564(20), 566(30), 606, 615(4), 650, 747
Ruch, J. V., 490(**47**), 492
Rucker, R. B., 565, 567, 569, 620, 650, 651, 652, 665, 666, 716
Rudall, K. M., 80, 83(22)
Ruddle, F. H., 487
Ruggeri, A., 136, 147(44), 164(44), 165, 244
Ruiz, H. A., 783
Ruoslahti, E., 507, 508, 518, 803, 804, 805(15), 806(12), 807(24), 808(5), 809(12, 16, 18, 23), 810(18, 41), 811(15, 27, 28), 813(24), 814(12), 816(12), 817(12, 41, 49, 50), 818(41, 43), 819(43), 820, 821(23, 41, 49), 822(41), 826(43), 828, 829, 831, 832(6), 833, 834(6), 837(6, 23), 838(15)
Ruska, C., 80
Ruska, H., 80
Russell, F., 500, 551
Russell, R. G. G., 442
Rutishauser, U., 514, 526(7)
Rutter, W. J., 218, 231, 733
Ryhänen, L., 251, 252(43), 274(43), 277(43), 280(43), 569
Ryvar, R., 475, 490(**15**), 491

S

Saba, T. M., 808, 821
Saccomani, G., 689
Sachar, L. A., 597
Sage, H., 5, 7(34), 8(5), 12(24, 34), 16, 18(24, 34), 20, 21(24), 22, 30(97), 52, 96, 99, 100(13), 120(1), 122(13, 48), 123(1, 48, 52), 124, 131, 160(17), 225(17), 226, 332, 373, 392(7), 394(7), 417, 495
Sainte-Marie, G., 828
Saito, H., 176, 180(17)
Saito, S., 379
Saito, T., 69, 75(11), 76(11), 95(11)
Sajdera, S. W., 36, 772, 774, 775(17), 779,

780(34), 781(78), 784(34), 786(34), 787(34), 797(17, 78), 798(78)
Sakai, R., 260
Sakakibara, S., 249, 250(25), 251, 260(25, 37), 261(37), 280(25, 56), 427, 428(14), 437(14), 453
Sakamoto, M., 445, 450
Sakamoto, S., 445, 450
Sakashita, S., 831, 832(6), 833, 834(6), 837(6), 838(15)
Sakato, K., 87(42), 89
Sakmann, B., 325
Saksela, O., 844
Salcedo, L. L., 561, 567(7), 569, 593, 596(36), 617, 622(2), 625, 636, 640, 642(9), 645(9), 649(9), 651, 717, 721, 728, 762, 765(6)
Salo, L., 252, 263(46)
Sanada, Y., 122
Sandberg, L. B., 562, 564(20), 569, 572, 573(12), 578, 592, 606, 615(4), 637, 650, 653, 655, 657, 659, 664, 665, 672, 673, 674, 675(10), 689(7), 704(10), 711(7), 712(7, 8, 9, 10), 714(7, 8, 10), 716, 728, 747, 749(27), 751
Sandell, L., 27, 28(128), 244
Sandson, J. I., 774, 786(35)
Sanes, J. R., 326
Santi, L., 807
Santoro, S. A., 509, 511(2)
Sasaki, M., 599
Sasse, J., 493, 495(91, 95)
Sato, T., 7
Sattler, C. A., 504, 505(13), 547, 549(9), 554(9)
Sattler, G. L., 547, 549(9), 551, 552(10), 554(9)
Savage, J. E., 316, 317(27), 563, 640, 649(7), 665
Savolainen, E.-R., 290
Saxén, L., 490(**46**), 492, 838
Scebat, L., 749
Schachman, H. K., 210
Schaefer, J., 175
Schafer, M. P., 387, 392(31), 393(31), 395(31)
Schapira, G., 364
Schauenstein, K., 489
Schecter, N. S., 420, 423(25)
Scheele, G., 600

Scheinman, J. I., 489, 490(**23**), 492, 838
Scherer, J., 747
Schessinger, J., 508
Scheuenstuhl, H., 473, 474(17)
Schimke, R. T., 734, 739
Schirmer, R. E., 707
Schleiter, G., 11
Schlesinger, D. H., 804, 805(7), 814(7)
Schlesinger, S., 354
Schlueter, R. J., 89
Schmidt, M. F. G., 348
Schmitt, F. O., 13, 128, 133(6), 134, 222
Schmitt, J. W., 349
Schnaar, R. L., 513
Schor, A. M., 500
Schor, S. L., 499, 500, 503
Schreiber, J. R., 777
Schrode, J., 774, 786(25), 795(25), 796(25)
Schroeder, W. A., 340, 342(8)
Schubart, A. F., 321
Schubert, M., 779
Schubotz, L. M., 281
Schulman, J. D., 777
Schumaker, V. N., 320
Schuppan, D., 98, 123(8), 486
Schwan, H. P., 200
Schwartz, D., 7, 14, 27, 38, 96, 196, 216(13)
Schwartz, E., 632
Schwartz, M. L., 396, 397(42)
Schwartz, N. B., 796
Schwarz, R. T., 348
Schweri, M. M., 437, 443(26)
Scocca, J., 357
Scott, D. B., 157, 158(80)
Scott, J. E., 787
Scott, R. D., 727
Sear, C. H. J., 26, 30(121), 99, 840(59), 849, 851
Seglen, P. O., 520, 522, 719
Segrest, J. P., 285, 398, 399(3), 408
Seifter, S., 295, 453, 455(2), 461(2), 579, 580, 582(29, 30), 584(29, 30), 593, 603(35), 629, 643
Sela, M., 473, 751
Seltzer, J. L., 424
Sen, A., 807, 821(32)
Sender, P. M., 851
Senior, R. M., 593, 596, 747
Seno, N., 771
Senyk, G., 476, 485

Seppä, H., 290
Serafini-Francassini, A., 160, 161(90), 572, 573, 577, 840(63), 850, 851(63)
Setsuiya, T., 87(42), 89
Seyer, J. M., 11, 41
Shapiro, R., 560(35), 565, 649, 650(30), 666, 668(14), 716, 734, 762, 764(7)
Shatton, J., 779
Sheehan, J. K., 775
Sheer, C. J., 473, 474(21), 497
Shellswell, G. B., 226, 490(**31, 34, 49**), 492
Sherman, M. I., 489, 490(**36**), 492
Shibata, T., 379
Shieh, J. J., 319
Shigeto, A., 498
Shimada, W., 606
Shinkai, H., 324
Shinomura, T., 781, 793(84), 795(84), 796(84)
Shinro, Y., 432
Shipp, D. W., 841
Shively, J. E., 816, 817(50), 820(50), 821(50)
Shotton, D. M., 589, 590
Showalter, A. M., 24
Shudo, K., 251, 260(37), 261(37)
Shulman, H. J., 774, 786(35)
Shuttleworth, A., 225(16) 226
Sicher, N., 597
Siegel, R. C., 34, 315, 316, 317(24, 26), 318(28), 433, 563, 565(22), 638, 639, 642, 647(4), 649(4), 666
Silbert, J. E., 769
Silescue, H., 176
Silkowitz, M. H., 22, 468, 475, 485
Silman, I., 325, 326(8), 328, 329(24), 486
Silver, F. H., 204, 210, 211, 217(36)
Silverman, D., 593
Simha, R., 193
Simmons, H. A., 578
Simons, K., 249, 250(24), 260(24), 267(24), 268(24, 67), 274(24)
Sims, T. J., 7, 11, 12(23), 13(51), 18(23)
Sinex, F. M., 563, 634
Singer, F. S., 399
Singen, S. J., 827
Skinner, M., 728, 739, 762, 765(6)
Slaevsky, E., 759
Sledge, C. R., 323
Slichter, C. P., 174, 176(4)
Sloan, B., 600

Smith, B. D., 11, 21(43), 24, 42(41), 43, 106, 107(34), 108(34), 109, 112(34), 120, 309
Smith, D. F., 289, 294, 302(122), 303(122), 304(122)
Smith, D. W., 560, 563(6), 565(6), 567, 568(31), 665, 673, 716, 717, 718(10), 721(11)
Smith, G. F., 581, 582(35)
Smith, I. C. P., 176, 180(17), 707
Smith, J. W., 160, 161(90), 572
Smith, K. K., 831(11), 832, 838(11)
Smith, R., 19, 64, 766
Smith, S. P., 326
Smolen, J. S., 498
Snider, G. L., 588, 605
Snider, R., 11
Snyderman, R., 504
Sobel, A., 320
Sobel, M. E., 231
Sobezak, E., 793
Soda, G., 181
Sodek, J., 30
Sokolovsky, M., 599
Solomon, E., 485, 487
Solomon, I., 175, 176, 178(5)
Solter, D., 831, 838(7)
Solursh, M., 775, 782(44)
Somorjai, R. L., 691
Sonneborn, J. H., 227, 234(33), 243(33)
Sorgente, N., 787
Soskel, N. T., 674, 675(10), 704(10), 712(10), 714(10)
Souchon, H., 320
Spackman, D. H., 387
Spaet, T. H., 512
Sparks, J. J., 790, 791(110), 792(110)
Sparks, T. F., 513
Speake, B. K., 350, 352(16)
Speakman, P. T., 222, 332
Spector, D., 732
Speziale, M., 776
Speziale, P., 776
Spiegelberg, H. L., 320
Spiess, H. W., 175, 176
Spina, M., 572, 573, 577
Spiro, M. J., 287, 288, 291, 294(117), 299(116, 117), 302(116, 117), 303(117), 304(116, 117), 349, 352(13)
Spiro, R. G., 7, 34, 78, 79, 95(18), 287, 288, 291, 294(117), 299(116, 117), 302(116,

117), 303(117), 304(116, 117), 343, 347, 349, 352(7, 13), 355(7), 783
Srere, P. A., 501, 507
Stanescu, V., 793, 794
Starcher, B. C., 566, 573, 597, 620, 630(7), 649, 680, 681(36), 689, 711(35, 36), 747, 759
Stark, M., 13, 24, 133, 468
Starkey, P. M., 589, 604
Stassen, F. L. H., 262, 271(60), 647, 648
Stathakis, N. E., 807, 813(26)
Stavitsky, A. B., 752
Steck, T. L., 234
Steffen, C., 475, 477, 483, 485, 498
Stein, W. H., 340, 404
Steinback, H., 665
Steiner, L. A., 323
Steinmann, B. U., 18(90), 19
Stejskal, E. O., 175
Stemberger, A., 806
Stenman, S., 487, 490(**48**), 491(**60**), 492
Stern, P., 572, 576(8)
Stern, R., 2220, 221(3), 244, 385, 386(22), 387(22)
Stevens, C. A., 589
Stevens, F. S., 458
Stingl, G., 508
Stinson, R. H., 145, 149(59), 164
Stirewalt, W. S., 393
Stirtz, T., 13, 370
Stokes, G., 190
Stoltz, M., 474, 475, 485, 486
Stone, P. J., 588, 590, 591, 594, 595(38), 597(38), 602, 603(22), 605
Storch, V., 80
Stoupy, J., 841, 845(18)
Straugh, L., 456
Strause, E. L., 26, 30(120)
Strickland, D. K., 835
Strickland, S., 777, 831(11), 832, 838(11)
Stricklin, G. P., 444, 445(36), 446, 447(36), 450
Strider, W., 786
Strohman, R. C., 732
Stroud, R. M., 322, 323(21)
Studier, F. W., 18, 59
Stuhlsatz, H. W., 776
Sugano, H., 675, 678(13), 681(13), 688(13), 692, 693(56), 694(56), 695(56), 696(56),
697(59), 698(56), 699(56), 701(56), 702(56), 705, 706(67), 707(59), 708(67)
Sullivan, C. E., 87(43), 88(43), 90, 91(43), 92(43), 93(43), 94(43), 95(43), 176, 177, 181(24, 25), 183(25), 676, 691(27), 712(27), 714(27)
Sullivan, K. A., 641, 642(12), 645(12), 647(12), 648(12), 649(12)
Summerhayes, I., 808
Summers, D. F., 353, 354(18)
Sundar Raj, C. V., 487
Sung, S. C., 233
Suzuki, E., 128, 141(9)
Suzuki, I., 769
Suzuki, 769, 781, 787, 790(96), 791(96), 793(84), 795(84), 796(84)
Swarm, R., 11, 12, 21(43), 34, 42(3), 44(3), 99, 105(21), 106(21), 107(21), 109, 110(21)
Sweeny, P. R., 145, 149(59), 164
Sweet, M. B. E., 790
Sweetman, F. R., 600
Swinney, H. L., 205
Sykes, B. C., 114, 565, 666, 672, 728, 748, 751
Szigeti, M., 845, 849(47), 851(47)
Szymanovicz, G., 379, 385(18), 397(18)

## T

Tabas, I., 354
Tabora, E., 348
Tachibana, Y., 355
Tacquet, A., 835
Tai, P. C., 238
Takahashi, S., 593, 603(35)
Takeichi, M., 514
Takeuchi, T., 599
Tanaka, M., 567
Tanford, C., 210, 420, 423(22, 23, 25), 683
Tang, L.-H., 774(42), 775, 781(42), 786(42), 798(48), 800(42)
Tannelle, C., 845
Tanzer, M. L., 11, 16, 22, 65, 67, 69(9), 72, 73, 74(15), 75(9), 76(9, 33), 77(14, 15), 78, 79, 83, 95(9), 134, 365, 369, 417, 454, 580
Tappel, A. L., 249, 252(20), 257(20), 258(20), 747
Tarentino, A. L., 355

Tarrab-Hazdai, R., 486
Tassin, J., 508
Tauber, S., 598, 600(52a)
Taubman, M. B., 473, 474(21), 497
Taylon P., 455, 461(17)
Temple, A., 498
Teng, M.-H., 804
Terato, K., 432
Terranova, V. P., 508, 838
Tesler, A., 633
Tewes, A., 328
Thesleff, I., 487, 498(**48**), 492
Thiery, J. P., 514, 526(7)
Thomas, J., 562, 572, 633, 747, 851
Thomas, M., 749
Thonar, E. M.-J., 772, 796(21), 799, 800(21)
Thurston, G. B., 202
Thyberg, J., 620, 799
Timmins, P. A., 148, 152(95), 158, 161
Timpl, R., 7, 8, 11, 12, 15(59), 16(47, 59), 17, 18(26), 22(47, 80), 23(80), 24(80), 25, 29, 30(116, 117), 31(145), 38, 42, 44(36), 49(44), 59, 83, 98, 99, 100(12, 16), 105(16), 106, 107(36) 108(36), 110, 111(42), 117(36), 118(36), 123(8), 127(12), 133, 136, 138, 162(52), 169(47), 196(14), 197, 215(14), 311, 369, 370(13), 411, 412, 415, 468, 469, 472, 473(1, 10), 474(10, 15, 19), 475, 476(15, 31), 477(16, 22), 478(1, 40, 44), 479(22, 26, 42, 44), 480(10, 22, 25), 482(26), 483(44, 51), 484(1, 2, 11, 15, 22, 44, 51), 485, 486(15, 22, 25, 42, 60), 487(1, 40, 47), 488(5, 16, 22, 40, 60, 72), 489(19, 22, 40, 46, 60, 72, 75), 490(**1**, **3**, **6**, **13**, **19**, **20**, **22**, **26**, **27**, **28**, **32**, **35**, **37**, **41**, **46**, **48**, **50**), 491(**27**, **28**, **52**, **53**, **55**, **58**), 492, 493(47, 71, 75, 87), 494(98), 495(68, 87, 91, 93), 496(104), 497(10, 26, 67, 79, 87, 104), 498(47, 112), 503, 514, 522(12), 649, 650(32), 654, 656, 806, 821, 831(9, 10), 832(3), 833(9), 834(3, 9, 13, 14, 16), 835(3, 12, 17, 18), 836(18), 837(3, 9, 10, 12, 13, 14, 23, 24), 838(3, 9, 10, 13, 14, 16, 26), 840(20), 841, 842
Tinker, D., 620, 650
Tinoco, I., 201
Tkacz, J. S.,, 772
Tocchetti, D., 153
Todaro, G. J., 807, 821(32)

Todd, E. W., 319, 322(1)
Todd, P., 418
Tom, K., 565, 567, 666, 716
Tomichek, E. A., 99, 100(15), 105(15)
Tomlin, S. G., 147, 149
Toole, B. P., 14, 42, 332
Torchia, D. A., 12, 40, 131, 133, 175, 176, 177, 178(22), 179(22), 180(21, 22), 181(24, 25), 183(23, 25), 676, 691(25, 26, 27), 712(25, 26, 27), 714(25, 26, 27), 775, 798(51), 799(52)
Torres, A. R., 674, 712(9)
Tosonev, I., 749
Toussaint, C., 77
Townes, A. S., 475
Trackman, P. G., 34, 315, 638, 642(5), 649, 650(5, 29)
Trapane, T. L., 674, 675, 692, 693(56), 694(56), 695(56), 696(56), 698(56), 699(56), 700, 701(56), 702(14, 56), 703(14), 705, 706(67), 708(67), 710(12), 711(12)
Traub, W., 5, 12(10), 21(10), 24(10), 97, 124, 128, 129(8), 154(8), 305, 373, 394(9)
Traverso, H. P., 573, 575(18), 845
Travis, J., 589, 590
Trelstad, R. L., 5, 11(7), 14, 42, 133, 135, 162(39), 204, 210, 211, 217(36), 332, 490(**41**, **43**), 492, 776
Trentham, D. E., 475
Trimble, R. B., 355
Trinkaus, J. P., 501
Trivedi, D., 789
Trus, B. L., 153
Tryggvason, K., 7, 16(28), 26(28), 42, 98, 105(6), 110(6), 111(6), 248, 251(10), 252(10, 36), 254(10, 34), 256(34), 257(36), 278(10, 34, 53), 279(10), 280(34, 36, 53), 281, 290, 387, 495
Tseng, L., 315, 649, 650(29)
Tseng, S. C. G., 385, 386, 387(22)
Tuderman, L., 29, 249, 252(19), 254(19), 256, 259(19), 260, 261(57), 262(19, 54), 265(19, 57), 266(57, 70), 267(19, 57, 70), 268(19, 70), 269(19, 70), 270(70), 271(62), 273(76, 78), 280(19, 57, 70), 281(91, 92), 282(91), 284(91, 92), 286, 290(106), 305, 309(5), 310(5), 312(5), 313(23), 314, 315(8)
Turino, G. M., 620, 630(6a), 748, 749

Turpeeniemi-Hujanen, T. M., 256, 274, 275(86, 87), 276(86, 87), 277(87), 280(52, 86, 87), 281(52, 93), 284(52, 93)
Turto, H., 719
Twumasi, D. Y., 589
Tyler, J. A., 444, 445, 448, 450(39), 451

## U

Udenfriend, S., 246, 247(3), 248(3), 249(3), 250(12, 23), 251(23), 252(15, 20), 257(20), 258(20), 260(3, 12, 23, 28), 262, 264(12), 265, 268(12, 69), 269(69), 270(3, 69), 271(59, 60, 69), 272(69), 273(69), 284(3), 372(2), 373, 464, 465, 845
Ui, N., 778, 780(71)
Uitto, J., 11, 16, 17, 24(78), 25, 27, 29(82), 99, 100(14), 104(14, 22), 114(30), 117, 118(47), 120, 124(30), 246, 247(2), 248(2), 249(2), 250(2), 251(2), 260(2), 271(2), 280(2), 286, 379, 385(17), 394(11), 568, 716(9), 717, 719(9), 720, 723(9), 727(9)
Ulreich, J. B., 317, 641
Upholt, W. B., 772, 800(20)
Urist, M. R., 504, 505(15)
Urry, D. W., 649, 650(29), 674, 675(5), 677(5), 678, 679(5, 15), 680(31), 681(36, 37, 40, 44), 682, 686, 687(40), 688, 689(15, 41), 691(20, 41), 692(17, 20), 693(50–57), 694, 695(17, 56), 696, 697(17, 59), 698, 699, 700(19, 20), 701, 702, 703(14, 17, 52), 704(17), 705(15, 50), 706(67), 707(59), 708(67), 709(11, 17, 30), 710, 711(5, 12, 15, 17, 31, 35, 36, 37, 40, 50–54), 712(18), 713, 715
Utiyama, H., 87, 89

## V

Vaes, G., 443, 449
Vaheri, A., 99, 100(12, 16), 105(16), 127(12), 252, 487, 488(72), 489(72, 75), 490(**32, 37, 48, 50**), 491(**60**), 492, 493(75, 87), 495(87, 93), 496(104), 497(87, 104), 514, 524(3), 803, 804, 806, 807(24), 808(1), 809, 813(24), 816, 817(49), 821(42, 49), 828(35), 833, 834(13, 14), 837(13, 14), 838(13, 14), 844

Vail, M. S., 6, 101
Valentine, T., 12, 34, 42(3), 44(3), 99, 105(21), 106(21), 107(21), 110(21), 831, 832(4)
Valic, M. I., 176
Valle, K.-J., 424
Vallet, B., 845
Vander Hart, D. L., 175, 177, 180(21)
van Eys, J., 340, 410
Vargas, L., 486, 497(67)
Varner, H. H., 277, 508
Varsson, B. I., 554
Vartio, T., 804, 844
Vater, C. A., 424, 433
Vaughan, D. W., 600
Vehri, A., 831(10), 832, 837(10), 838(10)
Veis, A., 7, 14, 27, 28(128), 36, 38, 89, 96, 129, 191, 196, 213, 214, 216(13), 217(12), 286, 772, 775(19), 798(19)
Vembu, D., 504
Venable, J. H., 136
Venkatachalam, C. M., 695, 697(59), 700, 705, 706(67), 707(59), 708(67), 715
Ventrella, G., 840(63), 850, 851(63)
Vergues, J. P., 831, 834(8), 835(8), 837(8)
Verhoeff, F. H., 574, 578
Vertel, B. M., 772, 800(20)
Vianden, G. D., 317
Vidal, G. P., 319
Vieillard, A., 841(18), 845(18)
Vigny, M., 325, 326(5)
Viljanto, J., 491(**59**), 492
Villiger, W., 320, 234(7)
Vlavoic, M., 606, 614(3), 615(3), 650
Vocaturo, A., 566, 620, 630(7)
Volanakis, J. F., 323
Volpin, D., 164, 165(103), 680, 681(41), 682(34), 686, 687(40), 689(41), 691(41), 704(33, 34), 711(33, 34, 37, 40)
Von Bassewitz, D. B., 133
von der Mark, K., 41, 468, 472, 473, 475, 485, 486(12), 487(12), 448(12), 489(12), 490(**10, 12, 20, 39, 40, 41**), 491, 492, 493, 495(95), 725
vonder Mark, K., 41, 468, 472, 473, 475, 477(12), 478(12), 485, 486(12), 487(12), 488(12), 489(12), 490(**8, 10, 12, 20, 39, 40, 41, 42, 43, 47**), 491, 492, 493, 494(101), 495(91, 92, 95, 101), 725

von Hippel, P. H., 178
Von Rudloff, E., 606
Voss-Foucart, M. F., 77
Voynow, P., 560(35), 565, 649, 650(30), 666, 668(14), 716, 743, 762, 764(7)
Vuento, M., 805, 809(16), 810(41), 817(41), 818(41), 821(41, 42), 822(41)

## W

Wachter, E., 23, 97, 99(4), 311
Wacker, W. B., 437, 443(26)
Waechter, C. J., 350
Waite, J. H., 73, 77(14), 454
Walker, W. H. C., 474, 482(28)
Wallace, B. J., 238
Wallach, D. F. H., 234
Wallach, J., 596
Walter, B. T., 533
Walter, P., 490(**17**), 491
Walton, A. G., 132, 204
Wampler, D. E., 22, 134
Wang, H., 747
Wang, S. K., 144
Wartiovaara, J., 487, 488(72), 489(72), 490(**37**), 492, 495, 831(10), 832, 837(10), 838(10)
Wasmuth, J. J., 227
Wastesson, Å., 776
Watkins, M. S., 326
Waugh, J. S., 175, 176(9)
Wautier, J. L., 320
Webb, G., 326, 329
Weber, K., 410, 411(2), 413, 414, 725, 727, 846
Weber, L., 488, 489(77), 490(**2**), 491
Weber, P., 821
Weeks, J. G., 440
Wegener, W. A., 215
Weinbaum, G., 597, 600, 748(27), 749(27), 756, 757(27)
Weinstein, E., 274
Weir, C. E., 747
Weiser, M., 841
Weislander, J., 799, 800(143)
Weisman, Z., 513
Weiss, J. B., 7, 19(33), 225(16), 226
Weiss, L., 502
Weiss, P. H., 398

Weissman, N., 560, 563(6), 565(6), 566, 657, 665, 673, 716
Weithmann, K. U., 287
Welgus, H. G., 429
Wells, P., 507
Welsh, C., 8, 12(36), 15(36), 42(40), 43, 47(40), 49(40)
Welsch, U., 80
Wendt, P., 24
Werb, Z., 596, 603(45)
Weser, U., 281
Wessells, N. K., 500
West, T. W., 418
Westermark, B., 776
Weston, J. A., 808, 814(40)
Whitaker, S., 561
White, B., 357
White, D. A., 350, 352(16)
White, R. R., 590
White, S. W., 145, 148(58), 151(77), 152(77), 155(58), 157, 158(77), 159(58), 160(58)
Whiting, A. H., 565, 651
Whitley, R. J., 599, 605(55)
Whitney, J. G., 387
Wick, G., 12, 15(59), 16(59), 29, 30, 31(145), 44, 83, 110, 111(42), 136, 138, 162(52), 169(47), 472, 473, 474(19), 475, 477(22), 478(40), 479(22, 42), 480(22), 484(22), 485, 486(22, 42, 60), 487(40), 488(5, 22, 40, 60), 489(19, 22, 40, 46, 60), 490(**1, 3, 13, 19, 22, 26, 27, 28, 35**), 491(**27, 28, 53, 55, 58**), 492, 493, 494(98), 498, 835, 837, 838(26)
Wiedemann, H., 11, 12, 15(59), 16(47, 59), 22(47), 44, 49(44), 83, 106, 107(36), 108(36), 110, 111(42), 117(36), 118(36), 133, 166(32), 490(**3**), 491
Wiestner, M., 28, 468, 491(**52**), 492, 493, 497, 838
Wight, T. N., 778
Wilczek, J., 28, 220, 221(3), 244, 788, 799, 837, 838(29)
Williams, C. A., 762
Williams, I. F., 26, 30(119), 99
Willings, S. R., 554
Winchester, R. J., 322
Winge, A. R., 630
Winter, K. K., 597
Winzler, R. J., 356, 403, 404(14), 405(14), 410(14)

Wirl, G., 440
Wisdom, B. J., Jr., 411, 418(6)
Witkop, B., 387
Wittnauer, L. P., 217
Wodak, S., 181
Woessner, D. E., 175, 176(7), 178(7), 179(7)
Woessner, J. F., 376, 380, 436
Wohllebe, M., 11
Wolfenstein-Todel, C., 786
Wolff, C., 147, 153(65)
Wolff, I., 475, 483, 841
Wolinsky, H., 617, 721
Wong, K., 473, 474(17)
Wood, G. C., 135
Wood, S., 428
Woodbury, R. G., 7, 12(24), 18(24), 21(24), 122, 123(52), 225, 417
Woodhead-Galloway, J., 132, 133(20), 134, 146(20), 149(93), 161, 162(20)
Woolley, D. E., 444, 445(35), 446(37), 448, 806
Worthington, C. R., 144, 147, 149
Wrann, M., 805, 809(16)
Wray, J. S., 151
Wright, G. P., 498
Wu, C., 289, 294, 302(122), 303(122), 304(122)
Wu, C. H., 490(**29**), 491(**29**), 492
Wu, S. C., 742, 743(19), 762
Wünsch, E., 461
Wyckoff, R. W. G., 159
Wylie, D. E., 526

## Y

Yaar, M., 837, 838(30)
Yaeger, J. A., 16, 22, 134
Yamada, K. M., 231, 507, 508, 803, 804, 805(7), 808(2, 13), 814(7, 40), 815
Yamagata, T., 790
Yamamoto, T., 24
Yamaoka, K., 201
Yamashina, I., 778, 780(71)
Yamashita, K., 355
Yanagishita, M., 777, 793(69), 794(69), 795(69)
Yang, F. C., 593, 603(35)
Yang, J., 499, 535, 554
Yankowski, R., 167
Yankeelov, J. A., 437, 443(26)
Yaoita, H., 490(**5**), 491, 493
Yasunobu, K. T., 319
Yau, M. C., 567
Yerganian, G., 227, 228(30)
Yonemasu, K., 322, 323(21), 324
Yoshida, E., 455, 461(12), 573
Yoshikane, K., 599
Yoshioko, K., 217
Yphantis, D. A., 210
Yu, S. Y., 593, 759
Yuan, L., 213, 214, 216
Yurchenco, P. D., 347, 348(9)
Yuspa, S. H., 838

## Z

Zardi, L., 807
Zeikus, R. D., 565
Zelmenis, G., 839, 844(1)
Zetter, B., 505
Zettner, A., 474, 481(29), 482(29)
Ziccardi, R. J., 319, 320
Ziegelmayer, G., 490(**35**), 492
Zimm, B. H., 215
Zimmerman, M., 590, 599, 603
Zinn, A. B., 348
Zizic, T. M., 498
Zuccarello, L. V., 81
Zupnik, J., 305, 312(4), 314(4)
Zweibaum, A., 839, 843(5), 844(5)

## Subject Index

### A

AB collagen, see Collagen, type V
ε-ACA, see ε-Aminocaproic acid
Acetic acid, in collagen extraction, 12
   0.5 $M$, 37–38
   with pepsin, 38
Acetylcholinesterase, 4, 325–339
   collagen-like domain, 334–335
   forms, 325
   intact tail subunits
     amino acid composition, 338–339
     isolation, 335–339
   in junctional basement membrane, 325–326
   pepsin-resistant fragments
     amino acid composition, 338–339
     isolation, 332–335
   145, isolation, by affinity chromatography, 328–331
   18 S, 325
     isolation, by affinity chromatography, 328–331
     structure, 326–327
Achilles tendon, see Tendon
Acridinium-linked affinity resin
   preparation, 328–329
   used to purify acetylcholinesterase, 328–331
Affinity chromatography
   isolation of acetylcholinesterase, 328–331
   for measurement of platelet-collagen adhesion, 512–513
   for purification of procollagens, 114–115
   for purification of vertebrate collagenase, 449–450
   zinc-chelate, purification of collagenase, 451
α-Aminoacetonitrile, 34
   administration, to chicks, and tropoelastin yield, 666–667
   inhibitor of lysyl oxidase, 563
ε-Aminocaproic acid, inhibitor of proteolysis, 35
β-Aminopropionitrile
   effect on insoluble elastin formation, 630–631
   inhibitor of lysyl oxidase, 563, 565, 638, 666
β-Aminopropionitrile fumarate, 33–34
   fumarate content, and tropoelastin yield, 666
Amnion, source of collagen, 12, 35
Amniotic membrane, collagens detected, 490
Annelids, see Cuticle collagen; species name
Annulus fibrosus, collagen fibril structure, 152, 161
Antibodies, conjugated, in localization of collagen
   for electron microscopy, 136–137
   for light microscopy, 137–138
$β_1$-Anticollagenase, 424
Aorta
   bovine, elastin, amino acid composition, 575
   collagen fibril structure, 164–165
   microfibrillar glycoprotein, 849
   pig, structural glycoprotein
     amino acid composition, 841
     apparent molecular weight, 840
   proteoglycans, 775–776
   structural glycoprotein, isolation, 851–852
*Ascaris*
   cuticle collagen, classification, 65–66
   *lumbricoides,* cuticle collagen
     amino acid composition, 88, 93
     extraction and purification, 89–91
     physical properties, 87, 92–93
     $R$ values, 95
     subunit, 87–88, 92–93

*suum,* basement membrane collagen, *R* values, 96
ASCC, *see* Cuticle collagen, acid-soluble
Ascorbate
 effect on insoluble elastin formation, 630–632
 role in hydroxylase reactions, 284
Ascorbic acid, *see* Ascorbate
Asparagine-linked glycosides, 346–360
Asymmetric particles
 solution behavior, 187–196
  hydrodynamic properties, 190–196
  relaxation processes, 196–209
   dielectric constant, 199
   dielectric loss, 199
   dielectric relaxation, 197–199
   electric birefringence, 200–204
   thermodynamic properties, 187–190
Atherosclerosis
 elastin degradation in, 588
 plaques, collagens detected in, 491
Azetidine-2-carboxylic acid, effect on insoluble elastin formation, 630–632

# B

Basement membrane
 glomerular
  isolation, 34–35
  proteoglycan, 778
  source of type IV collagen, 11
 isolation, 34
 laminin content, 838
 noncollagenous protein, 831
 proteolysis, inhibitors, 35
Basement membrane collagen, 65; *see also* Collagen, type IV
 molecular species, 6–7
 nomenclature, 6–7
 *R* values, 95–96
Blood vessel, *see also* Aorta
 brain, isolation, 35
 collagens detected, 490
 major, source of type III collagen, 11
 wall, source of collagen, 35
Bolton-Hunter reagent, $^{125}$I-labeled
 desmosine probe, 760–761
 in elastin radioimmunoassay, 753–754

Bone
 bovine dentine, collagen fibril structure, 152
 collagen fibril structure, 152, 157–158
 collagens detected, 490
 demineralized
  chicken leg, collagen fibril structure, 152, 158
  rat tibia, collagen fibril structure, 152, 158
  X-ray diffraction studies, 158
 developing, collagens detected, 490
 rat dentine, collagen fibril structure, 152
 source of type I collagen, 10, 35
Bovine serum albumin, molecular weight and collagen residues, 412–413
Brain microvessels, intact, isolation, 35
Bromelain, 747
*n*-Butanol, fractionation of tropoelastin, 653–654
4-Butyrobetaine hydroxylase, 247

# C

C1q, 319–324
 assay, 321
 chemical composition, 324
 collagen-like region, 4
 non-human, 323
 properties, 323–324
 purification procedure, 321–323
 serum concentration, 320
 stability, 323
 structure, 324
C1r, 319–320
C1s, 319–320
*Caenorhabditis elegans,* cuticle collagen, 90, 93
 amino acid composition, 88
 classification, 65–66
 *R* values, 95
Calcification, 157, 158
Calf, copper-deficient, isolation of soluble elastin from, 565
Calvaria
 chick
  collagen fibrils, preparation and $^{13}$C- and $^2$H-labeling, 177–178

## SUBJECT INDEX

embryo, recovery of procollagen
  mRNA from, 218–222
type V collagen from, $R$ values, 95
Carboxymethyl cellulose chromatography
  at acid pH, in purification of collagen, 45–47
  at alkaline pH, in purification of type V collagen, 47–49
  concave gradients, for rechromatography of peptides, 57–58
  of denatured collagen, 52–55
  linear gradients, for resolution of CNBr collagen peptides, 55–57
  for separation of type I procollagen and pC-collagen chains, 112
  of type IV collagen, 15
  of type V collagen, 15
Cartilage, *see also* Hyaline cartilage
  bovine
    nasal, collagen fibril structure, 152
    structural glycoprotein, apparent molecular weight, 840
  canine, structural glycoprotein, amino acid composition, 841
  collagen fibril structure, 152, 160–162
  ear
    elastin, amino acid composition, 575
    tropoelastin
      amino acid composition, 671
      isolated from lathyritic piglet, 671
    electron microscopy, 160–161
  mammalian articular, source of type II collagen, 10
  noncollagenous components, 130
  proteoglycan, 772–775
    content, 160
    extraction, 36
  rabbit knee articular, collagen fibril structure, 152
  type V collagen from, $R$ values, 95
  X-ray diffraction studies, 161–162
Cathepsin G, elastase activity, 589
Cell-collagen interaction, 499–503
  in cell growth and differentiation, 499–500
  influence of collagen organization on, 502–503

influence of collagen type on, 502
mediated by specific adhesion proteins, 503
requirement for specific attachment factors, 507–508
scanning electron microscopy analysis, 537–538, 541–544
transmission electron microscopy analysis, 538–540, 543–544
ultrastructural studies, 535–544
Cell culture
  biosynthesis of soluble elastin, 716–731
  biosynthetic matrices from, 530–535
  long-term, synthesis of soluble elastin, 720–723
  short-term
    cell characterization, 719
    incubation, 719–720
    isolation of cells for, 719
    synthesis of soluble elastin, 719–720
Cell-free synthesis
  of noninterstitial procollagen chains, 225–245
  product characterization, 234–235
  of protein
    directed by CHL cell components, 242–245
    reaction conditions, 232–234
    reaction parameters and characterization of products, 235–242
  preparation of components, 229–232
Cell-substrate adhesion, 530–535
  microexudate carpets, 530–531
    characterization, 532
    enzyme-digested, 531
  SDS carpet, preparation, 531
  urea carpet
    adhesion of single cells, 533–534
    characterization, 533
    collagenase-digested, 531
    preparation, 530–531
    trypsin-digested, 531
Cetylpyridinium chloride, precipitation of proteoglycan with, 787

Chick
    copper-deficient
        isolation of soluble elastin from, 565, 650–657
        nutritional conditioning, 651–652
        tissue homogenization, 652–653
    eye, collagens detected, 490
    lathyritic, isolation of soluble elastin from, 565, 667
    production of collagen antibodies, 475
Chinese hamster lung cells
    cell-free postnuclear extract (S30)
        preparation, 229
        procollagen chain synthesis, reaction parameters and characterization of products, 235–238
        reactions catalyzed by, 232–234
    growth, in culture, 229
    messenger RNA
        isolation, 230–232
        procollagen chain synthesis, reaction parameters and characterization of products, 240–243
        reactions catalyzed by, 234
    polysomes
        preparation, 229–231
        procollagen chain synthesis, reaction parameters and characterization of products, 238–240
        reactions catalyzed by, 233–234
    procollagen chains, synthesis, 225–245
CHL cells, see Chinese hamster lung cells
6-Chloronorleucine, 364, 366
Chondroblast, culture, synthesis of elastin, 630
Chondrocyte, collagen adhesion, 502, 508
Chondroitin sulfate, disaccharide repeat unit, 770
Chondronectin, 503
    collagen binding, 508
Chorion, source of collagen, 12, 35
Chromatographic techniques, see specific technique
Chymotrypsin, enhancement of elastin solubilization, 597
Chymotrypsinogen, molecular weight and collagen residues, 412–413
Clamworms, see *Nereis* sp.
*Clostridium histolyticum,* collagenase, see Collagenase, bacterial
Clostripain, contamination of commercial collagenase, 465
CM-cellulose chromatography, see Carboxymethyl cellulose chromatography
Colchicine, effect on collagen gel, 554
Collagen, see also Basement membrane collagen; Cuticle collagen; Interstitial collagen; Invertebrate collagen
    antibodies, 472–498; see also Antibodies, conjugated
        applications, 487–498
            in biosynthetic studies, 494–496
            in localization of collagen in tissues and cells, 487–494
            in quantitative determination of antigens, 497–498
        assays
            immunochemical, 479–481
            inhibition, 481–484
        purification, 476–479
        specificity, 484–487
    autoantibodies against, in disease, 498
    biosynthesis, 226–227
        measurement, using *C. histolyticum* collagenase, 471
        posttranslational enzymes in, 245–319; see also specific enzyme
        studies using antibodies, 494–496
    calf skin
        α chains, compared to standard globular proteins on SDS-gel electrophoresis, 415
        CNBr peptides
            electrophoretic mobility and number of residues, 417–419
            molecular weights and number of residues, 412–413
            SDS-gel electrophoretic analysis, 415–417
        categories, 65–66, 94–96

in cells and tissues, determination by
    hydroxyproline measurements,
    393–394
chain
  definition, 4
  precursors, 4; see also
    Preprocollagen
α-chains
  amino acid composition, 19–21
  CNBr cleavage products, 21–22
    electrophoresis profiles, 63–64
    resolution, 55–59
  nomenclature, 5–8
characterization, 18–22, 49–64
  by polyacrylamide gel
    electrophoresis, 59–64
chromatography, 15; see also specific
    chromatographic technique
cross-links
  aldimine, 362
  borohydride reduction, 361
  covalent, 360–372
  intermolecular, 360
  intramolecular, 360–361
    reduced, 364, 366
  keto-imine, 362
  mature, 360
    labeling, 371–372
    presumptive, isolation, 368–369
  quantitation, 368
  reduced components
    identification, 364–368
    isolation, 361–364
  reducible, 360
degradation
  measurement, 374, 380, 396–397
    by urinary hydroxylysine
      excretion, 399
    by urinary hydroxyproline
      excretion, 398–399
denatured, used for cell attachment,
  517
electrophoretic mobility, correlated
  with number of residues,
  411–413
extraction, 37–39
  in 0.5 M acetic acid
    insolubility in neutral salt
      solvents, 40
    recovery, 40
  with pepsin, recovery, 40–41
  in neutral salt solvent, initial
    purification, 39–40
  solvents, 12, 37–39
  use of limited proteolysis, 13–14
fibrils
  antibody-coated, inhibitors of
    platelet aggregation, 498
  $^{13}C$- and $^{2}H$-labeled, preparation,
    for NMR studies, 177–178
  D periodicity, 129–130, 136, 137
  diameter
    factors determining, 174
    related to carbohydrate
      content, 285
  native
    reprecipitated, 134–135
    reconstituted, for x-ray
      diffraction, 138–139
  NMR data
    on motion of polypeptide
      backbone, 178–183
    on motions of polypeptide side
      chains, 183–186
  NMR studies, 174–186
  radiolabeled
    assay for collagenase, 433–434
    film collagenase assay, 435–436
  structure
    in bone, 157–158
    in connective tissue disease,
      167–172
    in embryonic connective
      tissues, 167, 168
    in rat tail tendon, 146–154
    in type I-containing tissues,
      146–160
    in type II-containing tissues,
      152, 160–162
    in type III-containing tissues,
      162–166
    variations, in tendons, 154–157
  substrate for vertebrate
    collagenase, 433–436
  ultrastructural localization,
    493–494
  used for cell attachment studies,
    517–518
fibronectin binding, 806–807
  assay, 823–826

fibrous, see also Interstitial collagen
   characterization, 127–174
   D periodicity, 141
   electron microscopy, 131–137
      labeled with conjugated
         antibodies, 136–137
   light microscopy, 137–138
   preparation and investigation,
      131–146
   reprecipitated forms, 132–135
   thin sections, for electron
      microscopy, 135–136
fibrous-long-spacing, 134
gels; see also Collagen, substrates
   floating, 547, 551–552
      contraction during cell culture,
         552–554
      effects on cell culture, 549–550
   nylon mesh support, 554
   preparation, 546–547
H peptides, 369–370
heterogeneity, 4
   genetic, 3–4
   molecular, 3, 31–32
immunization with, 473–476
insect ejaculatory duct, $R$ values, 95
isotopic labeling
   in vitro, 426–427
   in vivo, 425–426
labeled, storage, 427
lattice, in corneal stroma, 776–777
mixture, of types I and III, resolution,
   44, 64
molecule, length, 128
native, monomers, used for cell
   attachment, 517
newly synthesized
   degradation, use of hydroxyproline
      measurements to quantify,
      396–397
   quantification by collagenase
      digestion, 394–396
nomenclature, 5–9
peptides
   chromatography, 55–59
   cross-linked
      containing reducible bonds,
         preparation and
         identification, 369–370
   isolation and characterization,
      370–372
   molecular weight, estimations,
      58–59
   polyacrylamide gel electrophoresis,
      63–64
precipitation, 14
   selective salt, 14–15
preparation, 10–15, 33–64
   tissue acquisition, 33–37
purification
   chromatographic techniques,
      44–49; see also specific
      technique
   denaturing conditions, 50–55
   initial steps, 39–41
   selective salt precipitation, 41–44
   for use as substrate in collagenase
      assay, 425
segment-long-spacing form, 133–134
selective salt precipitation
   from dilute acid solvent, 43
      NaCl concentrations for, 43
   from neutral salt solvent, 43
      NaCl concentrations for, 43
$R$ values, 66, 94–96
   calculation, 94
radiolabeled, binding to platelets, 512
reconstituted, preparation, 505
7S
   isolation, 44, 469
   precipitated from neutral salt
      solvent, NaCl concentration
      for, 43
   quantitation, 497
soluble, see also Asymmetric
   particles, solution behavior
   characterization, 186–217
      experimental considerations,
         215–217
      dipole moment, 197–200
      electric birefringence studies, 217
      equilibrium ultracentrifugation,
         210–215, 216
      light scattering techniques for,
         203–210
      molecular weight determinations,
         188–189
      monomer and model dimers
         calculated diffusion coefficients,
            191

calculated hydrodynamic
parameters, 194
calculated sedimentation
coefficients, 191
calculated translational
frictional coefficients,
190–191
excluded volume second virial
coefficients, 188–189
sedimentation techniques, 216–217
solutions
coating of plastic tissue culture
plates, 546
for tissue culture, preparation,
545–546
substrates
appearance
observed by scanning electron
microscopy, 540–541
observed by transmission
electron microscopy,
540–541
baby hamster kidney cells on,
ultrastructural studies,
541–544
cell adhesion on, 500–501
inhibitors, 501
for cell culture, 544–555
cell motility on, 501
cell separation from, 502
denatured collagen, 517
dried, 535
incubation with cells, 537
preparation, 536–537
requirement for adhesion
proteins, 503
fibroblast adhesion to, 503–508
assay, 505–506
characteristics, 506–507
ultrastructural studies,
539–540, 542–544
film, preparation, 504–505
for hepatocyte adhesion studies,
516–518
hydrated, 535
incubation with cells, 537
preparation, 537
native fibers, 517–518
native monomers, 517
preparation, 504–505

synthesis, measurement, 374, 380,
387, 391
tissue distribution, determination,
487–494
triple-helical conformation, 128
type I, 127; *see also* Interstitial
collagen
amino acid sequence, 128
antigenicity, 484–486
chains, 128
$\alpha$-chains
amino acid composition, 19–21
electrophoretic mobility, 63–64
cleavage by vertebrate
collagenase, 428–429
$\beta$ component, electrophoresis
profile, 64
$\gamma$ component, electrophoresis
profile, 64
denatured
CM-cellulose elution pattern,
53–54
molecular sieve
chromatography, 50–51
extracted in 0.5 $M$ acetic acid,
insolubility in neutral salt
solvents, 40
immunohistological localization,
489–493
nomenclature, 5–6
precipitated from neutral salt
solvent, NaCl concentration
for, 43
purification, 18
$R$ values, 95
reconstituted fibers, structure, 152,
159–160
segment-long-spacing, banding
pattern, 133, 162, 166
sources, 10, 35
tissues containing, collagen fibril
structure in, 146–160
type I-trimer, 6; *see also* Interstitial
collagen
precipitated from neutral salt
solvents, NaCl concentrations
for, 43
sources, 11
type II, 127; *see also* Interstitial
collagen
amino acid sequence, 128

antigenicity, 484–486
α chain, 128
  amino acid composition, 19–21
  CM-cellulose elution pattern, 53–54
  cleavage by vertebrate collagenase, 428–429
  denatured, electrophoresis profile, 63–64
  immunohistological localization, 489–493
  nomenclature, 5–6
  possible molecular species, 6
  precipitated from neutral salt solvent, NaCl concentration for, 43
  purification, 18
  $R$ values, 95
  reconstituted fibers, structure, 162
  segment-long-spacing, staining pattern, 133, 162
  sources, 10–11, 35
type III, 127; see also Interstitial collagen
  amino acid sequence, 128
  antigenicity, 484–486
  α chain, 128
    amino acid composition, 19–21
    CM-cellulose elution pattern, 53–54
    cleavage by vertebrate collagenase, 428–429
    denatured, electrophoresis profile, 63–64
    extracted in 0.5 $M$ acetic acid with pepsin, insolubility in neutral salt solvents, 40–41
    immunohistological localization, 489–493
    nomenclature, 5–6
    precipitated from neutral salt solvent, NaCl concentration for, 43
    purification, 18
    $R$ values, 95
    reconstituted fibers, structure, 166
    segment-long-spacing, staining pattern, 133, 166
    sources, 11, 35
    tissues containing, collagen fibril structure in, 162–166
type IV, 127, 225–226; see also Basement membrane collagen
  antigenicity, 486
  bovine, $R$ values, 96
  canine, $R$ values, 95
  α chains
    amino acid composition, 19–21
    CM-cellulose elution pattern, 53–55
  degradation, 451–452
  human, $R$ values, 96
  immunohistological localization, 489–493
  mouse, $R$ values, 96
  nomenclature, 6–7
  pepsin-resistant regions, 18
  pepsin-solubilized
    CM-cellulose chromatography, 46–47
    electrophoresis profile, 63–64
  precipitated from neutral salt solvent, NaCl concentration for, 43
  purification, 45
  $R$ values, 95–96
  sources, 11–12, 35
  viscosity modeling, 196
type V, 127, 226
  antigenicity, 486
  α chains
    amino acid composition, 19–21
    CM-cellulose elution profile, 54–55
    CNBr peptides, CM-cellulose elution profile, 56–57
  electrophoresis profile, 63–64
  immunohistological localization, 489–493
  molecular species, 6–8
  nomenclature, 6–8
  pepsin-solubilized, CM-cellulose chromatography, 48–49
  precipitated from neutral salt solvent, NaCl concentration for, 43
  purification, 18
    CM-cellulose chromatography, 47–49
    DEAE-cellulose chromatography, 49
  $R$ value, 66, 95

sources, 12, 35
types, 3–32
   characterization, 33–64
urinary metabolites, 398–410
vertebrate, $R$ values, 95–96
x-ray diffraction studies, 138–146
Collagenase
   bacterial, 453–471
      *Achromobacter iophagus,* 454
      Advance Biofactures
         digestion of acid-soluble collagen, 466–467
         purity, 465–466
         assays, 456–461
            Azocoll substrate, 456–457
            insoluble tendon collagen substrate, 457–458
            radioactive acid-soluble collagen substrate, 458–461
            synthetic substrates, 461
      Boehringer Mannheim, 465–466
      *Clostridium histolyticum,* 453, 454–456
         purification, 461–462
         substrate heterogeneity, 454–456
         used to isolate nonhelical regions of procollagen, 468
         used to measure collagen biosynthesis, 471
      commercial preparations, purity, 465–467
      contamination with nonspecific proteases, 456
         assay for, 463–467
      properties, 453–454
      protease-free, uses, 468–471
      S-200 purified CLSPA, 465–466
         digestion of acid-soluble collagen, 466–467
      as structural probe, 468–469
      in studies of hepatocyte-collagen adhesion, 515
      *Vibrio B-30,* 454
      Worthington CLSPA, 465–466
   catalysis, 424
   complexed to $\alpha_2$-macroglobulin, assay, using synthetic substrate, 439
   digestion of labeled collagen, to determine collagen content, 394–396
   inhibition, 424
   production, 424
   type IV, assay, 452
   vertebrate, 423–452
      assays, 428–439
         DNP-peptides for, 437–439
         using fibrillar substrate, 433–436
         plaque, using agarose-collagen, 437
         using soluble substrate, 429–433
         substrates, 425–428, 437–439
      cleavage of procollagens, 124
      diffuse fibril assay, 434–435
      extraction
         from cells, 441
         from tissues, 440–441
      human skin, amino acid analysis, 445–446
      isolation, 439–444
      latent, 439
         activation, 443–444
      microassay, using endogenous substrate, 436–437
      purification, 444–451
         special techniques, 449–451
         stability during, 446–447
      recovery
         from explant cultures, 441–442
         from monolayer cultures, 442–443
      specific activity, 445
      synthetic substrates, 427–428, 437–439
      tadpole, amino acid analysis, 445–446
      units of activity, 444
Collagenous components, separated by column chromatography, identification, 469–471
Collagenous proteins
   quantitation, 497–498
   size estimation, 410–423
   ultrastructural localization, 493–494
Complement, 319–320; *see also* C1q; C1r; C1s

Connective tissue
  collagen component, 127
  disease, collagen fibril structure in, 167–172
  embryonic, collagen fibril structure, 167, 168
  freeze-fracture preparation, for electron microscopy, 136
  isolation of procollagen messenger RNA from, 218–222
  noncollagenous components, 130
  total RNA, isolation, 732–733
  x-ray diffraction studies of axial fibril structure, 149
Copper, cofactor for lysyl oxidase, 565, 648–649
Cornea
  calf
    structural glycoprotein
      amino acid composition, 841
      apparent molecular weight, 840
  macular dystrophy, 777
  source of type V collagen, 12
  stroma, proteoglycan, 773, 776–777
Culticle collagen
  acid-soluble
    components, 75
    denatured, gel filtration, 69
    extraction, 66–69
    segment-long-spacing crystallites, 83
    subunits, purification, 69–75
  annelid
    native fibers, electron microscopy, 80–81
    reconstituted fibers, electron microscopy, 81
    segment-long-spacing crystallites, 81–84
  *Ascaris*
    extraction, 89–91
    properties, 92–93
    purification, 89–91
    subunit gelatin chains, preparation, 91–92
    subunits, properties, 92–93
  classification, 65–66, 94–96
  nematode, 89–94
    amino acid composition, 88
    classification, 65–66

    physical properties, 87
    $R$ values, 95
  *Nereis,* degradation, 454
  neutral pH salt-soluble
    components, 75
    extraction, 69–70
    segment-long-spacing crystallites, 83–84
  oligochaete
    amino acid composition, 88
    chemical characteristics, 88–89
    classification, 65–66
    extraction, 83–86
    properties, 86–89
    purification, 83–86
    $R$ values, 95
  polychaete
    classification, 65–66
    extraction, 66–69
    glycosylation, 78–79
    properties, 75–79
      chemical composition, 77–79
      physical, 75–77
    purification, 69–75
    $R$ values, 95
    stability, 77
    subunits, 75
      chemical composition, 77–79
      physical properties, 75–77
Cytochalasin, effect on collagen gel, 554

# D

$D$ period, 129–130, 132
  in bone, 149, 157
  in cartilage, 149, 161
  electron density distribution in, 145, 147
  in embryonic tissue, 167
  in skin, 149, 164–165
  in spleen reticulum, 149, 166
  in tendon, 129, 141, 146, 149, 154, 159
  in type III collagen, 166
  variations, in electron microscopy, 137
  in various connective tissues, 149
Dansyl chloride, labeling of mature cross-links, 372

DEAE-cellulose chromatography
 initial purification of collagen with, 41
 in resolving procollagens, 16–17
 of type II collagen, 15
 of type IV collagen, 15
 of type V collagen, 15, 49
Dehydrolysinonorleucine, in elastin, 580–582
Dentin, source of type I-trimer collagen, 11
Deoxycholate, in proteoglycan extraction, 782
Dermatan sulfate, 770
Dermatosparaxis
 collagen fibril structure in, 168–172
 collagens detected, 491
 procollagen quantitation in, 497
Dermis
 rabbit, structural glycoprotein
  amino acid composition, 841
  apparent molecular weight, 840
 source
  of type I collagen, 10, 35
  of type I-trimer collagen, 11
  of type III collagen, 11, 35
Descemet's membrane, source of type IV collagen, 11
Desmosine, 369; see also Elastin, cross-links
 antibodies, 759–762
 radioimmunoassay, 759–762
 pH optimum, 762
Diffusion coefficient
 rotational, 194
 translational, calculation, 191
DFP, see Diisopropyl fluorophosphate
Dihydroxylysinonorleucine, 362, 365, 367
 galactosyl derivatives, 365, 367–368
 glucosylgalactosyl derivatives, 365, 367–368
5,6-Dihydroxynorleucine, 364, 366
Diisopropyl fluorophosphate
 added to neutral salt solvent, 37
 inhibitor of proteolysis, 35, 37
α,α'-Dipyridyl
 effect on insoluble elastin formation, 630–632
 inhibitor
  of lysyl oxidase, 649

of proteolysis, 37
Disease, see Connective tissue, disease; specific disease

# E

Earthworm, see Cuticle collagen; species name
Edman degradation, use, for release of carboxy-terminal peptides from elastin desmosine cross-links, 606–607
Eel electric organ, acetylcholinesterase from, see Acetylcholinesterase, 18S
Ehlers-Danlos syndrome
 type IV
  collagen abnormalities in, 168
  type III procollagen quantitation, 497
 type VII, 29
Elastase, see also Elastolysis
 enhancement of type III collagen solubilization, 11
 inhibitors, measurement of activity, 604–605
 leukocyte
  cleavage of type IV collagen, 424
  degradation of type IV collagen, 451–452
  digestion of elastin, 747
 maleylated, 601–603
 mouse peritoneal macrophage, 589, 590, 603
 pancreatic, 588–589
  pH optimum, 600
 porcine
  digestion of insoluble elastin, 602, 747
  exhaustive digestion of elastin, 608
  properties, 590
 polymorphonuclear leukocyte, 588–589
  assay, sodium chloride concentration, 600
  properties, 590
Elastase-like activity, soluble substrates, 599–600
Elastic fibers
 components, 559
  amino acid composition, 560

morphology, variation, with tissue type, 578–579
Elastin
  amino acid composition, 560, 576–577
    from tissues of different ages, 561
  amorphous, 616
    ultrastructural analysis, 618–622
  antibody, reaction with soluble elastin, 728–729
  biosynthesis, 566–570
    in cell cultures, 568–569
    in organ cultures, 567
  cartilage, sequence data, 578
  characterization, 576–587
    electron microscopic, 578–579
    histochemical, 578–579
  coacervation, definition, 673
  Congo red, 593
  cross-linked peptides
    amino acid composition, 613
    isolation, 608–613
    separation from released single-chain peptides, 613–614
    two chains, simultaneous sequencing, 615
  cross-links, 561–564
    alanine enrichment, 563
    analysis, 585–586
    desmosine, 562–563
      amino acid composition, 564
      formation, 562–564, 633–634
      isolation of cross-linked peptides, 608–613
      photolysis in ultraviolet light, 585–587
      single-chain peptides released from, 563–564, 606–607
    formation, inhibition by copper deficiency, 565
    intermediates, lysine-derived, 635–636
    isodesmosine, 562
      formation, 564
    lysine-derived, 626–627, 635–636
      sodium borohydride reduction, 625–626
    precursors, 571
    ring, reduction, 582–585
  enzymic digestion, 747

  fibrous
    physical properties characterizing randomness, 711–716
    twisted filament model, 711–716
  hot alkali extraction, 572–573, 591
    effect on susceptibility to proteolysis, 603
  insoluble
    amino acid composition, and age of cell culture, 623
    antibodies, 744–759
    antiserum
      cross-reactivity between species, 758–759
      cross-reaction with tropoelastin, 744–745
      preparation, 747–751
      recovery of IgG fraction 750–751
      storage, 750
    autoclaving, 591
    biosynthesis, 615–637
      by cell culture, 617–632
      effects of various reagents, 630–632
      incorporation of soluble precursors during, 630
      in organ cultures, 632–636
    from cell culture
      amino acid analysis, 622–623
      preparation, 622–629
    chromophoric ligands, 593
    definition, 636–637
    isolation, 561
    labeled sodium borohydride reduction, 593
    lysine-pulsed, 625–628
    peptide bond hydrolysis, pH stat assay, 597–598
    preparation, for use as elastase substrate, 589–592
    proteolysis, 588–605
    pulse-chase procedures, 624–625
    radiolabeled, solubilization assay, 594–596
    sodium borohydride reduction, 625–626
    solubilization assays, 594–598
    structure, 606–615

# SUBJECT INDEX

isolation, 559, 571–576
  from elastic cartilage, 575
  from lung, 571–575
labeled sodium borodeuteride reduction, 579–580
labeled sodium borohydride reduction, 579
  effect on cross-linking compounds, 580–582
ligament, sequence data, 578
messenger RNA
  cell-free translation, 569–570, 731–743
    conditions, 734–739
    isolation, 732–734
microfibrillar glycoprotein, 616
  ultrastructural analysis, 618–622
peptide
  C-O chemical shift, solvent dependence, 695–701
  NH chemical shift
    solvent dependence, 700
    temperature dependence, 693–698
peptide aggregates
  structure, 680–691
  temperature-dependent characterization, 676
peptides
  repeat sequence
    conformational energy analyses, 704–708
    cyclic conformations with linear correlates, 692–693
    NMR characterization of torsion angles, 704–708
  soluble, characterization, 673–716
  structural characterization, detailed, 691–711
preparation
  effect on susceptibility to proteolysis, 603
  for use as elastase substrate, 589–593
purification
  using CNBr treatment, 592
  difficulty, 560
  with guanidine-HCl, 592
  to homogeneity, 608
radiolabeled, 593

reaction with [$^{14}$C]NaCN and $NH_3$, 580
  effect on cross-linking compounds, 585
single-chain peptides, released by Edman degradation, 606–615
  purification, 614
  sequencing, 614–615
solubilization, 744–746
solubilized
  antiserum
    cross-reactivity between species, 758–759
    preparation, 747–751
  competitive protein-binding radioimmunoassay, 752–756
  $^{125}$I-labeled, preparation, 752–754
  immunoassays, 751–758
  microtiter ELISA, 756–758
  passive hemagglutination assay, 751–752
soluble, 560, 565–566; *see also* Tropoelastin
  amino acid composition, 566
  biosynthesis, *in vitro*, 716–731
  Bolton-Hunter probes, 729–730
  coacervation, 669–671
  extraction, 723–725
  [$^{14}$C]hydroxyproline analysis, 728
  identification techniques, 725–731
  isolation
    from copper-deficient calf, 565
    from copper-deficient chicks, 565, 650–657
    from copper-deficient piglets, 565
    from lathyritic tissues, 665–673
    methods, 565
  molecular weight, 565
  preparation for use as elastase substrate, 592
  radioimmune displacement, 729–731
  radiolabeling with $^{128}$I, 729–730
  reaction with elastin-specific antibody, 728–729
  resistance to CNBr cleavage, 727
  SDS polyacrylamide gel electrophoresis and gel filtration, 725–727

solubility in aqueous alcohol
mixture, 727–728
structure, 559–566
secondary, NMR characterization, 693–700
tertiary, hydrophobic, and inverse temperature transition, 709–711
synthesis, via cell-free translation of elastin mRNA, 569–570
type I, 571
type II, 571
valylprolyl content, 577–578
α-Elastin, 592, 673
Bolton-Hunter labeled, competitive protein binding assay, 753–754
coacervates
circular dichroism, 689–691
electron microscopy, 680–687
coacervation, temperature profile, 677–678
lactoperoxidase labeled, competitive protein binding assay, 753
preparation, 744–746
β-Elastin, 592
κ-Elastin, 592
$κ_2$-Elastin, preparation, 746
Elastolysis
assays, 593–600
modulation, 600–603
by elastin ligands, 601–603
physical-chemical environment, 600
stimulation, 603
Electron microscopy, see also Scanning electron microscopy; Transmission electron microscopy
of annelid cuticle collagen, 80–83
characterization of rat tail tendon collagen fibrils, 146–147
of collagen labeled with conjugated antibodies, 136–137
of collagen fibrils
in cartilage, 160–161
in connective tissue disease, 169
negative staining of collagenous materials, 132
positive staining of collagenous materials, 132
of proteoglycan, 799
of reconstituted type I collagen fibers, 159

of skin, 163–164
structural preservation during, 137, 153–154
of suspensions of collagen fibrils
reprecipitated collagen, 132–135
tissues, 131–132
of thin sections of collagen, 135–136
*Electrophorus electricus,* electric organ, acetylcholinesterase from, see Acetylcholinesterase, 18S
ELISA, see Enzyme-linked immunoabsorbent assay
Embryo
chick, collagens detected, 490
mouse, collagens detected, 490
Emphysema, elastin degradation in, 588
Endo-β-N-acetylglucosaminidase D, 355
Endo-β-N-acetylglucosaminidase H, 355
Endothelial cells
culture, elastin synthesis in, 629–630, 721–723
isolation, 722–723
laminin synthesis, 837
Enzyme-linked immunoabsorbent assay
of elastin, 756–758
of fibronectin, 821, 823–826
Enzymes
extracellular, in collagen biosynthesis, 305–319
in hydroxyproline synthesis, 373
intracellular, in collagen biosynthesis, 245–304; see also specific enzyme
Epiphysial growth plate, source of type II collagen, 10
Epithelial cells, see also Hepatocytes
collagen adhesion, 502, 508
embedding, in collagen matrix, 554
laminin synthesis, 837
primary culture, on collagen gels, 544
Ethylenediaminetetraacetic acid
added to acetic acid, 37
added to neutral salt solvent, 37
inhibitor of proteolysis, 35, 37
N-Ethylmaleimide
added to acetic acid, 37
added to neutral salt solvent, 37
inhibitor of proteolysis, 35, 37
Extensions, definition, 97

## F

Fibroblast, 630
  adhesion to collagen substrate, 503–508
    fibronectin requirement, 503
  human skin
    on collagen substrates, ultrastructural studies, 539–540, 542–544
    structural glycoprotein, 849
  spreading, 524
Fibronectin, 503, 524, 803–831
  amino terminal sequence, 804
  amniotic fluid, electrophoretic mobility, 804–805
  antibodies
    affinity purification, 819
    against functional domains, 817
    monoclonal, 819–821
      reactivities, 820–821
    species-specific, 817–819
  antisera, 816–817
    absorption, 819
    interspecies cross-reactivity, 817–819
  assays, interference with, 821–822
  attachment of hepatocytes to, 514
  binding
    to actin, 807
    to collagen, 503–504, 506–508, 806–807
    to fibrinogen, 806
    to staphylococci, 807
  carbohydrate content, 804
  in cell attachment, 828–831
    assay, 828–831
  cell layer-derived, isolation, 814
  chemical characteristics, 803–806
  codistribution with collagen, 489
  from different sources, comparisons, 804–806
  enzyme immunoassay, 821, 823–826
  fibroblast, electrophoretic mobility, 804
  functional properties, 806–808
  gelatin-Sepharose affinity chromatography, 809–816
  immunochemical properties, 816–821
  insoluble, 803
  interaction
    with cell surfaces, 807
    with fibrin, 807
  isolation, 808–816
    from cell culture, 813–814
    from plasma, 812–813
  localization
    by immunofluorescence, 826–828
    *in vitro* staining, 826–827
    tissue staining, 828
  molecular weight, 804
  plasma
    antisera, 817
    compared to cell culture-derived, 808
    electrophoretic mobility, 804–805
    structural glycoprotein, apparent molecular weight, 840
  plasma level, 803
  preparation, for cell adhesion studies, 518–519
  quantitation, 821–826
  radioimmunoassay, 821–823
  serum levels, 813
  soluble, 803
  storage, 814–815
  substrates, 519
  in wound healing, 807–808
Ficin, 747
Fluorescein-elastin, 593
Follicular fluid, proteoglycan, 773, 777–778
Freeze fracture, preparation of connective tissue, 136

## G

$\beta$-Galactosidase, molecular weight and collagen residues, 412–413
$\beta$-1-Galactosyl-$O$-hydroxylysine, 399
  standard, preparation, 404–407
  urinary level, determination, 399
Galactosylhydroxylysl glucosyltransferase, 285
  catalytic properties, 301–304
  molecular properties, 297–299
  purification, 294–298
Gel chromatography, estimation of molecular weight of collagenous polypeptides, 420–423

# SUBJECT INDEX

Glial cells, proteoglycans, 776
Globulin, cold insoluble, see Fibronectin
Glomeruli, isolation, 35
Glucosylgalactose, structure, 284, 285
α-1,2-Glucosylgalactosyl-$O$-hydroxylysine, 399
    standard, preparation, 404–407
    urinary levels, determination, 403–410
Glycopeptide
    gel filtration, 352–355
    hydrolysis, 356–357
    partial characterization, 352–360
    sugar-labeled, from protein, preparation, 347–348
Glycoprotein
    cell surface, in hepatocyte-collagen adhesion, 514
    hydrolysis, 356–357
    microfibrillar
        extraction from elastic tissues, 849–850
        with lysine oxidase activity, extraction, 850–851
    structural, 839–852; see also Fibronectin; Laminin
        amino acid composition, 840–841
        apparent molecular weights, 840
        extraction
            procedures, 843–844
            from salt-insoluble stroma, 844–845
        fractionation and separation, 846–847
        isolation
            from marine sponges, 847–849
            procedures, principles, 842–843
            from pig, rabbit, and calf aortic media, 851–852
        reversible chemical modifications, to improve solubility and extractability, 846
        sugar content, 842
Glycosaminoglycan
    classes, 769
    contamination of fibronectin, 811
    definition, 769
    disaccharide repeat units, 769–770
    structure, variations, 769–770
Glycosides, see Asparagine-linked glycosides; Hydroxylysine glycosides; Hydroxylysyl glycosides
Goat, immunization with collagen, 475
Guanidinium thiocyanate, denaturation of tissue ribonuclease, 218, 221
Guinea pig, production of collagen antibodies, 474–476

# H

Heparan sulfate-heparin, disaccharide repeat unit, 770
Heparin-agarose, collagenase binding, 450
Hepatocyte
    cell-surface receptor, interaction with heparan sulfate, 778
    cytochrome P-450 levels on different substrates, 551–552
    incubation
        for determination of collagen adhesion, 521–522
        for determination of spreading, 524
    morphology, in culture on plastic and collagen-coated plates, 548–549
    plasma membrane proteoglycan, 773, 778, 782
    preparation, for attachment studies, 519–521
    seeding
        for determination of collagen attachment, 521
        for determination of spreading, 524
    spreading, 524–525
Hepatocyte-collagen adhesion, 513–529
    cell-cell adhesion, 526–528
    determination, 522–523
    effect of antibodies, 525–527
    lactate dehydrogenase activity as indicator, 522–523
Hepatoma cells, morphology, in culture on plastic versus collagen substrates, 549–550
Hexosylhydroxylysine, 365, 366
Hexosyllycine, 365, 366
Histidinohydroxymerodesmosine, 365, 366

SUBJECT INDEX 901

*Homo sapiens,* type IV collagen, R values, 96
Hyaline cartilage
 collagens detected, 490
 source of type II collagen, 10, 35
Hyaluronic acid, 769
 disaccharide repeat unit, 770
Hydrodynamic theory, 190–196
 translational motion, 190–192
  diffusion coefficient, 191
  sedimentation coefficient, 191
  translational frictional coefficient, 190–191
 viscosity, 192–196
  rotational diffusion coefficient, 194
  rotational frictional coefficient, 194–195
Hydroxyapatite, in bone, 157, 158
Hydroxylase, 246–248; *see also* specific enzyme
 activities, extraction from crude tissue or cell specimens, 251–252
 assays, 248–264
  using radioactively labeled protocollagen substrates, 252–259
  using synthetic peptide substrates, 260–262
  standardization, 258–259
  synthetic polypeptide substrates, 249–250
 properties, 264–284
  catalytic, 280–284
 purification, 264–284
 stimulation
  by bovine serum albumin, 249
  by catalase, 249
  by dithiothreitol, 249
  by thymol, 249
Hydroxylysine, 571
 structure, 246
 urinary levels, determination, 401–403
 as urinary marker for collagen degradation, 399
Hydroxylysine glycosides, 339–346
 Gal-Hyl
  residues, sequence analysis, 344–346
  structure, 339

Glc-Gal-Hyl
 conversion to Gal-Hyl, 342–343
 isolation, from sponge, 339–342
 residues, sequence analysis, 344–346
 structure, 339
in proteins and peptides, automatic analysis, 343–345
Hydroxylysinonorleucine, 362, 365, 367
 galactosyl derivative, 365, 367–368
 glucosylgalactosyl derivative, 365, 367–368
Hydroxylysyl galactosyltransferase, 285
 catalytic properties, 301–304
 molecular properties, 301
 purification, 299–301
Hydroxylysyl glycosides
 ratios, in collagens, 399
 urinary excretion, 399
 urinary levels, determination, 403–410
Hydroxylysyl glycosyltransferase, 284–287; *see also* specific enzyme
 activities
  assays, 287–294
   principles and comparison of methods, 287–290
   procedure, 291–294
  extraction from crude tissue or cell specimens, 290–291
 properties, 294–304
  catalytic, 301–304
 purification, 294–301
 unit of activity, 293–294
6-Hydroxynorleucine, 364, 366
*p*-Hydroxyphenylpyruvate hydroxylase, 247
Hydroxyproline
 in elastin, 571
 hydrolysis, 374–375
 measurements
  use in quantifying collagenase-sensitive labeled collagen, 394–396
  use in quantifying collagen in unlabeled cells and tissues, 393–394
  use in quantifying degradation of newly synthesized collagen, 396–397
 synthesis, 373–374

urinary levels, determination, 399–401
3-Hydroxyproline
  determination, 372–398
  high-voltage paper electrophoresis, 385–387, 397–398
  ion-exchange chromatography, 387–391, 397–398
  structure, 246
trans-3-Hydroxy-L-proline
  elution position, 390–391
  separation from cis epimer, 387
  separation from trans-4-hydroxy-L-proline, 387
4-Hydroxyproline, 246–247
  colorimetric reaction for, 375–380
  determination, 372–398
  high-voltage paper electrophoresis, 385–387, 397–398
  ion-exchange chromatography, 387–392, 397–398
  radiochemical assay, 380–385
  structure, 246
cis-4-Hydroxy-D-proline
  elution position, 390–391
  radiochemical assay, 385
cis-4-Hydroxy-L-proline
  colorimetric assay, 379
  elution position, 390–391
  radiochemical assay, 385
trans-4-Hydroxy-L-proline
  conversion to cis epimer during hydrolysis, 374–375
  elution position, 390–391
  separation from cis epimer, 387

## I

Interstitial collagen, 65, 225; see also Collagen
  $\alpha$ chains, amino acid composition, 19–21
  in connective tissue, 127
  mammalian, carbohydrate units, 284–285
  nomenclature, 6
  $R$ values, 95
Intervertebral disk, collagens detected, 490
Intestine, collagens detected, 490

Invertebrate collagen, see also Cuticle collagen; specific invertebrate
  preparation and characterization, 65–96
  $R$ values, 95
Ion-exchange chromatography
  of collagen, 45
  of denatured collagen, 51–55
  for separation of cuticle collagen subunits, 71
  for separation of type I procollagen $\alpha$ chains, 111–113
Isodesmosine, see Elastin, cross-links

## J

JMI cell line, see Hepatoma cells

## K

Keratan sulfate, 769
  disaccharide repeat unit, 770
  linkage region, 771
  synthesis, 771–772
Kidney
  collagens detected, 490
  mouse, tubulogenesis, collagens detected, 490

## L

Lactate dehydrogenase, activity, indicator of hepatocyte-collagen adhesion, 522–523
Laminin, 508, 831–838
  amino acid composition, 835
  biological activity, 838
  biosynthesis, 837
  carbohydrate content, 835
  circular dichroism studies, 835
  codistribution with type IV collagen, 489
  electrophoretic profile, 834
  fragment P1, 835–836
    immunology, 837
    isolation, 836–837
  fragment P2, 835–836
    immunology, 837
  immunological properties, 837
  molecular weight, 835

mouse tumor, apparent molecular weight, 840
purification, from EHS sarcoma, 832–833
solubilization, by proteases, 835–836
sources, 831
structure, 834–835
tissue distribution, 838
Lathyrism
and "blank" radioactivity in fibril assays, 425
enhancement of extractability of types I and II collagen, 10–11
enhancement of extractability of type IV collagen, 12
induction
in chicks, 667
experimental, to obtain tropoelastin, 665–673
in piglets, 667–668
inhibition of elastin cross-link formation, 565
in young rats, 33
Lathyrogen, 33–34; see also specific lathyrogen
inhibition of lysyl oxidase, 563, 565
Lens capsule
source of type IV collagen, 11
type IV collagen from, $R$ values, 96
Leukocyte elastase, see Elastase, leukocyte
Leupeptin, added to acetic acid, 37
Ligamentum nuchae, bovine elastin
amino acid composition, 575
antiserum, interspecies reactivity, 730–731
desmosine cross-linked, 564
microfibrillar glycoprotein, 849
with lysine oxidase activity, 850–851
structural glycoprotein, apparent molecular weight, 840
Light microscopy, of fibrous collagen, using labeled antibodies, 137–138
Light scattering
quasi-electric laser, 204–209
Rayleigh, 203, 209–211, 217
Liver
cirrhotic, source of collagen, 11, 35

collagens detected, 490
disease, antibodies against collagen in, 498
fibrosis, collagens detected, 491
*Locusta migratoria,* ejaculatory duct collagen, $R$ values, 95
*Lumbriconereis heteropoda,* cuticle collagen
chemical composition, 77–78
$R$ values, 95
stability, 77
*Lumbricus terrestris,* cuticle collagen
amino acid composition, 88
extraction and purification, 83–86
physical properties, 86–88
$R$ values, 95
Lung
collagens detected, 490
fibrosis, collagens detected, 491
monkey, structural glycoproteins, apparent molecular weights, 840
parenchyma
bovine, elastin, amino acid composition, 575
source of type III collagen, 11
structural glycoprotein, heterogeneity, 847
Lysate, rabbit reticulocyte, preparation, for elastin mRNA translation assay, 734
Lysinonorleucine, 364, 366
in elastin, 580–582
Lysyl hydroxylase, 247
activity assay, 251
using radioactively labeled protocollagen substrate, 256, 257
using synthetic peptide substrates, 260–262
activity unit, 261–262
ascorbate requirement, 284
catalytic properties, 280–284
intracellular location, 248
mechanism of reaction, 281–284
molecular properties, 276–278
molecular weight, 276–277
from porcine fetal skin, molecular properties, 277
purification, 274–277

sequence requirement for interaction, 247
Lysyl oxidase, 315–319, 637–650
  activity unit, 641
  amino acid composition, 647–648
  assay, 315–317
    enzyme-coupled spectrophotometric, 642
    optima, 645–647
    tritium release, 638–642
  cofactors, 648–649
  copper as cofactor, 565
  in elastin cross-link formation, 563
  inhibition, 563, 565, 665–666
    by $\beta$-aminopropionitrile, 631, 638
  molecular weight, 647–648
  pH optimum, 645–647
  properties, 645–650
  purification, 317–319, 642–647
  reaction catalyzed by, 637–638
  stability, 645
  substrate specificity, 649–650

## M

$\alpha_2$-Macroglobulin
  anti-elastase activity, measurement, 604–605
  inhibition of collagenase, 424
Manganese, role in glycosylation, 285
*Marphysa sanguinea,* cuticle collagen
  chemical composition, 77–78
  $R$ values, 95
  stability, 77
Mast cell protease
  cleavage of procollagen, 122–123
  cleavage products, 22
*Metridium dianthus,* structural glycoprotein, amino acid composition, 841
Microscopy, *see* Electron microscopy; Light microscopy; Scanning electron microscopy; Transmission electron microscopy
Molecular sieve chromatography
  of denatured collagen, 50–51
  for purification of procollagen, 113–114
  for resolution and purification of CNBr peptides, 58–59

Monosaccharides, analysis, 355–360
Mouse
  peritoneal macrophage elastase, *see* Elastase
  production of collagen antibodies, 474–476
Muscle
  collagens detected, 490
  developing, collagens detected, 490
Myosin, molecular weight and collagen residues, 412–413

## N

Nagarase, 747
Nasal septum, bovine, *see also* Cartilage
  source of type II collagen, 10
NEM, *see* $N$-Ethylmaleimide
Nematode, *see* Cuticle collagen; species name
*Nereis*
  *japonica,* cuticle collagen
    chemical composition, 77–78
    extraction, 67
    stability, 77
    $R$ values, 95
    subunits
      chemical composition, 77–79
      separation, 71–73
  *virens,* cuticle collagen
    chemical composition, 77–78
    extraction, 67
    glycosylation, 78–79
    native fibers, electron microscopy, 80–82
    $R$ values, 95
    segment-long-spacing crystallites, 83–84
    subunits
      chemical composition, 77–79
      separation, 71–73
Nerve, collagens detected, 490
Neutral salt solutions, in collagen extraction, 12
Neutral salt solvent, 37
Nightcrawler, *see* Cuticle collagen; *Lumbricus terrestris*
NMR, *see* Nuclear magnetic resonance
Notochord, lamprey, collagen fibril structure, 149, 152, 161

NSCC, see Cuticle collagen, neutral pH salt-soluble
Nuclear magnetic resonance
  $^{13}$C, characterization of proteoglycan, 798–799
  $^{13}$C and $^{2}$H, in solids, 174–176
  characterization of repeat sequence elastin peptide torsion angles, 704–708
  characterization of elastin structure, 674–675, 693–700
  characterization of elastin tertiary structure, and inverse temperature transition, 709–711
  study of collagen fibrils, 174–186
Nucleus pulposus, collagen fibril structure, 152, 161

## O

Oligochaete see Cuticle collagen; species name
Oligosaccharide
  endoglycosidase digestions, 355
  gel filtration, 352–355
  hydrolysis, 356–357
  partial characterization, 352–360
  sugar-labeled
    from lipid intermediates, preparation and characterization, 348–352
    from protein, preparation, 347–348
Orcein-elastin, 597
Organ culture
  biosynthesis of insoluble elastin, 632–636
  biosynthesis of soluble elastin, 717–731
  cartilage, 718
  chick aorta, 717–718
  incubation in, 717–718
  pig aorta, 718
  rapid method, 634–636
Organic acid solvent, for collagen extraction, 37–38
Osteogenesis imperfecta, 168, 172
Ovalbumin, molecular weight and collagen residues, 412–413

## P

*Panagrellus silusiae*, cuticle collagen, 90, 94
  subunit, physical properties, 87
Pancreatitis, elastin degradation in, 588
Papain, 747
p-Collagen, 9, 97
  chains, amino acid composition, 117–118
  type I, tissue source, 107
  type III, tissue source, 107
  type V, fα chains, 106
pC-collagen, 97
  definition, 9
  type II, tissue source, 107
D-Penicillamine, enhancement of collagen extractability, 34
DL-Penicillamine, 630
Pepsin
  added to acetic acid, for collagen extraction, 38
  in collagen extraction, 13–14
  digestion
    of elastin, 747
    of procollagen mRNA translation products, 223–225
Pepstatin
  added to acetic acid, 37
  added to neutral salt solvent, 37
Phenylmethanesulfonylfluoride
  added to acetic acid, 37
  inhibitor of proteolysis, 37
*Pheretima communissima*, cuticle collagen, physical properties, 87, 89
Phosphocellulose chromatography, for purification of acidic CNBr peptides, 58
Phosphorylase *a*, molecular weight and collagen residues, 412–413
Photodesmosine, 586
Piglet
  copper-deficient
    harvest of aortas, 661
    isolation of soluble elastin from, 565, 657–665
    monitoring, 659–661
    raising, 658–659
  lathyritic, isolation of soluble elastin from, 565, 667–668

# SUBJECT INDEX

Placenta
  human
    source of type IV collagen, 12
    type IV collagen from, $R$ values, 96
  source of type V collagen, 12
  type V collagen from, $R$ values, 95
  villi, source of collagen, 35
Platelet
  aggregation, see also Platelet-collagen adhesion
    blocked by collagen, 498
  binding to radiolabeled collagen, 512
Platelet-collagen adhesion, 509–513
  assay
    affinity chromatography, 512–513
    column filtration methods, 511–512
    membrane-filtration method, 509–511
    turbidimetric method, 512
  on collagen-coated surfaces, 513
PMSF, see Phenylmethanesulfonyl fluoride
pN-collagen, 97
  accumulation, in skin, 168
  definition, 9
  type II, structure, 25–26
  type III, 163
    tissue source, 107
Polyacrylamide gel electrophoresis, characterization of collagen
  continuous buffer system, 59–61
  discontinuous buffer system, 61–64
Polychaete, see Cuticle collagen; species name
Polychondritis, antibodies against collagen in, 498
Poly(L-proline)-agarose, preparation, 265–266
Preprocollagen, 4, 27–28
Procollagen
  adsorption chromatography, 17
  analysis by SDS-PAGE, 116–117
  antibodies, 472–498
  asparagine-linked glycosides, 346–360
  chains
    amino acid composition, 117–118
    noninterstitial, cell-free synthesis, 225–245

  purification, 111–116
  characterization, 22–27, 116–127
  CNBr cleavage, 120–122
  extraction
    in denaturing solvents, 17
    solvents, 16
  $^{125}$I-labeled, two-dimensional peptide mapping, 124–127
  identification, by SDS-PAGE, 98–99
  immunization with, 473–476
  intracellular processing, 245–246
  mast cell protease cleavage, 122–123
  messenger RNA
    interstitial
      preparation, 218–222
      translation, 222–225
      translation products, 222–225
        peptic digestion, 223–225
  nonhelical regions, isolation by bacterial collagenase, 468
  nomenclature, 9, 10
  pepsin cleavage, 119–120
  peptide mapping, 118–127
  peptides, quantitation, 497–498
  preparation, 15–17, 96–116
  processing, to collagens, 27–32
  proteinases, 305–315
  purification, 111–116
  segment-long-spacing, aggregates, 133–134
  selective salt precipitation, 16, 44
  sources, 16
    in vitro, 99–100
  structure, 22–25
  sugar-labeled, preparation, 347
  synthesized in vitro, isolation, 99–106
  from tissues, isolation, 106–111
  type I
    $\alpha$ chains, separation, 111–113
    CNBr cleavage, 120–121
    conversion to collagen, 28–29
    mast cell protease cleavage, 122–123
    messenger RNA translation products, 223, 225
    nomenclature, 9, 10
    pepsin treatment, 119
    sources
      in vitro, 99–100
      tissue, 107

structure, 22–25, 97
synthesized *in vitro,* isolation,
   101–104
from tissue, isolation, 106–109
type I-trimer, structure, 26
type II
   amino acid composition, 117–118
   conversion to collagen, 29
   messenger RNA translation
      products, 225
   pepsin treatment, 119
   physicochemical features, 25
   source
      *in vitro,* 99–100
      tissue, 107
   structure, 97
   synthesized *in vitro,* isolation, 104
   from tissue, isolation, 109–110
type III, 163
   amino acid composition, 118
   CNBr cleavage, 120–121
   mast cell protease cleavage,
      122–123
   pepsin treatment, 119–120
   processing, 29–30
   source
      *in vitro,* 99–100
      tissue, 107
   structure, 25–26, 97
   synthesized *in vitro,* isolation,
      101–104
   from tissue, isolation, 106–109
type IV
   amino acid composition, 117–118
   cleavage, by V-8 protease from *S.
      aureus,* 127
   human
      purification on CM-cellulose
         chromatography, 113
      *R* values, 96
   mast cell protease cleavage,
      122–123
   pepsin treatment, 120
   processing, 30–31
   sources, 16
      *in vitro,* 99–100
      tissue, 107
   structure, 26–27, 97–98
   synthesized *in vitro,* isolation, 105
   from tissue, isolation, 110–111

type V
   cleavage, by V-8 protease from *S.
      aureus,* 127
   pepsin treatment, 120
   purification, by sedimentation
      velocity centrifugation,
      115–116
   sources, *in vitro,* 100
   structure, 27
   synthesized *in vitro,* isolation,
      105–106
   vertebrate collagenase cleavage, 123,
      124
   viscosity modeling, 196
Procollagen C-proteinase
   assay, 305–312
      by gel electrophoresis, 311, 312
      rapid, 310–312
   properties, 306
   purification, 314–315
   substrates, sources, 305–307
Procollagen intermediates, *see also* p-
   collagens; pC-collagen; pN-collagen;
   Protocollagen
   preparation, 96–116
Procollagen N-proteinase
   assay, 305–312
      by gel electrophoresis, 310, 311
      rapid, 309–310
   properties, 306
   purification, 312–314
   substrates, sources, 305–307
Proelastin, 569–570
Proline
   hydroxylation, measurement, 471
   intracellular pool, specific activity,
      391–393
Prolyl hydroxylase, 373
Prolyl 3-hydroxylase, 247
   activity assay, 251
      using radioactively labeled
         protocollagen substrate, 256
   catalytic properties, 280–284
   intracellular location, 248
   molecular properties, 279–280
   purification, 278–279
   sequence requirement for interaction,
      248

Prolyl 4-hydroxylase
    activity assays, 249–250
        using radioactively labeled
            protocollagen substrate,
            255–258
        using synthetic peptide substrates,
            260–262
    activity unit, 261
    amino acid composition, 268, 270
    ascorbate requirement, 284
    catalytic properties, 280–284
    distribution, 247
    intracellular location, 248
    mechanism of reaction, 281–284
    molecular properties, 267–270
    molecular weight, 267
    protein, assay, 262–264
    purification, 264–269
    sequence requirement for interaction,
        247
    $\beta$-subunit-related protein,
        function, 270–271
        intracellular location, 271
        molecular properties, 273–274
        purification, 270–273
    subunits
        amino acid composition, 269–270
        isolation, 268–269
        tritiation, 262–263
$n$-Propanol, fractionation of tropoelastin,
    653–654
Propeptide
    definition, 97
    molecular weight, 97
Protease, elastolytic, 588; see also
    Elastase
Protease inhibitor
    in collagen extraction, 12–13, 37
    inhibition of proteolysis, in basement
        membrane, 35
    in proteoglycan extraction, 781
$\alpha_1$-Protease inhibitor, anti-elastase
    activity, measurement, 604–605
Proteoglycan, 769–800
    aorta, 775–776
    basement membrane, 778
    cartilage, 160–161, 772–775
        chondroitin sulfate-rich region,
            772–773

extraction, with guanidine-HCl,
    781
hyaluronic acid-binding region,
    773–774
    purification, 796–797
    interaction with hyaluronic acid,
        774–775
    keratan sulfate-rich region,
        772–773
    molecular weight, 774
    tissues with proteoglycans similar
        to, 775–776
in cell adhesion, 507
chemical characterization, 794–797
    core protein, 796–797
    glycosaminoglycan analyses,
        794–795
    oligosaccharide analyses, 795–796
in chondrocyte cultures, extraction,
    782
contamination of affinity-purified
    fibronectin, 811
corneal stroma, 773, 776–777
definition, 769–772
dermatan sulfate
    in corneal stroma, 776
    in follicular fluid, 777–778
electrophoresis in composite
    polyacrylamide-agarose gels,
    793–794
extraction, 36
    with associative solvents, 782–783
    from cartilage, 161
    from chondrosarcoma, 782–783
    with detergents, 782
    with guanidine HCl, 779–782
    preparation of tissue, 780–781
    procedures, 780–784
        associative conditions, 780
        development, 779–780
        disruptive, 779
        dissociative, 779–780
follicular fluid, 773, 777–778
glial cell, 776
heparan sulfate, in hepatocyte plasma
    membrane, 773, 778
    extraction, 782
in hepatocyte plasma membrane, 773,
    778, 782
immunology, 799–800

ion-exchange purification procedure, 788
isolation, from extracted tissue residue, 783–784
isopycnic CsCl density gradient centrifugation
associative conditions, 784–785
dissociative conditions, 785–787
direct gradients, 787
keratan sulfate, in corneal stroma, 776
linkage region, 769–771
link-proteins, apparent molecular weights, 840
molecular-sieve chromatography, 791–793
physical characterization, 797–799
analytical centrifugation, 797–798
electron microscopy, 799
light scattering, 798
nuclear magnetic resonance, 798–799
viscometry, 798
purification and separation, 784–794
in rat epiphysial cartilage aspirate, microanalytical investigation, 783
rate zonal-velocity centrifugation
associative conditions, 789–790
dissociative conditions, 790–791
smooth muscle cell, 776
in tendon, 160
Proteolysis
of basement membrane, inhibitors, 35
limited, in collagen extraction, 13–14
Protocollagen
purification, from chick embryo tendon, 252–254
substrate for hydroxylase activities, 248–251
radioactively labeled, 252–259
Pyridinoline, 369
Pyrimidine deoxyribonucleoside 2-hydroxylase, 247

## R

$R$ values, 65–66, 94–96
Rabbit
immunization, with elastin, 728, 729
production
of collagen antibodies, 474–475
of desmosine antibodies, 760
of elastin antibody, 728–729
of insoluble elastin antisera, 747–750
of tropoelastin antiserum, 762–763
Ragworms, see Nereis sp.
Rat, production of collagen antibodies, 475
Remazolbrilliant blue-elastin, 593
Renal tubules, isolation, 35
Reticulum
calf spleen, collagen fibril structure, 152
rat spleen, collagen fibril structure, 163–164
spleen
$D$ period, 149, 166
X-ray diffraction study, 164–165
Retinal microvessels, intact, isolation, 35
Rheumatoid arthritis, antibodies against collagen in, 498
Rhinodrilus fafneri, 89
Rhodamine-elastin, 593
Ribonuclease, tissue, denaturation, 218, 221
Roundworm, see Ascaris lumbricoides

## S

Salt
precipitation techniques, see Selective salt precipitation
solvents, see Neutral salt solvents
Sandworms, see Nereis sp.
Scanning electron microscopy, study of cell-collagen interactions, 537–538, 541–544
Scar tissue, collagens detected, 491
Scleroderma
collagens detected, 491
type III procollagen, quantitation, 497
SDS PAGE, see Sodium dodecyl sulfate-polyacrylamide gel electrophoresis
Sedimentation coefficient, calculation, 191
Sedimentation velocity, centrifugation, for purification of procollagens, 115–116

Segment-long-spacing
  banding pattern, 133
  crystallites
    from annelid cuticle collagen, 81–84
    staining patterns, 133
Selective salt precipitation, for recovery of collagen types, in tissue extracts, 41–44
SEM, see Scanning electron microscopy
Sheep, immunization with collagen, 475
Skin
  collagen fibril structure in, 152, 163–165
  calf, collagen fibril structure, 152
  cat, collagen fibril structure, 152
  chicken, collagen fibril structure, 152
  collagens detected, 490–491
  D period, 149, 164–165
  diseased, X-ray diffraction study of, 171–172
  embryonic, collagen composition, 167
  human, collagen fibril structure, 152
  lamb, collagen fibril structure, 152, 164
  mummified, collagens detected, 490
  noncollagenous components, 130
  rat, collagen fibril structure, 152, 165
  type V collagen from, $R$ values, 95
Smooth muscle cells
  culture
    structural glycoprotein, 850
    synthesis of insoluble elastin by, 617–629
      assessment, 618–629
    ultrastructural analysis, for elastin morphology, 618–622
  isolation, from rabbit or pig aorta, 721
  in vitro, isolation, 101
  proteoglycans, 776
  synthesis of laminin, 837
Sodium dodecyl sulfate
  carpet, for cell-substrate adhesion study, see Cell-substrate adhesion
  effect on elastolysis, 603
  hot solutions, extraction of soluble elastin with, 723–725
  polyacrylamide gel electrophoresis, 18–19
    analysis of procollagens, 116–117
    of annelid cuticle collagens, 71–75
    for size estimation of collagenous peptides, 411–419
    of soluble elastin, 725–727
Spleen, collagens detected, 490
Sponge, as source of hydroxylysine glycoside, 339–342
*Spongia officinalis*, structural glycoprotein
  amino acid composition, 841
  isolation, 847–849
*Staphylococcus aureus*
  immunoabsorbent
    in elastin immunoassay, 754–755
    in elastin immunoprecipitation, 729
    in tropoelastin immunoprecipitation, 740–742
  V-8 protease
    cleavage of procollagen, 127
    cleavage products, 22
Sternum, chick, source of type II collagen, 10
Subtilisin, 747
Swine, see Piglet

## T

Telopeptides, 119
  structure, 129–130
TEM, see Transmission electron microscopy
Tendon
  Achilles
    bovine
      axial fibril structure, 149–150, 155
      lateral fibril structure, 152, 157
      structural glycoprotein, 840–841
    rat
      axial fibril structure, 149–150, 154–155
      lateral fibril structure, 152, 155
  chicken leg
    axial fibril structure, 149–150, 155
    lateral fibril structure, 152, 157

chicken wing, collagen fibril
  structure, 157
collagens detected, 490
embryonic
  chick leg
    collagen fibril structure, 167
    source of protocollagen,
      252–254
    collagen composition, 167
  kangaroo tail
    axial fibril structure, 149
    lateral fibril structure, 152
  lamb, collagen fibril structure,
    149–150, 152, 155, 157, 164–165
  proteoglycan content, 160
  quail, collagen fibril structure, 155
  rabbit leg
    axial fibril structure, 149–150, 155
    lateral fibril structure, 155
  rat tail, collagen fibril structure in,
    146–154, 155
  source of type I collagen, 10, 35
  turkey leg
    axial fibril structure, 150, 155
    calcification, 157, 158
    lateral fibril structure, 152, 157
Thermodynamic theory, 187–190
  excluded volume, 188–189
Thermolysin, 747
Thymine 7-hydroxylase, 247
Tissue
  for collagen preparation
    acquisition, 33–34
    processing, 33–37
  dispersal, 36
  lathyritic, recovery of collagen during
    decalcification, 36
  mineralized, decalcification, 36
  sources of procollagens, 106–107
  vertebrate, used in collagen
    preparation, 35
Tooth, developing, collagens detected,
  490
*Torpedo californica,* fibronectin, 809, 817
Trachea, collagens detected, 490
Translational frictional coefficient,
  calculation, 190–191
Transmission electron microscopy, study
  of cell-collagen interactions,
  538–540, 543–544

$\varepsilon$-$N$-Trimethyllysine hydroxylase, 247
Tropocollagen, vertebrate, physical
  properties, 87
Tropoelastin, 560
  amino acid sequence, 566
  antibody
    in cell-free translation of mRNA,
      739
    preparation, 763–765
  antiserum, preparation, 762–765
  coacervates, electron microscopy of,
    680–687
  coacervation, 674, 676
    temperature profiles, 677–680
  from copper-deficient chick aorta
    alcohol fractionation, 653–654
    amino acid composition, 656
    ammonium sulfate precipitation,
      653–654
    extraction, 652–653
    isolation, 654–656
    recovery, from acetic acid extracts,
      655–656
  from copper-deficient piglet
    isolation, 657–665
    purity, 664–665
  detection, in cell-free translation of
    mRNA, 739–742
  hexapeptide, 674
  immunoprecipitation, from cell-free
    translation products, 739–742
  isolation, from lathyritic tissue,
    668–673
  molecular weight, 651
  nonequilibrium pH gradient
    electrophoresis, 739
  pentapeptide, 674
  peptide sequences, repeating, 674
  polyhexapeptide
    coacervation, temperature profile,
      679–680
    cyclododecapeptide analog,
      electron microscopy of,
      686–689
    electron microscopy of, 685–687
    model of $\beta$-spiral conformation,
      712–714
  polypentapeptide
    assigned NMR spectra, 693–694
    circular dichroism, 689–691

coacervation, temperature profile, 677–679
cyclic conformational correlate, 693–700
    crystal structure, 700–704
electron microscopy, 683, 686–687
model of $\beta$-spiral conformation, 712–714
    derived from cyclic correlate, 714–716
polypeptides, NMR characterization of torsion angles, 704–708
polytetrapeptide
    circular dichroism, 689–691
    coacervation, temperature profile, 677–679
    electron microscopy, 684, 686–687
    hydrophobic interactions, and inverse temperature transition, 709–711
    quantitation, in cell-free translation of mRNA, 742–743
    radioactive, incorporated into insoluble elastin during synthesis, 630
    tetrapeptide, 674
Tropoelastin a, 570
    isolation from lathyritic tissue, 672–673
    separation from tropoelastin b, 739
Tropoelastin b, 570
    amino acid composition, 671
    coacervation, 669–671
    electrophoretic separation from tropoelastin a, 736–739
    molecular weight, 669
Trypsin, enhancement of elastin solubilization, 597
Tumor
    EHS sarcoma
        laminin purification from, 832–833
        type IV collagen from, $R$ values, 96
        source of type IV procollagen, 99, 107, 110
    human HT 1080, source of type IV procollagen, 99
    murine
        basement membrane containing, source of type IV collagen, 12

polyoma-virus-induced, source of type I-trimer collagen, 11
smooth muscle cell, source of type III collagen, 11
Swarm chondrosarcoma, source of type II procollagen, 107, 109
Tunicamycin, inhibitor of keratan sulfate synthesis, 771–772

## U

Urea carpet, for cell-substrate adhesion study, *see* Cell-substrate adhesion
Uterus, wall, source of collagen, 35

## V

*Vibrio B-30* collagenase, 454
Visceral pleura, bovine, elastin, amino acid composition, 575
Viscometer, Zimm-Crothers rotating cylinder, 215–216
Viscosity, 192–196

## W

Wheat germ
    lysate, preparation, 232
    post-ribosomal supernatant fraction, preparation, 232

## X

X-ray diffraction
    advantages, 138–139
    applications to collagenous structures, 139–141
    data processing, 141–145
        calculation of integrated intensities, 141–143
        correction of integrated intensities, 143–145
    interpretation of data, 145–146
    specimen preparation, 138–139
    studies
        of cartilage, 161–162
        of demineralized bone, 158
        of fibers in connective tissue disease, 169–172

of rat tail tendon, 147–154
  equatorial diffraction, 151–154
  meridional diffraction, 147–151
of reconstituted type I collagen
  fibers, 159–160
of skin, 164–165, 170–171

# Z

Zone precipitation, chromatography, 15
Zwittergent 3-12, in proteoglycan extraction, 782
Zwitterionic detergent, in proteoglycan etraction, 782

191497